AF598428

Advances in Photosynthesis

VOLUME 4

Series Editor:

GOVINDJEE

Department of Plant Biology
University of Illinois, Urbana, Illinois, U.S.A.

Advances in Photosynthesis is an ambitious new book series seeking to provide a comprehensive and state-of-the-art account of photosynthesis research. Photosynthesis is the process by which higher plants, algae and certain species of bacteria transform and store solar energy in the form of energy-rich organic molecules. These compounds are in turn used as the energy source for all growth and reproduction in these organisms. As such, virtually all life on the planet ultimately depends on photosynthetic energy conversion. This series of multiauthored books spans topics from physics to agronomy, from femtosecond reactions to season long production, from the photophysics of reaction centers to the physiology of whole organisms, and from X-ray crystallography of proteins to the morphology of intact plants. The intent of this new series of publications is to offer beginning researchers, graduate students, and even research specialists a comprehensive current picture of the remarkable advances across the full scope of photosynthesis research.

The titles to be published in this series are listed on the backcover of this volume.

Oxygenic Photosynthesis: The Light Reactions

Edited by

Donald R. Ort
USDA/ARS, Photosynthesis Research Unit,
University of Illinois,
Urbana, Illinois, U.S.A.

and

Charles F. Yocum
Department of Biology/Chemistry,
University of Michigan,
Ann Arbor, Michigan, U.S.A.

Assistant Editor
Iris F. Heichel
University of Illinois

KLUWER ACADEMIC PUBLISHERS
DORDRECHT / BOSTON / LONDON

Library of Congress Cataloging-in-Publication Data

Oxygenic photosynthesis : the light reactions / edited by Donald R.
Ort and Charles F. Yocum.
p. cm. -- (Advances in photosynthesis ; v. 4)
Includes bibliographical references and index.
ISBN 0-7923-3683-6 (alk. paper)
1. Photosynthesis. I. Ort, Donald R. II. Yocum, Charles F.
III. Series.
QK882.0875 1996
581.1'3342--dc20 96-9020

ISBN 0-7923-3683-6 (HB)
ISBN 0-7923-3684-4 (PB)

Published by Kluwer Academic Publishers,
P.O. Box 17, 3300 AA Dordrecht, The Netherlands.

Kluwer Academic Publishers incorporates
the publishing programmes of
D. Reidel, Martinus Nijhoff, Dr W. Junk and MTP Press.

Sold and distributed in the U.S.A. and Canada
by Kluwer Academic Publishers,
101 Philip Drive, Norwell, MA 02061, U.S.A.

In all other countries, sold and distributed
by Kluwer Academic Publishers Group,
P.O. Box 322, 3300 AH Dordrecht, The Netherlands.

The camera ready text was prepared
by Lawrence A. Orr, Center for the Study of Early Events in Photosynthesis
Arizona State University, Tempe, Arizona 85287-1604, U.S.A.

The structures on the cover are, clockwise: ferredoxin, from D. B. Knaff;
Photosystem II supercore particle, from E. J. Boekema and coworkers; cytochrome *f*,
from W. A. Cramer and coworkers; and ATPase, from E. J. Boekema and coworkers

Artistic composition is by Joan Apperson, University of Illinois

Printed on acid-free paper

Printed in the Netherlands

Contents

Thylakoid Membranes: Synthesis and Assembly of Thylakoid Membranes

The Photosynthetic Apparatus: Oxygen Evolution

The Photosynthetic Apparatus: Photosystem II

The Photosynthetic Apparatus: Photosystem I

The Photosynthetic Apparatus: Components of Intersystem Electron Transfer

The Photosynthetic Apparatus: Coupling Factor

The Photosynthetic Apparatus: Light Harvesting Complexes

Molecular Biology/Genetics of the Photosynthetic Apparatus: The Chloroplast Genome

Molecular Biology/Genetics of the Photosynthetic Apparatus: The Nuclear Genome

Preface

Oxygenic Photosynthesis: The Light Reactions is the fourth volume to appear in the series *Advances in Photosynthesis* by Kluwer Academic Publishers. Following the precedent set by earlier books in this series, our goal as editors has been to provide beginning students and seasoned researchers alike with a comprehensive reference text devoted to the major topics that comprise the area of photosynthetic energy transduction in oxygenic eukaryotes. We have interceded as editors to write a brief prefatory chapter that provides a general, integrated outline of the major areas of oxygenic photosynthesis research as they exist at the present time. Our goal for this chapter is to provide a capsule summary of the book's contents that can be used by readers to gain a general view of the field before they delve into the various subject areas that are given specialized coverage in the text. In addition to the guidance offered in the introductory chapter, we have worked closely with the authors of individual chapters to provide extensive cross-referencing among chapters. Using these cross-references, an interested reader can readily pursue a particular topic in substantial depth if this is desired.

The book is organized so as to divide its contents into three general subject areas. These are (1) thylakoid membranes (structure, synthesis and assembly); (2) the photosynthetic apparatus (photosystems, interphotosystem components, coupling factor, and light-harvesting complexes); and (3) molecular biology/genetics of the photosynthetic apparatus (chloroplast and nuclear genomes). The first chapter of most of the subsections within each of the major topic areas is structured by its author(s) as a mini-review of the subject matter at hand while the following chapters contain detailed reviews. It is our hope that this organization will enhance the attractiveness and value of the text to students who are making their first acquaintance with the fascinating topics embraced by the text, as well as to faculty who participate in advanced courses on photosynthesis. All of the contributing authors were given wide latitude in reviewing and evaluating the present status of their areas of expertise. We believe that this strategy has produced a volume that will stand as a benchmark reference for some time to come.

Reading and editing the chapters for this book was a pleasant exercise. Time and again our attention was drawn to the enormous progress that has been made in understanding mechanisms and structures associated with oxygenic photosynthesis. Although this process, oxygen evolution coupled to $NADP^+$ reduction, is unavoidably complex on account of the number of components and reactions involved, readers of this book will quickly discern that questions surrounding structure and mechanism are being answered at an unprecedented pace. Crystal structures are replacing 'cartoon' representations of multi-subunit protein complexes, mechanistic details of reactions are under intense scrutiny, and the techniques of molecular biology are providing a new means of probing photosynthetic reactions at all levels, from the intact membrane to reaction centers.

The substantial leaps forward that characterize photosynthesis research today occur from a foundation laid by the research efforts of a generation of scientists who are the founders of this field. This book is in many ways a tribute to those scientists. Our own progress in this field was heavily influenced by contacts with these individuals, who were dedicated to advancing their field and to nurturing the next generation. We both profited immensely from our contacts with Norman Good, as a mentor (DRO) and supportive colleague in the early stages of a career (CFY), and it is to his memory that we dedicate this book.

Production of a book of this scope is an enormous undertaking. *Oxygenic Photosynthesis: The Light Reactions* owes its existence to the efforts of the authors who contributed to this book. Just as important are the two people who assisted us at all stages of editing, communication and assembly. As we point out in the introductory chapter, Iris Heichel assumed a major responsibility for checking our editing (a major duty in itself), assembling the index and acting in the most polite way possible to prevent the sorts of idiocies that academic scientists (in this case the editors) are famous for committing. Without Iris, this book would not exist. In turn, Iris and the editors would like to thank Larry Orr of the Center for the Study of Early Events in Photosynthesis at Arizona State University. Larry must be among the best-natured editorial production experts in the world. He managed to convert disks containing incomprehensible programs into lucid text, caught the errors we missed, and did much more to assure the professional as well as timely appearance of this book. Finally, we would like also to acknowledge Govindjee, not only for his leadership in developing the *Advances in Photosynthesis* series, but also for his work in the trenches, reading and commenting on each of the chapters in this book.

Donald R. Ort
Charles F. Yocum

Chapter 1

Electron Transfer and Energy Transduction in Photosynthesis: An Overview

D. R. Ort
Photosynthesis Research Unit, USDA/ARS and, Department of Plant Biology, University of Illinois, Urbana, IL 61801, USA

C. F. Yocum
Departments of Biology and Chemistry, The University of Michigan, Ann Arbor, MI 48109-1048, USA

Summary

The intent of this book is to provide readers with a comprehensive view of the activities, structure, and molecular biology of the components of the photosynthetic apparatus in chloroplast thylakoid membranes. The sections that follow provide coverage of major topics that represent foci of active research investigations. In this introductory chapter we provide a brief overview of thylakoid properties and a guide to our readers to appropriate chapters in the following sections of this book, where our colleagues have provided in-depth coverage of the topics outlined in this overview chapter. We outline the activities of the two photoreactions (Photosystems I and II) and discuss the electron carriers that mediate transfer of electrons between these light-driven enzyme systems. Thylakoid energy transduction is described in terms of the origins of the protonmotive force and its subsequent utilization by the chloroplast ATP synthase, CF_1. Finally, we provide some comments on the molecular biology and evolution of the eukaryotic photosynthetic apparatus.

Donald R. Ort and Charles F. Yocum (eds): Oxygenic Photosynthesis: The Light Reactions, pp. 1–9.

I. Introduction

The energy-transducing membranes of the chloroplast, called thylakoids, are structurally intricate and functionally very complex. The appressed morphology of these membranes is unusual, and the membrane-embedded protein complexes involved in energy capture, electron transfer and ATP synthesis are all comprised of multiple intrinsic and, in some cases, extrinsic proteins. However, this complexity is steadily yielding to structural and functional investigations as well as to systematic sequencing and mutagenesis of the genes that encode thylakoid-associated proteins. Such research approaches are providing numerous insights into the electron transfer system of oxygenic photosynthesis. The high resolution, 3-dimensional X-ray crystal structure of the reaction center of the bacterium *Rhodopseudomonas viridis* (Michel and Deisenhofer, 1988) has proved to be a valuable model (Rutherford, 1989) that has guided explorations of the structure of the PS II reaction center. The more recent progress towards resolution of the structure of the reaction center of PS I (Krauss et al., 1993; reviewed by Witt, Chapter 18), elucidation of the structure of the major light harvesting chlorophyll *a/b* binding protein (CAB) (Kühlbrandt et al., 1994), and of mitochondrial F_1 (Abrahams et al., 1994; see Boekema and Lücken, Chapter 26), along with the structures of ferredoxins (Ikemizu et al., 1994), ferredoxin-NADP reductase (Karplus and Bruns, 1994), plastocyanin (Colman et al., 1978), and cytochrome *f* (Martinez et al., 1994; reviewed by Martinez et al. in Chapter 22) provide researchers working on photosynthetic electron transfer and energy coupling reactions with a solid foundation of structural models on which to base biochemical, biophysical and genetic experiments. Indeed, where information about thylakoid protein complexes cannot yet be derived from crystallographic studies, data from those systems that are structurally-defined have been very productively coupled with the substantial reservoir of gene sequence data. The resulting combination of information has been applied to directed mutagenesis experiments that have proven to be extremely useful in identifying amino acids that participate in electron transfer processes (Debus et al., 1988). It is, however, an inevitable consequence of the complexity of the energy trapping and transducing protein complexes of thylakoid membranes that many important questions about individual components and about the intact membrane system remain unanswered. This, in turn, assures a productive and vital future for research on photosynthetic systems.

As in many other fields of biological research, expansion of the research literature in the field of photosynthesis has been extremely rapid. This can be overwhelming for both the seasoned researcher and the beginner who is just entering the field. The last comprehensive reviews of photosynthetic electron transfer reactions and energy conservation in intact thylakoid membranes were by Ort (1986) and Andreasson and Vanngard (1988). In this prefatory chapter, we will provide a brief, updated overview of thylakoid membrane activities as a means of guiding the reader to appropriate chapters in this book where more detailed information on a given topic may be found.

Abbreviations: CAB – chlorophyll a/b-binding protein; CF – coupling factor (ATP synthase); Cyt – cytochrome; PQ – plastoquinone; Pheo – pheophytin; PS – Photosystem; Q – quinone

II. Thylakoid Membranes: Energy Trapping and Electron Transfer Reactions

The chloroplast thylakoid membrane is unique among biological membranes in its structure and composition. Chapters 2 (by Staehelin and van der Staay) and 3 (by Wolfe and Hoober) outline the current view of these membranes, and set the stage for the contributions of Webber and Baker, Mustardy, Douce and Joyard and by Robinson. These Chapters (4–7) cover development, lipid composition, and the intracellular trafficking associated with biogenesis of thylakoids. The unusual 'stacking' behavior of these membranes, which leads to the distinctive appressed appearance resolved by transmission electron microscopy, as well as their lipid composition, primarily galactolipids rather than phospholipids (Webb and Green, 1991), clearly distinguish the thylakoid membrane bilayer from those of other organelles and from the cell plasma membrane.

The immediate end-products of reactions catalyzed by illuminated thylakoid membranes in the intact chloroplast are O_2, NADPH + H^+ and ATP. The latter two species are for the most part consumed to provide energy for the reduction of CO_2 in the surrounding chloroplast stroma. The key to utilization of thylakoid-generated reducing power is ferredoxin, an iron-sulfur protein whose activities are reviewed by Knaff in Chapter 17. The ATP produced by thylakoid energy coupling reactions is also consumed in a variety of reactions unrelated to carbon metabolism (for example organelle biogenesis and maintenance). The energy conversion and storage capabilities of chloroplast thylakoid membranes may be traced directly to the ability of chlorophyll, ligated to thylakoid protein components of the two photosystems, to trap radiant energy and convert it into redox energy. Stabilization of this energy, through its transfer to other protein complexes of the thylakoid membrane by secondary redox reactions, constitutes a major function of the thylakoid membrane.

A. Photosystems I and II

Differences in both function and polypeptide composition distinguish the two photosystems of thylakoid membranes, as described by Bricker and Ghanotakis and by Ki. Satoh for PS II (Chapters 8 and 11) and by Nechushtai, et al. for PS I in Chapter 15. At the same time, these photochemical enzyme systems display some remarkable similarities. Each of the reaction centers has an associated ensemble of chlorophyll molecules that, along with carotenoids, act as antenna systems that absorb light and transfer the resulting exciton energy to the reaction center chlorophyll a. In contrast to the bacterial reaction center, which binds only four bacteriochlorophyll a molecules, larger numbers of chlorophyll a antennae pigments are ligated to the polypeptides that comprise the PS II and PS I reaction centers. Chlorophylls a and b along with various carotenoids are also distributed among a family of polypeptides referred to as CAB proteins, which associate with the reaction center proteins and in doing so form a more extended antenna system. The Chapters (27–30) by Simpson and Knoetzel, Pichersky and Jansson, Melis, and Yamamoto and Bassi respectively, detail the structure, genetics, and dynamic properties of this interesting family of proteins.

In each photosystem the specially-ligated chlorophyll *a* molecules that form the reaction center primary donor are rapidly oxidized to a cation radical following transfer of exciton energy from the antenna pigments (van Grondelle et al., 1994). Time-resolved electronic absorption spectroscopy shows that the reaction centers of both photosystems initiate electron transfer, or charge-separation, reactions on a picosecond time-scale, which is in the same time frame as molecular vibrations. The result of charge separation is the creation of very large redox potential differences between the oxidized donor reaction center chlorophyll a cation radical species ($P680^+$ or $P700^+$) and their immediate acceptors ($Pheo^-$ or A_0^-). These redox potential differences represent the earliest steps in photosynthetic energy storage, and in addition are perhaps the largest known for biological systems (~1.8 V for PS II and ~1.5 V for PS I). In both photosystems, the initial charge-separated state is rapidly stabilized by a series of reactions, discussed by Malkin for PS I (Chapter 16) and by Satoh (Chapter 11) and Diner and Babcock (Chapter 12) for PS II. The energetically downhill electron transfer reactions that occur following charge separation effectively prevent recombination of the highly oxidizing reaction center chlorophyll molecules with highly reducing primary acceptor species. A simplified model for the 1-electron charge separation reactions and their approximate time constants is given in Table 1.

The oxidants (D^{+}) and reductants (A^-) shown in Eq. (4) of Table 1 are identified in Table 2; these species react rapidly with other redox carriers to form stable intermediates that produce energy transformations through the process of vectorial electron transfer reactions, discussed later in this chapter. In the case of PS II, a tetranuclear Mn cluster sequentially accumulates the oxidizing power used to liberate a molecule of O_2 from two molecules of H_2O ($E^{\circ\prime}$= +0.815 V), as reviewed by Britt in Chapter 9. In PS I, a sequence of iron-sulfur clusters, soluble ferredoxin and a flavoprotein catalyze reduction of $NADP^+$ ($E^{\circ\prime}$= –0.34 V), as described in Chapters 16 and 17 by Malkin and Knaff respectively.

Photosystems must couple the 1-electron charge separation events to multielectron redox chemistry of subsequent electron transfer reactions. On the acceptor side of PS I, this involves the light-driven 2-electron reduction of $NADP^+$ whereas on the donor side of Photosytem II the 4-electron oxidation of H_2O to molecular oxygen takes place. As shown in Table 2, these redox reactions are catalyzed by elaborate metalloprotein clusters containing either Fe (PS I) or Mn (PS II). Crystallization of PS I has opened the door to elucidation of its structure, as described by Witt (Chapter 18), and, as mentioned previously, the crystal structure of the bacterial reaction center has proven to be a useful framework for hypothetical models of the reaction center of

Table 1. Charge Separation Kinetics for Photosystems I and II*

Reaction	Approximate half-time	
	PS II	PS I
(1) $D P I A + h\nu \rightarrow D P^* I A$	< 1 ps	< 1 ps
(2) $D P^* I A \rightarrow D P^+ I^- A$	~ 20 ps	< 1–2 ps
(3) $D P^+ I^- A \rightarrow D P^+ I A^-$	~ 250 ps	~ 200 ns
(4) $D P^+ I A^- \rightarrow D^+ P I A^-$	~ 50–200 ns	< 1 ms

* D = Donor; P = Chlorophyll; P* = excited state reaction center chlorophyll; I = Intermediate Acceptor; A = Terminal Acceptor

PS II. Individual polypeptides and the interactions among them are obviously essential in providing the overall structure of the photosystem complexes. At a more subtle level, the folding pattern of reaction center polypeptides creates the binding domains for the optimal orientation of both organic and inorganic electron transfer cofactors, ensuring that rapid electron transfer can occur. The nature of these domains, clearly revealed in the bacterial reaction center crystal structure, is still obscure in the counterpart structures of oxygenic photosystems. However, the similarity in charge separation kinetics between the bacterial system and Photosystems I and II (Table 1) suggests that the fundamental processes that guide electron transfer from one electron transfer component to the next in all reaction center systems are probably the same.

Structural features likely to be shared among reaction centers would include spatial proximity of the participating redox components, as well as a deployment of these species with respect to one another so as to promote a vectorial charge-separation reaction whose ultimate product is charges of opposite sign at some distance from one another on opposing sides of the reaction center. This phenomenon creates an electrical potential that first appears within the reaction centers themselves. Ultimately, migration of oxidizing and reducing equivalents away from the photosystems, by way of interphotosystem electron transfer components, provides the means by which redox potential energy is converted into trans-membrane ion potentials that are utilized for ATP synthesis.

In spite of their similarities, the oxygenic photosystems also diverge in certain respects. As shown in Table 2, PS II utilizes quinone as the first stable electron acceptor while PS I employs a bound Fe/S cluster in this role. A number of interesting 'quirks' exist in PS II. As detailed in the review by Lavergne and Briantais (Chapter 14), energy-trapping and charge separation in this photoreaction are marked by heterogeneities that are just now becoming better understood. In addition, PS II contains tightly-bound cytochrome *b*-559. The ability to purify PS II (Berthold et al., 1981; Ikeuchi et al., 1985) and to employ genetic techniques to modify amino acid residues (Vermaas, 1993) is slowly lifting the veil of enigma that has surrounded this cytochrome. The current view of cytochrome *b*-559 is reviewed by Whitmarsh and Pakrasi in Chapter 13.

B. Interphotosystem Electron Transfer: Plastoquinone, the Cytochrome b_6f *Complex and Plastocyanin*

A schematic diagram of the thylakoid electron transfer sequence and its associated protein complexes is shown in Figure 1, which identifies the major membrane-embedded protein complexes. The small H_2O-soluble proteins, ferredoxin and plastocyanin, along with the lipid-soluble species, plastoquinone, constitute the mobile components of the thylakoid-associated energy transducing apparatus. The function of the interphotosystem electron carrier system is to deliver reducing equivalents generated by PS II to PS I, where charge-separation creates an oxidizing equivalent, $P700^+$ The difference in redox potentials between these two points in the oxygenic electron transfer chain is about 0.4 V, and electron transfer across this energy gap produces part of the protonmotive force that is utilized for ATP synthesis.

Table 2. Electron Transfer Intermediates Associated with the Photosystems

	Photosystem II	Photosystem I
D(onor)	Tetranuclear Mn Cluster; Tyrosine	Plastocyanin (Cu^{1+})
P(igment)	Chlorophyll *a* (P680)	Chlorophyll *a* (P700)
I(ntermediate)	Pheophytin *a*	Chlorophyll *a* (A_0) Menaquinone (A_1)
A(cceptor)	Plastoquinone (Q_A)	4 Fe/4 S Center (F_x)

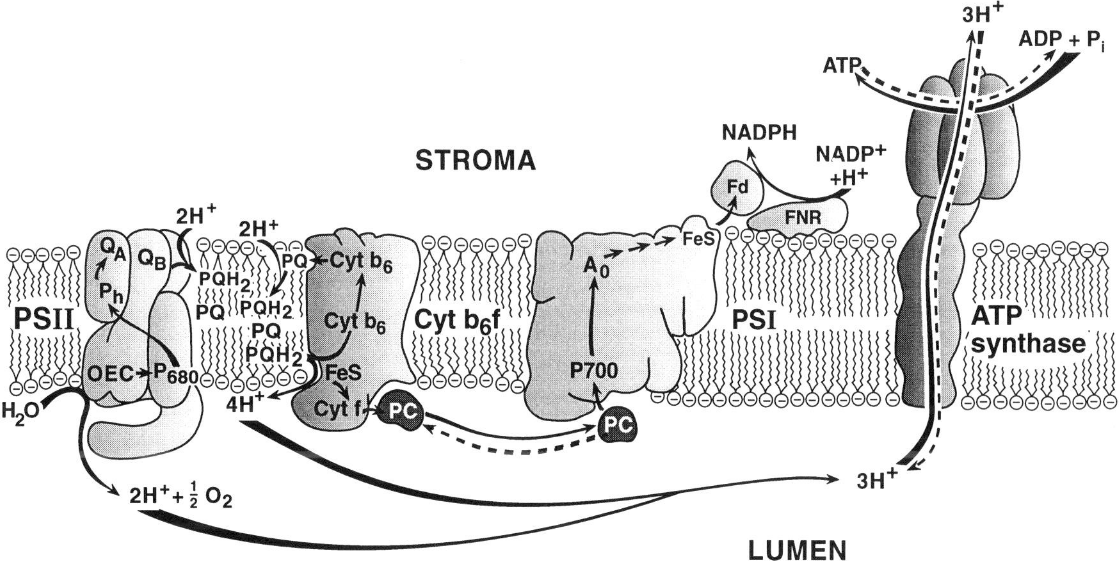

Fig. 1. Schematic diagram of the oxygenic electron transport chain in eukaryotic thylakoid membranes.

1. Plastoquinone Reduction/Oxidation

Generation of reduced plastoquinone by PS II initiates interphotosystem electron transfer. This reaction is an illustration of the coupling between 1-electron transfer at the reaction center and the subsequent shift to multielectron redox reactions. Single electrons generated by PS II photochemistry produce quinone anion radicals bound to protein sites, called Q_A and Q_B, in the reaction center. (Phenylurea and S-triazine herbicides cause electron transfer inhibition by displacing Q_B from its binding site, which is situated on the PS II reaction center protein known as D1.) The model for PS II-catalyzed plastoquinone reduction proposes a gated electron transfer sequence, as shown in the equations that follow (Crofts and Wraight, 1983):

$$Q_A\,Q_B + h\nu \rightarrow Q_A^-\,Q_B \quad (1)$$

$$Q_A^-Q_B \rightarrow Q_A\,Q_B^- \quad (2)$$

$$Q_A\,Q_B^- + h\nu \rightarrow Q_A^-\,Q_B^- \quad (3)$$

$$Q_A^-\,Q_B^- + 2\,H^+ \rightarrow Q_A\,Q_BH_2 \quad (4)$$

$$Q_A\,Q_BH_2 + PQ \rightarrow Q_A\,Q_B + PQH_2 \quad (5)$$

Rapid exchange of Q_BH_2 for an oxidized PQ molecule (Eq. (5) above) at the Q_B site completes the cycle and prepares the reaction center for another round of quinone reduction reactions. The PQH_2 generated by these reactions is deposited in a reservoir of reduced quinone that acts as the electron donor source for reactions catalyzed by the cytochrome b_6f

complex (described by Hauska et al. in Chapter 19), which in turn functions as the reductant of plastocyanin, a blue copper protein (Gross, Chapter 21). In sum, this sequence of reactions delivers electrons to $P700^+$ while the H^+ created by plastoquinone oxidation are conserved as part of the proton electrochemical potential used for ATP synthesis, as diagrammed in Fig. 1.

2. The Cytochrome b_6f *Complex and Plastocyanin*

The cytochrome b_6f complex and its counterparts in other membrane-bound electron transfer systems (the bc_1 complexes of bacterial plasma membranes and mitochondrial inner membranes) are comprised of a single *c*-type heme, 2 *b*-type hemes and an unusual Fe/S center (termed a 'Rieske' center after its discoverer), characterized by a positive, rather than negative, reduction potential. These complexes are universally involved in catalysis of quinol oxidation as well as reduction of a soluble electron transfer metalloprotein (plastocyanin or cytochrome *c*). As described by Hauska et al. in Chapter 19, the '*b-c*' complexes may also participate in gated quinone oxidation reactions, which are also known as 'Q-cycles'. Such oxidation reactions produce a 2-fold larger net translocation of H^+ than expected from standard quinone chemistry or from the simple concerted $2H^+/2\ e^-$ oxidations of the type described in Eqs. (1–4) above.

An additional feature of the cytochrome b_6f complex in thylakoid membranes is its probable function in catalysis of cyclic electron transfer reactions catalyzed by reduced soluble ferredoxin. Although the literature concerning cyclic electron transfer reactions has not yet arrived at consensus, it is apparent that in the presence of ferredoxin, the amount of ATP produced in vitro by thylakoid preparations is larger than can be accounted for on the basis of energy conservation associated with electron transfer involving a linear pathway from PS II through PS I to $NADP^+$. It is also clear that the ability of the cytochrome b_6f complex to catalyze diverse quinone oxidation reactions is central to thylakoid H^+-pumping activity.

Figure 1 shows diagramatically that bridging the spatial separation between the photosystems is a key function of the interphotosystem electron carriers. As noted in the chapters on thylakoid structure, PS II is confined largely to the appressed regions of thylakoid membranes, whereas PS I is found predominantly in the non-appressed, or stromal regions of these membranes. It is exceedingly unlikely that lateral mobility of the cytochrome b_6f complex in the thylakoid bilayer can account for measured rates of electron transfer between the photosystems, so either plastoquinone or plastocyanin (or both) must function as mobile carriers to shuttle electrons between the photosystems. It is of interest to note in this regard that about half of the membrane's complement of cytochrome b_6f complex is associated with the appressed regions of thylakoid membranes (Allred and Staehelin, 1986), where PS II is also found. This observation would suggest that plastocyanin, rather than PQH_2, plays the major role in long distance diffusion coupling electron transfer between the two photosystems (Whitmarsh, 1986).

As Gross describes in Chapter 21, plastocyanin is a member of the family of so-called 'blue' copper proteins, describing their color when copper is in the oxidized (2+) state. A relatively small protein (M_r of about 10,000), plastocyanin should be capable, on theoretical grounds, of possessing a mobility in accord with its proposed role as a diffusable electron carrier.

III. Energy Conservation In Thylakoid Membranes

Immediate products of light-induced electron transfer in thylakoid membranes are O_2, NADPH + H^+, and an electrochemical potential consisting of a membrane potential ($\Delta\Psi$) and a transmembrane pH differential (ΔpH). The negative component of the $\Delta\Psi$ resides on the stroma-facing surface of thylakoids while the thylakoid lumen is lower in pH than the surrounding external medium (Avron, 1978). The sum of these 2 components comprises the proton-motive force:

$$\Delta\mu_{H^+} = \Delta\Psi + 59\ \text{mV}\ \Delta\text{pH}$$

The relative insensitivity of chloroplast ATP synthesis in the steady-state to reagents that collapse $\Delta\Psi$ but not ΔpH suggests that under continuous illumination, most, if not all energy for ATP synthesis is derived from the chemical potential of the protons released into the thylakoid lumen as a result of electron transfer activity.

A. Proton Pumping and ATP/e_{2^-} Ratios

Mechanisms and sites of proton generation within the thylakoid membrane are discussed by Haumann and Junge (Chapter 10) and by Krishtalik and Cramer (Chapter 20). It may be that in the steady-state, only two reactions need be invoked to account for lumen acidification:

(1) $H_2O \rightarrow 1/2\ O_2 + 2\ e^- + 2\ H^+$ (PS II-catalyzed)

(2) $PQH_2 \rightarrow PQ + 2\ e^- + 2\ H^+$ (b_6f complex-catalyzed)

The sum of these processes predicts that for each electron pair transferred from H_2O to PS I, 4 H^+ will be accumulated that contribute to $\Delta\mu_{H^+}$. An extensive literature documents energy coupling associated both with H_2O oxidation and with oxidation of PQH_2 by PS I-catalyzed electron transfer. It has been shown that in the early stages of thylakoid energization, protons may not necessarily be deposited directly in the bulk aqueous phase of the lumen (Ort and Dilley, 1976).

The net yield of ATP from thylakoid electron transfer reactions has been the subject of extensive research. In isolated membranes where electron transfer can be forced into a 'pure' non-cyclic mode (for example through the presence of artificial acceptors such as $Fe(CN)_6^{3-}$ or methylviologen that rapidly oxidize the acceptor domain of PS I), measured efficiencies of energy transduction are on the order of 1.2 – 1.3 ATP synthesized per 2 e^- transferred (Saha et al. 1971). In thylakoids reconstituted with ferredoxin/$NADP^+$, higher ATP yields (about 1.6 ATP per 2 e^- transferred) have been measured. This discrepancy is taken to indicate that with ferredoxin present, cyclic electron transfer catalyzes additional H^+ pumping by way of the cytochrome b_6f complex to produce the 'extra' ATP detected in these experiments (Hosler and Yocum, 1985). Many questions remain about the significance of cyclic electron flow and about means by which thylakoid electron transfer may be regulated between cyclic and non-cyclic modes.

B. The ATP Synthase

Chapters 23–26 by McCarty, Richter and D.A. Mills, J.D. Mills, and by Boekema and Lücken discuss the structure and function of CF_1, the chloroplast ATP synthase or coupling factor, and CF_0, the membrane spanning portion of the holoenzyme. These enzyme systems are ubiquitously distributed in membrane systems where electron transfer reactions are coupled to ATP synthesis. Their structures are remarkably similar, and their subunit composition is similarly conserved. The critical subunits of CF_1 are designated α, β, γ, δ and ε. The enzyme is comprised of three copies each of α and β, and single copies of δ and ε. The intact enzyme is large, with a M_r of 400,000. As noted earlier, the crystal structure of mitochondrial F_1 is now available, and from this structure numerous analogies to CF_1 may be drawn (see Boekema and Lücken, Chapter 26). The CF_0 portion of the coupling apparatus is an oligomer comprised of four different intrinsic protein subunits that self-assemble in the membrane bilayer to form an H^+ conducting 'pore' as well as the site to which CF_1 binds.

α-β dimers comprise a trio of active sites of coupling factors; the exact mechanism by which these enzymes couple ATP synthesis to proton flow through CF_0 is not yet settled, but a popular mechanism proposed by Boyer (1989) suggests that each α/β dimer undergoes a series of subtle conformational shifts in the course of substrate binding and the subsequent steps involved in formation of the terminal phosphate bond of ATP and ATP release from the enzyme. The γ, δ and ε subunits carry out multiple functions, anchoring CF_1 to CF_0 and regulating the integrity of H^+ flow through the enzyme and the expression of its ATPase activity. In addition, the asymmetric positioning of the major subunits to the catalytic sites on the $\alpha\beta$ pair is proposed to account for sequentially altering nucleotide binding properties of the three sites.

The ATP synthases of oxygenic photosynthesis are distinguished from other such enzymes by the latency of their ATPase activities. Such ATP hydrolytic activity is present only at very low levels in CF_1, and ATPase activation requires imposition of a $\Delta\mu_{H^+}$ which can subsequently be stabilized by treatments that reduce a γ-subunit -S-S- bond. This aspect of the enzyme, discussed by McCarty in Chapter 23, points to an important difference between the ATP synthases of thylakoids and those of mitochondrial inner membranes and the plasma membrane of prokaryotes.

IV. Molecular Biology of Thylakoids

When DNA was discovered in chloroplasts and

mitochondria the stage was set for a number of pioneering investigations on the genetics of these organelles as well as for research into the mechanisms of organelle assembly. The observations in both organelles that the DNA content was insufficient to code for their protein and RNA complements posed a mystery that was solved by the discovery that the largest portion of the DNA codes for chloroplast and mitochondrial proteins reside on nuclear, rather than organellar, DNA. This discrepancy was resolved by the numerous investigations, summarized by Robinson in Chapter 7, that have documented the steps by which the products of nuclear genes are targeted into chloroplasts (Keegstra et al., 1989), and in some cases to the lumen of thylakoid membranes.

An additional level of sophistication in organelle assembly is revealed by the molecular analyses that have demonstrated that, in all cases, the DNA codes for individual components of a thylakoid membrane complex are split between the organelle and the nucleus. An example of this division of information is presented in Table 3 for many of the polypeptides that comprise PS II. Additional post-translational steps are necessary for many nuclear-encoded proteins; for example, CAB proteins must be imported into the chloroplast stroma, processed to their mature sizes, and inserted into the thylakoid membranes coincident with binding of chlorophyll, which is synthesized in the chloroplast. Not only must the assembly processes leading to multi-protein complexes be coordinated, but in addition, expression of both organellar and nuclear genes must be coordinated as well. This topic is addressed by Roell and Gruissem (Chapter 31-organellar gene expression) and by Gray (Chapter 33-nuclear gene expression).

The number of available sequences of chloroplast DNAs is now very large. These data have provided a basis for comparison of the sequences of individual genes across species lines as a way to identify regions of sequence similarity as well as regions where sequences differ. These data are now in routine use in the design of hypothetical structures for thylakoid membrane proteins, and have, in many instances, pointed to conserved amino acids whose function can be probed by site-directed mutagenesis. This is currently a very active research endeavor that started in cyanobacteria (the reader is referred to the volume edited by D. A. Bryant (1994) in this series for further information on cyanobacterial mutagenesis experiments). Owing to the greater complexity of eukaryotic systems, the mutagenesis approach has developed more slowly. Now, however, the ability to use biolistic techniques for chloroplast transformation has opened the door to such experiments, particularly in the unicellular green alga *Chlamydomonas*. Erickson (Chapter 32) reviews the current status of this approach, as well as its future prospects.

A strong impetus for sequencing the chloroplast genome has come from the utility of these data in research related to questions surrounding the evolution of oxygenic photosynthesis. This process is now estimated to have first appeared about 3.5×10^9 years ago (Awramik, 1992). Given its extremely long existence, it is remarkable to discover that comparisons of gene sequences among both eukaryotic and prokaryotic species reveal, for many key photosynthetic proteins, a very high degree of conservation in their respective gene sequences. Results of this approach to analyses of chloroplast evolution is discussed by Reith in Chapter 34, concluding what we hope is a comprehensive coverage of the major components and processes that comprise the most important set of biological reactions on earth.

Table 3. Coding Sites For Polypeptides of PS II

M.W.[1] (kD)	Function	Coding site
47 & 43	Reaction center chlorophyll *a* binding proteins	C
33	Stabilizes Mn binding	N
34 (D2)	Reaction center polypeptides	C
32 (D1)		C
28	Antenna pigment binding protein	N
25	CAB protein	N
23	Ca^{2+} retention	N
17	Cl^- retention	N
9	Cyt *b*-559 heme binding	C
4.5	Cyt *b*-559 heme binding	C

[1] Values determined by SDS-Polyacrylamide gel electrophoresis
[2] C: chloroplast DNA; N: nuclear DNA

Acknowledgments

The authors are deeply grateful to Iris Heichel for her extensive, highly professional assistance in editing the manuscripts in this book. We also gratefully acknowledge the United States Department of Agriculture National Research Initiative Competitive Grants Program and the National Science Foundation for their support of our research programs.

References

Abrahams JP, Leslie AGW, Lutter R, and Walker JE (1994) Structure at 2.8 Å resolution of F_1-ATPase from bovine heart mitochondria. Nature 370: 621–628

Allred DR and Staehelin LA (1986) Spatial organization of the cytochrome b_6-f complex within chloroplast thylakoid membranes. Biochim Biophys Acta 849: 94–103

Andreasson LE and Vanngard T (1988) Electron transport in Photosystems I and II. Ann Rev Plant Physiol Plant Mol Biol 39: 379–411

Avron M (1978) Energy transduction in photosynthesis. FEBS Lett 96: 225–232

Awramik SM (1992) The oldest records of photosynthesis. Photosyn Res 33: 75–79.

Berthold DA, Babcock GT and Yocum CF (1981) A highly resolved, oxygen-evolving Photosystem II preparation from spinach thylakoid membranes. EPR and electron-transport properties. FEBS Lett 134: 231–234

Boyer PD (1989) A perspective of the binding change mechanism for ATP synthesis. FASEB J 3: 2164–2178

Bryant DA (ed) (1994) The Molecular Biology of Cyanobacteria. 916 pp. Kluwer Academic Publishers, Dordrecht

Colman PM, Freeman HC, Guss JM, Murata M, Norris VA, Ramshaw JAM and Venkatappa MP (1978) X-ray crystal structure analysis of plastocyanin at 2.7 Å resolution. Nature 272: 319–324

Crofts, AR and Wraight C (1983) The electrochemical domain of photosynthesis. Biochim Biophys Acta 726: 149–183

Debus RJ, Barry BA, Sithole I, Babcock GT and McIntosh L (1988) Directed mutagenesis indicates that the donor to $P680^+$ in Photosystem II is tyrosine-161 of the D1 polypeptide. Biochemistry 27: 9071–9074

Hosler JP and Yocum CF (1985) Evidence for two cyclic photophosphorylation reactions concurrent with ferredoxin-catalyzed non-cyclic electron transport. Biochim Biophys Acta 808: 21–31

Ikemizu S, Bando M, Sato T, Morimoto Y, Tsukihara T and Fukuyama K (1994) Structure of [2Fe-2S] ferredoxin I from *Equisetum arvense* at 1.8 Å resolution. Acta Cryst D50: 167–174

Ikeuchi M, Yuasa M and Inoue Y (1985) Simple and discrete isolation of an O_2-evolving PS II reaction center complex retaining Mn and the extrinsic 33 kDa protein. FEBS Lett 185: 316–322

Karplus PA and Bruns CM (1994) Structure-function relations for ferredoxin reductase. J Bioenerg Biomemb 26: 89–99

Keegstra K, Olsen LJ and Theg S (1989) Chloroplastic precursors and their transport across the envelope membranes. Ann Rev Plant Physiol Plant Mol Biol 40: 471–501

Krauss N, Hinrichs W, Witt I, Fromme P, Pritzkow W, Dauter Z, Betzel C, Wilson KS, Witt HT and Saenger W (1993) Three-dimensional structure of system I of photosynthesis at 6 Å resolution. Nature 361: 326–331

Kühlbrandt W, Wang DN and Fujiyoshi Y (1994) Atomic model of plant light-harvesting complex. Nature 350: 130–134

Martinez SE, Huang D, Szczepaniak A, Cramer WA and Smith JL (1994) Crystal structure of chloroplast cytochrome *f* reveals a novel cytochrome fold and unexpected heme ligation. Structure 2: 95–105

Michel H and Deisenhofer J (1988) Relevance of the photosynthetic reaction center from purple bacteria to the structure of PS II. Biochemistry 27: 1–7

Ort DR (1986) Energy transduction in oxygenic photosynthesis: An overview of structure and mechanism. In: Staehelin LA and Arntzen C (eds) Photosynthesis III: Photosynthetic Membranes and Light Harvesting Systems, pp 143–196. Springer Verlag, New York

Ort DR and Dilley R (1976) Photophosphorylation as a function of illumination time I. Effects of permeant cations and permeant anions. Biochim Biophys Acta 449: 95–107

Rutherford AW (1989) Photosystem II, the oxygen-evolving enzyme. Trends Biochem Sci 14: 227–232

Saha S, Ouitrakul R, Izawa S and Good NE (1971) Electron transport and photophosphorylation in chloroplasts as a function of the electron acceptor. J Biol Chem 246: 3204–3209

van Grondelle R, Dekker JP, Gillbro T and Sundstrom V (1994) Energy transfer and trapping in photosynthesis. Biochim Biophys Acta 1187: 1–66

Vermaas WFJ (1993) Molecular-biological approaches to analyze Photosystem II. Ann Rev Plant Physiol and Plant Mol Biol 44: 457–482

Webb MS and Green BR (1991) Biochemical and biophysical properties of thylakoid acyl lipids. Biochim Biophys Acta 1060: 133–158

Whitmarsh J (1986) Mobile electron carriers in thylakoids. In: Staehelin LA and Arntzen C (eds) Photosynthesis III: Photosynthetic Membranes and Light Harvesting Systems, pp 508–527. Springer Verlag, New York

Chapter 2

Structure, Composition, Functional Organization and Dynamic Properties of Thylakoid Membranes

L. Andrew Staehelin and Georg W. M. van der Staay
Department of Molecular, Cellular and Developmental Biology, University of Colorado at Boulder, Boulder, CO 80309-0347, USA

Summary

Chloroplasts are semi-autonomous organelles comprised of two envelope membranes, an aqueous matrix known as stroma, and internal membranes called thylakoids. All of the light-harvesting and energy-transducing functions are located in the thylakoids, which form a physically continuous membrane system that encloses an aqueous compartment, the thylakoid lumen. With few exceptions thylakoids are differentiated into stacked grana and non-stacked stroma membrane regions. A model of the three-dimensional relationship between grana and stroma thylakoids is presented. The membrane continuum is formed by a lipid bilayer that contains unique types of lipids. The principal functions of thylakoids are the trapping of light energy and the transduction of this energy into the chemical energy forms, ATP and NADPH. During this process, water is oxidized and oxygen is released. These functions are performed by five large protein complexes: Photosystem I with bound antennae, Photosystem II with bound antennae, light-harvesting complex II, cytochrome b_6f, and ATP synthase. The roles of these complexes in photosynthetic electron transport and ATP synthesis are discussed. The differentiation of thylakoids into grana and stroma membrane regions is a morphological reflection of an underlying non-random distribution of the five complexes between the two types of membrane domains. The most prominent effect of

Donald R. Ort and Charles F. Yocum (eds): Oxygenic Photosynthesis: The Light Reactions, pp. 11–30.

membrane stacking is the physical segregation of most Photosystem II to stacked grana membranes, and of most Photosystem I to unstacked stroma membranes. The evolutionary roots and the functional implications of this non-random organization of thylakoid membrane components are discussed in some detail. The final section of this chapter describes how thylakoid membranes adapt to long-term and short-term changes in the light-environment. Long-term light changes cause alterations in the ratios of the different types of protein complexes in turn to optimize the use of available light energy. In contrast, short-term light changes modulate the organization of membrane components and serve primarily to protect the photosystems and only secondarily to optimize the turnover of the electron transport chain.

I. Structural Organization and Compartmentalization of Chloroplasts

The chloroplasts of higher plants are bounded by two envelope membranes that surround an aqueous matrix, the stroma, and the internal photosynthetic membranes, the thylakoids (Fig. 1). The principal function of the *envelope membranes* is to control the movement of metabolites, lipids and proteins into and out of chloroplasts. In addition, they serve as assembly/processing sites for membrane lipids, pigment molecules and proteins. Most of these transport and synthetic activities have been localized to the inner envelope membrane; the outer membrane appears to serve primarily as a physical barrier to large molecules between the cytoplasm and the inner envelope membrane (Douce and Joyard, 1979). Multi-subunit complexes that form bridges across both envelope membranes facilitate the transport of newly synthesized nuclear encoded proteins from the cytoplasm into the interior of the chloroplasts (de Boer and Weisbeek, 1993).

All light harvesting and energy-transducing functions of chloroplasts have been localized to the *thylakoid membranes*. Within each chloroplast, the thylakoids form a physically continuous, three-dimensional network that encloses a single, anastomosing chamber, the thylakoid lumen. The thylakoid network is comprised of two distinct types of membrane domains, the cylindrical stacks of appressed thylakoids, known as grana, and the interconnecting, single membranes called stroma thylakoids (Fig. 2). The end and side membranes of the grana stacks are also exposed to the stroma. A typical mature chloroplast may contain 40 to 60 grana stacks with diameters of 0.3 to 0.6 μm. The spatial relationship between grana and stroma thylakoids was initially determined by means of reconstructions of serially sectioned chloroplasts (e.g. Paolillo, 1970). More recently, freeze-fracture electron micrographs have provided direct support for the central ideas developed by the earlier studies. Figure 3 depicts a three-dimensional model of the spatial organization of a thylakoid network in which two stacks of grana thylakoids are intersected at an angle by parallel, evenly spaced stroma thylakoids that spiral up and around the stacks. Where each stroma thylakoid intersects with a grana thylakoid, the two types of thylakoids are interconnected through a necklike membrane bridge. Such bridges are clearly illustrated in Figs. 4 and 5. The number of stroma thylakoids that spiral up and around a given thylakoid is variable, with values of 2 to 12 reported in the literature. In Fig. 5, eight stroma thylakoids (arrowheads) can be seen to be attached to both of the centrally located stacks (PF_s, EF_s membrane domains) in a windmill type of arrangement. An explanation of how the fracture faces arise and how they are labeled is shown in Fig. 6. The formation of grana stacks during chloroplast development is discussed by Mustardy in Chapter 5.

The *chloroplast stroma* is defined as the aqueous compartment located between the inner envelope membrane and the thylakoids (Fig. 1). The main components of the stroma include: a) multiple (~300) copies of chloroplast DNA (Boffey and Leech, 1982), 70 S ribosomes, mRNAs and all other elements needed for protein synthesis (Jagendorf and Michales, 1990); b) carbon reduction cycle enzymes, particularly ribulose bisphosphate carboxylase-oxygenase (Süss et al., 1993); and c) enzymes

Abbreviations: Chl – chlorophyll; Cyt – cytochrome; DGDG – digalactosyldiacylglycerol; EF_s – exoplasmic fracture face of stacked membranes; EF_u – exoplasmic fracture faces of unstacked membranes; LHC – light-harvesting complex; MGDG – monogalactosyldiacylglycerol; P680 – special pair of chlorophylls in the reaction center of Photosystem II; P700 – special pair of chlorophylls in the reaction center of Photosystem I; PC – phosphatidylcholine; PF_s – protoplasmic fracture face of stacked membranes; PF_u – protoplasmic fracture face of unstacked membranes; PG – phosphatidylglycerol; PQ – plastoquinone; PS I – Photosystem I; PS II – Photosystem II; RC – reaction center; SQDG – sulfoquinovosyldiacylglycerol

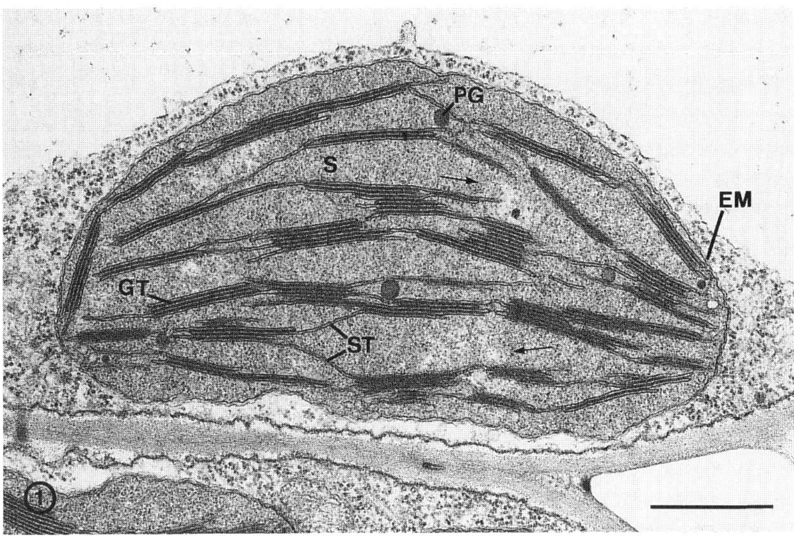

Fig. 1. Thin section electron micrograph of a tobacco chloroplast. Two envelope membranes (EM) delineate the chloroplast stroma (S), within which stacked grana thylakoids (GT) and unstacked stroma thylakoids (ST) can be recognized. Plastoglobuli (PG) and DNA containing regions (arrows) are also seen. Bar: 1 μm.

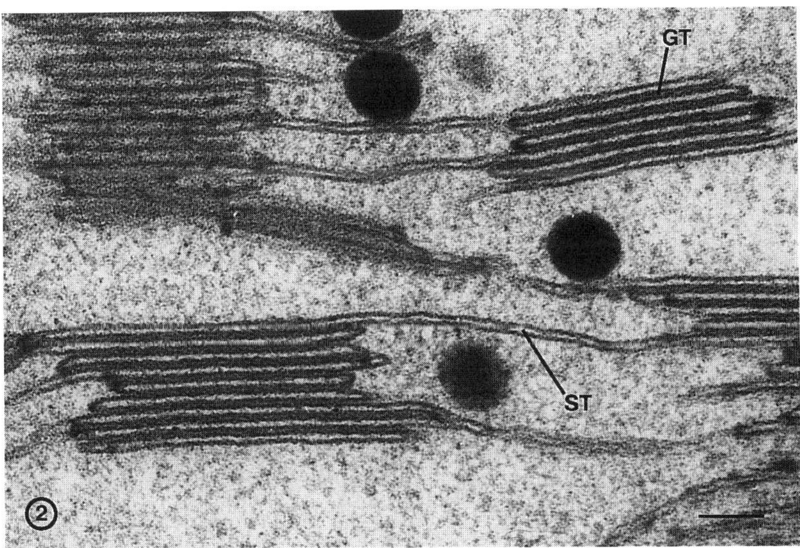

Fig. 2. Thin section through a portion of a spinach chloroplast illustrating interconnected stacked (grana; GT) and unstacked (stroma; ST) thylakoid membranes. Note the flattened, sac-like structure of the individual thylakoids. Bar: 0.1 μm.

involved in lipid, porphyrin, terpenoid, quinoid and aromatic compound synthesis (Somerville and Browse, 1991; Senge, 1993). Often starch granules and small (~50 nm), round and lipidic globuli, known as plastoglobuli are also seen. Plastoglobuli contain large amounts of carotenoids and plastoquinone and usually accumulate when development is arrested and during leaf senescence (Lichtenthaler, 1968; Dahlin and Ryberg, 1986).

II. Composition and Distribution of Acyl Lipids in Thylakoid Membranes

The acyl lipid composition of the thylakoid bilayer differs significantly from that of other plant membranes, a reflection of the prokaryotic origin of chloroplasts (Chapter 3). Not only are the two principal lipid types, monogalactosyldiacylglycerol (MGDG) and digalactosyldiacylglycerol (DGDG)

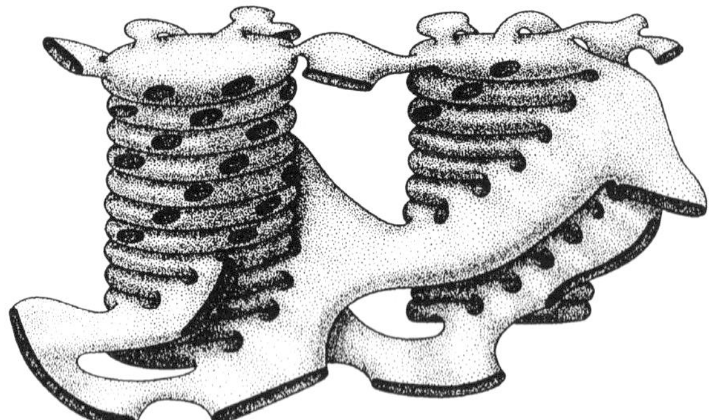

Fig. 3. Three-dimensional model of the spatial relationship between grana and stroma thylakoids. The drawing shows two grana stacks interconnected by parallel sets of unstacked stroma thylakoids that spiral up and around the stacks in a right-handed helical conformation. Where the stroma thylakoids intersect with the grana thylakoids, they are interconnected through necklike membrane bridges.

unique to thylakoids of oxygen-evolving photosynthetic organisms (Douce and Joyard, 1979; Gounaris and Barber, 1983; Webb and Green, 1991), but they also contain exceptionally large amounts of polyunsaturated fatty acids (Table 1). Sulfoquinovosyldiacylglycerol (SQDG) too is an acyl lipid that appears to be unique to photosynthetic membranes. Phosphatidylglycerol (PG) has been found in many types of lipid bilayer membranes, but the chloroplast versions of PG are special in that they contain a C16:1 *trans* (16:1^{t}) fatty acid tail (Table 1). Whether phosphatidylcholine (PC) is a genuine thylakoid lipid or a contaminant of other membranes has not been resolved unequivocally (Webb and Green, 1991).

The location of acyl lipids in thylakoid membranes can be defined both in terms of their distribution between inner and outer membrane leaflets and their distribution between grana and stroma thylakoids. Unlike the transmembrane organization of proteins, which is absolute for a given type of protein, the distribution of lipids is relative, and technically difficult to measure. This is because the biochemical methods used, e.g. degradation by specific lipases, usually perturb the bilayer structures that are being studied (Siegenthaler and Rawyler, 1986). There seems to be a general consensus that about 60% of the galactolipids (MGDG and DGDG) are associated with the outer leaflet and 40% with the inner leaflet of the thylakoid bilayer membrane. In contrast, about 70% of PG has been localized to the outer, and 95% of SQDG to the inner membrane leaflet (Sundby and Larsson, 1985; Sprague, 1987).

Only physical fractionation techniques can be employed in studies of the lateral distribution of lipids between grana and stroma thylakoid regions,

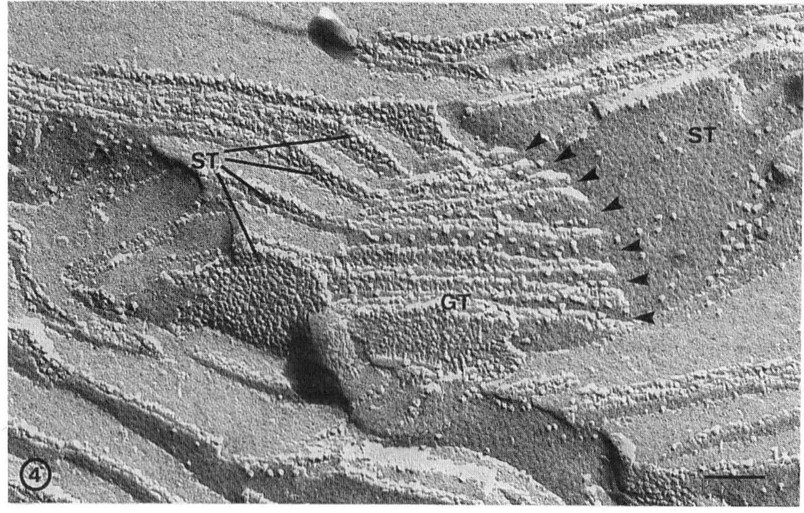

Fig. 4. Freeze-fracture electron micrograph of a single granum and associated stroma thylakoids from a pea chloroplast. Note the angle between the grana membranes (GT) and the stroma thylakoids (ST) on the right, which allows the stroma thylakoid to spiral up and around the granum (cf. Fig. 3). Where the stroma thylakoid crosses a grana thylakoid, the two membrane systems are continuous (arrowheads). On the left of the granum, five additional, evenly spaced stroma thylakoids (ST) are seen. Each one of these makes an angle with the grana thylakoids and is connected to the grana membranes as shown on the right. Bar 0.1 μm.

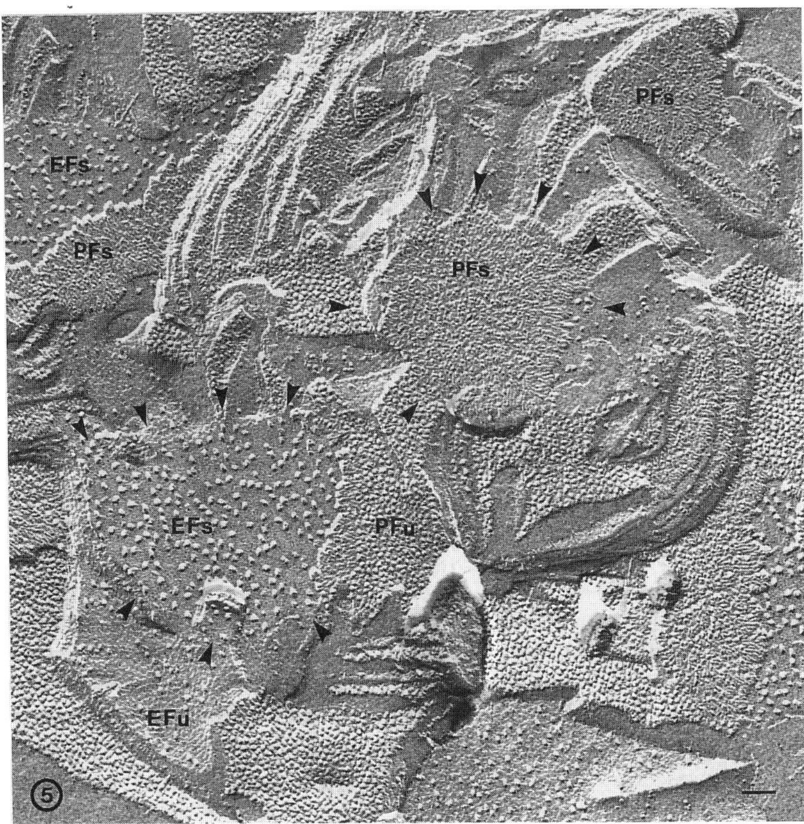

Fig. 5. Freeze-fractured pea thylakoids illustrating the relationship between grana and stroma thylakoids as seen in a face-on view. The *circularly arranged arrowheads* point to crossover/attachment sites between two grana (G) stacks and surrounding stroma thylakoids (ST). The angled stroma thylakoids resemble the blades of a conventional windmill, with the central grana membrane corresponding to the hub (cf. the top view of the model depicted in Fig. 3). The fracture faces EF_s and PF_s belong to stacked regions, the faces EF_u and PF_u to unstacked ones. Note the physical continuity of the stacked and unstacked regions. Bar 0.1 μm.

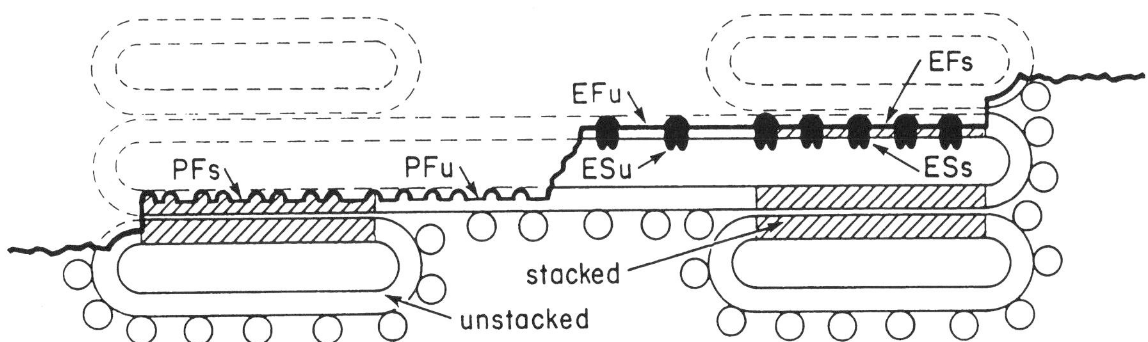

Fig. 6. Diagram illustrating how the faces EF_s, EF_u, PF_s and PF_u arise during freeze-fracture of thylakoid membranes. In addition, the true surfaces ES_s and ES_u , which are exposed by deep-etching, are shown. Compare with Figs. 4,5 and 8.

both because of the high degree of lateral mobility of thylakoid lipids, and because of the rapid, uncontrollable transfer of membrane lipids into detergent micelles. Murphy and Woodrow (1983) have estimated that the acyl lipids occupy about 14% of the surface area of appressed grana membranes and about 40% of the stroma membranes. The average ratio of MGDG to DGDG in grana thylakoids is ~2.5, and in stroma thylakoids 1.2.

Little definitive information is available on the

Table 1. The diacylglycerol lipids of thylakoid membranes

Name	% of total lipids[1]	Formal charge at pH 7	Preferential location(leaflet)	Major acyl lipid chains[2]
MGDG	46	0	outer	18:3/18:3 (92%); 18:3/16:0 (4%)
DGDG	29	0	outer	18:3/18:3 (70%); 18:3/16:0 (24%)
SQDG	10	(–)	inner	18:3/16:0 (62%); 18:2/16:0 (33%)
PG	11	(–)	outer	18:0/16:1^t (45%); 16:2/16:1^t (37%)
PC	4	(+ –)	?	18:3/16:0 (51%); 18:2/16:0 (34%)

The inner and outer leaflet refer to the luminal and stromal leaflets of the lipid bilayer.
MGDG, monogalactosyldiacylglycerol; DGDG, digalactosyldiacylglycerol; SQDG, sulfoquinovosyldiacylglycerol, PG, phosphatidylglycerol; PC, phosphatidylcholine.
[1] average values from measurements of spinach, tobacco and pea thylakoids
[2] average values from measurements of spinach, maize, oleander and soybean thylakoids
The numbers refer to the length and the number of double bonds of the two fatty acid chains associated with each lipid type (for example, *18:3* denotes a fatty acid with *18* carbon atoms and *3* double bonds). With the exception of trans D^3-hexadecenoic acid (16:1^t), all fatty acids in chloroplasts have the *cis* configuration at the double bonds.

functional significance of the different acyl lipid species in thylakoid membranes (reviewed by Webb and Green, 1991). The most convincing evidence for a functional role for a specific type of lipid has come from studies of LHC II. In particular, the formation of stable LHC II trimers appears to depend on the presence of PG containing C16:1^t fatty acids (Dubacq and Trémolières, 1983; Nußberger et al., 1993). The functional significance of this fact is that only trimeric LHC II (but not monomeric LHC II) is able to efficiently transfer excitation energy to PS II. This feature is exploited in cold-acclimated rye plants, where a decrease in PG 16:1^t causes a reversible uncoupling of energy transfer from LHC II to PS II, which appears to help protect PS II from photodamage under cold-stress conditions (Krupa et al., 1987). In temperature and salt-stressed thylakoids the acyl lipids can become reorganized into non-bilayer structures such as hexagonal II tubular configurations (Staehelin, 1986). In vitro studies with purified lipids have traced the origin of these configurations to MGDG (Murphy, 1986).

III. The Photosynthetic Electron Transport Chain: Components and Functional Organization

Thylakoids contain all of the functional elements needed for trapping and transducing light energy into the chemical energy forms, ATP and NADPH. The reactions are carried out by five types of membrane spanning protein complexes, associated cofactors and peripheral proteins as illustrated schematically in Fig. 7. Three of these complexes, Photosystem I (PS I/bound LHC I), Photosystem II (PS II/bound inner antennae) and the peripheral, light-harvesting complex II (LHC II), bind chlorophyll (Chl); the two others, cytochrome b_6f (Cyt b_6f) and the ATP synthase, also known as coupling factor (CF_0/CF_1), do not.

The first event depicted in Fig. 7 is the absorption of a photon by a Chl molecule in the LHC II antenna complex. Following its absorption, the photon's energy is conserved in the form of an exciton that 'migrates' by means of a resonance energy transfer mechanism among Chls to P680, the special pair of Chls that constitute the primary donor of PS II. In P680 the actual charge separation event leads to the production of a high energy electron that is then transferred into the electron transport chain. The main electron carriers in PS II include pheophytin, a bound plastoquinone (Q_A), and a dissociable plastoquinone. The lipid soluble PQ molecules (~7 per PS II complex) serve as mobile carriers between the PS II and the Cyt b_6f complexes. The Q_B-binding site serves essentially as a 'loading dock' for the sequential transfer of two electrons onto PQ, which, through the binding of two H^+, is converted to PQH_2. The lower affinity of PQH_2 for the Q_B site causes it to dissociate from PS II and enter the PQH_2 pool through which the electrons are transferred to Cyt b_6f. As shown in Fig. 7, the loading of electrons onto PQ is coupled to the binding of protons on the stromal side of the membrane and the subsequent 'unloading' of the electrons into Cyt b_6f with the discharge of the protons into the thylakoid lumen, thereby helping produce the electrochemical potential needed for ATP synthesis. A second proton-discharging site is the water-splitting (oxygen-evolving) enzyme

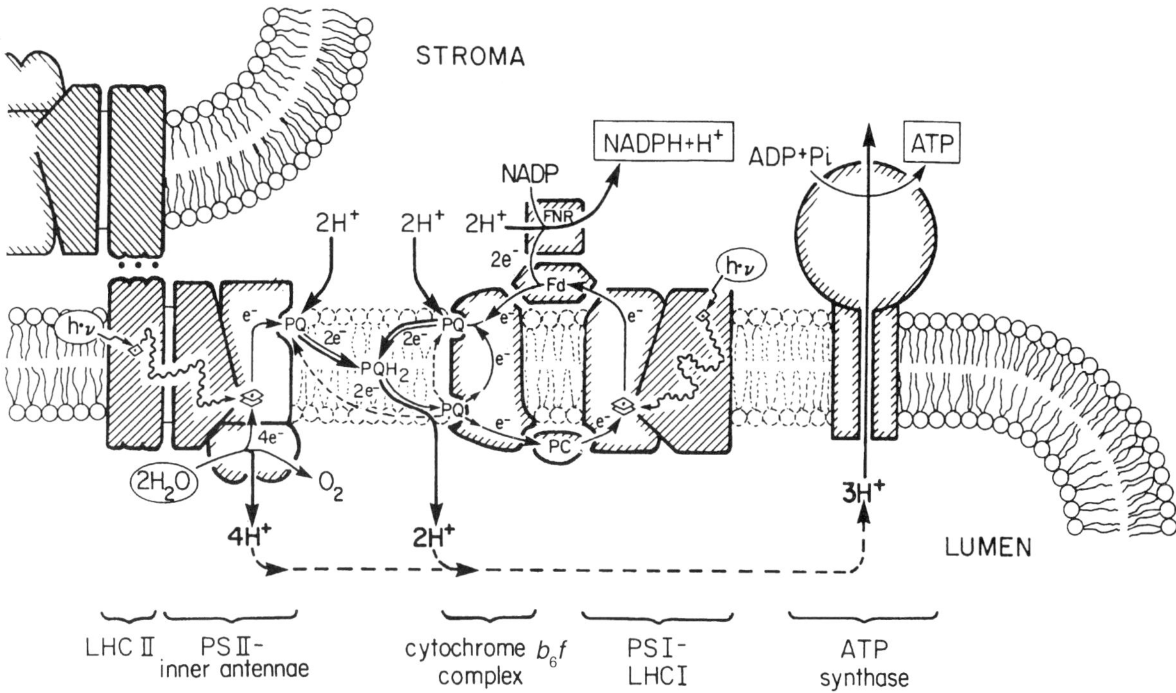

Fig. 7. Functional organization of the five protein complexes that couple photosynthetic electron transport to the transport of protons across the thylakoid membrane, the synthesis of ATP, and the reduction of NADP. The numbers indicate how many electrons or protons are involved in a given subreaction of the chain.
Thin line arrows: electron transfer reactions; *thick line arrows*: H^+-transport; *thin, dashed line arrows*: recycling of PQ; *thick dashed line arrows*: lumenal transport of H^+; *wavy arrows*: resonance energy transfer of exciton energy to reaction center chlorophylls. PQ: plastoquinone; PQH_2: reduced plastoquinone; Fd: ferredoxin; FNR: ferredoxin-NADP reductase; PC: plastocyanin.

complex associated with the lumenal side of PS II. The primary function of the oxygen-evolving complex is to provide replacement electrons for P680 to compensate for those donated to the electron transport chain.

Cyt b_6f has two PQ binding sites associated with the b_6 polypeptide (Fig. 7), the q-site that binds quinol and is located close to the lumenal surface, and the n-site that binds quinone and is closer to the stromal surface. The q-site serves as an electron 'unloading dock' for PQH_2, whereas the n-site functions as an electron 'loading dock' for PQ similar in function to the Q_B site on PS II. The first electron to be transferred from PQH_2 to Cyt b_6f in the q-site is immediately passed on to the Rieske iron-sulfur protein, Cyt *f* and then to plastocyanin, which carries the electrons to PS I. In contrast, the more energetic second electron is 'recycled' through the two Cyt b_6 hemes across the membrane to the n-site where it is used to reduce another PQ molecule. This quinone electron transport cycle, Q-cycle, increases the overall efficiency of the electron transport chain by translocating two additional protons for every two electrons that pass through the non-cyclic electron transport chain. This yields a total of six protons that are deposited into the thylakoid lumen for every two electrons that are released by splitting of one water molecule.

The energy from a second photon is needed to excite the P700 reaction center electron of PS I so that it can enter the second part of the electron transport chain. The oxidized PS I is rereduced by an electron provided by reduced plastocyanin. As the excited electron moves through PS I across the membrane to ferredoxin, it passes through a number of prosthetic groups including a Chl known as A_0, a bound phylloquinone, A_1, and the F_x and F_A/F_B iron sulfur centers. Ferredoxin serves at a very important regulatory site of the electron transport chain in that it controls the flow of electrons either back to Cyt b_6f to increase proton transport and thereby ATP synthesis (cyclic phosphorylation), or to the NADP reductase for the synthesis of reduced NADP.

The protons that accumulate in the thylakoid lumen

lead to the buildup of an electrochemical potential, whose energy is utilized by the CF_0/CF_1 complexes for the synthesis of ATP. In this scheme, the CF_0 subunit funnels the H^+-ions across the membrane to the catalytic CF_1 subunit, via the connecting stalk proteins.

A more detailed discussion of all of these events can be found in Section III of this book.

IV. Composition of the Five Complexes of the Thylakoid Membrane

Each of the five complexes shown in Fig. 7 is composed of multiple proteins and prosthetic groups. Table 2 lists the known protein components of each of the complexes together with their size, their gene name, their coding site (nuclear versus chloroplast) and their function.

PS II and PS I are both very large complexes each consisting of close to twenty different polypeptides. The Chl-binding reaction center and associated light-harvesting complex proteins all have a transmembrane disposition. In PS II, only the lumenally-located enzymes of the oxygen-evolving complex are peripheral proteins. In contrast, the peripheral proteins of the PS I complex are found on both sides of the membrane, with the iron-sulfur centers being associated with the stroma side polypeptides, and the lumenal proteins facilitating the binding of plastocyanin (Chapter 15).

The Cyt b_6f complex consists entirely of integral membrane proteins (Chapter 19). A set of transmembrane proteins also make up the CF_0 subunit of the ATP synthase, whereas the CF_1 subunit is composed entirely of peripheral proteins (Chapter 23).

Fig. 7 and Table 2 both show the inner antennae of PS II - CP29, CP26, CP24, CP22 - as part of the PS II complex. In contrast, LHC II (also called peripheral LHC II) is depicted as a separate complex. This distinction is based on results from freeze-fracture studies of thylakoid membranes of wild type and mutant plants. In particular, PS II-inner antenna and LHC II complexes have been shown to give rise to distinct and separate classes of freeze-fracture particles (Simpson, 1979; Staehelin and DeWit, 1984), and the two types of complexes can move independently of each other in the plane of the membrane (Kyle et al., 1983; Staehelin and Arntzen, 1983; Sundby and Andersson, 1985; Simpson, 1986; Dunahay and Staehelin, 1987). This contrasts with the PS I-LHC I system in which all of the LHC I polypeptides appear to be tightly bound to the PS I core (Boekema et al., 1994), similar to the inner antennae of PS II. Thus, the LHC I polypeptides are listed as part of the PS I complex in Table 2.

The PS II, PS I, Cyt b_6f and ATP synthase complexes are all composed of both chloroplast and nuclear-encoded gene products, (Table 2; also Chapters 4 and 7), which allows the nucleus to maintain control over the assembly of these complexes. In the case of the photosystems, the reaction center polypeptides (*psbA*, *psbD* and *psaA* and *psaB* gene products) are encoded in the chloroplast genome, but other critical components such as the proteins of the oxygen-evolving complex of PS II, and most of the iron-sulfur proteins of PS I are encoded in the nucleus. All of the Chl *a/b* light-harvesting proteins (CP29, CP26, CP24, CP22, LHC II, LHC I), are also coded by nuclear genes (*Lhcb*, *Lhca* genes) as discussed in greater detail in Chapter 28.

V. Lateral Distribution of Protein Complexes Between Grana and Stroma Membrane Regions

The differentiation of thylakoids into grana and stroma membrane regions is a morphological reflection of the underlying compositional and functional differences between these two types of membrane domains. These differences are readily apparent in freeze-fracture images of thylakoid membranes (Fig. 5) and have been further characterized by means of biochemical fractionation and immunolabeling techniques. A summary of typical differences in protein composition between grana and stroma thylakoids is shown in Table 3. The term 'typical' is used to highlight the fact that the indicated percentages are not absolute numbers; differences between sun and shade plant species, and between plants grown under different illumination conditions have been documented (Staehelin, 1986; Anderson et al., 1988; Simpson and von Wettstein, 1989; Melis, 1991). In addition, differences in values for some complexes have been reported depending on which biochemical or functional parameters are measured. For example, to the extent that grana and stroma PS II differ in functional properties (Melis, 1991), functional measurements of the distribution between grana and stroma thylakoids may produce somewhat different numbers than immunocytochemical studies of the distribution of specific PS II polypeptides (Melis, 1991; Olive and Vallon, 1991). Nevertheless, the reported values rarely differ by more than 10 percent, thus indicating that the compositional differences between grana and stroma thylakoids as shown in Table 3 are common to most higher plant chloroplasts studied to date.

Thylakoid stacking and grana formation are *not* needed for the functioning of the photosynthetic electron transport chain. This is most readily seen in Chl *b*-deficient chloroplast mutants that are unable to produce normal amounts of membrane stacks (Allen et al., 1988; T. G. Falbel and L. A. Staehelin, unpublished). In vitro unstacking of isolated thylakoids induced by low salt conditions has also been shown not to impair their ability to absorb light energy, carry out electron transport, synthesize ATP and reduce NADP. Instead, the evolutionary driving force for the lateral differentiation of thylakoids by means of membrane stacking apparently came from the functional need to spatially segregate PS II and PS I to achieve optimal performance (Staehelin and Arntzen, 1983; Trissl and Wilhelm, 1993). To this end, most PS II complexes and attendant light-harvesting systems are located in the appressed grana, and most PS I-LHC I complexes in the non-appressed stroma membrane regions (Table 3). The distribution of all of the other thylakoid proteins appears to follow their functional association with PS II and/or PS I (e.g. LHC II, Cyt b_6f, plastocyanin), or from morphological constraints arising from the size and shape of the complexes (e.g. ATP synthase). A consequence of this non-random organization is that regulation of the photosynthetic electron transport chain requires coordination of functionally interacting complexes that are physically separated by distances of up to 0.3 μm. This is achieved, in part, by the lateral redistribution of specific components between grana and stroma thylakoids (Table 3). The functional implications of these rearrangements are discussed in the section VIII.B of this Chapter.

In recent years a number of studies have provided suggestive evidence for the grana margins having a specific composition and serving specific functional roles (Albertsson et al., 1990; Wollenberger et al., 1994). The primary experimental support for these hypotheses has come from mechanically (French press, Yeda press) fragmented thylakoids fractionated by phase partition methods (Andreasson et al., 1988). Such studies have yielded a number of subfractions of thylakoids, that have been interpreted in terms of a concentric functional organization of subtypes of PS II complexes (PS IIα_1, PS IIα_2, PS IIα_3) and other components in grana and grana margin domains (Albertsson et al., 1990). Although it is clear that any transitional region between two different and dynamically interacting membrane domains will have transitional functional properties as thoughtfully discussed by Anderson (1989), the biochemical fractionation studies designed to characterize these domains appear to be too simplistic to yield reliable data. In particular, there is a basic assumption that the membranes do not develop a new distributional equilibrium of protein complexes following each round of traumatic mechanical shearing and fractionation. This assumption cannot be sustained in light of what is known about the dynamic and fluid nature of thylakoid membranes. Furthermore, electron microscopical and immunocytochemical staining of *intact* thylakoids and chloroplasts has failed to confirm many of the findings of these biochemical fractionation studies. In conclusion, the grana margins are transitional domains between grana and stroma thylakoids and obviously can be expected on statistical grounds to have a transitional composition of protein complexes and corresponding functional properties. However, there is no reliable evidence to support the idea that specific types of complexes preferentially partition into these membrane domains.

VI. Macromolecular Organization of Thylakoid Membranes: Correlative Freeze-fracture, Freeze-etch and Biochemical Studies

Freeze-*fracture* electron microscopy provides a means for visualizing the organization and spatial distribution of integral membrane protein complexes within the plane of bilayer membranes, and freeze-*etch* electron microscopy for studying the structural arrangements of peripheral membrane proteins (Chapman and Staehelin, 1986). As such these methods are invaluable tools for examining directly differences in composition between grana and stroma membrane regions, and for studying regulatory rearrangements of complexes in response to changes of environmental (e.g. light) parameters.

Freeze-fracturing at temperatures below –100 °C splits bilayer membranes along the central hydro-

Table 2. Protein components of the five complexes of the thylakoid membrane

Components (location)	Size (kDa)	Gene	Remarks
Photosystem II			
D1 (M)	32	*psbA* (C)	RC core, Q_B, pheo, Chl special pair
D2 (M)	34	*psbD* (C)	RC core, Q_A, pheo, Chl special pair
cyt $b_{559}\alpha$ (M)	9	*psbE* (C)	RC core, heme *b*
cyt $b_{559}\beta$ (M)	4	*psbF* (C)	RC core, heme *b*
CP47 (M)	47-51	*psbB* (C)	inner antenna, Chl *a*
CP43 (M)	43-47	psbC (C)	inner antenna, Chl *a*
PS II-H (M)	7.6	*psbH* (C)	phosphoprotein
PS II-I (M)	4.8	*psbI* (C)	RC core
PS II-J (?)	<5	*psbJ* (C)	open reading frame, no identified polypeptide
PS II-K (M)	4.3	*psbK* (C)	absent in purified oxygen-evolving core
PS II-L (M)	4.3	*psbL* (C)	in oxygen-evolving core
PS II-M (M)	3.8	*psbM* (C)	in oxygen-evolving core
PS II-N (M)	4.7	*psbN* (C)	in oxygen-evolving core
PS II-R (L)	10	*PsbR* (N)	membrane anchored; regulatory in O_2-evolution?
Oxygen-evolving complex			
PSII-O (OEE1) (L)	33	*OecA* or *PsbO* (N)	Mn^{++} stabilizing
PSII-P (OEE2) (L)	23	*OecB* or *PsbP* (N)	absent in cyanobacteria
PSII-Q (OEE3) (L)	16	*OecC* or *PsbQ* (N)	absent in cyanobacteria
inner Chl a/b antennae			
CP29 (M)	28	*Lhcb4* (N); 2 genes[1]	binds Chl *a* and *b*, lutein, neoxanthin , violaxanthin
CP26 (M)	27	*Lhcb5* (N)	binds Chl *a* and *b*, lutein, neoxanthin, violaxanthin
CP24 (M)	23	*Lhcb6* (N); 2 genes[1]	binds Chl *a* and *b*, lutein, neoxanthin (?), violaxanthin
CP22 (M)	21.7	*PsbS* (N) (*Lhcb7*?)	binds Chl *a* and *b*
LHC II			
LHC II-Type I (M)	25	*Lhcb1* (N); 8 genes[1]	LHC II binds 6-8 Chl *a* and 6 Chl *b* , 2 luteins, one
LHC II-Type II (M)	25	*Lhcb2* (N); 2 genes[1]	neoxanthin; DGDG and PG; PG is necessary for the
LHC II-Type III (M)	24	*Lhcb3* (N); 4 genes[1]	trimerization of LHC II
Cytochrome b_6f-complex			
cytochrome *f* (M)	31.3	*petA* (C)	heme *c*
cytochrome b_6 (M)	23.7	*petB* (C)	heme *b*
Rieske FeS (M)	18.8	*PetC* (N)	2Fe-2S
Subunit IV (M)	15.2	*petD* (C)	binds quinone
Subunit V (M)	4	*petE* (C)	
Photosystem I			
PS I-A [Ia] (M)	83	*psaA* (C)	P700, A_0 (Chl), A_1 (vitamin K_1), F_x [4Fe-4S], hetero-
PS I-B [Ib] (M)	82	*psaB* (C)	dimer binds about 100 Chl a, 12-15 [beta]-carotene
PS I-C [VII] (S)	8.9	*psaC* (C)	2 [4Fe-4S], terminal electron acceptors
PS I-D [II] (S)	17.9	*PsaD* (N)	ferredoxin docking
PS I-E [IV] (S)	9.7	*PsaE* (N)	ferredoxin docking?
PS I-F [III] (L?)	17.3	*PsaF* (N)	plastocyanin docking
PS I-G [V] (M)	10.8	*PsaG* (N)	found in higher plants only
PS I-H [VI] (S)	10.2	*PsaH* (N)	found in higher plants only
PS I-I [X] (M)	4.6	*psaI* (C)	
PS I-J [IX] (M)	3.3	*psaJ* (C)	
PS I-K [VIII] (M)	5.6	*PsaK* (N)	
PS I-L [V'] (M)	15.4	*PsaL* (N)	
PS I-M (M)	3.5	*psaM*	found in cyanobacteria only
PS I-N (M)	4.8	*psaN*	found in cyanobacteria only
PS I-O (M)	9.0	*PsaO* (N)	found in higher plants only
Ferredoxin (FD) (S)		*PetG* (N)	Fe-S center
FNR (S)	35.4	PetH (N)	ferredoxin-NADP reductase
FNR binding protein (S)	17.5	PetI (N)	
LHC I-Type I (M)	22	*Lhca1* (N); 2 genes[1]	LHC I proteins bind Chl *a* and *b* and some
LHC I-Type II (M)	23	*Lhca2* (N); 1 gene[1]	carotenoids
LHC I-Type III (M)	25	*Lhca3* (N); 1 gene[1]	
LHC I-Type IV (M)	22	*Lhca4* (N); 2 genes[1]	
Plastocyanin (L)	10.2	PetF (N)	mobile electron carrier, binds Cu

Table 2. Continued

Components (location)	Size (kDa)	Gene	Remarks
ATP synthase (Coupling Factor)			
CF_1 α (S)	55.4	*atpA* (C)	3 molecules per complex
CF_1 β (S)	53.9	*atpB* (C)	3 molecules per complex
CF_1 γ(S)	37[2]	*AtpC* (N)	1 molecule per complex
CF_1 δ (S)	27.7	*AtpD* (N)	1 molecule per complex
CF_1 ε (S)	14.7	*atpE* (C)	1 molecule per complex
CF_0 I (M)	20.9	*atpF* (C)	
CF_0 II (M)	14[2]	*AtpG* (N)	
CF_0 III (M)	8	*atpH* (C)	
CF_0 IV (M)	27	*atpI* (C)	

The most commonly used name for each component is given. Light-harvesting pigment-binding complexes/proteins are shown in bold letters. The location denotes whether this protein is membrane-spanning (M) or associated with the membrane at the stromal (S) or lumenal (L) surface. The size of each protein is based on its amino acid sequence; for different species, the size of the components can vary. The apparent molecular weight of proteins on denaturing gels can significantly deviate from the one deduced from the amino acid sequence. The name of the genes and their location, chloroplast (C) or nuclear (N) genome, is given. By convention, nuclear encoded genes are written with a capital letter. Functions or special characteristics of proteins are indicated.
Data were compiled from: Hennig and Herrmann, 1986; Nalin and Nelson, 1987; Simpson and von Wettstein, 1989; Almog et al., 1992; Bryant, 1992; Erickson and Rochaix, 1992; Green et al., 1992; Jansson et al., 1992; Bassi et al., 1993; Hope, 1993; Nußberger et al., 1993; Thornber et al., 1993; Vermaas, 1993; Funk et al., 1994; Jansson, 1994.
[1] in tomato; [2] molecular weight based on migration on gel

Table 3. Distribution of chloroplast membrane components between grana and stroma thylakoids under state 1 and state 2 conditions

	State 1		State 2	
Component	Grana	Stroma	Grana	Stroma
PS II	85	15	85	15
PS I	10	90	10	90
Cyt b_6f	50	50	30	70
LHC II	90	10	70	30
plastocyanin	40	60	55	45
ATP synthase	0	100	0	100

The numbers indicate the percentage of the component found associated with the grana and stroma membranes in state 1 and state 2 (LHC II phosphorylation).

phobic plane of the lipid bilayer continuum to produce two complementary fracture faces, designated EF and PF for Endoplasmic and Protoplasmic Fracture faces (Fig. 6). These fracture faces can be visualized by means of thin platinum/carbon replicas of the frozen surfaces. Since the integral protein complexes that span the bilayer are not split during the fracturing process, they are seen as 'particles' that rise above the smooth fracture face of the bilayer (Figs. 4–6). Each type of complex gives rise to a freeze-fracture particle that can be characterized in terms of its size, its particular behavior between the EF and PF faces, its lateral distribution between stacked (EF_s, PF_s) and unstacked (EF_u, PF_u) membrane regions, and its topological features on freeze-etch membrane Surfaces (e.g. ES_s, ES_u; Fig. 6).The differences in appearance of the EF_s, EF_u, PF_s and PF_u fracture faces shown in Fig. 5 reflect the different sets of protein complexes associated with the grana and stroma membrane regions. Since the EF and PF images constitute essentially two artificially produced views of what was originally one membrane, one has to sum the particles of two complementary fracture faces to obtain information on the full complement of protein complexes in a given type of membrane domain (EF_s and PF_s for grana thylakoids, EF_u and PF_u for stroma thylakoids). In contrast, the true surface view images (ES and PS, Fig. 8) allow investigators to see the actual complexes associated with the thylakoid surfaces. For example, Fig. 8 illustrates the organization of the oxygen-evolving complexes on the lumenal membrane surface.

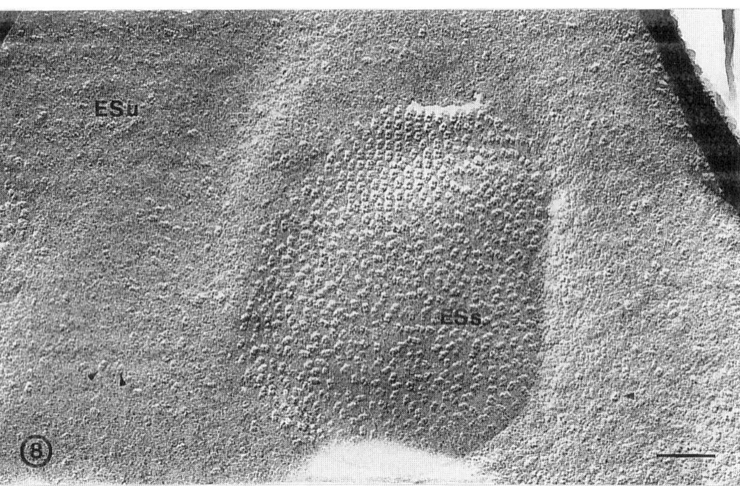

Fig. 8. Lumenal surface of a spinach thylakoid as seen by freeze-etch electron microscopy. The stacked membrane region (ES_s) in the center of the micrograph can be clearly distinguished from the surrounding unstacked (ES_u) due to the high density of large, tetrameric/multimeric particles. The ES_u areas contain relatively few such particles (*arrowheads*). Each tetrameric/multimeric particle corresponds to the oxygen-evolving enzyme complexes associated with the dimeric PS II complexes of grana thylakoids. Bar: 0.1 μm.

One of the primary concerns of researchers involved in freeze-fracture or freeze-etch electron microscope studies of thylakoid membranes has been the correlation of specific types of fracture face and membrane surface particles with the five functional complexes that have been identified (Staehelin and DeWit, 1984). The approaches used to determine the identity of freeze-fracture particles have included quantitative/correlative structural and biochemical studies of thylakoid membranes of plants grown under different illumination conditions (Armond et al., 1977; Staehelin et al., 1977), of many types of chloroplast mutants (Simpson, 1986; Olive and Vallon, 1991), and of purified and reconstituted protein complexes (McDonnel and Staehelin, 1980; Mörschel and Staehelin, 1983; Dunahay and Staehelin, 1985). In contrast, most investigations designed to identify specific surface complexes of thylakoid membranes have involved controlled buffer washes to selectively remove specific proteins, biochemical monitoring of the removed proteins, and re-attachment of the isolated proteins to the thylakoid surfaces (Miller and Staehelin, 1976, for CF_1; Seibert et al., 1987, for the oxygen-evolving complex). Table 4 and Fig. 9 summarize information concerning the dimensions of the functional complexes of thylakoid membranes and what types of freeze-fracture particles are produced by the different complexes.

A. Photosystem II Complexes

Only one type of protein complex appears to partition with the EF_s and EF_u faces, namely the PS II with its tightly bound inner antennae (CP29, CP26, CP24, CP22). However, the EF_s particles range in size from 8–16 nm and the EF_u particles from 8 to 11 nm (Staehelin et al., 1977). Based on many studies it now appears that fully functional, 'complete' PS II complexes give rise to the ~16 nm EF_s particles, and functional PS II particles that lack one or several of the inner antennae to 8–14 nm EF_s particles (Armond et al., 1977; Marquardt and Bassi, 1993). As discussed in greater detail elsewhere (Simpson, 1986; Olive and Vallon, 1991), the identity and composition of the EF_u particles is somewhat less certain. Based on their lumenal surface morphology, and on biochemical fractionation and immunocytochemical labeling studies, it appears that most EF_u particles also correspond to PS II complexes, but what causes them to associate with stroma rather than grana membranes is still a matter of debate. The much studied D1-protein repair cycle of PS II (Barbato et al., 1992; Aro et al., 1993) involves transfer of the damaged PS II complexes from a grana to a stroma membrane region, partial disassembly of the complex, replacement of the damaged D1 protein, reassembly, and finally return of the functional complex to a grana region. Based on this scheme and the increase in number of EF_u particles seen under conditions

Table 4. Size of complexes (in nm) of the thylakoid membrane

Complex	Computer averaged image of isolated complexes	Freeze-fracture images of: Particles in thylakoid membranes (location)	Freeze-fracture images of: Complexes reconstituted into liposomes
PS II monomer	12.3 × 7.5 (cyanobacteria)	10	10[1]
	11.9 × 7.2 (spinach)	6–12 (EF_u)	7.5[2]
PS II dimer	15.5 × 12 (cyanobacterium)	10 × 20 (EF)[1]	10 × 20[1]
	16.3 × 10 (spinach)	13–20 (EF_s)	n.d.
PS I monomer[3]	11 × 7	10–13 (PF_u)	10
PS I trimer[4]	20 × 20	?	n.d.
PS I-LHC I	16 × 12	10–16 (PF_u)	13
LHC II-trimer	7.3[5]	8 (PF_s; PF_u)	8
cyt b_6f-monomer	8.3 × 4.4	n.d.	n.d.
cyt b_6f-dimer	8.3 × 8.8	8–10 (PF_s; PF_u)	8.5
CF_0/CF_1	6.5 × 6.5 (CF_0)[6]	9–10 (PF_u)	9–10[7]

The sizes were measured after computer averaged imaging of isolated complexes, or after freeze-fracturing of whole thylakoid membranes or isolated complexes reconstituted into liposomes. Unless otherwise stated, the sizes of averaged images after correction for the detergent shell are from Boekema et al., 1994; the sizes of freeze-fracture particles are from Simpson, 1986; Staehelin, 1986; Olive and Vallon, 1991. Other data are from: [1] Mörschel and Schatz, 1987; [2] Sprague et al., 1985; [3] in cyanobacteria and chloroplasts from plants grown in intermittent light; [4] isolated from cyanobacteria, it is unclear whether PS I trimers exist in vivo; [5] Kühlbrandt et al., 1994; [6] Boekema et al., 1988; [7] Mörschel and Staehelin, 1983. n.d.: not determined.

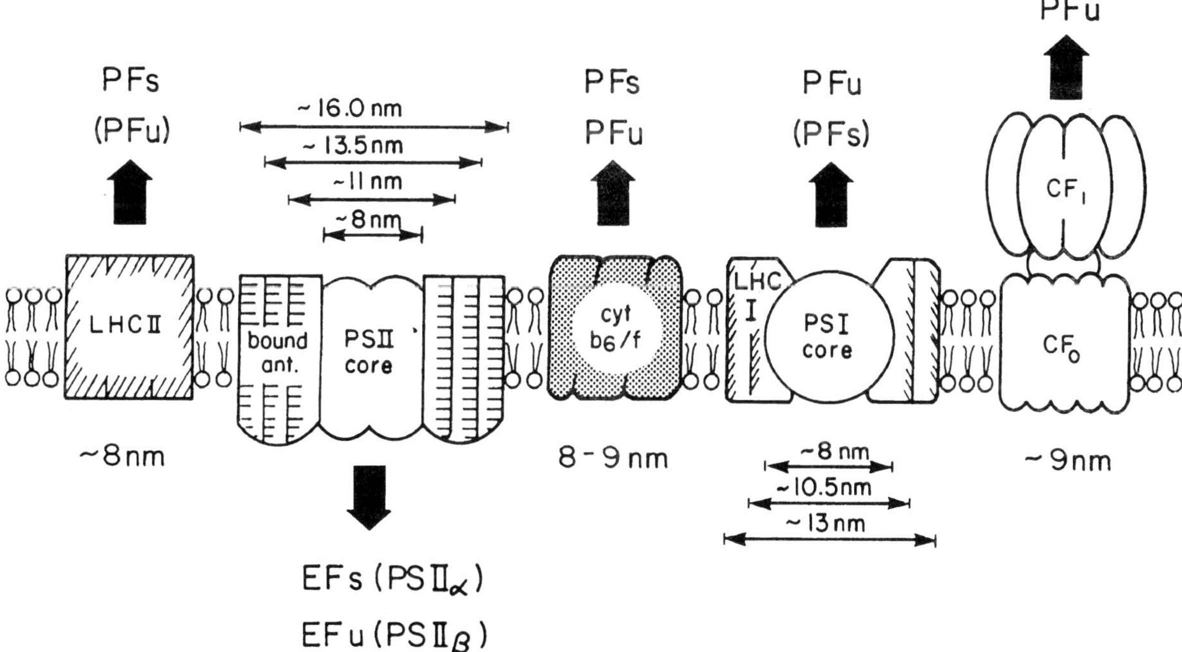

Fig. 9. Schematic diagram relating the five functional complexes of thylakoid membranes to different categories of freeze-fracture particles.

where the assembly of PS II complexes would be expected to be perturbed (e.g. Mn-deficient growth conditions; Simpson and Robinson, 1984), we postulate that most EF_u particles are structural equivalents of partially assembled and photochemically incompetent (e.g. PS IIβ/Q_B non-reducing; Melis, 1991) PS II complexes. What keeps incomplete PS II complexes in stroma thylakoids and allows fully functional ones to enter the stacked grana thylakoid regions is unknown, but may be related to the palmitoylation of D1 that seems to coincide with its migration from stroma to grana membranes (Mattoo and Edelmann, 1987).

Another important difference between grana and stroma membrane PS II complexes is that grana PS II complexes appear to be dimeric complexes and stroma PS II monomeric complexes. This was first demonstrated in a freeze-etch study that correlated the 'tetrameric' particle complexes seen on the lumenal ES_s surface of grana membranes (Fig. 8) with the enzymes of the oxygen-evolving complex (Seibert et al., 1987). After removal of the 33, 23 and 16 kDa oxygen-evolving complex proteins, the underlying PS II core complex was shown to be a dimer. Typically, the grana thylakoid regions contain between 1500 and 2000 dimeric PS II complexes per μm^2 based on freeze-fracture studies (Staehelin, 1986). When a D1 protein is damaged, the PS II dimers appear to fall apart, thereby allowing only the damaged PS II complex to transfer to the stroma membrane regions (Barbato et al., 1992). PS II dimer structures are also seen in sheets of purified and crystallized PS II complexes of spinach (Lyon et al., 1993; M. K. Lyon, personal communication). These two dimensional crystals are now being used to determine the arrangement of the subunits within the PS II complex.

B. Photosystem I Complexes

All PS I complexes partition with the PF faces during freeze fracturing. Two types of studies have produced correlative evidence linking 8–13 nm PF_u particles with PS I complexes; studies with PS I mutants (Simpson, 1982; 1986) and PS I reconstitution studies (Dunahay and Staehelin, 1985). The barley mutant viridis-n^{34} contains only 40% of the P700 Chl-protein (CP I), and viridis-zb^{63} has no detectable levels of this Chl-protein. Comparisons of freeze-fracture particles of the thylakoids of these mutants with those of wild type barley thylakoids demonstrated a loss of larger (10–13 nm) PF_u particles proportional to the loss of CP I. In the study in which purified PS I complexes were reconstituted into liposomes, the resulting particles also measured 10–13 nm.

In intermittent light plastids, which exhibit normal PS I activity but possess greatly reduced amounts of LHC I antennae (Marquardt and Bassi, 1993), only PF freeze fracture particles smaller than 10 nm are seen (Armond et al., 1977). Transfer of the intermittent light seedlings to continuous illumination, which triggers the accumulation of LHC I proteins, also leads to the formation of larger PF_u particles. This suggests that the larger 10–13 nm PF_u particles contain a 7–9 nm PS I core surrounded by LHC I pigment proteins (Armond et al., 1977). This model, based on freeze-fracture data, has received support from recent computer aided image analysis studies of individual PS I complexes with or without bound LHC I (Boekema et al., 1990), which show a PS I core surrounded by 8 LHC I measuring 12 × 16 nm in projection views. The slightly smaller size of the freeze-fracture particles suggests that not all PS I complexes contain a full complement of LHC I antennae.

C. Light-harvesting Complex II

LHC II, the peripheral light-harvesting antenna complex of PS II, gives rise to 8 nm PF_s and PF_u particles (Kyle et al. 1983; Staehelin, 1986). The recently published crystal structure of LHC II with a resolution of 3.4 Å confirms the idea that the 8 nm particles correspond to LHC II trimers, with each of the subunits containing 7 Chl *a*, 5 Chl *b* and 2 carotenoid molecules (Kühlbrandt et al., 1994; see also section III F: Light Harvesting Complexes). Following reconstitution of liposomes with purified LHC II, the liposomes can be induced to adhere to each other by the addition of physiological concentrations of mono- and divalent cations, thereby mimicking cation-mediated stacking of thylakoid membranes (McDonnel and Staehelin, 1980). This explains why most LHC II is found in the stacked grana membranes (Table 3). The LHC II complexes can move independent of the PS II complexes in thylakoid membranes and are redistributed in response to environmental stimuli (see below).

D. Cytochrome b_6f Complex

The Cyt b_6f complex consists of four major polypeptides and possibly a small hydrophobic polypeptide of ~4 kDa (Hauska, Chapter 19). Following reconstitution of this complex into phospholipid and galactolipid liposomes, freeze-fracture reveals particles with a diameter of 8–9 nm (Mörschel and Staehelin, 1983). Based on volumetric calculations, it was postulated that the particles correspond to Cyt b_6f dimers. More recent functional studies (Graan and Ort, 1986; Hope, 1993) and electron microscopical analysis of purified complexes

(Boekema et al., 1994), have provided further support for the hypothesis that the in vivo form of Cyt b_6f complex is a dimer. As listed in Table 3, Cyt b_6f appears to be fairly evenly distributed between grana and stroma thylakoids (Allred and Staehelin, 1986). Since virtually all EF particles can be accounted for by PS II complexes, the Cyt b_6f complex presumably gives rise to 8–9 nm PF_s and PF_u particles. Some recent studies of *Chlamydomonas reinhardtii* PS II mutants suggest that under certain conditions Cyt b_6f complexes might interact with non-complexed PS II polypeptides and give rise to a small subfraction of the EF_u particles (Olive and Vallon, 1991). To what extent this occurs in wild type thylakoids has yet to be determined.

E. CF_0/CF_1 ATP Synthase Complex

The lollipop shaped CF_1 complex was the first functional complex of thylakoids to be positively identified and localized (McCarty and Racker, 1966). The water soluble, globular CF_1 domain has a diameter of 10–12 nm and exhibits a six-fold rotational symmetry. Upon reconstitution of the CF_0/CF_1 complex into galactolipid liposomes, the CF_0 subunit gives rise to ~9.5 nm freeze-fracture particles (Mörschel and Staehelin, 1983). Immunolabeling and freeze-etch studies have demonstrated that the CF_0/CF_1 complexes are localized exclusively to non-stacked stroma thylakoids including the grana margins, with the CF_1 subunits protruding into the stroma (Miller and Staehelin, 1976). Since a large number of the PF_u particles fall into the 9–10 nm size range, it has not yet been possible to positively identify the CF_0 basepiece particles in freeze-fracture replicas of thylakoid membranes. Spinach thylakoids possess about 700 CF_0/CF_1 complexes per μm^2 of stroma membrane (Miller and Staehelin, 1976).

VII. Why do Thylakoid Membranes Form Grana Stacks?

Because grana stacks are such a prominent structural feature of Chl *b*–containing thylakoid membranes, the question of their functional significance remains a central question in photosynthesis research. The most prominent effect of membrane stacking appears to be the physical separation of PS II and PS I, with most PS II being sequestered in the stacked grana membranes and most PS I in the non–stacked stroma membranes (Table 3). However, grana formation is not required for the principle features of photosynthesis, since partially developed thylakoids that are largely devoid of stacked membranes are fully capable of electron transport and photophosphorylation (Boardman et al., 1978).

Two biophysical arguments for grana stack formation and the concomitant separation of PS II and PS I have been put forward by Staehelin and Arntzen (1983). The first is based on the fact that PS I contains the most long wavelength-absorbing pigment systems and in a well-connected pigment bed can draw excitation energy from PS II antenna systems (Butler, 1977). Thus, without the physical separation of the photosystems, PS I would unbalance excitation energy distribution within the pigment bed. The second argument was that PS I was also more efficient in exciton usage than PS II, as evidenced by the low room temperature fluorescence of PS I (Williams, 1977).

Trissl and Wilhelm (1993) have further analyzed these ideas in terms of the energetic and kinetic properties of PS II and PS I. Their conclusion is that the red-shifted antenna bed of PS I is of lesser importance than the need to physically separate a slow exciton trapping photosystem, PS II, from a ~3× faster and thus more efficient photosystem, PS I. This separation prevents quenching of PS II by PSI, and provides a mechanism for regulating the distribution of excitation energy between the two photosystems.

VIII. Alteration in Thylakoid Membrane Composition and Organization in Response to Changes in the Light Environment

Light is essential for plant growth. However, light is not a constant commodity that plants can draw upon in a predictable manner. Instead, plants must cope with light environments that vary by two orders of magnitude in irradiance. For example, shade plants on a rain forest floor often grow with less than 0.5% of the light irradiance of the top of the canopy (Björkmann and Ludlow, 1972). In addition, plants are subject to both momentary (e.g. cloud or leaf shadowing), diurnal and seasonal light changes. To survive and thrive under these constantly changing light environments, plants have evolved a number of short and long term molecular level adaptation strategies, which enable them to optimize the photosynthetic apparatus of each chloroplast for a given light environment. Long-term acclimation requires alterations in the composition of the photosynthetic membranes. In contrast, short-term adaptation involves spatial reorganization of existing components in the plane of the membrane.

A. Long-term Acclimation

Plants that grow in a shady habitat typically produce more thylakoids per chloroplast with a higher grana to stroma membrane ratio than high light plants. These morphological differences are a reflection of underlying molecular changes that optimize the *ratios* of the different functional complexes to an environment where light is limiting (Anderson et al., 1988; Melis, 1991). Thus, 'shade chloroplasts' produce proportionally more Chl-rich light-harvesting complexes (esp. LHC II, which leads to larger grana stacks) per reaction center complex than 'sun chloroplasts'. In addition, there are changes in the PS II to PS I ratios. The general effect of these light-harvesting/reaction center changes is to maximize the rate of turnover of the electron transport chain under low light conditions. This has been confirmed by measurements of the light intensity needed to saturate PS II and PS I activities, which are much lower in shade plants (Björkmann and Ludlow, 1972). Nevertheless, because the rate of electron transport in shade plants still tends to be lower than in sun chloroplasts, there are also compensatory decreases in electron transport complexes and mobile electron carriers (Cyt b_6f, plastoquinone, plastocyanin, ferredoxin), as well as in ATP synthase complexes. Conversely, the higher photosynthetic activities of sun chloroplasts are brought about by proportional increases in these complexes and carriers per unit Chl, since fewer Chl-rich light harvesting complexes are needed to sustain high rates of electron transport (Anderson et al., 1988).

Although most plants are genetically adapted to growing in either a shade or sun environment, within limits they also have the potential to adapt to varying light irradiances. There are some differences of opinion concerning the extent of light adaptability of plants and exact ratios of PS II to PS I complexes (see. e.g. Anderson et al., 1988; Lee and Whitmarsh, 1989). However, most of the discrepancies can be traced to the heterogeneity of the PS II centers (PS II α, PS II β; Ort and Whitmarsh, 1990) as well as differences between species.

Light adaptation is regulated at the cellular level as demonstrated in studies in which chloroplasts in the upper, palisade cells of a leaf have been shown to exhibit sun type characteristics, and those in the lower, spongy mesophyll cells have more shade type properties. However, when leaves are turned upside down for a week, the characteristics become reversed (Terashima and Inoue, 1985; Terashima and Takenaka, 1986).

The exact rate of conversion has not been determined for higher plant chloroplasts, but in the alga *Dunaliella* the adaptation from high to low light or low to high light occurs with a half-time of about a day (Kim et al., 1993). With the nature and extent of long-term acclimation of chloroplasts now firmly established, the question at hand is how does a plant determine the optimal thylakoid composition for its given growth habitat considering that there is no steady state light condition during any day of its life. One of the critical if not the most critical regulatory site of chloroplast development could be the Mg-chelatase complex that appears to serve as the 'master switch' for chlorophyll synthesis (T. G. Falbel and L. A. Staehelin, unpublished).

B. Short-term Adaptation

The photosynthetic apparatus of most leaves appears to be optimized for light conditions encountered mid-morning or mid-afternoon. When the light-irradiance exceeds these 'average' conditions, the problem is no longer to optimize the production of ATP and NADPH but to avoid photoinhibitory damage to the photosynthetic apparatus (Demmig-Adams and Adams, 1992; Horton et al., 1994). The short-term adaptive strategies mentioned here are designed primarily to prevent damage to the photosynthetic apparatus during these periods where light intensities exceed the photosynthetic capacity of the thylakoids. The ability of these mechanisms to regulate the rates of turnover of the two photosystems appears to be of secondary importance.

Three photoprotective regulatory mechanisms have been reported in the literature: the state 1-state 2 transitions (Williams and Allen, 1987), the xanthophyll-zeaxanthin cycle (Demmig-Adams, 1990), and the controlled aggregation of LHC II complexes (Horton and Ruban, 1992). Although all three mechanisms manifest themselves as distinct phenotypes, they should be viewed as synergistic and functionally related alterations of the thylakoid membrane with the ultimate goal of preventing photodestruction and possibly of optimizing the function of the photosynthetic electron transport chain. All three mechanisms are feedback controlled by the buildup of critical levels of intermediates in the electron transport chain (reduced PQ), or intermediates in ATP synthesis (the transmembrane ΔpH), with response times of one to several minutes.

State 1-state 2 transitions involve phosphorylation of a subset (up to 30%) of the LHC II complexes and the reversible migration of these complexes from grana to stroma membranes (Staehelin and Arntzen, 1983; Bennett, 1991; Allen, 1992). The primary goal of these changes is to reduce the antenna size of PS II. Secondary consequences with regulatory implications include a slight increase in the functional antenna size of PS I and changes in the partitioning of other electron transport chain components between grana and stroma thylakoids (Table 3). The sensor that activates the LHC II kinase appears to be associated with the Cyt b_6f complex, which responds to the redox potential of the PQ pool (Wollman and Lemaire, 1988; Gal et al., 1990; Bennett, 1991). Removal of the phosphate groups is mediated by a basal phosphatase. Other electron transport chain components that redistribute between the grana and stroma membrane regions during state 1-state 2 transitions include the Cyt b_6f complex (Vallon et al., 1991) and plastocyanin (Haehnel et al., 1989). Most likely, these rearrangements enhance the overall efficiency of the electron transport chain (Anderson, 1992).

The violaxanthin-zeaxanthin cycle and the LHC II aggregation mechanisms of photoprotection both appear to be triggered by a low lumenal pH (Demmig-Adams and Adams, 1992; Horton and Ruban, 1992). This activates the de-epoxidase enzyme, which converts violaxanthin via antheraxanthin to zeaxanthin, and also brings about a conformational change that engages the xanthophylls such that they can intercept and thermally dissipate excess light energy before it can cause damage to reaction centers (Gilmore and Yamamoto, 1993). The principle site of damage caused by excess light irradiance is PS II, where free-radical formation can cause damage either to the donor or the acceptor side of the reaction center (Barber and Andersson, 1992; Aro et al., 1993). Interestingly, zeaxanthin can not only deactivate singlet and triplet state Chls (Demmig-Adams and Adams, 1992), but also destroy oxygen radicals by an epoxidation reaction (Lichtenthaler and Schindler, 1992). At present there is little doubt about the ability of both mechanisms to protect by bringing about an increase in thermal dissipation of excitation energy in thylakoids, but how they work at the molecular level, and how LHC II is rearranged in the native membrane remains to be determined.

Acknowledgments

Thanks are due to Cris Cowan for critiquing the manuscript, and to Janet Meehl for help with the plates. Supported by NIH grant GM 22912 to LAS.

References

Albertsson PA, Andreasson E and Svensson P (1990) The domain organization of the plant thylakoid membrane. FEBS Lett 273: 36–40

Allen JF (1992) Protein phosphorylation in regulation of photosynthesis. Biochim Biophys Acta 1098: 275–335

Allen KD, Duysen ME and Staehelin LA (1988) Biogenesis of thylakoid membranes is controlled by light intensity in the conditional chlorophyll *b*-deficient CD3 mutant of wheat. J Cell Biol 107: 907–919

Allred DR and Staehelin LA (1986) Spatial organization of the cytochrome b_6f complex within chloroplast thylakoid membranes. Biochim Biophys Acta 849: 94–103

Almog O, Shoham G and Nechushtai R (1992) Photosystem I: composition, organization and structure. In: Barber J (ed) The Photosystems: Structure, Function and Molecular Biology, pp 443–469. Elsevier, Amsterdam

Anderson JM (1989) The grana margins of plant thylakoid membranes. Physiol Plant 76: 243–248

Anderson JM (1992) Cytochrome-b_6f complex - dynamic molecular organization, function and acclimation. Photosynth Res 34: 341–357

Anderson JM, Chow WS and Goodchild DJ (1988) Thylakoid membrane organisation in sun/shade acclimation. Austr J Plant Physiol 15: 11–26

Andreasson E, Svensson P, Weibull C and Albertsson P-A (1988) Separation and characterization of stroma and grana membranes-evidence for heterogeneity in antenna size of both Photosystem I and Photosystem II. Biochim Biophys Acta 936: 339–350

Armond PA, Staehelin LA and Arntzen CJ (1977) Spatial relationship of Photosystem I, Photosystem II, and the light-harvesting complex in chloroplast membranes. J Cell Biol 73: 400–418

Aro EM, Virgin I and Andersson B (1993) Photoinhibition of photosystem 2 - inactivation, protein damage and turnover. Biochim Biophys Acta 1143: 113–134

Barbato R, Friso G, Rigoni F, Vecchia FD and Giacometti GM (1992) Structural changes and lateral redistribution of Photosystem-II during donor side photoinhibition of thylakoids. J Cell Biol 119: 325–335

Barber J and Andersson B (1992) Too much of a good thing: Light can be bad for photosynthesis. Trends Biochem Sci 17: 61–66

Bassi R, Pineau B, Dainese P and Marquardt J (1993) Carotenoid-binding proteins of Photosystem II. Eur J Biochem 212: 297–303

Bennett J (1991) Protein phosphorylation in green plant chloroplasts. Ann Rev Plant Physiol Plant Mol Biol 42: 281–311

Björkman O and Ludlow MM (1972) Characterization of the light climate on the floor of a Queensland rainforest. Carnegie Inst Washington Yearbook 71: 85–94

Boardman NK, Anderson JM and Goodchild DJ (1978) Chlorophyll-protein complexes and structure of mature and developing chloroplasts. Curr Top Bioenerg 8: 36–109

Boekema EJ, Schmidt G, Gräber P and Berden JA (1988) Structure of the ATP synthase from chloroplasts and mitochondria. Z Naturforsch 43c: 219–225

Boekema E, Wynn RM and Malkin R (1990) The structure of spinach Photosystem I studied by electron microscopy. Biochim Biophys Acta 1017: 49–56

Boekema EJ, Boonstra AF, Dekker JP and Rögner M (1994) Electron microscopic structural analysis of Photosystem I, Photosystem II, and the cytochrome-b_6f complex from green plants and cyanobacteria. J Bioenerg Biomemb 26: 17–29

Boffey SA and Leech RM (1982) Chloroplast DNA levels and the control of chloroplast division in light-grown wheat leaves. Plant Physiol 69: 1387–1391

Bryant DA (1992) Molecular biology of Photosystem I. In: Barber J (ed) The Photosystems: Structure, Function and Molecular Biology, pp 501–549. Elsevier, Amsterdam

Butler WL (1977) Chlorophyll fluorescence: A probe for electron transfer and energy transfer. In: Trebst A and Avron M (eds) Photosynthesis I: Photosynthetic Electron Transfer and Photophosphorylation, pp 149–167. Springer-Verlag, Berlin

Chapman RL and Staehelin LA (1986) Freeze-fracture (-etch) electron microscopy. In: Aldrich HC and Todd WJ (eds) Ultrastructure Techniques for Microorganisms, pp 213–240. Plenum Publishing, New York

Dahlin C and Ryberg H (1986) Accumulation of phytoene in plastoglobuli of SAN-9789 (Norflurazon)-treated dark grown wheat. Physiol Plant 68: 39–45

de Boer D and Weisbeek P (1993) Import and routing of chloroplast proteins. In: Sundquist C and Ryberg M (eds) Pigment-Protein Complexes in Plastids, pp 311–334. Academic Press, San Diego

Demmig-Adams B (1990) Carotenoids and photoprotection in plants: A role for the xanthophyll zeaxanthin. Biochim Biophys Acta 1020: 1–24

Demmig-Adams B and Adams WW (1992) Photoprotection and other responses of plants to high light stress. Ann Rev Plant Physiol Plant Molec Biol 43: 599–626

Douce R and Joyard J (1979) Structure and function of the plastid envelope. Adv Bot Res 7: 1–116

Dubacq JP and Trémolières A (1983) Occurrence and function of phosphatidylglycerol containing Δ3-trans-hexadecenoic acid in photosynthetic lamellae. Physiol Vég 21: 293–312

Dunahay TG and Staehelin LA (1985) Isolation of Photosystem I complexes from octylglucoside/SDS solublized spinach thylakoids. Plant Physiol 78: 606–613

Dunahay TG and Staehelin LA (1987) Immunolocalization of the Chl *a/b* light harvesting complex and CP29 under conditions favoring phosphorylation and dephosphorylation of thylakoid membranes (state 1-state 2 transition). In: Biggins J (ed) Progress in Photosynthesis Research, Vol. 2, pp 701–704. Martinus Nijhoff, Dordrecht

Erickson JM and Rochaix JD (1992) The molecular biology of Photosystem II. In: Barber J (ed) The Photosystems: Structure, Function and Molecular Biology, pp 101–177. Elsevier, Amsterdam

Funk C, Schröder WP, Green BR, Renger G and Andersson B (1994) The intrinsic 22 kDa protein is a chlorophyll-binding subunit of Photosystem II. FEBS Lett 342: 261–266

Gal A, Hauska G, Herrmann R and Ohad I (1990) Interaction between light harvesting chlorophyll-*a/b* protein (LHCII) kinase and cytochrome b_6f complex. In vitro control of kinase activity. J Biol Chem 265: 19742–19749

Gilmore AM and Yamamoto HY (1993) Linear models relating xanthophylls and lumen activity to non-photochemical fluorescence quenching. Evidence that antheraxanthin explains zeaxanthin-independent quenching. Photosynth Res 35: 67–78

Gounaris K and Barber J (1983) Monogalactosyldiacylglycerol: The most abundant polar lipids in nature. Trends Biochem Sci 8: 378–381

Graan T and Ort DR (1986) Quantitation of 2,5-dibromo-3-methyl-6-isopropyl-p-benzoquinone binding sites in chloroplast membranes: Evidence for a functional dimer of the cytochrome b_6f complex. Arch Biochem Biophys 248: 445–451

Green BR, Shen DR, Aebersold R and Pichersky E (1992) Identification of the polypeptides of the major light-harvesting complex of Photosystem II (LHCII) with their genes in tomato. FEBS Lett 305: 18–22

Haehnel W, Ratajczak R and Robenek H (1989) Lateral distribution and diffusion of plastocyanin in chloroplast thylakoids. J Cell Biol 108: 1397–1405

Hennig J and Herrmann RG (1986) Chloroplast ATP synthase of spinach contains nine nonidentical subunit species, six of which are encoded by plastid chromosomes in two operons in a phylogenetically conserved arrangement. Mol Gen Genet 203: 117–128

Hope AB (1993) The chloroplast cytochrome *bf* complex - a critical focus on function. Biochim Biophys Acta 1143: 1–22

Horton P and Ruban AV (1992) Regulation of Photosystem II. Photosynth Res 34: 375–385

Horton P. Ruban AV and Walters RG (1994) Regulation of light harvesting in green plants. Plant Physiol 106: 415–420

Jagendorf AT and Michales A (1990) Rough thylakoids: translation on photosynthetic membranes. Plant Sci 71: 137–145

Jansson S (1994) The light-harvesting chlorophyll *a/b* -binding proteins. Biochim Biophys Acta 1184: 1–19

Jansson S, Pichersky E, Bassi R, Green BR, Ikeuchi M, Melis A, Simpson DJ, Spangfort M, Staehelin LA and Thornber JP (1992) A nomenclature for the genes encoding the chlorophyll *a/b*-binding proteins of higher plants. Plant Mol Biol Reporter 10: 242–253

Kim JH, Glick RE and Melis A (1993) Dynamics of photosystem stoichiometry adjustment by light quality in chloroplasts. Plant Physiol 102: 181–190

Krupa Z, Huner NPA, Williams JP, Maissan E and James DR (1987) Development at cold-hardening temperatures: The

structure and composition of purified rye light harvesting complex II. Plant Physiol 84: 19–24
Kühlbrandt W, Wang DN and Fujiyoshi Y (1994) Atomic model of plant light-harvesting complex by electron crystallography. Nature 367: 614–621
Kyle DJ, Staehelin LA and Arntzen CJ (1983) Lateral mobility of the light harvesting complex in chloroplast membranes controls excitation energy distribution in higher plants. Arch Biochem Biophys 222: 527–541
Lee W-J and Whitmarsh J (1989) Photosynthetic apparatus of pea thylakoid membranes: Response to growth light intensity. Plant Physiol 89: 932–940
Lichtenthaler HK (1968) Plastoglobuli and the fine structure of plastids. Endeavour 27: 144–149
Lichtenthaler HK and Schindler C (1992) Studies on the photoprotective function of zeaxanthin at high-light conditions. In: Murata M (ed) Research in Photosynthesis, Vol IV, pp 517–520. Kluwer Academic Publishers, Dordrecht
Lyon MK, Marr KM and Furcinitti PS (1993) Formation and characterization of two-dimensional crystals of Photosystem II. J Struct Biol 110: 133–140
Marquardt J and Bassi R (1993) Chlorophyll-proteins from maize seedlings grown under intermittent light conditions—their stoichiometry and pigment content. Planta 191: 265–273
Mattoo A and Edelman M (1987) Intramembrane translocation and posttranslational palmitoylation of the chloroplast 32-kDa herbicide-binding protein. Proc Natl Acad Sci USA 84: 1497–1501
McCarty RE and Racker E (1966) Effect of a coupling factor and its antiserum on photophosphorylation and hydrogen ion transport. Brookhaven Symp Biol 19: 202–214
McDonnel A and Staehelin LA (1980) Adhesion between liposomes mediated by the chlorophyll *a/b* light harvesting complex isolated from chloroplast membranes. J Cell Biol 84: 40–56
Melis A (1991) Dynamics of photosynthetic membrane composition and function. Biochim Biophys Acta 1058: 87–106
Miller KR and Staehelin LA (1976) Analysis of the thylakoid outer surface: Coupling factor is limited to unstacked membrane regions. J Cell Biol 68: 30–47
Mörschel E and Schatz GH (1987) Correlation of Photosystem II complexes with exoplasmatic freeze-fracture particles of thylakoids of the cyanobacterium *Synechococcus* sp. Planta 172: 145–154
Mörschel E and Staehelin LA (1983) Reconstitution of cytochrome b_6f and CF_0-CF_1 ATP synthetase complexes into phospholipid and galactolipid liposomes. J Cell Biol 97: 301–310
Murphy DJ (1986) Structural properties and molecular organization of acyl lipids of photosynthetic membranes. In: Staehelin LA and Arntzen CJ (eds) Photosynthesis III: Photosynthetic Membranes and Light-Harvesting Systems, pp 713–726. Springer-Verlag, Berlin
Murphy DJ and Woodrow IE (1983) Lateral heterogeneity in the distribution of thylakoid membrane lipid and protein components and its implication for the molecular organization of photosynthetic membranes. Biochim Biophys Acta 725: 104–112
Nalin CM and Nelson N (1987) Structure and biogenesis of chloroplast coupling factor CF_0/CF_1-ATP synthase. Curr Topics Bioenerg 15: 273–294
Nußberger S, Dörr K, Wang DN and Kühlbrandt W (1993) Lipid-Protein interactions in crystals of plant light-harvesting complex. J Mol Biol 234: 347–356
Olive J and Vallon O (1991) Structural organization of the thylakoid membrane: Freeze-fracture and immunocytochemical analysis. J Electron Microscopy Technique 18: 360–374
Ort DR and Whitmarsh J (1990) Inactive Photosystem II centers: A resolution of discrepancies in Photosystem II quantitation? Photosynth Res 23: 101–104
Paolillo D (1970) The three dimensional arrangement of integral lamellae in chloroplasts. J Cell Sci 6: 243–255
Seibert M, DeWit M and Staehelin LA (1987) Structural localization of the O_2-evolving apparatus to multimeric (tetrameric) particles on the lumenal surface of freeze-etched photosynthetic membranes. J Cell Biol 105: 2257–2265
Senge MO (1993) Recent advances in the biosynthesis and chemistry of the chlorophylls. Photochem Photobiol 57: 189–206
Siegenthaler PA and Rawyler A (1986) Acyl lipids in thylakoid membranes: Distribution and involvement in photosynthetic functions. In: Staehelin LA and Arntzen CJ (eds) Photosynthesis III: Photosynthetic Membranes and Light-Harvesting Systems, pp 693–705. Springer Verlag, Berlin
Simpson D (1979) Freeze-fracture studies on barley membranes III. Location of the light harvesting chlorophyll-protein. Carlsberg Res Commun 44: 305–336
Simpson D (1982) Freeze-fracture studies on barley plastid membranes *V. Viridis* n^{34}, a Photosystem I mutant. Carlsberg Res Commun 47: 215–225
Simpson D (1986) Freeze-fracture studies of mutant barley chloroplast membranes. In: Staehelin LA and Arntzen CJ (eds) Photosynthesis III: Photosynthetic Membranes and Light-Harvesting Systems, pp 665–674. Springer Verlag, Berlin
Simpson D and Robinson S (1984) Freeze-fracture ultrastructure of thylakoid membranes in chloroplasts from manganese-deficient plants. Plant Physiol 74: 735–741
Simpson DJ and von Wettstein D (1989) The structure and function of thc thylakoid mcmbranc. Carlsbcrg Rcs Comm 54: 55–65
Somerville C and Browse J (1991) Plant lipids: Metabolism, mutants, and membranes. Science 252: 80–87
Sprague SG (1987) Structural and functional consequences of galactolipids on thylakoid membrane organization. J Bioenerg Biomembr 19: 691–703
Sprague SG, Camm EL, Green BR and Staehelin LA (1985) Reconstitution of light-harvesting complexes and Photosystem II cores into galactolipid and phospholipid liposomes. J Cell Biol 100: 552–557
Staehelin LA (1986) Chloroplast structure and supramolecular organization of photosynthetic membranes. In: Staehelin LA and Arntzen CJ (eds) Photosynthesis III: Photosynthetic Membranes and Light-Harvesting Systems, pp 1–84. Springer-Verlag, Berlin
Staehelin LA and Arntzen CJ (1983) Regulation of chloroplast membrane function: Protein phosphorylation changes the spatial organization of membrane components. J Cell Biol 97: 1327–1337
Staehelin LA and DeWit M (1984) Correlation of structure and function of chloroplast membranes at the supramolecular level. J Cell Biochem 24: 261–269
Staehelin LA, Armond PA and Miller KR (1977) Chloroplast

membrane organization at the supramolecular level and its functional implications. Brookhaven Symp Biol 28: 278–315
Sundby C and Andersson B (1985) Temperature-induced reversible migration along the thylakoid membrane of Photosystem II regulates its association with LHC II. FEBS Lett 191: 24–28
Sundby C and Larsson C (1985) Transbilayer organization of the thylakoid membrane. Biochim Biophys Acta 813: 61–67
Süss K-H, Arkona C, Manteuffel R and Adler K (1993) Calvin cycle multienzyme complexes are bound to chloroplast thylakoid membranes of higher plants in situ. Proc Natl Acad Sci USA 90: 5514–5518
Terashima I and Inoue Y (1985) Palisade tissue chloroplasts and spongy tissue chloroplasts in spinach: Biochemical and ultrastructural differences. Plant Cell Physiol 26: 63–75
Terashima I and Takenaka A (1986) Organization of photosynthetic system of dorsiventral leaves as adapted to the irradiation from the adaxial side. In: Marcells R, Clijsters H and van Pouke M (eds) Biological Control of Photosynthesis, pp 219–230. Martinus Nijhoff, Dordrecht
Thornber JP, Peter GF, Morishige DT, Gomez S, Anandan S, Welty BA, Lee A, Kerfeld C, Takeuchi T and Preiss S (1993) Light harvesting in Photosystem I and Photosystem II. Biochem Soc Trans 21: 15–18
Trissl HW and Wilhelm C (1993) Why do thylakoid membranes from higher plants form grana stacks? Trends Biochem Sci 18: 415–419
Vallon O, Bulte L, Dainese P, Olive J, Bassi R and Wollman F-A (1991) Lateral redistribution of cytochrome b_6f complexes along thylakoid membranes upon state transitions. Proc Natl Acad Sci USA 88: 8262–8266
Vermaas W (1993) Molecular-biological approaches to analyze Photosystem II. Ann Rev Plant Physiol Plant Mol Biol 44: 457–481
Webb MS and Green BR (1991) Biochemical and biophysical properties of thylakoid acyl lipids. Biochim Biophys Acta 1060: 133–158
Williams WP (1977) The two photosystems and their interaction. In: Barber J (ed) Primary Process of Photosynthesis, pp 99–144. Elsevier, Amsterdam
Williams WP and Allen JF (1987) State 1-state 2 changes in higher plants and algae. Photosynth Res 13: 19–45
Wollenberger L, Stefansson H, Yu SG and Albertsson PA (1994) Isolation and characterization of vesicles originating from the chloroplast grana margins. Biochim Biophys Acta 1184: 93–102
Wollman F-A and Lemaire C (1988) Studies on kinase-controlled state transitions in Photosystem II and b_6f mutants from *Chlamydomonas reinhardtii* which lack quinone-binding protein. Biochim Biophys Acta 933: 85–94

Chapter 3

Evolution of Thylakoid Structure

Gregory R. Wolfe and J. Kenneth Hoober
Department of Botany and the Center for the Study of Early Events in Photosynthesis, Arizona State University, Tempe, Arizona 85287-1601, USA

Summary

Photosynthesis, an ancient process, originated among the earliest forms of life. Its broad distribution through at least half of the eubacterial phyla is an indication of this antiquity and, as stated by Woese (1987), the complexity of this process deems it unlikely that such a process arose on multiple occasions. This chapter summarizes and compares the structure of reaction centers of oxygenic photosynthetic organisms, purple bacteria, green bacteria and heliobacteria. Though there are two different types of contemporary photochemical reaction centers, recent comparative studies of structure and function revealed remarkable similarities that led to speculation of a common ancestor. A feature common to reaction centers is their association with a pigment bed that serves as a light-harvesting antenna. Evolutionary relationships of peripheral light-harvesting antenna complexes in chloroplasts also suggest a monophyletic origin of the organelle.

I. Introduction

Photosynthetic organisms were among the earliest life forms, and were present on earth 3,500 million years ago (Awramik, 1992; Blankenship, 1992). Evidence from stromatolites (fossilized microbial mats) suggests cyanobacteria-like organisms existed by this time. However, early photosynthetic mechanisms probably were much simpler than contemporary complex photosystems (Pierson and Olson, 1989). The origin of eukaryotes is often estimated at approximately 1,100 million years ago (Raven, 1970) though rare fossils of macroscopic algae from rocks dating back as far as 1,400 to 1,500 million years ago have been reported (Peat et al., 1978). Microbial eukaryotes were apparently rather abundant by 1000 million years ago (Awramik, 1992).

II. Photosystems

Photochemical reaction centers (RC) are integral membrane complexes comprising polypeptides and cofactors that catalyze the initial electron transfer

Donald R. Ort and Charles F. Yocum (eds): Oxygenic Photosynthesis: The Light Reactions, pp. 31–40.

step in conversion of light to chemical energy in photosynthetic organisms. Oxygenic photosynthesis involves the concerted action of two distinct RCs that are often thought to have evolved from different prokaryotic lineages (Blankenship, 1992). Comparative studies based on DNA sequences and biochemical and biophysical analyses have shown that all RCs can be categorized in two major groups. Type I, or 'iron-sulfur type' RCs, have iron-sulfur centers as electron acceptors that in the reduced form subsequently reduce $NAD(P)^+$. Examples of type I RCs include those found in *Chlorobiaceae* (green sulfur bacteria), *Heliobacteriaceae*, and PS I of plants and cyanobacteria. Type II RCs include those in purple bacteria and green filamentous bacteria (*Chloroflexus*), as well as PS II of oxygenic photosynthetic organisms. Type II RCs are 'quinone type' centers, in which (Bacterio)pheophytin and a pair of quinones serve as electron acceptors. This distinction between the two types of RCs is useful although probably not absolute. A feature common to both types is their association with a pigment bed that collects energy and transfers this energy to the RC.

A. Type II Reaction Centers

It is generally accepted that PS II and the RC of purple bacteria are evolutionarily related complexes. Cores of type II RCs are composed of two distinct subunits. Polypeptides D1 and D2 form the RC heterodimer in PS II and polypeptides L and M are the corresponding subunits in the RCs of purple bacteria. In PS II, D1 and D2 bind the primary electron donor (P680) and the primary acceptor (pheophytin). Q_A, a plastoquinone, functions as the secondary acceptor and is tightly bound to D2. Reduced Q_A subsequently transfers electrons to Q_B, a mobile plastoquinone that is loosely bound by D1. Whereas the semiquinone intermediate is tightly bound by D1, following double reduction and protonation of Q_B, the fully reduced hydroquinone dissociates from the RC and the Q_B site is refilled with a quinone from the plastoquinone pool in the membrane. There is also a nonheme iron coordinated by both core polypeptides. However, no native redox activity has been demonstrated for this cofactor. A similar organization of redox components and the nonheme iron occur on the L/M heterodimer of the purple bacterial RC (see Table 1).

Zurawski et al. (1982) first located and sequenced the *psb*A gene encoding the D1 polypeptide from spinach. Shortly thereafter, amino acid sequences of RC apoproteins from the purple bacteria *Rhodobacter capsulatus* (Youvan et al., 1984) and *Rhodopseudomonas sphaeroides* (now called *Rhodobacter sphaeroides*) (Williams et al., 1983, 1984) were reported and both groups noted sequence homology to D1. The sequence of D2 from spinach was determined and exhibited considerable homology to D1 (Alt et al., 1984; Rochaix et al., 1984). Residual homology between D1/D2 and L/M is 20–25% (Hearst and Sauer, 1984; Williams et al., 1984; Michel et al., 1986), although there are localized regions of higher homology (Williams et al., 1986). Hearst and Sauer (1984) speculated that highly conserved regions were likely quinone-binding sites of the RCs. The true relevance of highly conserved regions was not elucidated until Deisenhofer et al. (1985) reported the three-dimensional structure of the RC from *Rhodopseudomonas viridis* at atomic resolution. Analysis of X-ray diffraction patterns from RC crystals revealed that both subunits L and M contained 5 transmembrane helices in agreement with previous predictions from hydropathy plots (Williams et al., 1983, 1984; Youvan et al., 1984). Hydropathy plots of D1 and D2 sequences indicated that these also contain five transmembrane helices. Furthermore, crystallographic data for the *Rhodopseudomonas viridis* RC revealed that conserved amino acids, including histidine residues, bind the special pair Chls and the nonheme iron, and conserved phenylalanine and tryptophan residues provide the binding site for quinones (Deisenhofer et al., 1985; Deisenhofer and Michel, 1989, 1991). Confirmation of the structural analysis was provided by Allen et al. (1987a,b; 1988) with their presentation of crystallographic data on the RC from *Rhodobacter sphaeroides*.

Recently, various PS II preparations from higher plants have been crystallized (Fotinou et al., 1993; Holzenburg et al., 1993) and X-ray diffraction data may soon be available for direct comparison of three dimensional structures of PS II with purple bacterial RCs.

The significant sequence homology observed between the two subunits of type II RCs suggests that these distinct polypeptides arose via duplication of a

---

Abbreviations: BChl – bacteriochlorophyll; Chl – chlorophyll; LHC – light-harvesting complex; PS I – photosystem I; PS II – photosystem II; RC – reaction center

Table 1. Electron carriers in Type II and Type I reaction centers

Redox Component	Type II reaction centers		Type I reaction centers	
	Purple bacteria and *Chloroflexus*	PS II	Green sulfur bacteria and heliobacteria	PS I
Primary donor	BChl *a* dimer (P860)	Chl *a* dimer (P680)	BChl a dimer (P800 *Heliobacillus*; P840 *Chlorobium*)	Chl *a* dimer (P700)
Primary acceptor	BPheo	Pheophytin	BChl a	Chl *a* monomer (A_0)
Secondary acceptor	Ubiquinone or menaquinone (Q_A) (bound to M)	Plastoquinone (Q_A) (bound to D2)	Napthoquinone	Phylloquinone (A_1)
Tertiary acceptor	Ubiquinone (Q_B) (binds to L)	Plastoquinone (Q_B) (binds to D1)	Iron-sulfur centers (F_X, F_A, F_B)	Iron -sulfur centers (F_X, F_A, F_B)

single ancestral gene. It has often been assumed that D1 descended from L and D2 from M. This concept is premised primarily on the fact that the Q_B-binding site resides on D1 in PS II (Pfister et al., 1981) and on L in purple bacteria (Brown et al., 1984; de Vitry and Diner, 1984). Sequence analysis of the RC subunits does not support this hypothesis. D1 and D2 have greater identity to each other than to L or M, which suggests that two independent gene duplication events followed divergence of PS II and purple bacterial lineages (Williams et al., 1986).

A study by Coleman and Youvan (1993) provides further evidence supporting the hypothesis that contemporary heterodimeric type II RCs may have evolved from a homodimeric RC. They constructed a 'Q_AQ_A' RC mutant of *Rhodobacter capsulatus* in which the Q_B-binding region of the L subunit was replaced by a segment identical to the Q_A-binding region of the M subunit. This 'Q_AQ_A' mutant was not photosynthetically active. By selecting for photosynthetic competency, a strain derived from 'Q_AQ_A' was obtained that had a RC capable of supporting photosynthetic growth yet had two Q_A-binding sites and still lacked a Q_B-binding site. Two 'compensatory mutations' were found on the M subunit that apparently allowed photosynthetic capacity.

Though the similarities between PS II and bacterial type II RCs are numerous, there are several structural distinctions that should not be overlooked. Cytochrome *b*-559, comprising two subunits of 9 kDa and 4 kDa, occurs in eukaryotic PS II RCs but comparable subunits are not found in bacterial photosystems. Associated with the PS II RCs are core antenna complexes, CP43 and CP47. Furthermore, PS II is coupled to the oxygen evolving complex while the RC of anoxygenic purple and green filamentous bacteria are associated with sulfur oxidizing systems. Whereas the H_2O-oxidizing (reviewed by Rutherford, 1989; Renger, 1993) and H_2S-oxidizing systems (reviewed by Brune, 1989; 1995) may appear to carry out parallel reactions:

Cyanobacterial/plant:

$$2H_2O + CO_2 \overset{\text{light}}{\rightarrow} (CH_2O) + H_2O + O_2$$

Bacterial:

$$2H_2S + CO_2 \overset{\text{light}}{\rightarrow} (CH_2O) + H_2O + 2S$$

there is little similarity between the enzymology of the two systems. The oxygen evolving complex of PS II contains four manganese atoms as well as several extrinsic polypeptides situated on the lumenal side of thylakoid membranes and does not exist in green and purple photosynthetic bacteria. Thus it is improbable that these processes are evolutionarily related.

B. Type I Reaction Centers

There are sufficient similarities between PS I and RCs of green sulfur bacteria and heliobacteria to suggest a common lineage for all type I RCs (reviewed by Lockau and Nitschke, 1993). In PS I, the primary electron donor is P700, a Chl *a* dimer, and binding sites for the electron acceptors A_0 (a Chl *a* monomer), A_1 (phylloquinone), and F_X (a 4Fe-4S cluster) are generated by a heterodimeric structure comprised of *psa*A and *psa*B gene products (reviewed by Golbeck and Bryant, 1991; Golbeck, 1992). These genes from a variety of sources including higher plants (Fish et al., 1985), green algae (Kück et al., 1987) and cyanobacteria (Cantrell and Bryant, 1987) encode relatively large polypeptides with calculated molecular masses of around 82 kDa (Fish et al., 1985). There is approximately 45% amino acid homology between the genes of these polypeptides for a given species and the genes are highly conserved from cyanobacteria through higher plants with near 90% homology when conservative amino acid replacements are considered (Cantrell and Bryant, 1987). Hydropathy plot analysis of deduced amino acid sequences predict 11 transmembrane helices for each polypeptide (Fish et al., 1985). The three-dimensional structure of PS I from the cyanobacterium *Synechococcus* sp. at 6Å resolution was recently determined (Krauss et al., 1993). These latter data suggest that each RC apoprotein may span the membrane only nine times.

Closely associated with the PS I RC heterodimer is a polypeptide of ~8 kDa that serves as the apoprotein for terminal electron acceptors of PS I, F_A and F_B, both of which are 4Fe-4S centers. The sequence of the *psa*C gene encoding the F_A/F_B-binding polypeptide is highly conserved (near 90% identity) among higher plants and cyanobacteria (Golbeck and Bryant, 1991).

Redox components associated with PS I are also apparently present in heliobacteria (Nitschke et al., 1990b) and green sulfur bacteria (Nitschke et al., 1990a) (Table 1). Unlike PS I, the redox elements of *Chlorobium limicola* (Buttner et al., 1992a,b) and *Heliobacillus mobilis* (Liebl et al., 1993) are bound to a homodimeric RC. A single gene encoding a RC polypeptide with a deduced molecular mass of 82.24 kDa from *C. limicola* was characterized. No evidence was found for a second RC gene (Buttner et al., 1992a,b). Similarly, a single gene encoding a RC polypeptide of 68 kDa was found in *H. mobilis* (Liebl et al., 1993). Amino acid sequences of maize PsaA and PsaB show 16.3% and 16.8% identity with the RC polypeptide of *H. mobilis*. There is 15% and 14% identity between these two polypeptides and the *C. limicola* RC subunit. Hydropathy plots predict each of these RC polypeptides has 11 transmembrane helices. Furthermore, an amino acid identity of 17.4% exists between the RC polypeptides of *C. limicola* and *H. mobilis* (Liebl et al., 1993). The sequence and structural homology plus the similarity of redox components suggests that these Fe-S type RCs are descendants of a common ancestor. It is likely that *psa*A and *psa*B arose via gene duplication and that divergence of the PS I line from RCs of green sulfur bacteria occurred prior to the duplication event that led to heterodimeric cores.

The gene for the RC from *C. limicola* was isolated as part of an operon that also contained a second structural gene. This second gene encodes a polypeptide with a deduced mass of 23.87 kDa (Buttner et al., 1992a). This polypeptide is apparently related to the *psa*C gene. A similar gene has not been found in *H. mobilis*, although EPR experiments provided evidence for F_A and F_B centers (Nitschke et al., 1990b). Possibly an F_A/F_B-binding subunit is also present in the latter organism. If so, it will be of interest, from an evolutionary perspective, to determine whether this protein resembles the larger (24 kDa) subunit of *C. limicola* or the 8–9 kDa (PsaC) polypeptide that is highly conserved (identity 90%) in PS I of higher plants, algae and cyanobacteria.

C. Similarities between Type I and Type II Reaction Centers

Several structural and functional similarities between the two types of RCs led to speculation that a single ancestral RC gave rise to both quinone and iron-sulfur type RCs (Pierson and Olson, 1989; Nitschke and Rutherford, 1991; Blankenship, 1992; Golbeck, 1993; Vermaas, 1994). As discussed above, both types of RCs are composed of a hetero- or homodimeric protein that coordinates similar cofactors. Electrons flow from the (bacterio)Chl special pair to a Chl or Chl-like monomer, (bacterio)Chl *a* in type I and (bacterio)pheophytin in type II RCs, and subsequently to a pair of quinones (eg. A_1 in type I and Q_A/Q_B in type II). A single Fe atom bridges the two RC apoproteins in both types of RCs. Numerous hypotheses have been proposed based on these similarities that address the question of origin of RCs from a common ancestral form and how both RC types have come to occur in certain organisms.

Olson and Pierson (1987a,b; Pierson and Olson, 1989) suggested that the primordial RC was an Fe-S/quinone-containing complex, which was the progenitor to a primitive type I RC. Earliest type II RCs may have evolved from primitive type I RCs by loss of the Fe-S center, resulting in an organism containing two different types of RCs. Organisms with a single type of RC may then have arisen via selective loss of one RC or the other to give rise to either the green sulfur/heliobacterial or the purple bacterial lineages. An alternate proposal suggests an organism containing a primordial RC diverged into two distinct groups, a primitive Fe-S RC-containing prokaryote which gave rise to green bacteria and heliobacteria, and a prokaryote containing a primitive quinone type RC from which purple bacteria arose. The occurrence of both RC types within a single organism then would have arisen by fusion of these two lines (Mathis, 1990; Blankenship, 1992).

The recent characterization of the RC gene from *H. mobilis* (Liebl et al., 1993) provided important clues to the evolutionary origin of type I RCs. Perhaps even more intriguing are the implications concerning the origin of type II RCs. Examination of this gene sequence (Vermaas, 1994) revealed a 46-amino acid region that includes the putative sixth transmembrane helix of the RC apoprotein, which displays significant homology (72% homology and 33% identity) to a region that includes the sixth transmembrane helix of the CP47 apoprotein, a core antenna protein of PS II. Furthermore, the N-terminal 232 residues of the RC apoprotein is 22% identical to the CP47 apoprotein, significantly higher than the homology of the RC apoprotein to the corresponding region of PS I RC polypeptides. Vermaas (1994) proposed that a primordial Fe-S/quinone RC served as the progenitor for not only type I but also type II RCs. According to this hypothesis, the ancestral RC gene split to yield separate core antenna and RC complexes. Subsequent divergence and loss of the Fe-S complex would have given rise to the type II RC. As pointed out by Vermaas, such an hypothesis would imply that type II RC apoproteins should also bear at least residual sequence homology to the region including the last five helices of the apoproteins of type I RCs. Such homology is not obvious although limited sequence similarity was previously reported. Margulies (1991) reported that D2, L, and PS I-B all contain a conserved region of 29 amino acids with eight identical amino acids and four conservative replacements. A weaker alignment is seen among D1, L and PS I-B. A comparable region was not found in PS I-A. This conserved region is located in helix IV of D1, D2, L and M, and on helix IX of PS I-B and is therefore consistent with Vermaas' proposal that type II RC apoproteins were derived from the last five helices of type I RC apoproteins.

III. Light-Harvesting Complexes

The endosymbiotic theory for the origin of chloroplasts is generally accepted, though derivation of the numerous chloroplast types is not yet understood. It is a matter of debate whether the differentially pigmented plastids currently in existence had a monophyletic origin, i.e., all plastids derived from a single cyanobacterial-like ancestor (Cavalier-Smith, 1982), or a polyphyletic origin, i.e., multiple endosymbiotic events (Raven, 1970).

Algae are taxonomically segregated into three major groups based upon pigmentation of their plastids. 1) *Chlorophyta* contain Chl *a*/*b*-binding light-harvesting complexes (LHC). 2) *Chromophyta* have Chl *a*/*c*-binding LHCs. 3) *Rhodophyta* harvest light with Chl *a*-binding LHCs and phycobilisomes. Apoproteins of LHCs are typically 20–30 kDa in mass and the genes encoding these polypeptides reside in the nuclear genome (Green et al., 1991; Thornber et al., 1991). Immunological relatedness of LHCs from groups as phylogenetically distinct as Rhodophytes, Chromophytes and higher plants indicates conservation of primary structure during evolution (Wolfe et al., 1994). LHC genes were isolated from a variety of higher plants, and chlorophytic and chromophytic algae. Analysis of hydropathy plots based on deduced amino acid sequences predict that nearly all LHC polypeptides contain three transmembrane helices, which was confirmed by electron crystallography at a resolution of 6Å (Kühlbrandt and Wang, 1991) and now to a resolution of 3.4Å (Kühlbrandt et al., 1994). The first and third membrane-spanning domains and their preceding sequences are highly conserved among all LHC apoproteins. Furthermore, there is considerable similarity between the first and third helices. Other segments of these proteins, particularly the second helix and its flanking regions, have diverged considerably (Green et al., 1991).

Kim et al. (1992) characterized the *Psb*S gene that encodes a 22 kDa intrinsic polypeptide of PS II. The sequence of *Psb*S suggests that it is related to the LHC gene family, although hydropathy analysis suggests existence of four transmembrane helices. Like other members of the LHC family, the first and third helices display homology to each other. Interestingly, the second and fourth helices of *Psb*S are related to each other. Kim et al. (1992) speculated that a polypeptide containing four helices may have resulted from duplication of an ancestral gene encoding a polypeptide with two membrane spanning domains. A LHC with four transmembrane helices may have served as the prototype for the contemporary LHCs which arose by deletion of the fourth helix.

The evolutionary origin of LHCs remains enigmatic because of the absence of related polypeptides in photosynthetic prokaryotes. Functionally analogous BChl-binding antenna proteins from photosynthetic purple bacteria are small (5–7 kDa) (Zuber, 1985; Hawthornthwaite and Cogdell, 1991), traverse the membrane only one time and share no distinguishable sequence homology with LHC polypeptides. Similarly, antenna complexes from green sulfur bacteria are unrelated to LHCs (Hawthornthwaite and Cogdell, 1991).

A group of prokaryotic photosynthetic organisms known as Prochlorophytes are oxygenic, contain Chl *a*/*b*-protein complexes and lack phycobilisomes (Lewin, 1975, 1976). Discovery of this group led to speculations that an ancestral Prochlorophyte served as progenitor for green chloroplasts (Chisholm et al., 1988). However, biochemical evidence suggests that Prochlorophyte Chl *a*/*b*-binding proteins are not related to LHCs of eukaryotic chloroplasts. They are larger (30–36 kDa in mass) and apparently immunologically unrelated to their eukaryotic counterparts (Hiller and Larkum, 1985; Bullerjahn et al., 1990; Wolfe et al., 1994a). Sequence analysis of the Chl *a*/*b*-binding proteins from Prochlorophytes is necessary to ascertain whether these Chl-binding proteins share a common lineage with LHCs of eukaryotes.

Phycobilisomes serve as the major light-harvesting antenna in cyanobacteria and in red algae (Gantt, 1986; Glazer and Melis, 1987). Phycobiliproteins, the chromophore-binding proteins of the phycobilisome, are soluble proteins and are not related to LHC apoproteins. It was assumed that phycobilisome-containing organisms lack LHCs (Green et al., 1991). Wolfe et al. (1992; 1994b) reported that PS I from the red alga *Porphyridium cruentum* is structurally more similar to PS I of other eukaryotes than to PS I of cyanobacteria. Furthermore, the demonstration of coexistence of LHCs and phycobilisomes in red algae established the first clear link between organisms containing phycobilisomes and those containing the Chl-based LHCs (Wolfe et al., 1994a). Apoproteins of red algal LHCs bind Chl *a* and xanthophylls and are immunologically related to Chl *a*/*b*- and Chl *a*/*c*-binding LHCs. These data suggest the existence of LHCs prior to the divergence of major groups of eukaryotic algae.

It is presumed that cyanobacteria lack LHCs on the basis of analyses of relatively few species. Possibly, among the myriad of cyanobacterial species not yet examined, species similar to rhodophytes that contain both LHCs and phycobilisomes may exist. Nilsson et al. (1990) reported detecting a 22 kDa polypeptide in thylakoids from the cyanobacterium *Synechocystis* 6803 that was immunologically related to the intrinsic 22 kDa PS II subunit of spinach. Determination of the cyanobacterial gene sequence for comparison with *Psb*S and LHC sequences may prove quite interesting with regard to evolution of LHCs. At present there remains a paucity of evidence of LHC-related polypeptides in prokaryotes. If a prokaryotic chloroplast progenitor existed that contained both phycobilisomes and LHC-like polypeptides, chloroplasts containing only LHCs, such as those in chromophytes, chlorophytes and green plants, may then have evolved by the subsequent loss of phycobilisomes.

IV. Thylakoid Morphology

Photosynthetic organisms evolved significant differences in overall thylakoid morphology. In higher plants and the majority of chlorophytic algae, thylakoids are arranged into stacked regions (grana), and unstacked stromal lamellae. Lateral heterogeneity of thylakoid components in these organisms (Andersson and Anderson, 1980) is well accepted. PS II is localized in grana while PS I and ATPase predominate in stromal lamellae and are virtually excluded from granal stacks. Segregation of photosystems is believed to be an integral part of a strategy that evolved for regulation of energy distribution between the photosystems (Trissl and Wilhelm, 1993). Light conditions that favor excitation of PS II promote reduction of the plastoquinone pool. A membrane-bound kinase is activated under these conditions and subsequently phosphorylates LHCII. The resulting increase in charge density within the grana causes minor unstacking and migration of a portion of the phosphorylated LHCII to stromal thylakoids. The decrease in the cross-sectional area of antenna serving PS II is concomitant with an increase in PS I antenna size. This phenomenon of redistribution of excitation energy to maintain balance in photosynthetic electron transport is known as a state transition. Miller and Lyon (1985) rebuffed the hypothesis that segregation of the photosystems was advantageous based on the fact that red algae and cyanobacteria efficiently regulate energy distribution between the photosystems despite the apparent lack of thylakoid lateral heterogeneity. They suggested instead that while the phosphorylated state of LHCII may directly affect the degree of thylakoid stacking, phosphorylation-dependent migration of LHCs evolved to overcome problems inherent in a stacked membrane system.

Trissl and Wilhelm (1993) proposed that physical separation of PS I from PS II by grana formation represents a strategy that increases photosynthetic efficiency. PS I has faster trapping kinetics (~100 ps) in stabilization of charge separation than PS II (~300 ps). Furthermore, PS I antenna absorbs light of a longer wavelength than antenna of PS II. Therefore, if the photosystems were not separated, PS I would siphon off excitation energy from PS II and decrease the photochemical yield of PS II. Chromophytic algae and phycobilisome-containing organisms that have a more homogeneous distribution of photosystems apparently adopted alternative strategies to circumvent this problem. The relatively high ratio of PS II to PS I of 1.5 to 3.0 observed in chromophytic algae thus would reflect their way of compensating for the proximity of the two photosystems. In red algae and cyanobacteria, though there is significant spillover of energy from PS II to PS I (Biggins and Bruce, 1989), the phycobilisomes serve as the major light-harvesting antenna and are energetically coupled to PS II (Chereskin et al., 1985; Gantt, 1986). This arrangement directs sufficient excitation energy to PS II despite the siphoning of energy by PS I. Trissl and Wilhelm (1993) proposed that evolution of the various thylakoid morphologies displayed by extant photosynthetic oxygen-evolving species was driven by a common strategy: to segregate PS I, the 'fast photosystem' from PS II, the 'slow photosystem' to increase photosynthetic efficiency.

V. Conclusions

Our understanding of the evolution of photosynthesis has advanced immensely over the past decade as the result of comparative biophysical and biochemical studies of RCs and their associated antenna complexes and sequence analysis of the respective genes. It has been generally accepted for some time that the RC of purple bacteria and PS II from oxygenic organisms are evolutionarily related (Deisenhofer and Michel, 1991). Recent biochemical and molecular characterization of type I RCs from green sulfur bacteria (Nitschke et al., 1990a; Nitschke and Rutherford, 1991) and heliobacteria (Nitschke et al., 1990b; Trost et al., 1992) revealed similarities between these RCs and PS I that imply a common lineage. Sequence analysis of the RC apoproteins from heliobacteria has also revealed possible residual homology between this polypeptide and the PS II core antenna polypeptide CP47. These data led Vermaas (1994) to speculate that an ancestral type I RC may have given rise to a type II RC. Perhaps an ancestral RC gene split to yield genes encoding the separate RC and core antenna apoproteins found in contemporary PS II. Therefore it is not inconceivable that all RCs are derived from a common progenitor.

Recently there has been exciting progress towards discerning the possible evolutionary lineage of LHCs in photosynthetic eukaryotes. The discovery that groups as phylogenetically distinct as *Rhodophyta*, *Chlorophyta* and *Chromophyta* contain immunologically related LHCs strongly suggests a common origin for all chloroplasts (Wolfe et al., 1994a). The coexistence of LHCs and phycobilisomes within the Rhodophytes provides the first clear link between LHC- and phycobilisome-containing organisms (Wolfe et al., 1994a) and lends credence to the hypothesis that the progenitor of contemporary chloroplasts may have also had these two types of antenna systems. Chloroplasts containing only LHCs may then have originated via loss of the phycobilisome. Though there is some indication of the possible existence of LHC-like polypeptides in cyanobacteria (Nilsson et al, 1990), unequivocal evidence is lacking. Further investigation in this area to link LHCs to a group of prokaryotes should prove interesting.

This paper has focused primarily on the evolution of the pigmented components of photosynthetic membranes. Though the pigment-protein complexes are indeed fundamental in light harvesting and the initial steps in energy transduction, a more comprehensive review on the evolution of thylakoid structure should also include discussion of the major nonpigmented components such as the oxygen-evolving complex, the cytochrome b_6f complex and the chloroplast coupling factor. Other vital processes including carbon fixation and porphyrin biosynthesis were also apparently in existence in the earliest organisms and analysis of these processes should provide additional useful phylogenetic information.

References

Allen JP, Feher G, Yeates TO, Komiya H and Rees DC (1987a) Structure of the reaction center from *Rhodobacter sphaeroides* R-26: The cofactors. Proc Natl Acad Sci USA 84: 5730–5734

Allen JP, Feher G, Yeates TO, Komiya H and Rees DC (1987b) Structure of the reaction center from *Rhodobacter sphaeroides* R-26: The protein subunits. Proc Natl Acad Sci USA 84: 6162–6166

Allen JP, Feher G, Yeates TO, Komiya H and Rees DC (1988) Structure of the reaction center from *Rhodobacter sphaeroides* R-26: Protein-cofactor (quinones and Fe^{2+}) interactions. Proc Natl Acad Sci USA 85: 8487–8491

Alt J, Morris J, Westhoff P and Herrmann RG (1984) Nucleotide sequence of the clustered genes for the 44 kd chlorophyll *a* apoprotein and the '32 kd'-like protein of the photosystem II reaction center in the spinach plastid chromosome. Curr Genet 8: 597–606

Andersson B and Anderson JM (1980) Lateral heterogeneity in the distribution of chlorophyll-protein complexes of the thylakoid membranes of spinach chloroplasts. Biochem Biophys Acta 593: 427–440

Awramik SM (1992) The oldest records of photosynthesis. Photosynth Res 33: 75–89

Biggins J and Bruce D (1989) Regulation of excitation energy transfer in organisms containing phycobilins. Photosynth Res 20: 1–34

Blankenship RE (1992) Origin and early evolution of photosynthesis. Photosynth Res 33: 91–111

Brown AE, Gilbert CW, Guy R and Arntzen CJ (1984) Triazine herbicide resistance in the photosynthetic bacterium *Rhodopseudomonas sphaeroides*. Proc Natl Acad Sci USA 81: 6310–6314

Brune DC (1989) Sulfur oxidation by phototrophic bacteria. Biochim Biophys Acta 975: 189–221

Brune DC (1995) Sulfur compounds as photosynthetic electron donors. In: Blankenship RE, Madigan MT and Bauer CE (eds) Anoxygenic Photosynthetic Bacteria, pp 847–870. Kluwer Academic Publishers, Dordrecht

Bullerjahn GS, Jensen TC, Sherman DM and Sherman LA (1990) Immunological characterization of the *Prochlorothrix hollandica* and *Prochloron sp*. chlorophyll *a/b* antenna proteins. FEMS Microbiol Lett 67: 99–105

Buttner M, Xie D-L, Nelson H, Pinther W, Hauska G and Nelson N (1992a) Photosynthetic reaction center genes in green sulfur bacteria and in photosystem I are related. Proc Natl Acad Sci USA 89: 8135–8139

Buttner M, Xie D-L, Nelson H, Pinther W, Hauska G and Nelson N (1992b) The photosystem I-like P840-reaction center of green S-bacteria is a homodimer. Biochim Biophys Acta 1101: 154–156

Cantrell A and Bryant DA (1987) Molecular cloning and nucleotide sequence of the *psaA* and *psaB* genes of the cyanobacterium *Synechococcus* PCC 7002. Plant Mol Biol 9: 453–468

Cavalier-Smith T (1982) The origin of plastids. Biol J Linn Soc 17: 289–306

Chereskin BM, Clement-Metral JD and Gantt E (1985) Characterization of a purified photosystem II-phycobilisome particle preparation from *Porphyridium cruentum*. Plant Physiol 77: 626–629

Chisholm SW, Olson RJ, Zettler ER, Goericke R, Waterbury JB and Welschmeyer NA (1988) A novel free-living prochlorophyte abundant in the oceanic ephotic zone. Nature 334: 340–343

Coleman WJ and Youvan DC (1993) Atavistic reaction centre. Nature 366: 517–518

Deisenhofer J and Michel H (1989) The photosynthetic reaction centre from the purple bacterium *Rhodopseudomonas viridis*. EMBO J 8: 2149–2170

Deisenhofer J and Michel H (1991) High-resolution structure of photosynthetic reaction centers. Annu Rev Biophys Chem 20: 247–266

Deisenhofer J, Epp O, Miki K, Huber R and Michel H (1985) Structure of the protein subunits in the photosynthetic reaction centre of *Rhodopseudomonas viridis* at 3 Å resolution. Nature 318: 618–624

de Vitry C and Diner BA (1984) Photoaffinity labeling of the azidoatrazine receptor site in reaction centers of *Rhodopseudomonas sphaeroides*. FEBS Lett 167: 327–331

Fish LE, Kück U and Bogorad L (1985) Two partially homologous adjacent light-inducible maize chloroplast genes encoding polypeptides of the P700 chlorophyll a-protein complex of photosystem I. J Biol Chem 260: 1413–1421

Fotinou C, Kokkinidis M, Fritzsch G, Haase W, Michel H and Ghanotakis DF (1993) Characterization of a photosystem II core and its three-dimensional crystals. Photosynth Res 37: 41–48

Gantt E (1986) Phycobilisomes In: Staehelin LA and Arntzen CJ (eds) Encyclopedia of Plant Physiology: Photosynthesis III, Vol 19, pp 260–268. Springer-Verlag, Berlin

Glazer AN and Melis A (1987) Photochemical reaction centers: Structure, organization and function. Annu Rev Plant Physiol 38: 11–45

Golbeck JH (1992) Structure and function of photosystem I. Annu Rev Plant Physiol Plant Mol Biol 43: 293–324

Golbeck JH (1993) Shared thematic elements in photochemical reaction centers. Proc Natl Acad Sci USA 90: 1642–1646

Golbeck and Bryant (1991) Photosystem I. Curr Topics Bioenerg 16: 83–177

Green BR, Pichersky E and Kloppstech K (1991) Chlorophyll *a/b*-binding proteins: An extended family. Trends Biochem Sci 16: 181–186

Hawthornthwaite AM and Cogdell RJ (1991) Bacteriochlorophyll-binding proteins. In: Scheer H (ed) Chlorophylls, pp 493–528. CRC Press, Boca Raton

Hearst JE and Sauer K (1984) Protein sequence homologies between portions of the L and M subunits of reaction centers of *Rhodopseudomonas capsulata* and the Q_B-protein of chloroplast thylakoid membranes: A proposed relation to quinone-binding sites. Z Naturforsch 39c: 421–424

Hiller RG and Larkum AWD (1985) The chlorophyll-protein complexes of *Prochloron* sp. (Prochlorophyta). Biochim Biophys Acta 806: 107–115

Holzenburg A, Bewley MC, Wilson FH, Nicholson WV and

Ford RC (1993) Three-dimensional structure of photosystem II. Nature 363: 470–472

Kim S, Sandusky P, Bowlby NR, Aebersold R, Green BR, Vlahkis S, Yocum CF and Pichersky E (1992) Characterization of a spinach *psbS* cDNA encoding the 22 kDa protein of photosystem II. FEBS Lett 314: 67–71

Krauss N, Hinrichs W, Witt I, Fromme P, Pritzkow W, Dauter Z, Betzel C, Wilson KS, Witt HT and Saenger W (1993) Three-dimensional structure of system I of photosynthesis at 6Å. Nature 361: 326–330

Kück U, Choquet Y, Schneider M, Dron M and Bennoun P (1987) Structural and transcription analysis of two homologous genes for the P700 chlorophyll *a*-apoproteins in *Chlamydomonas reinhardtti*: Evidence for in vivo *trans*-splicing. EMBO J 6: 2185–2195

Kühlbrandt W and Wang DN (1991) Three-dimensional structure of plant light-harvesting complex determined by electron crystallography. Nature 350: 130–134

Kühlbrandt W, Wang DN and Fujiyoshi Y (1994) Atomic model of plant light-harvesting complex by electron crystallography. Nature 367: 614–621

Lewin RA (1975) Extraordinary pigment composition of a prokaryotic alga. Nature 256: 735–737

Lewin RA (1976) Prochlorophyta as a proposed new division of algae. Nature 261: 697–698

Liebl U, Mockensturm-Wilson M, Trost JT, Brune DC, Blankenship RE and Vermaas W (1993) Single core polypeptide in the reaction center of the photosynthetic bacterium *Heliobacillus mobilis*: Structural implications and relations to other photosystems. Proc Natl Acad Sci USA 90: 7124–7128

Lockau W and Nitschke W (1993) Photosystem I and its bacterial counterparts. Physiol Plant 88: 372–381

Margulies MM (1991) Sequence similarity between photosystem I and II. Identification of a photosystem I reaction center transmembrane helix that is similar to transmembrane helix IV of the D2 subunit of photosystem II and the M subunit of the non-sulfur purple and flexible green bacteria. Photosynth Res 29: 133–147

Mathis P (1990) Compared structure of plant and bacterial photosynthetic reaction centers. Evolutionary implications. Biochim Biophys Acta 1018: 163–167

Michel H, Weyer KA, Gruenberg H, Dunger I, Oesterhelt D and Lottspeich F (1986) The 'light' and 'medium' subunits of the photosynthetic reaction centre from *Rhodopseudomonas viridis*: Isolation of the genes, nucleotide and amino acid sequence. EMBO J 5: 1149–1158

Miller KR and Lyon MK (1985) Do we really know why chloroplast membranes stack? Trends Biochem Sci 10: 219–222

Nilsson F, Andersson B, and Jansson C (1990) Photosystem II characteristics of a constructed *Synechocystis* 6803 mutant lacking synthesis of the D1 polypeptide. Plant Molec Biol 14: 1051–1054

Nitschke W and Rutherford AW (1991) Photosynthetic reaction centres: Variations on a common theme? Trends Biochem Sci 16: 241–245

Nitschke W, Feiler U and Rutherford AW (1990a) Photosynthetic reaction center of green sulfur bacteria studied by EPR. Biochemistry 29: 3834–3842

Nitschke W, Setif P, Liebl U, Feiler U and Rutherford AW (1990b) Reaction center photochemistry of *Heliobacterium chlorum*. Biochemistry 29: 11079–11088

Olson JM and Pierson BK (1987a) Origin and evolution of photosynthetic reaction centers. Origins of Life 17: 419–430

Olson JM and Pierson BK (1987b) Evolution of reaction centers in photosynthetic prokaryotes. Int Rev Cytol 108: 209–248

Peat CJ, Muir MD, Plumb KA, McKirdy DM and Norvick MS (1978) Proterozoic microfossils from the Roper Group, Northern Territory, Australia. J Austr Geol Geophys 3: 1–17

Pfister K, Steinback KE, Gardner G and Arntzen CJ (1981) Photoaffinity labeling of an herbicide receptor protein in chloroplast membranes. Proc Natl Acad Sci USA 78: 981–985

Pierson BK and Olson JM (1989) Evolution of photosynthesis in anoxygenic photosynthetic prokaryotes. In: Cohen Y and Rosenberg E (eds) Microbial Mats: Physiological Ecology of Benthic Microbial Communities, pp 402–427 Am Soc Microbiol, Washington

Raven PH (1970) A multiple origin for plastids and mitochondria. Science 169: 36–40

Renger G (1993) Water cleavage by solar radiation - an inspiring challenge of photosynthesis research. Photosynth Res 38: 229–247

Rochaix J-D, Dron M, Rahire M and Malone P (1984) Sequence homology between the 32 kdalton and the D2 chloroplast membrane polypeptides of *Chlamydomonas reinhardii*. Plant Molec Biol 3: 363–370

Rutherford AW (1989) Photosystem II, the water-splitting enzyme. Trends Biochem Sci 14: 227–232

Thornber JP, Morishige DT, Anandan S and Peter GF (1991) Chlorophyll-carotenoid proteins of higher plant thylakoids. In: Scheer H (ed) Chlorophylls, pp 549–585. CRC Press, Boca Raton

Trissl H-W and Wilhelm C (1993) Why do thylakoid membranes from higher plants form grana stacks? Trends Biochem Sci 18: 415–419

Trost JT, Brune DC and Blankenship RE (1992) Protein sequences and redox titrations indicate that the electron acceptors in reaction centers from heliobacteria are similar to Photosystem I. Photosynth Res 32: 11–22

Vermaas WFJ (1994) Evolution of heliobacteria: Implications for photosynthetic reaction center complexes. Photosynth Res 41: 285–294

Williams JC, Steiner LA, Ogden RC, Simon MI and Feher G (1983) Primary structure of the M subunit of the reaction center from *Rhodopseudomonas sphaeroides*. Proc Natl Acad Sci USA 80: 6505–6509

Williams JC, Steiner LA, Feher G and Simon MI (1984) Primary structure of the L subunit of the reaction center from *Rhodopseudomonas sphaeroides*. Proc Natl Acad Sci USA 81: 7303–7307

Williams JC, Steiner LA and Feher G (1986) Primary structure of the reaction center from *Rhodopseudomonas sphaeroides*. Proteins: Structure Function Genetics 1: 312–325

Woese CR (1987) Bacterial evolution. Microbiol Rev 51: 221–271

Wolfe GR, Cunningham FX and Gantt E (1992) In the red alga *Porphyridium cruentum* photosystem I is associated with a putative LHCI complex. In: Murata N (ed) Research in Photosynthesis, pp 315–318. Kluwer Academic Publishers, Dordrecht

Wolfe GR, Cunningham FX, Durnford D, Green BR and Gantt E (1994a) Evidence for a common origin of chloroplasts with differently pigmented light-harvesting complexes. Nature 367: 566–568

Wolfe GR, Cunningham FX, Grabowski B, Gantt E (1994b) Isolation and characterization of Photosystem I and II from the red alga *Porphyridium cruentum*. Biochim Biophys Acta 1188: 357–366

Youvan DC, Bylina EJ, Alberti M, Begusch H and Hearst JE (1984) Nucleotide and deduced polypeptide sequences of the photosynthetic reaction-center, B870 antenna, and flanking polypeptides from *R. capsulata*. Cell 37: 949–957

Zuber H (1985) Structure and function of light-harvesting complexes and their polypeptides. Photochem Photobiol 42: 821–844

Zurawski G, Bohnert H, Whitfeld PR and Bottomley W (1982) Nucleotide sequence of the gene for the Mr 32,000 thylakoid membrane protein from *Spinacia oleracea* and *Nicotiana debneyi* predicts a totally conserved primary translation product of Mr 38,950. Proc Natl Acad Sci USA 79: 7699–7703

Chapter 4

Control of Thylakoid Membrane Development and Assembly

Andrew N. Webber
Department of Botany and Center for the Study of Early Events in Photosynthesis, Arizona State University, Tempe, AZ 85287-1601, USA

Neil R. Baker
Department of Biology, University of Essex, Colchester, CO4 3SQ, Essex, UK.

Summary

Development of photosynthetically competent thylakoid membranes requires the coordinated synthesis and assembly of a large number of polypeptides of both cytoplasmic and chloroplast origins. Chloroplast and nuclear gene expression has been the subject of intensive research in recent years and has led to detailed understanding of molecular aspects of the control of transcription, mRNA stability and translation. In the first part of this review we discuss recent advances in the understanding of the control of photosynthetic gene expression and how this may relate to the assembly of thylakoid protein complexes and processes by which stoichiometric levels of thylakoid protein complexes are achieved and maintained. In the second part we emphasize that the rate of chloroplast development and rapid establishment of a photochemically competent photosynthetic apparatus is an important determinant of plant productivity. Consideration is given to the appearance of photosynthetic activities during chloroplast development and how this may be perturbed by changes in the physical environment. It is concluded that a major gap in our understanding of the developmental

Donald R. Ort and Charles F. Yocum (eds): Oxygenic Photosynthesis: The Light Reactions, pp. 41–58.

biology of the thylakoid membrane lies between the detailed knowledge of the molecular processes of photosynthetic gene expression and the appearance of physiologically active chloroplast protein complexes in the membrane.

I. Introduction

The development of a photochemically competent chloroplast population is an obvious prerequisite for the attainment of photosynthetic competence in leaves. However, a factor that is often overlooked when evaluating the limitations to the photosynthetic productivity of plants is the rate of chloroplast development. It is becoming increasingly apparent that the photosynthetic productivity of many crops and natural vegetation can often be determined by environmental effects on the rate of development of the photosynthetic apparatus. Although the mature thylakoid has been a widely studied biological system and a detailed understanding of its components, structure and function now exist, knowledge of the factors determining the assembly and functional organization of the membrane's photosynthetic apparatus are relatively poor. Thylakoid membrane development is a highly complex process involving co-ordination and regulation of the synthesis of membrane components, their insertion and targeting into the membrane and organization into functional units. In this chapter we review the intrinsic factors determining the development of the thylakoid multimeric protein complexes and the attainment of photochemical competence, and consider the influence of environmental variables on this process.

II. Multimeric Protein Complexes

The PS I, PS II, cytochrome b_6f and ATP synthase complexes are all multimeric protein complexes containing from 4–6 polypeptides in the cytochrome b_6f complex to over 20 polypeptides in PS II. Individual polypeptides of these complexes are encoded by genes located in both the nucleus and the chloroplast (Fig. 1). Each complex contains a precise stoichiometric number of polypeptide subunits. Therefore, there must be precise coordination of the expression of the nuclear and plastid genes. In addition, thylakoid membrane protein complexes contain a large number of non-proteinaceous cofactors such as chlorophylls, pheophytins, quinones, hemes, manganese, iron and iron-sulfur clusters. These cofactors are also bound in precise stoichiometric amounts, and the synthesis and assembly of these cofactors must also be controlled, and may be involved in coordinating the synthesis and assembly of the apoprotein subunits. A further level of control must also exist to maintain the overall stoichiometry of the four major complexes in the thylakoid membrane, although this stoichiometry may vary depending on environmental variables, primarily light quality and quantity.

The synthesis and assembly of thylakoid membrane-protein complexes may be regulated at many stages of gene expression, and evidence exists for control of transcription, mRNA stability, translation and protein turn-over. In this section we will discuss how thylakoid membrane complex synthesis and assembly may be controlled. The chapter will focus on aspects that relate to control of thylakoid protein synthesis and assembly, and maintenance of thylakoid protein stoichiometry. Detailed reviews on mechanisms of nuclear (Chapter 33) and chloroplast (Chapter 31) gene expression and protein translocation (Chapter 7) are provided elsewhere in this book.

A. Transcription

Associated with chloroplast development are large changes in the accumulation of both nuclear and chloroplast encoded mRNAs for thylakoid proteins (Link, 1988). The steady state level of mRNAs from different chloroplast genes may vary depending on the relative transcription rate and message stability (reviewed by Mullet, 1988; 1993; Gruissem, 1989; Gruissem and Tonkyn, 1993). The accumulation of mRNA from different chloroplast genes, at the same developmental stage, can vary by as much as 900 fold in some cases (Rapp et al., 1992). These different levels of mRNA in the chloroplast are thought to be determined by transcription rate regulated primarily

Abbreviations: Cab – Chlorophyll *a/b* binding protein; CP43 – 43 kDa chlorophyll *a*; binding protein of PS II core antenna; CP47 – 47 kDa chlorophyll *a* binding protein of PS II core antenna; P680 – Primary donor of Photosystem II; P700 – Primary donor of Photosystem I; PMS – *N*-methylphenazonium methosulfate; PS I – Photosystem I; PS II – Photosystem II; UTR – untranslated region of mRNA

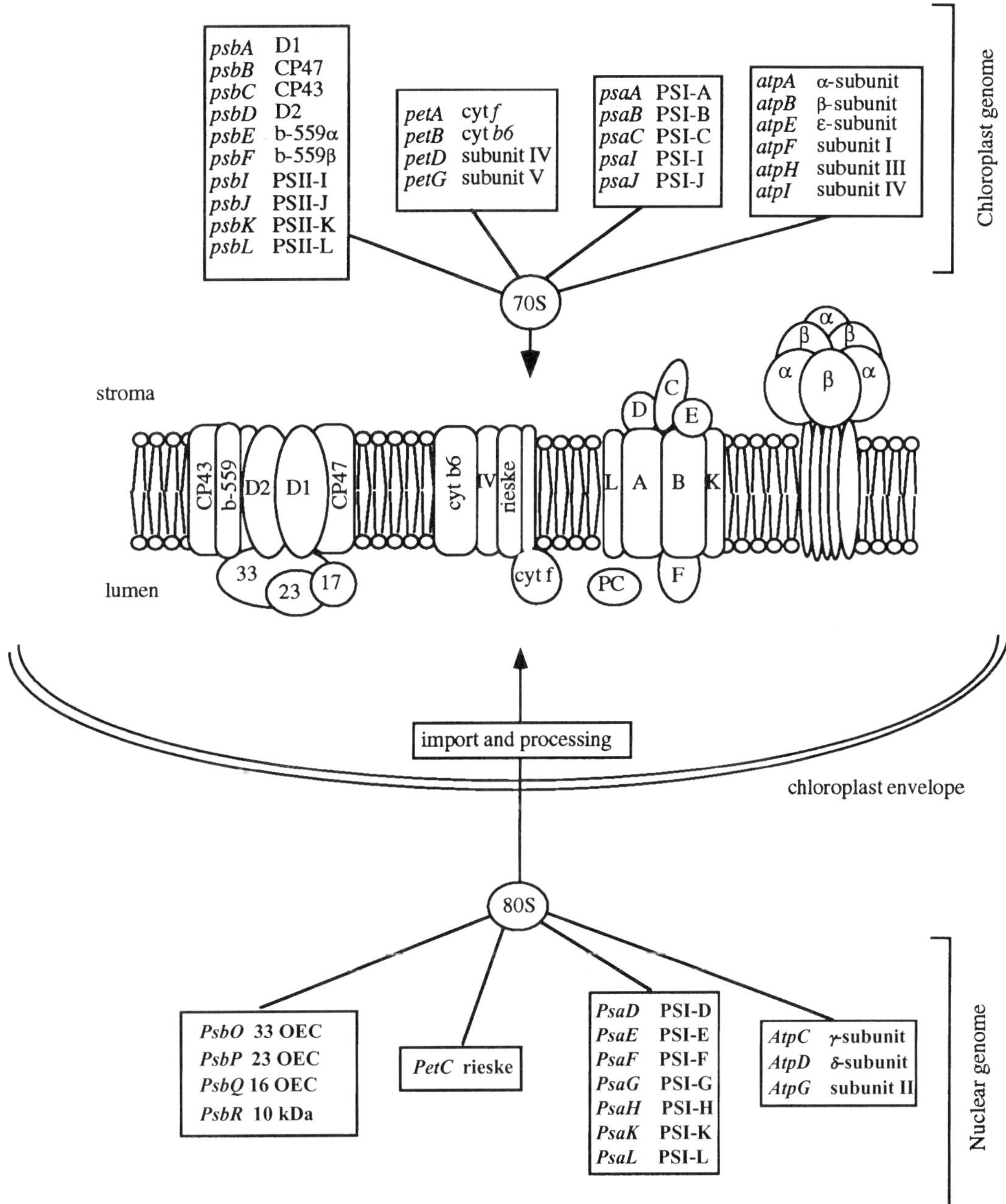

Fig. 1. Outline of the biosynthesis of thylakoid membrane protein complexes. All four major thylakoid complexes are multi-protein complexes encoded by both the nuclear and chloroplast genome. Transcripts from nuclear genes are translated on 80S ribosomes in the cytoplasm and the proteins post-translationally imported into the chloroplast. Chloroplast mRNAs for thylakoid proteins are translated on membrane bound 70S ribosomes and are thought to be co-translationally integrated into the membrane. In eukaryotic organisms, both PS II and PS I are associated with chlorophyll *a/b*-binding proteins (omitted from the diagram for clarity) which are encoded by the nuclear *cab* gene family.

by relative promoter strength (Gruissem and Tonkyn, 1993). The level of chloroplast transcripts may partially account for the relative stoichiometry of polypeptides within individual thylakoid protein complexes (Rapp et al., 1992), although it is clear that other levels of control must be involved to account for the steady state protein levels maintained. The most abundant mRNA for thylakoid proteins is from the *psbA* gene encoding polypeptide D1, which may reflect the requirement for high rates of synthesis of this protein. The maintenance of gene clusters in the chloroplast genome, whereby expression of several genes encoding thylakoid proteins are driven by shared promoters, e.g., the *psbB-psbH-petD-petB* (Hird et al., 1991) and *psbK-psbI-psbD-psbC* (Sexton et al., 1990b) operons, may help maintain transcript levels at a sufficient level to coordinate stoichiometric synthesis of the proteins. In particular, it is interesting that *psbA* is not co-transcribed which is consistent with the notion that the high transcript levels required to maintain D1 at stoichiometric levels would be excessive for more stable thylakoid polypeptides.

During the light-induced conversion of etioplasts to chloroplasts, mRNA accumulation from a number of chloroplast genes increases. In 4.5 day-old dark-grown barley seedlings, chloroplast mRNA from *psbA* and *psaA-psaB* genes are present (Klein and Mullet, 1987; Mullet and Klein, 1987). Upon irradiation there is an increase in *psbA* mRNA over a 24 h period, followed by a decline for a subsequent 72 h (Mullet and Klein, 1987). In the case of *psaA-psaB* mRNA, light produces a steady decline in the mRNA level (Klein and Mullet, 1987; Mullet and Klein, 1987). In spinach the situation is slightly different, mRNA from *psbA* and *atpB* increases upon irradiation of 3 day-old dark grown cotyledons. Whether the changes in mRNA accumulation found in barley and spinach reflect changes in transcription or mRNA stability has been the focus of much research (Deng and Gruissem, 1987; Mullet and Klein, 1987). In barley and spinach, rates of transcription and transcript decay have been analyzed in vitro using run-on chloroplast transcription assays. In both cases it was found that steady state mRNA levels did not always parallel the measured estimate of transcription, and it was concluded that post-transcriptional events had a significant influence on the accumulation of specific chloroplast mRNAs (Deng and Gruissem, 1987; Mullet and Klein, 1987; see section IIB). However, in the case of certain genes an increase in transcription rate has been observed, and it is becoming clear that transcription rate of some chloroplast genes can be modulated during chloroplast development and in response to light and light quality (Deng et al., 1989; Haley and Bogorad, 1990; Klein and Mullet, 1990; Christopher et al., 1992). Differences in the rate of transcription of various chloroplast genes during development and in response to light may in part be due to species differences and developmental stage of the plants being studied. Also, the complicated pattern of many chloroplast transcripts has made it difficult to determine accurately any variation in transcription rate of a specific gene. One of the more interesting features to emerge from studies of light-induced chloroplast mRNAs is that the light induction often involves the use of alternative chloroplast promoters (Haley and Bogorad, 1990; Sexton et al., 1990a; Christopher et al, 1992). For example, in 4.5 day-old dark-grown barley co-transcription of the *psbK-psbI-psbD-psbC* gene loci gives rise to at least 6 mRNAs when isolated mRNA is analyzed by northern hybridization with DNA probes specific for *psbD* or *psbC* (Gamble et al., 1988). Upon irradiation for 108 h the total mRNA from this transcript declined and was composed of only two mRNA species of 4.0 and 3.2 kb (Gamble et al., 1988). Thus only a subset of the transcripts from the *psbD-psbC* operon is maintained with irradiation. A combination of nuclease mapping and 5′ capping assays has shown that transcripts maintained in the light originate from a second promoter, located within the *psbD-psbC* cluster. This second promoter is virtually inactive in the dark, and is induced by blue light (Gamble and Mullet, 1989; Sexton et al., 1990a, Christopher and Mullet, 1994). Many chloroplast promoter sequences strongly resemble bacterial –10 and –35 elements suggesting that chloroplasts contain sigma factor-like components that aid correct transcription initiation (Helman and Chamberlin, 1988). In the case of the *psbK-psbI-psbD-psbC* of barley chloroplasts, DNA sequences upstream of the blue light induced transcript show no homology to typical chloroplast or *E. coli* promoter consensus sequences. However, inhibitor studies have shown that a nuclear encoded factor is required to activate this promoter (Gamble and Mullet, 1989). In prokaryotes, different sigma factors are used to initiate transcription of different genes that are induced under different environmental conditions, e.g., heat shock. It is possible that different chloroplast 'sigma factors' may be involved in activation of chloroplast promoters in response to light and perhaps other environmental variables. Tiller and Link (1993) isolated 'sigma factor-like' transcription factors from mustard etioplasts that are functionally distinct from those found in chloroplasts. A plastid extract from 4 h illuminated wheat plastids has been shown to contain a factor required for in vitro transcription of the *psbD-psbC* light-induced transcript (Wada et al., 1994).

During greening of etiolated leaves light increased

the rate of transcription of a number of nuclear genes encoding thylakoid proteins including *Cab*I, *Cab*II, *PsaD*, *PsbO*, *PsbP*, *PsbQ*, and *PsbR* (see Thompson and White, 1991; Chapter 33). The increased rate of transcription is mediated by phytochrome and/or the blue light receptor. Experiments with transgenic plants have allowed dissection of *cis*-acting sequences required for light, developmental and organ specific expression of several nuclear genes (Castresana et al., 1988; Manzara and Gruissem, 1988; Gidoni et al., 1989; Stockhaus et al., 1989; Lam and Chua, 1990). Trans-acting factors that interact with specific DNA sequences have been identified in plant extracts (see Gilmartin et al., 1990; Schindler and Cashmore, 1990). The mechanism of the signal transduction pathway by which light induces gene expression is being studied using a combination of genetic and biochemical approaches (reviewed by Thompson and White, 1991) and recent results implicate a role for calcium, calmodulin and G-proteins (e.g., Neuhaus et al., 1993). While phytochrome is the prominent photo-receptor in dark-grown leaves, green leaves contain much lower levels of type I phytochrome. The changes in relative levels of phytochrome and blue-light receptors in green leaves might imply a more important role for blue light in regulating gene expression in plants growing in the natural environment (Jenkins and Smith, 1985). This may be of special importance when considering mechanisms for changes in photosystem stoichiometry and light harvesting capabilities during adaptation to differing light environments.

The presence of a plastid factor controlling expression of nuclear genes encoding chloroplast proteins has been inferred from studies with carotenoid deficient mutants (Mayfield and Taylor, 1984; Taylor, 1989) and with plants treated with norflurazon, an inhibitor of carotenoid biosynthesis (Oelmuller and Mohr, 1986). In the absence of carotenoids, high light causes photobleaching resulting in disappearance of chlorophyll and many chloroplast proteins of both nuclear and plastid origin. Even though photooxidative damage was limited to the chloroplast, reduced levels of mRNA from *Cab*, *RbcS* and *PsbR* (ST-LS1) genes were observed (Batschauer et al., 1986; Burgess and Taylor, 1988; Stockhaus et al., 1989; Tonkyn et al., 1992). The nature of the chloroplast factor controlling this phenomenon is unknown, although it is thought to be a metabolic intermediate rather than a protein or RNA species (Taylor, 1989). Such a notion is consistent with reports that metabolites, such as acetate, sucrose and glucose (Sheen, 1990; Krapp et al., 1993), can modulate expression of nuclear genes such as *Cab* and *RbcS*. However, these metabolites generally repress expression of the nuclear genes, whereas the plastid factor is thought to be an activator that can be destroyed by photooxidation.

B. Stability of mRNA

Run-on transcription assays show that differential mRNA stability also plays an important role in determining the steady-state level of chloroplast mRNAs (Deng and Gruissem, 1987; Mullet and Klein, 1987). Biochemical and genetic evidence indicates that both *cis*-acting elements and *trans*-acting factors are involved in mRNA stabilization. It has been shown that both 5′ and 3′ untranslated regions can serve as RNA stabilizing elements (Stern and Gruissem, 1987; Sakamoto et al., 1993; Salvador et al., 1993). At the 3′ untranslated region inverted repeat sequences, that can fold into a stable stem-loop structure, help protect the mRNA from degradation by 3′ $\rightarrow$ 5′ exonuclease activity. The 3′ inverted repeats are important for stabilization of in vitro generated mRNAs added to chloroplast lysates (Stern et al., 1989; Adams and Stern, 1990) and have also been shown to be of importance to mRNA stability in vivo (Stern et al., 1991; Stern and Kindle, 1993). While it may be speculated that inverted repeats in the 5′ region also stabilize the mRNA, the *cis*-acting sequences involved have not been clearly identified. The presence of mRNA stabilizing elements at the 5′ and 3′ ends of chloroplast mRNAs implies that a rate limiting step in degradation is the initial endonucleic cleavage making the mRNA more susceptible to an exonuclease. Chloroplast fractions enriched in ribonuclease activity have revealed the presence of both exo- and endo-ribonucleases that possibly function in mRNA maturation and degradation in vivo (Hsu-Ching and Stern, 1991a).

Gel retardation assays and UV crosslinking have shown that several proteins interact with the chloroplast mRNA 3′ sequences (Stern et al., 1989). Some of the mRNA-binding proteins are nucleotide sequence specific and may serve to stabilize gene specific transcripts (Hsu-Ching and Stern, 1991b). Genes for several chloroplast RNA-binding proteins have been isolated and sequenced (Schuster and Gruissem, 1991; Mieszezak et al., 1992) revealing that each protein contains conserved motifs typical of RNA-binding proteins. A spinach chloroplast 28 kDa RNA-binding protein is induced by light, and may be important in modulating mRNA stability during chloroplast development (Schuster and Gruissem, 1991). Further studies of the 28 kDa ribonucleotide-binding protein, and similar studies with other mRNA-binding proteins, are required to determine if they are involved in regulating the differential accumulation of chloroplast RNAs in response to light, and other environmental and developmental signals.

Genes that may encode RNA-binding proteins have not been identified in any of the chloroplast genomes that have been entirely sequenced. It is most probable that control of mRNA stability within chloroplasts will be controlled by the expression of nuclear genes. In line with this, photosynthesis-deficient nuclear mutants of *Chlamydomonas reinhardtii* that are unable to accumulate transcripts from the *psbB* (Sieburth et al., 1991; Monod et al., 1992), *psbC* (Sieburth et al., 1991), *psbD* (Kuchka et al., 1989) and *atpB* (Drapier et al., 1992) genes have been isolated. Each mutant affects accumulation of only one specific transcript, suggesting that each chloroplast message requires at least one nuclear encoded factor for stabilization. One *C. reinhardtii* nuclear mutant, *nac2-26*, has provided evidence that 5′ mRNA sequences of *petD* may interact with the nuclear mRNA stabilizing factor (Kuchka et al., 1989; Rochaix, 1992). These nuclear-encoded factors are presumably related to the RNA-binding proteins identified in chloroplast extracts; however, more work is required to establish this as fact.

The stability of several chloroplast mRNAs has been observed to change in vivo with leaf age (Klaff and Gruissem, 1991) and chloroplast development (M. Kim et al., 1993). How selective stabilization of chloroplast mRNAs is regulated during chloroplast development in vivo is unknown. The role of inverted repeat sequences and specific mRNA binding proteins has been discussed above. It has also been demonstrated that translation rate and polysome association of chloroplast mRNAs can determine transcript stability in leaves (Klaff and Gruissem, 1991) and cells of *C. reinhardtii* (Rochaix et al., 1989; Drapier et al., 1992; Xu et al., 1993). Thus, in vivo, the stability of chloroplast mRNAs may be controlled by a complex interaction of a number of different and perhaps counteracting pathways.

In the case of nuclear genes, expression is thought to be controlled predominantly at the level of transcription. Regulation of stability of mRNA from nuclear genes has been reviewed by Gallie (1993). While nuclear encoded mRNA undergoes considerable post-transcriptional modification, only changes in stability of mRNA from *RbcS* (Ernst, et al., 1987; Shirley and Meagher, 1990; Thompson and Meagher, 1990) and *Fed-1* (encoding ferredoxin I; Elliot et al., 1989; Dickey et al., 1992) have been demonstrated to be light regulated.

C. Translation

1. Translational Control

In etiolated leaves, mRNAs encoding many thylakoid proteins are present. However, the polypeptides, particularly those involved in chlorophyll binding, do not accumulate until the leaves are exposed to light (Klein and Mullet, 1986;1987; Kreuz et al, 1986; Hird et al., 1991). For example, in barley, mRNA from *psaA-psaB* and *psbA* are present in dark-grown leaves, yet the apoproteins are absent (Klein and Mullet, 1987). In the *y-1* mutant of *C. reinhardtii*, which is unable to synthesize chlorophyll without light, chlorophyll-binding proteins are also absent in the dark (Malnoe et al., 1988; Herrin et al., 1992). In wheat, mRNA from the cotranscribed *psbB* and *psbH* genes, encoding the 47 kDa chlorophyll *a*-binding protein and the 10 kDa phosphoprotein of PS II, are present in etiolated leaves (Hird et al., 1991). While the 10 kDa phosphoprotein is present in etioplasts, the 47 kDa chlorophyll *a*-binding protein only accumulates after exposure to light (Hird et al., 1991). The latter demonstrates that there is not an overall block on chloroplast mRNA translation. Certain proteins, such as the 10 kDa phosphoprotein of PS II, are constitutively expressed in etioplasts and chloroplasts. Therefore, there is specific regulation of certain proteins such that they are either not translated in the dark, or are rapidly degraded in the etioplast.

Since the light-induced accumulation of thylakoid proteins is primarily associated with chlorophyll-binding proteins, it is thought that chlorophyll plays a key role in mediating chlorophyll-binding protein accumulation. Availability of chlorophyll could stabilize newly synthesized proteins, or somehow control translation. By pulse labeling of proteins in barley etioplast with 35S-methionine, Klein and Mullet (1986) were unable to detect the synthesis of chlorophyll apoproteins suggesting a block in translation or very rapid degradation of the newly synthesized protein. In barley it was found that mRNAs encoding the chlorophyll *a*-binding apoproteins are associated with polysomes in dark-grown leaves (Klein et al., 1987), and that following illumination there was no significant shift in the distribution of mRNAs in the polysome profile. These results suggest that in etioplasts elongation of the polypeptide chain is blocked at some stage(s), or that polypeptides are not released from the ribosome, rather than a block in translation initiation. Upon illumination, the onset of chlorophyll biosynthesis is thought to trigger continued synthesis of the chlorophyll-binding protein. The importance of chlorophyll synthesis in apoprotein accumulation has been directly tested in etioplasts where it was demonstrated that addition of chlorophyll biosynthesis precursors chlorophyllide-*a* and phytyl-pyrophosphate, which induce chlorophyll synthesis in the dark, results in chlorophyll-binding protein accumulation (Eichacker, et al., 1990). It is possible that binding of chlorophyll *a* to the apoprotein somehow facilitates translation and/or release of the protein from the ribosome complex. Kim et al. (1991) detected D1 protein translation intermediates associated with polysomes in pulse-labeled chloroplasts suggesting that a lack of chlorophyll might result in stalling or pausing of protein synthesis. However, recent work quantifying the abundance of translation initiation complexes, and the extent of translational run-off in the presence of lincomycin, indicates that chlorophyll availability has little influence on either translation initiation or elongation (Kim et al., 1994). It was concluded from these studies that chlorophyll binding to newly synthesized apoproteins leads rapidly to protein stabilization.

There is strong genetic evidence that nuclear-encoded factors are required for translation of chloroplast mRNAs. As with the factors involved in mediating mRNA stability, these factors are transcript specific, regulating the translation of individual thylakoid proteins. Nuclear mutants of *C. reinhardtii* have been characterized that are unable to translate chloroplast mRNAs from *psbC* (Rochaix et al., 1989), *psbD* (Kuchka et al., 1988), and *atpA* (Drapier et al., 1992). In each case, the native mRNAs are present at wild-type or slightly elevated amounts. A chloroplast suppressor of the nuclear F34 mutant (Rochaix et al., 1989), deficient in PS II, has been isolated and shown to be due to a single base change in the 3′ region of the *psbC* mRNA. The single base change may destabilize a potential stem loop sequence that is a target for a nuclear-encoded translational activator. Ribonuclear-binding proteins have been identified that bind to the 5′UTR of several chloroplast mRNAs (Danon and Mayfield, 1991). One of these proteins, a 47 kDa polypeptide that binds to the *psbA* 5′UTR, has been purified. The protein is present in low amounts in etioplasts and accumulates with chloroplast development. While light does not induce synthesis of the polypeptide, affinity for the *psbA* 5′ region is enhanced, suggesting that a post-translational modification may be involved in mediating translation of D1 (Danon and Mayfield, 1991). Recently, it has been demonstrated that reduced thioredoxin can reduce the *psbA* mRNA-binding proteins increasing their ability to bind to the 5′UTR (Danon and Mayfield, 1994). Less is known about the control of chloroplast translation in plants. A number of nuclear mutants of maize have been characterized that are deficient in cytochrome b_6f and PS I protein accumulation (Barken et al., 1986). The mRNA levels in many of the mutants are unaltered suggesting that the lesions occur at post-transcriptional stages. It remains to be seen if these mutants are impaired in translation of specific mRNAs, or whether other assembly processes are impaired.

In the future it will be important to dissect the relative importance of the various stages of translational control in regulating synthesis of thylakoid proteins. It will be important to distinguish between light effects on chlorophyll availability, which will rapidly stabilize apoproteins, versus regulation of *trans*-acting factors controlling translation initiation. Light has also been shown to increase the specific activity of an elongation factor, EF-G, in pea chloroplasts (Akkaya and Breitenberger, 1992). Further studies into the control of translation during chloroplast development will be an important future area of research as we begin to understand in more detail the components involved in initiation, elongation and termination in plastids.

Detectable amounts of Cab proteins of PS II are absent in etiolated leaves even though detectable levels of the mRNA are present (Bennett, 1981). This was shown to be due to rapid post-translation degradation of the protein (Bennett, 1981). In the absence of chlorophyll *b* only, as found in chlorophyll *b*-less mutants, a similar destabilization of the Cab protein is observed (e.g., Bellemare et al., 1982). Growth of *Lemna gibba* under intermittent red light, which allows synthesis of chlorophyll *a* but not chlorophyll *b*, has provided evidence that there may also be translational control of Cab protein biosynthesis (Slovin and Tobin, 1982).

2. Co-ordinated Translation

There is limited evidence that the synthesis of certain chloroplast proteins may be coordinated. This is best characterized in *C. reinhardtii*, where rapid pulse labeling of proteins is possible. A *C. reinhardtii* mutant, *thm24*, that cannot synthesize the β-subunit of the CF1 complex also exhibits a reduced rate of synthesis of the α-subunit (Drapier et al., 1992). Mutants with a reduced rate of synthesis of the α-subunit show an enhanced rate of synthesis of the β-subunit (Drapier et al., 1992). In the case of the PS I reaction center, frameshift mutants in *psaB* that do not produce the PsaB polypeptide also do not synthesize the PsaA protein (Girard-Bascou et al., 1987). However, the converse is not true since mutants unable to synthesize the PsaA protein still synthesize PsaB (Girard-Bascou et al., 1987; Takahashi et al., 1991), although the PsaB polypeptide is subsequently rapidly degraded. In the case of PS II, a mutant lacking D2 exhibits reduced rates of synthesis of both D1 and CP47 (Erickson et al., 1986; deVitry et al., 1989). In the absence of D1 synthesis, however, D2 protein synthesis is unaffected, although synthesis of CP47 is reduced (Jensen et al., 1986; Bennoun et al., 1986; de Vitry et al., 1989). The mechanism controlling coordinated synthesis of these polypeptides is unknown. Many of the chloroplast polypeptides are synthesized on membrane bound ribosomes, and membrane integration and assembly may occur as the protein chain is elongating. If assembly requires formation of a dimer, as in reaction center polypeptides, or close interaction with a second protein of the complex, then the lack of synthesis of one polypeptide may prevent efficient assembly of other proteins of the complex, and result in inhibition of translation or rapid degradation. However, it is interesting that translation of only one protein of the PS II or PS I heterodimer is tightly coupled to synthesis of the other. For PS II, rates of D1 synthesis apparently do not alter the rate of D2 synthesis and for PS I rates of synthesis of PsaA do not influence PsaB synthesis. In the case of PS II, the selective turnover of D1 under high light may require higher rates of synthesis of D1 than D2, and so tight coupling in rates of synthesis would result in wasteful synthesis of D2.

D. Post-translational Regulation

1. Assembly

Factors that control the assembly of thylakoid protein complexes are almost completely unknown. It is generally thought that absence of an individual polypeptide of the PS II, cytochrome b_6f and ATP synthase complex prevents the assembly of the remainder of the complex, and this seems also to be the case for several PS I proteins. The chlorophyll *a/b* light harvesting complexes associated with PS II and PS I assemble without the core polypeptides and vice versa. This decoupling may be important in regulating the light harvesting capabilities of the two reaction centers when grown under different light quantity and quality. Polypeptides of the oxygen evolving complex also have been reported to accumulate without PS II (de Vitry et al., 1989; Palomares et al., 1993).

Genetic transformation of cyanobacteria has allowed gene deletion approaches to study the requirement of specific polypeptides for the assembly of thylakoid proteins, in particular PS II (reviewed by Vermaas, 1993) and PS I (Golbeck, 1992). Recent advances in chloroplast transformation in *C. reinhardtii* (Boynton et al., 1988) have allowed site-directed and reverse genetic approaches to study assembly of PS II (Roffey et al, 1991), PS I (Takahashi et al., 1991; Webber et al., 1993; Cui et al., 1995), ATPase (Frasch WD, personal communication) and the cytochrome b_6f complex (Kuras and Wollman, 1994). In cyanobacteria, deletion of *psbA* (Nilsson et al., 1990), *psbD* or *psbB* genes (Vermaas et al., 1988) leads to a lack of assembly of the remainder of the major PS II polypeptides. In the absence of CP43, due to interruption of *psbC*, the remainder of the PS II complex is still able to assemble, and is capable of normal charge separation and electron flow (Rögner, et al., 1991). Deletion of either of the PS I reaction center genes, *psaA* and *psaB*, prevents the assembly of the remainder of the PS I reaction center in *Synechocystis* PCC 6803 (Smart et al., 1991). Deletion of *psaD* (Chitnis et al., 1989a), *psaE* (Chitnis et al., 1989b), *psaF* (Chitnis et al., 1991) and *psaL* (Chitnis et al., 1993) genes still allow photoheterotrophic growth of the mutants implying that PS I still assembles in these deletion mutants, although all subunits may not assemble normally in the absence of others (Cohen et al., 1993). In *Anabaena variabilis* ATCC 29413 (Mannan et al., 1991) deletion of the *psaC* gene also does not disrupt the assembly of PS I, although in *C. reinhardtii* similar disruption of *psaC* in chloroplasts results in destabilization of the remainder of the PS I complex (Takahashi et al., 1991). This latter observation suggests there are significant differences in the role of certain polypeptides in assembly of membrane-protein complexes in different organisms. As more studies are performed in *C. reinhardtii* it is likely that further differences in assembly requirements will be found between prokaryotic and eukaryotic organisms.

It has been suggested that the Rieske iron-sulfur protein plays a key role in assembly of the cytochrome b_6f complex. A mutant of *Lemna perpusilla* that is unable to synthesize the Rieske protein also does not accumulate cytochrome b_6f (Bruce and Malkin, 1991). However, cytochrome *f*, cytochrome b_6 and subunit IV are synthesized and, at least in the case of subunit IV, rapidly degraded (Bruce and Malkin, 1991). This argues that Rieske protein is essential for assembly and stability of the cytochrome b_6f complex, and thus control of Rieske protein synthesis could control the accumulation of the complex during thylakoid biogenesis. Similarly, without either the α, β or ε subunits (Lemaire and Wollman, 1989; Robertson et al., 1989; 1990) the remainder of the ATPase complex does not assemble although all subunits are synthesized. Again, rapid degradation of unassembled subunits is thought to occur. Taken together, these studies suggest that there must be concerted synthesis of the majority of the thylakoid proteins, for each complex, in order for correct assembly to occur. The mechanisms regulating this are unknown, but further studies of co-ordinated translation should begin to provide insights into how the synthesis of individual proteins is controlled so that stoichiometric quantities of each polypeptide are available for efficient assembly.

PS II (Ikeuchi and Inoue, 1988; Koike et al., 1989; Webber et al., 1989a,b), PS I (Ikeuchi et al., 1990) and the cytochrome b_6f complex (Haley and Bogorad, 1989; Schmidt and Malkin, 1993) contain a number of low molecular weight polypeptides of 10 kDa or less. These small polypeptides are generally encoded by chloroplast genes and are predicted to contain a hydrophobic membrane spanning region such that the N- and C-terminus are located on opposite sides of the thylakoid membrane. The function of these low molecular weight polypeptides is unknown. Deletion of *psbJ* (Lind et al., 1993) and *psbK* (Ikeuchi et al., 1991) genes in *Synechocystis* 6803 leads to reduced levels of PS II, although otherwise the complex operates normally. Therefore, it appears likely that these small polypeptides are important for either efficient assembly, or help stabilize the complex. It is interesting that, in contrast to most other thylakoid proteins, some of the small polypeptides, e.g., the 10 kDa phosphoprotein and cytochrome *b*-559, are present in thylakoid membranes in the absence of other PS II proteins. The 10 kDa phosphoprotein has been detected immunologically in etioplasts (Hird et al., 1991) when other proteins (particularly the chlorophyll binding proteins) do not accumulate. It is possible that these small polypeptides serve as a nucleus for assembly of other hydrophobic membrane proteins during thylakoid development.

2. Protein Turnover

There is a large amount of evidence for rapid, and sometimes selective, post-translational degradation of unassembled or damaged thylakoid-membrane proteins. These post-translational events may be important in determining the level to which thylakoid proteins accumulate, and perhaps function as a proof-reading mechanism to ensure that mis-assembled complexes do not accumulate. In PS II, the turn-over rate of the D1 protein is significantly faster than other protein components (Mattoo et al., 1984). Similar behavior has been reported for the cytochrome b_6f complex, where turn-over of cytochrome b_6 is slower than other components of the complex (Bruce and Malkin, 1991). These results appear to demonstrate that individual protein components can be selectively degraded, resynthesized and reinserted into the complex. In PS II, selective degradation and resynthesis of D1 (Chapter 2) is important because it becomes damaged by oxygen radicals produced during electron transfer, especially under conditions that promote photoinhibition. The significance of selective removal and insertion of proteins in complexes other than PS II is unknown.

Rapid degradation of unassembled thylakoid proteins suggests the presence of chloroplast proteases (reviewed by Vierstra, 1993). Proteases involved in post-translational modification of chloroplast proteins following import into the chloroplast have been intensively studied and are discussed in Chapter 7. Here, we will focus on proteases involved in degradation of thylakoid proteins. Protease enriched fractions, one of which requires ATP for activity, have been isolated from pea chloroplasts (Liu and Jagendorf, 1984; 1986; Malek et al., 1984). Further characterization of the pea chloroplast proteases revealed a neutral endopeptidase and three amino peptidases, based on their specificity for different amino acyl-β-naphthylamide substrates (Liu and Jagendorf, 1986). An interesting feature of the endopeptidase is that it is inhibited by dithiothreitol (Liu and Jagendorf, 1986) suggesting it may be under redox control. A protease purified from *C. reinhardtii y-1* membranes degrades newly synthesized Cab proteins, and other unidentified thylakoid membrane proteins (Hoober and Hughs, 1992). Inhibitor studies indicate that *C. reinhardtii* protease is different from commonly found proteases, and may contain an active histidyl moiety (Hoober and Hughs, 1992). A prolyl endopeptidase has also been isolated from spinach thylakoids that specifically degrades the 18 kDa polypeptide of the oxygen-evolving complex (Kuwabara, 1992). The scant literature suggests that there are multiple proteases involved in degradation of thylakoid membrane components. Future work will probably reveal numerous chloroplast proteases subject to various forms of control, as indicated by the finding that chloroplasts probably contain a protein related to the *E. coli* ATP dependent ClpA/P protease based on gene homology comparisons (Vierstra, 1993).

The selective turn-over of specific polypeptides (e.g., D1) further suggests that there are specific proteases that degrade individual polypeptides, or that proteins can be targeted for selected degradation. However, only one isolated protease (Kuwabara, 1992) has actually been reported to be specific for a single polypeptide (18 kDa protein of PS II). What targets a thylakoid protein for degradation is unknown. It was suggested that the prolyl endopeptidase from spinach recognizes higher-order structure of the 18 kDa oxygen-evolving complex protein (Kuwabara, 1992). Therefore, possibilities might include changes in protein conformation among assembled, unassembled and damaged forms of a polypeptide, or other post-translational modifications. Considering the importance of post-translational turn-over in thylakoid membrane assembly, our lack of understanding of the proteolytic events in thylakoids is a major gap in our knowledge.

E. Conclusions

There is evidence that all stages of gene expression are under tight control. Which of these stages are limiting to the production of functional thylakoid proteins and protein complexes remains a challenging question to address. Stern et al. (1991) have shown, by generating chloroplast mutants lacking the 3′ inverted repeat of *atpB*, that reduction in the level of mRNA from *atpB* by 90% results in approximately a 40% reduction in β-subunit production. *C. reinhardtii* cells containing reduced DNA copy numbers, due to inhibitor treatment, have reduced mRNA levels from *rbcL* and *atpA* and yet show normal rates of synthesis of the corresponding polypeptides (Hosler et al., 1989). Thus, mRNA levels may be in excess in the chloroplast, and events limiting to expression may be associated with control of translation and post-translational events. This conclusion is based on a very limited number of studies, and more work is required with different genes, and at varying stages of plastid development.

Assembly of thylakoid complexes may be limited by availability of individual polypeptide components. Cytochrome b_6f accumulation could be limited by availability of the nuclear encoded Rieske subunit, and thus regulation of *petC* expression could control cytochrome b_6f levels. PS II and PS I reaction center accumulation are more likely to be controlled by availability of chloroplast-derived subunits. Thus, control of PS II and PS I stoichiometry is ultimately likely to be controlled by rates of chloroplast protein synthesis and chlorophyll availability. Much of the evidence discussed previously indicates that chloroplast protein synthesis in turn is controlled by yet mostly unidentified products of nuclear genes. Recent advances in chloroplast transformation will allow the *cis*-acting sequences that are the targets of nuclear factors to be determined. Together with the analysis of the nuclear genes and their products, a more detailed understanding of the control of assembly will be obtained.

III. Photochemical Activities

Attainment of photosynthetic competence requires not only the synthesis and assembly of thylakoid complexes, but also their organization into functional entities. As is clear from the previous section, significant progress has been made in our understanding of the synthesis of chloroplast protein complexes. Unfortunately little is known about the molecular factors determining the functionality of assembled complexes. The majority of studies on the development of thylakoid activities are not recent and the majority are descriptive rather than analytical with respect to determinants of function. Here we review some of the important observations that may serve to provide a useful base for the resolution of the factors determining the appearance of membrane photochemical activities. Where appropriate, references will be made to previous, more extensive review articles rather than the numerous original papers being cited.

A. Electron Transport

Primary photochemical activities of the reaction centers of PS I and PS II would appear to develop within minutes of the synthesis and insertion of the component photosystem proteins into the membrane. The detection of variable chlorophyll *a* fluorescence, indicative of the light-induced reduction of Q_A by P680, from greening etiolated leaves within minutes of illumination demonstrates that photochemically active PS II reaction centers are present immediately, or very soon, after the insertion of the component proteins of PS II into the membrane (Baker, 1984). A single flash, which produces a single turnover of protochlorophyllide reductase in etiolated leaves, is sufficient to induce the development of PS II primary photochemistry; in 5-day-old etiolated barley leaves variable fluorescence can be detected within 25 min of exposure to a single 1 ms flash (Franck, 1993). There are many reports of significant levels of PS I-mediated electron transport in thylakoids isolated from etiolated leaves within minutes of illumination (Baker, 1984). These data imply the presence of P700, although spectroscopic evidence of the photo-oxidation of P700 has not been detected within such short times, presumably due to the very low P700 concentrations in the tissue (Baker and Butler, 1976; Baker, 1984). The rapid development of PS I and PS II primary photochemical activities in etiolated leaves is associated with the reduction of protochlorophyllide to chlorophyllide and the biosynthesis of chlorophyll-binding polypeptides (Franck, 1993). Accumulation of CP43 and CP47 proteins of PS II and the PsaB apoprotein has been shown to be triggered by the onset of chlorophyll synthesis in barley etioplasts (Eichacker et al., 1990).

Studies with etiolated leaves greened under intermittent irradiation show that factors other than the development of PS II primary photochemistry are associated with the appearance of water photooxidation (Baker, 1984). Leaves grown in intermittent light flashes develop PS II complexes that exhibit photoreduction of Q_A but not photo-oxidation of water. On exposure to continuous illumination, the ability to photooxidize water rapidly appears. The 16, 23 and 33 kDa proteins of the extrinsic oxygen-evolving complex have been detected in etioplasts (Ryrie et al., 1984; Liveanu et al., 1986) and their association with the PS II core complex occurs concomitantly with the appearance of the ability to photooxidize water (Ono et al., 1986), as does the incorporation of Mn into the complex and the appression of adjacent thylakoid vesicles (Baker, 1984).

In leaves grown in a natural diurnal light environment both water photooxidation and linear electron transport from PS II through PS I with associated proton pumping into the thylakoid lumen were detected at very early stages of chloroplast development (Webber et al., 1986), thus demonstrating the well co-ordinated development of the thylakoid photochemical apparatus during normal leaf biogenesis. This strict co-ordination of thylakoid development can be disrupted by subjecting leaves to different light regimes during plastid biogenesis, e.g., exposing dark-grown etiolated leaves and leaves grown in intermittent light to continuous illumination. Changes in temperature during chloroplast development can also break the co-ordinated development process (see section IV).

B. Photophosphorylation

Significant rates of N-methylphenazonium methosulfate (PMS)-mediated cyclic photophosphorylation have been observed in thylakoids isolated from etiolated leaves minutes after exposure of the leaves to continuous illumination and appear to be dependent upon the development of PS I activity. However, non-cyclic photophosphorylation has generally been detected only at much later stages of thylakoid development, presumably because the sustainable rate of linear electron transport at such early stages of development is considerably less than the PMS-mediated rate of cyclic electron flow around PS I (reviewed in Baker, 1984). Clearly, thylakoids contain active ATP synthase at very early stages of development and can use a light-generated proton electrochemical potential difference across the membrane to phosphorylate ADP. Although it has not been definitively demonstrated, primarily due to the technical difficulty, it is most likely that the appearance of photophosphorylation in leaves developing in a natural diurnal light regime is limited by the ability of the thylakoids to generate a sufficiently large proton electrochemical potential difference across the membrane and not by the absence of an active ATP synthase.

IV. Environmental Influences

Changes in the physical environment of a plant can have profound effects on the biogenesis of the photosynthetic apparatus. There are many reports of changes in nutrient and water status and exposure to pollutants modifying thylakoid composition and photosynthetic function; however, little is known of the mechanistic bases of such perturbations of thylakoid developmental biology. It is beyond the scope of this chapter to review comprehensively the relevant literature on this topic. Perhaps more relevant here is an anecdotal consideration of the effects of light and chilling temperatures on thylakoid biogenesis that demonstrate the complexity of interactions between thylakoid biogenesis and the physical environment.

The greening of etiolated leaves is triggered by light and involves controls on regulation of synthesis and assembly of the components of the thylakoid protein complexes. The photoreceptors known to be involved in this developmental response are phytochrome and protochlorophyllide; light converts protochlorophyllide to chlorophyllide, signal transduction is initiated, and the abundance of phytochrome and protochlorophyllide reductase mRNAs rapidly decrease (Thompson and White, 1991). It has been suggested that the initiation of synthesis and assembly of the photosystems following a single flash may only involve the photoconversion of protochlorophyllide, and not phytochrome, since these events are not inhibited by far red light (Franck, 1993).

The mechanisms of developmental responses of leaves growing in natural diurnal light regimes to changes in the light environment appear to be different from those of etiolated leaves, and care must be exercised in extrapolating information from greening etiolated systems to explain chloroplast development phenomena in natural light environments. For example, phytochrome is responsible for triggering the synthesis of mRNAs for LHCII polypeptides and the small subunit of Rubisco in greening etiolated pea leaves, but large changes in phytochrome photoequilibrium in light-grown pea leaves had no effect on the abundance of these mRNAs (Jenkins and Smith, 1985). Light-grown leaves appear to be more responsive to changes in blue than red light, and it has been suggested that phytochrome may only play a minor role in the light-induced regulation of chloroplast development in these leaves (Thompson and White, 1991).

Leaves of most species respond to increases in growth irradiance by increasing the capacity for linear electron transport per chlorophyll (Evans, 1996). Such acclamatory responses involve increases in the concentration of PS I, PS II, cytochrome b_6f and ATP synthase complexes relative to the antennae chlorophyll-protein complexes (Anderson, 1986). Enhancement of the far red region of the spectrum, as would occur due to shading by vegetation, results in an increase in the ratio of PS II:PS I complexes (Chow et al., 1990; McKiernan and Baker, 1991; J. H. Kim et al., 1993). Changes can also occur in the stoichiometry of light-harvesting antenna complexes and reaction centers (Anderson, 1986). Intensity changes alone apparently do not elicit such a change in reaction center stoichiometry or size of light-harvesting complex per reaction center in pea leaves (Lee and Whitmarsh, 1989). Such light-induced changes in the stoichiometry of thylakoid components have been implicated in the optimization of photosynthetic performance (Evans, 1996), and are important, but mechanistically poorly understood, acclamatory responses. The few studies that have investigated the molecular basis for photosystem stoichiometry adjustments have indicated that light quality can regulate chloroplast gene expression (Glick et al., 1986; Deng et al., 1989). Germinating spinach seedlings adapted to red light show a 10-fold decrease in *psaA* mRNA levels that correlates with decreased levels of PS I polypeptides (Deng et al., 1989). Run-on transcription assays demonstrate that the decrease in mRNA from *psaA* under red light growth conditions is a result of a decrease in *psaA* transcription activity and *psaA* transcript stability (Deng et al., 1989).

Temperature has wide ranging effects on the rate of chloroplast development and photosynthetic performance. However, the majority of these effects are attributable to the direct, and often differential, effects of temperature on metabolic reactions. Of particular interest in the context of thylakoid biogenesis is the effect of chilling temperatures on developing maize, a subtropical plant that is particularly sensitive to chilling. Reduction of temperature inhibits selectively the accumulation of some chloroplast, relative to nuclear, encoded thylakoid proteins in maize mesophyll cells (Nie and Baker, 1991) and suggests that chilling may inhibit preferentially the expression of chloroplast genes as has been found in rice (Hahn and Walbot, 1989; Maruyama et al., 1990). Although in vivo labeling studies indicated that the synthesis and/or assembly of chloroplast-encoded thylakoid membrane polypeptides in maize was more sensitive to low temperature than was the case for the nuclear-encoded thylakoid polypeptides, it also appeared that the chloroplast-encoded thylakoid proteins that were inserted into the membrane were rapidly broken down (Nie and Baker, 1990). More recent studies have shown that the stability of newly synthesized D1 in the thylakoid membranes of isolated maize chloroplasts was dependent upon both the temperature at which the plants were grown and on the temperature during the pulse-labeling period when the protein was being synthesized and assembled (Bredenkamp and Baker, 1994). The factors determining this phenomenon are not known, although it has been speculated that temperature-induced changes in the thylakoid lipid complement may be involved in modifying protein stability in the membrane (Nie and Baker, 1990). Immunocytological analyses of maize leaves growing at low temperature (Robertson et al. 1993; Nie et al. 1995) have demonstrated that the effects of temperature on thylakoid development are considerably more complex than biochemical and physiological analyses suggested. Heterogeneity of chilling effects was observed at the levels of individual proteins and between individual mesophyll cells. Some proteins were found to be affected considerably more than others as expected; however these effects were not uniform throughout the mesophyll cells with some cells showing dramatic decreases in protein levels compared to neighboring cells. A similar heterogeneity was found in the ability of cells to recover from the chilling perturbation when exposed to optimal growth temperature (Nie et al. 1995).

V. Concluding Remarks

It is apparent from this review that there have been tremendous advances in our knowledge of the molecular processes involved in the synthesis and assembly of the thylakoid protein complexes. However, remarkably little is known about the factors that regulate these processes and the appearance of physiological activity of the individual complexes in the membrane. The importance of chloroplast development and the impact of environmental variables on this process should not be underestimated in the context of plant productivity. It is to be hoped that as our knowledge of the fundamental processes involved in thylakoid biogenesis continues to expand that increasing effort will be focused on resolving the intriguing questions of what regulates thylakoid development and how the physical environment perturbs this regulation.

Acknowledgments

We thank the National Science Foundation and the National Research Initiatives Competitive Grants Program of USDA for support of some work cited in this chapter. This is publication number 232 from the Arizona State University Center for the Study of Early Events in Photosynthesis.

References

Adams CC and Stern DB (1990) Control of mRNA stability in chloroplasts by 3′ inverted repeats: Effects of stem and loop mutations on degradation of *psbA* mRNA in vitro. Nucl Acids Res 18: 6003–6010

Akkaya MS and Breitenberger CA (1992) Light regulation of protein synthesis factor EF-G in pea chloroplasts. Plant Mol Biol 20: 791–800

Anderson, JM (1986) Photoregulation of the composition, function and structure of thylakoid membranes. Annu Rev Plant Physiol 37: 93–136

Baker NR (1984) Development of chloroplast photochemical functions. In: Baker NR and Barber J (eds) Chloroplast Biogenesis, pp 207–252. Elsevier Science Publishers, Amsterdam

Baker NR and Butler WL (1976) Development of the primary photochemical apparatus of photosynthesis during greening of etiolated bean leaves. Plant Physiol 58: 526–529

Barkan A, Miles D and Taylor WC (1986) Chloroplast gene expression in nuclear, photosynthetic mutants of maize. EMBO J 5: 1421–1427

Batschauer A, Mösinger E, Kreuzz K, Dörr I and Apel K (1986) The implication of a plastid-derived factor in the transcriptional control of nuclear genes encoding the light-harvesting chlorophyll *a/b* protein. Eur J Biochem 154: 625–634

Bellemare G, Bartlett SG and Chua NH (1982) Biosynthesis of chlorophyll *a/b*-binding polypeptides in wild type and the chlorina f2 mutant of barley. J Biol Chem 157: 7762–7767

Bennett J (1981) Biosynthesis of the light-harvesting chlorophyll *a/b* protein. Polypeptide turnover in darkness. Eur J Biochem 118: 61–70

Bennoun P, Spierer-Herz M, Erickson J, Girard-Bascou J, Pierre Y, Delosme M and Rochaix J -D (1986) Characterization of Photosystem II mutants of *Chlamydomonas reinhardtii* lacking the *psbA* gene. Plant Mol Biol 6: 151–160

Boynton JE, Gillham NW, Harris EH, Hosler JP, Johnson AM, Jones AR, Randolph-Anderson, BL, Robertson D, Klein TM, Shark KB and Sanford JC (1988) Chloroplast transformation in *Chlamydomonas* with high velocity microprojectiles. Science 240: 1534–1538

Bredenkamp GJ and Baker NR (1994) Temperature-sensitivity of D1 protein metabolism in isolated *Zea mays* chloroplasts. Plant Cell Environ 17: 205–210

Bruce BD and Malkin R (1991) Biosynthesis of the chloroplast b_6f complex: Studies in a photosynthetic mutant of *Lemna*. Plant Cell 3: 203–212

Burgess DG and Taylor WC (1988) The chloroplast affects the transcription of a nuclear gene family. Mol Gen Genet 214: 89–96

Castresana C, Garcia-Luque I, Alonso E, Malik VL and Cashmore AR (1988) Both positive and negative regulatory elements mediate expression of a photoregulated CAB gene from *Nicotiana plumbaginifolia*. EMBO J 7: 1929–1936

Chitnis PR, Reilly PA and Nelson N (1989a) Insertional inactivation of the gene encoding subunit II of Photosystem I from the cyanobacterium *Synechocystis* sp. PCC 6803. J Biol Chem 264: 18381–18385

Chitnis PR, Reilly PA, Miedel MC and Nelson N (1989b) Structure and targeted mutagenesis of the gene encoding 8-kDa subunit of Photosystem I from the cyanobacterium *Synechocystis* sp. PCC 6803. J Biol Chem 264: 18374–18380

Chitnis PR, Purvis D and Nelson N (1991) Molecular cloning and targeted mutagenesis of the gene *psaF* encoding subunit III of Photosystem I from the cyanobacterium *Synechocystis* sp. PCC 6803. J Biol Chem 266: 20146–20151

Chitnis VP, Xu Q, Yu L, Golbeck JH, Nakamoto H, Xie D-L and Chitnis PR (1993) Targeted inactivation of the gene *psaL* encoding a subunit of Photosystem I of the cyanobacterium *Synechocystis* sp. PCC 6803. J Biol Chem 268: 11678–11684

Chow WS, Goodchild DJ, Miller C and Anderson JM (1990) The influence of high levels of brief or prolonged supplementary far-red illumination during growth on the photosynthetic characteristics, composition and morphology of *Pisum sativum* chloroplasts. Plant Cell Environ 13: 135–145

Christopher DA and Mullet JE (1994) Separate photosensory pathways co-regulate blue light/ultraviolet-A-activated *psbD-psbC* transcription and light-induced D2 and CP43 degradation in barley (*Hordeum vulgare*) chloroplasts. Plant Physiol 104: 1119–1129

Christopher DA, Kim M and Mullet JE (1992) A novel light-regulated promoter is conserved in cereal and dicot chloroplasts. Plant Cell 4: 785–798

Cohen Y, Chitnis VP, Nechushtai R and Chitnis PR (1993) Stable assembly of *PsaE* into cyanobacterial photosynthetic membranes is dependent on the presence of other accessory subunits of Photosystem I. Plant Mol Biol 23: 895–900

Cui L, Bingham SE, Kuhn M, Käss H, Lubitz W and Webber AN (1995) Specific mutagenesis of conserved histidines in the helix VIII domain of PsaB impairs assembly of the Photosystem I reaction center without altering spectroscopic characteristics of P700. Biochemistry 34: 1549–1558

Danon A and Mayfield SP (1991) Light regulated translational activators: Identification of chloroplast gene specific mRNA binding proteins. EMBO J 10: 3993–4001

Danon A and Mayfield SP (1994) Light-regulated translation of chloroplast messenger RNAs through redox potential. Science 266: 1717–1719

de Vitry C, Olive J, Drapier D, Recouvreur M and Wollman F-A (1989) Post-translational events leading to the assembly of Photosystem II protein complex: A study using photosynthesis mutants from *Chlamydomonas reinhardtii*. J Cell Biol 109: 991–1006

Deng X-W and Gruissem W (1987) Control of plastid gene expression during development: The limited role of transcriptional regulation. Cell 49: 379–387

Deng X-W, Tonkyn JC, Peter GF, Thornber JP and Gruissem W (1989) Post-transcriptional control of plastid mRNA accumulation during adaptation of chloroplasts to different light quality environments. Plant Cell 1: 645–654

Dickey LF, Gallo-Meagher M and Thompson WF (1992) Light regulatory sequences are located within the 5′ portion of the *Fed-1* message sequence. EMBO J 11: 2311–2317

Drapier D, Girard-Bascou J, and Wollman F-A (1992) Evidence for nuclear control of expression of the *atpA* and *atpB* chloroplast genes in *Chlamydomonas*. Plant Cell 4: 283–295

Eichacker LA, Soll J, Lauterbach P, Rüdiger W, Klein RR and Mullet JE (1990) In vitro synthesis of chlorophyll *a* in the dark triggers accumulation of chlorophyll *a* apoproteins in barley etioplasts. J Biol Chem 265: 13566–13571

Elliott RC, Dickey LF, White MJ and Thompson WF (1989) *cis*-Acting elements for light regulation of pea ferredoxin I gene expression are located within transcribed sequences. Plant Cell 1: 691–698

Erickson JM, Rahire M, Malnoe P, Girard-Bascou J, Pierre Y, Bennoun P and Rochaix J -D (1986) Lack of the D2 protein in a *Chlamydomonas reinhardtii psbD* mutant affects Photosystem II stability and D1 expression. EMBO J 5: 1745–1754

Ernst D, Pfeiffer F, Schefbeck K, Weyrauch C and Oesterhelt D (1987) Phytochrome regulation of mRNA levels of ribulose-1,5-bisphosphate carboxylase in etiolated rye seedlings (*Secale cereale*). Plant Mol Biol 10: 21–23

Evans JR (1996) Developmental constraints on photosynthesis: Effects of light and nutrition. In: Baker NR (ed) Photosynthesis and the Environment. Kluwer Academic Publishers, Dordrecht, in press

Franck F (1993) Photosynthetic activities during early assembly of thylakoid membranes. In: Sunqvist C and Ryberg M (eds) Pigment-protein Complexes in Plastids: Synthesis and Assembly, pp 365–381. Academic Press, Inc., San Diego

Gallie DR (1993) Posttranscriptional regulation of gene expression in plants. Annu Rev Plant Physiol Plant Mol Biol 44: 77–105

Gamble PE and Mullet JE (1989) Blue light regulates the accumulation of two *psbD-psbC* transcripts in barley chloroplasts. EMBO J 8: 2785–2794

Gamble PE, Sexton TB and Mullet JE (1988) Light dependent changes in *psbD* and *psbC* transcripts of barley chloroplasts: Accumulation of two transcripts maintains *psbD* and *psbC* translation capability in mature chloroplasts. EMBO J 7: 1289–1297

Gidoni D, Brosio P, Bond-Nutter D, Bedbrook J and Dunsmuir P

(1989) Novel *cis*-acting elements in petunia Cab gene promoters. Mol Gen Genet 215: 337–344

Gilmartin PM, Sarokin, L Memelink, J and Chua N-H (1990) Molecular light switches for plant genes. Plant Cell 2: 369–378

Girard-Bascou J, Choquet Y, Schneider M, Delosme M and Dron M (1987) Characterization of a chloroplast mutation in the *psaA*2 gene of *Chlamydomonas reinhardtii*. Curr Genet 12: 489–495

Glick RE, McCauley SW, Gruissem W and Melis A (1986) Light quality regulates expression of chloroplast genes and assembly of photosynthetic membrane complexes. Proc Natl Acad Sci USA 83: 161–170

Golbeck JH (1992) Structure and function of Photosystem I. Annu Rev Plant Physiol Plant Mol Biol 43: 293–324

Gruissem W (1989) Chloroplast RNA: Transcription and processing. In: Marcu A (ed) The Biochemistry of Plants, Vol. 15, pp 151–191. Academic Press, New York

Gruissem W and Tonkyn JC (1993) Control mechanisms of plastid gene expression. Cric Rev Plant Sci 12: 19–55

Hahn M and Walbot V (1989) Effects of cold-treatment on protein synthesis and mRNA levels in rice leaves. Plant Physiol 91: 930–938

Haley J and Bogorad L (1989) A 4 kDa maize chloroplast polypeptide associated with the cytochrome b_6f complex: subunit 5, encoded by the chloroplast *petE* gene. Proc Natl Acad Sci USA 86: 1534–1538

Haley J and Bogorad L (1990) Alternative promoters are used for genes within maize chloroplast polycistronic transcription units. Plant Cell 2: 323–333

Helmann JD and Chamberlin MJ (1988) Structure and function of bacterial sigma factors. Annu Rev Biochem 57: 839–872

Herrin DL, Battey JF, Greer K and Schmidt GW (1992) Regulation of chlorophyll apoprotein expression and accumulation: Requirements for carotenoids and chlorophyll. J Biol Chem 267: 8260–8269

Hird SM, Webber AN, Wilson R, Dyer TA and Gray JC (1991) Differential expression of the chloroplast genes for the 47 kDa chlorophyll *a*-protein and the 10 kDa phosphoprotein during chloroplast development in wheat. Current Genetics 19: 199–206

Hoober JK and Hughs MJ (1992) Purification and characterization of a membrane-bound protease from *Chlamydomonas reinhardtii*. Plant Physiol 99: 932–937

Hosler JP, Wurtz EA, Harris EH, Gillham NW and Boynton JE (1989) Relationship between gene dosage and gene expression in the chloroplast of *Chlamydomonas reinhardtii*. Plant Physiol 91: 648–655

Hsu-Ching C and Stern DB (1991a) Specific ribonuclease activities in spinach chloroplasts promote mRNA maturation and degradation. J Biol Chem 266: 24205–24211

Hsu-Ching C and Stern DB (1991b) Specific binding of chloroplast proteins in vitro to the 3′ untranslated region of spinach chloroplast *petD* mRNA. Mol Cell Biol 11. 4380–4388

Ikeuchi M and Inoue Y (1988) A new Photosystem II reaction center component (4.8 kDa protein) encoded by chloroplast genome. FEBS Lett 241: 99–104

Ikeuchi M, Hirano A, Hiyama T and Inoue Y (1990) Polypeptide composition of higher plant Photosystem I complex: Identification of *psaI*, *psaJ* and *psaK* gene products. FEBS Lett 263: 274–278

Ikeuchi M, Eggers B, Shen G, Webber A, Yu L, Hirano A, Inoue Y and Vermaas W (1991) Cloning of the *psbK* gene from *Synechocystis* sp. PCC 6803, and characterization of Photosystem II mutants lacking PS II-K. J Biol Chem 266: 11111–11115

Jenkins GI and Smith H (1985) Red: far-red ratio does not modulate the abundance of transcripts for two major chloroplast polypeptides in light-grown *Pisum sativum* terminal shoots. Photochem Photobiol 42: 679–684

Jensen KH, Herrin DL, Plumley G and Schmidt GW (1986) Biogenesis of Photosystem II complexes: Transcriptional, translational and post-translational regulation. J Cell Biol 103: 1315–1325

Kim J, Klein PG and Mullet JE (1991) Ribosomes pause at specific sites during synthesis of membrane-bound chloroplast reaction center protein D1. J Biol Chem 266: 14931–14938

Kim J, Eichacker LA, Rudiger W and Mullet JE (1994) Chlorophyll regulates accumulation of the plastid-encoded chlorophyll proteins P700 and D1 by increasing apoprotein stability. Plant Physiol 104: 907–916

Kim JH, Glick RE and Melis A (1993) Dynamics of photosystem stoichiometry adjustment by light quality in chloroplasts. Plant Physiol 102: 181–190

Kim M, Christopher DA and Mullet JE (1993) Direct evidence for selective modulation of *psbA*, *rpoA*, *rbcL* and 16S RNA stability during barley chloroplast development. Plant Mol Biol 22: 447–463

Klaff P and Gruissem W (1991) Changes in chloroplast mRNA stability during leaf development. Plant Cell 3: 517–529

Klein RR and Mullet JE (1986) Regulation of chloroplast-encoded chlorophyll-binding protein translation during higher plant chloroplast biogenesis. J Biol Chem 261: 11138–11145

Klein RR and Mullet JE (1987) Control of gene expression during higher plant chloroplast biogenesis: Protein synthesis and transcript levels of *psbA*, *psaA-psaB* and *rbcL* in dark-grown and illuminated barley seedlings. J Biol Chem 262: 4341–4348

Klein RR and Mullet JE (1990) Light-induced transcription of chloroplast genes: *psbA* transcription is differentially enhanced in illuminated barley. J Biol Chem 265: 1895–1902

Klein RR, Mason H and Mullet JE (1987) Light-regulated translation of chloroplast proteins. Transcripts of *psaA-psaB*, *psbA* and *rbcL* are associated with polysomes in dark-grown and illuminated barley seedlings. J Cell Biol 106: 289–302

Koike H, Mamada K, Ikeuchi M and Inoue Y (1989) Low-molecular-mass proteins in cyanobacterial Photosystem II: Identification of *psbH* and *psbK* gene products by N-terminal sequencing. FEBS Lett 244: 391–396

Krapp A, Hofmann B, Schafer C and Stitt M (1993) Regulation of the expression of *rbcS* and other photosynthetic genes by carbohydrates: A mechanism for the 'sink regulation' of photosynthesis? Plant Journal 3: 817–828

Kreuz K, Dehesh K and Apel K (1986) The light dependent accumulation of the P700 chlorophyll *a* protein of the Photosystem I reaction center in barley. Eur J Biochem 159: 459–467

Kuchka MR, Mayfield SP and Rochaix J -D (1988) Nuclear mutations specifically affect the synthesis and/or degradation of the chloroplast-encoded D2 polypeptide of Photosystem II in *Chlamydomonas reinhardtii*. EMBO J 7: 319–324

Kuchka MR, Goldschmidt-Clermont M, Van Dillewijn J and

Rochaix J -D (1989) Mutation at the *Chlamydomonas* nuclear NAC2 locus specifically affects stability of the chloroplast *psbD* transcript encoding polypeptide D2 of PS II. Cell 58: 869–876

Kuras R and Wollman F-A (1994) The assembly of cytochrome b_6/f complexes: An approach using genetic transformation of the green alga *Chlamydomonas reinhardtii*. EMBO J 13: 1019–1027

Kuwabara T (1992) Characterization of a prolyl endopeptidase from spinach thylakoids. FEBS Lett 300: 127–130

Lam E and Chua N -H (1990) GT-1 binding sites confer light-responsive expression in transgenic tobacco. Science 248: 471–474

Lee W-J and Whitmarsh J (1989) Photosynthetic apparatus of pea thylakoid membranes: Response to growth light intensity. Plant Physiol 89: 932–940

Lemaire C and Wollman F-A (1989) The chloroplast ATP synthase in *Chlamydomonas reinhardtii*. II. Biochemical studies on its biogenesis using mutants defective in photophosphorylation. J Biol Chem 264: 10235–10242

Lind LK, Shukla VK, Nyhus KJ and Pakrasi HB (1993) Genetic and immunological analysis of the cyanobacterium *Synechocystis* sp. PCC 6803 show that the protein encoded by the *psbJ* gene regulates the number of Photosystem II centers in thylakoid membranes. J Biol Chem 268: 1575–1579

Link G (1988) Photocontrol of plastid gene expression. Plant Cell Environ 11: 329–338

Liu X -Q, and Jagendorf AT (1984) ATP dependent proteolysis in pea chloroplasts. FEBS Lett 166: 248–252

Liu X -Q and Jagendorf AT (1986) Neutral peptidases in the stroma of pea chloroplasts. Plant Physiol 81: 603–608

Liveanu V, Yocum CF and Nelson N (1986) Polypeptides of the oxygen-evolving Photosystem II complex: Immunological detection and biogenesis. J Biol Chem 261: 5296–5300

Malek K, Bogorad L, Ayers A and Goldberg AL (1984) Newly synthesized proteins are degraded by ATP stimulated proteolytic process in isolated pea chloroplasts. FEBS Lett 166: 253–255

Malnöe P, Mayfield SP and Rochaix J -D (1988) Comparative analysis of the biogenesis of Photosystem II in the wild-type and *y-1* mutant of *Chlamydomonas reinhardtii*. J Cell Biol 106: 609–616

Mannan RM, Whitmarsh J, Nyman P and Pakrasi HB (1991) Directed mutagenesis of an iron sulfur protein of the Photosystem I complex in the filamentous cyanobacterium *Anabaena variablis* ATCC 29413. Proc Natl Acad Sci USA 88: 10168–10172

Manzara T and Gruissem W (1988) Organization and expression of the genes encoding ribulose 1,5 bisphosphate carboxylase in higher plants. Photosynth Res 16: 117–139

Maruyama S, Yatomi M and Nakumura Y (1990) Response of rice leaves to low temperature. 1. Changes in basic biochemical parameters. Plant Cell Physiol 31: 303–309

Mattoo AK, Hoffman-Falk H, Marder JB and Edelman E (1984) Regulation of protein metabolism: Coupling of photosynthetic electron transport to in vivo degradation of the rapidly metabolized 32 kDa protein of the chloroplasts membranes. Proc Natl Acad Sci USA 81: 1380–1384

Mayfield SP and Taylor WC (1984) Carotenoid deficient maize seedlings fail to accumulate light-harvesting chlorophyll *a/b* binding protein mRNA. Eur J Biochem 144: 79–84

McKiernan M and Baker NR (1991) Adaptation to shade of the light-harvesting apparatus in *Silene dioica*. Plant Cell Environ 14: 205–212

Mieszczak M, Klahre U, Levy JH, Goodall GJ and Filipowicz W (1992) Multiple plant RNA binding proteins identified by PCR: Expression of cDNAs encoding RNA binding proteins targeted to chloroplasts in *Nicotiana plumbaginifolia*. Mol Gen Genet 234: 390–400

Monod C, Goldschmidt-Clermont M and Rochaix J-D (1992) Accumulation of chloroplast *psbB* RNA requires a nuclear factor in *Chlamydomonas reinhardtii*. Mol Gen Genet 231: 449–459

Mullet JE (1988) Chloroplast development and gene expression. Annu Rev Plant Physiol Plant Mol Biol 39: 475–502

Mullet JE (1993) Dynamic regulation of chloroplast transcription. Plant Physiol 103: 309–313

Mullet JE and Klein RR (1987) Transcription and RNA stability are important determinants of higher plant chloroplast RNA levels. EMBO J 6: 1571–1579

Neuhaus G, Bowler C, Kern R and Chua N-H (1993) Calcium/calmodulin-dependent and -independent phytochrome signal transduction pathways. Cell 73: 937–952

Nie G-Y and Baker NR (1990) Low temperature perturbation of thylakoid protein metabolism during maize leaf development. In Baltscheffsky M (ed) Current Research in Photosynthesis, Vol. IV, pp 687–690. Kluwer Academic Publishers, Netherlands

Nie G-Y and Baker NR (1991) Modifications of thylakoid composition during development of maize leaves at low growth temperatures. Plant Physiol 95: 184–191

Nie G-Y, Robertson EJ, Fryer MJ, Leech RM and Baker NR (1995) Response of the photosynthetic apparatus in maize leaves grown at low temperature to increases in temperature. Plant Cell Environ 18: 1–12

Nilsson F, Andersson B and Jansson C (1990) Photosystem II characteristics of a constructed *Synechocystis* 6803 mutant lacking synthesis of the D1 polypeptide. Plant Mol Biol 14: 1051–1054

Oelmüller R and Mohr H (1986) Photooxidative destruction of chloroplasts and its consequences for expression of nuclear genes. Planta 167: 106–113

Ono T, Kajikawa H and Inoue Y (1986) Changes in protein composition and Mn abundance in Photosystem II particles on photoactivation of the latent O_2-evolving system in flash-grown leaves. Plant Physiol 80: 85–90

Palomares R, Herrmann RG and Oelmuller R (1993) Post-transcriptional and post-translational regulatory steps are crucial in controlling the appearance and stability of thylakoid polypeptides during the transition of etiolated tobacco seedlings to white light. Eur J Biochem 217: 345–352

Rapp JC, Baumgartner BJ and Mullet J (1992) Quantitative analysis of transcription and RNA levels of 15 barley chloroplast genes: Transcription rates and mRNA levels vary over 300-fold; predicted mRNA stabilities vary 30 fold. J Biol Chem 267: 21404–21411

Robertson D, Woessner JP, Gillham NW and Boynton JE (1989) Molecular characterization of two point mutants in the chloroplast *atpB* gene of the green alga *Chlamydomonas reinhardtii* defective in assembly of the ATP synthase complex. J Biol Chem 264: 2331–2337

Robertson D, Boynton JE and Gillham NW (1990) Cotranscription of the wild-type chloroplast *atpE* gene encoding the CF_1/CF_0

epsilon subunit with the 3′ half of the *rps7* gene in *Chlamydomonas reinhardtii* and characterization of frameshift mutations in *atpE*. Mol Gen Genet 221: 155–163

Robertson EJ, Baker NR and Leech RM (1993) Chloroplast thylakoid protein changes induced by low growth temperature in maize revealed by immunocytology. Plant Cell Environ 16: 809–818

Rochaix J-D (1992) Post-transcriptional steps in the expression of chloroplast genes. Annu Rev Cell Biol 8: 1–28

Rochaix J-D, Kuchka M, Mayfield S, Schirmer-Rahire M, Girard-Bascou J and Bennoun P (1989) Nuclear and chloroplast mutations affect the synthesis or stability of the chloroplast *psbC* gene product in *Chlamydomonas reinhardtii*. EMBO J 8: 1013–1021

Roffey RA, Golbeck JH, Hille CR and Sayre RT (1991) Photosynthetic electron transport in genetically altered Photosystem II reaction centers of chloroplasts. Proc Natl Acad Sci USA 88: 9122–9126

Rögner M, Chisholm DA and Diner BA (1991) Site-directed mutagenesis of the *psbC* gene of Photosystem II: Isolation and functional characterization of CP43-less Photosystem II core complexes. Biochemistry 30: 5387–5395

Ryrie IJ, Young S and Andersson B (1984) Development of the 33-, 23- and 16-kDa polypeptides of the photosynthetic oxygen-evolving system during greening. FEBS Lett 177: 269–273

Sakamoto W, Kindle KL and Stern DB (1993) In vivo analysis of *Chlamydomonas* chloroplast *petD* gene expression using stable transformation of β-glucuronidase translational fusions. Proc Natl Acad Sci USA 90: 497–501

Salvador ML, Klein U and Bogorad L (1993) 5′ sequences are important positive and negative determinants of the longevity of *Chlamydomonas* chloroplast gene transcripts. Proc Natl Acad Sci USA 90: 1556–1560

Schindler U and Cashmore AR (1990) Photoregulated gene expression may involve ubiquitous DNA binding proteins. EMBO J 9: 3415–3427

Schmidt CL and Malkin R (1993) Low molecular weight subunits associated with the cytochrome b_6f complexes from spinach and *Chlamydomonas reinhardtii*. Photosynth Res 38: 73–81

Schuster G and Gruissem W (1991) Chloroplast mRNA 3′ end processing requires a nuclear encoded RNA-binding protein. EMBO J 10: 1493–1502

Sexton TB, Christopher DA and Mullet JE (1990a) Light-induced switch in barley *psbD-psbC* promoter utilization: A novel mechanism regulating chloroplast gene expression. EMBO J 9: 4485–4494

Sexton TB, Jones JT and Mullet JE (1990b) Sequence and transcriptional analysis of the barley ctDNA region upstream of *psbD-psbC* encoding *trnK*(UUU), *rps16*, *trnQ*(UUG), *psbK*, *psbI*, and *trnS*(GCU). Curr Genet 17: 445–454

Sheen J (1990) Metabolic repression of transcription in higher plants. Plant Cell 2: 1027–1038

Shirley BW and Meagher RB (1990) A potential role for RNA turnover in the light regulation of plant gene expression: Ribulose-1,5-bisphosphate carboxylase small subunit in soybean. Nucleic Acids Res 18: 3377–3385

Sieburth LE, Berry-Lowe S, Schmidt GW (1991) Chloroplast RNA stability in *Chlamydomonas*: Rapid degradation of *psbB* and *psbC* transcripts in two nuclear mutants. Plant Cell 3: 175–189

Slovin JP and Tobin EM (1982) Synthesis and turnover of the light harvesting chlorophyll *a/b*-protein in *Lemna gibba* grown with intermittent red light: Possible translational control. Planta 154: 465–472

Smart LB, Anderson SL and McIntosh L (1991) Targeted genetic inactivation of the Photosystem I reaction center in the cyanobacterium *Synechocystis* sp. PCC 6803. EMBO J 10: 3289–3296

Stern DB and Gruissem W (1987) Control of plastid gene expression: 3′ inverted repeats act as mRNA processing and stabilizing elements, but do not terminate transcription. Cell 51: 1145–1157

Stern DB and Kindle KL (1993) 3′ end maturation of the *Chlamydomonas reinhardtii* chloroplast *atpB* mRNA is a two-step process. Mol Cell Biol 13: 2277–2285

Stern DB, Jones H and Gruissem W (1989) Function of plastid mRNA 3′ inverted repeats: RNA stabilization and gene-specific protein binding. J Biol Chem 264: 18742–18750

Stern DB, Radwanski ER and Kindle KL (1991) A 3′ stem/loop structure of the *Chlamydomonas* chloroplast *atpB* gene regulates mRNA accumulation in vivo. Plant Cell 3: 285–297

Stockhaus J, Schell J and Willmitzer L (1989) Identification of enhancer elements in the upstream region of the nuclear photosynthetic gene ST-LS1. Plant Cell 1: 805–813

Takahashi Y, Goldschmidt-Clermont M, Soen S -Y, Franzen LG and Rochaix J -D (1991) Directed chloroplast transformation in *Chlamydomonas reinhardtii*: Insertional inactivation of the *psaC* gene encoding the iron sulfur protein destabilizes Photosystem I. EMBO J 10: 2033–2040

Taylor WC (1989) Regulatory interactions between nuclear and plastid genomes. Annu Rev Plant Physiol Plant Mol Biol 40: 211–233

Thompson DM and Meagher RB (1990) Transcriptional and post-transcriptional processes regulate expression of RNA encoding the small subunit of ribulose-1,5-bisphosphate carboxylase differently in petunia and in soybean. Nucleic Acids Res 18: 3621–3629

Thompson WF and White MJ (1991) Physiological and molecular studies of light-regulated nuclear genes in higher plants. Annu Rev Plant Physiol Plant Mol Biol 42: 423–466

Tiller K and Link G (1993) Sigma-like transcription factors from mustard (*sinapis alba* L.) etioplast are similar in size to, but functionally distinct from, their chloroplast counterparts. Plant Mol Biol 21: 503–513

Tonkyn JC, Deng X -W and Gruissem W (1992) Regulation of plastid gene expression during photooxidative stress. Plant Physiol 99: 1406–1415

Vermaas W (1993) Molecular-biological approaches to analyze Photosystem II structure and function. Annu Rev Plant Physiol Plant Mol Biol 44: 457–481

Vermaas WFJ, Ikeuchi M and Inoue Y (1988) Protein composition of the Photosystem II core complex in genetically engineered mutants of the cyanobacterium *Synechocystis* sp. PCC 6803. Photosynth Res 17: 97–113

Vierstra RD (1993) Protein degradation in plants. Annu Rev Plant Physiol Plant Mol Biol 44: 385–410

Wada T, Tunoyama Y, Shiina T, and Toyoshima Y (1994) In vitro analysis of light-induced transcription in the wheat *psbD/C* gene cluster using plastid extracts from dark-grown and short-term-illuminated seedlings. Plant Pysiol 104: 1259–1267

Webber AN, Baker NR, Page CD and Hipkins MF (1986) Photosynthetic electron transport and establishment of an associated trans-thylakoid proton electrochemical gradient during development of the wheat leaf. Plant Cell Environ 9: 203–208

Webber AN, Hird SM, Packman LC, Dyer TA and Gray JC (1989a) A Photosystem II polypeptide is encoded by an open reading frame co-transcribed with genes for cytochrome *b*-559 in wheat chloroplast DNA. Plant Mol Biol 12: 141–151

Webber AN, Packman LC, Chapman DJ, Barber J and Gray JC (1989b) A fifth chloroplast-encoded polypeptide is present in the Photosystem II reaction centre complex. FEBS Lett 242: 259–262

Webber AN, Gibbs PB, Ward JB and Bingham SE (1993) Site-directed mutagenesis of the Photosystem I reaction center in chloroplasts: The proline-cysteine motif. J Biol Chem 268: 12990–12995

Xu R, Bingham SE and Webber AN (1993) Increased mRNA accumulation in a *psaB* frame-shift mutant of *Chlamydomonas reinhardtii* suggests a role for translation in *psaB* mRNA stability. Plant Mol Biol 22: 465–474

Chapter 5

Development of Thylakoid Membrane Stacking

László Mustárdy
Institute of Plant Biology, Biological Research Center, Hungarian Academy of Sciences, H-6701 Szeged, P.O.Box 521, Hungary

Summary

To fully understand function it is essential to have detailed information about structure. In mesophyll cells of higher plants, photosynthesis takes place in a granal thylakoid system where the two photosystems are spatially separated. Photosystem II (PS II) is located mainly in the stacked grana thylakoids while Photosystem I (PS I) is relegated to the unappressed, stromal exposed, portions of the thylakoid. This lateral heterogeneity in the deployment of photosynthetic complexes develops during thylakoid membrane biogenesis. The complexity of thylakoid membrane architecture, and the segregation of two photosystems within, place important conditions on efficient electron transfer.

The three-dimensional reconstitution of the thylakoid system, typical of chloroplasts from mesophyll cells of C_3 and C_4 plants, from a complete series of sections and the scanning electron microscopic observation, shows multiple helices of stroma lamellae around a central core of granum. Each of the helically wound stromal membranes is connected to each of the granum compartments via membrane fusions. In turn, neighboring grana are interconnected via fusion of stroma lamellae. This way, the entire thylakoid system in the chloroplasts constitutes a single unit where the intrathylakoidal space is continuous. Reconstitution of serial sections reveals that this continuity stems from the prothylakoidal system of proplastids and is retained throughout the process of chloroplast development.

Donald R. Ort and Charles F. Yocum (eds): Oxygenic Photosynthesis: The Light Reactions, pp. 59–68.

I. Introduction

In electron micrographs, higher plant chloroplasts appear as elongated vesicles filled with a matrix, or 'stroma,' surrounded by a pair of envelope membranes. Embedded within the stroma is the chlorophyll-containing lamellar system, known as thylakoid membranes (see Chapter 2). In a two-dimensional view, thylakoid membrane features appear to fall into two general size ranges. The smaller size thylakoids (generally around 0.5 μm) are stacked and termed grana. Along one side of a granum approximately every other thylakoid appears to extend into the stroma. These unstacked portions of the thylakoid membrane system, frequently called stroma lamellae, are elongated and appear to interconnect the grana. A three-dimensional view of the lamellar system can be developed from an ordered series of two-dimensional sections.

Menke (1960) developed the first model of the chloroplast thylakoid system in higher plants from early electron microscopic observations (Fig. 1a). He proposed that the lamellae extending from every second granum continued from the edge of the granum on each side thereby intersecting several grana. Weier et al. (1963), using $KMnO_4$ staining, suggested that, instead of flat sheets, the stroma lamellae was an anastomosing system of membranous channels interconnecting the grana. These investigators introduced new terminology for specific parts of the grana. Their proposed structures became known as frets (Fig. 1b); 'partition' was used to designate the two adjacent thylakoids of the granum and 'margin' to indicate the ends of granum thylakoid which are in direct contact with the stroma. A granum thylakoid was called 'compartment' and the electron transparent intrathylakoidal space was termed 'loculus'. This terminology is still in use (Fig. 1f). However, later investigations showed that the tubular model of stroma thylakoids was in error due to an artifact caused by $KMnO_4$ staining (Falk and Sitte, 1963). Subsequent models show lamellar sheets instead of tubes and in this way the meaning of the 'fret' was changed to describe the interconnection between the granum and the lamellar sheet. Heslop-Harrison (1963) observed that one stroma lamella can be connected to more than a single granum compartment. He suggested that a stroma thylakoid can intersect a granum at an angle to the planes of the granum compartments (Fig. 1c for a 3-dimensional model). A consequence of such an arrangement was that many, or perhaps all, loculi within the chloroplast were interconnected, indicating that the entire intrathylakoidal space may be a continuum in the lamellar system. This idea was further developed by Wehrmeyer (1964) who proposed that a single stroma lamella exists in a spiral or helical arrangement deployed around the outside of the granum (Fig. 1d). The idea of helical arrangement was further elaborated by Paolillo and Falk (1966) who established that more than one stroma lamella could be attached to a single compartment and the interconnections (i.e. frets) were found to be much narrower than proposed by Wehrmeyer. Paolillo's latest model (Fig. 1e) depicts large stroma lamellae connected to the cylindrical granum in a pattern of a right-handed helix (Paolillo, 1970). The model was further refined by experiments showing that stroma thylakoids cannot be envisioned as large sheets but are instead strips encircling the granum helically and interconnected via narrow fusions (Brangeon and Mustárdy, 1979).

II. Reconstitution of Three-dimensional Thylakoid Arrangement

In a section taken through the median portion of a granum, the connecting stroma lamellae run parallel to the plane of the granum compartment. Near the margin of the granum, the tilt in the adjacent lamellae to the granum compartment increases (Fig. 2). In a tangentially sectioned granum, where both granum and connecting stroma thylakoids can be seen, the oblique compartment-fret relationship is evident and shows the apparent tilt of the fret (inset, Fig. 2) suggesting a helical arrangement. However, a single two-dimensional view is inconclusive. A complete section-by-section analysis through the entire granum was required to obtain an accurate picture of overall granum-stroma lamellae spatial relationship.

Reconstructions from serial thin sections confirm the helical relationship and add substantial detail to the three dimensional structural model of thylakoids. In Fig. 3, the inversion of the fret tilt is most easily visualized by comparing the sections in front of and behind a given granum with a median section. This relationship clearly indicates a helical arrangement. Further evidence can also be found by following the extensions of stroma lamellae through sections. Three

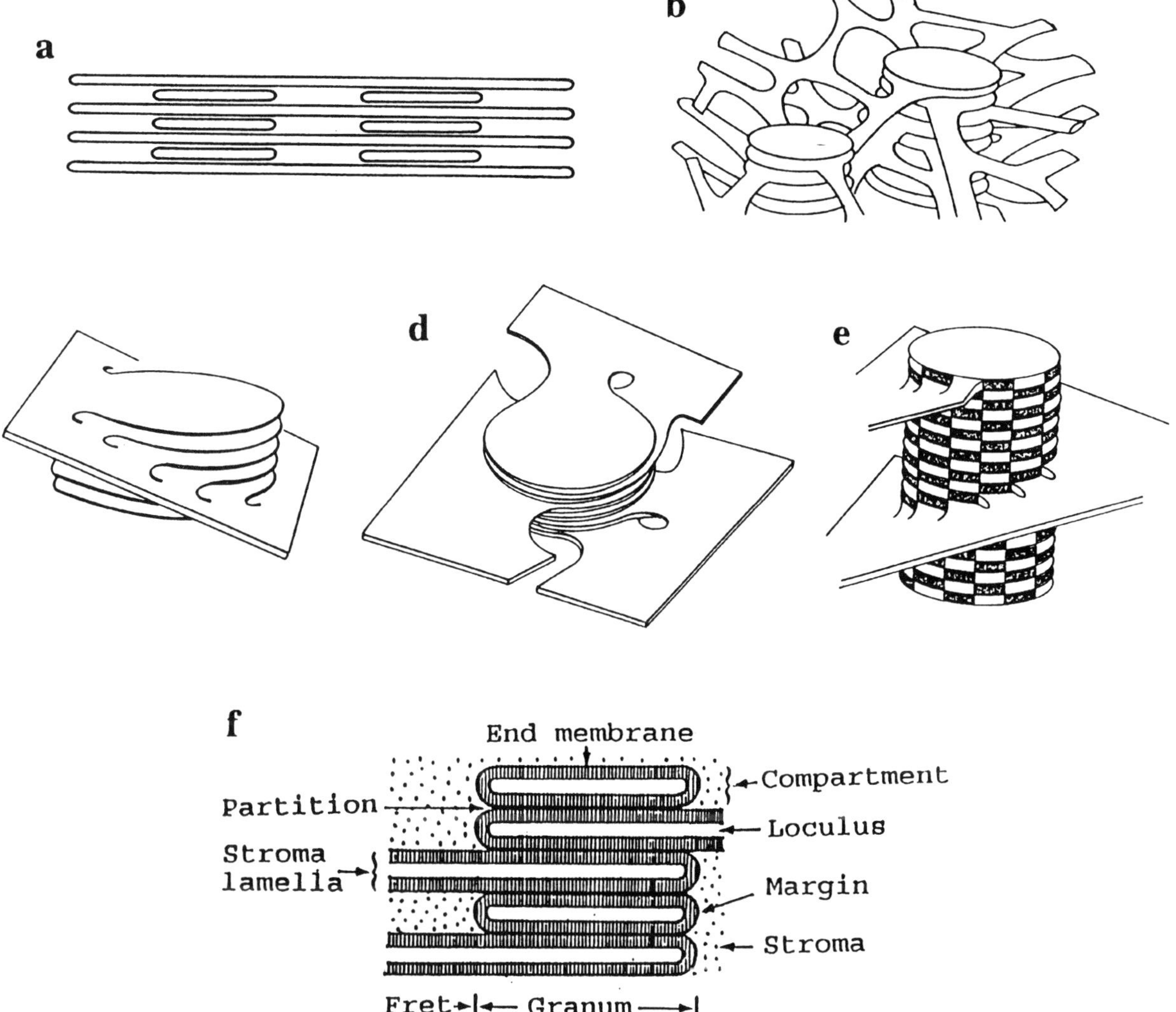

Fig. 1. Development of models on higher plant chloroplast lamellar arrangement. (Redrawn after different authors) a. First model proposed by Menke (1960) showing every second granum lamellae continues into the stroma and extends through several grana; b. Cylindrical grana are interconnected by an anastomosing system of lamellar channels (Weier et al., 1963); c. Stroma lamellae intersect the granum compartments at an angle connecting them together (Heslop-Harrison, 1963); d. Single stroma lamella interconnect the granum compartments in a helical manner (Wehrmeyer, 1964); e. Continuous, large helices of stroma lamellae around the granum (Paolillo, 1970); f. Terminology for different parts of the granum, proposed by Weier et al. (1965).

subsequent sections within the near-meridian region of two neighboring grana are illustrated in Fig. 4. Each connecting stroma lamella from the right margin of the granum shift upward, while those on the left margin shift downward.

The model developed from the analysis of reconstituted serial sections is shown in Fig. 5. It is an exact representation of two grana and their attending stroma lamellae. This direct reconstitution clearly shows an intergranal thylakoid system of multiple, helically arranged strips which are wound around the cylindrical granum.

This 3-dimensional structure is in line with the model proposed by Paolillo (1970). However, variations have been noted in the number of helices around a granum (more than eight can occur) and in the width of the slits at the periphery of each compartment. These irregularities in the junction between the consecutive compartments are averaged out and result in a constant tilt of about 22°.

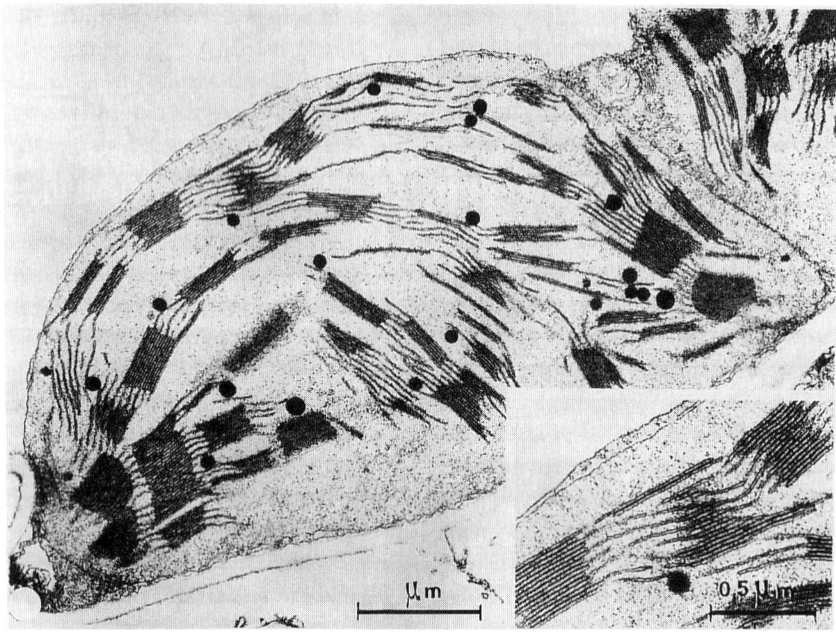

Fig. 2. Electron micrograph of mature *Lolium* chloroplast. Near midsection of a granum the grana-fret relationship is parallel. Oblique relationship of stroma membrane to granum is revealed in tangentially sectioned granum (inset) (Brangeon and Mustárdy, 1979).

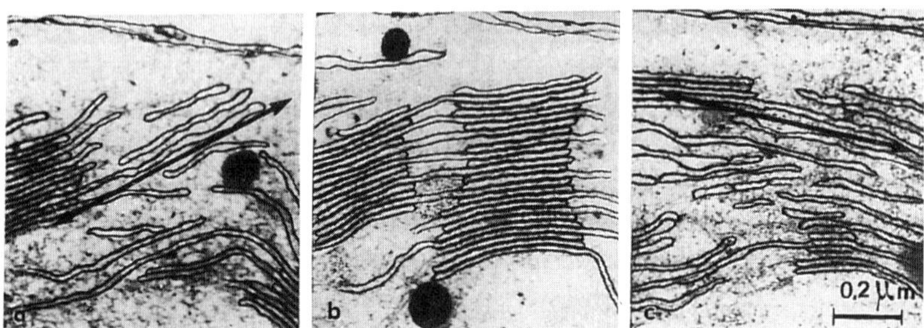

Fig. 3. Tilt of the stroma membrane with respect to granum illustrated by section a, in front of the granum; b, median section; and c, section behind the granum. The arrows show the tilts at reversed directions in front of and behind the granum (Brangeon and Mustárdy, 1979).

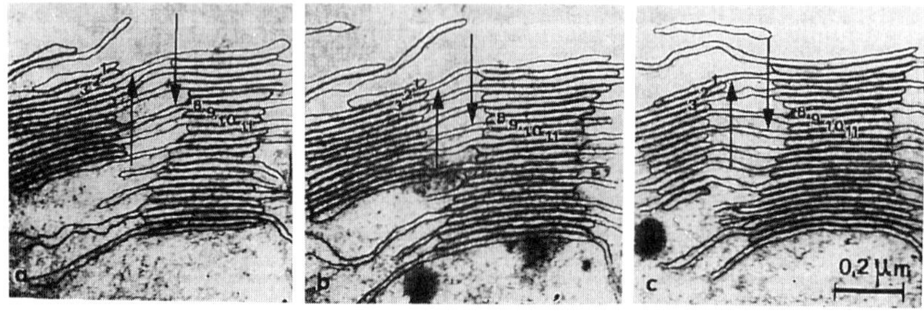

Fig. 4. Three serial sections in the near-median region of adjacent grana showing an upward shift of stroma membrane on the right and a downward shift on the left granal margin. The arrows point to the lamellae which join the granum at different levels (Brangeon and Mustárdy, 1979).

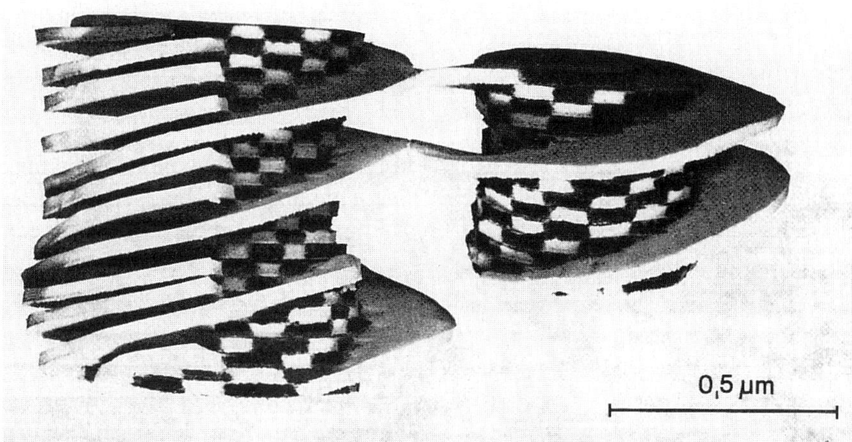

Fig. 5. Three-dimensional model of the thylakoid membrane from mature chloroplast reconstituted from serial sections. On the far left side the total number of helices are wound around the granum; Helices on the right side have been left off to show the alternating fret connections (light zones) and closed margins (darkened zones). Grana are interconnected via narrow fusions of helices at the same levels (Brangeon and Mustárdy, 1979).

Furthermore, the data from serial sections suggest that the interconnection of helices between neighboring grana is accomplished via heterogeneous fusion, with different widths rather than smooth helical continuities (large sheets) linking grana.

Helical arrangement of thylakoid membranes could also be demonstrated with scanning electron microscopy (Mustárdy and Jánossy, 1979). During sonication of isolated chloroplasts large parts of the stroma lamellae break off and the helical arrangement of the remaining narrow strips can be easily seen. The single thylakoids are arranged in a right-handed tilted spiral around the granal body (Fig. 6).

III. Ontogenetic Assembly of Thylakoids Viewed in Three-dimensions

Initial studies on chloroplast differentiation were for the most part based on electron microscopic analysis of individual chloroplast thin sections. Various strategies for investigating the different developmental stages of chloroplasts include studies during leaf ontogeny under seasonal changes, using leaves of different age from the same plant, tissues of different age from the same leaf, meristematic tissue of different age, and etioplasts during the greening process (see Kutik, 1985).

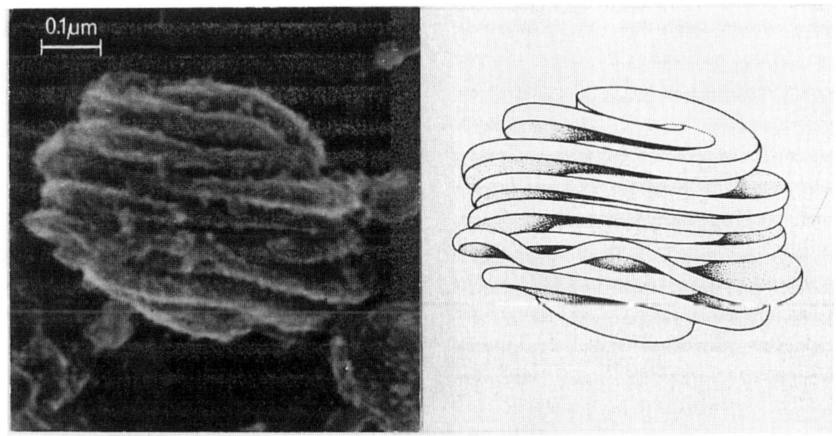

Fig. 6. Scanning electron micrograph of a granum isolated from mesophyll chloroplasts of maize and its schematic model. Narrow strips of remaining fret show helical arrangement (Mustárdy and Jánossy, 1979).

The spatial relationship of grana and stroma thylakoids during chloroplast differentiation can be most successfully studied by serial section analysis at different levels of ontogenesis. This approach was taken with *Lolium multiflorum* seedlings (Mustárdy and Brangeon, 1978; Brangeon and Mustárdy, 1979) in which four developmental stages of leaves were selected to follow the ontogenetic assembly of thylakoids by analysis and reconstitution of serial sections of developing chloroplasts.

A. Lamellar Formation

In meristematic tissues, the intrachloroplastic lamellae show pro-thylakoids of varying length distributed apparently at random within the proplastids (Fig. 7a). However, views of reconstituted serial sections reveal a single lamellar entity rather than separate and independent granal and stromal membrane features characteristic of fully developed chloroplasts. Thylakoids can be developmentally traced to a 'parent entity' which has apparently split and given rise to several branches of different length. In some cases, the pro-thylakoid form is flattened and fanlike, although apparently flexible and capable of changing orientation (Fig. 7b). In crossing this lamellar sheet, sharp cross-sectional views deteriorate into fuzzy tangential views. Some of the pro-thylakoids show a balloon- or barrel-like form with several marginal splits in the reconstituted serial sections (Fig. 7c). In cross section, these figures appear as roughly circular thylakoids with interruptions and/or as short separate lamellae, depending upon the section plane. A few rudimentary double stacks are present at this stage. Within one proplastid, bubble-like invaginations of the inner membrane of the envelope are evident (see inset Fig. 7a) but the large, flattened lamellae are never connected to the inner envelope. Although the possibility that these vesicles elaborate into thylakoid sheets was proposed by several authors (Hodge et al., 1956; Mühlethaler and Frey-Wyssling, 1959; Bradbeer et al., 1974), it has still not been shown convincingly.

B. Lamellar Orientation and Grana Initiation

The developing chloroplasts increase in size and become flattened. Development is accompanied by rapid thylakoid synthesis although the formation of photosynthetic units has been studied mainly during the greening process of etioplasts and not in proplastids (Bradbeer, 1981). It is generally the case

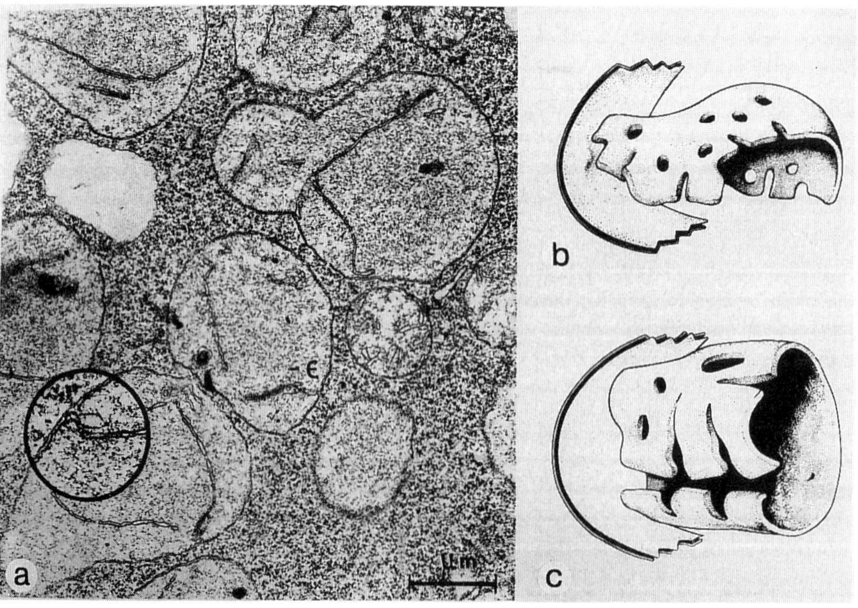

Fig. 7. Lolium proplastids from meristematic tissue (a) and two reconstituted 3-dimensional views of lamellae (b and c). Circular inset shows invagination of inner plastid envelope (Brangeon and Mustárdy, 1979).

that, in the newly formed thylakoids, most of the chlorophyll is initially present in the reaction centers. The light-harvesting complex, which facilitates stacking of membranes and grana formation (Mullet and Arntzen, 1980), associates with the reaction centers later in thylakoid development (see Chapter 4).

At this stage of chloroplast development, the newly synthesized thylakoids are disposed parallel to one another and to the equatorial plane of chloroplasts (Fig. 8a). In the partial serial section, aligned thylakoids appear as individual broad sheets often traversing the width of the plastid (Fig. 8b). However, when the sections cross in a right angle, lamellar connections are encountered at the plastid ends (Fig. 8a arrow-heads). During this differentiation step there is the appearance of multiple perforations dispersed throughout the lamellar sheets. This phenomenon has been seen in a variety of membrane systems, particularly in a rapidly expanding membrane (Dalton and Hagueneau, 1968). The interruptions give a beaded appearance to the thylakoids in the cross-section (Fig. 8a inset) but in the reconstituted view these are revealed as slits or tears (Fig. 8c). In this developmental stage several double or triple stacks can be observed which overlap on the top of and beneath the lamellar sheet (Fig. 8a, arrows). The three-dimensional reconstitutions show discs of varying diameter in the course of expansion and reveal an overgrowing phenomenon (simultaneous growth and appression). These reconstitutions sometimes disclose remnants of slits alongside the overlaps suggesting that the perforations may be the sites of grana initiation. This was also suggested by the biometric analysis, which showed that the frequency of holes in the thylakoid increase between stage 1 and 2. In stage 3, however, when granum initiations occurred, there was a rapid decrease in these lamellar interruptions that coincided with the increase in overlap frequency (Brangeon and Mustárdy, 1979). It is very likely that the edges of the lamellae are the growth points, while at the periphery of the holes, the overlapping growth pattern above and below the single lamellar sheet gives rise to stacking.

It is very probable that the appearance of light-harvesting complex at the growth points associated with PS II is a consequence of the high self-aggregation tendency of this protein, which thereby initiates the granum formation. The lateral-aggregation of PS II particles via adhesion of peripheral light-harvesting complexes can lead to

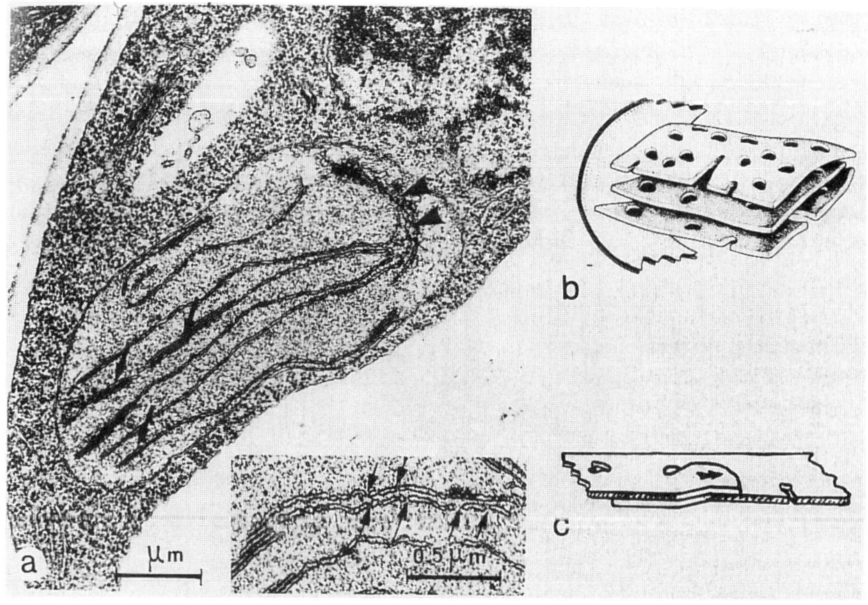

Fig. 8. Plastid profile from early stage of differentiation (a) showing parallel lamellar array with lamellar interruptions (inset-arrows) and nascent grana stacks (arrows). At the right end of the plastid lamellar connections are evident as 'parent entity' (arrow-heads). In spatial view these lamellae are independent through several sections (b) and the lamellar interruptions are revealed as slits or holes. (c) 3-dimensional reconstitution of a nascent granum (Brangeon and Mustárdy, 1979).

the formation of nearly homogeneous regions that are enriched in PS II while largely excluding PS I. These so-called 'macrodomains', through their head-to-head aggregations, ensure the maintenance of granal stacking (Garab et al., 1991; Barzda et al., 1994) and are responsible for the spatial separation of the two photosystems.

C. Granum Build-up, Multiplication and Integration of Helical Membrane

The parent lamellar sheet, interrupted in the vicinity of each nascent granum, constitutes a fret attachment at different levels to the stack. This arrangement maintains a continuity and consequently connects all grana that have been initiated on the same sheet. The 'insertion' of the stacked discs into the lamellar sheet introduces a twist in the lamellae surrounding the developing granum, an orientation that would be the primary helix around the granum. Three-dimensional reconstitution of developing grana often reveal incomplete discs at the end of the stacks that have only a single fret connection. This seems to indicate that the granum can be increased by 'overtopping' of the overgrown fret (Fig. 9b).

As the number of granum compartments increases, the fretwork is built into multiple layers. There are two types of configurations contributing to the fretwork increase. Firstly, the primary helix divides and directly gives rise to a secondary helix (Fig. 9c). Secondly, numerous thylakoid 'splittings' can be reconstituted during this stage (Fig. 9d). After ramifications the cleaved thylakoids extend and link into the existing helical system of neighboring grana or give an additional helix to the stack. This can happen not only in the same lamellar sheet but the growing thylakoid can also extend towards the lower (or upper) plane (Fig. 9a, arrows). From this developmental stage the parallel 'parent' lamellae have been replaced by a network which interconnects at different levels.

These multiple mechanisms—lamellar expansion, perforation, overgrowth, thylakoid splitting, branching, and bridging via fusion—guide the morphogenesis of the intrachloroplastic network formation and lead to the mature chloroplast (see Chapter 2).

IV. Possible Functional Role of Helical Arrangement

For the optimal energy distribution between the kinetically slow (PS II) and fast (PS I) photosystems, it may be necessary to physically separate the

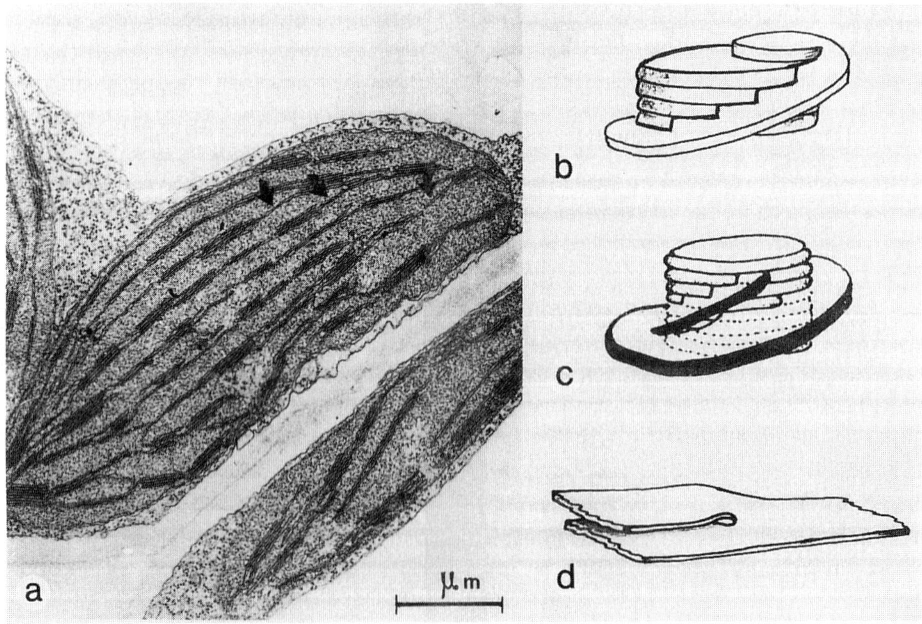

Fig. 9. Plastid profile from young chloroplast (a) showing fret connections between parent lamellar sheets (arrows). Three-dimensional reconstitutions of developing granum showing new compartment formation via overtopping (b), additional helix formation via splitting of primary helix (c) and thylakoid splitting on the lamellar sheet (d) (Brangeon and Mustárdy, 1979).

photosystems to decrease the probability of an exciton spilling over from PS II to PS I antenna pigments (Trissl and Wilhelm, 1993). On the other hand, for efficient energy transfer, an optimal spherical arrangement of the two photosystems is essential. This double requirement can be fulfilled by a helically arranged granal system.

Among the basic requirements for photosynthetic ATP synthesis, two may be relevant to a discussion of possible functional roles of the helical arrangement of thylakoids: i) the closed intrathylakoid space and ii) the vectorial flow of protons across the thylakoid membrane (Mitchell, 1961). In higher plant chloroplasts the proton accumulation in the granum is laterally separated from the site of utilization (stroma thylakoids) via the ATPase. The utilization of protons that accumulate within the granum is possible because the slits formed by fret connections ensure the continuity of the inner aqueous space. Since helices interconnect all granum compartments they allow coupling proton currents that might otherwise be formed between the different loculi of the granum. Such gradients can be formed as a result of the light gradient effect caused by self-shadowing (Osváth et al., 1994). A continuous intrathylakoid space thus ensures a homogeneous proton distribution in the loculus region of the entire chloroplast.

Another possibly important function of the helical arrangement involves the multiple connections of the two types of membranes facilitating the rapid diffusion of membrane components between the granum and stroma regions (Whitmarsh, 1986), e.g. the turnover of PS II reaction center protein D1 (Kim et al., 1994).

References

Anderson JM and Andersson B (1982) The architecture of photosynthetic membranes: Lateral and transverse organization. Trends Biochem Sci 7: 288–292

Barzda V, Mustárdy L and Garab G (1994) Size dependency of circular dichroism in macroaggregates of photosynthetic pigment-protein complexes. Biochemistry 33: 10837–10841

Bradbeer JW (1981) Development of photosynthetic function during chloroplast biogenesis. In: Hatch MD and Boardman NK (eds) The Biochemistry of Plants. Photosynthesis, Vol 8 pp 423–472. Academic Press, New York

Bradbeer JW, Ireland HMM, Smith JW, Rest J and Edge HJW (1974) Plastid development in primary leaves of *Phaseolus vulgaris*. VII. Development during growth in continuous darkness. New Phytol 73: 263–270

Brangeon J and Mustárdy L (1979) The ontogenetic assembly of intra-chloroplastic lamellae viewed in 3-dimension. Biol Cellulaire 36: 71–80

Dalton AJ and Hagueneau F (1968) The Membranes. Academic Press, N.Y., London

Falk H and Sitte P (1963) Zellfeinbau bei Plasmolyse. I. Der Feinbau der Elodea-Blattzellen. Protoplasma 57: 290–303

Garab G, Kieleczawa J, Sutherland JC, Bustamante C and Hind G (1991) Organization of pigment-protein complexes into macrodomains in thylakoid membranes of wild-type and chlorophyll *b*-less mutant of barley as revealed by circular dichroism. Photochem Photobiol 54: 273–281

Heslop-Harrison J (1963) Structure and morphogenesis of lamellar system in grana-containing chloroplasts. I. Membrane structure and lamellar architecture. Planta 60: 243–260

Hodge AJ, McLean JD and Mercer FV (1956) A possible mechanism for the morphogenesis of lamellar system in plant cells. J Biophys Biochem Cytol 2: 597–608

Kim J, Klein PG and Mullet JE (1994) Synthesis and turnover of photosystem II reaction center protein D1. Ribosome pausing increases during chloroplast development. J Biol Chem 269: 17918–17923

Kutik J (1985) Chloroplast development. In: Sestak Z (ed) Photosynthesis During Leaf Development, pp 51–75. Academia, Praha.

Menke W (1960) Das allgemeine Bauprinzip des Lamellarsystems der Chloroplasten. Experientia 16: 537–538

Mitchell P (1961) Coupling of phosphorylation to electron and hydrogen transfer by a chemi-osmotic type of mechanism. Nature 191: 144–148

Mullet JE and Arntzen CJ (1980) Stimulation of grana stacking in a model membrane system. Mediation by a purified light-harvesting pigment-protein complex from chloroplasts. Biochim Biophys Acta 589: 100–117

Mustárdy L and Brangeon J (1978) 3-Dimensional chloroplast infrastructure: Developmental aspects. In: Akoyunoglou G et al. (eds) Chloroplast Development, pp 489–494. Elsevier North-Holland Biomedical Press, Amsterdam-New York-Oxford

Mustárdy L and Jánossy AGS (1979) Evidence of helical thylakoid arrangement by scanning electron microscopy. Plant Sci Lett 16: 281–284

Mühlethaler K and Frey-Wissling A (1959) Entwicklung und Struktur der Proplastiden. J Biophys Biochem Cytol 6: 507–512

Osváth Sz, Meszéna G, Barzda V and Garab G (1994) Trapping magnetically oriented chloroplast thylakoid membranes in gel for electric measurements. J Photochem Photobiol B: Biol 26: 287–292

Paolillo DJ (1970) The three-dimensional arrangement of intergranal lamellae in chloroplasts. J Cell Sci 6: 243–255

Paolillo DJ and Falk RH (1966) The ultrastructure of grana in mesophyll plastids of *Zea Mays*. Amer J Bot 53: 173–180

Staehelin LA (1976) Reversible particle movements associated with unstacking and restacking of chloroplast membranes in vitro. J Cell Biol 71: 136–158

Trissl H-W and Wilhelm C (1993) Why do thylakoid membranes from higher plants form grana stacks? Trends Biochem Sci 18: 415–419

Wehrmeyer W (1964) Zur Klärung structurellen Variabilität der Chloroplastengrana des Spinats in Profil und Aufsicht. Planta 62: 272–293

Weier TE, Stocking CR, Thomson WW and Drever H (1963)

The grana as structural units in chloroplasts of mesophyll of *Nicotiana rustica* and *Phaseolus vulgaris*. J Ultrastruc Res 8: 122–143

Weier TE, Stocking CR, Bracker CE and Risler EB (1965) The structural relationships of the internal membrane system of in situ and isolated chloroplasts of *Hordeum vulgare*. Amer J Bot 52: 339–352

Whitmarsh J (1986) Mobile electron carriers in thylakoids. In: Staehelin LA and Arntzen CJ (eds) Encyclopedia of Plant Physiology, New series, Photosynthesis III, Photosynthetic Membranes and Light Harvesting Systems, pp 508–527. Springer-Verlag, Berlin

Chapter 6

Biosynthesis of Thylakoid Membrane Lipids

Roland Douce and Jacques Joyard
Laboratoire de Physiologie Cellulaire Végétale, URA CNRS n°576, Département de Biologie Moléculaire et Structurale, CEA-Centre d'Etudes Nucléaires de Grenoble et Université Joseph Fourier, 17 rue des Martyrs, F-38054 Grenoble-cedex 9, France

Summary

Plastid membranes (thylakoids as well as the two envelope membranes) contain specific polar lipids (galactolipids, sulfolipid, phospholipids), pigments (chlorophylls, carotenoids) and prenylquinones (plastoquinone, tocopherols). In this chapter, we describe our present understanding of the structure of these lipids and their distribution within chloroplasts, of the biosynthetic pathway for glycerolipids, pigments and prenylquinones and of the biochemical properties of the enzymes involved. The biosynthesis of plastid glycerolipids takes place in the inner envelope membrane, which is the site of assembly of fatty acids, glycerol and polar head groups

Donald R. Ort and Charles F. Yocum (eds): Oxygenic Photosynthesis: The Light Reactions, pp. 69–101.

(galactose, for galactolipids; sulfoquinovose, for sulfolipid and glycerol for phosphatidylglycerol). The inner envelope membrane contains all the enzymatic equipment for the biosynthesis of glycerolipids containing almost exclusively a C18/C16 diacylgycerol backbone. In contrast, the origin of plastid glycerolipids with a C18/C18 diacylglycerol backbone is still poorly understood. Then, fatty acids of the newly synthesized molecules are desaturated to form the polyunsaturated molecular species that are characteristic of plastid glycerolipids. Although the sequences of reactions involved in the biosynthesis of plastid prenyllipids (pigments and prenylquinones) have been thoroughly studied, little is known about the precise localization and properties of the different enzymes involved. There is a tight cooperation within chloroplasts between the stroma, the envelope membranes and the thylakoids to form the various pigments (chlorophyll and carotenoids) and prenylquinones (plastoquinone-9 and α-tocopherol), and many of the enzymes involved are in fact located in the inner envelope membrane. Because thylakoids represent the main proportion of the plastid membranes, they contain the largest amount of the plastid lipid constituents, therefore massive transport of lipid molecules from their site of synthesis (envelope membranes) to their site of accumulation (thylakoids) should take place during plastid development. The possible mechanisms that could be involved, i.e. vesicular transport, transfer of lipid monomers through the stroma either by facilitated transport or by spontaneous diffusion of free monomers, lateral diffusion of lipids between membranes at regions of direct intermembrane contact, will be discussed.

I. Introduction

The most common and best-known plastid is the chloroplast. Electron micrographs of green leaf cells show that chloroplasts present the three following major structural regions: a) a highly organized internal membrane network, formed of flat compressed vesicles and called thylakoids; b) an amorphous background rich in soluble proteins ($0.4\,g.ml^{-1}$) called the stroma; c) a pair of outer membranes known as the chloroplast envelope. Chloroplast membranes contain an astonishing variety of lipids including polar lipids (galactolipids, phospholipids, sulfolipid), pigments (chlorophylls, carotenoids) and prenylquinones (plastoquinone, tocopherols). As detailed throughout this chapter, generation of this diversity requires sophisticated metabolic pathways. The lipid compounds representing the end products of these pathways must bestow significant evolutionary advantages to the chloroplast membrane system in which they reside, implying particular functional roles for each component. Furthermore, the wide diversity of polar lipid species found in chloroplast membranes suggests a broader spectrum of functions for polar lipids than are currently appreciated. The object of this chapter is to provide an overview of lipid metabolism in chloroplasts at an advanced level. References to reviews will guide the reader to the current literature.

Abbreviations: ACP – acyl carrier protein; DAG – diacylglycerol; DGDG – digalactosyldiacylglycerol; LHC – light-harvesting complex; MGDG – monogalactosyldiacylglycerol; PA – phosphatidic acid; PC – phosphatidylcholine; PG – phosphatidylglycerol; SL – sulfolipid; SQ – sulfoquinovose; SQDG – sulfoquinovosyldiacylglycerol

II. Characterization and Biosynthesis of Chloroplast Glycerolipids

A. Structure and Distribution of Chloroplast Glycerolipids

Plastid membranes are characterized by the presence of large amounts of glycolipids: sulfolipid and galactolipids. The major plastid glycolipids are neutral lipids, i.e. galactolipids. They contain one or two galactose molecules attached to the *sn*-3 position of the glycerol backbone, corresponding to 1,2-diacyl-3-O-(β-D-galactopyranosyl)-*sn*-glycerol (or monogalactosyldiacylglycerol, MGDG) and 1,2-diacyl-3-O-(α-D-galactopyranosyl-(1 → 6)-O-β-D-galactopyranosyl)-*sn*-glycerol (or digalactosyldiacylglycerol, DGDG) (Fig. 1). MGDG was first isolated as a galactose-containing lipid from wheat flour by Carter et al. (1956), and from leaves by Sastry and Kates (1964). Galactolipids represent up to 80% of thylakoid membrane glycerolipids, from which MGDG constitute the main part (50%). A unique feature of galactolipids is their very high content of polyunsaturated fatty acids: in some species, up to 95% of the total fatty acids is linolenic acid (18:3). Therefore, the most abundant molecular species of galactolipids have 18:3 at both the *sn*-1 and *sn*-2

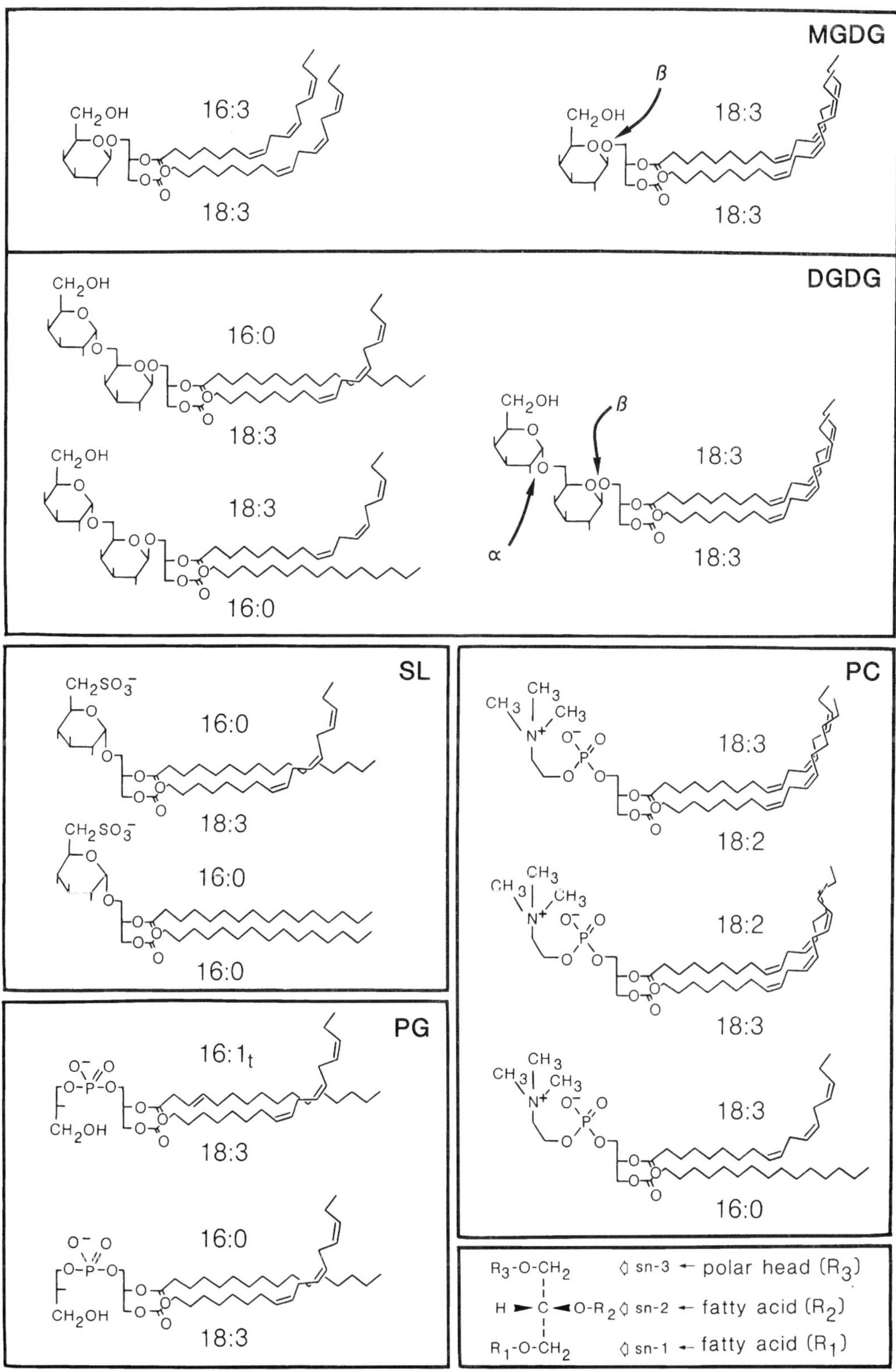

Fig. 1. Structure of the major plastid glycerolipids. Glycerolipids with C 16 fatty acids at the *sn*-2 position of the glycerol have a prokaryotic structure.

positions of the glycerol backbone. Some plants, such as pea, which have almost exclusively 18:3 in MGDG are called '*18:3 plants*'. Other plants, such as spinach, which contain large amounts of 16:3 in MGDG are called '*16:3 plants*' (Heinz, 1977). The positional distribution of 16:3 in MGDG is highly specific: this fatty acid is present at the *sn*-2 position of glycerol and is almost excluded from the *sn*-1 position. Therefore, different galactolipid molecular species with either C18 fatty acids at both *sn* positions or with C18 fatty acids at the *sn*-1 and C16 fatty acids at the *sn*-2 positions can be present in plastid membranes. The proportions of these two types of MGDG molecular species varies widely among plants (Heinz, 1977). The first structure is typical of '*eukaryotic*' lipids (such as phosphatidylcholine) and the second one corresponds to a '*prokaryotic*' structure, since it is characteristic of cyanobacterial glycerolipids (Heinz, 1977). In addition, this difference is also true of other glycerolipids since any membrane lipid containing C16 fatty acids at the *sn*-2 position of glycerol is considered as prokaryotic. The proportion of eukaryotic to prokaryotic molecular species is not identical in all glycerolipids from a given plant. For instance, although in spinach half of MGDG has the prokaryotic structure, this holds true for only 10–15% of DGDG (Bishop et al., 1985).

The most important sulfolipid found in higher plants is a 1',2'-diacyl-3'-O-(6-deoxy-6-sulfo-α-D-glucopyranosy)-*sn*-glycerol (sulfoquinovosyl-diacylglycerol or SQDG). The sulfonic residue at C6 of deoxyglucose (quinovose) carries a formal negative charge at physiological pH. This glycolipid is also found in cyanobacteria (Heinz, 1977). Analyses of the positional distribution of fatty acids (see for instance Siebertz et al., 1979) demonstrate that a significant proportion of SQDG in higher plants has a dipalmitoyl backbone. However, the major molecular species in SQDG contain both 16:0 and 18:3 fatty acids (Siebertz et al., 1979). In fact, two distinct structures can be observed in higher plants: they contain either 18:3/16:0 or 16:0/18:3 (Siebertz et al., 1979; Bishop et al., 1985). As discussed above, the structures having 16:0 at the *sn*-2 position (16:0/16:0 and 18:3/16:0) are typical of '*prokaryotic*' lipids whereas the last one (16:0/18:3), having 18:3 at the *sn*-2 position, is typical of '*eukaryotic*' lipids (see above). In a 16:3 plant, such as spinach, a higher proportion of SQDG (compared to the situation for MGDG) has a prokaryotic structure. In contrast, wheat, a 18:3 plant, contains almost exclusively SQDG with an eukaryotic structure (Bishop et al., 1985). All plastid membranes contain SQDG and, like MGDG, this glycolipid is present in the outer leaflet of the outer envelope membrane since it is accessible from the cytosolic face of isolated intact chloroplasts to specific antibodies (Billecocq et al., 1972).

Phosphatidylglycerol is a genuine constituent of all extraplastidial membranes, but it is present in low amounts (in general it represents only a few percent of the total glycerolipid content). In contrast, it is the major plastid phospholipid (Table 1). Phosphatidylglycerol represents (in spinach) about 7–10% of the total glycerolipid of thylakoids and of the envelope (outer as well as inner) membranes (Table 1). However, phospholipase *c* digestion of intact chloroplasts suggests that, in contrast to phosphatidylcholine, little phosphatidylglycerol is present in the outer leaflet of the outer membrane (Dorne et al., 1985). Plastid phosphatidylglycerol has a unique structure (Dorne and Heinz, 1989) since (a) it has a typical prokaryotic structure, with C16 fatty acid esterified at the *sn*-2 position of the glycerol backbone and (b) contains a unique *trans*-hexadecenoic acid ($16{:}1_t$) at the *sn*-2 position of the glycerol backbone (Haverkate and Van Deenen, 1965). No eukaryotic structure of phosphatidylglycerol (with C18 fatty acids at the *sn*-2 position of the glycerol backbone) was reported in chloroplasts (Bishop et al., 1985; Dorne and Heinz, 1989). Because of this uniqueness, specific functions for chloroplast phosphatidylglycerol have been proposed in plants. Although difficult to demonstrate unambiguously, this hypothesis is supported by the following observations. The chilling sensitivity of plants can be correlated with the occurrence of high-melting 16:0/16:0 and $16{:}0/16{:}1_t$ combinations in plastid phosphatidylglycerol (Murata et al., 1982; Bishop, 1986). Correlations have also been made between phosphatidylglycerol accumulation and the development of appressed membranes: for instance, mutants lacking PS II reaction centers were shown to be lacking phosphatidylglycerol (Garnier et al., 1990; Trémolières et al., 1991). Finally, Nussberger et al. (1993) have demonstrated that LHCII complex binds specifically with DGDG and PG, this phospholipid being intimately involved in trimer formation (Hobe et al., 1994) and is likely to be located at the subunit interface.

Phosphatidylcholine is only a minor constituent of plastid membranes (Table 1). In contrast, this

Table 1. Glycerolipid, pigment and prenylquinone composition of spinach chloroplast membranes

	Outer envelope membrane	Inner envelope membrane[a]	Total envelope membranes	Thylakoids[a]
Total polar lipids[b] (mg /mg protein)	2.5 – 3	1	1.2 – 1.5	0.6 – 0.8
Polar lipids (% of total)				
MGDG	17	55	32	57
DGDG	29	29	30	27
Sulfolipid	6	5	6	7
Phosphatidylcholine[a]	32	0	20	0
Phosphatidylglycerol	10	9	9	7
Phosphatidylinositol	5	1	4	1
Phosphatidylethanolamine	0	0	0	0
Total chlorophylls[c] (μg /mg protein)	*nd*	*nd*	0.1 – 0.3	160
Chlorophylls (% of total)				
Chlorophyll *a*	*nd*	*nd*	86	72
Chlorophyll *b*	*nd*	*nd*	14	28
Chlorophyll precursors[c] (Protochlorophyllide + chlorophyllide, μg /mg protein)	*nd*	*nd*	0.41	0 – 0.35
Total carotenoids[d] (μg /mg protein)	2.9	7.2	6 – 12	20
Carotenoids (% of total)				
β-Carotene	9	12	11	25
Violaxanthin	49	47	48	22
Lutein + zeaxanthin	16	23	21	37
Antheraxanthin	–	5	6	–
Neoxanthin	26	13	13	16
Total prenylquinones[e] (μg /mg protein)	4 – 12	4 – 11	4 – 11	4 – 7
Prenylquinones[e] (% of total)				
α-Tocopherol + α-tocoquinone	81	67	69	24
Plastoquinone-9 + plastoquinol	18	32	28	70
Phylloquinone K1	1	1	3	6

[a] The presence of phosphatidylcholine in thylakoids and inner envelope membrane preparations is due to a contamination by the outer envelope membrane (Dorne et al., 1985, 1990). Therefore, the polar lipid composition of thylakoids and inner envelope membrane was recalculated (data from Block et al., 1983a,b) to account for contamination of these membranes by the outer envelope membrane (see Joyard et al., 1991).
[b] Polar lipid composition of plastid membranes was obtained after thermolysin treatment of intact organelles (Dorne et al., 1982).
[c] Data are from Pineau et al. (1993).
[d] Data are from Jeffrey et al. (1975), Siefermann-Harms et al. (1978) and Block et al. (1983b).
[e] Data are from Lichtenthaler et al. (1981), Soll et al. (1980, 1985).
nd – not determined.

phospholipid represents about half of the glycerolipid content of extraplastidial membranes such as mitochondrial or peroxysomal membranes. The only plastid membrane containing significant amounts of phosphatidylcholine is the outer envelope membrane, with 30 to 35% of the glycerolipid content (Cline et al., 1981; Block et al., 1983b), compared to a few percent of the glycerolipid content of thylakoid preparations. The marked difference in the composition of the inner and outer membrane systems of chloroplast envelope has been established due to the development of reliable methods for the separation of these two structures (Cline et al., 1981; Block et al., 1983a). If the association between the two membrane systems is broken, the two membranes can be clearly resolved using density gradient centrifugation. The lower protein to lipid ratio in the outer membrane results in the banding of the outer membrane at a lower density (1.08 g.cm^{-3}) than the inner membrane (1.13 g.cm^{-3}). Phospholipase *c* digestion of intact chloroplasts provided evidence that chloroplast phosphatidylcholine is actually concentrated in the outer leaflet of the outer membrane (Dorne et al., 1985; Miquel et al., 1987). The outer envelope phosphatidylcholine is apparently of extraplastidial origin since (a) envelope membranes are unable to synthesize phosphatidylcholine (Joyard and Douce, 1976a) and (b) plastid and extraplastidial phosphatidylcholine have the same diacylglycerol structure (Siebertz et al., 1979). Therefore, mechanisms should exist for the transfer of phosphatidylcholine molecules from their site of synthesis (probably the endoplasmic reticulum) to the outer envelope membrane, but to date, no clear evidence for a precise mechanism is available. For instance, phospholipid transfer proteins have been demonstrated to mediate net in vitro transfer of phosphatidylcholine between plant cell membranes, but it is not yet clear whether such proteins are active in vivo since they have been demonstrated to have features of secreted proteins (Kader, 1990, 1993). Finally, using isolated intact chloroplasts in which envelope phosphatidylcholine was removed by phospholipase *c* treatment, Dorne et al. (1990) demonstrated that the presence of this phospholipid in thylakoid preparations is probably due to contaminating outer envelope membranes and therefore that thylakoids (and probably the inner envelope membrane) are devoid of phosphatidylcholine, like membranes from cyanobacteria. These observations raise the problem of the apparent lack of transmembrane diffusion in the outer envelope membrane: although it is a major component in the outer leaflet of the outer membrane, phosphatidylcholine is not redistributed to the thylakoids (Dorne et al., 1990). Again, the mechanism involved is entirely unknown.

Finally, Table 1 shows that all plastid membranes, and most notably the outer envelope membrane, are devoid of phosphatidylethanolamine, one of the major components (together with phosphatidylcholine) of extra-plastidial membranes.

B. Biosynthesis of Chloroplast Glycerolipids

Chloroplast polar lipids are synthesized exclusively for use in the biogenesis of plastid membranes since there does not appear to be any significant alternative fate for these lipids. Plastid envelope membranes play a central role in the biosynthesis of plastid glycerolipids (Douce, 1974; Joyard and Douce, 1977; Joyard et al., 1993), since they are the site of assembly of fatty acids, glycerol and polar head groups (galactose, for galactolipids; sulfoquinovose, for sulfolipid; and phosphorylglycerol, for phosphatidylglycerol). However, in vitro kinetics of acetate incorporation into chloroplast lipids have demonstrated that, like prokaryotes, isolated intact chloroplasts can synthesize glycerolipids containing almost exclusively a C18/C16 diacylgycerol backbone, but are *apparently* unable to catalyze the formation of phosphatidic acid and diacylglycerol with only C18 fatty acids. Therefore, it is likely that the formation of these two structures proceeds from two distinct pathways. We must point out, however, that the biochemical mechanisms that control the ratio of galactolipids, sulfolipid and phosphatidylglycerol in the membranes of plastids are unknown.

1. Origin of the Diacylgycerol Backbone

a. Diacylglycerol Biosynthesis in Chloroplast Envelope Membranes

The first enzyme of the glycerolipid biosynthetic pathway is a soluble glycerol-3-phosphate acyltransferase, closely associated with the inner envelope membrane, that catalyzes the transfer of oleic acid (18:1), from 18:1-ACP (acyl carrier protein) to the *sn*-1 position of glycerol (Frentzen et al., 1983; Frentzen, 1993) producing 1-oleoyl-*sn*-glycerol-3-phosphate (lysophosphatidic acid). Bertrams and Heinz (1981) first provided biochemical data for the purified chloroplast glycerol-3-phosphate acyltransferase from pea and spinach. Depending upon the plant species analyzed, the chloroplast stroma contains one to three isomeric forms of glycerol-3-phosphate acyltransferase (Bertrams and Heinz, 1981; Nishida et al., 1987; Douady and Dubacq, 1987). The sequences of the cDNAs corresponding to two isoforms of the squash chloroplast protein were determined by Ishizaki et al. (1988). Squash is a chilling-sensitive plant in which the chloroplast glycerol-3-phosphate acyltransferase incorporates a higher proportion of palmitic acid at the *sn*-1 position of the glycerol than in chilling-resistant plants, such as pea (see for instance Roughan, 1986). An oleate-selective (i.e. which discriminates efficiently against palmitoyl-ACP) acyl-ACP:*sn*-glycerol-3-phosphate acyltransferase has been purified from pea chloroplasts and its cDNA sequenced by Weber et al. (1991). Unfortunately, the comparison between the selective (pea) and non-selective (squash) acyltransferase does not provide a clue for recognizing any structural difference resulting in different selectivities (Weber et al., 1991). Interestingly, Kunst et al. (1988) characterized an *Arabidopsis thaliana* mutant, JB25, in which the chloroplast glycerol-3-phosphate acyltransferase activity was reduced to less than 4% of the activity of the wild type. This mutant is an interesting tool to investigate possible regulation of glycerolipid biosynthesis in higher plants (see below).

Lysophosphatidic acid is further acylated to form 1,2-diacyl-*sn*-glycerol-3-phosphate (phosphatidic acid) by the action of a 1-acylglycerol-3-phosphate acyltransferase (Joyard and Douce, 1977). In spinach chloroplasts, both the outer and the inner envelope membranes contain this acyltransferase (Block et al., 1983b), but in pea chloroplasts the enzyme is only found in the inner membrane (Andrews et al., 1985). Since lysophosphatidic acid used for this reaction is esterified at the *sn*-1 position, the enzyme will direct fatty acids, almost exclusively palmitic acid (16:0), to the available *sn*-2 position (Frentzen et al., 1983; Frentzen, 1993). The physiological significance of the outer envelope 1-acylglycerol-3-phosphate acyltransferase is totally unknown, as well as its specificity and selectivity for the different substrates.

Therefore, the two plastid acyltransferases have distinct specificities and selectivities for acylation of *sn*-glycerol-3-phosphate (Fig. 2). Together, they lead to the formation of phosphatidic acid with 18:1 fatty acid at the *sn*-1 and 16:0 fatty acid at the *sn*-2 positions of the glycerol backbone. This structure is typical of the so-called prokaryotic glycerolipids. In contrast, extraplastidial acyltransferases have distinct localization and properties (nature of the acyl donor, specificities, selectivities) as discussed by Frentzen (1986, 1993).

Phosphatidic acid, synthesized in envelope membranes, is further metabolized into either diacylglycerol or phosphatidylglycerol (Fig. 2). Diacylglycerol (1,2-diacyl-*sn*-glycerol) biosynthesis occurs in the envelope membrane owing to a membrane-bound phosphatidate phosphatase (Joyard and Douce, 1977; 1979) exclusively located on the inner envelope membrane (Block et al., 1983b; Andrews et al., 1985). In contrast to chloroplasts from 16:3 plants, those from 18:3 plants have a rather low phosphatidate phosphatase activity (Heinz and Roughan, 1983) and cannot deliver diacylglycerol fast enough to sustain the full rate of glycolipid synthesis. The same is true in non-green plastids from 18:3 plants (Alban et al., 1989b). These results may explain why 18:3 plants contain only small amounts of galactolipids and sulfolipid with C16 fatty acids at the *sn*-2 position, but contain phosphatidylglycerol (which is synthesized from phosphatidic acid) with such a structure (Figs. 2 and 3). It is not yet known whether the reduced level of phosphatidate phosphatase activity is due to lower expression (species-specific) of the gene coding for the enzyme or to the presence of regulatory molecules which control the activity of the enzyme.

The envelope phosphatidate phosphatase exhibits biochemical properties clearly different from similar enzymes described in the various cell fractions from animals or yeast (for reviews, see Bishop and Bell, 1988; Carman and Henry, 1989). Phosphatidate phosphatase activities described in extraplastidial compartments from plant tissues strongly resemble their animal counterpart and are very different from the envelope enzyme (Stymne and Stobert, 1987; Joyard and Douce, 1987). First, the envelope enzyme is tightly membrane-bound, whereas in yeast or animal cells the activity is recovered in both cytosolic and microsomal fractions. Furthermore, the pH optimum for the envelope enzyme is alkaline (9.0) and cations, such as Mg^{2+}, are powerful inhibitors of the enzyme. After solubilization, phosphatidate phosphatase remains sensitive to Mg^{2+} and to a wide range of other cations. Mn^{2+}, Cu^{2+} and Zn^{2+} were found to be the most potent inhibitors of the solubilized enzyme (Malherbe et al., 1995). In marked contrast, the yeast or animal phosphatidate phosphatases are active in the pH range of 5.5 to 7.5 and their activity is strongly dependent upon the presence of Mg^{2+} (for a review, see Joyard and Douce, 1987). Malherbe et al. (1992) demonstrated that diacylglycerol is a powerful inhibitor (apparent Ki $\cong$ 70 μM) of chloroplast envelope phosphatidate phosphatase whereas the affinity of this enzyme for its substrate phosphatidic acid is rather low (apparent Km 600 μM). In vivo, the steady state activity of phosphatidate phosphatase is sensitive to the diacylglycerol/phosphatidic acid molar ratio (Malherbe et al., 1992). In other words, increasing the ratio of diacylglycerol to phosphatidic acid in the inner envelope membrane resulted in a logarithmic increase in inhibition. Feedback inhibition of phosphatidate phosphatase (and consequently of galactolipid and sulfolipid synthesis) by diacylglycerol might lead to accumulation of phosphatidic acid and therefore will favor phosphatidylglycerol synthesis. Therefore, the rate of diacylglycerol formation is tightly related to the rate of its utilization by the envelope enzymes involved in galactolipids and sulfolipid biosynthesis.

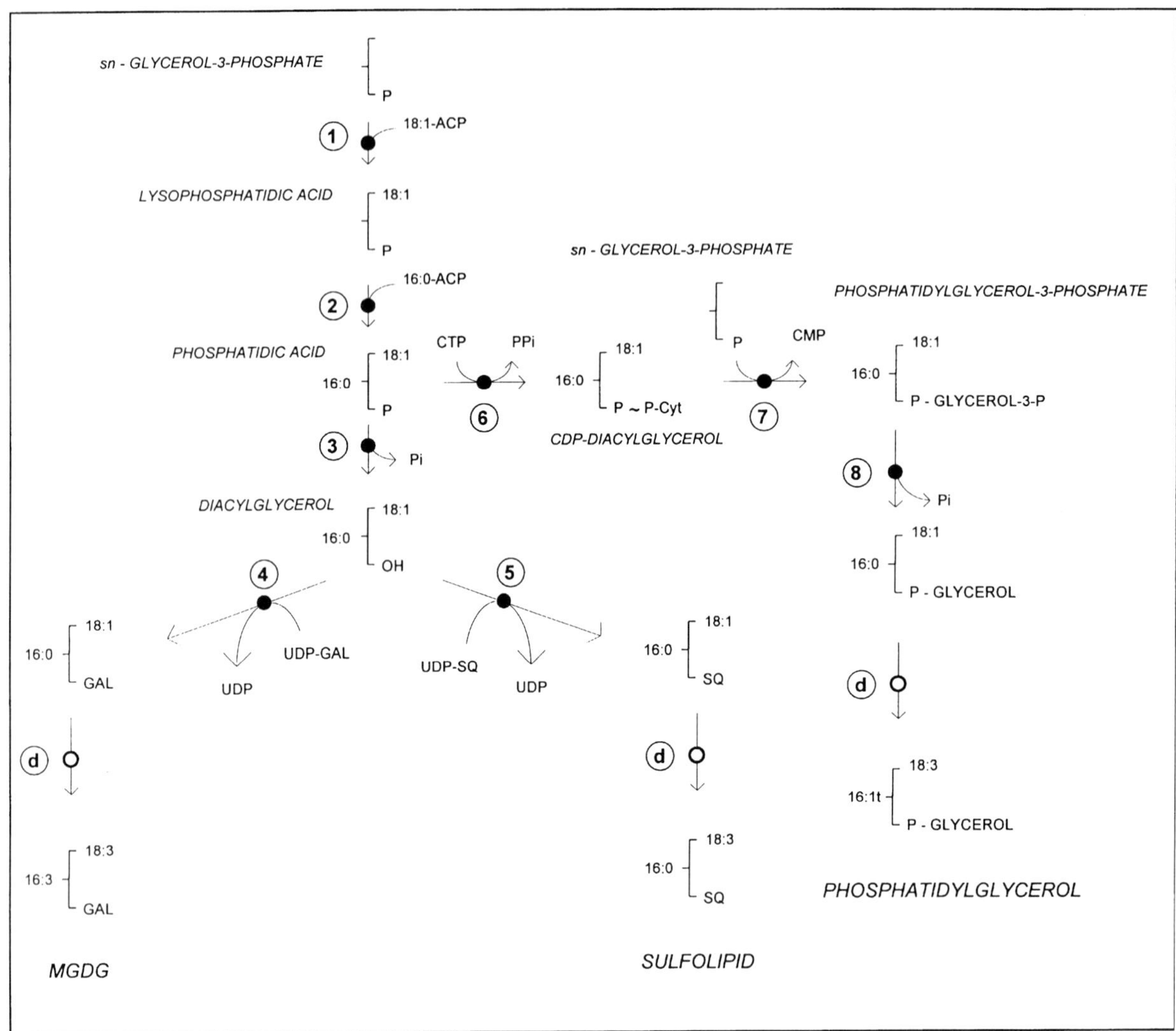

Fig. 2. Biosynthesis of plastid glycerolipids by plastid envelope membranes. The enzymes involved are: (1) glycerol-3-phosphate acyltransferase, (2) 1-acylglycerol-3-phosphate acyltransferase, (3) phosphatidate phosphohydrolase, (4) MGDG synthase, (5) sulfolipid synthase, (6) phosphatidate cytidyltransferase, (7) CDP-diacylglycerol-glycerol-3-phosphate 3-phosphatidyltransferase, (8) phosphatidylglycerophosphatase, (d) desaturases. This biosynthetic pathway leads to the synthesis of MGDG, sulfolipid and phosphatidylglycerol with unsaturated C 18 and C 16 fatty acids at the *sn*-1 and *sn*-2 positions of glycerol respectively.

b. Possible Origin of Dioleoylglycerol

As discussed above, the specificities of the envelope acyltransferases do not allow the formation of phosphatidic acid and diacylglycerol with exclusively C18 fatty acids. Prokaryotes are also unable to synthesize such structures. In vivo kinetics of acetate incorporation into chloroplast lipids suggest that phosphatidylcholine could provide the diacylglycerol backbone for eukaryotic plastid glycerolipids. The reader is referred to reviews by Heinz (1977), Douce and Joyard (1980), Roughan and Slack (1982) and Joyard and Douce (1987) for detailed presentations of the arguments in favor of this hypothesis. This hypothesis (Fig. 3) involves *(a)* the synthesis of phosphatidylcholine in extraplastidial membranes, *(b)* its transfer to the outer envelope membrane, *(c)* the formation of diacylglycerol and *(d)* the integration of diacylglycerol into MGDG or sulfolipid. Except during the initial steps, the plastid envelope membranes should play a central role in this pathway. However, the demonstration that this pathway indeed operates in plants has yet to be provided. The missing link is the conversion of phosphatidylcholine into

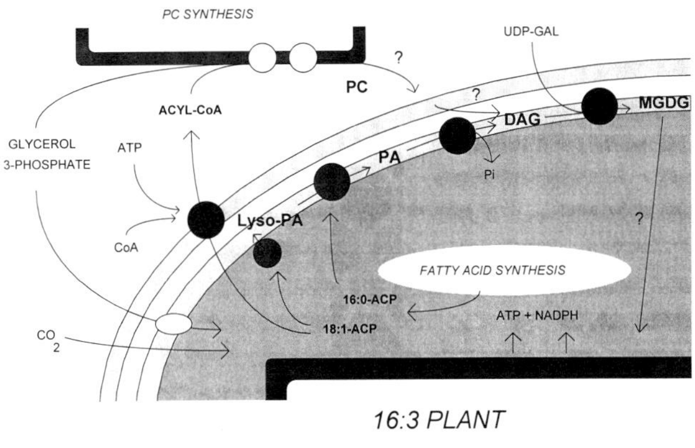

16:3 PLANT

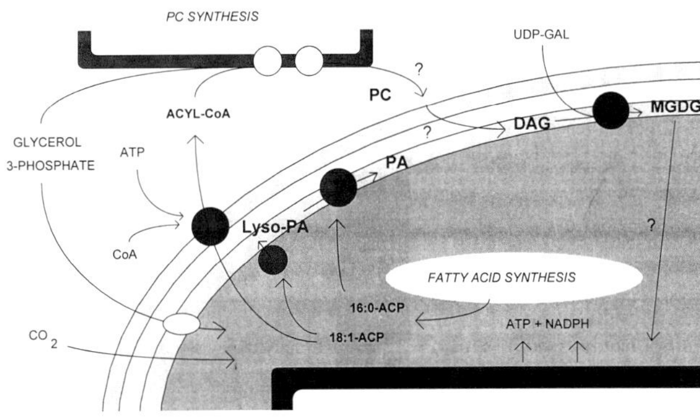

18:3 PLANT

Fig. 3. Biosynthesis of MGDG in 16:3- and 18:3-plants. The formation of MGDG with the 18:1/16:0 diacylglycerol backbone (prokaryotic structure) involves the Kornberg-Pricer pathway of the chloroplast envelope. This molecular species is then desaturated in envelope membranes and transported to the thylakoids (see Fig. 8). The formation of C18/C18 (eukaryotic structure) is not clearly established. In 16:3-plants, both pathways are generally active, whereas in 18:3-plants, only the latter pathway is operating. The lack of phosphatidate phosphatase in 18:3 plants prevents the formation (in the inner envelope membrane) of diacylglycerol, the substrate for MGDG synthase. The C18/C18 diacylglycerol used for MGDG synthesis in envelope membranes probably derives from phosphatidylcholine synthesized outside the chloroplast, in the endoplasmic reticulum. This MGDG molecule is then desaturated in envelope membranes and transported to the thylakoids (see Fig. 8).

diacylglycerol since envelope membranes apparently lack a phospholipase *c* activity which would produce diacylglycerol (Joyard and Douce, 1976b; Roughan and Slack, 1977; Siebertz et al., 1979).

2. Galactolipid Biosynthesis

a. MGDG Biosynthesis

The inner envelope membrane of spinach chloroplasts is characterized by the presence of a 1,2-diacylglycerol 3-β-galactosyltransferase (or MGDG synthase) which transfers a galactose from the water-soluble donor, UDP-galactose, to an hydrophobic acceptor molecule, diacylglycerol, to synthesize MGDG with the release of UDP (Douce, 1974). MGDG synthase, which is the first committed step specific for the synthesis of galactolipids, is concentrated in envelope membranes of all plastids and is therefore probably the best marker enzyme for envelope membranes (Douce, 1974). In spinach, this enzyme is located exclusively in the inner envelope membrane (Block et al., 1983b).

Covès et al. (1986, 1987) first described solubilization of MGDG synthase from spinach chloroplast envelope membranes and the development of a specific assay for the solubilized activity allowed partial purification of this enzyme (Covès et al., 1986). Teucher and Heinz (1991) and Maréchal et al. (1991) independently purified several hundred fold MGDG synthase activity from spinach chloroplast envelope, but in both cases, unambiguous characterization of the polypeptide associated with MGDG synthase activity was difficult. In particular, the final amount of enzyme was so low (in the range of μg when starting the purification from 100 mg envelope proteins), that further analyses of the protein were almost impossible. Electrophoretic analyses of the most purified fractions demonstrated that a polypeptide with a molecular mass around 20 kDa is probably associated with MGDG synthase activity (Teucher and Heinz 1991; Maréchal et al., 1991). These data also led to the conclusion that, despite its importance for chloroplast membrane biogenesis, MGDG synthase is only a minor envelope protein. This observation raised the question of the regulation of MGDG synthase activity within envelope membranes.

The apparent Km of MGDG synthase for UDP-gal is 100 μM (Covès et al., 1988; Maréchal et al., 1994a). However, no UDP-gal was detected in spinach chloroplast extracts by ^{31}P-NMR analyses (Bligny et al., 1990). In contrast, UDP-gal was characterized in perchloric extracts from whole spinach leaves, corresponding to concentrations in the cytosol of about 0.2 to 0.5 mM (Bligny et al., 1990). These values are high enough to sustain optimal rates of MGDG synthesis, suggesting that UDP-gal concentration is probably not a major regulatory factor of the biosynthetic pathway. Since MGDG synthase is located on the inner membrane of the spinach chloroplast envelope (Block et al., 1983b), and since this membrane is impermeable to UDP-gal (Heber and Heldt, 1981), there is no need for UDP-gal to accumulate in the plastid stroma if MGDG synthesis occurs on the outer surface of the inner envelope membrane, facing the intermembrane space which is connected to the cytosol by the presence of a pore protein in the outer envelope membrane (Flügge and Benz, 1984). If this hypothesis is true, then, the UDP-gal binding site of the envelope MGDG synthase could be localized at the outer surface of the inner envelope membrane.

Maréchal et al. (1994a) carried out kinetic experiments in mixed micelles containing the partially purified enzyme, the substrate (diacylglycerol) and the detergent (CHAPS), according to the 'surface dilution' kinetic model proposed by Deems et al. (1975). The dependence of kinetic parameters of MGDG synthase on the diacylglycerol mole fraction allows a comparison of the affinity of the enzyme for a wide range of diacylglycerol molecular species. Km values obtained ranged between 0.0089-mol fraction for dilinoleoylglycerol and dipalmitoylglycerol and 0.0666-mol fraction for distearoylglycerol, but the differences observed were not really related to the unsaturation of the molecule since the Km value for dilinoleoylglycerol was much lower than that (0.040-mol fraction) for dilinolenoylglycerol. Km values for dioleoylglycerol and for 1-oleoyl-2-palmitoylglycerol were in the average range, i.e. lower than 0.030-mol fraction. In addition, 1-oleoyl-2-palmitoylglycerol is produced in significant amounts by the envelope phosphatidate phosphatase. This explains why this typical prokaryotic structure is readily incorporated into MGDG. However, the best substrate for MGDG synthase was dilinoleoylglycerol. In mixed micelles, dilinoleoylglycerol concentration corresponding to an apparent Km value was as low as 1 molecule/micelle. The same result was obtained with isolated envelope vesicles (Maréchal et al., 1994b). One should question whether dilinoleoylglycerol could actually be present (or formed) in envelope membranes. As discussed above, it is assumed that MGDG molecules with the eukaryotic structure could derive from phosphatidylcholine which is synthesized on the endoplasmic reticulum and transferred to cell organelles such as chloroplasts. Interestingly, phosphatidylcholine, which (in chloroplasts) is concentrated in the outer leaflet of the outer envelope membrane (Dorne et al., 1985, 1990), contains significant amounts of 18:2 at both *sn*-1 and *sn*-2 position of the glycerol backbone (Siebertz et al., 1979), but the enzyme which could generate diacylglycerol from phosphatidylcholine, i.e. phospholipase *c*, has not been found in envelope membranes (see above). Therefore, if phosphatidylcholine is indeed the source of dilinoleoylglycerol in the outer membrane, only limited amounts would be delivered to the inner membrane where MGDG synthase is located. In addition, nothing is known about the possible mechanisms that could be involved in diacylglycerol transfer from the outer to the inner envelope membrane (spontaneous diffusion of free monomers, lateral diffusion of lipids between membranes at regions of direct intermembrane contact). Therefore, only the high affinity of the envelope MGDG synthase for dilinoleoylglycerol would explain the presence of C18 fatty acids at both *sn* positions of glycerol in MGDG.

MGDG molecules do not contain dipalmitoyl species (Siebertz et al., 1979). In addition, dipalmitoylglycerol is only synthesized in small amounts within envelope membranes by the enzymes of the Kornberg-Pricer pathway whereas 1-oleoyl-2-palmitoylglycerol is the major product (see above). Furthermore, Maréchal et al. (1994a) demonstrated that the envelope MGDG synthase has a low affinity for dipalmitoylglycerol (compared to that for 1-oleoyl-2-palmitoylglycerol). Therefore, the combination of these two major limitations probably explains why the biosynthesis of MGDG molecules containing dipalmitoylglycerol is unlikely. As we shall see latter, this is in contrast with SQDG biosynthesis.

The mechanism of MGDG synthase was also investigated with two-substrate kinetic studies at varied UDP-galactose molar concentrations and varied dioleoylglycerol surface concentrations (Maréchal et al., 1994a). The families of reciprocal plots obtained were shown to intersect at a single point of the 1/[substrate]-axis thus demonstrating that MGDG synthase was a sequential, either random or ordered, bireactant system, and not a ping-pong mechanism as proposed by Van Besouw and Wintermans (1979). Therefore, MGDG synthase has two distinct and independent substrate-binding sites, a hydrophilic one for UDP-galactose that is accessible from the outer surface of the intermembrane space, and a hydrophobic one for diacylglycerol (Maréchal et al., 1994a). Further investigations on the inhibition pattern of MGDG synthase (at varied UDP-galactose molar concentrations and varied dioleoylglycerol surface concentrations) by UDP, and on the effect of various inhibitors (NEM, etc.) demonstrated that in fact the enzyme mechanism (Fig. 4) is a random, sequential, bireactant system (Maréchal et al., 1994a, 1995).

b. DGDG Biosynthesis

The galactolipid:galactolipid galactosyltransferase, which is the only DGDG-forming enzyme clearly described in plastids to date, was characterized first in envelope membranes by Van Besouw and Wintermans (1978). It catalyzes an enzymatic galactose exchange between galactolipids with the formation of diacylglycerol, but in vitro, unnatural

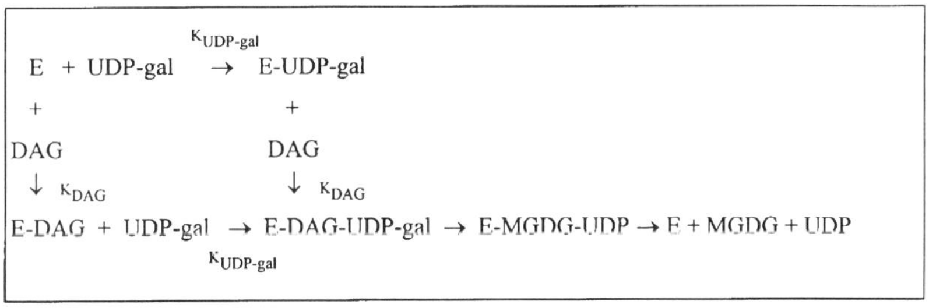

$$E + \text{UDP-gal} \xrightarrow{K_{\text{UDP-gal}}} \text{E-UDP-gal}$$

+ DAG $\downarrow K_{DAG}$ (from E) ; + DAG $\downarrow K_{DAG}$ (from E-UDP-gal)

$$\text{E-DAG} + \text{UDP-gal} \xrightarrow[K_{\text{UDP-gal}}]{} \text{E-DAG-UDP-gal} \rightarrow \text{E-MGDG-UDP} \rightarrow E + \text{MGDG} + \text{UDP}$$

Fig. 4. Mechanism of the envelope MGDG synthase.

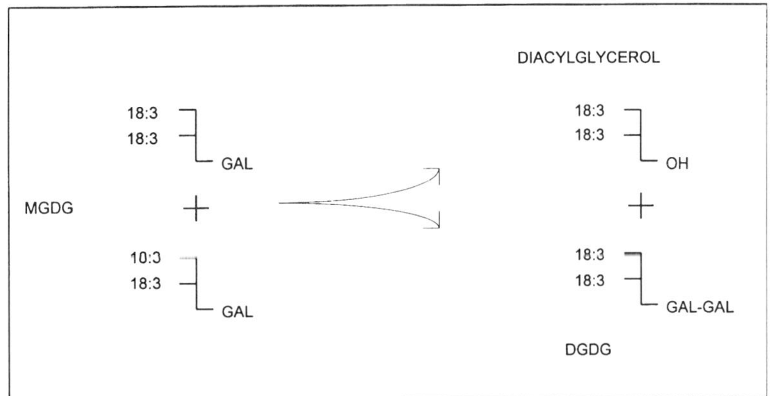

Fig. 5. Reactions catalyzed by the galactolipid:galactolipid galactosyltransferase of the outer envelope membrane.

galactolipids with more than two galactose residues can be synthesized (Fig. 5). This enzyme is located on the cytosolic side of the outer envelope membrane because it is destroyed during mild proteolytic digestion (with thermolysin) of intact chloroplasts (Dorne et al., 1982). Van Besouw and Wintermans (1978), Heemskerk and Wintermans (1987) and Heemskerk et al. (1990) have proposed that the galactolipid:galactolipid galactosyltransferase is indeed responsible for DGDG synthesis. For instance, Heemskerk et al. (1990) analyzed galactolipid synthesis in chloroplasts or chromoplasts from 8 species of 16:3 or 18:3 plants and found that digalactosyldiacylglycerol formation is never stimulated by UDP-gal or any other nucleoside 5'-diphosphodigalactoside; in all cases, DGDG formation was reduced by thermolysin digestion of intact organelles. In vitro, the galactolipid:galactolipid galactosyltransferase does not show strong specificity for any MGDG molecular species. However, if this enzyme is indeed the DGDG-synthesizing enzyme, it should discriminate in vivo between the various MGDG molecular species that are available since (a) the proportion of eukaryotic molecular species is higher in DGDG than in MGDG (Heinz, 1977, Bishop et al., 1985) and (b) DGDG contains 16:0 fatty acids (up to 10–15%) at both the *sn*-1 and *sn*-2 position and very little 16:3 (in 16:3 plants), whereas MGDG contains little 16:0, but (in 16:3 plants) 16:3 at the *sn*-2 position (Heinz, 1977).

Sakaki et al. (1990) proposed another physiological significance for the envelope galactolipid:galactolipid galactosyltransferase. Using ozone-fumigated spinach leaves, they demonstrated in vivo (Sakaki et al., 1990) that MGDG was converted into diacylglycerol by the galactolipid:galactolipid galactosyltransferase and then to triacylglycerol (by acylation with 18:3-CoA), owing to a diacylglycerol acyltransferase associated with envelope membranes (Martin and Wilson, 1984). Whether the galactolipid:galactolipid galactosyltransferase is involved in DGDG synthesis is the focus of current investigation in several laboratories and hopefully a definitive answer to this question should be available in the near future.

3. Sulfolipid Biosynthesis

Intact chloroplasts are able to incorporate SO_4^{2-} into SQDG (Haas et al. 1980; Kleppinger-Sparace et al., 1985; Joyard et al., 1986). Heinz et al. (1989) synthesized different nucleoside 5'-diphosphosulfoquinovoses and demonstrated that both UDP- and GDP-sulfoquinovose significantly increased SQDG synthesis by spinach chloroplasts and by isolated envelope membranes; UDP-sulfoquinovose was observed to be twice as active as the GDP derivative. Therefore, SQDG synthesis in envelope membranes is probably due to a UDP-sulfoquinovose:1,2-diacylglycerol 3-β-sulfoquinovosyltransferase or 1,2-diacylglycerol 3-β-sulfoquinovosyltransferase (SQDG synthase).

In plants containing SQDG molecular species with C18 fatty acids at the *sn*-2 position, the diacylglycerol used for SQDG synthesis could derive from eukaryotic lipids, as proposed for MGDG (see above). But this is only a small proportion of SQDG molecules, except in plants such as wheat or cucumber (Bishop et al. 1985). Because most SQDG molecules have a prokaryotic structure, they should derive from diacylglycerol molecules formed de novo in the inner envelope membrane by the enzymes of the Kornberg-Pricer pathway (Bishop et al. 1985; Joyard et al., 1986). Among the diacylglycerol molecular species formed within chloroplasts, dipalmitoylglycerol represents only a minor portion, but SQDG contains such a structure in significant amounts (Siebertz et al. 1979). Furthermore, the two enzymes involved in MGDG and SQDG synthesis compete for the same pool of diacylglycerol molecules (Joyard et al., 1986) and it is unclear how these two enzymes discriminate between the different molecular species to form MGDG and SQDG molecules with a distinct backbone. Envelope membranes loaded with 16:0/16:0 and/or 18:1/16:0 diacylglycerol and incubated in presence of UDP-gal or UDP-sulfoquinovose, incorporated 16:0/16:0 diacylglycerol with a much greater efficiency into SQDG than into MGDG whereas 18:1/16:0 diacylglycerol was incorporated into both MGDG and SQDG with almost the same efficiency (Seifert and Heinz, 1992). Therefore, by comparison with the kinetic studies of MGDG synthase (Maréchal et al., 1994a), one can propose that high affinity of SQDG synthase for 16:0/16:0 diacylglycerol explains its presence in SQDG. Experiments with partially purified SQDG synthase similar to those performed by Maréchal et al. (1994a) on MGDG synthase, would be an interesting test to confirm this hypothesis.

4. Phosphatidylglycerol Biosynthesis

Mudd and de Zacks (1981) first demonstrated the ability of intact chloroplasts to synthesize phosphatidylglycerol. Plastid phosphatidylglycerol is synthesized from phosphatidic acid formed by the inner envelope (Fig. 2), i.e. with C18 fatty acids at the *sn*-1 and C16 fatty acids at the *sn*-2 position of the glycerol, and the enzymes involved in its synthesis are localized in the inner envelope membrane (Andrews and Mudd., 1985). Therefore, in addition to glycerol-3-phosphate acyltransferase and 1-acylglycerol-3-phosphate acyltransferase, which catalyze the formation of phosphatidic acid, phosphatidylglycerol synthesis requires the coordinated functioning of the following three enzymes (Fig. 2): phosphatidate cytidyltransferase, CDP-diacylglycerol-glycerol-3-phosphate 3-phosphatidyl-transferase and phosphatidylglycerophosphatase, which catalyze the formation of CDP-diacylglycerol, 3-(3-*sn*-phosphatidyl)-*sn*-glycerol-1-phosphate and phosphatidylglycerol respectively. The structure of phosphatidylglycerol, with C16 fatty acids at the *sn* 2 position of the glycerol backbone, is typical of prokaryotic lipids (see above). Since, in chloroplasts, this phospholipid does not contain C18 fatty acids at the *sn*-2 position, its biosynthesis does not require the participation of eukaryotic lipids, as proposed for MGDG and to a lesser extent for SQDG (see above). In fact, chloroplasts from 16:3 and 18:3 plants both synthesize phosphatidic acid. In 16:3 plants, part of the phosphatidic acid is converted into diacylglycerol for galactolipids and sulfolipid synthesis, whereas in 18:3 plants, almost all phosphatidic acid formed is used for phosphatidylglycerol synthesis. It is clear that phosphatidate phosphatase is in a position to help to regulate the relative flux of phosphatidate to diacylglycerol (and thus to galactolipids and SQDG) versus the production of CDP-diacylglycerol to be used for PG synthesis.

C. Fatty Acid Desaturation

Glycerolipids that are synthesized on envelope membranes contain 16:0 and 18:1 fatty acids which must be desaturated to polyunsaturated fatty acids. This desaturation of C16 and C18 fatty acids occurs when esterified to MGDG (for a review, see Heinz; 1993). Roughan et al. (1979) first demonstrated that isolated intact spinach chloroplasts can synthesize MGDG containing polyunsaturated fatty acids. Heinz and Roughan (1983) further analyzed MGDG synthesized from [^{14}C]-acetate by isolated *Solanum nodiflorum* chloroplasts and found the complete series of radioactive fatty acids, from 16:0 to 16:3 at the *sn*-2 position and from 18:1 to 18:3 at the *sn*-1 position of the glycerol. Desaturation also takes place in non-green plastids (Alban et al., 1989b). Schmidt and Heinz (1990a,b) showed that pure envelope membranes could be used as a source of enzyme for the desaturation of oleic acid into linoleic acid. Catalase was required in the incubation medium, apparently to destroy the inhibitory hydrogen peroxide that was formed by autooxidation of ferredoxin (Heinz, 1993). A cDNA encoding a 40-kDa polypeptide from chloroplast envelope membranes that could possibly be involved in the n-6 (Δ^{12}) desaturase from spinach was isolated by Schmidt and Heinz (1992). The sequence data suggest a phylogenetic and functional relationship between this desaturase and the *des*A-coded protein, another Δ^{12} desaturase (Heinz, 1993).

In fact, the nature of the enzymes involved in vivo in the *cis*-desaturation of fatty acyl chains engaged in glycerolipids is not yet clearly established. Likewise, the biosynthesis of the unique *trans*-Δ^3-hexadecenoic acid component of chloroplast phosphatidylglycerol is still a mystery. Until recently, most of our knowledge of chloroplast desaturases was limited to the soluble components of the 18:0 to 18:1 desaturation system (ferredoxin, ferredoxin:NADP oxidoreductase, stearoyl-ACP desaturase). Because intact chloroplasts are able to catalyze desaturation of MGDG-linked unsaturated or monounsaturated fatty acids to polyunsaturated fatty acids (Roughan et al., 1979), the addition of electron transport inhibitors was expected to provide information about electron donors and transport components involved in this type of desaturation (Andrews et al., 1989). Indeed, fatty acid desaturation in newly synthesized MGDG by intact chloroplasts can be blocked by compounds that interfere with light-driven electron transport in thylakoids (Andrews et al., 1989). In the case of envelope-bound desaturase, preliminary in vitro evidence for the involvement of a ferredoxin:NADPH oxidoreductase (FNR) was obtained by the use of anti-FNR-IgG (Schmidt and Heinz, 1990a,b). Experiments with *Arabidopsis* chloroplasts (Norman et al., 1991) suggest that O_2 is the final electron acceptor whereas reduced ferredoxin ($E'_0 = -0.4$ v) is the source of electrons for the reduction of O_2 to H_2O ($E'_0 = -0.8$ v). Since ferredoxin delivers only one electron at a time, the desaturase has to oxidize two reduced ferredoxin, and store the first electron before the double bond is formed (Heinz, 1993). This is possible only in the presence of a complex electron transfer chain that has not yet been characterized in envelope membranes. Clearly, this should be a major goal in the future. In addition, the exact mechanism of the desaturation still remains to be understood. For instance, unambiguous evidence that the true electron donor for the envelope-bound fatty acid desaturase is indeed reduced ferredoxin is still lacking.

Another approach to study fatty acid desaturation is the use of mutants from cyanobacteria or *Arabidopsis*. Browse et al. (1985) have initiated a genetic approach to the analysis of lipid composition and synthesis by the isolation of a series of mutants of the small crucifer *Arabidopsis thaliana* with specific alterations in leaf fatty acid or glycerolipid composition. The mutants were isolated without selection by screening an EMS-mutagenized population of plants (about 10000 randomly chosen individuals) by gas chromatography of small leaf samples. Such investigations led to the characterization of several genes associated with membrane-bound fatty acid desaturases (Somerville and Browse, 1991; Slabas and Fawcett, 1992; Heinz, 1993). Most of the mutations with alterations in membrane lipid composition that have been analyzed cause the loss or the reduction in the amount of an unsaturated fatty acid and the corresponding accumulation of a less unsaturated precursor (Somerville and Browse, 1991; Heinz, 1993). Thus, it was inferred that the different *fad* mutants were defective in the desaturation of a glycerolipid-linked fatty acid. Most of these mutants had apparently a normal phenotype and their growth was almost the same as the wild type *Arabidopsis*, thus demonstrating that mutations that were selected were not lethal under the growth conditions that were selected. Several mutants affecting fatty acid desaturation in chloroplasts (*fad*A, B, C, D.) have been isolated and characterized. However, it is not yet clear whether the genes characterized from the different mutants correspond to the desaturases themselves or to some regulatory elements. Analysis of the effects of the mutations on membrane lipid composition complement the biochemical studies of leaf lipid metabolism and its regulation. Among the most interesting observations provided by the use of these mutants was the demonstration of the flexibility of the proportions observed between 16:3 and 18:3 fatty acid distribution. Wild type *Arabidopsis* is a typical 16:3 plant and therefore it is possible to follow the possible switching between MGDG molecular species whether or not they contain 16:3 fatty acids. Such analyses have been done first using the mutant JB 25 (a mutant deficient in chloroplast glycerol-3-phosphate acyltransferase) in which MGDG no longer contains any 16:3, but only 18:3. In addition, the MGDG content of the mutant is nearly the same as the wild type (Kunst et al., 1988). Since the acyltransferase involved in the biosynthesis of lysophosphatidic acid is strongly reduced, the normal pathway for the synthesis of glycerolipids having only the prokaryotic structure is less active and the activity of the enzymes catalyzing the synthesis of the eukaryotic structure is increased to compensate for the loss of prokaryotic glycerolipids (Kunst et al., 1988). This experiment and others done with *fad* mutants demonstrate the remarkable flexibility of glycerolipid biosynthesis in plants and provides an insight into the nature of the regulatory mechanisms involved.

III. Characterization and Biosynthesis of Chloroplast Prenyllipids

Plants, and especially their plastids, contain various classes of terpenoid lipids (prenyllipids) that are very different in their chemical structure, but which all have a common precursor deriving from mevalonic acid, isopentenyldiphosphate, a C5 isoprene unit (Ruzicka, 1953). The origin of mevalonic acid and isopentenyldiphosphate is not clearly established. However, Soler et al. (1993) recently characterized in the chloroplast envelope a specific carrier involved in the transport of isopentenyldiphosphate. The major prenyllipids in chloroplasts are pigments (chlorophylls, carotenoids) and prenylquinones (plastoquinone-9, phylloquinone and α-tocopherol). However, only the side chain of chlorophylls and prenylquinones derive from isopentenyldiphosphate, whereas the whole carotenoid structure derives from the isoprene unit. Interestingly, some of these plastid prenyllipids are essential vitamins: β-carotene (vitamin A), α-tocopherol (vitamin E) and phylloquinone (vitamin K_1). Reviews providing the basic general information on structure, distribution and biosynthesis of plant prenyllipids are given by Gray (1987), Rudiger and Schoch (1988), Goodwin and Britton (1988), Britton (1988), Kleinig (1989), Beale and Weinstein (1990) and Lichtenthaler (1993).

A. Distribution and Biosynthesis of Pigments

1. Chlorophylls

a. Distribution

Chlorophylls are the most conspicuous plastid pigment (Table 1). There are about $6.7 \cdot 10^8$ chlorophyll molecules per chloroplast of a typical C3 plant, representing 4% of the dry mass (estimated at about $20 \cdot 10^{-12}$ g) of a chloroplast (Lawlor, 1987). These metallo-tetrapyrroles are bound, together with carotenoids, to proteins associated within thylakoids with light-harvesting and photosystem proteins. Detailed presentations of the chlorophyll-carotenoid proteins (there are about 300 chlorophyll molecules per photosystem) found in chloroplasts were given by Cogdell (1988) and Lichtenthaler (1993) and chapters 27 through 30 of this volume. In contrast to thylakoids, envelope membranes are devoid of chlorophyll since only traces are found in the best envelope preparations (Jeffrey et al., 1974). As shown in Table 1, on average and on a protein basis, an envelope membrane preparation contains about 500 times less chlorophyll than thylakoids. Pineau et al. (1993) have separated by HPLC these minute traces of pigments found in envelope preparations. Because the chlorophyll *a* to *b* ratio in the purest envelope preparations was around 6, one can suggest that (a) the envelope fractions were almost devoid of chlorophyll *b* and (b) the traces of chlorophyll *a* that are found in envelope membrane fractions do not derive from a contamination by small thylakoid fragments enriched in plastoglobules (Pineau et al., 1993). The origin and significance of these traces is not yet clear, but they could correspond to a biosynthesis pool.

Pineau et al. (1986) demonstrated that envelope membranes from mature spinach chloroplasts contain low amounts of pigments with the absorption and fluorescence spectroscopic properties, and the behavior in polar/non polar solvents, of protochlorophyllide and chlorophyllide. As shown in Table 1, their concentration is in the range of 0.1 to 1.5 nmol/mg protein (Pineau et al., 1993). In contrast, protochlorophyllide and chlorophyllide are barely detectable in thylakoids: the molar ratio of protochlorophyllide and chlorophyllide to chlorophyll was up to 1000 times higher in envelope membranes than in thylakoids (Pineau et al., 1993). Interestingly, Hinterstoisser et al. (1988) and Peschek et al. (1989) have demonstrated that chlorophyll-free cytoplasmic membranes from the cyanobacterium *Anacystis nidulans* contain significant amounts of protochlorophyllide and chlorophyllide.

b. Biosynthesis

Chlorophyll is made of two parts with a distinct origin: (a) the porphyrin (i.e. chlorophyllide) that derives from (2)δ-aminolevulinic acid, and (b) a prenylalcohol, namely the diterpene phytol, that derives from isopentenyldiphosphate. In higher plants, (2)δ-aminolevulinic acid is formed from the intact carbon skeleton of glutamic acid, in a process requiring three enzymatic reactions and tRNA (Beale, 1990). We shall consider first the formation of the chlorophyllide part of the chlorophyll molecule. The sequence of the reactions involved in chlorophyllide biosynthesis has been thoroughly studied (for reviews, see Castelfranco and Beale, 1983; Beale and Weinstein, 1990). The first steps in chlorophyllide biosynthesis, namely those from (2)δ-aminolevulinic acid to protoporphyrin IX, occur in the soluble phase of chloroplasts (Smith and Rebeiz, 1979), but all the subsequent steps in protoporphyrin IX transformation are catalyzed by membrane-bound enzymes (Castelfranco and Beale, 1983; Beale and Weinstein, 1990). Since they are only present in plastids, these enzymes should be either localized in thylakoids or in envelope membranes. Some information is now available about the possible compartmentation, within chloroplasts, of the enzymes involved in chlorophyll biosynthesis.

In fact, the characterization of protochlorophyllide and chlorophyllide in envelope membranes (see above) raised the possibility for this membrane system to play a role in chlorophyll biosynthesis. Incubation of chloroplast envelope membranes under weak light in the presence of NADPH induced a progressive decrease in the level of fluorescence at 636 nm (attributed to protochlorophyllide) together with a parallel increase in fluorescence at 680 nm (attributed to chlorophyllide) due to the action of a protochlorophyllide reductase (Pineau et al., 1986; Joyard et al., 1990). Pineau et al. (1993) extended this observation and were able to demonstrate, using HPLC, that envelope membranes from fully developed chloroplasts catalyze the NADPH- and light-dependent conversion of endogenous protochlorophyllide into chlorophyllide. The rate of photoconversion observed in vitro was very low (up to 1 nmol protochlorophyllide converted/h/mg protein) but this probably does not reflect the in vivo rate of activity because of the low level of endogenous protochlorophyllide in native envelope membranes. Frank and Strzalka (1992) detected, in greening barley leaves, a photoactive protochlorophyllide reductase complex consisting of protochlorophyllide reductase, protochlorophyllide and NADPH. It is not yet known whether the envelope protochlorophyllide reductase and protochlorophyllide are indeed associated in the same photoactive structure, but this could be compatible with the levels of protochlorophyllide reductase and protochlorophyllide found in envelope membranes. Joyard et al. (1990) demonstrated that in mature spinach chloroplasts, a 37,000 Da-polypeptide from purified envelope membranes was immunodecorated by specific antibodies (anti-PCR) raised against oat protochlorophyllide reductase. In contrast, purified thylakoids, as well as stroma proteins, were not found to react with anti-PCR (Joyard et al., 1990). The envelope polypeptide reacting with anti-protochlorophyllide reductase antibody was not the major envelope E37 polypeptide (Block et al., 1992), but a very minor protein that can be separated from E37 only by two-dimensional polyacrylamide gel electrophoresis (Joyard et al., 1990). Protochlorophyllide reductase from mature spinach chloroplasts appears to be localized on the cytosolic side of the outer envelope membrane because (a) this polypeptide was susceptible to thermolysin digestion of isolated intact chloroplasts and (b) anti-PCR antibodies induced agglutination of isolated intact chloroplasts (Joyard et al., 1990). However, there is a marked difference between etiolated tissues and green tissues concerning the localization of protochlorophyllide reductase visualized by immunocytochemistry with anti-PCR antibodies. In etiolated leaf samples, gold particles are highly concentrated in prolamellar bodies of etioplasts, while in chloroplasts almost no gold particles can be detected (Shaw et al., 1985; Dehesh et al., 1986; Benli et al., 1991). In addition, it was also possible to visualize gold particles in the vicinity of the plasmalemma and in the cytosol (Dehesh et al., 1986). The significance of this unexpected distribution is not clearly understood, but this suggests that the direct measurement of protochlorophyllide reductase activity in conjunction with the determination of the presence of protochlorophyllide, are essential for proper enzyme localization. Therefore, in the absence of such assays in purified inner and outer envelope membranes, the occurrence of traces of protochlorophyllide reductase activity in membranes like the inner envelope membrane cannot be ruled out.

Molecular data concerning protochlorophyllide reductase are now available for several plant species, including monocots (oat, barley), dicots (*Arabidopsis*, pea) and gymnosperms (Schulz et al., 1989; Darrah et al., 1990; Benli et al., 1991; Spano et al., 1992a,b; Teakle and Griffiths, 1993). There is a high degree of conservation of reductase structure among these plant species, but some uncertainty still persists as to the identity of the NH_2-terminal residues of the protein (Teakle and Griffiths, 1993). Protochlorophyllide reductase is coded for by nuclear DNA and is synthesized, as a precursor of the 44 kDa protein in barley (Schulz et al., 1989), 41 kDa in oat (Darrah et al., 1990) and 43 kDa in pea and *Pinus* (Spano et al., 1992a,b). Protochlorophyllide reductase contains a relatively high content of basic amino acids, with arginine and lysine accounting for more than 12% of the total (Darrah et al., 1990). Teakle and Griffiths (1993) obtained a full-length cDNA clone for wheat protochlorophyllide reductase (from dark-grown tissues) and have demonstrated that isolated pea chloroplasts can import, process and target the mature reductase to the stromal side of thylakoid membranes. In contrast to the firm attachment of protochlorophyllide reductase to prolamellar bodies (Grevby et al., 1989), the interaction of the reductase with the thylakoid membrane was concluded to be at least in part electrostatic as low salt washing (with 50 mM KCl) can readily displace it (Teakle and Griffiths, 1993). Therefore, we are facing two series of clear-cut but apparently contradictory results. First, the import/processing experiments with intact pea chloroplasts suggest a thylakoid location for protochlorophyllide reductase whereas, in mature chloroplasts, only polypeptides from highly purified envelope membranes, and not from thylakoids, are immunodecorated with anti-PCR antibody (Joyard et al., 1990). In addition, the outer envelope proteins studied so far are synthesized in their mature form, without any transit peptide, in contrast with proteins which are destined for the stroma or thylakoids (Salomon et al., 1990; Li et al., 1991). Therefore, the presence of a transit peptide in the precursor analyzed by Teakle and Griffiths (1993) strongly suggests, but does not prove, a location of the mature protochlorophyllide reductase on internal membranes of plastids. In fact, only the enzyme from etiolated tissues has been purified and the most recent and reliable molecular data are related to cDNA from dark-grown plants (Teakle and Griffiths, 1993). No molecular data are available on protochlorophyllide reductase from green tissues. In addition, different isoenzymes for protochlorophyllide reductase have been described (Ikeuchi and Murakami, 1982), thus the possibility that distinct isoenzymes, with a distinct localization, could exist in etiolated and green tissues cannot be ruled out (see Dehesh et al., 1986; Benli et al., 1991). For instance, protein gel electrophoretic studies previously demonstrated that protochlorophyllide reductase in green plants is associated with one or a small group of immunologically related polypeptides varying in Mr between 36 and 38 kDa (Ikeuchi and Murakami, 1982; Dehesh et al., 1986; Spano et al, 1992b), whereas in prolamellar bodies, a predominant polypeptide of 38 kDa has been described (Selstam et al., 1987).

It is not yet known how many genes code for protochlorophyllide reductase and whether the presence of multiple genes could be responsible for the existence of different isoenzymes present in different membrane systems (outer envelope membrane, prolamellar body, etc.). This raises the question of whether a single protochlorophyllide reductase precursor could be transferred to different membranes or whether different precursors with (or without) distinct transit peptides could be synthesized in a plant cell. The different localization of protochlorophyllide reductase in etioplasts and chloroplasts also raises the question of the physiological significance of the envelope enzyme. One can question whether the presence of protochlorophyllide reductase in the envelope membrane is indeed relevant to the participation of this membrane system in chlorophyll biosynthesis. During barley leaf biogenesis in illuminated plants, chloroplasts accumulate about 10^6 PS I and PS II complexes containing $3.7 \cdot 10^8$ chlorophyll molecules during a 36 to 48 hr period (Klein and Mullet, 1987). Mature chloroplasts need chlorophyll to replace the molecules that are continuously destroyed by photooxidation. Therefore, there is a physiological requirement for continuous chlorophyll biosynthesis in chloroplasts. The regulation of this process is still unknown, but it is most likely that chlorophyll accumulation proceeds through continuous regeneration and phototransformation of the photoactive protochlorophyllide reductase complex (see above), as proposed by Frank and Strzalka (1992).

Matringe et al. (1992) analyzed the localization of protoporphyrinogen oxidase within mature chloroplasts (Fig. 6). This enzyme is of particular interest since it is the last step common to chlorophyll and heme biosynthesis before the pathways branch (for details, see Beale and Weinstein, 1990). In addition, it is the target enzyme for diphenylether-type herbicides, such as Acifluorfen. Direct assay of the enzymatic activity by spectrofluorimetry and analyses of the specific binding of the herbicide Acifluorfen gave clear evidence that protoporphyrinogen oxidase activity is associated with both envelope and thylakoid membranes (Matringe et al., 1992). On a protein basis, envelope membranes contain 10 times more Acifluorfen-binding sites than do thylakoids. These results suggest that envelope and thylakoid protoporphyrinogen oxidase activity could be associated with one of the two porphyrin synthesis pathways. In support for this notion, Matringe et al. (1994) demonstrated that ferrochelatase activity (the enzyme that chelates Fe on protoporphyrin IX for the biosynthesis of protoheme; for a review see Jones, 1968) was associated with thylakoids, and not with the envelope membranes, thus demonstrating that thylakoids are the site for protoheme biosynthesis in chloroplasts (Fig. 6). The possibility that envelope protoporphyrinogen oxidase catalyzes the first step of chlorophyll biosynthesis is supported by both direct and indirect evidence. To date, the envelope is the only chloroplast membrane system in which protochlorophyllide reductase activity has been directly and unambiguously demonstrated. In addition, the envelope membranes contain both the substrate and product of the reaction (Pineau et al., 1986, 1993; Joyard et al., 1990).

Other evidence for a possible role of envelope membranes in chlorophyll biosynthesis is only indirect. After protoporphyrinogen oxidase, the next enzyme in this pathway is Mg-chelatase (Fig. 6). Direct determination of Mg-chelatase activity in purified chloroplast membranes is very difficult, since after rupture of the envelope, only 10% of the initial activity found in chloroplasts can be recovered (Walker and Weinstein, 1991a). Walker and Weinstein (1991b) demonstrated that Mg-chelatase can be resolved into two forms: one soluble and another membrane-bound. In vivo, the enzyme is probably

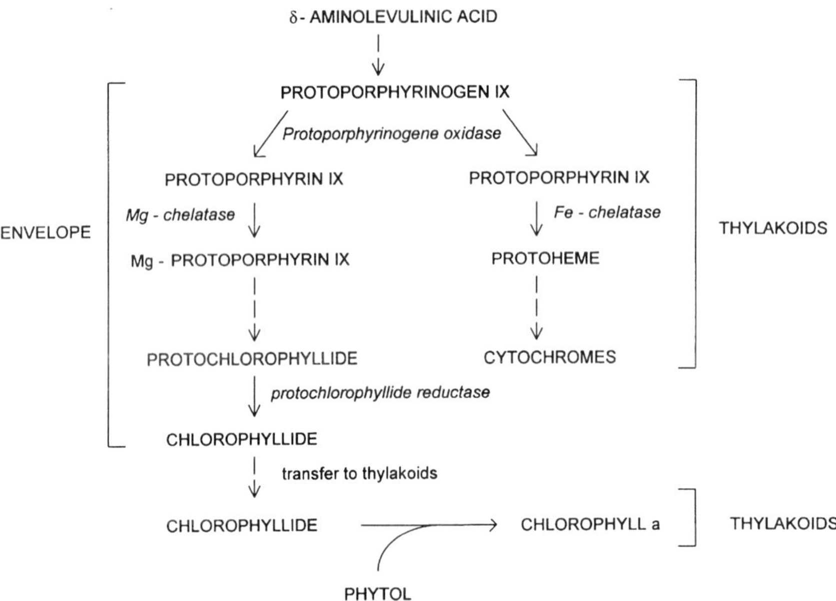

Fig. 6. Biosynthesis of chlorophyll and hemes in chloroplasts. Heme biosynthesis takes place in thylakoids. Chlorophyllide biosynthesis is probably associated with envelope membranes, but the last step, catalyzed by chlorophyll synthase, is associated with thylakoids.

membrane-bound because of the high stromal Mg^{2+} concentration. Fuesler et al. (1984) showed that in intact chloroplasts, Mg-chelatase was accessible to molecules unable to pass through the inner envelope membrane, thus suggesting that Mg-chelatase could be present in the envelope membranes. In addition, Walker and Weinstein (1991a) demonstrated that (a) chloroplast integrity is essential for a good enzymatic activity and (b) Mg-chelatase activity is greatest in the presence of 1.5 μM protoporphyrin IX. Since an hydrophobic substrate, like protoporphyrin IX, when added to a suspension of intact chloroplasts has little chance to reach thylakoids, the observations of Walker and Weinstein (1991a) provide indirect evidence for a localization of the enzyme on envelope membranes. Furthermore, Johanningmeier and Howell (1984) have shown that the level of Mg-protoporphyrin methyl esters can regulate the accumulation of light-induced cytosolic mRNA for light-harvesting chlorophyll *a/b* binding protein. Consequently, the authors postulated that such intermediates in chlorophyll synthesis could be located in the chloroplast envelope. Together, the experiments described above suggest a role of envelope membranes in the biosynthesis of chlorophyllide, i.e. the porphyrin structure of chlorophyll (Fig. 6). However, direct assays of all enzymes of the chlorophyllide biosynthetic pathway between protoporphyrinogen oxidase and protochlorophyllide reductase are necessary to confirm this hypothesis.

The biosynthesis of the second part of the chlorophyll molecule, i.e. the prenylalcohol phytol, also takes place in envelope membranes (Soll and Schultz, 1981; Soll et al. 1983; Soll, 1987). As we shall see below (Fig. 7), envelope membranes (associated with the stromal prenyltransferase) catalyze the condensation of isopentenyldiphosphate into geranylgeranyldiphoshate and to phytyldiphosphate (Block et al., 1980; Soll and Schultz, 1981; Soll et al. 1983; Soll, 1987) the biosynthetic precursor of phytol (Fig. 7). The final step of chlorophyll formation, i.e. the addition of phytyldiphosphate (or its precursor geranylgeranyldiphosphate) to chlorophyllide, is a reaction catalyzed by chlorophyll synthase (Fig. 6). This enzyme is specifically associated with thylakoids (Block et al., 1980). In vitro, chlorophyll synthase is able to use either geranylgeranyldiphosphate or phytyldiphosphate as a substrate, but there is some evidence that in vivo, chlorophyll is formed directly by assembly of chlorophyllide and phytyldiphosphate. This is in contrast with the situation in non-green plastids: chlorophyll synthase from oat etioplasts readily accepts geranylgeranyldiphosphate rather than phytyldiphosphate (for a review, see Lichtenthaler, 1993). The problem of how chlorophyll *b*, which

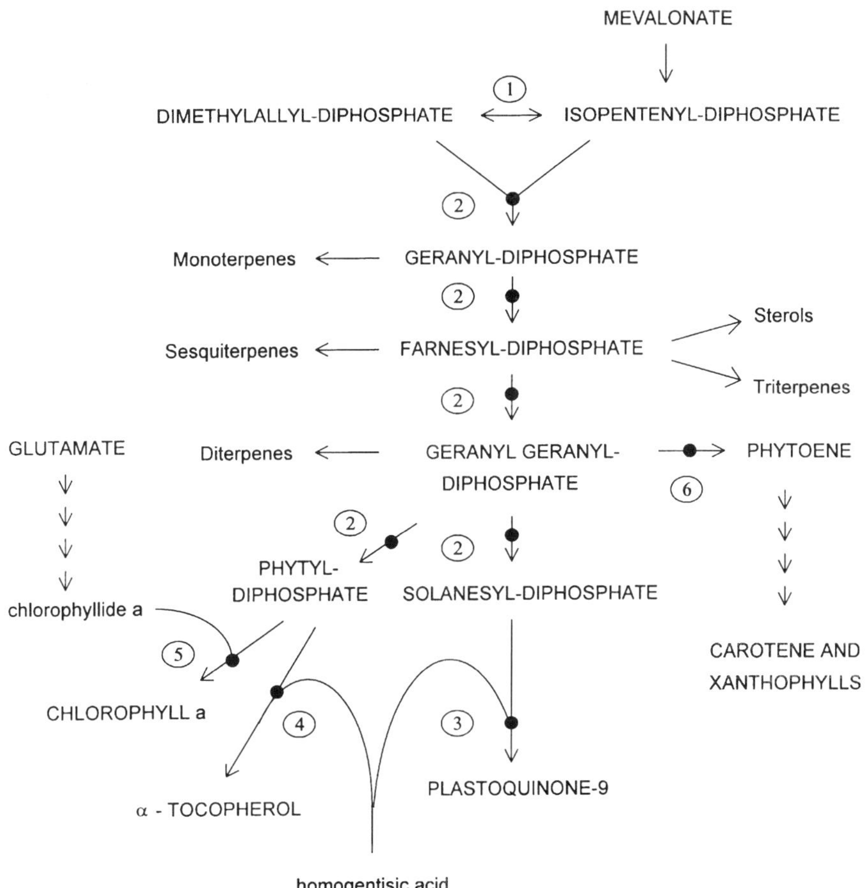

Fig. 7. Biosynthesis of plastid prenyllipids (carotenoids, side chains of chlorophyll and of prenylquinones). 1, isopentenyl-diphosphate isomerase; 2, prenyltransferase; 3, Solanesyl-diphosphate:homogentisate solanesyltransferase; 4, phytyl-diphosphate:homogentisate phytyltransferase; 5, chlorophyll synthase; 6, phytoene synthase.

bears an aldehyde group in place of the B-ring methyl substituent of the tetrapyrrole of chlorophyll *a*, is formed is not yet solved (Lichtenthaler, 1993).

Finally, little is known about the regulation of gene expression for enzymes involved in chlorophyll biosynthesis (Rüdiger and Schoch, 1988; Beale and Weinstein, 1990). The major control of chlorophyll biosynthesis is at the step of (2)δ-aminolevulinic acid formation (e.g. Ilag et al. 1994). For instance, addition of (2)δ-aminolevulinic acid to dark-grown seedlings leads to the accumulation of protochlorophyllide (Castelfranco et al., 1974). To study gene expression and regulation of chlorophyll synthesis, potentially interesting tools are presently available. For instance, albino mutants that are devoid of chlorophyll were expected to be useful for such studies. In fact, in most of these mutants, the primary block is in carotenoid biosynthesis whereas chlorophyll is synthesized, but then, due to massive photooxidative damage, chlorophyll is destroyed and fails to re-accumulate (Taylor, 1989). Chlorophyll-free chromoplasts from daffodil contain all the enzymes involved in the conversion of (2)δ-aminolevulinic acid to Mg-protoporphyrin IX monomethyl ester as well as the last enzyme of the pathway, chlorophyll synthase, involved in the transfer of the geranylgeranyl moiety to exogenous chlorophyllide *a*, but the formation of the isocyclic chlorophyll ring was not observed (Lützow and Kleinig, 1990). Thus, the genes for most of the membrane-bound enzymes involved in chlorophyll biosynthesis were not expressed in chromoplasts, in contrast with those for soluble enzymes. In general, etioplasts have been extensively used as working models for studies of chlorophyll biosynthesis. The main reasons are: (a) they contain high levels of protochlorophyllide and of protochlorophyllide reductase and (b) they are initially devoid of chlorophyll which greatly facilitates studies on chlorophyll biosynthesis. In consequence, very little is known about chlorophyll biosynthesis and its regulation in mature chloroplasts. In fact, the obvious differences in enzyme localization and specificities between chloroplasts and non-green plastids raise the question of a distinct set of isoenzymes and of specific regulatory processes for these two types of organelles.

2. Carotenoids

a. Distribution

All plastid membranes contain carotenoids. They represent about 0.9% of the dry mass of a chloroplast (Lawlor, 1987). As shown in Table 1, thylakoids as well as chloroplast envelope membranes contain significant amounts of carotenoids. Rather surprisingly, the presence of carotenoids in envelope membranes has been a subject of debate (see for instance Britton, 1988). However, envelope membranes are deep yellow (for a color picture of envelope membranes, see Neuburger et al., 1977), because the only colored pigments present in highly purified envelope preparations are carotenoids. In addition, unambiguous characterization of carotenoids as genuine constituents of envelope membranes from various plastid types has been made: envelope membranes from chloroplasts and from most non-green plastids have a very similar pigment composition, and this composition, as shown in Table 1, is distinct from that of thylakoids (Joyard et al., 1991). Violaxanthin is the major carotenoid in envelope membranes whereas thylakoids are richer in β-carotene (Jeffrey et al., 1974). The xanthophyll to β-carotene ratio is about 6 in chloroplast envelope membranes, but only 3 in thylakoids. Siefermann-Harms et al. (1978) described light-induced changes in the carotenoid levels of envelope membranes (see Table 1). The physiological significance of these modifications is still unclear, but there is probably no direct relationship with the light-induced changes of the carotenoid levels associated with the xanthophyll cycle in thylakoids (Siefermann-Harms, 1977). The xanthophyll cycle is probably related to the dissipation of excess light energy (Demmig-Adams and Adams, 1992). Both chloroplast envelope membranes contain carotenoids, the outer membrane being slightly richer in neoxanthin (Block et al., 1983b), but it is not yet known whether carotenoids are present as protein-pigment complexes, as in thylakoids (Siefermann-Harms, 1985). When present in protein-pigment complexes, carotenoids are not usually covalently bound to proteins, thus the procedures used to separate envelope membrane proteins involving detergents are not mild enough to preserve the integrity of such complexes if they do occur in envelope membranes (see however, Markwell et al., 1992). Therefore, carotenoids are generally released as 'free' carotenoids during membrane solubilization. This strongly limits our understanding of the physiological significance of envelope carotenoids, although a role (as a singlet-oxygen quencher) in preventing (photo)oxidative damage cannot be excluded. The specific distribution of carotenoids among the different protein-pigment complexes in thylakoids has been described extensively by Siefermann-Harms (1985), Cogdell (1988) and Lichtenthaler (1993). The main functions of carotenoids in thylakoids are their participation in light absorption processes, and their function in the dissipation of excited states. All these functions have been reviewed in detail by Cogdell (1988), Rau (1988), Lichtenthaler (1993) and Demmig-Adams and Adams (1992).

b. Biosynthesis

Bramley (1985), Jones and Porter (1986) and Britton (1988) have summarized early studies on the different enzymes involved in carotenoid biosynthesis. Most biochemical and molecular biology studies that have been reported on carotenoid biosynthesis in higher plants concern chromoplasts. In fact, these plastids are the most active in carotenogenesis because during the chloroplast-chromoplast transformation associated with fruit ripening, rapid and massive synthesis of carotenoids occurs (for instance of lycopene in tomatoes and capsanthin, in pepper). Carotenoid accumulation in chromoplasts occurs in several types of structures such as fibrils, crystals or membranes (Gunning and Steer, 1975). For instance, in *Narcissus pseudonarcissus*, carotenoids are present in a concentrically stacked multilayer (about 20 layers in *Narcissus*) of lipid-rich membranes deriving from the inner envelope membrane (Gunning and Steer, 1975). In other plants, such as bell pepper (*Capsicum annuum*) fruits, carotenoids overaccumulate in lipoprotein fibrils (Deruère et al., 1994).

Since carotenoids are tetraterpenoids, they are

made of isoprenoid units deriving from isopentenyl-diphosphate. The question of whether plastids can synthesize this precursor for further use in carotenoid biosynthesis or whether isopentenyldiphosphate has to be imported is still controversial (Kleinig, 1989; Goodwin, 1993), but all plastids are able to convert isopentenyldiphosphate into carotenoids. However, as discussed by Beyer and Kleinig (1990), carotenogenic enzymes constitute a delicate system which is easily disturbed upon loss of plastid integrity and that apparently works in a product channeling manner. Therefore, the use of cell-free systems to analyze carotenoid biosynthesis is made difficult by the loss of enzyme activity during the experiments. The first C40 carotenoid synthesized is 15-*cis*-phytoene, a colorless compound with three conjugated double bonds. Phytoene is formed from isopentenyl-diphosphate by condensation of two all-*trans* geranylgeranyl diphosphate molecules and it is the precursor for all desaturated and oxygenated carotenoids (Fig. 7). Phytoene desaturation leads to the formation of lycopene, a colored carotenoid with 11 double bonds, *via* the formation of various intermediates (phytofluene, ζ-carotene, neurosporene). Then, α-carotene and β-carotene are formed by cyclization of lycopene at both ends of the molecule. α-carotene is the precursor for lutein, lutein-epoxide and neoxanthin, whereas β-carotene leads to the formation of zeaxanthin, antheraxanthin, violaxanthin and neoxanthin. Once phytoene desaturation proceeds, the newly synthesized carotenoids are channeled into a sequence leading to the formation of all primary carotenoids that are normally found in plastids. It is not known, however, at which stage pathways for carotene and the different xanthophylls diverge and how the formation of the different carotenoids is regulated (Britton, 1988). In addition, many features of the biosynthesis remain to be solved, such as the role of oxygen, the interaction of integral membrane and of peripheral proteins, etc. The use of specific inhibitors or the modification of growth conditions allows accumulation of the different intermediates that, in normal conditions, are present in too low amounts to be detected in plants. For instance, phytoene desaturase is inhibited by iron deficiency (Pascal et al., 1995) and by various herbicides such as norflurazon, diflufenican, etc. whereas the formation of cyclic carotenoids is inhibited by CTPA, or chlorophenylthiotriethylamine, (Britton, 1988; Barry and Pallett, 1990; Lichtenthaler, 1993). Indeed, although such tools proved to be very useful in understanding carotenogenesis in plastids (Britton, 1988; Barry and Pallett, 1990; Lichtenthaler, 1993), the reactions involved are in general extremely complex and the inhibition can be indirect. For instance, in vivo, quinones and factors regulating the redox state of quinones may play a major role in phytoene desaturation, although in vitro molecular oxygen is the terminal acceptor (Mayer et al., 1990). Schulz et al. (1993) demonstrated that phytoene desaturase inhibition by the bleaching herbicide of the benzoyl cyclohexane-dione type was due to an inhibition of quinone biosynthesis (see below) rather than a direct effect on phytoene desaturation.

In chloroplasts, carotenoid biosynthesis and its regulation are not as well characterized as in chromoplasts, or as in fungi, bacteria or even algae. In general, the reaction mechanisms of the enzymes involved have been extensively analyzed and there is no reason to suspect that different enzymatic mechanisms are involved in chloroplasts (Britton, 1988; Goodwin, 1993). However, according to Britton (1988) at least three phases of carotenoid biosynthesis within chloroplasts should be considered: (a) the bulk synthesis that occurs during the initial construction of the photosynthetic apparatus, (b) the synthesis that continues in mature chloroplasts as part of turnover, and (c) the synthesis that occurs as a response to or consequence of changes in environmental conditions, especially light intensity. Therefore, carotenoid biosynthesis in chloroplasts does not follow the same regulatory processes as it does in etioplasts or in chromoplasts, although control by light is probably essential in most cases. In addition, the main difference between chloroplasts and other plastids is that carotenoids are localized and function in different pigment-protein complexes of thylakoid membranes (Siefermann-Harms, 1985). Therefore, a complete understanding of carotenoid biosynthesis in chloroplasts requires also that of chlorophylls and of the proteins that are associated with them (see for instance Plumley and Schmidt, 1987; Humbeck et al., 1989; Mullet et al., 1990; Herrin et al., 1992). In addition, the localization of the later stages of carotenoid synthesis within chloroplasts is not yet clearly established. Because thylakoids accumulate most of the plastid carotenoids in green tissues, it is generally assumed that thylakoids, and not envelope membranes, play the central role in carotenogenesis (Britton, 1988). However, there is some indirect evidence for the participation of envelope membranes in carotenoid biosynthesis. First, all plastids contain the soluble and membrane-bound enzymatic machinery for carotenoid synthesis, and since in some plastids, such as proplastids or amyloplasts, the only membrane system present is their envelope membranes, they should be a site of carotenoid synthesis. This view is supported by the observation that the carotenoid composition of all plastid envelope membranes analyzed so far is rather similar (Joyard et al., 1991). In addition, chromoplast membranes, which are assumed to derive from the inner envelope membrane, are very active in carotenoid biosynthesis (Beyer and Kleinig, 1990). However, all of this evidence is only indirect, but some more direct evidence for a role of envelope membranes in carotenoid biosynthesis is currently available. First, phytoene synthase and desaturase (Lütke-Brinkhaus et al., 1982) and zeaxanthine epoxidase (Costes et al., 1979) activities were demonstrated in envelope membranes from mature spinach chloroplasts. Recent investigations in our laboratory (unpublished) demonstrated that phytoene desaturase activity was indeed present in envelope membranes as well as in thylakoids, and that the activity was completely inhibited by diflufenican, a specific inhibitor of phytoene desaturase.

Despite the crucial roles of carotenoids in higher plants, such as photooxidative protection and participation in light-harvesting protein-pigment complexes, few carotenoid biosynthetic enzymes have been characterized and, at least partially, purified from higher plants (Beyer, 1987; Schmidt et al., 1989; Hugueney et al., 1992; Camara, 1993). Some plant genes involved in carotenoid biosynthesis have been cloned (Buckner et al., 1990; Kuntz et al., 1992; Bartley et al., 1992; Buckner and Robertson, 1993) but the molecular mechanisms regulating carotenogenesis have not yet been elucidated. Because of the difficulties encountered in biochemical and molecular studies of carotenogenic enzymes in higher plants, some of the most promising ways to reach this goal are the use of plant mutants defective in carotenoid accumulation and the use of carotenoid genes from either fungi or bacteria. As discussed by Taylor (1989), the pigmentation of plastids makes it easy to find mutants that block their development. Unfortunately, since carotenoids protect plastids against photooxidative damage, carotenoid deficiencies will cause a wide range of pleiotropic effects (Taylor, 1989). It is therefore difficult to analyze higher plant mutants devoid of carotenoids: exposure to light will kill the mutant plant (bleaching compounds are powerful herbicides) or at least will significantly affect its development (for instance, they grow only in limited light conditions, Taylor, 1989). Despite this major limitation, numerous albino mutants have been characterized (Taylor, 1989). For instance, maize mutants defective in carotenoid accumulation in the endosperm of their kernels have been analyzed at the genetic level: some of these mutants are devoid of β-carotene but they do accumulate precursors of β-carotene in both their leaves and endosperm (Buckner et al., 1990; Buckner and Robertson, 1993). Such mutants could be most valuable models for understanding carotenoid biosynthesis and its regulation in higher plants. In fact, the most significant progress on the characterization of genes for enzymes of the carotenoid biosynthetic pathway were obtained from fungi, cyanobacteria and photosynthetic bacteria (see for instance Armstrong et al., 1993; Bartley et al., 1993; Morelli et al., 1993, Sandmann, 1993; and for a review of cloned genes, Table 2 from Lichtenthaler, 1993).

B. Distribution and Biosynthesis of Prenylquinones

1. Distribution

Plastid prenylquinones, especially α-tocopherol and plastoquinone-9, are essential compounds for photosynthesis, mostly as singlet-oxygen quenchers or electron carriers between Q_A, Q_B of PS II and cytochrome b_6f complex. For instance, the principal function of α-tocopherol is to remove polyunsaturated fatty acid radical species generated during lipid peroxidation. Both envelope membranes contain plastoquinone-9, α-tocopherol (vitamin E) and phylloquinone (vitamin K_1) (Lichtenthaler et al., 1981; Soll et al., 1985). Although envelope membranes contain qualitatively the same prenylquinones as thylakoids, these compounds are present in different proportions. The α-tocopherol to plastoquinone-9 ratio is therefore much higher in the envelope fraction (about 2.3) than in the thylakoids (about 0.3). The role of envelope quinones is not yet clear, but they could be involved in fatty acid and/or carotenoid desaturation. For instance, phytoene desaturase which is present in chloroplast envelope membranes (Lütke-Brinkhaus et al., 1982; Holford et al., unpublished) uses molecular oxygen as the terminal electron acceptor, but quinone compounds are able to replace it in phytoene desaturation in

Narcissus pseudonarcissus chromoplasts (Mayer et al., 1990). However, the participation of envelope quinones in fatty acid desaturation has not yet been demonstrated. Finally, α-tocopherol, a singlet-oxygen quencher, has been shown to be chemically modified after several cycles of physical interaction with O_2 (Elstner, 1987), necessitating the replenishment of these molecules that takes place in envelope membranes.

2. Biosynthesis

The biosynthesis of plastid prenylquinones is located in chloroplasts (Bickel and Schultz, 1976; Hutson and Threlfall, 1980; Schultze-Siebert et al., 1987). The biosynthetic pathway for prenylquinones was established (Soll et al., 1980a,b; Pennock, 1983; Morris and Threlfall, 1983) from studies of their chemical synthesis (Mayer and Isler, 1971). Prenylquinones are formed by condensation of two compounds, a quinone ring that derives from shikimic acid, and a prenyl side chain that derives from isopentenyldiphosphate. The formation of the prenyl side chain involves the biosynthesis of geranylgeranyl diphosphate. Therefore, these initial steps are identical to those described for carotenoids and for the side chain of chlorophyll. However, phylloquinone and α-tocopherol contains a C_{20} phytyl chain whereas plastoquinone-9 contains a C_{45} solanesyl chain (Fig. 7). Although the prenyltransferase involved in geranylgeranyl diphosphate biosynthesis is soluble and localized in the chloroplast stroma (Block et al., 1980), the products of the reaction are used by various membrane-bound enzymes including prenyltransferases, methyltransferases and cyclases (Fig. 7). In chloroplasts, envelope membranes are the site of prenylquinone biosynthesis (Soll et al., 1980a; Fiedler et al., 1982; Soll et al., 1985): α-tocopherol and plastoquinone-9 synthesis involves a series of enzymes located in the inner envelope membrane (Soll et al., 1985). The precursor for both α-tocopherol and plastoquinone-9, homogentisic acid, is synthesized from 4-hydroxyphenylpyruvate (a product of the plastid shikimate pathway) by a 4-hydroxyphenylpyruvate dioxygenase. Fiedler et al. (1982) proposed that this enzyme is probably located in envelope membranes, but the possibility that peroxisomes can catalyze this reaction cannot be entirely ruled out (Bickel et al., 1978). Interestingly, and as discussed above, this enzyme is the target for a benzoyl cyclohexane-dione type bleaching herbicide (Schulz et al., 1993). One of the consequences of the inhibition of this enzyme is an in vivo quinone depletion in plants leading to an indirect accumulation of phytoene corresponding to an inhibition of phytoene desaturase (Schulz et al., 1993). Tocopherols are synthesized by condensation of homogentisic acid and a C_{20}-prenyl pyrophosphate to form 2-methyl-6-prenylquinol (Soll et al., 1980a; Fiedler et al., 1982). By a series of methylations and cyclization, the 2-methyl-6-prenylquinol gives rise successively to 2,3-dimethyl-6-prenylquinol, γ-tocopherol or γ-tocotrienol and finally α-tocopherol or α-tocotrienol (Soll et al., 1980a). The same pathway has been demonstrated in pepper (*Capsicum annuum*) chromoplasts (Camara et al., 1982). The other major chloroplast prenylquinone, plastoquinone-9, is also synthesized in the inner envelope membrane by condensation of homogentisic acid and solanesyl-pyrophosphate to form 2-methyl-6-solanesylquinol, which is methylated and oxidized to form successively plastoquinol-9 and plastoquinone-9 (Soll et al., 1985). In contrast, only the initial steps of phylloquinone synthesis, i.e. the condensation of 1,4-dihydroxy-2-naphtoate with phytyl-pyrophosphate, occurs in envelope membranes, whereas the methyltransferase which forms phylloquinol is probably associated with thylakoids (Fiedler et al., 1982).

The assay of prenylquinone biosynthetic enzymes is rather difficult since of all the precursors and cofactors used, only homogentisic acid, polyprenyl diphosphate and S-adenosylmethionine are water soluble (Soll, 1987). Despite this major experimental limitation, specific activities for the different enzymes ranging from 15 pmol/h/mg protein (for the methylation of γ-tocopherol to α-tocopherol) to almost 500 pmol/h/mg protein (for the methylation of 2-methyl-6-phytylquinol to 2,3-dimethyl-6-phytylquinol) were obtained in purified inner envelope membranes from spinach chloroplasts (Soll et al., 1985). With the same membrane fraction, plastoquinone-9 biosynthesis was assayed in a single step, leading to the formation of 15 pmol of plastoquinone-9 /h/mg protein (Soll et al., 1985). In all cases, the reactions were almost linear for at least one hour. Since envelope membranes were highly pure and prepared from intact chloroplasts devoid of any contaminating extraplastidial membranes, the role of chloroplast envelope membranes in plastoquinone-9 and α-tocopherol biosynthesis was believed to be clearly established (Soll et al., 1980a; Soll et al., 1985). Thus it was rather surprising that

Swiezewska et al. (1993) proposed that plastoquinone biosynthesis was in fact localized in Golgi membranes and that a specific transport and targeting system was required for plastoquinone transfer to chloroplasts. This conclusion was mostly based on observations that microsomal membranes were 20 times more active in prenylquinone biosynthesis than chloroplasts and that among the different cell membranes the Golgi apparatus had the highest specific activity (expressed as dpm/h/mg protein). However, these observations were hampered by several limitations. First, the activities reported were extremely low, and hardly detectable in most cases. In addition, activities in envelope membranes were not determined, and no marker enzymes for the plastid envelope were assayed in any of the fractions analyzed. The only plastid marker analyzed was chlorophyll and the marker enzyme used for Golgi apparatus was UDP-galactosyltransferase, despite the fact that envelope membranes are an active site of UDP-galactose incorporation because of MGDG synthesis (see above). Therefore, further work is needed before the proposal that plastoquinone is formed in the Golgi apparatus and then transported to the outer envelope membrane followed by further transfer to the thylakoids could be accepted.

Most enzymes involved in quinone biosynthesis that have been purified are in fact the soluble proteins (prenyltransferases, etc.) of the common pathway shared by all the different terpenoid compounds (carotenoids, side chains of chlorophyll and prenylquinones, etc.). In contrast, many of the membrane components and associated mechanisms remain to be deduced. D'Harlingue and Camara (1985) purified γ-tocopherol methyltransferase activity from *Capsicum* chromoplasts. They determined some biochemical parameters (molecular weight, 33,000; kinetic parameters, optimum pH, etc). Since this work, few attempts at solubilization and purification of membrane-bound enzymes (in fact only for γ-tocopherol methyltransferase activity) have been published (Ishiko et al., 1992; Shigeoka et al., 1992). In addition, almost nothing has been reported so far on the genes involved in quinone biosynthesis or the regulation of their expression despite their major importance in plant cell metabolism. Complementation studies using yeast mutants together with *Arabidopsis* cDNA libraries is a possible approach for cloning some plant genes involved in terpenoid biosynthesis, and it is highly likely that the finding of a specific inhibition of quinone biosynthesis by herbicides (Schulz et al., 1993) will stimulate research in the field.

IV. Transfer of Lipid Constituents between Envelope and Thylakoids

Thylakoid formation first requires the biosynthesis of proteins, glycerolipids, pigments, quinones, etc. in a tightly regulated manner. Indeed, the biosynthesis of chlorophyll apoproteins is linked to the ability of plastids to synthesize chlorophyll and carotenoids (see for instance Plumley and Schmidt, 1987; Humbeck et al., 1989; Mullet et al., 1990; Herrin et al., 1992) and inhibition of carotenoid biosynthesis can inhibit transport of nuclear-encoded proteins (Dahlin, 1993). In addition, during plastid biogenesis, transport of proteins and lipids is essential to build up thylakoids. For instance, in the case of nuclear encoded plastid proteins, there is considerable evidence (for reviews, see Keegstra et al., 1989; Smeekens et al., 1990; Soll and Alefsen, 1993) to suggest that translocation of thylakoid- (chlorophyll *a/b* binding protein, etc.) or lumen- (plastocyanin, 33 kDa oxygen-evolving protein, etc.) directed proteins may occur through the stroma as soluble proteins. These results strongly suggest that the proteins do not depart from the inner membrane by a bulk flow process (Fig. 8). In contrast to what is known for protein transport, in the case of membrane lipids our knowledge is still in its infancy. All the observations summarized in this chapter demonstrate that the inner envelope membrane is the site of synthesis for galactolipids, sulfolipid, phosphatidylglycerol, α-tocopherol, plastoquinone-9 and of some pigments such as chlorophyll and carotenoids. Thylakoids represent the main plastid membrane system in the cell and contain the largest amount of the plastid lipid constituents. For instance, in 1 m^2 of leaves, there are 2.5 g of chloroplast lipids (representing 20% of a chloroplast dry mass) and 0.5 g chlorophyll (representing 4% of a chloroplast dry mass). These lipids and pigments, synthesized in envelope membranes, are almost exclusively concentrated in the $4.1 \cdot 10^{-18}$ m^3 of the corresponding thylakoids (Lawlor, 1987). Therefore, massive transport of molecules (glycerolipids, pigments, prenylquinones) from their site of synthesis (envelope membranes) to their site of accumulation (thylakoids) should take

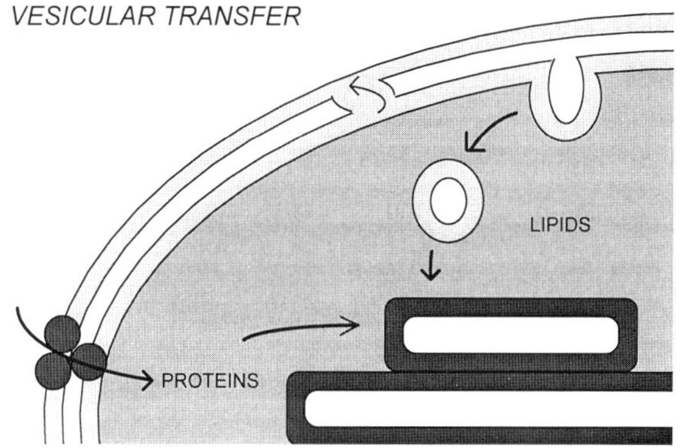

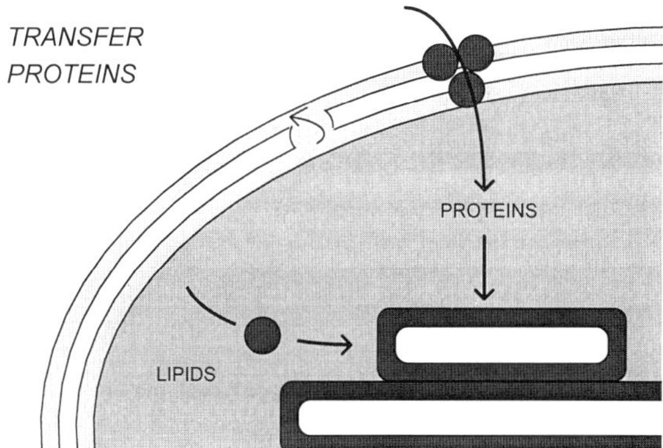

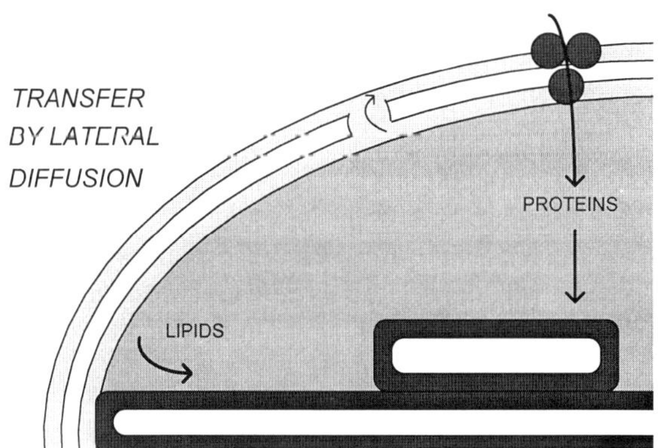

Fig. 8. Possible mechanisms for lipid transfer from their site of synthesis (the inner envelope membrane) to their site of accumulation (the thylakoids).

place during development (Fig. 8).

Joyard et al. (1980) investigated the distribution of radioactive glycerolipids between envelope and thylakoids from mature spinach chloroplasts labeled in vivo with $^{14}CO_2$. They demonstrated that envelope membranes contain a small lipid pool with a very high turnover, and that lipid export to thylakoids was rapid. Bertrams et al. (1981) detected labeled galactolipids in thylakoids only a few seconds after addition of UDP-galactose to chloroplasts. However, the rate of turnover sufficient to replace all the thylakoid MGDG which has been reported varied from 16 h (Morré et al., 1991) to 2.5 days (Rawyler et al., 1991). In fact, almost nothing is known about the mechanisms involved for the transfer of lipids. In addition, different strategies for lipid transfer from envelope membranes to thylakoids probably occur at various stages of chloroplast development. The process involved in mature chloroplasts is probably very different from that observed in young developing plastids. In the case of developing plastids, import of thylakoid membrane lipids en route to the thylakoids may occur through the stroma via vesicles derived from the inner membrane (Wellburn 1982), but this remains to be demonstrated. Fig. 8 summarizes some of the possible mechanisms which could be involved: vesicular transport, transfer of lipid monomers through the stroma either by protein-facilitated transport or by spontaneous diffusion of free monomers, lateral diffusion of lipids between membranes at regions of direct intermembrane contact, etc. Some of these possibilities have been investigated.

In their investigations, Rawyler et al. (1991) confirmed that galactolipid export was rapid and demonstrated that MGDG was more actively transported to thylakoids than DGDG. Because of this selectivity, Rawyler et al. (1991) dismissed the possibility that galactolipid transfer could be achieved by a vesicular transfer in mature chloroplasts. According to Rawyler et al. (1991), the transmembrane distribution of the newly inserted galactolipid molecules in the thylakoids suggests that the export consists of transient and partial fusions between stroma facing monolayers of both membranes and/or galactolipid transport by soluble proteins. They also demonstrated that once in the thylakoids, galactolipids were quickly diluted into the bulk of membrane galactolipids. Using intact chloroplasts incubated in the presence of [^{14}C]-acetate, [^{14}C]-glycerol-3-phosphate and UDP-[^{14}C]-galactose, Rawyler et al. (1994) followed both the transfer of various lipids (MGDG, DGDG and phosphatidylglycerol) into thylakoids as well as their transmembrane distribution. The latter was found to match closely the corresponding polar lipid distribution in thylakoids, regardless of plant age and species, incubation time and temperature. Rawyler et al. (1994) proposed that transient fusion between inner envelope and thylakoid membranes, rather than the participation of a lipid transfer protein, would allow lipid export by lateral diffusion and build the lipid asymmetry observed in thylakoids. In fact, the hypothesis of a galactolipid transport by stromal proteins was only supported by the discovery of a 30 kDa protein able to transfer galactolipids in a donor/acceptor vesicular system (Nishida and Yamada, 1986). However, almost ten years later, this observation has not been confirmed. In contrast, Morré et al. (1991) developed a cell-free system composed of labeled envelope vesicles and unlabelled thylakoids immobilized on nitrocellulose strips. They observed an ATP- and temperature-dependent transfer of radioactivity between donor and acceptor vesicles with no absolute requirement for a soluble protein carrier. According to Morré et al. (1991), these observations suggest a vesicular transfer. No cell membrane fractions except envelope membranes demonstrated any significant ability to transfer lipids to thylakoids. In fact, electron microscopy studies suggest that numerous vesicles, probably deriving from the inner envelope membrane, are formed during development (see for instance Carde et al., 1982). Since the inner envelope membrane and thylakoids do not differ in polar lipid composition (Cline et al., 1981, Block et al., 1983b), natural fusion of vesicles produced by the inner envelope membrane with growing thylakoids is possible. In addition, and as discussed above, MGDG can form hexagonal type II structures (rather than Lα bilayers) which could behave as intermediates in membrane fusion (Lindblom and Rilfors, 1989). Therefore, MGDG (which represents almost half of the lipid content of inner envelope vesicles and thylakoids) could favor fusions between plastid membranes (Quinn and Williams, 1983; Gounaris and Barber, 1983). Nothing is known about vesicle budding or how their selective fusion with thylakoids is programmed and catalyzed, nor is it known how vesicles deriving from the inner membrane 'select' their content while 'rejecting' the bulk constituents of the inner membrane from which they bud. If the vesicles deriving from the inner envelope represent rudimentary thylakoids, considerable modifications must take place following the initial step of invagination: vesicles deriving from the inner envelope membrane should be modified by both degradation of inner envelope proteins (unless they consist only of a lipid matrix) and addition of new polypeptides to acquire unique thylakoid characteristics. For instance, during development of young maize plastids (with a single prothylakoid) into mature chloroplasts, the activity of the two photosystems was detected only in thylakoids (even at early stages of development) but never in envelope membranes (Wrischer 1989). However, membrane synthesis never occurs de novo; a membrane always derives from a preexisting membrane and thylakoids are probably not an exception to this rule. Therefore, vesicles deriving from the inner envelope should be added to a preexisting membrane (pre-thylakoids), whose nature is totally unknown. The discrete templates for thylakoids can be either modified inner envelope membrane vesicles or genuine distinct prethylakoids. Little information is available to solve this problem, probably because the rules governing membrane identity, stability and function are largely unknown. Powerful in vitro systems for reconstitution of transport from the chloroplast envelope will undoubtedly also play a major role in understanding the regulation of lipid export from the inner envelope membrane. It is clear, therefore, that the inner membrane of the plastid envelope is both a stable and dynamic structure. The identity of envelope membranes is likely to be governed by a sophisticated set of structurally interactive nuclear-coded protein and polar lipids. While this assembly confers specific functional properties, it clearly does not impede the extensive flow of lipids mediated by carrier vesicles that continuously bud at the inner membrane surface. Many answers about both the biochemical and molecular details related to envelope structure and the dynamic of vesicle formation and reassociation (budding, targeting and fusion) will undoubtedly emerge from future studies.

V. Future Prospects

Although our knowledge of the lipid biosynthetic pathways has increased dramatically over the last 20 years, there are still a number of fundamental questions that remain unanswered. The identity of the enzyme systems that function as the pacemaker of polar lipids, pigments and prenylquinones is entirely unknown. Similarly, the mechanisms that operate to produce the observed distribution of glycerolipid polar head groups has not been uncovered, and the details of how glycerolipid, pigment and prenylquinone biosyntheses are coordinated during chloroplast division and differentiation remain to be established. For example, the regulatory points that control the divergence of 80% of the diacylglycerol to galactolipid biosynthesis rather than to sulfolipid production is a mystery. Finally, our present knowledge about the mechanism of hydrogen removal from phytoene or saturated fatty acyl chains by a sophisticated set of membrane-bound desaturases associated with unknown electron carriers, has not improved much. Cyanobacteria will facilitate the isolation of nonconditional as well as conditional mutants with lesions in the lipid biosynthetic pathway. Such a genetic approach promises to provide important information on the mechanisms and regulation of these pathways, an approach heretofore inaccessible in a eukaryotic organism.

Acknowledgments

We wish to acknowledge financial support of our work as a continuing program by the Commissariat à l'Energie Atomique (CEA) and the Centre National de la Recherche Scientifique (CNRS, URA 576).

References

Alban C, Dorne AJ, Joyard J and Douce R (1989a) [^{14}C]-acetate incorporation into glycerolipids from cauliflower proplastids and sycamore amyloplasts. FEBS Lett 249: 95–99

Alban C, Joyard J and Douce R (1989b) Comparison of glycerolipid biosynthesis in non-green plastids from sycamore (*Acer pseudoplatanus*) cells and cauliflower (*Brassica oleracea*) buds. Biochem J 259: 775–783

Andrews J and Mudd JB (1985) Phosphatidylglycerol synthesis in pea chloroplasts. Pathway and localization. Plant Physiol 79: 259–265

Andrews J, Ohlrogge J and Keegstra K (1985) Final steps of phosphatidic acid synthesis in pea chloroplasts occurs in the inner envelope membrane. Plant Physiol 78: 459–465

Andrews J, Schmidt H and Heinz E (1989) Interference of electron transport inhibitors with desaturation of monogalactosyl diacylglycerol in intact chloroplasts. Arch Biochem Biophys 270: 611–622

Armstrong GA, Hundle BS and Hearst JE (1993) Evolutionary conservation and structural similarities of carotenoid biosynthesis gene products from photosynthetic and nonphotosynthetic organisms. Meth Enzymol 214: 297–311

Barry P and Pallett KE (1990) Herbicidal inhibition of carotenogenesis detected by HPLC. Z Naturforsch 45: 492–497

Bartley GE, Viitanen P, Bacot KO and Scolnik PA (1992) A tomato gene expressed during fruit ripening encodes an enzyme of the carotenoid biosynthesis pathway. J Biol Chem 267: 5036–5039

Bartley GE, Kumle A, Beyer P and Scolnik PA (1993) Functional analysis and purification of enzymes for carotenoid biosynthesis expressed in photosynthetic bacteria. Meth Enzymol 214: 374–385

Beale SI (1990) Biosynthesis of the tetrapyrrole pigment precursor, δ-aminolevulinic acid, from glutamate. Plant Physiol 93: 1273–1279

Beale SI and Weinstein JD (1990) Tetrapyrrole metabolism in photosynthetic organisms. In: Dailey HA (ed) Biosynthesis of Heme and Chlorophylls, pp 287–391. McGraw-Hill Publishing Co, New York

Benli M, Schulz R and Apel K (1991) Effect of light on the NADPH-protochlorophyllide oxidoreductase of *Arabidopsis thaliana*. Plant Mol Biol 16: 615–625

Bertrams M and Heinz E (1981) Positional specificity and fatty acid selectivity of purified *sn*-glycerol-3-phosphate acyltransferase from chloroplasts. Plant Physiol 68: 653–657

Bertrams M, Wrage K and Heinz E (1981) Lipid labelling in intact chloroplasts from exogenous nucleotide precursors. Z Naturforsch 36: 62–70

Beyer P (1987) Solubilization and reconstitution of carotenogenic enzymes from daffodil chromoplast membranes using 3-[(3-cholamidopropyl)dimethylammonio]-1-propane sulfonate. Meth Enzymol 148: 392–400

Beyer P and Kleinig H (1990) On the desaturation and cyclization reactions of carotenes in chromoplast membranes. In: Krinsky NI, Mathews-Roth MM and Taylor RF (eds) Carotenoids: Chemistry and Biology, pp 195–206. Plenum Press, New York

Bickel H and Schultz G (1976) Biosynthesis of plastoquinone and β-carotene in isolated chloroplasts. Phytochem 15: 1253–1255

Bickel H, Palme L and Schultz G (1978) Incorporation of shikimate and other precursors into aromatic amino acids and prenylquinones of isolated spinach chloroplasts. Phytochem 17: 119–124

Billecocq A, Douce R and Faure M (1972) Structure des membranes biologiques: Localisation des galactosyldiglycérides dans les chloroplastes au moyen des anticorps spécifiques. CR Acad Sci Paris 275: 1135–1137

Bishop DG (1986) Chilling sensitivity in higher plants: The role of phosphatidylglycerol. Plant Cell Env 9: 613–616

Bishop DG, Sparace SA and Mudd JB (1985) Biosynthesis of sulfoquinovosyldiacylglycerol in higher plants: The origin of the diacylglycerol moiety. Arch Biochem Biophys 240: 851–858

Bishop WR and Bell RM (1988) Assembly of phospholipids into cellular membranes: Biosynthesis, transmembrane movement and intracellular location. Annu Rev Cell Biol 4: 579–610

Bligny R, Gardeström P, Roby C and Douce R (1990) ^{31}P NMR studies of spinach leaves and their chloroplasts. J Biol Chem 265: 1319–1326

Block MA, Joyard J and Douce R (1980) Site of synthesis of geranylgeraniol derivatives in intact spinach chloroplasts. Biochim Biophys Acta 631: 210–219

Block MA, Dorne AJ, Joyard J and Douce R (1983a) Preparation and characterization of membrane fractions enriched in outer and inner envelope membranes from spinach chloroplasts. I - Electrophoretic and immunochemical analyses. J Biol Chem 258: 13273–13280

Block MA, Dorne AJ, Joyard J and Douce R (1983b) Preparation and characterization of membrane fractions enriched in outer and inner envelope membranes from spinach chloroplasts. II - Biochemical characterization. J Biol Chem 258: 13281–13286

Block MA, Joyard J and Douce R (1992) Purification and characterization of E37, a major envelope protein. FEBS Lett 287: 765–769

Bramley PM (1985) The in vitro biosynthesis of carotenoids. Adv Lip Res 21: 243–279

Britton G (1988) Biosynthesis of carotenoids. In: TW Goodwin (ed) Plant Pigments, pp 133–182. Academic Press, London

Browse J, McCourt P and Somerville C (1985) A mutant of *Arabidopsis* lacking a chloroplast-specific lipid. Science 227: 763–765

Buckner B and Robertson DS (1993) Cloning of carotenoid biosynthetic genes from maize. Meth Enzymol 214: 311–323

Buckner B, Kelson TL and Robertson DS (1990) Cloning of the *yl* locus of maize, a gene involved in the biosynthesis of carotenoids. Plant Cell 2: 867–876

Camara B (1993) Plant phytoene synthase complex: Component enzymes, immunology and biogenesis. Meth Enzymol 214: 352–365

Camara B, Bardat F, Seye A, D'Harlingue A and Monéger R (1982) Terpenoid metabolism in plastids. Localization of α-tocopherol synthesis in *Capsicum* chromoplasts. Plant Physiol 70: 1562–1563

Carde JP, Joyard J, Douce R (1982) Electron microscopic studies of envelope membranes from spinach plastids. Biol Cell 44: 315–324

Carman GM and Henry SA (1989) Phospholipid biosynthesis in yeast. Annu Rev Biochem 58: 635–669

Carter HE, McCluer RH and Slifer ED (1956) Lipids of wheat flour. I. Characterization of galactosylglycerol components. J Am Chem Soc 78: 3735–3738

Castelfranco PA and Beale SI (1983) Chlorophyll biosynthesis: Recent advances and areas of current interest. Annu Rev Plant Physiol 34: 241–278

Castelfranco PA, Rich PM and Beale SI (1974) The abolition of lag phase in greening cucumber cotyledons by exogenous δ-aminolevulinic acid. Plant Physiol 53: 615–618

Cline K, Andrews J, Mersey B, Newcomb EH and Keegstra K (1981) Separation and characterization of inner and outer envelope membranes of pea chloroplasts. Proc Natl Acad Sci USA 78: 3595–3599

Cogdell R (1988) The function of pigments in chloroplasts. In: TW Goodwin (ed) Plant Pigments, pp 183–230. Academic Press, London

Costes C, Burghoffer C, Joyard J, Block MA and Douce R (1979) Occurrence and biosynthesis of violaxanthin in isolated spinach chloroplast envelope. FEBS Lett 103: 17–21

Covès J, Block MA, Joyard J and Douce R (1986) Solubilization and partial purification of UDP-galactose:diacylglycerol galactosyltransferase activity from spinach chloroplast envelope. FEBS Lett 208: 401–406

Covès J, Pineau B, Block MA, Joyard J and Douce R (1987) Solubilization and partial purification of chloroplast envelope proteins: Application to UDP-galactose:diacylglycerol galactosyltransferase. In Leaver C and Sze H (eds) Plant Membranes: Structure, Function, Biogenesis, pp 103–112. Alan R. Liss, New York

Covès J, Joyard J and Douce R (1988) Lipid requirement and kinetic studies of solubilized UDP-galactose:diacylglycerol galactosyltransferase activity from spinach chloroplast envelope membranes. Proc Natl Acad Sci USA 85: 4966–4970

D'Harlingue A and Camara B (1985) Plastid enzymes of terpenoid biosynthesis. Purification and characterization of γ-tocopherol methyltransferase from *Capsicum* chromoplasts. J Biol Chem 260: 15200–15203

Dahlin D (1993) Import of nuclear-encoded proteins into carotenoid-deficient young etioplasts. Physiol Plant 87: 410–416

Darrah PM, Kay SA, Teakle GR and Griffiths WT (1990) Cloning and sequencing of protochlorophyllide reductase. Biochem J 265: 789–798

Deems RA, Eaton BR and Dennis EA (1975) Kinetic analysis of phospholipase A2 activity towards mixed micelles and its implication for the study of lipolytic enzymes. J Biol Chem 250: 9013–9020

Dehesh K, van Cleve B, Ryberg M and Apel K (1986) Light-induced changes in the distribution of the 36,000-Mr polypeptide of NADPH-protochlorophyllide oxidoreductase within different compartments of barley (*Hordeum vulgare* L.) Planta 169: 172–183

Demmig-Adams B and Adams WW III (1992) Photoprotection and other responses of plants to high light stress. Annu Rev Plant Physiol Plant Mol Biol 43: 599–626

Deruère J, Römer S, d'Harlingue A, Backhaus RA, Kuntz M and Camara B (1994) Fibril assembly and carotenoid over-accumulation in chromoplasts: A model for supramolecular lipoprotein structures. The Plant Cell 6: 119–133

Dorne A-J and Heinz E (1989) Position and pairing of fatty acids in phosphatidylglycerol from pea leaf chloroplasts and mitochondria. Plant Sci 60: 39–46

Dorne A-J, Block MA, Joyard J and Douce R (1982) The galactolipid:galactolipid galactosyltransferase is located on the outer membrane of the chloroplast envelope. FEBS Lett 145: 30–34

Dorne A-J, Joyard J, Block MA and Douce R (1985) Localization of phosphatidylcholine in outer envelope membrane of spinach chloroplasts. J Cell Biol 100: 1690–1697

Dorne A-J, Joyard J and Douce, R (1990) Do thylakoids really contain phosphatidylcholine? Proc Natl Acad Sci USA 87: 71–74

Douady D and Dubacq J-P (1987) Purification of acyl-CoA:glycerol-3-phosphate acyltransferase from pea leaves. Biochim Biophys Acta 921: 615–619

Douce R (1974) Site of synthesis of galactolipids in spinach chloroplasts. Science 183: 852–853

Douce R and Joyard J (1980) Plant galactolipids. In: Stumpf PK (ed) The Biochemistry of Plants: Lipids: Structure and Function, Vol 4, pp 321–362. Academic Press, New York

Elstner EF (1987) Metabolism of activated oxygen species. In: Davies DD (ed) The Biochemistry of Plants: Biochemistry of Metabolism, Vol 11, pp 253–315. Academic Press, New York

Fiedler E, Soll J and Schultz G (1982) The formation of homogentisate in the biosynthesis of tocopherol and plastoquinone in spinach chloroplasts. Planta 155: 511–515

Flügge UI and Benz R (1984) Pore-forming activity in the outer

membrane of the chloroplast envelope. FEBS Lett 169: 85–89
Frank F and Strzalka K (1992) Detection of the photoactive protochlorophyllide-protein complex in the light during the greening of barley. FEBS Lett 309: 73–77
Frentzen M (1986) Biosynthesis and desaturation of the different diacylglycerol moieties in higher plants. J Plant Physiol 124: 193–209
Frentzen M (1993) Acyltransferases and triacylglycerol. In: Moore ST Jr (ed) Lipid Metabolism in Plants, pp 195–230. CRC Press, Boca Raton
Frentzen M, Heinz E, McKeon TA and Stumpf PK (1983) Specificities and selectivities of glycerol-3-phosphate acyltransferase and monoacylglycerol-3-phosphate acyltransferase from pea and spinach chloroplasts. Eur J Biochem 129: 629–636
Fuesler TP, Wong YS and Castelfranco PA (1984) Localization of Mg-chelatase and Mg-protoporphyrin IX monomethyl ester (oxidative) cyclase activities within isolated, developing cucumber cotyledons. Plant Physiol 75: 662–664
Garnier J, Wu B, Maroc J, Guyon D and Trémolières A (1990) Restoration of both an oligomeric form of the light-harvesting antenna CPII and a fluorescence state II-state I transition by D^3-*trans*-hexadecenoic acid-containing phosphatidylglycerol, in cells of a mutant of *Chlamydomonas reinhardtii*. Biochim Biophys Acta 1020: 153–162
Goodwin TW (1993) Biosynthesis of carotenoids: An overview. Meth Enzymol 214: 330–340
Goodwin TW and Britton G (1988) Distribution and analysis of carotenoids. In: TW Goodwin (ed) Plant Pigments, pp 61–132. Academic Press, London
Gounaris, K. and Barber, J.(1983) Monogalactosyldiacylglycerol: The most abundant polar lipid in Nature. Trends Biochem Sci 9: 378–381
Gray JC (1987) Control of isoprenoid biosynthesis in higher plants. Adv Bot Res 14: 25–91
Grevby C, Engdahl, Ryberg M and Sundqvist C (1989) Binding properties of NADPH-protochlorophyllide oxidoreductase as revealed by detergent and ion treatments of isolated and immobilized prolamellar bodies. Physiol Plant 77: 493–503
Gunning BES and Steer MW (1975) Ultrastructure and the Biology of Plant Cell, Edward Arnold: London
Haas R, Siebertz HP, Wrage K and Heinz E (1980) Localization of sulfolipid labeling within cells and chloroplasts. Planta 148: 238–244
Haverkate F and Van Deenen LLM (1965) Isolation and chemical characterization of phosphatidylglycerol from spinach leaves. Biochim Biophys Acta 106: 78–92
Heber U and Heldt HW (1981) The chloroplast envelope: Structure, function, and role in leaf metabolism. Ann Rev Plant Physiol 32: 139–168
Heemskerk JWM and Wintermans JFGM (1987) The role of the chloroplast in the leaf acyl-lipid synthesis. Physiol Plant 70: 558–568
Heemskerk JHW, Storz T, Schmidt RR and Heinz E (1990) Biosynthesis of digalactolsyldiacylglycerol in plastids from 16:3 and 18:3 plants. Plant Physiol 93: 1286–1294
Heinz E (1977) Enzymatic reactions in galactolipid biosynthesis. In: Tevini M and Licthenthaler HK (eds) Lipids and Lipid Polymers, pp 102–120. Springer Verlag, Berlin
Heinz E (1993) Biosynthesis of polyunsaturated fatty acids. In: Moore ST Jr (ed) Lipid Metabolism in Plants, pp 34–89. CRC Press, Boca Raton
Heinz E and Roughan PG (1983) Similarities and differences in lipid metabolism of chloroplasts isolated from 18:3 and 16:3 plants. Plant Physiol 72: 273–279
Heinz E, Schmidt H, Hoch M, Jung K-H, Binder H and Schmidt RR (1989) Synthesis of different nucleoside 5'-diphosphosulfoquinovoses and their use for studies on sulfolipid biosynthesis in chloroplasts. Eur J Biochem 184: 445–453
Herrin DL, Battey JF, Greer K and Schmidt GW (1992) Regulation of chlorophyll apoprotein expression and accumulation: Requirements for carotenoids and chlorophyll. J Biol Chem 267: 8260–8269
Hinterstoisser B, Missbichler A, Pineau B and Peschek G (1988) Detection of chlorophyllide in chlorophyll-free plasma membrane preparations from *Anacystis nidulans*. Biochem Biophys Res Comm 154: 839–846
Hobe S, Prytulla S, Külbrandt W and Paulsen H (1994) Trimerization and crystallization of reconstituted light-harvesting chlorophyll *a/b* complex. EMBO J 13: 3423–3429
Hugueney P, Römer S, Kuntz M and Camara B (1992) Characterization and molecular cloning of a bifunctional flavoprotein catalyzing the synthesis of phytofluene and zeta-carotene in *Capsicum* chromoplasts. Eur J Biochem 209: 399–407
Humbeck K, Römer S and Senger H (1989) Evidence for an essential role of carotenoids in the assembly of an active photosystem II. Planta 179: 242–250
Hutson KG and Threlfall DR (1980) Synthesis of plastoquinone-9 and phytylplastoquinone from homogentisate in lettuce chloroplasts. Biochim Biophys Acta 632: 630–648
Ikeuchi M and Murakami S (1982) Behavior of the 36,000-dalton protein in the internal membranes of squash etioplasts during greening. Plant Cell Physiol 23: 575–583
Ilag LL, Kumar AM and Söll D (1994) Light regulation of chlorophyll biosynthesis at the level of aminolevulinate formation in *Arabidopsis*. Plant Cell 6: 265–275
Ishiko H, Shigeoka S, Nakano Y and Mitsunaga T (1992) Some properties of γ-tocopherol methyltransferase solubilized from spinach chloroplasts Phytochem 31: 1499–1500
Ishizaki O, Nishida I, Agata K, Eguchi G and Murata N (1988) Cloning and nucleotide sequence of cDNA for the plastid glycerol-3-phosphate acyltransferase from squash. FEBS Lett 238: 424–430
Jeffrey SW, Douce R and Benson AA (1974) The carotenoids of the chloroplast envelope. Proc Natl Acad Sci USA 71: 807–810
Johanningmeier U and Howell SH (1984) Regulation of light-harvesting chlorophyll-binding protein mRNA accumulation in *Chlamydomonas reinhardtii*. Possible involvement of chlorophyll synthesis precursors. J Biol Chem 259: 13541–13549
Jones BL and Porter JW (1986) Biosynthesis of carotenes in higher plants. CRC Crit Rev Plant Sci 3: 295–324
Jones OTG (1968) Ferrochelatase of spinach chloroplasts. Biochem J 107: 113–119
Joyard J and Douce R (1976a) L'enveloppe des chloroplastes est-elle capable de synthétiser la phosphatidylcholine? CR Acad Sci Paris 282: 1515–1518
Joyard J and Douce R (1976b) Mise en évidence et rôle des diacylglycérols dans l'enveloppe des chloroplastes d'épinard. Biochim Biophys Acta 424: 126–131

Joyard J and Douce R (1977) Site of synthesis of phosphatidic acid and diacylglycerol in spinach chloroplasts. Biochim Biophys Acta 486: 273–285

Joyard J and Douce R (1979) Characterization of phosphatidate phosphohydrolase activity associated with chloroplast envelope membranes. FEBS Lett 102: 147–150

Joyard J and Douce R (1987) Galactolipid biosynthesis. In: Stumpf PK (ed) The Biochemistry of Plants: Lipids: Structure and Function, Vol 9, pp 215–274. Academic Press, New York

Joyard J, Douce R, Siebertz HP and Heinz E (1980) Distribution of radioactive lipids between envelopes and thylakoids from in vivo labeled chloroplasts. Eur J Biochem 108: 171–176

Joyard J, Blée E and Douce R (1986) Sulfolipid synthesis from $^{35}SO_4^{2-}$ and [1-^{14}C]-acetate in isolated intact spinach chloroplasts. Biochim Biophys Acta 879: 78–87

Joyard J, Block MA, Pineau B, Albrieux C and Douce R (1990) Envelope membranes from mature spinach chloroplasts contain a NADPH:protochlorophyllide reductase on the cytosolic side of the outer membrane. J Biol Chem 265: 21820–21827

Joyard J, Block MA and Douce R (1991) Molecular aspects of plastid envelope biochemistry. Eur J Biochem 199: 489–509

Joyard J, Block MA, Malherbe A, Maréchal E and Douce R (1993) Origin and synthesis of galactolipid and sulfolipid head groups. In: Moore ST Jr (ed) Lipid Metabolism in Plants, pp 231–258. CRC Press, Boca Raton

Kader JC (1990) Intracellular transfer of phospholipids, galactolipids and fatty acids in plant cells. Subcell Biochem 16: 69–111

Kader JC (1993) Lipid transport in plants. In: Moore ST Jr (ed) Lipid Metabolism in Plants, pp 309–336. CRC Press, Boca Raton

Keegstra K, Olsen LJ and Theg SM (1989) Chloroplastic precursors and their transport across the envelope membranes. Ann Rev Plant Physiol Plant Mol Biol 40: 471–501

Klein RR and Mullet JE (1987) Control of gene expression during higher plant chloroplast biogenesis. Protein synthesis and transcript levels of psbA, psaA-psaB, and rbcL in dark-grown and illuminated barley seedlings. J Biol Chem 262: 4341–4348

Kleinig H (1989) The role of plastids in isoprenoid biosynthesis. Annu Rev Plant Physiol Plant Mol Biol 40: 39–59

Kleppinger-Sparace KF, Mudd JB and Bishop DG (1985) Biosynthesis of sulfoquinovosyldiacylglycerol in higher plants: The incorporation of $^{35}SO_4$ by intact chloroplasts. Arch Biochem Biophys 240: 859–865

Kunst L, Browse J and Sommerville C (1988) Altered regulation of lipid biosynthesis in a mutant of *Arabidopsis* deficient in chloroplast glycerol-3-phosphate acyltransferase activity. Proc Natl Acad Sci USA 85: 4143–4147

Kuntz M, Römer S, Suire C, Hugueney P, Weil JH, Schantz R and Camara B (1992) Identification of a cDNA for the plastid located geranylgeranyl pyrophosphate synthase from *Capsicum annuum*: Correlative increase in enzyme activity and transcript level during fruit ripening. Plant J 21: 25–34

Lawlor DW (1987) Photosynthesis: Metabolism, Control and Physiology. Longman Scientific and Technical, Harlow, UK.

Li HM, Moore T and Keegstra K (1991) Targeting of proteins to the outer envelope membrane uses a different pathway than transport into chloroplasts. Plant Cell 3: 709–717

Lichtenthaler HK (1993) The plant prenyllipids, including carotenoids, chlorophylls, and prenylquinones. In: Moore ST Jr (ed) Lipid Metabolism in Plants, pp 427–470. CRC Press, Boca Raton

Lichtenthaler HK, Prenzel H, Douce R and Joyard J (1981) Localization of prenylquinones in the envelope of spinach chloroplasts. Biochim Biophys Acta 641: 99–105

Lindblom G and Rilfors L (1989) Cubic phases and isotropic structures formed by membrane lipids. Possible biological relevance. Biochim Biophys Acta 988: 221–256

Lütke-Brinkhaus F, Liedvogel B, Kreuz K and Kleinig H (1982) Phytoene synthase and phytoene dehydrogenase associated with envelope membranes from spinach chloroplasts. Planta 156: 176–180

Lützow M and Kleinig H (1990) Chlorophyll-free chromoplasts from daffodil contain most of the enzymes for chlorophyll synthesis in a highly active form. Arch Biochem Biophys 277: 94–100

Malherbe A, Block MA, Joyard J and Douce R (1992) Feedback inhibition of phosphatidate phosphatase from spinach chloroplast envelope membranes by diacylglycerol. J Biol Chem 267: 23546–23553

Malherbe A, Block MA, Douce R and Joyard J (1995) Solubilization and biochemical properties of phosphatidate phosphatase from spinach chloroplast envelope membranes. Plant Physiol Biochem 33: 149–161

Maréchal E, Block MA, Joyard J and Douce R (1991) Purification de l'UDP-galactose:1,2-diacylglycérol galactosyltransferase de l'enveloppe des chloroplastes d'épinard. C R Acad Sci Paris 313: 521–528

Maréchal E, Block MA, Joyard J and Douce R (1994a) Kinetic properties of monogalactosyldiacylglycerol synthase from spinach chloroplast envelope membranes. J Biol Chem 269: 5788–5798

Maréchal E, Block MA, Joyard J and Douce R (1994b) Comparison of the kinetic properties of MGDG synthase in mixed micelles and in envelope membranes from spinach chloroplast. FEBS Lett 352: 307–310

Maréchal E, Miége C, Block MA, Douce R and Joyard J (1995) The catalytic site of monogalactosyldiacylglycerol synthase from spinach chloroplast envelope membranes. Biochemical analysis of the structure and of the metal content. J Biol Chem 270: 5714–5722

Markwell J, Bruce BD and Keegstra K (1992) Isolation of a carotenoid-containing submembrane particle from the chloroplastic envelope outer membrane of pea (*Pisum sativum*). J Biol Chem 267: 13933–13937

Martin BA and Wilson RF (1984) Subcellular localization of triacylglycerol biosynthesis in spinach leaves. Lipids 19: 117–121

Matringe M, Camadro JM, Block MA, Joyard J, Scalla R, Labbe P and Douce R (1992) Localization within chloroplasts of protoporphyrinogen oxidase, the target enzyme for diphenylether-like herbicides. J Biol Chem 267: 4646–4651

Matringe M, Camadro JM, Joyard J and Douce R (1994) Localization of ferrochelatase activity within mature pea chloroplasts. J Biol Chem 269: 15010–15015

Mayer H and Isler O (1971) Synthesis of vitamin E. Meth Enzymol 18C: 241–248

Mayer MP, Beyer P and Kleinig H (1990) Quinone compounds are able to replace molecular oxygen as terminal electron acceptor in phytoene desaturation in chromoplasts of *Narcissus pseudonarcissus* L. Eur J Biochem 191: 359–363

Miquel M, Block MA, Joyard J, Dorne A-J, Dubacq J-P, Kader J-C and Douce R (1987) Protein-mediated transfer of phosphatidylcholine from liposomes to spinach chloroplast envelope membranes. Biochim Biophys Acta 937: 219–228

Morelli G, Neslon MA, Ballario P and Macino G (1993) Photoregulated carotenoid biosynthetic genes of *Neurospora crassa*. Meth Enzymol 214: 412–424

Morré DJ, Morré JT, Morré SR, Sundqvist C and Sandelius AS (1991) Chloroplast biogenesis. Cell-free transfer of envelope monogalactosylglycerides to thylakoids. Biochim Biophys Acta 1070: 437–445

Morris SR and Threlfall DR (1983) Synthesis of 2-methyl-6-phytyl-1,4-benzoquinone by a membrane preparation of the cyanobacterium *Anabaena variabilis*. Biochem Soc trans 11: 587–588

Mudd JB and de Zacks R (1981) Synthesis of phosphatidylglycerol by chloroplasts from leaves of *Spinacia oleracea* L. Arch Biochem Biophys 209: 584–591

Mullet JE, Klein PG and Klein RR (1990) Chlorophyll regulates accumulation of the plastid-encoded chlorophyll apoproteins CP43 and D1 by increasing apoprotein stability. Proc Natl Acad Sci USA 87: 4038–4042

Murata N, Sato N, Takahashi N and Hamazaki Y (1982) Compositions and positional distribution of fatty acids in phospholipids from leaves of chilling-sensitive and chilling resistant plants. Plant Cell Physiol 23: 1071–1079

Neuburger M, Joyard J and Douce R (1977) Strong binding of cytochrome *c* on the envelope of spinach chloroplasts. Plant Physiol 59: 1178–1181

Nishida I and Yamada M (1986) Semisynthesis of a spin-labeled monogalactosyl-diacylglycerol and its application in the assay for galactolipid transfer activity in spinach leaves. Biochim Biophys Acta 813: 298–306

Nishida I, Frentzen M, Ishizaki O and Murata N (1987) Purification of isomeric forms of acyl-[acyl-carrier-protein]:glycerol-3-phosphate acyltransferase from greening squash cotyledons. Plant Cell Physiol 28: 1071–1079

Norman HA, Pillai P and St John JB (1991) In vitro desaturation of monogalactosyldiacylglycerol and phosphatidylcholine molecular species by chloroplast homogenates. Phytochem 30: 2217–2222

Nussberger S, Dörr K, Wang DN and Külbrandt W (1993) Lipid-protein interactions in crystals of plant light-harvesting complex. J Mol Biol 234: 347–356

Pascal N, Block MA, Pallett KE, Joyard J and Douce R (1995) Inhibition of carotenoid biosynthesis in sycamore cells deprived of iron. Plant Physiol Biochem 33: 97–104

Pennock JF (1983) The biosynthesis of chloroplastic terpenoid quinones and chromanols. Biochem Soc trans 11: 504–510

Peschek G, Hinterstoisser B, Wastyn M, Kuntner O, Pineau B, Missbichler A and Lang J (1989) Chlorophyll precursors in the plasma membrane of a cyanobacterium, *Anacystis nidulans*. Characterization of protochlorophyllide and chlorophyllide by spectrophotometry, spectrofluorimetry, solvent partition and high performance liquid chromatography. J Biol Chem 264: 11827–11832

Pineau B, Dubertret G, Joyard J and Douce R (1986) Fluorescence properties of the envelope membranes from spinach chloroplasts. Detection of protochlorophyllide. J Biol Chem 261: 9210–9215

Pineau B, Gérard-Hirne C, Douce R and Joyard J (1993) Identification of the main species of tetrapyrrolic pigments in envelope membranes from spinach chloroplasts. Plant Physiol 102: 821–828

Plumley FG and Schmidt GW (1987) Reconstitution of chlorophyll *a/b* light-harvesting complexes: Xanthophyll-dependent assembly and energy transfer. Proc Natl Acad Sci USA 84: 146–150

Quinn PJ and Williams WP (1983) The structural role of lipids in photosynthetic membranes. Biochim Biophys Acta 737: 223–266

Rau W (1988) Functions of carotenoids other than in photosynthesis. In: TW Goodwin (ed) Plant Pigments, pp 231–255. Academic Press, London

Rawyler A, Meylan M and Siegenthaler PA (1991) Galactolipid export from envelope to thylakoid membranes in intact chloroplasts. I. Characterization and involvement in thylakoid asymmetry. Biochim Biophys Acta 1104: 331–341

Rawyler A, Meylan M and Siegenthaler PA (1994) (Galacto)lipid export from envelope to thylakoid membranes in intact chloroplasts. II. A general process with a key role for the envelope in the establishment of lipid asymmetry in thylakoid membranes. Biochim Biophys Acta 1233: 123–133

Roughan PG (1986) Acyl lipid synthesis by chloroplasts isolated from the chilling-sensitive plant *Amaranthus lividus* L. Biochim Biophys Acta 878: 371–379

Roughan PG and Slack CR (1977) Long-chain acyl-coenzyme A synthetase activity of spinach chloroplasts is concentrated in the envelope. Biochem J 162: 457–459

Roughan PG and Slack CR (1982) Cellular organization of glycerolipid metabolism. Annu Rev Plant Physiol 33: 97–132

Roughan PG, Mudd JB, McManus TT and Slack CR (1979) Linoleate and α-linolenate synthesis by isolated spinach (*Spinacia oleracea*) chloroplasts. Biochem J 184: 571–574

Rüdiger W and Schoch S (1988) Chlorophylls. In: TW Goodwin (ed) Plant Pigments, pp 1–59. Academic Press, London

Ruzicka L (1953) The isoprene rule and the biogenesis of terpenic compounds. Experientia 9: 357–396

Sakaki T, Kondo N and Yamada M (1990) Pathway for the synthesis of triacylglycerol from monogalactosyldiacylglycerols in ozone-fumigated spinach leaves. Plant Physiol 94: 773–780

Salomon M, Fisher K, Flügge UI and Soll J (1990) Sequence analysis and protein import studies of an outer chloroplast envelope polypeptide. Proc Natl Acad Sci USA 87: 5778–5782

Sandmann G (1993) Carotenoid analysis in mutants from *Escherichia coli* transformed with carotenogenic gene cluster and *Scenedesmus obliquus* mutant C-6D. Meth Enzymol 214: 341–347

Sastry PS and Kates M (1964) Lipid components of leaves. V. Galactolipids, cerebrosides, and lecithin of runner-bean leaves. Biochemistry 3: 1271–1280

Schmidt A, Sandmann G, Armstrong GA, Hearst JE and Böger P (1989) Immunological detection of phytoene desaturase in algae and higher plants using an antiserum raised against a bacterial fusion-gene construct. Eur J Biochem 184: 375–378

Schmidt H and Heinz E (1990a) Involvement of ferredoxin in

desaturation of lipid-bound oleate in chloroplasts. Plant Physiol 94: 214–220

Schmidt H and Heinz E (1990b) Desaturation of oleoyl groups in envelope membranes from spinach chloroplasts. Proc Natl Acad Sci USA 87: 9477–9480

Schmidt H and Heinz E (1992) n-6 desaturase from chloroplast envelopes: Purification and enzymatic characteristics. In: Cherif A, Miled-Daoud DB, Marzouk B, Smaoui A and Zarrouk M (eds) Metabolism, Structure and Utilization of Plant Lipids, pp 140–143. Cent Natl Ped Tunis, Tunisia

Schulz A, Ort O, Beyer P and Kleinig H (1993) SC-0051, a 2-benzoyl-cyclohexane-1,3-dione bleaching herbicide, is a potent inhibitor of the enzyme *p*-hydroxyphenyl pyruvate dioxygenase. FEBS Lett 318: 162–166

Schulz R, Steinmüller K, Klaas M, Forreiter C, Rasmussen S, Hiller C and Apel K (1989) Nucleotide sequence of a cDNA coding for the NADPH-protochlorophyllide oxidoreductase (PCR) of barley (*Hordeum vulgare* L.) and its expression in *Escherichia coli*. Mol Gen Genet 217: 355–361

Schulze-Siebert D, Homeyer U, Soll J and Schultz G (1987) Synthesis of plastoquinone-9, α-tocopherol and phylloquinone (vitamine K1) and its integration in chloroplast carbon metabolism of higher plants. In: Stumpf PK, Mudd JB and Nes WD (eds) The Metabolism, Structure, and Function of Plant Lipids, pp 29–36. Plenum Publishing Co, New York

Seifert U and Heinz E (1992) Enzymatic characteristics of UDP-sulfoquinovose:diacylglycerol sulfoquinovosyltransferase from chloroplast envelopes. Botanica Acta 105: 197–205

Selstam E, Widell A and Johannson BA (1987) A comparison of prolamellar bodies of wheat, Scots pine and Jeffrey pine. Pigment spectra and properties of protochlorophyllide reductase. Physiol Plant 70: 209–214

Shaw P, Henwood J, Oliver R and Griffiths T (1985) Immunogold localization of protochlorophyllide reductase in barley etioplasts. Eur J Cell Biol 39: 50–55

Shigeoka S, Ishiko H, Nakano Y and Mitsunaga T (1992) Isolation and properties of γ-tocopherol methyltransferase in *Euglena gracilis*. Biochim Biophys Acta 1128: 220–226

Siebertz HP, Heinz E, Linscheid M Joyard J and Douce R (1979) Characterization of lipids from chloroplast envelopes. Eur J Biochem 101: 429–438

Siefermann-Harms D (1977) The xanthophyll cycle in higher plants. In: Tevini M and Lichtenthaler HK (eds) Lipids and Lipid Polymers, pp 218–230. Springer Verlag, Berlin

Siefermann-Harms D (1985) Carotenoids in photosynthesis. I. Location in photosynthetic membranes and light harvesting functions. Biochim Biophys Acta 811: 325–355

Siefermann-Harms D, Joyard J and Douce R (1978) Light-induced changes of the carotenoid levels in chloroplast envelopes. Plant Physiol 61: 530–533

Slabas AR and Fawcett T (1992) The biochemistry and molecular biology of plant lipid biosynthesis. Plant Mol Biol 19: 169–191

Smeekens S, Weisbeek P and Robinson C (1990) Protein transport into and within chloroplasts. Trends Biochem Sci 15: 73–76

Smith BB and Rebeiz CA (1979) Chloroplast biogenesis. XXIV. Intrachloroplastic localization of the biosynthesis and accumulation of protoporphyrin IX, magnesium-proto-porphyrin monoester, and longer wavelength metalloporphyrins during greening. Plant Physiol 63: 227–231

Soler E, Clastre M, Bantignies B, Marigo G and Ambid C (1993) Uptake of isopentenyl diphosphate by plastids isolated from *Vitis vitifera* L. cell suspensions. Planta 191: 324–329

Soll J (1987) α-tocopherol and plastoquinone biosynthesis in chloroplast membranes. Meth Enzymol 148: 383–392

Soll J and Alefsen H (1993) The protein import apparatus of chloroplasts. Physiol Plant 87: 433–440

Soll J and Schultz G (1981) Phytol synthesis from geranylgeraniol in spinach chloroplasts. Biochem Biophys Res Commun 99: 907–912

Soll J, Douce R and Schultz G (1980a) Site of biosynthesis of α-tocopherol in spinach chloroplasts. FEBS Lett 112: 243–246

Soll J, Kemmerling M and Schultz G (1980b) Tocopherol and plastoquinone synthesis in spinach chloroplasts subfractions. Arch Biochem Biophys 204: 544–550

Soll J, Schultz G, Rüdiger W and Benz J (1983) Hydrogenation of geranylgeraniol. Two pathways exist in spinach chloroplasts. Physiol Plant 71: 849–854

Soll J, Schultz G, Joyard J, Douce R and Block MA (1985) Localization and synthesis of prenylquinones in isolated outer and inner envelope membranes from spinach chloroplasts. Arch Biochem Biophys 238: 290–299

Somerville C and Browse J (1991) Plant lipids: Metabolism, mutants, and membranes. Science 252: 80–87

Spano AJ, He Z and Timko MP (1992a) NADPH:protochlorophyllide oxidoreductases in white pine (*Pinus strobus*) and loblolly pine (*P. taeda*). Mol Gen Genet 236: 86–95

Spano AJ, He Z, Michel H, Hunt DF and Timko MP (1992b) Molecular cloning, nuclear gene structure, and developmental expression of NADPH:protochlorophyllide oxidoreductase in pea (*Pisum sativum* L.). Plant Mol Biol 18: 967–972

Stymne S and Stobart AK (1987) Triacylglycerol biosynthesis. In: Stumpf PK (ed) The Biochemistry of Plants: Lipids: Structure and Function, Vol 9, pp 175–214. Academic Press, New York

Swiezewska E, Dallner G, Andersson B and Ernster L (1993) Biosynthesis of ubiquinone and plastoquinone in the endoplasmic reticulum-Golgi membranes of spinach leaves. J Biol Chem 268: 1494–1499

Taylor WC (1989) Regulatory interactions between nuclear and plastid genomes. Annu Rev Plant Physiol Plant Mol Biol 40: 211–233

Teakle GR and Griffiths, WT (1993) Cloning, characterization and import studies on protochlorophyllide reductase from wheat (*Triticum aestivum*). Biochem J 296: 225–230

Teucher T and Heinz E (1991) Purification of UDP-galactose:diacylglycerol from chloroplast envelopes of spinach (*Spinacia oleracea* L.). Planta 184: 319–326

Trémolières A, Roche O, Dubertret G, Maroc J, Guyon D and Garnier J (1991) Restoration of thylakoid appression by D^3-*trans*-hexadecenoic acid-containing phosphatidylglycerol, in a mutant of *Chlamydomonas reinhardtii*. Relationship with the regulation of excitation energy distribution. Biochim Biophys Acta 1059: 286–292

Van Besouw A and Wintermans JFGM (1978) Galactolipid formation in chloroplast envelopes. I. Evidence for two mechanisms in galactosylation. Biochim Biophys Acta 529: 44–53

Van Besouw A and Wintermans JFGM (1979) The synthesis of galactolipids by chloroplast envelopes. FEBS Lett 102: 33–37

Walker CJ and Weinstein JD (1991a) Further characterization of the magnesium chelatase in isolated developing cucumber chloroplasts. Plant Physiol 95: 1189–1196

Walker CJ and Weinstein JD (1991b) In vitro assay of the chlorophyll biosynthetic enzyme Mg-chelatase: Resolution of the activity into soluble and membrane-bound fractions. Proc Natl Acad Sci USA 88: 5789–5793

Weber S, Wolter FP, Buck F, Frentzen M and Heinz E (1991) Purification and cDNA sequencing of an oleate-selective acyl-ACP:sn-glycerol-3-phosphate acyltransferase from pea chloroplasts. Plant Mol Biol 17: 1067–1076

Wellburn AR (1982) Bioenergetic and ultrastructural changes associated with chloroplast development. Int Rev Cytol 80: 133–191

Wrisher M (1989) Ultrastructural localization of photosynthetic activity in thylakoids during chloroplast development in maize. Planta 177: 18–23

Chapter 7

Targeting of Proteins Into and Across the Thylakoid Membrane

Colin Robinson
Department of Biological Sciences, University of Warwick, Coventry CV4 7AL, UK

Summary

The synthesis and assembly of photosynthetic proteins is a complex process in chloroplasts, since a variety of chloroplast- and nuclear-encoded proteins are targeted either into or across the thylakoid membrane. The insertion of chloroplast proteins is poorly understood, but in vitro reconstitution assays have provided a great deal of information on the pathways taken by imported thylakoid proteins. The import of these proteins from the cytosol is dependent on N-terminal presequences which serve to target the proteins across the double-membrane envelope, probably at contact sites between the two membranes. Most integral membrane proteins are subsequently targeted into the thylakoid by means of information located in the mature proteins, whereas cleavable targeting signals direct the translocation of hydrophilic lumenal proteins across the thylakoid membrane. Remarkably, this transport of lumenal proteins across the thylakoid membrane relies on at least two completely different mechanisms, and the available evidence strongly suggests that distinct translocases operate in the thylakoid membrane. One of these mechanisms appears to have been inherited from the cyanobacterial-type progenitor of the chloroplast, whereas the other may have evolved relatively recently in response to the appearance of novel photosynthetic proteins.

Donald R. Ort and Charles F. Yocum (eds): Oxygenic Photosynthesis: The Light Reactions, pp. 103–112.

I. Introduction

For many years, the thylakoid membrane has held promise as a fascinating system for the study of protein insertion/translocation. The processes of light capture, photosynthetic electron transport and photophosphorylation all require the targeting of proteins to specific locations within the thylakoid network, in order for the highly structured thylakoid membrane complexes to assemble correctly. The individual components of these complexes cover every possible permutation in terms of localization within the thylakoid network, since the soluble proteins are located on both the stromal and lumenal faces of the thylakoid membrane, and the hydrophobic proteins contain one, two or multiple membrane-spanning segments with widely varying orientations. And because the entire photosynthetic process is dependent on these components being in *exactly* the right place at the right time, there is little or no room for error in the targeting of these proteins into the thylakoids. These points alone suggested that the biogenesis of photosynthetic proteins was likely to be both complex and interesting in terms of the protein trafficking involved, but the overall process is of course made still more complex by the fact that, in the case of the chloroplast, the proteins originate from two different starting points.

The last nine or ten years have witnessed the cloning of genes encoding most of the prominent photosynthetic proteins, and within these sequences are many potential signals for the targeting of these proteins to the thylakoids. More recently, a variety of in vitro assays have been developed to examine in detail the mechanisms by which some of the proteins are targeted into or across the thylakoid membrane. In this article I will describe how this combination of structural and biochemical studies has led to important insights into the pathways and mechanisms involved in the targeting of proteins into the thylakoid membrane and lumen. Most importantly, these studies indicate that thylakoid protein insertion/translocation is likely to involve the operation of a far greater variety of mechanisms than had been envisaged previously, and the emerging story has interesting implications for the evolutionary origin of these mechanisms.

II. Sites of Synthesis of Thylakoid Proteins

A striking feature of all of the thylakoid protein complexes is that each consists of a mixture of chloroplast- and nuclear-encoded polypeptides. Figure 1 shows the organization of the photosystem II complex (PS II), which itself illustrates the complexity of protein traffic required for correct assembly of thylakoid complexes. The key reaction center proteins, D1 and D2, are both chloroplast-encoded, hydrophobic membrane proteins which are probably inserted co-translationally into the thylakoid membrane. The light-harvesting chlorophyll-binding protein associated with PS II (LHCPII) is similarly hydrophobic but is cytosolically-synthesized and thus has to be transported across the two envelope membranes and across the stromal phase. Different again, the 33, 23 and 16 kDa proteins of the oxygen-evolving complex (33K, 23K and 16K) are hydrophilic proteins which are synthesized in the cytosol and transported across all three chloroplast membranes (and the stroma) into the thylakoid lumen. The other complexes in the thylakoid membrane are similarly composed of a mixture of nuclear- and chloroplast-encoded subunits.

III. Insertion of Proteins into the Thylakoid Membrane

A. Import of Nuclear-encoded Thylakoid Membrane Proteins

1. Protein Translocation Across the Envelope Membranes

Numerous studies have examined the transport of proteins into chloroplasts, mostly using in vitro assays for the import of proteins by isolated intact chloroplasts, and the available evidence suggests that the known stromal and thylakoid proteins are all imported across the envelope membranes by a common mechanism. In each case, imported proteins are synthesized with a transient N-terminal presequence (or transit peptide) that serves to target the attached protein into the stroma by a posttranslational mechanism (Highfield and Ellis, 1978; Chua

Abbreviations: 33K, 23K, 16K – 33, 23, and 16 kDa proteins of the oxygen-evolving complex; ER – endoplasmic reticulum; hsp – heat shock protein; PC – plastocyanin; PS I – Photosystem I; PS II – Photosystem II; Rubisco – ribulose 1,5-bisphosphate carboxylase/oxygenase; SPP – stromal processing peptidase; TPP – thylakoidal processing peptidase

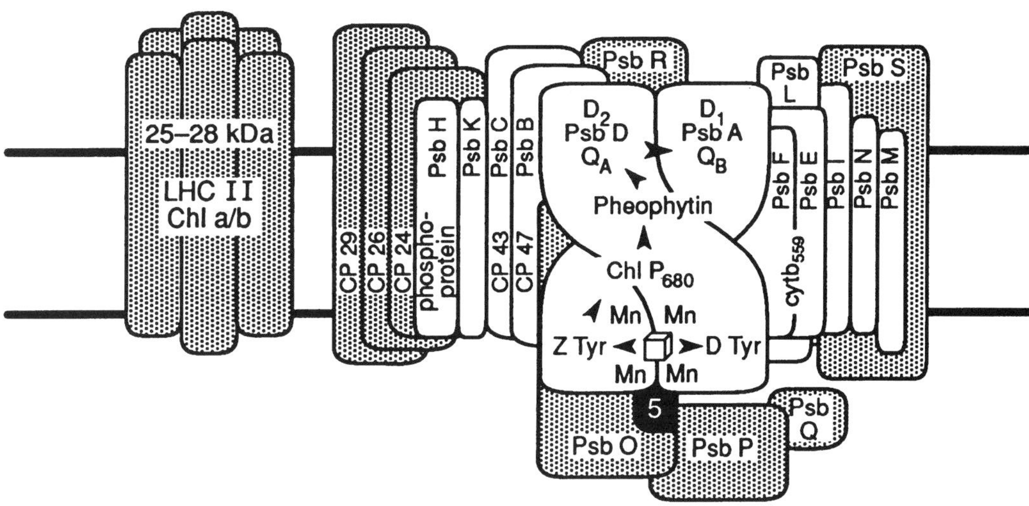

Fig. 1. Biogenesis of the photosystem II complex. The diagram illustrates the overall structure of the photosystem II complex from higher plants, together with the origins of the component polypeptides. Nuclear-encoded polypeptides are shaded; these include the lumenal 33, 23 and 16 kDa proteins of the oxygen-evolving complex (the *PsbO, P* and *Q* gene products) and the integral light-harvesting chlorophyll a/b-binding protein (denoted LHCII). Other proteins, including the key reaction center proteins (D1 and D2) are encoded and synthesized within the chloroplast.

and Schmidt, 1978). Numerous studies (e.g. van den Broeck et al., 1985) have confirmed that the presequences of stromal proteins are capable of targeting foreign proteins into the chloroplast, and it is therefore clear that most, if not all, of the essential targeting information is located within the presequence.

After synthesis on cytosolic ribosomes, the first recognized step in the import process involves the binding of the precursor protein to protease-sensitive receptors on the chloroplast envelope (Cline et al., 1985). This critical step is known to be highly specific (there is no evidence for significant mis-targeting of chloroplast proteins) but the molecular details are still vague. Interestingly, binding of precursor proteins to chloroplasts requires ATP (Olsen et al., 1989) suggesting that some form of active process is involved. Otherwise, attempts to explain the mode of action of the import receptors have met with little success, in part because the requisite features of the targeting signals have been so difficult to define. Stroma-targeting presequences are almost invariably hydrophilic and enriched in hydroxylated residues, with an N-terminal uncharged region followed by a positively charged section (von Heijne et al., 1989). However, they exhibit essentially no primary sequence similarity and there is no convincing evidence for the presence of significant secondary structure (von Heijne and Nishikawa, 1991). Accordingly, the critical targeting signals are something of an enigma.

A number of attempts have been made to identify the import receptors, with controversial results. Pain et al. (1988) used antiidiotypic antibodies to analyze the structures of the receptor molecules, and reported that a 36 kDa envelope protein was involved in binding the precursor of Rubisco small subunit. Cloning of this protein, however, showed that it was homologous to the phosphate translocator in the inner envelope membrane (Flügge et al, 1989; Schnell et al, 1990) and Flügge et al. (1991) have argued that this is the only function of this protein. Very recently, Perry and Keegstra (1994) have used a cross-linking approach to examine the early events in the import process, and showed that the precursor of Rubisco small subunit interacts with two proteins of about 85 and 75 kDa. Further studies, however, are required to confirm whether either (or both) of these proteins functions as an import receptor.

Another study (Guera et al., 1993) analyzed the conformation of chimeric precursor proteins following binding to the chloroplast, taking advantage of the fact that the cytosolic protein, dihydrofolate reductase, is known to fold particularly tightly under normal conditions. This study found that a chimeric construct, in which the ferredoxin presequence was linked to dihydrofolate reductase, was largely unfolded after binding to the chloroplast (the construct was unusually sensitive to proteolysis), suggesting that precursor proteins may be acted upon by a powerful 'unfoldase' activity prior to transport into the stroma. It is of course possible that the putative unfoldase activity *is* the receptor, and that unfolding is an integral part of the binding process; this intriguing possibility will hopefully be resolved in the near future.

The next major step on the import pathway involves the translocation of the bound precursor protein across the two envelope membranes. For many years it was thought likely that this process would take place at 'contact sites' between the outer and inner envelope membranes. Schnell and Blobel (1993) found that a construct consisting of the Rubisco small subunit presequence linked to Protein A, could be *partially* imported into isolated chloroplasts, showing that, under these conditions, the N-terminus of the construct was located in the stroma (the presequence had been removed) but the C-terminus was still accessible to external proteases. These data strongly suggest that precursor proteins are transported into the stroma at contact sites in an unfolded conformation (the precise extent of unfolding remains to be determined). Other details of the translocation mechanism, however, are scarce, although it is known that ATP is required for the translocation of proteins across the envelope membranes (Theg et al., 1989).

During or shortly after transport across the envelope membranes, the presequences are removed by a metal-dependent stromal processing peptidase, SPP (Robinson and Ellis, 1984). This enzyme is highly specific for imported precursors but, as with the chloroplast-targeting signals, there is little definitive information on the signals which are recognized by SPP because the processing sites are similarly devoid of significant primary sequence similarity. Oblong and Lamppa (1992) showed that purified pea SPP gives rise to two bands of about 140 kDa on SDS-PAGE gels, but it is as yet unclear whether the enzyme is in the form of a heterodimer.

There is also evidence that further protein-protein interactions are required before imported proteins are functional in the chloroplast stroma. Studies on the import of ferredoxin-NADP-reductase suggest that the imported protein interacts transiently with two distinct 'molecular chaperones': an Hsp form and Cpn60 (Tsugeki and Nishimura, 1993). Given the role of Cpn60 in assisting the folding of Rubisco large subunit (Hemmingsen et al, 1988) it is likely that Cpn60 is involved in mediating the correct folding of imported stromal proteins, although this has yet to be confirmed. The role of Hsp70 is currently unclear. Figure 2 depicts a current model for the overall import pathway utilized by cytosolically-synthesized proteins, based on the observed protein-protein interactions known to take place during import into the stroma.

2. Insertion of Proteins into the Thylakoid Membrane

Although the thylakoid membrane contains numerous cytosolically-synthesized integral membrane proteins, the vast majority of studies on the insertion of proteins into the thylakoid membrane have focused on a single protein, the light-harvesting chlorophyll-binding protein of PS II (LHCPII). This protein has a molecular mass of about 26 kDa, contains three membrane spanning helices, and is synthesized in the cytosol as a 32 kDa precursor protein. The presequence of this protein resembles that of Rubisco small subunit in overall structural terms, and there is clear evidence that it functions in targeting LHCPII only as far as the stroma; Lamppa (1988) showed that the small subunit presequence could effectively replace that of LHCPII. Signals specifying integration into the thylakoid membrane must, therefore, reside in the mature LHCPII sequence, although the nature of this information is as yet unclear. However, other details of the integration mechanism have emerged following the reconstitution of this event using chloroplast lysates: integration is dependent on the presence of at least one stromal protein factor and ATP (Cline, 1986; Chitnis et al., 1987). In addition, there are indications that the stromal factor may maintain LHCPII in an integration-competent form in the stroma, thereby preventing aggregation of this hydrophobic protein (Payan and Cline, 1991).

Many other integral thylakoid membrane proteins are imported from the cytosol, and the vast majority appear also to be synthesized with stroma targeting sequences (in that there is no indication of the bipartite presequence typically found in front of imported lumenal proteins; see below). Cai et al. (1993) showed by the use of chimeric constructs, that the 20 kDa apoprotein of the CP24 complex, like LHCPII, integrates into the thylakoid membrane by virtue of information contained in the mature protein sequence. However, it is as yet unclear whether the LHCPII import pathway is the 'standard' pathway for integral

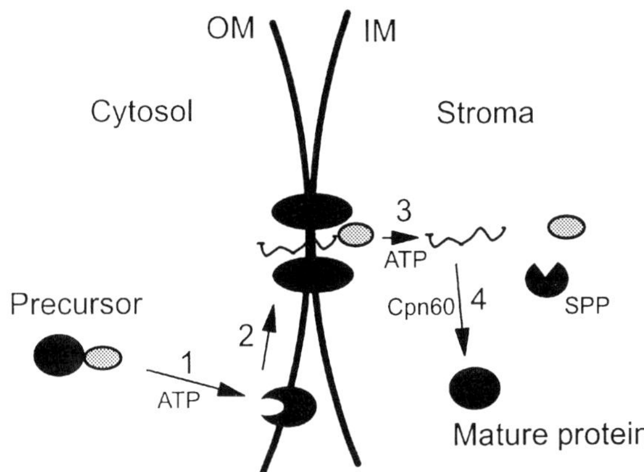

Fig. 2. Import of proteins into the chloroplast stroma. Nuclear-encoded chloroplast proteins are synthesized in the cytosol as precursors containing N-terminal presequences (grey oval). After synthesis, the presequence specifies binding to receptors on the chloroplast surface by an ATP-dependent process (step 1). During or shortly after binding, the precursor is unfolded and transported across the outer and inner envelope membranes (OM, IM) at contact sites between them (steps 2 and 3). Translocation requires stromal ATP, and probably takes place at proteinaceous pores, but details are unclear. Following translocation into the stroma, the presequence is removed by a specific stromal processing peptidase (SPP) and the chaperone molecule Cpn60 is believed to mediate refolding of the imported protein into a native form (step 4).

membrane proteins, because it is not known whether other proteins require stromal factors or ATP for integration into the thylakoid membrane.

B. Insertion of Chloroplast-Encoded Proteins

Although the chloroplast genome encodes about 20% of the total chloroplast protein complement, a high percentage of these proteins are constituents of thylakoid membrane photosynthetic complexes. Unfortunately, relatively little is known about the mechanisms by which these proteins are inserted into the thylakoid membrane, primarily because most expression studies have been carried out using intact chloroplasts, which for obvious reasons are of limited use for the analysis of insertion mechanisms. It is widely assumed, though not proven, that chloroplast-encoded proteins are inserted into the thylakoid membrane cotranslationally, and the insertion mechanisms may therefore differ fundamentally from those utilized by cytosolically-synthesized proteins. However, no information is currently available on the energetics of the insertion process or the role, if any, played by stromal proteins.

Further work is also required to elucidate the roles, if any, of the presequences that are found on a variety of chloroplast-encoded proteins. In the case of cytochrome *f*, which is synthesized with a 35-residue presequence in wheat and spinach (Willey et al, 1984; Alt and Herrmann, 1984), it is extremely likely that the presequence acts as a targeting signal directing partial translocation across the thylakoid membrane. These presequences resemble bacterial export, or 'signal' sequences, and it seems likely that pre-cytochrome *f* is transported into the thylakoid membrane by one of the mechanisms utilized for cytosolically-synthesized lumenal proteins (see below). However, other plastid-encoded proteins, for example subunits I and IV of the CF_0/CF_1 ATP synthase (Henning and Herrmann, 1986) are synthesized with short, hydrophilic presequences, completely unlike signal sequences in structural terms. The significance of these peptide extensions is unclear and it is also unclear whether these extensions are removed by specific processing enzymes.

IV. Translocation of Lumenal Proteins Across the Thylakoid Membrane

A. The Two-step Import Pathway for Nuclear-Encoded Lumenal Proteins

Relatively few thylakoid lumen proteins have been identified and characterized, and the list in higher plants currently consists of plastocyanin (PC), the 33, 23 and 16 kDa proteins of the PS II oxygen-evolving complex (33K, 23K, 16K), and PS I subunits N and F. Most biochemical studies on the import of lumenal proteins have focused on PC and the three PS II proteins. A two-step import pathway has emerged from a combination of in vitro import and processing experiments. The first phase of the import pathway involves the synthesis of the proteins as larger precursors and their import into the stroma, where they are processed to intermediate forms by SPP. Thereafter, the intermediates are transferred across the thylakoid membrane and processed to the mature size by a thylakoid processing peptidase, TPP (Hageman et al, 1986; Smeekens et al, 1986; James et al, 1989).

As might be expected from such an import pathway, the presequences of lumenal proteins contain two targeting signals in tandem, specifying the experimentally separable processes of 'envelope transfer' and 'thylakoid transfer' (Ko and Cashmore, 1989; Hageman et al, 1990). The envelope transfer signals are structurally and functionally equivalent to stroma-targeting presequence, being basic, hydrophilic and rich in serine and threonine. The characteristics of the second, 'thylakoid transfer' signals are very different. In each case, the signal contains a hydrophobic core region and short-chain residues, usually Ala, at the -3 and -1 positions, relative to the TPP cleavage site (von Heijne et al, 1989). These features are very reminiscent of 'signal' peptides that direct protein export in prokaryotes by the *sec*-dependent mechanism (Webb and Sherman, 1994), and the 33K thylakoid transfer signal has indeed been shown to be capable of directing protein export in *Escherichia coli* (Seidler and Michel, 1990). The reaction specificities of pea TPP and *E. coli* signal peptidase are also virtually identical (Halpin et al, 1989). On the basis of this series of observations it was widely assumed that the thylakoidal protein import mechanism was derived from a *sec*-type protein export mechanism in an ancestral, cyano-bacterial-type progenitor of the chloroplast.

B. Two Distinct Mechanisms for Thylakoid Protein Translocation

Although the studies outlined above convincingly demonstrated a two-step import pathway for lumenal proteins, none of these studies analyzed in detail the translocation of proteins across the thylakoid membrane, primarily because the in vitro import assays were carried out using intact chloroplasts. Subsequent studies were carried out using assays for the import of protein by isolated thylakoids, initially using an ATP-driven assay for the import of 33K by pea thylakoids (Kirwin et al., 1989) and subsequently using light-driven assays in which efficient import of all four of the above-mentioned lumenal proteins was demonstrated (Mould et al., 1991; Cline et al., 1992; Klösgen et al., 1992). These studies, together with more recent work, have unexpectedly shown that lumenal proteins fall into two clear groups in terms of their requirements for translocation across the thylakoid membrane: 23K and 16K are translocated by an apparently simple mechanism which is absolutely dependent on the thylakoid vesicle ΔpH, but which does not require the presence of soluble stromal factors or ATP (Mould and Robinson, 1991; Mould et al., 1991; Cline et al., 1992; Klösgen et al., 1992). In complete contrast, the translocation of 33K and PC is completely dependent on the presence of at least one soluble stromal protein and ATP, whereas the ΔpH is not a prerequisite (Hulford et al; 1994; Robinson et al, 1994).

Two different approaches indicate that these results probably reflect the operation of two distinct protein translocases in the thylakoid membrane. In the first, Cline et al. (1993) showed that in competition experiments using saturating concentrations of *E. coli*-expressed precursor proteins, 23K competed with 16K for translocation across the thylakoid membrane, and 33K competed with PC, but the two groups did not compete with each other. In a different study, Robinson et al. (1994) examined the translocation across the thylakoid membrane of a fusion protein, consisting of the 23K presequence linked to PC mature protein. They found that the construct behaved exactly as pre-23K in that translocation across the thylakoid membrane required a ΔpH, but not ATP or stromal factors, demonstrating that mature PC can be transported across the thylakoid membrane by either of two types of mechanism, depending on which presequence is present. These results strongly argue in favor of a two-translocase model for thylakoid protein biogenesis, although this has yet to be formally proven.

C. Why Two Translocation Mechanisms?

The most probable explanation for the data described above is that two protein translocation systems operate in the thylakoid membrane, and that the 23K/16K-system recognizes specific signals within a subset of lumenal protein presequences (no data are yet available on the translocation characteristics of the reciprocal types of construct, i.e. 33K/PC presequences in front of 23K or 16K, and so we do not yet know whether the 33K/PC system recognizes signals that are restricted within the presequences of 33K or PC). This finding is completely unexpected because all of the known lumenal proteins are synthesized with thylakoid transfer signals containing apparently similar hydrophobic core domains and TPP cleavage motifs. The transfer signals of 33K, 23K and PC are illustrated in Fig. 3, together with a *possible* structure of the 16K signal (the SPP cleavage site within pre-16K has yet to be identified). There are no genuinely clear-cut differences between the two types of transfer signals, but it is interesting that the charge distribution varies to some degree, with the 23K/16K signals possessing a patch of acidic residues in the N-terminal domain. However, it remains to be determined whether these (or any other) differences are functionally significant.

Assuming that two distinct translocases operate in the thylakoid membrane (this has yet to be formally proven) the critical question is simply: why? There is no obvious advantage to this arrangement, but there are some tantalizing clues as to how it might have arisen. Of the four lumenal proteins analyzed in detail, only two (33K and PC) are present in oxygenic cyanobacteria, where they are almost certainly transported into the thylakoid lumen by a *sec*-type mechanism (Kuwabara et al, 1987; Briggs et al, 1990), which in bacteria requires ATP (for SecA function) and the participation of a soluble factor, SecB (reviewed in Johnson et al, 1992). In chloroplasts the translocation of 33K and PC likewise requires soluble factors and ATP, consistent with the idea that transport is mediated by an ancestral, *sec*-related system. The presence of such a system in chloroplasts has also been inferred from the findings of *secA* and *secB* gene homologs in the plastid genomes of some red and chromophytic algae (Scaramuzzi et al., 1992; Valentin, 1993), and it seems likely that in higher plants the genes have been transferred to the nucleus. Finally, it was recently shown that azide, a potent inhibitor of SecA in bacteria, inhibits the translocation of 33K and PC (but not 23K or 16K) across the thylakoid membrane in chloroplasts (Knott and Robinson, 1994). It must be emphasized that there is as yet no *definite* proof of the involvement of Sec homologs in 33K and PC translocation in chloroplasts, but the circumstantial evidence appears overwhelming. Intriguingly, 23K and 16K are *not* present in cyanobacteria, and it is therefore possible that the evolution of the photosynthetic mechanism in chloroplasts has involved the acquisition of both new lumenal components and a new system to transport them across the thylakoid membrane. It must, however, be emphasized that further work is required to test whether this hypothesis stands up to more rigorous analysis, since only four lumenal proteins have been analyzed in significant detail to date.

D. A Novel Import Pathway for CFo subunit II

As if the complexities of thylakoidal protein insertion/translocation were not bewildering enough, recent studies on the biogenesis of CFo subunit II (CFoII) suggest that this protein is imported into the thylakoids by yet another pathway. Structurally, CFoII is unique in that it is an integral membrane protein (containing a single membrane-spanning section near the N-terminus) that is synthesized with a bipartite presequence apparently similar to those of lumenal proteins (Herrmann et al., 1993). This presequence can act as a lumen-targeting signal, having been shown capable of targeting plastocyanin into the lumen (unpublished data) and it was at first thought likely that CFoII was transported by the thylakoid protein translocation machinery, with the membrane-spanning section acting as a 'stop-transfer' signal to abort the translocation process. Recent studies on the integration of CFoII into isolated thylakoids, however, strongly suggest that this protein is in fact integrated into the thylakoid membrane by a completely novel mechanism. Remarkably, insertion into thylakoids takes place with high efficiency in the complete absence of stromal extracts, ATP or a ΔpH, and the insertion process is not affected by the presence of saturating concentrations of pre-23K (Michl et al., 1994). It is therefore possible that CFoII inserts into the thylakoid membrane by a

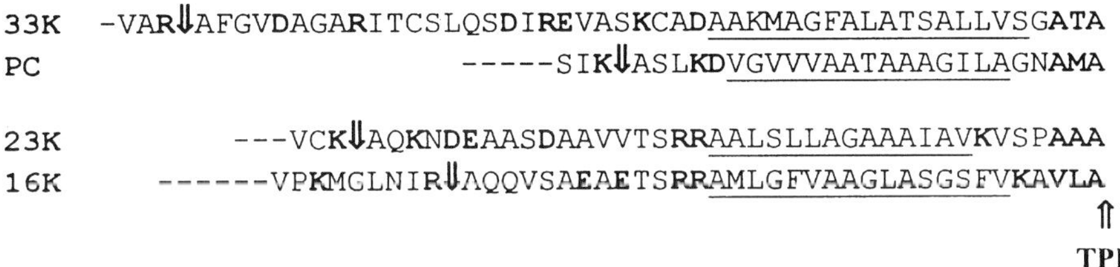

Fig. 3. Bipartite presequences of thylakoid lumen proteins. The figure illustrates the thylakoid transfer signals of wheat 33K, wheat 23K and spinach PC; the SPP cleavage sites are denoted by ⇓. The C-terminal section of the spinach 16K presequence is also given, together with a putative SPP cleavage site. H-regions are underlined, and charged residues and the TPP cleavage motif (final three residues) are given in bold.

spontaneous mechanism, and such a mechanism has been proposed on the basis of the observed structural similarities between CFoII and M13 procoat, which is known to insert into bacterial membranes by a spontaneous mechanism. The bipartite presequence of CFoII is thus believed to function in a completely different way than those of the known lumenal proteins. It will be interesting to determine whether any other thylakoid membrane proteins insert in a similar manner. Figure 4 depicts the known pathways for imported thylakoid proteins which have been elucidated, at least in part, using the in vitro import/ integration assays described above.

V. Concluding Remarks

The development of a variety of in vitro assays for the analysis of thylakoid protein biogenesis has led to an enormous increase in our understanding of the pathways by which these proteins are targeted into the thylakoid membrane. And there is no doubt about the main 'take-home' message from these studies, which is that the protein traffic involved in thylakoidal protein insertion/translocation is complex in the extreme. At least four apparently distinct pathways have been identified already, and there is a real likelihood that others will be elucidated in the near future, since only two integral membrane proteins have been studied in detail so far (LHCPII and CFoII) and the insertion mechanisms for these proteins could hardly be more different. It also appears very likely that chloroplast-encoded proteins may be cotranslationally inserted into the thylakoid membrane by additional mechanisms. A clue to one possible mechanism has emerged in recent work by Franklin and Hoffman (1993) who cloned a nuclear-encoded chloroplast homolog to the 54 kDa protein of the signal recognition particle involved in protein translocation across the endoplasmic reticulum and bacterial plasma membranes. The signal recognition particle is a ribonucleoprotein complex which, at least in eukaryotic systems, functions to arrest the translation of nascent chains until the ER translocation

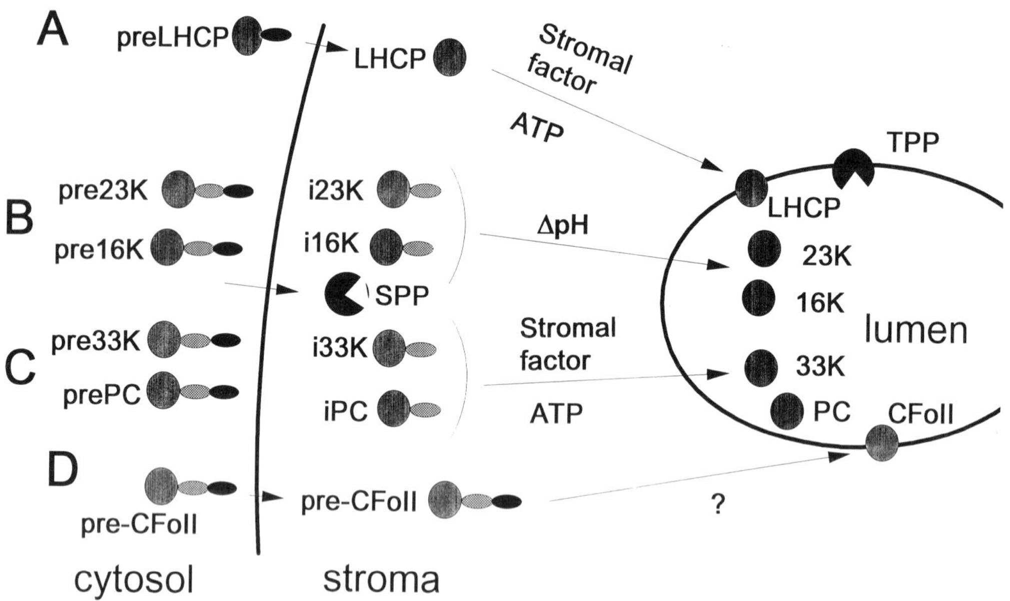

Fig. 4. Multiple pathways of thylakoidal protein insertion/translocation. All of the proteins shown are synthesized in the cytosol with N-terminal 'envelope transit' signals (black ovals), after which translocation across the envelopes probably occurs by a common mechanism. Integration of LHCPII into the thylakoid membrane (pathway A) is then mediated by information in the mature protein, and takes place by a mechanism involving the participation of stromal factors and ATP. Proteins destined for the thylakoid lumen are targeted across the thylakoid membrane by means of cleavable thylakoid transfer signals (grey ovals); translocation takes place by two distinct pathways (B and C) with differing requirements as shown. The integral membrane protein, CFoII (pathway D) is synthesized with a bipartite presequence but the lack of any identifiable integration requirements suggests that this protein may be integrated by a spontaneous mechanism.

machinery recognizes the N-terminal presequence and initiates transport across the membrane. At present the function of the chloroplast homolog is unclear, but there is a good possibility that it might function in the cotranslational insertion of proteins into the thylakoid membrane. It is also worth pointing out as an aside that the 54 kDa protein is unlikely to function in the absence of an RNA component, in which case such an RNA molecule is probably imported from the cytosol.

Despite the documented successes in unraveling the overall pathways taken by imported thylakoidal proteins, it has to be emphasized that the integration/translocation pathways are, for the most part, understood only in relatively superficial detail. None of the components that mediate these processes (for example, stromal factors, or the components of the translocation systems) have been purified or characterized in real detail, and this type of information is obviously required for a comprehensive understanding of the underlying mechanisms. This type of information is also particularly important for an understanding of the evolution of at least one of the thylakoidal protein translocation mechanisms, since there has so far been only supportive evidence, rather than conclusive proof, for the involvement of a *sec*-related mechanism in higher plant chloroplasts. In the same vein, it may also be worth keeping an eye out for a possible ΔpH-dependent, *sec*-independent mechanism for protein targeting into thylakoids in cyanobacteria, since it is by no means certain that this mechanism appeared relatively recently along with the 'newer' 23 kDa and 16 kDa proteins of the PS II oxygen-evolving complex.

References

Alt J and Herrmann RG (1984) Nucleotide sequence of the gene for pre-apocytochrome *f* in the spinach plastid genome. Curr Genet 8: 551–557

Briggs LM, Pecoraro VL, and McIntosh L (1990) Copper-induced expression, cloning and regulatory studies of the plastocyanin gene from the cyanobacterium *Synechocystis* sp. PCC 6803. Plant Mol Biol 15: 633–642

Cai D, Herrmann RG and Klösgen, RB (1993) The 20 kDa apoprotein of the CP24 complex of photosystem II: An alternative model to study import and intra-organellar routing of nuclear-encoded thylakoid proteins. Plant J 3: 383–392

Chitnis, PR, Nechushtai R and Thornber JP (1987) Insertion of the precursor of the light-harvesting chlorophyll *a/b* protein into the thylakoids requires the presence of a developmentally regulated stromal factor. Plant Mol Biol 10: 3–11

Chua N-H and Schmidt GW (1978) Post-translational transport into intact chloroplasts of a precursor to the small subunit of ribulose-1,5-bisphosphate carboxylase. Proc Natl Acad Sci USA 75: 6110–6114

Cline K (1986) Import of proteins into chloroplasts: Membrane integration of a thylakoid precursor protein reconstituted in chloroplast lysates. J Biol Chem 261: 14804–14810

Cline K, Werner-Washburne M, Lubben TH and Keegstra K (1985) Precursors to two nuclear-encoded chloroplast proteins bind to the outer envelope membrane before being imported into chloroplasts. J Biol Chem 260: 3691–3696

Cline K, Ettinger WF and Theg SM (1992) Protein-specific energy requirements for protein transport across or into thylakoid membranes. Two lumenal proteins are transported in the absence of ATP. J Biol Chem 267: 2688–2696

Cline K, Henry R, Li C and Yuan J (1993) Multiple pathways for protein transport into or across the thylakoid membrane. EMBO J 12: 1405–1412

Flügge U-I, Fischer K, Gross A, Sebald W, Lottspeich F and Eckerskorn C (1989) The triose phosphate-3-phosphoglycerate-phosphate translocator from spinach chloroplasts: Nucleotide sequence of a full-length cDNA clone and import of the in vitro synthesized precursor protein into chloroplasts. EMBO J 8: 39–46

Flügge U-I, Weber A, Fischer K, Lottspeich F, Eckerskorn C, Waegemann K and Soll J (1991) The major chloroplast envelope polypeptide is the phosphate translocator and not the protein import receptor. Nature 353: 364–367

Franklin AE and Hoffman NE (1993) Chloroplasts contain a homolog of the 54 kD subunit of the signal recognition particle. J Biol Chem 268: 22175–22180

Guera A, America T, van Waas M and Weisbeek P (1993) A strong protein unfolding activity is associated with the binding of precursor chloroplast proteins to chloroplast envelopes. Plant Mol Biol 23: 309–324

Hageman J, Robinson C, Smeekens S and Weisbeek P (1986) A thylakoid processing protease is required for complete maturation of the lumen protein plastocyanin. Nature 324: 567–569

Hageman J, Baecke C, Ebskamp M, Pilon R, Smeekens S and Weisbeek P (1990) Protein import into and sorting inside the chloroplast are independent processes. Plant Cell 2: 479–494

Halpin C, Elderfield PD, James HE, Zimmermann R, Dunbar B and Robinson C (1989) The reaction specificities of the thylakoidal processing peptidase and *Escherichia coli* leader peptidase are identical. EMBO J 8: 3917–3921

Hemmingsen SM, Woolford C, van der Vies SM, Tilly K, Dennis DT, Georgopoulos C, Hendrix RW, and Ellis RJ (1988) Homologous plant and bacterial proteins chaperone oligomeric protein assembly. Nature 339: 483–488

Hennig J and Herrmann RG (1986) Chloroplast ATP synthase of spinach contains nine nonidentical subunit species six of which are encoded by plastid chromosomes in two operons in a phylogenetically conserved arrangement. Mol Gen Genet 203: 117–128

Herrmann RG, Steppuhn J, Herrmann GS, and Nelson N (1993) The nuclear-encoded polypeptide CFoII from spinach is a real ninth subunit of chloroplast ATP synthase. FEBS Lett 326: 192–198

Highfield PE and Ellis RJ (1978) Synthesis and transport of the small subunit of chloroplast ribulose bisphosphate carboxylase. Nature 271: 420–424

Hulford A, Hazell L, Mould RM, and Robinson C (1994) Two distinct mechanisms for the translocation of proteins across the thylakoid membrane, one requiring the presence of a stromal protein factor and ATP. J Biol Chem 269: 3251–3256

James HE, Bartling D, Musgrove JE, Kirwin PM Herrmann RG and Robinson C (1989) Transport of proteins into chloroplasts: Import and maturation of precursors to the 33- 23- and 16-kDa proteins of the photosynthetic oxygen-evolving complex. J Biol Chem 264: 19573–19576

Johnson K, Murphy CK and Beckwith J (1992) Protein export in *E. coli*. Current Opinion in Biotechnology 3: 481–485

Kirwin PM, Meadows JW, Shackleton JB, Musgrove JE, Elderfield PD, Mould R, Hay NA and Robinson C (1989) ATP-dependent import of a lumenal protein by isolated thylakoid vesicles. EMBO J 8: 2251–2255

Klösgen RB, Brock IW, Herrmann RG and Robinson C (1992) Proton gradient-driven import of the 16kDa oxygen-evolving complex protein as the full precursor protein by isolated thylakoids. Plant Mol Biol 18: 1031–1034

Knott TG and Robinson C (1994) The SecA inhibitor azide reversibly blocks the translocation of a subset of lumenal proteins across the thylakoid membrane. J Biol Chem 269: 7843–7846

Ko K and Cashmore AR (1989) Targeting of proteins to the thylakoid lumen by the bipartite transit peptide of the 33kDa oxygen-evolving protein. EMBO J 8: 3187–3194

Kuwabara T, Reddy KJ and Sherman LA (1987) Nucleotide sequence of the gene from the cyanobacterium *Anacystis nidulans* R2 encoding the Mn-stabilising protein involved in photosystem II water oxidation. Proc Natl Acad Sci USA 84: 8230–8234

Lamppa GK (1988) The chlorophyll *a/b*-binding protein inserts into the thylakoids independent of its cognate transit peptide. J Biol Chem 263: 14996–14999

Michl D, Robinson C, Shackleton JB, Herrmann RG and Klösgen RB (1994) Targeting of proteins to thylakoids by bipartite presequences: CF_0II is imported by a novel third pathway. EMBO J 13: 1310–1317

Mould RM and Robinson C (1991) A proton gradient is required for the transport of two lumenal oxygen-evolving proteins across the thylakoid membrane. J Biol Chem 266: 12189–12193

Mould RM, Shackleton JB and Robinson C (1991) Transport of proteins into chloroplasts: Requirements for the efficient import of two lumenal oxygen-evolving complex proteins into isolated thylakoids. J Biol Chem 266: 17286–17289

Oblong JE and Lamppa GK (1992) Identification of two structurally related proteins involved in proteolytic processing of precursors targeted to the chloroplast. EMBO J 11: 4401–4409

Olsen LJ, Theg SM, Selman BR and Keegstra K (1989) ATP is required for the binding of precursor proteins to chloroplasts. J Biol Chem 264: 6724–6729

Pain D, Kanwar YS and Blobel G (1988) Identification of a receptor for protein import into chloroplasts and its localisation to envelope contact zones. Nature 331: 232–237

Payan LA and Cline K (1991) A stromal protein factor maintains the solubility and insertion competence of an imported thylakoid membrane protein. J Cell Biol 112: 603–613

Perry SE and Keegstra K (1994) Envelope membrane proteins that interact with chloroplastic precursor proteins. Plant Cell 6: 93–105

Robinson C and Ellis R J (1984) Transport of proteins into chloroplasts: Partial purification of a chloroplast protease involved in the processing of imported precursor polypeptides. Eur J Biochem 142: 337–342

Robinson C, Cai D, Hulford A, Hazell L, Michl D, Brock IW, Herrmann RG and Klösgen RB (1994) The presequence of a chimeric construct dictates which of two mechanisms is utilised for translocation across the thylakoid membrane: Evidence for the existence of two distinct translocation systems. EMBO J 13: 279–285

Scaramuzzi CD, Hiller RG and Stokes HW (1992) Identification of a chloroplast-encoded *secA* gene homologue in a chromophytic alga: Possible role in chloroplast protein translocation. Curr Genet 22: 421–428

Schnell DJ and Blobel G (1993) Identification of intermediates in the pathway of protein import into chloroplasts and their localisation to envelope contact sites. J Cell Biol 120: 103–115

Schnell DJ, Blobel G and Pain D (1990) The chloroplast import receptor is an integral membrane protein of chloroplast envelope contact sites. J Cell Biol 111: 1825–1838

Seidler A and Michel H (1990) Expression in *Escherichia coli* of the *psbO* gene encoding the 33 kd protein of the oxygen-evolving complex from spinach. EMBO J 9: 1743–1748

Smeekens S, Bauerle C, Hageman J, Keegstra K and Weisbeek P (1986) The role of the transit peptide in the routing of precursors toward different chloroplast compartments. Cell 46: 365–375

Theg SM, Bauerle C, Olsen LJ, Selman BR and Keegstra K (1989) Internal ATP is the only energy requirement for the translocation of precursor proteins across chloroplastic membranes. J Biol Chem 264: 6730–6736

Tsugeki R and Nishimura M (1993) Interaction of homologues of Hsp70 and Cpn60 with ferredoxin-NADP+ reductase upon its import into chloroplasts. FEBS Letts 320: 198–202

Valentin K (1993) SecA is plastid-encoded in a red alga: Implications for the evolution of plastid genomes and the thylakoid protein import apparatus. Mol Gen Genet 236: 245

Van den Broeck G, Timko MP, Kausch AP, Cashmore AR, Montagu MV and Herrera-Estrella L (1985) Targeting of a foreign protein to chloroplasts by fusion to the transit peptide from the small subunit of ribulose-1,5-bisphosphate carboxylase. Nature 313: 358–363

Von Heijne G and Nishikawa K (1991) Chloroplast transit peptides—the perfect random coil? FEBS Letts 278: 1–3

Von Heijne G, Steppuhn J and Herrmann R G (1989) Domain structure of mitochondrial and chloroplast targeting peptides. Eur J Biochem 180: 535–545

Webb R and Sherman LA (1994) The cyanobacterial heat-shock response and the molecular chaperones. In: Bryant DA (ed) The Molecular Biology of Cyanobacteria, pp 751–767. Kluwer Academic Publishers, Dordrecht

Willey DL, Howe CJ, Auffret AD, Bowman CM, Dyer TA and Gray JC (1984) Location and nucleotide sequence of the gene for cytochrome *f* in wheat chloroplast DNA. Mol Gen Genet 194: 416–422

Chapter 8

Introduction to Oxygen Evolution and the Oxygen-Evolving Complex

Terry M. Bricker
Dept. of Microbiology, Louisiana State University, Baton Rouge, LA 70803, USA.

Demetrios F. Ghanotakis
Department of Chemistry, University of Crete Iraklion, Crete, Greece

Summary

In this chapter, we will introduce Photosystem II and photosynthetic oxygen evolution. In this highly endergonic reaction, light energy is used to extract electrons from water, with the concomitant production of molecular oxygen, and to reduce plastoquinone to plastoquinol. The electron transport processes occurring during these reactions will be reviewed. The properties of the various subchloroplastic preparations which evolve oxygen will be discussed and summarized. The intrinsic proteins, which are required for oxygen evolution, and the extrinsic proteins, which act as enhancers of this reaction, will be examined from both structural and functional perspectives. Additionally, the roles of the cofactors associated with the photosystem (manganese, calcium and chloride) in the formation of the active site of the oxygen-evolving complex will be discussed. Finally, possible mechanisms for water oxidation will also be examined.

Donald R. Ort and Charles F. Yocum (eds): Oxygenic Photosynthesis: The Light Reactions, pp. 113–136.

I. Introduction

Photosystem II (PS II) functions as a light-driven, water-plastoquinone oxidoreductase. This enzyme catalyzes one of the most important and interesting biochemical reactions known, the oxidation of water to molecular oxygen. This reaction is important, because all of the oxygen in the earth's atmosphere is generated by this process, and interesting because of its extremely unfavorable energetics and apparently unique mechanism. PS II is a multi-protein complex which consists of three parts. Light energy is absorbed by light harvesting complexes that contain most of the pigments associated with PS II. Excitation energy is transferred from this antenna to the 'core' of the PS II complex where the primary photochemistry leading to charge separation takes place (Ghanotakis and Yocum, 1990). Finally, the photochemical reactions result in the accumulation of oxidizing equivalents in the oxygen-evolving complex; four oxidizing equivalents are used to convert two molecules of water into oxygen (Debus, 1992). In this chapter we will present our views of the current state of understanding of the structural and functional aspects of this important protein complex. This review is not intended to be encyclopedic; additional chapters in this volume will cover most of these topics in far more detail.

The electron transfer events occurring in PS II are summarized in Fig. 1. In plants, algae and cyanobacteria, light is trapped by and funneled through light-harvesting pigment-protein complexes (light-harvesting complex II or phycobilisomes) to the reaction center of PS II, which contains a special dimeric or possibly monomeric chlorophyll, P680 (Doring et al., 1969). After excitation, P680 is photooxidized and donates an electron to the primary acceptor of PS II, a protein-bound pheophytin (Klimov et al., 1980). This charge separation is stabilized by the transfer of an electron to Q_A and then to Q_B, protein-bound plastoquinones. The accumulation of two reducing equivalents on Q_B

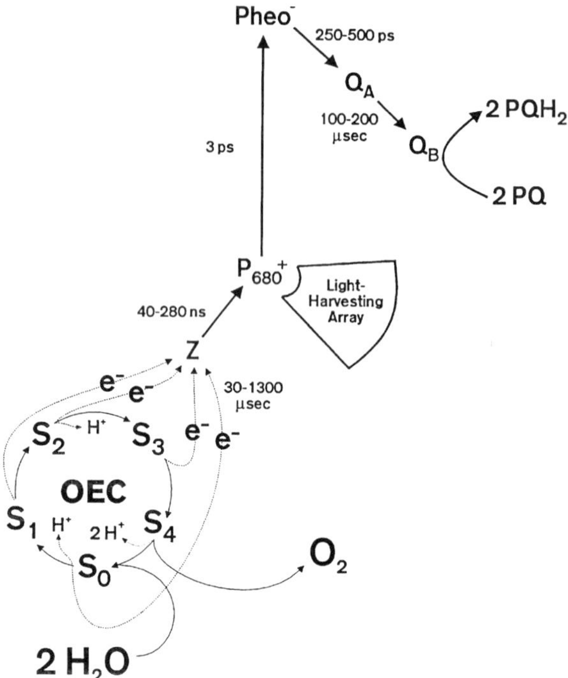

Fig. 1. Photosynthetic Electron Transport in Photosystem II.

leads to the formation of plastoquinol, which is released from PS II. $P680^+$ is reduced by TyrZ, the [161]Tyr of the D1 protein (Debus et al., 1988a and 1988b). $TyrZ^+$ gives rise to the EPR Signal II_{vf}, which exhibits ms decay kinetics (Babcock et al., 1976). The subsequent reduction of $TyrZ^+$ leads to the accumulation of an oxidizing equivalent in the oxygen-evolving complex. The accumulation of four oxidizing equivalents, in a manner consistent with the observed S-state transitions (Joliot et al., 1969; Kok et al., 1970), leads to the release of oxygen from the complex. During the accumulation of oxidizing equivalents, protons are released to the lumenal space of the thylakoid membrane (Saphon and Crofts, 1977; see also Chapter 10).

A number of ionic cofactors are associated with the oxygen-evolving complex. Several lines of evidence indicate that four protein-bound manganese ions are involved in the oxygen evolution reactions. These ions apparently function at the active site of the oxygen-evolving complex and probably exhibit a tetranuclear, trinuclear + one, or 'dimer of dimers' architecture (see Debus, 1992, for an in-depth discussion.). Two of these ions are chloride-sensitive and are released from the oxygen-evolving site in the absence of the extrinsic 33 kDa protein and high

Abbreviations: DTSP – dithiobis(succinimidyl propionate); EDC – 1-ethyl-3-[3-(dimethylamino)-propyl]carbodiimide; EPR – Electron Paramagnetic Resonance; ESEEM – Electron Spin Echo Envelope Modulation; EXAFS – Extended X-ray Absorption Fine Structure; LHCP – Light Harvesting Chlorophyll-protein; NHS-biotin – N-hydrosuccinimidobiotin; PS II – Photosystem II; Tris – Tris(hydroxymethyl)aminomethane; TyrD – [160]Tyr of the D2 protein; TyrZ – [161]Tyr of the D1 protein; XANES – X-ray Absorption Near Edge Structure

concentrations of chloride (>200 mM) (Kuwabara et al., 1985). The other two manganese ions are chloride-insensitive and are released by treatment with alkaline Tris or millimolar concentrations of hydroxylamine (Miyao and Inoue, 1991). The structural significance of this apparent heterogeneity in the manganese cluster is unclear. Additionally, two to three calcium ions are associated with PS II with one high affinity site and one to two sites with lower affinities (Boussac and Rutherford, 1988a,b). There is some evidence that the S-state transitions modulate the binding of calcium to these lower affinity sites. Finally, a number of chloride ions are also required for oxygen evolution (Homann, 1987). These either act as bridging ligands within the manganese cluster (Sandusky and Yocum, 1983) or screen positive charge equivalents on reaction center proteins (Coleman and Govindjee, 1987). Recently, EXAFS data indicated that there may be one chloride ion within the coordination sphere of manganese (Yachandra et al., 1993a). A more in-depth discussion of these inorganic cofactors is given in Section III of this chapter.

Comparisons of PS II with the elegant crystal structure of the purple bacterial reaction center (Deisenhofer et al., 1985) led to the formation of a number of hypotheses concerning the structure of PS II. Amino acid similarities between the L and M subunits of the bacterial reaction center and the D1 and D2 proteins of higher plant and cyanobacterial PS II led Trebst and Depka (1985) to suggest that D1 and D2 proteins bear the reaction center of PS II. Subsequently, a putative PS II reaction center complex was isolated from higher plant PS II membranes, by solubilization with Triton X-100, which contained the D1 and D2 proteins along with cytochrome *b*-559 (Nanba and Satoh, 1987) and the 4.8 kDa *psbI* gene product (Ikeuchi et al., 1989). This complex exhibited the spin-polarized triplet associated with PS II charge recombination (Okamura et al., 1987), pheophytin$^-$ formation (Danielius et al., 1987) and partial chain electron transport from diphenylcarbazide to silicomolybdate (Chapman et al., 1988). A similar complex was isolated using the detergent dodecyl maltoside and the chaotropic agent lithium perchlorate (Ghanotakis et al., 1989). The model that, in vivo, TyrZ, P680, pheophytin, Q_A and Q_B are located within the D1-D2-cytochrome *b*-559 complex is widely accepted (see Chapter 11). However, this core complex is severely damaged on both the oxidizing and reducing sides of PS II. It does not evolve oxygen, does not exhibit the EPR Signals II_s or II_f and contains variable (usually low) amounts of bound quinone. It is not clear whether the disruption of the reducing side is due to the relatively harsh conditions required for the removal of CP43 and CP47 or to the removal of these proteins. Although CP43 and CP47 function as the inner antennae of the PS II core, the involvement of these species in the regulation of quinone and manganese ion binding in PS II is likely.

The analogy drawn between the purple bacterial reaction center and PS II, while certainly very useful, is greatly oversimplified. Clearly, the purple bacteria do not evolve oxygen. Residues required for the ligation of the manganese cluster, calcium, chloride, and the extrinsic proteins of PS II are absent in this system. The number of chlorophyll molecules associated with the PS II reaction center is larger than the number associated with the purple bacterial reaction center. In the bacterial system, four bacteriochlorophylls are present per two bacteriopheophytins. In higher plant reaction center preparations, six to twelve chlorophylls are present per two pheophytins (Dekker et al., 1989; Kobayashi et al., 1990; Gounaris et al., 1990). These results indicate that the structural organization of the accessory chlorophylls in the vicinity of P680 is quite different from that observed in the bacterial reaction center. For a more detailed discussion of these and other differences between these two protein complexes, see van Mieghem and Rutherford (1993).

II. Photosystem II Preparations

The isolation of oxygen-evolving PS II membranes was a watershed event in the study of PS II (Berthold et al., 1981; Kuwabara and Murata, 1982; Yamamoto et al., 1982). PS II activities and signals, which could previously be examined only in relatively unstable thylakoid preparations and at low signal to noise ratios, can now be examined in very stable membrane preparations that are essentially devoid of PS I and highly depleted of the cytochrome b_6f complex. Using these oxygen-evolving membranes as starting material, numerous PS II submembrane (particle) preparations have been developed. Numerous biochemical and biophysical studies, which have taken advantage of the decreased chlorophyll and polypeptides per PS II reaction center present in these preparations, led to a better understanding of the structural and functional characteristics of this unique enzyme complex.

The biochemical manipulation of PS II membranes resulted in the purification of oxygen-evolving PS II submembrane complexes that are depleted of the main light-harvesting proteins and are referred to as PS II 'cores'. A great deal of effort was spent on isolation of a minimal PS II unit capable of stabilizing the primary charge separation at the reaction center and oxidizing water. Several groups have isolated oxygen-evolving PS II core complexes, most of which are depleted of the 23 and 17 kDa extrinsic polypeptides (Ikeuchi et al., 1985; Ikeuchi and Inoue 1986; Ghanotakis and Yocum 1986; Ghanotakis et al., 1987; Enami et al., 1989; Haag et al., 1990). However, most of these preparations also contain additional polypeptides, such as the 22 and 10 kDa species, which seem not to be required for photosynthetic water oxidation.

Preparations lacking the 23 and 17 kDa polypeptides exhibit high rates of light-induced oxygen evolution only in the presence of exogenous calcium and chloride. The proteins appear to concentrate these inorganic cofactors. The 23 and 17 kDa proteins also play a role in the structural architecture around the manganese complex and their release may result in an altered topology in its vicinity (Ghanotakis et al., 1984b,c,d; Tamura et al., 1986; Mei and Yocum, 1991). Procedures for the isolation of PS II core complexes which contain all three extrinsic proteins were recently reported (Enami et al., 1989; Mishra and Ghanotakis 1994). By using a non-ionic detergent, Mishra and Ghanotakis isolated a PS II core complex that retained the extrinsic 23 and 17 kDa polypeptides, but is depleted of the 22 and 10 kDa proteins present in the complex isolated by the solubilization of PS II membranes with octyl glucopyranoside at elevated ionic strength (Ghanotakis et al., 1987). The 10 and 22 kDa species are also present in the PS II core complexes isolated by Enami et al. (1989) and Ikeuchi et al (1985), who used heptyl thioglucopyranoside and octyl glucopyranoside for the solubilization of the PS II membranes. The 'pure complex' of Ikeuchi and Inoue (1986) is free of the 10 and 22 kDa polypeptides, but is also depleted of the 23 and 17 kDa extrinsic proteins. Bowden et al. (1991) demonstrated that by solubilizing the PS II membranes at pH 7.5, either with octyl glucopyranoside or heptyl thioglucoside, it is possible to remove the 22 and the 10 kDa polypeptides. These preparations, however, are less competent in oxygen evolution than those isolated at lower pH. The most highly purified PS II preparations capable of oxygen evolution consist of the chlorophyll-binding 43 kDa (CP43) and 47 kDa (CP47) polypeptides, two hydrophobic species with molecular weights of 34 and 32 kDa (called 'D2' and 'D1', respectively), cytochrome *b*-559, which consists of two polypeptides with molecular masses of 9 and 4.5 kDa (α and β subunits, respectively), the extrinsic 33 kDa protein, and three other smaller polypeptides (Ikeuchi et al., 1985; Ghanotakis et al., 1987).

III. The Proteins of Photosystem II

A. D1 and D2

The D1 and D2 intrinsic membrane protein components of PS II, which are encoded by the *psbA* and *psbD* genes, respectively, are the most intensively studied proteins in PS II (for a more in-depth discussion, see Chapter 11 of this volume). These proteins form the photochemical core of PS II. Because of the similarities of the primary electron transport processes between PS II and the purple non-sulfur bacteria and the strong local amino acid sequence homologies between these proteins and the L and M subunits of the bacterial reaction center, Trebst and Depka (1985) hypothesized that the D1 and D2 proteins form the reaction center of PS II. Strong support for this hypothesis was provided by the subsequent isolation of a D1-D2-cytochrome *b*-559-psbI core preparation (Nanba and Satoh, 1987; Fotinou and Ghanotakis, 1990). The PsbI protein and cytochrome *b*-559 are removed from this complex by additional detergent treatment to yield a D1-D2 complex capable of carrying out primary charge separation (Tang et al., 1990).

Structurally, the D1 and D2 proteins are very similar and appear to form a heterodimer that exhibits two-fold symmetry. Both of these proteins contain five transmembrane helices. Biochemical evidence supports the depicted topologies for both the D1 (Fig. 2) and the D2 proteins (Fig. 3) (Sayre et al., 1986; Geiger et al, 1987). The mature D1 protein contains 343 amino acid residues and is post-translationally cleaved after ^{344}A. The gene encoding the enzyme responsible for this C-terminal processing was cloned from *Synechocystis* 6803 (Anbudurai et al., 1994). The mature D2 protein is 352 amino acid residues long and is not C-terminally processed. After N-terminal formyl-methionine removal, both

Fig. 2. Topological Model of the D1 Protein. This consensus sequence is based on the thirty-eight species examined by Svensson et al. (1991). Conserved amino acid residues are indicated by their single letter code; open circles, unconserved residues; closed circles, conservatively replaced hydrophobic residues; open squares, conservatively replaced hydrophilic residues; - , D or E; + , R or K.

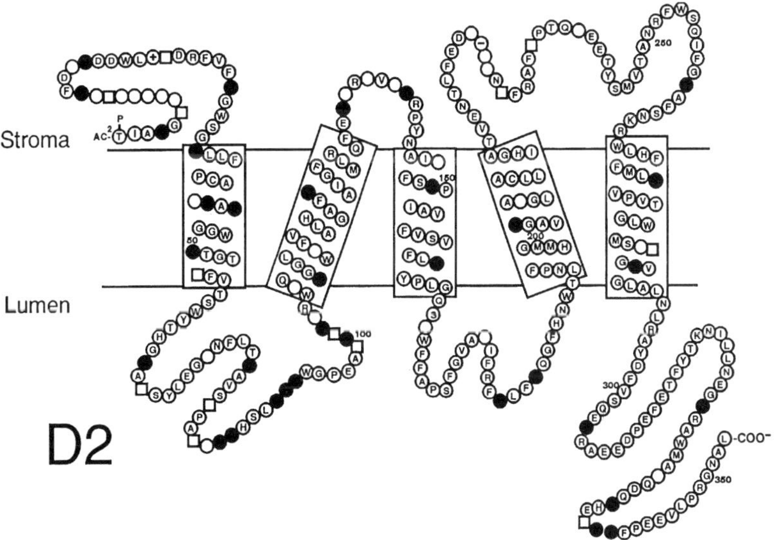

Fig. 3. Topological Model of the D2 Protein. This consensus sequence is based on the fifteen species examined by Svensson et al. (1991). Conserved amino acid residues are indicated by their single letter code; open circles, unconserved residues; closed circles, conservatively replaced hydrophobic residues; open squares, conservatively replaced hydrophilic residues; - , D or E; + , R or K.

proteins are acetylated and phosphorylated at ^{2}T (Michel et al., 1988) (it should be noted that ^{2}T is not conserved in *Euglena gracilis*). The enzymes responsible for these modifications have not been identified and the physiological roles of these alterations are unknown.

The D1-D2 heterodimer is hypothesized to bind a variety of cofactors. These include the chlorophylls of the primary donor P680, the primary acceptor pheophytin, Q_A, the non-heme iron, Q_B, and the radicals TyrZ$^+$ and TyrD$^+$. Four additional chlorophylls, the 'non-photochemical' pheophytin, and one or two β-carotenes also are bound (Kobayashi et al., 1990; Tang et al., 1990). It is also very likely that these proteins contribute ligands to the manganese, calcium and, possibly, chloride associated with the oxygen-evolving site. Additionally, binding sites for the extrinsic 33 kDa protein and possibly other extrinsic proteins of the oxygen-evolving complex may also be present on these reaction center proteins. Comparison of the amino acid sequences of the L and M proteins of the bacterial reaction center with those of the D1 and D2 proteins of PS II, coupled with the bacterial reaction center crystal structure and the evaluation of mutations in the D1 and D2 proteins has allowed the putative identification of binding sites for a number of these cofactors. The chlorophylls of P680 are probably bound to ^{198}H of both the D1 and D2 proteins. Q_A is hypothesized to be bound in a pocket of D2 that includes ^{215}H, ^{261}A and ^{254}W. The non-heme iron is probably bound to ^{215}H and ^{272}H of D1 and ^{215}H and ^{269}H of D2. Q_B is hypothesized to be bound in a pocket on D1 that includes ^{215}H, ^{262}Y, ^{263}A, ^{264}S, ^{255}F and ^{265}F (Trebst, 1986). Candidate residues for the binding of the four so-called 'accessory' chlorophylls in PS II include ^{118}H and ^{190}H of both the D1 and D2 proteins (Tang et al., 1990). Site-directed mutagenesis studies show that TyrZ$^+$, which gives rise to the EPR Signal II_{vf}, is ^{161}Y of D1 (Debus et al., 1988a). Similar studies revealed that TyrD$^+$, which gives rise to the EPR Signal II_s, is ^{160}Y of D2 (Debus et al., 1988b).

Unfortunately, it is much more difficult to identify residues on D1 and D2 that are involved in the binding of cofactors and proteins associated with the oxygen-evolving complex. This is because there are no analogous residues in the bacterial system. Models of the tetrameric manganese cluster (Yachandra et al., 1993b) indicate that at least 24 ligands may be required for cluster formation. Additional ligands are required for the calcium and chloride associated with PS II. Carboxylate and histidyl residues may be prime candidates for ligands within the manganese cluster (see below). Numerous site-directed mutagenesis studies have modified all of the lumenally exposed residues on both the D1 and D2 proteins that could serve as manganese ligands. Only a few changes have led to phenotypes consistent with the disruption of manganese (or calcium) ligation. These include ^{59}D, ^{61}D, ^{65}E, ^{170}D, ^{332}H, ^{337}H, ^{342}D, the terminal carboxylic acid group of ^{344}A of the D1 protein and ^{70}E of the D2 protein (see Debus, 1992, for a more in-depth discussion). Recently, Tang et al. (1994) presented direct ESEEM evidence that a histidyl residue(s) is ligated to manganese, at least in the S_2 state (see Chapter 9). The number of putative ligands from D1 and D2 are insufficient to provide all of the hypothesized ligands in the manganese cluster; other ligands may be provided by water or backbone carbonyls. Intrinsic proteins other than D1 and D2 (e.g. CP47, CP43, cytochrome *b*-559, psbI protein, etc.) may provide at least some manganese/calcium ligands.

B. Cytochrome b-*559*

Cytochrome *b*-559 is one of the most enigmatic components of PS II. This 'redox center without a function' has been proposed to participate in the assembly of PS II, oxygen evolution, protection against photoinhibition, cyclic electron transfer, etc. No unequivocal evidence supporting any of these proposed functions is available (for a more in-depth discussion, see Chapter 13).

Two redox forms of cytochrome *b*-559 are observed. The high potential form has an unusually high redox potential of +370 mV (Bendall, 1968). Most *b*-type cytochromes have redox potentials in the 0 to 100 mV range. The low potential form has a redox potential of 60–80 mV (Fan and Cramer, 1970; Hind and Nakatani, 1970). The low potential form is often associated with conditions or treatments that damage PS II.

Structurally, cytochrome *b*-559 is a novel *b*-type cytochrome apparently consisting of a heterodimer of two subunits: the 9 kDa, α subunit and the 4.5 kDa, β subunit (Fig. 4A). These subunits are encoded by the *psbE* and *psbF* genes, respectively. Each subunit contains a single transmembrane helix and a single histidyl residue. The histidyl residues of both subunits coordinate the heme of the cytochrome. The N-termini of both polypeptides are exposed at the stromal face of the membrane. An alternative hypothesis has been presented suggesting that the cytochrome consists of a pair of homodimers, $(\alpha)_2$ and $(\beta)_2$, each ligating a heme (Pakrasi and Vermaas, 1992).

Cytochrome *b*-559 is required for assembly of functional PS II. Deletion of the *psbE* (Pakrasi et al., 1991) or *psbF* (Pakrasi et al., 1990) genes in *Synechocystis* sp. PCC 6803 leads to the loss of PS II reaction center activity. Additionally, these mutants do not accumulate significant amounts of the D1 and D2 proteins. Site-directed mutagenesis experiments that alter the putative heme ligands ^{22}H in either the α or β subunits also lead to dramatic losses of D1, D2

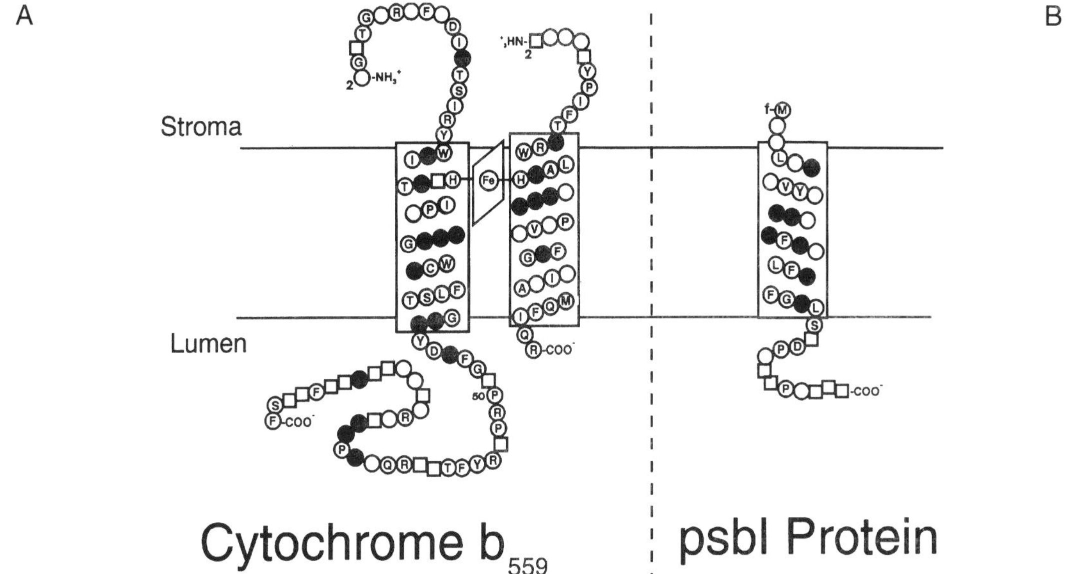

Fig. 4. Topological Model of (A) Cytochrome *b*-559 and (B) *psbI* gene product. The consensus sequence for cytochrome *b*-559 is based on the fifteen sequences examined by Cramer et al. (1993). The consensus sequence for the *psbI* gene product is based on the sequences of liverwort (Ohyama et al., 1986), tobacco (Shinozaki et al., 1986), wheat (Howe et al., 1988), *Anacystis nidulans* 6301 (Chen et al., 1990), rice (Hiratsuka et al., 1989), *Euglena gracilis* (Hallick et al., 1993) and the partial sequences from *Synechococcus vulcanus* (Ikeuchi et al., 1989) and pea (Webber et al., 1989). Conserved amino acid residues are indicated by their single letter code; open circles, unconserved residues; closed circles, conservatively replaced hydrophobic residues; open squares, conservatively replaced hydrophilic residues; - , D or E; + , R or K.

and of both subunits of the cytochrome (Pakrasi et al., 1991). Interestingly, truncation of the carboxyl terminus of the α subunit by 31 amino acid residues (Tae and Cramer, 1992) leads to an 80–90% loss of PS II centers without the loss of assembled cytochrome *b*-559. This result indicates that the C-terminus of the α-subunit is not required for the formation of functional cytochrome but is required for the assembly of functional and stable PS II centers.

C. psbI *Gene Product*

Little is known about the function of the PsbI protein. It is present in PS II core preparations and is removed by additional detergent treatment (Tang et al., 1990). The protein contains 31 amino acid residues and apparently one transmembrane α-helix (Fig. 4B). The N-terminus, which is not processed (Ikeuchi and Inoue, 1988a), is exposed at the stromal surface of the membrane (Tomo et al., 1993). Biochemical removal of this protein does not result in the release of pigments or other PS II cofactors and the resulting D1-D2 complex can carry out primary charge separation (Tang et al., 1990). It is difficult to evaluate the functional significance (if any) of the removal of the PsbI protein since, as noted previously, the D1-D2-cytochrome *b*-559-PsbI complex is functionally seriously impaired. Targeted mutagenesis of the *psbI* gene yields *Synechocystis* strains that assemble PS II and evolve oxygen (Pakrasi, H.B., personal communication).

D. CP47 and CP43

CP47 (CPa-1) and CP43 (CPa-2) are integral membrane protein components of PS II. They are encoded by the *psbB* and *psbC* genes, respectively. These chlorophyll proteins are interior transducers of excitation energy from the light-harvesting pigment proteins (LHCP in plants, the phycobilisomes in cyanobacteria and red algae) to the photochemical reaction center (Fig. 5). Additionally, a significant body of evidence indicates that these proteins possess other important functions in PS II.

CP47 from spinach is 508 amino acid residues long and is highly conserved. In the 15 species for which derived amino acid sequences are available, 80% of the amino acids are conserved or conservatively replaced. Hydropathy plot analysis indicates that this protein contains six putative transmembrane

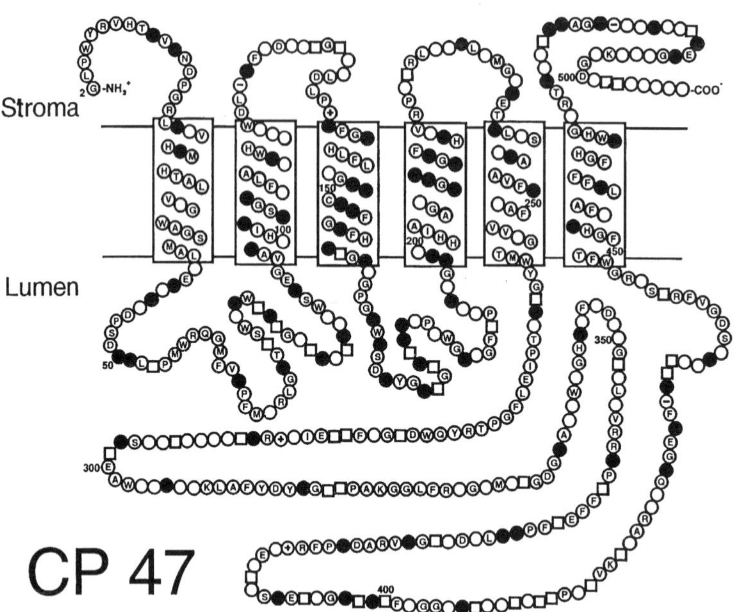

Fig. 5. Topological Model of CP47. The consensus sequence for CP47 is based on the sequences for the four species examined by Bricker (1990) as well as corn (Rock et al., 1987), rye (Bukharov et al., 1988), rice (Hiratsuka et al., 1989), barley (Andreeva et al., 1989), *Euglena gracilis* (Keller et al., 1989), *Anabaena* Sp. PCC 7120 (Lang and Haselkorn, 1989), wheat (Gray et al., 1990), *Oenothera hookeri* (Offermann-Steinhard and Herrmann, 1990), *Prochlorothrix hollandica* (Greer and Golden, 1991), *Chlamydomonas reinhardtii* (Berry-Lowe et al., 1992) and *Synechococcus* sp. PCC 7942 (Kulkarni et al., 1993). Conserved amino acid residues are indicated by their single letter code; open circles, unconserved residues; closed circles, conservatively replaced hydrophobic residues; open squares, conservatively replaced hydrophilic residues; - , D or E; + , R or K.

helices (Vermaas et al., 1987; Bricker, 1990) (see Fig. 5). A large (190 amino acid residue) extrinsic loop is one of the most intriguing structural features of CP47. Crosslinking experiments demonstrated that domains on this loop are easily crosslinked to the 33 kDa extrinsic protein of PS II (see below). This finding indicates that this extrinsic loop is located on the lumenal face of the thylakoid membrane. This result, which is consistent with the von Heijne and Gavel analysis (1988) of this protein sequence, strongly supports the predicted topology shown in Fig. 5.

Structurally, CP43 is very similar to CP47 (Fig. 6). This protein (461 amino acid residues) also appears to possess six transmembrane helices and a large (134 amino acid residue) lumenally exposed loop. It is also highly conserved; 85% of the residues are either identical or conservatively replaced. For CP43, direct evidence is available concerning the topology depicted in Fig. 6 (Sayre and Robel-Boerner, 1994). This immunological study, which examined the effects of proteases on inside-out and right side-out PS II preparations, essentially confirms the predicted topology of CP43.

The distribution of the conserved histidyl residues located in the transmembrane helices is another very interesting structural feature of these proteins. CP47 contains 12 residues while CP43 contains 8. These histidyl residues are clustered near both the stromal and lumenal membrane surfaces to give a distribution reminiscent of the positioning of the conserved histidyl residues in other light-harvesting chlorophyll proteins (Zuber et al., 1987). These residues are prime candidates for chlorophyll axial ligands. In a recent site-directed mutagenesis study, a number of these putative chlorophyll ligands in CP47 were modified to either tyrosyl or asparaginyl residues (Shen et al., 1993). Many of these mutants exhibited modest alterations in light harvesting efficiency and decreased PS II stability. Generally, mutants bearing tyrosyl residue substitutions were more severely affected than those bearing asparaginyl residue substitutions. These results are consistent with the altered histidyl residues being chlorophyll ligands. CP47 contains either 10–12 (Tang and Satoh, 1984; Barbato et al., 1991) or 20–25 bound chlorophyll *a* molecules (deVitry et al., 1984; Yamaguchi et al., 1988). Additional chlorophyll ligands could be

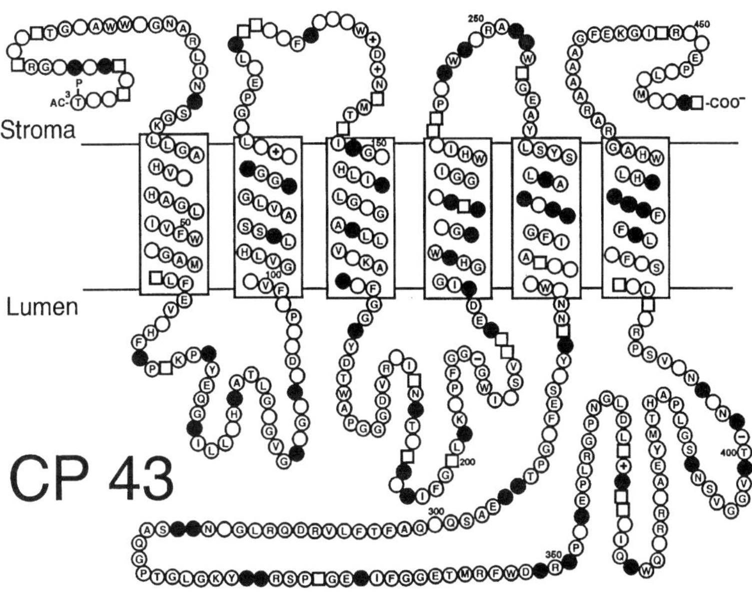

Fig. 6. Topological Model of CP43. The consensus sequence for CP47 is based on the sequences for the ten species examined by Bricker (1990) as well as barley (Reverdatto et al., 1988), rye (Bukharov et al., 1989), rice (Hiratsuka et al., 1989), and *Cyanidium caldarium* (Kessler, et al., 1992). Conserved amino acid residues are indicated by their single letter code; open circles, unconserved residues; closed circles, conservatively replaced hydrophobic residues; open squares, conservatively replaced hydrophilic residues; - , D or E; + , R or K.

provided by asparaginyl and glutamyl residues (Wechsler et al, 1985), carbonyl groups and water molecules (Tronrud et al., 1986). CP43 contains either 15 (Rögner et al., 1991) or 20–25 bound chlorophyll *a* molecules (deVitry et al., 1984).

It should be pointed out that CP43 is post-translationally modified. The first two amino acid residues are removed by a protease and the resulting N-terminal threonyl residue is acetylated and phosphorylated (Michel et al., 1988) (^{3}T, however, is not conserved in *Cyanidium caldarium* (Kessler et al., 1992)). The mechanisms and the physiological significance of these modifications have not been determined, and no similar modifications have been identified in CP47.

In their roles as chlorophyll proteins, CP47 and CP43 transfer excitation energy from the exterior antennae (LHCP or phycobilisomes) to the chlorophylls of the photochemical reaction center of PS II. It is unclear whether the path of excitation energy transfer is linear as shown in Fig. 7A or parallel as shown in Fig. 7B. Several very indirect lines of evidence favor a linear excitation energy transfer model. The 685 and 695 nm 77 K fluorescence emission bands of PS II appear to be associated with CP43 and CP47, respectively (Nakatani et al., 1984; Pakrasi et al., 1985). Excitation energy flows most efficiently from pigments with lower λ to those with higher λ fluorescence bands. CP43 appears to be more loosely associated with the PS II core complex than is CP47. Chaotropic agents (potassium thiocyanate and lithium perchlorate) or additional detergent treatments easily remove CP43, which yields a CP47-D1-D2-cytochrome *b*-559-PsbI complex (Akabori et al., 1988; Yamaguchi et al., 1988; Ghanotakis et al., 1989). Additionally, partially functional (non-oxygen evolving) PS II reaction centers can assemble in the absence of CP43 (Rögner et al., 1991), whereas this has not been observed in the absence of CP47. Finally, CP43 is required for the binding of LHCP to the CP47-D1-D2-Cytochrome *b*-559-PsbI complex and must be present to facilitate excitation energy transfer from the LHCP to the core complex (Bassi et al., 1987).

In addition to its role in energy transduction, numerous lines of evidence indicate that CP47 and CP43 perform other functions in PS II. Both proteins are required for the assembly of oxygen-evolving PS II centers and for the stability of PS II. In *Synechocystis*, insertional or deletion mutagenesis of the *psbB* gene yields a PS II$^-$ phenotype (Vermaas et al., 1988; Eaton-Rye and Vermaas, 1991) with no

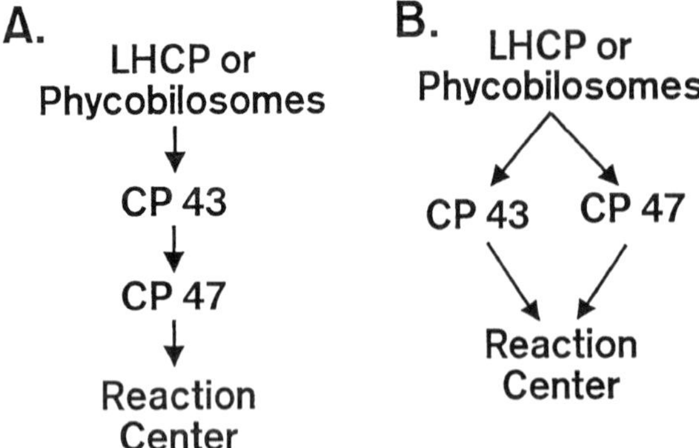

Fig. 7. Possible Routes of Excitation Energy Transfer from the Exterior Antenna to the Reaction Center of Photosystem II. A., linear model and B., parallel model. Indirect evidence, which is by no means conclusive, favors A.

immunologically detectable accumulation of the CP47, D1 or D2 proteins. Site-directed mutants of the *psbB* gene, which do not accumulate CP47, accumulate lower amounts of the D1 and D2 proteins (Shen et·al., 1993). These results led Vermaas and coworkers to suggest that CP47 is required for the assembly/stability of PS II. Insertional or deletion mutagenesis of the *psbC* gene also leads to a loss of functional PS II. Only about 10% of wild type PS II centers accumulate and these centers are not able to evolve oxygen (Rögner et al., 1991). Additionally, eight short deletions (7–11 residues) introduced into the large extrinsic loop of CP43 all lead to a loss of photoautotrophic growth, oxygen evolution, and marked declines in the amounts of other reaction center proteins (D1, D2 and CP47) (Kuhn and Vermaas, 1993).

CP47 is very closely associated with the oxygen-evolving complex of PS II. A monoclonal antibody (FAC2) was isolated that recognizes its antigenic determinant on CP47 only in the absence of the extrinsic 33 kDa protein *and* the chloride insensitive manganese ions associated with the oxygen-evolving site of PS II (Bricker and Frankel, 1987). The epitope for this antibody is in the domain ^{360}P-^{391}S (Frankel and Bricker, 1989), which is located in the large extrinsic loop of CP47.

A variety of protein crosslinkers are capable of crosslinking CP47 to the extrinsic 33 kDa protein in spinach PS II membranes. These include DTSP (Enami et al., 1987; Bricker et al., 1988), 2-iminothiolane (E. Camm, personal communication; C.B. Queirolo and T.M. Bricker, unpublished observations) and the water-soluble carbodiimide EDC (Bricker et al., 1988; Enami et al., 1991). EDC is particularly interesting since it crosslinks amino groups to carboxyl groups that are in van der Waals contact (Hackett and Strittmatter, 1984). Proteins crosslinked by this reagent are assumed to be interacting via a salt bridge. The domains on both the extrinsic 33 kDa protein and CP47 that are crosslinked with EDC have been mapped. The domain ^{364}E-^{440}D of the large extrinsic loop is crosslinked to the N-terminal domain (^{2}E-^{76}K) of the extrinsic 33 kDa protein (Odom and Bricker, 1992).

Additionally, the extrinsic 33 kDa protein shields lysyl residues located on CP47 from labeling with the amino group-modifying reagent NHS-biotin. Treatments that remove the extrinsic 33 kDa protein from spinach PS II membranes allow the specific labeling of CP47 with this reagent (Bricker et al., 1988). Recently, the biotinylated regions on CP47 were mapped (Frankel and Bricker, 1992). Two domains, ^{304}Lys-^{321}Lys and ^{389}Lys-^{419}Lys, both of which lie in the large extrinsic loop of CP47, are the only biotinylatable regions on this protein.

In PS II membranes, the extrinsic 33 kDa protein also protects CP47 from cleavage by trypsin (Bricker and Frankel, 1987). A study by Hayashi et al. (1993) showed that in an oxygen-evolving core preparation from which the extrinsic 33 kDa protein had been removed, cleavage of CP47 at ^{389}K by the endoproteinase Lys-C was strongly correlated with loss of oxygen-evolving capability and loss of the ability to rebind the extrinsic 33 kDa protein (Hayashi et al., 1993).

Finally, site-directed mutagenesis studies within the large extrinsic loop of CP47 indicate a possible role for this protein in water oxidation. Substitution of glycyl residues for the basic residue pair ^{384}R^{385}R in CP47 leads to a 50% loss of oxygen-evolving capability in vivo and a marked decrease in the stability of the oxygen-evolving complex in vitro (Putnam-Evans and Bricker, 1992). This mutation lies within the epitopic domain of the monoclonal antibody FAC2 and in the region which is EDC crosslinked to the extrinsic 33 kDa protein. It is possible that this mutation destabilizes the interaction of CP47 with the 33 kDa extrinsic protein. In another mutant, alteration of ^{448}R to ^{448}G (or ^{448}S, Q. Wu and T. M. Bricker, unpublished observations) yields another very interesting phenotype. Under chloride-limiting conditions (<20 μM), this mutant cannot grow autotrophically and exhibits a virtually complete loss of the ability to evolve oxygen. This defect is correlated with a loss of PS II centers and a decrease in the steady state amounts of D1 protein and CP47. Wild type strains do not exhibit these defects when grown under limiting chloride conditions. This is the first mutant identified in any system which exhibits a defect in the chloride requirement for PS II activity/stability (Putnam-Evans and Bricker, 1994). Deletion mutants within the large extrinsic loop of CP47 also exhibit defects in the oxygen-evolving complex (Eaton-Rye and Vermaas, 1991; Haag et al., 1993; Gleiter et al., 1994). In particular, the mutants Δ(A373-D380) and Δ(R384-V392) do not bind the 33 kDa extrinsic protein as tightly as does wild type; both the B and Q thermoluminescence bands, which arise from S_2 recombination with Q_A^- and Q_B^-, respectively, are shifted to higher temperatures and the oxygen evolution activities of these mutants are susceptible to both dark inactivation and photo-inactivation.

CP43 may also have functions other than light harvesting. As noted previously, all oxygen-evolving preparations isolated to date contain this protein. Biochemical studies originally indicated that the presence of CP43 was required for the functional association of Q_A with PS II (Akabori et al., 1988; Yamaguchi et al., 1988; Petersen et al., 1990). Mutagenesis studies in *Synechocystis* PCC 6803, however, indicate that Q_A is still functional in mutants lacking CP43 (Rögner et al., 1991). The relatively harsh treatments required to remove CP43 in the earlier studies may have damaged the Q_A binding site. Alternatively, Q_A may be more stable in the cyanobacterial system than in higher plants (Barry et al., 1994).

CP43 may function as a serine protease (Salter et al., 1992). Highly purified PS II preparations exhibit D1 degradation upon treatment with photoinhibitory light concentrations. Earlier studies showed that this degradation could be suppressed by inhibitors of serine proteases (Virgin et al., 1991; Shipton and Barber, 1992). Salter et al. (1992) demonstrated that a single 43 kDa band was labeled with radioactive diisopropyl fluorophosphate, a specific serine protease inhibitor, and presented a number of lines of evidence indicating that the radiolabeled band was CP43; however, they did not unequivocally demonstrate this by sequencing the labeled protein.

E. The 33, 24 and 17 kDa Extrinsic Proteins

It has been unequivocally demonstrated that PS II preparations containing only the intrinsic components of PS II evolve oxygen if supplied with high concentrations of calcium and chloride (Bricker, 1992). Thus, the extrinsic proteins of PS II enhance oxygen evolution at physiological concentrations of these inorganic cofactors. The 33 kDa protein, which is encoded by the *PsbO* gene, is found in all oxygenic organisms while the 24 and 17 kDa proteins, encoded by the *PsbP* and *PsbQ* genes, respectively, are absent in the cyanobacteria. See Shen and Inoue (1993), however, for a discussion of two proteins which may perform analogous functions in these organisms, and Wales et al. (1989a, 1989b) for analyses of the amino acid sequences of the *PsbO* and *PsbP* genes, respectively.

Of the three extrinsic proteins, the 33 kDa component appears to play the most central role in oxygen evolution rate enhancement. The removal of this protein from PS II preparations leads to a destabilization of the manganese cluster under low chloride (<100 mM) conditions (Miyao and Murata, 1984a,b; Kuwabara et al., 1985). Two of the four manganese ions become paramagnetically uncoupled from the cluster and are lost (Mavankal et al., 1986). At high concentrations of chloride (>100 mM) the manganese cluster is stable and oxygen evolution occurs at low rates (Bricker, 1992). Under these conditions, the S_2 and S_3 states appear to have longer lifetimes than in the presence of the 33 kDa protein and the $S_3 \rightarrow (S_4) \rightarrow S_0$ transition is retarded (Miyao et al., 1987; Vass et al., 1987).

In *Synechocystis* sp. PCC 6803, deletion of the *PsbO* gene, which encodes the 33 kDa protein, yields an autotrophic phenotype (Burnap and Sherman, 1991). This mutant evolves oxygen at about 45% of the rate observed in control strains. Additionally, the S_2 and S_3 states are stabilized, the $S_3 \rightarrow (S_4) \rightarrow S_0$ transition is retarded, and the miss factor is increased (Burnap et al., 1992). These findings correlate remarkably well with the in vitro studies cited above. Interestingly, in *Chlamydomonas reinhardtii*, a mutant (*Fud*44) which lacks the 33 kDa protein cannot assemble stable PS II centers (Mayfield et al., 1987), cannot grow autotrophically, and does not evolve oxygen. This difference may suggest that the assembly of PS II complexes differs dramatically in *Synechocystis* and *Chlamydomonas*.

Mutants lacking the 33 kDa protein cannot grow autotrophically at low calcium concentrations (Philbrick et al., 1991). Previously, it was suggested from sequence analysis that the 33 kDa protein contained a putative calcium-binding domain (Wales et al., 1989a; Yocum, 1991). ^{45}Calcium was shown to bind to a protein identified as the 33 kDa protein after SDS-PAGE and 'Western' transfer (Webber and Gray, 1989). The observed interaction, however, was rather weak. Finally, PS II membranes that have been depleted of the 33 kDa protein require a two-fold higher concentration of calcium for optimal oxygen-evolving activity (Bricker, 1992). These results may indicate a possible role for the 33 kDa protein in the concentration and/or binding of calcium in PS II.

While this protein obviously is an important component of the oxygen-evolving complex, relatively little is known of its structural organization. Studies examining the stoichiometry of this protein indicate that there are either one (Miyao and Murata, 1989; Enami et al., 1991) or two copies (Xu and Bricker, 1992) present per PS II reaction center. It has been shown that the reconstitution of oxygen evolution activity in NaCl-urea washed PS II membranes, which lack the 33 kDa extrinsic protein, saturates at about two copies of the 33 kDa protein per PS II reaction center (Betts et al., 1994). Additionally, the presence of an intramolecular disulfide bridge has been demonstrated (Camm et al., 1987; Tanaka and Wada, 1988). The secondary structure of the 33 kDa protein in solution has been examined by far-UV circular dichroism spectroscopy (Xu et al., 1994). The 33 kDa protein contains a large proportion of β-sheet (38%) and a relatively small amount of α-helix (9%).

The 33 kDa protein is tightly bound to intrinsic components of PS II and is closely associated with a number of intrinsic PS II components. Note that removal of the manganese cluster weakens the binding of the 33 kDa protein to PS II (Miyao and Murata, 1989; Kavelaki and Ghanotakis, 1991). Previously discussed evidence documents this association with the large extrinsic loop of CP47. The protein may also be associated with D1 (Sayre et al., 1986; Mei et al., 1989), D2 (Mei et al., 1989), CP43 (Isogai et al., 1985), cytochrome *b*-559 (Takahashi and Asada, 1991; Enami et al., 1992) and the PsbI protein (Enami et al., 1992). A number of PS II components probably provide binding sites for this protein. The N-terminal domain of the 33 kDa protein appears to be involved in these interactions. Eaton-Rye and Murata (1989) demonstrated that the N-terminal 16–18 residues are required for binding of the 33 kDa protein and the restoration of oxygen evolution. Additionally, Odom and Bricker (1992) have shown that residues in the N-terminal 7 kDa of the protein interact via charge pair interactions with the large extrinsic loop of CP 47.

Removal of the 24 kDa and 17 kDa extrinsic proteins greatly increases the amount of calcium and chloride required for optimal oxygen evolution by PS II. Removal of these proteins inhibits the rate of oxygen evolution by as much as 80% at low cofactor concentrations. The presence of the 24 kDa protein appears to be associated with a decrease in the calcium requirement (Miyao and Murata, 1984c; Waggoner and Yocum, 1987) while the presence of the 17 kDa protein may lower the requirement for chloride (Miyao and Murata, 1985). It is unclear if these proteins act to concentrate these cofactors or if their binding leads to conformational changes in the intrinsic proteins that favor high rates of oxygen evolution at low cofactor concentrations.

The removal of the 24 kDa protein leads to other interesting effects on PS II. The presence of the 24 kDa protein protects the manganese cluster from the effects of bulky reductants, such as hydroquinone (Waggoner and Yocum, 1987). Transmembrane effects also occur. Upon removal of the 24 kDa protein, the binding affinity for atrazine at the Q_B site is decreased (Rashid and Carpentier, 1990; C. F. Yocum, personal communication) and high potential cytochrome *b*-559 is converted to its low potential form (Briantais et al., 1985). These effects may be the result of an alteration in the native conformation of the transmembrane helices of intrinsic PS II components.

The 24 kDa protein requires the 33 kDa extrinsic protein for binding to PS II and the 17 kDa protein requires the 24 kDa protein for binding to PS II (Miyao and Murata, 1983; Miyao and Murata, 1989). It is unclear at this time if the 24 kDa protein actually binds to the 33 kDa component or if the presence of the 33 kDa protein induces a conformational change in the intrinsic components that leads to the binding of the 24 kDa protein to PS II. A similar ambiguity is associated with the binding of the 17 kDa protein in the presence of the 24 kDa component. The binding of both proteins to PS II requires their N-terminal domains (Kuwabara et al., 1986; Miyao et al., 1988). The association of both proteins with PS II is stabilized by the presence of the manganese cluster (Becker et al., 1985; Ono et al., 1986; Kavelaki and Ghanotakis, 1991).

F. The Low Mass Proteins—psbH, J, K, L, M, N, *and* R *Gene Products and Others*

A large number of low molecular mass proteins have been reported to be associated with PS II (summarized in Debus, 1992). Most of these proteins have been identified as either protein bands present in various PS II preparations or as open reading frames which are co-transcribed with known PS II components. While one can certainly form working hypotheses based on such circumstantial evidence, more rigorous biochemical and molecular analyses must be forthcoming. The demonstration that a protein is present in a PS II complex, while certainly a *necessary* requirement for the designation of a new PS II subunit, is not *sufficient*. Demonstration that biochemical removal and reconstitution reversibly modulates a known PS II activity (or affects stability of the complex) or targeted mutagenesis experiments demonstrating a modification of photosystem function and/or stability must be performed. A number of these putative subunits fulfill these more rigorous criteria. These are summarized in Table 1. A more detailed examination of these subunits appears in Barry et al. (1994).

G. Stoichiometry of the Proteins of Photosystem II

The generally accepted stoichiometry of the D1, D2, CP47 and CP43 proteins in PS II is 1:1:1:1. This result is based on labeling studies with green algae utilizing either [^{14}C]acetate (de Vitry et al., 1987) or [^{14}C]bicarbonate (Gounaris et al., 1987), and is virtually the only area of agreement on this topic. Either one or two copies of cytochrome *b*-559 are associated with PS II (see Chapter 13 for a more detailed discussion). The *psbI* gene product appears to be present in a 1:1:1 stoichiometry with the α and β subunits of cytochrome *b*-559 (Ikeuchi and Inoue, 1988a, 1988b). Thus, there are either one or two copies of this component. The 33, 24 and 17 kDa proteins of higher plants are thought to be present in a 1:1:1 stoichiometry (Murata et al., 1984; Andersson et al., 1984). The debate here is whether there are one or two copies of each of these proteins per oxygen-evolving PS II center. Reports based primarily on quantitation of Coomassie blue staining (Murata et al., 1984) or apparent stoichiometry of crosslinking (Enami et al., 1992) indicate the presence of one 33 kDa protein per PS II. Alternatively, a recent immunoquantitation study performed with both PS II membranes and an oxygen-evolving core preparation has indicated that there are two copies of this protein per PS II (Xu and Bricker, 1992). It is unlikely that a consensus will be reached on any of these questions until a crystal structure for a completely functional PS II is available (Adir et al., 1992; Fotinou and Ghanotakis, 1993).

III. Inorganic Cofactors of Photosystem II

A. Manganese

There is now a consensus that four manganese atoms are necessary for the photochemical conversion of water into oxygen (for a review, see Debus, 1992). Selective extraction experiments and a series of spectroscopic studies revealed a heterogeneity in the manganese population. Seventy-five percent of the total manganese is released by treatment of PS II with 0.8 M Tris at pH 8.5 or hydroxylamine (Yocum et al. 1981; Tamura and Cheniae, 1987). The same amount of manganese is also released by treatment of PS II membranes, first with 2M sodium chloride and subsequently with a reductant such as hydroquinone (Ghanotakis et al., 1984d). The main difference between the two treatments is that, while Tris releases all three extrinsic proteins, the high salt/reductant treatment results in a PS II system that has been depleted of most of its manganese and the 17 and 23 kDa polypeptides, but retains the extrinsic 33 kDa protein. Exposure of PS II membranes to 1.0 M calcium chloride results in depletion of all three extrinsic proteins but retention of the four manganese ions. When 1.0 M calcium chloride-treated PS II preparations are transferred to a low chloride medium they lose two of the four manganese ions (Ono and Inoue, 1985). This is evidence that one pair of manganese ions is more weakly bound than the other. It is not clear whether this heterogeneity is due to a different environment or different oxidation states of the various manganese ions.

Tamura and Cheniae (1987) have shown that, under certain conditions, it is possible to reassemble a functional manganese complex in PS II preparations that have been depleted of manganese. This process, which requires light, is known as photoactivation. The current model for the assembly of a tetranuclear manganese complex is the following: Photooxidation of a ligated Mn^{2+} to Mn^{3+} is the first step. Oxidation of the first manganese is followed by the binding of a second Mn^{2+} and the formation of a binuclear Mn^{3+}-Mn^{2+} intermediate. This intermediate is subsequently photooxidized to a stable Mn^{3+}-Mn^{3+} complex. The last step is the spontaneous binding of two more Mn^{2+} ions, a process that leads to the formation of the tetranuclear manganese complex. It is interesting that calcium and chloride are also required for the assembly of a functional manganese complex. The assembly of a stable manganese complex is followed by the spontaneous rebinding of the extrinsic 17 and 23 kDa polypeptides (Becker et al., 1985; Ono et al., 1986). The requirement for the manganese complex for productive binding of the 17 and 23 kDa species was also demonstrated by a series of in vitro extraction-reconstitution experiments (Kavelaki and Ghanotakis, 1991).

The catalytic role of manganese has been probed by various spectroscopic techniques. A thorough investigation of two EPR signals observed at liquid helium temperatures has provided information regarding the structure of the manganese complex in the S_2 state (see Chapter 9 for details). The first signal, the so-called 'multiline' signal, is centered at around $g = 2$ and is comprised of 18–20 lines spanning 1500–1800 G (Dismukes and Siderer, 1980). The second signal, the so called '$g = 4.1$' signal, shows a turning point at $g = 4.1$ and has a width of 320–360 G (Casey and Sauer, 1984; Zimmermann and Rutherford, 1986). Both signals are attributed to the S_2 state of the oxygen-evolving complex.

The redox changes in the manganese complex have also been probed by UV difference spectroscopy. Cycling through the various S-states is accompanied by UV changes that show period four oscillations. Although van Gorkom and co-workers (Dekker et al., 1984) have proposed that the UV changes observed during $S_0 \rightarrow S_1$, $S_1 \rightarrow S_2$ and $S_2 \rightarrow S_3$ are each due to the oxidation of one Mn(III) to Mn(IV), other groups have suggested alternative explanations (Lavergne, 1986; 1987; Saygin and Witt, 1987).

The structure of the manganese complex has been investigated by EXAFS studies. Although there is an agreement among the various groups regarding certain aspects of the manganese complex, there are still contradictory points which are probably due to the nature of the samples used for the EXAFS studies (Goodin et al., 1984; George et al., 1989; Penner-Hahn et al., 1990). The oxidation states of the manganese complex have been investigated by XANES measurements. Based on such measurements it has been proposed that, in the S_1 state, the manganese complex consists of two Mn(III) and two Mn(IV) ions (Yachandra et al., 1993a; Riggs et al., 1992)

There are various proposals regarding the localization of manganese-binding sites and the nature of the residues that bind manganese. The release of manganese along with the 33 kDa polypeptide during exposure of PS II to 0.8 M Tris at high pH was initially taken as evidence that the extrinsic 33 kDa protein is required for manganese binding. Subsequent experiments demonstrated that PS II membranes which had been depleted of the extrinsic 33 kDa protein by treatment with sodium chloride-urea showed almost 25% of the oxygen-evolving activity of control PS II membranes (Bricker, 1992). These results and the observation that depletion of the *Psb*O gene, which encodes the extrinsic 33 kDa protein from *Synechocystis*, did not destroy the capacity to evolve oxygen (Burnap and Sherman, 1991; Philbrick et al., 1991; Mayes et al., 1991) indicate that the release of the extrinsic 33 kDa protein along with manganese is probably a coincidental phenomenon related to ionic strength. Indeed, Kavelaki and Ghanotakis (1991) have shown that, upon removal of the manganese complex, the interaction of the extrinsic 33 kDa protein with the PS II core is altered and it is possible to release this protein by raising the ionic strength. Although the extrinsic 33 kDa protein is not necessary for manganese binding in PS II, there is strong evidence that the protein stabilizes and protects the manganese complex. We cannot exclude the possibility that the extrinsic 33 kDa protein provides ligands (or a ligand) to the manganese complex, and that upon removal of the protein, H_2O or other amino acid residues occupy the ligand sites on the metal. The structural changes that might accompany such a replacement could result in a modified manganese complex that maintains its catalytic activity but at a significantly diminished efficiency.

Based on studies with a mutant of the green alga *Scenedesmus obliquus* (the so-called LF-1 mutant), it has been proposed that the D1 protein is directly involved in manganese binding (Metz and Seibert, 1984; Metz et al., 1986). The LF-1 mutant, which contains the D1 protein in its unprocessed form, assembles PS II complexes that contain only a fraction of the manganese found in the wild type, and do not evolve oxygen. Nevertheless, the evidence that implicated the D1 protein in manganese binding could be considered circumstantial. The unprocessed D1 protein could, through an allosteric effect, alter the structure of other PS II proteins and thus prevent the assembly of a functional manganese complex.

Excess light excitation of oxygen-evolving photosynthetic membranes, which cannot be utilized for photochemical energy conversion, results in a loss of photosynthetic activity (Kyle, 1987). This phenomenon is referred to as photoinhibition. There is now a consensus that photoinhibition causes an impairment of electron transport capacity, and subsequently results in the degradation of the reaction center D1 polypeptide (Barber and Andersson, 1992; Prasil et al., 1992). A correlation between D1 degradation and loss of manganese during photoinhibition indicated that D1 may bind manganese (Virgin et al., 1988). Although manganese release could be due to D1 conformational changes that occur during photoinhibition, we cannot exclude the possibility of allosteric effects that alter the structure of other manganese-binding protein(s), and result in the destruction of the manganese complex. A series of experiments with PS II preparations, which had been modified either through light-induced iodination (Ikeuchi et al., 1988) or chemical modification (Tamura et al., 1989), also implicated the D1 polypeptide in manganese binding.

Various amino acid residues have been discussed as possible ligands to manganese. The properties of the manganese complex in PS II and the chemistry of model manganese compounds indicate that carboxylate residues, in combination with μ_2-oxo or μ_2-hydroxo bridges, may be ligands to manganese in the oxygen-evolving complex. Strong evidence suggesting that at least one ligand to manganese is a histidyl residue has been provided by ESEEM measurements on ^{14}N- and ^{15}N-labeled PS II membranes isolated from the cyanobacterium *Synechococcus elongatus* (DeRose et al., 1991). Direct evidence that the labeled amino acid residue is histidine has recently been provided (Tang et al., 1994).

B. Calcium and Chloride

Chloride depletion of PS II samples results in the inactivation of the oxygen-evolving complex. Addition of certain anions restores oxygen evolution activity; the effectiveness of the anions that have been examined follows the order chloride →bromide → iodide → nitrate (Kelly and Izawa, 1978). It has been proposed that chloride binds directly to manganese or to a site very close to manganese, affecting the redox properties of the manganese complex (Sandusky and Yocum, 1984; Brudvig et al., 1989). A charge- neutralizing role for chloride has also been proposed (Coleman and Govindjee, 1987; Homann, 1988). Although the current signal-to-noise ratio in the EXAFS data does not allow a definitive answer regarding the presence of chloride in the coordination sphere of manganese, there is some evidence that chloride might be a terminal ligand to manganese with a manganese-to-chloride distance of 2.3 Å (Yachandra et al., 1993b). Further studies are necessary in order to identify the chloride-binding site(s) within the oxygen-evolving complex.

Although the involvement of calcium at the oxidizing side of PS II has been known for a long time, its binding site(s) and mode of action remain unclear. Salt-washed PS II membranes, depleted of the extrinsic 17 and 23 kDa polypeptides, require relatively high concentrations of calcium for maximal oxygen evolution activity (Ghanotakis et al., 1984b; Miyao and Murata, 1984c). This has been attributed to the 23 kDa protein's property of enhancing the binding of calcium. Various estimates of the number of calcium ions per PS II necessary for oxygen evolution vary from 1–3 (Cammarata and Cheniae, 1987; Shen et al., 1988). The presence of a high affinity and one to two lower affinity binding sites has been suggested.

The calcium-binding sites have been investigated by using extraction-reconstitution experiments and selective substitution of calcium with other cations. Spectroscopic studies, such as low temperature EPR, room temperature kinetic EPR, EXAFS and optical spectroscopy, have provided information regarding its binding and interaction with other redox species at the oxidizing side of PS II. There is no agreement about which S-state transition is blocked in the absence of calcium (see Debus (1992) for a review). The controversy regarding this question is probably due to the different procedures used for calcium depletion.

Various cations compete with calcium for binding sites in PS II. Sodium, potassium, and cesium are weakly competitive with calcium, but they do not support oxygen evolution activity (see Yocum, 1991, for a review). Strontium can substitute for calcium; the steady state oxygen evolution rate of the strontium-containing PS II is low, and that has been attributed to a slowing of certain S-state transitions (Pistorius 1983; Pistorius and Schmid 1984; Ghanotakis et al., 1984a; Boussac and Rutherford, 1988a). In the presence of strontium the S_2 multiline signal exhibits a narrower hyperfine splitting (Boussac and Rutherford, 1988b). Replacement of calcium with various lanthanides inhibits oxygen evolution activity. Lanthanides also affect electron transport from TyrZ to $P680^+$; this effect depends on the ionic radius of the trivalent lanthanide and the pH of the system (Bakou et al., 1992; Bakou and Ghanotakis, 1993).

The various studies with PS II preparations depleted of calcium or in which calcium has been substituted with other cations, and the photoinactivation experiments described above, suggest a structural role for calcium. Binding of calcium could either provide the appropriate environment for a stable manganese complex, or affect its redox properties. The stabilizing role for calcium is supported by the observation that the cation protects the manganese complex against destruction that occurs upon exposure of salt-washed PS II to various reductants, and the ability of calcium to stabilize reduced forms of the oxygen-evolving complex (Mei and Yocum, 1991). A concanavalin A-type structure, in which calcium binds in close proximity to the manganese complex and affects its structure, was suggested by Ghanotakis et al. (1987). We cannot, however, exclude a catalytic role for calcium. An involvement of calcium in deprotonation that occurs during S-state transitions has been proposed (Boussac et al., 1990). Also, it has been suggested that calcium might bind substrate water that is exchanged with a chloride on the manganese complex immediately preceding water oxidation (Rutherford, 1989).

IV. The Mechanism of Water Oxidation

The cleavage of two molecules of water requires the extraction of four electrons from the oxygen-evolving complex. The period 4 oscillation in O_2 yield, discovered by Joliot et al. (1969), was interpreted as evidence of a linear oxidation reaction within the oxygen-evolving complex. Kok developed the S-state model on the basis of these results (Joliot and Kok, 1975). Since 1970 an abundance of experimental evidence has provided new information regarding the properties of this unique enzyme; unfortunately, the mechanism of water oxidation is still unknown.

Since a sequential, 4-electron oxidation of water is not favored thermodynamically, it is believed that the mechanism of water oxidation involves either a concerted 4-electron oxidation or a concerted 2-electron oxidation reaction leading to the formation of a 'peroxyl' intermediate. The absorption changes obtained by UV-difference spectroscopy appear to support the former model. The UV absorbance changes which accompany the $S_0 \rightarrow S_1$, $S_1 \rightarrow S_2$ and $S_2 \rightarrow S_3$ are dark-stable, with lifetimes approximating those of the corresponding S-states. During the $S_3 \rightarrow (S_4) \rightarrow S_0$ transition, a rapid collapse of the absorption occurs, which is interpreted to indicate the reduction of the accumulated higher oxidation states of manganese accompanying the oxidation of water to oxygen (Dekker et al., 1984). Further support for the concerted 4-electron oxidation model was also provided by a series of mass spectrometry experiments. Examination of the isotopic composition of O_2 released from PS II led to the proposal that water oxidation does not involve partially oxidized intermediates that may be stably bound in the S_2 or S_3 states (Radmer and Ollinger, 1986; Bader et al., 1987).

Based on the spectroscopic data and the properties of model compounds, a large number of mechanisms for water oxidation have been proposed. Although the elucidation of the exact mechanism of photo-oxidation of water will require more information regarding the structural aspects of the manganese complex and the role of the other inorganic cofactors, calcium and chloride, a few of the mechanisms that have been proposed are very interesting working models. Many proposed mechanisms involve the formation of an O-O bond between the oxygen atoms of μ_2-oxo bridges and/or terminal OH^- ligands of the manganese complex. On the basis of various di-μ_2-oxo bridged Mn(IV)-Mn(IV) dimers, which can either produce or disproportionate H_2O_2, Pecoraro (1992) has proposed an interesting mechanism of water oxidation. According to this model the S_2 state is assumed to consist of one Mn(III)-Mn(IV) dimer and one Mn(IV)-Mn(IV) dimer. The $S_2 \rightarrow S_3$ transition corresponds to the oxidation of an amino acid residue, whereas the $S_3 \rightarrow S_4$ transition corresponds to a Mn(III) to Mn(IV) oxidation. A molecule of H_2O_2, which is generated by one Mn(IV)-Mn(IV), reacts with the second Mn(IV)-Mn(IV) dimer to form O_2.

Although the oxidation of an organic species has been proposed to occur during the $S_2 \rightarrow S_3$ transition, the importance of this oxidation and the identification of the residue that is oxidized remains controversial. On the basis of a 130–164 Gauss-wide EPR signal observed in calcium-depleted PS II samples, a histidine residue may be oxidized during the $S_2 \rightarrow S_3$ transition (Boussac et al., 1990). Alternatively, the 130–164 G signal has been attributed to $TyrZ^+$, which is interacting with the modified S_2 state manganese complex (Hallahan et al., 1992). Since the oxidation of an amino acid residue during the $S_2 \rightarrow S_3$ transition is in conflict with UV data implicating a manganese oxidation during this transition, the involvement of a redox-active histidine during the oxidation of water remains speculative (see Britt, Chapter 9).

V. Conclusions

Despite significant progress in the biochemical and biophysical characterization of PS II, the unique structural and mechanistic properties of this membrane protein complex remain to be elucidated. The isolation of highly resolved oxygen-evolving PS II subcomplexes, the employment of site-directed mutagenesis, the development and refinement of sophisticated spectroscopic techniques and the crystallization of PS II core complexes (Fotinou et al., 1993) foretell major advances in the future.

Acknowledgments

Partial support for this work was provided by NSF Grant DMB 9304955 to T.M.B. The authors wish to thank Ms. Laurie K. Frankel for her help in preparing this manuscript.

References

Adir N, Okamura MY and Feher G (1992) Crystallization of the reaction center of Photosystem II. Biophysical J 61s: A101

Akabori K, Tsukamoto H, Tsukihara J, Nagatsuka T, Motokawa O and Toyoshima Y (1988) Disintegration and reconstitution of Photosystem II reaction center core complex. I. Preparation and characterization of three different types of subcomplexes. Biochim Biophys Acta 932: 345–357

Anbudurai PR, Tsafrir SM, Ohad I, Shestakov SV and Pakrasi HB (1994) The *ctpA* gene encodes the c-terminal processing protease for the D1 protein of the Photosystem II reaction center. Proc Natl Acad Sci USA 91: 8082–8086

Andersson B, Larsson C, Jansson C, Ljungberg U and Akerlund H-E (1984) Immunological studies on the organization of proteins in photosynthetic oxygen evolution. Biochim Biophys Acta 766: 21–28

Andreeva VA, Buryakova AA, Reverdatto SV, Chakhmakhcheva OG and Efimov VA (1989) Nucleotide sequence of the barley chloroplast DNA fragment containing *psbB-psbH-petB-petD* gene cluster. Nuc Acid Res 17: 2859–2860

Babcock GT, Blankenship RE and Sauer K (1976) Reaction kinetics for positive charge accumulation on the water side of chloroplast Photosystem II. FEBS Lett 61: 286–289

Bader KP, Thibault P and Schmid GH (1987) Studies on the properties of the S_3 state by mass spectrometry in the filamentous cyanobacterium *Oscillatoria chlaybea*. Biochim Biophys Acta 893: 564–571

Bakou A and Ghanotakis DF (1993) Substitution of lanthanides at the calcium site(s) in Photosystem II affects electron transport from tyrosine Z to $P680^+$. Biochim Biophys Acta 1141: 303–308

Bakou A, Buser C, Dandulakis G, Brudvig G and Ghanotakis DF (1992) Calcium binding site(s) of Photosystem II as probed by lanthanides. Biochim Biophys Acta 1099: 131–136

Barbato R, Race HL, Friso G and Barber J (1991) Chlorophyll levels in the pigment binding proteins of Photosystem II. A study based on the chlorophyll to cytochrome ratio in different Photosystem II preparations. FEBS Lett 286: 86–90

Barber J and Andersson B (1992) Too much of a good thing: Light can be bad for photosynthesis. Trends Biochem Sci 17: 61–66

Barry BA, Boerner RJ and dePaula JC (1994) The use of cyanobacteria in the study of the structure and function of Photosystem II. In: Bryant D (ed) The Molecular Biology of Cyanobacteria, pp 217–257. Kluwer Academic Publishers, Dordrecht

Bassi R, Hoyer-Hansen G, Barbato R, Giacometti GM and Simpson DJ (1987) Chlorophyll-proteins of the Photosystem II antenna complex. J Biol Chem 262: 13333–13341

Becker B, Callahan F and Cheniae G (1985) Photoactivation of NH_2OH-treated leaves: Reassembly of released extrinsic polypeptides and relegation of Mn into the polynuclear Mn catalyst of water oxidation. FEBS Lett 192: 209–214

Bendall DS (1968) Oxidation-reduction potentials of cytochromes in chloroplasts from higher plants. Biochem J 109: 46–47

Berry-Lowe SL, Johnson CH and Schmidt GW (1992) Nucleotide sequence of the *psbB* gene of *Chlamydomonas reinhardtii*. Plant Physiol 98: 1542

Berthold DA, Babcock GT and Yocum CF (1981) A highly resolved, oxygen-evolving Photosystem II preparation from spinach thylakoid membranes. FEBS Lett 13: 231–233

Betts S, Hachigian TM, Pichersky RE and Yocum CF (1994) Reconstitution of the spinach oxygen-evolving complex with recombinant Arabidopsis manganese-stabilizing protein. Plant Mol Biol 26:117–130

Boussac A and Rutherford AW (1988a) Nature of the inhibition of the oxygen-evolving enzyme of Photosystem II induced by NaCl washing and reversed by the addition of Ca^{+2} or Sr^{+2} FEBS Lett 236: 432–436

Boussac A and Rutherford AW (1988b) S-state formation after Ca^{+2} depletion in the Photosystem II oxygen-evolving complex. Chem Scripta 28A: 123–126

Boussac A, Zimmermann JL, Rutherford AW and Lavergne J

(1990) Histidine oxidation in the oxygen-evolving Photosystem II enzyme. Nature 37: 303–306

Bowden SJ, Hallahan BJ, Ruffle SV, Evans MCW and Nugent JHA (1991) Preparation and characterization of Photosystem II core particles with and without bound bicarbonate. Biochim Biophys Acta 1060: 89–96

Briantais J-M, Vernotte C, Miyao M, Murata N and Picaud M (1985) Relationship between O_2 evolution capacity and cytochrome *b*-559 high-potential form in Photosystem II particles. Biochim Biophys Acta 808: 348–351

Bricker TM (1990) The structure and function of CPa-1 and CPa-2 in Photosystem II. Photosyn Res 24: 1–13

Bricker TM (1992) Oxygen evolution in the absence of the 33 kDa manganese-stabilizing protein. Biochemistry 31: 4623–4628

Bricker TM and Frankel LK (1987) Use of a monoclonal antibody in structural investigations of the 49 kDa polypeptide of Photosystem II. Arch Biochem Biophys 256: 295–301

Bricker TM, Odom WR and Queirolo CB (1988) Close association of the 33 kDa extrinsic protein with the apoprotein of CPa-1 in Photosystem II. FEBS Lett 231: 111–117

Brudvig GW, Beck WF and dePaula JC (1989) Mechanism of photosynthetic water oxidation. Annu Rev Biophys Biophys Chem 18: 25–46

Bukharov AA, Kolosov VL and Zolotarev AS (1988) Nucleotide sequence of rye chloroplast DNA fragment encoding the *psbB* and *psbH* genes. Nuc Acid Res 16: 8737

Bukharov AA, Kolosov VL, Klezovich ON and Zolotarev AS (1989) Nucleotide sequence of rye chloroplast DNA fragment comprising *psbD*, *psbC* and *trnS* genes. Nuc Acid Res 17: 798

Burnap RL and Sherman LA (1991) Deletion mutagenesis in *Synechocystis* sp. PCC 6803 indicates that the Mn-stabilizing protein of Photosystem II is not essential for oxygen evolution. Biochemistry 30: 440–446

Burnap RL, Shen J-R, Jursinic PA, Inoue Y and Sherman LA (1992) Oxygen yield and thermoluminescence characteristics of a cyanobacterium lacking the manganese-stabilizing protein of Photosystem II. Biochemistry 31: 7404–7410

Camm EL, Green BR, Allred DR and Staehelin A (1987) Association of the 33 kDa extrinsic polypeptide (water splitting) with PS II particles: Immunochemical quantification of residual polypeptide after membrane extraction. Photosyn Res 13: 69–80

Cammarata K and Cheniae G. (1987) Studies on 17, 24 kD depleted Photosystem II membranes. I. Evidence for high and low affinity calcium sites in 17, 24 kD depleted PS II membranes from wheat versus spinach. Plant Physiol 84: 8577–8595

Casey JL and Sauer K (1984) EPR detection of a cryogenically photogenerated intermediate in photosynthetic oxygen evolution. Biochim Biophys Acta 767: 21–28

Chapman DJ, Gounaris K and Barber J (1988) Electron transport properties of the D1-D2-cytochrome *b*-559 Photosystem II reaction center. Biochim Biophys Acta 933: 423–431

Chen JC, Bing YM, Masakazu F and Sigiura M (1990) Nucleotide sequence of the *psb*I gene of the cyanobacterium, *Anacystis nidulans* 6301. Nucl Acid Res 18: 4017

Coleman WJ and Govindjee (1987) A model for the mechanism of chloride activation of oxygen evolution in Photosystem II. Photosyn Res 13: 199–223

Cramer WA, Tae G-S, Furbacher PN and Bottger M (1993) The enigmatic cytochrome *b*-559 of oxygenic photosynthesis. Physiol Plant 88: 705–711

Danielius RV, Satoh K, van Kan PJM, Plijter JM, Nuijs AM and van Gorkom HJ (1987) The primary reaction of Photosystem II in the D1-D2-cyt *b*-559 complex. FEBS Lett 213: 241–243

Debus RJ (1992) The manganese and calcium ions of photosynthetic oxygen evolution. Biochim Biophys Acta 1102: 269–352

Debus RJ, Barry BA, Sithole I, Babcock GT and McIntosh L (1988a) Directed mutagenesis indicates that the donor to $P680^+$ in Photosystem II is Tyr-161 of the D1 polypeptide. Biochemistry 27: 9071–9074

Debus RJ, Berry BA, Babcock GT and McIntosh L (1988b) Site specific mutagenesis identifies a tyrosine radical involved in the photosynthetic oxygen-evolving complex. Proc Natl Acad Sci USA 85: 427–430

Dedner N, Meyer HE, Ashyon C and Wildner GF (1988) N-terminal sequence analysis of the 8 kDa protein in *Chlamydomonas reinhardtii*. Localization of the phospho-threonine. FEBS Lett 236: 77–82

Deisenhofer J, Epp O, Miki K, Huber R and Michel H (1985) Structure of the protein subunits in the photosynthetic reaction center of *Rhodopseudomonas viridis* at 3 Å resolution. Nature 318: 618–623

Dekker JP, van Gorkom HJ, Wensink J and Ouwehand L (1984) Absorbance difference spectra of the successive redox states of the oxygen-evolving apparatus of photosynthesis. Biochim Biophys Acta 767: 176–179

Dekker JP, Bowlby NR and Yocum CF (1989) Chlorophyll and cytochrome *b*-559 content of the photochemical reaction center of Photosystem II. FEBS Lett 254: 150–153

DeRose VJ, Yachandra VK, McDermott AE, Britt RD, Sauer K and Klein MP (1991) Nitrogen ligation to manganese in the photosynthetic oxygen-evolving complex: Continuous-wave and pulsed EPR studies of Photosystem II particles containing ^{14}N or ^{15}N. Biochemistry 30: 1335–1341

de Vitry C, Wollmann F-A and Delepelaire P (1984) Function of the polypeptides of the Photosystem II reaction center in *Chlamydomonas reinhardtii*. Biochim Biophys Acta 767: 415–422

de Vitry C, Diner BA and Lemoine Y (1987) Chemical composition of Photosystem II reaction centers: Phosphorylation of PS II polypeptides. In: Biggins J (ed) Progress in Photosynthesis Research, Vol II, pp 105–108. Martinus Nijhoff, Dordrecht

Dismukes GC and Siderer Y (1980) EPR spectroscopic observation of a manganese center associated with water oxidation in spinach chloroplasts. FEBS Lett 121: 78–80

Doring G, Renger G, Vater J and Witt HT (1969) Properties of photoactive chlorophyll-*a*II in photosynthesis. Z Naturforsch 24b: 1139–1143

Eaton-Rye JJ and Murata N (1989) Evidence that the amino-terminus of the 33 kDa extrinsic protein is required for binding to the Photosystem II complex. Biochim Biophys Acta 977: 219–226

Eaton-Rye JJ and Vermaas WFJ (1991) Oligonucleotide-directed mutagenesis of *psbB*, the gene encoding CP47, employing a deletion strain of the cyanobacterium *Synechocystis* sp. PCC 6803 Plant Mol Biol 17: 1165–1177

Enami I, Satoh K and Katoh S (1987) Crosslinking between the 33 kDa extrinsic protein and the 47 kDa chlorophyll-carrying protein of the PS II reaction center core complex. FEBS Lett

226: 161–165

Enami I, Kamino K, Shen J-R, Satoh K and Katoh S (1989) Isolation and characterization of Photosystem II complexes which lack light harvesting chlorophyll *a/b* proteins but retain three extrinsic proteins related to oxygen evolution from spinach. Biochim Biophys Acta 977: 33–39

Enami I, Kaneko M, Kitamura N, Koike H, Sonoike K, Inoue Y and Katoh S (1991) Total immobilization of the extrinsic 33 kDa protein in spinach Photosystem II membrane preparations. Protein stoichiometry and stabilization of oxygen evolution. Biochim Biophys Acta 1060: 224–232

Enami I, Ohta S, Mitsuhashi S, Takahashi S, Ikeuchi M and Katoh S (1992) Evidence from crosslinking for a close association of the extrinsic 33 kDa protein with the 9.4 kDa subunit of cytochrome *b*-559 and the 38 kDa product of the *psbI* gene in oxygen-evolving Photosystem II complexes from spinach. Plant Cell Physiol 33: 291–297

Fan HN and Cramer WA (1970) The redox potential of cytochromes *b*-559 and *b*-563 in spinach chloroplasts. Biochim Biophys Acta 267: 375–382

Fotinou C and Ghanotakis DF (1990) A preparative method for the isolation of the 43 kDa, 47 kDa and the D1-D2-Cyt *b*-559 species directly from thylakoid membranes. Photosyn Res 25: 141–145

Fotinou C, Kokkinidis M, Haase M, Fritzch G, Michel H and Ghanotakis DF (1993) Characterization of a Photosystem II core and its three dimensional crystals. Photosyn Res 37: 41–48

Frankel LK and Bricker TM (1989) Epitope mapping of the monoclonal antibody FAC2 on the apoprotein of CPa-1 in Photosystem II. FEBS Lett 257: 279–282

Frankel LK and Bricker TM (1992) Interaction of CPa-1 with the manganese-stabilizing protein of Photosystem II: Identification of domains on CPa-1 which are shielded from N-hydroxysuccinimide biotinylation by the manganese-stabilizing protein. Biochemistry 31: 11059–11063

Geiger R, Brezborn R, Depka W, and Trebst A (1987) Site directed antisera to the D2 polypeptide subunit of Photosystem II. Z Naturforsch 42c: 491–498

George GN, Prince RC and Cramer SP (1989) The Mn site of the photosynthetic water-splitting enzyme. Science 243: 789–791

Ghanotakis DF and Yocum CF (1986) Purification and properties of an oxygen-evolving reaction center complex from Photosystem II membranes. FEBS Lett 197: 244–248

Ghanotakis, DF and Yocum CF (1990) Photosystem II and the oxygen-evolving complex. Ann Rev Plant Physiol Plant Mol Biol 41: 255–276

Ghanotakis DF, Babcock GT, and Yocum CF (1984a) Calcium reconstitutes high rates of oxygen evolution in polypeptide depleted Photosystem II preparations. FEBS Lett 167: 127–130

Ghanotakis DF, Topper JN, Babcock GT and Yocum CF (1984b) Water-soluble 17 and 23 kDa polypeptides restore oxygen evolution activity by creating a high-affinity binding site for Ca^{2+} on the oxidizing side of Photosystem II. FEBS Lett 170: 169–173

Ghanotakis DF, Babcock GT and Yocum CF (1984c) Structural and catalytic properties of the oxygen-evolving complex. Correlation of polypeptide and manganese release with the behavior of Z^+ in chloroplasts and a highly resolved preparation of the PS II complex. Biochim Biophys Acta 765: 388–398

Ghanotakis DF, Topper J and Yocum CF (1984d) Structural organization of the oxidizing side of Photosystem II. Exogenous reductants reduce and destroy the Mn-complex in Photosystem II membranes depleted of the 17 and 23 kDa polypeptides. Biochim Biophys Acta 767: 524–531

Ghanotakis DF, Demetriou DM and Yocum CF (1987) Isolation and characterization of an oxygen-evolving Photosystem II reaction center core preparation and a 28 kDa chl *a*-binding protein. Biochim Biophys Acta 891: 15–21

Ghanotakis DF, de Paula JC, Demetriou DM, Bowlby NR, Petersen J, Babcock GT and Yocum CF (1989) Isolation and characterization of the 47 kDa protein and the D1-D2-cytochrome *b*-559 complex. Biochim Biophys Acta 974: 44–53

Gleiter, HM, Haag, E, Shen, J-R, Eaton-Rye, JJ, Inoue, Y, Vermaas, WFJ and Renger, G (1994) Functional characterization of mutant strains of the cyanobacterium *Synechocystis* PCC 6803 lacking short domains within the large, lumen-exposed loop of the chlorophyll protein CP47 in Photosystem II. Biochemistry 33: 12063–12071

Goodin DB, Yachandra VK, Britt RD, Sauer K and Klein MP (1984) The state of manganese in the photosynthetic apparatus. 3 Light-induced changes in X-ray absorption (K-edge) energies of manganese in photosynthetic membranes. Biochim Biophys Acta 767: 209–216

Gounaris K, Pick U and Barber J (1987) Stoichiometry and turnover of Photosystem II polypeptides. FEBS Lett 211: 94–98

Gounaris K, Chapman DJ, Booth P, Crystall B, Giorgi LB, Klug DR, Porter G and Barber J (1990) Comparison of the D1/D2/cytochrome *b*-559 reaction center complex of Photosystem II isolated by two different methods. FEBS Lett 265: 88–92

Gray JC, Hird SM and Dyer TA (1990) Nucleotide sequence of a wheat chloroplast gene encoding the proteolytic subunit of an ATP-dependent protease. Plant Mol Biol 15: 947–950

Greer KL and Golden SS (1991) Nucleotide sequence of *psbB* from *Prochlorothrix hollandica*. Plant Mol Biol 17: 915–917

Haag E, Irrgang KD, Boekema EJ and Renger G (1990) Functional and structural analysis of Photosystem II core complexes from spinach with high oxygen evolution capacity. Eur J Biochem 189: 47–53

Haag, E, Eaton-Rye, JJ, Renger, G and Vermaas, WFJ (1993) Functionally important domains of the large hydrophilic loop of CP47 as probed by oligonucleotide-directed mutagenesis in *Synechocystis* sp. PCC 6803 Biochemistry 32: 4444–4453

Hackett CS and Strittmatter P (1984) Covalent crosslinking of the active sites of vesicle-bound cytochrome b_5 and NADH cytochrome b_5 reductase. J Biol Chem 259: 3275–3282

Hallahan BJ, Nugent JHA, Warden JT and Evans MCW (1992) Investigations of the origin of the 'S3' EPR signal from the oxygen-evolving complex of photosystem 2: The role of tyrosine Z. Biochemistry 31: 4562–4573

Hallick RB, Hong L, Drager RG, Favreau MR, Monfort A, Orsat B, Spielmann A and Stutz E (1993) Complete sequence of *Euglena gracilis* chloroplast DNA. Nuc Acid Res 21: 3537–3543

Hayashi H, Fujimura Y, Mohanty PS and Murata N (1993) The role of CP47 in the evolution of oxygen and the binding of the extrinsic 33-kDa protein to the core complex of Photosystem II as determined by limited proteolysis. Photosyn Res 36: 35–42

Hind G and Nakatani HY (1970) Determination of the

concentration and redox potential of chloroplast cytochrome *b*-559. Biochim Biophys Acta 216: 223–225

Hiratsuka J, Shimada H, Whittier R, Ishibashi T, Sakamoto M, Mori M, Kondo C, Honji Y, Meng B-Y, Li Y-Q, Kanno A, Nishizawa Y, Hirai A, Shinozaki K and Sugiura M (1989) The complete nucleotide sequence of the rice (*Oryza sativia*) chloroplast genome. Mol Gen Genet 217: 185–193

Homann PH (1987) The relations between the chloride, calcium and polypeptide requirements of photosynthetic water oxidation. J Bioenerg Biomem 19: 105–123

Homann PH (1988) The chloride and calcium requirement of photosynthetic water oxidation: Effects of pH. Biochim Biophys Acta 93: 1–13

Howe CJ, Barker RF, Bowman CM and Dyer TA (1988) Common features of three inversions in wheat chloroplast DNA. Curr Genet 13: 343–349

Ikeuchi M and Inoue Y (1986) Characterization of O_2 evolution by a wheat Photosystem II reaction center complex isolated by a simplified method: Disjunction of secondary acceptor quinone and enhanced Ca^{2+} demand. Arch Biochem Biophys 247: 97–107

Ikeuchi M, and Inoue Y (1988a) A new Photosystem II reaction center component (38 kDa protein) encoded by the chloroplast genome. FEBS Lett 241: 99–103

Ikeuchi M and Inoue Y (1988b) A new 38 kDa polypeptide intrinsic.to the Photosystem II reaction center as revealed by modified SDS-PAGE with improved resolution of the low-molecular-weight proteins. Plant Cell Physiol 29: 1233–1239

Ikeuchi M, Yuasa M and Inoue Y (1985) Simple and discrete isolation of an O_2-evolving PS II reaction center complex retaining Mn and the extrinsic 33 kDa protein. FEBS Lett 185: 316–322

Ikeuchi M, Koike H and Inoue Y (1988) Iodination of D1 (herbicide-binding protein) is coupled with photooxidation of $^{125}I^-$ associated with Cl-binding site in Photosystem II water oxidation system. Biochim Biophys Acta 932: 160–169

Ikeuchi M, Koike H, and Inoue Y (1989) N-terminal sequencing of Photosystem II low-molecular-mass proteins 5 and 3.1 kDa components of the O_2-evolving core complex from higher plants. FEBS Lett 242: 263–269

Ikeuchi M, Eggers B, Shen G, Webber A, Yu J, Hirano A, Inoue Y and Vermaas WFJ (1991) Cloning of the *psbK* gene from *Synechocystis* sp. 6803 and characterization of Photosystem II in mutants lacking PS II-K. J Biol Chem 266: 11111–11115

Isogai Y, Yamamoto Y and Nishimura M (1985) Association of the 33 kDa polypeptide with the 43 kDa component in Photosystem II particles. FEBS Lett 187: 240–243

Joliot P and Kok B (1975) Oxygen evolution in photosynthesis. In: Govindjee (ed), Bioenergetics of Photosynthesis, pp 387–411. Academic Press, London/New York

Joliot P, Barbieri G and Chabaud R (1969) Un nouveau modele des centres photochimique du systeme II. Photochem Photobiol 10: 309–329

Kavelaki K and Ghanotakis DF (1991) Effect of the manganese complex on the binding of the extrinsic proteins (17, 23 and 33 kDa) of Photosystem II. Photosynth Res 29: 149–155

Keller W, Weil JH and Nair CKK (1989) Nucleotide sequence of the *psbB* gene of *Euglena gracilis*. Plant Mol Biol 13: 723–725

Kelley PM and Izawa S (1978) The role of chloride ion in Photosystem II. I. Effects of chloride ion on Photosystem II electron transport and on hydroxylamine inhibition. Biochim Biophys Acta 502: 198–210

Kessler U, Maid U and Zetsche K (1992) An equivalent to bacterial *ompR* genes is encoded on the plastid genome of red algae. Plant Mol Biol 18: 777–780

Klimov VV, Dolan E and Ke B (1980) EPR properties of an intermediary electron acceptor (pheophytin) in Photosystem II reaction centers at cryogenic temperatures. FEBS Lett 112: 97–100

Kobayashi M, Maeda H, Watanabe T, Nakane H and Satoh Ki (1990) Chlorophyll *a* and *β*-carotene content in the D1/D2/ cytochrome *b*-559 reaction center complex from spinach FEBS Lett 260: 138–140

Kok B, Forbush B and McGloin M (1970) Cooperation of charges in photosynthetic oxygen evolution. Photochem Photobiol 11: 457–475

Kuhn MG and Vermaas WFJ (1993) Deletion mutations in a long hydrophilic loop in the Photosystem II chlorophyll-binding protein CP43 in the cyanobacterium *Synechocystis* sp. PCC 6803. Plant Mol Biol 23: 123–133

Kulkarni RD, Mueller UW and Golden SS (1993) Nucleotide sequence of *psbB* from *Synechococcus* sp. PCC 7942. Biochim Biophys Acta 1173: 329–332

Kuwabara T and Murata N (1982) Inactivation of oxygen evolution and concomitant release of three polypeptides in the Photosystem II particles of spinach chloroplasts. Plant Cell Physiol 23: 533–539

Kuwabara T, Miyao M, Murata T and Murata N (1985) The function of the 33 kDa protein in the oxygen evolution system studied by reconstitution experiments. Biochim Biophys Acta 806: 283–289

Kuwabara T, Murata T, Miyao M and Murata N (1986) Partial degradation of the 18 kDa protein of the photosynthetic oxygen-evolving complex: A study of a binding site. Biochim Biophys Acta 850: 146–155

Kyle DJ (1987) Photosynthetic oxygen evolution. In: Kyle DJ, Osmond CB and Arntzen CJ (eds) Topics in Photosynthesis,. Vol 9, pp 197–226. Elsevier, Amsterdam

Lang JD and Haselkorn R (1989) Isolation, sequencing and transcription of the genes encoding the Photosystem II chlorophyll-binding protein CP47 in the cyanobacterium *Anabaena* 7120. Plant Mol Biol 13: 441–446

Lavergne J (1986) Stoichiometry of the redox changes of manganese during the photosynthetic water oxidation cycle. Photochem Photobiol 43: 311–317

Lavergne J (1987) Optical difference spectra of the S-state transitions in the oxygen-evolving complex. Biochim Biophys Acta 894: 91–107

Lind LK, Shukla VK, Nyhus KJ and Pakrasi HB (1993) Genetic and immunological analyses of the cyanobacterium *Synechocystis* sp. 6803 show that the protein encoded by the *psbJ* gene regulates the number of Photosystem II centers in thylakoid membranes. J Biol Chem 268: 1575–1579

Mavankal G, McCain DC and Bricker TM (1986) Effects of chloride on paramagnetic coupling of manganese in calcium chloride-washed Photosystem II preparations. FEBS Lett 202: 235–239

Mayes SR, Cook KM, Self SJ, Zhang, Z and Barber J (1991) Deletion of the gene encoding the Photosystem II 33 kDa protein from *Synechocystis* sp. PCC 6803 does not inactivate water splitting but increases vulnerability to photoinhibition. Biochim Biophys Acta 1060: 1–12

Mayes SR, Dubbs JM, Vass I, Hideg E, Nagy L and Barber J (1993) Further characterization of the *psbH* locus of *Synechocystis* sp. 6803: Inactivation of *psbH* impairs Q_A to Q_B electron transport in Photosystem II. Biochemistry 32: 1454–1465

Mayfield SP, Bennoun P and Rochaix J-D (1987) Expression of the nuclear encoded OEE1 protein is required for oxygen evolution and stability of Photosystem II particles in *Chlamydomonas reinhardtii*. EMBO J 6: 313–318

Mei R and Yocum CF (1991) Calcium retards NH_2OH inhibition of O_2 evolution activity by stabilization of Mn^{2+} binding to Photosystem II. Biochemistry 30: 7836–7842

Mei R, Green JP, Sayre RT and Frasch WD (1989) Manganese-binding proteins of the oxygen-evolving complex. Biochemistry 28: 5560–5567

Metz JG and Seibert M (1984) Presence in Photosystem II core complexes of a 3 kDa polypeptide required for oxygen evolution. Plant Physiol 76: 829–832

Metz JG, Pakrasi HB, Seibert M and Arntzen (1986)Evidence for a dual function of the herbicide-binding D1 protein in Photosystem II. FEBS Lett 205: 269–273

Michel HP and Bennett (1987) Identification of the phosphorylation site of an 83 kDa protein from Photosystem II of spinach. FEBS Lett 212: 103–108

Michel HP, Hunt DF, Shabanowitz J and Bennett J (1988) Tandem mass spectrometry reveals that three Photosystem II proteins of spinach chloroplasts contain N-acetyl-O-phosphothreonine at their NH_2 termini. J Biol Chem 263: 1123–1130

Mishra R and Ghanotakis DF (1994) Selective extraction of CP26 and CP29 proteins without affecting the binding of the extrinsic proteins (17, 23, and 33 kDa) and the DCMU sensitivity of a Photosystem II core complex. Photosyn Res 42: 37–42

Miyao M and Inoue Y (1991) An improved procedure for photoactivation of photosynthetic oxygen evolution: Effect of artificial electron acceptors on the photoactivation yield of NH_2OH-treated wheat Photosystem II membranes. Biochim Biophys Acta 1056: 47–56

Miyao M and Murata N (1983) Partial disintegration and reconstitution of the photosynthetic oxygen evolution system. Binding of 24 kDa and 18 kDa polypeptides. Biochim Biophys Acta 725: 87–93

Miyao M and Murata N (1984a) Role of the 33 kDa polypeptide in preserving Mn in the photosynthetic oxygen evolution system and its replacement by chloride ions. FEBS Lett 168: 281–286

Miyao M and Murata N (1984b) Effects of urea on Photosystem II particles. Evidence for an essential role of the 33 kDa polypeptide in oxygen evolution. Biochim Biophys Acta 765: 253–257

Miyao M and Murata N (1984c) Calcium ions can be substituted for the 24 kDa polypeptide in photosynthetic oxygen evolution. FEBS Lett 168: 118–120

Miyao M and Murata N (1985) The chloride effect on photosynthetic oxygen evolution: Interaction of Cl^- with 18 kDa, 24 kDa and 33 kDa proteins. FEBS Lett 180: 303–308

Miyao M and Murata N (1989) The mode of binding of three extrinsic proteins of 33 kDa, 24 kDa and 18 kDa in the Photosystem II complex of spinach. Biochim Biophys Acta 977: 315–321

Miyao M, Murata N, Lavorel J, Maison-Peteri B, Boussac A and Etienne A-L (1987) Effects of the 33 kDa protein on the S-state transitions in photosynthetic oxygen evolution. Biochim Biophys Acta 890: 151–159

Miyao M, Fujimura Y and Murata N (1988) Partial degradation of the extrinsic 23 kDa protein of the Photosystem II complex of spinach. Biochim Biophys Acta 936: 465–474

Murata N, Miyao M, Omata T, Matsunami H and Kuwabara T (1984) Stoichiometry of components in the photosynthetic oxygen evolution system of Photosystem II particles prepared with Triton X-100 from spinach chloroplasts. Biochim Biophys Acta 765: 363–369

Nagatsuka T, Fukuhara S, Akabori K and Toyoshima Y (1991) Disintegration and reconstitution of Photosystem II reaction center core complex. II. Possible involvement of low-molecular-mass proteins in the functioning of Q_A in the PS II reaction center. Biochim Biophys Acta 1057: 223–231

Nakatani HY, Ke B, Dolan E, Arntzen CJ (1984) Identity of the Photosystem II reaction center polypeptide. Biochim Biophys Acta 765: 347–352

Nanba O and Satoh K (1987) Isolation of a Photosystem II reaction center consisting of D1 and D2 polypeptides and cytochrome *b*-559. Proc Natl Acad Sci USA 84: 109–112

Odom WR and Bricker TM (1992) Interaction of CPa-1 with the manganese-stabilizing protein of Photosystem II: Identification of domains crosslinked by 1-ethyl-3-[3(dimethylamino) propyl]carbo-diimide. Biochemistry 31: 5616–5620

Offermann-Steinhard K and Herrmann RG (1990) Nucleotide sequences of *psbB* and *psbH*, the plastid encoded genes for CP47 and the 10 kDa phosphoprotein of Photosystem II in *Oenothera hookeri* and *argillicola*. Nuc Acid Res 18: 6452

Ohyama K, Fukuzawa H, Kohchi T, Shirai H, Sano T, Sano S, Umesono K, Shiki Y, Takeuchi M, Chang Z, Aota Ş, Inokuchi H and Ozeki H (1986) Chloroplast gene organization deduced from complete sequence of liverwort *Marchantia polymorpha* chloroplast DNA. Nature 322: 572–573

Okamura MY, Satoh K, Isaacson RA and Feher G (1987) Evidence of the primary charge separation in the D1/D2 complex of Photosystem II from spinach: EPR of the triplet state. In: Biggins J (ed) Progress in Photosynthesis Research, Vol 1, pp 379–381. Martinus Nijhoff, Dordrecht

Ono T and Inoue Y (1985) S-state turnover in the O_2-evolving system of $CaCl_2$-washed Photosystem II particles depleted of three peripheral proteins as measured by thermoluminescence. Removal of 33 kDa protein inhibits S3 to S4 transition. Biochim Biophys Acta 806: 331–340

Ono T-A, Kajikawa H and Inoue Y (1986) Changes in protein composition and Mn abundance in Photosystem II particles on photoactivation of the latent O_2-evolving system in flash-grown wheat leaves. Plant Physiol 80: 85–90

Pakrasi HB and Vermaas WFJ (1992) Protein engineering of Photosystem II. In: Barber J (ed) The Photosystems: Structure Function and Molecular Biology, pp 231–258. Elsevier Science Publishers B.V., Amsterdam

Pakrasi HB, Riethman HC and Sherman LA (1985) Organization of the pigment proteins of the Photosystem II complex of the cyanobacterium *Anacystis nidulans* R2 Proc Natl Acad Sci USA 82: 6903–6907

Pakrasi HB, Nyhus K and Granok, H (1990) Targeted deletion mutagenesis of the beta subunit of cytochrome *b*-559 destabilizes the reaction center of Photosystem II. Z Naturforsch

45c: 423–429

Pakrasi HB, Ciechi PD and Whitmarsh J (1991) Site-directed mutagenesis of the heme axial ligands of cytochrome *b*-559 affects the stability of the Photosystem II complex. EMBO J 10: 1619–1627

Pecoraro VL (1992) Introduction to manganese enzymes. In: Pecoraro VL (ed) Manganese Redox Enzymes, pp 197–231. VCH Publishers, New York

Penner-Hahn JE, Fronko RM, Pecoraro VL, Yocum CF, Betts SD and Bowlby NR (1990) Structural characterization of the Mn sites in the photosynthetic oxygen-evolving complex using X-ray absorption spectroscopy. J Am Chem Soc 112: 2549–2557

Petersen J, Dekker JP, Bowlby NR, Ghanotakis DF, Yocum CF and Babcock GT (1990) EPR characterization of the CP47-D1-D2-cytochrome *b*-559 complex of Photosystem II. Biochemistry 29: 3226–3231

Philbrick JB, Diner BA and Zilinskas BA (1991) Construction and characterization of cyanobacterial mutants lacking the manganese-stabilizing protein of Photosystem II. J Biol Chem 266: 13370–13376

Pistorius EK (1983) Effects of Mn^{+2}, Ca^{+2} and chlorpromazine on Photosystem II of *Anacystis nidulans*. An attempt to establish a functional relationship of amino acid oxidase to Photosystem II. Eur J Biochem 135: 217–222

Pistorius EK and Schmid GH (1984) The effects of Mn^{+2} and Ca^{+2} on oxygen evolution and on the variable fluorescence yield associated with Photosystem II in preparations of *Anacystis nidulans*. FEBS Lett 171: 173–178

Prasil O, Adir N, and Ohad I (1992) Dynamics of Photosystem II: Mechanism of photoinduction and recovery process. In: Barber J (ed) Current Topics in Photosynthesis, Vol 11, pp 220–250. Elsevier, Amsterdam

Putnam-Evans C and Bricker TM (1992) Site-directed mutagenesis of the CPa-1 protein of Photosystem II: Alteration of the basic residue pair ^{384,385}R to ^{384,385}G leads to a defect associated with the oxygen-evolving complex. Biochemistry 31: 11482–11488

Putnam-Evans C and Bricker TM (1994) Site-directed mutagenesis of the CP47 protein of Photosystem II: Alteration of the basic residue ^{448}R to ^{448}G prevents the assembly of functional Photosystem II centers under chloride-limiting conditions. Biochemistry 33: 10770–10776

Radmer R and Ollinger O (1986) Do the higher oxidation states of the photosynthetic O_2-evolving system contain bound H_2O? FEBS Lett 195: 285–289

Rashid A and Carpentier R (1990) The 16 and 23 kDa extrinsic polypeptides and the associated Ca^{+2} and Cl^- modify atrazine interaction with the Photosystem II core complex. Photosynth Res 24: 221–227

Reverdatto SV, Andreeva AV, Buryahova AA, Chakhmakhcheva OG, and Efimov VA (1988) Nucleotide sequence of barley chloroplast *psbC* gene. Nuc Acid Res 17: 3996

Riggs PJ, Mei R, Yocum CF and Penner-Hahn JE (1992) Reduced derivatives of the manganese cluster in the photosynthetic oxygen-evolving complex. J Am Chem Soc 114: 10650–10651

Rock CD, Barkan A and Taylor WC (1987) The maize plastid *psbB-psbF-petB-petD* gene cluster – spliced and unspliced *petB* and *petD* RNAs encode alternative products. Curr Genet 12: 69–77

Rögner M, Chisholm DA and Diner B (1991) Site-directed mutagenesis of the *psbC* gene of Photosystem II: Isolation and functional characterization of CP43-less Photosystem II core complexes. Biochemistry 30: 5387–5395

Rutherford AW (1989) Photosystem II. The water splitting enzyme. Trends Biochem Sci 14: 227–232

Salter AH, Virgin I, Hagman Å and Andersson B (1992) On the mechanism of light-induced D1 protein degradation in Photosystem II core particles. Biochemistry 31: 3990–3998

Sandusky PO and Yocum CF (1983) The mechanism of amine inhibition of the photosynthetic oxygen-evolving complex. Amines displace functional chloride from a ligand site on manganese. FEBS Lett 162: 339–343

Sandusky PO and Yocum CF (1984) The chloride requirement for photosynthetic oxygen evolution. Analysis of the effects of chloride and other anions on the amine inhibition of the oxygen evolving complex. Biochim Biophys Acta 766: 603–611

Saphon S and Crofts T (1977) Protolytic reactions in Photosystem II: A new model for the release of protons accompanying the photooxidation of water. Z Naturforch 32c: 617–626

Saygin O and Witt HT (1987) Optical characterization of intermediates in the water-splitting enzyme system of photosynthesis—possible states and configurations of manganese and water. Biochim Biophys Acta 893: 452–469

Sayre RT and Wrobel-Boerner EA (1994) Molecular topology of the Photosystem II chlorophyll *a* binding protein, CP43: Topology of a thylakoid membrane protein. Photosyn Res 40: 11–19

Sayre RT, Andersson B and Bogorad L (1986) The topology of a membrane protein: The orientation of the 32 kDa Q_B-binding chloroplast thylakoid membrane protein. Cell 47: 601–608

Shen G, Eaton-Rye JJ and Vermaas WFJ (1993) Mutation of histidine residues in CP47 leads to a destabilization of the Photosystem II complex and to impairment of light energy transfer. Biochemistry 32: 5109–5115

Shen J-R and Inoue Y (1993) Binding and function of two new extrinsic components, cytochrome c_{550} and a 12 kDa protein, in cyanobacterial Photosystem II. Biochemistry 32: 1825–1832

Shen J-R, Satoh K and Katoh S (1988) Isolation of an oxygen evolving Photosystem II preparation containing only one tightly bound calcium atom from a chlorophyll *b*-deficient mutant of rice. Biochim Biophys Acta 936: 386–393

Shinozaki K, Ohme M, Tanaka M, Wakasugi T, Hayashida N, Matsubayashi T, Zaita N, Chunwongse J, Obokata J, Yamaguchi-Shinozaki K, Ohto C, Torazawa K, Meng BY, Sugita M, Dean H, Kamagishara T, Yamada K, Kasuda J, Takaiwa F, Kato A, Tohdoh N, Shimada H and Sugiura M (1986) The complete nucleotide sequence of the tobacco chloroplast genome: Its gene organization and expression. EMBO J 5: 2043–2049

Shipton CA and Barber J (1992) Characterization of photoinduced breakdown of the D1-polypeptide in isolated reaction centers of Photosystem II. Biochim Biophys Acta 1099: 85–90

Stockhaus J, Hofer M, Renger G, Westhoff P, Wydrzynski T and Willmitzer L (1990) Antisense RNA efficiently inhibits formation of the 10 kDa polypeptide of Photosystem II in transgenic potato plants: Analysis of the role of the 10 kDa protein. EMBO J 9: 3013–3021

Svensson B, Vass I and Styring S (1991) Sequence analysis of the

D1 and D2 reaction center proteins of Photosystem II. Z Naturforsch 46c: 765–776

Tae GS and Cramer WA (1992) Truncation of the COOH-terminal domain of the *psbE* gene product in *Synecocystis* sp. 6803: Requirements for Photosystem II assembly and function. Biochemistry 31: 4066–4073

Takahashi M-A and Asada K (1991) Determination of the molecular size of the binding site for the manganese-stabilizing protein of Photosystem II membranes. Biochim Biophys Acta 1059: 361–363

Tamura N and Cheniae G (1987) Photoactivation of the water-oxidizing complex in Photosystem II membranes depleted of Mn and extrinsic proteins. I. Biochemical and kinetic characterization. Biochim Biophys Acta 890: 179–193

Tamura N, Radmer R, Lantz S, Cammarata K and Cheniae GM (1986) Depletion of Photosystem II-extrinsic proteins. II. Analysis of the PS II/water-oxidizing complex by measurements of N,N,N′,N′-tetramethyl-*p*-phenylenediamine oxidation following an actinic flash. Biochim Biophys Acta 850: 369–379

Tamura N, Ikeuchi M and Inoue Y (1989) Assignment of histidine residues in D1 protein as possible ligands for functional manganese in photosynthetic water-oxidizing complex. Biochim Biophys Acta 973: 281–289

Tanaka S and Wada K (1988) The status of cysteine residues in the 33 kDa protein of spinach Photosystem II complexes. Photosyn Res 17: 255–266

Tang X-S and Satoh K (1984) Characterization of a 47-kilodalton chlorophyll-binding polypeptide isolated from a Photosystem II core complex. Plant Cell Physiol 25: 935–945

Tang X-S, Fushimi K and Satoh Ki (1990) D1-D2 complex of the Photosystem II reaction center from spinach. Isolation and partial characterization. FEBS Lett 273: 257–260

Tang X-S, Diner BA, Larsen BS, Gilchrist, Jr. ML, Lorigan GA and Britt RD (1994) Identification of histidine at the catalytic site of the photosynthetic oxygen-evolving complex. Proc Natl Acad Sci USA 91: 704–708

Tomo T, Enami I and Satoh K (1993) Orientation and nearest neighbor analysis of *psbI* gene product in the Photosystem II reaction center complex using bifunctional crosslinkers. FEBS Lett 323: 15–18

Trebst A (1986) The three-dimensional structure of the herbicide binding niche on the reaction center polypeptides of Photosystem II. Z Naturforsch 42c: 742–750

Trebst A and Depka B (1985) The architecture of Photosystem II in plant photosynthesis. Which polypeptide carries the reaction center of Photosystem II? In: Michel-Beyerle ME (ed) Antennas and Reaction Centers in Photosynthetic Bacteria-Interactions and Dynamics, pp 216–223. Springer-Verlag, Berlin

Tronrud DE, Schmidt MF and Matthews BW (1986) Structure and X-ray amino acid sequence of a bacteriochlorophyll-*a* protein from *Prosthecochloris aestuarii* refined at 1.9 Å resolution. J Mol Biol 188: 443–453

van Mieghem FJE and Rutherford AW (1993) Comparative spectroscopy of Photosystem II and purple bacterial reaction centres. Biochem Soc Trans 21: 986–991

Vass I, Ono T and Inoue Y (1987) Stability and oscillation properties of thermoluminescent charge pairs in the oxygen-evolving system depleted in Cl^- or the 33 kDa extrinsic protein. Biochim Biophys Acta 892: 224–235

Vermaas WFJ, Williams JGK and Arntzen CJ (1987) Sequencing and modification of *psbB*, the gene encoding the CP-47 protein of Photosystem II in the cyanobacterium *Synechocystis* 6803. Plant Mol Biol 8: 317–326

Vermaas WFJ, Ikeuchi M and Inoue Y (1988) Protein composition of the Photosystem II core complex in genetically engineered mutants of the cyanobacteriun *Synechocystis* PCC 6803 Photosyn Res 17: 97–113

Virgin I, Styring S and Andersson B. (1988) Photosystem II disorganization and manganese release after photoinhibition of isolated spinach thylakoid membranes. FEBS Lett 233: 408–412

Virgin I, Salter AH, Ghanotakis DF, and Andersson B (1991) Light-induced D1 degradation is catalyzed by a serine-type protease. FEBS Lett 287: 125–128.

von Heijne G and Gavel Y (1988) Topogenic signals in integral membrane proteins. Eur J Biochem 174: 671–678

Waggoner CM and Yocum CF (1987) Selective depletion of water-soluble polypeptides associated with Photosystem II. In: Biggins J (ed) Progress in Photosynthesis Research, Vol I, pp 685–688. Martinus Nijhoff, Dordrecht

Wales R, Newman BJ, Pappin D and Gray JC (1989a) The extrinsic 33 kDa polypeptide of Photosystem II is a putative calcium-binding protein and is encoded by a multi-gene family in pea. Plant Mol Biol 12: 439–451

Wales R, Newman BJ, Rose SA, Pappin D and Gray JC (1989b) Characterization of cDNA clones encoding the extrinsic 23 kDa polypeptide of the oxygen-evolving complex of Photosystem II. Plant Mol Biol 13: 573–582

Webber AN and Gray JC (1989) Detection of calcium binding by Photosystem II polypeptides immobilized onto nitrocellulose membrane. FEBS Lett 249: 79–82

Webber AN, Packman L, Chapman DJ, Barber J and Gray JC (1989) A fifth chloroplast encoded polypeptide is present in the Photosystem II reaction center complex. FEBS Lett 242: 259–262

Wechsler T, Suter F, Fuller RC and Zuber H (1985) The complete amino acid sequence of the bacteriochlorophyll c binding polypeptide from chlorosomes of the green photosynthetic bacterium *Chloroflexus aurantiacus*. FEBS Lett 181: 173–178

Xu QA and Bricker TM (1992) Structural organization of proteins on the oxidizing side of Photosystem II: Two molecules of the 33 kDa manganese-stabilizing protein per reaction center. J Biol Chem 267: 25816–25821

Xu QA, Nelson J and Bricker TM (1994) Secondary structure of the 33 kDa, extrinsic protein of Photosystem II: A far-UV circular dichroism study. Biochim Biophys Acta 1188: 427–431.

Yachandra VK, DeRose VJ, Latimer MJ, Mukerji, I, Sauer K and Klein MP (1993a) A structural model for the photosynthetic oxygen evolving manganese complex. Jpn J Appl Phys 32s: 523–526

Yachandra VK, DeRose VJ, Latimer MJ, Mukerji, I, Sauer K and Klein MP (1993b) Where plants make oxygen: A structural model for the photosynthetic oxygen-evolving manganese cluster. Science 260: 675–679

Yamaguchi N, Takahashi Y and Satoh K (1988) Isolation and characterization of a Photosystem II core complex depleted in the 43 kDa chlorophyll binding subunit. Plant Cell Physiol 29: 123–129

Yamamoto Y, Ueda T, Shinkai H and Nishimura M (1982) Preparation of oxygen-evolving Photosystem II subchloroplasts from spinach. Biochim Biophys Acta 679: 347–350

Yocum CF (1991) Calcium activation of photosynthetic water oxidation. Biochim Biophys Acta 1059: 1–15

Yocum CF, Yerkes CT, Blankenship RE, Sharp RR and Babcock GT (1981) Stoichiometry, inhibitor sensitivity, and organization of manganese associated with photosynthetic oxygen evolution. Proc Natl Acad Sci USA 78: 7507–7511

Zimmermann J-L and Rutherford AW (1986) Electron paramagnetic resonance properties of the S2 state of the oxygen evolving complex of Photosystem II. Biochemistry 25: 4609–4615

Zuber H, Brunisholz R and Sidler W (1987) Structure and function of light harvesting pigment protein complexes. In: Amesz J (ed) Photosynthesis, pp 233–271. Elsevier Press, Amsterdam

Chapter 9

Oxygen Evolution

R. David Britt
Department of Chemistry, University of California, Davis, CA 95616, USA

Summary

A unique capability of the Photosystem II (PS II) reaction center is the ability to extract electrons from water, producing molecular oxygen as a byproduct. Four photon-induced charge separations at the site of the chlorophyll moiety P680 couple sequentially to oxidation events at an Oxygen Evolving Complex (OEC), resulting in the formation of molecular oxygen. A tetranuclear manganese cluster is at the heart of the OEC. Recent biochemical and spectroscopic results have given new insights into the structure of this Mn cluster, its ligation to the PS II polypeptides, and the role of essential cofactors Ca^{2+} and Cl^-. Models for the oxygen evolution mechanism are discussed, including new models that assign a direct role in water splitting to the redox active tyrosine $Y_Z^{\bullet}$.

Donald R. Ort and Charles F. Yocum (eds): Oxygenic Photosynthesis: The Light Reactions, pp. 137–164.

I. Introduction and Overview

Two to three billion years ago a photosynthetic reaction center evolved that was functionally similar to current Photosystem II (PS II) reaction centers. This reaction center was capable of extracting electrons from water and releasing molecular oxygen as a byproduct. Organisms capable of oxygenic photosynthesis were able to utilize an inexhaustable supply of water as a source of biologically activated electrons and protons. The molecular oxygen released rapidly (by geological timescale standards) transformed earth's atmosphere from a composition of reduced gases to one rich in O_2, setting the stage for the evolution of the highly efficient respiration process. The focus of this chapter is to describe our current state of understanding of the process of photosynthetic oxygen evolution in higher plants and cyanobacteria. There has been rapid progress in research on oxygen evolution in recent years, but there are still many unanswered questions with regard to both the molecular structure of the oxygen evolving complex (OEC) and details of its function.

As described by Diner and Babcock (Chapter 12), light-induced charge separation in PS II occurs at the P680 chlorophyll moiety, which is bound at the interface of the D1/D2 heterodimer that forms the protein core of PS II. Rapid electron transfer from the excited state P680* pigment to an adjacent pheophytin is followed by slower electron transfer to a bound plastoquinone, Q_A. The electron is then transferred to another plastoquinone in a second quinone binding site, Q_B. After a second electron is sequentially transferred through Q_A to Q_B, the Q_B quinone takes up two protons from the stromal side of the thylakoid membrane and is released into the membrane as a reduced plastoquinol, to be replaced by an oxidized plastoquinone from a quinone pool in the membrane. The quinone cycle is closed in the cytochrome b_6f complex, where the quinol is oxidized to quinone, releasing the protons into the lumenal phase, helping to establish the proton gradient needed for ATP synthesis (see Chapter 23).

Abbreviations: ENDOR – Electron Nuclear Double Resonance; EPR – Electron Paramagnetic Resonance; ESE – Electron Spin Echo; ESEEM – Electron Spin Echo Envelope Modulation; ESE-ENDOR – Electron Spin Echo-Electron Nuclear Double Resonance; EXAFS – Extended X-ray Absorption Fine Structure; PS II – Photosystem II; OEC – Oxygen-Evolving Complex; TRIS – 2-amino-2-hydroxymethylpropane-1,3-diol; XANES – X-ray Absorption Near Edge Structure

The $P680^+$ chlorophyll cation formed by the initial electron transfer is a strong oxidant which rapidly oxidizes a nearby tyrosine designated Y_Z (see Chapter 12). This redox active tyrosine, residue 161 of the D1 polypeptide (*Synechocystis* numbering), is conventionally thought of as the lone electron transfer intermediate between $P680^+$ and the OEC. There is a clear need for such an electron transfer intermediate, as the slowest step in the oxygen evolution process occurs on the timescale of 1 ms, and therefore the quantum yield for oxygen evolution would be quite low in the absence of an electron transfer intermediate to stabilize charge separation with respect to fast charge recombination reactions. There is also a symmetry-related tyrosine designated Y_D, tyrosine 160 of the D2 polypeptide (again *Synechocystis* numbering), which is typically present as a dark-stable radical $Y_D^{\bullet}$, and which is bypassed in the fast electron transfer between the OEC and $P680^+$.

After its formation via electron transfer to $P680^+$, the tyrosyl radical $Y_Z^{\bullet}$ in turn oxidizes the OEC. The core of the OEC is a cluster of four Mn ions, with Ca^{2+} and Cl^- present as necessary cofactors. After four oxidation equivalents are transferred from the OEC, molecular oxygen is liberated, and the OEC resets to its most reduced state. At least some of the intermediate OEC oxidation steps occur within the Mn cluster, although it is also possible that one or more steps involve partial substrate oxidation or oxidation of a proximal amino acid. Details of the oxygen evolving cycle were accurately described by Kok, and the cycle involving five so-called S-state intermediates is referred to as the Kok cycle or the S-state cycle. There is continued debate about which state or states bind substrate water molecules in the OEC.

The conventional view of the role of the Mn cluster in oxygen evolution is two-fold, acting both to oxidize the substrate water molecules and to position them to facilitate the formation of the O_2 double bond. Recent proposals have suggested that tyrosine Y_Z also serves directly in the water-splitting process. In these models the transient $Y_Z^{\bullet}$ neutral radical operates to abstract protons or hydrogen atoms from water ligands to the Mn cluster. Thus the $Y_Z^{\bullet}$ component is proposed to be an integral part of the OEC, as well as the first component of the OEC to be oxidized by $P680^+$. In these models the OEC is therefore a metallo-radical enzyme rather than a metalloenzyme with radical mediated electron transfer.

II. Kinetics and Thermodynamics of Oxygen Evolution

A. Kinetics

Key kinetics experiments underpinning our current ideas about photosynthetic oxygen evolution were first performed some 25 years ago. Joliot et al. (1969) measured the release of oxygen during a series of short (≈10 μs) saturating flashes applied to previously dark-adapted suspensions of spinach chloroplasts or Chlorella. The result was a remarkable period-four oscillation of oxygen evolution vs. flash number. Figure 1 shows such a pattern as measured for spinach chloroplasts by Babcock (1973). The quantity of evolved oxygen is maximal after the third flash, with repeated maxima after successive four flash intervals. The oxygen evolution pattern eventually dephases and reaches a steady state value for each flash. These experiments clearly demonstrate that four independent photo-oxidation steps are required to oxidize water.

Kok et al. (1970) reproduced the flash pattern results of Joliot et al. (1969) and devised a kinetic model to explain the data. A current picture of the 'S-state' model is illustrated in Fig. 2. In the Kok S-state model each PS II reaction center acts as an independent unit in the oxidation of water. This is a key point: before the flash experiments it was conceivable that PS II reaction centers generated diffusable oxidants which acted to split water at some remote center independent of the site of charge separation. S-state transitions represent single electron oxidations of the OEC, caused by electron transfer through Y_Z to rereduce $P680^+$. The most reduced state, S_0, is formed after molecular oxygen is released from a previous cycle. Photon absorption and electron transfer from P680 lead to the formation of the highly oxidizing $P680^+$ cation, which oxidizes Y_Z to $Y_Z^{\bullet}$. In turn, $Y_Z^{\bullet}$ extracts an electron from the OEC, inducing the $S_0 \rightarrow S_1$ transition. Subsequent photon absorptions by P680 drive the $S_1 \rightarrow S_2$, $S_2 \rightarrow S_3$, and $S_3 \rightarrow S_4$ transitions. S_4 is an unstable activated complex which releases molecular oxygen on the timescale of 1 ms, resetting the cycle. One unusual feature of the cycle is the fact that maximal O_2 evolution occurs following the third flash of dark-adapted photosynthetic membranes rather than the fourth. This is a consequence of the fact that S_1 is the dark-stable state.

Electron transfer times through Y_Z have been measured by time-resolved optical and electron paramagnetic resonance spectroscopies. Figure 3 shows a summary of the electron transfer timescales in PS II as a function of S-state (Rutherford, 1989; Hansson and Wydrzynski, 1990). Electron transfers are relatively fast during the early S-state transitions, but slow appreciably as the OEC becomes more

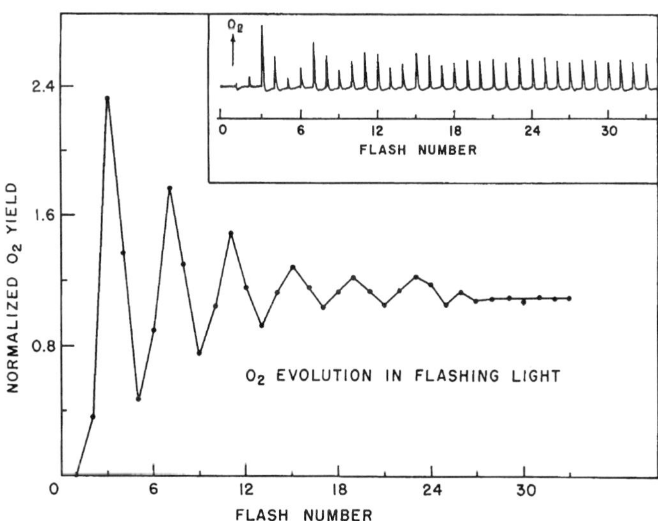

Fig. 1. The amplitude of molecular oxygen released during a train of 21 μs saturating light flashes applied at 1 s intervals to a dark-adapted spinach chloroplast suspension (Babcock, 1973). The inset shows kinetically resolved O_2 evolution traces, while the main plot displays amplitudes vs. flash number. This figure was kindly provided by Gerald T. Babcock.

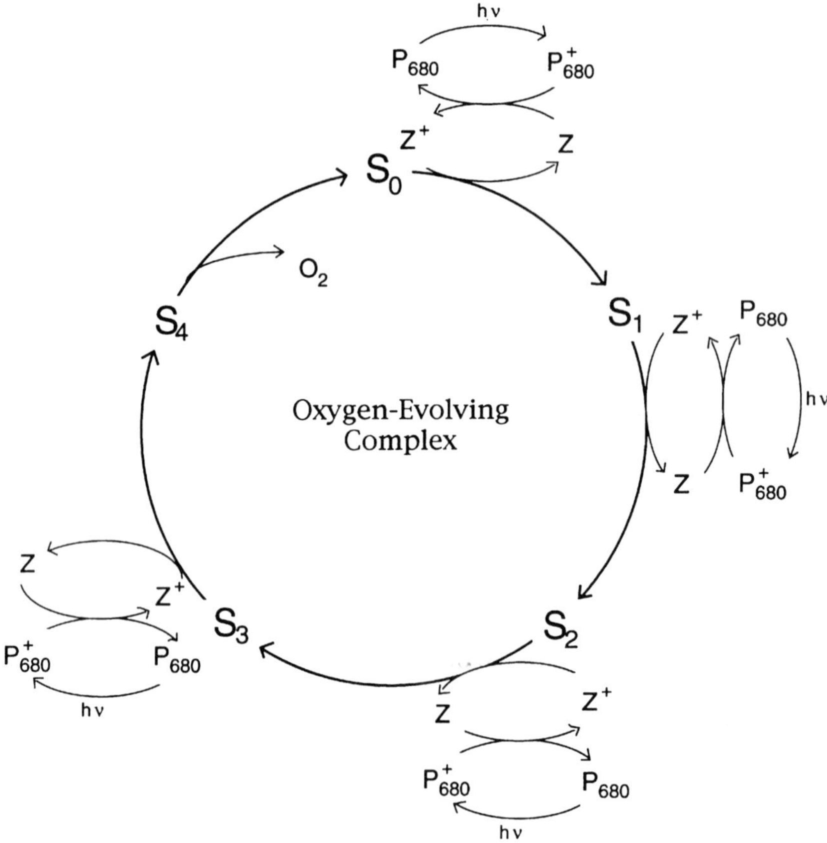

Fig. 2. The OEC S-state cycle introduced by Kok et al. (1970).

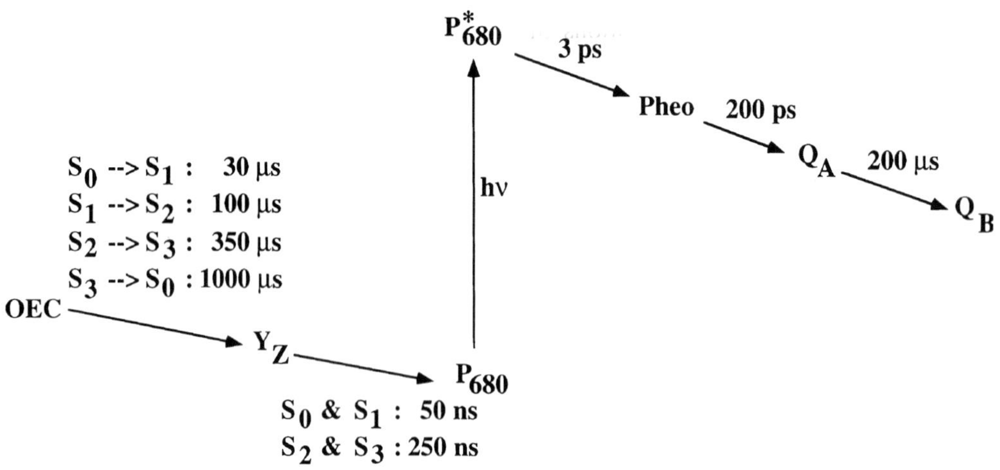

Fig. 3. Timescales of electron transfer in PS II (adapted from Rutherford, 1989).

oxidized. Oxygen release and the final electron transfer out of the OEC occur on the same 1 ms timescale. This leads to the possibility that the S_4 state is formally equivalent to $S_3 + Y_Z^{\bullet}$, and is one basis for recent models proposing that $Y_Z^{\bullet}$ functions to abstract protons or hydrogen atoms from substrate water.

B. Thermodynamics

From the flash dependence studies of oxygen evolution, we know that the four 1-electron oxidizing equivalents required to produce O_2 in the half-reaction $2H_2O \rightarrow O_2 + 4H^+ + 4e^-$ occur at the OEC level as the result of four sequential electron transfers, each reducing the photoxidized $P680^+$ species. P680 undergoes charge separation with photons of wavelengths as long as 680 nm. The energy of a 680 nm photon is 1.8 eV. The thermodynamic requirement for an oxygen evolving complex is clear from an examination of Fig. 4 (adapted from Radmer and Cheniae, 1977), which illustrates the redox potentials for the individual one-electron electrochemical oxidation steps on the path from H_2O to O_2. The first step, $H_2O \leftrightarrow HO + H^+ + e^-$, operates at a 2.3 V potential, so the driving force available from coupling to the energy of a 680 nm photon is clearly insufficient. On the other hand, the average potential for oxidizing H_2O to O_2 over four steps is only 0.81 V. The role of the OEC from a thermodynamic point of view is to bypass the high energy 1-electron intermediates via a series of lower energy oxidation intermediates that are energetically accessible through P680 photochemistry. For example, early S-state transitions are likely to occur as oxidation of Mn ions in the OEC. At later S-state transitions, the highly oxidized Mn ions can then carry out concerted water ligand oxidations of two or more electron equivalents, avoiding the high energy 1-electron transitions of Fig. 4.

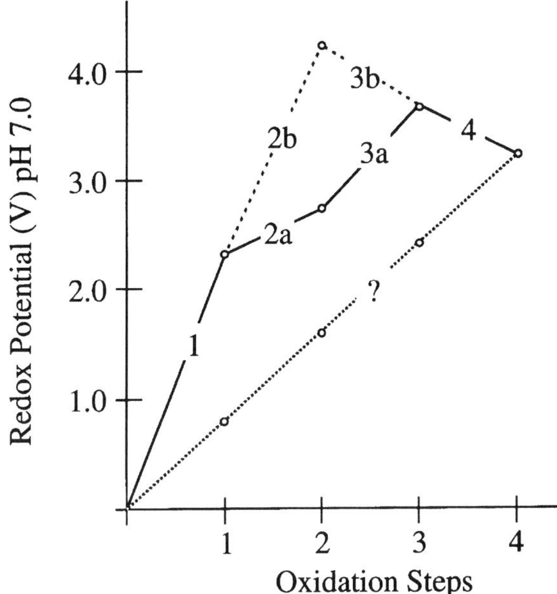

Fig. 4. Redox potentials of the individual redox couples for the full oxidation of water to molecular oxygen at pH 7.0 (adapted from Radmer and Cheniae, 1977). Two distinct pathways are shown. The solid line passes through a hydrogen peroxide intermediate. The dashed line employs a higher energy atomic oxygen intermediate. The oxidation-reduction couples indicated on the figure are:

(1) $H_2O \leftrightarrow HO + H^+ + e^-$
(2a) $HO + H_2O \leftrightarrow H_2O_2 + H^+ + e^-$
(3a) $H_2O_2 \leftrightarrow O_2^- + 2H^+ + e^-$
(4) $O_2^- \leftrightarrow O_2 + e^-$
(2b) $HO \leftrightarrow O + H^+ + e^-$
(3b) $O + H_2O \leftrightarrow O_2^- + 2H^+ + e^-$
(?) Path of minimum energy per step ($0.81 V/e^-$)

III. Structural Studies

A. Biochemical Refinement

Though experiments such as the O_2 flash oscillation studies were possible with chloroplast preparations or even whole unicellular organisms, a great advance in understanding oxygen evolution came with the advent of refined PS II preparations (see Chapter 8). These preparations give researchers the ability to quantitatively determine the cofactor composition of PS II and to perform spectroscopic experiments with high sensitivity and without contamination from other thylakoid membrane complexes. A major breakthrough was the 'BBY' preparation, where Berthold et al. (1981) isolated grana stack derived membrane sheets heavily enriched in PS II. Subsequent preparations have become more refined in terms of deletion of polypeptides not required for O_2 evolution. Details of the polypeptide composition of PS II and current views of individual polypeptide functions can be found in Chapters 8 and 11 and in other recent reviews (Debus, 1992; Yocum, 1994). As for the Mn ions and cofactors of the OEC, the current view of their stoichiometry is four Mn ions (Yocum et al., 1981), one Ca^{2+} ion (Ädelroth et al., 1995), and one Cl^- ion (Lindberg et al., 1990) (*vide infra*).

Although much has been learned about oxygen-evolution from the higher plant preparations described above, limitations are imposed by the difficulty of implementing modern molecular biology techniques such as site-directed mutagenesis in these complicated organisms. Cyanobacteria and the eukaryotic alga *Chlamydomonas* are more amenable to such methods

(Nixon et al., 1992). In particular, highly active oxygen-evolving PS II preparations have been devised for cyanobacteria (Noren et al., 1991; Tang and Diner, 1994).

Although a number of researchers have been working to obtain X-ray diffraction quality PS II crystals, progress to date has been slow. Therefore, most structural information about the OEC has come from spectroscopic investigations, with EPR and EXAFS playing dominant roles.

B. Electron Paramagnetic Resonance (EPR) Spectroscopy of the Manganese Cluster

Examination of the Kok cycle (Fig. 2) reveals the potential utility of EPR spectroscopy in studying the O_2-evolution process. EPR provides a powerful structural tool when paramagnetic species are present. The chlorophyll species $P680^+$ and tyrosines $Y_Z^{\bullet}$ and $Y_D^{\bullet}$ are paramagnetic radicals which have been extensively studied with EPR (see Chapter 12). For example, EPR studies of the dark-stable $Y_D^{\bullet}$ radical (Barry and Babcock, 1987) in cyanobacteria grown on deuterated tyrosine provided definitive evidence that $Y_D^{\bullet}$ and $Y_Z^{\bullet}$ are tyrosine radicals. Transition metals such as manganese may have unpaired *d*-orbital electrons and thus can be studied with EPR spectroscopy. The focus of this section is to describe EPR experiments that have led to the conclusion that the four Mn ions in PS II exist in a single tetranuclear Mn cluster.

A major breakthrough in understanding the role of manganese in oxygen-evolution came with the discovery of the $g = 2$ 'multiline' EPR signal by Dismukes and Siderer (1981). The pattern of hyperfine lines in this signal (Fig. 5) clearly indicated the presence of a multinuclear Mn cluster in the OEC. The ^{55}Mn nucleus is 100% naturally abundant. The EPR signal from a single Mn ion presents a total of six hyperfine lines arising from the six allowed m_I values of the spin $I = 5/2$ ^{55}Mn nucleus. In contrast, synthetic antiferromagnetically-coupled Mn(III)-Mn(IV) dinuclear clusters show approximately 16 hyperfine lines resulting from the superposition of EPR transitions arising from the 36 m_I combinations of the two ^{55}Mn nuclei in chemically inequivalent sites (Cooper et al., 1978). The multiline EPR spectrum bears some resemblance to the spectra from such mixed valence Mn dinuclear clusters, though it is broader and has several more resolved transitions. It was thus clear from the detection of the multiline EPR signal that there was minimally a dinuclear Mn cluster present in the OEC. The additional transitions and extra linewidth were suggestive of a higher nuclearity origin for the signal, with a tetranuclear cluster the prime candidate.

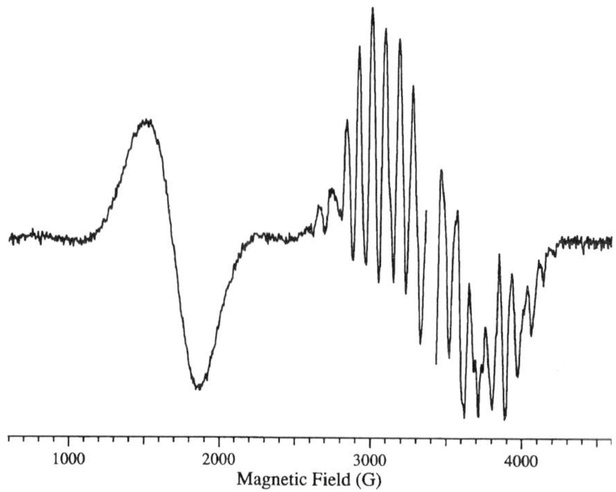

Fig. 5. The S_2-state manganese cluster EPR signals. The 'multiline' EPR signal consists of >16 partially-resolved ^{55}Mn hyperfine lines centered at the $g = 2$ region of the spectrum (3400 G for this figure obtained with 9.5 GHz microwave frequency). The $Y_D^{\bullet}$ tyrosine radical signal occurs at $g = 2$. It is offscale for the instrument gain used to record the multiline signal and is excised in the figure. The Mn $g = 4.1$ signal is the broad unstructured feature centered at 1700 G. This (195 K illumination – dark) difference spectrum was obtained with spinach PS II particles in the presence of 1.0 mM methanol and 400 mM sucrose, which gives appreciable amplitudes to both multiline and $g = 4.1$ signals. This figure was kindly provided by Dee Ann Force and Gary Lorigan.

The flash dependence of the multiline signal showed that it arises from the OEC in the S_2-state (Dismukes and Siderer, 1981). Another OEC signal in the $g = 4.1$ region of the spectrum (Fig. 5) was soon detected (Casey and Sauer, 1984; Zimmermann and Rutherford, 1984). This $g = 4.1$ signal extends over a field range of approximately 1000 G and shows no ^{55}Mn hyperfine structure in unoriented samples. Its origin in terms of S-states was initially the subject of debate, but later flash dependence studies showed that the $g = 4.1$ signal also originates from the S_2-state (Zimmermann and Rutherford, 1986). As discovered by Casey and Sauer (1984), the $g = 4.1$ signal is formed by lower temperature continuous illumination than is the multiline signal, with a maximal yield at 140 K vs. 195 K. However in PS II samples warmed to 195 K, the $g = 4.1$ signal converts into the multiline signal. Also, the $g = 4.1$ signal is stabilized by certain cryoprotectants such as sucrose (Zimmermann and Rutherford, 1986) and by various inhibitors of oxygen evolution such as F^- and small amines (Beck and Brudvig, 1986, 1988). The major question, what was the molecular origin of the two distinct S_2-state signals, appeared to have two possible answers.

The two signals could arise from two separate centers in the OEC (Hansson et al, 1987). The multiline signal clearly arises from a cluster with two or more Mn nuclei. The $g = 4.1$ signal could arise from an isolated Mn(IV) monomer, as S = 3/2 Mn(IV) ions are known to give rise to signals in the $g = 4$ region. The conversion of signal forms upon warming PS II samples from 140 K to 195 K could be explained by electron transfer between the two centers at the higher temperature, and the changes in relative signal intensities with different cryoprotectants or inhibitor incubation could be due to shifts in redox potentials and subsequent electron transfer between the centers. Alternatively, both signals could arise from different spin states of the same center, with the multiline signal arising from an S =1/2 state and the $g = 4.1$ signal arising from a higher spin state such as S = 3/2 or S = 5/2 (Zimmermann and Rutherford, 1986; de Paula et al., 1986). In this common cluster model, the 140 K to 195 K conversion between $g = 4.1$ and multiline signal forms would result from a conformational change in the cluster upon warming. The cryoprotectant and inhibitor sensitivity would also be triggered by conformational changes, perhaps including forming or replacing bridges between Mn ions by ligand molecules.

Some early temperature dependence studies pointed to an excited state origin for the multiline signal (de Paula and Brudvig, 1985; de Paula et al., 1986), which appeared to be an important result because a dinuclear mixed valence Mn cluster would certainly have an S = 1/2 ground state. However it became clear with further experiments that the multiline is really a ground state signal (Hansson et al., 1987; Britt et al., 1992, Lorigan and Britt, 1994), a result consistent with either a dinuclear or tetranuclear origin. A breakthrough came from EPR studies of oriented PS II membranes in which the $g = 4.1$ signal was stabilized by treatment with inhibitory levels of ammonia (Kim et al., 1990, 1992). For the unique orientation where the membrane normal was oriented parallel to the magnetic field, the $g = 4.1$ signal was itself shown to have 'multiline' character, with at least 16 ^{55}Mn hyperfine lines present. This was important for two reasons. First, this showed without question that the $g = 4.1$ signal arises from Mn, and in particular, a multinuclear Mn cluster. Second, the $g = 4.1$ signal had been shown to be a ground state signal (Hansson et al., 1987), but a ground state of an antiferromagnetically-coupled mixed valence Mn dinuclear cluster would have spin S = 1/2 and could not give rise to a signal in the $g = 4$ region. Thus the $g = 4.1$ signal was shown to arise minimally from a trinuclear center. Given that the multiline signal arises minimally from a dinuclear center, the separate center model was clearly ruled out because it would require five Mn ions rather than the four Mn ions present in PS II. Thus the common cluster/spin conversion model was proven to be the correct one, with the cluster being either trinuclear or tetranuclear. A tetranuclear cluster was highly favored because there was no spectroscopic evidence for a separate mononuclear Mn ion once the $g = 4.1$ signal was shown to arise from a cluster. This line of evidence for a tetranuclear cluster was reinforced by multifrequency EPR studies by Haddy et al. (1992) that favored an S = 5/2 origin of the $g = 4.1$ signal, which is inconsistent with an isolated Mn(IV) ion.

Another line of evidence for a tetranuclear Mn cluster has come from EPR spectral simulations of the multiline signal. There have been a number of recent simulations of the signal using four Mn hyperfine couplings to achieve the correct number of lines and overall spectral widths (Bonvoisin et al., 1992; Kusunoki, 1992; Zheng and Dismukes, 1992). Perhaps the most impressive of these is that of Zheng and Dismukes (1992 and personal communication), who not only simulate the normal multiline, but also an altered form accessible through ammonia binding (Beck et al., 1986; Britt et al., 1989). The two different signals are best modeled with similar anisotropically coupled Mn(III) and Mn(IV) ions, but with different exchange coupling pathways between them. These simulations invoke an unusual $d_{x^2-y^2}$ unpaired spin ground state for Mn(III), which could result from Mn(III) being five-coordinate in the S_2-state. However, the simulations require a 3Mn(III)-1Mn(IV) configuration for S_2, an oxidation state assignment inconsistent with the 1Mn(III)-3Mn(IV) assignments made by X-ray absorption spectroscopy (*vide infra*). Furthermore, a recent successful multifrequency simulation of the multiline EPR signal has been obtained using only a dinuclear Mn cluster model (Åhrling and Pace, 1995). However, exceedingly large hyperfine anisotropies and large ^{55}Mn quadrupolar couplings are required to achieve the dinuclear simulation. Moreover, invoking a dinuclear origin for the multiline signal forces one to add an additional non-Mn paramagnetic component (for example a histidine radical as proposed by Boussac et al., 1990) to another dinuclear Mn cluster to match the experimental $g = 4.1$ results.

The problem with conventional EPR in fully resolving this controversy is the fact that many of the EPR transitions overlap and much information is therefore lost. Even in a dinuclear cluster only 16 peaks are resolved out of a total of 36 transitions. The situation with a tetranuclear cluster is far worse, as the EPR signal will be composed of 1296 transitions! Our most recent approach to address the overall nuclearity of Mn centers in PS II has been to utilize electron spin echo - electron nuclear double resonance (ESE-ENDOR) to measure the ^{55}Mn nuclear spin transitions of the cluster (Randall et al., 1995). The number of ENDOR transitions increases only linearly with the number of nuclei coupled to the unpaired electron magnetization, making adequate resolution possible even in multinuclear clusters.

Figure 6 shows the experimental ^{55}Mn ENDOR and EPR spectra of the $g = 2$ multiline signal, along with simulations using dinuclear, trinuclear, and tetranuclear models. The problem with the dinuclear and trinuclear models is that the experimental hyperfine couplings revealed in the ENDOR data are too small to give the overall linewidth to the multiline EPR signal with only two or three coupled nuclei. However the tetranuclear model gives good fits to both types of magnetic resonance spectra. The *simultaneous* constraint of simulating both EPR and ENDOR data can only be met with a tetranuclear cluster model. Thus the magnetic resonance data demonstrate that the 4 Mn ions of the OEC are arranged in a single tetranuclear Mn cluster. The actual structure of the cluster cannot be determined from the current EPR information; fortunately, important structural information has also been contributed by EXAFS spectroscopy.

C. X-ray Spectroscopy of the Manganese Cluster

The advent of modern synchrotron radiation technology has made it feasible to perform X-ray spectroscopy on dilute non-crystalline biological samples. One powerful X-ray spectroscopic technique is Extended X-ray Absorption Fine Structure (EXAFS) spectroscopy. In the EXAFS experiment, X-ray generated photoelectrons which backscatter off the electron charge density of adjacent atoms give rise to a modulation in the X-ray absorption cross section as a function of the electron de Broglie wavelength, which depends on the excess X-ray photon energy above the electron ionization energy. The modulation of the X-ray absorption can be interpreted to reveal adjacent electron density as a function of radius from the parent atom. The maximum range from transition metal atoms that can be probed by EXAFS spectroscopy of dilute metalloenzymes is typically on the order of 4 Å. Because each element has a distinct ionization energy (to date the 'K' ionization of the Mn 1*s* core electrons has been primarily used for OEC studies), the EXAFS results are uniquely specific to individual elements. Thus for examining the local environment of Mn ions of the OEC cluster, EXAFS spectroscopy cleanly focuses on the active site only, with no background effects from the rest of the protein (assuming of course there is no additional Mn in the preparations). However, EXAFS cannot distinguish the individual environment of each Mn ion of the cluster. Rather, EXAFS spectroscopy reveals the summation of electron distributions surrounding all four Mn ions.

The Kok cycle state best characterized by Mn EXAFS is the dark-stable S_1-state (Fig. 7a). A consistent element of S_1-state Mn EXAFS data recorded by all investigators is the presence of a strong backscatter at 2.7 Å (Yachandra et al., 1986a; Yachandra et al., 1987; George et al, 1989; Penner-Hahn et al., 1990; MacLachlan et al., 1992; DeRose et al., 1994). The selectivity in the origin of the photoelectron is not matched in the identification of the backscatterer: only an approximate measure of the atomic number of the backscatterer can be obtained. However the 2.7 Å backscatterer data are consistent with a first-row transition metal such as Mn at this distance, and 2.7 Å is the distance between Mn atoms in di-μ-oxo bridged Mn(III)Mn(IV) complexes (Sauer et al., 1992). Approximate quantification of the 2.7 Å backscattering amplitude favors one such backscatterer per Mn, consistent with a pair of such dinuclear clusters. There is also a peak at approximately 3.3 Å (George et al., 1989; Penner-Hahn et al., 1990; DeRose et al., 1994; Latimer et al., 1995). The EXAFS data in this region are consistent with the presence of another Mn backscatterer at this distance, and it is appealing to assign this to Mn because we know from EPR spectroscopy that the

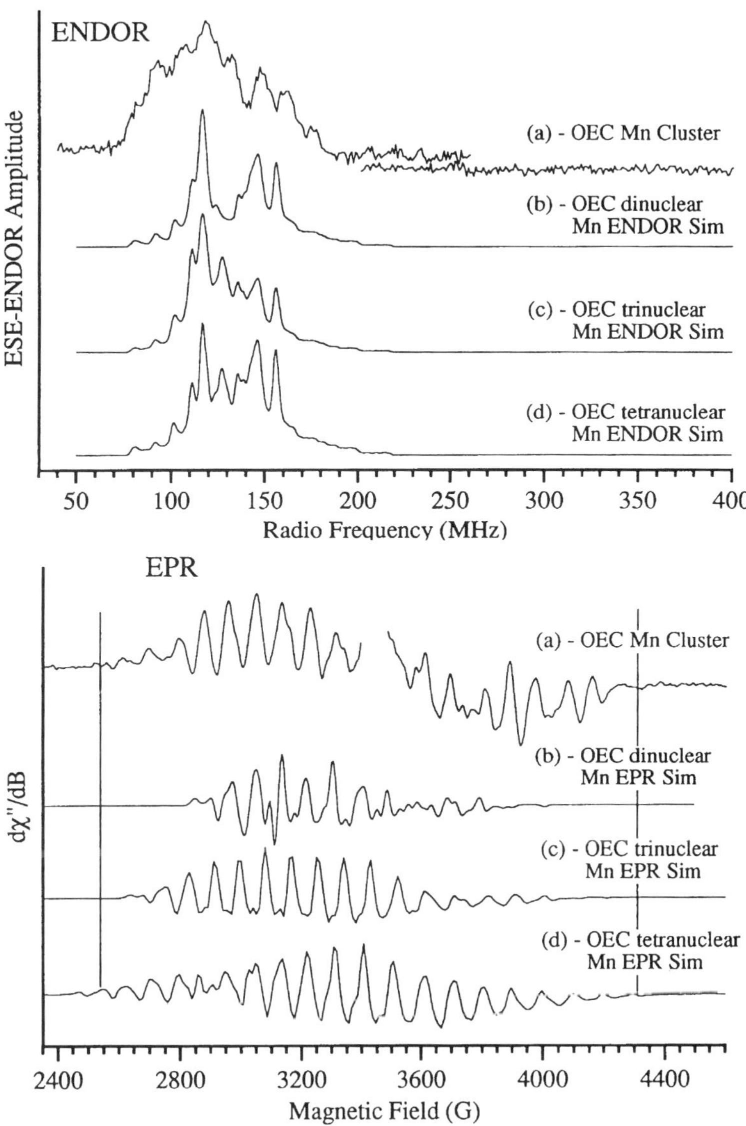

Fig. 6. Experimental ENDOR (trace (a), top panel) and EPR (trace (a), bottom panel) spectra of the $g = 2$ multiline signal, along with simulations using dinuclear (b), trinuclear (c), and tetranuclear (d) models for the Mn cluster. Only the tetranuclear model provides adequate simulations (Sim) to both ENDOR and EPR spectra. The figure utilizes ENDOR data and simulations published by Randall et al. (1995).

Mn ions are arrayed in a tetranuclear complex. However there is some evidence that there may be some contribution from Ca^{2+} to this peak, as Latimer et al. (1995) have reported that this peak increases in amplitude when Ca^{2+} is replaced by Sr^{2+}, which is a stronger backscatterer. However, this assignment is controversial, as Riggs, Yocum, and Penner-Hahn (personal communication) have not observed such changes in the peak amplitude following Sr^{2+} reconstitution. The shortest distance peak is from oxygen and nitrogen ligands to Mn. These light backscatterers are in the 1.8–2.0 Å range (George et al., 1989; Penner-Hahn et al., 1990; MacLachlan et al., 1992; DeRose et al., 1994). It is not possible to distinguish between oxygen and nitrogen with the EXAFS data, and this peak contains a sum of all ligands to all four Mn ions, including both bridging and terminal ligands. It is clear from the EXAFS data that there are no sulfur ligands to the cluster. EXAFS studies of oriented PS II membranes show some degree of anisotropy in backscattering amplitudes, providing additional constraints in modeling the

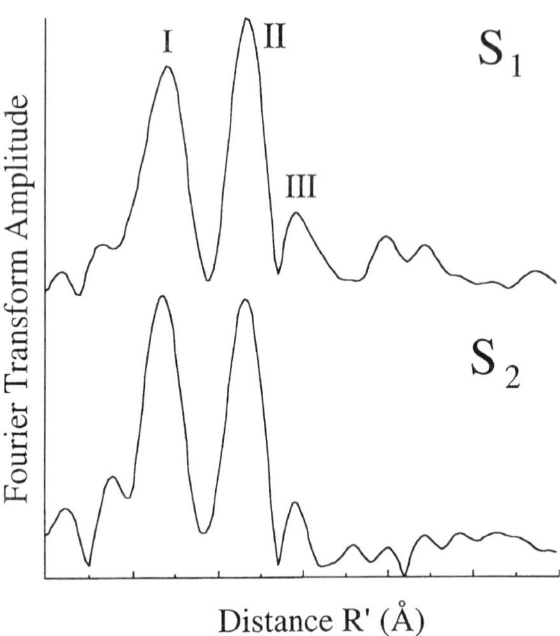

Fig. 7. Mn K-edge EXAFS Fourier transforms for the S_1 (a) and S_2 (b) states of a spinach PS II preparation. This figure is adapted from DeRose et al., (1994)

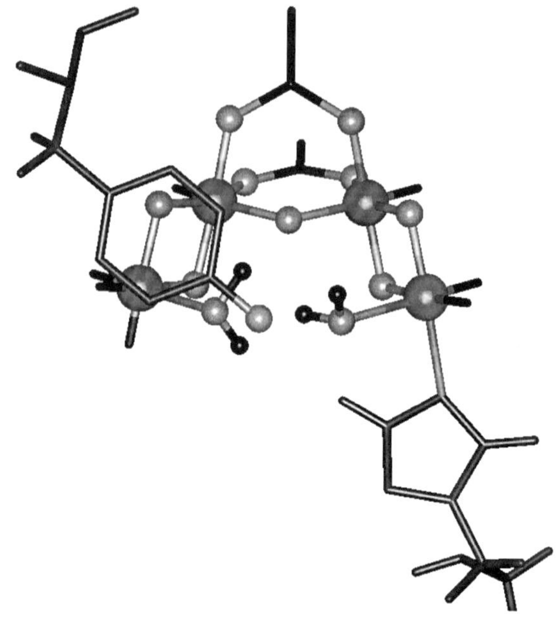

Fig. 8. A model for the OEC Mn cluster and its immediate environment. The core tetranuclear Mn cluster is adapted from the 'dimer of dimers' model of Yachandra et al. (1993), which is based on constraints provided by Mn EXAFS. This cluster is composed of a pair of di-μ-oxo bridged dinuclear Mn cores linked together via an oxo bridge and a pair of protein-provided carboxylato bridges. A histidine ligand (D1:his-190 is employed in this model) is provided as demonstrated by Tang et al. (1994). The redox-active tyrosine Y_Z is in close proximity as shown by Gilchrist et al. (1995). Water substrate molecules are positioned as terminal ligands to the outermost Mn ions of the cluster. This proposed OEC stucture is part of a larger PS II model (see Fig. 11) constructed by Dr. Jeffrey Peloquin based on homology with the *Rhodobacter* reaction center structure (Komiya et al, 1988).

geometry of the Mn cluster (George et al., 1989; Mukerji et al., 1994). In particular, such anisotropy rules out highly symmetric structures such as Mn_4O_4 cubane-like clusters (Brudvig and Crabtree, 1986).

The constraints provided by EXAFS have not provided a unique structure for the Mn cluster. However there are a number of possible Mn cluster structures that have been proposed based on the EXAFS constraints (George et al, 1989; DeRose et al., 1994). One that has received much recent attention is a 'dimer of dimers' model described by Klein and coworkers (Yachandra et al., 1993). A representation of the Klein model is included in Fig. 8, which shows a model for the OEC and its immediate environment. Two di-μ-oxo bridged dinuclear Mn 'dimers' are linked via a μ-oxo di-μ-carboxalato bridge. The planes of the two dinuclear Mn cores are tilted out into a 'jawlike' structure. It is important to emphasize that this is by no means a unique structure, but one of a number of tetranuclear clusters consistent with the constraints provided by Mn EXAFS.

The S_2-state EXAFS of PS II samples advanced by 195 K illumination to the form of the S_2 state that gives rise to the multiline EPR signal show very small changes with respect to the S_1 data (Fig. 7b) (DeRose et al., 1994; Mukerji et al., 1994). There appear to be no major structural rearrangements of the tetranuclear complex upon the $S_1 \rightarrow S_2$ transition. A slightly larger difference in the EXAFS data is observed when the OEC is advanced by 130 K illumination to the S_2 form that gives rise to the g = 4.1 signal (Liang et al., 1994). In this case one of the Mn-Mn distances is increased to approximately 2.85 Å. Such small structural differences seen in the Mn cluster advanced to S_2 at the different illumination temperatures appear consistent with the cluster conformational changes required to switch between a spin 3/2 or 5/2 ground state for the g = 4.1 signal form and a spin 1/2 ground state for the multiline signal form. Preliminary EXAFS data obtained on the S_3-state of the OEC show major differences consistent with large structural changes on the $S_2 \rightarrow S_3$ transition (MP Klein and VK Yachandra, personal communication). There appears to be a greater heterogeneity in Mn-Mn distances in the S_3 state.

Unfortunately it has not been possible to trap the evanescent S_4 state in order to perform EXAFS on the final activated complex.

High-resolution X-ray absorption spectra obtained at the absorption edge itself can reveal information about the oxidation state of a transition metal such as Mn, as higher oxidation states require additional photon energy to achieve photoionization of the core electrons. In this manner, X-ray Absorption Near Edge Structure (XANES) spectroscopy has been used to attempt to determine the oxidation states of the Mn ions of the OEC. The consensus view appears to be that the cluster contains two Mn(III) and two Mn(IV) ions in the dark-stable S_1 state of the Kok cycle (Riggs et al., 1992; Yachandra et al., 1993). Shifts in Mn K-edge absorption on the $S_1 \rightarrow S_2$ transition were first reported some years ago (Goodin et al, 1984; Yachandra et al., 1987), and these were interpreted as direct evidence of Mn oxidation in this transition. This result along with the appearance of Mn EPR signals following the $S_1 \rightarrow S_2$ transition leave little doubt that there is Mn oxidation in this step. If the S_1 state has an oxidation state configuration of 2Mn(III)-2Mn(IV), then this implies that the oxidation state configuration of the S_2 state is 1Mn(III)-3Mn(IV). However, this conflicts with the 3Mn(III)-1Mn(IV) assignment preferred by Zheng and Dismukes (1992 and personal communication) on the basis of their S_2-state multiline simulations. To a large degree the XANES oxidation state assignments rely on comparisons with model compounds, and one could argue that if Mn(III) in the OEC is five-coordinate and the models used in edge-energy calibration are all six-coordinate, then this could be a source of error in the XANES OEC oxidation state assignment (Dismukes, personal communication). However the Klein group has argued, based on the small amplitude of the 'pre-edge' (1s $\rightarrow$ 3d) transitions, that the Mn ions of the OEC are octahedrally coordinated. Another line of evidence against a 3Mn(III)-1Mn(IV) assignment for S_2 comes from XANES studies of reduced derivatives of the OEC (Riggs et al., 1992). For example, reduction of PS II with NH_2OH results in a two-flash retardation of the O_2 evolution cycle (Bouges, 1971). The XANES spectrum of Mn(II) is quite distinctive from the XANES of Mn(III) and Mn(IV), so XANES spectroscopy of reduced derivatives should be sensitive to the presence of stoichiometric quantities of Mn(II). The XANES spectrum of NH_2OH-reduced PS II can be fit with no more than one Mn(II) ion (Riggs et al., 1992). In fact, a 4Mn(III) model fits the data well. As this state is three oxidation state equivalents reduced with respect to S_2, it is difficult to rationalize this observation with a 3Mn(III)-1Mn(IV) assignment for S_2. Thus, the oxidation assignments for the Mn cluster in different S-states remain controversial.

Additional controversy concerns the question of whether Mn is oxidized in the $S_2 \rightarrow S_3$ transition. Ono et al. (1992) measured the XANES spectra of Mn in the OEC as a function of flash number. Mn absorption edge shifts were reported that were consistent with Mn oxidation during the $S_0 \rightarrow S_1$, $S_1 \rightarrow S_2$, and $S_2 \rightarrow S_3$ transitions, with a large rereduction of Mn upon cycling back to S_0. This conflicted with earlier Klein group results which showed no evidence of Mn absorption edge shifts upon forming ≈60% S_3 through cryogenic illumination of samples where the acceptor-side nonheme Fe was oxidized to Fe(III) (Guiles et al, 1990). This result had raised the possibility that Mn is not oxidized during the $S_2 \rightarrow S_3$ transition, but instead the oxidation is centered on some other species within the OEC such as an amino acid residue. Recently the Klein group has performed its own flash dependent XANES studies, and again sees no clear evidence for Mn edge shifts in the $S_2 \rightarrow S_3$ transition (Andrews et al., 1995). Even if the experimental controversy is resolved, it is unclear whether these results can be taken as reliable indicators of Mn oxidation during this transition, as there seems to be appreciable cluster reorganization as monitored by EXAFS in the $S_2 \rightarrow S_3$ transition, and the edge shifts are only an accurate measure of Mn oxidation if the core geometry and ligation of the cluster is unchanged.

D. UV and IR Difference Spectroscopies

EPR and EXAFS spectroscopies share a common advantage in studying a metal-based catalytic site in a large complex system such as Photosystem II: they probe only the local structure about metal atoms or clusters. Other spectroscopic techniques such as UV or IR absorption spectroscopies suffer from having added contributions from organic cofactors, as well as peptide bonds and amino acid side-chains in the case of IR spectroscopy. Nevertheless there has been significant progress in utilizing these methods.

Much of the focus of UV spectroscopy on the OEC has been to examine Mn oxidation state changes during the different S-state transitions, utilizing absorption changes in the 300 nm region thought to be associated with ligand-to-metal charge transfer band changes upon oxidation of Mn ions (Dekker et al., 1984a,b,c; Lavergne, 1986,1987; Saygin and Witt, 1987; Renger and Hanssum, 1988; Dekker 1992). These are difficult experiments, due to the small amplitudes of the UV absorption differences, overlapping spectral changes from the quinone acceptor molecules, and uncertainty about the precise S-state composition of dark-adapted membranes and the degree to which the S-states are synchronously advanced with subsequent flashes (see Debus 1992 for a detailed discussion). As a result, there has been a great deal of conflict in the literature, particularly concerning the $S_0 \rightarrow S_1$ transition. This transition is particularly challenging experimentally, because four flashes are required in a native system to observe it, and appreciable S-state dephasing will occur after this many flashes (see Dekker, 1992, for details). There is more consensus in experiments and interpretation of the $S_1 \rightarrow S_2$ and $S_2 \rightarrow S_3$ transitions, the results of which appear to support Mn(III) oxidation to Mn(IV) in these steps. One caveat concerns the $S_2 \rightarrow S_3$ transition, which some XANES studies suggested does not involve Mn oxidation. UV difference studies in inactivated Ca^{2+}-depleted PS II membranes were interpreted to favor histidine oxidation in this step (Boussac et al., 1990). However our recent ENDOR and ESEEM studies have shown this assignment to be incorrect (*vide infra*), so invoking histidine oxidation in either the $S_0 \rightarrow S_1$ transition (Dekker, 1992) or the $S_2 \rightarrow S_3$ transition in the native system (Boussac et al., 1990) is not experimentally supported. Even though there are a number of controversial issues in the details of these UV difference studies, it is important not to overlook the general trend of Mn oxidation during the lower S-state transitions and rereduction of Mn during the $S_3 \rightarrow S_4 \rightarrow S_0$ transitions, consistent with the idea that there is some component of concerted water oxidation during these steps.

To date the bulk of IR studies of PS II have focussed on the organic cofactors (for example, MacDonald et al., 1993), but there have been a few recent studies investigating the OEC directly. For example Berthomieu and Boussac (1995) have utilized FTIR to support the UV spectroscopy evidence that histidine is oxidized in Ca^{2+}-depleted PS II membranes. Noguchi et al. (1995) have performed FTIR studies to characterize amino acid ligation to the Mn cluster, presenting data in support of a carboxylate bridge between Mn and Ca^{2+} which is disrupted by Ca^{2+}-depletion. Although IR shares the disadvantage with optical spectroscopy of sampling the entire PS II complex, an advantage it shares with EPR is the ability to selectively manipulate the spectral features through the use of isotopic labeling. One can anticipate that many interesting results will come from coupling selective labeling of amino acids (as in Barry and Babcock, 1987; Tang et al., 1994) with the use of difference FTIR methods.

IV. Manganese Cluster Ligation

A tetranuclear Mn cluster will have 24 ligands if each of the Mn atoms is octahedrally coordinated. In such a cluster, a sizeable number of these ligands will form bridges between Mn ions. For example, in the Klein model (Yachandra et al., 1993) there are 14 ligand sites occupied by bridging ligands. Some of the bridges in this and other structural models are 'inorganic' oxygens. Acetato bridges can be provided by carboxyl donating amino acids such as glutamate and aspartate, and exogenous carboxyls can also bind to biological metal clusters. The terminal ligation positions not taken up by bridging ligands can come from a variety of sources: amino acid ligands, substrate and cofactor molecules, and when present, inhibitor molecules. Additionally, both bridging and terminal ligation may change during the course of the catalytic Kok cycle.

A. Amino Acid Ligands

One powerful approach to assign ligands to the Mn cluster is selective mutagenesis of amino acids such as histidine, aspartate, and glutamate that commonly provide ligands to metals. As described earlier, such mutagenesis studies have been performed primarily in *Synechocystis* (Diner et al., 1991; Debus, 1992; Nixon and Diner, 1992; Nixon et al., 1992; Pakrasi and Vermaas, 1992; Boerner et al., 1992; Barry et al., 1994; Chu et al., 1994a,b, 1995a,b). A complication to the site-directed mutagenesis approach is that the slowing or disabling of oxygen evolution in a mutant does not prove that the mutated amino acid was a ligand to the Mn cluster. Alternative possibilities include alteration of protein tertiary structure around the OEC, destabilization of binding of the required Ca^{2+} ion, alteration of Mn(II) binding sites required in photoassembly of the cluster, disruption of fast electron transfer between the Mn cluster and $Y_Z^{\bullet}$, and disruption of proton transfer pathways needed to transfer protons to the lumenal phase. Possible ligand 'hot spots' are those that eliminate or impair photoautotrophic growth, such as the carboxy terminus of D1 (ala-344), D1 residues asp-59, asp-61, glu-65, asp-170, glu-189, his-190, his-332, glu-333, his-337, and asp-342 and D2 residue glu-70. Further complicating the situation is the possibility that there could be mutations of ligands to the cluster which could be compensated for by the binding of exogenous ligands or alternative amino acid residues that can be repositioned through small protein conformational changes. Thus, mutagenesis experiments need to be coupled with extensive biophysical experimentation to completely characterize the mutated system. Such characterization is being aggressively pursued, now that all likely D1/D2 ligands have been mutagenized. Much in the way of mutant characterization can be done with chlorophyll fluorescence kinetics assays in whole cells (for example, Chu et al, 1994a,b, 1995a,b). More detailed characterizations will have to be performed with high resolution structural techniques such as EPR, FTIR, and EXAFS on isolated PS II particles from mutants. It is likely that these experiments will provide detailed information on amino acid ligation to the Mn cluster.

There has been appreciable progress in ligand characterization using spectroscopic methods to study wild-type PS II preparations. As described earlier, FTIR evidence has been presented to support carboxylate ligation to the Mn cluster (Noguchi et al., 1995). In principal, EXAFS spectroscopy could be useful for characterization of amino acid ligands to the Mn cluster (George et al., 1989; Penner-Hahn et al., 1990; MacLachlan et al., 1992; DeRose et al., 1994). The major difficulty is that the backscattering amplitudes from nitrogen and oxygen ligands are too similar to be resolved. In smaller clusters and mononuclear metal sites, backscattering from carbons of the histidine imidazole group can give a characteristic signature to histidine ligation. However, an additional problem with the OEC cluster is that the EXAFS data include contributions from all ($\leq$24) ligands to the four Mn ions, and small features associated with one or two ligands can be masked. It is not even clear whether backscattering from the close bridging O groups can be distinguished from backscattering from slightly more distant terminal O/N ligands. However, as noted previously, it is clear from EXAFS that there are no sulfur ligands (although one would certainly not propose, on chemical grounds, to place sulfurs in such a highly oxidizing environment).

The most definitive spectroscopic insights into Mn ligation have come from magnetic resonance methods. The PS II reaction center is far too large for high resolution solution NMR methods to be applied. However, detection of superhyperfine couplings to magnetic nuclei of ligands provides a path for EPR spectroscopy to determine cluster ligation. The problem is that the Mn-ligand couplings are so small that their presence is masked by the inhomogeneous broadening of the Mn EPR signals (Britt and Klein, 1992). Fortunately, this drawback to basic EPR spectroscopy can be overcome with double resonance methods such as ENDOR and electron spin echo envelope modulation (ESEEM), allowing one to study the nuclear spin transitions of magnetic nuclei near the Mn cluster. The information achieved is akin to that obtained with protein NMR spectroscopy, with the advantage that the spectra are simplified by the fact that only those nuclei with appreciable magnetic couplings to the Mn cluster are observed. Nitrogen ligation has been investigated by both ENDOR (Tang et al., 1993) and ESEEM (Britt et al., 1989; DeRose et al., 1991; Zimmermann et al., 1993). ESEEM is a pulsed EPR technique that is particularly useful in examining couplings to quadrupolar nuclei such as the spin $I = 1$ ^{14}N nucleus (Britt, 1995). In an ESEEM experiment, one increments the time between resonant microwave pulses in an electron spin echo (ESE) experiment. If the microwave pulses drive semi-allowed nuclear spin transitions in addition to the fully-allowed electron spin transitions, induced nuclear spin coherences give rise to a modulation of the ESE amplitude in the 'envelope' gathered over the incremented time range. The required semi-allowed transitions are typically strong for ^{14}N ligands, and this results in relatively deep modulation, making ESEEM quite sensitive for searching for nitrogen ligands. The ^{14}N spectrum is obtained by Fourier analysis of the envelope modulation pattern. Recently, the ESEEM method was used to definitively demonstrate that there is at least one histidine ligand to the Mn cluster (Tang et al., 1994). ^{14}N modulation from a nitrogen ligand to the cluster was found to disappear upon selective ^{15}N-labeling of histidine in PS II. The possible histidine ligands are defined by the mutagenesis experiments: D1 histidines 190, 332, and 337.

The combination of selective isotopic labeling with high resolution magnetic resonance spectroscopy is a powerful structural tool. Selective mutagenesis can be employed along with such experiments. This approach will also be essential in obtaining detailed information with vibrational spectroscopies such as FTIR and Raman. A combination of such approaches will almost certainly allow researchers to more fully determine the protein coordination of the Mn cluster in the near future.

B. Water Binding

The synchronization of the Kok cycle is defined by O_2 release during the $S_4 \rightarrow S_0$ transition. However, it has been more difficult to identify the Kok cycle transitions during which substrate waters bind to the OEC. One limiting view would have two waters binding in Mn ligation sites cleared upon the release of O_2 in the $S_4 \rightarrow S_0$ transition. In this case, the water ligands to Mn would be present in the most reduced state, S_0. Alternatively, the OEC may exclude water from the catalytic site during early S states to prevent formation of unwanted oxidation byproducts. Thus, the other limiting view would be to have waters bind to Mn only upon the $S_3 \rightarrow S_4$ transition, with an immediate concerted four-electron oxidation to form O_2. Of course the binding can occur at one or two intermediate S-state transitions as well. If water binds before the $S_3 \rightarrow S_4$ transition, the mechanistic question of whether water oxidation occurs sequentially or concertedly arises. Much of our current knowledge about water binding and oxidation comes from mass spectrometry and EPR studies.

Mass spectrometry provides a measure of water exchange rates. For example, the isotopic composition of evolved O_2 can be measured as a function of time after injecting isotopically labeled water into samples poised in different stages of the Kok cycle by previous flashes. Radmer and Ollinger (1986) demonstrated by diluting $H_2{}^{18}O$-incubated samples with unlabeled water following one or two preflashes that there are no non-exchangeable substrate waters bound through the S_3 state of the Kok cycle: essentially no $^{18}O_2$ was evolved. However the unlabeled water incubation time was 1 min, so bound waters with exchange rates $k_{ex} >> 10^{-2}\ s^{-1}$ would be recorded as freely exchangeable by this approach. Radmer and Ollinger posited that any oxidized water intermediates would exchange with water at a slower rate, and therefore concluded there was no water oxidation up to the S_3-state. These results were considered strong support for the model where water oxidation occurs as a fully concerted 4-electron oxidation during the $S_4 \rightarrow S_0$ transition. Recently, similar experiments have been performed by Messinger et al. (1995) with much shorter mixing times (≈30 ms). These experiments focused on the S_3-state. The result of their experiments and data analysis was that one substrate water exchanges in the S_3-state with a rate of approximately $k_{ex} = 2\ s^{-1}$, while the other exchanges with a rate $k_{ex} > 40\ s^{-1}$ or binds only in the S_4-state. The exchange rate of the slower exchanging water is far too fast to have been measured in the earlier experiments of Radmer and Ollinger. If present, the faster water exchanges too fast to be measured even with the improved time resolution obtained by Messinger et al. This experiment makes it clear that at least one substrate water is bound by the S_3-state, and that if two substrate waters are bound, they are not bound symmetrically. There are many viable explanations for the possible asymmetry in exchange times. Messinger et al. discuss the possibility that the slowly exchanging site corresponds to a terminal Mn(IV)=O group. However, their data in no way rule out the possibility that there are two H_2O ligands to Mn in the S_3-state that have different exchange rates due to differences in Mn oxidation states or Mn coordination geometry. An exchange rate of $k_{ex} = 2\ s^{-1}$ is very slow compared to the exchange rate of water bound to Mn(II) ($k_{ex} = 2 \times 10^7\ s^{-1}$), but it is exceedingly fast compared to the exchange for Cr(III) ($k_{ex} = 2 \times 10^{-6}\ s^{-1}$), which is isoelectronic with Mn(IV) (see Lincoln and Merbach (1995) for a comprehensive review of water exchange kinetics of solvated metal ions). Also, for a given oxidation state of a specific metal ion, the kinetics of water exchange can vary over several orders of magnitude depending on the other ligands to the metal ion.

In contrast to mass spectrometry, EPR spectroscopy can be used to examine binding of fast-exchanging ligands through detection of superhyperfine interactions between the electron spin of the metal ion and a magnetic nucleus of a ligand. Water binding to the S_2 state can be studied with the Mn cluster poised in the state giving rise to the $g = 2$ multiline signal. Because the multiline signal can only be detected at cryogenic temperatures, these studies involve incubating samples in isotopically enriched water for some period of time, then freezing to cryogenic temperatures to perform the spectroscopy. The non-magnetic ^{16}O nucleus of water can be replaced by the spin $I=5/2$ ^{17}O nucleus, and both spin $I=1/2$ ^1H and $I=1$ ^2H can be studied in the hydrogen sites of water. An early ^{17}O-labeled water study showed very subtle line broadening in the multiline signal after incubation of PS II particles under light in ^{17}O-labeled water (Hansson et al., 1986). This result was taken as evidence of water-derived ligand binding by the S_2 state. However, a concern about this study is the possibility that ^{17}O was exchanged into non-substrate oxygen sites during the long incubation time. No consistent line narrowing of the multiline EPR signal has been observed following ^2H_2O-incubation of PS II particles (Yachandra et al., 1986b).

More details about water binding have come from ENDOR and ESEEM experiments on the multiline EPR signal. An ^1H ENDOR study by Tang et al. (1993) concluded that there are no protons coupled as strongly as would be expected for those of water bound to Mn. However, Tang et al. could not exclude the possibility that they were unable, with the limited ENDOR signal amplitude, to observe broad anisotropic ENDOR signals from such strongly coupled protons. This indeed appears to be the case based on our ^2H ESEEM and ^1H ESE-ENDOR measurements.

We showed several years ago (Britt et al., 1990) that exchangeable deuterons in sites in close proximity to the Mn cluster could be detected by ESEEM of ^2H_2O-exchanged PS II particles. Very deep ^2H modulation of the ammonia-altered multiline signal was determined to have a contribution from ^2H nuclei of the ammonia-derived Mn ligand (*vide infra*). Comparable modulation depth was observed in ^2H_2O-exchanged native PS II membranes showing the normal form of the multiline signal. This was considered evidence for water or hydroxyl ligation to Mn by the S_2-state. Because the ^2H-modulation was observed in PS II particles advanced to the S_2-state by illumination at cryogenic temperatures, it was suggested that this ligation was also present in the S_1-state, as ligand exchange is unlikely at 195 K. No change in modulation was observed after 'annealing' the sample to 0 °C to allow for facile ligand exchange, and this result indicated that no additional ligand binding occurs during the $S_1 \rightarrow S_2$ transition. These experiments have been repeated recently with better signal-to-noise, and the same effects are observed. Spectral simulations support the assignment of direct ligation of water or hydroxyl ligands to Mn in the lower S-states. Figure 9 shows the 2-pulse ESEEM results for the S_2-state multiline signal following ^2H_2O exchange. The time-domain pattern in (a) is simulated with contributions from proximal deuterons. The simulation is not unique, but we do require multiple deuterons at dipolar coupled distances on the order of 2.5–2.7 Å to simulate the depth and rapid modulation damping of the time domain data, and such short distances are in the range expected for water or hydroxyl ligands to Mn. Recent ^1H ESE-ENDOR results confirm the presence of such strongly-coupled exchangeable protons at the Mn cluster of the OEC (not shown).

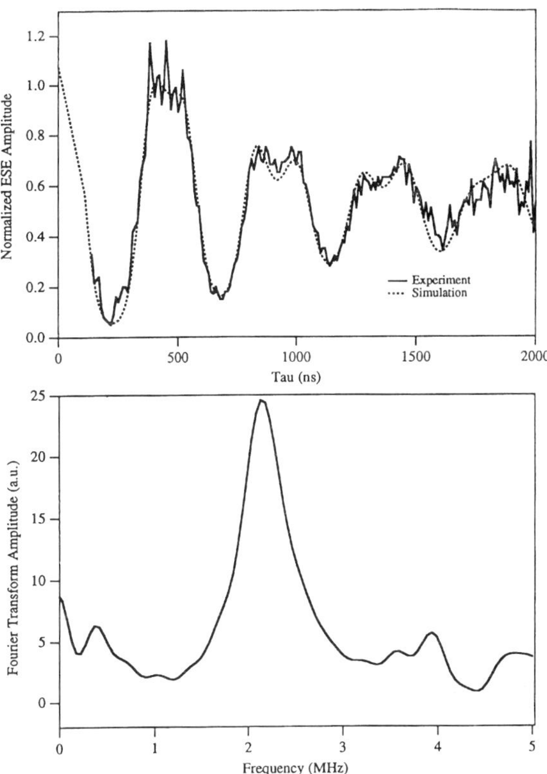

Fig. 9. (a) The time-domain 2-pulse ESEEM pattern of the S_2 multiline signal from the Mn cluster of PS II following ^2H_2O exchange. The dashed line is a simulation of this time-domain pattern. The dominant contributions to the dashed line simulation arise from two deuterons at a dipolar-coupled distance of 2.5 Å and an additional deuteron at a distance of 2.7 Å. Effects from more distant deuterons out to 7.4 Å are included. (b) The Fourier Transform of (a) showing the large peak at the ^2H Larmor frequency of 2.16 MHz. These unpublished data are kindly provided M. Lane Gilchrist.

C. Inhibitor and Cofactor Binding

It has long been known that Ca^{2+} is an essential cofactor for photosynthetic oxygen evolution (for detailed reviews see Ghanotakis and Yocum (1990), Yocum (1991), and Debus (1992)). Ca^{2+} binding and exchange is significantly affected by the presence of the small extrinsic polypeptides (see Chapter 8). Studies of Ca^{2+} binding utilizing radioactive ^{45}Ca showed that a single Ca^{2+} is required for oxygen evolution (Ädelroth et al., 1995).

Because Ca^{2+} is required for oxygen evolution, it is appealing to consider Ca^{2+} to be within the OEC, intimately associated with the Mn cluster. As previously described, EXAFS evidence for Ca^{2+} in close proximity to Mn has been presented (Latimer et al., 1995), but this result is controversial. FTIR evidence for a Ca^{2+} carboxylate-bridged to Mn has been presented (Noguchi et al., 1995). More definitive evidence for a link between Ca^{2+} and the Mn cluster comes from EPR studies, which clearly demonstrate that Ca^{2+} depletion affects the magnetic properties of the Mn cluster. The normal $g = 2$ multiline EPR signal is lost in Ca^{2+}-depleted PS II reaction centers and is replaced by a 'dark-stable' multiline EPR signal, which has approximately 26 partially resolved ^{55}Mn hyperfine lines with spacing on the order of 55 G (Boussac et al., 1989). Upon physiological temperature illumination, this EPR signal is replaced by a broadened radical signal that has been assigned to the S_3 state (Boussac et al., 1989) (but *vide infra*). Sr^{2+} reconstitution provides some reactivation of O_2 evolution in Ca^{2+}-depleted PS II particles (Ghanotakis et al., 1984), and the S_2-state of such Sr^{2+}-reconstituted samples shows an altered form of the S_2 multiline EPR signal and a stabilization of the $g = 4.1$ EPR signal (Boussac and Rutherford, 1988a). Moreover, Ca^{2+}-binding is greatly weakened upon treatments that deplete PS II of manganese (Tamura and Cheniae, 1988; Ädelroth et al., 1995), and the binding of Ca^{2+} is S-state dependent (Boussac and Rutherford, 1988b), with weaker binding in the S_3-state. However, removal of the one Ca^{2+} required for O_2 evolution also effects an enormous slowdown of acceptor side (Q_A to Q_B) electron transfer (Dekker et al., 1984d; Andréasson et al., 1995). Given the relatively large distance from the OEC to the acceptor side quinones, it is clear that removal of this Ca^{2+} has very long range effects. Perhaps the essential Ca^{2+} is moderately distant from the Mn cluster, but serves in a key structural role, stabilizing a protein conformation necessary for both S-state turnover and fast electron transfer between Q_A and Q_B.

Even if it is very close to the Mn cluster, the role of Ca^{2+} may be purely structural. In PS II particles depleted of the extrinsic peptides, Ca^{2+} acts to slow NH_2OH inhibition of oxygen evolution without slowing Mn reduction to Mn^{2+} (Mei and Yocum, 1991), a result suggesting that Ca^{2+} serves to stabilize the Mn ligand environment. However it is possible the Ca^{2+} is more actively involved in water oxidation. Electron transfer from Y_Z to $P680^+$ is slowed in Ca^{2+}-depleted PS II particles after the S_3-state is formed (Boussac et al., 1992). Perhaps Ca^{2+} is also required to balance electrostatic potentials within the OEC, possibly through modulating S-state dependent deprotonation or anion binding. Along the latter line, Rutherford (1989) described a possible role for Ca^{2+} acting as a site to deliver substrate water in exchange for Cl^- during the $S_3 \rightarrow S_4$ transition. A similar idea, that Ca^{2+} acts to exclude water from premature oxidation, has been forwarded by Tso et al. (1991). However this is not supported by our ^2H ESEEM work supporting water-binding in the earlier S-states. Moreover, as described above, Ca^{2+} does not block NH_2OH access to the Mn cluster (Mei and Yocum, 1991).

The Ca^{2+} binding site is quite specific: only Sr^{2+} gives partial restoration of O_2 evolution in Ca^{2+}-depleted PS II preparations (Ghanotakis et al., 1984). On the other hand, the requirement for Cl^- is more catholic: Cl^- can be replaced in Cl^--depleted membranes by Br^- with good rates of turnover, and to a lesser degree by I^- and NO_3^- (Kelley and Izawa, 1978). The F^- anion, on the other hand is a potent inhibitor. The majority of experiments investigating the role of Cl^- in oxygen evolution have examined the effect of adding Cl^- or other anions to PS II or thylakoid membranes initially depleted of Cl^-, such as in the aforementioned experiments of Kelley and Izawa (1978). The effects of chloride on amine inhibition of oxygen evolution have been studied intensively by Sandusky and Yocum (1984a,b, 1986). One amine inhibition site (called the Type 2 site by Sandusky and Yocum) is accessible only to ammonia and is not competitive with Cl^-. The other site of amine inhibition (Type I) is competitive with respect to Cl^- concentration. High concentrations of amines such as ammonia, Tris, methylamine, 2-amino-2-ethylpropanediol, and *tert*-butylamine lead to inhibition at this site under low Cl^- conditions. The increase in the binding affinity of these inhibitors with basicity across the series led Sandusky and Yocum to propose that this site is a metal binding site, probably representing direct ligation to the Mn cluster. The competitiveness of this site with Cl^-

argues that Cl^- ligates to Mn in non-inhibited samples, which opens a manifold of possibilities in which Cl^- plays a direct role in the water oxidation process, as well as for Cl^- participating structurally, for example as a Cl^--bridge between Mn ions. An alternative possibility is that Cl^- binding is associated with nearby amino acid residues (Homann, 1985, 1988a,b; Coleman and Govindjee, 1987). Chloride bound to amino acids in the vicinity of the Mn cluster could serve in either a structural or a catalytic role.

Direct structural evidence concerning the proximity of Cl^- to the Mn cluster is lacking at this time. Penner-Hahn et al. (1990) argued that the detection of one Cl ligand per four Mn ions would require EXAFS data far superior to those obtained to date. However the Klein group has stated that the difference in EXAFS spectra of cyanobacteria grown on Br^- vs. Cl^- were consistent with one halide ligand to Mn (Yachandra et al., 1991; Klein et al., 1993). More recently, DeRose et al. (1995) have examined small Mn EXAFS changes upon F^- inhibition, which they argue are consistent with the effects of F^- binding to Mn. However all of these halide substitution effects on the Mn EXAFS are very subtle and do not provide unambiguous assignment of halide ligation to the Mn cluster. Both the ^{35}Cl (76% natural abundence) and ^{37}Cl (24% natural abundence) nuclei are magnetic with nuclear spin $I = 3/2$. There are no multiline EPR signal changes observed upon substitution of Br^- for Cl^- (Yachandra et al., 1986b), although there is probably no information content in this experiment since inhomogenous broadening of the multiline EPR signal would mask any weak superhyperfine interactions. No ENDOR signals from coupled ^{35}Cl and ^{37}Cl have been reported, but this could be due to the large quadrupolar couplings of these nuclei, which could easily broaden any ENDOR signals beyond current detection capabilities.

It is true however, that just as for Ca^{2+}, depletion of Cl^- has a dramatic effect on the Mn cluster magnetic properties. Cl^--depleted PS II reaction centers show no multiline EPR signal upon illumination (Ono et al., 1986). However there is a stabilized S_2-like state formed following a single flash of Cl^--depleted PS II particles, as addition of Cl^- *after* the flash leads to multiline signal formation in the dark. Apparently the OEC does undergo advancement in the absence of Cl^-, but does not achieve the Mn cluster electronic configuration that gives rise to the multiline EPR signal. Flash dependent XANES studies of Cl^--depleted PS II particles provide support for Mn oxidation after one flash, with no Mn oxidations on subsequent flashes (Ono et al., 1995). Competitive inhibition by F^- causes a loss of the multiline EPR signal with a concomitant increase in the $g = 4.1$ EPR signal (Casey and Sauer, 1984; Ono et al., 1987; Beck and Brudvig, 1988; DeRose et al, 1995). In addition, amine binding to the Cl^- competitive site described by Sandusky and Yocum leads to stabilization of the $g = 4.1$ EPR signal (Beck and Brudvig, 1986, 1988).

It is clear that ammonia does bind directly to Mn in the ammonia inhibition site (Type 2) that is not competitive with Cl^-. Under ammonia inhibiting conditions with high Cl^- concentration, the multiline signal is the dominant S_2 signal, but its lineshape is significantly altered (Beck et al., 1986; Beck and Brudvig, 1986, 1988). ESEEM experiments performed on the ammonia-altered multiline EPR signal using both $^{14}NH_3$ and $^{15}NH_3$ showed that an NH_3-derived ligand binds directly to the Mn cluster during the $S_1 \rightarrow S_2$ transition (Britt et al., 1989). Comparison of ^2H ESE modulation in control and ammonia-treated samples incubated in ^2H_2O support the idea that ammonia displaces a water ligand upon binding to Mn in this Type 2 site (Britt et al., 1990). Analysis of the ^{14}N quadrupolar interaction led to the postulate that the ammonia-derived ligand is present as a NH_2 bridge between Mn ions (Britt et al., 1989). The change in multiline EPR lineshape may be the result of a change in the overall pattern of magnetic exchange interactions within the cluster caused by the formation of this new bridge, and recent EXAFS results reporting a change in a Mn-Mn distance are consistent with formation of such a bridge (Dau et al., 1995). Thus the ammonia binding site that is not competitive with Cl^- has been fairly well characterized. No ENDOR or ESEEM evidence has been provided for amine binding to Mn in the Cl^- competitive Type 1 site. Such results would be particularly important in giving insights into Cl^- as well as amine binding, but progress here has lagged due to the relative difficulty of performing double resonance spectroscopy on the high spin $g = 4.1$ EPR signal that is stabilized by amine binding to this site.

Much of the work described above points to a direct role of Cl^- in the water splitting chemistry. However, some results indicate that Cl^- serves only in a structural role. Lindberg et al. (1990, 1993) have used radioactive ^{36}Cl, which provides a direct measure of Cl^- binding, to study slow Cl^- exchange into and out of fully intact PS II. These studies characterize a slow exchange (halftime of several hours) binding site that saturates with 1 Cl^-/PS II. This slow exchange site is disrupted by removal of the 17 and 24 kDa extrinsic proteins with NaCl washing, as well as other treatments that also remove the 33 kDa protein. However, Cl^- occupation at this site does not appear to be correlated with rates of O_2 evolution: there is still appreciable O_2 evolution (>20% of control rates) with this site empty. Lindberg et al. determined that, if Cl^- is required at other fast exchanging sites, the binding must be extremely tight, with a dissociation constant no greater than 1 μM. This value is orders of magnitude smaller than those reported in previous studies, such as that of Kelley and Izawa (1978), which examined reconstitution of oxygen evolution in Cl^--depleted thylakoid membranes. This raises the possibility that such studies using Cl^--depleted membranes are examining alleviation of a lesion induced by the very processes used to Cl^--deplete the membranes (Wydrzynski et al., 1990). Thus Cl^- may not be directly involved in the Kok cycle in intact PS II membranes, but Cl^--depletion induces structural changes that lead to loss of activity, and addition of high levels of Cl^- either reverses the structural changes or somehow compensates for them. The slow exchange site defined by Lindberg et al. clearly plays a structural role in stabilizing the Mn complex, as oxygen evolution and the ability to form the $g = 2$ EPR signals decreases over a period of hours as this site is depopulated by dialysis against Cl^--free buffer. However, the degradation of oxygen evolution and loss of EPR signals is prevented by inclusion of 25 μM Cl^- in the dialysis buffer.

Despite much progress in characterizing the contribution of the cofactors Ca^{2+} and Cl^- to the water-splitting reactions, we see that it is still uncertain whether these ions are specifically involved in the water splitting chemistry, or whether they merely serve as crucial structural elements within the OEC. One can therefore anticipate continued research into the functions of both Ca^{2+} and Cl^-.

V. Water Oxidation Mechanisms

As our knowledge of the structure of the OEC has evolved, a number of models have been proposed for how the water splitting chemistry proceeds. This section reviews two types of mechanistic proposals. Discussed first are possible mechanisms by which a tetranuclear Mn cluster could produce molecular oxygen from water. Then new models involving tyrosine $Y_Z^{\bullet}$ directly in water oxidation are described.

A. Metal Cluster Mechanisms

The photosynthetic water oxidation reaction has attracted much attention from inorganic chemists interested in the biological activity of manganese (reviews include Pecoraro, 1988, 1992; Wieghardt, 1989, 1994; Brudvig et al., 1991; Armstrong, 1992; Pecoraro et al., 1994, 1995a,b). As evidence for a tetranuclear Mn cluster at the heart of the OEC gathered, models emerged for how a tetranuclear cluster could carry out the O_2-evolution chemistry. These models invoke structural changes within the tetranuclear cluster in the Kok cycle to force oxidation of water and the formation of molecular oxygen. I think of these as 'transformer toy' models because of the way they convert between different cluster forms in the course of the Kok cycle.

One such model was described by Brudvig and Crabtree (1986). In this model, the tetranuclear Mn cluster exists in a 'cubane-like' Mn_4O_4 structure in the lower S-states S_0, S_1, and S_2. Such a cubane structure is analogous to the well characterized Fe_4S_4 clusters. In the Brudvig and Crabtree model, two water derived ligands (O^{2-} or OH^-) bind into bridging positions during the $S_2 \rightarrow S_3$ transition, forming a Mn_4O_6 'adamantane-like' structure. The Mn oxidation during the $S_3 \rightarrow S_4$ transition then triggers a structural rearrangement, with oxo-bridges swinging together to form the O-O bond, the subsequent release of oxygen, and finally a return to the Mn_4O_4 cubane structure.

Christou and Vincent (1987) have proposed a different mechanism, but with similar flavor, based on their synthetic model chemistry. In their 'double-pivot' model the Mn cluster exists in a Mn_4O_2 'butterfly' configuration in the S_0 and S_1 states, with two 'wing-tip' Mn ions bound to the oxygens of a Mn_2O_2 planar core. Substrate waters bind to the wing-tip position Mn ions during either the $S_0 \rightarrow S_1$ or $S_1 \rightarrow S_2$ transition. These deprotonate during the $S_2 \rightarrow S_3$ transition, forming a Mn_4O_4 cubane structure in the S_3 state as the two wing-tip Mn ions pivot inward. On the $S_3 \rightarrow S_4$ transition, the formation of the O-O bond is triggered by bringing the two oxygens on one face of the cubane together, O_2 is eliminated, and the Mn cluster returns to the Mn_4O_2 'butterfly' configuration.

The presence of the very symmetric Mn clusters

utilized in the Brudvig and Crabtree and the Christou and Vincent models has not been supported by EXAFS, which favor a less symmetric structure such as displayed in Fig. 8.

In these types of models, the water oxidation chemistry is completely driven by changes in structure of the tetranuclear cluster, with a great deal of bond rearrangement at different steps of the cycle. Amino acid residues from the protein do not enter directly into the chemistry. In dramatic contrast, recent models for the water oxidation chemistry place great emphasis on a specific amino acid residue, the redox active tyrosine Y_Z, and consider photosynthetic water oxidation to occur through a metallo-radical mechanism.

B. Metallo-Radical Mechanisms

Recent proposals invoking a direct role of tyrosine $Y_Z^{\bullet}$ in oxygen evolution are in stark contrast to the cluster chemistry models described above (Gilchrist et al., 1995; Hoganson et al., 1995; Tommos et al., 1995). In these new metalloradical models, the Mn cluster serves to position and acidify substrate water molecules, with electron and proton transfers from the substrate waters driven by the tyrosine radical $Y_Z^{\bullet}$.

A major basis for placing $Y_Z^{\bullet}$ directly in the mechanism of water oxidation comes from evidence from pulsed EPR spectroscopy that the Mn cluster is very close to this redox active tyrosine (Gilchrist et al., 1995). These pulsed EPR measurements were carried out to identify the molecular origin of a broad radical EPR signal that is cryogenically trapped following illumination of Ca^{2+}-depleted PS II particles (Boussac et al., 1989). The flash dependence of the signal amplitude indicates that it originates from the S_3 state (Boussac et al., 1990). In Ca^{2+}-depleted PS II particles, the radical has a linewidth of approximately 160 G. Other treatments that block oxygen evolution, such as acetate or fluoride incubation, lead to similar signals upon illumination, with varying linewidths depending on details of the treatment and the resulting extrinsic polypeptide composition (Baumgarten et al., 1990; Andréasson and Lindberg, 1992; Boussac et al., 1992; Hallahan et al., 1992; Maclachlan and Nugent, 1993). Boussac et al. (1990) examined the UV absorption changes associated with the formation of the broad radical in Ca^{2+}-depleted preparations. These results were interpreted as favoring an oxidized histidine as the origin for this signal. The unusual breadth was attributed to the effect of magnetic coupling to the Mn cluster. The histidine assignment was reinforced by FTIR difference spectra correlated with radical formation (Berthomieu and Boussac, 1995). A potential problem in such experiments is that they are sensitive to UV or IR changes originating throughout the complex PS II preparation. For example, the UV changes could be due to the Mn cluster itself, and changes in the FTIR spectra could arise from a variety of sources, including secondary changes from a histidine ligand to Mn as the Mn ion is oxidized. The histidine assignment was challenged by Hallahan et al. (1992), whose CW EPR experiments were interpreted to favor a $Y_Z^{\bullet}$ origin of the broad spectrum. However the experimental procedure utilized in this study was criticized by Boussac and Rutherford (1992), who argued that the trapped $Y_Z^{\bullet}$ signal reported by Hallahan et al. was really an increase in the $Y_D^{\bullet}$ signal due to a decrease in microwave power saturation properties. It has also been suggested, based on the average g-value of the radical, that it arises from a partially oxidized substrate molecule (Kusunoki, 1995). Furthermore, it also seemed possible that the radical could have an acceptor-side origin.

We first used pulsed EPR spectroscopy to address the molecular origin of the broad radical in Ca^{2+}-depleted PS II using the ESE-ENDOR technique (Gilchrist et al., 1995). An advantage of this approach is that only protons of the radical itself can contribute to the ^{1}H ENDOR spectrum outside a narrow frequency band about the ^{1}H Larmor frequency. The result of this study showed that the proton ENDOR pattern is identical with those obtained for biological tyrosine radicals, and in stark contrast with the spectrum of a histidine radical similar to that utilized as a model in the UV difference spectrum study. ^{1}H ENDOR peaks were observed at the same frequencies of the 2,6 and 3,5 ring protons of tyrosine radicals, as well as at a higher frequency matching the strongly coupled β proton in photoaccumulated $Y_Z^{\bullet}$ of Mn-depleted PS II particles. However, in this first experiment, we were unable to observe a pure $Y^{\bullet}_Z$ spectrum in the Mn-depleted spinach PS II membranes because of the presence of the stable $Y_D^{\bullet}$ radical. More recently, the ESE-ENDOR experiment has been performed on $Y_Z^{\bullet}$ trapped in a Mn-depleted PS II particle from a *Synechocystis* mutant where the $Y_D^{\bullet}$ radical signal was deleted by converting the D2 tyrosine-160 to a phenylalanine residue (Britt et al., 1995). The ^{1}H ENDOR obtained on this sample was essentially identical to that obtained for the Ca^{2+}-

depletion radical, giving strong support to the $Y^{\bullet}_Z$ assignment to the broad radical. Conclusive assignment of a tyrosine origin of the broad radical signal is provided by an ESEEM study of the broad radical generated by acetate treatment in PS II particles prepared from *Synechocystis* grown on deuterated tyrosine (Fig. 10). Trace (a) shows the frequency domain illuminated-annealed ESEEM spectrum of the broad radical signal. The ratio of ESEEM patterns obtained for a deuterated tyrosine sample and a control sample is employed to isolate only those modulation changes due to the incorporation of ^2H in the nonexchangeable hydrogen positions of tyrosine. The peaks are at similar frequencies to those observed following a parallel ratio procedure for tyrosine $Y^{\bullet}_D$ (Trace (c)). The ESEEM assigned to the broad radical does not arise from contamination from the $Y^{\bullet}_D$ tyrosine. The broad radical ESEEM is taken 70 G upfield from the high field edge of the $Y^{\bullet}_D$ signal, and there is no ^2H modulation evident in the annealed data set (Trace (b)) even though the $Y^{\bullet}_D$ signal amplitude is essentially unchanged following the annealing step. The ^2H modulation features of the broad radical are also comparable to those observed for $Y^{\bullet}_Z$ cryotrapped in the $Y^{\bullet}_D$-less mutant grown on specifically-deuterated tyrosine (Tommos et al., 1995). The observed retardation of the electron transfer rate from Y_Z to $P680^+$ in Ca^{2+}-depleted PS II particles (Andréasson et al., 1995) provides an explanation of how $Y^{\bullet}_Z$ is trapped by fast freezing after illumination in such preparations.

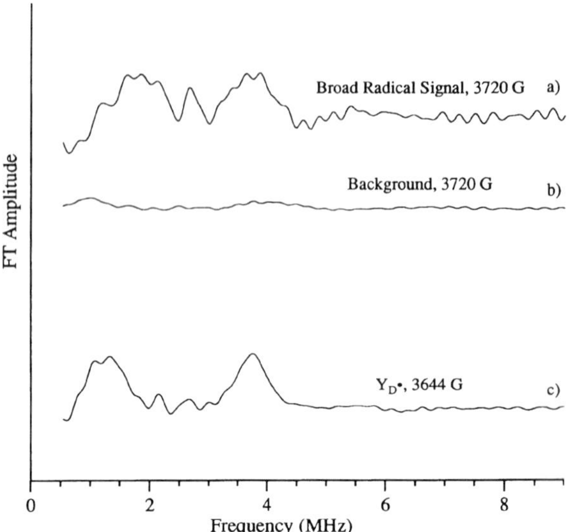

Fig. 10. ESEEM of the broad radical signal formed by illumination after acetate treatment of *Synechocystis* PS II membranes grown with deuterated tyrosine: (a) the frequency domain illuminated-annealed ESEEM spectrum of the broad radical signal. The ratio of ESEEM patterns obtained for a deuterated tyrosine sample and a control sample is employed to isolate only those modulation changes due to the incorporation of ^2H in the nonexchangeable hydrogen positions of tyrosine. (b) Ratioed ESEEM of the PS II sample after annealing to remove the broad radical signal. (c) Ratioed ESEEM of $Y^{\bullet}_D$. The ESEEM spectra in (a) and (b) were recorded 70 G upfield from the high field edge of this $Y^{\bullet}_D$ signal.

The large EPR linewidth of the $Y^{\bullet}_Z$ signal in the Ca^{2+}-depleted PS II preparation indeed appears to arise from a magnetic interaction with the Mn cluster, which is still present after Ca^{2+}-depletion ($Y^{\bullet}_Z$ trapped in the Mn-depleted sample shows a normal tyrosine radical linewidth). We were able to accurately model the linewidth and lineshape using a dipolar coupling model with a Mn tyrosine distance of only 4.5 Å. There may be some isotropic exchange component to the coupling, but that would also mandate a short Mn-tyrosine distance to broaden the tyrosine EPR spectrum to match the observed linewidth. Thus it is clear that the C_2 symmetry of PS II is broken in the placement of the Mn cluster: it is located exceedingly close to Y_Z. A speculative schematic for the overall architecture of PS II is shown in Fig. 11. This model is based on the homology of PS II with the purple bacteria reaction center. The chlorophyll pigments and the quinones are positioned in direct analogy to the purple bacteria. The symmetry related tyrosines Y_D and Y_Z are shown in positions corresponding to M arg164 and L arg135, respectively. The Mn cluster (shown in detail in Fig. 8) is positioned to be ligated by D1 his190 (replacing L tyr164), which positions the cluster in close proximity to Y_Z as indicated by our pulsed EPR results.

The proximity of the Mn cluster to tyrosine $Y^{\bullet}_Z$ raises the possibility that the radical acts as more than a simple electron transfer component. The PS II tyrosine radicals are neutral: the phenolic oxygen deprotonates as the tyrosine is oxidized (Babcock et al., 1989). Therefore tyrosine radical-mediated proton transfer is occurring in close proximity to the Mn cluster where water oxidation occurs, and it is therefore appealing to couple tyrosine based proton transfer directly into the water oxidation process.

One possible role of $Y^{\bullet}_Z$ as a strong proton donor arises from the work of Pecoraro and coworkers (1995b), who suggested that an oxidized amino acid residue, such as tyrosyl or histidyl, could protonate the oxo bridges of a relatively stable $Mn(IV)_2O_2$ unit,

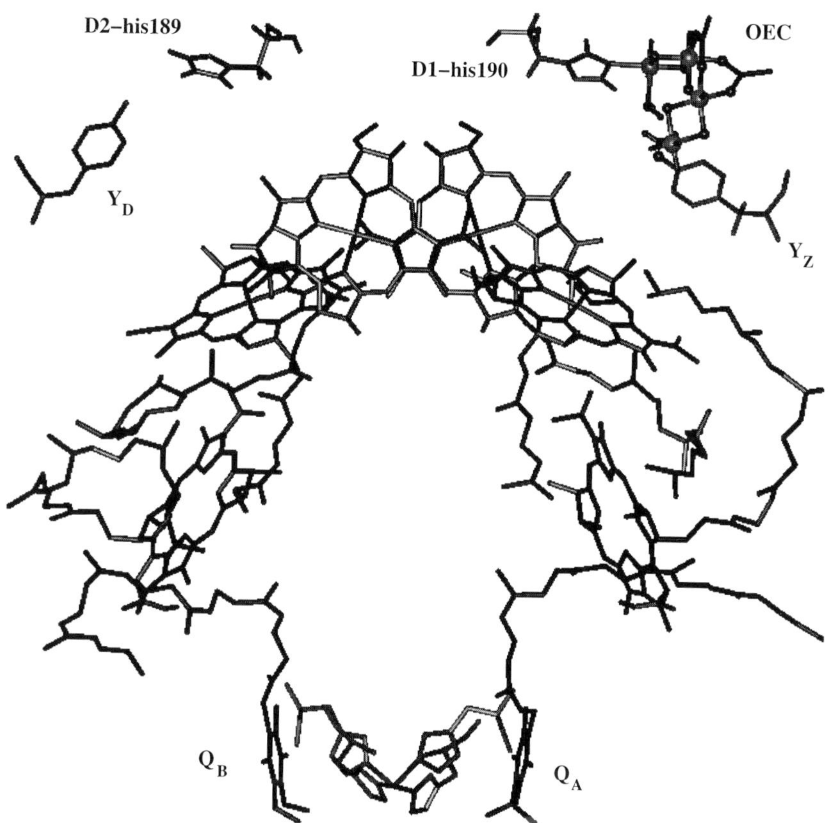

Fig. 11. A model for PS II. The organic cofactors are placed in homology with the *Rhodobacter* reaction center structure (Komiya et al, 1988). Amino acid substitutions are: L arg135 → D1 tyr161 (Y_Z); M arg164 → D2 tyr160 (Y_D); L tyr164 → D1 his190; and M his193 → D2 his189. The tetranuclear Mn cluster is placed so it is ligated to D1 his-190, which allows it to be positioned adjacent to tyrosine Y_Z. The Mn cluster region is shown to larger scale as Fig. 8. This model was constructed by Dr. Jeffrey Peloquin.

triggering a reduction to Mn(III) that begins the substrate oxidation process via a hydrogen peroxide intermediate formed at that dinuclear Mn core. We have proposed a model in which Y_Z acts as a photogenerated base in the water oxidation chemistry (Fig. 12). In this model, we start with substrate water bound to Mn of the OEC cluster in a specific S-state S_n. After the initial charge transfer, P680 is present as a cation radical $P680^+$. Electron transfer oxidizing Y_Z and re-reducing $P680^+$ (A) triggers the phenolic proton transfer to a proximal base *B* (B), leaving the Y_Z tyrosine in a neutral radical form $Y_Z^{\bullet}$ (C). Protons have been determined to be released into the lumenal phase on the same timescale as Y_Z oxidation (Haumann and Junge, 1994), so proton transfer from base B to the lumen is also illustrated at this step. In the schematic model of the OEC displayed in Fig. 8, one could envision D1 his190 serving as this proximal base. Subsequent electron transfer from the Mn cluster to $Y_Z^{\bullet}$ advances the Mn cluster to the S_{n+1}-state, leaving Y_Z as a deprotonated tyrosine (D). This strong tyrosine base abstracts a proton from a water acidified by its ligation to a high valence Mn ion of the Mn cluster (E), leaving a hydroxo ligand to the Mn cluster (F). The sequence is then repoised to act during the next S-state transition.

It is not clear if such a proton abstraction would occur at every S-state transition. For example, early transitions could occur with aqua ligand deprotonation to hydroxo ligands merely through the increase in Lewis acidity of the oxidized Mn cluster. In this case, the tyrosine radical assisted deprotonation would be required only in later transitions, for example to deprotonate hydroxo ligands. Indeed the best evidence for tyrosine radical involvement in water oxidation occurs in the final step of the Kok cycle, where oxygen release occurs on the same timescale as $Y_Z^{\bullet}$ rereduction (Babcock et al., 1976). In the

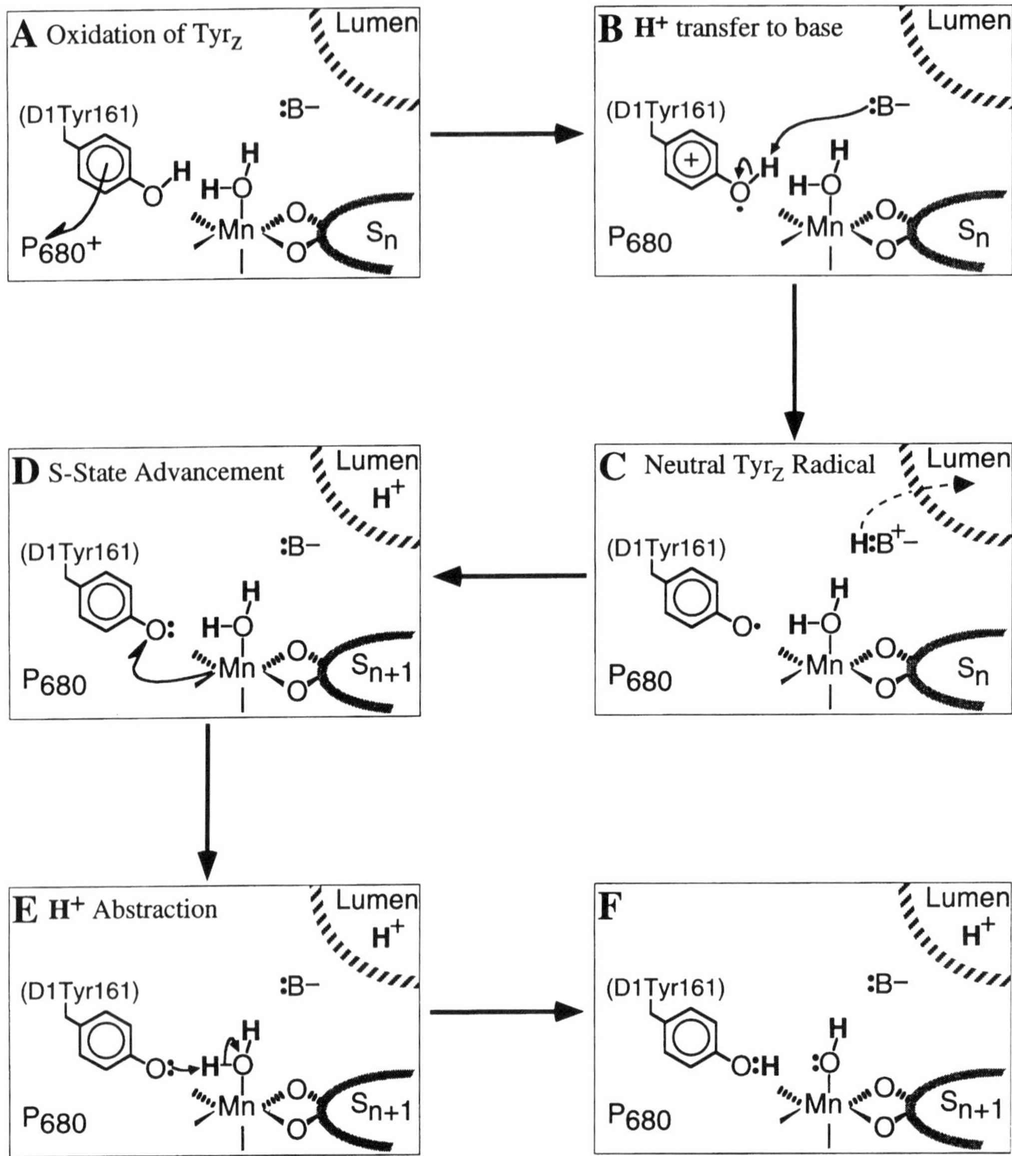

Fig. 12. A metalloradical mechanism for water oxidation, with tyrosine Y_Z acting as a photogenerated base, extracting protons as well as electrons from the Mn cluster and substrate waters. This figure was modified from Gilchrist et al. (1995) by David Randall.

metalloradical formulation, one could identify the formal S_4 state as $S_3+Y_Z^{\bullet}$, with O_2 formation triggered by the final deprotonation caused by the photogenerated tyrosine base.

A variant of this model would replace the sequential electron and proton transfers in steps (A-B) and (D-E) with single hydrogen-atom transfers. Gerald Babcock and coworkers have proposed such a hydrogen-atom transfer mechanism for oxygen evolution chemistry (Tommos et al., 1995; Hoganson et al., 1995). Along with our result that the Mn cluster is proximal to $Y_Z^{\bullet}$, a key observation leading to their hydrogen abstraction model was an ^{2}H ESEEM measurement of modestly large disorder in the dihedral angle between the $Y_Z^{\bullet}$ ring normal and the C-H bond to the strongly-coupled β methylene proton (Tommos et al., 1995). The ESEEM measurement was performed for $Y_Z^{\bullet}$ cryotrapped in Mn-depleted $Y_D^{\bullet}$-less mutants as described above. It was proposed that this angle distribution reflected a high degree of flexibility of the phenoxyl sidechain of this radical, which did not seem consistent with what would be predicted for a pure electron transfer component. By definition, a hydrogen-atom transfer requires strongly-coupled motion of the electron and proton. Since $Y_Z^{\bullet}$ is oxidized by $P680^+$ and reduced by the Mn cluster, considerable flexibility in the Y_Z tyrosine sidechain position would be necessary if there is a strong correlation between electron and proton motion as required in the hydrogen-atom transfer formulation.

One concern with the ^2H ESEEM measurement of disorder in the tyrosine ring angle is that it was obtained in Mn-depleted PS II particles rather than in PS II with an intact oxygen-evolving complex. However similar disorder appears to exist in our ^1H ESE-ENDOR measurements of Ca^{2+}-depleted PS II (Gilchrist et al., 1995). We showed, using ^2H ESE-ENDOR, that the hydrogen bonding to $Y_Z^{\bullet}$ in Mn-depleted PS II samples is quite disordered (Force et al., 1995). The disorder in the β methylene dihedral angle relative to the tyrosine ring may be due to this lack of well defined hydrogen bonding. A well-ordered hydrogen bond formed off the CO bond axis would act to form an energy barrier for rotation of the ring from its minimum energy position. The hydrogen bond disorder may simply be due to the disruption of the $Y_Z^{\bullet}$ environment induced by the removal of the proximal Mn cluster or extraction of Ca^{2+}. One could imagine that in the intact system, the $Y_Z/Y_Z^{\bullet}$ hydrogen bonding may be well-defined, but toggles between two sites for oxidation and reduction steps (Hoganson et al., 1995; Diner et al., 1995; Force et al., 1995), and this site switching would require appreciable flexibility.

In the full Kok cycle model discussed by Gerald Babcock and coworkers (Hoganson et al., 1995), S_0 consists of a pair of magnetically coupled dinuclear Mn clusters, one with an H_2O ligand and one with a Cl^- ligand. Two successive hydrogen atom transfers from the H_2O ligand site result in an O ligand at S_2, with another H_2O ligand binding in the site previously occupied by Cl^- in the S_2 state. This second H_2O ligand is the site of two more hydrogen atom transfers, with O_2 formed from the resulting two O ligands at S_4. The details of this model are very speculative. For example there is no clear evidence that the differential water exchange rates reported by Messinger et al. (1995) result from the presence of an O ligand in S_2 and S_3, or that Cl^- actually plays such a precise role in S-state turnover as indicated in this model. However the major intent of the model is quite interesting, the idea that hydrogen atom abstraction occurs at every step of the Kok cycle. One potential problem with this is that as a result, each oxidation event occurs as a ligand oxidation, and even if the oxidation equivalents are eventually relocated to Mn, transiently the OEC is forced to pass through high energy water oxidation intermediates such as illustrated in Fig. 4. We prefer the more conservative view that the tyrosine radical acts via sequential electron and proton transfers, possibly only in the later S-state transitions as described above. It will be interesting to see the results of future experiments designed to test if indeed $Y_Z^{\bullet}$ is directly involved in water oxidation, and if so, whether the electron and proton transfers are sequential or occur as hydrogen atom transfers, and whether $Y_Z^{\bullet}$ is actually involved in each S-state transition.

Acknowledgments

I thank Lane Gilchrist, David Randall, Jeffrey Peloquin, Gary Lorigan, and Dee Force for sharing data from our laboratory and for many useful discussions about the water-splitting chemistry. I have also had many interesting recent conversations with Gerald Babcock, Bruce Diner, Xiao-Song Tang, Gary Brudvig, Vittal Yachandra, Melvin Klein, Victoria DeRose, Richard Debus, Michael Baldwin, Bridgette Barry, Vincent Pecoraro, William Armstrong, James Penner-Hahn, Kurt Warncke, Charles Dismukes, William Rutherford, and Charles Yocum about these topics. Our own research is supported by grants from the United States Department of Agriculture, the National Science Foundation, and the National Institutes of Health.

References

Ädelroth P, Lindberg K and Andréasson LE (1995) Studies of Ca^{2+} binding in spinach Photosystem II using $^{45}Ca^{2+}$. Biochemistry 34: 9021–9027

Andréasson LE and Lindberg K (1992) The inhibition of photosynthetic oxygen evolution by ammonia probed by EPR. Biochim Biophys Acta 1100: 177–183

Andréasson LE, Vass I and Styring S (1995) Calcium ion depletion modifies the electron transfer on both donor and acceptor sides in Photosystem II from spinach. Biochim Biophys Acta 1230: 155–164

Andrews JC, Cinco R, Dau H, Latimer MJ, Liang W, Roelofs TA, Rompel A, Sauer K, Yachandra VK and Klein MP (1995) Photosynthetic water oxidation—structural insights to the catalytic manganese complex. Physica B 209: 657–659

Åhrling K and Pace RJ (1995) Simulation of the S_2 state multiline electron paramagnetic resonance signal of Photosystem II: A multifrequency approach. Biophys J 68: 2081–2090

Armstrong WH (1992) Polynuclear manganese complexes as models for the Photosystem II water oxidation catalyst. In: Pecoraro VL (ed) Manganese Redox Enzymes, pp 261–286. VCH Publishers, New York.

Babcock GT (1973) Kinetics and intermediates in photosynthetic oxygen evolution. Ph. D. thesis, Department of Chemistry, University of California, Berkeley

Babcock GT, Blankenship RE and Sauer K (1976) Reaction kinetics for positive charge accumulation on the water side of chloroplast Photosystem II. FEBS Lett 61: 286–289

Babcock GT, Barry BA, Debus RJ, Hoganson CW, Atamian M, McIntosh L, Sithole I and Yocum CF (1989) Water oxidation in Photosystem II: From radical chemistry to multielectron chemistry. Biochemistry 28: 9557–9565

Barry BA and Babcock GT (1987) Tyrosine radicals are involved in the photosynthetic oxygen-evolving system. Proc Natl Acad Sci USA 84: 7099–7103

Barry BA, Boerner RJ and de Paula JC (1994) The use of cyanobacteria in the study of the structure and function of Photosystem II. In: Bryant D (ed) The Molecular Biology of Cyanobacteria. pp 217–257. Kluwer Academic Publishers, Dordrecht

Baumgarten M, Philo JS and Dismukes GC (1990) Mechanism of photoinhibition of photosynthetic water oxidation by chloride depletion and fluoride substitution: Oxidation of a protein residue. Biochemistry 29: 10814–10822

Beck WF and Brudvig GW (1986) Binding of amines to the O_2-evolving center of Photosystem II. Biochemistry 25: 6749–6756

Beck WF and Brudvig GW (1988) Ligand-substitution reactions of the O_2-evolving center of Photosystem II. Chemica Scripta 28A: 93–98

Beck WF, de Paula JC and Brudvig GW (1986) Ammonia binds to the manganese site of the O_2-evolving complex of Photosystem II in the S_2 state. J Am Chem Soc 108: 4018–4022

Berthold DA, Babcock GT and Yocum CF (1981) A highly resolved, oxygen-evolving Photosystem II preparation from spinach thylakoid membranes. FEBS Lett 134: 231–234

Berthomieu C and Boussac A (1995) Histidine oxidation in the S_2 to S_3 transitions probed by FTIR difference spectroscopy in the Ca^{2+}-depleted Photosystem-II: Comparison with histidine radicals generated by UV-irradiation. Biochemistry 34: 1541–1548

Boerner RJ, Nguyen AP, Barry BA and Debus RJ (1992) Evidence from directed mutagenesis that aspartate 170 of the D1 polypeptide influences the assembly and/or stability of the manganese cluster in the photosynthetic water-splitting complex. Biochemistry 31: 6660–6672

Bonvoisin J, Blondin G, Girerd JJ and Zimmermann JL (1992) Theoretical study of the multiline EPR signal from the S_2 state of the oxygen evolving complex of Photosystem II: Evidence for a magnetic tetramer. Biophys J 61:1076–1086

Bouges B (1971) Action de faibles concentrations d'hydroxylamine sur emission d'oxygene des algues *Chlorella* et des chloroplasts d'épinards. Biochim Biophys Acta 234: 103–112

Boussac A and Rutherford AW (1988a) Nature of the inhibition of the oxygen-evolving enzyme of Photosystem II induced by NaCl washing and reversed by the addition of Ca^{2+} or Sr^{2+} Biochemistry 27: 3476–3483

Boussac A and Rutherford AW (1988b) Ca^{2+} binding to the oxygen evolving enzyme varies with the redox state of the Mn cluster. FEBS Lett 236: 432–436

Boussac A and Rutherford AW (1992) The origin of the split S_3 EPR signal in calcium-depleted Photosystem II: Histidine versus tyrosine. Biochemistry 34: 7441–7445

Boussac A, Zimmermann JL and Rutherford AW (1989) EPR signals from modified charge accumulation states of the oxygen-evolving enzyme in Ca^{2+}-deficient Photosystem II. Biochemistry 28: 8984–8989

Boussac A, Zimmermann JL, Rutherford AW and Lavergne J (1990) Histidine oxidation in the oxygen-evolving Photosystem-II enzyme. Nature 347: 303–306

Boussac A, Sétif P and Rutherford AW (1992) Inhibition of tyrosine Z photooxidation after formation of the S_3-state in Ca^{2+}-depleted and Cl^--depleted photosystem-II. Biochemistry 31: 1224–1234

Britt RD (1995) Electron spin echo methods in photosynthesis research. In: Amesz J and Hoff AJ (eds) Biophysical Techniques in Photosynthesis, pp 235–253. Kluwer Academic Publishers, Dordrecht

Britt RD and Klein MP (1992) Electron spin echo envelope modulation studies of mixed valence manganese complexes: Applications to the catalytic manganese cluster of photosynthetic oxygen evolution. In: Bagguley DMS (ed) Pulsed Magnetic Resonance: NMR, ESR, and Optics, a Recognition of E. L. Hahn, pp 361–376. Clarendon Press, Oxford

Britt RD, Zimmermann JL, Sauer K and Klein MP (1989) Ammonia binds to the catalytic manganese of the oxygen-evolving complex of Photosystem II. Evidence by electron spin-echo envelope modulation spectroscopy. J Am Chem Soc 111: 3522–3532

Britt RD, DeRose VJ, Yachandra VK, Kim DK, Sauer K and Klein MP(1990) Pulsed EPR studies of the manganese center of the oxygen-evolving complex of Photosystem II. In: Baltscheffsky M (ed) Current Research in Photosynthesis, Vol I, pp 769–772, Kluwer Academic Publishers, Dordrecht

Britt RD, Lorigan GA, Sauer K, Klein MP, and Zimmermann JL (1992) The $g = 2$ multiline EPR signal of the S_2 state of the photosynthetic oxygen-evolving complex originates from a ground spin state. Biochim Biophys Acta 1040: 95–101

Britt RD, Randall DW, Ball JA, Gilchrist ML, Force DA, Sturgeon BE, Lorigan GA, Tang XS, Diner BA, Klein MP, Chan MK and Armstrong WH (1995) Electron spin-echo endor studies of the tyrosine radicals and the manganese cluster of Photosystem II. In: Mathis P (ed) Photosynthesis: From Light to Biosphere, Vol II, pp 223–228. Kluwer Academic Publishers, Dordrecht

Brudvig GW and Crabtree RH (1986) Mechanism for photosynthetic O_2 evolution. Proc Natl Acad Sci USA 83: 4586–4588

Brudvig GW, Thorp HH and Crabtree RH (1991) Probing the mechanism of water oxidation in Photosystem II. Acc Chem Res 24: 311–316

Casey J and Sauer K (1984) EPR detection of a cryogenically photogenerated intermediate in photosynthetic oxygen evolution. Biochem Biophys Acta 767: 21–28

Christou G and Vincent JB (1987) The molecular 'double-pivot' mechanism for water oxidation. Biochem Biophys Acta 895: 259–274

Chu HA, Nguyen AP and Debus RJ (1994a) Site-directed Photosystem II mutants with perturbed oxygen-evolving properties. 1. Instability or inefficient assembly of the manganese cluster in vivo. Biochemistry 33: 6137–6149

Chu HA, Nguyen AP and Debus RJ (1994b) Site-directed Photosystem II mutants with perturbed oxygen-evolving properties. 2. Increased binding or photooxidation of manganese in the absence of the extrinsic 33-kDa polypeptide in vivo. Biochemistry 33: 6150–6157

Chu HA, Nguyen AP and Debus RJ (1995a) Amino acid residues that influence the binding of manganese or calcium to Photosystem II. 1. The lumenal interhelical domains of the D1 polypeptide. Biochemistry 34: 5839–5858

Chu HA, Nguyen AP and Debus RJ (1995b) Amino acid residues that influence the binding of manganese or calcium to Photosystem II. 2. The carboxy-terminal domain of the D1 polypeptide. Biochemistry 34: 5859–5882

Coleman WJ and Govindjee (1987) A model for the mechanism of chloride activation of oxygen evolution in Photosystem II. Photosynth Res 13: 199–223

Cooper SR, Dismukes GC, Klein MP and Calvin M (1978) Mixed valence interactions in di-μ-oxo bridged manganese complexes. Electron paramagnetic resonance and magnetic susceptibility studies. J Am Chem Soc 100: 7248–7252

Dau H, Andrews JC, Roelofs TA, Latimer MJ, Liang W, Yachandra VK, Sauer K and Klein MP (1995) Structural consequences of ammonia binding to the manganese center of the photosynthetic oxygen-evolving complex: An X-ray absorption spectroscopy study of isotropic and oriented Photosystem II particles. Biochemistry 34: 5274–5287

Debus RJ (1992) The manganese and calcium ions of photosynthetic oxygen evolution. Biochim Biophys Acta 1102: 269–352

Dekker JP (1992) Optical studies on the oxygen-evolving complex of Photosystem II. In: Pecoraro VL (ed) Manganese Redox Enzymes, pp 85–101. VCH Publishers, New York.

Dekker JP, Van Gorkom HJ, Brok M and Ouwehand L (1984a) Optical characterization of Photosystem II donors. Biochim Biophys Acta 764: 301–309

Dekker JP, Van Gorkom HJ, Wensink J and Ouwehand L (1984b) Absorbance difference spectra of the successive redox states of the oxygen-evolving apparatus of photosynthesis. Biochim Biophys Acta 767: 1–9

Dekker JP, Plijter JJ, Ouwehand L and Van Gorkom HJ (1984c) Kinetics of manganese redox transitions in the oxygen-evolving apparatus of photosynthesis. Biochim Biophys Acta 767: 176–179

Dekker JP, Ghanotakis DF, Plijter JJ, Van Gorkom HJ and Babcock GT (1984d) Kinetics of the oxygen-evolving complex in salt-washed Photosystem II preparations. Biochim Biophys Acta 767: 515–523

de Paula JC and Brudvig GW (1985) Magnetic properties of manganese in the photosynthetic O_2-evolving complex. J Am Chem Soc. 107: 2643–2648

de Paula JC, Beck WF and Brudvig GW (1986) Magnetic properties of manganese in the photosynthetic O_2-evolving complex. 2. Evidence for a manganese tetramer. J Am Chem Soc 108: 4002–4009

DeRose VJ, Yachandra VK, McDermott AE, Britt RD, Sauer K and Klein MP (1991) Nitrogen ligation to manganese in the photosynthetic oxygen-evolving complex: Continuous wave and pulsed EPR studies of Photosystem II particles enriched with ^{14}N and ^{15}N isotopes. Biochemistry 30: 1335–1341

DeRose VJ, Mukerji I, Latimer MJ, Yachandra VK, Sauer K and Klein MP (1994) Comparison of the manganese oxygen-evolving complex in Photosystem II of spinach and *Synechocystis* sp. with multinuclear manganese model compounds by X-ray absorption spectroscopy. J Am Chem Soc 116: 5239–5249

DeRose VJ, Latimer MJ, Zimmermann JL, Mukerji I, Yachandra VK, Sauer K and Klein MP (1995) Fluoride substitution in the Mn cluster from Photosystem II: EPR and X-ray absorption spectroscopy studies. Chem Phys 194: 443–459

Diner BA, Nixon PJ and Farchaus JW (1991) Site-directed mutagenesis of photosynthetic reaction centers. Curr Opin Struct Biol 1: 546–554

Diner BA, Tang XS, Zheng M, Dismukes GC, Force DA, Randall DW and Britt RD (1995) Environment and function of the redox active tyrosines of Photosystem II. In: Mathis P (ed) Photosynthesis: From Light to Biosphere, Vol II, pp 229–234. Kluwer Academic Publishers, Dordrecht

Dismukes GC and Siderer Y (1981) Intermediates of a polynuclear manganese center involved in photosynthetic oxidation of water. Proc Natl Acad Sci USA 78: 274–278

Force DA, Randall DW, Britt RD, Tang XS and Diner BA (1995) ^2H ESE-ENDOR study of hydrogen bonding to the tryosine radicals $Y_D^{\bullet}$ and $Y_Z^{\bullet}$ of Photosystem II. J Am Chem Soc 117: 12643–12644

George GN, Prince RN and Cramer SP (1989) The manganese site of the photosynthetic water-splitting enzyme. Science 243: 789–791

Ghanotakis DF and Yocum CF (1990) Photosystem II and the oxygen-evolving complex. Annu Rev Plant Physiol Plant Mol Biol 41: 255–276

Ghanotakis DF, Babcock GT and Yocum CF (1984) Calcium reconstitutes high rates of oxygen evolution in polypeptide depleted Photosystem II preparations. FEBS Lett 167: 127–130

Gilchrist ML, Ball JA, Randall DW and Britt RD (1995) Proximity of the manganese cluster of Photosystem II to the redox active tyrosine Y_Z. Proc Natl Acad Sci USA 92: 9545–9549

Goodin DB, Yachandra VK, Britt RD, Sauer K and Klein MP (1984) State of manganese in the photosynthetic apparatus. 3. Light-induced changes in X-ray absorption (K-edge) energies of manganese in photosynthetic membranes. Biochim Biophys Acta 767: 209–216

Guiles RD, Zimmermann JL, McDermott AE, Yachandra VK, Cole JL, Dexheimer SL, Britt RD, Wieghardt K, Bossek U, Sauer K and Klein MP (1990) The S_3 state of Photosystem II: Differences between the structure of the manganese complex in the S_2 and S_3 states determined by X-ray absorption spectroscopy. Biochemistry 29: 471–485

Haddy A, Dunham WR, Sands RH and Aasa R (1992) Multifrequency EPR investigations into the origin of the S_2-state signal at $g = 4.1$ of the O_2-evolving complex. Biochem Biophys Acta 1099: 25–34

Hallahan BJ, Nugent JHA, Warden JT and Evans MCW (1992) Investigation of the origin of the 'S_3' EPR signal from the oxygen-evolving complex of Photosystem 2: The role of tyrosine Z. Biochemistry 31: 4562–4573

Hansson Ö and Wydrzynski T (1990) Current perceptions of Photosystem II. Photosynth Res 23: 131–162

Hansson Ö, Andréasson LE and Vänngård T (1986) Oxygen from water is coordinated to manganese in the S_2 state of Photosystem II. FEBS Lett 195: 151–154

Hansson Ö, Aasa R and Vänngård T (1987) The origin of the multiline and $g = 4.1$ electron paramagnetic resonance signals from the oxygen-evolving system of Photosystem II. Biophys J 51: 825–832

Haumann M and Junge W (1994) Extent and rate of proton release by photosynthetic water oxidation in thylakoids—electrostatic relaxation versus chemical production. Biochemistry 33: 864–872

Hoganson CW, Lydakis-Simantiris N, Tang XS, Tommos C, Warncke K, Babcock GT, Diner BA, McCracken J and Styring S (1995) A hydrogen-atom abstraction model for the function of Y_Z in photosynthetic oxygen-evolution. Photosyn Res 46: 177–184.

Homann PH (1985) The association of functional anions with the oxygen-evolving center of chloroplasts. Biochim Biophys Acta 809: 311–319

Homann PH (1988a) The chloride and calcium requirement of photosynthetic water oxidation: Effects of pH. Biochim Biophys Acta 934: 1–13

Homann PH (1988b) Structural effects of Cl^- and other anions on the water oxidizing complex of chloroplast Photosystem II. Plant Physiol. 88: 194–199

Joliot P, Barbieri G and Chabaud R (1969) Un nouveau modèle des centres photochimique du systèm II. Photochem Photobiol 10: 309–329

Kelley PM and Izawa S (1978) The role of chloride ion in Photosystem II: I. Effects of chloride ion on Photosystem II electron transport and on hydroxylamine inhibition. Biochim Biophys Acta 502: 198–210

Kim DH, Britt RD, Klein MP and Sauer K (1990) The $g = 4.1$ EPR signal of the S_2 state of the photosynthetic oxygen evolving complex arises from a multinuclear Mn cluster. J Am Chem Soc 112: 9389–9391

Kim DH, Britt RD, Klein MP and Sauer K (1992) The manganese site of the photosynthetic oxygen-evolving complex probed by EPR spectroscopy of oriented Photosystem II membranes: The $g = 4$ and $g = 2$ multiline signals. Biochemistry 31: 541–547

Klein MP, Sauer K and Yachandra VK (1993) Perspectives on the structure of the photosynthetic oxygen evolving manganese complex and its relation to the Kok cycle. Photosyn Res 38: 265–277

Kok B, Forbush B and McGloin M (1970) Cooperation of charges in photosynthetic O_2 evolution – I. A linear four step mechanism. Photochem Photobiol 11: 457–475

Komiya H, Yeats TO, Rees DC, Allen JP and Feher G (1988) Structure of the reaction center from *Rhodobacter sphaeroides* R–26 and 2.4.1: Symmetry relations and sequence comparisons between different species. Proc Natl Acad Sci USA 85: 9012–9016

Kusunoki M (1992) A new paramagnetic hyperfine structure effect in manganese tetramers. The origin of 'multiline' EPR signals from an S_2 state of a photosynthetic water-splitting enzyme. Chem Phys Lett 197: 108–116

Kusunoki M (1995) EPR evidence for the primary water oxidation step upon the $S_2 \rightarrow S_3$ transition in the Joliot-Kok cycle of plant Photosystem II. Chem Phys Lett 239: 148–157

Latimer MJ, DeRose VJ, Mukerji I, Yachandra VK, Sauer K and Klein MP (1995) Evidence for the proximity of calcium to the manganese cluster of Photosystem II: Determination by X-ray absorption spectroscopy. Biochemistry 34: 10898–10909

Lavergne J (1986) Stoichiometry of the redox changes of manganese during the photosynthetic water oxidation cycle. Photochem Photobiol 43: 311–317

Lavergne J (1987) Optical-difference spectra of the S-state transitions in the photosynthetic oxygen-evolving center. Biochim Biophys Acta 894: 91–107

Liang W, Latimer MJ, Dau H, Roelofs TA, Yachandra VK, Sauer K and Klein MP (1994) Correlation between structure and magnetic spin state of the manganese cluster in the oxygen-evolving complex of Photosystem II in the S_2 state: Determination by X-ray absorption spectroscopy. Biochemistry 33: 4923–4932

Lincoln SF and Merbach AE (1995) Substitution reactions of solvated metal ions. In: Sykes AG (ed) Advances in Inorganic Chemistry, Vol 42, pp 2–88. Academic Press, San Diego

Lindberg K, Wydrzynski, T, Vänngård T and Andréasson LE (1990) Slow release of chloride from ^{36}Cl-labeled Photosystem II membranes. FEBS Lett 264: 153–155

Lindberg K, Vänngård T and Andréasson LE (1993) Studies of the slowly exchanging chloride in Photosystem II. Photosynth Res 38: 401–408

Lorigan GA and Britt RD (1994) Temperature-dependent pulsed electron paramagnetic resonance studies of the S_2 state multiline signal of the photosynthetic oxygen-evolving complex. Biochemistry 33: 12072–12076

MacDonald GM, Bixby KA and Barry BA (1993) A difference Fourier-transform infrared study of two redox-active tyrosine residues in Photosystem II. Proc Natl Acad Sci USA 90: 11024–11028

MacLachlan DJ and Nugent JHA(1993) Investigation of the 'S_3' EPR signal from the oxygen-evolving complex of Photosystem 2: The effect of inhibition of oxygen evolution by acetate. Biochemistry 32: 9772–9780

MacLachlan DJ, Hallahan BJ, Ruffle SV, Nugent JHA, Evans MCW, Strange RW and Hasnain SS (1992) An e.x.a.f.s. study of the manganese O_2-evolving complex in purified Photosystem II membrane fractions: The S_1 and S_2 states. Biochem J 285: 569–576

Mei R and Yocum CF (1991) Calcium retards NH_2OH inhibition of O_2 evolution activity by stabilization of Mn^{2+} binding to Photosystem II. Biochemistry 30: 7837–7842

Messinger J, Badger M and Wydrzynski T (1995) Detection of *one* slowly exchanging substrate water molecule in the S_3 state of Photosystem II. Proc Natl Acad Sci USA 92: 3209–3213

Mukerji I, Andrews JC, DeRose VJ, Latimer MJ, Yachandra VK, Sauer K and Klein MP (1994) Orientation of the oxygen-evolving manganese complex in a Photosystem II membrane preparation: An X-ray absorption spectroscopy study. Biochemistry 33: 9712–9721

Nixon PJ and Diner BA (1992) Aspartate 170 of the Photosystem II reaction center polypeptide D1 is involved in the assembly of the oxygen-evolving manganese cluster. Biochemistry 31: 942–948

Nixon PJ, Chisholm DA and Diner BA (1992) Isolation and functional analysis of random and site-directed mutants of Photosystem II. In: Shewry P and Gutteridge S (eds) Plant Protein Engineering, pp 93–141. Cambridge University Press, Cambridge

Noguchi T, Ono T and Inoue Y (1995) Direct detection of a carboxylate bridge between Mn and Ca^{2+} in the photosynthetic oxygen-evolving center by means of Fourier transform infrared spectroscopy. Biochim Biophys Acta 1228: 189–200

Noren GH, Boerner RJ and Barry BA (1991) EPR characterization of an oxygen-evolving Photosystem II preparation from the transformable cyanobacterium *Synechocystis* 6803. Biochemistry 30: 3943–3950

Ono T, Zimmermann JL, Inoue Y and Rutherford AW (1986) EPR evidence for a modified S-state transition in chloride-depleted Photosystem II. Biochim Biophys Acta 851: 193–201

Ono T, Nakayama H, Gleiter H, Inoue Y and Kawamori A (1987) Modification of the properties of S_2 state in photosynthetic O_2-evolving center by replacement of chloride with other anions. Arch Biochem Biophys 256: 618–624

Ono TA, Noguchi T, Inoue Y, Kusunoki M, Matsushita T and Oyanagi H (1992) X-ray detection of the period-four cycling of the manganese cluster in the photosynthetic water oxidizing enzyme. Science 258: 1335–1337

Ono T, Noguchi T, Inoue Y, Kusunoki M, Yamaguchi H and Oyanagi H (1995) XANES spectroscopy for monitoring intermediate reaction states of Cl^--depleted Mn cluster in photosynthetic water oxidation enzyme. J Am Chem Soc 117: 6386–6387

Pakrasi JB and Vermaas WFJ (1992) Protein engineering of Photosystem II. In: Barber J (ed) The Photosystems: Structure, Function, and Molecular Biology. pp 231–257. Elsevier, Amsterdam

Pecoraro VL (1988) Structural proposals for the manganese centers of the oxygen evolving complex: An inorganic chemist's perspective. Photochem Photobiol 48: 249–264

Pecoraro VL (1992) Structurally diverse manganese coordination complexes: From voodoo to oxygenic photosynthesis. In: Pecoraro VL (ed) Manganese redox enzymes, pp 197–231. VCH Publishers, New York

Pecoraro VL, Baldwin M and Gelasco A (1994) Interaction of manganese with dioxygen and its reduced derivatives. Chem Rev 94: 807–826

Pecoraro VL, Gelasco A and Baldwin M (1995a) Reactivity and mechanism of manganese enzymes. A modeling approach. Adv Chem Ser 246: 265–301

Pecoraro VL, Gelasco A and Baldwin M (1995b) Modeling the chemistry and properties of multinuclear manganese enzymes. In: Kessissoglou DP (ed) Bioinorganic Chemistry, pp 287–298. Kluwer Academic Publishers, Dordrecht

Penner-Hahn JE, Fronko RM, Pecoraro VL, Yocum CF, Betts SD and Bowlby NR (1990) Stuctural characterizaiton of the manganese sites in the photosynthetic oxygen-evolving complex using X-ray absorption spectroscopy. J Am Chem Soc 112: 2549–2557

Radmer R and Cheniae G (1977) Mechanisms of oxygen evolution. In: Barber J (ed) Primary Processes of Photosynthesis, pp 303–348. Elsevier, Amsterdam.

Radmer R and Ollinger O (1986) Do the higher oxidation states of the photosynthetic O_2-evolving system contain bound H_2O? FEBS Lett 195: 285–289

Randall DW, Sturgeon BE, Ball JA, Lorigan GA, Chan MK, Klein MP, Armstrong WH and Britt RD (1995) ^{55}Mn ESE-ENDOR of a mixed valence Mn(III)Mn(IV) complex: Comparison with the Mn cluster of the photosynthetic oxygen-evolving complex. J Am Chem Soc 117: 11780–11789

Renger G and Hanssum B (1988) Studies on the deconvolution of flash-induced absorption changes into the difference spectra of individual redox steps within the water-oxidizing enzyme system. Photosynth Res 16: 243–259

Riggs PJ, Mei R, Yocum CF and Penner-Hahn JE (1992) Reduced derivatives of the manganese cluster in the photosynthetic oxygen-evolving complex. J Am Chem Soc 114: 10650–10651

Rutherford AW (1989) Photosystem II, the water-splitting enzyme. TIBS 14: 227–232

Sandusky PO and Yocum CF (1984a) The mechanism of amine inhibition of the photosynthetic oxygen-evolving complex: Amines displace functional chloride from a ligand site on manganese. FEBS Lett 162: 339–343

Sandusky PO and Yocum CF (1984b) The chloride requirement for photosynthetic oxygen evolution: Analysis of the effects of chloride and other anions on amine inhibition of the oxygen-evolving complex. Biochim Biophys Acta 766: 603–611

Sandusky PO and Yocum CF (1986) The chloride requirement for photosynthetic oxygen evolution: Factors affecting nucleophilic displacement of chloride from the oxygen-evolving complex. Biochim Biophys Acta 849: 85–93

Sauer K, Yachandra VK, Britt RD and Klein MP (1992) The photosynthetic water oxidation complex studied by EPR and X-ray absorption spectroscopy. In: Pecoraro VL (ed) Manganese Redox Enzymes, pp 141–175. VCH Publishers, New York.

Saygin Ö and Witt HT (1987) Optical characterization of intermediates in the water-splitting enzyme system of photosynthesis-possible states and configurations of manganese and water. Biochim Biophys Acta 893: 452–469

Tamura N and Cheniae GM (1988) Photoactivation of the water oxidizing complex: The mechanisms and general consequences to Photosystem II. In: Stevens SE and Bryant D (eds) Light Energy Transduction in Photosynthesis: Higher Plants and Bacterial Models, pp 227–242. American Society of Plant Physiologists, Rockville

Tang XS and Diner BA (1994) Biochemical and spectroscopic characterization of a new oxygen-evolving Photosystem II core complex from the cyanobacterium *Synechocystis* PCC 6803. Biochemistry 33: 4594–4603

Tang XS, Sivaraja M and Dismukes GC (1993) Protein and substrate coordination to the manganese cluster in the photosynthetic water oxidizing complex: ^{15}N and ^{1}H ENDOR spectroscopy of the S_2 state multiline signal in the thermophilic cyanobacterium *Synechococcus elongatus*. J Am Chem Soc 115: 2382–2389

Tang XS, Diner BA, Larsen BS, Gilchrist ML, Lorigan GA and Britt RD (1994) Identification of histidine at the catalytic site of the photosynthetic oxygen-evolving complex. Proc Natl Acad Sci USA 91: 704–708

Tommos C, Tang XS, Warncke K, Hoganson CW, Styring S, McCracken J, Diner BA and Babcock GT (1995) Spin-density distribution, conformation, and hydrogen bonding of the redox-active tyrosine Y_Z in Photosystem II from multiple electron magnetic-resonance spectroscopies: Implications for photosynthetic oxygen evolution. J Am Chem Soc 117: 10325–10335

Tso J, Sivaraja M and Dismukes GC (1991) Calcium limits substrate accessibility or reactivity at the manganese cluster in photosynthetic water oxidation. Biochemistry 30: 4734–4739

Wieghardt K (1989) The active centers in manganese-containing metalloproteins and inorganic model complexes. Angew Chem 101: 1179–1198

Wieghardt K (1994) A structural model for the water-oxidizing manganese cluster in Photosystem II. Angew Chem 106: 765–768

Wydrzynski T, Baumgart F, MacMillan F and Renger G (1990) Is there a direct chloride cofactor requirement in the oxygen-evolving reactions of Photosystem II? Photosyn Res 25: 59–72

Yachandra VK, Guiles RD, McDermott AE, Britt RD, Dexheimer SL, Sauer K and Klein MP (1986a) The state of manganese in the photosynthetic apparatus. 4. Structure of the manganese complex in Photosystem II studied using EXAFS spectroscopy. The S_1 state of the O_2-evolving Photosystem II complex from spinach. Biochem Biophys Acta 850: 324–332

Yachandra VK, Guiles RD, Sauer K and Klein MP (1986b) The state of manganese in the photosynthetic apparatus. 5. The choloride effect in photosynthetic oxygen evolution. Is halide coordinated to the EPR-active manganese in the O_2-evolving complex? Studies of the substructure of the low-temperature multiline EPR signal. Biochem Biophys Acta 850: 333–342

Yachandra VK, Guiles RD, McDermott AE, Cole JL, Britt RD, Dexheimer SL, Sauer K and Klein MP (1987) Comparison of the structure of the manganese complex in the S_1 and S_2 states of the photosynthetic O_2-evolving complex: An X-ray absorption spectroscopy study. Biochemistry 26: 5974–5981

Yachandra VK, DeRose VJ, Latimer MJ, Mukerji I, Sauer K and Klein MP (1991) A structural model of the oxygen evolving manganese cluster. Photochem Photobiol 53: Supp 98S

Yachandra VK, DeRose VJ, Latimer MJ, Mukerji I, Sauer K and Klein MP (1993) Where plants make oxygen: A structural model for the photosynthetic oxygen-evolving manganese cluster. Science 260: 675–679

Yocum CF (1991) Calcium activation of photosynthetic water oxidation. Biochim Biophys Acta 1059: 1–15

Yocum CF (1994) Photosynthetic oxygen evolution. In: Nriagu JO and Simmons MS (eds) Environmental Oxidants, pp 1–30. John Wiley, New York.

Yocum CF, Yerkes CT, Blankenship RE, Sharp RR and Babcock GT (1981) Stoichiometry, inhibitor sensitivity and organization of manganese associated with photosynthetic oxygen evolution. Proc Natl Acad Sci USA 78: 7507–7511

Zheng M and Dismukes GC (1992) Photosynthetic water oxidation: What have we learned from the multiline EPR signals? In: Murata N (ed) Research in Photosynthesis, pp 305–308. Kluwer Academic Publishers, Dordrecht.

Zimmermann JL and Rutherford AW (1984) EPR studies of the oxygen-evolving enzyme of Photosystem II. Biochim Biophys Acta 767:160–167

Zimmermann JL and Rutherford AW (1986) Electron paramagnetic resonance properties of the S_2 state of the oxygen evolving complex of Photosystem II. Biochemistry 25: 4609–4615

Zimmermann JL, Boussac A and Rutherford AW (1993) The manganese center of oxygen-evolving and Ca^{2+}-depleted Photosystem II: A pulsed EPR spectroscopy study. Biochemistry 32: 4831–4841

Chapter 10

Protons and Charge Indicators in Oxygen Evolution

Michael Haumann and Wolfgang Junge
Abt. Biophysik, FB Biologie/Chemie, Universität Osnabrück, D-49069 Osnabrück, Germany

Summary

Driven by four quanta of light, Photosystem II evolves one molecule of dioxygen and releases four protons into the lumen of thylakoids. The pattern of proton release over the four reaction steps was considered as diagnostic for the partial reactions with two molecules of water. A long accepted pattern with integer proton over electron ratios, which was frequently used to model these reactions, is no longer valid. Instead, a rather large variability of the pattern as function of the material (thylakoids, IML-thylakoids, BBY-membranes and core particles) and,

Donald R. Ort and Charles F. Yocum (eds): Oxygenic Photosynthesis: The Light Reactions, pp. 165–192.

in some materials, of the pH has been observed. Proton release can occur in approximately 10 microseconds. It precedes the electron transfer from the manganese cluster to Y_Z^+, and probably reflects an electrostatic response of amino acid residues to the positive charge. Only upon the dark-transition, $S_4 \rightarrow S_0$, and in thylakoids, a partial protolytic reaction is detected which is geared by the slower electron transfer in a few milliseconds from manganese to Y_Z^+. Depending on the total number of bases formed during the preceding rapid steps, $S_0 \Rightarrow S_1 \Rightarrow S_2 \Rightarrow S_3 \Rightarrow S_4$, protons are released or taken up. This is interpreted as the net result of the constant proton production from water and the concomitant resetting of pKs of a variable number of bases upon the deposition of electrons from water into the manganese center. The kinetic and thermodynamic role of these bases in water oxidation is subject to speculation.

In another seemingly consistent theory the net charge of the catalytic center, calculated as the sum of electron abstraction and proton release, was considered as the direct source of the observed electrochromic bandshifts and different rates of the reduction of $P680^+$. Recent data show that these charge indicators do not straightforwardly follow the above variations of the pattern of proton release. Imperfect electrostatic compensation is indeed expected if proton release occurs from a different location than that of the electron hole in the manganese center. This calls for a new interpretation of 'net charge transients' which may reveal topological features of the storage sites of electron holes in the catalytic center.

I. Introduction

Four protons must be released during the tetravalent oxidation of two molecules of water to molecular oxygen. The release of these protons into the thylakoid lumen, together with the uptake of protons during the reduction of bound quinones at the stromal side of Photosystem II, is one source of the transmembrane protonmotive force which drives the synthesis of ATP.

The complete reaction cycle of the water oxidase is composed of four redox transitions. Each is powered by one quantum of light. The oxidized primary electron donor of Photosystem II, $P680^+$, oxidizes a tyrosine residue on the D1-protein, named Y_Z which in turn oxidizes a tetra-manganese cluster, Mn_4 (see Andersson and Styring, 1991; Renger and Wydrzynski, 1991; Debus, 1992; Rutherford et al., 1992; Yocum, 1992). This is formally described in terms of the Kok-scheme (Joliot and Kok, 1975) which involves five states, S_i (i = 0 – 4), of increasing oxidizing power: $S_0 \Rightarrow S_1 \Rightarrow S_2 \Rightarrow S_3 \Rightarrow S_4 \rightarrow S_0$. Herein the four double arrows denote the light reactions and the thin arrow the consecutive dark reaction that causes the burst of oxygen and closes the cycle. Excitation of Photosystem II with a series of short flashes of light clocks the catalytic center through these states, and since S_1 is usually most stable in the dark, the production of oxygen is observed upon the third flash, which continues with a damped oscillation of period of four. In contrast with this behavior, proton release is more evenly distributed over the transitions. The research on this topic (see Lavergne and Junge (1993) for a recent review) was, of course, motivated by the hope that the pattern of proton release might be diagnostic for the still cryptic reaction sequence of (bound) water. A particular pattern of proton release, namely 1:0:1:2, over the four transitions starting from S_0 has emerged from studies in several laboratories over almost ten years (Fowler, 1977; Saphon and Crofts, 1977; Bowes and Crofts, 1981; Wille and Lavergne, 1982; Förster and Junge, 1985a). This pattern was seemingly supported by measurements on two supposed indicators of the electrostatic net charge in the catalytic center, namely, certain electrochromic bandshifts in the blue and the red spectral region (Velthuys, 1981, 1988; Saygin and Witt, 1985) and the rate of reduction of $P680^+$ by Y_Z (Brettel et al., 1984; Meyer et al., 1989). Broadly speaking, their oscillatory behavior could be attributed to a pattern of net charge jumps of

Abbreviations: ATP – adenosine triphosphate; BBY – open PS II-enriched membrane fragments; BSA – bovine serum albumin; CAB – chlorophyll *a/b* binding proteins; DCBQ – dichloro-p-benzoquinone; DCCD – N,N'-Dicyclohexylcarbodiimide; DCMU – 3-(3,4-dichlorophenyl)-1,1-dimethylurea; DNP-INT – dinitrophenolether of iodonitrotoluol; EDTA – ethylenediamine-tetraacetic acid; ENDOR – electron nuclear double resonance; EPR – electron paramagnetic resonance; Fe – the non-heme iron of PS II; IML – intermittent light; LHCII – light harvesting complex of PS II; MES – 2-(N-morpholino)ethanesulfonic acid; Mn_4 – the tetranuclear manganese cluster of the water oxidase; P680 – primary electron donor of PS II; Pheo – pheophytin; PS I – Photosystem I; PS II – Photosystem II; Q_A – primary bound quinone acceptor of PS II; Q_B – secondary quinone acceptor of PS II; S_i – the redox states of the water oxidase; Y_D – tyrosine-160 of the D2 subunit of PS II; Y_Z – tyrosine-161 of the D1 subunit of PS II

0:+1:0:−1. This was understood in terms of the accepted pattern of proton release at that time of 1:0:1:2, considering that each transition abstracts one electron from the catalytic center (Brettel et al., 1984). This particular pattern of proton release has been used to hypothesize detailed reaction schemes (Renger and Govindjee, 1985; Kretschmann et al., 1991) for a stepped reaction with bound water.

This seemingly consistent theory has faded away for several reasons, which are the subject of this article. Gernot Renger (1987) proposed that one portion of proton release might be attributable to an electrostatic response of peripheral amino acid residues. A more rigorous approach to measurements of proton release and the assessment of the Kok-parameters has shown that the pattern of proton release can vary greatly as a function of the pH and of the protein periphery of the catalytic center (Lübbers and Junge, 1990; Wacker et al., 1990; Jahns et al., 1991, 1992; Rappaport and Lavergne, 1991; Jahns and Junge, 1992b), whereas the redox cycle progressed normally (Lübbers et al., 1993a). This has focused attention on other sources of protons than water itself. Inorganic chemists have disclosed the acid/base properties of μ-oxo bridges in response to the oxidoreduction of their manganese atoms (Thorp et al., 1989; Machanda et al., 1991, 1992). Measurements of proton release at high time resolution have revealed that the greater portion of proton release on any transition can occur at the level of oxidized Y_Z (Förster and Junge, 1985a; Haumann and Junge, 1994a). It is an open question whether this is caused by Y_Z proper or by an electrostatic response of ionogenic amino acid side groups, which are exposed to the electric field of the positive charge on Y_Z^+. An electrostatic response would account for the variability of proton release as a function of the pH and the protein periphery. But where are the chemically-produced water protons? Is the pattern of proton release more or less arbitrary? It is also conceivable that some bases in the vicinity of the catalytic center are essential for the final reaction with water to yield molecular oxygen. It has been proposed that the concerted transfer of protons and electrons may serve to lower the free energy barrier of an otherwise kinetically unfavorable multielectron transfer (Krishtalik, 1986, 1989, 1990; see Chapter 20). Finally, is the proposed strict correlation between supposed indicators of the net charge in the center and proton release still valid? If there are indications that it is not, are the electrochromic bandshifts perhaps indicators of internal electron transfer between different manganese atoms of the cluster?

The history of this problem, including a great deal of the recent data, and speculations have been reviewed recently (Lavergne and Junge, 1993). For the sake of the consistency of this article some portions of this review by Jerome Lavergne and one of us (WJ) have to be repeated. Here, we emphasize the then-unpublished material plus our most recent results.

II. Materials and Techniques

Some progress is owed to the comparison of various oxygen evolving materials, namely *thylakoids*, thylakoids with reduced contents of LHC-proteins, so called *IML-thylakoids*, PS II-enriched open membrane fragments, so called *BBY membranes,* and *core particles*. We start with a description of their technical advantages and drawbacks mainly in studies on proton release. This is followed by a digression on the spectroscopic detection of protolytic reactions at high time resolution, a summary of how to define the Kok-parameters by measurements of UV-absorption transients, and a description of the analysis of electrochromic absorption changes.

A. Comparison of Oxygen Evolving Materials

Different oxygen evolving materials have specific advantages and disadvantages in this context. The signal-to-noise ratio in flash spectrophotometric studies is reciprocal to the number of chlorophyll molecules per oxygen evolving reaction center. The respective numbers are as follows: thylakoids (500–700) (Melis and Anderson, 1983), IML-thylakoids (120–150) (M. Haumann, unpublished), BBY membranes (250–300) (Berthold et al., 1981) and core particles (50–120) (Schatz and Witt, 1984; Ghanotakis et al., 1987a; van Leeuwen et al., 1991). Core particles are the best system in this respect. The discrimination, mainly in the UV, of intrinsic absorption transients, which are attributable to the donor side of Photosystem II against those from the acceptor side, is a problem in all systems (see below). The discrimination of protolytic transients is perhaps most reliable in thylakoids and IML-thylakoids since the events at the donor side are spatially separated from those at the acceptor side by the rather proton impermeable thylakoid membrane. With the F-

ATPase blocked, the relaxation time of a flash-induced pH difference is longer than 10 s (Junge et al., 1986).

1. Thylakoids

Thylakoids are usually prepared from spinach or pea leaves. They contain typically 500–700 chlorophyll molecules per oxygen evolving reaction center, the full electron transport chain with PS I, and the ATPase (Melis and Anderson, 1983). Two types of approaches have been chosen to detect the release of protons by water oxidation in thylakoids: (a) The difference of pH transients in the suspending medium is recorded with and without a protonophore added. The protonophore (uncoupler) accelerates the relaxation of the pH-difference across the thylakoid membrane. Either glass electrodes (Fowler, 1977) or hydrophilic pH-indicating dyes (Saphon and Crofts, 1977) have been used for detection of pH-transients in the medium. The latter are practically selective for pH-transients in the medium because of the very large volume ratio (> 1000:1) of the medium over the lumen in thylakoids (Polle and Junge, 1986a;b). Both the glass electrode (intrinsic rise time >10 ms) and the indicator dyes, which detect protons from the lumen only after their passage across the membrane, have limited time resolution. (b) A particular amphiphilic dye, neutral red, has been used as an indicator for pH-transients in the lumen of thylakoids (Ausländer and Junge, 1975). Compared to hydrophilic pH-indicators, it has a much greater sensitivity for pH-transients in the lumen. This is due to its adsorption at the membrane-water interfaces which causes an effective 'upconcentration' by two to three orders of magnitude in the lumen (Junge et al., 1979; Hong and Junge, 1983). The dye originally responds to events at both interfaces, the lumen- and the stroma-oriented sides. Selectivity for lumenal events is obtained by the addition of a non-permeant buffer, like bovine serum albumin, in order to quench the pH transients in the medium. With this dye the different time windows of proton release by water oxidation (≤1 ms) and by plastoquinol oxidation (about 10 ms) have been detected (Ausländer and Junge, 1975) and it has been shown that some components of proton release are faster than electron transfer from Mn to Y_Z^+ (Förster and Junge, 1985a). However, the kinetic competence of this indicator dye is flawed in stacked thylakoids, as pointed out by Lavergne and Rappaport (1990). The rationale is that bovine serum albumin has no access to the very narrow (5 nm) partitions between stacked membranes, whereas neutral red has. For the time span of the lateral relaxation of the alkalinization jump, which is created by proton uptake from the partitions at the reducing side of PS II, the response of neutral red is mixed with contributions from the lumen and the partition compartments. This relaxation along the membrane surface (not across the membrane) takes about 100 ms (Polle and Junge, 1986b). The slowness is theoretically understood by diffusion/reaction of protons in a domain (the partition) which contains a high concentration of *fixed* buffering groups (Junge and Polle, 1986; Junge and McLaughlin, 1987). The addition of mobile buffers accelerates this relaxation (Polle and Junge, 1989), as theoretically expected. This problem of neutral red is greatly diminished, if not virtually absent, in unstacked thylakoids in the presence of a non-permeating buffer at high concentration (Jahns et al., 1991). In this material, proton release into the lumen by water oxidation has been resolved down to a half-rise time of 13 μs (Jahns et al., 1992). The intrinsic response of neutral red to events at *both sides* of the membrane has also been used for a high time resolution of proton uptake at the bound quinone-iron system, Q_AFeQ_B. Transient signals from neutral red recorded without and with addition of bovine serum albumin yielded a half-rise time of 230 μs for proton uptake upon the reduction of Q_AFe^{III} (Haumann and Junge, 1994b).

2. IML-thylakoids

IML-thylakoids are isolated from e.g. pea seedlings that are grown under intermittent light (about 10 days under a repetitive cycle of 2 min light and 118 min darkness). This regime impairs the synthesis of chlorophyll *b* and drastically diminishes the amount of light-harvesting CAB proteins (Argyroudi-Akoyunoglou and Akoyunoglou, 1970; Tzinas et al., 1987). Accordingly, the ratio of chlorophyll molecules per electron transport chain is decreased to about 150 (Haumann, unpublished). These membranes are intrinsically unstacked (Day et al., 1984). They contain both photosystems and the ATPase. IML-thylakoids share with ordinary thylakoids the spatial separation of the protolytic reactions at the donor and the acceptor side of PS II. The above mentioned kinetic problems with neutral red are absent (Jahns and Junge, 1992a,b).

3. BBY-membranes

BBY-membranes are prepared by mildly solubilizing the stroma lamellae by Triton X-100 (Berthold et al., 1981). The remaining membranous fraction is highly enriched in PS II and nearly devoid of PS I. The membranes are still partially stacked but not vesicular. Stacking is disadvantageous because it limits the kinetic competence of pH-indicating dyes (see Section IIB). The non-vesicular structure is also disadvantageous because any pH-indicating dye responds to the protolytic reactions *on both sides of PS II* which complicates the isolation of proton release caused by water oxidation.

The group of Jérôme Lavergne relied on using 20 μM DCBQ as a highly efficient electron acceptor (Rappaport and Lavergne, 1991), that interacts with Q_A (Oettmeier et al., 1987; Renger et al., 1988). Their rationale was based on the assumption that the reduction of DCBQ causes proton uptake without any oscillations as function of the flash number. Oscillations of proton release at the donor side cause an oscillation of the net pH-jumps around the baseline which is defined by the compensation of proton release and proton uptake. They calibrated the average extent of proton release under repetitive flashes by using hexacyanoferrate(III) as electron acceptor, which does not bind protons upon reduction. This advantageous property of hexacyanoferrate(III) is hampered in experiments with dark adapted material. During dark adaptation (typically for 20 min) hexacyanoferrate(III) oxidizes the non-heme iron (Diner et al., 1991) and the reduction of the non-heme iron upon the first flash causes the extra-uptake of protons at the acceptor side. This is clearly documented in the work from the group of Gernot Renger on BBY-membranes (Renger et al., 1987; Wacker et al., 1990). Surprisingly, evidence for extra-uptake of protons is also documented in experiments without any added electron acceptor (see Fig. 31 in Wacker, 1993). This may be understood by the oxidation of the non-heme iron by O_2 even in the absence of other oxidants (Petrouleas and Diner, 1986; Diner et al., 1991). The oxidation of the non-heme iron is pH-dependent (–60 mV per pH-unit, (Kok and Velthuys, 1977)) and therefore interferes differently at different pH values. The published patterns of proton release in BBY-membranes differ strongly between the groups of Jerome Lavergne, Gernot Renger and our group (see Section IIIB). We have the impression that neither the extent of proton uptake, which involves the non-heme iron, nor the complete absence of binary oscillations in the presence of DCBQ are adequately characterized.

4. Core Particles

Core particles with oxygen evolving capacity are usually prepared from BBY-membranes. The preparations according to the various original recipes (Ghanotakis et al., 1984, 1987a; Rögner et al., 1987; Haag et al., 1990; van Leeuwen et al., 1991; Kirilovsky et al., 1992) differ in their protein and pigment composition and aggregation state. The original preparation from spinach according to Ghanotakis et al. (1984, 1987a) (adaptable to pea (Lübbers et al., 1993a)) is prepared with octylglucoside. The preparation made in our laboratory (Lübbers et al., 1993a) largely contains the full set of extrinsic 'oxygen evolution enhancing' proteins (17, 23, 33 kDa). The stability of the more highly oxidized states of the catalytic center is in the time range of minutes as for thylakoids and BBY-membranes. One major disadvantage of this preparation in experiments on proton release is the aggregation of about 20 reaction centers (determined by coherent light scattering (Lübbers, 1993)). As in stacked thylakoids the diffusion/reaction of protons through these aggregates limits the time resolution of pH-indicating dyes. The preparation of van Leeuwen et al. (1991), on the other hand, which is prepared with β-dodecyl maltoside, is monodisperse and rise times of proton release down to 10 μs have been observed (Bögershausen and Junge, 1995) (see Fig. 3B). This preparation still contains the extrinsic protein of 33 kDa, but it has lost the 17 and 23 kDa polypeptides. The higher oxidation states decay much more rapidly, in seconds (van Leeuwen et al., 1992b). What, on first sight, appears as a disadvantage for a properly flash-synchronized clocking of the catalytic center through its oxidation states, can be advantageous. As the dark equilibrium state, with a predominance of S_1, forms within seconds after firing of the previous flash group, the averaging of repetitively induced transients which start from the dark adapted state is feasible at a repetition frequency of 0.1 Hz. The interval between flashes in a group has to be sufficiently short, about 100 ms (Bögershausen and Junge, 1995). With all the other oxygen evolving materials the dark equilibria are not reached before about 15 min of incubation in the dark. Thus, the averaging of repetitive transients with the *same* sample is impractical, and instead, one uses the stepped flow of large amounts of material through the measuring cell, with each sample exposed only to a single group of flashes.

B. Rates of Protolytic Reactions and Their Spectroscopic Detection

According to the fundamental work of Eigen (1963), protolytic reactions *in an aqueous environment* are among the most rapid reactions outside of photochemistry. The protonation of a base in water can be diffusion controlled with a rate constant ranging from 10^{10} to 10^{11} M^{-1} s^{-1}. The rate constant of the off-reaction, k_-, is related to the constant of the on-reaction, k_+, by the dissociation constant K_0 of the acid: $k_- = K_0 \cdot k_+$. The validity, in principle, of Eigen's concepts for acid/base reactions at the surface of proteins and membranes has been experimentally evaluated and theoretically extended by Gutman and coworkers (see Gutman and Nachliel, 1990, for a review). There are, in principle, three pathways of proton transfer between a donor and an accepting indicator dye, namely through protolysis, through hydrolysis, and through direct collision (Eigen, 1963). The pathway that dominates is determined by the pH and the concentration of the donor and the acceptor. In most practical applications with proteins and membranes the concentration of fixed buffering groups is large. This determines the magnitude of the pH-jump upon protolysis of the donor.

The response of a pH-indicating dye, I, to a sudden injection of protons into water (in the limit of a small pH-perturbation) is very fast. The effective rate constant is $k_{eff} = k_+ \cdot [I^-]$. With a conservative estimate for k_+, namely 10^{10} M^{-1} s^{-1}, and with a concentration of the basic form of the dye of $[I^-]$ = 1 mM, the effective rate constant, k_{eff}, is 10^7 s^{-1} and the relaxation time is 100 ns. If the dye concentration is one hundred-fold lower, typically 10 μM, as in spectrophotometric experiments, the response time is still short, 10 μs.

How rapid is proton release into water? If an acid *in contact with water* (with $pK_{red} > pH$) suddenly acquires a new, lower pK ($pK_{ox} < pH$), e.g. because of the deposition of a positive charge in its vicinity, the rate constant of deprotonation is given by $k_{eff} \cong k_+ \cdot K_{ox}$, which implies that the logarithm of the relaxation rate is linearly related to the new pK of the respective group. With $k_+ = 10^{10}$ M^{-1} s^{-1}, as before, and $pK_{ox} = 4$, the effective rate constant is 10^6 s^{-1}, equivalent to a relaxation time of 1 μs (or 10 μs if $pK_{ox} = 5$). However, an acid group in *direct contact* with the electrolyte solution can only experience a pK-shift by a 'chemical' mechanism (see below). A pK-shift by electrostatic interactions with a newly created positive charge is expected for groups *within* the protein dielectric. For the latter situation the assumption of a diffusion-controlled protonation is no longer valid and both the on- and the off-reaction are delayed by an activation barrier.

In the above kinetic estimates we addressed the spontaneous deprotonation of an acid (into water) and the response of an indicator dye to a small pH-jump in water. A dye, I, may also react with an acid, XH, through direct collision, according to the following reaction scheme (Since this case applies to neutral red, we here assume that the protonated form of the indicator, I, is cationic):

$$XH + I \underset{k_-}{\overset{k_+}{\leftrightarrow}} X^- + IH^+$$

The ratio of the rate constants is related to the ratio of the respective dissociation constants as follows:

$$k_+ / k_- = K_x / K_i$$

where the subscripts refer to the acid and the dye.

A treatment of this reaction in the limit of small perturbations (i.e. relaxation theory) yields a single relaxation rate constant:

$$k = k_+ \cdot ([XH]+[I]) + k_- \cdot ([X^-] + [IH^+])$$

If the dye concentration is in large excess over the one of the acid donor group and if the pH is not too far away from the pK of the dye, the relaxation rate constant, k, is (see Eigen, 1963):

$$k = [I]_{total} \cdot k_- \cdot (K_x + [H^+]) / (K_i + [H^+])$$

where k_- denotes the reverse rate constant. Under these conditions the relaxation rate is proportional to the total concentration of the dye, and it depends on the pH and on the respective pK-values of the donor group and the acceptor dye. As k_- is usually much smaller than the rate constant of the diffusion controlled protonation (it is expected to lie in the range of 10^7 to 10^8 M^{-1} s^{-1}) this pathway becomes effective only at higher dye concentrations. Assuming $K_x \cong [H^+] \cong K_i$ and a total dye concentration of 1mM, the relaxation rate constant ranges between 10^4 to 10^5 s^{-1}. As will be shown, the direct transfer of protons has been observed only with a particular dye that is adsorbed to the membrane surface and thereby upconcentrated in the vicinity of the donor groups.

The theoretically expected high time resolution of indicator dyes is not met, however, if the indicator is kept far from the primary groups, even more so if proton transfer to the indicator dye involves diffusion steps together with transient buffering reactions. This can delay the pulse response of the dye by several orders of magnitude (Junge and Polle, 1986b; Junge and McLaughlin, 1987). This has been observed for proton uptake by PS II in stacked thylakoid membranes (Polle and Junge 1986a, 1989) and with aggregated core particles (Lübbers and Junge, 1990). The relaxation of a pH-transient as described by Fick's second law is then approximately described by an effective diffusion coefficient, according to Eq. (19) in Junge and McLaughlin (1987):

$$D^{\mathrm{eff}} \cong D_H \cdot \left(\frac{2.3[H^+]}{\beta_{tot}} \right) + D_{OH} \cdot \left(\frac{2.3 \cdot 10^{-14}}{\beta_{tot} \cdot [H^+]} \right) + \sum_i D_i \cdot \frac{\beta_i}{\beta_{tot}}$$

where the symbols D denote the diffusion coefficients of protons (sub H), hydroxyl anions (OH) and of the various species of mobile buffers (i). The β-terms denote the differential buffering capacity at the given pH of the mobile buffers (*i*) and of all buffers (*tot*), i.e. fixed plus mobile species.

C. Oxygen Evolution and Absorption Transients in the UV as Indicators for the Stepped Progression Through the S-states

In dark adapted material, i.e. with the catalytic center of the water oxidase largely synchronized in state S_1, the pattern of oxygen evolution as a function of flash number reflects the stepped progression towards higher oxidation states (Joliot and Joliot, 1968; Joliot and Kok, 1975; Kok and Velthuys, 1977). Its measurement by a platinum electrode requires a relatively high concentration of material and often a different composition of the suspending medium than in spectrophotometric experiments. Thus, it is convenient to assay the progression of the water oxidase by absorption transients in the UV attributed to the oxidoreduction of the manganese cluster (Dekker et al., 1984c). Both indicators have specific advantages and disadvantages that are described below.

1. Oxygen Evolution

The pattern of oxygen evolution as a function of flash number was the first and, still is, the only truly reliable indicator for the progression through the oxidation states. It is usually described by only three parameters (the Kok-triple), namely: (1) the relative population of the most stable state, S_1, in the dark (s_1), (2) the proportion of photochemical double hits (β), and (3) the proportion of misses (α) (Kok et al., 1970; Forbush et al., 1971). The common restriction to only three parameters is based on the implicit assumption that the photochemical performance of PS II is independent of the redox state, both at the donor side (oscillating with period of four) and at the acceptor side (oscillating with period of two). This has been qualified by several authors (Lavorel, 1976, 1991, 1992; Thibault, 1978; Meunier, 1993; Shinkarev and Wraight, 1993).

When deconvoluting the pattern of proton release as a function of flash number to yield the contribution of each of the four transitions, most authors have relied on the three parameter model. Even if one accepts the, perhaps minor, deficiencies compared to using multi-parameter models, there are still two other problems: (1) The deconvolution parameters (s_1, α and β) inferred from patterns of oxygen evolution result from the competent subset of PS II reaction centers. Measurements of UV/VIS-absorption transients and pH transients may be complicated by contributions from *all centers*, both competent and incompetent in oxygen evolution (see the following section). (2) Patterns of oxygen evolution are often recorded under chemical conditions (e.g. much higher concentration of reaction centers, no redox buffers and mediators) that differ from those typical for optical experiments. The second problem is related to the design of the oxygen electrode. Historically, the oscillating pattern of oxygen evolution was first detected with bare platinum electrodes in direct contact with the suspending medium of the photosynthetic material (Joliot and Joliot, 1968). Redox mediators in the medium give rise to large artifact signals. This is not a problem with algae and only a small problem with isolated thylakoids that contain the plastoquinol pool which allows for several turnovers of PS II in the absence of an added electron acceptor. It is a big problem with PS II core particles that have lost the secondary quinone acceptor. To avoid this complication our group has used a Clark-type oxygen electrode with a large platinum surface covered by a thin Teflon sheet. With this method we obtained more regular patterns of oxygen evolution, even with core particles (Lübbers et al., 1993a). The major advantage, however, is that oxygen evolution can now be measured in samples with a composition similar to those that are used for flash spectrophotometry. After deposition of the photosynthetic material on the Clark-electrode by centrifugation we observed half-rise times of oxygen evolution of about 1 ms with thylakoids, BBY-membranes, and PS II core particles (M. Haumann, unpublished). Even this rapid rise time may be limited by the diffusion of oxygen to the platinum surface (Lavorel, 1992).

2. UV-absorption Transients

The progression of the water oxidase with a period of four produces periodic absorption transients in the near UV that have been ascribed to the oxidation-reduction of manganese (Mathis and Haveman, 1977; Velthuys, 1981; Dekker et al., 1984a,b; Schatz and van Gorkom, 1985; Renger and Weiss, 1986; Witt et al., 1986; Lavergne, 1987, 1991). An unequivocal interpretation of these transients, however, has not been achieved, because of the contributions of the oxidation-reduction of the bound quinones (Q_A, Q_B) and added quinones. Reaction centers which are incompetent in oxygen evolution (Lavergne and Leci, 1993) and only perform a single turnover are another complication. Therefore, the UV-transients generated by the first flash have often been excluded from the analysis (Lavergne, 1991; Lavergne and Leci, 1993). Still, the spectra attributed to the four redox transitions differed between laboratories (Dekker et al., 1984c; Witt et al., 1986; Lavergne, 1991; van Leeuwen et al., 1992). Only recently the data from two groups (Lavergne, 1991; van Leeuwen et al., 1992a) seem to converge. It is now agreed that transition $S_1 \Rightarrow S_2$ gives rise to the largest absorption transient that is probably due to a Mn(III)/Mn(IV) transition (Dekker et al., 1984c) in agreement with the multiline EPR-signal attributable to S_2 (Dismukes and Siderer, 1981; Haddy et al., 1989) and the results from other techniques (de Groot et al., 1986; Srinivasan and Sharp, 1986; McDermott et al., 1988; Styring and Rutherford, 1988; Guiles et al., 1990; Sauer et al., 1992). $S_2 \Rightarrow S_3$ gives rise to a smaller and slightly differently shaped transient. Again, this is tentatively attributed to a Mn(III)/Mn(IV) transition (Dekker et al., 1984c). The extent of the absorption transient attributable to transitions $S_0 \Rightarrow S_1$ is still under debate. Whereas the group of J. Lavergne (1991) claims that its extent is negligible, others find it significant (Dekker et al., 1984c; van Leeuwen et al., 1993).

The rise time of the UV transients has been interpreted to indicate the transfer time of electrons from the manganese cluster to Y_Z^+. Decker et al. (1984a) determined half rise times of 30, 110, 350, and 1300 ms for the four transitions from $S_0 \Rightarrow S_1 \Rightarrow S_2 \Rightarrow S_3 \Rightarrow S_4$ ($\rightarrow S_0$) (see also Gerken et al., 1987; Renger and Hanssum, 1992; Van Leeuwen et al., 1992). They were confirmed for the slower transients by time resolved EPR of Y_Z^+ (Babcock et al., 1976; Boska and Sauer, 1984; Bock et al., 1988; Hoganson and Babcock, 1988). Recently, the rise time of $S_0 \Rightarrow S_1$, has been debated. This transient is particularly difficult to access because it appears only upon the fourth flash in a series where the mixing of states by photochemical double hits and misses is large. In contradiction with the claims of other groups, Rappaport et al., (1994) have reported a rise time of about 250 μs. This discrepancy is unsettled.

D. Local Electrochromism and the Reduction Rate of P680⁺ as Indicators of Electrostatic Transients

1. Local Electrochromism

Electrochromism is the small shift of pigment absorption bands in response to a strong electric field. The difference spectra plus and minus the electric field are characterized by antisymmetric peaks resembling the first derivative of the respective absorption band. The electrochromic response to more localized electric fields should not be confused with electrochromism in response to the trans-membrane field (Junge and Witt, 1968; Emrich et al., 1969). In BBY-membranes from spinach and in core particles from *Synechococcus*, oscillations with a period of four as a function of the flash number have been observed in the Soret- and the Q_y-band of chlorophyll *a*. According to their spectral shape they are interpreted as local electrochromism in response to the net charge of the catalytic center (Saygin and Witt, 1985; Lavergne, 1987, 1991; Velthuys, 1988). The net charge is changed by an amount resulting from the difference of electron abstraction and proton release. It has been proposed that P680, the primary electron donor of PS II, may be the source of local electrochromism (Schatz and Van Gorkom, 1985). The major problem around 430 nm is the clear separation of local electrochromism from absorption transients of other origin in the same wavelength region. Absorption transients in the blue spectral region are a superimposition of manganese-, tyrosine-, and P680 oxidation-reduction, all of which, at least kinetically, oscillate with a period of four, and induce local electrochromism of chlorophyll *a*. A further component, oscillating with period of two, is the oxidation-reduction of Q_A, which causes electrochromic shifts of pheophytin (Velthuys, 1988). The spectra around 430 nm as determined for the oxidation-reduction of Q_A, Y_Z and manganese on transition $S_1 \Rightarrow S_2$ are shown in Fig. 1. Attempts to isolate the electrochromic components attributable to manganese and Y_Z have been based on differences recorded at two wavelengths, namely 440 nm and, depending on the author either 420, 424, or 428 nm. The precision of this approach may be hampered by unequal contributions to these wavelengths of Q_A/Q_A^- (and pheophytin electrochromism, see Fig. 1) and of P680/P680$^+$ (Gerken et al., 1989). These complications are even more severe at microsecond time resolution. Another way to separate the electrochromic shifts attributable to the donor side from absorption transients of the acceptor side is to record electrochromic transients at 443 nm in dark adapted material and to subtract transients obtained at the same wavelength which result from repetitive excitation (Haumann et al., 1994). Again, this is based on the assumption that binary oscillations at the acceptor side are absent in the presence of an efficient electron acceptor/mediator couple like hexacyanoferrate(III)/DCBQ.

There is only partial agreement about the spectral shape of local electrochromism in the blue that is attributable to chlorophyll *a*. The largest transient is observed on transition $S_1 \Rightarrow S_2$ (see Fig. 1) and another one of similar extent but opposite direction occurs on transition $S_3 \Rightarrow S_4 \rightarrow S_0$. The extent of $S_2 \Rightarrow S_3$ is negligible (Lavergne, 1991; van Leeuwen et al., 1993; Rappaport et al., 1994). But again, the extent of the transient which is induced by transition $S_0 \rightarrow S_1$ is under contention. Whereas Lavergne's group

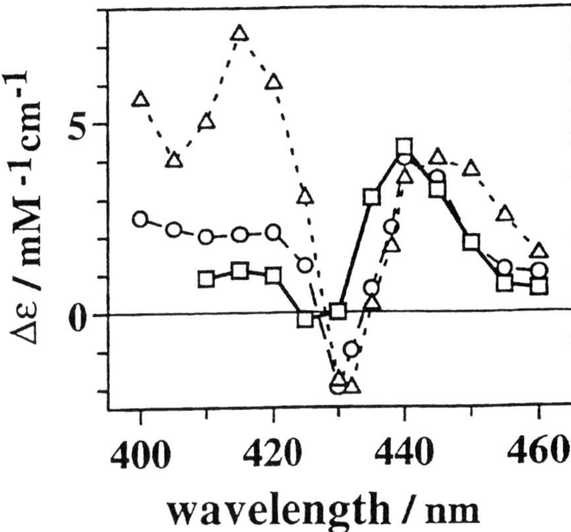

Fig. 1. Flash-induced difference spectra in the blue region which resemble local electrochromic band shifts in response to the following redox couples (the data points are taken from the respective source in parenthesis): Q_A–Q_A^- (triangles, (Dekker et al., 1984a)), Y_Z^+–Y_Z (circles, (Dekker et al., 1984a)), S_2–S_1 (squares, (Velthuys, 1988)). Note the nearly equal extents of all three spectra around 440 nm.

(Lavergne, 1991; Rappaport et al., 1994) reported that the electrochromism is smaller in magnitude than on transition $S_1 \Rightarrow S_2$, van Gorkom's group (van Leeuwen et al., 1993) found no significant extent on $S_0 \Rightarrow S_1$. This discrepancy is unresolved.

The group of Horst Witt studied core particles from *Synechococcus* sp. and recorded local electrochromism in two other wavelength regions, at 514 nm with dominant contributions of carotenoids (Saygin and Witt, 1984), and around 680 nm, the Q_y-band region of chlorophyll *a* (Saygin and Witt, 1984, 1985, 1987; see also Velthuys, 1988). Redox transients of manganese and of Q_A in the narrow sense are absent in both wavelength regions. Electrochromism of pheophytin in response to Q_A^- is present at least around 680 nm (see Fig. 2 in Velthuys, 1988).

2. The Rate of Reduction of P680$^+$

The rate of reduction of P680$^+$ is most conveniently recorded at 820 nm to minimize fluorescence artifacts (Haveman and Mathis, 1976). Recently, it has been recorded at high time resolution in the Soret region (van Leeuwen et al., 1993). The rate has been shown to oscillate depending on the redox transition of the water oxidase. Witt and coworkers (Brettel et al., 1984; Meyer et al., 1989) used oxygen evolving reaction centers from *Synechococcus* sp. and found a half rise time of 20 ns for the reactions starting from states S_0 and S_1 and a biphasic rise with half times of 50 ns and 250 ns for centers starting from S_2 and S_3. The slowing has been interpreted as arising from Coulombic attraction of the electron, which travels from Y_Z to $P680^+$, by a positive surplus charge in the manganese center (Brettel et al., 1984).

III. Extents and Rates of Proton Release During the Redox Transitions

The checkered history of proton release and the arguments leading to a reinterpretation have been presented in detail elsewhere (Lavergne and Junge, 1993). In brief, until 1990 there was apparent agreement among several laboratories (Saphon and Crofts, 1977; Fowler, 1977; Bowes and Crofts, 1981; Wille and Lavergne, 1982; Förster and Junge, 1985a), except for Hope and Morland (1979), on a given integer pattern of the proton-over-electron stoichiometry over the four redox transitions $S_0 \Rightarrow S_1 \Rightarrow S_2 \Rightarrow S_3 \Rightarrow S_4 \rightarrow S_0$, namely 1:0:1:2. In the earlier experiments thylakoids were the only oxygen evolving material used and the pH was not varied systematically. In 1990 it became apparent that the supposed 'standard' pattern was not met in other materials such as BBY-membranes (Lavergne and Rappaport, 1990) and core particles (Lübbers and Junge, 1990; Wacker et al., 1990). In BBY-membranes proton release varied greatly as a function of the pH (Rappaport and Lavergne, 1991). In this chapter we present new data on the *rate of proton release* from water oxidation as a function of the material, the pH and the type and the concentration of pH-indicating dyes used to monitor the pH-transients. In light of the kinetic properties, the recent data on the *extent* of proton release are analyzed for the source of protons, namely from water, from ligands to the manganese center, or from more peripheral amino acid residues.

A. Rates as a Function of Material, Indicators, Mobile Buffers and pH

So far, proton release has been time resolved in a direct way only by using pH-indicating dyes, either neutral red (Förster and Junge, 1985a; Jahns et al., 1992; Haumann and Junge, 1994a) or hydrophilic dyes such as cresol red (Bögershausen and Junge, 1995). Rapid protolytic reactions have been indirectly inferred from transients of local electrochromism (Lavergne et al., 1992; Rappaport et al., 1994). Time resolution by dyes was restricted to *unstacked thylakoids* (with neutral red) and to monodisperse core particles. In all other materials, including BBY-membranes, stacking or aggregation has prevented a rapid dye response (see IIB). The rate of proton release from water oxidation is of interest as it may allow the discrimination of protolytic events caused by a newly created positive charge on P680, by Y_Z and the manganese cluster. The respective transient times of electron transfer range approximately between 10 and 100 ns for $Y_Z \rightarrow P680^+$, depending on the redox state S_i (Brettel et al., 1984; Meyer et al., 1989) and 50 to 1000 μs for $S_i \rightarrow Y_Z^+$ (Dekker et al., 1984a; Renger and Hanssum, 1992; Rappaport et al., 1994). This is illustrated in Fig. 2. The former reactions are outside of, but the latter ones are within, the time domain which is accessible to pH-indicating dyes.

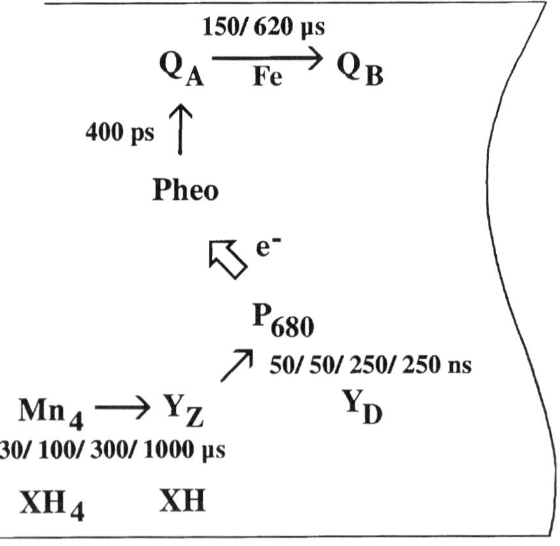

Fig. 2. Sequence of electron transfer steps in Photosystem II. The half-rise times are taken from the respective reference in parenthesis: $Q_A^- \rightarrow Q_B / Q_A^- \rightarrow Q_B^-$ (Bowes and Crofts, 1980); Pheophytin$^- \rightarrow Q_A$ (Trissl et al., 1987); $Y_Z \rightarrow P680^+$ ($S_0 \Rightarrow S_1$ to $S_3 \Rightarrow S_4 \rightarrow S_0$) (Schlodder et al., 1984); $Mn_4 \rightarrow Y_Z^+$ (Dekker et al., 1984b). At least 5 deprotonable groups (XH) are supposed to exist in the vicinity of manganese-Y_Z.

1. Different Mechanisms of Proton Transfer from the Donor Side of PS II to Dyes in Thylakoids and Core Particles—Studies Under Repetitive Flashes

a. Thylakoids—Collisional Proton Transfer

The rate of proton transfer from water oxidation to neutral red in unstacked thylakoids has been studied under excitation with repetitive flashes of light, i.e. with equal contributions from all four transitions of the water oxidase (Haumann and Junge, 1994a). Proton release from the oxidation of plastoquinol was blocked by DNP-INT. More than 80% of the total extent of proton transfer to neutral red occurred as a fast phase with a rate proportional to the dye concentration. Its half-rise time ranged from 170 μs (5 μM neutral red) over 35 μs (27 μM) to 12 μs (100 μM). This is documented in Fig. 3A. Such a concentration dependence is indicative of a bimolecular reaction between the indicator dye and acid groups. It is noteworthy that the effective concentration of neutral red in the thylakoid lumen is 2 to 3 orders of magnitude larger than the average concentration in the suspension because of its adsorption at the membrane surface (Hong and Junge, 1983). Any additional mobile buffer is expected to further increase the rate. Indeed, the addition of imidazole accelerated the transfer of protons to neutral red (see Fig. 5 in Haumann and Junge (1994a)).

The very rapid transient at high concentrations of neutral red ($\rightarrow$ 12 μs) implies that *the major portion of proton release can be triggered by* Y_Z^+. The result at low concentration, on the other hand, implies that *the normal appearance of protons in the lumen is rather slow* (> 300 μs).

b. Core Particles—Protolysis

In reaction center core particles prepared according to Ghanotakis et al. (1987a) proton release by water oxidation has revealed an apparent rise time of about 100 ms (Lübbers and Junge, 1990). This long time is caused by aggregation (see Section IIA4). The preparation by the method of van Leeuwen et al. (1991), on the other hand, is probably monodisperse. Proton release as rapid as 10 μs was observed at acid pH (Fig. 3B) (Bögershausen and Junge, 1995). In contrast with the situation in thylakoids, the rate is nearly independent of the concentration and the type (chemical nature and pK) of the added hydrophilic indicator. In the case of core particles proton transfer is not caused by a collision of the dye with the donor groups but rather by protolysis followed by the diffusion controlled protonation of fixed buffers and the dye. As the dye can respond very rapidly to a sharp pH-step it almost certainly reports the time course of a spontaneous deprotonation act. In parallel

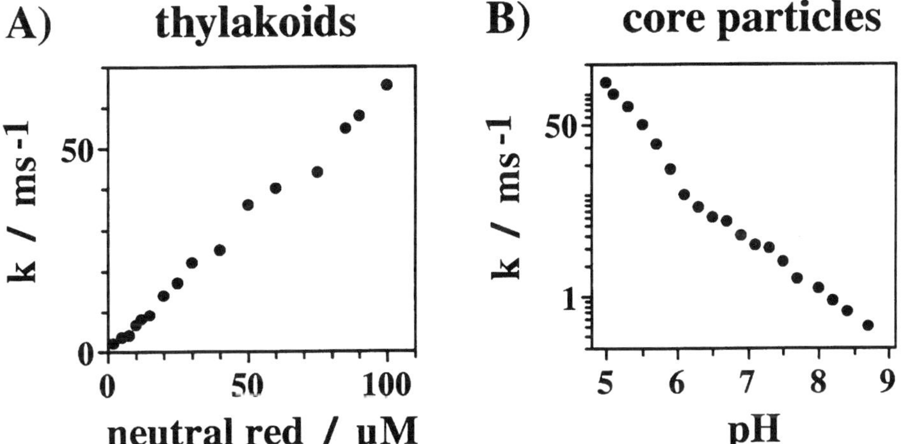

Fig. 3. Rates of proton release under repetitive flash excitation A) in unstacked thylakoids (20 μM of chlorophyll) as function of the neutral red concentration in the medium (adapted from Haumann and Junge, 1994a) B) in PS II core particles (3 μM of chlorophyll) prepared according to van Leeuwen et al. (1991) as function of the pH recorded with bromocresol purple (5 < pH < 7.5) and cresol red (7 < pH <8.7) (Bögershausen and Junge, 1995). Note the logarithmic ordinate scale in B.

with the above described experiments we checked the actual response time of the dye by nanosecond laser pulses applied to a similar sample that additionally contained 'caged protons' (4-formyl-6-methoxy-3-nitrophenoxyacetic acid). The half-rise time as recorded at pH 6.3 (with 40 μM MES) by 15 μM bromocresol purple was 600 ns (Lakomiak, Drevenstedt and Junge, unpublished).

As documented in Fig. 3B the rate increased in proportion to the free proton concentration in the suspension in the pH-range from 6 to 5, with half rise times of 70 μs at pH 6 down to 7 μs at pH 5 (Bögershausen and Junge, 1995). In the pH-domain from 8 to 6 the rise was less steep and the lowest rate at pH 8.6 was about 500 s^{-1}. A linear decrease of the log-rate of protolysis as a function of the pH is expected if there are several acid groups involved, and the major contribution comes from those groups with a pK equal or close to the pH (see Section IIB).

We summarize these data: (1) The concentration dependence of proton transfer to neutral red *in thylakoids* indicates a collisional mechanism between the (locally highly concentrated) dye and the proton releasing groups. (2) *In core particles*, the response of hydrophilic dyes indicates, on the contrary, the spontaneous protolysis of acids. Several acid groups with different pK values seem to be involved. (3) *In both materials* the greater portion of proton release can be faster than the electron transfer from the manganese cluster to Y_Z^+ under certain conditions, namely at high concentrations of neutral red in thylakoids and at acid pH in core particles. Here, protons are released primarily at the level of Y_Z^+. Under other conditions, proton release appears to lag behind electron transport. As will be shown below, the former or the latter behavior is without effect on the oscillatory progress of water oxidation. (4) *In both materials*, but under other conditions, proton release can also occur more slowly and after the completion of the electron transfer between the manganese cluster and Y_Z^+. This is true for thylakoids (no neutral red added) and core particles (alkaline pH).

2. The Kinetics of Proton Release in Dark-adapted, Unstacked Thylakoids

The first detailed study of the kinetics of proton release on the successive S-transitions was published by Förster and Junge (1985a) using neutral red. The main findings are as follows: There is a 1.2 ms phase accompanying the $S_3 \Rightarrow S_4 \rightarrow S_0$ reaction that is compatible with the rate of electron transfer from manganese to Y_Z^+ in this transition. However, there is also a faster component (200 μs), which indicates that some proton transfer precedes the reduction of Y_Z^+. The second flash, which mainly promotes the step $S_2 \Rightarrow S_3$, causes proton release rising with a half time of 200 μs and therewith somewhat faster than electron transfer between the manganese center and Y_Z (300 μs, (Dekker et al., 1984a)). The rapid components of proton release in these two transitions have been attributed to the rapid dissociation and rebinding of protons during the transient oxido-reduction of Y_Z, caused by the prosthetic group proper, or by adjacent amino acid side chains. This interpretation agrees with observations in Tris-treated thylakoids that are unable to oxidize water but still carry out the reduction of $P680^+$ by Y_Z. Proton release and proton uptake follow the oxidation and reduction of Y_Z (Renger and Voelker, 1982; Förster and Junge, 1984; Conjeaud and Mathis, 1986).

These studies were extended by using unstacked thylakoids and thereby avoiding the kinetic problems of the neutral red technique when applied to stacked thylakoids as in the above cited work. Figure 4 shows the response of neutral red at a rather high concentration, (45 μM) to a nanosecond flash from a Ruby laser that followed zero to three priming Xenon flashes (for experimental details, see Haumann and Junge, 1994). One remarkable feature is the presence of a fast rising phase in *all* transitions. At pH 7.4 the respective half-rise times range between 30 μs (1st flash) and 50 μs (4th flash), at pH 6.3 between 50 μs (1st) and 80 μs (2nd flash) (see Table II in Haumann and Junge, 1994). In all transitions, except perhaps $S_0 \Rightarrow S_1$, which is dominant on the fourth flash, proton transfer to neutral red precedes electron transfer from the manganese cluster to Y_Z^+. The fast phases are accelerated when the concentration of neutral red is increased. As documented in Fig. 7 in Haumann and Junge (1994), a half-rise time of 13 μs results for the first flash at 75 μM neutral red and 1 mM imidazole. It is noteworthy that the extent of the initial transfer of protons at the level of Y_Z^+ is not modified when the positive charge is transferred more slowly to the manganese cluster, e.g. in about 300 μs upon $S_2 \Rightarrow S_3$. The absence of a 300 μs-phase is immediately obvious from inspection of the respective traces (2nd flash) in Fig. 4. The absence during the first flash of components with 25, 50 and 100 μs rise or decay is evident from inspection of

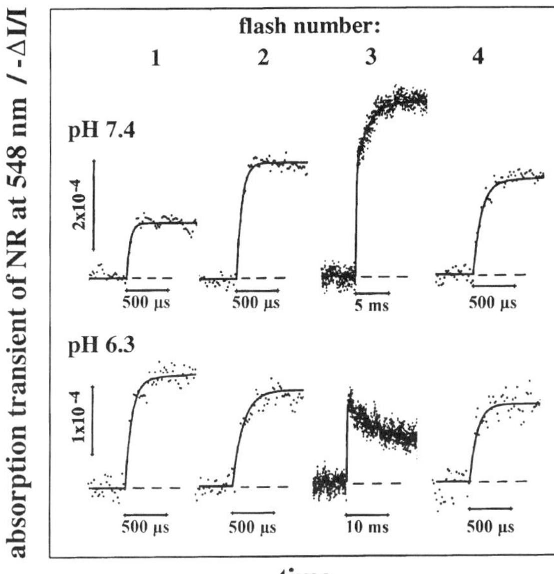

Fig. 4. Time-resolved proton release after the first four flashes in dark-adapted unstacked thylakoids recorded with 45 μM neutral red. A) upper row at pH 7.4 B) lower row at pH 6.3. The flashes were provided by a Q-switched Ruby-laser. Note the different time scale for flash no. 3 at both pH-values (points experimental, lines fit by two exponentials) (Haumann and Junge, 1994a).

Fig. 7 in Haumann and Junge (1994a). The fast proton transfer may be caused by the deprotonation of tyrosine or by an electrostatic response of other amino acids. In both cases it is all but trivial that the relocation of the positive charge from Y_Z^+ to manganese, which according to structural modeling implies the passage of an electron over at least 1, if not 1.5 nm (Svensson et al., 1990; Ruffle et al., 1992) does not cause any alteration of the extent of proton release. There are two ways of visualizing this behavior. As illustrated in Fig. 2 different acidic groups may be responsible for the response to the oxidation of tyrosine (XH) and of the manganese center (XH_4). It is conceivable that the passage of an electron from manganese to tyrosine resets the pK of XH and causes the rebinding of a proton, which, however, remains undetected, because it is synchronized with proton release by XH_4. This push-and-pull mechanism around Y_Z has been proposed previously (Förster and Junge, 1985a). Alternatively, it is conceivable that the manganese-tyrosine system behaves as an electrostatic entity. The electrostatic implications of this concept are further discussed in Section IV. At this point we maintain that there are at least five acidic residues that react to a positive charge on Y_Z-manganese by proton transfer to neutral red.

Another remarkable feature (Förster and Junge, 1985a) is the slow proton release, mainly at the third flash. Its half-rise time is about 1200 μs at pH 7.4 and its decay-time about 5000 μs at pH 6.3 (Haumann and Junge, 1994a). The rise of the slow phase is independent of the concentration of neutral red. Moreover, its Arrhenius activation energy, about 20 kJ/mol, differs from the about 40 kJ/mol of the rapid phases (Haumann, unpublished). An activation energy of 20 kJ/mol has been determined for the UV-transients associated with transition $S_3 \Rightarrow S_4 \rightarrow S_0$ (Renger and Hanssum, 1992). The properties of the slow phase, namely the appearance on the 'oxygen producing' third flash, a millisecond rise, the activation energy, and the independence of neutral red, identify this phase as geared by the intrinsic oxygen liberating chemical reaction during transition $S_4 \Rightarrow S_0$.

It is most remarkable, however, that the slow phase indicates *proton release* at pH 7.4 but *proton uptake* at pH 6.3 (Haumann and Junge, 1994a. We have tentatively interpreted this observation based on the assumption that the reaction with two molecules of water to yield dioxygen is fully concentrated in the final dark reaction $S_4 \rightarrow S_0$ (see Fig. 5): The four photochemical transitions, $S_0 \Rightarrow S_1 \Rightarrow S_2 \Rightarrow S_3 \Rightarrow Y_Z^+$, create a certain number of bases. These protons appear in less than 100 μs if neutral red is present. The subsequent reaction with water, i.e. during the transition $S_3Y_Z^+(\Rightarrow S_4) \rightarrow S_0$, produces four electrons and four protons. The electrons cancel the positive charges in the catalytic center and thereby the pK values of the bases are reset. The protons are used either to reprotonate the bases in response to the resetting of their pKs or they are released into the lumen. The observed behavior during the slow transition $S_3Y_Z^+(\Rightarrow S_4) \rightarrow S_0$ (mainly on the third flash of light) differs depending on the total number of these bases, which is a function of the pH. If the number is 4 the production of protons and the rebinding of protons cancel each other. If this number is smaller than four, for example 3.4 as at pH 7.4, the slow release of 0.6 proton is expected. This reflects a fraction of the chemically produced protons from water. If the number of bases is larger than four, e.g. 4.5 as at pH 6.3, the resetting of the pKs produces an overshoot of proton uptake. The slow phase of proton release at alkaline pH (to the amount of 0.6 H^+ at pH 7.4) is the only instance where chemical protons

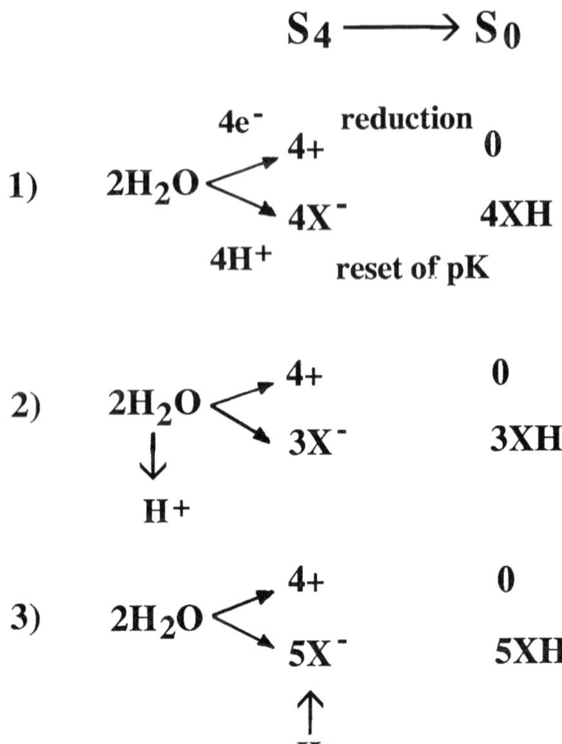

Fig. 5. Model for the concomitant electron and proton transfer during the oxygen producing dark-reaction $S_4 \rightarrow S_0$. The reduction of the Mn-cluster by four electrons is coupled to the resetting of the pK's of bases X^- which have been created in the four rapid transitions, $S_0 \Rightarrow S_1 \Rightarrow S_2 \Rightarrow S_3 \Rightarrow S_4$. 1) We assume that four bases are created and then reprotonated by the four protons from two water molecules. No proton release with the characteristic half-rise time of about 1 ms of transition $S_4 \rightarrow S_0$ is expected, as observed in core particles. 2) Only 3 bases X^- are created. It follows the net release of 1 proton with a half-rise time of about 1 ms. This resembles the observed behavior in thylakoids at alkaline pH. 3) 5 bases are created. The resetting of their pKs leads to the net uptake of 1 proton. This resembles the observed behavior in thylakoids at acidic pH.

from water oxidation are obvious and clearly kinetically labeled (Fig. 5). In dark adapted core particles where we have obtained high kinetic resolution of proton release the number of rapidly formed bases seems to be 4 at any pH between 5.5 and 7.5. There is no evidence for an extra release of protons with ms-half-rise times on the third flash (Haumann et al., 1994; Bögershausen and Junge, 1995).

Before closing this segment it is worth recalling that in thylakoids there is no *rapid* proton release in the absence of neutral red. Instead, proton release occurs rather slowly, in more than 300 μs. An important question is whether or not operations on the velocity of proton transfer affect the velocity or the sequence of electron transfer reactions (see Section IV).

B. Extent of Proton Release as a Function of the Material and the pH

More than ten years of work in several laboratories led to the conclusion that the pattern of proton release as a function of the state transitions $S_0 \Rightarrow S_1 \Rightarrow S_2 \Rightarrow S_3 \Rightarrow S_4 \rightarrow S_0$ was 1:0:1:2 (Saphon and Crofts, 1977; Fowler, 1977; Bowes and Crofts, 1981; Wille and Lavergne, 1982; Förster and Junge, 1985a). Only one laboratory maintained a 1:1:1:1 pattern (Hope and Morland, 1979). All experiments were then carried out on thylakoids. It is now apparent that in dark adapted material the extent of proton release from water oxidation greatly differs both as a function of flash number and of the material. The first to note that the pattern differed from the long accepted classical one and that it varied as a function of the pH were Lavergne and Rappaport (1990). They studied BBY-membranes. At the same time two groups reported the absence of any oscillations of proton release in core particles (Lübbers and Junge, 1990; Wacker et al., 1990). The variability of the pattern of proton release as a function of the material and the pH is documented in Fig. 6 (adapted from Lavergne and Junge, 1993). The original data on the extent of proton release as a function of flash number have been deconvoluted to yield the extent as a function of the redox transition by using the Kok-parameter triple resulting from measurements of oxygen evolution and/or UV-absorption transients.

The stoichiometric pattern of proton release *in thylakoids* (Haumann and Junge, 1994a) is illustrated in Fig. 6A. The most pronounced features are: 1) The only marginally pH-dependent proton yield of 1 for the transition $S_2 \Rightarrow S_3$, a feature which is shared by all other materials (see below). 2) A decrease from 1 to less than 0.5 starting at pH 6.7 and increasing towards alkaline pH for the transition $S_0 \Rightarrow S_1$. 3) A mutually compensating behavior of the respective proton yields for $S_1 \Rightarrow S_2$ and $S_3 \Rightarrow S_4 \rightarrow S_0$, with the former dominating at pH 5.8–6.6 (up to 2 protons compared to 0.3 for $S_3 \Rightarrow S_4 \rightarrow S_0$) and the latter around pH 7.5 (1.5 vs. 0.5 protons). The crossover point where both transitions release one proton is at pH 6.7. The above pH-dependence is corroborated by an alternative technique, with hydrophilic pH-indicating dyes

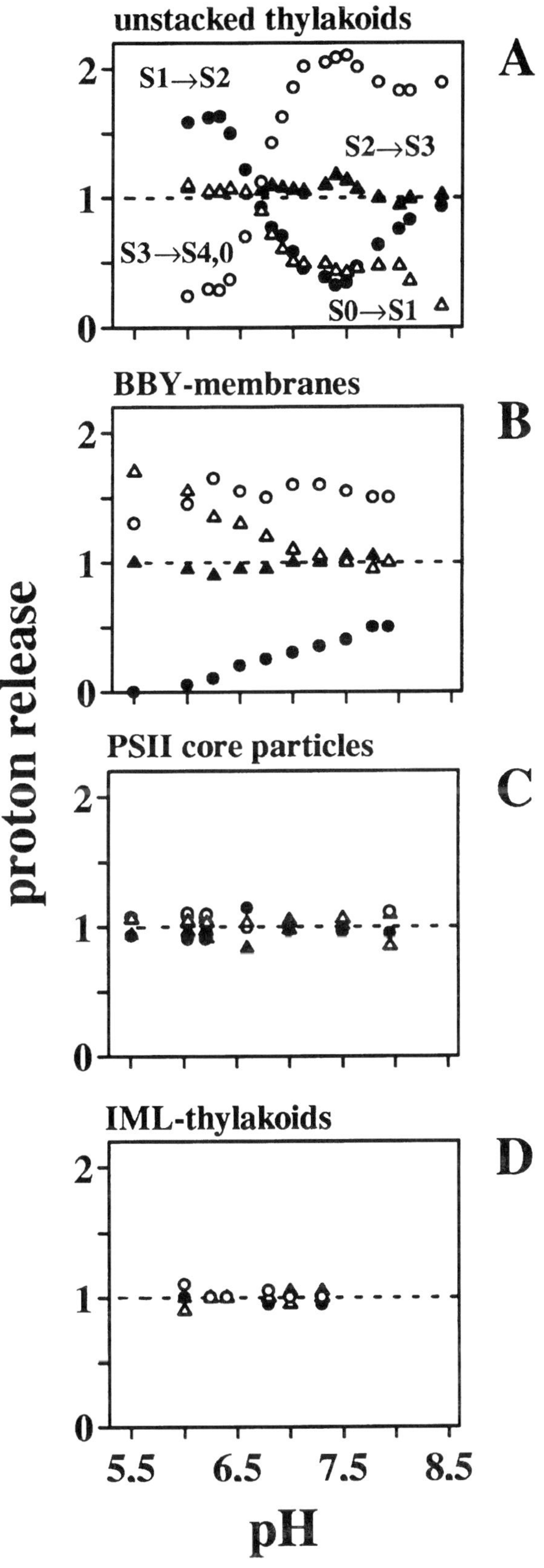

(instead of neutral red) and with and without the added protonophore nigericin (Haumann and Junge, 1994a). These results strongly deviate from the pH-dependence of proton release published for BBY-membranes (Fig. 6B) and core particles (Fig. 6C).

The pattern in *BBY-membranes* (Rappaport and Lavergne, 1991) is shown in Fig. 6B. A constant release close to 1 is obtained for $S_2 \Rightarrow S_3$. For $S_0 \Rightarrow S_1$, the release becomes greater than 1 below pH 7. Upon the $S_1 \Rightarrow S_2$ transition, the stoichiometric coefficient remains below 1, rising from 0 below pH 6 to about 0.5 proton at pH 7.5. The same authors have tried to consolidate these results into a formal model, involving at least four deprotonating groups, undergoing pK shifts in response to the oxidation steps of the charge-storing system (Rappaport and Lavergne, 1991). A set of four pK's for each S-state implies that 16 pK values are potentially involved. Among these pK's, only three lie within, or close to, the accessible pH range and could be estimated (the others correspond to groups that are fully protonated or unprotonated). One pK was originally close to 6.0 in the presence of S_0, and it shifted to below 4.5 when S_1 was formed. Another pK was 8.2 in S_1 and shifted to 7.2 in S_2.

An apparent pK emerging from such an analysis is not necessarily attributable to a particular residue. Since the newly created positive charge (e.g. on manganese) may interact electrostatically with several ionogenic residues, which, in turn interact with each other; the apparent pKs may be collective parameters of the ensemble. They may be calculated then by standard methods of statistical mechanics as shown for the related problem of proton uptake at the reducing side of bacterial reaction centers (Takahashi and Wraight, 1990; Beroza et al., 1991).

It is disturbing, however, that the reported pH-dependent patterns of proton release in BBY-membranes vary greatly among authors (compare Fig. 4 in Rappaport and Lavergne (1991) and Fig. 35

Fig. 6. Patterns of proton release by water oxidation as a function of the pH. The original patterns as function of the flash number were deconvoluted to yield the extent at any of the four transitions, namely $S_0 \Rightarrow S_1$ (open triangles), $S_1 \Rightarrow S_2$ (closed circles), $S_2 \Rightarrow S_3$ (closed triangles) and $S_3 \Rightarrow S_4 \rightarrow S_0$ (open circles). Four different types of material, all capable of oxygen evolution were used; from top to bottom: A: unstacked thylakoids (Haumann and Junge, 1994a), B: BBY-membranes (Rappaport and Lavergne, 1991), C: PS II core particles (Lübbers, 1993), and D: IML-thylakoids (Jahns and Junge, 1992b). This figure is reproduced from Lavergne and Junge, 1993).

in Wacker (1993)). Whether these differences are caused by different preparation techniques or are mainly attributable to the poorly characterized and superimposed proton uptake at the acceptor side is not clear.

The pattern of proton release in *core particles* is shown in Fig. 6C. The first reports that suggested a featureless pattern of proton release in core particles (Wacker et al., 1990; Lavergne and Rappaport, 1990) were obtained with preparations according to Ghanotakis et al. (1987a) and Haag et al. (1990). Proton release was constant from the second flash on, whereas the net production after the first flash was much decreased (see Fig. 2 in Lübbers and Junge (1990) and Fig. 3 in Wacker et al. (1990)). The diminution was caused by the superimposition of proton uptake due to the photoreduction, on the first flash, of the oxidized non-heme iron at the acceptor side. The true pattern of proton release was therefore approximately 1:1:1:1. In core particles prepared according to Ghanotakis et al. (1987a) a scan of the pattern showed very little variation over the pH-range from 5.5 to 7.5 (Lübbers et al., 1993a). The deconvolution of the raw data on proton release to yield the distribution of protons as a function of the redox transition has been based on both the pattern of oxygen evolution and the pattern of UV-transients. In parallel experiments and by the same analysis the pattern of proton release was determined again in thylakoids. The comparison between both systems corroborated the absence of oscillations of proton release in this core particle (Lübbers et al., 1993a). The same holds true in the core particle of van Leeuwen et al. (1991) in experiments where the spacing between flashes in each group is short enough (typically 100 ms) to overcome the more rapid deactivation of higher S-states (Jahns et al., 1992, Bögershausen and Junge, 1995).

Another example of a less structured pattern is observed in thylakoids prepared from pea plants grown under a regime of intermittent light with long dark periods (10 days of 2 min light followed by 2 h dark), so called *IML-thylakoids*. These growth conditions severely impair the synthesis of chlorophyll *b* with the consequence that most of the light-harvesting proteins around Photosystem II are lacking (Jahns and Junge, 1992a). IML-thylakoids are intrinsically unstacked, so that the kinetic problem of the neutral red technique is absent. In this material, the pattern of oxygen release oscillates conventionally in the pH-range from 6 to 7.5. The pattern of proton release is approximately 1:1:1:1 and its pH-dependence, which is illustrated in Fig. 6D, is much flatter than in control thylakoids (Jahns and Junge, 1992b, 1993).

It is obvious that the pattern of proton release as a function of the redox transition and the pH differs considerably between different materials. It differs not only between core particles and thylakoids, with a richer protein periphery in the latter, but also between thylakoids and BBY-membranes, which are both membranous and expected to have the complete repertoire of PS II proteins. The data on IML-thylakoids suggest that the light-harvesting LHCII proteins may be involved in the modulation of the pattern of proton release (Jahns and Junge, 1992a,b, 1993). This is in line with observations that controlled trypsination converts a structured release into a more featureless pattern in BBY-membranes (Renger et al., 1987). The influence of peripheral proteins on the pattern of proton release may involve particular amino acid residues of these proteins, but it is equally conceivable that it is caused by different electrostatic screening (see Section IV).

How the variability of the pattern of proton release affects the stepped progress of water oxidation has to be questioned. This has been addressed in a comparative study with thylakoids and core particles carried out at three pH-values, 5.5, 6.5 and 7.5 (Lübbers et al., 1993a). The Kok parameter triple was determined from oxygen evolution and, in parallel, from UV-transients. The Kok-parameters differed between both systems. S_0 was more highly populated in core particles that showed more misses and fewer double hits. When comparing situations with very different proton release, e.g. thylakoids at pH 5.5 and 6.5, however, the differences between the patterns of oxygen release and UV-transients were very small. The same holds true when comparing thylakoids with core particles. *Thus, there is no evidence for any feedback of proton liberation on the progress of water oxidation in the investigated pH domain.*

C. Proton Release and Indicators of the Net Charge

The extent of electrochromic bandshifts was studied under excitation of dark adapted core particles from *Synechococcus* sp. with a group of light flashes. Saygin and Witt (1984, 1985) monitored both the green and the red spectral region. They found that the extent of the bandshift oscillated with a period of four and with the following pattern as a function of the state transition ($S_0 \Rightarrow S_1 \Rightarrow S_2 \Rightarrow S_3 \Rightarrow S_4 \rightarrow S_0$): 0:+1:0:−1 (Saygin and Witt, 1985). The extent of proton release was not recorded in parallel, but the authors assumed that the then accepted pattern of proton release of 1:0:1:2 (see Section IIIB) was also valid in their material. They concluded that the oscillation of electrochromism was caused by net charge jumps, which in turn were reflecting the 'intrinsic' proton release. The pattern of the net charge for each of the states S_0, S_1, S_2 and S_3 was then 0, 0, +1, +1. This concept gained credence from experiments on the rate of reduction of $P680^+$ (Brettel et al., 1984). The half-transfer time oscillated as follows: 23 ns, 23 ns, 50 / 260 ns, 50 / 260 ns. This was attributed to the above proposed net charge pattern over the states S_0, S_1, S_2 and S_3 (see Witt et al., 1986 for a review).

When reinvestigating the pattern of absorption transients in the UV and the blue spectral region in BBY-membranes, Lavergne (1991) found a probable electrochromic component that oscillated in a different way, but in agreement with the pattern of proton release in this material (Rappaport and Lavergne, 1991). More than one proton was released on $S_0 \Rightarrow S_1$ below pH 7. This led to a negative net charge in S_1 and, as expected, an inversion of the electrochromic bandshift was observed during $S_0 \Rightarrow S_1$. Oscillations of electrochromism and protons were investigated over a wider pH-range, and were found to vary in a complementary way (Rappaport and Lavergne, 1991). Whereas the correlation between electrochromism and 'intrinsic proton shifts' was only postulated in the above cited work, it was established here by an observed proton release.

This seemingly coherent picture was flawed when such investigations were carried out with core particles from higher plants. Van Gorkom's group (van Leeuwen et al., 1992) showed an inversion of the direction of the electrochromic bandshift similar to Lavergne's group when comparing $S_0 \Rightarrow S_1$ with $S_1 \Rightarrow S_2$. They also reported similar oscillations to those observed by Witt's group for the rate of reduction of $P680^+$ (van Leeuwen et al., 1992). At the same time our group reported the absence of oscillating proton release in these core particles (Jahns et al., 1992). This was contradictory to the previously postulated and, in BBY-membranes, observed coherence between proton release and electrochromic bandshifts.

Van Leeuwen (1993) corroborated this discrepancy by scanning the oscillations of the electrochromic bandshifts in the blue as function of the pH. He found a pH-dependence similar to that reported for BBY-membranes (Rappaport and Lavergne, 1991), but, in agreement with our data, no pronounced oscillations of proton release. Taking into account that the most drastic variations of the pattern of proton release as function of the pH are observed in thylakoids (Haumann and Junge, 1994), we recently compared electrochromism with proton release in thylakoids, IML-thylakoids, BBY-membranes and two different core particles. Broadly speaking, the patterns of electrochromism as a function of flash number were very much alike, whereas the patterns of proton release varied. In part, this is documented in Fig. 7 (see Haumann et al., 1994). There was no inversion of the sign of electrochromism between the first and the third flash under conditions where the extent of proton release upon the third flash varied from less to greater than one. Our electrochromic patterns in the blue spectral region, documented in Fig. 7, are similar to the pattern recorded in the green and the red spectral region for core particles from *Synechococcus* sp. (Saygin and Witt, 1984, 1985a,b, 1987). Accordingly, these oscillations are not the proposed straightforward indicator of the net charge as inferred from electron abstraction and proton release. As the spectral features in all wavelength domains are typical for an electrochromic bandshift it is worth asking for the exact nature of the *internal charge shifts* they may indicate (see Section IV).

The group of J. Lavergne studied electrochromic bandshifts at microsecond time resolution (Rappaport et al., 1994). They used dark adapted BBY-membranes and recorded electrochromic transients in the blue spectral region together with absorption transients of manganese at 295 nm. At a wavelength difference of 440–424 nm they observed a very rapid and unresolved rise of absorption that was followed by a rapid decay in micro- to milliseconds onto a level which oscillated, as known from their previous work at lower time resolution. This was interpreted to indicate the very rapid formation of Y_Z^+ followed by the motion of the positive charge further away from the probe molecule, possibly $P680^+$. The contribution of the four S-state transitions is interpreted as follows: During three transitions, namely $S_1 \Rightarrow S_2 \Rightarrow S_3 \Rightarrow S_4 \rightarrow S_0$, the microsecond phase of electrochromism showed half times similar to the electron transfer from manganese to Y_Z^+ (given in parenthesis): at pH

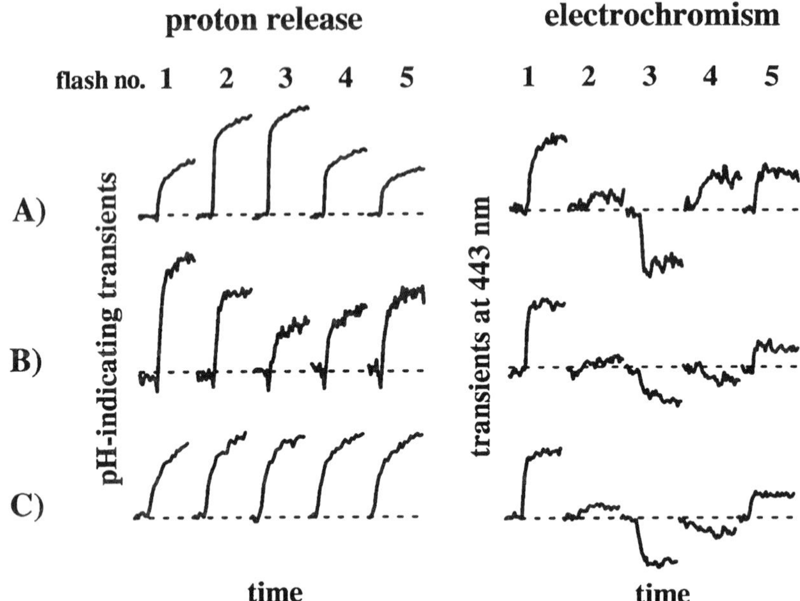

Fig. 7. Proton release and local electrochromism as induced by five flashes of light. A) thylakoids at pH 6.5. B) thylakoids at pH 5.5. C) PS II core particles prepared according to Ghanotakis et al. (1987a) at pH 6.5. Proton release (left-hand panel) was recorded with neutral red (+/– dye) in thylakoids and bromocresol purple (+/– dye) in core particles. Local electrochromism was recorded at 443 nm as the difference of transients obtained with dark-adapted material and transients obtained under repetitive flash-excitation. Both preparations were dark-adapted for 20 min. 100 μM DCBQ plus 200 μM hexacyanoferrate(III) were used as electron acceptors.

6.5 55 μs (50 μs) – 200 (200–290) – 1200 (1200). This is in accord with the above interpretation. During $S_0 \Rightarrow S_1$ the respective half-rise times were 225 μs (250 μs), again in agreement with each other. The transfer time of electrons between manganese and Y_Z^+, 250 μs, was much longer than the about 30 and 70 μs previously reported by two other laboratories (Dekker et al., 1984a; Renger and Hanssum, 1992). This particular transition is least well characterized due to the progressive mixing of states by double hits and misses during a flash sequence. This is apparent from the considerable discrepancies between the associated UV spectra obtained by different investigators (compare Fig. 9 in Lavergne (1991) with Fig. 1 in Rappaport et al. (1994) and with Fig. 4 in van Leeuwen et al. (1992)).

Most interesting was the result with respect to $S_3 \Rightarrow S_4 \rightarrow S_0$. Superimposed on the slow, 1200 μs decay of electrochromism was a more rapid phase of 30 μs (35% of total). The rapid phase correlated with a lag phase of the electron transfer (Lavergne et al., 1992; Rappaport et al., 1994). The authors tentatively interpreted this phase as a rapid proton shift (release or internal displacement)(see also Lavergne and Junge, 1993).

To *directly* correlate the kinetics of electrochromism and proton release we performed parallel experiments with thylakoids and core particles. The stepped progress over the redox states and the transfer time of electrons from manganese to Y_Z^+ was studied in the UV. We used conditions where proton release occurred in about 25 μs, i.e. at high concentration of neutral red in thylakoids and at acid pH in core particles. In both systems the transients at a wavelength of 443 nm were adequately described by the respective transfer time ($t_{1/2}$) of the electron from manganese to Y_Z^+ (in thylakoids about 40:75:220:1100 μs for $S_0 \Rightarrow S_1 \Rightarrow S_2 \Rightarrow S_3 \Rightarrow S_4 \rightarrow S_0$) whereas proton release occurred more rapidly, at 25 μs, as mentioned. In both systems a minor fast phase was discernible upon $S_3 \Rightarrow S_4 \rightarrow S_0$, whose half-decay time, about 100 μs, was still significantly longer than the rise time of proton release (Haumann et al., 1994).

Our combined results on the extent and rate of the electrochromic transients imply that the *observable proton release* causes only small, if any, electrochromism. This is in line with data on core particles showing oscillations of the reduction rate of $P680^+$ (van Leeuwen et al., 1992) under conditions of constant proton release.

D. Chemically Modified Materials

1. Chloride Depletion of PS II

The chloride anion is an essential cofactor of water oxidation. In its absence the normal progress towards the higher oxidation states (i. e. oxygen evolution) is blocked, but it is unclear where and how (Debus, 1992; Yocum, 1992). It is believed by some authors that the first flash produces a modified state S_2^* (Ono et al., 1986a,b, 1987) whose further conversion to S_3^* is blocked (Itoh et al., 1984; Theg et al., 1984), although the oxidation of Y_Z proceeds normally (Ono et al., 1986a,b). On the contrary, other authors have interpreted their data to show that the transition $S_2^* \rightarrow S_3^*$ is still possible but that the following transition is blocked (Sinclair, 1984; Homann et al., 1986; Boussac et al., 1992; Boussac and Rutherford, 1994). A new EPR-signal in Cl^--depleted material has been attributed to an organic radical within a distance of 1 nm from the manganese cluster (Baumgarten et al., 1990). This led to the suggestion that a histidine-residue may be the newly oxidized species in state S_3^* (Boussac et al., 1990, 1992; see Chapter 9). Our group has studied proton release, the rate of P680$^+$ reduction and transmembrane electrochromism in Cl^- depleted thylakoids (Lübbers et al., 1993b). In agreement with Ono et al. (1987) we found a fast reduction of P680$^+$ upon flashes 1 and 2, but a much slower reduction from flash number 3 on. The half-rise times of the biphasic decay under repetitive excitation were 70 and 200 μs. The reduction of P680$^+$ was caused by a backreaction from the acceptor side of PS II, as evident from the parallel decay of the transmembrane potential. Proton release was absent from the second flash on (Lübbers et al., 1993b). Under conditions (70 μM neutral red plus 1 mM imidazole) where proton release occurred in about 10 μs in untreated thylakoids there was no indication for even transient proton release on flash no. 2 in Cl^--depleted material (Lübbers, 1993). Thus, the deprotonation around Y_Z^+ is prevented although the formation of Y_Z^+ is still evident from the very rapid reduction of P680$^+$.

The absence of deprotonation has been indirectly inferred (Lavergne, 1991) from the difference spectrum on transition $S_2^* \Rightarrow S_3^*$ in the blue in calcium-depleted PS II membranes (Boussac et al., 1990). The extent of the difference has been attributed to proton release (see the foregoing section). In the latter material a significant peak around 440 nm was apparent which was absent in non-depleted material. This was interpreted as indicating the appearance of an uncompensated surplus charge in state S_3^* (Lavergne, 1991).

Even without Cl^--depletion, transition $S_2 \Rightarrow S_3$ is unique in the sense that it lacks variation of proton release as a function of the material and the pH. In Section IV we discuss the possibility that proton release may originate from a more intrinsic group of the catalytic center. The absence of proton release during the corresponding transition in Cl^--depleted material, concomitant with the block of any further stable oxidation of the center, is compatible with a more pivotal role of this particular deprotonation step.

2. Hydroxylamine

Hydroxylamine and hydrazine are considered to be 'water analogues' that interact with the catalytic manganese center. Hydroxylamine at μM-concentration causes a shift of the burst of oxygen in a series of flashes from flash no. 3 to flashes 4, 5 or 6 (Bennoun and Joliot, 1969; Beck and Brudvig, 1988; Messinger et al., 1991). The resetting of the S-transitions caused by hydroxylamine can be observed in oxygen evolution, donor side transients in the UV and the red spectral region equally well (Kretschmann et al., 1991; Kretschmann and Witt, 1993). Hydroxylamine reduces the normally dark-stable state S_1 to states S_0, S_{-1} or even S_{-2} (Kretschmann and Witt, 1993). In thylakoids the concentration dependence seemed to suggest the cooperative and reversible binding of 2–3 molecules of hydroxylamine to the Mn-cluster and a light dependent redox reaction (Förster and Junge, 1985b, 1988). On the contrary, studies with PS II complexes have established an irreversible reduction of the Mn-cluster in the dark (Messinger and Renger, 1990; Messinger et al., 1991; Riggs et al., 1993), whereby N_2 is liberated (Radmer and Ollinger, 1982; Renger et al., 1990; He et al., 1991). Two protons are released upon the first flash in hydroxylamine-treated samples (Förster and Junge, 1985b, 1986). The half-rise was about 2 ms, whereas on the subsequent flashes it was much shorter (about 200 μs) (Förster and Junge, 1985b). These experiments were performed with stacked thylakoids. The delay of proton release upon the first flash is now understood by the superimposition of events from the acceptor side (see Section IIA1). A repetition of these experiments with unstacked thylakoids has confirmed the extent, namely about 2 protons per PS II on the first flash, but with a much shorter rise-time (about 350 μs, M. Haumann and W. Junge, unpublished).

3. The Effect of DCCD

In a certain concentration window, treatment with the hydrophobic diimide DCCD (N,N′-Dicyclohexylcarbodiimide) diminishes the proton pumping activity of PS II without an equivalent effect on oxygen evolution. The extent of proton release by water oxidation and of proton uptake by quinone reduction decreases and a rapid, transmembrane electrogenic back-reaction becomes apparent (Jahns et al., 1988). The interpretation of this result was that a covalent modification by DCCD closes the normal outlet for protons from water oxidation into the lumen. Instead, these protons are apparently channeled across the membrane to the site of plastoquinone reduction. Because all reaction steps of the water oxidase are equally affected there seems to be a common outlet for all protons from water oxidation (Jahns and Junge, 1989). Sequencing studies with ^{14}C-labeled DCCD have identified lumenal parts of chlorophyll *a/b* binding proteins (CAB-proteins) as targets of the modification (Jahns and Junge, 1992a,b). Further attempts to identify any specific member of the CAB family as responsible for the DCCD effects on PS II have been unsuccessful (Jahns and Junge, 1992b, 1993). It is conceivable that certain CAB-proteins, aside from their antenna function, shield the site of proton production by PS II and channel protons into the lumen.

4. Proton Sequestering Domains

The idea that protons from water oxidation and also from plastoquinol oxidation are not always directly released into the lumen, but are instead transiently trapped in membrane sequestered domains has been promoted mainly by Richard Dilley (for a review see Dilley, 1991). Peter Homann′s group has studied the influence of the metastable proton pool on the release of functional chloride from PS II and on the stability of the manganese cluster (Theg and Homann, 1982). Our group has found that the supposed domains activated by low uncoupler concentrations trap up to 6 protons (Theg and Junge, 1983). Titration of the domain capacity as a function of pH revealed an apparent pK of about 7.5 and a Hill-coefficient of 2 (Theg and Junge, 1983). The shifting of the pK toward acidity under oxidizing conditions has been interpreted as attributable to the location of the domain buffering groups in the vicinity of a redox component, whose oxidation creates a positive electric field at these groups (Polle and Junge, 1986c). The same study also showed that the domain storing capacity is doubled if one half of the PS II centers is blocked by DCMU. This implies the delocalization of the domains over at least two PS II-complexes.

IV. Structural and Mechanistic Considerations

Protons from the cleavage of water. Two extreme cases have been discussed in the literature. (1) Water binding occurs only after completion of three if not four oxidation steps of the catalytic center, $S_0 \Rightarrow S_1 \Rightarrow S_2 \Rightarrow S_3 \Rightarrow S_4$. This theory seems to be supported by studies with isotopically labeled water, which show that the isotopic composition of dioxygen is determined by the one present in bulk water during state S_3 (Radmer and Ollinger, 1980; Bader et al., 1993). It is a convincing argument, however, that this observation provides no proof for this theory because of possible oxygen exchange between bound water derivatives and bulk water (Debus, 1992). (2) Water derivatives are bound to the manganese cluster and deprotonated stepwise during all transitions. This view has been promoted mainly by the group of Horst Witt (Kretschmann et al., 1991; Kretschmann and Witt, 1993) based on experiments on local electrochromism in the presence of hydroxylamine.

Since protons can be released at the level of the intermediate electron carrier Y_Z^+, their appearance in the medium yields little information on the nature of water intermediates present in each oxidation state. The only evidence for *chemically produced protons* has been observed in thylakoids at slightly alkaline pH, namely a millisecond component upon the third flash which reflects transition $S_4 \rightarrow S_0$. The extent of this slow component is small, only about 0.5 to 1.0 protons per PS II. It probably represents protons produced in a final reaction with water. The equivalent of this phase at acid pH is a fractional slow proton *uptake*. This led us to postulate the rapid creation of between 3.5 and at least 4.5 bases during the four rapid oxidations of the catalytic center, $S_0 \Rightarrow S_1 \Rightarrow S_2 \Rightarrow S_3 \Rightarrow S_4$ (Haumann and Junge, 1994a). The resetting of their pKs upon electron transfer from bound water causes the rebinding of protons which largely but not completely masks the chemical production. We did not observe proton release with the respective time course of the electron transfer from the manganese cluster to Y_Z^+ during any other transition. This is compatible with the notion that only one transition, namely $S_3 \Rightarrow S_4 \rightarrow S_0$, involves a reaction with bound water, but it does not prove it. Some of the 3.5 to 4.5 bases which are created before the last reaction step may serve a function that has been postulated by Lev Krishtalik (1986, 1990). He has argued that the final reaction with water may be thermodynamically and kinetically more favorable if there is one step involving the concerted transfer of at least two electrons from water to the manganese cluster. The activation barrier of this step could be lowered by the cotransfer of one if not two protons to bases formed during the preceding oxidation steps.

Protons from the oxidation of Y_Z It is striking that proton release already occurs at the level of the intermediate carrier Y_Z^+; whether directly or not by tyrosine is an open question. The behavior of the symmetrically placed but functionally more inert, stable tyrosine, Y_D, on the D_2-branch of PS II is better characterized (Eckert and Renger, 1988; Prince, 1988; Babcock et al., 1989; Vass et al., 1990; Barry, 1993). Its oxidation does not cause the deprotonation of tyrosine proper, but it polarizes the hydrogen bond with a neighboring histidine (Rodriguez et al., 1987; Hoganson and Babcock, 1992). Recent ENDOR data by Mino and Kawamori (1994) reveal a similar local arrangement of protons surrounding Y_Z^+ and Y_D^+. Our data on proton release in thylakoids reveal the rapid (i.e. Y_Z^+-related) release of less than one proton per PS II molecule during $S_1 \Rightarrow S_2$ at slightly alkaline pH, the release of about one proton during the other transitions and also on the same transition at more acidic pH. Because of this variability and the above cited Y_Z/Y_D-symmetry we tend to assume that the rapid proton release associated with the formation of Y_Z^+ is due to an electrostatic response of neighboring amino acid residues rather than to tyrosine itself.

The observed variability of the pattern of proton release as a function of the redox transition, of the pH and the particular material is most readily understood if the variation around the mean of 1 proton per transition per PS II is attributed to an electrostatic response of amino acid residues which are located between Y_Z-manganese and the boundary of PS II with the electrolyte solution. The best candidates are acid carboxylic groups (solution pK around 4) and basic guanidinium groups (solution pK around 11) whose pK is shifted toward neutrality in the more hydrophobic environment of the protein. When these groups are exposed to the electric field originating from a new positive charge on Y_Z or on the manganese cluster they undergo an acid directed pK-shift that causes their deprotonation. If one acid group is deprotonated and thus charged, its electric field acts on all the other acid groups. If the mutual interaction is comparable to the one between any particular group and Y_Z-manganese the system of acid/bases behaves as one coupled entity with collective pKs. These can be calculated from the Boltzmann sum over the various microstates (for an example see Beroza et al., 1991).

Our data on the sum over the rapid jumps of proton release show that at least five acid/base groups are required to account for the maximum observed release of 4.5 protons in thylakoids up to state S_4. A total of only four groups was used by Rappaport and Lavergne (1991) for a simulation of their data on BBY-membranes. With a minimum of five groups arranged at reasonable distances, i.e. 2–8 Å from bulk water and 5–15 Å from-Y_Z manganese, we could simulate all the observed pH-dependencies (D. Cherepanov, W. Drevenstedt and W. Junge, unpublished). Without an atomic structure of PS II, however, these simulations serve only illustrative purposes. The flattening of the pattern of proton release in core particles and IML-thylakoids both as a function of the redox transition and of the pH can be understood as follows: The abstraction of some peripheral proteins exposes a greater number of ionogenic groups to the aqueous phase (Rappaport and Lavergne, 1991). The involvement of many groups with pK-values spread out over pH is compatible with the pH-dependent rate of proton release which we observed in core particles (Bögershausen and Junge, 1995). The alternative is an increase of the coulombic interaction between a few groups such that the deprotonation of any group suppresses the deprotonation of any other one.

The variability of the pattern of proton release without any obvious influence on the progress of water oxidation indicates that at least some of the deprotonated groups are not directly linked (as substrate or as ligands) to the manganese cluster. One possible exception is transition $S_2 \Rightarrow S_3$ (see Section IIID1).

Non-integer stoichiometries of protons released

over electrons abstracted from the catalytic center imply the partial deprotonation of an acid group, for example with pK 6. At a pH of 6 the group is half protonated on the average. Concerning the following electron transfer step, which is induced by a flash, there are two different situations, depending on the protolytic relaxation time, τ. If we take the pK of the respective group as 6 and assume a diffusion controlled rate constant of 10^{10} M^{-1} s^{-1} (as for the on-reaction of a group in contact with water) we expect a relaxation time of 100 μs. For events that happen in the time domain below 100 μs the ensemble of reaction centers is heterogeneous, some carry the protonated and others the deprotonated group, so the ensemble average is e.g. 0.5. One straightforwardly expects the heterogeneity to affect the reduction of $P680^+$ by Y_Z in nanoseconds. This may account for the observed biphasic reduction in about 50 and 250 ns upon $S_2 \Rightarrow S_3$ and $S_3 \Rightarrow S_4 \rightarrow S_0$. An influence of the protonation state on the reaction rate between Y_Z and $P680^+$ is established (Meyer et al., 1989). Moreover, recent work from our institute on transient photovoltages (Pokorny et al., 1994) shows that the membrane-normal component of the distance between Y_Z and P680 is about one tenth of the distance between P680 and Q_A, if the same dielectric constant is assumed along both stretches. This orientation is particularly sensitive to charge displacements by deprotonation. On a time scale much larger than τ the time averaging smears out the exchange between the two protonation states, 0 and 1, to yield the partially protonated state.

Is there any influence of the protonation state on the rate of electron transfer from the manganese cluster to Y_Z^+? Our data give no evidence for such an effect. The patterns of oxygen evolution and of the UV absorption transients are not very different under conditions where the pattern of proton release varies between 1 : 1.5 : 1 : 0.5 over 1 : 1 : 1 : 1 to 0.3 : 0.4 : 1.2 : 2.1 (see Fig. 4 in Lübbers et al., 1993 and Fig. 1 in Haumann et al., 1994). Even more meaningful is the observation that the rise and decay kinetics of UV-transients are independent of the rate of proton transfer to neutral red (in thylakoids) or to a hydrophilic buffer (in core particles). This is astounding since the difference between the midpoint potentials of Y_Z and P680, and between the manganese cluster and Y_Z are small, namely 110 mV (Metz et al., 1989) and about 55–170 mV (Vass and Styring, 1991) respectively. A small effect of the protonation state on the rate of electron transfer to Y_Z^+ is only expected if proton release is caused by more remote groups which undergo a small pK shift under the action of the electric field of a newly deposited positive charge on Y_Z-manganese. A voltage jump of about 120 mV, which is induced at the position of a given acid group, causes a pK-shift of 2 units. This shift can imply the release of about 0.8 protons. The symmetry of Coulomb's law calls for an oppositely directed voltage jump of similar extent (about −120 mV) at the position of the primary charge. This is all but negligible in comparison with the about 100 mV drop of the redox potential between these cofactors. It is therefore not self-evident that there should be no effect on the redox equilibria, or at least on the rates of electron transfer between the manganese cluster and Y_Z^+. One interesting possibility is that Y_Z-manganese is embedded in a pocket or cleft with a greater dielectric constant than that of the surrounding protein matrix. Such a structural constraint is not unlikely if one considers the polar or even ionic ligands of the manganese cluster and the well established vicinity of Ca^{2+} and Cl^- (Debus, 1992; Sauer et al., 1992; MacLachlan et al., 1992). Under this condition the electric potential jump which is caused by the deprotonation of a remote group, would be very similar at both cofactor positions and also at positions in between. This minimizes the influence of proton release on the rate constant of electron transfer between manganese and Y_Z^+. Computer models of the structure of PS II show a more polar environment of Y_Z as compared with that of Y_D (Svensson et al., 1991).

Proton release and indicators of the electric field strength. A seemingly consistent picture which has emerged from the older literature (see Section IID) was that the sum of electron abstraction and proton release during each of the transitions $S_i \Rightarrow S_{i+1}$ defines the net charge jump of the catalytic center. The pattern of net charge jumps is reflected by oscillating patterns a) of electrochromic bandshifts and b) of the rate of reduction of $P680^+$. Not only the terminology has to be qualified. The term 'net charge of the catalytic center' may be misleading. One should take into account that the loci of a newly created positive charge, e.g. on manganese, and of proton abstraction, from an acid, can be remote from each other and at different distances from the protein/electrolyte boundary. This implies that the electrostatic compensation of the newly created positive charge by the release of one proton from another site is incomplete. We calculated a rather invariant pattern of electrochromism in parallel with a pronounced pattern of proton release by peripheral amino acids under the assumption of a rather polar environment of Mn-Y_Z (D. Cherepanov et al., unpublished).

The observed absence of electrochromic bandshifts during transitions $S_0 \Rightarrow S_1$ and $S_2 \Rightarrow S_3$ then may indicate an almost perfect electrostatic compensation which would be expected if proton release occurs from the center of the manganese cluster. For transition $S_2 \Rightarrow S_3$ this notion is compatible with the release of one proton per PS II independent of the pH ranging between 5.5 and 7.75, and of the material, namely thylakoids, IML-thylakoids, BBY-membranes, and core particles.

On the other hand, there are electrochromic bandshifts and shifts of the reduction rate of $P680^+$ upon transitions $S_1 \Rightarrow S_2$ and $S_3 \Rightarrow S_4 \rightarrow S_0$. Their extent is rather constant although proton release varies as a function of the pH and the material. At higher time resolution we found no correlation between the rate of electrochromic bandshifts and the rate of proton transfer to neutral red (in thylakoids) and to the aqueous phase (in core particles). Thus, the net charge is decoupled from proton release during $S_1 \Rightarrow S_2$ and $S_3 \Rightarrow S_4 \rightarrow S_0$.

Because of this decoupling from proton release, a quantitative analysis of the electrochromic response that utilizes 'tagging' of the pigments by polarized spectroscopy is expected to provide information on the distances between the charges and the innermost pigments and on the orientation of the latter. This is currently under investigation.

Acknowledgments

We wish to thank our colleagues Dr. Karin Lübbers, Oliver Bögershausen and Wolfgang Drevenstedt for their experimental contributions and many discussions, Hella Kenneweg for excellent technical assistance, and Drs. Jérôme Lavergne (Paris), Armen Mulkidjanian (Osnabrück/Moscow), Dimitri Cherepanov and Lev Krishtalik (Moscow) for sharing with us their ideas.

This work was financially supported by the Deutsche Forschungsgemeinschaft (Project A2-SFB 171), the Land Niedersachsen and the Fond der Chemischen Industrie.

References

Andersson B and Styring S (1991) Photosystem-II - Molecular organization, function, and acclimation. In: Lee CP (ed) Current Topics in Bioenergetics, Vol 16, pp 161–181. Academic Press, New York

Argyroudi-Akoyunoglou JH and Akoyunoglou G (1970) Photoinduced changes in the chlorophyll *a* to chlorophyll *b* ratio in young bean plants. Plant Physiol 46: 247–249

Ausländer W and Junge W (1975) Neutral red, a rapid indicator for pH changes in the inner phase of thylakoids. FEBS Lett 59: 310–315

Babcock GT, Blankenship RE and Sauer K (1976) Reaction kinetics for positive charge accumulation on the water side of chloroplast Photosystem II. FEBS Lett 61: 286–289

Babcock GT, Barry BA, Debus RJ, Hoganson CW, Atamian M, McIntosh L, Sithole I and Yocum CF (1989) Water oxidation in Photosystem II: From radical chemistry to multielectron chemistry. Biochemistry 28: 9557–9565

Bader KP, Renger G and Schmid GH (1993) A mass spectrometric analysis on the water splitting reaction.. Photosynth Res 38: 355–361

Barry BA (1993) The role of redox-active amino acids in the photosynthetic water-oxidizing complex. Photochem Photobiol 57: 179–188

Baumgarten M, Philo JS and Dismukes GC (1990) Mechanism of photoinhibition of photosynthetic water oxidation by Cl- depletion and F- substitution: Oxidation of a protein residue. Biochemistry 29: 10814–10822

Beck WF and Brudvig GW (1988) Resolution of the paradox ammonia and hydroxylamine as substrate analogues for the water-oxidation reaction catalyzed by Photosystem II. J Am Chem Soc 110: 1517–1525

Bennoun P and Joliot A (1969) Etude de la photooxidation de l'hydroxylamine par les chloroplasts d'epinards. Biochim Biophys Acta 189: 85–94

Beroza P, Fredkin DR, Okamura MY and Feher G (1991) Protonation of interacting residues in a protein by a Monte-Carlo method - Application to lysozyme and the photosynthetic reaction center of *Rhodobacter-sphaeroides*. Proc Natl Acad Sci USA 88: 5804–5808

Berthold DA, Babcock GT and Yocum CF (1981) A highly resolved, oxygen-evolving Photosystem II preparation from spinach thylakoid membranes. FEBS Lett 134(2): 231–234

Bock CH, Gerken S, Stehlik D and Witt HT (1988) Time resolved EPR on Photosystem II particles after irreversible and reversible inhibition of water cleavage with high concentrations of acetate. FEBS Lett 227: 141–146

Bögershausen O and Junge W (1995) Rapid proton transfer under flashing light at both functional sides of dark-adapted Photosystem II particles. Biochim Biophys Acta 1230: 177–185

Boska M and Sauer K (1984) Kinetics of EPR signal II in chloroplast Photosystem II. Biochim Biophys Acta 765: 84–87

Boussac A and Rutherford W (1994) Electron transfer events in chloride-depleted Photosystem II. J Biol Chem 269: 12462–12467

Boussac A, Zimmermann JL, Rutherford AW and Lavergne J (1990) Histidine oxidation in the oxygen-evolving photosystem-

II enzyme. Nature 347: 303–306

Boussac A, Setif P and Rutherford AW (1992) Inhibition of tyrosine Z photooxidation after formation of the S_3 state in Ca^{2+}-depleted and Cl^--depleted Photosystem II. Biochemistry 31: 1224–1233

Bowes JM and Crofts AR (1980) Binary oscillations in the rate of reoxidation of the primary acceptor of Photosystem II. Biochim Biophys Acta 590: 373–384

Bowes JM and Crofts AR (1981) The role of pH and membrane potential in the reactions of Photosystem II as measured by effects on delayed fluorescence. Biochim Biophys Acta 637: 464–472

Brettel K, Schlodder E and Witt HT (1984) Nanosecond reduction kinetics of photooxidized chlorophyll-$_{aII}$ (P-680) in single flashes as a probe for the electron pathway, H^+-release and charge accumulation in the O_2-evolving complex. Biochim Biophys Acta 766: 403–415

Conjeaud H and Mathis P (1986) Electron transfer in the photosynthesis membrane. Influence of pH and surface potential on the P 680 reduction kinetics. Biophys J 49: 1215–1221

Day DA, Ryrie IJ and Fuad N (1984) Investigations of the role of the main light-harvesting chlorophyll-protein complex in thylakoid membranes. Reconstruction of depleted membranes from intermittent-light-grown plants with the isolated complex. J Cell Biol 97: 163–172

De Groot A, Plijter JJ, Evelo R, Babcock GT and Hoff AJ (1986) The influence of the oxidation state of the oxygen-evolving complex of Photosystem II on the spin-lattice relaxation time of signal II as determined by electron spin-echo spectroscopy. Biochim Biophys Acta 848: 8–15

Debus RJ (1992) The manganese and calcium ions of photosynthetic oxygen evolution. Biochim Biophys Acta 1102: 269–352

Dekker JP, Plijter JJ, Ouwenand L and Van Gorkom HJ (1984a) Kinetics of manganese redox transitions in the oxygen evolving apparatus of photosynthesis. Biochim Biophys Acta 767: 176–179

Dekker JP, Van Gorkom HJ, Brok M and Ouwehand L (1984b) Optical characterization of Photosystem II electron donors. Biochim Biophys Acta 764: 301–309

Dekker JP, Van Gorkom HJ, Wensink J and Ouwehand L (1984c) Absorbance difference spectra of the successive redox states of the oxygen-evolving apparatus of photosynthesis. Biochim Biophys Acta 767: 1–9

Dilley RA (1991) Energy coupling in chloroplasts - A calcium-gated switch controls proton fluxes between localized and delocalized proton gradients. In: Slayman CL (ed) Current Topics in Bioenergetics, Vol 16, pp 265–318. Academic Press, New York

Diner BA, Petrouleas V and Wendoloski JJ (1991) The iron-quinone electron-acceptor complex of Photosystem II. Physiol Plant 81: 423–436

Dismukes GC and Siderer Y (1981) Intermediates of a polynuclear manganese center involved in photosynthetic oxidation of water. Proc Natl Acad Sci USA 78: 274–278

Eckert HJ and Renger G (1988) Temperature dependence of P680+ reduction in oxygen-evolving PS II membrane fragments at different redox states Si of the water oxidizing system. FEBS Lett 236: 425–431

Eigen M (1963) Protonenübertragung, Säure-Base-Katalyse und enzymatische Hydrolyse.Teil I: Elementarvorgänge. Angew Chem 12s: 489–588

Emrich HM, Junge W and Witt HT (1969) Further evidence for an optical response of chloroplast bulk pigments to a light-induced electrical field in photosynthesis. Z Naturforsch B 24: 1144–1146

Forbush B, Kok B and McGloin MP (1971) Cooperation of charges in photosynthetic O_2 evolution—II. Damping of flash yield oscillation, deactivation. Photochem Photobiol 14: 307–321

Förster V and Junge W (1984) Protolytic reactions at the donor side of PS II: Proton release in tris-washed chloroplasts with $t_{1/2} = 100\mu s$. Implications for the interpretation of the proton release pattern in untreated, oxygen-evolving chloroplasts. In: Sybesma M (ed) Advances in Photosynthetic Research, Vol 1, pp 305–308. Martinus Nijhoff, The Hague

Förster V and Junge W (1985a) Stoichiometry and kinetics of proton release upon photosynthetic water oxidation. Photochem Photobiol 41: 183–190

Förster V and Junge W (1985b) Interaction of hydroxylamine with the water-oxidizing enzyme investigated via proton release. Photochem Photobiol 41: 191–194

Förster V and Junge W (1985c) Cooperative and reversible action of three or four hydroxylamine molecules on the water-oxidizing complex. FEBS Lett 186: 153–157

Förster V and Junge W (1986) On the action of hydroxylamine, hydrazine and their derivatives on the water-oxidizing complex. Photosynth Res 9: 197–210

Förster V and Junge W (1988) Protolytic reactions of the photosynthetic water oxidase in the absence and in the presence of added ligands. Chem Scr 28A: 111–116

Fowler CF (1977) Proton evolution from Photosystem II. Stoichiometry and mechanistic considerations. Biochim Biophys Acta 462: 414–421

Gerken S, Brettel K, Schlodder E and Witt HT (1987) Direct observation of the immediate electron donor to chlorophyll-aII^+ ($P\text{-}680^+$) in oxygen-evolving Photosystem II complexes. Resolution of nanosecond kinetics in the UV. FEBS Lett 223: 376–380

Gerken S, Dekker JP, Schlodder E and Witt HT (1989) Studies on the multiphasic charge recombination between chlorophyll aII^+ ($P\text{-}680^+$) and plastoquinone Q_A^- in Photosystem II complexes. Ultraviolet difference spectrum of Chl-aII^+/Chl-aII. Biochim Biophys Acta 977: 52–61

Ghanotakis DF, Babcock GT and Yocum CF (1984) Structural and catalytic properties of the oxygen-evolving complex. Correlation of polypeptide and manganese release with the behavior of Z^+ in chloroplasts and a highly resolved preparation of the PS II complex. Biochim Biophys Acta 765: 388–398

Ghanotakis DF, Demetriou DM and Yocum CF (1987a) Purification of an oxygen evolving Photosystem II reaction center core preparation. In: Biggins J (ed) Progress in Photosynthesis Research, Vol 1, pp 681–684. Martinus Nijhoff, Dordrecht

Ghanotakis DF, Waggoner CM, Bowlby NR, Demetriou DM, Babcock GT and Yocum CF (1987b) Comparative structural and catalytic properties of oxygen-evolving Photosystem II preparations. Photosynth Res 14: 191–199

Guiles RD, Zimmermann JL, McDermott AE, Yachandra VK, Cole JL, Dexheimer SL, Britt RD, Wieghardt K, Bossek U,

Sauer K and Klein MP(1990) The S3 state of Photosystem II: Differences between the structure of the manganese complex in the S2 and S3 states determined by X-ray absorption spectroscopy. Biochemistry 29: 471–485

Gutman M and Nachliel E (1990) The dynamic aspects of proton transfer processes. Biochim Biophys Acta 1015: 391–414

Haag E, Irrgang KD, Boekema EJ and Renger G (1990) Functional and structural analysis of Photosystem II core complexes from spinach with high oxygen evolution capacity. Eur J Biochem 189: 47–53

Haddy A, Aasa R and Andreasson LE (1989) S-band EPR studies of the S2-state multiline signal from the photosynthetic oxygen-evolving complex. Biochemistry 28: 6954–6959

Haumann M and Junge W (1994a) Extent and rate of proton release by photosynthetic water oxidation in thylakoids: Electrostatic relaxation versus chemical production. Biochemistry 94: 864–872

Haumann M and Junge W (1994b) The rates of proton uptake and electron transfer at the reducing side of Photosystem II in thylakoids. FEBS Lett 347: 45–50

Haumann M, Bögershausen O and Junge W (1994) Photosynthetic oxygen evolution: Net charge transients as inferred from electrochromic bandshifts are independent of proton release into the medium. FEBS Lett 355: 101–105

Haveman J and Mathis P (1976) Flash-induced absorption changes of the primary donor of Photosystem II at 820 nm in chloroplasts inhibited by low pH or tris-treatment. Biochim Biophys Acta 440: 346–355

He P, Bader KP and Schmid GH (1991) Mass spectrometric analysis of N_2-formation induced by the oxidation of hydrazine and hydroxylamine in flash illuminated thylakoid preparations of the filamentous cyanobacterium *Oscillatoria-chalybea*. Z Naturforsch C 46: 629–634

Hoganson CW and Babcock GT (1988) Electron-transfer events near the reaction center in O_2-evolving Photosystem II preparations. Biochemistry 27: 5848–5855

Hoganson CW and Babcock GT (1992) Protein-tyrosyl radical interactions in Photosystem II studied by electron spin resonance and electron nuclear double resonance spectroscopy: Comparison with ribonucleotide reductase and in vitro tyrosine. Biochemistry 31: 11874–11880

Homann PH, Gleiter H, Ono T and Inoue Y (1986) Storage of abnormal oxidants 'S1', 'S2' and 'S3' in photosynthetic water oxidases inhibited by Cl- removal. Biochim Biophys Acta 850: 10–20

Hong YQ and Junge W (1983) Localized or delocalized protons in photophosphorylation? On the accessibility of the thylakoid lumen for ions and buffers. Biochim Biophys Acta 722: 197–208

Hope AB and Morland A (1979) Proton translocation in isolated spinach chloroplasts after single-turnover actinic flashes. Aust J Plant Physiol 6: 1–16

Itoh S, Yerkes CT, Koike H, Robinson HH and Crofts AR (1984) Effects of chloride depletion on electron donation from the water-oxidizing complex to the Photosystem II reaction center as measured by the microsecond rise of chlorophyll fluorescence in isolated pea chloroplasts. Biochim Biophys Acta 766: 612–622

Jahns P and Junge W (1989) The protonic shortcircuit by DCCD in Photosystem II. A common feature of all redox transitions of water oxidation. FEBS Lett 253: 33–37

Jahns P and Junge W (1990) Dicyclohexylcarbodiimide-binding proteins related to the short circuit of the proton-pumping activity of Photosystem II. Identified as light-harvesting chlorophyll-*a/b*-binding proteins. Eur J Biochem 193: 731–736

Jahns P and Junge W (1992a) Thylakoids from pea seedlings grown under intermittent light: Biochemical and flash-spectrophotometric properties. Biochemistry 31: 7390–7397

Jahns P and Junge W (1992b) Proton release during the four steps of photosynthetic water oxidation: Induction of 1:1:1:1 pattern due to lack of chlorophyll *a/b* binding proteins. Biochemistry 31: 7398–7403

Jahns P and Junge W (1993) Another role of chlorophyll a/b binding proteins of higher plants: They modulate protolytic reactions associated with Photosystem II. Photochem Photobiol 57: 120–124

Jahns P, Polle A and Junge W (1988) The photosynthetic water oxidase: Its proton pumping activity is short-circuited within the protein by DCCD. EMBO J 7: 589–594

Jahns P, Lavergne J, Rappaport F and Junge W (1991) Stoichiometry of proton release during photosynthetic water oxidation: A reinterpretation of the response of Neutral red leads to a non-integer pattern. Biochim Biophys Acta 1057: 313–319

Jahns P, Haumann M, Bögershausen O, and Junge W (1992) Water oxidation: 1:1:1:1 proton-over-electron stoichiometry in CAB-protein depleted thylakoids and PS II core particles. In: Murata N (ed) Progress in Photosynthesis Research, Vol 2, pp 333–336. Kluwer, Dordrecht

Joliot P and Joliot A (1968) A polarographic method for detection of oxygen production and reduction of Hill reagent by isolated chloroplasts. Biochim Biophys Acta 153: 625–634

Joliot P and Kok B (1975) Oxygen evolution in photosynthesis. In: Govindjee (ed) Bioenergetics of Photosynthesis, pp 387–412. Academic Press, New York

Junge W and McLaughlin S (1987) The role of fixed and mobile buffers in the kinetics of proton movement. Biochim Biophys Acta 890: 1–5

Junge W and Polle A (1986) Theory of proton flow along appressed thylakoid membranes under both non-stationary and stationary conditions. Biochim Biophys Acta 848: 265–273

Junge W and Witt HT (1968) On the ion transport system of photosynthesis—Investigation on a molecular level. Z Naturforsch 23b: 244–254

Junge W, Ausländer W, McGeer AJ and Runge T (1979) The buffering capacity of the internal phase of thylakoids and the magnitude of the pH changes inside under flashing light. Biochim Biophys Acta 546: 121–141

Junge W, Schönknecht G and Förster V (1986) Neutral red as an indicator of pH transients in the lumen of thylakoids - some answers to criticism. Biochim Biophys Acta 852: 93–99

Kirilovsky DL, Boussac A, Van Mieghem FJE, Ducruet JM, Setif P, Yu J, Vermaas WFJ and Rutherford AW (1992) Oxygen-Evolving Photosystem II preparation from wild type and Photosystem II mutants of *Synechocystis sp.* PCC 6803. Biochemistry 31: 2099–2107

Kok B and Velthuys B (1977) Present status of the O_2 evolution model. In: Castellani A (ed) Research in Photobiology, pp

111–119. Plenum Press, New York

Kok B, Forbush B and McGloin M (1970) Cooperation of charges in photosynthetic O_2 evolution - I. A linear four-step mechanism. Photochem Photobiol 11: 457–475

Kretschmann H and Witt HT (1993) Chemical reduction of the water splitting enzyme system of photosynthesis and its light-induced reoxidation characterized by optical and mass spectroscopical measurements: A basis for the estimation of the states of redox active manganese and of water in the quarternary oxygen-evolving S-state cycle. Biochim Biophys Acta 1144: 331–345

Kretschmann H, Pauly S and Witt HT (1991) Evidence for a chemical reaction of hydroxylamine with the photosynthetic water splitting enzyme-S in the dark—possible states of manganese and water in the S-cycle. Biochim Biophys Acta 1059: 208–214

Krishtalik LI (1986) Energetics of multielectron reactions. Photosynthetic oxygen evolution. Biochim Biophys Acta 849: 162–171

Krishtalik LI (1989) Energetics of photosynthetic oxygen evolution. Biofizika 34: 883–886

Krishtalik LI (1990) Activation energy of photosynthetic oxygen evolution: An attempt at theoretical analysis. Bioelectrochem Bioenerg 23: 249–263

Lavergne J (1987) Optical-difference spectra of the S-state transitions in the photosynthetic oxygen-evolving complex. Biochim Biophys Acta 894: 91–107

Lavergne J (1991) Improved UV-visible spectra of the S-transitions in the photosynthetic oxygen-evolving system. Biochim Biophys Acta 1060: 175–188

Lavergne J and Junge W (1993) Proton release during the redox cycle of the water oxidase. Photosynth Res 38: 279–296

Lavergne J and Leci E (1993) Properties of inactive Photosystem II centers. Photosynth Res 35: 323–343

Lavergne J and Rappaport F (1990) On the stoichiometry of proton release by the oxygen-evolving system. In: Baltscheffsky M (ed) Current Research in Photosynthesis, pp 873–876. Kluwer, Dordrecht

Lavergne J, Blanchard-Desce M and Rappaport F (1992) Oxidation and deprotonation reactions in the Kok cycle. In: Murata N (ed) Research in Photosynthesis, Vol 2, pp 273–280. Kluwer, Dordrecht

Lavorel J (1976) Matrix analysis of the oxygen evolving system of photosynthesis. J Theor Biol 57: 171–185

Lavorel J (1991) On the proper use of sigma analysis in the study of flash-induced oxygen evolution. Photosynth Res 27: 15–18

Lavorel J (1992) Determination of the photosynthetic oxygen release time by amperometry. Biochim Biophys Acta 1101: 33–40

Lübbers K (1993) Photosynthetische Wasseroxidation: Eigenschaften von Reaktionszentren und der Einfluss des Kofaktors Chlorid. Thesis, Universität Osnabrück, Germany

Lübbers K and Junge W (1990) Is the proton release due to water oxidation directly coupled to events at the manganese centre? In: Baltscheffsky M (ed) Current Research in Photosynthesis, Vol I, pp 877–880. Kluwer, Dordrecht

Lübbers K, Haumann M and Junge W (1993a) Photosynthetic water oxidation under flashing light: Oxygen release, proton release and absorption transients in the near UV—a comparison between thylakoids and a reaction center core preparation. Biochim Biophys Acta 1183 : 210–214

Lübbers K, Drevenstedt W and Junge W (1993b) Chloride depletion of photosynthetic water oxidase. FEBS Lett 336: 304–308

Machanda R, Thorp HH, Brudvig GW and Crabtree RH (1991) Proton coupled electron transfer in high-valent oxomanganese dimers: Role of the ancillary ligands. Inorg Chem 30: 4 94–497

Machanda R, Thorp HH, Brudvig GW and Crabtree RH (1992) An unusual example of multiple proton-coupled electron transfer in a high-valent oxomanganese dimer [(phen)2 Mn(III) (O)2 Mn(IV) (phen)2](Clo4)3;(phen= 1,10-Phenanthrolin). Inorg Chem 31: 4040–4041

MacLachlan DJ, Nugent JHA and Evans MCW (1992) XAS and EPR studies of cofactor depleted spinach PS II. In: Murata N (ed) Research in Photosynthesis, Vol 2, pp 373–376. Kluwer, Dordrecht

Mathis P and Haveman J (1977) Analysis of absorption changes in the ultraviolet related to charge-accumulating electron carriers in Photosystem II of chloroplasts. Biochim Biophys Acta 461: 167–181

McDermott AE, Yachandra VK, Guiles RD, Cole JL, Dexheimer SL, Britt RD, Sauer K and Klein MP (1988) Characterization of the manganese O2-evolving complex and the iron-quinone acceptor complex on Photosystem II from a thermophilic cyanobacterium by electron paramagnetic resonance and X-ray absorption spectroscopy. Biochemistry 27: 4021–4031

Melis A and Anderson JM (1983) Structural and functional organization of the photosystems in spinach chloroplasts. Antenna size, relative electron-transport capacity, and chlorophyll composition. Biochim Biophys Acta 724: 473–484

Messinger J and Renger G (1990) The reactivity of hydrazine with Photosystem II strongly depends on the redox state of the water-oxidizing system. FEBS Lett 277: 141–146

Messinger J, Pauly S and Witt HT (1991) The flash pattern of photosynthetic oxygen evolution after treatment with low concentrations of hydroxylamine as a function of the previous S_1/S_0-ratio —further evidence that NH_2OH reduces the water oxidizing complex in the dark. Z Naturforsch C 46: 1033–1038

Metz JG, Nixon PJ, Rögner M, Brudvig GW and Diner BA (1989) Directed alteration of D1 polypeptide of Photosystem II: Evidence that Tyrosine-161 is the redox component, Z, connecting the oxygen-evolving complex to the primary electron donor, P680. Biochemistry 28: 6960–6969

Meunier PC (1993) Oxygen evolution by Photosystem II: The contribution of backward transitions to the anomalous behavior of double-hits revealed by a new analysis method. Photosynth Res 36: 111–118

Meyer B, Schlodder E, Dekker JP and Witt HT (1989) O_2 evolution and Chl a_{II}^+ (P-680$^+$) nanosecond reduction kinetics in single flashes as a function of pH. Biochim Biophys Acta 974: 36–43

Mino H and Kawamori A (1994) Microenvironments of tyrosine D^+ and tyrosine Z^+ in Photosystem II studied by proton matrix ENDOR. Biochim Biophys Acta 1185: 213–220

Oettmeier W, Masson K and Dostatni R (1987) Halogenated 1,4-benzoquinones as irreversible binding inhibitors of photosynthetic electron transport. Biochim Biophys Acta 890: 260–269

Ono T, Conjeaud H, Gleiter H, Inoue Y and Mathis P (1986a) Effect of preillumination on the P-680$^+$ reduction kinetics in chloride-free Photosystem II membranes. FEBS Lett 203: 215–219

Ono T, Zimmermann JL, Inoue Y and Rutherford AW (1986b) EPR evidence for a modified S-state transition in chloride-depleted Photosystem II. Biochim Biophys Acta 851: 193–201

Ono T, Zimmermann JL, Inoue Y and Rutherford AW (1987) Abnormal S_2 state formed in chloride depleted Photosystem II as revealed by manganese EPR multiline signal. In: Biggins J (ed) Progress in Photosynthesis Research, Vol I, pp 652–656. Martinus Nijhoff, Dordrecht

Petrouleas V and Diner BA (1986) Identification of Q_{400}, a high-potential electron acceptor of Photosystem II, with the iron of the quinone-iron acceptor complex. Biochim Biophys Acta 849: 264–275

Pokorny A, Wulf K and Trissl H-W (1994) An electrogenic reaction associated with the re-reduction of P680 by tyr Z in Photosystem II. Biochim Biophys Acta 1184: 65–70

Polle A and Junge W (1986a) The slow rise of the flash-light-induced alkalization by Photosystem II of the suspending medium of thylakoids is reversibly related to thylakoid stacking. Biochim Biophys Acta 848: 257–264

Polle A and Junge W (1986b) The slow rate of proton consumption at the reducing side of Photosystem I is limited by the rate of redox reactions of extrinsic electron acceptors, but not by a diffusion barrier for protons. Biochim Biophys Acta 848: 274–278

Polle A and Junge W (1986c) Transient and intramembrane trapping of pumped protons in thylakoids. The domains are delocalized and redox-sensitive. FEBS Lett 198: 263–267

Polle A and Junge W (1989) Proton diffusion along the membrane surface of thylakoids is not enhanced over that in bulk water. Biophys J 56: 27–31

Prince RC (1988) Tyrosine radicals. Trends Biochem Sci 13: 286–289

Radmer R and Ollinger O (1980) Isotopic composition of photosynthetic O_2 flash yields in the presence of $H_2{}^{18}O$ and $HC^{18}O_3^-$. FEBS Lett 110: 57–61

Radmer R and Ollinger O (1982) Nitrogen and oxygen evolution by hydroxylamine-treated chloroplasts. FEBS Lett 144: 162–166

Rappaport F and Lavergne J (1991) Proton release during successive oxidation steps of the photosynthetic water oxidation process: Stoichiometries and pH dependence. Biochemistry 30: 10004–10012

Rappaport F, Blanchard-Desce M and Lavergne J (1994) Kinetics of electron transfer and electrochromic change during the redox transitions of the photosynthetic oxygen-evolving complex. Biochim Biophys Acta 1184: 178–192

Renger G (1987) Mechanistic aspects of photosynthetic water cleavage. Photosynthetica 21: 203–224

Renger G and Govindjee (1985) The mechanism of photosynthetic water oxidation. Photosynth Res 6: 33–55

Renger G and Hanssum B (1992) Studies on the reaction coordinates of the water oxidase in PS II membrane fragments from spinach. FEBS Lett 299: 28–32

Renger G and Voelker M (1982) Studies on the proton release pattern of the donor side of system II: Correlation between oxidation and deprotonization of donor D_1 in Tris-washed inside-out thylakoids. FEBS Lett 149: 203–207

Renger G and Weiss W (1986) Studies on the nature of the water oxidizing enzyme. III. Spectral characterization of the intermediary redox states in the water-oxidizing enzyme system Y. Biochim Biophys Acta 850: 184–196

Renger G and Wydrzynski T (1991) The role of manganese in photosynthetic water oxidation. Biol Met 4: 73–80

Renger G, Wacker U and Voelker M (1987) Studies on the protolytic reactions coupled with water cleavage in Photosystem II membrane fragments from spinach. Photosynth Res 13: 167–184

Renger G, Hanssum B, Gleiter H, Koike H and Inoue Y (1988) Interaction of 1,4-benzoquinones with Photosystem II in thylakoids and Photosystem II membrane fragments from spinach. Biochim Biophys Acta 936: 435–446

Renger G, Bader KP and Schmid GH (1990) Mass spectroscopic analysis of N_2 formation by flash-induced oxidation of hydrazine and hydroxylamine in normal and Tris-treated tobacco chloroplasts. Biochim Biophys Acta 1015: 288–294

Riggs PJ, Mei R, Yocum CF and Penner-Hahn JE (1993) Characterization of the manganese site in the photosynthetic oxygen evolving complex: The effect of hydroxylamine and hydroquinone on the X-ray absorption spectra. Jpn J Appl Phys 32: 527–529

Rodriguez ID, Chandrashekar TK and Babcock GT (1987) Endor characterization of H_2O/D_2O exchange in the D^+Z^+ radical in photosynthesis. In: Biggins J (ed) Progress in Photosynthesis Research, pp 471–474. Martinus Nijhoff, Dordrecht

Rögner M, Dekker JP, Boekema EJ and Witt HT (1987) Size, shape and mass of the oxygen-evolving Photosystem II complex from the thermophilic cyanobacterium *Synechococcus* sp. FEBS Lett 219: 207–211

Ruffle SV, Donelly D, Blundell, TL and Nugent JHA (1992) Computer modeling of the Photosystem II core proteins. In: Murata N (ed) Research in Photosynthesis, Vol 2, pp 191–194. Kluwer, Dordrecht

Rutherford AW, Zimmerman JL and Boussac A (1992) Oxygen evolution. In: Barber J (ed) The Photosystems: Structure, Function and Molecular Biology, pp 179–229. Elsevier, Amsterdam

Saphon S and Crofts AR (1977) Protolytic reactions in Photosystem II: A new model for the release of protons accompanying the photooxidation of water. Z Naturforsch C: Biosci 32C: 617–626

Sauer K, Yachandra WK, Britt RD, et al. (1992) The photosynthetic water oxidation complex studied by EPR and X-ray absorption spectroscopy. In: Pecoraro VL (ed) Manganese Redox Enzymes, pp 141–175. VCH, New York

Saygin Ö and Witt HT (1984) On the change of the charges in the four photoinduced oxidation steps of the water-splitting enzyme system S. Optical characterization at O_2-evolving complexes isolated from *Synechococcus*. FEBS Lett 176: 83–87

Saygin Ö and Witt HT (1985) Evidence for the electrochromic identification of the change of charges in the four oxidation steps of the photoinduced water cleavage in photosynthesis. FEBS Lett 187: 224–226

Saygin Ö and Witt HT (1987) Optical characterization of intermediates in the water-splitting enzyme system of photosynthesis—possible states and configurations of manganese and water. Biochim Biophys Acta 893: 452–469

Schatz GH and Van Gorkom HJ (1985) Absorbance difference spectra upon charge transfer to secondary donors and acceptors in Photosystem II. Biochim Biophys Acta 810: 283–294

Schatz GH and Witt HT (1984) Extraction and characterization of oxygen-evolving Photosystem II complexes from a thermophilic cyanobacterium *Synechococcus* spec.. Photobiochem Photobiophys 7: 1–14

Schlodder E, Brettel K, Schatz GH and Witt HT (1984) Analysis of the Chl-a_{II}^{+} reduction kinetics with nanosecond time resolution in oxygen-evolving Photosystem II particles from *Synechococcus* at 680 and 824 nm. Biochim Biophys Acta 765: 178–185

Shinkarev VP and Wraight CA (1993) Oxygen evolution in photosynthesis: From unicycle to bicycle. Proc Natl Acad Sci USA 90: 1834–1838

Sinclair J (1984) The influence of anions on oxygen evolution by isolated spinach chloroplasts. Biochim Biophys Acta 764: 247–252

Srinivasan AN and Sharp RR (1986) Flash-induced enhancements in the proton NMR relaxation rate of Photosystem II particles: Response to flash trains of 1–5 flashes. Biochim Biophys Acta 851: 369–376

Styring SA and Rutherford AW (1988) The microwave power saturation of SII_{slow} varies with the redox state of the oxygen-evolving complex in Photosystem II. Biochemistry 27: 4915–4923

Svensson B, Vass I, Cedergren E and Styring S (1990) Structure of donor side components in Photosystem II predicted by computer modeling. EMBO J 9: 2051–2059

Svensson B, Vass I and Styring S (1991) Sequence analysis of the D1 and D2 reaction center proteins of photosystem-II. Z Naturforsch C 46: 765–776

Takahashi E and Wraight CA (1990) A crucial role for Asp(L213) in the proton transfer pathway to the secondary quinone of reaction centers from *Rhodobacter sphaeroides*. Biochim Biophys Acta 1020: 107–111

Theg SM and Homann PH (1982) Light-, pH- and uncoupler-dependent association of chloride with chloroplast thylakoids. Biochim Biophys Acta 679: 221–234

Theg SM and Junge W (1983) The effect of low concentrations of uncouplers on the detectability of proton deposition in thylakoids. Evidence for subcompartmentation and preexisting pH differences in the dark. Biochim Biophys Acta 723: 294–307

Theg SM, Jursinic PA and Homann PH (1984) Studies on the mechanism of chloride action on photosynthetic water oxidation. Biochim Biophys Acta 766: 636–646

Thibault P (1978) A new attempt to study the oxygen evolving system of photosynthesis: Determination of transition probabilities of a state i. J Theor Biol 73: 271–284

Thorp H, Sasneskhi JE, Brudvig GW and Crabtree RH (1989) Proton coupled electrons transfer in [(bpy) 2 Mn(O) 2 Mn(bpy)2]$^{3+}$. J Am Chem Soc 3: 9249–9250

Trissl H-W, Breton J, Deprez J and Leibl W (1987) Primary electrogenic reactions of Photosystem II as probed by the light gradient method. Biochim Biophys Acta 893: 305–319

Tzinas G, Argyroudi-Akoyunoglou JH and Akoyunoglou A (1987) The effect of the dark interval in intermittent light on thylakoid development: Photosynthetic unit formation and light harvesting protein accumulation. Photosynth Res 14: 241–258

Van Leeuwen P (1993) The redox cycle of the oxygen evolving complex of Photosystem II. Thesis, University of Leiden, The Netherlands

Van Leeuwen PJ, Nieveen MC, van de Meent EJ, Dekker JP and Van Gorkom HJ (1991) Rapid and simple isolation of pure Photosystem II core and reaction center particles from spinach. Photosynth Res 28: 149–153

Van Leeuwen PJ, Heimann C, Dekker JP, Gast P and van Gorkom HJ (1992a) Redox changes of the oxygen evolving complex in Photosystem II core particles as studied by UV spectroscopy. In: Murata N (ed) Research in Photosynthesis, Vol 2, pp 325–326. Kluwer, Dordrecht

Van Leeuwen PJ, Heimann C, Kleinherenbrink FAM, et al. (1992b) Kinetics of electron transport on the donor side of spinach photosystem core particles. In: Murata N (ed) Research in Photosynthesis, Vol 2, pp 341–344. Kluwer, Dordrecht

Van Leeuwen PJ, Heimann C, Gast P, Dekker JP and Van Gorkom HJ (1993) Flash-induced redox changes in oxygen-evolving spinach Photosystem II core particles. Photosynth Res 38: 169–176

Vass I and Styring S (1991) pH-dependent charge equilibria between tyrosine-D and the states in Photosystem II. Estimation of relative midpoint redox potentials. Biochemistry 30: 830–839

Vass I, Deak Z, Jegerschoeld C and Styring S (1990) The accessory electron donor tyrosine-D of Photosystem II is slowly reduced in the dark during low-temperature storage of isolated thylakoids. Biochim Biophys Acta 1018: 41–46

Velthuys BR (1981) Spectrophotometric studies on the S-state transition of Photosystem II and of the interactions of its charged donor chains with lipid soluble anions. In: Akoyunoglou G (ed) Photosynthesis II. Electron Transport and Photophosphorylation, pp 75–85. Balaban International Science Services, Philadelphia

Velthuys BR (1988) Spectroscopic characterization of the acceptor state Qa- and the donor state S2 of Photosystem II of spinach in the blue, red and near-infrared. Biochim Biophys Acta 933: 249–257

Wacker U (1993) Untersuchengen der Kopplung von H^{+}-Reaktionen und Elektronentranfervorgaengen im Photosystem II hoeherer Pflanzen. Thesis, Technische Universität Berlin, Germany

Wacker U, Haag E and Renger G (1990) Investigation of pH-change-patterns of photosystem-II membrane fragments from spinach. In: Baltscheffsky M (ed) Current Research in Photosynthesis, pp 869–872. Kluwer, Dordrecht

Wille B and Lavergne J (1982) Measurement of proton translocation in the thylakoids under flashing light using a spin-labelled amine. Photobiochem Photobiophys 4: 131–144

Witt HT, Schlodder E, Brettel K and Saygin Ö (1986) Reaction sequences from light absorption to the cleavage of water in photosynthesis. Routes, rates and intermediates. Photosynth Res 10: 453–471

Yocum CF (1992) The calcium and chloride requirements for photosynthetic water oxidation. In: Pecoraro VL (ed) Manganese Redox Enzymes, pp. 71–83. VCH, New York

Chapter 11

Introduction to the Photosystem II Reaction Center—Isolation and Biochemical and Biophysical Characterization

Kimiyuki Satoh
Department of Biology, Okayama University, Okayama 700, Japan

Summary

This chapter provides a brief summary of developments in research on the isolation, structure and function of a pigment-protein complex, which is now called the Photosystem II reaction center (PS II RC). The complex contains D1 and D2 proteins, cytochrome *b*-559 and the *psbI* gene product, presumably in an equimolar ratio. Two types of preparations, with different pigment stoichiometry, have been prepared from a variety of organisms, which are highly functional and stabilized; one contains 6 chlorophyll *a* and 2 β-carotene, the other contains 4 chlorophyll a and 1 β-carotene, per 2 pheophytin a molecules.

Because many constituents concerned with light-harvesting functions and secondary electron transport are depleted, or are nonfunctional, in the isolated complex, the RC can carry out a very limited number of functional activities of Photosystem II. This greatly facilitated detailed spectroscopic studies of the energy and electron transfer reactions.

However, there is an argument among different groups concerning the rate of primary photochemistry; i.e., does it occur at 3 ps or 21 ps? On the other hand, the separated charges on the primary radical pair, the primary donor (P680) and the pheophytin acceptor, are unstable in the isolated complex, because of the absence of the primary quinone acceptor, and thus they recombine with a half-time of 30–40 ns, forming the triplet state of P680 with a high yield. The structure and molecular interactions of P680 in the isolated PS II RC are discussed in this article, based on the EPR and FTIR properties of the triplet state of P680, formed by radical pair recombination.

Donald R. Ort and Charles F. Yocum (eds): Oxygenic Photosynthesis: The Light Reactions, pp. 193–211.

I. Introduction

The early processes occurring in Photosystem II (PS II) of oxygenic photosynthesis, catalyzed by higher plants, as well as algae and cyanobacteria, can be divided into several parts shown in the following reaction sequences: (1) Absorption of light quanta by distal and proximal antennae to form electronically excited states of pigments; (2) Trapping of excitation energy in the antenna system by a special chlorophyll a (Chl *a*) molecule(s) constituting the primary electron donor (P680) in the reaction center (RC); (3) Ejection of an electron from the singlet excited state of P680 and transfer to a pheophytin *a* (Pheo *a*) molecule; and (4) Stabilization of the separated charges on the primary radical pair, $P680^{+}Pheo\ a^{-}$, by the operation of secondary electron transport systems, i.e., on the reducing side, the electron proceeds from reduced Pheo *a* to plastoquinone (PQ) molecules, referred to as Q_A and Q_B, in a special environment, and on the oxidizing side, an electron is supplied to oxidized P680 from a water molecule, via a tyrosine residue on a constituent protein subunit (D1 protein), through the function of a manganese-cluster. A conceptual model for the organization of PS II, in terms of biochemical preparation, as well as the outline of the electron transport chain in this system, is given in Figs. 1 and 2.

In general, the RC in photosynthetic systems can be defined as a minimum unit capable of photochemical charge separation between the primary electron donor and the primary electron acceptor, followed by stabilization of separated charges by secondary electron transport reactions. In purple bacteria, a complex consisting of L, M and H subunits follows this criterion (Deisenhofer et al., 1985; Chang et al., 1986; Allen et al., 1987). However, for PS II of oxygenic photosynthesis, this kind of preparation has not yet been isolated. Instead, there are several such biochemical preparations of PS II as shown in Fig. 1. One type contains a full complement of PS II components which include large numbers of antenna pigments associated with both the peripheral (outer or bulk) antenna, such as light-harvesting chlorophyll a/b proteins, and the proximal (inner or core) antenna complexes (Satoh, 1993b). This type of preparation consists of membrane fragments, often called 'PS II particles' or more specifically 'BBY- or KM-particles' using the names of the individuals who first prepared

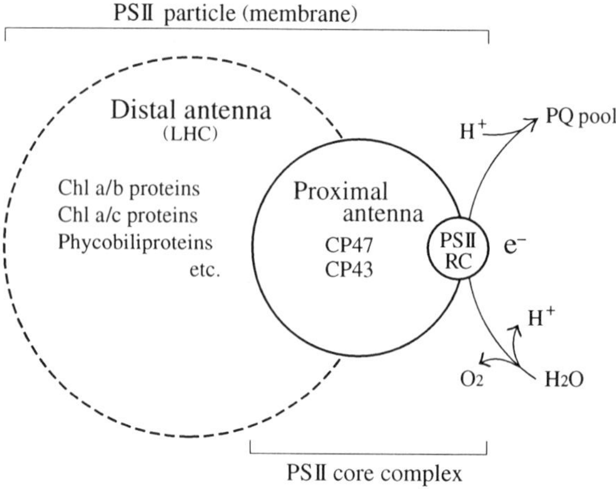

Fig. 1. A conceptual model for the organization of Photosystem II: Antenna systems and electron transport.

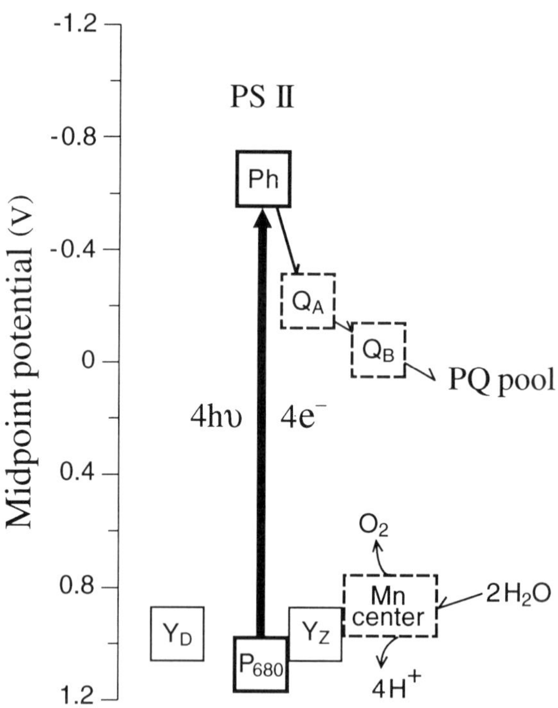

Fig. 2. Electron transport chain in PS II: The isolated PS II RC contains neither quinone acceptors (Q_A, Q_B) nor a manganese center.

Abbreviations: ADMR – absorbance-detected magnetic resonance; β-Car – β-carotene; BChl – bacteriochlorophyll; Chl – chlorophyll; Cyt *b*-559 – cytochrome *b*-559; EPR – electron paramagnetic resonance; FTIR – Fourier transform infrared; P680 – primary electron donor of PS II; Pheo – pheophytin; PQ – plastoquinone; PS I – Photosystem I; PS II – Photosystem II; Q_A – primary quinone acceptor; Q_B – secondary quinone acceptor; RC – reaction center

these materials (Berthold et al., 1981; Kuwabara and Murata, 1982). The other type of PS II preparation is the PS II core complex as described by Satoh and Butler (1978). It is a pigment-protein complex with a particle size of 400–500 kilodaltons (kDa), and contains six major polypeptide subunits of a hydrophobic character (Satoh, 1982, 1983, 1985; Satoh et al., 1983), which are encoded by chloroplast DNA in eukaryotic organisms (Ohyama et al., 1986; Shinozaki et al., 1986). Many other components, mostly of lower molecular weight, are also found in this type of preparation (Ghanotakis and Yocum, 1990; Ikeuchi, 1992). The core complex containing four manganese atoms per RC and retaining a peripheral protein of nuclear origin (*PsbO* gene product or 33-kDa manganese stabilizing protein) has also been isolated (Ikeuchi et al., 1985; Satoh et al., 1985; Tang and Satoh, 1985; Ghanotakis and Yocum, 1986; Yamada et al., 1987; Tang and Diner, 1994; see chapter 8). At present, this type of pigment-protein complex is the smallest structural unit capable of water oxidation by PS II. The PS II core complex contains 40–60 Chl *a* and about 10 β-carotene (β-Car) molecules per RC, but no chlorophyll *b* or xanthophyll. Most of the pigments in the core complex are bound to the two protein subunits that are encoded by *psbB* and *psbC* genes, the holocomplexes of which are often called CP47 and CP43 (CPa-1 and CPa-2), respectively (see Bricker, 1990). A photochemically active pigment-protein complex of PS II depleted of CP43 was also isolated and characterized (for example, Ghanotakis et al., 1989; Ghanotakis and Yocum, 1990).

The PS II complex depleted in both of the proximal antenna proteins has been isolated (Nanba and Satoh, 1987). This protein-pigment complex retains the activity of primary charge separation of PS II (Danielius et al., 1987; Nanba and Satoh, 1987; Okamura et al., 1987; Takahashi et al., 1987). However, in this preparation, the separated charges cannot be stabilized, since components responsible for secondary and tertiary electron transport, i.e., quinone acceptors (Q_A and Q_B), tyrosine donors (Y_Z and Y_D) and the manganese-cluster for the water splitting reaction, are not present (quinones) or are not fully functional (tyrosines). This chapter will address the isolation and some biochemical and biophysical characterizations of this protein-pigment complex, which is now called PS II RC, although this complex does not satisfy the criterion for a RC discussed above. The description will be focused on some specific aspects of the PS II RC, such as the structure of P680. For more details, see recent reviews by Satoh (1988, 1989, 1993a,b) and by Seibert (1993).

II. Isolation

In 1986, Nanba and Satoh succeeded in isolating a pigment-protein complex of PS II depleted in both of the proximal antenna components, CP47 and CP43, that retained photochemical activity, namely primary charge separation. This type of preparation has since been isolated from a wide variety of organisms including cyanobacteria, algae and other higher plants (Barber et al., 1987; Akabori et al., 1988; Demetriou et al., 1988; Gounaris et al., 1988, 1989; Dekker et al., 1989; Ghanotakis et al., 1989; Braun et al., 1990; Fotinou and Ghanotakis, 1990; Montoya et al., 1991).

Since the PS II RC is a membrane-embedded intrinsic protein complex, the procedures used for its isolation require extraction with a detergent. The purification, from extracts of thylakoid membranes, usually was conducted successfully in the presence of a low concentration of detergent using standard techniques developed for water-soluble proteins, such as ion-exchange chromatography, isoelectric focusing (Fig. 3), gel filtration, or sucrose density gradient centrifugation. Triton X-100 was used initially as the detergent for extraction and purification (Nanba and Satoh, 1987). However, this detergent was shown to cause difficulties in that it can modify spectroscopic properties of the complex, such as absorption and fluorescence spectra, and fluorescence lifetime or energy transfer processes (Seibert et al., 1988), as well as photochemical properties, under certain conditions (see Seibert, 1993). Although the use of Triton X-100 is highly advantageous in extracting the complex from thylakoid membranes, some improvements in the preparation have been made and it is now possible to provide highly functional and stabilized materials by replacing Triton X-100 with digitonin (Satoh and Nakane, 1989), dodecyl-β-D-maltoside (Chapman et al., 1988; Seibert et al., 1988; McTavish et al., 1989; Yruela et al., 1994), or other nonionic detergents (McTavish et al., 1989), after extraction and complete or partial chromatographic purification in the presence of a low concentration of Triton. The PS II RC was solubilized from the PS II core complex, with (Ikeuchi et al., 1985; Ghanotakis and Yocum, 1986) or without (Akabori et al., 1988; Dekker et al., 1989; Ghanotakis

Fig. 3. Isoelectric focusing of PS II RC (pH range, 4.5–6.0).

et al., 1989) Triton X-100. Akabori et al. (1988) used octyl-β-D-glucoside for solubilization, followed by sucrose density gradient centrifugation; Ghanotakis et al.(1989) used dodecyl-β-D-maltoside with a chaotropic reagent, $LiClO_4$, for solubilization, followed by ion-exchange column chromatography. A preparative method of purification directly from thylakoids using dodecylmaltoside and $LiClO_4$ was also reported (Fotinou and Ghanotakis, 1990). Preparations of PS II RC obtained using these different methods, however, exhibited essentially similar pigment and protein compositions (however, see Satoh, 1993a), as well as similar photochemical properties.

The isolated PS II RC described here is active in terms of primary charge separation, measured either by steady-state photoaccumulation of reduced primary acceptor (Pheo a^-) under continuous illumination in the presence of dithionite and methyl viologen (Nanba and Satoh, 1987) (see Fig. 8) or by direct optical detection of the formation of oxidized primary donor ($P680^+$) and Pheo a^- upon flash excitation (Danielius et al., 1987; Takahashi et al., 1987). Formation of the triplet state of P680 ($^3P680$) with a high yield also was observed in these preparations, either by optical spectroscopy (Takahashi et al., 1987), or by EPR spectroscopy, which detects the spin polarization characteristic of a radical-pair precursor (Okamura et al., 1987; Telfer et al., 1988; Ghanotakis et al., 1989) (see Fig. 9). The isolated complex contains no functional PQ, irrespective of the type of detergent used at the solubilization and purification stages (however, see later).

The purified PS II RC contains D1 and D2 proteins, α and β subunits of Cyt *b*-559 and the *psbI* gene product (Nanba and Satoh, 1987; Ikeuchi and Inoue, 1988; Webber et al., 1989). It also contains a number of cofactors important in the primary processes of PS II such as Chl *a* and β-Car, as well as a cytochrome *b*-559 (Cyt *b*-559) heme; this will be discussed in the following sections.

III. Chemical Composition and Polypeptide Sequence

A. Protein Subunits and Their Primary Sequences

Figure 4 shows the polypeptide profile of the isolated PS II RC, analyzed by sodium dodecyl sulfate-polyacrylamide gel electrophoresis (SDS-PAGE). The two polypeptides of about 30 kDa are named D1 and D2, respectively, as shown in the figure. The origin of the name 'D' comes from their 'diffuse' migration patterns in early SDS-PAGE experiments that used resolving gels containing no urea (see, Satoh, 1982; Satoh et al., 1983). Cyt *b*-559 consists of two subunits, i.e., an α-subunit of about 10 kDa, and a β-subunit of about 4 kDa, *psbE* and *psbF* gene products, respectively (see Cramer et al., 1986). A small polypeptide of about 4.8 kDa present in the complex was identified as the *psbI* gene product (Ikeuchi and Inoue, 1988; Webber et al., 1989). The D1 and D2 proteins tend to form dimeric aggregates in SDS-PAGE, with a size of about 60 kDa, as shown in Fig. 4.

The stoichiometry among the five subunits of the PS II RC prepared from spinach was analyzed, based on the amino acid composition of the isolated pigment-protein complex and the deduced primary sequence of each subunit, taken together with their experimentally determined N-terminal and C-terminal sequences (Satoh, 1992). The conclusion derived from this analysis is that the five polypeptides are present in equimolar ratios in the complex, although some arguments still exist concerning the copy number of Cyt *b*-559 in its native state.

The D1 and D2 proteins are homologous in amino acid sequence to the L and M subunits of the purple bacteria RC, for which the recent X-ray crystallographic structure determination (Deisenhofer et al., 1985; Chang et al., 1986; Allen et al., 1987) has provided a picture at atomic resolution of the detailed organization of the pigments and proteins. The amino acid sequence homology, supported by the fact that these two photosystems, i.e., PS II and the purple bacterial RC, are similar in their functional organization, especially on the reducing side, has

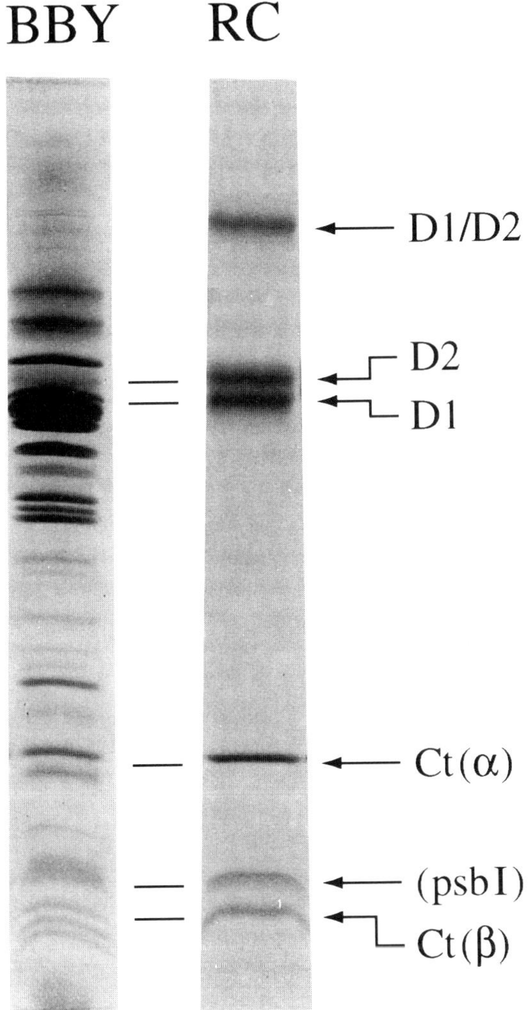

Fig. 4. Polypeptide subunits of PS II RC analyzed by SDS-PAGE: BBY, PS II membrane; RC, PS II RC

given strong support to the assumption that the structure of PS II RC is analogous to that of its purple bacteria counterpart and for constructing models for the molecular organization of PS II RC (Trebst, 1986; Svensson et al., 1991) (Fig. 4).

To date, probably more than 45 genes for the D1 protein (*psbA*), and more than 15 genes for the D2 protein (*psbD*), have been sequenced from a wide variety of organisms, ranging from cyanobacteria to higher plants (Svensson et al., 1991). The variations in deduced amino acid sequences are very small. By rough calculation, it can be said that more than 60–70% of the amino acids in the D1 and D2 proteins have remained the same throughout the long evolutionary time-span from cyanobacteria to higher plants; the lowering of the % homology value originates mostly from two sources; (1) the sequence homology in specific domains (modules) of proteins: for example, that for the N-terminal 25 amino acids for both D1 and D2 proteins (Fig. 5) and that for the C-terminal 8–16 amino acids for the D1 protein (Fig. 6), is low as compared with the other parts of the same proteins; and (2) there are large variations in the amino acid sequences for only one deduced sequence of a D1 protein, the putative product of the *psbAI* gene of *Synechocystis* PCC 6803: this is one out of three *psbA* genes in this organism which is not transcribed under conditions so far examined (Mohamed et al., 1993).

In both D1 and D2 proteins, the highest degree of sequence conservation is found in the putative membrane spanning D helix and the CD helices according to the molecular organization proposed by Trebst (1986). These regions are considered to be important in furnishing the site for primary photochemistry of PS II since the amino acids, supposedly important in coordinating co-factors of PS II RC, are almost completely conserved among species; e.g., the putative axial His ligands to the magnesium atoms of the primary donor Chl (His-198 for both D1 and D2 proteins) and to the ferrous-iron in the putative quinone (Q_A and Q_B)-iron acceptor complex (His-215 and His-272 on the D1 protein and His-215 and His-269 on the D2 protein) are conserved in all *psbA* genes; the amino acids such as Trp-254 on the D2 protein and Phe-255 on the D1 protein, both of which are proposed to interact with Q_A and Q_B quinones, respectively, are conserved in all species so far analyzed; the Tyr residue at position 161 on the D1 protein, which serves as the immediate donor to P680 and is responsible for EPR signal II (Signal IIvf (f)) (Debus et al., 1989), is conserved in all species so far reported, and, in a symmetrical way, there is another conserved Tyr which is responsible for the dark stable EPR signal II (Signal IIs) residing on the D2 protein as well (Debus et al., 1988). These are just a few, typical, examples of the conservation of amino acids in the subunits of PS II RC. For additional details, see chapter 8 by Bricker and Ghanotakis.

Another well-documented aspect of the gross organization of D1 and D2 proteins, based on the primary sequences, is that the amino acid sequences of proteins that are supposedly extruded into the lumenal space of thylakoids are highly conserved; e.g., polypeptide loops between the A and B helices

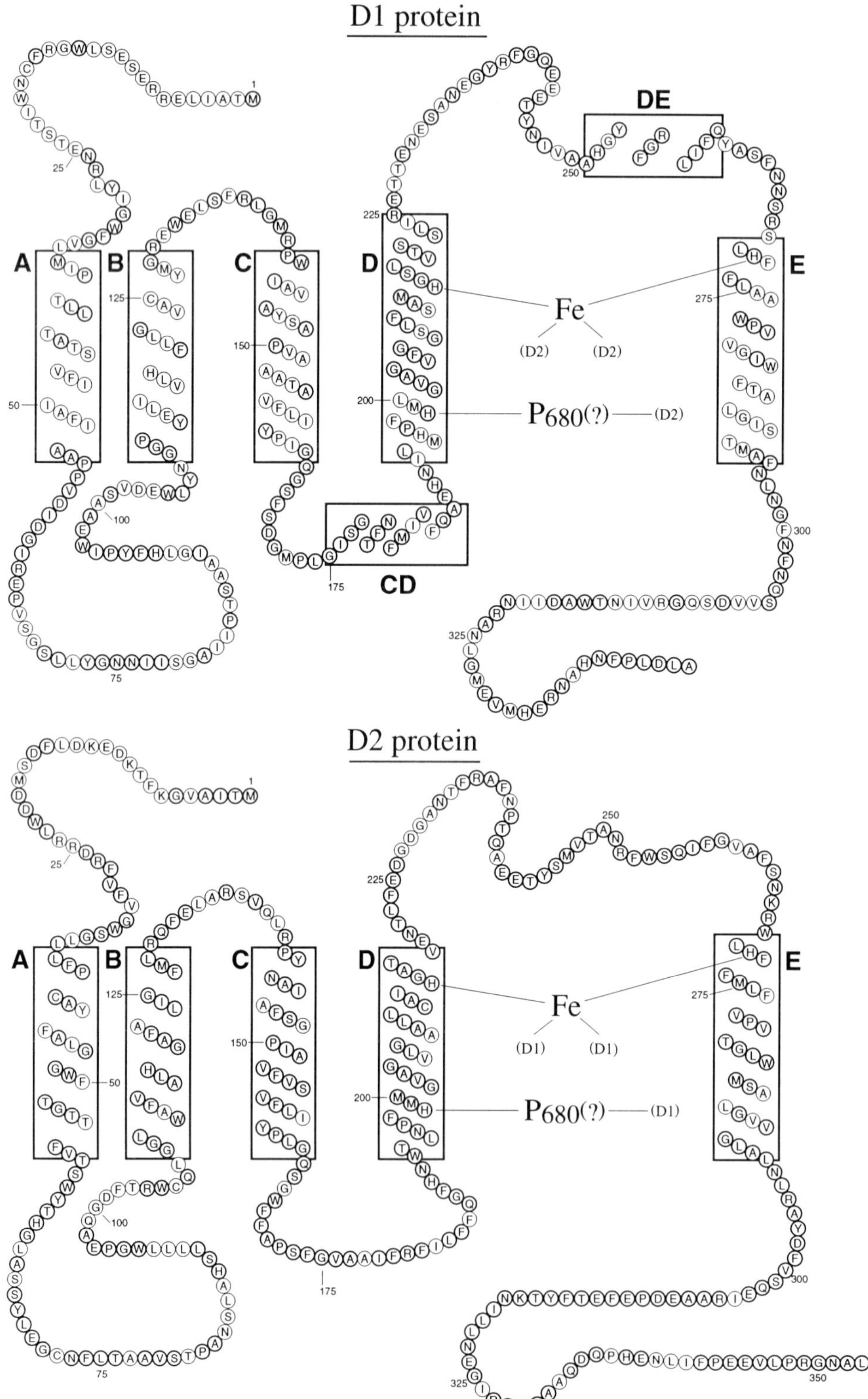

Fig. 5. Two-dimensional folding model of D1 and D2 proteins in PS II RC, as originally proposed by Trebst (1987). Bold letters correspond to the conserved amino acid residues in the deduced sequences of D1 and D2 proteins.

and between the C and D helices, as well as the C-terminal extension from the E helices, except for the C-terminal 8–16 amino acids of the D1 protein mentioned above (Fig. 6); this part of the D1 protein, but not of the D2 protein, is known to be cleaved immediately after translation by a protease present in the lumenal space (Taylor et al., 1988; Inagaki et al., 1989). Thus, it is not present in the mature protein. These observations suggest that the lumenal parts of these two proteins play an important role in the structure and function of PS II RC, possibly in organizing the water-splitting machinery or ligating the manganese cluster. The importance of the C-terminus and lumenal loops of proteins for manganese-binding is supported by a mutation study of *Scenedesmus obliquus* (Metz et al., 1986), as well as by recent site-directed mutagenesis experiments using *Synechocystis* PCC 6803 (Debus, 1992; Nixon and Diner, 1992). The structure and function of the other subunits of PS II RC are discussed in Chapter 8.

A preliminary result suggested that a complex consisting of D1 and D2 proteins, but depleted in Cyt *b*-559 and the *psbI* gene product, carries all the pigments in the PS II RC and furnishes the site for primary charge separation in PS II (Tang et al., 1990). Consequently, Cyt *b*-559 and the *psbI* gene product seem not to be essential for the occurrence of primary photochemistry in PS II. These components, on the other hand, are supposed to have important functions in the stable assembly of PS II RC in vivo.

B. Pigments and Redox Cofactors

Absorption spectrum. The PS II RC contains only Chl *a*, Pheo *a*, β-Car and Cyt *b*-559 heme. These components absorb in the visible region, so the electronic absorption spectrum of the RC shows peaks at: 665–685 nm and 550–650 nm due to Pheo *a* and Chl *a*, 545 nm due to Pheo *a*, 460–530 nm due to β-Car, 435 nm due to Chl *a* and 415 nm due to Pheo *a* and Cyt *b*-559 (Fig. 7). When the sample is reduced by dithionite, the contribution of Cyt *b*-559 in the spectrum becomes apparent at 559 nm (α-band). The assignment of absorption contributions from Chl *a* other than P680 and Pheo *a*, both on active and inactive branches of the electron transport chain, is a subject of much debate at present. This is particularly true for assigning the position of two Pheo *a* bands; either 672 nm (Montoya et al., 1993), 676 nm (Otte et al., 1992; van Kan et al., 1990), or 680–682 nm (Nanba and Satoh, 1987; Braun et al., 1990; Yruela et al., 1994) for Pheo *a* on the active branch.

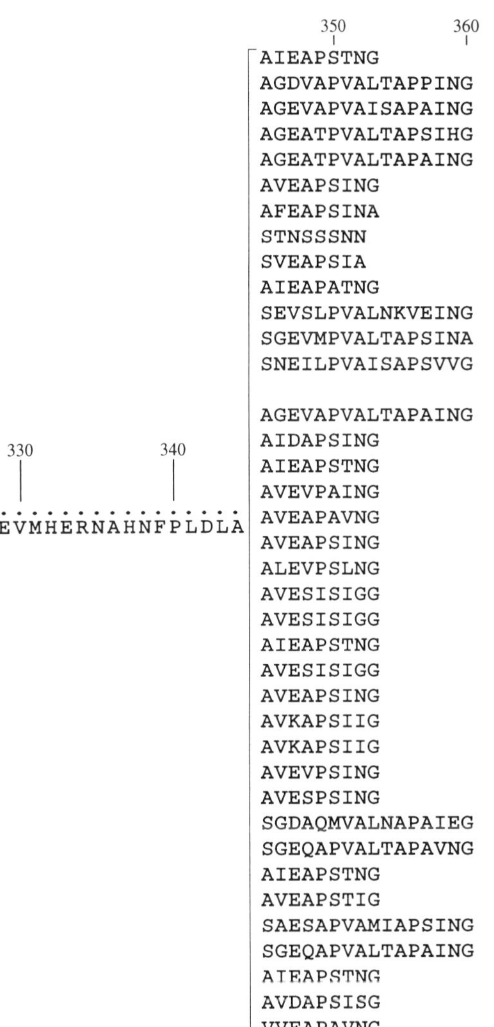

Fig. 6. Some examples of the deduced C-terminal sequence of D1 protein. The amino acids between positions 329 and 344 are conserved in every sequence so far analyzed, except for the *psbAI* gene product of *Synechocystis* PCC 6803, which is eliminated in this comparison.

Pigment stoichiometry. PS II RC preparations are now widely accepted to contain two Pheo *a* per RC; the primary photochemical reaction requires only one out of two Pheo *a* in the RC, as demonstrated by comparing the molar ratio between the chemically estimated and photochemically reduced amounts of pigment, using appropriate molar absorption

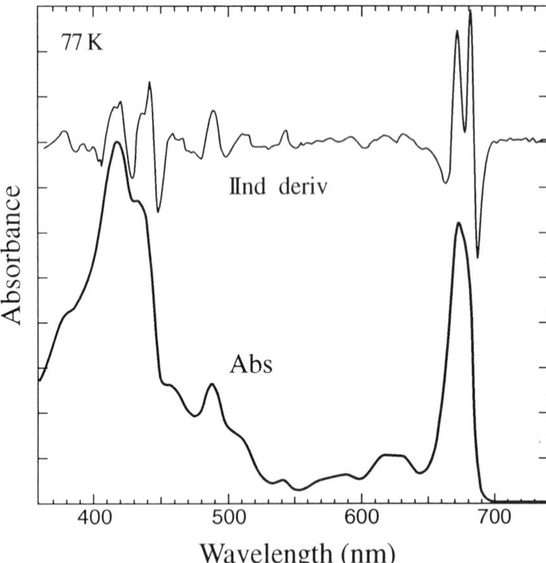

Fig. 7. Absorption spectrum of the PS II RC at 77 K and its 2nd derivative.

coefficients (Nanba and Satoh, 1987). The number of Chl *a* and *β*-Car per 2 Pheo *a* (per RC) in the isolated RC complex, reported in the literature, varies between 2–8 and 0.5–1 respectively (see, Satoh, 1993a). However, typical preparations of PS II RC so far analyzed can be classified into two types in terms of pigment stoichiometry. In one type of preparation, the Chl *a* to Pheo *a* ratio is 2, consistent with its purple bacteria counterpart (Barber et al., 1987; Montoya et al., 1993). The other type of preparation, which seems to be more native than the other, judging from the procedure for preparation, contains two additional Chl *a* and one additional *β*-Car and thus the Chl *a* to Pheo *a* ratio is 3 and the *β*-Car to Pheo *a* ratio is 1 (Kobayashi et al., 1989). Therefore, the stoichiometric ratio of Chl *a* to Pheo *a* of 2, in the former type of preparation, can be explained by a selective pigment release caused by detergent treatment, although, in principle, the original number of pigments associated with the PS II RC in its native state is difficult to interpret based on these biochemical analyses using detergents (see Satoh, 1993a).

As mentioned, the isolated PS II RC complex contains virtually no PQ-9, which serves as the secondary and tertiary electron acceptor in PS II. A preparation of PS II RC that contains one stoichiometric amount of PQ per 2 Pheo *a* has been reported (Akabori et al., 1988). However, the quinone in this preparation is not functional in terms of charge stabilization, so it can be assumed that the PQ is bound non-specifically to the protein.

The number of Cyt *b*-559 in the PS II RC complex with 6 Chl *a* and 2 Pheo *a* was estimated to be one per RC using a milli-molar absorption coefficient of Cramer et al. (1986). However, there are arguments about the number of Cyt *b*-559 hemes associated in the PS II RC in vivo (Miyazaki et al., 1989).

Other components. The presence of at least one non-heme iron in the PS II RC, as in its purple bacteria counterpart, was reported for a preparation from pea (Barber et al., 1987). However, quantitation of metals in most other preparations revealed that PS II RC contains only one iron which should correspond to the heme-iron of Cyt *b*-559 present in the complex (Satoh, unpublished). The isolated PS II RC complex contains no manganese and thus the complex is inactive in oxygen evolution, although some of the manganese-binding sites are predicted to be present on the D1 and D2 proteins (see later). As mentioned previously, the D1 subunit contains a conserved Tyr-residue at position 161, which serves as the secondary electron donor in PS II, as well as a conserved Tyr-161 on the D2 protein. However, these components seem not to be fully active, possibly because of modifications in the organization of the RC due to detergent action during preparation (see later).

IV. Functional Aspects of PS II RC Organization—Energy and Electron Transfer Processes

Because many constituents necessary for light-harvesting functions and secondary electron transport are depleted, as previously mentioned, isolated PS II RC can carry out a very limited number of functional activities. This, on the other hand, has greatly facilitated detailed spectroscopic studies of energy and electron transfer in PS II, although precautions had to be taken to account for the modification caused by detergent action (van Mieghem, 1993).

A. Energy Transfer and Trapping

In the purple bacteria RC, the energy transfer processes are relatively simple, since the energy level of the primary donor BChl dimer is at least 150 meV below those of the other accessory pigments.

This situation results in unidirectional transfer of excitation energy to the primary donor in approximately 100 fs, followed by BPheo reduction at the rate of about $(3\ ps)^{-1}$ at room temperature. In contrast, in the PS II RC the energy levels of Q_Y transitions for Chl *a* and Pheo *a* overlap in such a way that these pigments are separated by no more than about 30 meV and the excited singlet state of P680 is not likely to be a deep trap for excitation energy at room temperature. This may result in extensive reversible energy transfers between pigments, upon excitation, which may not have physiological significance. There have been several reports of energy transfer processes in the isolated PS II RC with lifetimes ranging from several tens to hundreds of ps, using spectroscopic techniques such as time resolved fluorescence spectroscopy (Mimuro et al., 1988; Govindjee et al., 1990; Roelofs et al., 1991), hole-burning spectroscopy (Tang et al., 1990, 1991) and ultra-fast transient absorption spectroscopy (Wasielewski et al., 1989; Durrant et al., 1992; Hastings et al., 1992; Durrant et al., 1993). Some of the results obtained in the earlier analyses were interpreted as arising from energy transfer processes that are not associated with charge separation (see Seibert, 1993). Using hole-burning spectroscopy, Tang et al. (1990, 1991) determined that the decay kinetics of the Chl absorption around 670 nm and that of Pheo *a*, which are 12 and 50 ps at 4.2 K, respectively, possibly was due to energy transfer to P680. These lifetimes are two or three orders of magnitude slower than those reported for the purple bacteria RC and thus it is argued that the so-called trapping limited model, in which there is a rapid equilibration of excitation energy between the antenna pigments and P680, followed by the observed trapping of the excitation energy by radical pair formation, may not be appropriate for the PS II RC.

On the other hand, two faster kinetic components with lifetimes of 400–600 fs and 3.5 ps after possible direct excitation of P680 at 695 nm were also reported (Hastings et al., 1992), which are assignable to the decay of a delocalized P680 excited singlet state. Furthermore, Durrant et al. (1992), exciting the PS II RC at both 612 and 694 nm, concluded from difference transient absorption changes that the excitation energy equilibrates between the majority of pigments at a rate of $(100 \pm 50\ fs)^{-1}$ and proposed that it is appropriate to describe the kinetics within the PS II RC in terms of the trapping limited model. The authors, basing their conclusion on the quantitative analysis of this process, suggested that the initial P680 excited singlet state is delocalized over at least two chlorins and that this delocalization lasts for at least 200 fs, suggesting that P680 is a dimeric state of Chl (see later). Further analysis with improved temporal resolution is needed in order to cover the gaps among different dynamic processes taking place in the PS II RC, such as vibrational relaxation, energy transfer and energy trapping.

B. Primary Photochemistry

The primary photochemistry that takes place in the isolated PS II RC is a charge separation between P680 and Pheo *a*, which results in the formation of a biradical state. This state is formed in approximately 100 ps in PS II particles retaining their antenna complexes (Nuijs et al., 1986; Schatz et al., 1988). However, when P680 is excited via exciton transfer from the antenna system, the rate of charge separation is expected to be limited by the trapping of excitation energy by the RC. In isolated PS II RC, where no associated light harvesting complexes are present, excitation has been shown to result in the formation of a biradical in less than 25 ps with a near unity quantum yield (Danielius et al., 1987; Booth et al., 1991). There are some direct measurements of the absorption transients of pigments in the isolated PS II RC and there is a controversy concerning the time for charge separation (see Seibert, 1993).

Wasielewski et al. (1989), exciting at 610 nm and probing at 674 nm at room temperature (277 K), have measured a rate of $(3.0 \pm 0.6\ ps)^{-1}$ for the appearance of the oxidized state of P680, as observed by the absorption increase at 820 nm, and assigned this to primary charge separation between P680 and Pheo *a*. The time constant for decay of the lowest singlet excited state of P680, monitored at 650 nm, was 2.6 ± 0.6 ps; this coincided with the appearance of oxidized P680 observed at 820 nm, within experimental error. An estimate of the lifetime of excited states from photochemical hole-burning studies is 1.8–2.1 ps at 4.2 K (Tang et al., 1991). These data are roughly consistent with the 2.6 ± 0.6 ps lifetime obtained from transient absorption measurements at 4 °C, if the cooling effect, i.e., increased electronic coupling, etc., is taken into account (see Seibert, 1993).

On the other hand, Durrant et al. (1992), working with a temporal resolution of about 160 fs, employing sufficiently low excitation levels to avoid multiphoton processes, have observed at 295 K a 21-ps decaying component. The process(es) responsible for this component was found to produce bleaching of a Pheo ground state absorption band at 545 nm and the simultaneous appearance of a Pheo anion band at 460 nm, as well as of a negative peak at 681 nm, assignable to the formation of the $P680^+Pheo^-$ biradical; the relative amplitudes of these peaks are 49 (at 681 nm): 1 (at 545 nm) : 6.7 (at 460 nm), essentially in agreement with those from steady-state observations in Fig. 8 (Nanba and Satoh, 1987; Hastings et al., 1992). A lower limit of 60% of the final Pheo reduction was reported to occur at this rate when excited at 694 nm (Durrant et al., 1992). The rate of oxidation of P680 in the isolated PS II RC, monitored by loss of Chl stimulated emission, also supported the conclusion that P680 oxidation primarily occurs with an effective rate of $(21\ ps)^{-1}$ (Durrant et al., 1993). Thus, at this moment, there is a controversy over whether the charge separation occurs in 3 ps or 21 ps in the PS II RC. The critical points in this argument center on how to eliminate contributions of the rate-limiting processes of energy transfer and relaxation in interpreting the data. Since, at room temperature, excitons are in equilibrium between pigments, only the average trapping time would be observed by experiment. This limits the apparent rate of charge separation.

No evidence for the presence of an intermediate species in electron transport, as may exist in the purple bacteria RC (Lauterwasser et al., 1991), has been obtained at this stage of analysis of the PS II RC.

C. Biradical Recombination

Because of the absence of a primary quinone acceptor, biradicals formed by primary charge separation are unstable in the isolated PS II RC. Therefore, a recombination of the primary radical pair takes place either to the excited singlet state, via a radical pair mechanism to the triplet state, or directly to the ground state by a non-radiative mechanism. The half-time of the recombination in the isolated PS II RC was estimated to be 30–40 ns (Danielius et al., 1987; Takahashi et al., 1987). Similar results were obtained by observing the kinetics of recombination fluorescence (Mimuro et al., 1988; Seibert et al., 1988; Booth et al., 1990; Govindjee et al., 1990; Roelofs et al., 1991). The half-time of biradical recombination in the isolated PS II RC, however, is significantly longer than the 1–6 ns and 20–30 ns

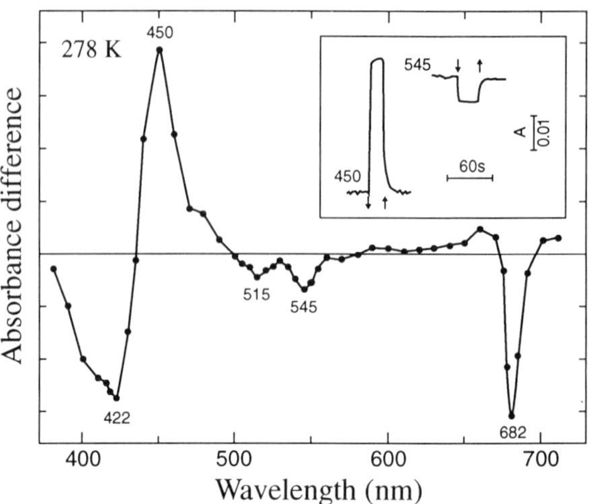

Fig. 8. Photoaccumulation of pheophytin anion in the PS II RC under steady-state illumination in the presence of dithionite and methyl viologen (from Nanba and Satoh, 1987): Inset, kinetic traces at 450 and 545 nm.

reported by others for more intact systems. This discrepancy is most probably caused by one or more of the following: variations in antenna size, the presence of charge on Q_A, and the presence of Fe^{2+}/Q_A^- interactions (Hansson et al., 1988; Gulyaev, 1990; Roelofs et al., 1991). Some studies have suggested that the rate of recombination may be more complex, possibly due to variation in the conformational states of PS II RCs, multiple relaxations of metastable radical pair state, or equilibration between singlet and triplet radical pair states etc.. (Booth et al., 1990, 1991; Roelofs and Holzwarth, 1990; Freiberg et al., 1994).

D. Formation of Triplet States

Recombination between $P680^+$ and $Pheo^-$ results in the formation of the triplet state of P680, with a temperature-dependent quantum yield, (23% at 276 K; 80% at 10 K), which can be detected either by optical spectroscopy (Takahashi et al., 1987) or by EPR spectroscopy (Okamura et al., 1987) (Fig. 8). The low yield of the primary radical pair in intact PS II preparations was attributed to the presence of Q_A^-, which causes a repulsive coulombic interaction with the radical pair. When Q_A is doubly reduced, it presumably loses its negative charge by protonation, resulting in an increase in triplet yield (van Mieghem et al., 1991; van Mieghem, 1993). Thus, complete removal of quinone acceptors in the PS II RC results in formation of the P680 triplet with a high yield

(Takahashi et al., 1987) (Fig. 9).

PS II RC contains 1–2 *β*-Car, and a triplet state of Car was observed with a half-time of about 16 ps (with a yield of 3%), which is faster than the decay time of $P680^+Pheo^-$ (Takahashi et al., 1987). The *β*-Car in the isolated PS II RC is in all-trans form, as shown by resonance Raman spectroscopy (Fujiwara et al., 1987), and would not be expected to quench the triplet state of P680. Thus the shorter rise-time compared with radical pair recombination is consistent with this interpretation.

E. Other Redox Reactions Associated with PS II RC

Cytochrome b-559. Isolated PS II RC contains an additional redox component, Cyt *b*-559, which is present in the oxidized form and does not respond to illumination. However, in a quinone-reconstituted sample, Cyt *b*-559 is reduced rapidly in the light, either by steady-state illumination or by flash excitation (Satoh et al., 1990). Under this condition, the reduced cytochrome recombines with $P680^+$ with a half-time of 2 ms (Satoh et al., 1990). However, the meaning of this light reaction in vitro is not very well understood. The functioning of Cyt *b*-559 in the photoprotection of PS II RC is a current issue of discussion (see Chapter 13).

Secondary electron donor (Y_Z). Radical pair recombination between $P680^+$ and $Pheo^-$ is much faster than electron donation to $P680^+$ from Y_Z, which is in the 1 s time range when oxygen evolution is inactivated (Mathis et al., 1989). Thus this reaction would not be expected to occur under normal conditions in the isolated PS II RC. When DBMIB or SiMo, which serve as the electron acceptor for PS II, is present a 5-*μ*s transient absorption decay phase appears as observed at 820 nm or at 430 nm that has been attributed to electron donation from Y_Z (Mathis et al., 1989; Takahashi et al., 1989). However, this kinetic species is detected in only about 5% of the centers. On the other hand, Nugent et al. (1989), using EPR spectroscopy, observed a signal attributable to the formation of Y_Z^+ in isolated PS II RC. Again, the fraction of the signal is very small, suggesting that a large part of Y_Z is inactive, possibly because of modifications to protein structure (van Mieghem, 1993), although the Tyr residue is apparently present in the complex.

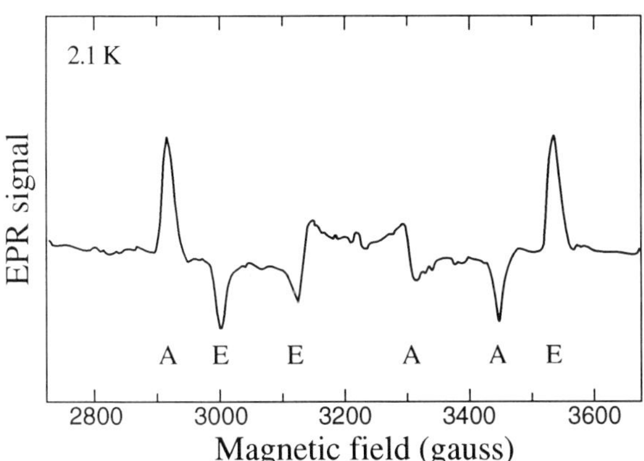

Fig. 9. Formation of the spin-polarized triplet state of P680 in PS II RC (from Okamura et al., 1987).

Reactivity of artificial electron donors and acceptors. The isolated PS II RC can oxidize artificial electron donors such as 1,5-diphenylcarbazide, Mn, NH_2OH, and KI, as well as reduce SiMo and 2,6-dichlorophenol indophenol dye in a dichlorophenyldimethylurea (DCMU) sensitive fashion (Chapman et al., 1988). However, the degree of reactivity depends largely on the preparations and on reaction conditions.

V. Structural Aspects of Organization—Structure of the Primary Donor (P680)

As mentioned previously, there is extensive sequence homology between the L and M subunits of the purple bacteria RC and the D1 and D2 proteins of PS II RC (Trebst, 1986; Michel and Deisenhofer, 1988). This homology is also apparent around the binding site of the primary donor. Thus, it is generally believed that the structure of P680 is analogous with that of its purple bacteria counterpart. However, one critical difference between P680 and the primary donor of the purple bacteria RC is that the former generates a high oxidizing potential (about +1 V), approximately twice that of the purple bacteria; this oxidizing potential is crucial for almost all organisms on this planet since it drives the water-splitting reaction which gives rise to oxygen evolution. Thus, although the major polypeptides of PS II RC are homologous with those of purple bacteria, some modifications must exist to allow the generation and stabilization of this high oxidizing potential at the binding site of primary electron donor. From this view point, it is interesting to note that the structure of PS II RC is shown to have some unique characteristics when compared to those of the bacterial RC (Rutherford, 1986; Babcock, 1987; Renger, 1987; Hansson and Wydrzynski, 1990; Svensson et al., 1990).

The primary donor of the purple bacteria RC is known to form a dimeric structure called the special pair, in which a pair of BChl molecules are symmetrically arranged in parallel, overlapping each other in ring I of the BChl molecule (Deisenhofer et al., 1985; Allen et al., 1987); this arrangement causes a strong exciton coupling, resulting in a large splitting of the far-red absorption bands.

The overall similarity between PS II RC and its purple bacteria counterpart, both in the sequence of constituent subunits mentioned above and in the functional organization i.e., reaction rates and components involved, and in particular, the fact that the His ligands for the primary donor in the purple bacteria RC are conserved in the D1 and D2 proteins of PS II RC (Trebst, 1986; Sayre et al., 1986; Michel and Deisenhofer, 1988) naturally lead to a model in which P680 is a dimer of Chl *a*, oriented perpendicular to the plane of thylakoid membranes as in purple bacteria.

In accordance with this proposal, spectroscopic measurements using isolated PS II RC suggested that P680 consists of two Chl molecules that show exciton interaction, and are responsible for both the main long-wavelength absorption band at about 680 nm and a shoulder at 684 nm, although these wavelengths vary somewhat among preparations (Tetenkin et al., 1989; van Kan et al., 1990; Tang et al., 1990, 1991; Braun et al., 1990). However, the maximum possible interaction (exciton coupling) energy, estimated in these studies, between RC pigments in PS II (142 cm^{-1}) or potential P680 dimer components (85–100 cm^{-1}) are much lower than the 600–1500 cm^{-1} range observed for the purple bacteria special pair (Knapp et al., 1985; Scherer and Fischer, 1987). These results were interpreted to suggest that the geometry of the special pair of PS II RC, if it is present, is quite different from that of purple bacterial photosystems, e.g., the chlorin rings are farther apart or rotated so as to result in minimization of pair interactions between the two Chls.

On the other hand, Davis et al. (1979) showed that the optical, EPR, and redox properties of monomeric Chl *a* could be varied, depending on the solvent used for measurement, and suggested the possibility that P680 is a ligated Chl *a* monomer whose characteristics were controlled by the protein environment. Values of zero-field splitting parameters derived from the spectrum of the P680 triplet (Fig. 9) are identical to those for the in vitro monomeric Chl triplet state, and not smaller, as would be expected for a dimer (Rutherford et al., 1981; Hoff, 1986). The small Stark effect (Losche et al., 1988) also favors this proposal. The observation of bleaching of a single keto C=O band upon formation of the P680 triplet state in the resonance Raman spectrum (Moënne-Locoz et al., 1990) supports a Chl *a* monomer structure for P680.

Additional support for the monomeric nature of triplet P680 was provided by an EPR orientation measurement (Rutherford, 1985; van Mieghem et al., 1991). For the EPR spectrum of triplet P680 in the oriented PS II RC, the angle of the pigment which carries the triplet state was analyzed and it was concluded that the orientation maximum for the Z peak was tilted approximately 30° out of the plane of the membrane, as demonstrated for oriented PS II particles (Rutherford, 1985; van Mieghem et al., 1991). These observations show that the structure of P680 is significantly different from that of its purple bacteria counterpart, in spite of the fact that the preserved His ligands in the L and M subunits are conserved in the D1 and D2 proteins. Thus, the existing results can be interpreted to support one of the following possibilities: (1) One out of two PS II Chls that are structural analogs of the BChl special pair which carries the triplet state is rotated from 90° to 30° with respect to the membrane plane; (2) P680 is a Chl monomer, structurally equivalent to one of the BChl monomers in the purple bacterial RC (accessory BChl) but is in-plane rotated 45° compared with the bacterial case (van Mieghem et al., 1991); or (3) P680 is a dimer as in the purple bacterial case, but triplet transfer takes place and the triplet state resides on a Chl *a* monomer in the isolated PS II RC which is equivalent to one of the BChl monomers in the purple bacteria RC. However, in discussing the monomeric or dimeric structure of P680, the definition has to be specific about the molecular species (P680, *P680, P680$^+$, ^{3}P680), dynamic time range of the state, degree of coupling, etc.. In fact, Den Blanken et al. (1983) have presented ADMR evidence based on comparisons with Chl *a* monomers and dimers that indicates that P680 is dimeric in its singlet ground state but that the triplet is localized on a monomeric Chl.

In order to elucidate the structure of P680,

molecular interactions in the isolated PS II RC were investigated by Noguchi et al. (1993), by detecting light-induced FTIR difference spectra upon formation of the triplet state (Fig. 10). The ground and triplet state difference spectrum of P680 was analyzed by comparing it with that of pure Chl *a* in different organic solvents. The negative peaks at 1669 and 1707 cm^{-1} accompanied by the positive peaks at 1627 and 1659 cm^{-1} in the triplet difference spectrum in vivo were assigned to the keto C=O stretching mode; the feature at 1669 cm^{-1} is hydrogen-bonded, while the other, absorbing at 1707 cm^{-1} is free in a highly nonpolar environment. The appearance of these two pairs of bands originating from different molecular environments, indicates that P680 has a dimeric structure analogous to that of the purple bacteria primary donor. From the band positions of the keto C=O (1627–1707 cm^{-1} region) and carbomethoxy C=O (1716–1723 cm^{-1} region) stretches, the hydrogen-bonding properties of these two Chl *a* molecules (P_1 and P_2) were deduced to be asymmetrical. In one Chl *a* molecule (P_1), both the keto and carbomethoxy C=O groups form hydrogen bonds, while in the other Chl *a* molecule (P_2), the keto C=O is not hydrogen-bonded whereas the carbomethoxy C=O group is hydrogen bonded. The temperature dependence of the intensity ratios of the keto C=O bands revealed that the triplet state is equilibrated between the two Chl *a* molecules, with or without hydrogen-bonding at the keto C=O, with an energy gap of 8.4 ± 0.7 meV (Noguchi et al., 1993). This implies that most of the triplet population is localized on one Chl *a* molecule at low temperature, e. g., 86% at 80 K, in which both of the C=O groups, keto and carbomethoxy, are hydrogen-bonded (P_1). The crystallographic data for hydrogen-bond interactions in the purple bacteria RC (Deisenhofer et al., 1985; Allen et al., 1987), show that, in *Rhodopseudomonas viridis* the keto C=O of P_L (BChl mainly associated with L-subunit) is hydrogen-bonded to Thr L248, and no hydrogen-bonding occurs for keto C=O of P_M (BChl mainly associated with M-subunit). The closest amino acid residue to the keto group is Ile M282; in *Rhodobacter sphaeroides*, keto C=O groups of both P_L and P_M are not H-bonded and the closest amino acids are Met L248 and Ile M284, respectively. In view of the similarities (sequence and folding) between D1/D2 and L/M, we can assume that the amino acid residues proximal to the keto C=O groups in P680 are as follows; the residues in the D1 and D2 subunits in PS II corresponding to the

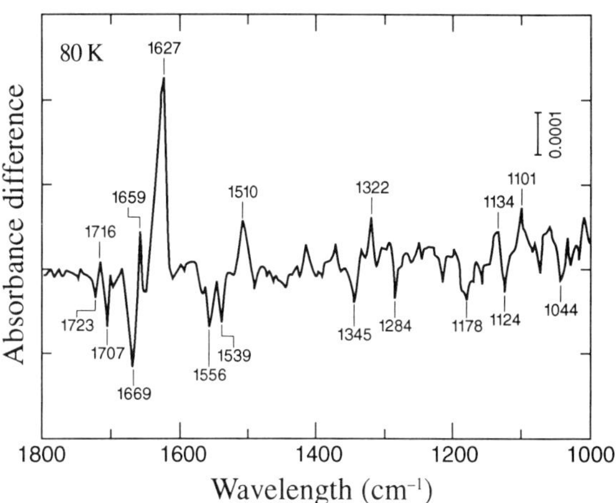

Fig. 10. FTIR absorption difference spectrum of PS II RC upon formation of the triplet state of P680 (from Noguchi et al., 1993).

bacteria counterparts adjacent to the keto C=O groups for both P_1 and P_2 Chls are Ile D1-290 and Val D2-287, respectively, and neither residue is capable of hydrogen-bonding (Fig. 11). However, in D1, there are two hydrogen-bonding amino acids adjacent to Ile D1-290, i.e., Ser D1-291 and Thr D1-292, while no hydrogen-bonding amino acids are found around Val 287 in the D2 protein, consistent with the conclusion, derived from the FTIR measurement, that the keto C=O of one out of the two Chls is located in a nonpolar environment. Thus, it can be speculated that the most probable candidate for the amino acid residue that forms a hydrogen bond with one of the two keto C=O groups in P680 is either Ser D1-291 or Thr D1-292, or both. On the other hand, in the carbomethoxy C=O region, a single negative band appeared at 1723 cm^{-1} responding to P680 triplet formation, and this frequency value indicates hydrogen-bond formation. Based on the sequence homology and the crystallographic data for hydrogen-bonding, it is predicted that Thr D1-286 and Ser D2-283 are the hydrogen-bonding partners of the two carbomethoxy C=O groups of the two Chls of P680 (Noguchi et al., 1993). Figure 12 presents a model of P680 and its hydrogen-bonding interactions with apoproteins that is obtained by summarizing the above discussion. The model also includes previous results on the orientation of the P680 triplet that show that the triplet Z axis is tilted by about 30° to the membrane (van Mieghem et al., 1991; van der Vos et al., 1992). This deviation in the orientation is speculated to be caused by the fact that the hydrogen-bonding amino acid on the D1 protein for the keto C=O group is shifted by one or two residues from Ile

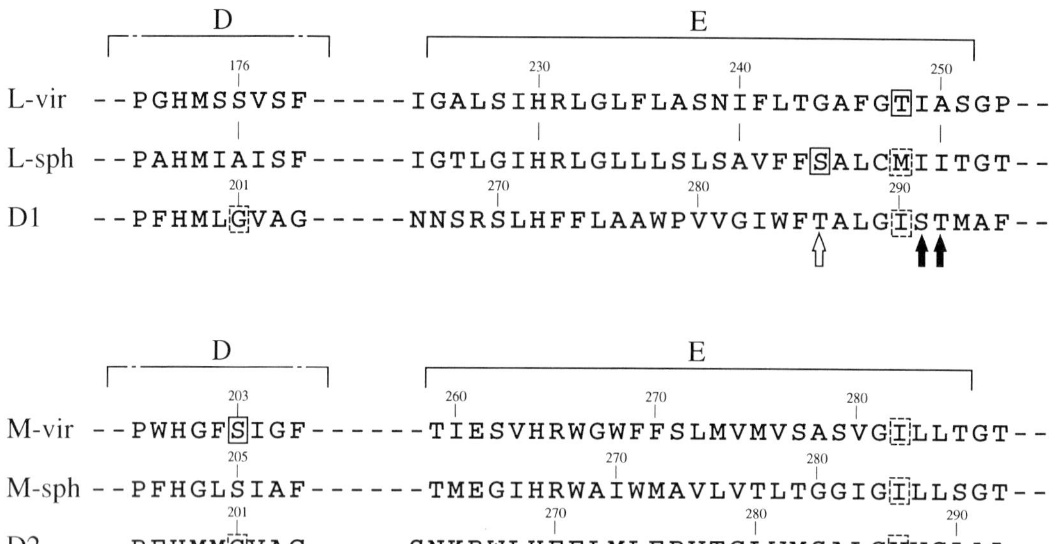

Fig. 11. Partial sequence comparison between L and M subunits of purple bacteria RC and D1 and D2 proteins of spinach: L-vir and M-vir, L and M subunits of *Rhodopseudomonas viridis*; L-sph and M-sph, L and M subunits of *Rhodobacter spheroides*.

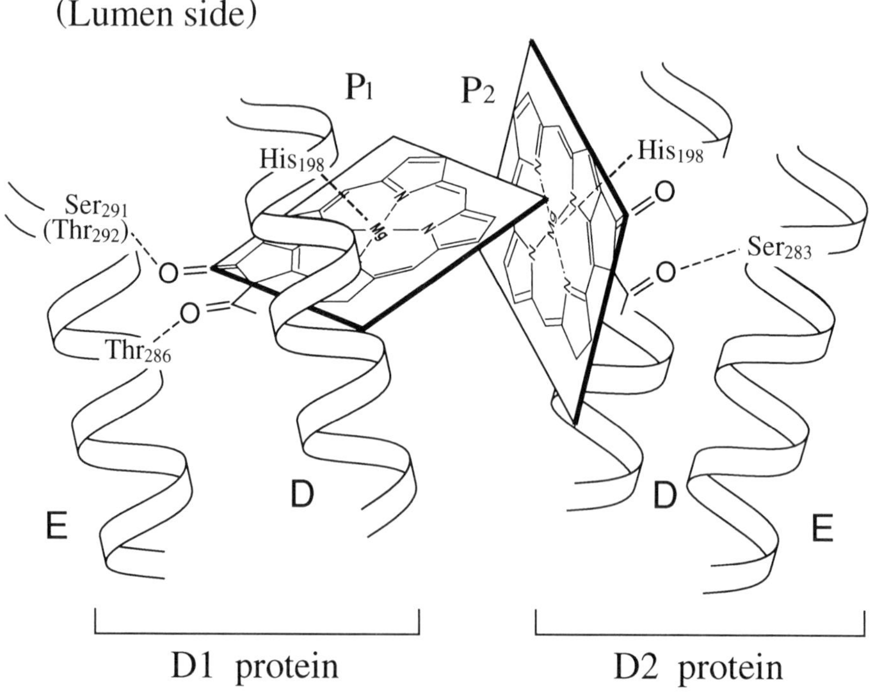

Fig. 12. Structural model of P680 in PS II RC (modified from Noguchi et al., 1993).

D1-290, as discussed above, which corresponds to the residue hydrogen-bonding to the keto C=O in purple bacteria.

As a consequence of the arrangement of the two Chl *a* (P_1 and P_2 in Fig. 10) discussed above, the dimeric structure of P680 appears considerably different from that of its purple bacteria counterpart. The Chl planes are no longer oriented parallel to each other, and the angle between them is so large that the overlap between the macrocycles will be much smaller. Thus, it is readily apparent that this configuration would cause a weak exciton coupling, resulting in small band splitting in the red absorption spectrum. It is also likely that this asymmetric structure, as well as the hydrogen-bonding characteristics of the Chl dimer might be the reasons why P680 has a higher redox potential than other primary donors, since hydrogen-bonding interactions to the keto C=O position of BChl are believed to be important for the redox-potential of the primary donor in the case of *Rhodobacter sphaeroides* (Murchison et al., 1992). The interactions may also cause the triplet localization.

VI. Perspectives

Identification of the PS II RC by successful isolation of a minimum unit of pigment-protein complex, capable of charge separation between P680 and Pheo *a*, has greatly facilitated detailed analysis of the structure, function and biodynamics of PS II. This includes spectroscopic characterizations of energy and electron transfer reactions in primary processes, biochemical and molecular genetic analysis of the structure and assembly of the PS II complex, as well as detailed investigations of co-factor binding to the protein subunits by site-directed mutagenesis, etc.

Arguments, however, have not been settled concerning the number of Chl *a* and *β*-Car associated with the RC or the rate of primary charge separation in PS II. Successful reconstitution of quinone acceptors, Q_A and Q_B, which are depleted in the isolated RC complex, has not been achieved on the reducing side. On the other hand, a large part of the secondary electron donor (Y_Z) seems to be inactive on the oxidizing side, possibly because of modifications to protein structure, although the Tyr-161 residue is apparently present in the isolated complex. Thus, development of other refined preparations of the PS II RC will be necessary in order to analyze the structure and function.

The area of crystallographic analysis for PS II is left behind, compared to that for PS I and light-harvesting complexes, as well as for purple bacterial photosystems. Therefore, the preparation of crystals of PS II RC, or of more intact complexes of PS II, suitable for X-ray crystallographic study is urgently needed in order to elucidate unique structures, such as P680 and the Mn-cluster. In order to understand the dynamic nature of organization in PS II, elucidation of the molecular mechanism of light-regulated turn-over of D1 subunit of the PS II RC will be another target of interest in future studies.

References

Akabori K, Tsukamoto H, Tsukihata J, Nagatsuka T, Motokawa O and Toyoshima Y (1988) Disintegration and reconstitution of Photosystem II reaction center core complex. I. Preparation and characterization of three different types of subcomplex. Biochim Biophys Acta 932: 345–357

Allen JP, Feher G, Yeates TO, Komiya H and Rees DC (1987) Structure of the reaction center from *Rhodobacter sphaeroides* R-26: The co-factors. Proc Natl Acad Sci USA 84: 5730–5734

Babcock GT (1987) The photosynthetic oxygen-evolving process. In: Amesz J (ed) New Comprehensive Biochemistry, pp 125–128. Elsevier, Amsterdam

Barber J, Chapman DJ and Telfer A (1987) Characterization of a PS II reaction centre isolated from the chloroplasts of *Pisum sativum*. FEBS Lett 220: 67–73

Berthold DA, Babcock GT and Yocum CF (1981) A highly resolved, oxygen-evolving Photosystem II preparation from spinach thylakoid membranes. EPR and electron-transport properties. FEBS Lett 134: 231–234

Booth PJ, Crystall B, Giorgi LB, Barber J, Klug, DR and Porter G (1990) Thermodynamic properties of D1/D2/cytochrome *b*-559 reaction centres investigated by time-resolved fluorescence measurements. Biochim Biophys Acta 1016: 141–152

Booth PJ, Crystall B, Ahmad I, Barber J, Porter G and Klug DR (1991) Observation of multiple radical pair states in Photosystem 2 reaction centers. Biochemistry 30: 7573–7586

Braun P, Greenberg BM and Scherz A (1990) D1-D2-cytochrome *b*-559 complex from the aquatic plant *Spirodela oligorrhiza*: Correlation between complex integrity, spectroscopic properties, photochemical activity, and pigment composition. Biochemistry 29: 10376–10387

Bricker TM (1990) The structure and function of CPa-1 and CPa-2 in Photosystem II. Photosynth Res 24: 1–13

Chang C-H, Tiede DM, Tang J, Smith U, Norris JR and Schiffer M (1986) Structure of *Rhodobacter sphaeroides* R-26 reaction center. FEBS Lett 205: 82–86

Chapman DJ, Gounaris K and Barber J (1988) Electron-transport properties of the isolated D1-D2-cytochrome *b*-559 Photosystem II reaction center. Biochim Biophys Acta 933: 423–431

Cramer WA, Theg SM and Widger WR (1986) On the structure and function of cytochrome *b*-559. Photosynth Res 10: 393–403

Danielius RV, Satoh K, van Kan PJM, Plijter JJ, Nuijs, AM and van Gorkom HJ (1987) The primary reaction of Photosystem II in the D1-D2-cytochrome *b*-559 complex. FEBS Lett 213: 241–244

Davis MS, Forman A and Fajer J (1979) Ligated chlorophyll cation radicals: Their function in Photosystem II of plant photosynthesis. Proc Natl Acad Sci USA 76: 4170–4174

Debus RJ (1992) The manganese and calcium ions of photosynthetic oxygen evolution. Biochim Biophys Acta 1102: 269–352

Debus RJ, Barry BA, Babcock GT and McIntosh L (1988) Site-directed mutagenesis identifies a tyrosine radical involved in the photosynthetic oxygen-evolving system. Proc Natl Acad Sci USA 85: 427–430

Debus RJ, Barry BA, Sithole I, Babcock GT and McIntosh L (1989) Directed mutagenesis indicates that the donor to $P680^+$ in Photosystem II is Tyr-161 of the D1 polypeptide. Biochemistry 26: 9071–9074

Deisenhofer J, Epp O, Miki K, Huber R and Michel H (1985) Structure of the protein subunits in the photosynthetic reaction centre of *Rhodopseudomonas viridis* at 3Å resolution. Nature 318: 618–624

Dekker JP, Bowlby NR and Yocum CF (1989) Chlorophyll and cytochrome *b*-559 content of the photochemical reaction center of Photosystem II. FEBS Lett 254: 150–154

Demetriou C, Lockett CJ and Nugent JHA (1988) Photochemistry in the isolated Photosystem II reaction center core complex. Biochem J 252: 921–924

Den Blanken HJ, Hoff AJ, Jongenelis APJM and Diner BA (1983) High-resolution triplet-minus-singlet absorbance difference spectrum of Photosystem II particles. FEBS Lett 157: 21–27

Durrant JR, Hastings G, Joseph DM, Barber J, Porter G and Klug DR (1992) Subpicosecond equilibration of excitation energy in isolated Photosystem II reaction centers. Proc Natl Acad Sci USA 89: 11632–11636

Durrant JR, Hastings G, Joseph DM, Barber J, Porter G and Klug DR (1993) Rate of oxidation of P680 in isolated photosystem 2 reaction centers monitored by loss of chlorophyll stimulated emission. Biochemistry 32: 8259–8267

Fotinou C and Ghanotakis DF (1990) A preparative method for the isolation of the 43 kDa, 47 kDa and the D1-D2-Cyt *b*559 species directly from thylakoid membranes. Photosynth Res 25: 141–145

Freiberg A, Timpmann K, Moskalenko AA and Kuznetsova NY (1994) Pico- and nanosecond fluorescence kinetics of Photosystem II reaction centre and its complex with CP47 antenna. Biochim Biophys Acta 1184: 45–53

Fujiwara M, Hayashi H, Tasumi M, Kanaji M, Koyama Y and Satoh K (1987) Structural studies on a Photosystem II reaction-center complex consisting of D-1 and D-2 polypeptides and cytochrome *b*-559 by resonance Raman spectroscopy and high-performance liquid chromatography. Chem Lett 2005–2008

Ghanotakis DF and Yocum CF (1986) Purification and properties of an oxygen evolving reaction center complex from Photosystem II membranes: A simple procedure utilizing a nonionic detergent and elevated ionic strength. FEBS Lett 197: 244–248

Ghanotakis DF and Yocum CF (1990) Photosystem II and oxygen-evolving complex. Annu Rev Plant Physiol 41: 255–276

Ghanotakis DF, de Paula JC, Demetriou DM, Bowlby NR, Petersen J, Babcock GT and Yocum CF (1989) Isolation and characterization of the 47 kDa protein and the D1-D2-cytochrome *b*-559 complex. Biochim Biophys Acta 974: 44–53

Gounaris K, Chapman DJ and Barber J (1988) The interaction between the 33 kDa manganese-stabilising protein and the D1/D2 cytochrome *b*-559 complex. FEBS Lett 234: 374–378

Gounaris K, Chapman DJ and Barber J (1989) Isolation and characterization of a D1/D2/cytochrome *b*-559 complex from *Synechocystis* 6803. Biochim Biophys Acta 973: 296–301

Govindjee, van de ven M, Preston C, Seibert M and Gratton E (1990) Chlorophyll a fluorescence lifetime distributions in open and closed Photosystem II reaction center preparations. Biochim Biophys Acta 1015: 173–179

Gulyaev BA (1990) Organization of Pigment-Protein Complexes and Transfer of Excitation Energy in Photosynthetic Membranes. Thesis. Moscow State University, Moscow.

Hansson Ö and Wydrzynski T (1990) Current perceptions of Photosystem II. Photosynth Res 23: 131–162

Hansson Ö, Duranton J and Mathis P (1988) Yield and lifetime of the primary radical pair in preparations of Photosystem II with different antenna size. Biochim Biophys Acta 932: 91–96

Hastings G, Durrant JR, Barber J, Porter G and Klug DR (1992) Observation of pheophytin reduction in photosystem two reaction centers using femtosecond transient absorption spectroscopy. Biochemistry 31: 7638–7647

Hoff AJ (1986) Triplets: Phosphorescence and magnetic resonance. In: Govindjee, Amesz J and Fork DC (eds) Light Emission by Plants and Bacteria, pp 225–265. Academic Press, Orlando

Ikeuchi M (1992) Subunit proteins of Photosystem II. Bot Mag Tokyo 105: 327–373

Ikeuchi M and Inoue Y (1988) A new Photosystem II reaction center component (4.8 kDa protein) encoded by chloroplast genome. FEBS Lett 241: 99–104

Ikeuchi M, Yuasa M and Inoue Y (1985) Simple and discrete isolation of an O_2-evolving PS II reaction center complex retaining Mn and the extrinsic 33 kDa protein. FEBS Lett 185: 316–322

Inagaki N, Fujita S and Satoh K (1989) Solubilization and partial purification of a thylakoidal enzyme of spinach involved in the processing of D1 protein. FEBS Lett 246: 218–222

Knapp EW, Fischer SF, Zinth W, Sander M, Kaiser W, Deisenhofer J and Michel H (1985) Analysis of optical spectra from single crystals of *Rhodopseudomonas viridis* reaction centers. Proc Natl Acad Sci USA 82: 8463–8467

Kobayashi M, Maeda H, Watanabe T, Nakane H and Satoh K (1990) Chlorophyll a and β-carotene content in the D1/D2/cytochrome *b*-559 reaction center complex from spinach. FEBS Lett 260: 138–140

Kuwabara T and Murata N (1982) Inactivation of photosynthetic oxygen evolution and concomitant release of three polypeptides in the Photosystem II particles of spinach chloroplasts. Plant Cell Physiol 23: 533–539

Lauterwasser C, Finkele U, Scheer H and Zinth W (1991) Temperature dependence of the primary electron transfer in photosynthetic reaction centers from *Rhodobacter sphaeroides*. Chem Phys Lett 183: 471–477

Lösche M, Satoh K, Feher G and Okamura MY (1988) Stark effect in PS II RCs from spinach. Biophys J 53: 270a

Mathis P, Satoh K and Hansson Ö (1989) Kinetic evidence for the function of Z in isolated Photosystem II reaction centers. FEBS Lett 251: 241–244

McTavish H, Picorel R and Seibert M (1989) Stabilization of isolated Photosystem II reaction center complex in the dark and in the light using polyethylene glycol and an oxygen-scrubbing system. Plant Physiol 89: 452–456

Metz JG, Pakrashi HR, Seibert M and Arntzen CJ (1986) Evidence for a dual function of the herbicide-binding D1 protein in Photosystem II. FEBS Lett 205: 269–294

Metz JG, Nixon PJ, Rogner M, Brudvig GW and Diner BA (1989) Directed alteration of the D1 polypeptide of Photosystem II: Evidence that tyrosine-161 is the redox component, Z, connecting the oxygen-evolving complex to the primary electron donor, P680. Biochemistry 28: 6960–6969

Michel H and Deisenhofer J (1988) Relevance of the photosynthetic reaction center from purple bacteria to the structure of Photosystem II. Biochemistry 27: 1–7

Mimuro M, Yamazaki I, Itoh S, Tamai N and Satoh K (1988) Dynamic fluorescence properties of D1-D2-cytochrome *b*-559 complex isolated from spinach chloroplasts: Analysis by means of the time-resolved fluorescence spectra in picosecond time range. Biochim Biophys Acta 933: 428–486

Miyazaki A, Shina T, Toyoshima Y, Gounaris K and Barber J (1989) Stoichiometry of cytochrome b-559 in Photosystem II. Biochim Biophys Acta 975: 142–147

Möenne-Locoz P, Robert B and Lutz M (1990) Structure of the primary reactants in Photosystem II: Resonance Raman studies of D1D2 particles. In: Baltscheffsky M (ed) Current Research in Photosynthesis, Vol I, pp 423–426. Kluwer Academic Pub, Dordrecht

Mohamed A, Eriksson J, Osiewacz HD and Jansson C (1993) Differential expression of the *psbA* genes in the cyanobacterium *Synechocystis* 6803. Mol Gen Genet 238: 161–168

Montoya G, Yruela I and Picorel R (1991) Pigment stoichiometry of a D1-D2-Cyt *b*559 complex isolated from the higher plant *Beta vulgaris*. FEBS Lett 283: 255–258

Montoya G, Cases R, Yruela I and Picorel R (1993) Spectroscopic characterization of two forms of the D1-D2-cytochrome *b*559 complex from sugar beet. Photochem Photobiol 58: 724–729

Murchison HA, Alden RG, Allen JP, Peloquin JM, Taguchi AKW, Woodbury NW and Williams JC (1992) Mutations designed to modify the environment of the primary electron donor of the reaction center from *Rhodobacter sphaeroides*; Phe to Leu at L167 and His to Phe at L168. Biochemistry 31: 11029–11037

Nabedryk E, Andrianambinintsoa S, Berger G, Leonhard M, Mäntele W and Breton J (1990) Characterization of bonding interactions of the intermediary electron acceptor in the reaction center of Photosystem II by FTIR spectroscopy. Biochim Biophys Acta 1016: 49–54

Nanba O and Satoh K (1987) Isolation of a Photosystem II reaction center consisting of D-1 and D-2 polypeptides and cytochrome *b*-559. Proc Natl Acad Sci USA 84: 109–112

Nixon PJ and Diner BA (1992) Aspartate 170 of the Photosystem II reaction center polypeptide D1 is involved in the assembly of the oxygen-evolving manganese cluster. Biochemistry 31: 942–948

Noguchi T, Inoue Y and Satoh K (1993) FTIR studies on the triplet state of P680 in the Photosystem II reaction center: Triplet equilibrium within a chlorophyll dimer. Biochemistry 32: 7186–7195

Nugent JHA, Telfer A, Demetriou C and Barber J (1989) Electron transfer in the isolated Photosystem II reaction centre complex. FEBS Lett 255: 53–58

Nuijs AM, van Gorkom HJ, Plitjer JJ and Duysens LNM (1986) Primary-charge separation and excitation of chlorophyll *a* in Photosystem II particles from spinach as studied by picosecond absorbance-difference spectroscopy. Biochim Biophys Acta 848: 167–175

Ohyama K, Fukuzawa H, Kohchi T, Shirai H, Sana T, Sano S, Umesono K, Shiki Y, Takeuchi M, Chang Z, Aota S, Inokuchi H and Ozeki H (1986) Chloroplast gene organization deduced from complete sequence of liverwort *Marchantia polymorpha* chloroplast DNA. Nature 322: 572–574

Okamura MY, Satoh K, Isaacson RA and Feher J (1987) Evidence of the primary charge separation in the D1-D2 complex of Photosystem II from spinach; EPR of the triplet state. In: Biggins J (ed) Progress in Photosynthesis Research, Vol 1, pp 379–381. Nijhoff Publishers, Dordrecht

Otte SCM, van der Vos R and van Gorkom HJ (1992) Steady state spectroscopy at 6 K of the isolated Photosystem II reaction centre: Analysis of the red absorption band. J Photochem Photobiol 15: 5–14

Renger G (1987) Biological exploitation of solar energy by photosynthetic water splitting. Angew Chem Int Ed Engl 26: 643–660

Roelofs TA and Holzwarth AR (1990) In search of a putative long-lived relaxed radical pair state in closed Photosystem II. Biophys J 57: 1141–1153

Roelofs TA, Gilbert M, Shuvalov VA and Holzwarth AR (1991) Picosecond fluorescence kinetics of the D_1-D_2-cyt-*b*-559 Photosystem II reaction center complex. Energy transfer and primary charge separation processes. Biochim Biophys Acta 1060: 237–244

Rutherford AW (1985) Orientation of EPR signals arising from components in Photosystem II membranes. Biochim Biophys Acta 807: 189–201

Rutherford AW (1986) How close is the analogy between the reaction centre of Photosystem II and that of purple bacteria? Biochem Soc Trans 14: 15–17

Rutherford AW, Paterson DR and Mullet JE (1981) A light-induced spin-polarized triplet detected by EPR in Photosystem II reaction centers. Biochim Biophys Acta 635: 205–214

Satoh K (1982) Fractionation of thylakoid-bound chlorophyll-protein complexes by isoelectric focussing. In: Edelman M, Hallick RP and Chua N-H (eds) Methods in Chloroplast Molecular Biology, pp 845–856. Elsevier Biomedical, Amsterdam

Satoh K (1983) Photosystem II reaction center complex purified from higher plants. In: Inoue Y, Crofts AR, Govindjee, Murata N, Renger G and Satoh K (eds) The Oxygen Evolving System of Photosynthesis, pp 27–38. Academic Press, Tokyo

Satoh K (1985) Protein-pigments and Photosystem II reaction center. Photochem Photobiol 42: 845–853

Satoh K (1988) Reality of P680 chlorophyll protein—

Identification of the site of primary photochemistry in oxygenic photosynthesis. Physiol Plant 72: 209–212

Satoh K (1989) Molecular organization in Photosystem II reaction center. In: Norris JR (ed) Photochemical Energy Conversion, pp 238–250. Elsevier Biomedical, Amsterdam

Satoh K (1992) Organization of the Photosystem II reaction center. In: Argyroudi-Akoyunoglou JH (ed) Regulation of Chloroplast Biogenesis, pp 375–382. Plenum Press, New York

Satoh K (1993a) Isolation and properties of the Photosystem II reaction center. In: Deisenhofer J and Norris JR (eds) The Photosynthetic Reaction Center, Vol I, pp 289–318. Academic Press, San Diego

Satoh K (1993b) Molecular organization of the photochemical apparatus of oxygenic photosynthesis. In: Shima A et al. (eds) Frontiers of Photobiology, pp 3–11. Elsevier Sci Pub, Amsterdam

Satoh K and Butler WL (1978) Low temperature spectral properties of subchloroplast fractions purified from spinach. Plant Physiol 61: 373–379

Satoh K and Nakane H (1989) Refined purification and characterization of the D1-D2 reaction center of Photosystem II In: Baltscheffsky M (ed) Current Research in Photosynthesis, Vol 1, pp 271–274. Kluwer Academic Publishers, Dordrecht

Satoh K, Nakatani HY, Steinback KE, Watson J and Arntzen CJ (1983) Polypeptide composition of a Photosystem II core complex; Presence of a herbicide-binding protein. Biochim Biophys Acta 724: 142–150

Satoh K, Ohno T and Katoh S (1985) An oxygen-evolving complex with a simple subunit structure —'a water-plastoquinone oxidoreductase'— from the thermophilic cyanobacterium *Synechococcus* sp. FEBS Lett 180: 326–330

Satoh K, Hansson Ö and Mathis P (1990) Charge recombination between stabilized $P680^+$ and reduced cytochrome b-559 in quinone-reconstituted PS II reaction center. Biochim Biophys Acta 1016: 121–126

Sayre RT, Andersson B and Bogorad L (1986) The topology of a membrane protein: The orientation of the 32 kd Qb-binding chloroplast thylakoid membrane protein. Cell 37: 601–608

Schatz GH, Brock H and Holzwarth AR (1988) Kinetic and energetic model for the primary processes in Photosystem II. Biophys J 54: 397–405

Scherer POJ and Fischer SF (1987) Model studies to low-temperature optical transitions of photosynthetic reaction centers. II. *Rhodobacter sphaeroides* and *Chloroflexus aurantiacus*. Biochim Biophys Acta 891: 157–164

Seibert M (1993) Biochemical, biophysical, and structural characterization of the isolated Photosystem II reaction center complex. In: Deisenhofer J and Norris JR (eds) The Photosynthetic Reaction Center, Vol I, pp 319–356. Academic Press, San Diego

Seibert M, Picorel R, Rubin AB and Connolly JS (1988) Spectral, photophysical, and stability properties of isolated Photosystem II reaction center. Plant Physiol 87: 303–306

Shinozaki K, Ohme M, Tanaka M, Wakasugi T, Hayashida N, Matsubayashi T, Zaita N, Chunwongse J, Obokata J, Yamaguchi-Shinozaki K, Ohto C, Torazawa K, Meng BY, Sugita M, Deno H, Kamogashira T, Yamada K, Kusuda J, Takaiwa F, Katoh A, Tohdoh N, Shimada H and Sugiura M (1986) The complete nucleotide sequence of the tobacco chloroplast genome: Its gene organization and expression. EMBO J 5: 2043–2049

Svensson B, Vass I, Cedergren E and Styring S (1990) Structure of donor side components in Photosystem II predicted by computer modelling. EMBO J 9: 2051–2059

Svensson B, Vass I and Styring S (1991) Sequence analysis of the D1 and D2 reaction center proteins of Photosystem II. Z Naturforsch 46c: 765–776

Takahashi Y, Hansson Ö, Mathis P and Satoh K (1987) Primary radical pair in the Photosystem II reaction centre. Biochim Biophys Acta 893: 49–59

Takahashi Y, Satoh K and Itoh S (1989) Silicomolybdate substitutes for the function of a primary electron acceptor and stabilizes charge separation in Photosystem II reaction center complex. FEBS Lett 255: 133–138

Tang D, Jankowiak R, Seibert M, Yocum CF and Small GJ (1990) Excited-state structure and energy-transfer dynamics of two different preparations of the reaction center of Photosystem II: A hole-burning study. J Phys Chem 94: 6519–6522

Tang D, Jankowiak R, Seibert M and Small GJ (1991) Effects of detergent on the excited state structure and relaxation dynamics of the Photosystem II reaction center: A high resolution hole burning study. Photosynth Res 27: 19–29

Tang X-S and Diner BA (1994) Biochemical and spectroscopic characterization of a new oxygen-evolving Photosystem II core complex from the cyanobacterium *Synechocystis* PCC 6803. Biochemistry 33: 4594–4603

Tang X-S and Satoh K (1985) The oxygen-evolving Photosystem II core complex. FEBS Lett 179: 60–64

Tang X-S, Fushimi K and Satoh K (1990) D1-D2 complex of the Photosystem II reaction center from spinach: Isolation and partial characterization. FEBS Lett 273: 257–260

Taylor MA, Packer JCL and Bowyer JR (1988) Processing of the D1 polypeptide of the Photosystem II reaction centre and photoactivation of a low fluorescence mutant (LF-1) of *Scenedesmus obliquus*. FEBS Lett 237: 229–233

Telfer A, Barber J and Evans MCW (1988) Oxidation-reduction potential dependence of reaction centre triplet formation in the isolated D1/D2/cytochrome *b*-559 Photosystem II complex. FEBS Lett 232: 209–213

Tetenkin VL, Gulvaev BA, Seibert M and Rubin AB (1989) Spectral properties of stabilized D1/D2/cytochrome *b*-559 Photosystem II reaction center complex. Effects of Triton X-100, the redox state of pheophytin, and β-carotene. FEBS Lett 250: 459–463

Trebst A (1986) The topology of plastoquinone and herbicide binding peptides of Photosystem II in the thylakoid membrane. Z Naturforsch 41c: 240–245

van der Vos R, van Leeuwen PJ, Braun P and Hoff AJ (1992) Analysis of the optical absorbance spectra of D1-D2-cytochrome *b*-559 complexes by absorbance-detected magnetic resonance. Structural properties of P680. Biochim Biophys Acta 1140: 184–198

van Kan PJM, Otte SCM, Kleinherenbrink FAM, Nieveen MC, Aartsma TJ and van Gorkom HJ (1990) Time-resolved spectroscopy at 10 K of the Photosystem II reaction center: Deconvolution of the red absorption band. Biochim Biophys Acta 1020: 146–152

van Mieghem F (1993) Photochemistry and structural aspects of the Photosystem II reaction centre. Thesis Univ. Wageningen

van Mieghem FJE, Satoh K and Rutherford AW (1991) A chlorophyll tilted 30° relative to the membrane in the

Photosystem II reaction centre. Biochim Biophys Acta 1058: 379–385

Wasielewski MR, Johnson DG, Seibert M and Govindjee (1989) Determination of the primary charge separation rate in isolated Photosystem II reaction centers with 500-fs time resolution. Proc Natl Acad Sci USA 86: 524–528

Webber AN, Packman L, Chapman DJ, Barber J and Gray JC (1989) A fifth chloroplast-encoded polypeptide is present in the Photosystem II reaction centre complex. FEBS Lett 242: 259–262

Yamada Y, Tang X-S, Itoh S and Satoh K (1987) Purification and properties of an oxygen-evolving Photosystem II reaction-center complex from spinach. Biochim Biophys Acta 891: 129–137

Yruela I, van Kan PJM, Müller MG and Holzwarth AR (1994) Characterization of a D1-D2-cyt *b*-559 complex containing 4 chlorophyll *a*/2 pheophytin *a* isolated with the use of $MgSO_4$. FEBS Lett 339: 25–30

Chapter 12

Structure, Dynamics, and Energy Conversion Efficiency in Photosystem II

Bruce A. Diner

Central Research and Development Department, Experimental Station, E. I. Du Pont de Nemours and Company, Wilmington, Delaware 19880-0173, USA

Gerald T. Babcock

Department of Chemistry, Michigan State University, East Lansing, Michigan 48824, USA

Donald R. Ort and Charles F. Yocum (eds): Oxygenic Photosynthesis: The Light Reactions, pp. 213–247.

Summary

In the absence of an X-ray crystallographic structure for Photosystem II (PS II), the reaction centers of the purple non-sulfur photosynthetic bacteria have been serving as surrogate models for the probing of structure-function relationships in PS II. The folding of the polypeptide subunits is thought to be very similar between the two types of reaction centers and the pheophytins and quinones which function as electron acceptors occupy highly conserved binding sites. These redox cofactors also operate at very nearly the same oxidation-reduction potentials in both systems.

The need to oxidize water as the electron donor in PS II has driven the evolutionary divergence in what was undoubtedly a common ancestral reaction center, resulting in a greater conservation of the excitation energy in the stable charge-separated products, 68% for PS II and 35% for the purple bacterium, *Rb. sphaeroides*. Despite the homologies in pigment composition and in primary protein structure, the primary electron donor of Photosystem II, P680, appears to be a much more weakly coupled dimer of chlorophylls than is its bacterial homologue, resulting in a shallower trap for excitation energy and a higher photochemical conversion efficiency in PS II. Strong evidence also exists that the C2 symmetry of the bacterial special pair chlorophyll is not conserved in PS II and that the spin density of the oxidized donor is much closer to the redox active tyrosine Y_Z (D1-Tyr161) than to Y_D (D2-Tyr160), accounting for the much more rapid oxidation of Y_Z by $P680^+$. Despite the location of these tyrosines in homologous stretches of sequence of the D1 and D2 polypeptides, their environments and their interactions with neighboring residues are clearly different. These interactions may influence the way in which proton transfer is coupled to electron transfer, accounting in part for the contrasting redox behavior of Y_Z and Y_D.

As similar as are the binding domains of the primary and secondary quinone electron acceptors, Q_A and Q_B of PS II with those of the purple bacteria, there are marked differences in the pathways for protonation of the Q_B quinol and in the coordination environment of the non-heme iron located midway between the two quinones. Two of the iron coordination positions in PS II are readily exchangeable and are able to accommodate a wide range of small coordinating molecules, including bicarbonate. The nature of the bound molecule influences the coupling of proton transfer to interquinone electron transfer.

I. Introduction

A. Overview of PS II Function

Photosytem II uses photons in the visible region of the spectrum to produce a charge separation between the reaction-center chlorophyll complex, P680, and the quinone acceptor, Q_A. The oxidized product $P680^+$ is among the strongest oxidants generated in biological systems and has a reduction potential (~1.2V vs. SHE) sufficient to oxidize water to molecular oxygen; the protons and electrons made available in this process are ultimately used to reduce CO_2 to more complex organic molecules in the carbon-fixing reactions of photosynthesis. Water is a thermodynamic sink, hence its ubiquity, and evolution of the capability to use H_2O as a reducing substrate for PS II photochemistry allowed photosynthesis to proliferate.

The oxidizing capacity generated by the photo-

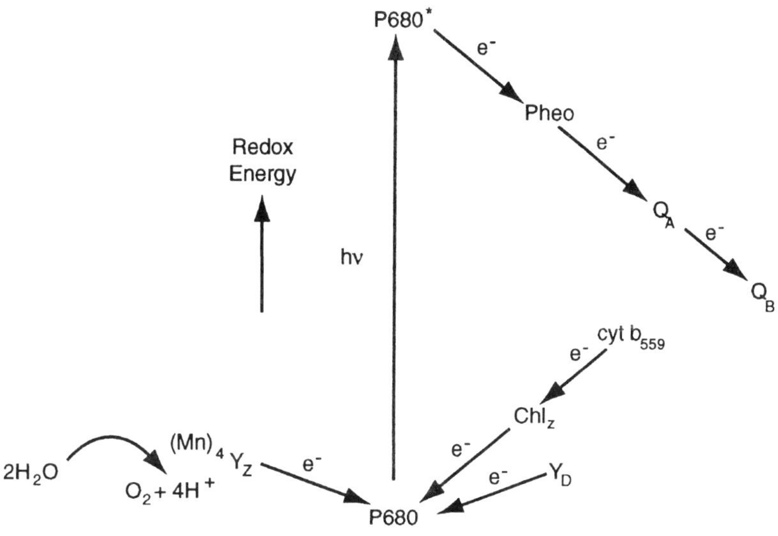

Fig. 1. Sequential arrangement of the electron transfer components of Photosystem II on a relative energy scale. The arrows indicate the direction of electron flow.

production of P680$^+$ is harnessed effectively to the oxidation of water; spurious electron-transfer reactions that would oxidize essential components in the D1/D2 heterodimer that provides the protein core of PS II are minimized, but not eliminated (for a review of photoinhibition in PS II see Prásil et al., 1992). The principal pathway for the physiologically productive reduction of P680$^+$ proceeds through a redox-active tyrosine, Y_Z, which has been identified as Tyr161 in the D1 polypeptide, and an ensemble of four manganese ions that oxidize substrate water. Figure 1 summarizes this pathway, as well as the reducing reactions that transfer an electron from photoexcited P680 through a pheophytin intermediate to Q_A. On the reducing side, secondary electron transfers and proton-uptake reactions eventually produce the reduced, protonated Q_BH_2 cofactor, which then dissociates from the PS II reaction center to shuttle reducing equivalents through the cytochrome b_6f complex to Photosystem I and, ultimately, to the carbon-fixing dark reactions.

In addition to the primary electron-transfer pathway to P680$^+$ that produces water oxidation, there are at least three other redox active cofactors in PS II that can be oxidized by the reaction-center chlorophyll complex. These are shown in Fig. 1 as the redox-active tyrosine, Y_D, identified as Tyr160 (cyanobacterial numbering, 161 for spinach numbering) on the D2 polypeptide, a chlorophyll that has been designated Chl_Z, and a low-spin, bis-histidine coordinated b-type cytochrome, Cyt *b*-559. As discussed in more detail, these redox centers act under various conditions to provide reducing equivalents to P680$^+$.

The protein components in Photosytem II have been extensively characterized. In addition to the D1/D2 heterodimer, two large chlorophyll-binding polypeptides, CP47 and CP43, are isolated with the reaction-center complex and are essential for water oxidizing activity. A number of smaller polypeptides,

Abbreviations: ADMR – absorbance detected magnetic resonance; BChl – bacteriochlorophyll; BPheo – bacteriopheophytin; Cyt – cytochrome; D1 and D2 – polypeptides of the Photosystem II reaction center; ENDOR – electron nuclear double resonance; EPR – electron paramagnetic resonance; ESEEM – electron spin echo envelope modulation; FTIR – Fourier transform Infrared; OEC – oxygen evolving complex; P680 – primary electron donor of PS II; P870 – primary electron donor of *Rb. sphaeroides* reaction centers; Pheo – pheophytin; PS II – Photosystem II; Q_A – primary quinone electron acceptor; Q_B – secondary quinone electron acceptor; RC – reaction center; SHE – standard hydrogen electrode; Sn – states of the OEC where 'n' corresponds to the number of stored oxidizing equivalents; Y_D – redox-active tyrosine 160 (cyanobacterial numbering) of the D2 polypeptide; Y_Z – redox-active tyrosine 161 of the D1 polypeptide

with molecular weights in the 10–35 kDa range, also occur. Three of these, with MW = 17, 23 and 33 kDa, are extrinsic, membrane associated polypeptides and are important for the stable binding of the manganese, calcium and chloride cofactors that are essential to water oxidation. The heme of Cyt *b*-559 is bound into the reaction center by a heterodimeric set of low molecular weight polypeptides, designated as the α and β subunits. These have molecular weights less than 10 kDa, and each has a single membrane spanning region that contains the ligating histidine. For a more detailed discussion of the PS II polypeptide composition and function, the reader is referred to review articles by Erickson and Rochaix (1992) and by Seibert (1993), and to Chapters 8, 9 and 11.

B. Thermodynamic Efficiency in PS II Relative to Bacteria: The Evolution of Weak Coupling

There are strong similarities between the organization of the primary photochemistry in Photosystem II and in the reaction centers from purple photosynthetic bacteria. These have been discussed in detail elsewhere (e. g., Rutherford, 1986; Michel and Deisenhofer, 1988; Renger, 1992; Seibert, 1993), are considered briefly below, and are discussed in Chapter 11. These include sequence homologies between the PS II core polypeptides D1 and D2 and the corresponding L and M polypeptides in bacterial reaction centers, the occurrence of conserved histidines in D1 and D2 that are likely to ligate the chlorophylls that comprise P680, sequence homologies in the $Q_AFe^{2+}Q_B$ region, and the demonstration of an intermediate pheophytin acceptor. The striking C_2 symmetry that occurs in the bacterial systems is preserved as a structural theme in PS II and extends beyond the primary photochemistry, as demonstrated by the identification of the symmetry-related redox-active tyrosines, Y_Z and Y_D in D1 and D2 (Debus et al., 1988 a,b; Vermaas et al., 1988; Metz et al., 1989; Babcock et al., 1989; Barry, 1993, Koulougliotis et al., 1995).

At the same time it is apparent that there are significant differences between the two classes of reaction centers. To date, the emphasis in discussing these differences has been on structural and sequence discrepancies that have emerged. Thus, for example, the histidines that ligate the accessory chlorophylls in the bacterial reaction centers have no sequence analogs in D1 and D2 and the orientation of the primary donor triplet axis in PS II differs substantially from that of the triplet axis in the bacterial protein (Rutherford, 1985; Satoh, 1993).

More fundamentally, the two types of reaction centers differ functionally and thermodynamically. The bacterial center, of course, does not split water, but perhaps ultimately of more importance is the fact that the PS II reaction center is a much more efficient thermodynamic machine than are reaction centers of the purple non-sulfur photosynthetic bacteria. Figure 2 illustrates the thermodynamic efficiencies and highlights the differences. In Photosystem II, quanta at 680 nm (= 1.84 eV) drive light-induced charge separation to produce the initial charge-separated and relatively stabilized $P680^+ Q_A^-$ pair. The midpoint potential of $P680^+$ has not been measured but has been estimated at +1.12 V; for the Q_A/Q_A^- couple, the operating potential is in the range of –0.13 V (see below). Thus, of the 680 nm photon (1.84 eV), 1.25 eV is converted into a charge-separated pair, which represents an efficiency of 68%. In *Rb. sphaeroides* reaction centers, 870 nm (= 1.44 eV) photons produce the $P870^+ Q_A^-$ pair, which corresponds to a redox energy difference of 0.50 V (e. g., Warncke and Dutton, 1993). Calculating the efficiency of this process yields a value of 35%. Thus, the PS II reaction center has evolved so that its initial charge-separation efficiency, under no load conditions, is more than twice that of the bacterial system under similar conditions. Although this circumstance may change somewhat in the physiological situation—for example, the prevailing membrane potential, which opposes charge separation in both reaction centers, is significantly larger in bacteria than in oxygenic systems—the above considerations show striking thermodynamic differences between bacterial and higher plant reaction centers.

The structural and physical bases for these functional differences have arisen within structures that have retained a good deal of overall similarity. Although the details of the underlying mechanisms by which this has occurred cannot be specified precisely now, a theme that emerges from recent investigations of PS II is that the strong coupling between pigments and reactants in the bacterial reaction center is much weaker in the PS II photochemical core. Thus, exciton coupling in the primary donor complex in PS II is significantly less pronounced than in bacteria (van Gorkom and

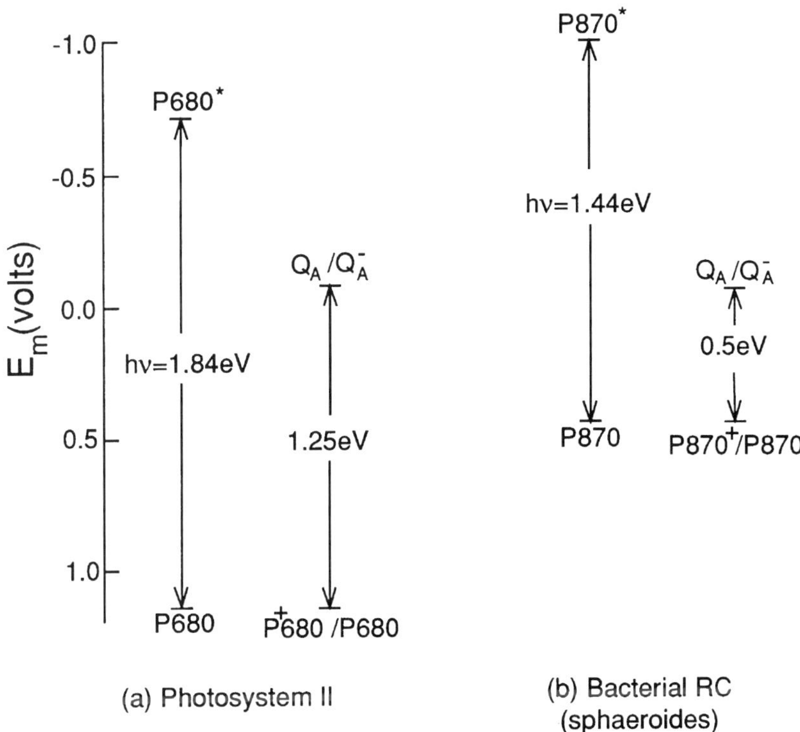

Fig. 2. Schematic representation of the relationship between photon energy ($h\nu$) and $P^{+}_{\lambda max}$ Q_A^- energy differences in (a) Photosystem II reaction centers and (b) *Rb. sphaeroides* photosynthetic bacterial reaction centers.

Schelvis, 1993), and the remarkable hole-burning and Stark spectroscopic effects that have been observed in bacteria are less striking in Photosystem II (Lösche et al., 1988; Tang et al., 1990; Jankowiak and Small, 1993; Boxer, 1993). In reaction-center complexes, the primary donor absorption band is split much more strongly from those of the accessory pigments in bacteria than in oxygenic reaction centers (see below). Overall, the weaker coupling that occurs in PS II produces a situation in which both the energy-transfer and charge-separation processes are thermodynamically driven much less strongly than in the bacteria. By arranging a set of these weakly driven processes in series, good overall thermodynamic efficiency is still achieved but with substantially less degradation of the photon energy. Because the photosynthetic apparatus is essentially a free energy conversion device, the evolution of the weak coupling strategy represents a significant advance in effectiveness. The weaker coupling, however, makes the system more complicated to probe experimentally; the spectral distinctions between antenna, trap, and acceptor are substantially reduced and severe spectral congestion results. Thus, work to understand the details of PS II reaction center function has been more difficult and, at present, is marked by considerable uncertainty and controversy. This is largely a consequence of the structural and functional departures from the bacterial reaction center that have occurred to increase energy conversion efficiency in PS II and to provide an oxidant sufficiently strong to oxidize water. In the sections that follow, we first review aspects of the physical structure of PS II and then consider aspects of its spectroscopy and function.

II. Biochemical and Pigment Organization of Photosystem II

A. Resolution of the Reaction Center Complex Polypeptides

The biochemical resolution of PS II has been accomplished over the past 15 years. Although crude reaction-center preparations had been prepared from thylakoid membranes earlier, the isolation of O_2-evolving PS II membranes by Yocum and co-workers marked the beginning of a period in which the underlying principles of the architecture of the reaction center and its associated antenna were understood. Several laboratories have been involved in this research (reviewed in Ghanotakis and Yocum, 1990; Satoh, 1993) and their efforts have shown that the PS II integral protein assembly can be dissociated sequentially. The light-harvesting chlorophyll complex (LHC) can be stripped away to yield a complex comprising the reaction-center polypeptides D1 and D2, the Cyt *b*-559 polypeptides, the chlorophyll-containing CP47 and 43 species, and several interesting, but less well-characterized, low-molecular-weight subunits. These preparations retained the peripheral 33 kDa polypeptides and O_2 evolving capacity. Removal of CP43 released the extrinsic polypeptide and both O_2 evolution and Q_A and Q_B were lost in these CP47/D1/D2/Cyt *b*-559 preparations. Rögner et al. (1991), however, used genetic and biochemical techniques to show that PS II reaction centers are assembled in vivo with functional Q_A in the absence of CP43. Nanba and Satoh (1987) were the first to isolate a D1/D2/Cyt *b*-559 complex that retained $P680^+Pheo^-$ charge separation (see Chapter 11). These initial PS II reaction-center preparations were unstable, however. Procedures were later developed by Seibert (McTavish et al., 1989), Yocum (Dekker et al., 1989; Ghanotakis et al., 1989), and Barber (Booth et al., 1991), and their co-workers that provided more robust D1/D2/Cyt *b*-559 reaction-center complexes; the stoichiometry and functional relevance of the pigment complement that occurs in these reaction centers has become an issue of considerable interest and debate.

B. Pigment Composition of the PS II Reaction Center

The isolated D1/D2/Cyt *b*-559 PS II RC contains chlorophyll *a*, pheophytin *a*, β-carotene, and the protoheme group of the cytochrome; the Q_A and Q_B quinones, however, are removed by all isolation procedures developed thus far (Satoh, 1993; Seibert, 1993). The stoichiometries of these components have been a matter of controversy and values ranging from four to greater than ten chlorophylls per PS II have been reported. If one assumes that there are two pheophytins per P680, then values of one β-carotene, one cytochrome *b*-559, and four to six chlorophylls generally result. As the bacterial reaction center contains four bacteriochlorophyll per special pair, the higher chlorophyll values found in PS II are problematic. This is especially so in view of the fact that the milder isolation procedures noted above tend to give chlorophyll stoichiometries closer to six. Despite variations in isolation procedure, chlorophyll content, and optical absorption characteristics (see Fig. 3), however, the primary charge-separating photochemistry in the various PS II reaction-center preparations is similar, which could suggest that the lower stoichiometry may represent the limiting chlorophyll content. The isolated reaction center has fluorescence components that have been attributed to pigments inactive in primary charge separation (e.g., Schelvis et al., 1994), which also supports this view. Recently, Chang et al. (1994) have proposed on the basis of optical and hole-burning experiments that the limiting chlorophyll stoichiometry is, in fact, four and that the higher values result from residual CP47 contamination and from the presence of a 684 nm-absorbing linker pigment, the latter of which is easily modified in the presence of Triton X-100 to produce a 670 nm absorbing chromophore. A recent contrasting view is given by Eijckelhoff and Dekker (1995) who find six chlorophyll *a* per two pheophytin in all reaction center preparations analyzed, and attribute the lower stoichiometry observed by others to errors in extinction coefficients. The analogy between the bacterial reaction center and PS II may yet hold for four of the chlorophyll *a*, but with weaker excitonic coupling of the bacterial special pair homologues. Likely binding sites for the extra two chlorophylls in PS II are D1-His118 and D2-His117 (cyanobacterial numbering, Schelvis et al., 1994) for which there are no bacterial homologues.

III. Absorption Characteristics and Organization of Pigments in Photosystem II

A. Optical Properties

Figure 3 shows the long-wavelength, optical-absorption spectra of D1/D2/Cyt *b*-559 reaction-center preparations that contain six (*a*), five (*b*), or four (*c*) chlorophylls; the spectra were recorded at 4.2 K. Several features are noteworthy in these spectra (Braun et al., 1990, van Kan et al., 1990; Otte et al., 1992; Chang et al., 1994; van Grondelle et al., 1994). First, the Q_y region is composed of two bands with absorption maxima close to 680 and 670 nm. The former has contributions from the lower-energy P680 exciton component (see below), from one of the two pheophytins, and from the putative 684 nm linker chromophore described above. The latter has contributions from the accessory chlorophylls in the reaction center; the second pheophytin contributes in the region between 670 and 680 nm. Second, these spectra show that, as the chlorophyll content of the RC preparation increases, the wavelength region between the 670 and 680 nm peaks in the resolved, four chlorophyll preparation fills in, and, as expected, the absorbance, normalized to the Pheo *a* Q_x-band at 543 nm, increases with the chlorophyll content. The suggestion, however, that much of the increased Q_y absorbance in the six chlorophyll preparations is due to contaminating CP47 runs counter to the biochemical evidence (e.g., Dekker et al., 1989). Either loosely bound accessory chlorophylls (possibly bound at D1-His118 and D2-His117) or 684 nm absorbers not removed by the milder dodecyl maltoside extraction are likely to be the origin(s) of the chlorophyll and absorbance increase in the higher chlorophyll stoichiometry preparations. Third, and as opposed to the bacterial reaction centers, the extent of the Q_y absorption region is significantly compressed in PS II—the width of the total absorbance envelope in Fig. 3 is ~ 500 cm^{-1} at half height. Accordingly, spectral resolution of the contributions of the individual chromophores is markedly diminished. In *Rhodobacter sphaeroides* reaction centers, for example, the splitting between the absorbance maxima of the special pair and the accessory pigments is ~1,000 cm^{-1}; the pheophytins are split further, to ~1,500 cm^{-1}, from the special pair absorbance (for recent reviews of the bacterial reaction center, see Deisenhofer and Norris, 1993, Volume 2).

The spectral congestion in the PS II Q_y region, as shown in Fig. 3, has made spectroscopic and kinetic investigations of the reaction-center structure and dynamics considerably more difficult than in the bacteria. Nonetheless, weakened exciton coupling in the special pair (see below) and the resulting shallower trap in PS II appears to be a key aspect of the evolutionary strategy whereby the photochemical conversion efficiency of the oxygenic systems was increased relative to the purple bacterial reaction centers.

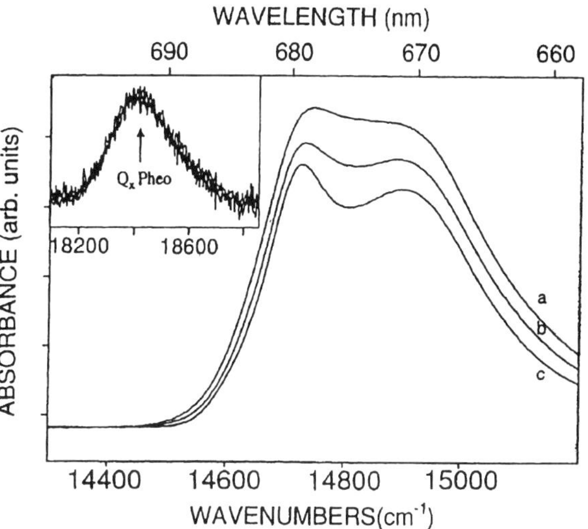

Fig. 3. 4.2 K absorption spectra in the Qy absorption region of (a) 6 chlorophyll a/RC, (b) 5 chlorophyll a/RC, and (c) 4 chlorophyll a/RC Photosystem II reaction center preparations. The spectra were normalized to the same intensity of the Q_x band of Pheo *a* (inset). From Chang et al., 1994, with permission.

B. P680 Spectroscopy and Structure

Resolution of the contribution(s) of P680 to the optical spectra in Fig. 3 and of its structure are central problems in PS II. As noted above, the spectral overlap in the Q_y absorption region and uncertainties in the pigment stoichiometry in PS II have provided significant obstacles to resolving these issues. At present, both monomer and dimer models for P680 are available in the literature; moreover, the orientation of the chlorophylls in dimer models and of the pigment planes in both monomer and dimer structures, with respect to the membrane, are under debate. Figure 4 summarizes proposed structures that have been presented for P680. Figure 4a is based on a strict analogy between the bacterial RC and PS II and suggests an approximately parallel orientation of the two macrocycles with the interdimer axis parallel to the membrane plane. On the basis of spectral data discussed below, this structure can almost certainly be eliminated. Figure 4b preserves the basic bacterial structure but suggests that P680 is more monomeric and localized on one of the accessory chlorophylls in the reaction center. The genesis of this model is the observation that, at low temperature, the P680 triplet

P680 Structural Models

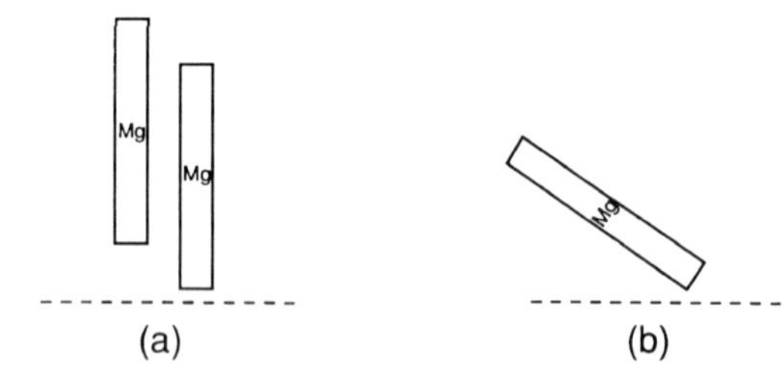

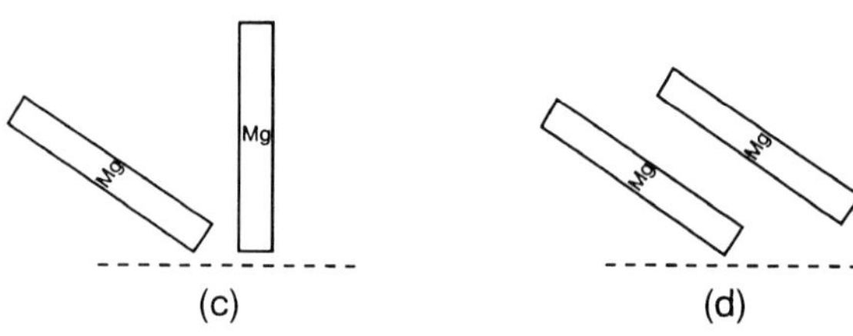

Fig. 4. Schematic representation of various structures that have been proposed for the photochemically active P680 species in Photosystem II. The dotted line in each part of the figure represents the inner plane of the membrane. See text for details.

is localized (Rutherford, 1985; van der Vos et al., 1992) and that its triplet axis system has an orientation very similar to that expected for the monomeric accessory bacteriochlorophyll in *Rhodopseudomonas viridis* or *Rhodobacter sphaeroides*, i. e., that the ring plane is tilted ~30° with respect to the membrane plane (van Mieghem et al., 1991). This structure has been supported by spectroscopic data indicating that the spin-polarized triplet in PS II is located on P680 (Durrant et al., 1990) and by absorbance detected magnetic resonance studies, which were interpreted to indicate a monomeric structure for P680 with its Q_y transition dipole oriented at close to the magic angle (54° ± 8°) with respect to the membrane normal (van der Vos et al., 1992). This model also has the important consequence of rationalizing the distinction between the redox active tyrosines in PS II, as an off-axis location for P680 places it closer to one of the two tyrosines. Figure 4c is a variation on the original van Mieghem et al. (1991) model of Fig. 4b that was proposed by Noguchi et al. (1993) on the basis of variable-temperature FTIR experiments on the P680 triplet state. In the vibrational spectroscopic work, localization of the triplet to a single chlorophyll was observed at low temperature ($T \leq 80$ K); but, above these temperatures, equilibration of the triplet over two macrocycles was detected. Low-temperature localization is consistent with both EPR and ADMR results, which were obtained at liquid-helium temperatures; higher-temperature delocalization suggests a dimer structure such as that in Fig. 4c. Figure 4d has been proposed recently by van Gorkom and co-workers (van Gorkom and Schelvis, 1993; Schelvis et al., 1994; see also van Grondelle et al., 1994) and involves a dimeric structure with monomeric Q_y transition moments that are close to parallel (15–21°, Kwa et al. 1994) and oriented at close to the magic angle with respect to their connecting axis; the physical separation of the two monomers in the dimer is postulated to be greater than typical van der Waals distances (see also Braun et al., 1990; Carbonera et al., 1994). This model is discussed in more detail below.

The evolution of postulated structures in Fig. 4 for P680 reflects a deepening understanding of the spectroscopy of the reaction center; the tilted dimer in Fig. 4d accounts for most of the data currently available and also retains the potential to rationalize the difference in Y_Z and Y_D oxidation kinetics. Thus, the dimer structure is suggested by the occurrence of conserved histidines in D1 and D2 that are the analogs of the special pair ligating residues in the bacterial reaction center (Michel and Deisenhofer, 1988; Satoh, 1993). Exciton coupling in P680, indicative of a true dimeric structure, has been consistently suggested by circular dichroism measurements (Tetenkin et al., 1989; Braun et al., 1990; Otte et al., 1992); recent hole-burning work by Small and co-workers (Jankowiak et al., 1989; Chang et al., 1994) and polarized site selection excitation spectroscopy by Kwa et al. (1994) support the CD results. These methods suggest that the upper exciton state occurs at ~667 nm and that it carries less than 5% of the absorption intensity of P680. As with the bacterial reaction center, the polarization of the upper 667 nm exciton component is opposite to that of the lower P680 exciton component, indicating the occurrence of C_2 symmetry. That the lower exciton component (680–684 nm), i. e., P680, carries most of the dimer oscillator strength is also consistent with the time-resolved measurements of Durrant et al. (1992a,b) and of Schelvis et al. (1994), both of whom noted enhanced oscillator strength of P680 at early times in the excitation and charge-separation process (see also Kwa et al., 1992). These results indicate, then, a dimer structure with a small (~300 cm^{-1}) exciton coupling. The weaker exciton coupling, relative to bacterial reaction centers, in this model for P680 is achieved by the spatial relationships of the monomer Q_y transition dipoles and the intermonomer connecting axis and by a larger distance between the two monomers. The absence of linear dichroism in the ADMR experiments of van der Vos et al. (1992) suggests an orientation with the P680 Q_y transition dipole oriented at the magic angle, with respect to the membrane normal, as shown in Fig. 4d. The triplet data, indicating localization to half the dimer at low temperature, is accounted for by the FTIR data showing activated delocalization over the dimer at higher temperatures (Noguchi et al., 1993). Hydrogen bonding of the two monomers, as detected in the FTIR work, shows that, while both carbomethoxy groups are hydrogen bonded, the keto group of only one of the two is involved in such an interaction; the other monomer keto is free. This dichotomy in hydrogen bonding provides a plausible structural basis for the temperature effects on triplet delocalization.

The work thus far on P680 structure and dynamics shows that significant differences occur with respect to the bacterial reaction center. The exciton coupling in the dimer in PS II is significantly smaller: 300 cm^{-1} in P680, as opposed to 1900 and 1300 cm^{-1} in *Rp. viridis* and *Rb. sphaeroides*, respectively. Recent optically detected magnetic-resonance data (Carbonera et al., 1994) have been interpreted to indicate that the exciton coupling in P680 is even less than 300 cm^{-1}. Compared to the bacterial reaction center, the reduced exciton coupling in PS II decreases the red-shift of the special pair dimer substantially relative to the accessory chlorophylls. For *Rb. sphaeroides*, for example, the accessory bacteriochlorophyll at 800 nm absorbs photons that are 1,000 cm^{-1} higher in energy than those absorbed by the P870 dimer; for PS II, P680 is split by only 200 cm^{-1} from the 670 nm absorbing accessory chromophores. Both of these phenomena, i.e., decreased exciton coupling and a reduced accessory chlorophyll/special pair energy difference, reflect the weak coupling scenario that appears to be operational in oxygenic reaction centers. Stark effects are also significantly reduced in PS II (Lösche et al., 1988), relative to bacteria (Boxer, 1993), and the low-frequency marker mode (125–145 cm^{-1}) progression that occurs in bacterial reaction centers has not been observed in analogous hole-burning experiments on PS II (Chang et al., 1994). If the interpretation of the marker mode as reflecting an inter-monomer vibration in the bacterial reaction center is correct, then the absence of this mode in PS II also suggests a greater distance separation between the two monomers in PS II (Fig. 4d).

Despite these differences between oxygenic and bacterial reaction centers, there are also striking similarities. First, and most obvious, is the homology that occurs between the L and M polypeptides in the bacteria and D1 and D2 in PS II. Second, as discussed above, the pigment composition of PS II appears to mimic that of the bacteria. Third, the data now available clearly show that the dimeric special pair motif is preserved to some extent in PS II, at least with respect to singlet excited-state optical characteristics. In this regard, Durrant et al. (1995) have proposed a model for P680 in which the pigment molecules are largely localized as in the bacterial reaction centers but where, instead of a strongly coupled dimer, the excitonic coupling is best described as a weakly coupled multimer. Delocalization of the triplet state over the multimer also appears to occur, although localization has been established at low temperature. The extent to which the hole is delocalized in $P680^+$ remains to be determined. While preserving the dimer, however, its orientation in PS II shows a marked departure from the C_2 symmetry axis. Fourth, the strong coupling to low-frequency protein phonon modes that occurs in the bacterial special pair is preserved in the PS II P680 dimer; likewise, the weaker phonon coupling to accessory pigments in bacteria carries over to the PS II reaction center. Although the full significance of the strong phonon coupling behavior remains to be established, it is likely to be important for facilitating efficient and rapid charge separation upon photoexcitation. Quite clearly, the lessons that have been learned in the bacteria special pair will continue to be useful in comparison and in contrast to Photosystem II.

C. The Accessory Chlorophylls and Pheophytins

To date, the major focus in PS II has been on establishing the structure and dynamics of P680, as described above; the accessory chlorophylls and the pheophytins have received less attention. The situation with respect to the non-dimer pigments has also been clouded by uncertainties in pigment stoichiometries, as noted in the previous section. Nonetheless, some important information on these pigments is available.

If the 4 chlorophyll stoichiometry for the PS II

reaction center inferred from recent hole-burning and optical spectroscopic studies is correct, then PS II, like the bacteria, contains two accessory chlorophylls and two pheophytins. The former contribute to the 670 nm band in the optical absorption spectra of Fig. 3. The optical properties of the pheophytins are somewhat more obscure, although recent data indicate that one of the two pheophytins absorbs near 676 nm and the other near 680 nm (Breton, 1990; Tang et al., 1990; van Kan et al., 1990; Otte et al., 1992; van der Vos et al., 1992; van Gorkom and Schelvis, 1993). As with the bacteria, only one of the two pheophytins is active in electron transfer and forms the charge-separated $P680^+ Pheo^-$ state. Photo-accumulation data indicate that the active Pheo is the 676 nm absorber (van Kan et al., 1990). Data from several spectroscopies (Lubitz et al., 1989; Moënne-Loccoz et al., 1989; Nabedryk et al., 1990; Kwa et al., 1992) and from site-directed mutagenesis (Nixon, 1994) indicate that the binding site for this pigment has several analogies to that of the active bacteriopheophytin in the bacterial reaction center. A strict analogy to the bacterial RC for both of the PS II pheophytins, however, appears to be precluded by linear dichroism data that indicate oppositely signed LD features for these two chromophores (Breton, 1990; van der Vos et al., 1992; van Gorkom and Schelvis, 1993).

The bacterial analogy for the binding of the two accessory chlorophylls appears to break down even further. In the bacterial RC, conserved histidines in the L and M subunits that bind the accessory bacteriochlorophylls do not have analogs in the PS II reaction center polypeptides (Michel and Deisenhofer, 1988). Results from polarized fluorescence spectroscopy indicate that the average orientations of the PS II chlorophylls are markedly different from those of the bacterial accessory pigments (Kwa et al., 1992). van Gorkom and Schelvis (1993) have argued that a close association of the accessory pigments in PS II with the P680 dimer is unlikely, as too small a separation distance would allow the accessory chlorophylls to compete effectively with the Y_Z donor in reducing the highly oxidizing $P680^+$ photoproduct. By considering the distance dependence of both energy and electron-transfer processes, they conclude that a separation of ~23 Å allows an effective antenna function while minimizing the photooxidation process. Modulation of the redox properties of the accessory chlorophylls by the local protein environment may alleviate the potential for photooxidation to some extent and allow somewhat closer approach of the accessory pigments. Within this model, conserved histidines at positions 118 and 117 in D1 and D2 (cyanobacterial numbering, respectively) are postulated as likely candidates for accessory chlorophyll binding sites (Schelvis et al., 1994).

IV. Primary Photochemistry

A. Excitation Energy Transfer and Trapping

The absorption of a photon in the pigment bed associated with PS II produces an energy transfer process that culminates in charge separation as follows:

$$\text{(antenna)}^* \underset{k_{-1}}{\overset{k_1}{\rightleftarrows}} \text{(P680 Pheo)}^* \underset{k_{-2}}{\overset{k_2}{\rightleftarrows}} \text{P680}^+ \text{ Pheo}^- \tag{1}$$

where the asterisk indicates an excited-state chromophore. There has been an extended effort to evaluate the rate and equilibrium constants associated with this energy-transfer and trapping process, and two extreme models can be envisioned (see van Grondelle et al., 1994, for an excellent, recent review of energy-transfer processes in oxygenic and anoxygenic photosynthesis). If the four rate constants indicated in Eq. (1) are comparable, then extensive delocalization of the photon over the pigment array, including P680, will occur and exciton localization to produce charge separation will result in a shallowly trapped $P680^+ Pheo^-$ species. This process reflects a trap-limited situation in which the excitation visits the primary photoreactants several times before stable charge separation occurs. If, on the other hand, k_1, k_{-1}, and k_{-2} are small compared to k_2, then exciton migration in the pigment bed limits charge separation, the charge-separated state is deeply trapped, and the process reflects a situation in which charge separation is limited by exciton diffusion in the antenna.

Owing to the nearly isoenergetic electronic properties of antenna, P680 and pheophytin in the complex ensemble that comprises the PS II pigment array, the trap-limited situation would, *a priori*, be expected to be the case. An isoenergetic pigment bed is an additional manifestation of the weak coupling scenario described above; that is, such a situation does not produce significant degradation of the photon energy prior to its use in producing charge separation.

Such a weak-coupling, trap-limited model makes two strong predictions. (1) The rate of charge separation should be independent of the excitation wavelength, as rapid equilibration of the exciton over the pigment ensemble should occur. The fast equilibration and (relatively) slow trapping is expected to make the identity of the original absorber immaterial to the eventual charge-separation process. (2) The experimentally observed rate of charge separation should be a fraction of the total number (N) of pigment molecules that are associated with each PS II reaction center according to the relation k_{cs}(observed) = k_2/N, where k_{cs}(observed) is the experimentally detected charge-separation rate. These predictions are expected to hold at room temperature; observed behavior may deviate as the temperature is lowered and the thermal energy in the system decreases, as noted by van Grondelle et al. (1994).

Until recently, the behavior of PS II in isolated D1/D2/Cyt *b*-559 preparations was unclear with respect to these predictions. Initial work from Holzwarth and co-workers on PS II reaction centers that retained a significant antenna, 40-80 chlorophylls, showed that the fluorescence decay was biphasic, with half-times of 80–100 ps and 520 ps (Schatz et al., 1987; 1988; see also Leibl et al., 1989; Holzwarth, 1989). They explained these results within the context of a trap-limited model, which they termed an exciton/radical pair equilibrium model, by assuming equilibration of the excitation over the pigment array in times short relative to charge separation. Within the context of this model, they extracted a value for k_2 of (2.7 ps)$^{-1}$, which is close to the value now generally accepted for the intrinsic rate of charge separation (see below). Later work by Durrant et al. (1992a,b) strongly supported the weak-coupling, trap-limited model used in the fluorescence work by demonstrating excitation delocalization in the PS II reaction-center on the 100 fs time scale. They pointed out that this result is entirely consistent with the nearly isoenergetic pigment array in PS II. They noted, in fact, that exciton equilibration in the isolated bacterial reaction center is still fast with respect to the charge-separation process, despite the fact that the stronger coupling in the bacterial RC leads to a six-fold increase in the energy difference between accessory and special pair pigments relative to PS II, as discussed above.

Despite the support for the weak-coupling, trap-limited model provided by the fluorescence and absorption measurements above and recent fluorescence measurements in D1/D2/CP47 complexes (Freiberg et al., 1994), there are a number of counter indications now in the literature. Hole-burning experiments by Small and co-workers (Jankowiak et al., 1989; Tang et al., 1990) suggested that equilibration occurred on much longer time scales, and the data were interpreted to indicate exciton transfer from the accessory chlorophylls and from pheophytin to 680 nm absorbing chromophores in ~12 ps and ~50 ps, respectively. Similarly, fluorescence work has been consistent in identifying a decay phase, in some cases, the fastest decay phase (Mimuro et al., 1988; Roelofs et al., 1993), that occurs in ~20 ps. The interpretation of this phase as reflecting energy transfer from 670 nm absorbers to 680 nm absorbers has also been consistent. Most of these experiments, however, have been carried out at low temperatures (liquid nitrogen and lower) and the slower equilibration times could reflect a combination of the decreased available thermal energy, spurious (non-charge-separating) long wavelength traps, and inhomogeneous broadening of the pigment ensemble in PS II, as noted in the recent review by van Grondelle et al. (1994).

Nonetheless, the 20 ps fluorescence phase has been observed at higher temperatures in both fluorescence and absorption studies and most recently Schelvis et al. (1994) have provided a serious challenge to the weak-coupling, trap-limited model by showing excitation wavelength dependent charge-separation kinetics in D1/D2/Cyt *b*-559 preparations that contain six chlorophylls. Excitation at wavelengths to the red of the 680 nm absorption produced $P680^+ Pheo^-$ in 3 ps; excitation of accessory pigments in the 670 nm region led to 30 ps charge-separation kinetics.

These results, taken together, have led to a confused situation as to the energy-transfer and trapping processes that occur in Photosystem II: although the trap-limited model is intellectually appealing and has strong experimental support, there are several experiments now in the literature that run counter to its predictions. A resolution of this conundrum has been suggested by Durrant et al. (1992b), Roelofs et al. (1993), and, most recently, by Chang et al. (1994). Each of these sets of workers proposed that contaminating pigments, not closely and, by implication, not physiologically associated with the PS II reaction center, are present in preparations that contain more than four chlorophylls per P680. Durrant et al. (1992b) pointed out that in earlier studies from other labs as well as their own, energy-transfer processes on the tens to hundreds of ps time scale had been observed. In arguing for sub-ps excitation equilibration in PS II, they asserted that the slower events arose from a minority of chlorins, implicitly contaminants, present to varying degrees in PS II preparations. Roelofs et al. (1993) reached a similar, but more forceful, conclusion in attributing the tens of ps energy-transfer component in their fluorescence data to the two additional chlorophylls present in their six chlorophyll/P680 preparations. Chang et al. (1994) have now explicitly stated the idea that the functional core of PS II consists of four chlorophylls/P680 (see above) and that the two additional chlorophylls that are present in preparations isolated by the currently favored methods lead to unphysiological photophysical processes.

To summarize, the current state of our understanding of energy transfer and trapping in PS II is in flux. Models that favor either rapid or slow excitation transfer are available in the literature. Nonetheless, on the basis of (a) the behavior of the bacterial systems, where exciton equilibration is fast relative to charge separation, (b) the much more nearly isoenergetic array of pigments in PS II relative to the bacterial reaction centers, (c) recent experimental data showing fast exciton equilibration under certain experimental conditions, and (d) the reassessment of the pigment composition of the PS II reaction center, it appears as if fast equilibration, trap-limited kinetics are operational in the oxygenic reaction center as well.

B. Primary Charge Separation

Many of the same factors that have complicated interpretation of energy transfer in PS II have also made an unambiguous identification of the charge-separation kinetics to produce $P680^+ Pheo^-$ difficult. As indicated above, resolution of these difficulties appears to be emerging, and the charge-separation kinetics can be interpreted within this model.

Initial estimates of the charge-separation time came from work on fluorescence decay kinetics in partially resolved cyanobacterial preparations that retained a sizable antenna (40–80 chlorophylls). This work, from which a $P680^+ Pheo^-$ formation time of 2 ps was estimated, was discussed above. This time course was supported by both the room- and low-temperature absorption measurements of Wasielewski et al. (1989) and by the hole-burning experiments reported by Chang et al. (1994). Roelofs et al. (1991, 1993) and Schelvis et al. (1994) have reported time-resolved fluorescence and optical absorption data that also indicated a primary charge-separation rate of $(2–3\ ps)^{-1}$. Taken together, these results strongly favor a reaction time course for radical pair formation that is analogous to that for the bacterial reaction centers. As van Gorkom and Schelvis (1993) point out, these rapid kinetics are necessary to insure high quantum efficiency for the energy-conserving forward reaction. In these studies, however, 20–25 ps phases that could be attributed to primary charge separation were consistently observed. For example, Wasielewski et al. (1989) originally detected a 3 ps charge-separation phase but noted, as well, a ~20 ps phase that was assigned to a relatively slow energy transfer process; the 20-ps phase detected in fluorescence studies has been discussed above. Thus, the majority view of the primary process in PS II is that charge separation occurs on the 3 ps time scale; the 20 ps phase has been thought by most workers to reflect energy transfer.

Klug and co-workers, in a carefully done series of measurements, however, have made several important observations on primary processes in PS II (Durrant et al., 1990, 1992a,b, 1993; Hastings et al., 1992; see also McCauley et al., 1992). These include the following: a) reduction of a significant fraction ($\gtrsim$50%) of the Pheo acceptor, with a 20 ps time course; b) energy equilibration within 100 fs between shorter and longer wavelength absorbing pigments in isolated PS II reaction centers (see above), and c) exciton localization in the P680 absorber in the sub-ps time range. The reaction centers used in these measurements typically contained six chlorophylls per P680. While there is support for the latter two observations, the 20 ps charge-separation time clearly runs counter to the majority of the measurements of the charge-separation time.

As with the energy-transfer work, some clarification of this situation now seems to be emerging. Two recent observations are particularly relevant to measurements of the charge-separation process. First, as noted above, Durrant et al. (1992b), Roelofs et al. (1993), and Chang et al. (1994) have suggested that PS II reaction centers contain minimally four chlorophylls; they postulate that the additional two that commonly occur in the current generation of PS II preps are only loosely associated with the core pigments and do not participate in rapid energy equilibration. Second, Holzwarth and co-workers (1994) and Wasielewski and co-workers (Wiederrecht et al., 1994) have pointed out that photoselection processes in time-resolved PS II work occur when pump and probe beam polarizations are not held at the magic angle relative to each other. In particular, Wiederrecht et al. (1994) have shown that 20 ps phases in their preparations are emphasized when the two beams are parallel to one another. In the earlier work reported by the Klug group, parallel polarization and six chlorophyll/P680 preparations were used. This group has recently countered both objections by using magic angle transient absorption spectroscopy (Klug et al., 1995) and five chlorophyll/P680 preparations (Vacha et al., 1995) and obtaining substantially similar results for primary charge separation $(21\ ps)^{-1}$ and a reduced amplitude for a distinct $(27\ ps)^{-1}$ component attributed to energy transfer.

To conclude this section, it appears that a reasonably consistent view of PS II energy-transfer and charge-separation processes is now emerging. Following photon absorption, rapid exciton equilibration occurs within ~100 fs. P680 operates as an effective, but shallow, trap and has an absorption cross section roughly twice that of monomer chlorophyll; following its initial excitation, however, exciton localization occurs within ~400 fs. Charge separation at room temperature produces $P680^+\ Pheo^-$ on the 3 ps time scale. However, at room temperature an intrinsic time constant of 3 ps for charge separation may be extended to an experimentally observed 21 ps by exciton equilibration over multiple chromophores. The number of pigment molecules implicated in such a multimer may vary between preparations and thereby modify the observed rate of charge separation.

C. Charge Stabilization

Charge separation in PS II reaction center preparations produces the radical pair state, $P680^+\ Pheo^-$ (Danielius et al., 1987). In the normal course of events, this species undergoes further redox processes to produce stable products that can be used in water oxidation and in supplying electrons to Photosystem I. Forward electron transfer from $Pheo^-$ to Q_A proceeds in the 100's of ps time range (Renger, 1992) and the redox-active tyrosine Y_Z reduces $P680^+$ on the 10–500 ns time scale. Thus, in less than 1 μs, the charge separation is stabilized as the Y_Z^+ P680 Pheo Q_A^- species (see below). Prior to these stabilizing electron transfers, however, dynamic events occur in the initial charge-separated state.

Because the $P680^+\ Pheo^-$ radical pair is shallowly trapped, repopulation of the P680 exciton state occurs with fairly high probability. Deeper trapping of the charge separation only occurs upon electron transfer to Q_A. Thus, we can extend Eq. (1) above by focusing on the equilibria involving the radical pair state

$$(\text{P680 Pheo})^* \underset{k_{-2}}{\overset{k_2}{\rightleftharpoons}} \text{P680}^+\ \text{Pheo}^- \xrightarrow{k_3} \text{P680}^+\ \text{Pheo}\ Q_A^- \tag{2}$$

This model, which has been elaborated by Holzwarth and co-workers as the radical pair/exciton model (Schatz et al., 1987, 1988; Holzwarth, 1989) predicts biphasic fluorescence decay kinetics; one phase corresponds to the charge separation and the other to the decay of the recombination repopulation of the excited state. Attempts to implement this interpretation in PS II have been complicated by multiexponential fluorescence decays and by uncertainties as to how to interpret the relevance of the various kinetic phases of the emission. The 20 ps phase discussed above is an example. Nonetheless, the model can be used to make estimates of the free energy drop in forming the initial charge-separated state. Values of 50–125 meV have been obtained in the room-temperature to liquid-nitrogen-temperature range (Roelofs et al., 1991). These are substantially less than the ~170 meV drop that occurs upon forming the initial radical pair state (P^+BPheo^- (Renger, 1992)), in photosynthetic bacterial reaction centers, as expected from the greater thermodynamic efficiency of the oxygenic reaction centers.

The decay of the $P680^+\ Pheo^-$ radical pair state is multiphasic and occurs on the 10s of ns time scale in the absence of Q_A (Danielius et al., 1987; Takahashi et al., 1987; van Mieghem et al., 1992; Liu et al., 1993). In isolated PS II reaction centers, a substantial fraction of these radical pairs rephase to form the spin polarized triplet state that has been discussed above. Before considering this process, however, we turn to the oxidizing side of PS II, where the redox active tyrosine residue, Y_Z, reduces the $P680^+$ species.

V. Secondary Electron Transfer

A. Secondary Electron Donors, Y_D and Y_Z

1. Oxidation by P680⁺

Two of the redox components that reduce the oxidized primary electron donor ($P680^+$) are the redox active tyrosines Y_Z (Boska et al., 1983; Gerken et al., 1988; Debus et al., 1988a,b; Metz et al., 1989) and Y_D (Debus et al., 1988a; Vermaas et al., 1988; Buser et al., 1990). For recent reviews see Debus (1992), Babcock (1993), Barry (1993) and Hoganson and Babcock (1994). Y_D is probably located in a cul de sac interacting only with P680 (Debus et al., 1988b; Metz et al., 1989; Buser et al., 1990; Vass and Styring, 1991). Y_Z is most likely the immediate oxidant to the Mn cluster responsible for photosynthetic water oxidation (Babcock et al. 1976; Dekker et al., 1984a; Vos et al., 1991; Rappaport et al., 1994), though an intermediate has recently been proposed based on the kinetic predictions of a new structural model for the PS II donor side (Un et al., 1994).

2. Y_D and Y_Z are Tyrosines

Arguments identifying Y_Z and Y_D as tyrosines originate from many sources. The first evidence was provided by Barry and Babcock (1987), who showed that specific labeling of the cyanobacterium *Synechocystis* 6803 with perdeuterated tyrosine produced a loss of proton hyperfine coupling and consequent narrowing of the characteristic EPR signal IIs arising from Y_D•. A similar isotopic labeling experiment has been carried out for Y_Z• with detection at room temperature (Boerner and Barry , 1993), and also at low temperature in a Y_D-less mutant (Tang et al., 1996) with a similar result. A spectroscopic comparison of Y_D•, following in vivo incorporation of specifically deuterated tyrosine, with the corresponding in vitro model tyrosine radicals has further solidified identification of a biological tyrosyl radical and contributed to an interpretation of the partially resolved hyperfine components of the in vivo EPR spectrum of Y_D• (Barry et al., 1990). The optical spectrum for Y_Z•-Y_Z was reported by a number of groups in non-oxygen evolving PS II core complexes (Diner and de Vitry; 1984, Dekker et al., 1984b). The same optical (Gerken et al., 1988) and EPR (Hoganson and Babcock, 1988) difference spectra were observed in non-oxygen-evolving and in oxygen evolving complexes despite the much faster oxidation of Y_Z by $P680^+$ in the latter. These data confirmed that the same species was reducing $P680^+$ in the presence and absence of the OEC. A comparison of the optical difference spectrum Y_Z•-Y_Z with that of Tyr-Tyr in vitro also supported the tyrosyl radical assignment (Gerken et al., 1988).

3. Localization of Y_D and Y_Z

a. Directed Mutagenesis

Debus et al. (1988a) and Vermaas et al. (1988) proposed that the two redox active tyrosines were symmetrically located on either side of the two-fold rotational symmetry axis that characterizes the reaction centers of the purple non-sulfur photosynthetic bacteria and that is thought to exist in PS II as well (Michel and Deisenhofer, 1988). Identification of two tyrosines, D1-Tyr161 and D2-Tyr160 (cyanobacterial numbering, D2-Tyr161 in higher plants), toward the lumenal ends of the third transmembrane helices of polypeptides D1 and D2 in largely conserved regions of nine amino acid residues (Table 1), favored a two-fold symmetry model and suggested a common origin for these tyrosine containing sequences. Debus and coworkers (1988a,b) targeted these symmetrically localized tyrosines for directed mutagenesis, replacing each with a non-reactive Phe. Mutants at D1-Tyr161 and D2-Tyr160 were also constructed by Metz and coworkers (1989) and by Vermaas et al. (1988) in independent parallel efforts. The mutations resulted in a loss of Y_Z• function for D1-Tyr161Phe and a loss of Y_D• function for D2-Tyr160Phe. In the case of the mutant strain, D1-Tyr161Phe, with Y_D oxidized, the oxidizing equivalent went no further than $P680^+$, disappearing by charge recombination with Q_A^- ($t_{1/2}$ = 1 ms, Metz et al., 1989). These are negative experiments in the sense that they entail the disappearance of a signal, but they are consistent with the tyrosyl radical origin of Y_Z and Y_D and provide a likely localization for these residues. Replacement of D2-Tyr160 with Trp (Boerner et al., 1993) and D1-Tyr161 with Cys, Met or His (P. J. Nixon and B. A. Diner, unpublished) has given no clear evidence for oxidation of the substituted residue.

b. Dipolar Coupling between Y_D• /Y_Z• and the Non-heme Fe(II)

Further support for the localization of Y_Z and Y_D has

come from measurements of dipolar coupling of the respective tyrosyl radicals with the non-heme iron of the iron-quinone complex of PS II (Bosch et al., 1991, Hirsh and Brudvig, 1993). Using the dipolar interaction between the homologous non-heme Fe(II) of the bacterial reaction centers and $P870^+$ as a reference, Hirsh and Brudvig (1993) estimated by saturation recovery measurements the distance between the PS II non-heme iron and Y_D• to be 37 ± 5 Å. A similar saturation recovery experiment in Mn-depleted core complexes of the mutant D2-Tyr160Phe lacking Y_D and with Y_Z• trapped at low temperature (Koulougliotis et al., 1995) gives a similar estimate for the distance between the non-heme iron and Y_Z•. These observations confirm the symmetric localization of tyrosines Y_Z and Y_D around the putative C2 symmetry axis of the PS II reaction center that is extrapolated from bacterial reaction centers. According to the crystallographic structures of the bacterial reaction centers, the guanidinium group carbon of the amino acid residues L-Arg135 and M-Arg165 of *Rb. sphaeroides* or the guanidinium group carbon of the amino acid residue L-Arg135 and the imidazole group of M-His162 of *Rp. viridis* are 37Å from the non-heme iron. That L-Arg135 and M-Arg165 (M-His162) are the positional homologues of D1-Tyr161 and D2-Tyr160, respectively, from primary and secondary structural comparisons, also confirms the assignment of Z and D, respectively, to these tyrosines. Again, using the coordinates of the bacterial reaction center structure, the distance between the positional homologues of Y_Z and Y_D, L-Arg135 to M-Arg165 (M-His162), respectively, are 39 or 44 Å for *Rb. sphaeroides* and 32 or 39 Å for *Rp. viridis*, depending on whether the measurement is made between the side chain reference points (as above) or between the alpha carbons, respectively.

c. Dipolar Coupling between Y_Z• and $P680^+$

The linewidth of a transient EPR signal attributed to $P680^+$ has been studied by two groups under conditions where the lifetime of Y_Z• has been increased to allow simultaneous generation of the two radicals. Bock et al. (1988) observed two kinetic components for the reduction of $P680^+$ with $t_{1/2}$ of 170 and 500 μs having linewidths of 8G and 14G, respectively. Why these authors observed two components is unclear, though the slower of the two apparently matches a major component of charge recombination with Q_A^- observed optically. These linewidths are to be compared to 7.9G in the absence of Y_Z• but probably in the presence of Y_D•. Hoganson and Babcock (1988) observed a broadening of $P680^+$ to 9G of a component with a half lifetime of 150–200 μs. The broadening in both cases probably arises from the effect of Y_Z• on the local magnetic field at $P680^+$. This is consistent with a distance ranging from 10–15 Å, with the results of Bock et al. (1988) favoring the shorter end of this range. This distance is less than half of that estimated above for the Y_Z-Y_D distance. Because Y_Z and Y_D are equidistant from the likely two-fold symmetry axis passing through the non-heme iron of the reaction center, these measurements imply that $P680^+$ is located on the Y_Z-side of the symmetry axis. Current views of the structure of P680 (see text and Fig. 4, above) are able to accommodate this conjecture.

d. Dipolar Coupling Between Y_Z• and the OEC

Dipolar coupling between the OEC and Y_Z• was indicated by a difference in the power saturation characteristics of Y_Z• in the presence and absence of the Mn cluster (Warden et al. 1976, Yocum and Babcock, 1981). This Y_Z•-OEC distance has been estimated to be greater than 10 Å because of a lack of broadening of Y_Z• by the OEC (Hoganson and Babcock, 1988), although the strength of this interaction depends not only on distance but on the g-tensor orientation and anisotropy, hyperfine interactions, spin-lattice relaxation rates, spin density distribution, and zero-field splittings of the paramagnetic centers. A shorter distance, possibly as close as 4.5 Å, has been proposed (Gilchrist et al., 1995), based on the assignment to Y_Z• of the 160G S_3-split signal, observed upon Ca^{2+} depletion (Boussac et al. 1989; see Chapter 9). A comparison of the microwave power saturation characteristics of Y_Z• and Y_D• with the Mn cluster present showed that Y_Z• required greater microwave power to saturate (Warden et al., 1976; Babcock et al., 1989). With the reservation that the S-state distribution of the OEC was probably not identical for the power saturation measurements of these two radicals, these observations have been interpreted as implying a closer proximity of Y_Z• to the Mn cluster. Un and coworkers (1994) have placed the OEC equidistant from the two radicals, arguing that the difference in power saturation of Y_Z• and Y_D•, from Babcock et al. (1989), translates into practically the same distance (factor of only 1.26 closer for Y_Z) based on an r^{-6} power dependence, where r is the distance between the centers of spin density. At the other extreme, Gilchrist et al. (1995) estimate a Y_Z•-

OEC distance of 4.5 Å based on pulsed ENDOR measurements and identify Y_Z• with the 'split S_3 signal' in Ca^{2+}-depleted PS II particles.

The influence of the oxidation states of the OEC on the spin-lattice relaxation times of the EPR signal of Y_D• has been examined by the electron spin echo technique between 5 and 30K (Evelo et al., 1989c) and by measurements of the power saturation between 8 and 70K (Styring and Rutherford, 1988). Y_D• relaxes most slowly in the S_1 state and about equally in the S_2 and S_3 states. Relaxation enhancement by S_0 versus S_2, S_3 depends on the temperature. A relaxation enhancement for these states is interpreted as reflecting the presence of a mixed valence Mn cluster in the OEC. The distance between the OEC and Y_D• has been estimated at 29–43 Å by Evelo et al. (1989c), at 24–27 Å by Kodera et al. (1992a) and at 25–35 Å by Un et al. (1994).

e. Photovoltage and Electroluminescence

Additional information on the localization of Y_Z comes from photovoltage measurements across the thylakoid membrane that have now been extended into the 10–100 ns time scale (Pokorny et al., 1994). Flash excitation of O_2-evolving BBY (Berthold et al., 1981) membranes in the S_1 state of the OEC produces electron transfer from Y_Z to P680$^+$ that contributes an electrogenic step ($\tau = 29 \pm 5.6$ ns) equal in dielectrically weighted amplitude to 16% of the charge separated state, P680$^+$$Q_A^-$. Assuming a homogeneous dielectric, this amplitude corresponds to a distance along the membrane normal of 4 Å. The photovoltage technique, presently limited to the ≤100ns time domain, cannot detect the slower steps associated with the reduction of Y_Z• and proton release to the lumen. However, measurements of chlorophyll luminescence induced by electric field pulses (Vos et al., 1991) indicate a further electrogenicity associated with oxidation of the OEC by Y_Z equal to 5% of the total transmembrane charge separation or just under 2 Å projected along the membrane normal, assuming location of the OEC in the 35 Å hydrophobic membrane core (Pokorny et al., 1994). While the photovoltage measurements can measure only the fastest step associated with the oxidation of Y_Z in the S_1 state, the electroluminescence measurements are made over a sufficiently long time period to be able to resolve all of the electrogenic steps associated with each of the S-states, including proton release from the OEC.

f. Dipolar Coupling of Y_D• to Dy^{+3}

Dipolar relaxation enhancement studies by Innes and Brudvig (1989) using paramagnetic lanthanide ions (Dy^{+3}-EDTA) have led to the estimation of Y_D• at 26 Å from the stromal surface, 41 Å from the lumenal surface in the presence of the extrinsic polypeptides of PS II and 27 Å from this surface following Tris washing, which removes the 33, 23 and 17 kDa extrinsic polypeptides. A similar study by Isogai et al. (1990) estimated these same distances to be at 36, 28 and 20 Å, respectively. These EPR studies place Y_D• and, by symmetry, Y_Z• much deeper into the membrane from the lumenal surface than do the electroluminescence measurements. Likewise, little delocalized electrochromicity (carotenoid band shift) associated with oxidation of and deprotonation of the OEC has been detected. The distances obtained from electrochromic and electroluminescence techniques are dielectrically weighted, and could correspond to a greater physical separation of Y_Z from the OEC and/or the OEC from the lumenal space, if the intervening distance were to be a region of high dielectric constant .

4. EPR and ENDOR Spectra of Y_D• and Y_Z•

a. Fitting of Spectroscopic Parameters

Hoganson and Babcock (1992) have collated much of the spectroscopic information on Y_D•, from X-band orientation studies (Rutherford, 1985), EPR and ENDOR spectra after isotopic labeling (Barry et al. 1990, Hoganson and Babcock, 1992) as well as Q-band measurements (Brok et al., 1985). Using literature values of 2.0074, 2.0044 and 2.0023 for the g-tensor principal components, g_x, g_y and g_z, respectively (Brok et al., 1985, Gulin et al., 1992 and Un et al., 1994) and the proton hyperfine coupling constants determined by ENDOR, they were able to simulate well the literature spectra, assigning spin densities to the ring carbons in an odd-alternant fashion, and to the phenolic oxygen as well. These numbers have now been further refined by deuterium ESEEM spectroscopic analysis of specifically deuterated Y_D• in *Synechocystis* 6803 (Warncke et al., 1994) to give C1 (0.37), C2,6 (–0.06), C3,5 (0.25), C4-O (0.25) and C_β (0). The angles between the β-methylene protons and the C1 p_z-orbital are 68° and 52° (Warncke et al., 1994). Rigby et al. (1994) have calculated similar spin densities for several other species by using high resolution ENDOR techniques. The differences between Y_D• and other tyrosyl radicals described in the literature (e.g. ribonucleotide reductase, Hoganson and Babcock, 1992) concern primarily the C1 and the C4-O spin densities and the orientation of the β-methylene protons. Simulation of the orientation spectra indicate that Y_D• is oriented such that the membrane normal is 75 ± 10° away from the g_z axis, perpendicular to the plane of the ring, and 18–20° away from the long axis of the ring (essentially superimposed on g_x) passing through C1, C4 and the phenolic O. These orientation studies cannot distinguish between the phenolic oxygen pointing toward the lumenal or stromal surfaces. Modeling of the PS II reaction center by Svensson et al. (1990) better accommodates D2-Tyr160 with the phenolic O pointing toward the lumen.

Although cyanobacterial and higher plant PS II reaction centers resemble each other, there are differences in the EPR and ENDOR (Hoganson and Babcock, 1992, 1994) spectra of Y_D• that arise principally from differences in the hyperfine coupling of the β-methylene protons (Rigby et al., 1994). However, mutation of D2-Gln164 to Leu in *Synechocystis* 6803 gives rise to an EPR spectrum for Y_D• that more closely resembles that of spinach (Tommos et al., 1993). Mutations at D2-Met159 (Vermaas et al., 1988) and D2-Pro161 (Tommos et al., 1994) adjacent to the Y_D tyrosine also perturb the EPR spectral properties of Y_D•.

Kodera et al. (1992b) have recently trapped tyrosine Y_Z at 253K. Proton ENDOR spectroscopy indicates that the ring plane of Y_Z• is rotated around the C1-C_β bond with respect to Y_D•, altering the hyperfine coupling of the β-methylene protons. This conclusion is consistent with differences in the room temperature EPR lineshapes of Y_Z• and Y_D• reported by Boerner and Barry (1993). Tang et al. (1996), in an independent effort, have succeeded in trapping Y_Z• in a mutant of *Synechocystis* 6803 lacking tyrosine Y_D. The low temperature EPR (150K) and proton ENDOR (135K) spectra of the trapped Y_Z• are clearly different from those of Y_D•. Transient ENDOR and deuterium ESEEM spectra on isotopically labeled Y_Z• have now been collected in the Y_D•-less mutant (Tommos et al., 1995). Very small differences are observed for the hyperfine coupling constants of the ring protons and for the spin densities of Y_D• and Y_Z•. One difference, however, is the presence in the Y_D• spectrum of a well-defined $A_\perp$ component of the hyperfine tensor of an exchangeable proton with an axial line shape (hyperfine frequency 3.1 MHz, Tang et al., 1996). This feature has been assigned to a hydrogen-bonded proton (Rodriguez et al., 1987, Evelo et al., 1989b, Tang et al., 1993, 1996). This signal is broadened in the Y_Z• spectrum and is more difficult to observe (Tang et al., 1996, Tommos et al. 1995, see below). The position of the β-methylene protons is also different, showing in the case of Y_Z• a stronger hyperfine coupling of one of the β-methylene protons and a greater distribution of angles between the β-methylene protons and the ring plane (Tommos et al., 1995). Mino and Kawamori (1994), however, report low temperature proton ENDOR spectra of Y_Z• and Y_D• that are quite similar between 10 and 20 MHz. In this case, Y_Z• and Y_D• were both present so relaxation phenomena at low temperature or incomplete subtraction of the contribution of Y_D• may have enhanced the contribution of the Y_D• ENDOR spectrum to that attributed to Y_Z•.

b. Y_D• and Y_Z• are Neutral Radicals

Based on a comparison with model tyrosines, the zero crossing point (g_{iso}) of 2.0046 for Y_Z• and Y_D• (Table 1) has been interpreted as indicating that they are deprotonated during oxidation to give the neutral radical (Barry and Babcock, 1987, Evelo et al., 1989a). A protonated radical would be expected to show a g-value of 2.0032 (Dixon and Murphy, 1976). Such deprotonation is not surprising given the pKa of –2 of the tyrosyl radical (Dixon and Murphy, 1976). The ESEEM and proton ENDOR of Y_D• indicate the presence of a magnetically coupled proton, attributed to a hydrogen bond, that undergoes exchange in D_2O (Rodriguez et al., 1987, Evelo et al., 1989b, Tang et al., 1993). Measurements of the D/H exchange rate, using proton ENDOR to follow the fate of the exchangeable proton (hyperfine coupling constant, 3.1 or 3.5 MHz, depending on where measured), show the $t_{1/2}$ to be on the order of 8 h at 0 °C (Tang et al., 1993). Slow exchange suggests that the proton released by oxidation of Y_D is likely to remain localized in the vicinity of the radical so as to rejoin Y_D• upon its rereduction. An ESE-ENDOR experiment on Y_Z• indicates the presence of an exchangeable proton(s) (Diner et al., 1995, Force et al., 1995) after D/H exchange. This proton(s) shows a dispersion of hydrogen bond strengths, whose average resembles that of Y_D•. Furthermore, effects of D/H exchange upon the matrix

Table 1. Summary of the kinetic and thermodynamic properties of redox active tyrosines Y_Z and Y_D

Property	D	Z
Location	D2-161[(1)]	D1-160[(1)]
Local amino acid sequence	D2-SQGLPYMLFVS[(1)]	D1-GQGIPYILFVA[(1)]
Oxidation by $P680^+$ ($t_{1/2}$)	10 ms, pH 5–7.5[(2)]	23 ns (S_0, S_1), pH 7.5[(3)] 50–260 ns (S_2, S_3), pH 7.5[(3)] 1.4–32 μs (–OEC, pH 9 to 4)[(4)]
Oxidation by OEC (S_2, S_3)($t_{1/2}$)	~1 s (pH 7.6)[(5)]	
Kzp ($ZP680^+ = Z^+P680$)	NA	29 (S_0), pH 7.5[(3)] 29 (S_1), " [(3)] 2 (S_2), " [(3)] 2 (S_3), " [(3)] 18–600 (–OEC, pH 5–8.5)[(2, 17)]
E'_0 (pH 7)	0.72–0.76 V[(6,7)]	~1.1 V (+OEC) ($E_{mP680+/P680)} = 1.12$V) 0.97–1.05 V (–OEC), "
Reduction by OEC ($t_{1/2}$)	(a) 5–40 min (S0), pH 8–4.5[(7)] (b) 500 min (S0), pH 4.7–7.2[(7)] 30 min (S–1), pH 7.2[(8)] 13 min (S0, S-2), pH 7.2[(8)]	≤3 μs to 350 μs (S_0), pH 6-6.5[(10–13)] 55–110 μs (S_1), pH 6–6.5[(9, 10, 13)] 250–350 μs (S_2), pH 6–6.5[(9, 10, 13)] 1.2–1.3 ms (S_3), pH 6–6.5[(9, 10, 13)]
Back reaction with Q_A^- (–OEC) ($t_{1/2}$)		10 ms–500 ms (pH 5–8.5)[(2, 17)]
Reduction by Mn^{2+} (–OEC) ($t_{1/2}$)		6×10^6 to $4–4.8 \times 10^7$ M^{-1} s^{-1} (pH 6)[(14, 15)]
Protonation state	Neutral radical ($g_{iso} = 2.0046$)[(1)]	Neutral radical ($g_{iso} = 2.0046$)[(1)]
Accessibility to solvent (D_2O/H_2O exchange)	Extremely slow[(1)] ($t_{1/2}$ ~8 h, pD = 7.5)	Slow on time scale of e^- transfer[(13, 16)]

[(1)] see text; [(2)] Buser et al., 1990; [(3)] Brettel et al., 1984; [(4)] Conjeaud and Mathis, 1980; [(5)] Babcock and Sauer, 1973; [(6)] Boussac and Etienne, 1984; [(7)] Vass and Styring, 1991; [(8)] Messinger and Renger, 1993; [(9)] Babcock et al., 1976; [(10)] Dekker et al., 1984a; [(11)] Vos et al., 1991; [(12)] van Leuwen et al., 1993; [(13)] Rappaport et al., 1994; [(14)] Hoganson and Babcock, 1989; [(15)] Diner and Nixon, 1992; [(16)] Diner, 1986; [(17)] Yerkes et al., 1983.

region of the Y_Z• ENDOR spectrum have also been interpreted to indicate ready accessibility of solvent to Y_Z (Tang et al., 1996, Hoganson et al., 1995).

A local electrochromic shift ($\lambda_{max} = 440$ nm, $\lambda_{min} = 430$ nm) is associated with the oxidation of Y_Z in the presence (Rappaport et al., 1994) and absence (Diner and de Vitry, 1984, Dekker et al. 1984b) of the OEC. A similar band shift has now been observed for Y_D• (Diner et al., 1995). Although of smaller amplitude, this same shift is observed for the S state transition, $S_1 \rightarrow S_2$, where less than one proton is released per center (Rappaport and Lavergne, 1991). This signal is indicative, therefore, of the presence of a bound charge on the donor side of PS II, associated with the oxidation of Y_Z (Diner, 1986, Rappaport et al., 1994). Were the proton freed, upon the oxidation of Y_Z and Y_D, to be released to the lumen, then, contrary to what is observed, the oxidation and reduction of Y_Z and Y_D would no longer show a local electrochromic shift. Consequently, the phenolic proton released upon the oxidation of Y_Z or of Y_D remains in close proximity to the oxidized radical.

An alternative view, particularly with the Mn cluster intact, is that electrochromic shifts arise from local charge rearrangements (e.g. Haumann et al., 1994). In this interpretation, phenolic proton release from Y_Z into the lumen could occur and the observed electrochromism would be attributed to charge rearrangements in the protein matrix.

c. Identity of the Proton Acceptor

The exchangeable proton coupled to Y_D• was originally reported to have $A_\perp$ and $A_\parallel$ components of the hyperfine tensor with hyperfine frequencies of 3.5 and 7 MHz (Rodriguez et al., 1987), although a reinvestigation of the proton ENDOR now seems to indicate the absence of an exchangeable component at 7 MHz (Tang et al., 1993 and M. P. Espe and G. T. Babcock, unpublished). A proton ENDOR spectrum reaffirming this assignment (Mino et al., 1993) may have been contaminated by a narrow radical that appears during the lengthy incubation in D_2O (Rodriguez et al., 1987). The exchangeable proton has been suggested by two groups (Rodriguez et al., 1987; Evelo et al., 1989b) as arising from a proton, hydrogen bonded to the phenolic proton. An early proposal of Debus et al. (1988a) and modeling studies by Svensson et al. (1990) and by Ruffle et al. (1992) suggest D1-His190 and D2-His189 (cyanobacterial numbering, D2-His190 in higher plants) as being potential proton acceptors and hydrogen bond partners with Y_Z and Y_D, respectively. While the models indicate that the distance between the Tyr-His pair in D1 may be too great for a hydrogen bond, there is insufficient precision to eliminate this possibility. As a test of these His-Tyr interactions, two groups have examined the spectroscopic and kinetic consequences of site-directed mutations at these His residues. Replacement of D2-His189 by Leu and Tyr (Tommos et al. 1993) and by Asp, Gln and Leu (Tang et al., 1993) results in a narrowing of the Y_D• EPR signal. While the EPR spectra of the Asp, Gln and Leu mutants differ with regard to their resolved hyperfine structure, the proton ENDOR spectra of these same mutants all show the loss of the exchangeable proton with axial line shape at 3.1 MHz (Tang et al., 1993). These results indicate that there is a close interaction between D2-His189 and D2-Tyr160 and that this histidine is implicated in a hydrogen bond to Y_Z•. Using a point-dipole approximation and employing a spin density of 0.25 on the oxygen (Rigby et al., 1994; Warncke et al., 1994) one obtains an oxygen-proton distance of 1.85 Å, consistent with the proton serving as a hydrogen bond.

Mutations similar to those mentioned above have been made at the homologous D1-His190. Here, unlike the case for Y_D•, there is no apparent alteration of the Y_Z• EPR spectrum (Tang et al., 1993, 1996, Roffey et al., 1994a). The 3.1 MHz hyperfine component arising from an exchangeable proton coupled to Y_D• is greatly broadened in the proton and ESE-ENDOR and cw-ENDOR spectra of Y_Z• (Force et al., 1995; Tang et al., 1996), making it possible that D1-His190 is involved in a hydrogen bond to Y_Z•. However, the rate of oxidation of Y_Z following charge separation is slowed by a factor of 200 in Mn-depleted mutant core complexes giving a $t_{1/2}$ of about 2 ms and lowering the flash yield of Y_Z oxidation to ~1/3 that of wild type (Diner et al., 1991a; P. J. Nixon et al., unpublished). Here, D1-His190 might serve as a proton acceptor upon formation of Y_Z•, the absence of which could slow the deprotonation step required for the oxidation of Y_Z. Pokorny et al. (1994) argue against D1-His190 acting as a proton acceptor based on photovoltage measurements that measure a dielectrically weighted distance between Y_Z and P680 projected onto the membrane normal. However, not knowing exactly where $P680^+$ is located leaves the question open. Bernard et al. (1995) have examined the Y_Z•-Y_Z FTIR difference spectrum in *Synechocystis* 6803 wild type and a mutant at D1-His190 and conclude that on the time scale of their measurements, D1-His190 is not protonated and Y_Z• is not hydrogen bonded. However, conclusions concerning assignments of the vibrational bands in the Y_Z•-Y_Z difference spectra remain controversial and therefore tentative because of possible interference from either Q_A^--Q_A and Chl^+-Chl.

5. Kinetics of Oxidation and Reduction of Y_Z

a. Oxidation of Y_Z (OEC intact)

Delosme (1971) found that the fluorescence yield was markedly quenched in the S_2 and S_3 states of the OEC during microsecond flash excitation of dark-adapted *Chlorella* cells. Such quenching reflects the enhanced concentration of $P680^+$, known to be a quencher of fluorescence (Butler et al., 1973). It was first pointed out by Bouges-Bocquet (1980) that this enhanced fluorescence quenching in the S_2 and S_3 states is a reflection of a lowered equilibrium constant, Kzp, ($Y_ZP680^+ = Y_Z$•(H^+)P680) in these states. According to Delosme (1971), fluorescence yields measured during the second and third actinic flashes given to dark-adapted *Chlorella* cells are nearly equal. Furthermore, the fluorescence yield of these are about two-thirds of that of the fifth flash, which should contain a high concentration of $S_0 + S_1$. By setting the fifth flash concentration of Y_ZP680^+ to near zero and the quenching non-linear with $P680^+$ concentration with a probability (p) of energy transfer of 0.5 (Joliot and Joliot, 1964), about 20% of the centers in the S_2 and S_3 states should contain Y_ZP680^+ during the flash lifetime. This concentration of Y_ZP680^+ implies an equilibrium constant, Kzp, of about 4.

Brettel et al. (1984) approached this problem from another angle and found that the rate of oxidation of Y_Z by $P680^+$ varies with the oxidation state of the Mn cluster, showing at pH 7.5 a $t_{1/2}$ of 23 ns in the S_0 and S_1 states and equal contributions of phases of 50 and

260 ns in the S_2 and S_3 states. It is known that oxidation of the OEC from the S_1 to the S_2 state is accompanied by a release of 0–0.5 protons, depending on the pH (Rappaport and Lavergne, 1991; for review, see Lavergne and Junge, 1993; see also Chapter 10) that results in an increase in the net charge of the OEC. Brettel and coworkers (1984) suggested that this net positive charge of the S_2 and S_3 states slows oxidation of Y_Z and decreases Kzp. These authors estimated Kzp at 29 for states S_0 and S_1 and 2.2 for states S_2 and S_3 at pH 7.5 (Table 1). These kinetics were further examined as a function of pH by Meyer et al. (1989) who observed components of 20, 40 and 280 ns attributed to the accumulation of 0, 1 and 2 positive charges, respectively, on the Mn cluster. In agreement with the above estimates for Kzp, Rappaport et al. (1994) have found that the flash-induced difference, $\Delta A_{440\ nm} - \Delta A_{424\ nm}$ is larger by 35% in states $Y_Z{\bullet}S_0$ and $Y_Z{\bullet}S_1$ than it is in $Y_Z{\bullet}S_2$ and $Y_Z{\bullet}S_3$ measured 10 μs after a light flash at pH 6.5. This difference is attributed to 15% of the centers in the latter two states as having $P680^+$ (Kzp = 5–6, Jérôme Lavergne, personal communication). All of the above observations are also consistent with the appearance of microsecond components in the reduction kinetics of $P680^+$ that arise from the reduction of $Y_Z{\bullet}$ in the S_2 and S_3 states (Schlodder et al., 1985). The behavior of Kzp as a function of flash number forms the basis of an analysis by Shinkarev and Wraight (1993) of the 'misses' observed in oxygen evolution arising from charge recombination and from centers containing either $P680^+$ or Q_A^-, all governed by PS II donor and acceptor side equilibria. All of these observations are consistent with the model of Brettel et al. (1984) in which the accumulated charges on the OEC govern the donor side equilibria. While not the only explanation, should the coulombic interpretation be correct, it would imply that the dielectrically weighted distance from the OEC to Y_Z is less than it is to P680.

b. Oxidation of Y_Z (Mn-depleted OEC)

Oxidation of Y_Z by $P680^+$ is slowed by loss of the Mn cluster to between 1.4 and 32 μs ($t_{1/2}$) over the pH span of 9.0 to 4.0, respectively (Conjeaud and Mathis, 1980, Table 1). The equilibrium constant for Kzp is also increased to between 18–600 for the pH range 5.0 to 8.5, respectively (Yerkes et al., 1983, Buser et al., 1990). Judging from the likely effect of charge of the OEC on Kzp in the presence of the Mn cluster, it is probable that the net charge of the region formerly occupied by the Mn cluster is more negative, owing to free carboxyl groups normally involved in metal coordination, and consequently more favorable to the stabilization of $Y_Z{\bullet}$, particularly at alkaline pH. The oxidation rate would normally be expected to increase with a decrease in the Em of $Y_Z{\bullet}/Y_Z$. The observed slowing may reflect a small increase in the Y_Z-$P680^+$ distance, giving a factor of ten slowing of the rate per 1.7 Å (Moser et al., 1992). Alternatively, the slowing could reflect an altered interaction between Y_Z and D1-His190 (e.g. shift in pKa) that occurs upon the loss of the Mn cluster.

c. Reduction of $Y_Z{\bullet}$ (OEC Intact)

The rate of reduction of $Y_Z{\bullet}$ by the OEC has recently undergone a reevaluation from earlier EPR (Babcock et al., 1976) and optical measurements (Dekker et al., 1984a). Rappaport et al. (1994) report rates of $Y_Z{\bullet}$ reduction by the OEC of 200–250 μs, 55 μs, 250–300 μs and 1.2 ms for the transitions $S_0 \rightarrow S_1$, $S_1 \rightarrow S_2$, $S_2 \rightarrow S_3$ and $S_3 \rightarrow S_0$, respectively (Table 1). The first rate is probably the most contested; values range from 30 μs to ≤3 μs in optical measurements (Dekker et al., 1984; van Leeuwen et al., 1993, respectively). A possible source for confusion is that the $S_0 \rightarrow S_1$ transition is generally measured on the fourth flash, where it needs to be deconvoluted from the contributions of the other S states. The absence of slow kinetics, observed by EPR, for this transition (Babcock et al., 1976) may stem from magnetic coupling between the Mn cluster and $Y_Z{\bullet}$ making it more difficult to detect the latter. The very fast rate proposed by van Leeuwen et al. (1993) for $S_0 \rightarrow S_1$ is attributed by Rappaport et al. (1994) to an overestimation of the extinction coefficients for S_1-S_0 at the wavelengths used for their measurement. Furthermore, Rappaport et al. (1994) suggest that a 350 μs component detected by Vos et al. (1991) in a study of electroluminescence associated with $S_0 \rightarrow S_1$ could be consistent with their slower rate.

Rappaport et al. (1994) also report a lag of 30–50 μs in the reduction of $Y_Z{\bullet}$ in the time range of the partial decay of a local electrochromic shift (440–424 nm) following the formation of $Y_Z{\bullet}S_3$ and $Y_Z{\bullet}S_0$ (pH>6.5). These authors propose that Y_Z oxidation induces a deprotonation in the vicinity of the Mn cluster, rather than in the vicinity of $Y_Z{\bullet}$, thus lowering the activation energy for oxidation of the OEC and allowing the oxidation of the OEC to proceed. This phenomenon is therefore a further indicator of the electrostatic interactions that likely occur between the components on the donor side of the PS II reaction center. They also imply that Y_Z and the OEC are close to each other. Further support of this proximity comes from the attribution of the 'S_3 split signal' (see below) to Y_Z•, broadened at low temperature by the location of the Mn cluster possibly as close as 4.5 Å away (Gilchrist et al., 1995).

6. Kinetics and Reduction Potentials of Y_Z and Y_D

Y_Z• is not the only tyrosyl radical to interact with the OEC, although it is the only one of the two to do so directly. Y_D•, acting through P680, is able to oxidize the S_0 state to S_1 with a rate of 6–9 × 10^{-4} s^{-1} (Styring and Rutherford, 1987, Messinger and Renger, 1993, Table 1). Following reduction of the OEC by hydrazine to the S_{-1} and S_{-2} states, such oxidation by Y_D• occurs with rates of 4 × 10^{-4} and 9 × 10^{-4} s^{-1}, respectively, (Messinger and Renger, 1993). Y_D is able to reduce both S_2 and S_3 by one equivalent with a $t_{1/2}$ of ~1 s (Babcock and Sauer, 1973; Velthuys and Visser, 1975). In the absence of the Mn cluster its rate of oxidation for $Y_D \rightarrow P680^+$ is, unlike Y_Z, pH independent from pH 5.0 to 7.5 with a rate constant of about 60 s^{-1} (Buser et al. 1992). Vass and Styring (1991) and Boussac and Etienne (1984) have estimated the midpoint potential of Y_D•/Y_D at 720–760 mV, far less than the Em for Y_Z•/Y_Z of 1.03 V for S-states S_0 and S_1 and 1.10 V for states S_2 and S_3 based on the estimates of Kzp above and an Em of 1.12 V for $P680^+$/P680 (Klimov and Krasnovskii, 1981). The reasons for the higher reduction potential for Y_Z•/Y_Z than Y_D•/Y_D are at present unclear. The immediate environment of Y_Z has been proposed, based on modeling of the reaction center (Svensson et al. 1990), to be more hydrophilic than that of Y_D. This proposal is now supported by experimental data showing H/D exchange of matrix protons in the vicinity of Y_Z• (Hoganson et al., 1995; Tang et al., 1996). While this proposal is consistent with the greater accessibility to Y_Z• of hydrophilic donors like Mn^{2+} (see below) or ferrocyanide (Dekker et al., 1984b), the fact that the phenolic proton, lost upon oxidation of Y_Z, remains localized may indicate that these hydrophilic donors are still maintained at a distance, particularly as the rates of electron transfer are not particularly fast ($t_{1/2}$ of first order oxidation of Mn^{2+} oxidation is ~400 μs, Hoganson and Babcock 1989). Because the proton, released upon oxidation of Y_D and Y_Z, is sequestered, the pKa of the proton acceptor becomes a determinant of the reduction potential of the tyrosyl radical as do the charges located in the immediate environment. D1-His190 may serve as both a proton acceptor to Y_Z (see above) as well as a ligand to the Mn cluster (all site-directed mutants at this site are either inactive or severely impaired for oxygen evolution, Roffey et al., 1994a; Chu et al. 1995; P. D. Nixon, X. -S. Tang and B. A. Diner, unpublished). Such coordination would be expected to lower the pKa of D1-His190 and at the same time raise the Em of Y_Z•/Y_Z.

7. Other Organic Radicals on the Donor Side of PS II

Barry and coworkers have recently observed an organic radical, M^+, upon room temperature illumination of non oxygen-evolving core complexes isolated from the mutants D1-Tyr161Phe (Noren and Barry, 1992) and D2-Tyr160Phe and Trp (Boerner et al., 1993), that does not resemble a typical tyrosyl powder EPR spectrum, showing a 1:3:3:1 lineshape. However, a narrowing of the EPR signal upon labeling with specifically deuterated tyrosine of mutant D2-Tyr160Phe supports the assignment to at least a modified tyrosyl radical (Boerner and Barry, 1994) or possibly to a tyrosyl radical superimposed on another paramagnetic component. The observation of this signal is probably preparation dependent as this signal is not observed in PS II core complexes isolated by a different procedure (Metz et al., 1989; Tang and Diner, 1994; Tang et al., 1996) from independent isolates of the same mutants.

Under conditions of Ca^{+2} depletion, a split signal, with a linewidth of 160G (Boussac et al. 1989) is observed in the g = 2 region upon illumination of centers in the S_2 state. The signal, commonly referred to as the 'S_3 split signal,' has been proposed to arise from an organic radical coupled to the S_2 state of the OEC (S_2X^+). Under conditions of Cl^- depletion similar signals are observed in this S_2X^+ state with linewidths of 90G (Boussac et al., 1992), 160G (F^- treatment) Baumgarten et al., 1990). Acetate treatment (MacLachlan and Nugent, 1993) produces a 230 G signal for S_2X^+. This signal, assumed to be the same species in all cases, has been variously attributed to either a histidine (or OH-histidine) radical (Boussac et al., 1990; Berthomieu and Boussac, 1995) or to Y_Z• itself (Hallahan et al., 1992; Deak et al. 1994;

Gilchrist et al.. 1995), broadened by interaction with a nearby paramagnetic center, presumably the OEC in the S_2 state. The experiments of Hallahan et al. (1992) and Deak et al. (1994) have, however, been criticized by Boussac and Rutherford (1995) as being unable to distinguish between a photoaccumulation of Y_Z• and an accelerated relaxation by the S_2X^+ state of a partially microwave power saturated Y_D•.

Boussac et al. (1992) have reported that under conditions that produce S_2X^+, Y_Z is no longer oxidizable. However, if the split signal reflects the presence of Y_Z•, then no further oxidation would be possible. Alternatively, whatever is responsible for the broadening of the split signal may also broaden the Y_Z• signal making it difficult to detect. Rutherford and coworkers favor the histidine model and suggest its possible role as an intermediate between Y_Z• and the Mn cluster, operating even under normal physiological conditions where Ca^{2+} and Cl^- are still present (Boussac and Rutherford, 1994). On the other hand, histidine (Em ~1.55 V, Isied, 1991) would not normally be expected to be oxidizable by $P680^+$ and replacement of D1-Tyr161 by His does not show oxidation of the substituted residue (P. J. Nixon and B. A. Diner, unpublished). While proponents of the assignment of the split signal to a tyrosyl radical have been strengthened by the similarity in the pulsed ENDOR spectra of S_2X^+ and of Y_Z• (Gilchrist et al., 1995), there remain nagging questions that keep the controversy alive.

Thermoluminescence emission from Tris-washed thylakoids has been examined by illumination at –20 °C, cooling to –50 °C, and then detecting light emission upon warming. Ono and Inoue (1991) have attributed the A_T emission band at –20 °C to a back-reaction between Q_A^- and oxidized histidine and not to Q_A^- and Y_Z•. Their arguments are based on the sensitivity of the A_T band to diethylpyrocarbonate (Ono and Inoue, 1991), a reagent known to modify histidine, and to differences in the sensitivity to DCMU of Y_Z• and the charge pair responsible for the A_T band (Koike et al., 1986). A mutation introduced at D1-His190 (D1-His190Phe) has also been reported to eliminate the A_T band. While this His could be the oxidized species, a mutation at this site also lowers the quantum yield of Y_Z oxidation (Diner et al., 1991a; Tang et al. 1993; Roffey et al., 1994a; Chu et al., 1995; P. J. Nixon et al., unpublished) to the point where Y_Z• is no longer generated during the –20 °C preillumination of the thermoluminescence experiment. Debus (1992) has argued that the conditions used for generation and inhibition of the A_T band do not provide compelling proof of histidine involvement (except perhaps indirectly, e.g. D1-His190, see above) and could as well be consistent with recombination of Q_A^- with Y_Z•. Indeed, the recent demonstration by Kodera and collaborators (1992b) of oxidation of Y_Z upon illumination at –20 °C would tend to support this view. However, in a recent report, Kramer et al. (1994) have examined the A_T band emission in *Chlamydomonas reinhardtii* PS II-enriched membranes bearing mutations at D1-His195. These authors (Roffey et al., 1994b) report that the mutation D1-His195Asp decreases the equilibrium constant for Kzp 50-fold at pH 7.6 without shifting the peak temperature of the A_T band detected at –16.2 °C. Roffey et al. (1994b) conclude, based on these observations, that Y_Z• is not a participant in the recombination leading to the A_T thermoluminescence. If the position of the A_T thermoluminescence band were determined primarily by the activation energy for reduction of Y_Z•, and the activation energy were large compared to ΔG_{ZP}, then the A_T band position might be expected to be relatively insensitive to K_{ZP}. Here too, the question of the source of the oxidizing equivalent remains open.

8. Photoactivation of the OEC

Tamura and Cheniae (1987) have proposed a model for the assembly of the Mn cluster that requires at least two steps of photoactivation. The first step generates an unstable intermediate that is further oxidized and stabilized by the second. Two additional Mn^{2+} are subsequently bound to give a mixed valence tetranuclear Mn cluster. Additional light reactions are thought to be required before this cluster reaches the S_0 state. Hoganson and Babcock (1989), Nixon and Diner (1992) and Diner and Nixon (1992) have shown that after Mn depletion of PS II-enriched membranes and core complexes Mn^{2+} can act as a donor to Y_Z•, implicating Y_Z• as the immediate oxidant of the Mn^{2+} that are incorporated into the cluster. Oxidation of Mn^{2+} is largely second order at pH 5.7–6 with second order rate constants ranging from 6×10^6 $M^{-1} s^{-1}$ (Hoganson and Babcock,1989) to 4–4.8×10^7 $M^{-1} s^{-1}$ (Diner and Nixon, 1992, Table 1). This oxidation becomes at least partially first order at pH 7.5 showing a 400 μs phase (Hoganson and Babcock,1989). Eleven site-directed mutations introduced at D1-Asp170 result in strains that vary in their ability to assemble the Mn cluster and consequently to evolve oxygen (Nixon and Diner, 1992). The second order rate constant for Mn^{2+} oxidation in the D1-Asp170Ser mutant, completely inactive for oxygen-evolution, is decreased to $5.5 \times 10^5\ M^{-1}\ s^{-1}$, further linking the oxidation of Mn^{2+} by Y_Z• to the assembly of the Mn cluster. Boerner et al. (1992) have shown that one of the least active of these mutants for oxygen evolution, D1-Asp170Asn, has reduced levels of Mn bound per reaction center. D1-Asp170, only nine amino acid residues away from D1-Tyr161 (Y_Z), is therefore implicated in a binding site that positions, for oxidation by Y_Z•, at least the first Mn^{2+} that ultimately goes into the assembled Mn cluster.

Tyrosine Y_D•, most likely acting through P680 and Y_Z, has been shown to oxidize S_0 to S_1 (Styring and Rutherford, 1987). These authors also propose that Y_D• could help maintain the Mn cluster in a more oxidized and therefore more stable state. Y_D• could also be potentially involved in photoactivation of the OEC, in steps that have reduction potentials below 750 mV. Such an oxidation would be of particular value where an unstable intermediate could be oxidized to a stable product by Y_D•, without having to wait for a second photoevent. Consistent with such a role are the results of Messinger and Renger (1993) who have shown that Y_D• can oxidize states S_0, S_{-1} and S_{-2} of the OEC.

B. Primary and Secondary Quinone Electron Acceptors, Q_A and Q_B

1. Q_A is a Plastoquinone

The primary quinone electron acceptor, Q_A, is a plastoquinone-9, normally participating in one-electron redox chemistry involving the quinone and semiquinone states. The electron on Q_A^- is normally transferred to plastoquinone, Q_B, which by contrast, can exist under physiological conditions not only in these redox states but in the quinol form as well. Reduction of Q_A to the quinol form requires extremely reducing conditions or extensive illumination (see below). The chemical identification of Q_A was first based on the resemblance of the light generated Q_A^--Q_A difference spectrum to the in vitro difference spectrum of the plastosemiquinone anion radical minus plastoquinone (Stiehl and Witt, 1969; van Gorkom 1974) and later to the quantification of the number of plastoquinone-9 per center in PS II core complexes, functional for the reduction of Q_A but lacking Q_B (de Vitry et al., 1986; Takahashi and Satoh, 1987). The plastosemiquinone remains in the anionic form even down to pH 4 (Pulles et al., 1976), indicating either an inaccessibility of the semiquinone to a proton donor of the appropriate pKa, or a pKa of the semiquinone depressed by its environment to <<4 from a pKa of 5.9 in aqueous solution (Rich and Bendall, 1980). Oddly, the generation of the anion radical appears not to engender proton uptake at pH 7 at the stromal surface of PS II, at least not in the presence of DCMU where the electron goes no further than Q_A (Fowler and Kok, 1974).

2. Q_A Binding Site

Modeling of the Q_A binding site in PS II (Michel and Deisenhofer, 1988; Diner et al., 1991b; Ruffle et al., 1992), based on the crystallographic structure of the Q_A binding site in the photosynthetic bacteria (Michel et al., 1986; Allen et al., 1988), indicates substantial homology between the two. Residues D2-Thr217, Phe252 and Trp253 (cyanobacterial numbering, plus 1 for spinach) are homologues of M-Thr220, Phe249 and Trp250 of *Rp. viridis*, respectively. D2-His214 (cyanobacterial numbering), is homologous to M-His217 and probably shares with this residue a role in non-heme iron coordination and hydrogen bonding to the Q_A quinone. Recent evidence in favor of a histidine providing a hydrogen bond to Q_A^- has been obtained through ESEEM detection of nitrogen coupling to Q_A^- using PS II preparations in which Q_A^- has been magnetically uncoupled from the non-heme iron (Atashkin et al., 1995). However, Tang et al. (1995), using isotopic labeling, have found no evidence for histidyl nitrogen coupling to Q_A^-. M-Ala258 which provides another hydrogen-bond to the Q_A quinone through its peptide bond may be conserved in D2-Ala260 (cyanobacterial numbering) or replaced by the peptide bond of another residue. ENDOR mesurements of Q_A^- uncoupled from the non-heme iron by treatment with trypsin (Macmillan et al. 1990) indicate the presence of an asymmetry in the hydrogen bonds like that previously reported for Q_A^- in reaction centers of *Rb. sphaeroides* (Feher et al., 1986).

Unlike Q_A of the reaction centers of *Rb. sphaeroides*, where it has been straightforward to extract ubiquinone and to replace it with a wide variety of quinones and quinone analogues (see for example Gunner et al., 1986), efforts to extract plastoquinone and to replace Q_A of PS II with plastoquinone or other quinones have been less successful. Many attempts at extraction of Q_A and subsequent reconstitution have either lacked the appropriate spectroscopic controls to show that the quinone actually went back into the Q_A site or resulted in an irreversible loss of quinone photoreduction. The detection of local electrochromic shifts of pheophytin (540–550 nm, van Gorkom, 1974), an EPR doublet arising from $Pheo^-$ (Klimov et al., 1980), or of FTIR vibrational bands (Araga et al., 1993), associated with the formation of the semiquinone anion are good indicators that the quinone has returned to its binding site. Such controls are essential, particularly in light of a report by Warncke and Dutton (1991) in bacterial reaction centers of direct electron transfer from the pheophytin anion to a quinone located outside the Q_A site and presumably at the surface of the reaction center. Using the electrochromic shift as a control, Diner et al. (1988) found that Q_A could be replaced by exchange in the presence of detergent at elevated temperature, but that extraction and reconstitution with the same quinones did not work. The reasons for the irreversible loss of Q_A photoreduction upon quinone extraction imply either irreversible denaturation of the center or the extraction of a cofactor required for function. The latter explanation has been strengthened recently with a report of reconstitution in PS II core complexes of Q_A function that required the addition of both lipid and the PsbH, PsbL and 4.1 kD subunits of PS II (Araga et al., 1993). These observations open the possibility of one of these small subunits being located close to the Q_A binding site.

3. Spectroscopic Properties of Q_A^-

The EPR spectrum of Q_A^- of PS II is broadened by an exchange interaction with the non-heme iron (Nugent et al., 1981, Rutherford and Zimmermann, 1984) probably located at a distance of ~7 Å (edge to edge) by analogy with the bacterial reaction centers (Deisenhofer et al., 1984). Two characteristic *g*-values have been observed for $Q_A^-Fe(II)$ over the pH range 6–9 (Rutherford and Zimmermann, 1984). A value of $g = 1.82$ with a high field feature at $g = 1.67$ is observed at acid pH (<7). Upon the addition of formate a very much enhanced signal is observed at $g = 1.82$ (Vermaas and Rutherford, 1984). This form is most similar to the $Q_A^-Fe(II)$ EPR signal observed in reaction centers of *Rb. sphaeroides*. Also, following formate treatment, the Mössbauer spectrum of the non-heme iron (Petrouleas et al., 1992) changes from an isomer shift and quadrupole splitting of 1.19 mm s^{-1} and 2.1–2.9 mm s^{-1}, respectively, to values of 1.18 mm s^{-1} and 2.07 mm s^{-1}, respectively, close to those observed for the reaction centers of *Rb. sphaeroides* (IS = 1.16 mm/s, QS = 2.15 mm/s). These similarities imply that with formate treatment the inner coordination sphere of the iron resembles more closely that of the bacteria. At pH>7, and probably with bicarbonate bound, an alternative $g = 1.9$ signal is observed for $Q_A^-Fe(II)$ with a high field feature at $g = 1.64$. The $g = 1.9$ form has been associated with a lowered $E_{m,7}$ for the Fe(III)/Fe(II) couple (~400 mV) of the non-heme iron, oxidizable by ferricyanide (Petrouleas and Diner, 1987). Formate or acid treatment, both of which induce a $g = 1.82$ $Q_A^-Fe(II)$ EPR signal, have been reported (Zimmermann and Rutherford, 1986, Diner and Petrouleas, 1987a) to result in an increase in the $E_{m,7}$ of the iron (>500 mV), making it, like the non-heme iron of the bacterial reaction centers, less likely to be oxidized by ferricyanide. There is, however, some recent controversy on this point. Stemler and Jursinic (1993) observe no effect of 100 mM formate on the $E_{m,7.2}$ of the Fe(III)/Fe(II) couple, in contrast with Deligiannakis et al. (1994), who observe an increase in the $E_{m,6.2}$ in the presence of 100 mM formate.

An additional PS II acceptor side EPR signal has been observed at $g = 1.66$ (Corrie et al., 1991). This signal has been attributed to the state $Q_A^-Fe(II)Q_B^-$, a state not EPR detectable in the bacterial reaction centers (Wraight, 1978; Rutherford and Evans, 1979).

4. Q_B Binding Site

The binding site of Q_B shows substantial homology to that of the reaction centers of the purple non-sulfur photosynthetic bacteria. D1-Leu218, Tyr254, Phe255 and Ser264 are the PS II homologues of the *Rps. viridis* L-Leu193, Tyr215, Phe216, and Ser233 (Michel and Deisenhofer, 1988; Diner et al., 1991a; Ruffle et al., 1992). Many of the inhibitors that block photosynthetic electron transport bind to the Q_B binding pocket. The above mentioned homology is strengthened by the observation of homologous herbicide resistance mutations in the purple bacterial and PS II reaction centers (see Oettmeier, 1992 for a review). The Q_B quinone is hydrogen bonded to L-His190 and L-Ser223 in the reaction centers of *Rb. sphaeroides* (Allen et al. 1988) and *Rps. viridis* (Michel and Deisenhofer, 1988). The likely PS II homologues of these residues are D1-His215 and D1-Ser264. Mutation of the latter residue to glycine raises the dissociation constant, Kq, of the Q_B binding site for plastoquinone by a factor of ten (Taoka and Crofts, 1990). L-Ser223 and L-Asp213 (Paddock et al., 1990; Takahashi and Wraight, 1990) on the one hand and L-Glu212 (Paddock et al., 1989) on the other, have been implicated in providing the first and second protons, respectively, for the protonation of the semiquinone and the quinol: $Q_B^- + H(1)^+ = Q_BH$; $Q_BH + e^- + H(2)^+ = Q_BH_2$ in the reaction centers of *Rb. sphaeroides*. The PS II homologues of L-Glu212 or L-Asp213 have not yet been identified, though one possibility is D1-His252. The protonation of D1-His252 has been proposed to be coupled to formation of Q_B^- (H. H. Robinson and A. R. Crofts, personal communication). Indeed, mutation of this residue to Leu or Gly lowers by a factor of ten the apparent equilibrium constant, $K'_{AB(1)}$ for the reaction $Q_A^- + PQ \rightleftharpoons Q_A^-Q_B \rightleftharpoons Q_AQ_B^-$ (Diner et al., 1991a,b). It is possible though that both the dissociation constant, Kq, and the intrinsic equilibrium constant, $K_{AB(1)}$ for reaction AB(1): $Q_A^-Q_B \rightleftharpoons Q_AQ_B^-$, related to $K'_{AB(1)} = K_{AB(1)}/(1 + Kq/[PQ])$, are lowered by these mutations. In addition, replacement of D1-Ser264 by glycine (Taoka and Crofts, 1990), unlike its bacterial homologue (L-Ser233Ala, Paddock et al., 1990), has surprisingly little effect on the forward rate of reaction AB(2): $Q_A^-Q_B^- + H^+ \rightleftharpoons Q_AQ_BH^-$, implying that D1-Ser264 does not play an essential role in the protonation of the Q_B quinol in PS II.

A large number of benzo-, naptho- and anthra-quinones have been demonstrated to bind to the Q_B site as demonstrated by competition with endogenous plastoquinone (Oettmeier, 1992), by competition with labeled herbicide inhibitors (Oettmeier, 1992) and by the ability to oxidize the non-heme iron of PS II (see below) either as the quinone in the case of the tetrahalogenated p-benzoquinones (Koike et al., 1993) or as the semiquinone as in the cases of p-benzoquinone and phenyl-p-benzoquinone, formed by oxidation of Q_A^- (Zimmermann and Rutherford, 1986, Petrouleas and Diner, 1987).

One novel consequence of the diversity of molecules that bind to the Q_B site is the ability to combine an electron acceptor with an inhibitor molecule. Bocarsly and Brudvig (1992) have designed a (dimethylureido) phenyl nitroxide, a stereo analogue of the herbicide chloroxuron. This species binds tightly to the Q_B site, accepting only one electron. Upon continuous illumination at 250K in the presence of this acceptor, only two oxidizing equivalents are generated by the reaction center allowing synchronized generation of the S_3 state, a feat otherwise difficult to perform in a dense EPR sample.

The endogenous plastoquinol bound to the Q_B binding site as Q_BH_2 has recently been implicated as a reductant of oxidized Cyt *b*-559, a cytochrome tightly associated with the PS II reaction center (Buser et al., 1992). The redox couples Cyt *b*-559$^+$/Cyt *b*-559 and Q_B^-/Q_BH_2 were shown to interact in a pH-dependent redox equilibrium, implying a role for Q_B in a slow electron transfer cycle around PS II.

5. Coordination of the Non-heme Iron by Small Molecules

The non-heme iron is most likely coordinated by D1-His215, D1-His272, D2-His214 and D2-His268 (cyanobacterial numbering) by analogy with the purple bacterial reaction centers (Michel and Deisenhofer, 1988). The $g = 1.9$ and 1.82 EPR signals, mentioned above, have been suggested (Corrie et al., 1991) to correspond to a state with bicarbonate bound and a bicarbonate-free state, respectively, though this point has not been unequivocally established. See Blubaugh and Govindjee (1988) for a review of bicarbonate effects on PS II. Michel and Deisenhofer (1988) and van Rensen et al. (1988) have proposed that bicarbonate might replace the M-Glu232 (*Rp. viridis* numbering) as a ligand to the non-heme iron, occupying the remaining one or two coordination positions. Supportive of this assignment is the observation of a reversible decrease of the quadrupole splitting of 0.7 mm s^{-1} in the Mössbauer spectrum of the non-heme iron upon replacement of bicarbonate with formate, indicating a change in the inner coordination sphere of the iron (Diner and Petrouleas, 1987a). Probably the strongest evidence thus far that bicarbonate is actually coordinated to the non-heme iron, comes from the demonstration that NO can bind to the iron and that NO is displaced by bicarbonate (Diner and Petrouleas, 1990). NO does not bind to the non-heme iron of bacterial reaction centers, probably because of the stability of glutamate coordination. In PS II, a large $g = 4$, X-band EPR signal arising from the $S = 3/2$, Fe(II)-NO adduct disappears reversibly when either Q_A or Q_B is in the semiquinone form, consistent with the signal arising from the NO adduct to the non-heme iron of the iron-quinone complex and the formation of a magnetically coupled integral spin system (Petrouleas and Diner , 1990). The loss of the $g = 4$ signal upon addition of bicarbonate, implies a displacement of NO and a competition between NO and bicarbonate for coordination to the non-heme iron. In the presence of bicarbonate, the forward rate of reaction AB(1): $Q_A^-Q_B \rightleftharpoons Q_AQ_B^-$ is faster ($t_{1/2} = 100–200\ \mu s$) than that for reaction AB(2): $Q_A^-Q_B^- + H^+ \rightleftharpoons Q_AQ_BH^-$ ($t_{1/2} = 300–500\ \mu s$) (Robinson and Crofts, 1983). One consequence of NO binding is the slowing of forward electron transfer that is considerably more marked for reaction AB(2) than it is for reaction AB(1) (Diner and Petrouleas, 1990). This observation matches that observed upon bicarbonate displacement by formate (Eaton-Rye and Govindjee, 1988), and has been interpreted as indicating a role for bicarbonate in the protonation of the quinol form of Q_B (Diner et al., 1991a; Xu et al., 1991). The kinetics of electron transfer on the PS II acceptor side, upon dissociation of bicarbonate, resemble those observed in reaction centers of *Rb. sphaeroides* upon replacement of L-Asp213 with Asn (Takahashi and Wraight, 1990). These kinetics were also interpreted in terms of an inhibition of quinol protonation in reaction AB(2). The mechanism of action of bicarbonate is at present unknown, though its binding is likely to modify the pKas of nearby residues implicated in proton transfer.

Two residues, D2-Lys264 and D2-Arg265, probably located one turn of an α-helix above D2-His268, one of the four putative histidine ligands to the non-heme iron, are both required for tight binding of bicarbonate (Diner et al., 1991a). Two other residues, D2-Arg233 and Arg251, have also been implicated, though more weakly, in bicarbonate binding (Cao et al., 1991). Whether bicarbonate binds in a mono- or bidentate fashion has not been established. However, the absence of a PS II homologue of D2-Glu232, the conservation of the four bacterial histidine ligands, and the ability of NO and formate, which displace bicarbonate (Stemler and Murphy, 1985), to bind at the same time (Diner and Petrouleas, 1990), imply the availability of two exchangeable coordination positions in the first coordination sphere of the non-heme iron. A wide variety of molecules can also bind in these positions, including, in addition to NO, formate, and bicarbonate, CN^- (Koulougliotis et al., 1993) and a variety of carboxylate anions (acetate, glycolate, glyoxylate, and oxalate, Deligiannakis et al. 1994). These molecules shift the midpoint potential of the Fe(III)/Fe(II) couple and the observed g-values of the Q_A^--Fe(II) EPR signal. The most marked of these is CN^- which induces a $g = 1.98$ signal, attributed to the binding of two CN^- (Koulougliotis et al., 1993). Recently, Sanakis et al. (1994) have found that at high CN^- concentration and at alkaline pH, the Q_A^- EPR signal moves to $g = 2.0045$ (linewidth 9.5G), indicating that the semiquinone is no longer magnetically coupled to the iron (Feher and Okamura, 1978, Klimov et al., 1980). Mössbauer data indicate, however, that the iron is still bound but in a diamagnetic form. Sanakis and coworkers (1994) propose that these conditions result in the binding of 3 or 4 CN^- per iron. The associated increase in the crystal-field splitting converts the iron from high ($S = 2$) to low spin ($S = 0$). Surprisingly, the ligand exchange has very little effect on the rates of Q_A/Q_B electron transfer, indicating that while the coordination environment of the iron has been modified, that of the quinones is barely perturbed. The only previous observations in PS II of magnetically uncoupled Q_A^- have come from a report of outright extraction of the non-heme iron (Klimov et al., 1980) and of trypsinized PS II membrane fragments (Macmillan et al., 1990).

6. Oxidation of the Non-heme Iron

A high potential electron acceptor of PS II with an $E_{m,7}$ of 400 mV (Ikegami and Katoh, 1973) was shown to be the non-heme iron operating as redox couple Fe(III)/Fe(II) (Petrouleas and Diner, 1986). The Em shows a pH-dependence of –60 mV/pH between pH 6-8 implying that a proton is released upon oxidation (Bowes et al., 1979; Petrouleas and Diner, 1986). Oxidation of the iron can be observed in the dark using oxidants such as ferricyanide (Petrouleas and Diner, 1986) or in the light by reduction-induced oxidation involving high potential semiquinones in the Q_B binding site (Zimmermann and Rutherford, 1986; Petrouleas and Diner, 1987). The Fe(III) can oxidize Q_A^- with a $t_{1/2}$ of 7 μs at pH 6.5 at room temperature (Diner and Petrouleas, 1987a). No oxidation of the non-heme iron has been observed in bacterial reaction centers in the presence of ferricyanide, implying that the Fe(III)/Fe(II) redox couple has a much higher midpoint potential (>>500 mV) in these centers or that the oxidation is extremely slow. Two groups (Diner and Petrouleas, 1987a; Aasa et al., 1989) have succeeded in modeling the EPR spectra of the non-heme Fe(III). Resonances at $g = 8.15 \pm 0.05$ and at $g = 5.63 \pm 0.03$ are attributed to g_x

of the lowest Kramer's doublet (±1/2) and g_z of the middle Kramer's doublet (±3/2), respectively. These assignments give, for the octahedral symmetry of the coordination environment of the iron, an E/D of 0.11 in the range 0 (axial) to 0.33 (rhombic). A temperature study of these two resonances gives a zero-field splitting, D, of 2–3K (Diner and Petrouleas, 1987a; Aasa et al., 1989). The binding of quinones (Petrouleas and Diner, 1987), DCMU and o-phenanthroline (Itoh et al., 1986; Diner and Petrouleas, 1987b) to the Q_B site makes the iron coordination geometry more axial as indicated by the decrease of g_x toward 6. In a reciprocal fashion, oxidation of the non-heme iron decreases the affinity of o-phenanthroline and DCMU for the Q_B binding site (Wraight, 1985). An orientation study (Deligiannakis et al., 1992) of the Fe(III) places the principal axes of the g-tensor, g_x and g_z in the membrane plane and g_y perpendicular to the plane. The two exchangeable coordination positions, probably occupied by oxygens, and the two nitrogen ligands, belonging to the histidines closest to Q_A and Q_B, were assigned to the x-y plane.

7. Redox Properties of Q_A

The $E_{m,7}$ of the redox couple Q_A/Q_A^- has been measured in equilibrium redox titrations to be in the range –5 to –15 mV with pH dependencies of close to, but also less than, –60 mV/pH unit (Knaff, 1975, Horton and Croze, 1979). A pKa for Q_A^- of 8.9 has been measured by Knaff (1975) above which the Em of the Q_A/Q_A^- couple was found to be –130 mV. In the presence of DCMU, there is no proton uptake observed associated with the reduction of Q_A (Fowler and Kok, 1974), nor was proton uptake observed on the time scale of electron transfer from Q_A to Q_B (Auslander and Junge, 1974). It is likely, therefore, that proton uptake does not occur on the time scale of interquinone electron transfer, and that –130 mV is the operating potential of Q_A/Q_A^-.

Evans et al. (1985) have titrated ($E_{m,10}$ = –430 mV) the appearance of the spin polarized triplet arising from charge recombination between P680$^+$ and Pheo$^-$. While these authors attributed this component to an acceptor between Pheo and Q_A, van Mieghem et al. (1989) have provided an alternative interpretation of double reduction of Q_A. In their model the Q_A^- anion would, for electrostatic reasons, raise the energy of P680$^+$Ph$^-$, decreasing its transient concentration during the lifetime of the excitation energy within PS II. This model is consistent with an increase in the standard free energy of the P680$^+$Ph$^-$ state with Q_A single reduction as proposed by Schatz et al. (1988). Upon the formation of the uncharged, Q_AH_2, with illumination under reducing conditions, P680$^+$Ph$^-$ would decrease in energy. Its yield and that of the spin polarized triplet would both increase. Recently, however, van Mieghem (1993) and coworkers have been forced to modify this model as they have found the yield of P680$^+$Ph$^-$ and of ^{3}P680 to be nearly the same with Q_A singly and doubly reduced. The unaltered P680 triplet yield with Q_A double reduction is also consistent with the observations of Liu et al. (1993). Van Mieghem (1993) explains the earlier inability to observe the spin polarized triplet in the presence of Q_A^- as the modulation of its lifetime, and not its yield, with kinetic components (τ) of 2.6 and 20 μs in the presence of Q_A^- and 1.4 and 4.5 ms in the presence of Q_AH_2. In addition, the lifetime of the primary radical pair increased with kinetic components of 26 and 130 ns in the presence of Q_A^- and 72 and 290 ns in the presence of Q_AH_2. The reasons for the increased lifetimes with the double reduction of Q_A are not at present known (but see below).

Van Mieghem (1993) has found that double reduction of Q_A followed by reoxidation of Q_AH_2 by ferricyanide, results in a form of the reaction center where Q_A can still be singly photoreduced but where substantial double reduction of Q_A can now occur in the dark even with ascorbate; long incubation with dithionite and benzyl viologen are no longer required. This new state is also suggested to be associated with the loss of oxidizable non-heme iron, an interpretation that is consistent with the observation of the destabilization of semiquinone in the Q_A site and the facile reduction of Q_A to the quinol form following removal of the iron in bacterial reaction centers (Dutton et al., 1978). A role for the non-heme iron in stabilizing the semiquinone form of Q_A is the only clearly demonstrated role for the metal that has emerged, although a structural anchor around which the reaction center assembles is also possible. Whether a loss of the iron is responsible for the altered lifetimes of the radical pair state and of the spin polarized triplet observed above remains to be established. Double reduction of Q_A has also been proposed as one of the possible mechanisms for the photoinhibition of the PS II reaction center, resulting in a dissociation of the quinol from its binding site (Styring et al., 1990). Firm evidence of this conclusion is still lacking, and loss of the iron could be another explanation.

Acknowledgments

Several of the ideas expressed in this review emerged during the weekly photosynthesis group meetings in East Lansing. GTB thanks the participants (N. Bowlby, G. Deinum, C. Essenmacher, M. Gardner, Y. Gindt, C. Hoganson, D. Kreszowski, N. Lydakis, M. Mac, W. Shi, C. Tommos, K. Warncke, and C. Yocum) for their critical and incisive comments. We appreciate having received preprints and/or helpful discussion from Drs. G. W. Brudvig, J. P. Dekker, J. Lavergne, P. J. Nixon, V. Petrouleas, A. W. Rutherford, G. Small, S. Styring, X.-S. Tang, H. J. van Gorkom, M. Wasielewski, and C. F. Yocum. Research support for the work in photosynthesis in the authors' laboratories is provided by the Central Research and Development Dept. of the E. I. Du Pont de Nemours Co. (BAD), by the USDA NRI/CRGP Photosynthesis and Respiration Project (BAD and GTB) and by NIH GM37300 (GTB).

References

Aasa R, Andréasson L-E, Styring S and Vänngård T (1989) The nature of the Fe(III) EPR signal from the acceptor-side iron in photosystem II. FEBS Lett 243: 156–160

Allen JP, Feher G, Yeates TO, Komiya H and Rees DC (1988) Structure of the reaction center from *Rhodobacter sphaeroides* R-26: Protein-cofactor (quinones and Fe^{2+}) interactions. Proc Natl Acad Sci USA 85: 8487-8491

Araga C, Akabori K, Sasaki J, Maeda A, Shiina T and Toyoshima Y (1993) Functional reconstitution of the primary quinone acceptor, Q_A, in the Photosystem II core complexes. Biochim Biophys Acta 1142: 36–42

Atashkin AV and Kawamori A, Kodera Y, Kuroiwa S and Akabori K (1995) An electron spin echo envelope modulation study of the primary acceptor quinone in Zn-substituted plant photosystem II. J Chem Phys 102: 5583–5588

Ausländer W and Junge W (1974) The electric generator in the photosynthesis of green plants: II. Kinetic correlation between protolytic reactions and redox reactions. Biochim Biophys Acta 357: 285–298

Babcock GT (1993) Proteins, radicals, isotopes, and mutants in photosynthetic oxygen evolution. Proc Natl Acad Sci USA 90: 10893–10895

Babcock GT and Sauer K (1973) Electron paramagnetic resonance signal in spinach chloroplasts. I. Kinetic analysis for untreated chloroplasts. Biochim Biophys Acta 325: 483–503

Babcock GT, Blankenship RE and Sauer K (1976) Reaction kinetics for positive charge accumulation on the water side of chloroplast Photosystem II. FEBS Lett 61: 286–289

Babcock GT, Barry BA, Debus RJ, Hoganson CW, Atamian M, McIntosh L, Sithole I and Yocum CF (1989) Water oxidation in Photosystem II: From radical chemistry to multielectron chemistry. Biochem 28: 9557–9565

Barry B (1993) The role of redox active amino acids in the photosynthetic water-oxidizing complex. Photochem Photobiol 57: 179–188

Barry BA and Babcock GT (1987) Tyrosine radicals are involved in the photosynthetic oxygen-evolving system. Proc Natl Acad Sci USA 84: 7099–7103

Barry BA, El-Deeb MK, Sandusky PO and Babcock GT (1990) Tyrosine radicals in Photosystem II and related model compounds. J Biol Chem 265: 20139–20143

Baumgarten M, Philo JS and Dismukes GC (1990) Mechanism of photoinhibition of photosynthetic water oxidation by Cl^- depletion and F^- substitution: Oxidation of a protein residue. Biochem 29: 10814–10822

Bernard MT, MacDonald GM, Nguyen AP, Debus RJ and Barry BA (1995) A difference infrared study of hydrogen bonding to the Z• tyrosyl radical of Photosystem II. J Biol Chem 270: 1589–1594

Berthold DA, Babcock GT and Yocum CF (1981) A highly resolved, oxygen-evolving Photosystem II preparation from spinach thylakoid membranes. FEBS Lett 134: 231–234

Berthomieu C and Boussac A (1995) Histidine oxidation in the S2 to S3 transition probed by FTIR difference spectroscopy in the Ca^{2+}-depleted Photosystem II: Comparison with histidine radicals generated by UV irradiation. Biochem 34: 1541–1548

Blubaugh D and Govindjee (1988) The molecular mechanism of the bicarbonate effect at the plastoquinone reductase site of photosynthesis. Photosynth Res 19: 85–128

Bocarsly JR and Brudvig GW (1992) Turnover control of Photosystem II: Use of redox-active herbicides to form the S3 state. J Am Chem Soc 114: 9762–9767

Bock CH, Gerken, Stehlik D and Witt HT (1988) Time resolved EPR on Photosystem II particles after irreversible and reversible inhibition of water cleavage with high concentrations of acetate. FEBS Lett 227: 141–146

Boerner RJ and Barry BA (1993) Isotopic labeling and EPR spectroscopy show that a tyrosine residue is the terminal electron donor, Z, in manganese-depleted Photosystem II preparations. J Biol Chem 268: 17151–17154

Boerner RJ and Barry BA (1994) EPR evidence that the M^+ radical, which is observed in three site-directed mutants of Photosystem II, is a tyrosine radical. J Biol Chem 269: 134–137

Boerner RJ, Nguyen AP, Barry BA and Debus RJ (1992) Evidence from directed mutagenesis that the aspartate 170 of the D1 polypeptide influences the assembly and/or stability of the manganese cluster in the photosynthetic water-splitting complex. Biochem 31: 6660–6672

Boerner RJ, Bixby KA, Nguyen AP, Noren GH, Debus RJ and Barry BA (1993) Removal of stable tyrosine radical D^+ affects the structure of redox properties of tyrosine Z in manganese-depleted Photosystem II particles from *Synechocystis* 6803. J Biol Chem 268: 1817–1823

Booth PJ, Crystall B, Ahmad I, Barber J, Porter G and Klug DR (1991) Observation of multiple radical pair states in Photosystem 2 reaction centers. Biochem 30: 7573–7586.

Bosch MK, Evelo RG, Styring S, Rutherford AW and Hoff AJ (1991) ESE relaxation measurements in Photosystem II. The influence of the reaction center non-heme iron on the spin-lattice relaxation of Tyr D·. FEBS Lett 292: 279–283

Boska M, Sauer K, Buttner W and Babcock GT (1983) Similarity of EPR Signal IIf rise and $P680^+$ decay kinetics in Tris-washed chloroplasts Photosystem II preparations as a function of pH. Biochim Biophys Acta 722: 327–330

Bouges-Bocquet B (1980) Kinetic models for the electron donors of Photosystem II of photosynthesis. Biochim Biophys Acta 594: 85–103

Boussac A and Etienne A-L (1984) Midpoint potential of Signal II (slow) in Tris-washed Photosystem-II particles. Biochim Biophys Acta 766: 576–581

Boussac A and Rutherford AW (1994) Electron transfer events in chloride-depleted Photosystem II. J Biol Chem 269: 12462–12467

Boussac A and Rutherford AW (1995) Does the formation of the S3-state in Ca^{2+}-depleted Photosystem II correspond to an oxidation of Tyrosine Z detectable by cw-EPR at room temperature? Biochim Biophys Acta 1230: 195–201

Boussac A, Zimmermann J-L and Rutherford AW (1989) EPR signals from modified charge accumulation states of the oxygen evolving enzyme in Ca^{2+}-deficient Photosystem II. Biochem 28: 8984-8989

Boussac A, Zimmermann J-L, Rutherford AW and Lavergne J (1990) Histidine oxidation in the oxygen-evolving Photosystem-II enzyme. Nature 347: 303–306

Boussac A, Sétif P and Rutherford AW (1992) Inhibition of Tyrosine Z photooxidation after formation of the S3 state in Ca^{2+}-depleted and Cl^--depleted Photosystem II. Biochem 31: 1224–1234

Bowes JM, Crofts AR and Itoh S (1979) A high potential acceptor for Photosystem II. Biochim Biophys Acta 547: 327–335

Boxer SG (1993) Photosynthetic reaction center spectroscopy and electron transfer dynamics in applied electric fields. In: Deisenhofer J and Norris JR (eds) The Photosynthetic Reaction Center, Vol. II, pp 179–220. Academic Press, San Diego

Braun P, Greenberg BM and Scherz A (1990) D1-D2-cytochrome *b*-559 complex from the aquatic plant *Spirodela oligorrhiza*: Correlation between complex integrity, spectroscopic properties, photochemical activity, and pigment composition. Biochem 29: 10376–10387

Breton J (1990) Orientation of the pheophytin primary electron acceptor and of the cytochrome *b*-559 in the D1-D2 Photosystem II reaction center. In: Jortner J and Pullman B (eds) Perspectives in Photosynthesis, pp 23–28. Kluwer Academic Publishers, Dordrecht

Brettel K, Schlodder E and Witt HT (1984) Nanosecond reduction kinetics of photooxidized chlorophyll-a_{II} (P-680) in single flashes as a probe for the electron pathway, H^+-release and charge accumulation in the O_2-evolving complex. Biochim Biophys Acta 766: 403–415

Brok M, Ebskamp FCR and Hoff AJ (1985) The structure of the secondary donor of Photosystem II investigated by EPR at 9 and 35 GHz. Biochim Biophys Acta 809: 421–428

Buser CA, Thompson LK, Diner BA and Brudvig GW (1990) Electron-transfer reactions in manganese-depleted Photosystem II. Biochem 29: 8977-8985

Buser CA, Diner BA and Brudvig GW (1992) Photooxidation of cytochrome *b*-559 in oxygen-evolving Photosystem II. Biochem 31: 11449–11459

Butler WL, Visser JWM and Simons HL (1973) The kinetics of light-induced changes of C-550, cytochrome *b*-559 and fluorescence yield in chloroplasts at low temperature. Biochim Biophys Acta 292: 140–151

Cao J, Vermaas WFJ and Govindjee (1991) Arginine residues in the D2 polypeptide may stabilize bicarbonate binding in photosystem II of *Synechocystis* sp. PCC 6803. Biochim Biophys Acta 1059: 171–180

Carbonera D, Giacometti G and Agostini G (1994) A well resolved ODMR triplet minus singlet spectrum of P680 from PS II particles. FEBS Lett 343: 200–204

Chang HC, Jankowiak R, Reddy NRS, Yocum CF, Picorel R, Seibert M and Small, GJ (1994) On the question of the chlorophyll *a* content of the Photosystem II reaction center. J Phys Chem 98: 7725–7735

Chu H-A, Nguyen AP and Debus RJ (1995) Amino acid residues that influence the binding of manganese or calcium to Photosystem II. 1. The lumenal inter-helical domains of the D1 polypeptide. Biochem 34: 5839–5858

Conjeaud H and Mathis P (1980) The effect of pH on the reduction kinetics of P-680 in Tris-treated chloroplasts. Biochim Biophys Acta 590: 353–359

Corrie AR, Nugent JHA and Evans MCW (1991) Identification of EPR signals from the states $Q_A^{\cdot -}Q_B^{\cdot -}$ and $Q_B^{\cdot -}$ in Photosystem II from *Phormidium laminosum*. Biochim Biophys Acta 1057: 384–390

Danielius RV, Satoh K, van Kan PJM, Plijter JJ, Nuijs AN and van Gorkom HJ (1987) The primary reaction of Photosystem II in D1-D2-cytochrome *b*-559 complex. FEBS Lett 213: 241–244

de Vitry C, Carles C and Diner BA (1986) Quantitation of plastoquinone-9 in Photosystem II reaction center particles. FEBS Lett 196: 203–206

Deak, Z, Vass, I and Styring, S (1994) Redox interaction of Tyrosine-D with the S-states for the water-oxidizing complex in intact and chloride-depleted Photosystem II. Biochim Biophys Acta 1185: 65–74

Debus RJ (1992) The manganese and calcium ions of photosynthetic oxygen evolution. Biochim Biophys Acta 1102: 269–352

Debus RJ, Barry BA, Babcock GT and McIntosh L (1988a) Site-directed mutagenesis identifies a tyrosine radical involved in the photosynthetic oxygen-evolving system. Proc Natl Acad Sci USA 85: 427–430

Debus RJ, Barry BA, Sithole I, Babcock GT and McIntosh L (1988b) Directed mutagenesis indicates that the donor to $P680^+$ in Photosystem II is Tyrosine–161 of the D1 polypeptide. Biochem 27: 9071–9074

Deisenhofer J and Norris JR (1993) The Photosynthetic Reaction Center. Structure, Spectroscopy and Photochemistry, Vols I and II. Academic Press, New York

Deisenhofer J, Epp O, Miki K, Huber R and Michel H (1984) X-ray structure analysis of a membrane protein complex: Electron density map at 3Å resolution and a model of the chromophores of the photosynthetic reaction center from *Rhodopseudomonas viridis*. J Mol Biol 180: 385–398

Dekker JP, Plijter JJ, Ouwehand L and van Gorkom HJ (1984a) Kinetics of manganese redox transitions in the oxygen-evolving apparatus of photosynthesis. Biochim Biophys Acta 767: 176–179

Dekker JP, van Gorkom HJ, Brok M and Ouwehand L (1984b) Optical characterization of Photosystem II electron donors. Biochim Biophys Acta 764: 301–309

Dekker JP, Bowlby NR and Yocum CF (1989) Chlorophyll and cytochrome *b*-559 content of the photochemical reaction center of photosystem II. FEBS Lett 254: 150–154

Deligiannakis Y, Tsekos N, Petrouleas V and Diner BA (1992) Orientation dependence of the Fe^{2+}-NO and Fe^{3+} EPR signal associated with the non-heme iron of Photosystem II. Biochim Biophys Acta 1140: 163–168

Deligiannakis Y, Petrouleas V and Diner BA (1994) Binding of carboxylate anions at the non-heme Fe(II) of PS II. I. Effects on the $Q_A^-Fe^{2+}$ and Q_AFe^{3+} EPR spectra and the redox properties of the iron. Biochim Biophys Acta 1188: 260–270

Delosme R (1971) New results about chlorophyll fluorescence 'in vivo'. In: Forti G (ed.) Proceedings IInd International Congress on Photosynthesis. pp 187–195, Dr. W. Junk, Den Haag

Diner BA (1986) The reaction center of Photosystem II. In: Staehelin LA and Arntzen CJ (eds.) Photosynthesis III: Photosynthetic Membranes and Light-harvesting Systems. pp 422–436. Springer-Verlag, Berlin

Diner BA and de Vitry (1984) Optical spectrum and kinetics of the secondary electron donor, Z, of Photosystem II. In: Sybesma C (ed.) Advances in Photosynthesis Research, Vol I, pp 407–411. Martinus Nijhiff/Dr. W. Junk, The Haag

Diner BA and Nixon PJ (1992) The rate of reduction of oxidized redox active-tyrosine, Z^+, by exogenous Mn^{2+} is slowed in a site-directed mutant, at aspartate 170 of polypeptide D1 of Photosystem II, inactive for photosynthetic oxygen evolution. Biochim Biophys Acta 1101: 134–138

Diner BA and Petrouleas V (1987a) Q_{400}, the non-heme iron of the Photosystem II iron-quinone complex. A spectroscopic probe of quinone and inhibitor binding to the reaction center. Biochim Biophys Acta 895: 107–125

Diner BA and Petrouleas V (1987b) Light-induced oxidation of the acceptor-side Fe(II) of Photosystem II by exogenous quinones acting through the Q_B binding site. II. Blockage by inhibitors and their effects on the Fe(III) EPR spectra. Biochim Biophys Acta 893: 138–148

Diner BA and Petrouleas V (1990) Formation by NO of nitrosyl adducts of redox components of the Photosystem II reaction center. II. Evidence that HCO_3^-/CO_2 binds to the acceptor-side non-heme iron. Biochim Biophys Acta 1015: 141–149

Diner BA, de Vitry C. and Popot J-L (1988) Quinone exchange in the Q_A binding site of Photosystem II reaction center core preparations isolated from *Chlamydomonas reinhardtii*. Biochim Biophys Acta 934: 47–54

Diner BA, Nixon PJ and Farchaus JW (1991a) Site -directed mutagenesis of photosynthetic reaction centers. Curr Op Struct Biol 1: 546–554

Diner BA, Petrouleas V and Wendoloski JJ (1991b) The iron-quinone electron-acceptor complex of photosystem II. Physiol Plant 81: 423–436

Diner BA, Tang X-S, Zheng M, Dismukes GC, Force DA, Randall DW and Britt RD (1995) Environment and function of the redox active tyrosines of Photosystem II. In: Mathis P (ed) Photosynthesis: From Light to Biosphere, Vol II, pp 229–234. Kluwer Academic Publishers, Dordrecht

Dixon WT and Murphy D (1976) Determination of the acidity constants of some phenol radical cations by means of Electron Spin Resonance. J Chem Soc, Faraday Trans II 72: 1221–1230

Durrant JR, Giorgi IB, Barber J, Klug DR and Porter G (1990) Characterisation of triplet states in isolated photosystem II reaction centres: Oxygen quenching as a mechanism for photodamage. Biochim Biophys Acta 1017: 167–175

Durrant JR, Hastings G, Hong Q, Barber J, Porter G and Klug DR (1992a) Determination of P680 singlet state lifetimes in photosystem two reaction centres. Chem Phys Lett 188: 54–60

Durrant JR, Hastings G, Joseph DM, Barber J, Porter G and Klug DR (1992b) Subpicosecond equilibration of excitation energy in isolated Photosystem II reaction centers. Proc Natl Acad Sci USA 89: 11632–11636

Durrant JR, Hastings G, Joseph DM, Barber J, Porter G and Klug DR (1993) Rate of oxidation of P680 in isolated Photosystem 2 reaction centers monitored by loss of chlorophyll stimulated emission. Biochem 32: 8259-8267

Durrant JR, Klug, DR, Kwa SLS, van Grondelle, R, Porter, G and Dekker, JP (1995) A multimer model for P680, the primary electron donor of photosystem II. Proc Natl Acad Sci USA 92: 4798–4802

Dutton PL, Prince RC and Tiede DM (1978) The reaction center of photosynthetic bacteria. Photochem Photobiol 28: 939–949

Eaton-Rye J and Govindjee (1988) Electron transfer through the quinone acceptor complex of Photosystem II after one or two actinic flashes in bicarbonate-depleted spinach thylakoid membranes. Biochim Biophys Acta 934: 47–54

Eijckelhoff, C and Dekker JP (1995) Determination of the pigment stoichiometry of the photochemical reaction center of Photosystem II. Biochim Biophys Acta 1231: 21–28

Erickson JM and Rochaix J-D (1992) The molecular biology of photosystem II. In: Barber J (ed) The Photosystems: Structure, Function and Molecular Biology, Topics in Photosynthesis, Vol 11, pp 101–177, Elsevier Science Publishers, Amsterdam

Evans MCW, Atkinson YE and Ford RC (1985) Redox characterisation of the Photosystem II electron acceptors. Evidence for two electron carriers between pheophytin and Q. Biochim Biophys Acta 806: 247–254

Evelo RG, Dikanov SA and Hoff AJ (1989a) Electron spin echo envelope modulation (ESEEM) studies of the tyrosyl radical D• of plant Photosystem II. Chem Phys Lett 157: 25–30

Evelo RG, Hoff AJ, Dikanov SA and Tyryshkin AM (1989b) An ESEEM study of the oxidized electron donor of plant Photosystem II: Evidence that D• is a neutral tyrosine radical. Chem Phys Lett 161: 479–484

Evelo RG, Styring S, Rutherford AW and Hoff AJ (1989c) EPR relaxation measurements of Photosystem II reaction centers: Influence of S-state oxidation and temperature. Biochim Biophys Acta 973: 428–442

Feher G and Okamura MY (1978) Chemical composition and properties of reaction centers. In Clayton R (ed.) The Photosynthetic Bacteria pp 349–386, Plenum, New York

Feher G, Isaacson RA, Okamura MY and Lubitz W (1986) ENDOR of semiquinones in RC's from *Rhodopseudomonas sphaeroides*. In: Michel-Beyerle M (ed) Springer Series in Chemical Physics Vol 42 pp 174–189, Springer-Verlag, Berlin

Force DA, Randall DW, Britt RD, Tang X-S and Diner BA (1995) 2H ESE-ENDOR study of hydrogen bonding to the tyrosine radicals Y_D• and Y_Z• of Photosystem II. J Am Chem Soc 117: 12643–12644

Fowler CF and Kok B (1974) Proton evolution associated with the photooxidation of water in photosynthesis. Biochim Biophys Acta 357: 299–307

Freiberg A, Timpmann K, Moskalenko AA and Kuznetsova NY (1994) Pico- and nanosecond fluorescence kinetics of

Photosystem II reaction centre and its complex with CP47 antenna. Biochim Biophys Acta 1184: 45–53

Gerken S, Brettel K, Schlodder E and Witt HT (1988) Optical characterization of the immediate electron donor to chlorophyll a_{II}^{+} in O_2-evolving photosystem II complexes. Tyrosine as possible electron carrier between chlorophyll a_{II} and the water-oxidizing manganese cluster. FEBS Lett 237: 69–75

Ghanotakis DF and Yocum CF (1990) Photosystem II and oxygen-evolving complex. Ann Rev Plant Physiol 41: 255–276

Ghanotakis DF, de Paula JC, Demetriou DM, Bowlby NR, Petersen J, Babcock GT and Yocum CF (1989) Isolation and characterization of the 47 kDa protein and the D1-D2-cytochrome *b*-559 complex. Biochim Biophys Acta 974: 44–53

Gilchrist ML, Ball JA, Randall DW and Britt RD (1995) Proximity of the manganese cluster of Photosystem II to the redox active tyrosine Y_Z. Proc Natl Acad Sci USA 92: 9545–9549

Gulin VI, Dikanov SA, Tsvetko YU, Evelo RG and Hoff A (1992) Very high frequency (135 GHz) EPR of the oxidized primary donor of the photosynthetic bacteria *Rb. sphaeroides* R-26 and *Rps. viridis* and of Y_D• (signal II) of plant Photosystem II. J Pur Appl Chem 64, 903–906

Gunner MR, Robertson DE and Dutton PL (1986) Kinetic studies on the reaction center protein from *Rhodopseudomonas sphaeroides*: The temperature and free energy dependence of electron transfer between various quinones in the Q_A site and the oxidized bacteriochlorophyll dimer. J Phys Chem 90: 3783–3795

Hallahan BJ, Nugent JHA, Warden JT and Evans MCW (1992) Investigation of the origin of the 'S3'EPR signal from the oxygen-evolving complex of Photosystem 2: The role of Tyrosine Z. Biochem 31: 4562–4573

Hastings G, Durrant JR, Barber J, Porter G and Klug DR (1992) Observation of pheophytin reduction in Photosystem Two reaction centers using femtosecond transient absorption spectroscopy. Biochem 31: 7638–7647

Haumann M, Bögershausen O and Junge W (1994) Photosynthetic oxygen evolution: Net charge transients as inferred from electrochromic bandshifts are independent of proton release into the medium. FEBS Lett 355: 101–105

Hirsh DJ and Brudvig GW (1993) Long-range electron spin-spin interactions in the bacterial photosynthetic reaction center. J Phys Chem 97: 13216–13222

Hoganson CW and Babcock GT (1988) Electron-transfer events near the reaction center in O_2-evolving Photosystem II preparations. Biochem 27: 5848–5855

Hoganson CW and Babcock GT (1989) Redox cofactor interactions in Photosystem II: Electron spin resonance spectrum of $P680^{+}$ is broadened in the presence of Y_Z^{+}. Biochem 28: 1448–1454

Hoganson CW and Babcock GT (1992) Protein-tyrosyl radical interactions in Photosystem II studied by Electron Spin Resonance and Electron Nuclear Double Resonance Spectroscopy: Comparison with ribonucleotide reductase and in vitro tyrosine. Biochem 31: 11874–11880

Hoganson CW and Babcock GT (1994) Photosystem II. In: Sigel H and Sigel A (eds), Metal Ions in Biological Systems, Vol. 30, 'Metalloenzymes involving amino acid-residues and related radicals'. Marcel Dekker, New York

Hoganson CW, Lydakis-Simantiris N, Tang X-S, Tommos C, Warncke K, Babcock GT, Diner BA McCracken J, Styring S (1995) A hydrogen-atom abstraction model for the function of Y_Z in photosynthetic oxygen evolution. Photosynth Res 46: 177–184

Holzwarth, A. R. (1989) Applications of ultrafast laser spectroscopy for the study of biological systems. Q Rev Biophys 22: 239–326

Holzwarth, A. R., Muller, M. G., Gatzen, G., Hucke, M., Griebenow, K. (1994) Ultrafast spectroscopy of the primary electron and energy transfer processes in the reaction center of photosystem II. J Lumin 60–61: 497–502

Horton P and Croze E (1979) Characterization of two quenchers of chlorophyll fluorescence with different midpoint oxidation-reduction potentials in chloroplasts. Biochim Biophys Acta 545: 188–201

Ikegami I and Katoh S (1973) Studies on chlorophyll fluorescence in chloroplasts II. Effect of ferricyanide on the induction of fluorescence in the presence of 3-(3,4-dichlorophenyl)-1,1-dimethylurea. Plant Cell Physiol 14: 829-836

Innes JB and Brudvig GW (1989) Location and magnetic properties of the stable tyrosine radical in Photosystem II. Biochem 28: 1116–1125

Isied SS (1991) Electron transfer reactions in metalloproteins. In: Sigel H and Sigel A (eds.) Metal Ions in Biological Systems, Vol. 27 pp 1–56, Marcel Dekker, New York

Isogai Y, Itoh S and Nishimura M (1990) Location of D^{+} and distribution of surface charges in Photosystem II. Biochim Biophys Acta 1017: 204–208

Itoh S, Tang X-S and Satoh K (1986) Interaction of the high spin Fe atom in the Photosystem II reaction center with the quinones, Q_A and Q_B in purified oxygen-evolving PS II reaction center complex and in PS II particles. FEBS Lett 205: 275–281

Jankowiak R and Small GJ (1993) Spectral hole burning: A window on excited state electronic structure, heterogeneity, electron-phonon coupling, and transport dynamics of photosynthetic units. In: Deisenhofer J and Norris JR (eds) The Photosynthetic Reaction Center, Vol. II, pp 133–177. Academic Press, San Diego

Jankowiak R, Tang D, Small GJ and Seibert M (1989) Transient and persistent hole burning of the reaction center of Photosystem II. J Phys Chem 93: 1649–1654

Joliot A and Joliot P (1964) Etude cinétique de la réaction photochimique libérant l'oxygène au cours de la photosynthèse. C R Acad Sc Paris 258: 4622–4625

Klimov VV and Krasnovskii AA (1981) Pheophytin as the primary electron acceptor in Photosystem 2 reaction centers. Photosynthetica 15: 592–609

Klimov VV, Dolan E, Shaw ER and Ke B (1980) Interaction between the intermediary electron acceptor (pheophytin) and a possible plastoquinone-iron complex in photosystem II reaction centers. Proc Natl Acad Sci USA 77: 7227–7231

Klug DR, Rech T, Joseph M, Barber J, Durrant JR and Porter G (1995) Primary processes in isolated Photosystem II reaction centres probed by magic angle transient absorption spectroscopy. Chem Phys 194: 433–442

Knaff DB (1975) The effect of pH on the midpoint oxidation-reduction potentials of components associated with plant Photosystem II. FEBS Lett 60: 331–335

Kodera Y, Takura K and Kawamori A (1992a) Distance of P680 from the manganese complex in Photosystem II studied by time resolved EPR. Biochim Biophys Acta 1101: 23–32

Kodera Y, Takura K, Mino H and Kawamori A (1992b) Pulsed EPR study of tyrosine-Z^+ in Photosystem II. In: Murata N (ed) Research in Photosynthesis. Vol II, pp 57–60, Kluwer Academic Publishers, Dordrecht

Koike H, Siderer Y, Ono T-A and Inoue Y (1986) Assignment of thermoluminescence to $S_3Q_A^-$ charge recombination: Sequential stabilization of S_3 and Q_A^- by a two-step illumination at different temperatures. Biochim Biophys Acta 850: 80-89

Koike H, Kashino Y and Satoh K (1993) Interactions of halogenated benzoquinones with the non-heme iron (Q_{400}) in Photosystem II. Z Naturforsch 48c: 168–173

Koulougliotis D, Kostopoulos T, Petrouleas V and Diner BA (1993) Evidence for CN^- binding at the PS II non-heme Fe^{2+}. Effects on the EPR signal for $Q_A^-Fe^{2+}$ and on Q_A/Q_B electron transfer. Biochim Biophys Acta 1141: 275–282

Koulougliotis D, Tang X-S, Diner BA and Brudvig GW (1995) Spectroscopic evidence for the symmetric location of tyrosines D and Z in Photosystem II. Biochem 34: 2850–2856

Kramer DM, Roffey RA, Govindjee and Sayre RT (1994) The A_T thermoluminescence band from *Chlamydomonas reinhardtii* and the effects of mutagenesis of histidine residues on the donor side of the Photosystem II D1 polypeptide. Biochim Biophys Acta 1185: 228–237

Kwa SLS, Newell WR, van Grondelle R, and Dekker J (1992) The reaction center of Photosystem II studied with polarized fluorescence spectroscopy. Biochim Biophys Acta 1099: 193–202

Kwa S LS, Eijckelhof C, van Grondelle R, and Dekker JP (1994) Site-selection spectroscopy of the reaction center complex of Photosystem II. 2. Identification of the fluorescing species at 4K, J Phys Chem 98: 7712–7716

Lavergne J and Junge W (1993) Proton release during the redox cycle of the water oxidase. Photosynth Res 38: 279–296

Leibl W, Breton J, Deprez J and Trissl H-W (1989) Photoelectric study on the kinetics of trapping and charge stabilization in oriented PS II membranes. Photosynth Res 22: 257–275

Liu B, Napiwotzki A, Eckert H-J, Eichler HJ and Renger G (1993) Studies on the recombination kinetics of the radical pair $P680^+Pheo^-$ in isolated PS II core complexes from spinach. Biochim Biophys Acta 1142: 129–138

Lösche M, Satoh K, Feher G, and Okamura MY (1988) Stark effect in PS II RCs from spinach. Biophys J 53: 270a

Lubitz W, Isaacson RA, Okamura MY, Abresch EC, Plato M, and Feher G (1989) ENDOR studies of the intermediate electron acceptor radical anion $I^{\bullet-}$ in Photosystem II reaction centers. Biochim Biophys Acta 977: 227–232

MacLachlan DJ and Nugent JHA (1993) Investigation of the S3 electron paramagnetic resonance signal from the oxygen-evolving complex of Photosystem 2: Effect of inhibition of oxygen evolution by acetate. Biochem 32: 9772–9780

Macmillan F, Gleiter H, Renger G, Lubitz W (1990) EPR/ENDOR studies of plastoquinone anion radical in Photosystem II ($Q_A^{\bullet-}$) and in organic solvents. In: Baltscheffsky M (ed) Current Research in Photosynthesis, Vol I, pp 531–534, Kluwer Academic Publishers, Dordrecht

McCauley, S. W., Baronavski, A. P., Rice, J. K., Ghirardi, M. L., and Mattoo, A. K. (1992) A search for subpicosecond absorption components in Photosystem II reaction centers. Chem Phys Lett 198: 437–442

McTavish H, Picorel R, and Seibert, M (1989) Stabilization of isolated Photosystem II reaction center complex in the dark and in the light using polyethylene glycol and an oxygen-scrubbing system. Plant Physiol 89: 452–456

Messinger J and Renger G (1993) Generation, oxidation by the oxidized form of the tyrosine of polypeptide D2, and possible electronic configuration of the redox states S0, S-1 and S-2 of the water oxidase in isolated spinach thylakoids. Biochem 32: 9379–9386

Metz JG, Nixon PJ, Rögner M, Brudvig GW and Diner BA (1989) Directed alteration of the D1 polypeptide of PS II: Evidence that Tyrosine-161 is the redox component, Z, connecting the oxygen-evolving complex to the primary electron donor, P680. Biochem 28, 6960–6969

Meyer B, Schlodder E, Dekker JP and Witt HT (1989) O_2 evolution and Chl a_{II}^+ (P680+) nanosecond reduction kinetics in single flashes as a function of pH. Biochim Biophys Acta 974: 36–43

Michel H and Deisenhofer J (1988) Relevance of the photosynthetic reaction center from purple bacteria to the structure of Photosystem II. Biochem 27: 1–7

Michel H, Epp O and Deisenhofer J (1986) Pigment-protein interactions in the photosynthetic reaction centre from *Rhodopseudomonas viridis*. EMBO J 5: 2445–2451

Mimuro M, Yamazaki I, Itoh S, Tamai N, and Satoh K (1988) Dynamic fluorescence properties of D1-D2-cytochrome *b*-559 complex isolated from spinach chloroplasts: Analysis by means of the time-resolved fluorescence spectra in picosecond time range. Biochim Biophys Acta 933: 478–486

Mino H and Kawamori A (1994) Microenvironments of tyrosine D^+ and tyrosine Z^+ in Photosystem II studied by proton matrix ENDOR. Biochim Biophys Acta 1185: 213–220

Mino H, Satoh J-I, Kawamori A, Toriyama K and Zimmermann J-L (1993) Matrix ENDOR of tyrosine D+ in oriented Photosystem II membranes. Biochim Biophys Acta 1144: 426–433

Moënne-Loccoz, P, Robert, B, and Lutz, M (1989) A resonance Raman characterization of the primary electron acceptor in Photosystem II. Biochem 28: 3641–3645

Moser CC, Keske JM, Warncke, K, Farid RS and Dutton PL (1992) Nature of biological electron transfer. Nature 355: 796-802

Nabedryk E, Andrianambinintsoa S, Berger G, Leonhard M, Mäntele W, and Breton J (1990) Characterization of bonding interactions of the intermediary electron acceptor in the reaction center of Photosystem II by FTIR spectroscopy. Biochim Biophys Acta 1016: 49–54

Nanba O and Satoh K (1987) Isolation of a Photosystem II reaction center consisting of D-1 and D-2 polypeptides and cytochrome b-559. Proc Natl Acad Sci USA 84: 109–112

Nixon PJ (1994) Using site-directed mutagenesis to probe pigment/protein interactions within the photosystem two complex. Abstract BBSRC Second Robert Hill Symposium on Photosynthesis, Imperial College, London 4/11–13/94

Nixon PJ and Diner BA (1992) Aspartate 170 of the Photosystem II reaction center polypeptide D1 is involved in the assembly of the oxygen-evolving manganese cluster. Biochem 31: 942–948

Noguchi T, Inoue Y, and Satoh K (1993) FT-IR studies on the triplet state of P680 in the Photosystem II reaction center: Triplet equilibrium within a chlorophyll dimer. Biochem 32: 7186–7195

Noren GH and Barry BA (1992) The YF161D1 mutant of *Synechocystis* 6803 exhibits an EPR signal from a light-induced Photosystem II radical. Biochem 31: 3335–3342

Nugent JHA, Diner BA and Evans MCW (1981) Direct detection of the electron acceptor of Photosystem II. Evidence that Q is an iron-quinone complex. FEBS Lett 124: 241–244

Oettmeier W (1992) Herbicides of Photosystem II. In: Barber J (ed.) The Photosystems: Structure, Function and Molecular Biology, pp 349–408, Elsevier, Amsterdam

Ono T-A and Inoue Y (1991) Biochemical evidence for histidine oxidation in photosystem II depleted of the Mn-cluster for O_2-evolution. FEBS Lett 278: 183–186

Otte SCM, van der Vos R, and van Gorkom HJ (1992) Steady state spectroscopy at 6 K of the isolated Photosystem II reaction centre: Analysis of the red absorption band. J Photochem Photobiol B: Biol. 13: 5–14

Paddock ML, Rongey SH, Feher G and Okamura MY (1989) Pathway of proton transfer in bacterial reaction centers: Replacement of Glutamic acid 212 in the L subunit by glutamine inhibits quinone (secondary acceptor) turnover. Proc Natl Acad Sci USA 86: 6602–6606

Paddock ML, McPherson PH, Feher G and Okamura MY (1990) Pathway of proton transfer in bacterial reaction centers: Replacement of Serine-L223 by alanine inhibits electron and proton transfers associated with reduction of quinone to dihydroquinone. Proc Natl Acad Sci USA 87: 6803–6807

Petrouleas V and Diner BA (1986) Identification of Q_{400}, a high-potential electron acceptor of Photosystem II with the iron of the quinone-iron acceptor complex. Biochim Biophys Acta 849: 264–275

Petrouleas V and Diner BA (1987) Light-induced oxidation of the acceptor-side Fe(II) of Photosystem II by exogenous quinones acting through the Q_B binding site. I. Quinones, kinetics and pH-dependence. Biochim Biophys Acta 893: 126–137

Petrouleas V and Diner BA (1990) Formation by NO of nitrosyl adducts of redox components of the Photosystem II reaction center. I. NO binds to the acceptor-side non-heme iron. Biochim Biophys Acta 1015: 131–140

Petrouleas V, Sanakis Y, Deligiannakis Y and Diner BA (1992) The non-heme Fe(II) of PS II: (1) Binding of new carboxylate anions, (2) Study of two Mössbauer components. In Murata A (ed.) Research in Photosynthesis Vol. II, 119–122, Kluwer Academic Publishers, The Netherlands

Pokorny A, Wulf K and Trissl HW (1994) An electrogenic reaction associated with the re-reduction of P680 by Tyr Z in Photosystem II. Biochim Biophys Acta 1184: 65–70

Prásil O, Adir N and Ohad I (1992) Dynamics of photosystem II: Mechanism of photoinhibition and recovery processes. In: Barber J (ed) The Photosystems: Structure, Function and Molecular Biology, Topics in Photosynthesis, Vol. 11, pp 295–348, Elsevier Science Publishers, Amsterdam

Pulles MPJ, van Gorkom HJ and Verschoor GAM (1976) Primary reactions of Photosystem II at low pH. 2. Light-induced changes of absorbance and electron spin resonance in spinach chloroplasts. Biochim Biophys Acta 440: 98–106

Rappaport F and Lavergne J (1991) Proton release during successive oxidation steps of the photosynthetic water oxidation process: Stoichiometries and pH dependence. Biochem 30: 10004–10012

Rappaport F, Blanchard-Desce M and Lavergne J (1994) Kinetics of electron transfer and electrochromic change during the redox transitions of the photosynthetic oxygen-evolving complex. Biochim Biophys Acta 1184: 178–192

Renger G, (1992) Energy transfer and trapping in Photosystem II. In: Barber J (ed) The Photosystems: Structure, Function and Molecular Biology, Topics in Photosynthesis, Vol. 11, pp 45–99. Elsevier Science Publishers, Amsterdam

Rich PR and Bendall DS (1980) The kinetics and thermodynamics of the reduction of cytochrome c by substituted p-benzoquinols in solution. Biochim Biophys Acta 592: 506–518

Rigby SEJ, Nugent JHA and O'Malley PJ (1994) The dark stable tyrosine radical of Photosystem 2 studied in three species using ENDOR and EPR spectroscopies. Biochem 33: 1734–1742

Robinson HH and Crofts AR (1983) Kinetics of the oxidation-reduction reactions of the photosystem II quinone acceptor complex, and the pathways for deactivation. FEBS Lett 153: 221–226

Rodriguez ID, Chandrashekar TK and Babcock GT (1987) ENDOR characterization of H_2O/D_2O exchange in the D^+Z^+ radical in photosynthesis. In: Biggins J (ed.) Progress in Photosynthesis Vol. I, pp 471–474, Martinus Nijhoff Publishers, Dordrecht

Roelofs TA, Gilbert M, Shuvalov VA and Holzwarth AR (1991) Picosecond fluorescence kinetics of the D_1-D_2-Cyt *b*-559 Photosystem II reaction center complex. Energy transfer and primary charge separation processes. Biochim Biophys Acta 1060: 237–244

Roelofs TA, Kwa SLS, van Grondelle R, Dekker JP and Holzwarth AR (1993) Primary processes and structure of the Photosystem II reaction center: II. Low-temperature picosecond fluorescence kinetics of a D_1-D_2-Cyt *b*-559 reaction center complex isolated by short Triton exposure. Biochim Biophys Acta 1143: 147–157

Roffey R, van Wijk K, Sayre R and Styring S (1994a) Spectroscopic characterization of Tyrosine-Z in Histidine 190 mutants of the D1 protein in Photosystem II (PS II) in *Chlamydomonas reinhardtii*. J Biol Chem 269: 5115–5121

Roffey RA, Kramer DM, Govindjee and Sayre RT (1994b) Lumenal side histidine mutations in the D1 protein of Photosystem II affect donor side electron transfer in *Chlamydomonas reinhardtii*. Biochim Biophys Acta 1185: 257–270

Rögner M, Chisholm DA and Diner BA (1991) Site-directed mutagenesis of the *psbC* gene of Photosystem II: Isolation and functional characterization of CP–43-less Photosystem II core complexes. Biochem 30: 5387–5395

Ruffle SV, Donnelly D, Blundell TL and Nugent JHA (1992) A three-dimensional model of the Photosystem II reaction centre of *Pisum sativum*. Photosynth Res 34: 287–300

Rutherford AW (1985) Orientation of EPR signals arising from components in Photosystem II membranes. Biochim Biophys Acta 807: 189–201

Rutherford AW, (1986) How close is the analogy between the reaction centre of photosystem II and that of purple bacteria? Biochem Soc Trans. 14: 15–17

Rutherford AW and Evans MCW (1979) A high potential semiquinone-iron type EPR signal in *Rhodopseudomonas viridis*. FEBS Lett 100: 305–308

Rutherford AW and Zimmermann J-L (1984) A new EPR signal attributed to the primary plastosemiquinone acceptor in Photosystem II. Biochim Biophys Acta 767: 168–175

Sanakis Y, Petrouleas V and Diner BA (1994) Cyanide binding at the non-heme Fe^{2+} of the iron-quinone complex of Photosystem II: At high concentrations, cyanide converts the

Fe^{2+} from high (S = 2) to low (S = 0) spin. Biochem 33: 9922–9928

Satoh, K., (1993) Isolation and properties of the Photosystem II reaction center. In: Deisenhofer J and Norris JR (eds) The Photosynthetic Reaction Center, Vol. I, pp 289–318, Academic Press, San Diego

Schatz GH, Brock H, Holzwarth AR (1987) Picosecond kinetics of fluorescence and absorbance changes in Photosystem II particles excited at low photon density. Proc Natl Acad Sci USA 84: 8414

Schatz GH, Brock H and Holzwarth AR (1988) Kinetic and energetic model for the primary processes in Photosystem II. Biophys J 54: 397–405

Schelvis JPM, van Noort PI, Aartsma TJ and van Gorkom HJ (1994) Energy transfer, charge separation and pigment arrangement in the reaction center of Photosystem II. Biochim Biophys Acta 1184: 242–250

Schlodder E, Brettel K and Witt HT (1985) Relation between microsecond reduction kinetics of photooxidized chlorophyll-a_{II} (P-680) and photosynthetic water oxidation. Biochim Biophys Acta 808: 123–131

Seibert M (1993) Biochemical, biophysical, and structural characterization of the isolated Photosystem II reaction center complex. In: Deisenhofer J and Norris JR (eds) The Photosynthetic Reaction Center, Vol. I, pp 319–356. Academic Press, San Diego

Shinkarev VP and Wraight C (1993) Oxygen evolution in photosynthesis: From unicycle to bicycle. Proc Natl Acad Sci USA 90: 1834–1838

Stemler A and Jursinic PA (1993) Oxidation-reduction potential dependence of formate binding to Photosystem II in maize thylakoids. Biochim Biophys Acta 1183: 269–280

Stemler A and Murphy JB (1985) Bicarbonate-reversible and irreversible inhibition of Photosystem II by monovalent anions. Plant Physiol 77: 974–977

Stiehl HH and Witt HT (1969) Quantitative treatment of the function of plastoquinone in photosynthesis. Z Naturforsch 24b: 1588–1598

Styring SA and Rutherford AW (1987) In the oxygen-evolving complex of Photosystem II the S_0 state is oxidized to the S_1 state by D+ (Signal II_{slow}). Biochem 26: 2401–2405

Styring SA and Rutherford AW (1988) The microwave power saturation of SII_{slow} varies with the redox state of the oxygen-evolving complex in Photosystem II. Biochem 27: 4915–4923

Styring S, Virgin I, Ehrenberg A and Andersson B (1990) Strong light photoinhibition of electron transport in Photosystem II. Impairment of the function of the first quinone acceptor, Q_A. Biochim Biophys Acta 1015: 269–278

Svensson B, Vass I, Cedergren E and Styring S (1990) Structure of donor side components in Photosystem II predicted by computer modeling. EMBO J 9: 2051–2059

Takahashi E and Wraight CA (1990) A crucial role for Asp^{L213} in the proton transfer pathway to the secondary quinone of reaction centers from *Rhodobacter sphaeroides*. Biochim Biophys Acta 1020: 107–111

Takahashi Y and Satoh K (1987) Quantitation of plastoquinone and functional electron carriers in the photosystem II reaction center complex. In Biggins J (ed.) Progress in Photosynthesis Research, Vol. II, pp 73–76, Martinus Nijhoff, Dordrecht

Takahashi Y, Hansson Ö, Mathis P, and Satoh K (1987) Primary radical pair in photosystem II reaction centre. Biochim Biophys Acta 893: 49–59

Tamura N and Cheniae G (1987) Photoactivation of the water oxidizing complex in Photosystem II membranes depleted of Mn and extrinsic proteins. I. Biochemical and kinetic characterization. Biochim Biophys Acta 890: 179–194

Tang D, Jankowiak R, Seibert M, Yocum CF and Small GJ (1990) Excited-state structure and energy-transfer dynamics of two different preparations of the reaction center of Photosystem II: A hole-burning study. J Phys Chem. 94: 6519–6522

Tang X-S and Diner BA (1994) Biochemical and spectroscopic characterization of a new oxygen-evolving PS II core complex from the cyanobacterium, *Synechocystis* PCC 6803. Biochem 270: 225–235

Tang X-S, Chisholm DA, Dismukes GC, Brudvig GW and Diner BA (1993) Spectroscopic evidence from site-directed mutants of *Synechocystis* PCC6803 in favor of a close interaction between Histidine 189 and redox-active Tyrosine 160, both of polypeptide D2 of the Photosystem II reaction center. Biochem 32: 13742–13748

Tang, X-S, Peloquin JM, Lorigan GA, Britt RD and Diner BA (1995) The binding environment of the reduced primary quinone electron acceptor, Q_A^- of PS II. In: Mathis P (ed) Photosynthesis: From Light to Biosphere, Vol I, pp 775–778. Kluwer Academic Publishers, Dordrecht

Tang X-S, Zheng M, Chisholm DA, Dismukes GC and Diner BA (1996) Investigation of the differences in the local protein environments surrounding tyrosine radicals, Y_Z and Y_D, in Photosystem II using wild type and the D2-Tyr160Phe mutant of *Synechocystis* 6803. Biochemistry 35: 1475–1484

Taoka S and Crofts AR (1990) Two-electron gate in triazine resistant and susceptible *Amaranthus hybridus*. In: Baltscheffsky M (ed) Current Research in Photosynthesis, Vol. I, pp 547–550. Kluwer Academic Publishers, Dordrecht

Tetenkin VL, Gulyaev BA, Seibert M and Rubin AB (1989) Spectral properties of stabilized D1/D2/cytochrome *b*-559 Photosystem II reaction center complex. FEBS Lett 250: 459–463

Tommos C, Davidsson L, Svensson B , Madsen C, Vermaas, W and Styring S (1993) Modified EPR spectra of the $Tyrosine_D$ radical in Photosystem II in site-directed mutants of *Synechocystis* sp. PCC 6803: Identification of side chains in the immediate vicinity of $Tyrosine_D$ on the D2 protein. Biochem 32: 5436–5441

Tommos C, Madsen C, Styring S and Vermaas W (1994) Point-mutations affecting the properties of $tyrosine_D$ in Photosystem II. Characterization by isotopic labeling and spectral simulation. Biochem 33: 11805–11813

Tommos C, Tang, X-S, Warncke, K, Hoganson, CW, Styring, S., McCracken, J, Diner, BA and Babcock, GT (1995) Spin-density distribution, conformation and hydrogen bonding of the redox-active tyrosine, Y_Z, in Photosystem II from multiple electron magnetic-resonance spectroscopies: Implications for photosynthetic oxygen evolution. J Am Chem Soc 117: 10325–10335

Un S, Brunel L-C, Brill T, Zimmermann J-L and Rutherford AW (1994) Angular orientation of the stable tyrosyl radical within Photosystem II by high field 245 GHz EPR. Proc Natl Acad Sci USA 91: 5262–5266

Vacha F, Joseph DM, Durrant JR, Telfer A, Klug DR, Porter G and Barber J (1995) Photochemistry and spectroscopy of a five-chlorophyll reaction center of Photosystem II isolated using a Cu affinity column. Proc Natl Acad Sci USA 92: 2929–2933

van der Vos R, van Leeuwen PJ, Braun P and Hoff AJ (1992) Analysis of the optical absorbance spectra of D1-D2-cytochrome *b*-559 complexes by absorbance-detected magnetic resonance. Structural properties of P680. Biochim Biophys Acta 1140: 184–198

van Gorkom HJ (1974) Identification of the reduced primary electron acceptor of Photosystem II as a bound semiquinone anion. Biochim Biophys Acta 347: 439–442

van Gorkom HJ and Schelvis JPM (1993) Kok's oxygen clock: What makes it tick? The structure of P680 and consequences of its oxidizing power. Photosynth Res 38: 297–301

van Grondelle R, Dekker JP, Gillbro T and Sundström V (1994) Energy transfer and trapping in photosynthesis. Biochim Biophys Acta 1187: 1–65

van Kan PJM, Otte SCM, Kleinherenbrink FAM, Nieveen MC, Aartsma TJ and van Gorkom HJ (1990) Time-resolved spectroscopy at 10 K of the Photosystem II reaction center; deconvolution of the red absorption band. Biochim Biophys Acta 1020: 146–152

van Leeuwen PJ, Heiman C, Gast P, Dekker JP and van Gorkom HJ (1993) Flash-induced redox changes in oxygen-evolving spinach Photosystem II core particles. Photosynth Res 38: 169–176

van Mieghem F (1993) Photochemistry and structural aspects of the Photosystem II reaction centre. Thesis, University of Wageningen

van Mieghem F, Nitschke W, Mathis P and Rutherford AW (1989) The influence of the quinone-iron electron acceptor complex on the reaction centre photochemistry of Photosystem II. Biochim Biophys Acta 977: 207–214

van Mieghem FJE, Satoh K and Rutherford AW (1991) A chlorophyll tilted 30° relative to the membrane in the Photosystem II reaction centre. Biochim Biophys Acta 1058: 379–385

van Mieghem FJE, Searle GFW, Rutherford AW and Schaafsma TJ (1992) The influence of the double reduction of Q_A on the fluorescence decay kinetics of Photosystem II. Biochim Biophys Acta 1100: 198–206

van Rensen JJS, Tonk WJM and de Bruijn SM (1988) Involvement of bicarbonate in the protonation of the secondary quinone electron acceptor of Photosystem II via the non-haem iron of the quinone iron acceptor complex. FEBS Lett 226: 347–351

Vass I and Styring S (1991) pH-dependent charge equilibria between tyrosine-D and the S-states in Photosystem II. Estimation of relative midpoint potentials. Biochem 30: 830–839

Velthuys BR and Visser JWM (1975) The reactivation of EPR signal II in chloroplasts treated with reduced dichlorophenol-indophenol: Evidence against a dark equilibrium between two oxidation states of the oxygen evolving system. FEBS Lett 55: 109–112

Vermaas WFJ and Rutherford AW (1984) EPR measurements on the effects of bicarbonate and triazine resistance in the acceptor side of Photosystem II. FEBS Lett 175: 243–248

Vermaas WFJ, Rutherford AW and Hansson O (1988) Site-directed mutagenesis in Photosystem II of the cyanobacterium *Synechocystis* sp. PCC 6803: Donor D is a tyrosine residue in the D2 protein. Proc Natl Acad Sci USA 85: 8477–8481

Vos MH, van Gorkom HJ and van Leeuwen PJ (1991) An electroluminescence study of stabilization reactions in the oxygen-evolving complex of Photosystem II. Biochim Biophys Acta 1056: 27–39

Warden JT, Blankenship RE and Sauer K (1976) A flash photolysis ESR study of Photosystem II Signal IIvf, the physiological donor to P-680$^+$. Biochim Biophys Acta 423: 462–478

Warncke K and Dutton PL (1991) Function of exotic primary electron acceptors in the reaction center protein. Biophys J 59: 146a

Warncke K and Dutton PL (1993) Influence of Q_A site redox cofactor structure on equilibrium binding, in situ electro-chemistry, and electron-transfer performance in the photo-synthetic reaction center protein. Biochem 32: 4769–4779

Warncke, K, McCracken J and Babcock GT (1994) Structure of the Y_D tyrosine radical in Photosystem II as revealed by ^2H Electron Spin Echo Envelope Modulation (ESEEM) spectroscopic analysis of hydrogen hyperfine interactions. J Am Chem Soc 116: 7332–7340

Wasielewski MR, Johnson DG, Seibert M and Govindjee (1989) Determination of the primary charge separation rate in isolated Photosystem II reaction centers with 500-fs time resolution. Proc Natl Acad Sci USA 86: 524–528

Wiederrecht GP, Seibert M, Govindjee and Wasielewski MR (1994) Femtosecond photodichroism studies of isolated Photosystem II reaction centers. Proc Natl Acad Sci USA 91: 8999–9003

Wraight CA (1978) Iron-quinone interactions in the electron acceptor region of bacterial photosynthetic reaction centers. FEBS Lett 93: 283–288

Wraight CA (1985) Modulation of herbicide-binding by the redox state of Q_{400}, an endogenous component of Photosystem II. Biochim Biophys Acta 809: 320–330

Xu C, Taoka S, Crofts AR and Govindjee (1991) Kinetic characteristics of formate/formic acid binding at the plastoquinone reductase site in spinach thylakoids. Biochim Biophys Acta 1098: 32–40

Yerkes CT, Babcock GT and Crofts AR (1983) A Tris-induced change in the midpoint potential of Z, the donor to photosystem II, as determined by the kinetics of the back reaction. FEBS Lett 158: 359–363

Yocum CF and Babcock GT (1981) Amine-induced inhibition of photosynthetic oxygen evolution. FEBS Lett 130: 99–102

Zimmermann JL and Rutherford AW (1986) Photoreductant-induced oxidation of Fe^{2+} in the electron acceptor complex of Photosystem II. Biochim Biophys Acta 851: 416–423

Chapter 13

Form and Function of Cytochrome *b*-559

John Whitmarsh
Photosynthesis Research Unit, USDA/Agricultural Research Service, Department of Plant Biology, University of Illinois, 1201 W. Gregory Drive, Urbana, IL 61801, USA

Himadri B. Pakrasi
Department of Biology, Box 1137, Washington University, St. Louis, MO 63130, USA

Summary

Cytochrome *b*-559 is an integral part of all Photosystem II reaction centers. The cytochrome is composed of two polypeptides that are linked by a single heme. The linkage is created by two histidine residues that serve as axial ligands for the protoheme. Each polypeptide provides one histidine. The structure of cytochrome *b*-559 is not known, nor is its organization in the Photosystem II reaction center. Although genetic and biochemical data show that cytochrome *b*-559 is composed of two different subunits, α and β, it is not known if the native protein is a heterodimer ($\alpha\beta$) or two homodimers (α_2 and β_2). The function of cytochrome *b*-559 is also a mystery, although it is clear that the cytochrome is not involved in the primary electron transfer reactions of Photosystem II. Among its more perplexing features is the occurrence of two different thermodynamic forms, one high potential and one low potential, that differ by 300 mV. Furthermore, although cytochromes typically function as electron carriers, cytochrome *b*-559 is a reluctant electron donor or acceptor. Under physiological

Donald R. Ort and Charles F. Yocum (eds): Oxygenic Photosynthesis: The Light Reactions, pp. 249–264.

conditions its detectable light-induced turnover is negligible. When conditions are selected to promote a significant light-induced turnover, the rate of electron transfer to or from the cytochrome is slow compared to the rate of photosynthetic electron transport. The sluggish behavior of cytochrome *b*-559 has prompted suggestions that it is involved in secondary electron transport that serves to protect Photosystem II against excess light. This chapter focuses on the structural organization of cytochrome *b*-559 in Photosystem II and current models to explain its role in photosynthesis.

I. Introduction

Cytochrome *b*-559 is a heme protein that is an essential component of all Photosystem II reaction centers. If the cytochrome is not present in the membrane, a stable PS II reaction center cannot be formed. Although the structure and function of Cyt *b*-559 remain to be discovered, it is known that the cytochrome is not involved in the primary enzymatic activity of PS II, which is the transfer of electrons from water to plastoquinone. As such, Cyt *b*-559 is one of a handful of redox components in PS II with no known function. Why PS II reaction centers contain redox components that are not involved in the primary enzymatic reactions is a puzzling question. The answer may be found in the unusual chemical reactions occurring in PS II and the fact that the reaction center operates at a very high power level. Photosystem II is an energy transforming enzyme that must switch between various high energy states which involve the creation of the powerful oxidants required for removing electrons from water and the complex chemistry of plastoquinone reduction which is strongly influenced by protons. In saturating light a single reaction center can have an energy throughput of 600 eV/s (equivalent to 60,000 kW per mole of PS II). Operating at such a high power level results in damage to the reaction center. Current research focuses on the possibility that Cyt *b*-559 may protect PS II from inopportune oxidation or reduction reactions that could cause irreversible protein damage when photosynthesis is driven by excess light. To discover the role of Cyt *b*-559 we need to understand its form and function in PS II. Fortunately, recent insights into the composition, structure and function of PS II have brought this possibility within reach. Here we describe the physical features of Cyt *b*-559 and discuss recent models to describe its role in photosynthesis. Previous reviews of Cyt *b*-559 include those by Cramer and Whitmarsh (1977), Bendall (1982), Bendall and Rolfe (1987), Cramer et al. (1993) and Whitmarsh et al. (1994).

Abbreviations: Cyt *b*-559LP – low potential form of Cyt *b*-559; Cyt *b*-559HP – high potential form of cyt b-559; msp – manganese stabilizing protein

II. Cytochrome *b*-559 Structure

A. Genes and Polypeptides

Cytochrome *b*-559 has two polypeptide subunits, α and β, that are encoded by two genes, *psbE* and *psbF*. In chloroplasts (Herrmann et al., 1984) and in cyanobacteria (Pakrasi et al., 1988) the two genes are adjacent and form part of an operon that includes two other genes, *psbL* and *psbJ*. This is not the case in the alga *Chlamydomonas reinhardtii* in which the two genes encoding the α and β subunits occupy different locations on the chromosome (Mor et al., 1994). In the cyanobacterium *Synechocystis* 6803 deletion of the entire *psbEFLJ* operon or deletion of the individual genes *psbE* or *psbF* leads to a destabilization of the PS II reaction center complex (Pakrasi et al., 1988, 1990; Shukla et al., 1992a). In these mutants the reaction center protein D2 is not integrated into the thylakoid membrane in a stable form. However, significant amounts of the PS II proteins, D1, CP47 and CP43, are found in the membranes of the mutant cells. It seems that the presence of the Cyt *b*-559 protein is required for the stable assembly of the PS II reaction center.

B. Polypeptide Sequences

Figure 1 shows a comparison of the derived amino acid sequences of the α and the β subunits of Cyt *b*-559 from a number of prokaryotic and eukaryotic organisms. Analysis of the amino acid sequences of the α and the β subunits of Cyt *b*-559 indicate that each polypeptide has one membrane spanning alpha-helical domain (Herrmann et al., 1984; Pakrasi et al., 1988). EPR and Raman spectroscopic analysis demonstrate that the two axial ligands of the heme cofactor are nitrogen atoms provided by two histidine residues (Babcock et al., 1985). Because the α- and

A

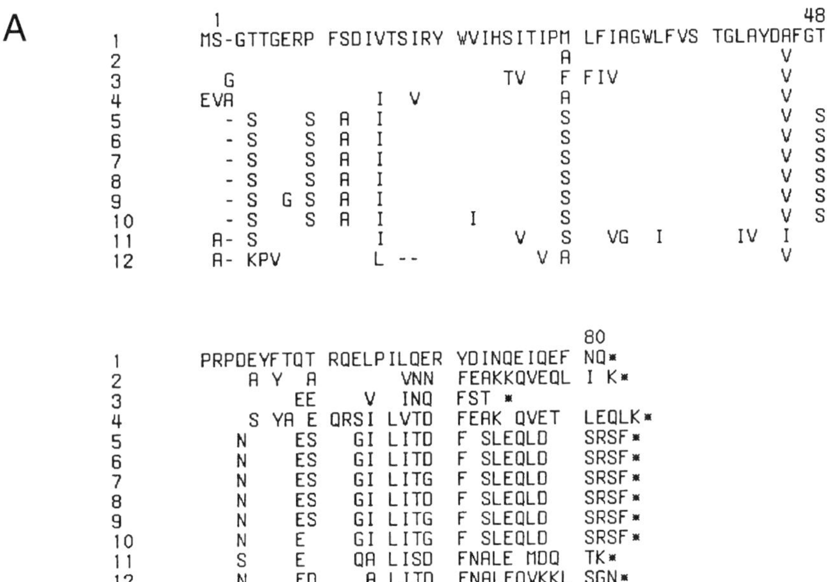

```
     1                                                   48
1   MS-GTTGERP FSDIVTSIRY WVIHSITIPM LFIAGWLFVS TGLAYDAFGT
2                                  A                  V
3     G                       TV   F FIV              V
4   EVA            I  V            A                  V
5     - S    S  A  I               S                  V  S
6     - S    S  A  I               S                  V  S
7     - S    S  A  I               S                  V  S
8     - S    S  A  I               S                  V  S
9     - S  G S  A  I               S                  V  S
10    - S    S  A  I       I       S                  V  S
11   A- S          I           V   S   VG  I      IV  I
12   A- KPV        L --          V A                  V


                                     80
1   PRPDEYFTQT RQELPILQER YDINQEIQEF NQ*
2       A Y  A       VNN  FEAKKQVEQL I K*
3           EE    V  INQ  FST *
4       S YA E QRSI LVTD  FEAK QVET  LEQLK*
5      N    ES   GI LITD  F SLEQLD   SRSF*
6      N    ES   GI LITD  F SLEQLD   SRSF*
7      N    ES   GI LITG  F SLEQLD   SRSF*
8      N    ES   GI LITD  F SLEQLD   SRSF*
9      N    ES   GI LITG  F SLEQLD   SRSF*
10     N    E    GI LITG  F SLEQLD   SRSF*
11     S    E    QA LISD  FNALE MDQ  TK*
12     N    ED    A LITD  FNALEQVKKL SGN*
```

Fig. 1A. Comparison of the sequence of the α subunit of Cyt *b*-559 in various organisms. Blank spaces in the sequences signify residues that are identical to that in *Synechocystis* 6803 (line 1). (-) indicates a gap introduced for optimal alignment of sequences. The first line shows residue numbers of the mature α subunit polypeptide in *Synechocystis* 6803. For alignments of similar sequences, the PILEUP and PRETTY programs of the Genetics Computer Group software package were used (Devereux et al., 1984). References: (1) *Synechocystis* 6803 (Pakrasi et al., 1988), (2) *Anabaena* 29413 (Pakrasi, unpublished), (3) Cyanelle (Cantrell and Bryant, 1988), (4) *Synechococcus vulcanus* (Koike and Pakrasi, unpublished), (5) Barley (Krupinska and Berry-Lowe, 1988), (6) Maize (Haley and Bogorad, 1990), (7) Spinach (Herrmann et al., 1984), (8) Wheat (Hird et al., 1986), (9) *Oenothera hookeri* (Carrillo et al., 1986), (10) Pea (Willey and Gray, 1989), (11) *Euglena* (Cushman et al., 1988), (12) *Chlamydomonas reinhardtii* (Mor et al., 1995).

B

```
      1                                             43
1.  -MATQNPNQP VTYPIFTVRW LAVHTLAVPS VFFVGAIAAM QFIQR*
2.  MTSGN I                        T    L    S       *
3.  -  MN       S           I AIGI A    I S T        *
4.  MTSNTPNQE   S         V        T I  L            *
5.  ----MTIDRT --           I G    T    L S S        *
6.  ----MTIDRT --             G    T    L S S        *
7.  ----MTIDRT --           I G    T  S L S S        *
8.  ----MTIDRT --           I G    T    L S S        *
9.  ----MTIDRT --             G    T  S L S S        *
10. ----MTIDRT --             G    T  S L S S        *
11. ----MTT KD TR             A  I T    L S S        *
12. - T KKSAEV LV   P       I GI   T I  L   T        *
```

Fig. 1B. Comparison of the sequence of the β subunit of Cyt *b*-559 in various organisms. Blank spaces in the sequences signify residues that are identical to that in *Synechocystis* 6803 (line 1). The first line shows residue numbers of the mature β subunit polypeptide in *Synechocystis* 6803.

β-subunit each has only one histidine residue, located in the alpha helix, it is presumed that Cyt *b*-559 is composed of two alpha helices linked together by two histidines that act as ligands for the heme. Similar heme-crosslinked membrane bound alpha helical domains are present in a number of membrane bound *a*- and *b*-type cytochromes (Esposti, 1989). Site directed replacement of either the αHis or the βHis to a Leu residue results in complete loss of PS II activity in *Synechocystis* 6803 (Pakrasi et al., 1991). Like the *psbE* and *psbF* deletion mutants, these His to Leu mutants lack the D2 protein, whereas CP47, CP43 and msp are present in significant amounts. These data indicate that proper coordination of the heme cofactor in Cyt *b*-559 is required for the stability of the PS II reaction center.

C. Heme Environment

The heme in Cyt *b*-559 is found in at least two distinct thermodynamic forms, one with a high midpoint potential (330–400 mV) and the other with a low midpoint potential (20–80 mV) (Cramer and Whitmarsh, 1977; Rich and Bendall, 1980). Although both forms coexist in normally functioning thylakoid membranes, neither form has been strictly correlated with any functional feature of the PS II complex. In addition, some membrane preparations have intermediate redox forms (e.g., Horton and Croze, 1977; Ortega et al., 1988). The Cyt *b*-559 heme is found both in grana and stroma membranes. Only the low potential form of the cytochrome is found in isolated stromal membranes (Vallon et al., 1987; Tae et al., 1993). McNamara and Gounaris (1995) show that a grana membrane preparation can be isolated that contains only the high potential form. It is noteworthy that the distribution of high and low potential Cyt *b*-559 between grana and stromal membranes correlates with the distribution of active and inactive PS II reaction centers, prompting the suggestion that the low potential form may be associated with inactive PS II centers and the high potential form with active centers (Chylla et al., 1987).

Experiments using thylakoid membranes show that various treatments convert the high potential form to the low potential form (e.g., heat, trypsin treatment, incubation at alkaline pH, treatment with detergents, and high salt buffers (Cramer and Whitmarsh, 1977)). However, treatments that convert the low potential form to the high potential form are rare (Matsuda and Butler, 1983; Whitford et al., 1984). Matsuda and Butler (1983) reported that reconstitution of PS II in the presence of lipids can restore at least a fraction of low potential Cyt *b*-559 to the high potential form. Recent work by Shuvalov and coworkers (1994) reveals an extremely low potential form of Cyt *b*-559 in highly purified PS II reaction center preparations. At pH 7.2 the extra low potential heme has a midpoint potential of –45 mV. At pH 9.4 the extra low potential heme exhibits two forms, one at a midpoint potential of +40 mV and the other at –220 mV. The absorption peak of the very low potential form shifts to 562 nm, which the authors attribute to replacement of a histidine axial ligand by a hydroxyl ion.

The structural features of the Cyt *b*-559 protein that determine the midpoint potential of the heme, in particular the unusual high potential form, are not understood. Babcock et al. (1985) have proposed a model to account for the presence of the two redox forms as well as their interconvertibility. As mentioned above, the heme cofactor is coordinated between the N-atoms of two His residues. EPR experiments indicate that for both the high and low potential form of the cytochrome, the plane of the heme is perpendicular to that of the membrane (Crowder et al., 1982). Babcock et al. (1985) suggested that the plane of the imidazole ring of one of the ligand His residues is twisted with respect to that of the other one in the high potential form of this protein. Rotation of the first imidazole back to the plane of the second one releases torsional energy, resulting in the transformation of the cytochrome to its low potential form. However, it should be noted that Walker et al. (1986) have argued that this transformation would be insufficient to account for such a large change in midpoint potential. Recently, Krishtalik et al. (1993) have offered a different explanation to account for the anomalously high potential form. They propose that the strong dipole electric field contributed by the α- and β-subunit alpha helices and the low dielectric constant of the heme environment can explain the high midpoint potential. According to this hypothesis exposure of the heme to the external water environment results in the low potential form of the cytochrome. It has also been shown that the conformation stabilizing the high potential form requires the ligation of calcium at the lumenal surface of the photosystem (McNamara and Gounaris, 1995).

III. Cytochrome *b*-559 Forms in Photosystem II

A. Role in Assembly

Two lines of evidence show that Cyt *b*-559 plays a key role in the biogenesis and stabilization of the PS II complex. First is the observation that in etiolated tissues of higher plants all of the PS II polypeptides are absent, except the Cyt *b*-559 protein (Herrmann et al., 1991). When the etiolated plants are placed in the light and start to green, other PS II polypeptides appear in the thylakoid membrane, ultimately forming functional PS II complexes. These observations have led to the suggestion that Cyt *b*-559 forms a nucleation center for the biogenesis of PS II. The second, and more compelling line of evidence is based on experiments using targeted mutations in the genes encoding the subunit polypeptides of Cyt *b*-559. Deletion of the *psbEFLJ* operon, as well as the individual *psbE* and *psbF* genes result in the absence of functional PS II complexes in the cyanobacterium *Synechocystis* 6803 (Pakrasi et al., 1988, 1990). In particular, the steady state level of the D2 protein is severely decreased in the thylakoid membranes of these mutant strains (Shukla et al., 1992a). Furthermore, when the heme ligand His residue in either the α- or β-subunit is changed to Leu, a residue that cannot bind a heme cofactor, assembly of the PS II reaction center is severely affected (Pakrasi et al., 1991). Tae and Cramer (1992) have created mutants of *Synechocystis* 6803 in which the C-terminal lumen-exposed hydrophilic domain of the α-subunit has also been truncated by the introduction of translational stop codons at various positions in the *psbE* gene. Analysis of these mutants led them to the conclusion that the C-terminal 31 residues of the α subunit of Cyt *b*-559 are important for the assembly of PS II, but not for its function. In summary, the genetic studies show that the α and β-subunits must be present and that the heme cofactor must be properly coordinated by both the α-His and β-His for the stable assembly of functional PS II complexes.

B. Polypeptide Composition

In the absence of crystallographic data the most productive approach to predict the organization of Cyt *b*-559 in the membrane has been hydropathy analysis combined with genetic, biochemical and spectroscopic data. This approach led Cramer and coworkers to propose that Cyt *b*-559 is a dimeric structure composed of the α and β subunits. The model is based on the observation that each α and β subunit has one membrane-spanning alpha helix and one histidine residue (Herrmann et al., 1984). This information and spectroscopic measurements which demonstrated that each of the two axial ligands of the heme are histidine residues (Babcock et al., 1985), prompted Cramer and coworkers to propose that the Cyt *b*-559 is a heterodimeric protein in which the heme cofactor links the α and β subunits by histidine ligands as shown in Fig. 2A.

It should be noted that the evidence favoring a heterodimeric model is not overwhelming, and it is possible that the cytochrome exists as two homo-dimers (α_2) and (β_2) as shown in Figs. 2B and C. Moskalenko et al. (1992) have examined the polypeptide organization of PS II reaction center preparations by chemical cross-linking, followed by gel analysis. Antibodies to the α-subunit recognized a band corresponding to an α_2 homodimer, whereas the band corresponding to the $\alpha\beta$ hetero-dimer was absent or extremely weak. It should be noted that the amino acid composition of the β subunit makes it less likely to cross-link. However, this observation shows that two α subunits are close neighbors, which can be interpreted to support two b-hemes per reaction center and even a homodimeric model. Alternatively, it is possible that in the detergent preparation two PS II centers could come in close contact, leading to α-α crosslinking.

C. Orientation of the α- and β-Subunits in the Thylakoid Membrane

Cramer and coworkers have provided convincing evidence that the α subunit of Cyt *b*-559 is oriented with its amino terminal end located in the stromal phase and its carboxyl end in the lumenal phase. Using spinach thylakoid membranes they showed that the carboxyl end of the α subunit is accessible to protease digestion in inside out vesicles, but not in right side out vesicles (Tae et al., 1988). Additional evidence supporting this orientation is provided by immuno-gold labeling and electron microscopy showing that the carboxyl end of the α-subunit is accessible to antibodies in inside out vesicles, but not in right side out vesicles (Vallon et al., 1989). If the heterodimeric model is correct, then the heme of Cyt *b*-559 would be near the stromal phase, about 8 Å from the end of the alpha helix. Cramer et al. (1993) have pointed out that the heterodimeric model shown in Fig. 2A is supported by the fact that the $\alpha\beta$ heterodimer obeys the rule for transmembrane orientation of integral proteins proposed by von Heijne (1986). The rule states that integral proteins tend to insert into membranes in a manner that minimizes the amount of positive charge that must be translocated across the membrane. The rule has been shown to apply to thylakoid membrane proteins (Gavel et al., 1991). In the case of the α-subunit, the predicted orientation would place its carboxyl end in the stromal phase, contradicting the biochemical evidence described above. However, the $\alpha\beta$ heterodimer considered as a single protein has more positive charge on the amino end of the complex, which would place the amino terminal end in the stromal phase. This consideration led Cramer et al.

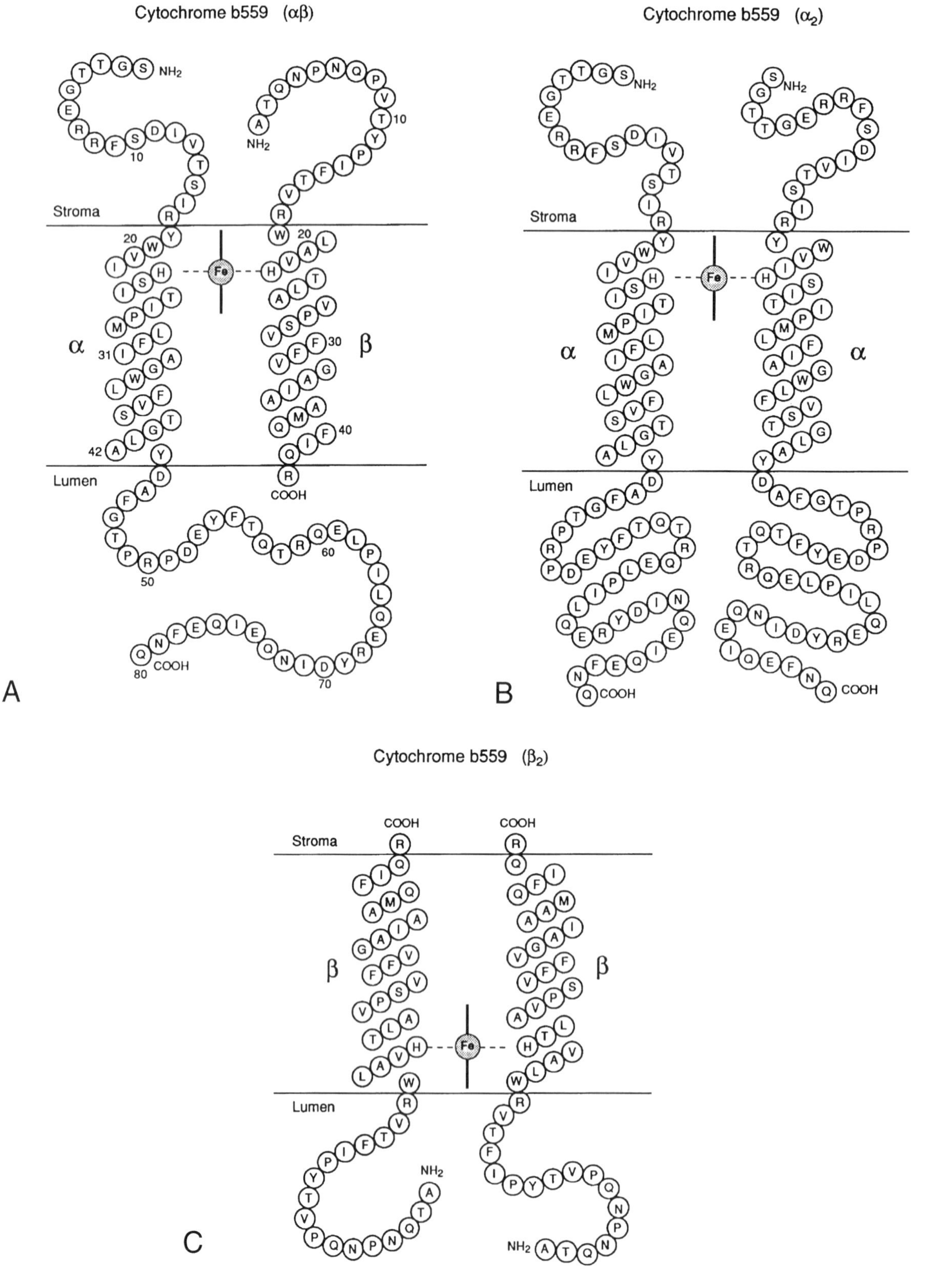

Cytochrome b559 ($\alpha\beta$)

Stroma

Lumen

α β

A

Cytochrome b559 (α_2)

Stroma

Lumen

α α

B

Cytochrome b559 (β_2)

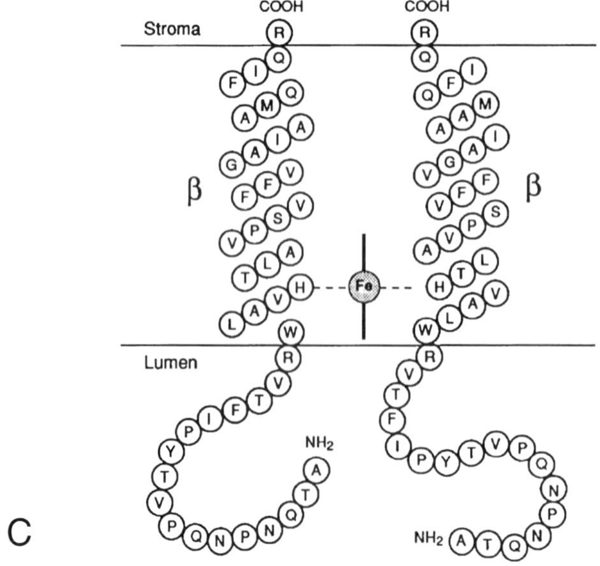

Fig. 2. Possible composition and orientation of Cyt *b*-559 polypeptides in PS II.

(1993) to propose that the insertion of Cyt *b*-559 into the membrane involves the $\alpha\beta$ heterodimer. It should be noted that if this argument is applied to Cyt *b*-559 in *Synechococcus vulcanus* or *Chlamydomonas reinhardtii*, the von Heijne rule predicts the opposite orientation of an $\alpha\beta$ heterodimer due to the positively charged lysine residues on the carboxyl end of the α subunit (Fig. 1).

Tae and Cramer (1994) have investigated the accessibility of the β-subunit amino terminal end to proteolysis by trypsin in membranes isolated from spinach and *Synechocystis* 6803. In *Synechocystis* 6803 the amino terminal end was modified by the introduction of an epitope that reacted with antibody raised against the Cyt *b*-559 α-subunit from spinach. Using Western analysis, they showed that the amino terminal end of the β-subunit of *Synechocystis* 6803 was accessible to trypsin, whereas the manganese stabilizing protein of PS II and the carboxyl terminal end of the α-subunit, both known to be located on the lumen side of the membrane, were resistant to trypsin. Based on these data, these authors have argued that the β-subunit of Cyt *b*-559 is oriented with its amino terminal end in the stromal phase. If this is correct, it means that for the $\alpha\beta$, or β_2 model of Cyt *b*-559, the heme would be located near the stromal side of the membrane. It is important to note that the interpretation of the data depends on the assumption that the isolated membranes are vesicular, right-side out, and that trypsin cannot penetrate to the inner aqueous phase. Whereas the data support this assumption, the fact that there is some trypsin digestion of the manganese stabilizing protein, but at a slower rate than of Cyt *b*-559, indicates that trypsin can infiltrate the lumen. In summary, the data cited above support, but do not prove, that the amino terminal end of the β-subunit is located in the stromal phase and the possibility remains that it may be oriented in the opposite direction, placing the heme near the lumenal phase.

D. One or Two Hemes per Reaction Center?

Understanding the organization of Cyt *b*-559 in PS II requires knowing the number of Cyt *b*-559 hemes in a functional reaction center. Unfortunately, there is disagreement in the literature concerning this question. Reports of two Cyt *b*-559 hemes per PS II reaction center are common (Vermeglio and Mathis, 1974b; Lam et al., 1983; Murata et al., 1984; Whitford et al., 1984; Whitmarsh and Ort, 1984; Yamamoto et al., 1984; Briantais et al., 1985; de Paula et al., 1985; Dekker et al., 1989; Lee and Whitmarsh, 1989; Shuvalov et al., 1989, 1994; Haag et al., 1990; Boerner et al., 1992, 1993; MacDonald et al., 1994), as are reports of one Cyt *b*-559 heme per reaction center (Ford and Evans, 1983; Ghanotakis et al., 1984; Yerkes and Crofts, 1984; Franzen et al., 1986; Barber et al., 1987; Yamagishi and Fork, 1987; Gounaris et al., 1989, 1990; Miyazaki et al., 1989; Barbato et al., 1991; Buser et al., 1992b; MacDonald et al., 1994; Tang and Diner, 1994). Determining the number of Cyt *b*-559 hemes in PS II is a difficult problem, which is illustrated by the fact that many researchers have papers in both of the groups cited above. The controversy is due in part to problems inherent in comparing data from different preparations that include isolated thylakoid membranes, detergent preparations of PS II-enriched membranes, and purified PS II reaction centers. Van Leeuwen et al. (1991) showed that the number of Cyt *b*-559 hemes per reaction center decreases as the PS II preparation is exposed to added detergent. This observation is particularly relevant to the work of Buser et al. (1992b), in which they determined the Cyt *b*-559 heme content of a PS II-enriched preparation to be 1 heme/PS II. In fact, convincing measurements showing one Cyt *b*-559 heme per PS II are invariably based on detergent preparations (Franzen et al., 1986; Miyazaki et al., 1989; Buser et al., 1992b). However, not all detergent preparations yield a single heme per reaction center. Recent work by MacDonald et al. (1994) provides evidence that in PS II reaction center preparations isolated from the cyanobacterium *Synechocystis* 6803, there are 1.5–2.0 Cyt *b*-559 hemes per PS II. However, in membrane fragments isolated from spinach they find one heme per reaction center. One explanation of these data is that detergent can selectively remove one Cyt *b*-559 from PS II. This would suggest that one of the Cyt *b*-559 hemes is more readily destabilized by detergent, a notion that is supported by the observation that Cyt *b*-559 heme shows a differential accessibility to ferricyanide (Selak et al., 1984).

There are two additional problems in establishing an accurate heme content of PS II. One is that there is no agreement on the extinction coefficient for Cyt *b*-559. Most determinations of the concentration of Cyt *b*-559 depend on absorption spectra of the reduced minus oxidized form of the cytochrome and application of the Beer-Lambert Law, which requires knowledge of the correct extinction coefficient.

However, there is considerable variation in the literature on the extinction coefficient for Cyt *b*-559. For the reduced minus oxidized extinction coefficient for the wavelength pairs 559 nm minus 577 nm the data of Garewal and Wasserman (1975) indicate a value of 16 mM^{-1} cm^{-1} (see Cramer and Whitmarsh, 1977), whereas Cramer et al. (1986) reported that the extinction coefficient for the wavelength pairs 560 minus 575 nm is 21.5 mM^{-1} cm^{-1} and for the wavelength pairs 560 nm minus 570 nm the extinction coefficient is 17.5 mM^{-1} cm^{-1}. The data of Miyazaki et al. (1989) give a value of 24.6 mM^{-1} cm^{-1} (estimated from Fig. 2D). A critical factor in determining the extinction coefficient is creating the pyridine hemochrome from the cytochrome. We have found that although this can be done easily for myoglobin, the reaction is less satisfactory for Cyt *b*-559. In particular, the pyridine hemochrome spectrum created using Cyt *b*-559 does not appear to be pure. Furhop and Smith (1975) state that the ratio of the α-band peak at 556 nm to the trough at 539 nm should be 3.47. Lower values indicate degradation or contamination of the product and therefore draw into question the accuracy of the method. In our hands this ratio is nearer three (V. McNamara and J. Whitmarsh, unpublished observations), as are the apparent ratios in other published reports . This is an important point that we think should be reexamined, particularly because the pyridine hemochrome spectra are rarely published.

The second problem is that many estimates of the heme content of PS II depend on an accurate determination of the concentration of the PS II reaction centers which is difficult, in part, because PS II exists in the membrane in different forms. Many PS II assays detect only active centers, whereas a significant fraction of PS II centers are inactive (e.g., Chylla and Whitmarsh, 1989). For example, Tang and Diner (1994) have presented data showing one Cyt *b*-559 per reaction center in core preparations isolated from *Synechocystis* 6803, whereas MacDonald et al. (1994) and Boerner et al. (1993) argue that the correct stoichiometry is 2:1. The difference between these two groups is in the quantitation of PS II. They agree on the amount of Cyt *b*-559 per chlorophyll. Although there is no supportive evidence, the possibility should be considered that some forms of PS II, for example inactive centers, may contain one Cyt *b*-559 heme per reaction center, whereas fully functional centers may contain two hemes per reaction center.

In considering the heme stoichiometry of PS II, it is interesting to consider the observation of Moskalenko et al. (1992) that two α subunits of Cyt *b*-559 are close enough together to allow cross linking. The simplest interpretation of this observation is that two α subunits are close neighbors in a single reaction center. Since both the α and β subunits contribute histidine ligands, this observation supports the view that there are two hemes per reaction center. However, as mentioned above it is possible that α-subunits from two different reaction centers could cross react.

At present it is fair to say that the number of Cyt *b*-559 hemes in a functional PS II reaction center remains a cloudy issue.

E. Relation to Other Photosystem II Proteins

One of the smallest photoactive PS II preparations contains the α and β subunits of Cyt *b*-559 as well as the D1, D2 and PsbI proteins, implying that these five polypeptides are closely associated in the PS II complex. Chemical cross-linking experiments by Enami et al. (1992) indicated that in a purified O_2-evolving PS II preparation from spinach, the α subunit of Cyt *b*-559 is within 11 angstroms of the extrinsic 33 kDa msp. This is consistent with the observation by Tae et al. (1988) that the carboxy-terminal lumen exposed domain of the α subunit is shielded by msp. In an independent study, Moskalenko and coworkers (1992) used a number of chemical cross-linkers to demonstrate that in two different highly resolved PS II preparations, the α subunit can form cross linked products with D1, D2 and the β subunit of Cyt *b*-559. As mentioned earlier, the Cyt *b*-559 deletion mutant strains do not accumulate the D2 protein, implying that these two proteins are structurally associated in PS II (Shukla et al., 1992a). The fact that Cyt *b*-559 can be photooxidized at low temperature by $P680^+$ within a few ms (Butler et al., 1973; Thompson and Brudvig, 1988; Vermeglio and Mathis, 1974a) indicates that the cytochrome is near P680. (For a discussion of intraprotein transfer rates versus distance see Moser et al., 1992). Figure 3 shows a bird's-eye view of the PS II reaction center core that is consistent with these putative associations. The circles represent the transmembrane alpha helices of the indicated proteins. The arrangement of the proteins is speculative and is guided by the structure of the purple bacteria reaction center (for a bird's-eye view of the reaction center of *Rhodobacter sphaeroides* see Norris and Schiffer, 1990).

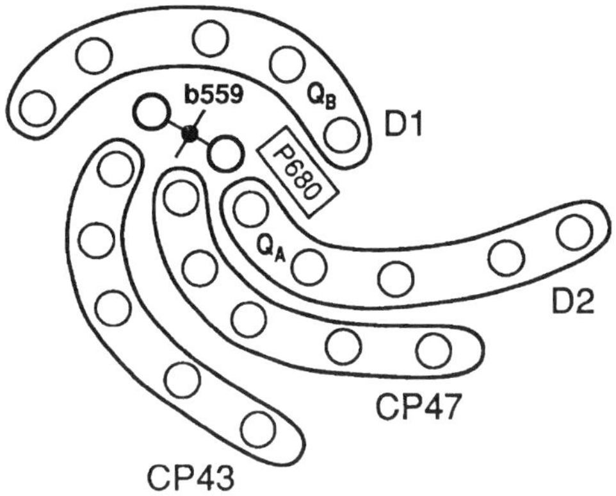

Fig. 3. Cartoon view of PS II alpha helices viewed from the stroma aqueous phase (perpendicular to the membrane). The arrangement shown here is guided by the structure of the purple bacteria reaction center and biochemical and biophysical data relating to Cyt *b*-559 discussed in the text.

IV. Cytochrome *b*-559 Function

A. Background

Understanding the functional role of Cyt *b*-559 has proven difficult largely because its light-induced turnover under physiological conditions is slow compared to other redox carriers known to be involved in photosynthetic electron transfer (e.g., Whitmarsh and Cramer, 1977, 1978; Canaani and Havaux, 1990; Klughammer et al., 1990). In fact, under most experimental conditions little or no Cyt *b*-559 can be observed to undergo light-induced redox activity. Despite the apparent inactivity of Cyt *b*-559 under conditions in which electrons are traveling rapidly through PS II, the cytochrome has been invoked to play a key role in numerous side reactions, most often in cyclic electron transport around PS II. The putative cycle is usually proposed to protect the reaction center from damage due to high light intensities, by dissipating excess energy that would otherwise damage the reaction center or a neighboring reaction center (e.g., Heber et al., 1979; Falkowski et al., 1986; Thompson and Brudvig, 1988; Canaani and Havaux, 1990; Rees and Horton, 1990; Whitmarsh and Chylla, 1990; Buser et al., 1992a; McNamara and Gounaris, 1992). For these mechanisms to offer effective protection, cyclic electron transport must operate at rates comparable to linear electron transport. At present there is no direct evidence that Cyt *b*-559 can turn over at such high rates. Not all cyclic models are designed to dissipate excess energy. Arnon and Tang (1988) have suggested that Cyt *b*-559 participates in a redox-linked proton pump that contributes to the proton electrochemical gradient across the photosynthetic membrane. Examination of Fig. 4, which illustrates several of the electron pathways that Cyt *b*-559 has been proposed to accommodate, shows that there is little consensus in the literature concerning the behavior of the cytochrome, let alone its function. Cramer et al. (1993) have provided a useful comparison of the functional roles of various b-cytochromes, although there are no obvious role models for understanding the enigmatic Cyt *b*-559.

Additional roles include the suggestion by Cramer et al. (1986) that Cyt *b*-559 contributes to the photoactivation of the water splitting Mn cluster by oxidizing Mn^{2+}. Cramer and coworkers have also investigated the possibility that the cytochrome plays a role in fatty acid desaturation (Cramer et al., 1990). A fresh idea from Ananyev and coworkers (1994) suggests that the high potential form of Cyt *b*-559 is involved in a superoxide dismutase activity. Recent results point to a role for Cyt *b*-559 in photoprotection not as a dissipater of excess energy, but rather as a safety valve to dissipate a potentially damaging radical that may be formed in the normal functioning of PS II. These models have the advantage of explaining the slow and limited turnover of Cyt *b*-559 observed under physiological conditions and are discussed below.

B. Role in Photoprotection

Photosystem II is susceptible to damage by visible light (reviewed by Prasil et al., 1992 and Aro et al., 1993). The light-induced inhibition of PS II is a multistep process, starting with photoinactivation of electron transfer reactions in PS II that eventually leads to the degradation of one of the two reaction center core proteins (D1) and to a lesser extent degradation of the other core protein (D2). Some experiments show that the initial site of damage is on the reducing side of PS II, whereas other experiments, often done using material in which the water oxidizing system has been disconnected from primary photochemistry, show that the primary site of damage is on the oxidizing side of PS II. Such results have prompted two models of photoinhibition—those that focus on damage done by highly oxidizing cation

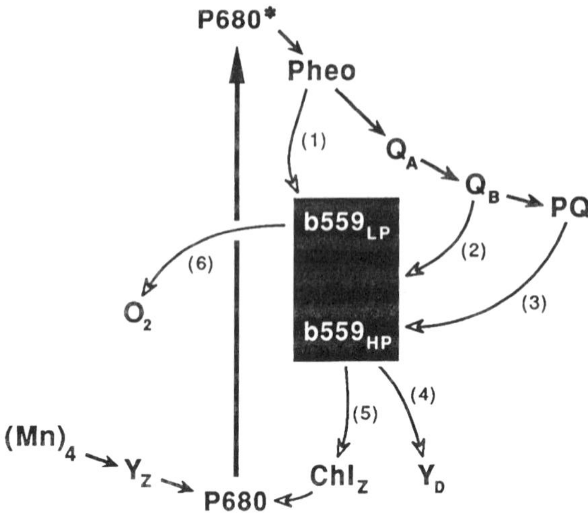

Fig. 4. Electron pathways suggested in the literature for Cyt *b*-559. (1) (Nedbal et al., 1992; Barber and De Las Rivas, 1993; Whitmarsh et al., 1994), (2) (Buser et al., 1992a), (3) (Whitmarsh and Cramer, 1978), (4) (Vass et al., 1990), (5), (6) (Whitmarsh et al., 1994).

radicals and those that focus on damage done by highly reducing anion radicals. There is controversy concerning which type of photoinhibition may be significant under physiological conditions. In part, the controversy is due to differences in experimental approaches. However, as discussed by Aro et al. (1993) and Prasil et al. (1992), the different pathways demonstrated in vitro may actually occur in plants under different stress conditions .

1. Cyt b-559 as an Electron Acceptor

A redox component can exert strong control over the sensitivity of PS II to photoinhibition (Nedbal et al., 1992). Changing the redox state of a one-electron component from the reduced to the oxidized state causes the rate of photoinhibition in continuous light to decrease by an order of magnitude. In this work photoinhibition was monitored under anaerobic conditions at various ambient redox potentials by measuring the impact of various doses of visible light on the rate of water oxidation and by the variable fluorescence decrease associated with the impairment of PS II activity. The data fit a one-electron Nernst equation with a midpoint potential of 28 ± 18 mV (pH 7.5) which is nearly pH independent over the range from 5.5 to 8.0. The results are the same in the presence or absence of DCMU and the loss of water oxidation capacity was irreversible.

These results reveal what appears to be a protective electron transport pathway in PS II that can severely retard the rate of photoinhibition. One model that provides a molecular mechanism to account for photoinhibition focuses on the reducing anion radicals formed in the sequential reactions leading to the reduction of plastoquinone (e.g., van Mieghem et al., 1989; Setlik et al., 1990; Styring et al., 1990; Vass et al., 1992). In this model the trigger for photoinhibition is the creation of stable $Pheo^-$, which can occur when Q_A is reduced and PS II is in the state $P680^+ Pheo^- Q_A^-$. Although this state typically decays by a back reaction between $P680^+$ and $Pheo^-$, there is a low probability that the primary donor $P680^+$ will be reduced by Y_Z, creating the stable state ($P680\ Pheo^- Q_A^-$) (Rutherford and Zimmerman, 1984; Klimov et al., 1985). Since reduced pheophytin is a strong reductant, it can reduce Q_A^-. The doubly reduced Q_A^{2-} is proposed to be protonated, and then to dissociate from PS II (van Mieghem et al., 1989; Styring et al., 1990; Vass et al., 1992), triggering subsequent steps that result in the removal of the D1 polypeptide.

Redox control of photoinhibition is most simply explained in the context of the above model, in which photoinhibition is initiated by Q_A^{2-}. The key idea is that in normally active PS II centers an alternative electron transfer pathway exists, which drains electrons from $Pheo^-$, thereby preventing the formation of Q_A^{2-}. The pathway depends on a redox component with an operating midpoint potential of 20 mV or somewhat higher being in the oxidized state. When the component is poised in the reduced state, the protective pathway is blocked, which allows reduction of Q_A^- by $Pheo^-$, leading to photoinhibition. Comparison of the redox behavior of PS II prosthetic groups indicates that the low potential form of Cyt *b*-559 is the most likely candidate for the redox component controlling the rate of photoinhibition (Nedbal et al., 1992). This model assumes that Cyt *b*-559 can be reduced by $Pheo^-$. Barber and De Las Rivas (1993) have provided evidence for this pathway in a PS II reaction center preparation. They demonstrated the full reduction of low potential Cyt *b*-559 by pheophytin in continuous light. Similar behavior has been observed in PS II enriched membrane preparations and thylakoid membranes under anaerobic conditions (Whitmarsh et al., 1994; Poulson et al., 1995). The reduction of Cyt *b*-559 in continuous light occurred in the presence or absence of DCMU in both PS II enriched membranes and thylakoid membranes, which is consistent with reduction of the cytochrome by pheophytin. Furthermore, the data show that once Cyt *b*-559 becomes reduced, the rate of photoinhibition increases dramatically. These observations support two key features of the photoprotection scheme. First, that as the low potential form Cyt *b*-559 is reduced, photoinhibition increases (Whitmarsh et al., 1994; Poulson et al., 1995), and second, that low potential Cyt *b*-559 can be reduced by pheophytin (Barber and De Las Rivas, 1993; Whitmarsh et al., 1994; Poulson et al., 1995).

Although several molecular species have been shown to oxidize Cyt *b*-559 under various conditions, an endogenous oxidant under physiological conditions has not been identified. Within the context of a photoprotective model operating under reducing conditions, plastoquinone, molecular oxygen or a redox component on the donor side of PS II could serve as an oxidant. The notion that free plastoquinone could act as an oxidant is supported by the observation of an interaction between free plastoquinone and high potential Cyt *b*-559 in thylakoid membranes (Whitmarsh and Cramer, 1978). These data raise the intriguing possibility that plastoquinone may interact directly with Cyt *b*-559 via a quinone binding site on PS II other than the Q_A or Q_B sites. Such a binding site could account for observations of a second low affinity inhibitor binding site on PS II (e.g., Jursinic et al., 1991).

A consequence of proposing that the photoprotective redox component is low potential Cyt *b*-559 is that it requires the presence of at least one low potential heme per reaction center. At present, it is not known if there is one low potential Cyt *b*-559 heme per reaction center endogenously present in thylakoid membranes. Alternatively, high potential Cyt *b*-559 may be reversibly transformed to the low potential form in the light (Nedbal et al., 1992; Barber and De Las Rivas, 1993). In fact, Styring et al. (1990) have observed such a conversion, but it is not clear if the conversion precedes or is concurrent with loss of water oxidation or whether it is reversible.

The photoprotective model described here accounts for the slow turnover of Cyt *b*-559 observed under physiological conditions. Since the reduction of Cyt *b*-559 depends on the stable reduction of pheophytin and the photoaccumulation of reduced Pheo is a low quantum yield process (Klimov et al., 1977), the rate of electron transfer through Cyt *b*-559 can be slow, and still provide protection against production of Q_A^{2-}. In fact, a slow rate of electron transfer through Cyt *b*-559 would be required to ensure the high quantum yield normally observed in PS II (Nedbal et al., 1992; Whitmarsh et al., 1994; Poulson et al., 1995).

2. Cyt b-559 as an Electron Donor

Some models of photoinhibition suggest that the events which initiate photodamage are on the donor-side of PS II, within the sequential reactions leading to water oxidation (Prasil et al., 1992; Aro et al., 1993). Damage to PS II is envisioned to be a consequence of the highly oxidizing radicals Y_z^+ or $P680^+$, which can oxidize nearby chromophores or residues (Blubaugh et al., 1991; van der Bolt and Vermaas, 1992). Within this scheme, Brudvig and coworkers (Thompson and Brudvig, 1988; Buser et al., 1992a) have proposed that high potential Cyt *b*-559 provides a protective electron pathway in PS II by reducing the damaging cation radicals. Photooxidation of high potential Cyt *b*-559 in PS II can be observed when the main electron donation pathway from water to P680 is impaired by low temperatures (Erixon and Butler, 1971), ADRY agents (Heber et al., 1979; Yerkes and Crofts, 1984; Samson and Fork, 1992) or by inhibition of the oxygen-evolving complex (Buser et al., 1990). Butler et al. (1973) observed similar kinetics of cytochrome *b*-559 photooxidation and fluorescence induction at 77 K preceded by C550 photoreduction, indicating that under these conditions the fluorescence increase was not due to the photoreduction of Q_A but to the reduction of $P680^+$ by Cyt *b*-559. The auxiliary nature of the electron donation to $P680^+$ by Cyt *b*-559 was clearly demonstrated at cryogenic temperatures by the inverse relationship between the amount of photooxidized Cyt *b*-559 and the amplitude of the EPR S_2 state multiline signal (Thompson and Brudvig, 1988).

If Cyt *b*-559 is oxidized, illumination of PS II at low temperatures leads to the oxidation of a reaction center auxiliary chlorophyll molecule, Chl_z (Visser et al., 1977). Thompson and Brudvig (1988) demonstrated that Cyt *b*-559 and Chl_z are sequential electron donors to $P680^+$. Assuming simultaneous Cyt *b*-559 photooxidation (observed in non-O_2-evolving PS II) and Cyt *b*-559 photoreduction (characteristic of O_2-evolving PS II), these authors propose that Cyt *b*-559 is involved in cyclic electron flow that may protect PS II reaction centers from photoinhibition by Chl_z^+. In this model, the first event of photoinhibition is the irreversible modification of Chl_z causing a loss of efficiency in energy transfer from the chlorophyll antenna to the reaction center. Other researchers, studying photoinhibition in PS II membranes that have been inactivated on the donor side by treatment with hydroxylamine, proposed that $P680^+$ rather than Chl_z^+ is the main cation radical responsible for photodamage (Blubaugh et al., 1991). Under the experimental conditions described by this group, decoupling of the antenna chlorophyll from the reaction center complex was not observed.

C. Role in Water Oxidation

In most cases treatments that convert the high potential form of Cyt *b*-559 to the low potential form result in the loss of water oxidation, a correlation that has led to suggestions that the cytochrome is involved in water oxidation (e.g., Butler, 1978). For example, Matsuda and Butler (1983) found that reconstitution of low-activity PS II preparations in liposomes stimulates O_2 evolution and at the same time converts some Cyt *b*-559LP to Cyt *b*-559HP. However, the redox potential of Cyt *b*-559HP makes it an unlikely acceptor in the electron-transfer chain between water and P680. Furthermore, trypsin treated thylakoid membranes, in which all of cytochrome *b*-559 is converted into the low potential form, are still capable of water oxidation (Völker et al., 1986). In addition, experiments with subthylakoid PS II preparations show that it is possible to retain high O_2 evolution activity, even when Cyt *b*-559 is present only in its low potential form. For example, Briantais et al. (1985) showed that BBY PS II membrane preparations treated with 1 M $CaCl_2$, in which Cyt *b*-559 is converted to the low potential form, can mediate high rates of water oxidation.

As mentioned earlier, the carboxy terminal domain of the α subunit of Cyt *b*-559 is exposed to the thylakoid lumen. Since Cyt *b*-559 is intimately associated with the reaction center proteins D1 and D2, as well as msp, it is possible that the lumen exposed region of the cytochrome may influence the water oxidizing mechanism. Tae and Cramer (1992) investigated the role of the hydrophilic C-terminal region of the α-subunit of Cyt *b*-559 by creating mutants lacking the last 12, 22 and 31 amino acids in *Synechocystis* 6803. On the basis of the PS II activity in the R50-stop mutant, they conclude that the C-terminal 31 residues of the α-subunit are not required for normal water oxidation activity. However, site specific modifications of the 37 amino acid long lumen exposed portion of the α subunit in *Synechocystis* 6803 indicate that this part of the polypeptide may have some role in the water oxidation reactions. Oligonucleotide directed deletion of various domains in the C terminal portion of this polypeptide show that the region between Asp52 and Ile71 has a significant influence on oxygen evolution activity (Shukla et al., 1992b). In particular, site specific modification of Arg68 to Leu results in the formation of a PS II complex that is vulnerable to light, presumably due to donor side photoinhibition. In summary, there is no evidence that Cyt *b*-559 plays a redox role in water oxidation, however, it may play a structural role in maintaining an active complex or it could possibly play a role in the assembly of the manganese cluster.

Acknowledgments

We thank Vince McNamara for helpful discussions. The preparation of this article was supported in part by NRICGP/USDA (JW) and NIH (HP).

References

Ananyev G, Renger G, Wacker U and Klimov V (1994) The photoproduction of superoxide radicals and superoxide dismutase activity of Photosystem II. The possible involvement of cytochrome *b*-559. Photosynth Res 41: 327–338

Arnon DI and Tang GMS (1988) Cytochrome *b*-559 and proton conductance in oxygenic photosynthesis. Proc Natl Acad Sci USA 85: 9524–9528

Aro E-M, Virgin I and Andersson B (1993) Photoinhibition of Photosystem II. Inactivation, protein damage and turnover. Biochim Biophys Acta 1143: 113–134

Babcock GT, Widger WR, Cramer WA, Oertling WA and Metz JG (1985) Axial ligands of chloroplast cytochrome b-559: Identification and requirement for a heme-cross-linked polypeptide structure. Biochemistry 24: 3638–3645

Barbato R, Race HL, Friso G and Barber J (1991) Chlorophyll levels in the pigment-binding proteins of Photosystem II. FEBS Lett 286: 86–90

Barber J and De Las Rivas J (1993) A functional model for the role of cytochrome *b*-559 in the protection against donor and acceptor side photoinhibition. Proc Natl Acad Sci USA 90: 10942–10946

Barber J, Chapman DJ and Telfer A (1987) Characterisation of a PS II reaction centre isolated from the chloroplasts of *Pisum sativum*. FEBS Lett 220: 67–73

Bendall DS (1982) Photosynthetic cytochromes of oxygenic

organisms. Biochim Biophys Acta 683: 119–151

Bendall DS and Rolfe SA (1987) Characterization of chloroplast cytochromes. Meth Enzym 148: 259–273

Blubaugh DJ, Atamian M, Babcock GT, Golbeck JH and Cheniae GM (1991) Photoinhibition of hydroxylamine-extracted Photosystem II membranes: Identification of the sites of photodamage. Biochemistry 30: 7586–7597

Boerner RJ, Nguyen AP, Barry BA and Debus RJ (1992) Evidence from directed mutagenesis that aspartate 170 of the D1 polypeptide influences the assembly and/or stability of the manganese cluster in the photosynthetic water-splitting complex. Biochemistry 31: 6660–6672

Boerner RJ, Bixby KA, Nguyen AP, Noren GH, Debus RJ and Barry BA (1993) Removal of stable tyrosine radical D^+ affects the structure or redox properties of tyrosine Z in Manganese-depleted Photosystem II particles from *Synechocystis* 6803. J Biol Chem 268: 1817–1823

Briantais JM, Vernotte C, Miyao M, Murata N and Picaud M (1985) Relationship between oxygen evolution capacity and cytochrome *b*-559 high-potential form in Photosystem II particles. Biochim Biophys Acta 808: 348–351

Buser CA, Thompson LK, Diner BA and Brudvig GW (1990) Electron-transfer reactions in manganese-depleted Photosystem II. Biochemistry 29: 8977–8985

Buser CA, Diner BA and Brudvig GW (1992a) Photooxidation of cytochrome *b*-559 in oxygen-evolving Photosystem II. Biochemistry 31: 11449–11459

Buser CA, Diner BA and Brudvig GW (1992b) Reevaluation of the stoichiometry of cytochrome *b*-559 in Photosystem II and thylakoid membranes. Biochemistry 31: 11441–11448

Butler WL (1978) On the role of cytochrome *b*-559 in oxygen evolution in photosynthesis. FEBS Lett 95: 19–25

Butler WL, Visser JWM and Simons HL (1973) The kinetics of light-induced changes of C-550, cytochrome *b*-559 and fluorescence yield in chloroplasts at low temperature. Biochim Biophys Acta 292: 140–151

Canaani O and Havaux M (1990) Evidence for a biological role in photosynthesis for cytochrome *b*-559—a component of Photosystem II reaction center. Proc Natl Acad Sci USA 87: 9295–9299

Cantrell A and Bryant D (1988) Nucleotide sequence of the genes encoding cytochrome *b*-559 from the cyanelle genome of *Cyanophora paradoxa*. Photosynth Res 16: 65–71

Carrillo N, Seyer P, Tyagi A and Herrmann RG (1986) Cytochrome *b*-559 genes from *Oenothera hookeri* and *Nicotiana tabacum* show a remarkably high degree of conservation as compared to spinach. Curr Genet 10: 619–624

Chylla RA and Whitmarsh J (1989) Inactive Photosystem II complexes in leaves: Turnover rate and quantitation. Plant Physiol 90: 765–772

Chylla RA, Garab G and Whitmarsh J (1987) Evidence for slow turnover in a fraction of Photosystem II complexes in thylakoid membranes. Biochim Biophys Acta 894: 562–571

Cramer WA and Whitmarsh J (1977) Photosynthetic cytochromes. Ann Rev Plant Physiol 28: 133–172

Cramer WA, Theg SM and Widger WR (1986) On the structure and function of cytochrome *b*-559. Photosynth Res 10: 393–403

Cramer WA, Furbacher PN, Szczepaniak A and Tae G-S (1990) The chloroplast b cytochromes: Crosslinks, topography, and functions. In: Baltscheffsky M (eds) Current Research in Photosynthesis, Vol 3, pp 221–230. Kluwer Academic Publishers, Dordrecht

Cramer WA, Tae G-S, Furbacher PN and Böttger M (1993) The enigmatic cytochrome *b*-559 of oxygenic photosynthesis. Physiologia Plantarum 88: 705–711

Crowder MS, Prince RC and Bearden A (1982) Orientation of membrane-bound cytochromes in chloroplasts detected by low-temperature EPR spectroscopy. FEBS Lett 144: 204–208

Cushman JC, Christopher DA, Little MC, Hallick RB and Price CA (1988) Organization of the *psbE*, *psbF*, orf38, and orf42 Gene Loci on the *Euglena gracilis* chloroplast genome. Curr Genet 13: 173–180

de Paula JC, Innes JB and Brudvig GW (1985) Electron transfer in Photosystem II at cryogenic temperatures. Biochemistry 24: 8114–8120

Dekker JP, Bowlby NP and Yocum CF (1989) Chlorophyll and cytochrome *b*-559 contents of the photochemical reaction center of Photosystem II. FEBS Lett 254: 150–154

Devereux J, Haeberli P and Smithies O (1984) A comprehensive set of sequence analysis programs for the VAX. Nucl Acid Res 12: 387–394

Enami I, Ohta S, Mitsuhashi S, Takahashi S, Ikeuchi M and Katoh S (1992) Evidence from crosslinking for a close association of the extrinsic 33 kDa protein with the 9.4 kDa subunit of cytochrome *b*-559 and the 4.8 kDa product of the *psbI* gene in oxygen-evolving Photosystem II complexes from spinach. Plant Cell Physiol 33: 291–297

Erixon K and Butler WL (1971) The relationship between Q, C-550 and Cytochrome *b*-559 in photoreactions at –196° in chloroplasts. Biochim Biophys Acta 234: 381–389

Esposti MD (1989) Prediction and comparison of the haem-binding sites in membrane haemoproteins. Biochim Biophys Acta 977: 249–265

Falkowski PG, Fujita Y, Ley A and Mauzerall C (1986) Evidence for cyclic electron flow around Photosystem II in *Chlorella pyrenoidosa*. Plant Physiol 81: 310 312

Ford RC and Evans MCW (1983) Isolation of a photosystem 2 complex from higher plants with highly enriched oxygen evolution activity. FEBS Lett 254: 159–164

Franzen LG, Styring S, Etienne AL, Hansson O and Bernotte C (1986) Spectroscopic and functional characterization of a highly oxygen evolving Photosystem II reaction center complex from spinach. Photobiochem Photobiophys 13: 15–28

Furhop JH and Smith KM (1975) Laboratory methods. In: Smith KM (eds) Porphyrins and Metalloporphyrins, pp 757–869. Elsevier Scientific Publishing Co., Amsterdam

Garewal HS and Wasserman AR (1975) Triton X-100-4M urea as an extraction medium for membrane proteins. I. Purification of chloroplast cytochrome *b*-559. Biochemistry 13: 4063–4071

Gavel Y, Steppuhn J, Herrmann R and von Heijne G (1991) The positive-inside rule applies to thylakoid membrane proteins. FEBS Lett 282: 41–46

Ghanotakis DF, Babcock GT and Yocum CF (1984) Structural and catalytic properties of the oxygen-evolving complex. Correlation of polypeptide and manganese release with the behavior of Z^+ in chloroplasts and a highly resolved preparation of the PS II complex. Biochim Biophys Acta 765: 388–398

Gounaris K, Chapman DJ and Barber J (1989) Isolation and

characterisation of a D1/D2/cytochrome *b*-559 reaction centre complex of photosystem two isolated by two different methods. Biochim Biophys Acta 973: 296–301

Gounaris K, Chapman DJ, Booth P, Crystall B, Giorgi LB, Klug DR, Porter G and Barber J (1990) Comparison of the D1/D2/ cytochrome *b*-559 reaction centre complex of photosystem two isolated by two different methods. FEBS Lett 265: 88–92

Haag E, Irrgang K-D, Boekema EJ and Renger G (1990) Functional and structural analysis of Photosystem II core complexed from spinach with high oxygen evolution capacity. FEBS Lett 47–53

Haley J and Bogorad L (1990) Alternative promoters are used for genes within maize chloroplast polycistronic transcription units. Plant Cell 2: 323–333

Heber U, Kirk MR and Boardman NK (1979) Photoreactions of cytochrome *b*-559 and cyclic electron flow in Photosystem II of intact chloroplasts. Biochim Biophys Acta 546: 292–306

Herrmann RG, Alt J, Schiller B, Widger WR and Cramer WA (1984) Nucleotide sequence of the gene for apocytochrome *b*-559 on the spinach plastid chromosome: Implications for the structure of the membrane proteins. FEBS Lett 176: 239–244

Herrmann RG, Oelmüller R, Bichler J, Schneiderbauer A, Steppuhn J, Wedel N, Tyagi AK and Westhoff P (1991) The thylakoid membrane of higher plants: Genes, their expression and interaction. In: Herrmann RG and Larkins BA (eds) Plant Molecular Biology, Vol 2, pp 411–427. Plenum, New York

Hird SM, Willey DL, Dyer TA and Gray JC (1986) Location and nucleotide sequence of the gene for cytochrome *b*-559 in wheat chloroplast DNA. Mol Gen Genet 203: 95–100

Horton P and Croze E (1977) The relationship between the activity of chloroplast Photosystem II and the midpoint oxidation-reduction potential of cytochrome *b*-559. Biochim Biophys Acta 462: 86–101

Jursinic PA, McCarthy SA, Bricker TM and Stemler A (1991) Characteristics of two atrazine-binding sites that specifically inhibit Photosystem II function. Biochim Biophys Acta 1059: 312–322

Klimov VV, Klevanik AV and Shuvalov VA (1977) Reduction of pheophytin in the primary light reacton of Photosystem II. FEBS Lett 82: 183–186

Klimov VV, Shuvalov VA and Heber U (1985) Photoreduction of pheophytin as a result of electron donation from the water-splitting system to Photosystem II reaction centers. Biochim Biophys Acta 809: 345–350

Klughammer C, Kolbowski J and Schreiber U (1990) LED array spectrophotometer for measurement of time resolved difference spectra in the 530–600 nm wavelength region. Photosynth Res 25: 317–327

Krishtalik LI, Tae G-S, Cherepanov DA and Cramer WA (1993) The redox properties of cytochromes *b* imposed by the membrane electrostatic environment. Biophys J 65: 184–195

Krupinska K and Berry-Lowe S (1988) Characterization and in vitro expression of the cytochrome *b*-559 genes in barley. I. Localization and sequence of the genes. Carlsberg Res Commun 53: 43–55

Lam E, Baltimore B, Ortiz W, Chollar S, Melis A and Malkin R (1983) Characterization of a resolved oxygen-evolving Photosystem II preparation from spinach thylakoids. Biochim Biophys Acta 724: 210–221

Lee WJ and Whitmarsh J (1989) The photosynthetic apparatus of pea thylakoid membranes: Response to growth light intensity. Plant Physiol 89: 832–840

MacDonald GM, Boerner RJ, Everly RM, Cramer WA, Debus RJ and Barry BA (1994) Comparison of cytochrome *b*-559 content in Photosystem II complexes from spinach and *Synechocystis* species PCC 6803. Biochemistry 33: 4393–4400

Matsuda H and Butler WL (1983) Restoration of high-potential cytochrome *b*-559 in Photosystem II particles in liposomes. Biochim Biophys Acta 725: 320–324

McNamara V and Gounaris K (1992) Electron cycling via cytochrome *b*-559 in Photosystem II core particles under photoinhibitory conditions. In: Murata N (eds) Research in Photosynthesis, Vol 2, pp 89–92. Kluwer Academic Publishers, Dordrecht

McNamara V and Gounaris K (1995) Granal Photosystem II complexes contain only the high potential form of cytochrome *b*-559 which is stabilized by the ligation of calcium. Biochim Biophys Acta 1231: 289–296

Miyazaki A, Shina T, Toyoshima Y, Gounaris K and Barber J (1989) Stoichiometry of cytochrome *b*-559 in Photosystem II. Biochim Biophys Acta 975: 142–147

Mor TS, Ohad I, Hirschberg J and Pakrasi HB (1995) An unusual organization of the genes encoding cytochrome *b*-559 in *Chlamydomonas reinhardtii*: *psbE* and *psbF* genes are separately transcribed from different regions of the plastid chromosome. Mol Gen Genet 246: 600–604

Moser CC, Keske JM, Warncke K, Farid RS and Dutton PL (1992) Nature of biological electron transfer. Nature 355: 796–802

Moskalenko AA, Barbato R and Giacomette GM (1992) Investigation of the neighbor relationships between Photosystem II polypeptides in the two types of isolated reaction centres (D1/D2/cyt*b*-559 and CP47/D1/D2/cyt*b*-559 complexes). FEBS Lett 314: 271–274

Murata N, Miyao M, Onata T, Matsunami H and Kuwabara T (1984) Stoichiometry of components in the photosynthetic oxygen evolution system of Photosystem II particles prepared with Triton X-100 from spinach chloroplasts. Biochim Biophys Acta 765: 363–369

Nedbal L, Samson G and Whitmarsh J (1992) Redox state of a one-electron component controls the rate of photoinhibition of Photosystem II. Proc Natl Acad Sci USA 89: 7929–7933

Norris JR and Schiffer M (1990) Photosynthetic reaction centers in bacteria. Chem Eng News July, 1990: 22–37

Ortega JM, Hervás M and Losada M (1988) Redox and acid-base characterization of cytochrome *b*-559 in Photosystem II particles. Eur J Biochem 171: 449–455

Pakrasi HB, Williams JGK and Arntzen CJ (1988) Targeted mutagenesis of the *psbE* and *psbF* genes blocks photosynthetic electron transport: Evidence for a functional role of cytochrome *b*-559 in Photosystem II. EMBO J 7: 325–332

Pakrasi HB, Nyhus KJ and Granok H (1990) Targeted deletion mutagenesis of the β-subunit of cytochrome *b*-559 protein destabilizes the reaction center of Photosystem II. Z Naturforsch 45c: 423–429

Pakrasi HB, Ciechi PD and Whitmarsh J (1991) Site directed mutagenesis of the heme axial ligands of Cytochrome *b*-559 affects the stability of the Photosystem II complex. EMBO J 10: 1619–1627

Poulson M, Samson G and Whitmarsh J (1995) Evidence that cytochrome *b*-559 protects Photosystem II against photoinhibition. Biochemistry 34: 10932–10938

Prasil O, Adir N and Ohad I (1992) Dynamics of Photosystem II: Mechanism of photoinhibition and recovery processes. In: Barber J (eds) The Photosystems: Structure, Function and Molecular Biology, pp 295–348. Elsevier, Amsterdam

Rees D and Horton P (1990) The mechanisms of changes in Photosystem II efficiency in spinach thylakoids. Biochim Biophys Acta 1016: 219–227

Rich P and Bendall DS (1980) The redox potentials of the b-type cytochromes of higher plant chloroplasts. Biochim Biophys Acta 591: 153–161

Rutherford AW and Zimmerman JL (1984) A new EPR signal attributed to the primary plastosemiquinone acceptor in Photosystem II. Biochim Biophys Acta 767: 168–175

Samson G and Fork D (1992) Simultaneous photoreduction and photooxidation of cytochrome *b*-559 in Photosystem II treated with carbonylcyanide-m-chlorophenylhydrazone. Photosynth Res 33: 203–212

Selak ME, Koch-Whitmarsh BE and Whitmarsh J (1984) Evidence for a heterogeneous population of high potential cytochrome *b*-559 in the thylakoid membrane. In: Sybesma C (eds) Advances in Photosynthesis Research, Vol 1, pp 493–496. Martinus Nijhoff/Dr. Junk Publishers

Setlik I, Allakhverdiev SI, Nedbal L, Setlikova E and Klimov VV (1990) Three types of Photosystem II photoinactivation. I. Damaging processes on the acceptor side. Photosynth Res 23: 39–48

Shukla VA, Stanbekova GE, Shestakov SV and Pakrasi HB (1992a) The D1 protein of the Photosystem II reaction center complex accumulates in the absence of D2: Analysis of a mutant of the cyanobacterium *Synechocystis sp.* PCC 6803 lacking cytochrome *b*-559. Mol Microbiol 6: 947–956

Shukla VK, Anbudurai PR, Wu R and Pakrasi HB (1992b) Site-directed modifications of the *psbE* and *psbJ* genes in *Synechocystis* sp. PCC 6803. In: Murata N (eds) Research in Photosynthesis, Vol 3, pp 409–412. Kluwer Academic Publishers, Dordrecht

Shuvalov VA, Heber U and Schreiber U (1989) Low temperature photochemistry and spectral properties of a photosystem 2 reaction center complex containing the proteins D1 and D2 and two hemes of Cyt *b*-559. FEBS Lett 258: 27–31

Shuvalov VA, Schreiber U and Heber U (1994) Spectral and thermodynamic properties of the two hemes of the D1D2 cytochrome *b*-559 complex in spinach. FEBS Lett 337: 226–230

Styring S, Virgin I, Ehrenberg A and Andersson B (1990) Strong light photoinhibition of electron transport in Photosystem II. Impairment of the function of the first quinone acceptor, Q_A. Biochim Biophys Acta 1015: 269–278

Tae G-S and Cramer WA (1992) Truncation of the COOH-Terminal domain of the psbE gene product in *Synechocystis* sp. PCC 6803: Requirements for the Photosystem II assembly and function. Biochemistry 31: 4066–4074

Tae G-S and Cramer WA (1994) Topography of the heme prosthetic group of cytochrome *b*-559 in the Photosystem II reaction center. Biochemistry 33: 10060–10068

Tae G-S, Black MT, Cramer WA, Vallon O and Bogorad L (1988) Thylakoid membrane protein topography: Trans-membrane orientation of the chloroplast cytochrome *b*-559 *psbE* gene product. Biochemistry 27: 9075–9080

Tae G-S, Everly RM, Cramer WA, Madgwick SA and Rich PR (1993) On the question of the identity of cytochrome *b*-560 in thylakoid stromal membranes. Photosynth Res 36: 141–146

Tang XS and Diner BA (1994) Biochemical and spectroscopic characterization of a new oxygen-evolving Photosystem II core complex from the cyanobacterium *Synechocystis* PCC 6803. Biochemistry 33: 4594–4603

Thompson LK and Brudvig GW (1988) Cytochrome *b*-559 may function to protect Photosystem II from photoinhibition. Biochemistry 27: 6653–6658

Vallon O, Høyer-Hansen G and Simpson DJ (1987) Photosystem II and cytochrome *b*-559 in the stroma lamellae of barley chloroplasts. Carlsberg Res Commun 52: 405–421

Vallon O, Tae G-S, Cramer WA, Simpson D, Hoyer-Hanson G and Bogorad L (1989) Visualization of antibody binding to the photosynthetic membrane: The transmembrane orientation of cytochrome *b*-559. Biochim Biophys Acta 975: 132–141

van der Bolt F and Vermaas W (1992) Photoinactivation of Photosystem II as studied with site-directed D2 mutants of the cyanobacterium *Synechocystis* sp. PCC 6803. Biochim Biophys Acta 1098: 247–254

van Leeuwen PJ, Nieveen MC, van de Meent EJ, Dekker JP and van Gorkom J (1991) Rapid and simple isolation of pure Photosystem II core and reaction center particles from spinach. Photosynth Res 28: 149–153

van Mieghem FJE, Nitschke W, Mathis P and Rutherford AW (1989) The influence of the quinone-iron electron acceptor complex on the reaction centre photochemistry of Photosystem II. Biochim Biophys Acta 977: 207–214

Vass I, Deak Z, Jegerschöld C and Styring S (1990) The accessory electron donor tyrosine-D of Photosystem II is slowly reduced in the dark during low-temperature storage of isolated thylakoid membranes. Biochim Biophys Acta 1018: 41–46

Vass I, Styring S, Hundal T, Koivuniemi A, Aro EM and Andersson B (1992) Photoinhibition of Photosystem II—Stable reduced Q_A species promote chlorophyll triplet formation. Proc Natl Acad Sci USA 89: 1408–1412

Vermeglio A and Mathis P (1974a) Light-induced absorbance changes at –170 °C with spinach chloroplasts: Charge separation and field effect. Biochim Biophys Acta 368: 9–17

Vermeglio A and Mathis P (1974b) Light-induced absorption changes in spinach chloroplasts/ a comparative study at –50° and –170 °C. In: Avron M (eds) Proceedings of the Third International Congress on Photosynthesis, pp 323–334 Elsevier Scientific Publishing Company, Amsterdam

Visser JWM, Rijgersberg CP and Gast P (1977) Photooxidation of chlorophyll in spinach chloroplasts between 10 and 180 K. Biochim Biophys Acta 460: 36–46

Völker M, Renger G and Rutherford AW (1986) Effects of trypsin upon EPR signals arising from components of the donor side of Photosystem II. Biochim Biophys Acta 851: 424–430

von Heijne G (1986) The distribution of positively charged residues in bacterial inner membrane proteins correlates with the trans-membrane topology. EMBO J 5: 3021–3027

Walker FA, Huynh GH, Scheidt WR and Osvath SR (1986) Models of Cyt *b*. The effects of axial ligand plane orientation on electron paramagnetic resonance and Mössbauer spectra of low spin ferrihemes. J Amer Chem Soc 108: 5288–5297

Whitford D, Gounaris K and Barber J (1984) Studies on cytochrome *b*-559 of the higher plant thylakoid membranes. In: Sybesma C (ed) Advances in Photosynthetic Research, Vol

1, pp 497–500. Martinus Nijhoff/Dr. Junk Publishers, The Hague

Whitmarsh J and Chylla R (1990) Hypothesis: In leaves inactive PS II complexes are converted during illumination to an active form that drives an electron cycle and quenches fluorescence. Plant Physiol 93S: 142

Whitmarsh J and Cramer WA (1977) Kinetics of the photoreduction of cytochrome *b*-559 by Photosystem II in chloroplasts. Biochim Biophys Acta 460: 280–289

Whitmarsh J and Cramer WA (1978) A pathway for the reduction of cytochrome *b*-559 by Photosystem II in chloroplasts. Biochim Biophys Acta 501: 83–93

Whitmarsh J and Ort DR (1984) Stoichiometries of electron transport complexes in spinach chloroplasts. Arch Biochem Biophys 231: 378–389

Whitmarsh J, Samson G and Poulson M (1994) Photoprotection in Photosystem II—the role of cytochrome *b*-559. In: Baker N and Bower J (eds) Photoinhibition of Photosynthesis—From Molecular Mechanisms to the Field, pp 75–93. BIOS Scientific Publishers, Ltd., Oxford

Willey DL and Gray JC (1989) Two small open reading frames are cotranscribed with the pea chloroplast genes for the polypeptides of cytochrome *b*-559. Curr Genet 15: 213–220

Yamagishi A and Fork DC (1987) Photoreduction of Q_A, Q_B, and cytochrome *b*-559 in an oxygen-evolving Photosystem II preparation from the thermophilic cyanobacterium *Synechococcus* sp. Arch Biochem Biophys 259: 124–130

Yamamoto Y, Tabata K, Isogai Y, Nishimura M, Okayama S, Matsuura K and Itoh S (1984) Quantitative analysis of membrane components in a highly active O_2-evolving Photosystem II preparation from spinach chloroplasts. Biochim Biophys Acta 767: 493–500

Yerkes CT and Crofts AR (1984) A mechanism for ADRY induced photo-oxidation of cytochrome *b*-559. In: Sybesma C (ed) Advances in Photosynthesis Research, Vol 1, pp 489–492. Marinus Nijhoff/Dr. Junk Publishers, Dordrecht

Chapter 14

Photosystem II Heterogeneity

Jérôme Lavergne
Institut de Biologie Physico-Chimique, 13 rue Pierre et Marie Curie, 75005 Paris, France

Jean-Marie Briantais
Laboratoire d'Ecologie Végétale, Université de Paris-Sud, Bât. 362, 91405 Orsay, France

Summary

Two main topics are addressed in this chapter: 'static heterogeneities' of PS II, as they appear in standard dark-adapted material, and 'dynamic heterogeneities' possibly involved in the non-photochemical quenching processes that modulate the steady-state yield of PS II. Three types of static heterogeneities and possible correlations between them are discussed: (i) Granal and stromal PS II. A fraction, around 10–15%, of the PS II complex is found in stroma lamellae. (ii) PS II α and β. A fraction, around 35%, of PS II (β) appears to have a smaller antenna size and to be organized in isolated units, in contrast with the major part of PS II (α). (iii) A fraction, around 15%, of PS II centers are blocked on the acceptor side (non Q_B-transferring) and thus inactive with regard to oxygen evolution. A unifying model has been proposed by Melis (1985, 1991), wherein stromal PS II, PS IIβ and inactive centers represent essentially the same sub-population of PS II, assumed to reflect a

Donald R. Ort and Charles F. Yocum (eds): Oxygenic Photosynthesis: The Light Reactions, pp. 265–287.

dynamic stock in the biosynthetic turnover of the PS II complex. The various correlations implied by this model are reexamined and evaluated in light of the currently available data, and alternative interpretations are discussed. It is argued that inactive centers belong to PS IIα and that, on the other hand, stromal PS II centers are active. The antenna size of stromal PS II is probably consistent with their belonging to PS IIβ, but the amount of the latter exceeds significantly that of stromal PS II: it is suggested that a significant part of PS IIβ may be located in the grana margins. The concept of non-photochemical quenching 'qN' covers three different contributions. 'qT', that appears at low irradiance levels, is interpreted as a 'state 2 transition', involving detachment of a fraction of LHCII from PS II α and thus probably increasing the β-fraction. 'qE', controlled by the lumenal pH, is the major contribution to non-photochemical quenching at physiological irradiances. Most of the available evidence supports its interpretation as due to a dissipation pathway at the antenna level. The alternative mechanism of a formation of inactive centers of the quenching sink type does not account for the results obtained in vivo in normal materials, but seems to prevail in LHCII-deficient material. At over-saturating intensities, the 'qI' quenching reflects photoinhibition associated with inactivation of PS II centers in a quenching sink state. Significant formation of non Q_B-transferring centers does not take place as a result of photodegradation or of blocking the synthesis of the PS II complex. At physiological irradiances, there is no evidence that a sub-population of damaged PS II centers could be ascribed to insufficient synthetic turnover of the PS II complex.

I. Introduction

The subject of this review covers potentially a large body of phenomena and interpretations which are at least as heterogeneous as PS II itself. We thus feel unable, keeping within the space limits allocated to this chapter, to aim at any exhaustive review. Rather, we shall attempt to focus on some current debates, describe their main lines and give our (admittedly biased) opinion on them. There have been few previous reviews specifically devoted to this subject, but an excellent one is by Black et al. (1986). Related topics have, however, been dealt with in a number of reviews, such as Diner (1986), Govindjee (1990), Diner et al. (1991) and Melis (1991).

Why has there been more work devoted to PS II heterogeneity than, say, on PS I or b_6f heterogeneity? Part of the answer is trivial: PS II has been more studied than other (however respectable) components of the photosynthetic apparatus, perhaps mostly because specific signals (chlorophyll fluorescence and oxygen evolution) are easier to detect. This is not the whole story, however: there are reasons to believe that PS II may be more heterogeneous than other components. One is that the PS II complex comprises a polypeptide (D1) with a remarkably high turnover rate, implying rapid (photo-) degradation and synthesis, producing a flux that may be expected to result in different subpopulations. Related to this aspect is the fact that the donor side of PS II has to handle highly oxidizing intermediates that may be self-deleterious. It should also be mentioned that the PS II complex is heterogeneously distributed between the appressed and non-appressed membrane regions of the thylakoid membrane (see Chapter 2 by Staehelin), whereas PS I is strictly confined to the non-appressed regions. By this token, the cytochrome b_6f complex, which appears in both membrane regions, may also be suspected of heterogeneity and, indeed, some suggestions of this kind have been put forward (Olive et al., 1986; Vallon et al., 1991). Another hint for PS II heterogeneity is that some of its core polypeptides have been reported to undergo a regulated phosphorylation (see the review by Allen, 1992).

A topic that will be omitted from our coverage is that of an *internal* heterogeneity of acceptors in the reaction center. Models involving two parallel acceptors that could be reduced in two successive photoacts have been hypothesized in the past, such as the Q_1/Q_2 model of Joliot and Joliot (1979, 1981a)

Abbreviations: BBY – granal membrane preparation according to Berthold et al. (1981); DCBQ – 2,6-dichloro-p-benzoquinone; DCMU – 3-(3,4-dichlorophenyl)-1,1-dimethylurea; DCIP – 2,6-dichlorophenol-indophenol; DMBQ – 2,6-dimethyl-p-benzoquinone; F – level of chlorophyll fluorescence [F_o: level with all centers open; F_{pl}: plateau reached after the first phase of the induction kinetics (weak light, no DCMU); F_s: steady-state level; F_m: level with all centers closed]; LHCII – light-harvesting chlorophyll-protein complex II; PpBQ – phenyl-p-benzoquinone; PS – Photosystem

or the related 'B type/non-B type' model of Lavergne (1982a,b). Concerning the latter, it should be emphasized that there is no relation, except for a nomenclature ambiguity, with the inactive, non PQ-transferring centers that have arisen in later literature: the 'non-B type' acceptors were believed to transfer electrons to a secondary acceptor in a DCMU-sensitive reaction, whereas the 'inactive centers' are intrinsically blocked (see below). The Q_1/Q_2 or B/non-B models did not imply heterogeneous populations of PS II as an essential feature. Almost no further work in this area has been reported since the early 1980's and, clearly, these models would need serious reevaluation in the light of current knowledge on PS II. We shall also leave aside some other suggestions for PS II heterogeneities (e.g. Joliot, 1974; Delrieu and Rosengard, 1993) and only marginally allude to information derived from analyses of components of the chlorophyll fluorescence decay after a picosecond flash (see Holzwarth, 1991, for a review).

In structuring this chapter, we will employ a tentative distinction between two classes of PS II heterogeneity, based on experimental procedures:

(i) 'Static heterogeneity' present in dark-adapted material. The main topics here are: heterogeneity correlated with the grana/stroma structure; heterogeneity of antenna size; heterogeneity with respect to the ability of sustaining electron transfer.

(ii) 'Dynamic heterogeneity' that is implied when considering the functional changes of the photosynthetic apparatus under prolonged illumination. The quantum yield of PS II is decreased under such conditions, due to photoprotective and regulatory mechanisms or to photoinhibition and we shall discuss to what extent these processes may be ascribed to the development of PS II heterogeneities.

II. Static Heterogeneities: Basic Data

A. Appressed and Non-appressed Membrane Regions

The organization of thylakoid membranes in green plant chloroplasts involves stacked grana and unstacked stroma lamellae. In green algae, the grana stacking is not found, but nevertheless, a large fraction of the membranes are appressed (see Chapter 2 by Staehelin and van der Staay). Important heterogeneity is found for the protein content of the appressed and non-appressed regions. Most of the PS II complex is included in the former (while PS I is completely excluded from it). However, a fraction of PS II has been reported to be present in the unappressed stroma lamellae regions. Values of about 10–15% emerge from a number of biochemical assays of the PS II polypeptides: 10–20% in Andersson and Anderson (1980), 10% in Vallon et al. (1985), 12% in Callahan et al. (1989), and <7% in Vallon et al. (1987). Similar figures were found for various PS II activities measured in preparations of stroma lamellae (Anderson and Melis, 1983, Mäenpää et al., 1987; Albertsson et al., 1990; Henrysson and Sundby, 1990).

This heterogeneous distribution of the PS II complex may either reflect a partition equilibrium of a unique entity or a true biochemical difference (e.g. modification of polypeptides). Although the latter situation seems more plausible, there is little biochemical evidence for this as yet. The only difference in polypeptide composition reported between granal and stromal PS II complexes concerns a decreased amount in the latter, with respect to the core polypeptides, of the minor chlorophyll-binding proteins CP26 and CP29 (Vallon et al., 1987; Bassi et al., 1988). It has been suggested by Mattoo et al. (1989) that phosphorylation of the D1 and D2 polypeptides or acylation of D1 might play a role in the spatial separation of PS II between grana and stroma in the same way as for LHCII. There are, however, indications of phosphorylation heterogeneities even in the grana fraction. Four subpopulations of granal PS II core complexes were separated by electrofocusing by Giardi et al. (1991). Functional alterations of the acceptor side caused by phosphorylation have been proposed (see the discussion by Michel et al., 1988 and references therein), such as a decreased light-saturated electron transfer rate (Horton and Lee, 1984, Hodges et al., 1987). In spite of the similar polypeptide content, an important difference between granal and stromal PS II was reported by Dainese et al. (1992b). From a combined study of sucrose gradient ultracentrifugation, electron microscopy and Deriphat-PAGE analyses, these authors concluded that granal PS II is organized in a dimeric form, in agreement with Peter and Thornber (1991a), whereas stromal PS II was concluded to be monomeric.

It has been suggested that finer structural

heterogeneity should be taken into account, beyond grana/stroma partitioning. According to Albertsson and co-workers (1990), the grana margins constitute a third specific membrane domain (see Wollenberger et al., 1994, for a recent investigation on a margin fraction).

B. PS IIα and β

The distinction between α and β PS II was proposed by Melis and Homann (1975, 1976) based on an analysis of chlorophyll fluorescence induction kinetics in the presence of DCMU, which blocks electron transfer beyond Q_A. The fluorescence yield is not, in general, a linear measurement of the amount of closed centers (see Lavorel and Etienne, 1977; Dau, 1994, for reviews), but it is linearly correlated with the photochemical yield (rate of center closing) to a very good approximation (Bennoun and Li, 1973). Thus, the kinetics of Q_A photoreduction can be obtained through integration ('complementary area') of the fluorescence curve (Murata et al., 1966; Malkin and Kok, 1966). The direct fluorescence induction kinetics display a slow 'tail' accounting for typically 15% of the amplitude of variable fluorescence. Melis and Homann ascribed this phase to a population 'β' of PS II centers with lower photochemical efficiency than the major 'α' population. When analyzing the kinetics of the complementary area, the slow phase appears as an exponential that can be subtracted in order to retrieve the kinetics of α-centers. The latter are responsible for the sigmoidal shape of the fluorescence induction. This phenomenon was analyzed by Joliot and Joliot (1964) as reflecting an increase of the cross-section of open traps during the photo-reduction process, due to a probability for transferring excitons from closed to open centers through a connecting antenna bed (see also Joliot et al., 1973). Conversely, the exponential character of the β-kinetics suggests little or no connectivity of these particular photosynthetic units (see the discussion by Lavergne and Trissl, 1995).

1. The Mainstream Perception

The interpretation developed by Melis and his coworkers may be summarized as follows: PS IIβ consists of units (i.e. one center and its antenna) isolated from each other and from PS IIα with regard to exciton migration. The number of light-harvesting chlorophylls per center (antenna size) is 2–3-fold smaller than in PS IIα. The intrinsic trapping and fluorescence properties of the α or β centers are similar so that the relative amounts of both types of centers can be estimated from their contributions to the fluorescence area, and the antenna sizes from the rates of Q_A photoreduction.

The correctness of this model is supported by the work of Melis and Duysens (1979), which shows a good correlation between the photoreduction kinetics deduced from the complementary area of fluorescence or from UV absorption changes of Q_A. The slower, exponential β kinetics were observed in both cases to a similar relative extent, confirming the interpretation of the tail of the fluorescence curve in terms of Q_A reduction as well as the validity of the fluorescence area procedure. This result was extended by Melis and Schreiber (1979) to the absorption change monitoring Q_A reduction (C-550).

The agreement between the fluorescence area method and detection of the photoreduction of Q_A through absorption changes is understandable only if the intrinsic trapping efficiency of the β-centers is similar to that of α-centers. A lower quantum yield of β centers, be it due to a more shallow trap or to rapid recombination losses, would increase the fluorescence area with respect to the amount of stabilized charges (Dau, 1994, Lavergne and Trissl, 1995). Independent confirmation of this point was important, since the detection in thylakoids of pure PS II absorption changes, uncontaminated by other signals, is not an easy problem and the accuracy of the results of Melis and Duysens or Melis and Schreiber could be questioned on such grounds. A first aspect is the stabilization efficiency of the two types of centers, expressed by the efficiency of a saturating short flash. This experiment was first done by Melis and Duysens (1979) who reported that a single turnover flash removed 70% of the induction area ascribed to β-centers, compared with 98% for α-centers. By itself, this effect would predict a 1.4 times slower photoreduction of PS IIβ and also a 30% overestimate of the β fraction ('β_{max}') through the area method, but does not suffice to account for the 2–3 times slower photochemical rate of PS IIβ. Higher flash yields were later reported by Melis and Thielen (1980) and Thielen and van Gorkom (1981a), who found an efficiency of more than 95% for PS IIβ. On the other hand, a quite different result was found by Doschek and Kok (1972), who had noticed the biphasic character of the fluorescence induction curves and reported an extremely small efficiency of a saturating flash for removing the slow phase ('tail'). However, the induction curves shown by Doschek and Kok have somewhat unusual features, such as an extremely slow 'β-phase' which would presumably yield a considerable area (several fold that of the fast phase) and may not correspond to the usual β-phenomenon. Possible clues to the origin of these features are the rather low intensity used in these experiments, which may not compete efficiently enough with recombination, and the use of a far-red preillumination that may create a large fraction of centers with the Q_B pocket occupied by a plastosemiquinone before DCMU is added (see below). Nevertheless, even if the amount of flash-inefficient centers appears exceptionally high in the data of Doschek and Kok, it seems clear that this phenomenon is present to some extent in more typical materials. Joliot and Joliot (1977) reported that one short saturating flash raises the fluorescence to 90–93% of the total amplitude. About 20 flashes are required to destroy the remaining quenchers. The relative amplitude of the variable fluorescence involved in this phenomenon (in the range of 2–5% after correcting for the effect of 'misses' occurring on the first flash in normal centers) is too low to account entirely for the tail of the induction kinetics, but it may nevertheless have a large impact on area estimates, and the result of Thielen and van Gorkom (1981a) is surprising in this respect.

Besides the stabilization efficiency, Thielen and van Gorkom (1981a) gave an overall picture of the fluorescence and trapping parameters of α and β centers that validates the applicability of the fluorescence area method. A similar conclusion emerges from the analysis of picosecond fluorescence data by Roelofs et al. (1992). These authors found significant differences between α and β centers with regard to the detailed rate constants governing the 'exciton-radical pair equilibrium', but the quantum yields for photochemistry and variable fluorescence turn out to be similar. Interestingly, however, Melis and Ow (1982) reported a situation where the similarity between fluorescence parameters of the two fractions breaks down. These authors confirmed the agreement of the kinetics of fluorescence area and 320 nm absorption changes (reduction of Q_A) in the presence of divalent cations, but found a dramatic decorrelation in their absence (unstacked thylakoids). According to their analysis, the true amounts of α and β centers remain unchanged, but the fluorescence properties of the α-centers appear modified in the absence of divalent cations (this effect is in fact most probably due to quenching of PS II fluorescence by PS I).

It should be emphasized that the possibility that PS IIβ might correspond to a fraction of PS II subjected to a quenching by PS I is not consistent with the evidence showing quantitative agreement between the fluorescence area and absorption changes of Q_A^-. Such a competition for exciton trapping should decrease the complementary fluorescence area with respect to the amount of centers.

Although the β-phenomenon controls a small relative amplitude (approx. 15%) of the total variable fluorescence, its weight in terms of complementary area is much larger. A rather broad range of values has been reported for the fraction of β to total PS II centers: e.g. 60% (Percival et al., 1984), 45–35% (Melis and Homann, 1976, Melis and Duysens, 1979; Thielen and van Gorkom, 1981a), and 20–25% (Melis, 1991). Independently from the area method or absorption change kinetics, Roelofs et al. (1992) estimated a fraction of 50% β-centers from their analysis of fluorescence lifetimes. This scatter may reflect true physiological fluctuations or different intactness of the material (see Schreiber and Pfister, 1982), but it seems also likely that some of the complicating features that we describe later are the origin of significant distortions (i.e. sensitivity of the area estimate to the determination of the asymptote of the fluorescence induction, incomplete DCMU inhibition).

Consistent with the view that PS IIβ represents a set of isolated units with a specific antenna, distinct action spectra (Thielen et al., 1981; Percival et al., 1984) and emission spectra (Brearley and Horton, 1984) have been reported, showing that PS IIα is enriched in chlorophyll b with respect to PS IIβ (the difference in emission spectra was not, however, confirmed by Hodges and Barber, 1986). Accordingly, the smaller antenna size has been ascribed to the absence in PS IIβ of 'peripheral LHCII' (Melis, 1991).

At this stage, it can be imagined that the α/β heterogeneity is simply an antenna feature, involving no intrinsic heterogeneity of the reaction center. This view has been advocated by Percival et al. (1984). Nevertheless, there are several pieces of evidence that β-reaction centers have distinct characteristics. The recombination rate of these centers in the presence of DCMU was found (Melis and Homann,

1976, Thielen and van Gorkom, 1981a) to be markedly slower than that of α-centers (half-times of, respectively, 20 s and 1–3 s). Thus, a second fluorescence induction experiment carried out after a dark delay of less than, say, 10 s, shows a distinct relative enrichment (with respect to the first, dark-adapted trace) in fast, α-type contribution (Doschek and Kok, 1972; Joliot et al., 1973, Melis and Homann, 1976). Other characteristics of β-centers have been described: (i) Thielen and van Gorkom (1981b) reported that β-centers did not show the binary flash oscillations reflecting the two-electron gate Q_B. (ii) It has been suggested that the Q_A of β-centers has a markedly higher mid-point potential (120 mV) than that of α centers (Horton, 1981; Thielen and van Gorkom, 1981b). However, this view is not easily reconciled with other redox titration data of the PS II acceptor which do not reveal the presence of the 120 mV component (see the discussion by Black et al., 1986; a review of titration data on Q_A is given by Diner et al., 1991). It is known that the plastoquinone pool exerts a non-photochemical quenching effect in its oxidized state (Delosme, 1967; Vernotte et al., 1979), that should titrate around + 120 mV and may have been erroneously ascribed to Q_A-β (see Hodges and Barber, 1983).

2. Discrepancies

A basic requirement for applying meaningfully the complementary area method is the *photochemical* character of the fluorescence curve (illumination intensity-limited kinetics). This character was established long ago by Joliot (1965) for the induction curve as a whole (with no special attention to the β-phase) and was also verified by Delosme (1967) in the high intensity range. Surprisingly, the available data (to our knowledge) *do not* substantiate a photochemical behavior for the β-phase. This question was investigated by Melis and Homann (1975) who confirmed the proportionality of the rate of the α-phase to the illumination intensity, whereas the rate of the β-phase appeared only weakly intensity-dependent. In this paper, the authors declared that they were unable to reach a definite conclusion because of insufficient experimental accuracy. In a recent work, Hsu and Lee (1991a) found no intensity dependence for the β phase (or for a still slower phase, denoted γ, by these authors) and suggest limitation by a dark reaction. If this were true, the quantitation of β centers from their relative fluorescence area would lose any relevance as would the interpretation of the exponential kinetics in terms of separate units.

Several authors reported results that do not question the existence of an heterogeneous β fraction of PS II, but do affect the interpretation or quantitation. A first problem is technical and concerns the sensitivity of the complementary area method to the determination of the asymptote of the fluorescence induction (Bell and Hipkins, 1985). A mathematical treatment was proposed for overcoming this difficulty (Hsu et al., 1989), but, clearly, the problem remains that any small distortion of the slower phase of the induction will considerably affect the complementary area. This may require some caution regarding the multiplication of heterogeneous phases beyond α/β, claimed to derive from careful analysis of the kinetics of the complementary area (γ-phase according to Hsu et al., 1989; Hsu and Lee, 1991a; γ- and δ-phases according to Sinclair and Spence, 1990). There is, however, an independent indication for the involvement of a slower γ-phase, namely the existence of a fraction of quenchers whose destruction requires a large number of saturating flashes (Joliot and Joliot, 1977). To interpret this finding, these authors favored the idea of a low trapping efficiency in a fraction of centers, but the alternative possibility of a multi-turnover filling of a pool cannot be excluded. This phenomenon cannot account entirely (or even predominantly) for the slow tail of the induction kinetics. Its amplitude, approx. 2–5% of the amplitude of variable fluorescence, is smaller than found for the β phase, and the flash inefficiency also imposes a slower rate than found for the β phase. Both features, however, would fit the γ phase introduced by Hsu and coworkers (Hsu et al., 1989; Hsu and Lee, 1991a). Obviously, the area analysis becomes inappropriate for this phase and would grossly overestimate the number of centers involved due to their low trapping efficiency (even if this was corrected for, the antenna size still remains unknown).

According to several authors (Schreiber and Pfister, 1982; Brearley and Horton, 1984), the β-phenomenon involves centers with a lower affinity for binding DCMU. This was challenged by Horvath et al. (1984) who observed a decrease of the amplitude of variable fluorescence or of the 320 nm absorption change when increasing the DCMU concentration above 50 μM (whereas the F_0 level remained unaffected) and concluded that the inhibitor had some disruptive effect affecting preferentially the β centers. This quenching effect did not occur, however, in the Schreiber and Pfister experiments in a concentration range where large modifications of the β_{max} were observed. The interpretation of Horvath et al. also does not account for the findings of Brearley and Horton (1984). Furthermore, Hodges and Barber (1986) could not reproduce the effect reported by Horvath et al. on the fluorescence levels, but did find a specific sensitivity of the rate and area of the β-phase to the DCMU concentration. When varying the inhibitor concentration from 0.2 to 100 μM, the rate of the β-phase was increased by 60% and its area decreased by 70%. Studies of binding of radio-actively-labeled inhibitors to PS II have detected two sites, of high and low affinity (McCarthy et al., 1988, Chow et al., 1990). The possible relation with the α/β heterogeneity is not obvious, however, since McCarthy et al. concluded that both sites belonged to the same centers (on the D1 and D2 polypeptides, respectively). In our hands (J. Lavergne, unpublished) the affinity of the β-centers for DCMU seems indeed lower than that of α-centers, so that at subsaturating DCMU concentration (in the presence of 10 μM DCBQ added to suppress the fluorescence rise associated with photoreduction of the plastoquinone pool), pure α-kinetics are observed for the fraction of inhibited centers. Increasing the DCMU concentration first gives rise to a very slow 'β-phase' (presumably multi-turnover), which becomes faster at saturating DCMU concentrations, but nevertheless remains distinct from the α-phase. Similar observations were reported by Hodges and Barber (1986). In a way, these findings are reminiscent of the behavior of centers with a plastosemiquinone bound at site Q_B before adding DCMU. Due to the strong binding of the semiquinone, these centers have an approximately 20-fold diminished affinity to inhibitors of the Q_B pocket (Lavergne, 1982c) and their photoreduction (accompanying inhibitor binding) is expected to imply a multiturnover process. However, the β-phenomenon is still observed at DCMU concentrations above 10 μM that should be saturating even for centers initially in the $Q_AQ_B^-$ state. Schreiber and Pfister (1982) and Hodges and Barber (1983) have proposed that the β-phenomenon could simply be ascribed to a multi-turnover process due to a leak through the DCMU block. Clearly, if a multiturnover process of some kind is involved, the complementary area method would lead to a gross overestimate of β-centers. The 'DCMU-leak' hypothesis would require additional assumptions to account for the actual rate of the β-phase. The secondary fluorescence rise phase accompanying the reduction of the plastoquinone pool in the absence of DCMU is much slower than the β-phase and is still slowed when a fraction of the centers is inhibited (since a common pool is accessible to several centers). Thus, the DCMU-leak possibility does not weaken in the least the α/β heterogeneity, since it implies that β-centers have a diminished sensitivity to DCMU *and* a specific acceptor pool of very small capacity. The appropriate experimental criterion for deciding whether a multiturnover process is involved is to test the efficiency of a saturating flash for removing the fluorescence area of the β phase. As discussed above, it appears that most of the β phase is indeed suppressed by a one flash preillumination and that the remaining 'flash-inefficient' quenchers are probably responsible for the third, slower, γ-phase of small amplitude, described by Hsu et al. (1989) and Hsu and Lee (1991a).

As a tentative conclusion to this section, we feel that substantial support exists for the mainstream perception of PS IIβ: isolated units with an antenna of smaller size, enriched in chlorophyll *a*; reaction centers with slower recombination rates and modified mechanisms of secondary electron transfer. Among issues where clarification would be welcome, we would list: confirmation of the higher mid-point potential of Q_A; more detailed understanding of the DCMU concentration dependence; demonstration of the photochemical character of the β-phase; analysis of the degree of coupling between the antenna feature (lack of peripheral LHCII) and reaction center specificities.

C. Inactive PS II Centers

There are many ways by which a PS II center may be inactive or may have a decreased activity and, since several of these possibilities have been proposed to occur, a nomenclature effort might be worthwhile. We propose:

- *Quenching sink* for centers that efficiently trap excitons but do not stabilize charge separation (e.g. centers where a fast recombination occurs due to a blocked reduction of $P680^+$).

- *Inactive closed* for centers that do not efficiently trap excitons (e.g. centers blocked in the dark with a reduced Q_A).

- *Non Q_B-transferring* for centers able to carry

out the $S_1Q_A \rightarrow S_2Q_A^-$ transition, but unable to transfer electrons to the secondary quinone acceptor (e.g. centers inhibited by DCMU).

Pool-blocked for centers that are capable of multiple turnover and plastoquinone reduction, but become blocked due to the reduction of their acceptor pool (insufficient rate of plastoquinol reoxidation).

From different lines of evidence, there is agreement that a fraction of PS II centers belong to the non Q_B-transferring type, behaving in most respects like DCMU-inhibited centers. The reoxidation of Q_A^- by Q_B (or Q_B^-) in normal centers occurs with a $t_{1/2}$ in the 100–600 μs range (Joliot et al., 1971; Robinson and Crofts, 1983), and may be monitored through the decay of the fluorescence yield following a single turnover flash (or from turnover experiments). While this fast phase accounts for the major fraction of variable fluorescence, slower phases are observed in the few ms to minutes ranges (Forbush and Kok, 1968, Joliot et al., 1971). The slowest of these phases is due to the decay of state S_2 of the water oxidase: indeed, states S_2Q_A and S_3Q_A happen to be slightly less good quenchers than states S_0Q_A or S_1Q_A (Joliot et al., 1971). On the other hand, the 1–2 s decay phase is typical of the recombination $S_2Q_A^- \rightarrow S_1Q_A$ in DCMU-inhibited centers and suggests the presence of non Q_B-transferring centers. Support for this interpretation comes from all measurements of the turnover of photochemical activity of PS II, such as: absorption changes monitoring charge separation through the delocalized electrochromic signal (Chylla et al., 1987; Lavergne, 1987, 1991) or signals directly associated with the oxidized donor or reduced acceptor (Dekker et al., 1984; Lavergne 1987, 1991). These experiments were done with different materials (thylakoids, BBY membranes, algal cells). In all cases, a specific additional absorption contribution was observed on the first flash given to dark-adapted material, that took several seconds to recover. The flash-induced change associated with inactive centers was measured in this manner and shown to be identical with the ($S_2Q_A^-$ - S_1Q_A) spectrum of normal centers (Lavergne, 1991), thus implying a functional water oxidase (at least up to the S_2 state). Another similarity between inactive and DCMU-blocked centers is the inhibition of the recombination reaction observed when added hydroxylamine is used as an artificial donor to PS II (Lavergne and Leci, 1993).

Quantitative estimates for the fraction of non Q_B-transferring centers range from 10 to 35% (for reasons explained below, we exclude the larger estimates derived from the use of DCBQ). Part of this scatter may, here again, come from true physiological differences (especially in algae, where Lavergne (1987) reported large variations depending on culture batches), but also, as discussed below, some of the methods used may be responsible for systematic errors. This quantitation has been a subject of debate in the context of establishing the stoichiometry of PS II to PS I centers: when measuring the amount of PS II from a single turnover experiment on dark-adapted material, the additional contribution of inactive centers is present, whereas these centers do not contribute steady-state activities, such as O_2-evolution, which is relevant for the functional balance of the photosynthetic chain (see, e.g., Jursinic and Dennenberg, 1989).

Variable fluorescence, as a tool for characterizing inactive centers, has the advantage of being a pure photosystem II signal, but several complicating features interfere. The contribution from inactive centers may be estimated from the rapid small phase of fluorescence induction observed in the absence of DCMU (F_o-F_{pl} in the nomenclature of Forbush and Kok, 1968). The actinic intensity must be weak enough to keep the photochemical rate of active centers slow with respect to their recovery rate. Reasonable kinetic separation of this phase and the subsequent rise associated with pool reduction can be achieved by adding an acceptor (i.e. ferricyanide, or better, DCBQ - but see below for the debate on this acceptor). Besides inactive centers, two other contributions to the F_o-F_{pl} phase are expected. The first one is the S-state dependent quenching described above: starting from the low fluorescence state S_1, the illumination will bring about 50% of the more fluorescent states S_2 and S_3 (Lavergne and Leci, 1993, Hsu, 1993): according to Lavergne and Leci, this contribution accounts for about 35% of the F_o-F_{pl} rise. As pointed out by Hsu (1992, 1993), some contribution may also be expected from the $Q_A^-Q_B \rightleftharpoons Q_AQ_B^-$ equilibrium. The equilibrium constant is about 20 at pH 7 (Diner, 1977) and becomes lower when the pH is increased (Lavergne, 1982c, Robinson and Crofts, 1984). Thus, if no Q_B^- is present in the dark-adapted state, about 5% Q_A^- at pH 7 (or 20% at pH 8) should appear after one flash because of this mechanism (two-fold less at steady-state). The first contribution can be eliminated by using hydroxylamine to suppress the S-system and the second one can be minimized to some extent by lowering the pH. When these (significant) corrections are made, the quantitation of inactive centers from the F_o-F_{pl} fluorescence rise still depends on an assumption about the antenna connectivity of active and inactive centers. If inactive PS II consists of isolated units, then under reasonable assumptions, the F_o-F_{pl} amplitude or complementary area should yield their relative amount (amplitude or area are equivalent since the kinetics are exponential to a good approximation). If, on the other hand, these centers share the main α-antenna with active centers, the amplitude or area underestimates the amount of centers by a factor of about 2.5 related to the connectivity of PS IIα (Lavergne and Leci, 1993). As discussed below, the exponential kinetics give no clue in this case for choosing between either possibility.

Quantitation of inactive PS II from fluorescence data is thus not straightforward and absorption change measurements may be preferred. Specific difficulties, however, are also encountered in this case, mainly the unambiguous distinction between PS II- and PS I-driven signals. The use of material deprived of PS I centers eliminates this problem, but may at the same time introduce other artifacts (biochemical preparations of PS II-enriched material may degrade active centers to an inactive state; cells of mutant algae generally require a treatment with benzoquinone which might also introduce artifactual degradation, see Lavergne, 1987, 1991; Lavergne and Leci, 1993).

According to several authors (Graan and Ort, 1986; Cao and Govindjee, 1990; Nedbal et al., 1991), the non Q_B-transferring centers can be reactivated by providing them with a low concentration of certain halogenated quinones (DCBQ), restoring their ability to sustain electron flow and oxygen-evolution. Graan and Ort reported a 40% enhancement of steady-state electron flow to ferricyanide when adding 20 μM DCBQ. This observation disagrees with reports by Dennenberg et al. (1986) or Chow et al. (1991) who found no enhancement upon DCBQ addition. It was also reported by Dekker et al. (1984) and Lavergne (1991) that 100 μM additions of DCBQ in BBY membranes did not affect the UV-visible absorption changes associated with the centers that become blocked after the first flash in a train. Lavergne and Leci (1993) similarly concluded that DCBQ does not cause any reactivation of non Q_B-transferring centers based on a study of flash-induced absorption changes in algae devoid of PS I centers and on fluorescence experiments using spinach or pea thylakoids. Besides a general quenching of the fluorescence levels, that proved to be approximately of the Stern-Volmer type, no specific quenching by DCBQ of the F_o-F_{pl} rise, induced by continuous illumination or by one single turnover flash could be detected, at variance with the previous report by Cao and Govindjee (1990). Two non-exclusive interpretations for Graan and Ort's observation have been proposed as an alternative to inactive center reactivation. Jursinic and Dennenberg (1988) reported that significant double turnovers could occur in the presence of DCBQ when using flashes of more than 5 μs duration. This double-hitting was detected by oxygen-evolution yield (or a shift in the flash oscillations of delayed fluorescence) but, surprisingly, did not appear in the 515 nm field-indicating absorption change. On the other hand, Lavergne and Leci found no double turnovers (using flashes of 2 μs width, thus not contradicting Jursinic and Dennenberg), but confirmed that DCBQ enhances by 25–30% the steady-state electron transport rate with respect to that observed with PS I acceptors (ferricyanide) or other quinones (DMBQ). The proposed interpretation was that the 'inactive centers' characterized in this manner were not of the non Q_B-transferring type, but rather of the pool-blocked type. Indeed, the fluorescence induction of thylakoids supplied with ferricyanide or DMBQ remains biphasic, with an initial single turnover phase (F_o-F_{pl}) contributed by the non Q_B-transferring centers and a slower multi-turnover phase (F_{pl}-F_{pl}'), indicative of a pool reduction (see Melis and Anderson, 1983, for a similar observation). Addition of 10–15 μM DCBQ totally suppresses this secondary rise. Thus, Lavergne and Leci suggested that ferricyanide or DMBQ could not, at least in the 100 μM range, maintain the plastoquinone pool fully oxidized and envisaged the possibility of a bottleneck at the cytochrome b_6f complex. A simpler hypothesis is a stoichiometric excess of (active) PS II over PS I, as proposed by Melis and Anderson (1983). It is possible that some of the discrepancies in this field (about DCBQ and about the photoreduction kinetics of inactive centers, see below) arise from the confusion of the two fluorescence phases, F_o-F_{pl}-F_{pl}': indeed, if one misses for some reason the first phase and ascribes (wrongly) the second one to non Q_B-transferring centers, it is natural to conclude that inactive centers are reactivated by DCBQ and that their photochemical rate is slow

(β-type). It is, however, easy to distinguish experimentally between these two phenomena by using single turnover flashes: the F_{pl} level is reached upon the first flash, while development of the F_{pl}-F_{pl}' phase requires 10–20 flashes.

III. Possible Correlations

In the foregoing, we have attempted to give a separate account of three basic definitions of PS II (static) heterogeneity: membrane location, light-harvesting efficiency and activity. However, much of the current literature is devoted to the overlap between these features. A unifying model proposed by Melis (1985, 1991) actually condenses the three heterogeneities into one, equating (in wild type, mature chloroplasts) stromal PS II to PS IIβ and to non Q_B-transferring centers. This view accommodates a general theory of PS II turnover (photodegradation-migration to stroma-repair-migration back to grana). In the following sections, we examine pairwise each of these relationships. A discussion of PS II biosynthetic turnover will be given in Section IV-B.

A. Does PS IIβ = Non Q_B-transferring centers?

Analyzing the F_o-F_{pl} fluorescence rise observed under weak illumination in the presence of an acceptor (ferricyanide), Melis (1985) assigned this phenomenon to inactive centers and reported similar kinetics (both for the rate and exponential character) with the β- phase of the fluorescence induction in the presence of DCMU. This prompted a positive answer to the question discussed in this section. Support for this came from the work of Chylla and Whitmarsh (1990) who studied the saturation curves of inactive and active centers through the flash-induced field-indicating absorption change. These authors reported a two-fold lower light-harvesting efficiency for non Q_B-transferring centers compared with active PS II.

The slow (β-type) rate found by Melis for the F_o-F_{pl} was not confirmed by Hsu and Lee (1991b), nor by Lavergne and Leci (1993) who obtained a *shorter* half-time than for the induction in the presence of DCMU (dominated by the fast α-phase). A theoretical analysis led the latter authors to conclude that a shorter half-time was expected from the simple assumption that non Q_B-transferring centers were a minor component of PS IIα, sharing a common antenna with active centers (this holds when comparing the F_o-F_{pl} rise to the *fluorescence* kinetics of the α-phase, due to its sigmoidal shape; if one considers the Q_A kinetics obtained from the area, the opposite ordering of half-times is observed). This view also accounts for the exponential kinetics of the F_o-F_{pl} rise: as shown by simulations (and intuitively straightforward), the sigmoidal character reflecting the increase in cross-section caused by 'reflection' of excitons by closed centers becomes significant only when a substantial fraction of closed centers builds up. The *ca* 15% inactive centers immersed among a majority of permanently open (active) traps can hardly influence each other's cross-section. On the other hand, if a fraction of formerly active centers could be blocked in a non-quenching state, this model would predict an increased photochemical rate for inactive centers. This is precisely what was reported in the paper of Briantais et al. (1988) (see Section IV-B), where a large fraction of active centers were either destroyed or converted to an inactive-closed state by treating a photoautotrophic culture of *Chlamydomonas* cells with chloramphenicol. An additional argument for a common antenna and similar photochemical efficiency of inactive and α-centers was given by Lavergne and Leci: in a plot of the fluorescence level against the total fraction of closed centers (measured from absorption changes), the datapoint corresponding to inactive centers was found to lie on the hyperbolic curve generated by closing active centers, in two algal strains with different LHC content.

Again, this controversy should—and can—be settled experimentally. Presently we can only propose the following (one-sided) leads. As suggested above, the discrepancy concerning the fluorescence induction kinetics of inactive centers may have its origin in the confusion of a single turnover process (F_o-F_{pl}) and a multi-turnover one (F_{pl}-F_{pl}'). Experimental procedures making use of single turnover flashes should allow a clear-cut discrimination on this issue. The experiment of Chylla and Whitmarsh (1990) suffers from no ambiguity in this respect. These authors compared saturation curves of the field-indicating absorption change caused by a short flash, either fired on dark-adapted material (all centers open), or following a saturating flash (inactive centers closed). The energy dependence of inactive centers, obtained as the difference between the two sets of data, indicated a twofold smaller absorption cross-section compared with active centers. A difficulty and possible source of error in this procedure is to eliminate (or keep rigorously constant on a series of two flashes) the contribution of PS I to the field-indicating change.

B. Do Non Q_B-transferring Centers = Stromal PS II?

Several authors have studied the PS II activity of stroma lamellae purified by various means (using detergents or mechanical fractionation). The activity of PS II was assayed by oxygen-evolution, generally using PpBQ as an acceptor (Mäenpää et al., 1987; Andreasson et al., 1988; Henrysson and Sundby, 1990). The rates obtained imply that most of the PS II complex present in these membranes was active under such conditions. Henrysson and Sundby studied the efficiency of various quinone acceptors and found no particular specificity for the halogenated species: the highest electron transfer rate was obtained with PpBQ, and the efficiency of DMBQ, pBQ or DCIP was greater than or equal to that of DCBQ. Thus, even accepting Graan and Ort's view that inactive centers are specifically reactivated by DCBQ, this hardly fits the view of stromal PS II as non Q_B-transferring. The low efficiency of ferricyanide or methylviologen photoreduction reported by Henrysson and Sundby is probably due to loss or inactivation of plastocyanin.

On the other hand, the available data on granal membrane fractions suggest as large a fraction of inactive centers as in whole thylakoids. This appears from flash-induced absorption changes in BBY preparations (Dekker et al., 1984, Lavergne 1991) but also from fluorescence results in preparations that do not involve detergents and should therefore be less contaminated with artifactual inactive centers (Delrieu and Rosengard, 1993; Yu et al., 1993). We thus conclude that most of the available evidence does not support the identification of non Q_B-transferring centers with stromal PS II. This does not imply, however, that stromal PS II centers are identical to active, granal centers. According to Vallon et al. (1987), DCMU has a weak inhibitory efficiency in stromal PS II. This is consistent with the binding study of Wettern (1986). In a study of fluorescence induction in stroma lamellae (R. Bassi and J. Lavergne, unpublished) we also found little inhibition by DCMU and, in the absence of inhibitor, a half-time of about 20 ms for the reoxidation of Q_A^- after a flash, about 50-fold slower than in normal centers. We believe that stromal PS II may be the origin for the small (2–4%) DCMU-resistant oxygen-evolution reported by Joliot and Joliot (1981a,b).

C. Does Stromal PS II = PS IIβ?

There is presently little biochemical evidence that the PS II complex present in stroma lamellae is associated with a reduced amount of chlorophyll-proteins, except the finding of a sub-stoichiometric amount of the CP26 and CP29 proteins (Vallon et al., 1987; Bassi et al., 1988). The stromal fraction of LHCII, on the other hand, exceeds that of PS II, but is believed to serve PS I (Melis and Anderson, 1983), perhaps because the absence of CP29 and CP26 removes the docking site on the PS II complex (Peter and Thornber, 1991b; Dainese et al., 1992b). Measurements of the photochemical efficiency of stromal PS II support the idea of a smaller (by a factor of 2–3) antenna size with respect to granal PS II (Anderson and Melis, 1983, Andreasson et al., 1988; Henrysson and Sundby, 1990). It should be noted that competition from PS I for exciton trapping, rather than a smaller antenna size may account for the lower photochemical efficiency of stromal PS II. At any rate, these data fit with the almost universally accepted view that stromal PS II is of the β-type. Further support comes from reports showing that granal membrane fractions are enriched in fast, α-type PS II centers (Anderson and Melis, 1983; Lam et al., 1983, Andreasson et al., 1988), although the data of Hodges and Barber (1986) disagree on this point. As discussed above, there is some evidence for a lower DCMU binding affinity both in stromal and β-PS II, which is consistent with both populations being identical.

Whereas there is little doubt that stromal PS II is of the β-type, it is debatable whether this contribution can account for the total amount of the β-phenomenon. The relative amount of PS II centers found in stroma lamellae ranges from 2–20% (with most data in the narrower 10–15% range; the lower values may be specific to cereals, maize and barley, see Bassi et al., 1988) and seems definitely too low to account for the typical fraction of 35% PS IIβ (range of reported values: 20–60%). Published data on fluorescence induction from stroma lamellae preparations are contradictory: in some of them (Anderson and Melis, 1983, Wollenberger et al., 1994), the stromal contribution to the amplitude of variable fluorescence appears sufficient to account for most of the β-phase observed in thylakoids, whereas other reports (Vallon et al., 1987; Henrysson and Sundby, 1990; R. Bassi and J. Lavergne, unpublished) suggest that it could account for only a few percent of the total variable fluorescence in thylakoids, thus much less than the 15% typical β-amplitude. A smaller variable fluorescence per stromal PS II unit may be due to quenching by PS I (however, as explained earlier, such a quenching is not consistent with the 'mainstream view' of PS IIβ). If the amount of stromal PS II is significantly below that of PS IIβ, and if we accept the absence of the latter from the grana core, this suggests that a substantial fraction of PS IIβ may be located in the grana margins. Some support for this view emerges from the recent study of Wollenberger et al. (1994) on vesicles originating from the grana margins.

IV. Dynamic Heterogeneities

A. Non-photochemical Quenching

1. Basic Information

The quantum yield of photosynthetic oxygen evolution (Φ_{O_2}) (or of CO_2 assimilation in the absence of photorespiration), at steady-state, is a decreasing function of the illumination intensity. A number of studies have shown that non-photochemical quenching ('q_N') in PS II is a significant factor in this reversible reduction in photochemical efficiency (Weis and Berry, 1987; Horton and Hague, 1988; Genty et al., 1989, 1990a; Dan and Hansen, 1990). Figure 1 illustrates the development of this quenching in vivo during a transition from the dark-adapted state to illuminated steady-state, and its relaxation kinetics when the actinic light is turned off and replaced by a weak far-red light. Using the 'light doubling' method introduced by Bradbury and Baker (1984) and Schreiber et al. (1986), the maximum fluorescence level F_m is probed at various times by imposing a brief intense pulse that transiently closes all PS II centers in the state Q_A^-. The quenching process does not concern only the closed state of the centers, but the open state as well, as may be seen from the decreased F_o level assayed by replacing the actinic light by the weak far-red beam that restores all PS II acceptors to an oxidized state.

A useful expression has been derived (Genty et al., 1989) relating the quantum yield of PS II ($\Phi_{PS\,II}$) to the fluorescence yields F and F_m measured at any stage of an induction process (or F_s and F_m' at steady-

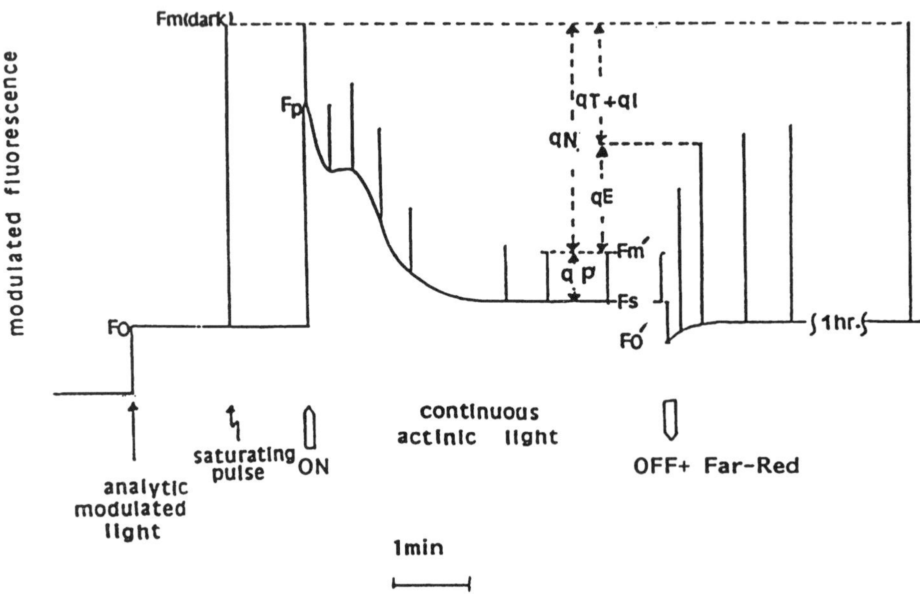

Fig. 1. Fluorescence induction with light of physiological intensity in a leaf, using the 'light doubling' technique. An analytical modulated beam allows detection of the variations of the chlorophyll fluorescence yield induced by the continuous actinic beam. Brief saturating pulses monitor the F_m level, decreasing as qN develops, and show the extent of photochemical quenching qP. When the actinic illumination is switched off and replaced by dim far red light, relaxations of the non-photochemical quenching on F_o and F_m can be followed.

state for some irradiance level) as illustrated in Fig. 1. $\Phi_{PS\ II}$ can be estimated as the ratio $(F_m - F)/F_m$. Schematically, this expression takes into account both the losses which affect the photochemical efficiency of an open trap, and the degree of openness of the centers (a proportionality constant should be inserted if significant losses occur in closed traps, Lavergne and Trissl, 1995). This relation predicts that at steady-state, when linear electron flow through both photosystems is limited by the same kinetic step, one should have $V_{O_2} = I\ \Phi_{PS\ II} = I\ (F_m - F)/F_m$ where I denotes the rate of photon absorption by PS II. Experimentally, this linear relationship has been verified for the steady-state values of V_{O_2} and $\Phi_{PS\ II}$ over a broad range of irradiances, with the exception of dim light (Van Wijk and Van Hasselt, 1990; Edwards and Baker, 1993; Hormann et al., 1994). Concerning the nature of the quenching process qN, this finding eliminates, in the intensity range where linearity is obeyed, two possibilities: (i) a removal of part of the PS II antenna (as in state 1 to state 2 transitions), that would decrease I, and (ii) the establishment of a diverted electron transfer pathway with respect to linear electron flow across the two photosystems. An example of this could be a cycle around PS II, provided its rate is slow enough to allow a significant contribution of the cycling centers to the variable fluorescence. Two major alternative possibilities remain open, namely an increased quenching in the antenna or the conversion of a fraction of centers into quenching sinks.

Non-photochemical quenching is, quantitatively, a large effect. In dark-adapted leaves $(F_m - F_o)/F_m$ is around 0.8. It declines two-fold under steady-state illumination levels that saturate photosynthesis (Genty et al., 1990b). As a result of this regulation process, PS II centers are poised between the open and closed state, but somewhat closer to the open state, which means that at physiological irradiances, PS II adjusts its yield close to, or slightly below, the rate-limiting step of electron withdrawal from its acceptors.

Non-photochemical quenching covers different phenomena. The time-course of the dark-relaxation of qN can be decomposed into, at least, three exponential components (Horton and Hague, 1988) that were denoted qT, qE, qI. These contributions have also been correlated with the irradiance range at which they saturate: qT in low light, qE at an intensity which saturates photosynthesis and qI at higher intensities. We will now briefly discuss each of these contributions and examine to what extent they may involve a dynamic heterogeneity of PS II.

2. Effects of the Various Non-photochemical Quenchings on PS II

a. qT Quenching

qT quenching occurs in dim light and its half-time of dark-relaxation is about five minutes. Most of the evidence concerning qT supports its interpretation as a state 2 transition (Horton and Hague, 1988; Hodges et al., 1989). This term denotes the process (see Allen, 1992; and Chapter 29 by Melis in this book, for reviews) by which a subpopulation of LHCII becomes phosphorylated, dissociates from PS II and migrates towards the non-appressed membrane region. Decreased fluorescence levels F_o and F_m are observed as a result of the state 2 transition, implying that the migrating LHCII does not rebind to stromal PS II (at least if the latter is isolated from PS I), but probably serves as a PS I antenna. Regulation by state transitions seems predominant in unicellular green algae and cyanobacteria (with, however, different mechanisms), whereas it is a relatively minor factor in higher plants.

A strong argument for ascribing qT to a state 2 transition is the finding that the dark relaxation of qT is blocked by NaF, a phosphatase inhibitor (Horton and Hague, 1988; Hodges et al., 1989). This interpretation is also consistent with the non-linearity of V_{O_2} vs. $(F_m - F)/F_m$ specifically observed under dim light. In this intensity range, the intrinsic yield of PS II, as measured by $(F_m - F)/F_m$, decreases much less than V_{O_2} (van Wijk and van Hasselt, 1990; Edwards and Baker, 1993), suggesting a diminished overall cross-section of the PS II antenna. The possibility that this deviation, first observed in leaves, could arise from optical factors (the fluorescence measurement selects the upper cell layers, while the oxygen evolution measurement does not) is ruled out by the similar observation recently made in a chloroplast suspension by Hormann et al., (1994).

A possible discrepancy arises from reports showing that in barley mutants (Lokstein et al., 1993) and in intermittent light-grown pea leaves (Briantais, 1994), both lacking LHCII, a qN component has very similar dark relaxation kinetics to that observed for qT in plants possessing LHCII. However, as concluded by Walter and Horton (1991), the kinetic analysis of qN dark-relaxation may not be accurate enough for discriminating the various components of qN in vivo.

Accepting the interpretation of qT as a state 2 transition, its impact on PS II heterogeneity is expected to be expressed as a decrease of the antenna size of granal PS II, possibly resulting in creation of PS IIβ, either in the appressed region or in stroma lamellae if a migration of the PS II complex results from the dissociation of LHCII.

b. qE Quenching

The component of qN called qE (E for energization) can be assayed in vitro by its elimination with uncouplers, or in vivo by its rapid dark-relaxation with a half-time of around 10 seconds (Horton and Bowyer, 1990; Krause and Weis, 1991). In contrast to unicellular green algae (Falkowski et al., 1986) and cyanobacteria (Demmig-Adams and Adams 1992), qE is the major qN component in higher plants at physiological irradiances. According to Krause and Behrend (1986), Ögren (1991) and van Wijk and van Hasselt (1993), qE provides protection from photoinhibition. qE occurs in response to the build-up of the thylakoid ΔpH or more specifically to the acidification of the thylakoid lumen (Briantais et al., 1979). qE is saturated at the same intensity as photosynthesis (Horton and Hague, 1988). It lowers the intrinsic photochemical efficiency of PS II, $(F_m - F_0)/Fm$, up to two-fold in most plants (Genty et al., 1990b). The molecular mechanism of qE is still controversial (see Walter and Horton, 1993 for a recent discussion). Two main models emerge from the literature, with opposite answers to our question: does qE imply the development of a functional PS II heterogeneity?

Weis and Berry (1987) have proposed that energization of the thylakoid membrane induces the formation of inactive centers of the quenching sink type. The quenching could result from accumulation of centers in the $P680^+Q_A^-$ state (Weis and Lechtenberg, 1989; Schreiber and Neubauer, 1990) or from rapid cyclic electron flow around PS II (Horton and Lee, 1985; Schreiber and Neubauer, 1987; Ramm and Hansen, 1993). Thus, qE in vivo would be the same phenomenon as the non-photochemical quenching induced by chemical acidification of isolated thylakoids and PS II-enriched preparations (Krieger and Weis, 1990; Krieger et al., 1992). According to results of Krieger and Weis (1993) acidification of the thylakoid lumen induces a reversible dissociation of Ca^{++} from the water splitting enzyme. The resulting inhibition of the donor side of PS II is expected to block chlorophyll fluorescence at the F_0 level (Butler et al., 1973). In agreement with this view, Krieger et al. (1992) observed that acidification of the lumen, either by changing the pH of the medium or by building a ΔpH through ATP hydrolysis, did not affect the F_0 level or the associated average fluorescence lifetime. This is in contrast, however, with the marked quenching of F_0 accompanying qE in vivo (see Fig. 1).

As discussed below, most studies of qE in vivo hardly seem consistent with the hypothesis of a conversion of active centers to quenching sinks. We believe, however, that a special case should be made for leaves lacking LHCII, such as in intermittent-light-grown peas or the *chlorina* f2 barley mutant. A significant qN quenching, that resembles qE with regard to its dark relaxation kinetics (Lokstein et al., 1993; Briantais, 1994) can be developed in such materials (it is however smaller than normal qE and gives rise to only a small decrease of F_0). However, at variance with leaves that contain a normal LHCII complement (Genty et al., 1990a), it was observed (Briantais, 1994) that non-photochemical quenching did not affect the photochemical efficiency of the PS II traps, as measured by the saturation curves of the flash-induced increase of fluorescence monitoring reduction of Q_A. This rules out an antenna quenching and suggests that in such materials, qN may involve conversion of centers to a quenching sink type. Since chloroplasts lacking LHCII have no or very few grana stacks and display no excitonic connection between PS II units (Arntzen et al., 1976), there is no reason for quenching sink units to influence the photochemical efficiency of active units.

In leaves with a normal antenna complement, a number of results favor the hypothesis that qE corresponds to an homogeneous enhancement of non-radiative losses in the PS II antenna. As already emphasized (see Fig. 1), a quenching of F_0 is observed (Bilger and Schreiber 1986; Rees et al., 1990; Genty et al., 1990a,c; see also Demmig-Adams 1990). This is accompanied by a decrease of the average fluorescence lifetime (Genty et al., 1992) . This diminution of F_0 is perfectly consistent with the decreased photochemical efficiency for the reduction of Q_A (Genty et al., 1990a). Genty et al. (1992) also reported that in chloroplasts which develop a qE quenching of F_0 and F_m similar to that observed in leaves, the dark relaxation of qN is not affected by the addition of a strong reductant (dithionite). This suggests that the accumulation of $P680^+$ is not involved in the manifestation of qE. Moreover, maximal qN in these chloroplasts was observed in the presence of ascorbate (Genty et al., 1992), a reducing mediator which reverses non-photochemical quenching induced by chemical acidification (Crofts and Horton, 1991; Rees et al., 1992). Direct evidence showing that qE does not involve a significant conversion of PS II centers to inactive sinks has been recently obtained by Genty (personal communication) who found that in protoplasts isolated from barley leaves, the same oxygen yield (per single turnover flash and per open trap) was measured in the presence and absence of non-photochemical quenching.

Horton et al., (1992) and Ruban and Horton (1992) proposed that qE in vivo results from an aggregation of LHCII triggered by the photoinduced ΔpH. Indeed, aggregation of isolated LHCII induces changes in its fluorescence emission spectra that are very similar to those accompanying qE in leaves (Ruban et al., 1993). Furthermore, antimycin A, a potent inhibitor of qE (Oxborough and Horton, 1987), inhibits aggregation of isolated LHCII (Horton et al., 1992). The quenching resulting from this aggregation would be enhanced by zeaxanthin (Noctor et al., 1991; Horton et al., 1992). Numerous data indicate a correlation between qE and the light-induced deepoxidation of violaxanthin to zeaxanthin (Demmig-Adams, 1990; Bilger and Björkman, 1990; Noctor et al., 1991; Gilmore and Yamamoto, 1991; 1993; Thayer and Björkman, 1992). According to Peter and Thornber (1991b), xanthophylls are located exclusively in the LHCII complex (including the minor complexes *a* and *e* that are present in the *chlorina* f2 mutant of barley). A special role for these minor complexes in the deepoxidation process has been suggested by Dainese et al. (1992a).

Therefore, at least in LHCII-containing PS II units (PS IIα), at physiological irradiances, an homogeneous quenching developing in the LHCII antenna and competing with photochemistry and fluorescence appears at the present time the most probable explanation of qE. On the other hand, conversion of centers to quenching sinks is observed in vitro as a result of lumen acidification, or in vivo in materials lacking LHCII. The reason why this process appears overridden by antenna quenching in normal material in vivo is not clear, but it may be suggested that the precise role of qE could be to prevent the accumulation of strongly oxidizing radicals on the PS II donor side that would otherwise result from excessive acidification of the lumen.

c. qI Quenching

This component of qN, called qI, (I for irreversible) was characterized by a slow dark-relaxation ($t_{1/2} > 30$ min). According to Horton and Hague (1988) it becomes large at irradiances which over-saturate photosynthesis. Consequently, it has been often interpreted as a photoinhibition, whereby degradation of PS II becomes faster than repair or new synthesis. Therefore, qI would induce a PS II heterogeneity (active/inactive centers), as discussed in the next section dealing with photodegradation of PS II. However, qI, as defined from its slow dark relaxation, may not be due solely to photoinhibition. According to Horton and Hague (1988), qI also occurs at physiological irradiances, and its intensity dependence displays two waves with an inflection near the saturation of photosynthesis. It has been suggested that qI may contain a contribution from a long-lasting component of qE that could result from hydrolysis in the dark of the ATP accumulated during the illumination, maintaining a ΔpH. Recently Ruban et al. (1993) proposed that qI is a quenching due to zeaxanthin, associated with the non aggregated form of LHCII and in which epoxidation would be slow. Ruban and Horton (1995) observed an instant relaxation of most of the qI upon nigericin addition. Nevertheless, at over-saturating irradiances there is no doubt that at least a part of qI is due to photoinhibition which will create a population of PS II centers of the quenching sink type.

B. Effects of Photodegradation of PS II

In the presence of intense light that over-saturates photosynthesis, Demmig and Björkman (1987) observed in leaves a parallel decrease, with the time of illumination, of the maximum quantum yield of oxygen evolution and of the maximum quantum yield of PS II $(F_m - F_o)/F_m$. Since the photosynthetic capacity was also diminished by this high light treatment it can be concluded that inactive centers were created. As a quenching is associated with this phenomenon, these inactive centers should be of the quenching sink type. In agreement with the creation of quenching sinks, rather than increased antenna losses, the quenching affects F_m but not F_o [except when the high irradiances are performed at low temperatures where F_0 is *increased* (Somersalo and Krause, 1989; Kirilovsky et al., 1990)]. A current debate in the field of photoinhibition concerns the question whether the primary photodamaging step occurs on the acceptor or donor side (see reviews by Kyle, 1987; Cleland, 1988; Critchley, 1988 and Ohad et al., 1994).

Several studies support the acceptor-side involvement, suggesting that photoinhibition is associated with a degradation of the 32 kDa D1 protein triggered by the formation of semiquinone anion radicals. Since DCMU and atrazine prevent photoinhibition (Mattoo et al., 1984) it was proposed that the initial damage is caused by Q_B^- in its niche, resulting in a decreased plastoquinone binding and creation of inactive, non Q_B-transferring centers. This interpretation was supported by the thermoluminescence data of Ohad et al. (1988). During the initial stages of a photoinhibitory treatment of *Chlamydomonas* cells, these authors observed a shift towards a lower temperature of the thermoluminescence peak, that may correspond to a relative increase of $S_2Q_A^-$ recombination at the expense of $S_2Q_B^-$ recombination. However, enhancement of non Q_B-transferring centers could only be a very transitory stage, quickly converted into quenching sinks, if one wants to account for the quenching of fluorescence routinely observed at room-temperature after a high light treatment. Moreover, in spinach leaves submitted to a 1 h photoinhibitory treatment, Briantais et al. (1992) observed only a decrease of the amplitude of the $S_2Q_B^-$ recombination which was not accompanied by an increase of the thermoluminescence peak associated with $S_2Q_A^-$.

In our opinion, the view that Q_B^- is responsible for the triggering of the photoinhibitory process encounters serious objections. First, a large fraction of centers have a bound Q_B^- in the dark-adapted state in vivo, at least in algae, where typically half of the centers are in this state (Wollman, 1978). In dark-adapted isolated thylakoids, about 20% Q_B^- is routinely found. If this semiquinone was deleterious to its niche, this process should appear already in the dark. Furthermore, the light regime that should maintain a maximum amount (50%) of Q_B^- at steady-state is weak illumination, and this does not fit with the intensity range where photoinhibition becomes a significant process. In recent work, Ohad and coworkers (Keren et al., 1995) developed a new interpretation of the role of Q_B^-, which does not suffer from the above objections. Photodegradation would be due to some probability for triplet chlorophyll and singlet oxygen formation during recombination of S_2 or S_3 with Q_B^-.

A different view, although involving also the acceptor side, is that the initial event in photoinhibition is a double reduction of Q_A, possibly leading to detachment of the primary quinone acceptor (Styring et al., 1990; Vass et al., 1993). This mechanism could transform the center into a quenching sink, involving transient photoreduction of pheophytin and its decay through a triplet state.

On the other hand, many reports suggest that photoinhibited PS II centers are blocked on the donor side (Cleland et al., 1986; Theg et al., 1986; Jegerschold et al., 1990). From their studies on photodegradation of D1 in chloride-depleted thylakoids, Jegerschold et al. (1990) proposed that D1 breakage is triggered by an accumulation of tyrosine Y_Z^+ and $P680^+$ which are potent oxidizing species. In these chloride-depleted thylakoids the sensitivity to photoinhibition is intensified. A similar observation was reported previously by Callahan et al. (1986) for chloroplasts inhibited on the donor side. This increased sensitivity of donor side-damaged centers is probably the origin of the observation by Greer and Laing (1988) and Le Gouallec et al. (1991) that a moderate light intensity, unable to induce photoinhibition by itself, totally blocks the recovery of a previously photoinhibited leaf. Even the primary charge separation may be blocked in photoinhibited PS II centers, as suggested by Demeter et al. (1985) who observed a parallel decrease of pheophytin and Q_A reduction both in *Chlamydomonas* cells and in spinach chloroplasts pretreated in strong light.

Therefore over-saturating light seems to create inactive PS II centers of the quenching sink type associated with degradation of D1, which account for at least a portion of qI quenching. Under physiological irradiances where products of PS II photoconversion are rapidly consumed, the phenomena reported above may be very minor (see the discussion of the 'low light syndrome' by Ohad et al., 1994). Continuous biosynthetic turnover of PS II must take place, but the repair or synthesis is probably sufficiently efficient to keep the amount of quenching sink centers at a low level. The detailed incidence of the successive steps involved in the photodegradation-repair cycle with regard to the trapping properties of the PS II centers is far from clear, in our opinion. The experiment described by Briantais et al. (1988) was designed to investigate the impact of blocking the repair-synthesis process. The translation of proteins encoded by the chloroplastic genome was inhibited by adding chloramphenicol to cells of *Chlamydomonas* under photoautotrophic conditions. This caused a decrease of PS II centers contributing to variable fluorescence, which was accelerated when raising the illumination intensity. Unexpectedly, however, the inactive centers thus created by chloramphenicol addition were neither of the quenching sink, nor non Q_B-transferring types, but of the closed type (or simply destroyed). The amount of non Q_B-transferring centers remained unaffected and these centers appeared to be protected from the photodegradation process. More generally, there is no evidence, to our knowledge, that inactive centers of this type play a role in the turnover of the PS II complex. On the other hand, we have as yet no explanation for the puzzling finding that the inhibition of chloroplast protein synthesis causes an accumulation of closed centers whereas quenching sink centers accumulate during 'normal' photoinhibition.

VI. Conclusions

In concluding this chapter, it would have been gratifying to the reader, and authors as well, to bring simplicity out of disorder and provide some unifying picture rationalizing the different aspects of PS II heterogeneity. Unfortunately, we have not really been able to achieve such a goal. Our opinion is that the unifying scheme proposed by Melis encounters serious objections: non Q_B-transferring centers seem predominantly a component of PS IIα; PS IIβ seems basically made of active centers having a smaller antenna size (although functional specificities of these centers are likely); no clear connection appears, at least under physiological conditions, between PS II heterogeneity and the turnover of the PS II complex.

The simplest view for PS II β is that it represents a fraction of the PS II complexes that are not associated with peripheral antenna, just reflecting a stoichiometric excess of the core complex over non-phosphorylated LHCII. The quantitation of PS II β has suffered a significant decrease from the moment it was hypothesized that these centers were inactive and belonged exclusively to stroma lamellae. The early estimates, around 35%, seem closer to most experimental data. If one accepts this figure, then it would be hard to account for the experimental quantum yield of photosynthesis if this fraction of PS II were inactive. Similarly, the population of PS II found in stroma lamellae does not suffice to account for such a large fraction. One may *speculate* that dissociation of LHCII induces a number of consequences for the PS II complex, such as conversion from dimer to monomer, modification of acceptor side properties, migration towards the periphery of the grana: but, although plausible, none of these propositions is presently substantiated by significant experimental support.

Concerning non Q_B-transferring centers, they probably represent a fraction of about 15% (in plants) whose negative impact on PS II efficiency will be diminished by their belonging to the connected α-antenna. Most of the available evidence does not support the idea that they represent a dynamic stock of damaged centers resulting from PS II turnover. A less appealing, but perhaps yet more tenable, hypothesis (Lavergne and Leci, 1993) would be a steric mismatch, due to the dense packing of membrane proteins in the grana, obstructing the accessibility of the Q_B niche in a fraction of centers. On the other hand, the fact that the underlying turnover flux of the PS II complex does not seem to entail a sizable stock of non-functional centers, in spite of the sensitivity of this system to photodegradation, demonstrates the efficiency of this process.

Dynamic regulations of the activity of PS II play a major role under physiological conditions. State transitions (in algae) or qE quenching (in higher plants) dramatically affect the photochemical yield of PS II. Both processes imply essentially modifications of the size and deactivation properties of the antenna rather than inactivation of the centers. This increase of non-radiative losses in the LHCII complex is tuning the delivery of excitons to PS II in response to the level of available secondary acceptors.

Acknowledgments

The authors wish to thank Dr. Michael Hodges for critical reading of the manuscript. This work was supported by the Centre National de la Recherche Scientifique.

References

Albertsson PÅ, Andreasson E and Svensson P (1990) The domain organization of the plant thylakoid membrane. FEBS Lett 273: 36–40

Allen JF (1992) Protein phosphorylation in regulation of photosynthesis. Biochim Biophys Acta 1098 : 275–335

Anderson JM and Melis A (1983) Localization of different photosystems in separate regions of chloroplast membranes. Proc Natl Acad Sci USA 80: 745–749

Andersson B and Anderson JM (1980) Lateral heterogeneity in the distribution of chlorophyll-protein complexes of the thylakoid membranes of spinach chloroplasts. Biochim Biophys Acta 593: 427–440

Andreasson E, Svensson P, Weibull C and Albertsson PÅ (1988) Separation and characterization of stroma and grana membranes—evidence for heterogeneity in antenna size of both photosystem I and photosystem II. Biochim Biophys Acta 936: 339–350

Arntzen CJ, Armond PA, Briantais JM, Burke JJ and Novitzky WP (1976) Dynamic interactions among structural components of the chloroplasts membrane, in chlorophyll proteins, reaction centers and photosynthetic membranes. Brookhaven Symposia in Biology 28: 316–337

Bassi R, Giacometti G and Simpson DJ (1988) Characterisation of stroma membranes from *Zea Mays* L. chloroplasts. Carlsberg Res Commun 53: 221–232

Bell DH and Hipkins MF (1985) Analysis of fluorescence induction curves from pea chloroplasts. Photosystem II reaction centre heterogeneity. Biochim Biophys Acta 807: 255–262

Bennoun P and Li YS (1973) New results about the mode of action of DCMU in spinach chloroplasts. Biochim Biophys Acta 292: 162–168

Berthold DA, Babcock GT and Yocum CF (1981) A highly resolved oxygen-evolving photosystem II preparation from spinach thylakoid membranes. EPR and electron transport properties. FEBS Lett 134: 231–234

Bilger W and Björkman O (1990) Role of xanthophyll cycle in photoprotection elucidated by measurements of light induced absorbance changes, fluorescence and photosynthesis in leaves of *Hedera canariensis*. Photosynth Res 25: 173–185

Bilger W and Schreiber U (1986) Energy dependent quenching of dark level chlorophyll fluorescence in intact leaves. Photosynth Res 10: 303–308

Black MT, Brearley TH and Horton P (1986) Heterogeneity in chloroplast photosystem II. Photosynth Res 8: 193–207

Bradbury M and Baker NR (1984) A quantitative determination of photochemical and non-photochemical quenching during the slow phase of the chlorophyll fluorescence induction curve of bean leaves. Biochim Biophys Acta 765: 275–281

Brearley T and Horton P (1984) Properties of photosystem IIα and photosystem IIβ in spinach chloroplasts. In: Sybesma C (ed) Advances in Photosynthesis Research, Vol I, pp 433–436. Martinus Nijhoff/Dr W. Junk Publishers, The Hague

Briantais JM (1994) Light-harvesting chlorophyll *a-b* complex requirement for regulation of photosystem II photochemistry by non-photochemical quenching. Photosynth Res 40: 287–294

Briantais JM, Vernotte C, Picaud M and Krause GH (1979) A quantitative study of the slow decline of chlorophyll *a* fluorescence in isolated chloroplasts. Biochim Biophys Acta 548: 128–138

Briantais JM, Cornic G and Hodges M (1988) The modification of chlorophyll fluorescence of *Chlamydomonas reinhardtii* by photoinhibition and chloramphenicol addition suggests a form of photosystem II less susceptible to degradation. FEBS Lett 236: 226–230

Briantais JM, Ducruet JM, Hodges M and Krause H (1992) The effects of low temperature acclimation and photoinhibitory treatments on photosystem 2 studied by thermoluminescence and fluorescence decay kinetics. Photosynth Res 31: 1–10

Butler WL, Visser JWM and Simons HL (1973) The kinetics of light-induced changes of C550, cytochrome *b*559 and fluorescence yield in chloroplasts at low temperature. Biochim Biophys Acta 292: 140–151

Callahan FE, Becker DW and Cheniae GM (1986) Studies on the photoactivation of the water oxidizing enzyme II. Characterization of weak light photoinhibition of PS II and its light-induced recovery. Plant Physiol 82: 261–269

Callahan FE, Wergin WP, Nelson N, Edelman M and Mattoo AK (1989) Distribution of thylakoid proteins between stromal and granal lamellae in *Spirodela*. Plant Physiol 91: 629–635

Cao J and Govindjee (1990) Chlorophyll *a* fluorescence transient as an indicator of active and inactive Photosystem II in thylakoid membranes. Biochim Biophys Acta 1015: 180–188

Chow WS, Hope AB and Anderson JM (1990) A reassessment of the use of herbicide binding to measure photosystem II reaction centres in plant thylakoids. Photosynth Res 24: 109–113

Chow WS, Hope AB and Anderson JM (1991) Further studies on quantifying photosystem II in vivo by flash-induced oxygen yield from leaf discs. Aust J Plant Physiol 18: 397–410

Chylla RA and Whitmarsh J (1990) Light saturation response of inactive photosystem II reaction centers in spinach. Photosynth Res 25: 39–48

Chylla RA, Garab G and Whitmarsh J (1987) Evidence for slow turnover in a fraction of Photosystem II complexes in thylakoid membranes. Biochim Biophys Acta 894: 562–571

Cleland RE (1988) Molecular events of photoinhibitory inactivation in the reaction center of photosystem II. Aust J Plant Physiol 15: 135–150

Cleland RE, Melis A and Neale PJ (1986) Mechanism of photoinhibition : Photochemical reaction center inactivation in system II of chloroplasts. Photosynth Res 9: 79–88

Critchley C (1988) The molecular mechanism of photoinhibition—facts and fiction. Aust J Plant Physiol 15: 27–41

Crofts J and Horton P (1991) Dissipation of excitation energy by photosystem II particles at low pH. Biochim Biophys Acta 1058: 187–193

Dainese P, Marquardt J, Pineau B and Bassi R (1992a) Identification of violaxanthin and zeaxanthin binding proteins in maize photosystem II. In: Murata (ed) Research in Photosynthesis, Vol I, pp 287–290, Kluwer Academic Publishers, Dordrecht

Dainese P, Santini C, Ghiretti-Magaldi A, Marquardt J, Tidu V, Mauro S, Bergantino E and Bassi R (1992b) The organization of pigment-proteins within photosystem II. In Murata N (ed) Research in Photosynthesis, Vol II, pp 13–20. Kluwer Academic Publishers, Dordrecht

Dan H and Hansen UP (1990) A study of the energy-dependent quenching of chlorophyll fluorescence by means of photoacoustic measurements. Photosynth Res 25: 269–278

Dau H (1994) Molecular mechanisms and quantitative models of variable photosystem II fluorescence. Photochem Photobiol 60: 1–23

Dekker JP, van Gorkom HJ, Wensink J and Ouwehand L (1984) Absorbance difference spectra of the successive redox states of the oxygen-evolving apparatus of photosynthesis. Biochim Biophys Acta 767: 1–9

Delosme R (1967) Etude de l'induction de fluorescence des algues vertes et des chloroplastes au début d'une illumination intense. Biochim Biophys Acta 143: 108–128

Delrieu MJ and Rosengard F (1993) Events near the reaction center in O_2 evolving PS II enriched thylakoid membranes: The presence of an electric field during the S_2 state in a population of centers. Photosynth Res 37: 205–215

Demeter S, Rozza ZS, Vass I and Sallai A (1985) Thermoluminescence study of charge recombination in photosystem II at low temperature. I Characterization of the Zv and A thermoluminescence bands. Biochim Biophys Acta 809: 369–378

Demmig B and Björkman O (1987) Comparison of the effect of excessive light on chlorophyll fluorescence (77 K) and photon yield of O_2 evolution in leaves of higher plants. Planta 171: 171–184

Demmig-Adams B (1990) Carotenoids and photoprotection : A role for the xanthophyll zeaxanthin.Biochim Biophys Acta 1020: 1–24

Demmig-Adams B and Adams WW (1992) Photoprotection and other responses of plants to high light stress. Annu Rev Plant Physiol Plant Mol Biol 43: 599–626

Dennenberg RJ, Jursinic PA and McCarthy S (1986) Intactness of the oxygen-evolving system in thylakoids and photosystem II particles. Biochim Biophys Acta 852: 222–233

Diner BA (1977) Dependence of the deactivation reactions of photosystem II on the redox state of plastoquinone pool A varied under anaerobic conditions. Equilibria on the acceptor side of photosystem II. Biochim Biophys Acta 460: 247–258

Diner BA (1986) The reaction center of photosystem II. In Staehelin LA and Arntzen CJ (eds) Photosynthesis III, Encyclopedia of Plant Physiology, Vol 19, pp 422–436. Springer Verlag, Berlin

Diner BA, Petrouleas V and Wendoloski JJ (1991) The iron-quinone electron-acceptor complex of Photosystem II. Physiol Plant 81: 423–436

Doschek WW and Kok B (1972) Photon trapping in photosystem II of photosynthesis. Biophys J 12: 832–838

Edwards GE and Baker NR (1993) Can CO_2 assimilation in maize leaves be predicted accurately from chlorophyll fluorescence analysis? Photosynth Res 37: 89–102

Falkowski PG, Wyman K, Ley AC and Mauzerall DC (1986) Relationship of steady-state photosynthesis to fluorescence in eucaryotic algae. Biochim Biophys Acta 849: 183–192

Forbush B and Kok B (1968) Reaction between primary and secondary electron acceptors of photosystem II of photosynthesis. Biochim Biophys Acta 162: 243–253

Genty B, Briantais JM and Baker NR (1989) The relationship between quantum yield of photosynthetic electron transfer and quenching of chlorophyll fluorescence. Biochim Biophys Acta 990: 87–92

Genty B, Harbinson J, Briantais JM and Baker NR (1990a) The relationship between non-photochemical quenching of chlorophyll fluorescence and the rate of photosystem 2 photochemistry in leaves. Photosynth Res 25: 249–257

Genty B, Harbinson J, Briantais JM and Baker NR (1990b) The relationship between the relative quantum efficiencies of photosystems in leaves. Efficiency of PS2 in relation to non-photochemical quenching. In: Baltcheffsky M (ed) Current Research in Photosynthesis, Vol IV, pp 365–368. Kluwer Academic Publishers, Dordrecht

Genty B, Wonders J and Baker NR (1990c) Non-photochemical quenching of F_0 in leaves is emission wavelength dependent: Consequences for quenching analysis and its interpretation. Photosynth Res 26: 133–139

Genty B, Goulas Y, Dimon B, Peltier G, Briantais JM and Moya I (1992) Modulation of efficiency of primary conversion in leaves, mechanisms involved at PS2. In: Murata N (ed) Research in Photosynthesis, Vol IV, pp 603–610, Kluwer Acad Publishers, Dordrecht

Giardi MT, Rigoni F, Barbato R and Giacometti GM (1991) Relationship between heterogeneity of PS II in grana particles in vitro and phosphorylation. Biochem Biophys Res Comm 176: 1298–1305

Gilmore A and Yamamoto HY (1991) Zeaxanthin formation and energy-dependent fluorescence quenching in pea chloroplast under artificially mediated linear and cyclic electron transport. Plant Physiol 96: 635–643

Gilmore AM and Yamamoto HY (1993) Linear models relating xanthophylls and lumen acidity to non-photochemical fluorescence quenching. Evidence that antheraxanthin explains zeaxanthin-independent quenching. Photosynth Res 35: 67–78

Govindjee (1990) Photosystem II heterogeneity: The acceptor side. Photosynth Res 25: 151–160

Graan T and Ort DR (1986) Detection of oxygen-evolving photosystem II centers inactive in plastoquinone reduction. Biochim Biophys Acta 852: 320–330

Greer DH and Laing WA (1988) Photoinhibition of photosynthesis in intact kiwifruit (*Actinida deliciosa*) leaves : Effect of light during growth on photoinhibition and recovery. Planta 175: 355–363

Henrysson T and Sundby C (1990) Characterization of photosystem II in stroma thylakoid membranes. Photosynth Res 25: 107–117

Hodges M and Barber J (1983) Analysis of chlorophyll fluorescence induction kinetics exhibited by DCMU-inhibited thylakoids and the origin of α and β centers. Biochim Biophys Acta 848: 239–248

Hodges M and Barber J (1986) Analysis of chlorophyll fluorescence induction kinetics exhibited by DCMU-inhibited thylakoids and the origin of α and β centres. Biochim Biophys Acta 848: 239–246

Hodges M, Boussac A and Briantais JM (1987) Thylakoid membrane protein phosphorylation modifies equilibrium between photosystem II quinone electron acceptors. Biochim Biophys Acta 894: 138–145

Hodges M, Cornic G and Briantais JM (1989) Chlorophyll fluorescence from spinach leaves. Resolution of non-photochemical quenching. Biochim Biophys Acta 974: 289–293

Holzwarth AR (1991) Excited-state kinetics in chlorophyll systems and its relationship to the functional organization of the photosystems. In H Scheer (ed) Chlorophylls, pp 1125–1151, CRC Press, Boca Raton

Hormann H, Neubauer C and Schreiber U (1994) On the relationship between chlorophyll fluorescence quenching and the quantum yield of electron transport in isolated thylakoids. Photosynth Res 40: 93–106

Horton P (1981) The effect of redox potential on the kinetics of fluorescence induction in pea chloroplasts. I. Removal of the slow phase. Biochim Biophys Acta 635: 105–110

Horton P and Bowyer JR (1990) Chlorophyll fluorescence transients. In: Harwood JL and Bowyer JR (eds) Methods in Plant Biochemistry, Vol 4, pp 259–296. Acad Press, New York

Horton P and Hague A (1988) Studies on the induction of chlorophyll fluorescence in barley protoplasts. IV Resolution of non-photochemical quenching. Biochim Biophys Acta 932: 107–115

Horton P and Lee P (1984) Phosphorylation of chloroplast thylakoids decreases the maximum capacity of photosystem II electron transfer. Biochim Biophys Acta 767: 563–567

Horton P and Lee P (1985) Phosphorylation of chloroplast membrane proteins partially protects against photoinhibition. Planta 165: 37–42

Horton P, Ruban AV, Rees D, Pascal AA , Noctor GD and Young AJ (1992) Control of the light-harvesting function of chloroplast membrane by aggregation of the LHCII chlorophyll protein complex. FEBS Let 292: 1–4

Horvath G, Droppa M and Melis A (1984) Herbicide action on photosystem II in spinach chloroplasts: Concentration effect on PS II α and PS II β. Photobiochem Photobiophys 7: 249–256

Hsu BD (1992) The active photosystem II centers can make a significant contribution to the initial fluorescence rise from F_0 to F_i. Plant Science 81: 169–174

Hsu BD (1993) Evidence for the contribution of the S-state transitions of oxygen evolution to the initial phase of fluorescence induction. Photosynth Res 36: 81–88

Hsu BD and Lee JY (1991a) A study on the fluorescence induction curve of the DCMU-poisoned chloroplast. Biochim Biophys Acta 1056: 285–292

Hsu BD and Lee JY (1991b) Characterization of the photosystem II centers inactive in plastoquinone reduction by fluorescence induction. Photosynth Res 27: 143–150

Hsu BD, Lee YS and Jang YR (1989) A method for analysis of fluorescence induction curve from DCMU-poisoned chloroplasts. Biochim Biophys Acta 975: 44–49

Jegerschold C, Virgin I and Styring S (1990) Light-dependent degradation of the D1 protein in photosystem II is accelerated after inhibition of the water splitting reaction. Biochemistry 29: 6179–6186

Joliot A (1974) Effect of low temperature (–30 to –60° C) on the reoxidation of the photosystem II primary acceptor in the presence and absence of DCMU. Biochim Biophys Acta 357: 439–448

Joliot A and Joliot P (1964) Etude cinétique de la réaction photochimique libérant l'oxygène au cours de la photosynthèse. CR Acad Sci Paris 258: 4622–4625

Joliot P (1965) Etudes simultanées des cinétiques de fluorescence et d'émission d'oxygène photosynthétique. Biochim Biophys Acta 102: 135–148

Joliot P and Joliot A (1977) Evidence for a double hit process in photosystem II based on fluorescence studies. Biochim Biophys Acta 462: 559–574

Joliot P and Joliot A (1979) Comparative study of the fluorescence yield and of the C550 absorption change at room temperature. Biochim. Biophys. Acta 546: 93–105

Joliot P and Joliot A (1981a) Characterization of photosystem II centers by polarographic, spectroscopic and fluorescence methods. In Akoyunoglou G (ed) Photosynthesis III. Structure and Molecular Organization of the Photosynthetic Apparatus, pp 885–899. Balaban International Science Services, Philadelphia PA

Joliot P and Joliot A (1981b) Double photoreactions induced by a laser flash as measured by oxygen emission. Biochim Biophys Acta 638: 132–140

Joliot P, Joliot A, Bouges B and Barbieri B (1971) Studies of system II photocenters by comparative measurements of luminescence, fluorescence, and oxygen emission. Photochem Photobiol 14: 287–305

Joliot P, Bennoun P and Joliot A (1973) New evidence supporting energy transfer between photosynthetic units. Biochim Biophys Acta 305: 317–328

Jursinic PA and Dennenberg RJ (1988) Enhanced oxygen yields caused by double turnovers of photosystem II induced by dichlorobenzoquinone. Biochim Biophys Acta 934: 177–185

Jursinic PA and Dennenberg RJ (1989) Measurement of stoichiometry of photosystem II to photosystem I reaction centers. Photosynth Res 21: 197–200

Keren N, Gong H and Ohad I (1995) Oscillation of reaction center II D1 protein degradation *in vivo* induced by repetitive light flashes: Correlation between the level of RC II Q_B^- and protein degradation in low light. J Biol Chem 270: 806–814

Kirilovsky DL, Vernotte C and Etienne AL (1990) Protection from photoinhibition by low temperature in *Synechocystis* 6714 and in *Chlamydomonas reinhardtii*: Detection of an intermediary state. Biochemistry 29: 8100–8106

Krause GH and Behrend U (1986) ΔpH-dependent chlorophyll fluorescence quenching indicates a mechanism of protection against photoinhibition of chloroplasts. FEBS Lett 200: 298–302

Krause GH and Weis E (1991) Chlorophyll fluorescence and photosynthesis: The basics. Annu Rev Plant Physiol Plant Mol Biol 42: 313–349

Krieger A and Weis E (1990) pH-dependent quenching of chlorophyll fluorescence in isolated PS II particles: Dependence on the redox potential. In: Baltscheffsky M (ed) Current Research in Photosynthesis, Vol VI, pp 563–566. Kluwer Academic Publishers, Dordrecht

Krieger A and Weis E (1993) The role of calcium in the pH dependent control of photosystem II. Photosynth Res 37: 117–130

Krieger A, Moya I and Weis E (1992) Energy-dependent quenching of chlorophyll a fluorescence : Effect of pH on stationary fluorescence and picosecond relaxation kinetics in thylakoid membranes and photosystem II preparations. Biochim Biophys Acta 1102: 167–176

Kyle DJ (1987) The biochemical basis for photoinhibition of photosystem II. In: Kyle DJ, Osmond CB and Artnzen CJ (eds), Photoinhibition, pp 197–226. Elsevier Publishers, Amsterdam

Lam E, Baltimore B, Ortiz W, Chollar S, Melis A and Malkin R (1983) Characterization of a resolved oxygen-evolving photosystem II preparation from spinach thylakoids. Biochim Biophys Acta 724: 201–211

Lavergne J (1982a) Two types of primary acceptors in chloroplasts photosystem II. I. Different recombination properties. Photobiochem Photobiophys 3: 257–271

Lavergne J (1982b) Two types of primary acceptors in chloroplasts photosystem II. II. Reduction in two successive photoacts. Photobiochem Photobiophys 3: 273–285

Lavergne (1982c) Mode of action of dichlorophenyldimethylurea. Evidence that the inhibitor competes with plastoquinone for binding to a common site on the acceptor side of photosystem II. Biochim Biophys Acta 682: 345–353

Lavergne J (1987) Optical difference spectra of the S-state transitions in the photosynthetic oxygen-evolving complex. Biochim Biophys Acta 894: 91–107

Lavergne J (1991) Improved UV-visible spectra of the S-transitions in the photosynthetic oxygen-evolving system. Biochim Biophys Acta 1060: 175–188

Lavergne J and Leci E (1993) Properties of inactive photosystem II centers. Photosynth Res 35: 323–343

Lavergne J and Trissl HW (1995) Theory of fluorescence induction in Photosystem II: Derivation of analytical expressions in a model including exciton radical pair equilibrium and restricted energy transfer between photosynthetic units. Biophys J 68: 2474–2492

Lavorel J and Etienne AL (1977) In vivo chlorophyll fluorescence. In Barber J (ed) Primary Processes of Photosynthesis, pp 203–268, Elsevier, Amsterdam

Le Gouallec JL, Cornic G and Briantais JM (1991) Chlorophyll fluorescence and photoinhibition in a tropical rainforest understory plant. Photosynth Res 27: 135–142

Lokstein H, Härtel H, Hoffmann P and Renger G (1993) Comparison of chlorophyll fluorescence quenching in leaves of wild type with a chlorophyll-*b*-less mutant of barley (*Hordeum vulgare* L.) J Photochem Photobiol B: Biol 19: 217–225

Mäenpää P, Andersson B and Sundby C (1987) Difference in sensitivity to photoinhibition between photosystem II in the appressed and non-appressed thylakoid regions. FEBS Lett 215: 31 36

Malkin S and Kok B (1966) Fluorescence induction studies in isolated chloroplasts. I Number of components involved in the reaction and quantum yields. Biochim Biophys Acta 126: 413–432

Mattoo AK, Hoffman-Falk H, Marder JB and Edelman M (1984) Regulation of protein metabolism : Coupling of photosynthetic electron transport to in vivo degradation of the rapidly metabolized 32 kilodalton protein of chloroplast membrane. Proc Natl Acad Sci USA 81: 1380–1384

Mattoo AK, Marder JB and Edelman M (1989) Dynamics of the photosystem II reaction center. Cell 56: 241–246

McCarthy S, Jursinic P and Stemler A (1988) Atrazine binding sites of photosystem II. Plant Physiol 86S: 46

Melis A (1985) Functional properties of photosystem IIβ in spinach chloroplasts. Biochim Biophys Acta 808: 334–342

Melis A (1991) Dynamics of photosynthetic membrane composition and function. Biochim Biophys Acta 1058: 87–106

Melis A and Anderson JM (1983) Structural and functional organization of the photosystem in spinach chloroplasts. Antenna size, relative electron-transport capacity, and chlorophyll composition. Biochim Biophys Acta 724: 473–484

Melis A and Duysens LNM (1979) Biphasic energy conversion kinetics and absorbance difference spectra of photosystem II of chloroplasts. Evidence for two different PS II reaction centers. Photochem Photobiol 29: 373–382

Melis A and Homann PH (1975) Kinetic analysis of the fluorescence induction in 3-(3,4-dichlorophenyl)-1,1-dimethylurea poisoned chloroplasts. Photochem Photobiol 21: 431–437

Melis A and Homann PH (1976) Heterogeneity of the photochemical centers in system II of chloroplasts. Photochem Photobiol 23: 343–350

Melis A and Ow RA (1982) Photoconversion kinetics of chloroplast photosystems I and II. Effect of Mg^{2+}. Biochim Biophys Acta 682: 1–10

Melis A and Schreiber U (1979) The kinetic relationship between the C-550 absorbance change, the reduction of Q (ΔA_{320}) and the variable fluorescence yield change in chloroplasts at room temperature. Biochim Biophys Acta 547: 47–57

Melis A and Thielen APMG (1980) The relative absorption cross-sections of photosystem I and photosystem II in chloroplasts from three types of *Nicotiana tabacum*. Biochim Biophys Acta 589: 275–286

Michel H, Hunt DF, Shabanowitz J and Bennett J (1988) Tandem mass spectrometry reveals that three photosystem II proteins of spinach chloroplasts contain N-acetyl-O-phosphothreonine at their NH_2 termini. J Biol Chem 263: 1123–1130

Murata N, Nishimura M and Takamiya A (1966) Fluorescence of chlorophyll in photosynthetic systems. II Induction of fluorescence in isolated spinach chloroplasts. Biochim Biophys Acta 120: 20–33

Nedbal L, Gibas C and Whitmarsh J (1991) Light saturation curves show competence of the water splitting complex in inactive photosystem II reaction centers. Photosynth Res 30: 85–94

Noctor G, Rees D, Young A and Horton P (1991) The relationship between zeaxanthin, energy-dependent quenching of chlorophyll fluorescence and the transthylakoid pH-gradient in isolated chloroplasts. Biochim Biophys Acta 1057: 320–330

Ögren E (1991) Prediction of photoinhibition of photosynthesis from measurements of fluorescence quenching components. Planta 184: 538–544

Ohad I, Koike H, Shochat S and Inoue Y (1988) Changes in the properties of reaction centers during the initial stages of photoinhibition as revealed by thermoluminescence measurements. Biochim Biophys Acta 993: 288–298

Ohad I, Keren N, Zer H, Gong H, Mor TS, Gal A, Tal S and Domovich Y (1994) Light-induced degradation of the photosystem II reaction centre D1 protein *in vivo*: An integrative approach. In: Baker NR and Bowyer JR (eds) Photoinhibition of Photosynthesis, From Molecular Mechanisms to the Field, pp 161–178. Environmental Plant Biology series, Davies (ed), Bios Scientific Publishers, Oxford

Olive J, Vallon O, Wollman F-A, Recouvreur M and Bennoun P (1986) Studies of the cytochrome b_6/f complex. II. Localization of the complex in the thylakoid membranes from spinach and *Chlamydomonas reinhardtii* by immunocytochemistry and freeze-fracture analysis of b_6-f mutants. Biochim Biophys Acta 851: 239–248

Oxborough K and Horton P (1987) Characterization of the effects of antimycin A upon the high energy state quenching of chlorophyll fluorescence (qE) in spinach and pea chloroplasts. Photosynth Res 12: 119–128

Percival MP, Webber AN and Baker NR (1984) Evidence for the role of the light-harvesting chlorophyll *a/b* protein complex in photosystem II heterogeneity. Biochim Biophys Acta 767: 582–589

Peter GF and Thornber JP (1991a) Biochemical evidence that the higher plant photosystem II core complex is organized as a dimer. Plant Cell Physiol 32: 1237–1250

Peter GF and Thornber JP (1991b) Biochemical composition and organization of higher plant photosystem II light harvesting pigment proteins. J Biol Chem 266: 16745–16754

Ramm D and Hansen UP (1993) Can charge recombination as caused by pH dependent donor side limitation in PS2 account for high-energy state quenching? Photosynth Res 35: 97–100

Rees D, Noctor GD and Horton P (1990) The effect of high-energy-state excitation quenching on maximum and dark level chlorophyll fluorescence yield. Photosynth Res 25: 199–211

Rees D, Noctor GD, Ruban AV, Crofts J, Young AJ and Horton P (1992) pH dependent chlorophyll fluorescence quenching in spinach thylakoids from light treated or dark adapted leaves. Photosynth Res 31: 11–19

Robinson HH and Crofts AR (1983) Kinetics of the oxidation-reduction reactions of the photosystem II quinone acceptor complex, and the pathway for deactivation. FEBS Lett 151: 221–226

Robinson HH and Crofts AR (1984) Kinetics of proton uptake and the oxidation-reduction reactions of quinone acceptor complex of PS II from pea chloroplasts. In: Sybesma C (ed) Advances in Photosynthesis Research, Vol 1, pp 477–480. Martinus Nijhoff/Dr W Junk Publishers, The Hague

Roelofs TA, Lee CH and Holzwarth AR (1992) Global target analysis of picosecond chlorophyll fluorescence kinetics from pea chloroplasts. A new approach to the characterization of the primary process in photosystem II α- and β-units. Biophys J 61: 1147–1163

Ruban AV and Horton P (1992) Mechanism of ΔpH-dependent dissipation of absorbed excitation energy by photosynthetic membranes. I—Spectroscopic analysis of isolated light-harvesting complexes. Biochim Biophys Acta 1102: 30–38

Ruban AV and Horton P (1995) An investigation of the sustained component of nonphotochemical quenching of chlorophyll fluorescence in isolated chloroplasts and leaves of spinach. Plant Physiol 108: 721–726

Ruban AV, Young AJ and Horton P (1993) Induction of non-photochemical energy dissipation and absorbance changes in leaves. Plant Physiol 102: 741–750

Schreiber U and Neubauer C (1987) The polyphasic rise of chlorophyll fluorescence upon onset of strong continuous illumination : Partial control by the photosystem II donor side and possible ways of interpretation. Z Naturforsch 42C: 1255–1264

Schreiber U and Neubauer C (1990) O_2-dependent electron flow, membrane energization and the mechanism of non-photochemical quenching of chlorophyll fluorescence. Photosynth Res 25: 279–293

Schreiber U and Pfister K (1982) Kinetic analysis of the light-induced chlorophyll fluorescence rise curve in the presence of dichlorophenyldimethylurea. Dependence of the slow-rise component on the degree of chloroplast intactness. Biochim Biophys Acta 680: 60–68

Schreiber U, Schliwa U and Bilger W (1986) Continuous recording of photochemical and non photochemical chlorophyll fluorescence quenching with a new type of modulation fluorimeter. Photosynth Res 10: 51–62

Sinclair J and Spence SM (1990) Heterogeneous photosystem 2 activity in isolated spinach chloroplasts. Photosynth Res 24: 209–220

Somersalo S and Krause GH (1989) Photoinhibition at chilling temperature, fluorescence characteristics of unharded and cold acclimated spinach leaves. Planta 177: 409–416

Styring S, Virgin I, Ehrenberg A and Andersson B (1990) Strong light photoinhibition of electron transport in photosystem II. Impairment of the function of the first quinone acceptor, Q_A. Biochim Biophys Acta 1015: 269–278

Thayer SS and Björkman O (1992) Carotenoid distribution and deepoxidation in thylakoids pigment-protein complexes from cotton leaves and bundle sheath cells of maize. Photosynth Res 33: 213–226

Theg SM, Filar LJ and Dilley RA (1986) Photoinactivation of chloroplasts already inhibited on the oxidizing side of photosystem II. Biochim Biophys Acta 849: 104–111

Thielen AMPG and van Gorkom HJ (1981a) Energy transfer and quantum yield in photosystem II. Biochim Biophys Acta 637: 439–446

Thielen AMPG and van Gorkom HJ (1981b) Redox potentials of electron acceptors in photosystems II α and II β. FEBS Lett 129: 205–209

Thielen AMPG, van Gorkom HJ and Rijgersberg CP (1981) Chlorophyll composition of photosystems IIα, IIβ and I in tobacco chloroplasts. Biochim Biophys Acta 635: 121–123

Vallon O, Wollman FA and Olive J (1985) Distribution of intrinsic and extrinsic subunits of the PS II protein complex between appressed and non-appressed regions of the thylakoid membrane: An immunocytochemical study. FEBS Lett 183: 245–250

Vallon O, Hoyer-Hansen G and Simpson DJ (1987) Photosystem II and cytochrome *b*-559 in the stroma lamellae of barley chloroplasts. Carlsberg Res Com 52: 405–421

Vallon O, Bulté L, Dainese P, Olive J, Bassi R and Wollman F-A (1991) Lateral redistribution of cytochrome b_6-f complexes along thylakoid membranes upon state transitions. Proc Natl Acad Sci 88: 8262–8266

Van Wijk KJ and Van Hasselt PR (1990) The quantum efficiency of photosystem II and its relation to non-photochemical quenching of chlorophyll fluorescence: The effect of measuring and growth temperature. Photosynth Res 25: 233–240

Van Wijk KJ and Van Hasselt PR (1993) Photoinhibition of photosystem II in vivo is preceded by down-regulation through light-induced acidification of the lumen: Consequence for the mechanism of photoinhibition in vivo. Planta 189: 359–368

Vass I, Gatzen G and Holzwarth AR (1993) Picosecond time-resolved fluorescence studies on photoinhibition and double reduction of Q_A in photosystem II. Biochim Biophys Acta 1183: 388–396

Vernotte C, Etienne AL and Briantais JM (1979) Quenching of the system II chlorophyll fluorescence by the plastoquinone pool. Biochim Biophys Acta 545: 519–527

Walter RG and Horton P (1991) Resolution of components of non-photochemical chlorophyll fluorescence quenching in barley leaves. Photosynth Res 27: 121–133

Walter RG and Horton P (1993) Theoretical assessment of

alternative mechanisms for non-photochemical quenching of PS II fluorescence in barley leaves. Photosynth Res 36: 119–139

Weis E and Berry J (1987) Quantum efficiency of PS II in relation to energy dependent quenching of chlorophyll fluorescence. Biochim Biophys Acta 283: 259–267

Weis E and Lechtenberg D (1989) Fluorescence analysis during steady state photosynthesis. Philos Trans R Soc London Biol Sci 233: 253–268

Wettern M (1986) Localization of the 32,000 Dalton chloroplast protein pools in thylakoids: Significance in atrazine binding. Plant Science 43: 173–177

Wollenberger L, Stefansson H, Yu SG and Albertsson Pè (1994) Isolation and characterization of vesicles originating from the chloroplast grana margins. Biochim Biophys Acta 1184: 93–102

Wollman FA (1978) Determination and modification of the redox state of the secondary acceptor of photosystem II in the dark. Biochim Biophys Acta 503: 263–273

Yu SG, Bjorn G and Albertsson Pè (1993) Characterization of a non-detergent PS II-cytochrome *b/f* preparation (BS) Photosynth Res 37: 227–236

Chapter 15

Introduction to Photosystem I: Reaction Center Function, Composition and Structure

Rachel Nechushtai, Amir Eden, Yuval Cohen and Judith Klein
Botany Department, The Hebrew University of Jerusalem, Jerusalem 91904, Israel

Donald R. Ort and Charles F. Yocum (eds): Oxygenic Photosynthesis: The Light Reactions, pp. 289–311.

Summary

Photosystem I (PS I), which is situated mainly in the non-stacked, stromal lamellae regions of the thylakoids, functions as a plastocyanin (PC)-ferredoxin (Fd) oxidoreductase. In eukaryotic photosynthetic organisms (algae and plants) the PS I complex is composed of a light-harvesting complex (LHCI) and a core component (CCI). While the LHCI serves as an accessory antenna to harvest light and funnel its energy to the CCI, it is the latter which contains the reaction center chlorophyll (P700), the photosynthetic pigments and the five different electron carriers (A_0-A_4) that perform the electron transfer process occurring in PS I.

The present chapter describes the evolution of the Photosystem I reaction center (PS I-RC) from green sulfur bacteria to higher plants. It summarizes the current knowledge of the composition of PS I-RC in different photosynthetic organisms, the functions of its different subunits and their association with different cofactors. In addition to the characterization of individual components, information on the overall three dimensional structure of the entire PS I-RC complex and its different compartments, derived from electron microscopy and X-ray crystallography, is presented.

The current resolution of the structural data permit neither the accurate assignment of the relations between all the different protein subunits within the PS I complex, nor their orientation in the thylakoid membranes. Hence, recent topological studies, tryptic treatments, epitope mapping and cross-linking experiments revealing the location and organization of different subunits of PS I, with respect to one another and to the thylakoids, are described.

Since we believe that a different insight into the understanding of the PS I-RC organization can be obtained from biogenesis and assembly studies, special attention is given to a description of the available information on the multi-step process of PS I assembly.

I. Introduction

A. Photosystem I (PS I) in Thylakoid Membranes—General Background

Photosystem I (PS I) is one of the four multisubunit complexes of the thylakoid membranes that interact in a concerted manner to carry out the photosynthetic electron transfer from water to $NADP^+$. The PS I complex, situated mainly in the non-stacked regions of the thylakoids (the stromal lamellae), functions at the reducing end of the electron transfer chain as a plastocyanin (PC) - ferredoxin (Fd) oxidoreductase (Scheller and Moller, 1990; Chitnis and Nelson, 1991; Golbeck and Bryant, 1991; Golbeck, 1993).

In eukaryotic photosynthetic organisms, algae and plants, the PS I complex is composed of a light-harvesting complex (LHCI) and a core component (CCI) (Thornber, 1986). The CCI and LHCI complexes together form the holo-PS I. The LHCI serves as an accessory antenna to harvest light and funnel its energy to the reaction center, P700, located in CCI. The CCI complex contains, in addition to P700, the photosynthetic pigments and all the electron carriers needed to carry out the electron transfer occurring in PS I (Scheller and Moller, 1990; Chitnis and Nelson, 1991; Golbeck and Bryant, 1991; Golbeck, 1993).

The holo PS I complex was found to contain 15–20 polypeptide subunits, 100–200 chlorophyll *a* molecules per P700 and was found to have a Chl *a/b* ratio greater than 5 (Mullet et al., 1980; Nechushtai et al., 1985; Chitnis and Nelson, 1991). A schematic representation of the holo-PS I complex is given in Fig. 1.

B. Electron Transfer in PS I—from Plastocyanin (PC) to Ferredoxin (Fd)

Electron transfer from PC, a copper protein situated on the lumenal side of the thylakoid membrane, to the stromal-located iron-sulfur protein, Fd, is mediated by the PS I core complex. The primary electron donor of the CCI complex, P700, was first

Abbreviations: CAB – chlorophyll *a/b* binding protein; CCI – core complex I; Chl – chlorophyll; EDC – N-ethyl-3-(3-dimethylaminopropyl) carbodiimide; EPR – electron paramagnetic resonance; EXAFS – extended X-ray absorption fine structure; Fd – ferredoxin; FNR – ferredoxin NADP reductase; LHC – light-harvesting complex; NHS-biotin – biotin N-hydrosuccinimide ester; orf – open reading frame; PC – plastocyanin; PsaA-N – the polypeptide subunits of PS I; *psaA-N* – the corresponding genes encoding the PS I subunits; PS I – Photosystem I; PS I-RC – Photosystem I reaction center; RC – reaction center

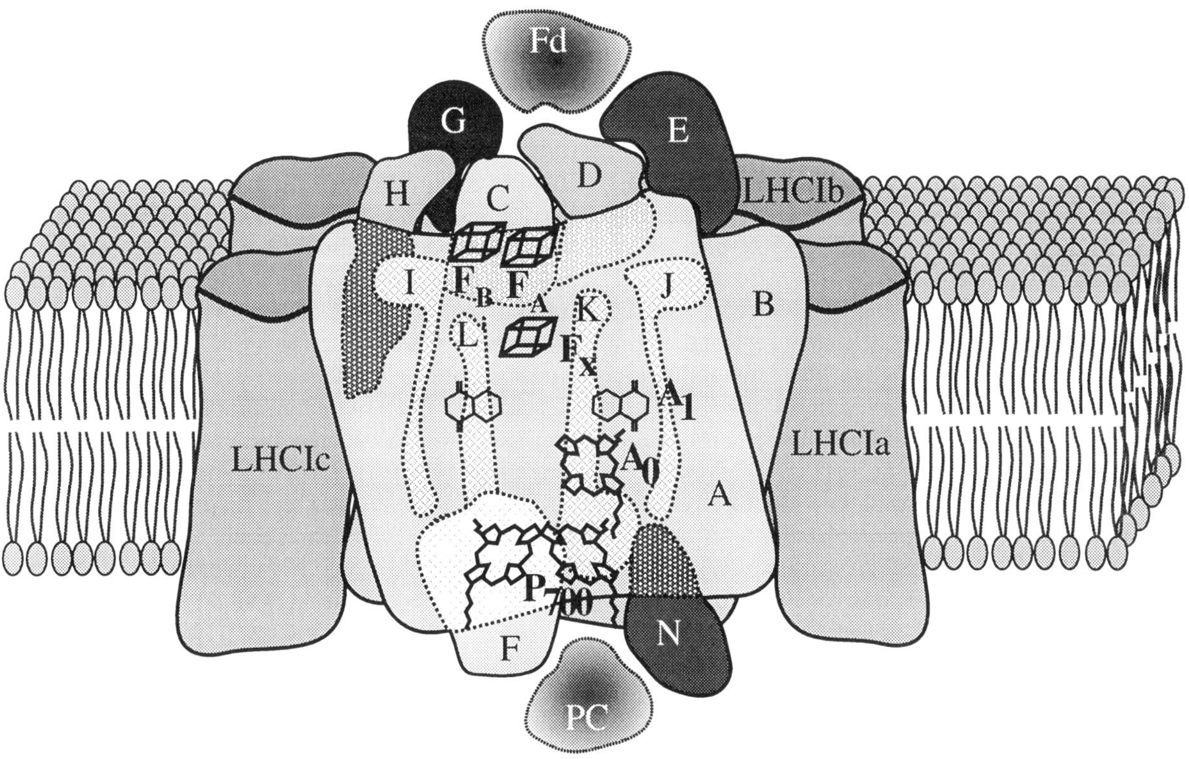

Fig. 1. A schematic representation of the photosystem I complex in the thylakoid membrane. The letters represent the different subunits according to the nomenclature given in Table 1. The different electron carriers of the complex are schematically illustrated.

identified by B. Kok to be a Chl *a* molecule (Kok, 1957,1961). The recently published structure of PS I indicates that P700 is a Chl *a* dimer (Krauss et al., 1993). Upon absorption of a photon, charge separation occurs and an electron is transferred to a primary acceptor A_0, a Chl *a* monomer (Furrer and Thurnauer, 1983; Brettel and Sétif, 1987) from which it proceeds to the intermediate acceptor A_1, which is a phylloquinone (vitamin K_1) (Furrer and Thurnauer, 1983; Brettel et al., 1986; Iwaki et al., 1992). From A_1 the electron is transferred to the intermediate acceptor A_2 (known also as F_x). Mössbauer spectroscopy, EXAFS and biochemical analysis of the iron and sulfur content of PS I, as well as the recent three dimensional structure of the complex indicate that A_2 is a [4Fe-4S] cluster. Similarly, the subsequent electron acceptors A_3 and A_4 (F_A and F_B, respectively) are [4Fe-4S] clusters (Cammack and Evans, 1975; Nugent et al., 1981; Golbeck, 1987). Because they have similar redox potentials, A_3 and A_4 were suggested to function in parallel rather than in a sequential manner (Malkin, 1982). Although the findings of different mutagenesis studies support this view (Zhao et al., 1992), the recent structural observations strongly favor sequential electron transfer (Guigliarelli et al., 1993; Krauss et al., 1993). A more detailed description of the cofactors, the kinetics and the overall electron transfer process occurring in PS I, is given in Chapters 16 and 18.

C. Evolution of the PS I Complex, from Green Sulfur Bacteria to Higher Plants

Knowledge collected in recent years suggests that the PS I complex has evolved from the reaction center of green sulfur bacteria (Nitschke et al., 1990; Nitschke and Rutherford, 1991). When *Chlorobium limicola* membranes were studied by EPR, three signals were detected. These were attributed to iron-sulfur centers acting as electron acceptors in the bacterial reaction center. Based on their spectroscopic similarities to signals in the PS I reaction center (PS I-RC), the three signals of *Chlorobium* were designated as F_B, F_A and F_X. Moreover, the orientation dependence of these EPR signals and the magnetic interaction between the reduced forms of F_B and F_A were found to be remarkably similar to those in PS I (Broadhurst et al., 1986; Nitschke et al., 1990).

Further evidence to support the described similarity has come from the isolation of genes encoding the reaction center of green sulfur bacteria (Buttner et al., 1992a,b; Liebl et al., 1993). The genes encoding the reaction center of the anoxygenic photosynthetic bacterium *Heliobacillus mobilis* (Liebl et al., 1993) showed a high percentage of sequence identity to *psaA* and *psaB*, the genes encoding the large subunits of PS I, in a cysteine-containing loop, which is the putative binding site of the iron-sulfur center F_X, and in a preceding hydrophobic region (Liebl et al., 1993). The putative binding elements for the primary donor, P840, in *Chlorobium limicola* showed similarities to the P700-binding region of PS I. The same holds true for the regions which bind the PS I acceptors A_0 and A_1 and their analogs in *Chlorobium* (Buttner et al., 1992a,b). Since, in both of these green sulfur bacteria, efforts to find a second, similar, gene coding another reaction center protein failed, a homodimeric organization of the reaction center in these bacteria was suggested. In this respect, the green sulfur bacterial reaction center differs from the PS I-RC of cyanobacteria, algae and plants and from any other reaction center.

Recently a gene encoding a 22 kDa protein was isolated from *Chlorobium limicola* and was proposed to be homologous to *psaC* and to contain the homologues of the F_A and F_B centers (Buttner et al., 1992a; Illinger et al., 1993). The latter indicated that all the electron carriers found in PS I exist in the ancestor reaction centers of green sulfur bacteria. Probably, upon evolution to cyanobacteria additional polypeptide subunits were evolved as part of the PS I-RC (Golbeck, 1994). Most of these were identified as structural accessory subunits, some of which facilitate the electron transfer process occurring in PS I, but do not take an active role in it. The majority of these subunits are highly conserved throughout evolution from cyanobacteria to higher plants, especially PsaA, PsaB and PsaC which together contain all the electron transfer components of the complex (Bryant, 1986). Upon the transition from prokaryotes to eukaryotes, some of the genes encoding the accessory subunits were transferred to the cell nucleus (still preserving a high degree of homology) and the light-harvesting complexes, LHCI, appeared .

II. The Composition and Function of PS I Components

A. The Light-Harvesting Complex, LHCI

Mullet et al. (1980) in their seminal work showed that PS I of pea is associated with an accessory antenna that contains Chl *b*, which they termed LHCI. LHCI is a minor pigment-binding component, which binds 5–10% of the thylakoid membrane pigment content, in contrast to LHCII which is the most abundant chlorophyll-protein complex and contains 40–50% of the thylakoid's pigments. LHCI complexes were isolated and characterized from a variety of different plants and algae, e.g., pea, spinach (Mullet et al., 1980; Malkin, 1982; Lam et al., 1984) *Lemna gibba* (Nechushtai et al., 1987), *Vicia faba*, barley (Machold et al., 1979), maize (Vainstein et al., 1989) cucumber (Iwasaki et al., 1991) and *Chlamydomonas reinhardtii* (Bassi et al., 1992). The common features found for LHCI complexes are: i) Their Chl *a*/Chl *b* ratio is greater than 3.5; ii) They contain polypeptide subunits with apparent molecular weights of 17–26 kDa; and iii) Their fluorescence emission occurs at a longer wavelength ($\lambda > 700$ nm) than that of LHCII (λ about 680 nm). Bassi et al. (1992) claimed that the long wavelength fluorescence emission typical of LHCI, correlates with changes in the aggregation state of its polypeptide components rather than with the presence of a specific protein.

Vainstein et al. (1989) isolated and characterized three components of LHCI from maize, and designated them as LHCIa, LHCIb, LHCIc. The recent characterization of LHCI of barley (Knoetzel et al., 1992) provided a new nomenclature based on the low-temperature fluorescence emission of each sub-LHCI complex; the different LHCI complexes were designated as LHCI-730, LHCI-680A and LHCI-680B. The LHCI-730 complex was shown to be composed of 21 kDa (encoded by LHCI-type IV gene) and 22 kDa (encoded by LHCI-type I) polypeptides. Each of the LHCI-680 complexes had a single apoprotein. LHCI-680A consisted of a 25 kDa polypeptide (encoded by LHCI-type III) and LHCI-680B of a 23 kDa polypeptide (encoded by LHCI-type II). LHCI-680B was associated with the non-pigmented PsaE subunit of CCI. This may indicate that PsaE functions in binding of this antenna to the reaction center (Knoetzel et al., 1992).

All apoproteins of LHCI are encoded by genes located in the cell nucleus. Their primary structure was deduced using recombinant DNA techniques (Thornber et al., 1991). The major studies on LHCI genes were performed in tomato from which the different genes encoding the LHCI proteins were isolated and characterized (Schwartz et al., 1991; Green and Pichersky, 1994).

A cDNA encoding the precursor to a major 20 kDa thylakoid polypeptide of *Chlamydomonas reinhardtii* (P22), previously assigned as a LHCI component, was characterized in a recent study (Hwang and Herrin, 1993). The primary sequence and predicted topology of P22 also exhibited features characteristic of light-harvesting chlorophyll *a/b*-binding proteins from higher plants. Sequence comparisons indicate that P22 has significantly greater identity with the Type-I LHCI protein of tomato than any other LHC protein. This result suggests that the divergence of LHC proteins into classes found in higher plants may have occurred early in evolution, prior to the separation of green algae and plants.

Comparison of the amino acid sequences of the polypeptides LHCI, LHCII, CP29 and CP24 confirms that all CAB proteins share three regions of very high similarity: the first and third transmembrane helices and the stroma-exposed region preceding them (Chitnis and Thornber, 1988; Thornber et al., 1991; Green and Pichersky, 1994). However, the regions between these conserved motifs as well as the N-terminus of LHCI apoprotein appear to be PS I-specific (Schwartz et al., 1991). Additional information on CAB proteins may be found in chapters 27–29.

B. The Core Complex (CCI), the Photosystem I Reaction Center (PS I-RC)

The core complex of PS I (CCI), often referred to as the PS I reaction center (PS I-RC), is the component of PS I where primary photochemical charge separation and electron transfer processes occur. The first active PS I-RC complex capable of photo-oxidizing PC and photoreducing Fd and $NADP^+$ was isolated from Swiss chard by Bengis and Nelson (1975). This complex was initially reported to contain six polypeptide subunits, 40–65 Chl *a* and 1–2 β-carotene molecules. In subsequent studies, other procedures yielded similar PS I-RC complexes with similar activity (Mullet et al., 1980; Nechushtai et al., 1987; Chitnis and Nelson, 1991). However, as a result of the use of modified polyacrylamide gel composition, additional polypeptide subunits were identified as an integral part of CCI. Currently, it is believed that PS I-RC is comprised of at least ten different polypeptides, approximately 100 Chl *a* molecules, several β-carotenes, a pair of vitamin K_1 molecules and three [4Fe-4S] clusters (Malkin, 1987; Golbeck, 1987; Golbeck and Bryant, 1991; Scheller and Moller, 1990; Chitnis and Nelson, 1991). In the following we describe the current knowledge of the different PS I-RC subunits and their association with the different PS I cofactors which carry out electron transfer in the complex. The discussion is divided according to the three functional domains of the PS I-RC complex: i) the *integral core* of the complex; ii) the *reducing* and iii) the *oxidizing parts* of the PS I-RC.

Since detailed information on the subunits of PS I-RC has been reviewed (Scheller and Moller, 1990; Chitnis and Nelson, 1991; Golbeck and Bryant, 1991; Golbeck, 1994), only a brief description of the different subunits with emphasis on their functional role is given. Greater attention is given to more recent studies.

1. The Integral Membrane Part of the PS I-RC

The integral, membrane embedded core of the PS I-RC is composed of two major components: i) the two core subunits, PsaA and PsaB, which contain all the pigments of the complex and most of its electron carriers; and ii) several low molecular weight subunits that contain transmembranal segments.

a. The Core Reaction Center Subunits (PsaA and PsaB)

PsaA and PsaB, the two high molecular weight subunits that are very similar in their apparent molecular weight, ranging between 58 to 70 kDa, form the membranal core of PS I-RC. The genes encoding these apoproteins, located in the chloroplast genome, have been cloned and sequenced from prokaryotic and eukaryotic organisms (Bryant, 1991). Their deduced amino acid sequences revealed the following structural features: i) PsaA and PsaB contain 11 putative transmembranal helices, each 19–25 residues long; ii) the majority of charged residues are present in the inter helical domain; iii) charged residues are mainly present in the extra-membranal loops; and iv) a high number of histidine residues was found in the transmembranal helices (Fish et al., 1985a,b). These histidine residues were proposed to be involved in coordinating the chlorophyll molecules associated with PS I-RC (Kirsch et al., 1986). The cysteine residues in PsaA and PsaB are also of special interest, since they were proposed to provide the ligands to the bound iron-sulfur cluster F_x. These residues along with those speculated to be involved in the Leucine zipper motif suggested to be present in PsaA and PsaB (Baker and Malkin, 1990; Kossel et al., 1990; Webber and Malkin, 1990), have become a preferred target for site-directed mutagenesis (Smart and McIntosh, 1993; Webber et al., 1993).

The first targeted inactivation of PsaA was reported by Smart et al. (1991). A kanamycin resistance gene was introduced into the unicellular cyanobacterium *Synechocystis* sp. PCC 6803 to disrupt the native *psaA* gene. This successful transformation paved the way for the use of the cyanobacterium for site-directed mutagenesis in *psaA* of the PS I core. The same research group also reported the inactivation of *psaB*, which resulted in disruption of stable PS I assembly, while PS II assembled normally in these cells. Expression of the *psaA* gene was not affected by *psaB* inactivation, but the PsaA protein was not detected, indicating that a stable PsaA homodimer could not form. The ability to inactivate *psaB* made its site-directed mutagenesis feasible (Smart and McIntosh, 1993). Indeed, a site-directed conversion of cysteine-565 to serine (C565S) in PsaB of *Synechocystis* sp. PCC 6803 was performed. The latter did not affect the accumulation of PS I-RC subunits or the low-temperature photoreduction of the terminal electron acceptors F_A and F_B. The PS I-RC isolated from this C565S mutant exhibited, according to EPR analysis, a [3Fe-4S] cluster, identified as having photoreduction characteristics similar to F_x (Warren et al., 1993).

The *psaB* gene was also investigated in the green alga *Chlamydomonas reinhardtii* . A chloroplast PS I mutation was complemented with a wild type *psaB* gene to restore photosynthetic competence. The mutation was found to be a single base pair deletion resulting in a reading frame shift and premature termination of the polypeptide. The identification of the mutation proved to be useful for in vivo studies of site-directed mutations in the *C. reinhardtii psaB* gene (Bingham et al., 1991). This system has been used to introduce specific amino acid changes into PsaB of the *C. reinhardtii* PS I-RC. Plasmids containing mutated copies of the *psaB* gene were introduced into the chloroplast genome. When the highly conserved proline-cysteine motif suggested to be important in coordinating the [4Fe-4S] iron-sulfur center F_x was mutated, the cysteine was found to be essential for assembly of the PS I-RC but no identifiable function was found for the adjacent proline (Webber et al., 1993).

b. The Low Molecular Weight Integral Membrane Subunits of PS I-RC

Additional low molecular weight subunits were detected in PS I-RC preparations. Smith et al. (1991) were the first to report the primary structure of a very low molecular weight integral membrane protein associated with PS I-RC of plants. The nucleotide sequence of the *orf40* gene of pea chloroplast DNA encoded a hydrophobic, membrane-spanning polypeptide identified as PsaI, a 4 kDa subunit of PS I-RC.

Ikeuchi and coworkers (Ikeuchi et al., 1990,1991; Ikeuchi and Inoue, 1991) used high resolution gel electrophoresis to resolve nine of the eleven subunits of *Anabaena variabilis* PS I-RC. Four of these proteins had molecular masses of 6.8, 5.2, 4.8 and 3.5 kDa. The 3.5 kDa subunit corresponded to PsaI of higher plants. The 6.8 kDa and 4.8 kDa proteins were identified as gene products of *psaK* and *psaJ*, respectively, and the 5.2 kDa protein was homologous to a 4.8 kDa subunit of PS I of the thermophilic cyanobacterium *Synechococcus vulcanus*, suggesting that this protein is a component of PS I in cyanobacteria. When the PsaI protein was cloned from the cyanobacterium *A. variabilis*, the nucleotide sequence encoded a 46 amino acid polypeptide (Sonoike et al., 1992). Since the 11 N-terminal residues were absent in the mature PsaI polypeptide, it was concluded that they constituted a cleavable presequence of the PsaI protein. The authors suggested that this presequence directs the protein to the thylakoid lumen. The presence of low molecular weight subunits in *A. variabilis* PS I-RC was confirmed by utilizing sodium dodecyl sulfate-urea-polyacrylamide gel electrophoresis and N-terminal amino acid sequencing of the fractionated proteins (Nyhus et al., 1992). Similar results were found for the cyanobacterium *Synechococcus* sp. (Muhlenhoff et al., 1993) and for cucumber PS I-RC complexes (Iwasaki et al., 1991).

Another low molecular weight protein whose gene was cloned and characterized was PsaK from barley (Kjaerulff et al., 1993), *Chlamydomonas reinhardtii*

(Franzen et al., 1989a) and spinach (Tittgen et al., 1986). The protein is encoded by a nuclear gene and is synthesized as a precursor that is post-translationally imported into the chloroplast. Analysis of its primary sequence indicates that PsaK has two hydrophobic regions predicted to be membrane-spanning alpha-helices.

An interesting suggestion arose from the comparison of PsaK and PsaG. The two proteins in barley have significant similarity to each other. Moreover when the barley PsaK and PsaG were compared with the reported PsaK sequence from *Synechococcus vulcanus*, the degree of similarity to both of the barley genes was equal, suggesting that the ancestral gene has been duplicated in a chloroplast, but not in a cyanobacterial progenitor (Kjaerulff et al., 1993).

The specific function of PsaI, PsaJ or PsaK is unknown since no mutants of these subunits are available. However based on the low resolution PS I structure, it was suggested that PsaI/J/K form the connecting domains between three monomers to form the trimeric PS I-RC (Krauss et al., 1993).

The recently discovered role for PsaL may, however, contradict this suggestion. The PsaL subunit, first described in barley, was proposed to have a molecular mass of 22.2 kDa (209 amino acids) as a precursor and 18 kDa as the mature form (Okkels et al., 1991). In spinach the *psaL* gene encodes a precursor polypeptide of 24 kDa (216 residues) and a mature protein of about 18.8 kDa (169 residues). Hydropathy analysis suggests that the polypeptide contains two transmembrane segments (Flieger et al., 1993). In the cyanobacterium *Synechocystis* sp. PCC 6803 the *psaL* gene was found to encode a protein of 16.6 kDa. The deduced amino acid sequence is homologous to PsaL of barley (Chitnis et al., 1993). A PsaL-less mutant was segregated, and the stable mutant strains exhibited photoautotrophic growth similar to the wild type. However, the PS I-RC isolated from this mutant was in the monomeric form only (Chitnis et al., 1993). This result implies that it is the PsaL subunit which is responsible for the formation of trimers of isolated PS I-RC in cyanobacteria (Chitnis and Chitnis, 1993; Chitnis et al., 1993).

2. The Peripheral Subunits of the PS I-RC

In addition to the hydrophobic integral membrane subunits of PS I-RC, two peripheral domains, a stromal-facing one and a lumenal-facing one, are present in the PS I-RC complex. These domains also represent the reducing and oxidizing sites of the complex.

a. The Reducing-Stromal Side of PS I-RC

Three subunits, PsaC, PsaD, PsaE, have been shown to be involved in the organization and stabilization of the reducing site of PS I-RC, as well as to contain the components involved in electron transfer to Fd and FNR.

PsaC has been identified as the subunit that binds the two final electron acceptors of PS I, F_A and F_B (Oh-oka et al., 1987). It was found to be a highly conserved chloroplast-encoded protein of 81 amino acids. PsaC was shown to contain nine cysteine residues, eight of which are involved in binding the two [4Fe-4S] clusters F_A and F_B (Golbeck, 1987; Golbeck and Bryant, 1991) Several models have been proposed for the structure of PsaC based on the similarity to the 2[4Fe-4S] bacterial ferredoxins (Oh-oka et al., 1987; Dunn and Gray, 1988), whose structure was solved by X-ray crystallography. Some of the cysteines binding the different clusters were altered by site-directed mutagenesis and the effect on electron transfer was studied. Following overproduction in *E. coli* of the mutated PsaC proteins, denoted C14D and C51D, the iron-sulfur clusters were inserted in vitro, and the reconstituted proteins were rebound to the P700-F_X core of *Synechococcus* sp. PCC 6301. Based on the photoreduction of F_A and F_B it was concluded that each cluster can be reduced independently by F_x (Zhao et al., 1992). These results suggest a parallel electron flow from F_x to either F_A or F_B. On the other hand, structural evidence obtained using two different methods favored a sequential, rather than parallel, electron flow. The first was the study of the structural properties of the iron-sulfur centers of PS I-RC of *Synechocystis* 6803 (Guigliarelli et al., 1993). This study indicated that the F_A-F_B direction is close to the membrane normal and therefore supported a sequential electron transfer mechanism. A similar conclusion can be drawn from the low resolution structure of PS I elucidated in a crystallographic study (Krauss et al., 1993). In this structure different distances from F_x were measured for the two F_A/F_B centers. Although both distances were in a range permitting electron transfer, the fact that one of the clusters was found to be 7Å closer to F_x strongly supports a sequential electron transfer from F_x to the closest cluster and then to the furthest cluster.

From $F_{A/B}$, electrons proceed to soluble Fd. However, the interaction between Fd and PS I is still poorly understood. The fact that Fd accepts electrons from F_A/F_B present in PsaC implies that the two proteins should come in contact with each other. However, the physical association between PsaC and Fd has not yet been demonstrated. On the other hand several studies have demonstrated that the zero-length cross-linker EDC stabilizes the association of Fd with PsaD (Zanetti and Merati, 1987; Zilber and Malkin, 1988; Wynn et al., 1989). Moreover, a *Synechocystis* sp. PCC 6803 mutant, in which the gene *psaD* was specifically inactivated, was shown to require glucose for normal growth (Chitnis et al., 1989b). In this mutant it appeared that, in the absence of PsaD, PS I lost the ability to donate electrons to Fd. This phenotype is consistent with the proposal that PsaD acts as a docking site for Fd. However, the possibility that the PS I dysfunction arises from a disturbance in the proper assembly of the other subunits of the reducing side cannot be ruled out. Previously PsaD has been suggested to serve as a template for the assembly of the entire PS I-RC (Nechushtai and Nelson, 1985; Chitnis et al., 1989b). It may very well be that in the absence of PsaD, the other components of the reducing side are not properly organized.

Information of great value on the assembly and function of the reducing site was obtained with the reconstitution system developed by Golbeck and Bryant (1991). In this system it was shown that reconstitution of the reducing site is absolutely dependent on reinsertion of the iron-sulfur clusters into the PsaC apoprotein and on the presence of the PsaD protein (Zhao et al., 1990). The [Fe-S] clusters in PsaC are necessary for the rebinding of subunits PsaC, PsaD and PsaE to the core subunits PsaA and B (Li et al., 1991a). Absence of PsaD affected the reconstitution of PsaC (Li et al., 1991b).

Results in vivo also show the importance of PsaC for the creation of a stable PS I complex: insertional inactivation of the *psaC* gene in *Chlamydomonas reinhardtii* caused a faster turnover of the PS I complex (Takahashi et al., 1991). On the other hand, P700 oxidation measured in a PsaC deficient *Anabaena variabilis* strain indicated stable assembly of PS I without PsaC (Mannan et al., 1991). One should bear in mind however that in a different cyanobacterium, *Synechocystis* sp. PCC 6803, two genes encoding PsaC were identified (Steinmuller, 1992).

Another subunit suggested to play a role in the reducing site is PsaE. Reconstitution of PsaE to the PS I complex induced the restoration of electron transport. Hence, PsaE was proposed to have a role in stabilizing the interaction between PsaC and the PS I core allowing electron flow from F_X to F_A/F_B (Weber and Strotmann, 1993).

Involvement of PsaE in the binding of Fd to PS I was also suggested (Rousseau et al., 1993). It was previously shown that a PsaE deleted mutant can grow autotrophically (Chitnis et al., 1989a). However, when PS I particles were isolated from this mutant and direct photoreduction of Fd was investigated by flash absorption spectroscopy, it was found that, relative to the wild type, PS I lacking PsaE exhibited a decreased rate of Fd reduction by a factor of at least 25. After reassociation of the purified PsaE polypeptide to the mutated PS I, the original rate of electron transfer was recovered (Chitnis et al., 1989a).

Topological studies and site-directed mutagenesis indicate that the N- and C-terminus of PsaE are required for the correct integration of PsaE into PS I. Mutations in these regions lead to phenotypes identical to mutants in which the entire PsaE protein was deleted. When modifications were made in the predicted stroma-exposed sequences, they did not impair PsaE integration, and Fd photoreduction was not significantly affected (Rousseau et al., 1993). Moreover, experiments measuring competition between Fd reduction and the $P700^+$-$(F_A/F_B)^-$ back reaction indicated that removal of PsaE resulted in inhibition of electron transport to Fd (Sonoike et al., 1993). Taken together, these data support a structural role for PsaE in the correct organization of the reducing site of PS I-RC.

Recent results suggest an additional role for PsaE. Following chemical cross linking, PS I particles were isolated in a complex with FNR. It was found that PsaE is the PS I subunit that interacts with FNR (Andersen et al., 1992). Moreover, the inhibition of electron transport to $NADP^+$ by an antibody raised against PsaE (Weber and Strotmann, 1993) resulted from the interference of binding of PsaE and FNR. The inactivation of the *psaE* gene from the cyanobacterium *Synechococcus* sp. PCC 7002 did not affect growth under normal conditions. However, the mutants grew more slowly under conditions which favor cyclic electron transport and could not grow at all under photoheterotrophic conditions. It was concluded that PsaE plays a role in cyclic electron transport in cyanobacteria (Zhao et al., 1993). Another study performed with the cyanobacterium *Synecho-*

cystis sp. PCC 6803 indicated that the PsaE subunit is required for $NADP^+$ photoreduction via ferredoxin, but not via flavodoxin (Xu et al., 1994b).

In summary, all these results lead to a generally accepted model in which the reducing site of PS I-RC is composed mainly of PsaC, PsaD and PsaE subunits. Probably all three are necessary for the proper structure and stability of this part of the complex, which is responsible for the interaction of PS I with Fd and FNR.

b. The Oxidizing-Lumenal Side of PS I-RC

The peripheral-lumenal domain of PS I-RC is the site of oxidative activity of the PS I complex, i.e., the photooxidation of PC, the secondary electron donor of PS I. Earlier biochemical and biophysical studies indicated that PsaF facilitates the electron transfer from PC to PS I (Bengis and Nelson, 1977; Haehnel et al., 1980; Nechushtai and Nelson, 1981a). The involvement of PsaF received support from crosslinking studies which indicated that PsaF is the subunit of PS I that associates with PC (Hippler et al., 1989; Wynn et al., 1989).

The genes encoding PsaF have been cloned and characterized from spinach (Steppuhn et al., 1988), *Chlamydomonas reinhardtii* (Franzen et al., 1989b), *Synechocystis* sp. PCC 6803 (Chitnis et al., 1991) and *Flaveria trinervia* (Lotan et al., 1993b). In all species *psaF* encodes a precursor protein with a transit sequence typical of proteins targeted to the thylakoid lumen. The amino acid sequence of PsaF, deduced from the gene sequence, reveals two moderately hydrophobic regions in this protein. One of these, a region of 17 hydrophobic amino acids in the C-terminal part of the protein, was found to be especially conserved in all PsaF proteins studied. This conserved region was speculated to be important primarily for the interaction of PsaF with the PsaA and B hydrophobic subunits (Lotan et al., 1993b).

The interaction of PC with the PS I-RC is not yet fully understood. Spectroscopic measurements of $P700^+$ reduction revealed that two types of PC associations with PS I exist: A very fast phase in which $t_{1/2} \cong 12$–$14\mu s$ and a second, with longer life-time $t_{1/2} \cong 200\ \mu s$ (Haehnel, 1986). The fast rate was attributed to a PC association in close proximity to PS I, whereas the slower one was assigned to another PC population (Haehnel et al., 1980,1986; Haehnel, 1982). The presence of PsaF was essential for the fast electron transfer.

In cyanobacteria, two donors, PC and cytochrome *c*-553, were shown to donate electrons to PS I (Laudenbach et al., 1990; Medina et al., 1993). In *Anabaena PCC 7119,* PC and Cyt c_6 were shown to have similar association constants for complex formation with spinach PS I (Medina et al., 1993).

Specific residues involved in the PC-PS I association were recently identified on PC. PC has been modified by site-directed mutagenesis at two separate electron-transfer sites: at a hydrophobic patch, Leu12 was converted to Glu and at an acidic patch, Tyr83 was converted to His. While the latter mutant showed a two-fold slower intracomplex electron transfer to PS I-RC, the actual affinity with PS I was unaffected. However, in the Leu12-Glu mutant there was a drastic decrease in the association with PS I. It was therefore concluded that the hydrophobic patch of PC is more important for association with and for electron transfer to PS I (Nordling et al., 1991). Plastocyanin structure and function is discussed in more detail in Chapter 21.

Haehnel et al. (1994) provided additional information on the amino acid residues on PC that are responsible for the interaction with PS I by characterizing the electron transfer process from PC to PS I. When Gly10 or Ala90 were replaced by the bulky hydrophobic Leu, formation of the PC-PS I complex was abolished. Hence, they concluded that these two residues, which are part of the conserved flat hydrophobic surface around the copper ligand His87, are part of two crucial structures required for both docking of PC on PS I and for efficient electron transfer from PC to $P700^+$. When Tyr83 was substituted by a Leu, a faster electron transfer to $P700^+$ was detected. The authors' conclusion, that this residue is not involved in electron transfer from PC to PS I, is in good agreement with the previous suggestion of Nordling et al. (1991).

In summary, the current suggestion for PC-PS I association and electron transfer is that of a four-step mechanism: i) first, long-range electrostatic interactions between the negative patch of PC and positive residues on PsaF are formed; ii) in a second stage, a tight docking of the flat hydrophobic patch of PC with a hydrophobic site, probably comprised of the two core subunits PsaA and B, occurs; iii) these two association steps allow for fast electron transfer (PC to $P700^+$) with a $t_{1/2}$ of 12–14μs; and iv) a facilitated release of oxidized PC takes place (Haehnel et al., 1994).

A cDNA clone encoding a 15 kDa subunit of

barley PS I-RC has been isolated. The protein has a 9.5 kDa apparent molecular weight on denaturing SDS-PAGE and has been designated PsaN. Its corresponding gene *psaN* appears to encode a hydrophilic, extrinsic protein with no predicted membrane-spanning regions. Its transit peptide, 60 residues of about 5.7 kDa, contains a predicted hydrophobic α-helix, suggesting that the protein is routed into the thylakoid lumen. Thus, PsaN may be the second known lumenal protein component, composing, together with PsaF, the oxidizing site of PS I (Knoetzel and Simpson, 1993). Table 1 summarizes the current information on the function and characteristics of the different subunits composing the PS I-RC.

III. The Organization and Structure of Photosystem I

A. PS I Structure—Deduced from Electron Microscopy and X-ray Crystallographic Studies

Two methods that have been used to elucidate the structure of PS I are negative staining electron microscopy (EM) and X-ray crystallography. While the former provides valuable information on the global shape and size of different PS I complexes (Boekema et al., 1987; Bottcher et al., 1992; Hefti et al., 1992; Ford and Cochrane, 1993), the latter permits a more accurate and detailed knowledge of secondary and tertiary structural elements of the reaction center core subunits and of the components carrying out the electron transfer process (Krauss et al., 1993).

The EM study of spinach holo-PS I revealed that the complex has a disk shape which is 16x12x6.8nm in size (Boekema et al., 1991,1994). Considering the molecular weight of the LHCI subunits and the 12 pigment molecules attached, Boekema and his coworkers predicted that 8 copies of LHCI are bound to the spinach PS I-RC complex. This conclusion is in good agreement with the stoichiometric calculations that resulted from the labeling experiments of the holo-PS I complex of *Chlamydomonas reinhardtii* (Schuster et al., 1988).

Negative staining EM studies performed on the PS I-RC were mainly conducted with complexes isolated from cyanobacteria. These predicted the complex to be either an ellipsoid of 18×8 nm (Williams et al., 1983), or to have a disk shape with a diameter of about 19 nm and a thickness of 6 nm (Boekema et al., 1987). The trimeric form of the cyanobacterial PS I-RC was measured to be a triangular disk of $18 \times 6 \times 6$ nm and the monomeric PS I-RC has been described as a pear-shape of $15 \times 9 \times 6$ nm (Ford and Holzenburg, 1988; Ford et al., 1988). The latter findings support results from freeze-fracture EM studies on spinach (Dunahay and Staehelin, 1985) and barley thylakoids (Simpson, 1982). Based on their results in the cyanobacterial PS I-RC, Ford and his coworkers (Ford and Holzenburg, 1988; Ford et al., 1988) suggested that the PS I-RC lies with its shortest axis across the thylakoid membrane. Moreover, since trimers were not observed in the membranes, the authors proposed that the formation of the trimeric form occurs as a result of the detergent environment and the extraction procedure. The proposed sizes of the monomeric complex predicted in the EM studies correlates fairly well with the size and arrangement of the monomeric PS I-RC of *Mastigocladus laminosus*, based on the packing and sizes of the unit cell of hexagonal crystals obtained from the monomeric PS I-RC (Almog et al., 1991).

Other EM studies have provided additional details on specific parts of the PS I-RC complex. Ordered two-dimensional arrays of PS I, reconstituted from detergent solubilized PS I-RCs and phospholipids, were analyzed by EM and digital image processing. These indicated that the reaction centers have a ridge of 2.5 nm height projecting from one side of the membrane while their other side was rather flat and exhibited a shallow, central indentation (Ford et al., 1990). Hefti et al. (1992), in their detailed study performed with PS I complexes in membranes under different ionic strength, confirmed the proposed asymmetry. Moreover, they showed that the complex's aggregation state remains in the monomeric form under all conditions tested (Hefti et al., 1992).

Additional evidence for the asymmetric structure of PS I came from the EM study performed on *Synechococcus sp*. PS I-RC. These studies (Bottcher et al., 1992) indicated that on one side, presumably the stromal side, there is a 3 nm high ridge. This ridge was speculated to be composed of the PsaC, PsaD and PsaE subunits. The other side, presumably the lumenal side, is flat, but has a 3 nm deep indentation in the center. This indentation was proposed to be a region in which the two large subunits (PsaA and PsaB) are partly separated (Bottcher et al., 1992). Kruip and coworkers (Kruip et al., 1993) found that comparison of averaged top and side view projections of PS I-RC and PS I-RC depleted of PsaC, D and E

Table 1. Polypeptide composition of PSI-RC

Subunit	Gene	Molecular weight	Bound Cofactors	Functions	Properties
			~100 chlorophyll *a*		
PsaA (Ia, PSI-A)	*psaA*	83.0	P_{700} chlorophyll *a* dimer		
			~12–16 β carotenes	Light-harvesting, charge separation and electron transfer.	Transmembrane proteins with 9–11 transmembranal helices
PsaB (Ib, PSI-B)	*psaB*	82.4	2 phylloquinones		
			[4Fe-4S] cluster (Fx)		
PsaC (VII, PSI-C)	*psaC*	8.9	2 [4Fe-4S] clusters (F_A,F_B)	Electron donation to Fd	peripheral on *s*-side
PsaD (II, PSI-D)	*psaD*	17.5–18 (18–22)	None	Fd-docking and confining a stable assembly of PSI-RC	peripheral on *s*-side
PsaE (IV, PSI-E)	*psaE*	7.7–10.8 (14–16)	None	FNR/Fd docking, cyclic e^- transfer and binding of LHCI to CCI	peripheral on *s*-side
PsaF (III, PSI-F)	*psaF*	17.3–18 (18–19)	None	PC-docking	peripheral on *l*-side, two hydrophobic regions
PsaG (?, PSI-G)	*psaG*	10–10.8	None	?	peripheral on *s*-side
PsaH (VI, PSI-H)	*psaH*	10.2–11	None	?	peripheral on *s*-side
PsaI (IX, PSI-I)	*psaI*	~4	None	?	transmembrane helix
PsaJ (VIII, PS I-J)	*psaJ*	~5.1	None	?	transmembrane helix
PsaK (?, PSI-K)	*psaK*	~8.4	None	?	intrinsic membrane protein
PsaL (V, PSI-L)	*psaL*	18 (14)	None	In cyanobacteria - responsible for trimerization	two putative transmembrane helices
PsaN	*psaN*	15 (9.5)	None	?	peripheral on *l*-side

Alternative names for each subunit are given in parentheses. The range of molecular weights predicted from deduced primary sequences from different sources is given. Apparent weights observed in SDS-PAGE are indicated in parentheses. *s* = stromal side, *l* = lumenal side

show that the height of the complex is reduced by 2.5–3.3 nm upon the removal of the stromal-facing subunits. For further details on the available knowledge on PS I structure derived from EM studies of cyanobacteria and higher plant PS I complexes, we refer the reader to articles by Ford and Cochrane (1993) and Boekema and coworkers (1994).

While EM studies yielded valuable information concerning the overall three dimensional structure of PS I complexes and to some extent on different parts of the complex, these studies did not (and could not) provide detailed high resolution structural data. Such information could be obtained by X-ray crystallography and indeed the most detailed structural data available to date on PS I comes from the three dimensional structure of *Synechococcus* sp. PS I-RC determined at 6Å resolution (Krauss et al., 1993). In this study, 28 α-helices and 45 Chl *a* molecules were located and assigned. Eighteen of the α-helices were identified as residing in PsaA and PsaB, nine on each. The two core subunits were found to relate to each other by a two-fold rotation axis that passes through the Fe-S cluster F_x. In addition to F_x, the location and the spatial arrangement of the other electron transfer components were determined. For further details on this PS I structure, we refer the reader to Chapter 18.

There is no doubt that the 6Å structure provided new information on the PS I complex; however the low resolution precluded a more accurate assignment of the different protein subunits within the complex and their relation to the thylakoid membrane. Hence, studies that provide information on the topology of different PS I-RC subunits are of great value and importance.

B. Topology of the PS I Components in the Thylakoid Membranes

The use of tryptic treatments has made it possible to

predict the location of different subunits of PS I with respect to the thylakoid membranes. These treatments in combination with theoretical predictions based on primary sequence analysis (von Heijne et al., 1989) enabled the determination of whether a subunit is located on the stromal or the lumenal side of the thylakoid membrane. Moreover, these methods have made it possible to characterize which of the subunits, and even what parts of the polypeptide chain, are peripheral or integral, i.e., embedded within the membrane bilayer.

Attempts to expose the LHCI apoproteins to tryptic digestion (Ortiz et al., 1985; Herrin et al., 1987) indicated that they are highly resistant to proteases. These findings support the theoretical predictions based on computer analysis of the LHCI apoprotein primary sequences that suggested the LHCI apoproteins to be integral membrane proteins (Hoffman et al., 1987; Pichersky et al., 1988; Green and Pichersky, 1994). Similar methods indicated that the core proteins of PS I-RC (PsaA and PsaB) are transmembranal (Andersson and Haehnel, 1982; Fish et al., 1985a; Kirsch et al., 1986; Cantrell and Bryant, 1987). This was further confirmed by the three-dimensional structure of the PS I-RC complex (Krauss et al., 1993). However, topological studies indicated that considerable parts of the PsaA and PsaB sequences extend into the stroma and the lumen (Sadewasser and Sherman, 1981; Ortiz et al., 1985; Kirsch et al., 1986). The detailed study by Zilber and Malkin (1992), which utilized several different proteases, indicated that the N-terminal halves of PsaA and PsaB can be cleaved by pronase E to generate 47, 45, 26 and 24 kDa fragments. In another study attempting to identify regions of PsaA that are exposed to the surrounding medium by using antibodies to synthetic peptides, it was shown that residues 413–421 of PsaA are exposed on the stromal surface of the membrane, and that the accessibility of this region is enhanced by NaSCN treatment, which removes extrinsic polypeptide subunits. This treatment also exposed a trypsin-cleavage site which may lie just after the residues of the 413–421 region. Immunogold labeling also indicates that residues 371–379 and 497–505 are exposed on the lumenal surface of the thylakoids (Vallon and Bogorad, 1993). All studies suggested that the F_x center, like the N-termini of the core proteins, is located on the stromal surface of the thylakoids.

Earlier studies had indicated that PsaC, D, E and H are extrinsic peripheral membrane proteins located on the stromal side of the thylakoids (Ortiz et al., 1985; Enami et al., 1986,1987; Oh-Oka et al., 1989; Golbeck and Bryant, 1991). More recent and detailed topological studies have enabled the identification of specific exposed residues or regions of these stromal facing subunits of PS I-RC. Zilber and Malkin (1992) found that only limited cleavage sites are accessible for proteolytic digestion of PsaD and concluded that the N-terminal domain of PsaD is exposed to the stroma, whereas the bulk of the protein is protected. Since Fd could be crosslinked to trypsin-treated thylakoids containing the 19 kDa PsaD fragment, it was concluded that the 15 amino acids at the N-terminus are not required for the association with Fd. Further supporting evidence for this conclusion comes from the fact that this N-terminal region is missing in PsaD of cyanobacteria (Chitnis et al., 1989b). The in situ topological studies conducted on PsaD using antibody binding and control proteolysis also indicated that the N-terminal domain of PsaD—up to Lys 15—is exposed to the medium (Lagoutte and Vallon, 1992). Partial mapping of exposed epitopes indicated that in addition to the N-terminal domain, another region—between Met 74 and Met 140—including the most basic amino acid clusters, is also accessible.

Proteolytic treatment of PsaE revealed that the N-terminal domain of this protein is also exposed to the stroma (Lagoutte and Vallon, 1992; Zilber and Malkin, 1992). Lagoutte and coworkers showed that this region extends up to Glu 15. They also suggested that another sequence—probably located after Met 39 in the second half of the PsaE protein is exposed (Lagoutte and Vallon, 1992; Rousseau et al., 1993). Topological mapping and site-directed mutagenesis indicated that specific residues in the N- and C-terminus of PsaE are required for correct integration and orientation of the protein in the thylakoids. Alterations of residues in other regions of the protein that are stroma-exposed did not impair protein integration into the membrane or photoreduction of Fd (Rousseau et al., 1993).

Studies by Q. Xu and P. R. Chitnis (personal communication), in which mapping of PsaD and E was performed in *Synechocystis* by identifying NHS-biotinylation sites and protease accessible sites, showed that the C- terminus of both PsaD and E are exposed to the stroma. Moreover, the PS I complex could assemble in vivo in PsaD C-terminal deletion mutants.

Information on the membrane topology of PsaH,

L and K came from the study of Zilber and Malkin (1992). While they confirmed the previous suggestion (Tjus and Andersson, 1991) that PsaH is a stromal-exposed protein accessible for degradation, they also showed that PsaL and PsaK are protected from extensive proteolysis. PsaL was found to have a protease cleavage site in the loop between the second and third proposed transmembranal spans, indicating that this loop is stromal-exposed. The sensitivity of PsaK to proteolysis by thermolysin and pronase E was somewhat surprising. From theoretical predictions, PsaK was expected to be mostly hydrophobic and hence protected. However, since the protein was cleaved in intact thylakoids and its mass decreased significantly, it was concluded that a central part is exposed on the stromal side of the membranes. Thus, two possible models for the PsaK membrane topology were proposed: In the first, there is one anchoring transmembrane span near the C-terminus leaving the N-terminal domain exposed to the stroma; alternatively, two transmembrane domains exist, with the second spanning the membrane near the N-terminus of the PsaK and only the intervening loop exposed to the stroma (Zilber and Malkin, 1992).

Crosslinking experiments have also provided information on the association between different PS I-RC subunits. The close interaction of PsaC and D was reconfirmed (Armbrust et al., 1994). Crosslinking experiments with glutaraldehyde indicated the formation of the following additional pairs: PsaD-PsaE, PsaE-PsaF and PsaL-PsaD (T. S. Armbrust, P. R. Chitnis and J. A. Guikema, personal communication). Interactions between PsaD and PsaL gained further support in a study which examined the PS I-RC in PsaD-less and PsaL-less strains of *Synechocystis* sp. PCC 6803 (Xu et al., 1994a). In a similar system, the interactions between PsaJ and PsaF were also observed (Xu et al., 1994c).

C. Formation of the Photosystem I Complex in the Thylakoid Membranes—Biogenesis and Assembly Studies

1. Different Steps of the Biogenesis Process

Studies on the biogenesis and assembly of the different PS I components provide an additional insight into the understanding of the process of PS I-organization in thylakoid membranes. Similar to other photosynthetic complexes, the biogenesis and assembly of PS I is a multi-step process. In eukaryotes, algae and plants, where about half of the PS I subunits are encoded in the cell nucleus, and the other half in the plastid genome, the multi-step biogenesis process takes place in two intracellular compartments, the chloroplast and the cytoplasm. Hence, a high degree of coordination is required between these different compartments in each step of the process. Biogenesis begins with transcription and translation of the various polypeptide subunits, which take place both in the chloroplast (for the chloroplast-encoded subunits) and in the cytoplasm (for the nuclear-encoded subunits). The polypeptide subunits synthesized in the cytoplasm are made as precursors having a leader (transit) sequence at their amino-terminus (Archer and Keegstra, 1990; de Boer and Weisbeek, 1992). These precursors are post-translationally imported into chloroplasts in an energy dependent process (requiring ATP) (Cline et al., 1985; Flugge and Hinz, 1986; Pain and Blobel, 1987; Theg et al., 1989; Pilon et al., 1992), probably via receptors situated in the chloroplast's envelope membranes (Pain et al., 1988; Schnell et al., 1990,1991). Within the plastids the precursors enter the thylakoid membrane. They are processed to their mature form and associate with their cofactors (pigments) and with the other subunits to form the fully active complex. The order and mechanism of these events are not yet fully understood. Several experimental systems, mainly in vitro , are now being used in order to try to dissect and follow the different stages of these complex processes and thereby determine the temporal sequence of the assembly of the PS I complex found in the thylakoid membrane. In addition to the discussion that follows, the reader is referred to Chapter 7.

2. Experimental Systems Used to Follow the Steps of the PS I Assembly Process

Different experimental systems, both in vivo and in vitro, have been used in the biogenesis studies of the PS I complex. The in vitro studies mainly utilized the import and the *in thylakoido* systems. The former, mainly used in studies of PS I assembly in eukaryotic organisms, utilizes isolated intact chloroplasts and/or etiochloroplasts to follow different steps in the assembly and integration of in vitro produced thylakoid proteins (Chua and Schmidt, 1979; Grossman et al., 1980). The latter is a more specific system to follow the integration and organization of the different components in the thylakoids themselves. The *in thylakoido* system, which was also termed the insertion (or integration) system, utilizes isolated thylakoids, rather than whole plastids, to follow integration and assembly on the membranal level (Cline, 1986; Chitnis et al., 1987). This system was used in studies of PS I assembly in eukaryotes (Cohen et al., 1992; Cohen and Nechushtai, 1992; Adam and Hoffman, 1993; Nielsen et al., 1994) as well as prokaryotes (Chitnis and Nelson, 1992a; Cohen et al., 1993).

Different in vivo experimental systems were also used to follow the biogenesis and assembly of PS I. Of great value is the system that utilizes mutants to study not only the function, but also the stability of different proteins in the thylakoid membranes. Moreover, in this system the influence of a specific subunit of PS I on the binding of other subunits and co-factors and on the formation of the entire active complex can be studied (Chitnis et al., 1989a, 1991, 1993; Takahashi et al., 1991; Mannan et al., 1991; Smart et al., 1991; Chitnis and Nelson, 1992a,b; Cohen et al., 1993; Rousseau et al., 1993; Warren et al., 1993; Smart and McIntosh, 1993). Another in vivo system, in which the events of assembly of the PS I complex were explored, was that of greening, i.e., during leaf development (Bredenkamp and Baker, 1988) and/or during light mediated chloroplast development (Nechushtai and Nelson, 1985; Herrmann et al., 1985; Vainstein et al., 1989; Lotan et al., 1993a).

3. In vitro Studies on the Assembly of PS I Subunits

a. The Integral Membrane Subunits of PS I

PsaA and PsaB, which form the main hydrophobic core of PS I, are translated on thylakoid-bound ribosomes and co-translationally inserted into the membrane. Several in vitro studies, with translation systems using lysed etioplasts, revealed that the insertion of these proteins into the membrane, and their stability within it are dependent on the presence of pigments. The PsaA and PsaB proteins accumulated only when chlorophyll precursors were provided (Eichacker et al., 1990,1992). However, pulse/chase translation assays showed that synthesis of chlorophyll does not result in increased chlorophyll apoprotein stability. It was therefore suggested that the initiation of translation of these proteins is controlled by chlorophyll-dependent regulation (Eichacker et al., 1992).

The assembly of an additional integral membrane component of PS I, PsaK, was analyzed in vitro. Pre-PsaK polypeptide was imported into intact chloroplasts, whereas pre-PsaK protein lacking 7 amino acid residues (Met-Ala-Ser-Gln-Leu-Ser-Ala) at the N-terminal end of the transit peptide, failed to be imported (Kjaerulff et al., 1993). It was concluded that the entire transit peptide is important for PsaK import into the chloroplast. Within the chloroplast, the mature protein was exclusively located in the thylakoid membrane (Kjaerulff et al., 1993).

b. Stromal-facing Peripheral PS I-RC Subunits

The stromal-facing peripheral subunits of PS I do not have hydrophobic domains that transverse the membrane. These proteins are either directly associated with the lipids or they interact specifically with integral and other peripheral membrane proteins in a manner that stabilizes their presence in the membrane. There have been several attempts to characterize the integration and assembly of the stromal-facing PS I proteins in the thylakoid membranes of plants and cyanobacteria.

The integration and assembly of PsaD were studied in intact plastids and in isolated thylakoids. The results indicated that both the precursor and mature forms of the protein integrated into the thylakoid membranes. However, when import into intact plastids was performed in the presence of an inhibitor of the stromal processing enzyme, only the precursor form could be localized in the thylakoid membranes (Cohen et al., 1992).

In contrast to the insertion of integral membrane proteins, neither the stromal fraction nor ATP were required for the integration of PsaD into the thylakoids of higher plants (Cohen et al., 1992) and cyanobacteria (Chitnis and Nelson, 1992a). Similarly, the insertion into thylakoids of PsaE did not require ATP, proton motive force, or stromal/cyanobacterial cytoplasmic factors (Chitnis and Nelson, 1992a; Y. Cohen and R. Nechushtai, unpublished). The assembly of PsaD and PsaE was found to be specifically into the PS I complex. No association of the newly integrated labeled protein subunit with any other membranal complex was observed (Cohen et al., 1992; Chitnis and Nelson, 1992a). Moreover, pre-PsaD could also assemble into isolated PS I complexes. Upon the addition of the stromal fraction, the precursor, bound to an isolated PS I complex, could be correctly processed by the stromal processing peptidase (Cohen and Nechushtai, 1992). The results suggest the following mechanism for the assembly of PsaD: First, the precursor form is inserted into the membrane and then the processing event takes place; i.e., processing follows the insertion and the binding of the precursor protein to the PS I complex. Although the precursor form is tightly bound to the PS I complex, it does not dissociate upon detergent extraction and the PS I isolation procedure (Cohen et al., 1993); removal of the transit peptide is required to allow proper interaction with other PS I subunits. Probably, the processing step induces a conformational change in the newly integrated protein and the latter becomes protected from proteolytic digestion (Y. Cohen and R. Nechushtai, unpublished). This proposed model for the different steps in the organization of the PS I-PsaD in the thylakoid membranes is schematically presented in Fig. 2.

A similar two-step assembly, in which only the last step is resistant to proteolysis, was reported for an integral PS I-protein, the LHCI chlorophyll *a/b* binding protein - CAB-7 (Adam and Hoffman, 1993). For this protein it was found that following its insertion into the thylakoids, CAB-7 binds pigments and assembles into the LHCI-680 form. In this state the protein is still susceptible to proteolysis. Only following the second assembly stage, in which the LHCI-680 assembles into the holo-PS I, does the CAB-7 protein become resistant to proteolytic digestion (Adam and Hoffman, 1993). A mutant of the *cab*-7 gene was generated to arrest protein assembly in the intermediate stage. This mutant indeed demonstrated that binding of pigments does not necessarily confer the final stabilization of the protein in the membrane; rather, the latter is induced only after the second step of the assembly occurs, i.e., after the binding to holo-PS I (Adam and Hoffman, 1993).

c. The Assembly of Lumenal-Facing Peripheral PS I-RC Subunits

The assembly of one of the lumenal facing subunits, PsaN protein, has been studied. This subunit is synthesized with a two-domain transit peptide, which first targets the protein into the chloroplast and then translocates it across the thylakoid membrane into the lumen (Knoetzel and Simpson, 1993). The translocation of PsaN across the thylakoid membranes does not require the presence of stromal factors or nucleoside-tri-phosphates (NTPs), but is dependent on the ΔpH across the membrane (Nielsen et al., 1994). The translocation pathway of PsaN seems to be similar to those of the 23 and 16 kDa proteins of the oxygen evolving complex (OEC), and not to those of PC and the 33 kDa polypeptide of OEC (Robinson et al., 1993; Hulford et al., 1994). However, although PsaN transit peptide has a two-domain presequence, no intermediate processing form was detected. It was hence concluded that it is the full precursor (pre-PsaN) that translocates both across the chloroplast envelope and across the thylakoid membrane, where it is directly processed by the thylakoid processing peptidase to yield the mature PsaN form (Nielsen et al., 1994).

4. The Interactions between Different Subunits Play a Major Role in Stabilizing Newly Integrated PS I Proteins

The studies on PsaD, PsaE and CAB-7 indicate that the interactions with other subunits of the PS I complex play an extremely important role in their assembly process, probably by confining the newly integrated proteins to stable conformations. The importance of interactions with other subunits of newly integrated PsaD and PsaE was more profound in studies that utilized thylakoids isolated from *Synechocystis* sp. mutants that lacked specific PS I subunits. The assembly of PsaD into thylakoids of a PsaD-less mutant was found to be more efficient. However, when the assembly of PsaE into the PsaD-less mutant was examined, it was found to be significantly less efficient (Chitnis and Nelson, 1992a; Cohen et al., 1993). While in thylakoids isolated from wild type cells the newly assembled PsaE was resistant both to washes by chaotropic agents and to proteolytic digestion, in membranes of the PsaD-less mutant the newly associated PsaE was not accessible to proteolytic digestion, but could, however, be washed from the thylakoids with NaBr (Cohen et al., 1993). When the membranes used for insertion were isolated from a mutant lacking PsaF and PsaJ, the PsaE that associated with them could be removed by both NaBr and proteolytic treatments (Cohen et al., 1993). These results suggest that PsaD and PsaJ interact with PsaE and stabilize its assembly in the PS I complex. This is in good agreement with the earlier studies that showed that PsaD has an effect on binding and/or stabilizing the association of other PS I

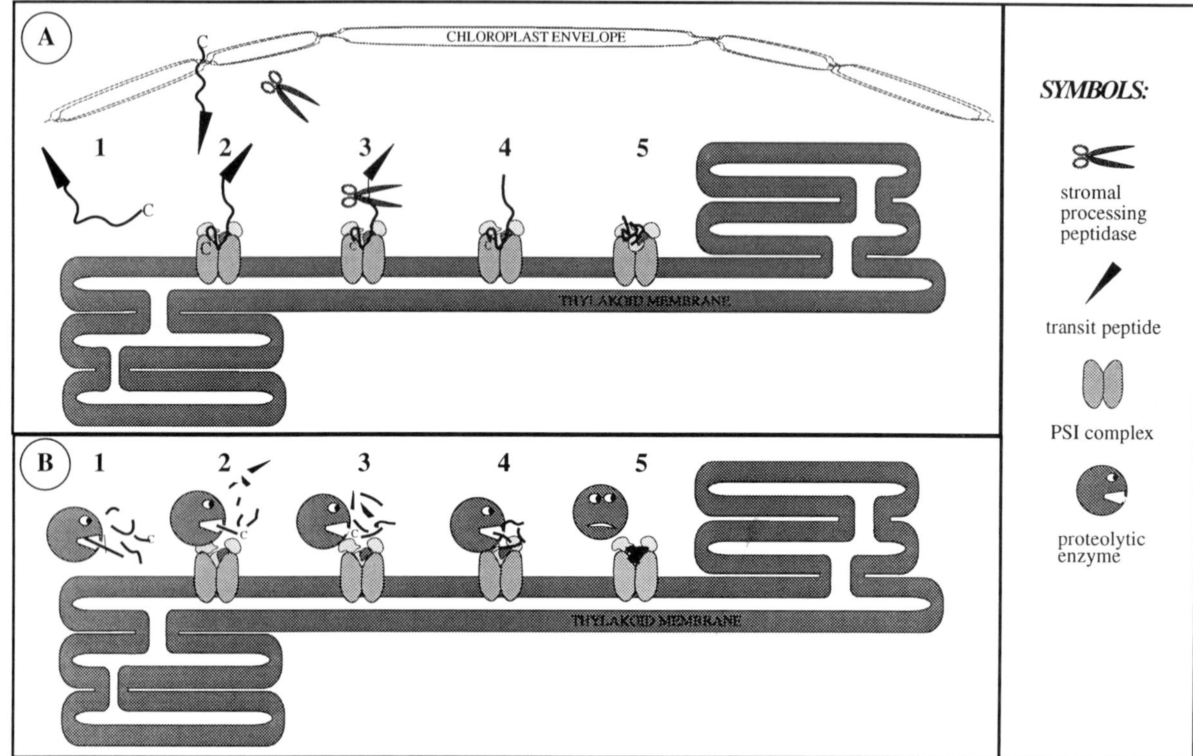

Fig. 2. A model for the assembly of PsaD into the Photosystem I complex in the thylakoid membrane. Panel A represents successive stages in the integration and assembly of PsaD. Panel B describes the sensitivity/resistance of the protein to proteolytic digestion in the different stages of the assembly process.

subunits in the complex (Nechushtai and Nelson, 1981b; Chitnis et al., 1989b). Supporting evidence for the importance of PsaD in the assembly of other subunits in the PS I complex comes from two other findings: i) PsaD was shown to be the first stromal-facing subunit that accumulates during greening of etiolated seedlings (Nechushtai and Nelson, 1985; Herrmann et al., 1985; Lotan et al., 1993a); and ii) in a reconstitution system, the importance of PsaD for stabilizing the binding of PsaC to the complex was also demonstrated (Li et al., 1991b).

In summary, the in vitro studies on the integration and assembly of PS I stromal-facing peripheral proteins strongly suggest that their stable assembly in the thylakoid membranes depends on their binding to other integral and/or peripheral proteins of the PS I complex. Modification, and even minor changes or complete absence of the binding/anchoring subunits, influence the capability of these proteins to bind and assemble in a stable manner in the membrane. However, despite the high homology of analog subunits from different organisms, a specific recognition mechanism controls the stable integration of the protein *only* to the appropriate membrane. This is probably the explanation for the finding that PsaD of cyanobacteria failed to assemble into thylakoids of higher plants (Chitnis and Nelson, 1992a).

5. In vivo Studies on the Assembly of the PS I Complex

When the biogenesis and assembly of PS I-RC were followed in vivo during the greening of etiolated seedlings, a step-by-step appearance and accumulation of the PS I-RC subunits was detected in leaves (Nechushtai and Nelson, 1985; Herrmann et al., 1985) and in thylakoid membranes (Lotan et al., 1991,1993a). While the PsaA and PsaB core subunits could be detected in dark grown seedlings, the nuclear-encoded peripheral subunits appeared sequentially upon exposure to light, with PsaD being the first to accumulate. It was found that upon exposure of etiolated seedlings to continuous white light, the mRNA, which was detected at low levels, accumulated in a specific pattern. In contrast, the PsaD protein subunit could not be detected in the etiolated seedlings; at least four hours of illumination were required to allow its accumulation. A pulse of red light induced the expression of the *psaD* and *psaE* mRNA, but the polypeptide could not be detected in the membranes unless the seedlings were re-exposed to light (white) for at least two hours. The results indicate that a dual regulatory mechanism in which both the level of mRNA and the presence of light control the accumulation of the PsaD and PsaE polypeptides (Lotan and Nechushtai, 1993; Lotan et al., 1993a). These studies also confirmed earlier reports that suggested that phytochrome is the photoreceptor involved in regulating the expression of nuclear-encoded PS I proteins (Oelmuller et al., 1989). Oelmuller and his co-workers showed the involvement of a blue-light photoreceptor (Oelmuller and Kendrick, 1991) as well as a blue-UVA-light photoreceptor (Palomares et al., 1991), in regulating the expression of PS I proteins.

When the expression of the *psaD*-mRNA and protein were followed in *Nicotiana sylvestris*, it was found that two copies of the *psaD* gene are present in this plant genome. The expression of these genes was found to vary during leaf development. While the ratio of *psaDb* to *psaDa* mRNA increased, the relative abundance of the corresponding proteins decreased over the same developmental period. These results indicate that differential regulation mechanisms control *psaDa* and *psaDb* expression at both the mRNA and protein levels during leaf development (Yamamoto et al., 1993). PsaD is the first subunit to accumulate during the greening process. However, one should note that the step by step assembly of PS I is confined to developing plastids. In steady-state conditions, pulse labeling experiments showed that the different subunits were similarly labeled (Nechushtai et al., 1981; Nechushtai and Nelson, 1981b; Chitnis and Nelson, 1992b). Hence, the steady state assembly of PS I either follows a mechanism different than the assembly in developing leaves, or in steady state conditions most of the assembly of the different subunits occurs into pre-existing complexes, and not into a de-novo synthesized complex (Chitnis and Nelson, 1992b).

A different system for following the expression of PS I subunits was that of *Chlamydomonas reinhardtii*, where the expression of chloroplast-encoded core PS I proteins (PsaA and PsaB) was followed during a 24 h cell cycle (12 h light, 12 h dark). It was found that *psaA* and *psaB* are maximally transcribed at the beginning of the light period. Transcription was induced before the onset of illumination by a light-independent mechanism. When the transcript level of the different genes was determined during the cell cycle, the *psaB* mRNAs were found to be most abundant in the first 6 hours of the light period and to decrease in the dark to about 15% of the maximal level. The abundance of *psaA* mRNA showed less variation and was maximal around the middle of the cell cycle. Since the results indicate similar transcriptional patterns, the differential steady state levels of these chloroplast transcripts appeared to be regulated at the post-transcriptional level (Leu et al., 1990).

The assembly of other nuclear-encoded PS I proteins, such as those of the LHCI components, were analyzed in relation to that of the PS I-RC subunits during greening of etiolated maize seedlings. It was found that the PS I-RC components accumulated prior to the appearance of LHCI protein (Bredenkamp and Baker, 1988; Vainstein et al., 1989). In maize, accumulation of the 17kDa protein preceded the appearance of the 24 and 21 kDa subunits of LHCI (Vainstein et al., 1989). As for the core components, the light requirement for the increase in the LHCPI transcript level is not specific for one of the light-dependent signal transduction chains; i.e., blue- as well as red-light photoreceptors were found to regulate it (Palomares et al., 1991).

As described above, in vivo systems in which different subunits of PS I were specifically disrupted have provided valuable information on the assembly of the complex. While the removal of many of the *Synechocystis* sp. PCC 6803 PS I-RC proteins, like PsaE, PsaF and PsaL, did not affect its function and assembly (Chitnis et al., 1989a,1991,1993), the removal of several other protein subunits affected the assembly of PS I-RC. The insertional inactivation of the *psaC* gene in *Chlamydomonas reinhardtii* induced a faster turnover of the entire PS I complex (Takahashi et al., 1991). Inactivation of the *Synechocystis psaD* gene resulted in a strain that has reduced P700 activity, and although it contains normal amounts of the PsaA-PsaB core subunits, the amounts of the low molecular weight subunits of PS I-RC were significantly reduced (Chitnis et al., 1989b). When the gene for either PsaA or PsaB core proteins was inactivated, the PS I complex could not assemble at all (Smart et al., 1991; Smart and McIntosh, 1993;

Webber et al., 1993). Although these subunits are very homologous to each other, when PsaB was absent a homo-dimer of two PsaA subunits was not formed (Smart and McIntosh, 1993). In the latter mutant, the peripheral PS I-RC subunits, like PsaC and PsaD could not be detected (Smart et al., 1991).

Acknowledgments

This work was supported by the following grants: BARD (United States-Israel Binational Agricultural Research and Development Fund) Grant No. IS-2191-92RC; GIF (German-Israeli Foundation for Scientific Research and Development) Grant No. I-259-153.03/92 and the Israel Science Foundation (administered by the Israel Academy of Science and Humanities) Grant No. 470/93-1. We thank Mrs. J. Goldstein for helpful suggestions.

References

Adam Z and Hoffman NE (1993) Biogenesis of Photosystem I light-harvesting complex: Evidence for a membrane intermediate. Plant Physiol 102: 35–43

Almog O, Shoham G, Michaeli D and Nechushtai R (1991) Monomeric and trimeric forms of Photosystem I reaction center of *Mastigocladus laminosus*: Crystallization and preliminary characterization. Proc Natl Acad Sci USA 88: 5312–5316

Andersen B, Scheller HV and Moller BL (1992) The PS I-E subunit of Photosystem I binds ferredoxin:$NADP^+$ oxido-reductase. FEBS Lett 311: 169–173

Andersson B and Haehnel W (1982) Location of Photosystem I and Photosystem II reaction centers in different thylakoid regions of stacked chloroplast. FEBS Lett 146: 13–17

Archer EK and Keegstra K (1990) Current views on chloroplast protein import and hypotheses on the origin of the transport mechanism. J Bioenerg Biomem 22: 789–810

Armbrust TS, Odom WR and Guikema JA (1994) Structural analysis of Photosystem I polypeptides using chemical crosslinking. J Exp Zool 269: 205–211

Baker AN and Malkin R (1990) Photosystem I reaction center proteins contain leucine zipper motifs—a proposed role in dimer formation. FEBS Lett 264: 1–4

Bassi R, Soen SY, Frank G, Zuber H and Rochaix JD (1992) Characterization of chlorophyll *a/b* proteins of Photosystem I from *Chlamydomonas reinhardtii*. J Biol Chem 267: 25714–25721

Bengis C and Nelson N (1975) Purification and properties of Photosystem I reaction center from chloroplasts. J Biol Chem 250: 2783–2788

Bengis C and Nelson N (1977) Subunit structure of chloroplast Photosystem I reaction center. J Biol Chem 252: 4564–4569

Bingham SE, Xu RH and Webber AN (1991) Transformation of chloroplasts with the *psaB* gene encoding a polypeptide of the Photosystem I reaction center. FEBS Lett 292: 137–140

Boekema EJ, Dekker JP, Van Heel MG, Rogner M, Saenger W, Witt I and Witt HT (1987) Evidence for a trimeric organization of the Photosystem I complex from the thermophilic cyanobacterium *Synechococcus* sp. FEBS Lett 217: 283–286

Boekema EJ, Wynn RM and Malkin R (1991) The structure of spinach Photosystem I studied by electron microscopy. Biochem Biophys Acta 1017: 49–56

Boekema EJ, Boonstra AF, Dekker JP and Rögner M (1994) Electron microscopic structural analysis of Photosystem I, Photosystem II and cytochrome b_6/f complex from green plants and cyanobacteria. J Bioener Biomemb 26: 17–29

Bottcher B, Graber P and Boekema E (1992) The structure of photosystem-I from the thermophillic cyanobacterium *Synechococcus* sp. determined by electron-microscopy of 2-dimensional crystals. Biochim Biophys Acta 1100: 125–136

Bredenkamp GJ and Baker NR (1988) The changing contribution of LHCI to Photosystem I activity during chloroplast biogenesis in wheat. Biochim Biophys Acta 934: 14–21

Brettel K and Sétif P (1987) Magnetic-field effects on primary reaction in Photosystem I. Biochim Biophys Acta 893: 109–112

Brettel K, Sétif P and Mathis P (1986) Flash-induced absorption changes in Photosystem I at low temperature: Evidence that the electron acceptor A, is vitamin K_1. FEBS Lett 203: 220–224

Broadhurst RW, Hoff AJ and Hore PJ (1986) Interpretation of polarized electron paramagnetic resonance signal of plant Photosystem I. Biochim Biophys Acta 852: 106–111

Bryant DA (1986) The cyanobacterial photosynthetic apparatus: Comparison to those of higher plants and photosynthetic bacteria. In: Platt T and Li WKW (eds) Photosynthetic Picoplankton Can Bull Fish Aquat Sci, Vol 214, pp 423–500.

Bryant DA (1991) Molecular biology of Photosystem I. In: Barber J (ed) The Photosystems: Structure, Function and Molecular Biology, pp 501–551. Elsevier, Amsterdam

Buttner M, Xie DL, Nelson H, Pinther W, Hauska G and Nelson N (1992a) Photosynthetic reaction center genes in green sulfur bacteria and in photosystem 1 are related. Proc Natl Acad Sci USA 89: 8135–8139

Buttner M, Xie DL, Nelson H, Pinther W, Hauska G and Nelson N (1992b) The Photosystem I-like P840-reaction center of green S-bacteria is a homodimer. Biochim Biophys Acta 1101: 154–156

Cammack R and Evans MCW (1975) EPR spectra of iron-sulfur protein in dimethylsulfoxide solution: Evidence that chloroplast Photosystem I particles contain 4Fe-4S centers. Biochem Biophys Res Commun 67: 544–549

Cantrell B and Bryant DA (1987) Molecular cloning and nucleotide sequence of the *psaA* and *psaB* genes of the cyanobacterium *Synechococcus* sp. PCC 7002. Plant Mol Biol 9: 453–468

Chitnis VP and Chitnis PR (1993) PsaL is required for the formation of Photosystem I trimers in the cyanobacteria *Synechocystis* sp. PC 6803. FEBS Lett 336: 330–334

Chitnis PR and Nelson N (1991) Photosystem I. In: Bogorad L and Vasil IK (eds) The Photosynthetic Apparatus: Molecular Biology and Operation, pp 178–224. Academic Press, San Diego

Chitnis PR and Nelson N (1992a) Assembly of two subunits of

the cyanobacterial Photosystem I on the n-side of thylakoid membranes. Plant Physiol 99: 239–246

Chitnis PR and Nelson N (1992b) Biogenesis of Photosystem I: The subunit PsaE is important for the stability of PS I complex. In: Argyroudi-Akoyunoglou J (ed) Regulation of Chloroplast Biogenesis, pp 277–282. Plenum Press, New York

Chitnis PR and Thornber JP (1988) The major light-harvesting complex of Photosystem II: Aspects of its molecular and cell biology. Photosynth Res 16: 41–63

Chitnis PR, Nechushtai R and Thornber JP (1987) Insertion of the precursor of the light-harvesting chlorophyll *a/b*-binding protein into the thylakoids requires the presence of a developmentally regulated stromal factor. Plant Mol Biol 10: 3–12

Chitnis PR, Reilly PA, Miedel MC and Nelson N (1989a) Structure and targeted mutagenesis of the gene encoding 8-kDa subunit of Photosystem I of the cyanobacterium *Synechocystis* sp. PCC 6803. J Biol Chem 264: 18374–18380

Chitnis PR, Reilly PA and Nelson N (1989b) Insertional inactivation of the gene encoding subunit II of Photosystem I of the cyanobacterium *Synechocystis* sp. PCC 6803. J Biol Chem 264: 18381–18385

Chitnis PR, Purvis D and Nelson N (1991) Molecular cloning and targeted mutagenesis of the gene *psaF* encoding subunit III of Photosystem I from the cyanobacterium *Synechocystis* sp. PCC 6803. J Biol Chem 266: 20146–20151

Chitnis VP, Xu Q, Yu L, Golbeck JH, Nakamoto H, Xie DL and Chitnis PR (1993) Targeted inactivation of the gene *psaL* encoding a subunit of Photosystem I of the cyanobacterium *Synechocystis* sp. PCC 6803. J Biol Chem 268: 11678–11684

Chua N-H and Schmidt GW (1979) Transport of proteins into mitochondria and chloroplast. J Cell Biol 81: 461–483

Cline K (1986) Import of proteins into chloroplasts: Membrane integration of a thylakoid precursor protein reconstituted in chloroplast lysates. J Biol Chem 261: 14804–14810

Cline K, Werner-Washburne M, Lubben TH and Keegstra K (1985) Precursors to two nuclear-encoded chloroplast proteins bind to the outer envelope membrane before being imported into chloroplasts. J Biol Chem 260: 3691–3696

Cohen Y and Nechushtai R (1992) Assembly and processing of subunit II (PsaD) precursor in the isolated Photosystem I complex. FEBS Lett 302: 15–17

Cohen Y, Steppuhn J, Yalovsky S, Herrman RG and Nechushtai R (1992) Insertion and assembly of the precursor of subunit II into the Photosystem I complex may precede its processing. EMBO J 11: 79–85

Cohen Y, Chitnis VP, Nechushtai R and Chitnis PR (1993) Stable assembly of PsaE into cyanobacterial photosynthetic membranes is dependent on the presence of other accessory subunits of Photosystem I. Plant Mol Biol 23: 895–900

de Boer DA and Weisbeek PJ (1992) Chloroplast protein topogenesis: Import, sorting and assembly. Biochim Biophys Acta 1071: 221–253

Dunahay TG and Staehelin LA (1985) Isolation of Photosystem I complexes from octyl glucoside/sodium dodecyl sulfate solubilized spinach thylakoids. Plant Physiol 78: 606–613

Dunn PPJ and Gray JC (1988) Nucleotide sequence of the *frxB* gene in wheat chloroplast DNA. Nucl Acids Res 16: 348

Eichacker LA, Soll J, Lauterbach P, Rudiger W, Klein RR and Mullet JE (1990) In vitro synthesis of chlorophyll *a* in the dark triggers accumulation of chlorophyll *a* apoproteins in barley etioplasts. J Biol Chem 265: 13566–13571

Eichacker L, Paulsen H and Rudiger W (1992) Synthesis of chlorophyll *a* regulates translation of chlorophyll *a* apoproteins P700, CP47, CP43 and D2 in barley etioplasts. Eur J Biochem 205: 17–24

Enami I, Ohta H and Katho S (1986) Topographical studies on subunit polypeptides of the Photosystem I reaction center complex in the thylakoid membrane of the thermophilic cyanobacterium *Synechococcus* sp. Plant Cell Physiol 27: 1395–1405

Enami I, Ohta H and Miyaoka T (1987) Cross linking studies on the membrane topography of Photosystem I reaction center complex in *Synecococcus* sp. Plant Cell Physiol 28: 101–111

Fish LE, Kuck U and Bogorad L (1985a) Analysis of the two partially homologous P700 chlorophyll *a* proteins of maize Photosystem I: Predictions based on the primary sequences and features shared by other chlorophyll proteins. In: Steinback KE, Bonitz S, Arntzen CJ and Bogorad L (eds) Molecular Biology of the Photosynthetic Apparatus, pp 111–120. Cold Spring Harbor Laboratory, Cold Spring, New York

Fish LE, Kuck U and Bogorad L (1985b) Two partially homologous adjacent light-inducible maize chloroplast genes encoding polypeptides of the P700 chlorophyll *a* protein complex of Photosystem I. J Biol Chem 260: 1413–1421

Flieger K, Oelmuller R and Herrmann RG (1993) Isolation and characterization of cDNA clones encoding a 18.8 kDa polypeptide, the product of the gene *psaL*, associated with Photosystem I reaction center from spinach. Plant Mol Biol 22: 703–709

Flugge UI and Hinz G (1986) Energy dependence of protein translocation into chloroplasts. Eur J Biochem 160: 563–570

Ford RC and Cochrane MA (1993) Cyanobacterial Photosystem I structure. Biochem Soc Trans 21: 19–21

Ford RC and Holzenburg A (1988) Investigation of the structure of trimeric and monomeric Photosystem I reaction centre complexes. EMBO J 7: 2287–2293

Ford RC, Pauptit RA and Holzenberg A (1988) Structural studies on improved crystals of the Photosystem I reaction center from *Phormidium laminosum*. FEBS Lett 238: 385–389

Ford RC, Hefti A and Engel A (1990) Ordered arrays of the Photosystem I reaction centre after reconstitution: Projections and surface reliefs of the complex at 2 nm resolution. EMBO J 9: 3067–3075

Franzen L-G, Frank G, Zuber H and Rochaix J-D (1989a) Isolation and characterization of cDNA clones encoding five subunits of Photosystem I from the green alga *Chlamydomonas reinhardtii*. In: Baltscheffsky M (ed) Current Research in Photosynthesis, Vol 3, pp 617–621. Kluwer Academic Publishers, Dordrecht

Franzen L-G, Frank G, Zuber H and Rochaix J-D (1989b) Isolation and characterization of cDNA clones encoding Photosystem I subunits with molecular masses 11.0, 10.0 and 8.4 kDa from *Chlamydomonas reinhardtii*. Mol Gen Genet 219: 137–144

Furrer R and Thurnauer MC (1983) Resolution of signals attributed to Photosystem I primary reactants by time resolved EPR at K band. FEBS Lett 153: 399–403

Golbeck JH (1987) Structure, function and organization of the Photosystem I reaction center complex. Biochim Biophys Acta 895: 167–204

Golbeck JH (1993) Shared thematic elements in photochemical

reaction centers. Proc Natl Acad Sci USA 90: 1642–1646

Golbeck JH (1994) Photosystem I in cyanobacteria. In: Bryant DA (ed) The Molecular Biology of Cyanobacteria, pp 319–360. Kluwer Academic Publishers, Dordrecht

Golbeck JH and Bryant DA (1991) Photosystem I. Current Topics in Bioenergetics 16: 83–177

Green BR and Pichersky E (1994) Chl *a/c* light-harvesting antenna proteins from helix and four helix ancestors. Photosyn Res 39: 149–162

Grossman AR, Bartlett SG and Chua N-H (1980) Energy-dependent uptake of cytoplasmically-synthesized polypeptides by chloroplasts. Nature 285: 625–628

Guigliarelli B, Guillaussier J, More C, Sétif P, Bottin H and Bertrand P (1993) Structural organization of the iron-sulfur centers in *Synechocystis* 6803 Photosystem I. EPR study of oriented thylakoid membranes and analysis of the magnetic interactions. J Biol Chem 268: 900–908

Haehnel W (1982) On the functional organization of electron transport from plastoquinone to Photosystem I. Biochim Biophys Acta 682: 245–257

Haehnel W (1986) Plastocyanin. In: Staehelin LA and Arntzen CJ (eds) Encyclopedia of Plant Physiology, Photosynthesis III, Vol 19, pp 547–559. Springer Verlag, Berlin

Haehnel W, Propper A and Krause H (1980) Evidence for complexed plastocyanin as the immediate electron donor of P-700. Biochim Biophys Acta 593: 384–399

Haehnel W, Jansen T, Gause K, Klosgen RB, Stahl B, Michel D, Huvermann B, Karas M and Herrmann RG (1994) Electron transfer from plastocyanin to Photosystem I. EMBO J 13: 1028–1038

Hefti A, Ford RC, Miller M, Cox RP and Engel A (1992) Analysis of the structure of Photosystem I in cyanobacterial thylakoid membranes. FEBS Lett 296: 29–32

Herrin DL, Plumley FG, Ikeuchi M and Michaels AS (1987) Chlorophyll antenna proteins of Photosystem I: Topology, synthesis, and regulation of the 20-kDa subunit of *Chlamydomonas* light-harvesting complex of Photosystem I. Arch Biochem Biophys 254: 397–408

Herrmann RG, Westhoff P, Alt J, Tittgen J and Nelson N (1985) Thylakoid membrane proteins and their genes. In: van Vloten-Doting L, Groot GSP and Hall TC (eds) Molecular Form and Function of the Plant Genome, pp 233–256. Plenum, Amsterdam

Hippler M, Ratajczak R and Haehnel W (1989) Identification of the plastocyanin binding subunit of Photosystem I. FEBS Lett 250: 280–284

Hoffman NE, Pichersky E, Malik VS, Castresana C, Ko K, Darr SC and Cashmore AR (1987) A cDNA clone encoding a Photosystem I protein with homology to Photosystem II chlorophyll *a/b*-binding polypeptides. Proc Natl Acad Sci USA 84: 8844–8848

Hulford N, Hazell L, Mould RM and Robinson C (1994) Two distinct mechanisms for the translocation of proteins across the thylakoid membranes, one requiring the presence of a stromal protein factor and nucleotide triphosphates. J Biol Chem 269: 3251–3256

Hwang S and Herrin DL (1993) Characterization of a cDNA encoding the 20-kDa Photosystem I light-harvesting polypeptide of *Chlamydomonas reinhardtii*. Curr Genet 23: 512–517

Ikeuchi M and Inoue Y (1991) Two new components of 9 and 14 kDa from spinach Photosystem I complex. FEBS Lett 280: 332–334

Ikeuchi M, Hirano A, Hiyama T and Inoue Y (1990) Polypeptide composition of higher plant Photosystem I complex: Identification of *psaI*, *psaJ* and *psaK* gene products. FEBS Lett 263: 274–278

Ikeuchi M, Nyhus KJ, Inou Y and Pakrasi HB (1991) Identities of four low-molecular-mass subunits of the Photosystem I complex from *Anabaena variabilis* ATCC 29413: Evidence for the presence of the *psaI* gene product in a cyanobacterial complex. FEBS Lett 287: 5–9

Illinger N, Xie DL, Hauska G and Nelson N (1993) Identification of the subunit carrying FeS centers A and B in the P840 center preparation of *Chlorobium limicola*. Photosyn Res 38: 111–114

Iwaki M, Takahashi M, Shimada K, Takahashi Y and Itoh S (1992) Photoaffinity labeling of the phylloquinone-binding polypeptides by 2-azidoanthraquinone in Photosystem I particles. FEBS Lett 312: 27–30

Iwasaki Y, Ishikawa H, Hibino T and Takabe T (1991) Characterization of genes that encode subunits of cucumber PS I complex by N-terminal sequencing. Biochim Biophys Acta 1059: 141–148

Kirsch W, Seyer P and Herrmann RG (1986) Nucleotide sequence of the clustered genes for two P700 chlorophyll *a* apoproteins of the Photosystem I reaction center and the ribosomal protein S14 of the spinach plastid chromosome. Curr Genet 10: 843–855

Kjaerulff S, Andersen B, Nielsen VS, Moller BL and Okkels JS (1993) The PS I-K subunit of Photosystem I from barley (*Hordeum vulgare* L.). Evidence for a gene duplication of an ancestral PS I-G/K gene. J Biol Chem 268: 18912–18916

Knoetzel J and Simpson DJ (1993) The primary structure of a cDNA for PsaN, encoding an extrinsic lumenal polypeptide of barley Photosystem I. Plant Mol Biol 22: 337–345

Knoetzel J, Svendsen I and Simpson DJ (1992) Identification of the Photosystem I antenna polypeptides in barley. Isolation of three pigment-binding antenna complexes. Eur J Biochem 206: 209–215

Kok B (1957) Absorption changes induced by the photochemical reaction of photosynthesis. Nature 179: 583–584

Kok B (1961) Partial purification and determination of oxidation-reduction potential of the photosynthetic chlorophyll complex absorbing at 700 nm. Biochim Biophys Acta 48: 527–533

Kossel H, Dory I, Igloi G and Maier R (1990) A leucine-zipper motif in Photosystem I. Plant Mol Biol 15: 497–499

Krauss N, Hinrichs W, Witt I, Fromme P, Pritzkow W, Dauter Z, Betzel C, Wilson KS, Witt HT and Saenger W (1993) Three-dimensional structure of system I of photosynthesis at 6 Å resolution. Nature 361: 326–331

Kruip J, Boekema EJ, Bald D, Boonstra AF and Rögner M (1993) Isolation and structural characterization of monomeric and trimeric Photosystem I complexes (P700.FA/FB and P700.FX) from the cyanobacterium *Synechocystis* PCC 6803. J Biol Chem 268: 23353–23360

Lagoutte B and Vallon O (1992) Purification and membrane topology of PS I-D and PS I-E, two subunits of the Photosystem I reaction center. Eur J Biochem 205: 1175–1185

Lam E, Ortiz W, Mayfield S and Malkin R (1984) Isolation and characterization of a light-harvesting chlorophyll *a/b* protein complex associated with PS I. Plant Physiol 74: 650–655

Laudenbach DE, Herbert SK, McDowell C, Fork DC, Grossman AR and Straus NA (1990) Cytochrome *c*-553 is not required for photosynthetic activity in the cyanobacterium *Synechococcus*. Plant Cell 2: 913–924

Leu S, White D and Michaels A (1990) Cell cycle-dependent transcriptional and post-transcriptional regulation of chloroplast gene expression in *Chlamydomonas reinhardtii*. Biochim Biophys Acta 1049: 311–317

Li N, Warren PV, Golbeck JH, Frank G, Zuber H and Bryant DA (1991a) Polypeptide composition of the Photosystem I complex and the Photosystem I core protein from *Synechococcus* sp. PCC 6301. Biochim Biophys Acta 1059: 215–225

Li N, Zhao J, Warren PV, Warden JT, Bryant DA and Golbeck JH (1991b) PsaD is required for the stable binding of PsaC to the Photosystem I core protein of *Synechococcus* sp. PCC 6301. Biochemistry 30: 7863–7872

Liebl U, Mockensturm WM, Trost JT, Brune DC, Blankenship RE and Vermaas W (1993) Single core polypeptide in the reaction center of the photosynthetic bacterium *Heliobacillus mobilis*: Structural implications and relations to other photosystems. Proc Natl Acad Sci USA 90: 7124–7128

Lotan O and Nechushtai R (1993) The light-regulated biogenesis of subunit V (PsaG) or the Photosystem I reaction center. In: Murata N (ed) Research in Photosynthesis, Vol I, pp 65–68. Kluwer Academic Publishers, Dordrecht

Lotan O, Cohen Y, Yalovsky S, Michaeli D and Nechushtai R (1991) Characterization of the sequential light-regulated assembly of Photosystem I core complex. In: Argyroudi-Akoyunoglou H (ed) Regulation of Chloroplast Biogenesis, pp 269–276. Plenum Publishing Corporation, New York

Lotan O, Cohen Y, Michaeli D and Nechushtai R (1993a) High levels of Photosystem I subunit II (PsaD) mRNA result in the accumulation of the PsaD polypeptide only in the presence of light. J Biol Chem 268: 16185–16189

Lotan O, Streubel M, Westhoff P and Nechushtai R (1993b) Subunit III (Psa-F) of Photosystem I reaction center of the C_4 dicotyledon *Flaveria trinervia*. Plant Mol Biol 21: 573–577

Machold O, Simpson DJ and Moller B (1979) Chlorophyll-proteins of thylakoids from wild-type and mutants of barley (*Hordeum vulgare* L.). Carlsberg Res Commun 44: 235–254

Malkin R (1982) Photosystem I. Ann Rev Plant Physiol 33: 455–479

Malkin R (1987) Photosystem I. In: Barber J (ed) The Light Reactions. Topics in Photosynthesis Research, Vol 8, pp 495–525. Elsevier Science Publishers, Amsterdam

Mannan RM, Whitmarsh J, Nyman P and Pakrasi HB (1991) Directed mutagenesis of an iron-sulfur protein of the Photosystem I complex in the filamentous cyanobacterium *Anabaena variabilis* ATCC 29413. Proc Natl Acad Sci USA 88: 10168–10172

Medina M, Diaz A, Hervas M, Navarro JA, Gomez-Moreno C, Delarosa MA and Tollin G (1993) A comparative laser-flash absorption spectroscopy study of *Anabaena* PCC 7119 plastocyanin and cytochrome c_6 photooxidation by Photosystem I particles. Eur J Biochem 213: 1133–1138

Mühlenhoff U, Haehnel W, Witt H and Herrmann RG (1993) Genes encoding eleven subunits of Photosystem I from the thermophilic cyanobacterium *Synechococcus* sp. Gene 127: 71–78

Mullet JE, Burke JJ and Arntzen CJ (1980) Chlorophyll proteins of Photosystem I. Plant Physiol 65: 814–822

Nechushtai R and Nelson N (1981a) Photosystem I reaction centers from *Chlamydomonas* and higher plant chloroplasts. J Bioenerg Biomembranes 13: 295–306

Nechushtai R and Nelson N (1981b) Purification properties and biogenesis of *Chlamydomonas reinhardtii* Photosystem I reaction center. J Biol Chem 256: 11624–11628

Nechushtai R and Nelson N (1985) Biogenesis of Photosystem I reaction center during greening. Plant Mol Biol 4: 377–384

Nechushtai R, Nelson N, Mattoo A and Edelman M (1981) Site of synthesis of subunits to Photosystem I reaction center and the proton-ATPase in spirodela. FEBS Lett 125: 115–119

Nechushtai R, Nelson N, Gonen O and Levanon H (1985) Photosystem I reaction center from *Mastigocladus laminosus*. Correlation between reduction state of the iron-sulfur centers and the triplet formation mechanisms. Biochim Biophys Acta 807: 35–43

Nechushtai R, Peterson CC, Peter GF and Thornber JP (1987) Purification and characterization of a light-harvesting chlorophyll-*a/b* protein of Photosystem I of *Lemna gibba*. Eur J Biochem 164: 345–350

Nielsen VS, Mant A, Knoetzel J, Moller BL and Robinson C (1994) Import of barley Photosystem I subunit N into the thylakoid lumen is mediated by a bipartite presequence lacking an intermediate processing site. J Biol Chem 269: 3762–3766

Nitschke W and Rutherford AW (1991) Photosynthetic reaction centres: Variations on a common structural theme? Trends Biochem Sci 16: 241–245

Nitschke W, Feiler U and Rutherford AW (1990) Photosynthetic reaction center of green sulfur bacteria studied by EPR. Biochemistry 29: 3834–3842

Nordling M, Sigfridsson K, Young S, Lundberg LG and Hansson O (1991) Flash-photolysis studies of the electron transfer from genetically modified spinach plastocyanin to Photosystem I. FEBS Lett 291: 327–330

Nugent JHA, Moller BL and Evans MCW (1981) Comparison of the EPR properties of Photosystem I iron-sulphur centres A and B in spinach and barley. Biochim Biophys Acta 634: 249–255

Nyhus KJ, Ikeuchi M, Inoue Y, Whitmarsh J and Pakrasi HB (1992) Purification and characterization of the Photosystem I complex from the filamentous cyanobacterium *Anabaena variabilis* ATCC 29413. J Biol Chem 267: 12489–12495

Oelmuller R and Kendrick RE (1991) Blue light is required for survival of the tomato phytochrome-deficient aurea mutant and the expression of four nuclear genes coding for plastidic proteins. Plant Mol Biol 16: 293–299

Oelmuller R, Kendrick RE and Briggs WR (1989) Blue-light mediated accumulation of nuclear-encoded transcripts coding for proteins of the thylakoid membrane is absent in the phytochrome-deficient aurea mutant of tomato. Plant Mol Biol 13: 223–232

Oh-oka H, Takahashi Y, Wada K, Matsubara H, Ohyama K and Ozeki H (1987) The 8 kDa polypeptide in Photosystem I is a probable candidate of an iron-sulfur center protein coded by the chloroplast gene *frxa*. FEBS Lett 218: 52–54

Oh-Oka H, Takahashi Y and Matsubara H (1989) Topological considerations of the 9-kDa polypeptide which contains centers A and B, associated with the 14- and 19-kDa polypeptides in the Photosystem I complex of spinach. Plant Cell Physiol 30: 869–875

Okkels JS, Scheller HV, Svendsen I and Moller BL (1991)

Isolation and characterization of a cDNA clone encoding an 18-kDa hydrophobic Photosystem I subunit (PS I-L) from barley (*Hordeum vulgare* L.). J Biol Chem 266: 6767–6773

Ortiz W, Lam E, Chollar S, Munt D and Malkin R (1985) Topography of the protein complexes of the chloroplast thylakoid membrane. Studies of Photosystem I using a chemical probe and proteolytic digestion. Plant Physiol 77: 389–397

Pain D and Blobel G (1987) Protein import into chloroplasts requires a chloroplast ATPase. Proc Natl Acad Sci USA 84: 3288–3292

Pain D, Kanwar YS and Blobel G (1988) Identification of a receptor for protein import into chloroplasts and its localization to envelope contact zones. Nature 331: 232–237

Palomares R, Herrmann RG and Oelmuller R (1991) Different blue-light requirement for the accumulation of transcripts from nuclear genes for thylakoid proteins in *Nicotiana tabacum* and *Lycopersicon esculentum*. J Photochem Photobiol B 11: 151–162

Pichersky E, Tanksley SD, Piechulla B, Stayton MM and Dunsmuir P (1988) Nucleotide sequence and chromosomal location of *Cab-7*, the tomato gene encoding chlorophyll *a/b*-binding polypeptide of Photosystem I. Plant Mol Biol 11: 69–71

Pilon M, De Kruijff B and Weisbeek PJ (1992) New insights into the import mechanism of the ferredoxin precursor into chloroplasts. J Biol Chem 267: 2548–2556

Robinson C, Klosgen RB, Herrmann RG and Shackleton JB (1993) Protein translocation across the thylakoid membrane—a tale of two mechanisms. FEBS Lett 325: 67–69

Rousseau F, Sétif P and Lagoutte B (1993) Evidence for the involvement of PS I-E subunit in the reduction of ferredoxin by Photosystem I. EMBO J 12: 1755–1765

Sadewasser DA and Sherman LA (1981) Internal and external membrane proteins of the cyanobacterium, *Synechococcus cedrorum*. Biochim Biophys Acta 640: 326–340

Scheller HV and Moller BL (1990) Photosystem I polypeptides. Physiologia Plantarum 78: 484–494

Schnell DJ, Blobel G and Pain D (1990) The chloroplast import receptor is an integral membrane protein of chloroplast envelope contact sites. J Cell Biol 111: 1825–1838

Schnell DJ, Blobel G and Pain D (1991) Signal peptide analogs derived from two chloroplast precursors interact with the signal recognition system of the chloroplast envelope. J Biol Chem 266: 3335–3342

Schuster G, Nechushtai R, Ferreira PCG, Thornber JP and Ohad I (1988) Structure and biogenesis of *Chlamydomonas reinhardtii* Photosystem I. Eur J Biochem 177: 411–416

Schwartz E, Shen D, Aebersold R, McGrath JM, Pichersky E and Green BR (1991) Nucleotide sequence and chromosomal location of *Cab11* and *Cab12*, the genes for the fourth polypeptide of the Photosystem I light-harvesting antenna (LHCI). FEBS Lett 280: 229–234

Simpson DJ (1982) Freeze-fracture studies on barley plastid membranes V. Viridis-n 34, a Photosystem I mutant. Carlberg Res Commun 47: 215–225

Smart LB and McIntosh L (1993) Genetic inactivation of the *psaB* gene in *Synechocystis* sp. PCC 6803 disrupts assembly of Photosystem I. Plant Mol Biol 21: 177–180

Smart LB, Anderson SL and McIntosh L (1991) Targeted genetic inactivation of the Photosystem I reaction center in the cyanobacterium *Synechocystis* sp. PCC 6803. EMBO J 10: 3289–3296

Smith AG, Wilson RM, Kaethner TM, Willey DL and Gray JC (1991) Pea chloroplast genes encoding a 4 kDa polypeptide of Photosystem I and a putative enzyme of C1 metabolism. Curr Genet 19: 403–410

Sonoike K, Ikeuchi M and Pakrasi HB (1992) Presence of an N-terminal presequence in the PsaI protein of the Photosystem I complex in the filamentous cyanobacterium *Anabaena variabilis* ATCC 29413. Plant Mol Biol 20: 987–990

Sonoike K, Hatanaka H and Katoh S (1993) Small subunits of Photosystem I reaction center complexes from *Synechococcus* elongatus. II. The *psaE* gene product has a role to promote interaction between the terminal electron acceptor and ferredoxin. Biochim Biophys Acta 1141: 52–57

Steinmuller K (1992) Identification of a second *psaC* gene in the cyanobacterium *Synechocystis* sp. PCC6803. Plant Mol Biol 20: 997–1001

Steppuhn J, Hermans J, Nechushtai R, Ljungberg U, Thummler F, Lottspeich F and Herrmann RG (1988) Nucleotide sequence of cDNA clones encoding the entire precursor polypeptides for subunits IV and V of Photosystem I reaction center from spinach. FEBS Lett 237: 218–224

Takahashi Y, Goldschmidt CM, Soen SY, Franzen LG and Rochaix JD (1991) Directed chloroplast transformation in *Chlamydomonas reinhardtii*: Insertional inactivation of the *psaC* gene encoding the iron sulfur protein destabilizes Photosystem I. EMBO J 10: 2033–2040

Theg SM, Bauerle C, Olsen LJ, SElman BR and Keegstra K (1989) Internal ATP is the only energy requirement for the translocation of precursor proteins across chloroplastic membranes. J Biol Chem 264: 6730–6736

Thornber JP (1986) Biochemical characterization and structure of pigment-proteins of photosynthetic organisms. In: Staehelin LA and Arntzen CJ (eds) Encyclopedia of Plant Physiol., Vol 19, pp 98–142, Springer-Verlag, Berlin

Thornber JP, Morishige DT, Anandan S and Peter GF (1991) Chlorophyll-carotenoid-proteins of higher plant thylakoids. In: Scheer H (ed) The Chlorophylls, pp 549–585. CRC Press, Boca Raton

Tittgen J, Hermans J, Steppuhn J, Jansen T, Jansson C, Andersson B, Nechushtai R, Nelson N and Herrmann RG (1986) Isolation of cDNA clones for fourteen nuclear encoded thylakoid proteins. Mol Gen Genet 204: 258–265

Tjus SE and Andersson B (1991) Extrinsic polypeptides of spinach Photosystem I. Photosynth Res 27: 209–219

Vainstein A, Peterson CC and Thornber JP (1989) Light-harvesting pigment-proteins of Photosystem I in maize: Subunit composition and biogenesis. J Biol Chem 264: 4058–4062

Vallon O and Bogorad L (1993) Topological study of PS I-A and PS I-B, the large subunits of the photosystem-I reaction center. Eur J Biochem 214: 907–915

von Heijne G, Steppuhn J and Herrmann RG (1989) Domain structure of mitochondrial and chloroplast targeting peptides. Eur J Biochem 180: 535–545

Warren PV, Smart LB, McIntosh L and Golbeck JH (1993) Site-directed conversion of cysteine-565 to serine in PsaB of Photosystem I results in the assembly of [3Fe-4S] and [4Fe-4S] clusters in Fx. A mixed-ligand [4Fe-4S] cluster is capable of electron transfer to FA and FB. Biochemistry 32: 4411–4419

Webber AN and Malkin R (1990) Photosystem I reaction-centre

proteins contain leucine zipper motifs. A proposed role in dimer formation. FEBS Lett 264: 1–4

Webber AN, Gibbs PB, Ward JB and Bingham SE (1993) Site-directed mutagenesis of the Photosystem I reaction center in chloroplast. The proline-cysteine motif. J Biol Chem 268: 12990–12995

Weber N and Strotmann H (1993) On the function of subunit PsaE in chloroplast Photosystem I. Biochim Biophys Acta 1143: 204–210

Williams RC, Glazer AN and Lundell DJ (1983) Cyanobacterial Photosystem I: Morphology and aggregation behavior. Proc Natl Acad Sci USA 80: 5923–5926

Wynn RM, Luong C and Malkin R (1989) Maize Photosystem I. Identification of the subunit which binds plastocyanin. Plant Physiol 91: 445–449

Xu Q, Armbrust TS, Guikema JA and Chitnis PR (1994a) Organization of Photosystem I polypeptides: A structural interaction between the PsaD and PsaL subunits. Plant Physiol 106: 1057–1063

Xu Q, Jung YS, Chitnis VP, Guikema JA, Golbeck JH and Chitnis PR (1994b) Mutational analysis of Photosystem I polypeptides in *Synechocystis* sp. PCC 6803. Subunit requirements for the reduction of $NADP^+$ mediated by ferredoxin and flavodoxin. J Biol Chem 269: 21512–21518

Xu Q, Odom WR, Guikema JA, Chitnis VP and Chitnis PR (1994c) Targeted deletion of *psaJ* from the cyanobacterium *Synechocystis* sp. PCC 6803 indicates structural interactions between PsaJ and PsaF. Plant Mol Biol 26: 291–302

Yamamoto Y, Tsuji H and Obokata J (1993) Structure and expression of a nuclear gene for the PS I-D subunit of Photosystem I in *Nicotiana sylvestris*. Plant Mol Biol 22: 985–994

Zanetti G and Merati G (1987) Interaction between Photosystem I and ferredoxin: Identification by chemical cross-linking of the polypeptide which binds ferredoxin. Eur J Biochem 169: 143–146

Zhao JD, Warren PV, Li N, Bryant DA and Golbeck JH (1990) Reconstitution of electron transport in Photosystem I with PsaC and PsaD proteins expressed in *Escherichia coli*. FEBS Lett 276: 175–180

Zhao J, Li N, Warren PV, Golbeck JH and Bryant DA (1992) Site-directed conversion of a cysteine to aspartate leads to the assembly of a [3Fe-4S] cluster in PsaC of Photosystem I. The photoreduction of FA is independent of FB. Biochemistry 31: 5093–5099

Zhao J, Snyder WB, Muhlenhoff U, Rhiel E, Warren PV, Golbeck JH and Bryant DA (1993) Cloning and characterization of the *psaE* gene of the cyanobacterium *Synechococcus* sp. PCC 7002: Characterization of a *psaE* mutant and over-production of the protein in *Escherichia coli*. Mol Microbiol 9: 183–194

Zilber A and Malkin R (1988) Ferredoxin cross-links to a 22 kDa subunit of Photosystem I. Plant Physiol 88: 810–814

Zilber AL and Malkin R (1992) Organization and topology of Photosystem I subunits. Plant Physiol 99: 901–911

Chapter 16

Photosystem I Electron Transfer Reactions— Components and Kinetics

Richard Malkin
Department of Plant Biology, University of California, 111 Koshland Hall, Berkeley, CA 94720, USA

Summary

All oxygenic photosynthetic organisms contain a membrane protein complex, known as Photosystem I, that catalyzes a light-induced transfer of electrons from reduced plastocyanin to ferredoxin. This complex contains several bound electron carriers that are involved in the initial charge separation and stabilization. These carriers include the reaction center chlorophyll, P700, which undergoes oxidation in the light, and a number of compounds that serve as electron carriers in what is now believed to be a sequence of electron transfer events: a monomeric chlorophyll *a* molecule is the initial electron acceptor, a phylloquinone molecule is the second electron acceptor and a series of bound iron-sulfur clusters then function as terminal electron acceptors. The properties of these electron carriers and recent evidence supporting their proposed roles are described and an attempt is made to present a unified picture of our understanding of the role of these individual components. The kinetic sequence of the electron transfer events is then considered in detail. After the discussion of these bound electron carriers, the reduction of ferredoxin by the Photosystem I complex is described and the role of this soluble electron carrier in catalyzing a cyclic transfer of electrons around Photosystem I is considered.

Donald R. Ort and Charles F. Yocum (eds): Oxygenic Photosynthesis: The Light Reactions, pp. 313–332.

I. Introduction

The PS I complex found in all oxygen-evolving photosynthetic organisms carries out a light-induced electron transfer from a reduced high-potential electron donor, such as plastocyanin in most eukaryotic organisms, to a low potential electron acceptor, ferredoxin. Electrons passed to ferredoxin ultimately are used for the reduction of $NADP^+$ although other routes for the reoxidation of reduced ferredoxin are also available in the chloroplast, as described in Chapter 17. The components involved in this series of reactions are localized in an integral protein complex that has been extensively studied in recent years, and several recent review articles have considered various topics related to the structure and function of this complex (Golbeck and Bryant 1991; Chitnis and Nelson 1991; Golbeck 1992, 1994; Sétif 1992). Chapter 15 in this book focused on structural aspects of the PS I complex, including a discussion of the function of specific protein subunits associated with this complex. The emphasis in this chapter will be more functional in terms of the electron carriers in PS I; the identity of the carriers in the photosystem will first be described and they will then be considered in the context of the electron transport pathway shown below:

$$\mathrm{P700A_0A_1(FeS)} \rightarrow \mathrm{P700^+A_0^-A_1(FeS)}$$

$$\rightarrow \mathrm{P700^+A_0A_1^-(FeS)} \rightarrow \mathrm{P700^+A_0A_1(FeS)^-}$$

It is important to stress at the outset of this discussion that a high resolution three dimensional structure of the PS I complex, comparable in resolution to that of the bacterial reaction center complex, is not yet available although several groups are working on this problem and Witt and co-workers (Krauss et al., 1993) have described a 6Å resolution structure of a cyanobacterial complex (see Chapter 18). While this work has given some overall indication of the geometry of the complex, it has not yet provided the molecular details necessary to define and identify electron transfer carriers.

Abbreviations: A_1 – The second electron acceptor in the PS I complex; A_0 – The first electron acceptor in the PS I complex; DABS – Diazonium benzene sulfonate; DBMIB – 2,5-dibromo-3-methyl-6-isopropylbenzoquinone; EDC – N-ethyl-3-(3-dimethylaminopropyl) carbodiimide; E_m – Midpoint redox potential; EPR – electron paramagnetic resonance; ESP EPR – electron spin polarized electron paramagnetic resonance; F_A – Iron-sulfur center A; F_B – Iron-sulfur center B; F_X – Iron-sulfur center X; PS I – Photosystem I; PS II – Photosystem II

II. Electron Carriers in PS I

A. P700

The reaction center chlorophyll of PS I, known as P700, is characterized by a bleaching at approximately 430 and 700 nm after the photochemical oxidation reaction (Kok 1956, 1957). The absorbance changes in the blue and red regions of the visible spectrum that accompany P700 oxidation are shown in Fig. 1A and 1B. Consistent with these absorption changes is the identification of the pigment as chlorophyll *a* and the photochemical reaction as a photooxidation based on early studies with model compounds which showed similar absorbance changes upon oxidation (for a review of this early literature, see Sétif and Mathis 1986). The photooxidation of P700 occurs with a high quantum yield in far-red light and is independent of temperature. While the chemical structure of P700 is not known with certainty due to the absence of a high-resolution structure for the PS I complex, a variety of biophysical measurements have supported the concept that P700 is a dimer of chlorophyll *a* molecules, similar in structure to the special pair found in the photosynthetic bacterial reaction center (Sétif and Mathis 1986; Moënne-Loccoz et al., 1990; see also Prisner et al., 1993). Some experimental data, however, have been interpreted in terms of a monomeric chlorophyll *a* molecule serving as this reaction center chlorophyll (O'Malley and Babcock, 1984). The E_m of the center is approximately +500 mV (Sétif and Mathis, 1980) but there are variations in this estimated value that may arise from the effects of detergents on the pigment and this may have led to lower values being reported in the literature for the E_m in different PS I preparations.

B. A_0

The electron acceptor A_0 is the first component that has been identified as undergoing photoreduction as P700 is photooxidized. This carrier has been characterized by EPR and optical absorption spectroscopy, but limitations in both these techniques have precluded a conclusive identification of the chemical nature of this compound. For example, in the case of optical experiments, overlapping

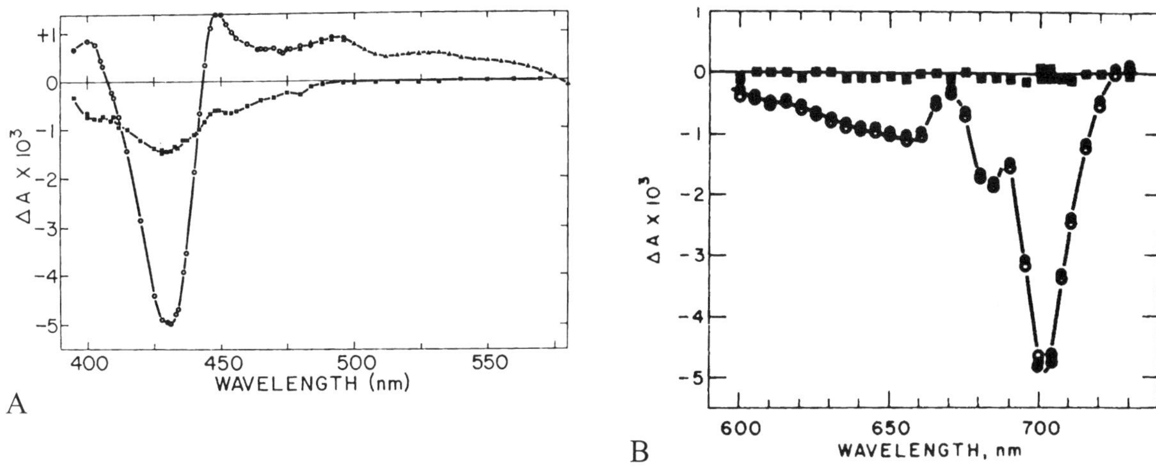

Fig. 1. The difference absorbance spectra of P700 and P430. (A) Difference spectrum of P700 and P430 from 390–580 nm. (B) Difference spectrum of P700 and P430 from 600 to 730 nm. The light-induced spectra were obtained in a spinach PS I preparation. Open circles represent the P700 difference absorbance spectrum and squares represent the P430 absorbance difference spectrum. From Hiyama and Ke (1971b).

absorbance changes associated with P700 oxidation, P700 triplet state and antenna chlorophyll absorbance changes have complicated the assignment of specific A_0 spectral properties.

The initial characterization of A_0 was based on the observation of an EPR signal at cryogenic temperatures in PS I samples poised at low potentials in a study by Shuvalov et al. (1979). The observed EPR signal, however, overlapped the signal from $P700^+$ and thus a conclusive assignment of the A_0^- signal could not be made although these workers did attribute the signal to a monomeric form of chlorophyll *a*. Subsequent work by Evans and co-workers utilized low temperature illumination (Bonnerjea and Evans, 1982). This group later developed an experimental protocol of illumination at approximately 200K to resolve some of these overlapping signals (Smith et al., 1987). An EPR spectrum attributable to the A_0 acceptor molecule was observed in PS I preparations under these illumination conditions although there was still an overlap of that signal with another electron acceptor, A_1. A signal that has EPR parameters consistent with an assignment as a monomeric form of chlorophyll *a* (*g*-value of 2.0025 and linewidth of approximately 13.5 gauss) was identified in this series of measurements (Mansfield and Evans, 1985; Smith et al., 1987). These EPR parameters may be compared with a *g*-value of 2.0029 for a monomeric form of chlorophyll a^- in solution reported by Fujita et al. (1978).

Early optical experiments in which the A_0^- component was accumulated in the light at physiological temperatures showed a bleaching at 692 nm and a substantial shoulder near 670 nm under conditions where secondary electron acceptors were reduced prior to illumination (Shuvalov et al., 1986; Nuijs et al., 1986), consistent with the photoreduction of a monomeric chlorophyll *a* molecule. It should be pointed out that other photochemical reaction centers, such as the PS II complex and the purple sulfur bacterial reaction center complex, contain either a pheophytin or bacteriopheophytin molecule as their initial electron acceptor, analogous in function to A_0 in PS I. However, careful analyses of purified PS I complexes have shown the absence of pheophytin in these preparations, leading to the conclusion that pheophytin cannot be the initial electron acceptor in PS I. The overall features of these PS I absorbance changes and their similarity to the difference spectrum presented by Fujita et al. (1978) in a study of model compounds led to the proposal that reduction of A_0 in PS I was associated with the reduction of chlorophyll *a*.

More recent time-resolved kinetic studies have continued to examine the optical features of the A_0 acceptor in PS I. The importance of this work is that these measurements were carried out under physiological conditions, that is, without the pre-reduction of any of the terminal electron acceptors prior to flash activation. Thus, negative charges in the electron acceptor complex that might influence the kinetics and absorbance properties of A_0 were absent.

Holzwarth et al. (1993) have recently reported picosecond kinetic studies of absorbance changes associated with the primary reactants in a PS I complex isolated from cyanobacteria. These measurements were made with low intensity flashes and showed that there were no absorbance changes in the 40 ps to nanosecond time domain that could be associated with the reduction of an early electron acceptor. They considered that the most likely explanation for this observation was that, although the initial acceptor was reduced rapidly, it was reoxidized even more rapidly (see below). Thus, the reduced acceptor was decaying faster than it was formed and transient absorbance changes associated with the photoreduction were impossible to detect. In a related series of papers by Hastings et al. (1994a,b), the identity of the initial PS I electron acceptor in cyanobacteria was again examined. In this case, a cyanobacterial mutant that contained only the PS I complex was used for spectral analysis and both high and low intensity flashes were used to activate the system. With low intensity flashes, Hastings et al. (1994a) confirmed the findings of Holzwarth et al. (1993) that there were no absorbance changes that could be identified with the photoreduction of an early electron acceptor. However, when high intensity flashes were used, absorbance changes associated with A_0 reduction could be resolved and these were analyzed in membranes under physiological conditions as well as under strongly reducing conditions to reduce terminal electron acceptors. Hastings et al. (1994b) found that the spectrum of the reduced A_0 molecule showed a maximum bleaching at 686–688 nm and was very similar to a spectrum reported by Fujita et al. (1978). A similar spectrum had been reported by Mathis et al. (1988) in the ns time range in a PS I preparation lacking A_1, and as shown in Fig. 2A in the blue region of the spectrum and Fig. 2B in the red region of the spectrum, the nanosecond spectrum is one of the better resolved A_0^- spectra and gives a clear definition of the bleaching of this pigment. In earlier studies by Swarthoff et al. (1982) as well as Nuijs et al. (1986) and Shuvalov et al. (1986), a component that showed a bleaching at 670 nm was assigned to A_0, but it is likely that this absorbance change is linked to the reduction of another chlorophyll *a* molecule near A_0 that can be reduced under the strongly reducing photoaccumulation conditions employed.

In the blue region of the spectrum, Shuvalov et al. (1979) reported a double peak on reduction of A_0 at

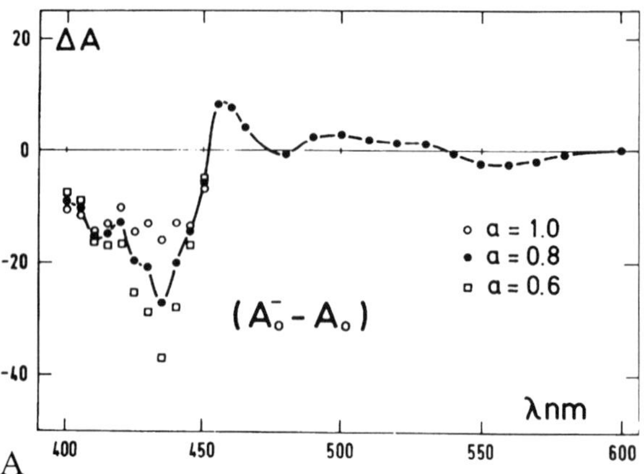

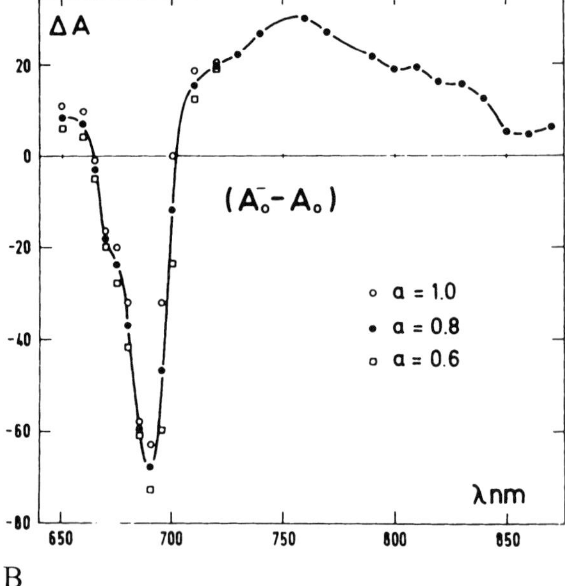

Fig. 2. The difference absorbance spectrum of A_0. (A) Difference spectrum attributed to A_0^- minus A_0 in the 400–600 nm region. (B) Difference spectrum attributed to A_0^- minus A_0 in the 650–850 nm region. Spectra were obtained in the ns time range in a PS I complex lacking A_1. From Mathis et al. (1988).

430 and 455 nm at cryogenic temperatures poised at low redox potentials. Warren et al. (1993) also showed a double peak in the blue region that was assigned to the photoreduction of A_0 in a PS I complex that was devoid of all the iron-sulfur centers. In the latter case, the doublet was at 412 and 438 nm, but the difference between this spectrum and the earlier one is most likely related to different experimental conditions. Studies with model compounds by Fujita et al. (1978) also showed that chlorophyll anion radicals have similar spectral features, and this comparative

approach has strengthened the assignment of A_0^- as a chlorophyll *a* anion. Interestingly, the spectrum of a component detected by Swarthoff et al. (1982) and assigned to the A_0 component does not show this double peak in the blue but rather a single band at 420 nm in addition to the decrease at 670 nm noted above. The ns spectrum shown in Fig. 2A also has a double peak in the blue region. Taken in conjunction with the EPR data, these results lead to the conclusion that a monomeric form of chlorophyll *a* serves as A_0 in PS I.

C. A_1

The identity of the second electron acceptor in PS I, known as A_1, has been a difficult experimental problem to solve although the majority of evidence indicates that phylloquinone (vitamin K_1) fills this role. Early evidence suggesting a function for quinones in the PS I complex was based on photoaccumulation studies of reduced electron acceptors at cryogenic temperatures by Evans and co-workers, as described above in the characterization of A_0. For example, an EPR signal with spectral parameters consistent with those of a semiquinone anion radical was observed after photoreduction experiments at 215K (Smith et al., 1987; Mansfield et al., 1987b). Optical kinetic studies also showed absorbance changes in the ultraviolet and visible spectral region that were consistent with such an assignment, but it should be mentioned that in both of these cases there is considerable overlap from spectral features of other electron transfer carriers, and this made definitive assignments based on these techniques complicated.

A more definitive spectral feature of the A_1 electron acceptor has come from studies of a characteristic electron spin polarized EPR signal (ESP EPR) by Stehlik and co-workers (Bock et al., 1989) as well as by Thurnauer and co-workers (Snyder et al., 1991). The ESP EPR signal has been directly attributed to the radical pair composed of $P700^+$ and reduced phylloquinone by employing both chemical reduction techniques and resolution and reconstitution of phylloquinone from PS I preparations (Rustandi et al., 1990; Sieckman et al., 1991; Rustandi et al., 1992). In the former case, conditions under which A_1 was doubly reduced led to a loss of the ESP EPR signal and in subsequent studies, it was shown that the ESP EPR signal was absent after solvent extraction of PS I under conditions that removed phylloquinone and that the signal could be restored by the addition of either protonated or deuterated phylloquinone (Rustandi et al., 1990). In the latter case, a very dramatic narrowing of the line width of the ESP EPR signal was observed, indicating that phylloquinone was not only required for the signal but also was a direct contributor to the species giving rise to the signal.

The presence of phylloquinone in PS I preparations from both plants and cyanobacteria has been confirmed by several groups (Schoeder and Lockau, 1986; Malkin, 1986; Sétif et al., 1987). It is particularly important to note from this work that the PS I reaction center core complex, known as CPI, which only contains the two high molecular mass subunits, still contains a substantial amount of tightly bound phylloquinone as well as A_0, the other early electron acceptor of PS I. While the presence of phylloquinone in all preparations analyzed is significant, it is also important to mention the stoichiometry of this component: two molecules of phylloquinone per P700 have generally been found in relatively intact PS I preparations. The two phylloquinone molecules in the PS I complex from spinach show differing extractabilities with respect to organic solvents: one is readily removed by hexane while the second is more tightly bound and can only be removed under more drastic conditions (Malkin, 1986; Biggins and Mathis, 1988). The removal of the more loosely bound phylloquinone molecule did not impair photochemical reduction of $NADP^+$ by PS I, a finding that suggests that electron transfer between the two quinones is not an essential part of the electron transfer sequence in the PS I acceptor complex, in analogy with the Q_A/Q_B acceptor system of purple photosynthetic bacteria and PS II.

Experiments on the resolution and reconstitution of phylloquinone from PS I complexes have given a clearer picture of the required role of the tightly bound quinone in PS I electron transfer reactions. Biggins and Mathis (1988) were able to extract both quinones, using a hexane plus methanol medium, and to demonstrate that activities which were lost were subsequently restored by the readdition of purified phylloquinone. Iwaki and Itoh (1989) were also able to extract both quinones using ether extractions, but in this work, back-reaction kinetics were used to monitor the reconstitution process, rather than more direct measurements of $NADP^+$ reduction. These differing measurements have led to some confusion as to whether only phylloquinone is able to restore activity to a fully extracted system or whether a wider range of quinones can restore activity. For example, Itoh and Iwaki (1991) found a number of different quinones of vastly different structure that could restore the presumed suppression of charge separation between $P700^+$ and A_0^-, although it is not clear from their results whether the added quinone was rebinding to the reaction center complex in its native binding site or if the quinone was being reduced at a more distant site. An examination of a similar problem by Biggins (1990) led to the suggestion that only phylloquinone would restore the photochemical reduction of P430, an optical component associated with the terminal Fe-S electron acceptors, and $NADP^+$ reduction. In Biggins' work, the finding that only phylloquinone would restore $NADP^+$ reduction is significant and suggests this assay may be a more accurate measure of the native electron transport pathway than measurements that utilize back-reaction kinetics. However, it is difficult to directly compare these two conflicting reports because the two groups are using different procedures for the removal of the bound quinones and they then employ different spectral measurements to assay their reconstituted activities. An examination of the ESP EPR signal (Rustandi et al., 1992) in resolved and reconstituted preparations showed that only a very limited number of quinones were able to reconstitute the ESP EPR signal, which suggests that quinones other than phylloquinone were not specifically bound in the A_1 site. These results support the view of the Biggins group that there is a rather high specificity for A_1 binding in this site, as opposed to the view of Itoh and co-workers. If one is seeking a unified viewpoint from this work at this time, it would appear to be that the removal of phylloquinone inhibits various photochemical activities in PS I, including $NADP^+$ reduction, and that under well-defined conditions, $NADP^+$ photoreduction activity can be restored by the addition of phylloquinone although there may be other quinone binding sites in the vicinity of the A_1 site.

Another complication in these studies has come from a comparison of photochemical activities at physiological temperatures and at cryogenic temperatures. There is now documentation from several groups that electron transfer from P700 to the bound Fe-S clusters can occur at cryogenic temperatures in the absence of phylloquinone, a result that is in contradiction to the assignment of a required role for this component as A_1 (see Sétif et al., 1987 for one such example, but also Mansfield et al., 1987a for a different set of results). Such PS I preparations from which the quinone has been fully extracted are, however, inactive in $NADP^+$ reduction at physiological temperatures, and this leads to the conclusion that the reactions observed at cryogenic temperatures may occur by non-physiological pathways. A series of results using UV irradiation to destroy quinones may be related to these observations. Several groups have reported that UV irradiation, resulting in the destruction of phylloquinone, did not substantially inhibit a number of PS I electron transfer reactions, including $NADP^+$ reduction (Ziegler et al., 1987; Palace et al., 1987; Biggins et al., 1989). If one accepts the premise that this treatment specifically affects the quinones in PS I, these results would contradict the above assignment of phylloquinone as A_1. However, there are many unresolved problems in these UV inactivation studies, such as the nature of the photochemical products after inactivation and the absence of any reconstitution of the inactivated samples by added quinones. Without the ability to reconstitute activity after a specific inactivating treatment, the specificity of the treatment remains open to question.

D. F_X

The first iron-sulfur center that accepts electrons in PS I is known as F_X. The chemical nature and functioning of this center has been clarified in recent years and there is now substantial agreement on many of the unusual properties of this center, particularly those relating to the structure of the center in PS I.

The EPR properties of F_X are unusual. The EPR signal from this reduced center, with *g*-values of approximately 2.04, 1.88 and 1.78 (see Fig. 3), is highly temperature sensitive, being barely detectable at temperatures greater than 10K. The linewidth, g_{av} and microwave power saturation properties are also substantially different from those of most other low potential iron-sulfur centers. The EPR properties of the center were difficult to define in relatively intact PS I complexes because the conditions necessary to observe the F_X signal also resulted in a reduction of the F_A and F_B centers, with overlapping EPR signals. Preparations of PS I complexes, which retained F_X in the absence of F_A and F_B, have led to a cleaner identification of spectral properties associated with the former center (Parrett et al., 1989).

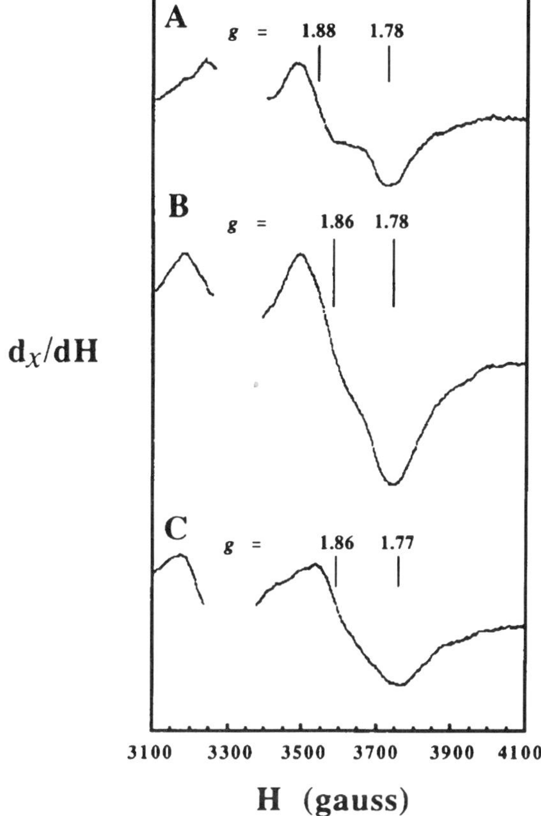

Fig. 3. The EPR spectrum of F_X in a PS I core complex. (A) Light minus dark difference spectrum of F_X after freezing in the dark and subsequent illumination at 7.5 K. (B) Light minus dark difference spectrum of F_X illuminated during freezing. (C) The spectrum of F_X with (A) subtracted from (B). From Parrett et al. (1989).

The E_m of F_X is also unusual in that it is the lowest of any iron-sulfur cluster characterized to date. The first estimation of the E_m, by Chamorovsky and Cammack (1982), was based on the magnitude of the high-field EPR signal and gave a value of –705 mV. A subsequent determination by Golbeck and co-workers (Parrett et al., 1989) utilizing a more indirect assay, the back-reaction kinetics for P700, gave a value of –670 mV at room temperature. It should be noted that the E_m of F_X shows a shift of approximately 60 mV to the positive when centers F_A and F_B are absent from the preparation.

The chemical nature of the F_X center was unclear until a number of biochemical approaches led to a consistent picture of this center as a [4Fe-4S] cluster that is coordinated to two different PS I subunits. Early evidence leading to this conclusion was based on analyses of the non-heme iron and acid-labile sulfide content of various PS I preparations. Intact preparations that have F_X, F_A and F_B contain approximately 12 moles of both non-heme iron and acid-labile sulfide per P700 (Høj and Møller, 1986; Parrett et al., 1990). Because the F_A and F_B clusters are known to be [4Fe-4S] clusters, this leaves a 'residual' of [4Fe-4S] for the F_X cluster. The preparation of a PS I core complex containing only F_X and devoid of F_A and F_B by Golbeck and co-workers (Golbeck et al., 1987a; Parrett et al., 1989) resulted in a direct determination of the amount of non-heme iron and acid-labile sulfide associated with F_X, and this determination showed approximately 4 moles of labile sulfide associated with the F_X center. However, this overall stoichiometry could indicate that F_X contains 2 [2Fe-2S] centers or 1 [4Fe-4S] cluster. Early evidence on this point was contradictory (Golbeck et al., 1987b; McDermott et al., 1987).

The finding that the PS I core complex, which contains only the two high molecular mass subunits of PS I, known as the PsaA and PsaB subunits, binds an intact F_X cluster puts serious chemical constraints on the nature of this center. In order to accommodate a single [4Fe-4S] cluster by what is considered to be normal ligation through cysteine residues, four cysteine sulfur atoms are required. If the F_X cluster actually contained two [2Fe-2S] centers, a minimum of eight cysteine residues would be required to coordinate these two centers since each [2Fe-2S] cluster also requires four cysteine ligands. However, the PsaA and PsaB subunits do not contain a sufficient number of cysteine residues to meet the latter requirements (Fish et al., 1985). The PsaA subunit of maize does contain four cysteine residues, while the PsaB subunit from the same organism contains only 2 cysteine residues (Fish et al., 1985; see also Golbeck and Bryant, 1991 for a complete review of the PsaA and PsaB sequences). In the former case, only three of the cysteines are conserved in other PsaA sequences, making it unlikely that this single subunit could provide all of the cysteines required to coordinate the F_X cluster. Thus, it has been proposed on the basis of this overview analysis that the F_X cluster is a [4Fe-4S] cluster that is ligated to two cysteine residues provided by PsaA and 2 cysteine residues provided by the PsaB subunit. Biophysical measurements are consistent with this model (McDermott et al., 1989; Petrouleas et al., 1989). This leads to a model in which F_X is formed by coordination to two different subunits, rather than a single polypeptide chain as is the case for most known iron-sulfur proteins. This heterodimeric structure, which includes one PsaA and one PsaB subunit, becomes the simplest core complex of PS I since this core complex is also known to bind P700 and the two other early electron acceptors, A_0 and A_1, found in PS I.

Supporting evidence for a heterodimeric structure for the PS I core has come from analyses of subunit stoichiometries in various PS I complexes. While an initial analysis with a cyanobacterial PS I complex gave a value of 4 (PsaA + PsaB) subunits per complex (Lundell et al., 1985), subsequent reinvestigation by several groups has shown that only a single PsaA subunit and a single PsaB subunit is present per PS I complex (Bruce and Malkin, 1988a,b; Scheller et al., 1989). The presence of single copies of the high molecular mass subunit is also consistent with electron microscopy studies of the overall structure of the complex, and, specifically, with the dimensions of the complex estimated by these techniques and estimations of the minimal molecular mass of the complex (Boekema et al., 1990).

E. F_A and F_B

All of the electron acceptors previously discussed (P700, A_0, A_1 and F_X) are associated with the PS I intrinsic core subunits, PsaA and PsaB, while the terminal electron acceptors in the complex, known as F_A and F_B, are bound to an extrinsic low molecular mass subunit. However, before this association was known, extensive analysis of the F_A and F_B centers was done in various PS I complexes isolated after detergent treatment and fractionation.

The first bound electron acceptors identified in PS I were the two Fe-S centers now known as F_A and F_B (Malkin and Bearden, 1971). These centers were initially characterized on the basis of their low temperature EPR signals. Photoreduction experiments at cryogenic temperatures resulted in the appearance of reduced F_A with EPR *g*-values of 2.05, 1.94 and 1.86 (Fig. 4). An additional center, with EPR *g*-values of 2.05, 1.92 and 1.89, was also identified in this work. Center F_A is generally the center that appears in photochemical reduction experiments at low temperatures, although experimental conditions can be established where a substantial portion of the F_B is also photoreduced. One additional feature of the EPR signals of F_A and F_B is the shift in the *g*-value of F_A when F_B is reduced.

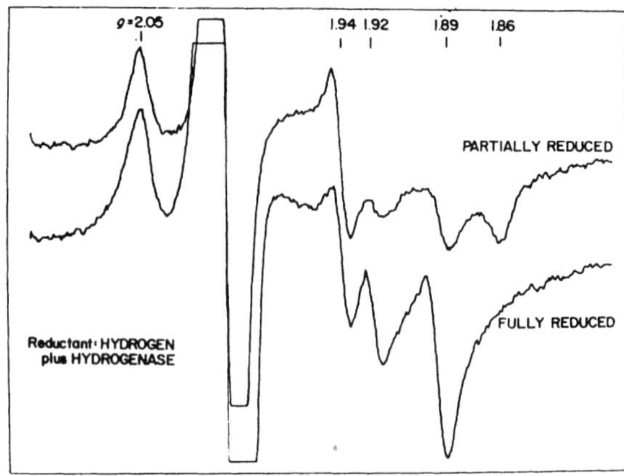

Fig. 4. EPR spectra of F_A and F_B in spinach thylakoids. The reductant in this experiment was hydrogen in the presence of the enzyme hydrogenase.

Under what may be called fully reduced conditions, the resonance line at $g = 1.86$ of F_A shifts to $g = 1.89$ to give a composite signal with four *g*-values: 2.05, 1.94, 1.92 and 1.89. This behavior had been previously noted for bacterial-type ferredoxins that contained 2 [4Fe-4S] clusters and led to the proposal that both F_A and F_B might be [4Fe-4S] clusters located in a single protein molecule.

At approximately the same time as the initial EPR characterization of F_A and F_B was being made, Hiyama and Ke (1971a,b) were analyzing absorbance changes associated with the PS I electron acceptor complex. This work led to the identification of P430, a component associated with the reduced electron acceptors of PS I. The absorbance change associated with P430 reduction is shown in Fig. 1A. A series of investigations has shown that P430 is probably the optical manifestation of iron-sulfur centers F_A and F_B (Ke and Beinert, 1973) although there are some workers who maintained that P430 is specifically associated with F_X (Hiyama and Fork, 1980). It should be mentioned that because all of these iron-sulfur clusters are now believed to contain [4Fe-4S] clusters, their absorbance properties would be expected to be very similar, and it is likely that the so-called P430 absorbance change arises from a combination of the photoreduction of all of these centers and that a clear assignment with any one cluster, based on optical properties, will depend on the exact experimental conditions used. For example, in PS I core preparations which are lacking F_A and F_B, the P430 absorbance change can be solely associated with F_X, and in this case, this overall absorbance change is very similar to that defined as the P430 change associated with F_A and F_B (see for example Parrett et al., 1989) (see Fig. 5 for the absorbance changes associated with F_X reduction and compare them with those in Fig. 1A). The E_m values for F_A and F_B have been determined in various PS I complexes and values of approximately –530 mV and –580 mV are now generally accepted (Ke et al., 1973; Evans et al., 1974). These midpoint potentials should be contrasted with the values determined for these centers in the isolated iron-sulfur protein, as described below.

The characterization of an isolated iron-sulfur protein by Oh-oka and colleagues (1987) and by Wynn and Malkin (1988) led to the identification of this protein as the product of the *psaC* gene in the chloroplast genome. This is the only extrinsic subunit that has been found to contain any of the electron carriers known to function in the PS I complex. The *psaC* gene was shown to encode a protein of approximately 9 kDa that corresponds to one of the low molecular mass subunits in the PS I holocomplex. The amino acid sequence of PsaC shows a typical ferredoxin-like sequence with two regions of clustered cysteine residues: CXXCXXCXXXCP. Such a sequence is common to all [8Fe-8S] containing ferredoxins that have been shown to contain two [4Fe-4S] clusters. The isolated protein contains two [4Fe-4S] clusters, and the midpoint redox potentials of these clusters have been determined as –470 and –560 mV(Oh-oka et al., 1991). The EPR spectrum of the centers in the isolated protein are considerably broadened when compared with the centers in the intact PS I complex and it has not been possible to correlate the signals in the isolated protein with those of centers F_A and F_B in situ. The isolated protein is considerably more labile with respect to oxygen than is the subunit in the PS I complex, allowing for the possibility that additional subunits in the PS I complex interact with the PsaC subunit, leading to a protection of the clusters from oxygen.

III. Kinetics of PS I Electron Transfer Reactions

Extensive studies of the kinetics of PS I electron transfer processes have been carried out for many years. Because of the technical limitations of accurately measuring the extremely rapid kinetics of the forward electron transfer reactions, much of this work has stressed kinetics of back reactions between reduced acceptors and $P700^+$. This work has been summarized by Golbeck and Bryant (1991) and Sétif (1992) and will not be stressed here. Rather, this presentation will focus on the kinetic studies detailing the forward electron transfer reactions through the carriers that have been described in detail in the preceding section and will emphasize recent results that have attempted to define the kinetic sequence of events in the PS I complex.

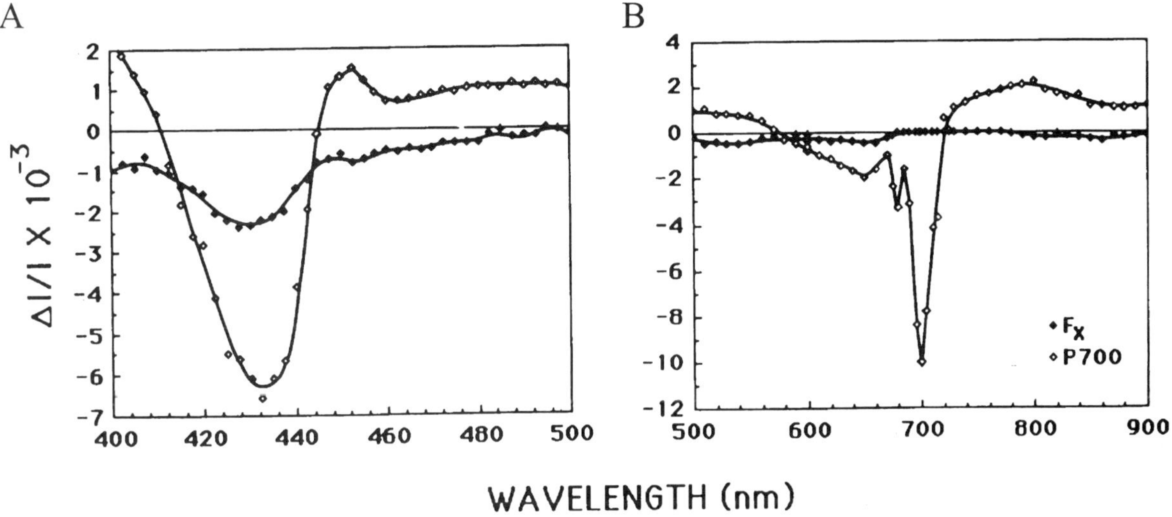

Fig. 5. Absorbance difference spectra of P700 and F_X in a PS I core complex. (A) Difference spectra for P700 and F_X in the 400–500 nm region. (B) Difference spectrum for P700 and F_X in the 500–900 nm region. From Parrett et al. (1989).

A. The Primary Charge Separation: P700$^+$ and A_0^- Formation

A determination of the kinetics of the charge separation event in PS I involves an understanding of the kinetics of photon absorption by PS I antenna, the transfer of this absorbed energy to the reaction center and the final charge separation event. In the case of PS I, since no reaction center complexes that are devoid of antenna are available, the determination of radical pair formation is complicated by energy transfer events, which in this particular case occur on a relatively slow time scale and produce absorption changes in the region where radical pair formation would be expected to be detected. Early reports of the kinetics of radical pair formation gave values in the 35–40 ps time range and also of 13.7 ps (Fenton et al., 1979; Wasielewski et al., 1987) for the appearance of P700$^+$ although these measurements may have been limited by pulse duration and there was less appreciation of some of the multiple components involved as early electron acceptors. Subsequent to this work, a number of laboratories have reinvestigated that trapping time in the PS I reaction center, using both absorbance and fluorescence measurements to define the individual processes that occur.

Measurements by several groups (Trissl et al., 1993; Holzwarth et al., 1993; Hastings et al., 1994a,b) have supported a much more rapid time of formation of P700$^+$$A_0^-$ in PS I. These picosecond and subpicosecond measurements have yielded an intrinsic time constant of 1–2 ps for radical pair formation, but they have also shown that trapping in PS I occurs by what is called a trapping-limited model. In this model, energy transfer between pigment molecules is a relatively slow process and radical pair formation is limited by the decay of the excitons in the PS I antenna system. What this model predicts in experimental terms is that the actual appearance of the charged separated state, represented by P700$^+$ and A_0^-, will only occur after excitons migrate through the antenna system to the reaction center site for ultimate charge separation. The antenna lifetime has been reported to be approximately 28 ps using low intensity picosecond flashes by Hastings et al. (1994a) although this value may vary depending on the particular preparation used because of differences in antenna size and organization. In fact, this value is in good agreement with measurements of other groups (Owens et al., 1987; Turconi et al., 1993). As will be discussed below in greater detail, Hastings et al. (1994b) have also reported a time constant of approximately 20 ps for the reoxidation of A_0^- under the same conditions that they measured a 28 ps formation time.

B. The Reoxidation of A_0^-

As described above, the kinetics of the A_0 reactions have been difficult to resolve because the chemical nature of this acceptor remains poorly defined in that absorbance bands associated solely with this acceptor have not been identified. If we accept that A_0 is likely to be a monomeric form of chlorophyll *a*, showing absorbance changes at approximately 686 nm, we can discuss early and more recent kinetic studies of this carrier.

In early picosecond measurements of PS I kinetics, Fenton et al. (1979) and Wasielewski et al. (1987) estimated that the first electron acceptor in PS I was reoxidized in approximately 40 ps. However, in this work, measurements of A_0 were indirect since the chemical identity of this carrier was not clear. Subsequently, Shuvalov et al. (1986) reinvestigated this question and came to the conclusion that the reduced primary acceptor was reoxidized in approximately 32 ps. It should be stressed that in this latter work, a 670 nm absorbing component was believed to be associated with the reduction of A_0, but this assignment is probably incorrect. However, even if the 670 nm component represents another carrier that is monitoring the state of A_0, there is some validity to the measured time constant.

In a more recent reconsideration of the kinetics of trapping and electron transfer in PS I, the use of both high and low intensity excitations has provided a clearer definition of kinetic constants associated with A_0 reduction (see above) and reoxidation. Several groups have noted that with low intensity flashes, accumulation of A_0^- is low, and this has hindered identification (see for example Shuvalov et al., 1986 and Holzwarth et al., 1993). Hastings et al. (1994b) confirmed this finding but were able to use high-intensity flashes to characterize A_0 more fully. Their work indicated that A_0^- is indeed oxidized with very fast kinetics, and a value of 21 ps was estimated. Under comparable conditions, the transfer of excitations from PS I antenna to the reaction center was found to be rate-limiting in terms of radical pair formation, the latter occurring in 1.5 ps. Excitation transfer to the reaction center was found to occur in

28 ps. Thus, under low excitation conditions, the A_0^- molecule is being oxidized in 21 ps while it is being formed in approximately 28 ps. It has been calculated by Hastings et al. (1994b) that a transient population of only 30% of A_0^- would accumulate, leading to problems in detectability by optical means.

From these early and more recent measurements, it is clear that A_0^- reoxidation in PS I occurs in the 20–30 ps time range. This time constant is to be contrasted with estimates of the rate of reoxidation of the analogous acceptor in PS II and photosynthetic reaction center complexes. In the latter cases, there is substantial agreement among a number of groups that reduced pheophytin *a* or bacteriopheophytin *a* is reoxidized in approximately 200 ps. While models that describe PS I and the bacterial-type reaction center complexes generally show strong analogy, this almost ten-fold difference in the rate of reoxidation between the PS I initial electron acceptor and acceptors in the so-called pheophytin-type reaction center remains to be explained on a structural basis.

C. Kinetics of A_1 Oxidation and F_X Reduction

The kinetics of A_1 have also been difficult to define and there remains some controversy concerning the rates of electron transfer steps involving this component. As noted above, based on absorbance changes of A_0, it has been concluded that A_0^- reoxidation occurs in approximately 20–30 ps, and this would then correspond to the reduction time for A_1. However, spectroscopic measurements of A_1 reduction have not been made to identify this time constant. In contrast, several groups have measured the reoxidation kinetics for A_1 using optical spectroscopy. For example, Mathis and Sétif (1988) reported absorbance changes in PS I complexes from spinach isolated with the detergent Triton X-100. Measurements were made at 370 nm in order to monitor phylloquinone reduction. This wavelength has been found to be a maximum in the spectrum of the phylloquinone anion in studies of phylloquinone in organic solvents. However, the applicability of this wavelength for studies of this electron carrier in a pigment-protein complex, such as the PS I complex, is still unclear. At 370 nm, there was a rapid absorbance increase, presumably due to the reduction of phylloquinone, followed by a biphasic decay, the primary decay component occurring with a half-time of approximately 15 ns. Since no absorbance changes associated with P700 could be observed in the ns time range and the kinetics were not altered after the addition of dithionite to reduce the bound iron-sulfur centers, it appeared unlikely that the 15 ns electron transfer step could involve a back reaction between a reduced acceptor, such as an Fe-S center and $P700^+$. Mathis and Sétif concluded that the 15 ns electron transfer was from A_1^-, phylloquinone, and the next bound electron acceptor, F_X. However, in another investigation of the same problem by Brettel (1988), using a cyanobacterial PS I complex, absorbance changes between 325 and 495 nm were measured, where P700, phylloquinone and the bound iron-sulfur centers absorb. Brettel measured a rapid absorbance increase in the 350–400 nm spectral range, which was followed by a bleaching centered at 387 nm with a half time of approximately 200 ns. This was interpreted in terms of an electron transfer where A_1 was reduced in less than 5 ns and subsequently, an electron was transferred from A_1^- to the first iron-sulfur acceptor, F_X, in approximately 200 ns. The disagreement between these first two sets of results is striking and no simple explanation for the disparity in results was presented by either group.

Following this work, two additional groups have attempted to measure the kinetics of A_1^- oxidation and the presumed forward electron transfer to F_X. Firstly, Warden (1990) measured the kinetics of F_X reduction at 445 nm in a cyanobacterial PS I preparation that had been depleted of F_A and F_B. F_X reduction occurred within 5 ns, suggesting that A_1 oxidation would occur with similar kinetics, but there was no indication of either a 200 ns or a 15 ns kinetic component. However, when A_1 was measured directly at 380 nm, a wavelength presumed in all these experiments to be specific for phylloquinone redox reactions, there was no correlation with the observed rapid kinetic changes determined at 445 nm for the iron-sulfur cluster. Thus, Warden has raised the question of whether the absorbance changes in the far UV actually monitor the redox state of phylloquinone since there does not seem to be any correlation between phylloquinone oxidation and F_X reduction.

This situation became more complicated after a recent report of Sétif and Brettel (1993) who showed that the kinetics of A_1^- oxidation, again measured at 370 nm, were dependent on the particular PS I preparation used. For example, reduced A_1 decayed significantly faster in a spinach PS I preparation as compared with a cyanobacterial preparation (biphasic decay of 25 and 150 ns in the spinach preparation and 200 ns in the cyanobacterial preparation). By prereducing iron-sulfur centers F_A and F_B prior to flash activation, Sétif and Brettel concluded that the fast kinetic component in the spinach PS I preparation represented the rate of reoxidation of A_1^- while the slower rate was more difficult to define. Although these workers considered the possibility that this slower rate represented reduction of centers F_A and F_B, it is not clear if this is a process that involved F_X. While further studies will be required to clarify the kinetics observed in this work, these results point out that different PS I complexes may show different kinetics for these terminal steps, and that comparisons are difficult to make between the results from different groups.

While the above-described optical studies have not given a clear answer for the rate of oxidation of A_1^-, an independent analysis has been carried out of the kinetics of the same reactions in PS I. Using ESP EPR techniques at room temperature, Bock et al. (1989) identified two different EPR signals, one arising from the $P700^+A_1^-$ state and the second arising from $P700^+$ alone. There is apparently no contribution to the ESP EPR signal from the bound iron-sulfur centers, so it is probable that the second signal actually originates from $P700^+F_X^-$, with the reduced electron acceptor not detected by the specific EPR technique used. It was found in this work with both spinach and cyanobacterial PS I complexes, and confirmed in similar measurements by Thurnauer and Norris (1980), that the disappearance of the first signal agrees with the appearance of the second, indicating a sequential series of events. Bock et al., concluded that the approximately 260 ns time constant that was measured for this conversion represents the transfer of an electron as shown below:

$$P700^+A_1^-F_X \rightarrow P700^+A_1F_X^-$$

Since the measurements of Bock et al. (1989) do not depend on assignment of spectral bands to individual acceptors, these results give an independent assessment of this electron transfer sequence. However, the limitation of the ESP EPR method is that events faster than approximately 50 ns are not detected, so that more rapid electron transfer events could not be observed.

It is difficult to draw conclusions about the precise rate of the reaction discussed above in detail because of the variability in the results that have been obtained by numerous groups. A current view based on all the evidence is that electron transfer to F_X occurs no slower than approximately 200 ns and that more rapid processes may also occur, depending on the specific PS I complex used.

D. Measurements of Fe-S Cluster Reduction

The complexities in determining the kinetics of A_1 reduction and reoxidation are based primarily on the spectral features of this carrier. The absorbance changes for the reduction of the quinone to the semiquinone state overlap substantially the absorbance changes associated with several other electron carriers in the PS I complex, such as P700 and A_o. The problems associated with measurements of the iron-sulfur clusters, F_X, F_A and F_B, are even more extreme in this regard. It is generally accepted that broad absorbance changes in the 430 nm region, denoted P430, are associated with the reduction of the bound iron-sulfur clusters (see Fig. 1A and 1B) (Hiyama and Ke, 1971b), but it is not clear which of the three clusters contribute under what conditions (see for example, Hiyama and Fork, 1980). It is likely that all the clusters are contributing since they all contain [4Fe-4S] centers and would be expected to have similar absorbance properties (for example, compare Figs. 1 and 5).

It has been possible to resolve F_X, F_A and F_B using low temperature EPR techniques, but this does not offer the opportunity of measuring kinetics and, in fact, there is recent evidence that indicates that pathways for electron transfer may exist at low temperatures that are not present at physiological temperatures (see below). This leads to the unsatisfying conclusion that we may never be able to resolve the individual electron transfer steps involving the terminal Fe-S acceptors in PS I.

In an early study of the kinetics of P430, Ke (1972) reported absorbance changes that occurred in less than 100 ns in the 430 nm spectral region after flash activation. Since both P700 and P430 show an absorbance decrease at 430 nm, it was not possible to resolve the two carriers, and Ke concluded that P430 was photoreduced in less than 100 ns as P700 was oxidized. In a preparation from cyanobacteria that lacked F_A and F_B, Warden (1990) also reported kinetics of the absorbance changes at 430 nm, arising from P700 and F_X in this case, and showed that F_X must be reduced in less than 5 ns. Whether the difference in these two kinetic values represents the rate of electron transfer between F_X and the terminal acceptors, F_A and F_B, is not clear, but the results do indicate electron transfer to the Fe-S clusters in the nanosecond time domain.

E. The Role of F_A and F_B in Non-cyclic Electron Transfer to Ferredoxin/NADP$^+$

An unresolved problem concerning the terminal electron acceptors in PS I is related to the function of the two [4Fe-4S] centers found in the PsaC subunit. Early EPR evidence indicated that F_A was the terminal acceptor that was photoreduced at cryogenic temperatures and little F_B accumulated under comparable conditions (Malkin and Bearden, 1971). When the midpoint potentials of the two centers were determined, this appeared reasonable as F_B has a more negative E_m value than F_A and the former could serve as a final sink for electrons at cryogenic temperatures. However, there were also different PS I preparations that showed a substantial amount of F_B photoreduction at cryogenic temperatures (Cammack et al., 1979; Hootkins et al., 1981) and additives, such as glycerol, were found to affect the amounts of F_A and F_B that accumulated after low temperature illumination (Evans and Heathcote, 1980). Thus, a simple picture of a series relationship between these two carriers appeared no longer tenable. In addition to this experimental complexity, there has also been a general conceptual problem concerning the two terminal acceptors in PS I. This has focused on the possible role of both or only one center in NADP$^+$ reduction. The focus of this discussion has been on whether both centers function in series to transfer electrons to ferredoxin, or whether some type of branched electron transfer sequence occurs:

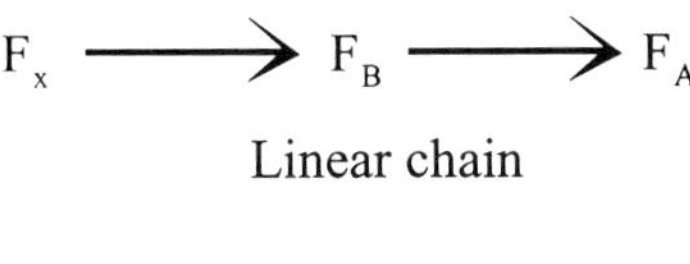

Linear chain

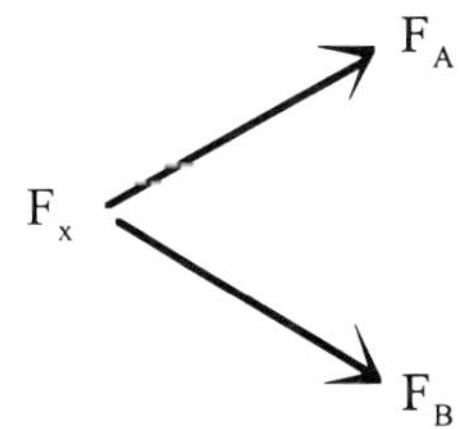

Branched Chain

Several different approaches have been used to attempt to define the relationship of F_A and F_B and the involvement of these centers in ferredoxin/NADP$^+$ reduction. One general approach has been to modify the PS I complex so that there has been a specific alteration of one of the two Fe-S centers. While the specificity of this approach may seem unreliable, there have been a number of chemical procedures that have modified F_B and have allowed for an analysis of the ensuing changes in activity. The second approach has been a molecular genetic one, using the techniques of site-directed mutagenesis to specifically alter one or both centers.

Golbeck and Warden (1982) were the first to report that one of the two terminal Fe-S centers could be altered. In this work, treatment of a PS I complex with increasing amounts of urea and ferricyanide (oxidative denaturation) led to a loss of F_B while F_A and F_X underwent much less destruction. Measurements of the EPR signals of the respective centers at cryogenic temperatures indicated that F_A could still be photoreduced after a substantial loss of F_B, and these workers argued for a branched electron transfer chain, as shown above. Unfortunately, there were no measurements of room temperature electron transfer reactions.

A different modification procedure was used by Malkin (1984) in studying the effect of the diazonium, DABS, on PS I. Modification with DABS led to a loss of NADP$^+$ photoreduction at physiological temperatures, but methyl viologen reduction was unaffected. Examination of the EPR signals of the PS I terminal acceptors showed that F_B was specifically modified again, as with urea-ferricyanide treatment, but in this case, the modified membranes showed a substantial inhibition of F_A photoreduction at cryogenic temperatures. The simplest model to explain these observations is that a linear electron transfer sequence occurs, with both F_B and F_A being required for the photoreduction of ferredoxin/NADP$^+$ and F_B being obligately required for F_A photoreduction.

One limitation of the above-described chemical modification procedures is that they may produce secondary changes in the PS I complex. One experimental approach for eliminating this possibility would be to reverse the modification procedure, but neither of the above procedures can be reversed. It has been found that treatment of PS I with mercurials, such as $HgCl_2$, leads to Fe-S alteration and that these changes can be reversed by subsequent reconstitution of the Fe-S centers. The initial observations using this approach were made by Sakurai and co-workers in 1987 (Kojima et al., 1987a,b) who reported that mercurial treatment specifically inhibited $NADP^+$ photoreduction with little effect on methyl viologen reduction. Since it is well established that mercurials react effectively with Fe-S clusters, this led to an examination of the bound Fe-S centers using EPR and the finding that, as previously described, F_B was being specifically altered by mercurial treatment. However, even though F_B was altered and photoreduction activity to $NADP^+$ at physiological temperatures was inhibited, F_A photoreduction at cryogenic temperatures was unaffected by this treatment. In contrast to the other treatments described, Inoue et al. (1992) reported that incubation of modified preparations with Na_2S, $FeCl_3$ and 2-mercaptoethanol led to a reconstitution of F_B, based on an EPR analysis. Inoue and Malkin (1994) have been able to confirm these findings. A concomitant restoration of $NADP^+$ reduction activity was also found after reconstitution of F_B in the PS I complex. One would argue from these results that F_B is an obligate carrier in PS I in the transfer of electrons to ferredoxin/$NADP^+$ but that F_B is not required for the photoreduction of F_A, at least at cryogenic temperatures. These apparently conflicting interpretations can be reconciled if one considers other measurements of the Fe-S centers that have been carried out at cryogenic temperatures that indicate alternative pathways of electron transfer may exist under these non-physiological conditions.

In experiments in which the phylloquinone electron acceptor has been removed from PS I by solvent extraction, resulting in a loss of $NADP^+$ reduction activity, it has been found that the terminal Fe-S centers can still be photoreduced at cryogenic temperatures (Sétif et al., 1987). The fact that physiological electron transfer reactions are inhibited in a reversible manner by this treatment would argue that major changes in the PS I complex have not occurred and that it is most likely that alternative pathways for electron transfer other than those that exist at physiological temperatures can occur at cryogenic temperatures. A similar effect has been noted in experiments where mutagenesis of F_X in a cyanobacterial PS I complex has produced a redox active F_X center, which is thermodynamically incapable of reducing F_A and F_B, yet the latter centers still can be photoreduced at cryogenic temperatures (Warren et al., 1993). On the basis of these additional studies, it would appear that measurements of electron transfer pathways at cryogenic temperatures are less reliable than those at physiological temperatures and that the ability to reconstitute a pathway that is functional under physiological conditions adds an additional positive control to these types of analyses.

The use of molecular genetics to study the F_A and F_B centers in the PS I complex has added a new and elegant approach to structure-function problems. The overall approach is based on the earlier observations of Golbeck and Bryant and their co-workers (see Golbeck and Bryant, 1991; Golbeck, 1992) that the Fe-S clusters in the PsaC subunit can be reconstituted after this protein was overexpressed in *E. coli*. Subsequent to the reconstitution, it was also possible to rebind the PsaC subunit to a PS I core complex containing the electron transfer carriers between P700 and F_X (Parrett et al., 1990). This reconstituted complex was fully competent in PS I electron transfer reactions. By molecular manipulation of the *psaC* gene, it has been possible to introduce specific mutations in the protein, altering the Fe-S centers. For example, Zhao et al. (1992) showed that a conversion of cysteine ligands to aspartate ligands was possible, that the proteins were assembled with Fe-S centers, and that it was possible to obtain a [3Fe-4S] cluster at either the F_A or F_B binding site. Subsequent electrochemical analyses of the redox potentials of the [4Fe-4S] clusters in these mutated proteins showed that they retained the redox properties of the native F_A or F_B cluster, that is, in a PsaC subunit containing a [3Fe-4S] center at the F_A site, the F_B potential was approximately –580 mv, while in a PsaC subunit with a [3Fe-4S] cluster at the F_B site, the F_A potential was approximately –515 mV (Yu et al., 1993a). Of particular interest to the functional problem concerning the F_A and F_B clusters was the finding that the presence of a non-functional F_B did not inhibit the photoreduction of F_A at cryogenic temperatures (Zhao et al., 1992). This result confirms others discussed above that indicate that F_B is not an obligatory carrier in the pathway of electrons to F_A, although we must also accept that the pathway at cryogenic temperatures may not be representative of processes at physiological temperatures. Interestingly, in these studies, it was also found that there was no enhanced reduction of F_B in a mutated protein that contained an altered F_A cluster, a finding that is not consistent with a simple linear chain where F_B precedes F_A as an electron carrier. Unfortunately, the critical measurements of room temperature reactions involving ferredoxin/NADP+ have not been reported with these mutated PsaC proteins so it is not yet possible to draw conclusions about the involvement of each of the Fe-S clusters in these reactions, allowing for a comparison of the results obtained by the chemical modification procedures described above.

F. Ferredoxin Reduction

Chloroplast ferredoxin serves as the first soluble electron carrier reduced through PS I. As described above, electrons are transferred through the bound electron carriers, terminating in the two Fe-S clusters localized on the PsaC subunit, to ferredoxin, although the role of the F_A and F_B clusters in this process remains to be resolved.

The interaction of ferredoxin with the PS I complex has been studied by a number of techniques. Using a biochemical approach, Zanetti and Merati (1987) and Zilber and Malkin (1988) cross-linked ferredoxin to spinach PS I complexes using the zero-length cross-linker, EDC. Using immunological techniques to identify specific cross-linked products, it was shown that ferredoxin cross-linked to a 22 kDa subunit, now known as the PsaD subunit, in the PS I complex. The interaction between ferredoxin and PsaD apparently is electrostatic, involving carboxyl groups on the soluble protein, since modification of these carboxyls with glycine ethyl ester led to a loss of $NADP^+$ photoreduction and a concomitant loss of cross-linking to the PsaD subunit (unpublished results). In these first experiments, it was also noted that a second cross-linked product was formed between ferredoxin and another PS I subunit, but the identity of the second subunit was not confirmed. In a reinvestigation of this problem, Andersen et al. (1992) found that the PsaE subunit of approximately 18 kDa also cross-linked to ferredoxin. Both the PsaD and PsaE subunits were localized to the stromal side of the thylakoid membrane in experiments using specific antibodies to each subunit, and this orientation would agree with their interaction with ferredoxin, a known stromal protein. What has emerged from these studies is a definition of two specific subunits in the PS I complex, PsaD and PsaE, that function as 'docking' proteins. In this regard, it is believed that these subunits act to hold ferredoxin in proximity to PsaC to facilitate electron transfer between the two proteins. The PsaC subunit is apparently more buried since it does not react with a specific antibody under conditions where PsaD and PsaE are reactive. It has been proposed that PsaD and PsaE actually shield PsaC, not permitting stromal contact by the latter, although there must still be a contact point or area between PsaC and ferredoxin. This organization of the PsaD and PsaE subunits relative to PsaC and ferredoxin is also consistent with reconstitution studies of Golbeck and Bryant and their groups (Li et al., 1991).

If we accept that complex formation between ferredoxin and PS I is the first step in the electron transfer sequence and that specific PS I subunits are involved in orienting the ferredoxin molecule, the next step in the sequence would be the actual electron transfer event. Hervas et al. (1992) reported the first measurements of the kinetics of the electron transfer from PS I to ferredoxin. Using an absorbance change at 480 nm to monitor ferredoxin reduction (Fig. 6, top), the kinetics were found to be non-linear with respect to protein concentration (Fig. 6, bottom). This has been interpreted as described above: complex formation precedes electron transfer. The second order rate constant for the reduction of ferredoxin was found to be 140–180 s^{-1}, indicating ferredoxin reduction occurs in the 5–10 ms time range (Fig. 6, bottom). This rate constant was sensitive to ionic strength, in agreement with the previous observations that suggested that an electrostatic interaction between ferredoxin and PS I occurs. It is also important to point out that the kinetics of all the electron transfer reactions that have been previously described in PS I occur in the sub-nanosecond time domain, but that the final step involving the interaction of a soluble carrier with the integral membrane complex is several orders of magnitude slower than this and the latter probably represents the rate-limiting electron transfer step in the overall PS I catalyzed reaction sequence.

Electron transfer reactions involving ferredoxin reduced by PS I, other than the cyclic electron transport sequence, will be discussed in greater detail in Chapter 17.

IV. Cyclic Electron Transfer and Cyclic Phosphorylation

The reactions that have been described thus far document the non-cyclic transfer of electrons through PS I to ferredoxin and on to $NADP^+$. In addition to this non-cyclic pathway, it has been known for over 25 years that PS I is also involved in a cyclic electron transport pathway that yields only ATP as a product.

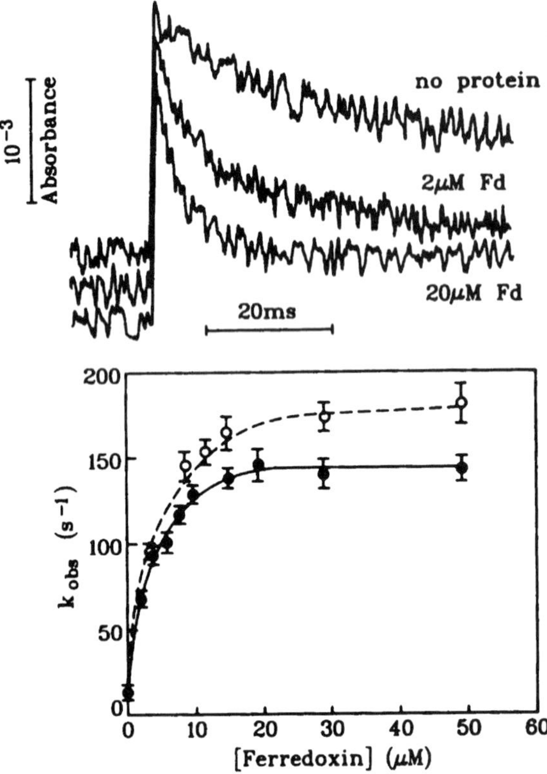

Fig. 6. Flash-induced absorbance changes associated with ferredoxin reduction. Absorbance transients were measured at 480 nm in a PS I complex from spinach with or without the addition of spinach or algal ferredoxin. Open circles: algal ferredoxin; closed circles: spinach ferredoxin. From Hervas et al. (1992).

The general features of the cyclic pathway are that it involves only PS I, based on measurements of ATP synthesis in monochromatic light and inhibitor sensitivity, and that a redox carrier or cyclic mediator, was required. In a series of studies by Arnon and co-workers (1967, 1975), evidence was presented that ferredoxin served as the endogenous catalyst for cyclic phosphorylation, although non-physiological mediators can function in this process. While the in vitro activity of cyclic phosphorylation is readily documented, there has been much more controversy as to whether this reaction occurs in vivo, although evidence with intact chloroplasts under conditions where CO_2 fixation was occurring did provide strong support for the concept that cyclic phosphorylation produces ATP required for CO_2 assimilation (Schürmann et al., 1971).

The mechanism of this PS I-mediated cyclic pathway and the concomitant synthesis of ATP remains unclear. As described above, inhibitors have been used to define the pathway and to eliminate any role for Photosystem II in the overall sequence of reactions. Inhibitors of the cytochrome b_6f complex, such as DBMIB, have been particularly useful in these studies since they have provided strong evidence that electron flow through the cytochrome complex is required for cyclic phosphorylation. On the basis of these findings, it is generally accepted that ferredoxin, reduced via PS I, can donate electrons either directly or indirectly to the cytochrome b_6f complex, and that subsequent transfer of electrons through the cytochrome complex can be used to establish a proton gradient leading to the synthesis of ATP.

A more controversial inhibitor used in these studies has been antimycin A, a well-established inhibitor of the cytochrome bc_1 complex of the mitochondrion. Antimycin A has been found to specifically inhibit cyclic phosphorylation while having little effect on non-cyclic electron transport and phosphorylation (Arnon et al., 1967). However, the antimycin concentration required for inhibition is higher than that required for the cytochrome bc_1 complex, and this has complicated interpretations. A simple view might be that the effect of antimycin is through the cytochrome b_6f complex, analogous to its behavior in the mitochondrial electron transport system, but studies with the isolated chloroplast cytochrome b_6f complex have shown it to be relatively insensitive to antimycin A (Hauska et al., 1983).

Because of the apparent insensitivity of the cytochrome b_6f complex to antimycin, Moss and Bendall (1984) considered the possibility that the antimycin site in chloroplasts was outside the cytochrome complex and provided the first evidence in support of this idea. They proposed the existence of a ferredoxin-plastoquinone oxidoreductase which functioned to transfer electrons from reduced ferredoxin to oxidized plastoquinone. The quinol formed in this reaction could then be reoxidized by the cytochrome b_6f complex. The behavior of antimycin would be to inhibit the ferredoxin-plastoquinone reductase activity rather than the cytochrome b_6f complex activity. Subsequent work by Cleland and Bendall (1992) provided a fuller description of the enzymatic activity of the reductase by direct measurements of the kinetics of ferredoxin reduction by PS I and its subsequent reoxidation by the presumed reductase. While the concept of such an enzyme is attractive in terms of offering a biochemical explanation for the unique aspects of the cyclic pathway in chloroplasts, molecular characterization of the elusive ferredoxin-plastoquinone reductase is still lacking and the nature of the interaction of this enzyme with PS I as well as with the cytochrome complex is unclear.

As indicated above, the biochemistry of the cyclic phosphorylation pathway in chloroplasts remains unclear. In the case of cyanobacteria, a molecular genetic approach has provided some new evidence that indicates one specific PS I subunit is required for this pathway. By studying a *psaE* deletion mutant of a cyanobacterium, Yu et al. (1993b) were able to demonstrate that the PsaE subunit is an essential component of the cyclic pathway. Interestingly, other groups, such as Andersen et al. (1992) and Rousseau et al. (1993), have also provided evidence for a role for this subunit in the non-cyclic pathway, so there remains a disagreement on the specificity of the role of PsaE, although the experimental evidence in both cases is different and the reactions have been studied in different organisms. However, the molecular genetic approach is a new experimental technique that will continue to be used to examine the properties of the cyclic pathway in organisms that can be easily manipulated at the molecular level.

Acknowledgments

The research from the author's laboratory described in this chapter was supported by a grant from the National Science Foundation. Thanks to Dr. L. Feldman for aid with the preparation of photographs for this chapter.

References

Andersen B, Koch B and Scheller HV (1992) Structural and functional analysis of the reducing side of Photosystem I. Physiol Plant 84: 154–161

Arnon DI and Chain RK (1975) Regulation of ferredoxin-catalyzed photosynthetic phosphorylation. Proc Natl Acad Sci USA 72: 4961–4965

Arnon DI, Tsujimoto HY and McSwain BD (1967) Ferredoxin and photosynthetic phosphorylation. Nature 214: 562–566

Biggins J (1990) Evaluation of selected benzoquinones, napthoquinones and anthraquinones as replacements for phylloquinone in the A_1 acceptor site of the Photosystem I reaction center. Biochemistry 29: 7259–7264

Biggins J and Mathis P (1988) Functional role of vitamin K1 in Photosystem I of the cyanobacterium *Synechocystis* 6803. Biochemistry 27: 1494–1500

Biggins J, Tanguay NA and Frank HA (1989) Electron transfer reactions in Photosystem I following vitamin K1 depletion by ultraviolet irradiation. FEBS Lett 250: 271–274

Bock CH, van der Est AJ, Brettel K and Stehlik D (1989) Nanosecond electron transfer kinetics in Photosystem I as obtained from transient EPR at room temperature. FEBS Lett 247: 91–96

Boekema EJ, Wynn RM and Malkin R (1990) The structure of spinach Photosystem I studied by electron microscopy. Biochim Biophys Acta 1017: 49–56

Bonnerjea J and Evans MCW (1982) Identification of multiple components in the intermediary electron acceptor complex of Photosystem I. FEBS Lett 148: 313–316

Brettel K (1988) Electron transfer from A_1^- to an iron-sulfur center with $t_{1/2}$ = 200 ns at room temperature in photosystem I. FEBS Lett 239: 93–98

Bruce BD and Malkin R (1988a) Subunit stoichiometry of the chloroplast Photosystem I complex. J Biol Chem 263: 7302–7308

Bruce BD and Malkin R (1988b) Structural aspects of photosystem I from *Dunaliella salina*. Plant Physiol 88: 1201–1206

Cammack R, Ryan MD and Stewart AC (1979) The epr spectrum of iron-sulphur centre B in Photosystem I in *Phormidium laminosum*. FEBS Lett 107: 422–426

Chamarovsky SK and Cammack R (1982) Oxidation-reduction potential of iron-sulphur centre X in Photosystem I. Photobiochem Photobiophys 4: 195–200

Chitnis PR and Nelson N (1991) Photosystem I. In: Bogorad L and Vasil IK (eds), The Photosynthetic Apparatus: Molecular Biology and Operation, pp 177–224. Academic Press, New York

Cleland RE and Bendall DS (1992) Photosystem I cyclic electron transport: Measurement of ferredoxin-plastoquinone reductase activity. Photosynth Res 34: 409–418

Evans MCW and Heathcote P (1980) Effects of glycerol on the redox properties of the electron acceptor complex in spinach Photosystem I particles. Biochim Biophys Acta 590: 89–96

Evans MCW, Reeves SG and Cammack R (1974) Determination of the oxidation-reduction potential of the bound iron-sulfur proteins of the primary electron acceptor complex of Photosystem I in spinach chloroplasts. FEBS Lett 49: 111–114

Fenton JM, Pellin MJ, Govindjee and Kaufmann KJ (1979) Primary photochemistry of the reaction center of Photosystem I. FEBS Lett 100: 1–4

Fish LE, Kuck U and Bogorad L (1985) Two partially homologous adjacent light inducible chloroplast genes encoding polypeptides of the P700 chlorophyll *a* protein complex of Photosystem I. J Biol Chem 260: 1413–1421

Fujita I, Davis MS and Fajer J (1978) Anion radicals of pheophytin and chlorophyll *a*: Their role in the primary charge separations of plant photosynthesis. J Am Chem Soc 100: 6280–6282

Golbeck J (1992) Structure and function of Photosystem I. Ann Rev Plant Physiol Plant Mol Biol 43: 293–324

Golbeck JH (1994) Photosystem I in cyanobacteria. In: Bryant DA (ed) The Molecular Biology of Cyanobacteria, pp 319–360. Kluwer Academic Publishers, Dordrecht

Golbeck J and Bryant D (1991) Photosystem I. Curr Topics in Bioenergetics 16: 83–177

Golbeck JH and Warden JT (1982) Electron spin resonance

studies of the bound iron-sulfur centers in Photosystem I. Photoreduction of center A occurs in the absence of center B. Biochim Biophys Acta 681: 77–84

Golbeck JH, McDermott AE, Jones WK and Kurtz DM (1987a) Evidence for the existence of [2Fe-2S] as well as [4Fe-4S] clusters among F_A, F_B and F_X. Implications for the structure of the Photosystem I reaction center. Biochim Biophys Acta 891: 94–98

Golbeck JH, Parrett KG, Mehari T, Jones KL and Brand JJ (1987b) Isolation of the intact Photosystem I reaction center core containing P700 and iron-sulfur center F_X. FEBS Lett 228: 268–272

Hastings G, Kleinherenbrink FAM, Lin S and Blankenship RE (1994a) Time resolved fluorescence and absorption spectroscopy of Photosystem I. Biochemistry 33: 3185–3192

Hastings G, Kleinherenbrink FAM, Lin, S, McHugh TJ and Blankenship RE (1994b) Observation of the reduction and reoxidation of the primary electron acceptor in Photosystem I. Biochemistry 33: 3193–3200

Hauska G, Hurt E, Gabellini N and Lockau W (1983) Comparative aspects of quinol-cytochrome *c*/plastocyanin oxidoreductases. Biochim Biophys Acta 726: 97–133

He W-Z and Malkin R (1994) Reconstitution of iron-sulfur center B of Photosystem I damaged by mercuric chloride. Photosynth Res 41: 381–388

Hervas M, Navarro JA and Tollin G (1992) A laser flash spectroscopic study of the kinetics of electron transfer from spinach Photosystem I to spinach and algal ferredoxins. Photochem Photobiol 56: 319–324

Hiyama T and Fork D (1980) Kinetic identification of Component X as P430: A primary electron acceptor of Photosystem I. Arch Biochem Biophys 199: 488–496

Hiyama T and Ke B (1971a) A new photosynthetic pigment, 'P430': Its possible role as the primary electron acceptor of Photosystem I. Proc Natl Acad Sci USA 68: 1010–1013

Hiyama T and Ke B (1971b) A further study of P430: a possible primary electron acceptor of Photosystem I. Arch Biochem Biophys 147: 99–108

Høj PB and Møller BL (1986) The 110 kDa reaction centre protein of Photosystem I, chlorophyll *a*-protein 1, is an iron-sulphur protein. J Biol Chem 261: 14292–14300

Holzwarth AR, Schatz G, Brock H and Bittersmann E (1993) Energy transfer and charge separation kinetics in Photosystem I. Part I: Picosecond transient absorption and fluorescence study of cyanobacterial Photosystem I particles. Biophys J 64: 1813–1826

Hootkins R, Malkin R and Bearden AJ (1981) EPR properties of Photosystem I iron-sulfur centers in the halophilic alga, *Dunaliella parva*. FEBS Lett 123: 229–234

Inoue K, Kusumoto N and Sakurai H (1992) Some properties of iron-sulfur centers in F_B-destroyed and F_B-reconstituted PS I particles. In: Murata N (ed) Research in Photosynthesis vol I, pp 577–580. Kluwer, Dordrecht

Itoh S and Iwaki I (1991) Full replacement of the function of the secondary electron acceptor phylloquinone (vitamin K1) by non-quinone carbonyl compounds in green plant photosystem I photosynthetic reaction centers. Biochemistry 30: 5339–5346

Iwaki M and Itoh S (1989) Electron transfer in spinach Photosystem I reaction center containing benzo-, naptho- and anthraquinones in place of phylloquinone. FEBS Lett 256: 11–16

Ke B (1972) The rise time of photoreduction, difference spectrum and oxidation-reduction potential of P430. Arch Biochem Biophys 152: 70–77

Ke B and Beinert H (1973) Evidence for the identity of P430 and chloroplast-bound iron-sulfur protein. Biochim Biophys Acta 305: 689–693

Ke B, Hansen RE and Beinert H (1973) Oxidation-reduction potentials of bound iron-sulfur proteins of Photosystem I. Proc Natl Acad Sci USA 70: 2941–2945

Kojima Y, Hiyama T and Sakurai H (1987a) Effects of mercurials on iron-sulfur centers of Photosystem I of *Anacystis nidulans*. In: Biggins J (ed) Progress in Photosynthesis Research, Vol 2, pp 57–60. Martinus Nijhoff, Dordrecht

Kojima Y, Niinomi Y, Tsuboi S, Hiyama T and Sakurai H (1987b) Destruction of Photosystem I iron-sulfur centers of spinach and *Anacystis nidulans* by mercurials. Bot Mag Tokyo 100: 243–253

Kok B (1956) Preliminary notes on the reversible absorption change at 705 μm in photosynthetic organisms. Biochim Biophys Acta 22: 399–401

Kok B (1957) Absorption changes induced by the photochemical reaction of photosynthesis. Nature 179: 583–584

Krauss N, Hinrichs W, Witt I, Fromme P, Pritzkow W, Dauter Z, Betzel C, Wilson KS, Witt HT and Saenger W (1993) Three-dimensional structure of system I of photosynthesis at 6Å resolution. Nature 361: 326–331

Li N, Zhao J, Warren PV, Warden JT, Bryant DA and Golbeck JH (1991) PsaD is required for stable binding of PsaC to Photosystem I core protein of *Synechococcus* sp . PCC 6301. Biochemistry 30: 7863–7872

Lundell DJ, Glazer AN, Melis A and Malkin R (1985) Characterization of a cyanobacterial PS I complex. J Biol Chem 260: 646–654

Malkin R (1984) Diazonium modification of Photosystem I. A specific effect on iron-sulfur center B. Biochim Biophys Acta 764: 63–69

Malkin R (1986) On the function of two vitamin K1 molecules in the Photosystem I electron acceptor complex. FEBS Lett 208: 343–346

Malkin R and Bearden AJ (1971) Primary reactions of photosynthesis: Photoreduction of a bound chloroplast ferredoxin at low temperatures as detected by EPR spectroscopy. Proc Natl Acad Sci USA 68: 16–19

Mansfield RW and Evans MCW (1985) Optical difference spectrum of the electron acceptor A_0 in Photosystem I. FEBS Lett 190: 237–241

Mansfield RW, Hubbard JAM, Nugent JHA and Evans MCW (1987a) Extraction of electron acceptor A_1 from pea Photosystem I. FEBS Lett 220: 74–78

Mansfield RW, Nugent JHA and Evans MCW (1987b) ESR characteristics of Photosystem I in deuterium oxide: Further evidence that electron acceptor A_1 is a quinone. Biochim Biophys Acta 894: 515–523

Mathis P and Sétif P (1988) Kinetic studies on the function of A_1 in the Photosystem I reaction center. FEBS Lett 237: 65–68

Mathis P, Ikegami I and Sétif P (1988) Nanosecond flash studies of the absorption spectrum of the Photosystem I primary acceptor A_0. Photosynth Res 16: 203–210

McDermott AE, Yachandra VK, Guiles RD, Britt RD, Dexheimer SL, Sauer K and Klein MP (1987) Characterization of the Mn-containing O_2 evolving complex from the cyanobacterium

Synechococcus using epr and X-ray absorption spectroscopy. In: Biggins J (ed) Progress in Photosynthesis Research, vol I, pp 565–568. Martinus Nijhoff, Dordrecht

McDermott AE, Yachandra VK, Guiles RD, Sauer, K, Parrett KG and Golbeck JH (1989) An EXAFS structural study of F_X, the low potential Fe-S center in Photosystem I. Biochemistry 28: 8056–8059

Moënne-Loccoz P, Robert B, Ikegami I and Lutz M (1990) Structure of the primary donor in Photosystem I: A resonance Raman study. Biochemistry 29: 4740–4746

Moss DA and Bendall DS (1984) Cyclic electron transport in chloroplasts. The Q-cycle and the site of action of antimycin. Biochim Biophys Acta 767: 389–395

Nuijs AM, Shuvalov VA, van Gorkom HJ, Plijter JJ and Duysens LNM (1986) Picosecond absorbance difference spectroscopy on the primary reactions and the antenna-excited states in Photosystem I particles. Biochim Biophys Acta 850: 310–318

Oh-oka H, Takahashi Y, Wada K, Matsubara H, Ohyama K and Ozeki H (1987) The 8 kDa polypeptide in Photosystem I is a probable candidate of an iron-sulfur protein coded by chloroplast gene *frx* A. FEBS Lett 218: 52–54

Oh-oka H, Itoh S, Saeki K, Takahashi Y and Matsubara H (1991) F_A/F_B protein from spinach Photosystem I complex: Isolation in a native state and some properties of the iron-sulfur centers. Plant Cell Physiol 32: 11–17

O'Malley PJ and Babcock GT (1984) Electron nuclear double resonance evidence supporting a monomeric nature for $P700^+$ in spinach chloroplasts. Proc Natl Acad Sci USA 81: 1098–1101

Owens TG, Webb SP, Mets L, Alberte RS and Fleming GR (1987) Antenna size dependence of fluorescence decay in the core antenna of Photosystem I: Estimates of charge separation and energy transfer rates. Proc Natl Acad Sci USA 84: 1532–1536

Palace GP, Franke JE and Warden JT (1987) Is phylloquinone an obligate electron carrier in Photosystem I? FEBS Lett 215: 58–62

Parrett KG, Mehari T, Warren PG and Golbeck JH (1989) Purification and properties of the intact P700 and Fx-containing Photosystem I core protein. Biochim Biophys Acta 973: 324–332

Parrett KG, Mehari T and Golbeck JH (1990) Resolution and reconstitution of the cyanobacterial Photosystem I complex. Biochim Biophys Acta 1015: 341–352

Petrouleas V, Brand JJ, Parrett KP and Golbeck JH (1989) A Mossbauer analysis of the low potential iron-sulfur center in Photosystem I. Spectroscopic evidence that F_X is a 4Fe-4S cluster. Biochemistry 28: 8980–8983

Prisner TF, McDermott AE, Un S, Norris JR, Thurnauer MC and Griffin RG (1993) Measurement of the *g*-tensor of the $P700^{+\cdot}$ signal from deuterated cyanobacterial Photosystem I particles. Proc Natl Acad Sci USA 90: 9485–9488

Rousseau F, Sétif P and Lagoutte B (1993) Evidence for the involvement of PS I-E subunit in the reduction of ferredoxin by Photosystem I. EMBO J 12: 1755–1765

Rustandi RR, Snyder SW, Feezel LL, Michalski TJ, Norris JR and Thurnauer MC (1990) Contribution of vitamin K1 to the electron spin polarization in spinach Photosystem I. Biochemistry 29: 8030–8032

Rustandi RR, Snyder SW, Biggins J, Norris JR and Thurnauer MC (1992) Reconstitution and exchange of quinones in the A_1 site of Photosystem I. An electron spin polarization paramagnetic resonance study. Biochim Biophys Acta 1101: 311–320

Scheller HV, Svendsen I, Møller BL (1989) Subunit comparison of Photosystem I and identification of center X as a 4Fe-4S iron-sulfur center. J Biol Chem 264: 6929–6934

Schoeder HU and Lockau W (1986) Phylloquinone copurifies with the large subunits of Photosystem I. FEBS Lett 199: 23–27

Schürmann P, Buchanan BB and Arnon DI (1971) Role of cyclic phosphorylation in photosynthetic carbon dioxide assimilation by isolated chloroplasts. Biochim Biophys Acta 267: 111–124

Sétif P (1992) Energy transfer and trapping in Photosystem I. In: Barber J (ed), The Photosystems: Structure, Function and Molecular Biology, pp 471–499. Elsevier, Amsterdam

Sétif P and Brettel K (1993) Forward electron transfer from phylloquinone A_1 to iron-sulfur centers in spinach Photosystem I. Biochemistry 32: 7846–7854

Sétif P and Mathis P (1980) The oxidation-reduction potential of P700 in chloroplast lamellae and subchloroplast particles. Arch Biochem Biophys 204: 477–485

Sétif P and Mathis P (1986) Photosystem I reaction center and its primary electron transfer reactions. In: Staehelin LA and Arntzen CJ (eds) Encyclopedia of Plant Physiology, New Series, volume 19, Photosynthesis III, pp 476–486. Springer-Verlag, Berlin

Sétif P, Ikegami I and Biggins J (1987) Light-induced charge separation in Photosystem I at low temperature is not influenced by vitamin K1. Biochim Biophys Acta 894: 146–156

Shuvalov VA, Dolan E and Ke B (1979) Spectral and kinetic evidence for two early electron acceptors in Photosystem I. Proc Natl Acad Sci USA 76: 770–773

Shuvalov VA, Nuijs AM, van Gorkom HJ, Smit HWJ and Duysens LNM (1986) Picosecond absorbance changes upon selective excitation of the primary electron donor P-700 in Photosystem I. Biochim Biophys Acta 850: 319–323

Sieckman I, van der Est A, Bottin H, Sétif P and Stehlik D (1991) Nanosecond electron transfer kinetics in Photosystem I following substitution of quinones for vitamin K1 as studied by time resolved EPR. FEBS Lett 284: 98–102

Smith NS, Mansfield RW, Nugent, JHA and Evans MCW (1987) Characterisation of electron acceptors A_0 and A_1 in cyanobacterial Photosystem I. Biochim Biophys Acta 892: 331–334

Snyder SW, Rustandi RR, Biggins J, Norris, JR and Thurnauer MC (1991) Direct assignment of vitamin K1 as the secondary acceptor A_1 in Photosystem I. Proc Natl Acad Sci USA 88: 9895–9896

Swarthoff T, Gast P, Amesz J and Buisman HP (1982) Photoaccumulation of reduced primary electron acceptors of Photosystem I of photosynthesis. FEBS Lett 146: 129–132

Thurnauer MC and Norris JR (1980) An electron spin echo phase shift in photosynthetic algae. Possible evidence for dynamic radical pair interactions. Chem Phys Lett 76: 557–561

Trissl H-W, Hecks B and Wulf K (1993) Invariable trapping times in Photosystem I upon excitation of minor long-wavelength-absorbing pigments. Photochem Photobiol 57: 108–112

Turconi S, Schweitzer G and Holzwarth AR (1993) Temperature dependence of picosecond fluorescence kinetics of a cyanobacterial Photosystem I particle. Photochem Photobiol

57: 113–119

Warden JT (1990) Nanosecond spectroscopy of early reduced electron transfer in Photosystem I. Iron-sulfur cluster F_X is reduced within 5 nanoseconds. In: Baltscheffsky M (ed) Current Research in Photosynthesis, Vol. II, pp 635–638. Kluwer, Dordrecht

Warren PV, Golbeck JH and Warden JT (1993) Charge recombination between $P700^+$ and A_1^- occurs directly to the ground state of P700 in a Photosystem I core devoid of F_X, F_B and F_A. Biochemistry 32: 849–857

Warren PV, Smart LB, McIntosh L and Golbeck JH (1993) Site directed conversion of cysteine-565 in *psaB* of Photosystem I results in the assembly of [3Fe-4S] and [4Fe-4S] clusters in F_X. A mixed-ligand [4Fe-4S] cluster is capable of electron transfer to F_A and F_B. Biochemistry 32: 4411–4419

Wasielewski MR, Fenton JM and Govindjee (1987) The rate of formation of $P700^+$-A_0^- in Photosystem I particles from spinach as measured by picosecond transient absorption spectroscopy Photosynth Res 12: 181–190

Wynn RM and Malkin R (1988) Characterization of an isolated chloroplast membrane Fe-S protein and its identification as the Photosystem I Fe-S_A/Fe-S_B. FEBS Lett 229: 293–297

Yu L, Zhao J, Lu W, Bryant DA and Golbeck JH (1993a) Characterization of the [3Fe-4S] and [4Fe-4S] clusters in unbound *psaC* mutants C14D and C51D. Midpoint potentials of the single [4Fe-4S] clusters are identical to F_A and F_B in bound PsaC of Photosystem I. Biochemistry 32: 8251–8258

Yu L, Zhao J, Muhlenhoff U, Bryant DH and Golbeck JH (1993b) PsaE is required for in vivo cyclic electron flow around Photosystem I in the cyanobacterium *Synechococcus* sp. PCC 7002. Plant Physiol 103: 171–180

Zanetti G and Merati G (1987) Interaction between Photosystem I and ferredoxin. Eur J Biochem 169: 143–146

Zhao J, Li N, Warren PV, Golbeck JH and Bryant DA (1992) Site-directed conversion of a cysteine to aspartate leads to the assembly of a [3Fe-4S] cluster in PsaC of Photosystem I. The photoreduction of F_A is independent of F_B. Biochemistry 31: 5093–5099

Ziegler K, Lockau W and Nitschke W (1987) Bound electron acceptors of Photosystem I. Evidence against the identity of redox center A_1 with phylloquinone. FEBS Lett 217: 16–20

Zilber A and Malkin R (1988) Ferredoxin cross-links to a 22 kD subunit of Photosystem I. Plant Physiol 88: 810–814

Chapter 17

Ferredoxin and Ferredoxin-Dependent Enzymes

David B. Knaff
Department of Chemistry and Biochemistry, and Institute for Biotechnology, Texas Tech University, Lubbock, Texas 79409-1061, USA

Summary

Ferredoxin, a M_r = 11 kDa protein that contains a single low potential [2Fe-2S] cluster, serves as the ultimate acceptor of electrons from Photosystem I. In addition to serving as the electron donor for $NADP^+$ reduction, a reaction catalyzed by the FAD-containing, membrane-associated enzyme ferredoxin:$NADP^+$ oxidoreductase, ferredoxin also serves as the electron donor to a number of soluble enzymes involved in nitrogen metabolism, sulfur metabolism and the regulation of carbon metabolism. This chapter describes the current state of our knowledge of the structure of ferredoxin and of these ferredoxin-dependent enzymes, of the biosynthesis of these proteins and of the mode of interaction between ferredoxin and its electron-accepting, reaction-partner proteins.

Donald R. Ort and Charles F. Yocum (eds): Oxygenic Photosynthesis: The Light Reactions, pp. 333–361.

I. Introduction

The ultimate destination for essentially all of the electrons removed from water by the action of Photosystem II and the oxygen-evolving complex during oxygenic photosynthesis is the soluble iron-sulfur protein ferredoxin. The actual reduction of ferredoxin is accomplished by the Photosystem I reaction center through a series of reactions that have been described in the preceding chapter. In some cyanobacteria and algae, the low molecular mass, FMN-containing protein flavodoxin replaces ferredoxin, particularly when these organisms are grown under iron-limiting conditions (Rogers, 1987; Mayhew and Tollin, 1992). However, this chapter will not deal with flavodoxin in any detail.

In addition to the non-cyclic flow of electrons from water to ferredoxin that requires both Photosystem I and Photosystem II, ferredoxin also participates in a cyclic electron flow that involves only Photosystem I and is coupled to ATP formation. Cyclic electron flow has been discussed in the preceding chapter, so this chapter will focus only on ferredoxin itself and on four enzymes which serve as acceptors of electrons from the ferredoxin reduced during non-cyclic electron flow: Ferredoxin:NADP$^+$ oxidoreductase (EC 1.18.1.2, hereafter abbreviated as FNR); ferredoxin:nitrite oxidoreductase (EC 1.7.7.1, hereafter referred to as nitrite reductase); glutamate synthase (EC 1.4.7.1, called glutamine:oxoglutarate aminotransferase or GOGAT in some references) and ferredoxin:thioredoxin oxidoreductase (hereafter referred to as thioredoxin reductase and abbreviated FTR).

II. Ferredoxin

A. Structure and Redox Properties

The ferredoxins found in oxygenic photosynthetic organisms (i.e., in cyanobacteria, algae and plants) are small, soluble, acidic proteins (M_r = ca. 11 kDa, pI's ranging from below 3.0 to approximately 4.0) that contain a single [2Fe-2S] cluster as the prosthetic group (Cammack et al., 1977; Matsubara et al., 1980; Matsubara and Hase, 1983; Cammack et al., 1985; Knaff and Hirasawa, 1991). Each iron in the cluster is coordinated by four sulfur ligands: Two inorganic sulfides that bridge the irons, and two cysteinyl sulfurs, with the four sulfurs arranged in an approximately tetrahedral array around each iron (Fig. 1, taken from Holden et al., 1994). Amino acid sequences are known for a large number of ferredoxins and the positions of the four cysteines that serve as ligands to the cluster irons are highly conserved (Matsubara et al., 1980; Matsubara and Hase, 1983; Knaff and Hirasawa, 1991; Barker et al., 1992; Holden et al., 1994). Typical oxidation-reduction midpoint potential (E_m) values for these ferredoxins are approximately –420 mV (Cammack et al., 1977), making the reduced forms of the proteins among the strongest soluble reductants found in nature.

The [2Fe-2S] cluster in oxidized ferredoxin contains two antiferromagnetically coupled, high-spin ($S = 5/2$) ferric irons (cluster spin state, $S = 0$) and can accept one electron, yielding a cluster in reduced ferredoxin (cluster spin state, $S = 1/2$) that contains one high-spin ferric iron antiferromagnetically coupled to one high-spin ferrous ($S = 2$) iron (Palmer, 1973; Sands and Dunham, 1975; Fu et al., 1992a). The iron and inorganic sulfide constituents can be removed from the purified protein with relative ease and the cluster can then be reconstituted into the apoprotein in vitro with good yields (Bayer et al., 1967). This has made it possible for nuclei with nuclear moments (e.g., ^{57}Fe and the sulfur analogs ^{77}Se and ^{80}Se) to be incorporated into the cluster, which in turn made it possible to obtain a better understanding of the electronic structure of the cluster and to characterize the interaction of the unpaired electron present in the reduced cluster with the inorganic sulfides. Some delocalization of electron density from the reduced cluster to the sulfur, β-CH_2 and α-CH of the ligating cysteines has also been shown to occur (Dunham et al., 1971; Salmeen and Palmer, 1972; Skjeldal et al., 1991).

The fact that the electron taken up by the cluster on reduction is substantially localized on one of the two iron atoms implies that the two iron atoms in the cluster are not in equivalent environments. It appears likely that the iron (usually designated Fe1) ligated to Cys^{41} and Cys^{46} (the numbering used here refers to the ferredoxin isolated from the vegetative cells of the cyanobacterium *Anabaena* 7120) is predominantly ferrous iron in the reduced protein, while the iron (usually designated Fe2) ligated to Cys^{49} and

Abbreviations: FNR – ferredoxin:NADP$^+$ oxidoreductase; FTR – ferredoxin:thioredoxin oxidoreductase

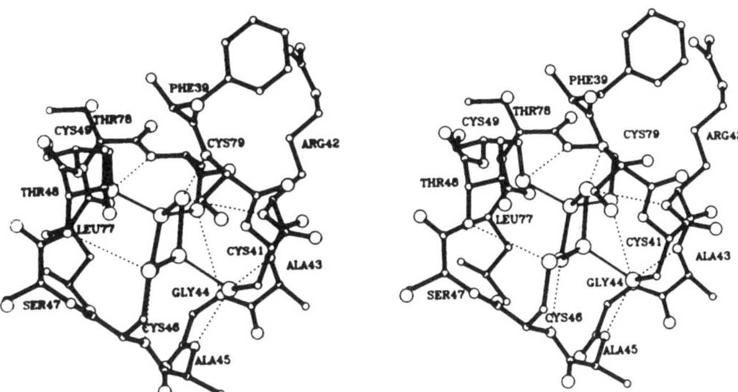

Fig. 1. Stereo view of the [2Fe-2S] binding pocket of the ferredoxin isolated from vegetative cells of *Anabaena* 7120. Amino acids within approximately 3.5 Å of the cluster are shown. Dashed lines indicate possible hydrogen bonds between backbone amide hydrogens and the sulfurs of the cluster or of three of the cysteinyl ligands, 41, 46 and 49. There are no apparent hydrogen bonds to the sulfur of the fourth liganding cysteine, Cys^{79}. Taken from Holden et al., 1994.

Cys^{79} is predominantly ferric (Tsukihara et al., 1990; Rypniewski et al., 1991; Skjeldal et al., 1991; Fu et al, 1992a). One would expect that the cluster iron located in the microenvironment with the highest relative dielectric constant to have the higher E_m value (Kassner and Yang, 1977). The fact that Fe1 is more exposed to the solvent (Tsukihara et al., 1990; Rypniewski et al., 1991) is thus consistent with this iron being predominantly reduced in reduced ferredoxin.

X-ray crystal structures are known for four cyanobacterial ferredoxins and for one plant ferredoxin (Tsukihara et al., 1981,1990; Rypniewski et al., 1991; Jacobson et al., 1993; Ikemizu et al., 1994). Figure 2 (taken from Rypniewski et al., 1991) shows the major secondary structure features of the 98 amino acid-long ferredoxin isolated from vegetative cells of *Anabaena* 7120. Four strands of β-pleated sheet are present, with strands A (residues 1–9) and B (residues 14–20) antiparallel to one another, strands C (residues 51–54) and D (residues 87–91) parallel and strand A and D parallel to each other. Two α-helices (residues 26–32 and 68–73) and a helical turn at the carboxyl terminus (residues 94–98) are also present. Ferredoxins isolated from two other cyanobacteria, *Spirulina platensis* (Tsukihara et al., 1981), and *Aphanothece sacrum* (Tsukihara et al., 1990), and from the higher plant, *Equisetum arvense* (Ikemizu et al., 1994) have quite similar overall structures. In all four of these ferredoxins the [2Fe-2S] cluster is located near the outer edge of the molecule in a loop that extends as much as 9 Å from the major portion of the protein (In the vegetative *Anabaena* 7120 ferredoxin, this loop runs from Pro^{38} through Ala^{50}). This cluster-binding loop is stabilized by a number of conserved electrostatic interactions, illustrated in Fig. 3 (taken from Holden et al., 1994) in schematic form for the vegetative *Anabaena* 7120 ferredoxin.

Fig. 2. Ribbon drawing of the ferredoxin isolated from vegetative cells of *Anabaena* 7120. Strands of β-pleated sheet are shown as arrows, α-helices are shown as coils and the [2Fe-2S] cluster is represented by a ball-and-stick model. Taken from Rypniewski et al., 1991.

Of interest is the observation that the ferredoxin found in the nitrogen-fixing heterocysts of *Anabaena* 7120, which is distinct from that found in the vegetative cells of the same cyanobacterium (Schrautemeier and Böhme, 1985; Böhme and Schrautemeier, 1987a; 1987b; Böhme and Haselkorn,

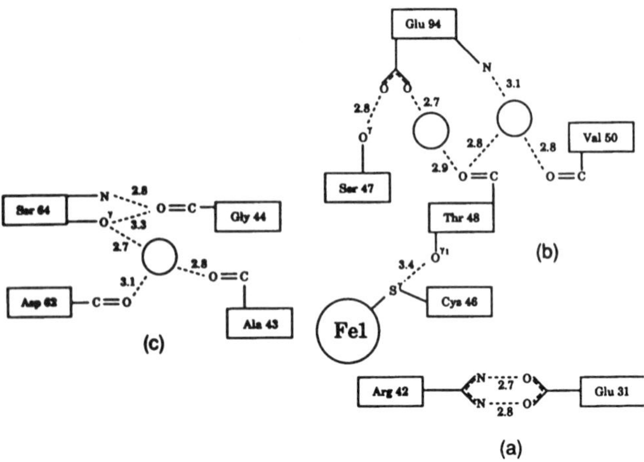

Fig. 3. Schematic representation of conserved electrostatic interactions that stabilize the [2Fe-2S] binding loop in the ferredoxin isolated from vegetative cells of *Anabaena* 7120. (a) Salt bridge between Glu^{31} and Arg^{42}; (b) Hydrogen bonding network between Ser^{47}, Thr^{48}, Val^{50} and Glu^{94}; (c) Interactions involving Ser^{64}. The large circles represent ordered water molecules. The numbers above the dotted lines are distances in Å. Taken from Holden et al., 1994.

1989), has a very similar structure to that of the *Anabaena* 7120 vegetative ferredoxin (Jacobson et al., 1993) even though the two proteins have different amino acids at 48 of the 98 positions (Alam et al., 1986; Böhme and Haselkorn, 1989). In fact, the *Anabaena* 7120 heterocyst ferredoxin differs at 4 of the 22 positions that appear to be most highly conserved in the [2Fe-2S] cluster-containing ferredoxins found in oxygenic phototrophs. One of these changes, substitution at position 42 of a histidine in the *Anabaena* heterocyst ferredoxin (Böhme and Haselkorn, 1989) for the arginine usually found at this position (Matsubara and Hase, 1983) results in an alteration of the hydrogen-bonding pattern in the region near the [2Fe-2S] cluster. In the *Anabaena* vegetative ferredoxin, the guanidino group Arg^{42} is involved in three hydrogen bonds with the carboxylate groups of Asp^{28} and Glu^{31} (See Fig. 3) but in the heterocyst protein, His^{42} forms only a single nitrogen to oxygen hydrogen bond to Asp^{28} and does not interact with Glu^{31} (Jacobsen et al., 1993). Despite these differences, the α-carbon positions of the two *Anabaena* ferredoxins can be superimposed with a root-mean-square value of 1.0 Å, or with a root-mean-square value of 0.45 Å if the two regions that differ most are excluded (Jacobson et al., 1993).

The distribution of charged residues in ferredoxins for which structures are known is asymmetric. Fig. 4C (taken from De Pascalis et al., 1994), which illustrates this for a model of spinach ferredoxin 1 constructed by fitting the sequence for the spinach protein to the *A. sacrum* structure, shows two regions of negative surface potential surrounding the [2Fe-2S] cluster. Very similar asymmetric, bilobal negative surface potentials were calculated for the ferredoxins from *A. sacrum*, *S. platensis* and the vegetative cells of *Anabaena* 7120 (De Pascalis et al., 1993). A dipole moment of 377 Debye has been calculated for the oxidized form of spinach ferredoxin 1, using a structure obtained by fitting the sequence of the spinach protein to the known structure of the ferredoxin from the cyanobacterium *A. sacrum* (De Pascalis et al., 1993). The negative end of the dipole moment vector lies near the iron-sulfur cluster and approximately midway between the two negatively charged domains (Fig. 4C). It has been proposed (See below) that the negative end of this strong molecular dipole moment plays a key role in steering ferredoxin in the initial stages of docking between the iron-sulfur protein and enzymes that utilize ferredoxin as an electron donor (De Pascalis et al., 1993; De Pascalis et al., 1994).

B. Biosynthesis

The genes for [2Fe-2S] ferredoxins from several oxygenic phototrophs have been cloned and sequenced (Smeekens et al., 1985; Alam et al., 1986; Reith et al., 1986; van der Plas et al., 1986a; 1986b; Dobres et al., 1987; Wedel et al., 1988; Böhme and Haselkorn, 1989; Somers et al., 1990; Vorst et al., 1990; Rogers et al., 1992). In plants and algae, apoferredoxins are encoded by nuclear genes and synthesized in the cytosol as precursor proteins that

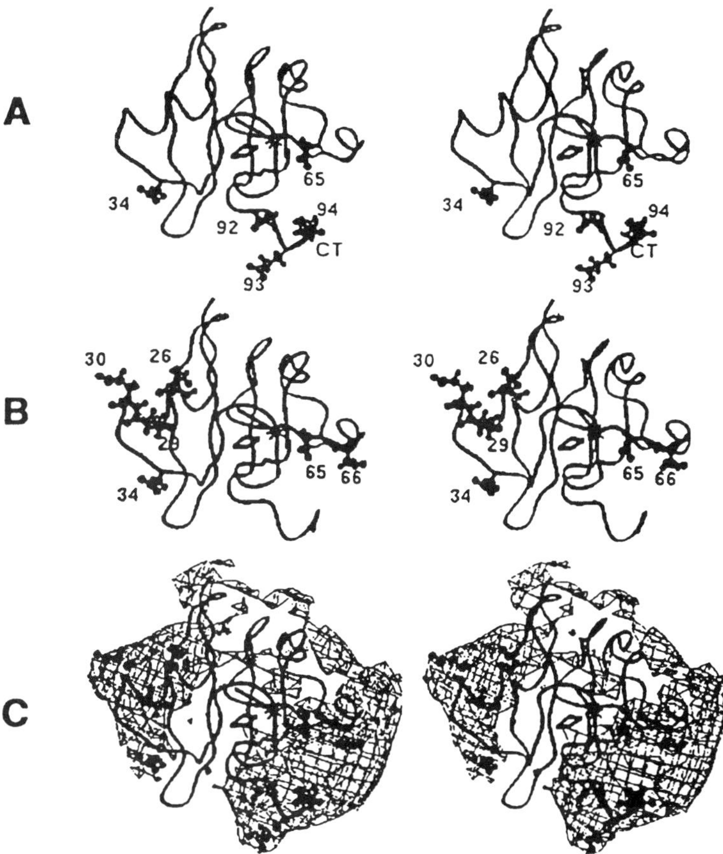

Fig. 4. Stereo ribbon model of the α-carbon trace of ferredoxin from *Aphanothece sacrum* to which the sequence of spinach ferredoxin 1 has been fitted. The [2Fe-2S] cluster is indicated by a diamond and the dipole moment vector is indicated by a straight line with the negative pole marked by an asterisk. CT marks the approximate location of the carboxy-terminus of spinach ferredoxin (Ala^{97}), which is missing in the *A. sacrum* ferredoxin. (A) Residues protected against chemical modification by complex formation with thioredoxin reductase. (B) Residues protected in the complex between ferredoxin and spinach FNR. (C) The calculated negative electrostatic surface potential of ferredoxin at a contour level of –15 kJ mol^{-1}. Taken from De Pascalis et al., 1994.

are larger than the mature forms, containing an amino-terminal extension referred to as the transit peptide or leader sequence (Huisman et al., 1978). The cyanobacterial apoferredoxins are not synthesized with leader sequences, but in some cases the initiating methionine is post-translationally removed (Alam et al., 1986; Reith et al., 1986; van der Plas et al., 1986a; van der Plas, 1986b; Böhme and Haselkorn, 1989). The leader sequences found in the pre-ferredoxins of photosynthetic eukaryotes provide the information necessary to target the pre-ferredoxin for post-translational translocation across the chloroplast envelope to its final location in the chloroplast stromal space (Smeekens et al., 1986; Smeekens et al., 1987; Elliott et al., 1989a; Pilon et al.,1990,1992a). During or immediately after translocation, the leader sequence is cleaved by a stromal protease (Robinson and Ellis, 1984). See Chapter 7, Sec. II for additional material on protein import.

Binding and translocation of ferredoxin into chloroplasts (for reviews see Keegstra and Olsen, 1989; de Boer and Weisbeek, 1991) requires the hydrolysis of stromal ATP (Cline et al., 1985a; Flügge and Hinz, 1986; Pain and Blobel, 1987; Olsen et al., 1989; Theg et al., 1989). Neither a membrane potential (Flügge and Hinz, 1986; Pain and Blobel, 1987; Theg et al., 1989) nor any soluble cytoplasmic factors (Lubben et al., 1989; Pilon et al., 1990, 1992a,b; Takahashi et al., 1990b) are required for the complete processing of pre-ferredoxin. It appears likely that the translocation of pre-ferredoxin involves a high affinity receptor protein located in the outer chloroplast envelope (Pilon et al., 1992b) and that this receptor, present at a level of approximately 2,000 per chloroplast (Friedman and Keegstra, 1989), is also involved in the uptake of the precursor of other proteins targeted to the chloroplast, such as the small subunit of Rubisco (Perry et al., 1991; Schnell et al., 1991; Pilon et al., 1992b).

In plants, coordination of iron and sulfide by apoferredoxin to yield the holoprotein occurs after the pre-apoferredoxin has been imported into the chloroplast stroma and the leader sequence cleaved (Li et al., 1990; Takahashi et al., 1990a; 1990b; Pilon et al., 1992a). It has been demonstrated, using both intact (Takahashi et al., 1986) and lysed chloroplasts (Takahashi et al., 1990a; 1990b), that the inorganic sulfide required for cluster assembly can be derived from cysteine in a reaction that requires NADPH (Takahashi et al., 1990b). Although assembly of the cluster into apoferredoxin to yield the holoprotein in vitro does not require an energy source (Bayer et al., 1967), cluster assembly appears to be stimulated by ATP hydrolysis in vivo (Takahashi et al., 1990a; 1990b). There is no requirement for light per se in cluster assembly (Takahashi et al., 1990a; 1990b; Suzuki et al., 1991) and cluster assembly can even occur in etioplasts that lack a functional photosynthetic electron transfer chain (Suzuki et al., 1991).

Most, and perhaps all, oxygenic phototrophs contain more than one form of ferredoxin and, in the case of higher plants, some ferredoxin forms are most abundant in non-photosynthetic tissues (Takahashi et al., 1983; Sakihama and Shin, 1987; Hirasawa et al., 1988a; Kimata and Hase, 1989; Wada et al., 1989; Somers et al., 1990; Green et al., 1991; Hase et al., 1991). Not only are there tissue-specific distributions of the different ferredoxin forms in some plants, but the relative abundances of the different ferredoxins vary during development of plants (Wada et al., 1985; Kimata and Hase, 1989; Wada et al., 1989; Green et al., 1991; Hase et al., 1991). The temporal dependence of the relative abundance of the different ferredoxin isoforms arises from the fact that the amount of the leaf-specific form(s) increases upon illumination of etiolated tissue while the amount of the ferredoxin isoform(s) detected in all organs is not affected by light (Takahashi et al., 1983; Hase et al., 1991).

A major component of the light-dependent regulation of the expression of *Fed*A, the gene encoding the major leaf form of ferredoxin, has been shown to result from increased mRNA levels in the light and to involve phytochrome (Kaufman et al., 1986; Dobres et al., 1987; Caspar and Quail, 1993). In both pea (Gallo-Meagher et al., 1992) and *Arabidopsis thaliana* (Vorst et al., 1990; Caspar and Quail, 1993), evidence suggests that the upstream promoter is involved in both light-regulation and tissue specificity, i.e, the promoter is involved in the light-induced increase in *Fed*A mRNA levels and in producing higher expression in aerial tissue than in roots. Additional light-responsive regulatory elements have been demonstrated to lie within the transcribed region of the pea gene (Elliott et al., 1989a; 1989b; Dickey et al., 1992; Gallo-Meagher et al., 1992). While the effect of upstream light-responsive elements appears to be most pronounced during the initial induction of gene expression in pea, the internal elements appear to play the more prominent role in modulating expression once the gene is already active (Gallo-Meagher et al., 1992). It is not yet clear whether these effects result from increased rates of mRNA synthesis or from increased mRNA stability (Dickey et al., 1992). No evidence for the presence of *cis*-acting, light-regulatory elements within the 5′ untranslated leader region of the *Arabidopsis Fed*A gene have been detected, although elements within this region have been shown to be required for full expression of the gene in both the light and the dark (Caspar and Quail, 1993). It should also be mentioned that regulation of *Fed*A gene expression in *Arabidopsis thaliana* appears to also involve a post-transcriptional process and an as yet uncharacterized 'plastidic factor' (Vorst et al., 1993).

It is known that at least two ferredoxin-dependent enzymes involved in relatively early steps in nitrogen metabolism, nitrite reductase and glutamate synthase, are found in non-photosynthetic plant tissues. Presumably the role of the ferredoxin(s) found in these tissues is to serve as an electron donor to enzymes that require reduced ferredoxin as an electron donor. In the case of some nitrogen fixing cyanobacteria, different physiological roles for two different ferredoxins found in the same organism have been clearly established. Separate ferredoxins are present in the vegetative cells, where oxygenic photosynthesis occurs, and in the heterocysts, specialized cells lacking Photosystem II, where nitrogen fixation occurs (Schrautemeier and Böhme, 1985; Böhme and Schrautemeier, 1987a; 1987b; Böhme and Haselkorn R, 1988). In *Anabaena variabilis*, the two ferredoxins are equally effective electron donors for $NADP^+$ photoreduction and both serve effectively as electron donors to nitrogenase

(Schrautemeier and Böhme, 1985). However, only the heterocyst ferredoxin is able to effectively couple electrons released by the action of glucose-6-phosphate dehydrogenase to nitrogenase (Schrautemeier and Böhme, 1985; Böhme and Schrautemeier, 1987b). It has been shown that the interaction of the *Anabaena* heterocyst ferredoxin with nitrogenase appears to require positively charged groups on ferredoxin (Schmitz et al., 1993). This is in contrast to the interaction of spinach leaf ferredoxin with the spinach leaf enzymes FNR, nitrite reductase and glutamate synthase, where negatively charged groups on ferredoxin are essential for optimal protein:protein interaction (see below).

III. Ferredoxin:NADP$^+$ Oxidoreductase (FNR)

A. Structure and Redox Properties

FNR is a monomeric protein that contains a single non-covalently bound FAD as its sole prosthetic group (Carrillo and Vallejos, 1987; Knaff and Hirasawa, 1991; Zanetti and Aliverti, 1991; Karplus and Bruns, 1994). In chloroplasts, the physiological role of FNR is to catalyze the reduction of NADP$^+$ to NADPH, using two molecules of reduced ferredoxin to supply the two electrons required for the reaction. In non-photosynthetic plant tissues and perhaps also in heterocysts of cyanobacteria, the physiological role of FNR is likely to involve catalysis of the reverse reaction so that NADPH generated by reactions of the pentose phosphate pathway can be used to reduce ferredoxin and thus provide an electron donor for reductive reactions of nitrogen and sulfur metabolism. In cyanobacteria, FNR may also function as a respiratory NADPH dehydrogenase (Scherer et al., 1988).

In chloroplasts, FNR is bound to the stromal side of the thylakoid membrane in stroma-exposed lamellae (Carrillo and Vallejos, 1987; Zanetti and Aliverti, 1991), but it can be released from the membrane with relative ease and most studies of the protein are carried out with the soluble form. The activity of the membrane-bound form of the enzyme appears to be controlled by a number of regulatory mechanisms (Carrillo and Vallejos, 1987), but this aspect of FNR is beyond the scope of this chapter. The mature proteins in plants have M_r values near 35 kDa (Carrillo and Vallejos, 1987; Knaff and Hirasawa, 1991; Zanetti and Aliverti, 1991). In some cyanobacteria, the mature proteins are considerably larger, with M_r near 45 kDa (Schluchter and Bryant, 1992). In the best characterized case of a large cyanobacterial FNR, that from *Synechococcus* PCC 7002 (Schluchter and Bryant, 1992), the protein contains a domain similar to the 35 kDa FNR's found in plants and a 85 amino acid-long extension at the amino-terminus that may be involved in anchoring it to the thylakoid surface (See below). Earlier reports that FNR's from cyanobacteria all had M_r = ca. 35 kDa may have resulted from the susceptibility of the larger form to proteolysis during purification (Schluchter and Bryant, 1992).

Complete amino acid sequences are available for FNR's from three cyanobacteria, from a protist and from leaves of four higher plants (Karplus and Bruns, 1994; Aoki et al., 1994). A complete sequence is also available for the FNR from one non-photosynthetic tissue, rice root (Aoki and Ida, 1994), and partial sequences are available for FNR's from other non-photosynthetic tissues (Morigasaki et al., 1990a; Green et al., 1991). The plant leaf FNR's are quite similar to one another in sequence, but differ significantly from the cyanobacterial enzymes and from rice root FNR (Karplus and Bruns, 1994). For example, the spinach and pea enzymes are 87% identical, while the spinach and *Synechococcus* enzymes are only 52% identical (Karplus and Bruns, 1994). Interestingly, the FNR's from rice leaf and rice root show only 49% identity (Aoki et al., 1994). Despite these differences among FNR's, sequence alignments have allowed identification of well-conserved regions within the FNR family (Karplus and Bruns, 1994).

A three-dimensional structure for the oxidized form of spinach leaf FNR at 2.2 Å resolution has been reported (Karplus et al., 1991) and the structure has recently been refined to 1.7 Å (Bruns and Karplus, 1995). The structure of fully reduced spinach FNR is also known at 2.0 Å resolution and shows only very small differences from the structure of the oxidized protein (Karplus and Bruns, 1994; Bruns and Karplus, 1995). Crystals that diffract to better than 2 Å resolution have been reported for FNR isolated from the cyanobacterium *Anabaena* PCC 7119 (Serre et al.,1991) and the structure, which was predicted to be similar to that of the spinach enzyme (Medina et al., 1993), has recently been solved (Serre et al., 1994). As the structure of the *Anabaena* FNR is indeed quite similar to that of the spinach enzyme and because of the limitations of space, the discussion below will focus on the structure of the spinach enzyme.

Figure 5 shows a ribbon diagram of oxidized spinach FNR, starting at residue 20. The first 19 residues are not visible and are thus likely to be disordered. Interestingly, the first eight amino acids of the *Anabaena* FNR are not visible in the electron density map and are also likely to be disordered (C Gómez-Moreno, personal communication). The structure consists of two domains, the first of which runs from residue 20 through residue 161. This FAD-binding domain consists of a six-stranded antiparallel β-barrel with a peripheral two-stranded antiparallel hairpin turn that leads into a single, eight amino acid long, α-helix. The second domain, which runs from residue 162 to residue 314 and contains a central five-stranded parallel β-sheet with six associated helices, provides the binding site for a single $NADP^+$. Residues 20 through 27 and 150 through 161 are in extended conformations and occupy portions of the interface between the two domains.

The FAD prosthetic group, which is in an extended conformation, is bound outside the β-barrel, with the riboflavin portion in a pocket between two strands of the β-barrel (Fig. 5). The 'front' (*re*-face) of the flavin isoalloxazine ring faces towards the $NADP^+$-binding domain and is partially covered by the parallel side chain of Tyr^{314}. Tyr^{95} forms a hydrogen bond with the 3′-hydroxyl group of the FAD ribityl moiety and,

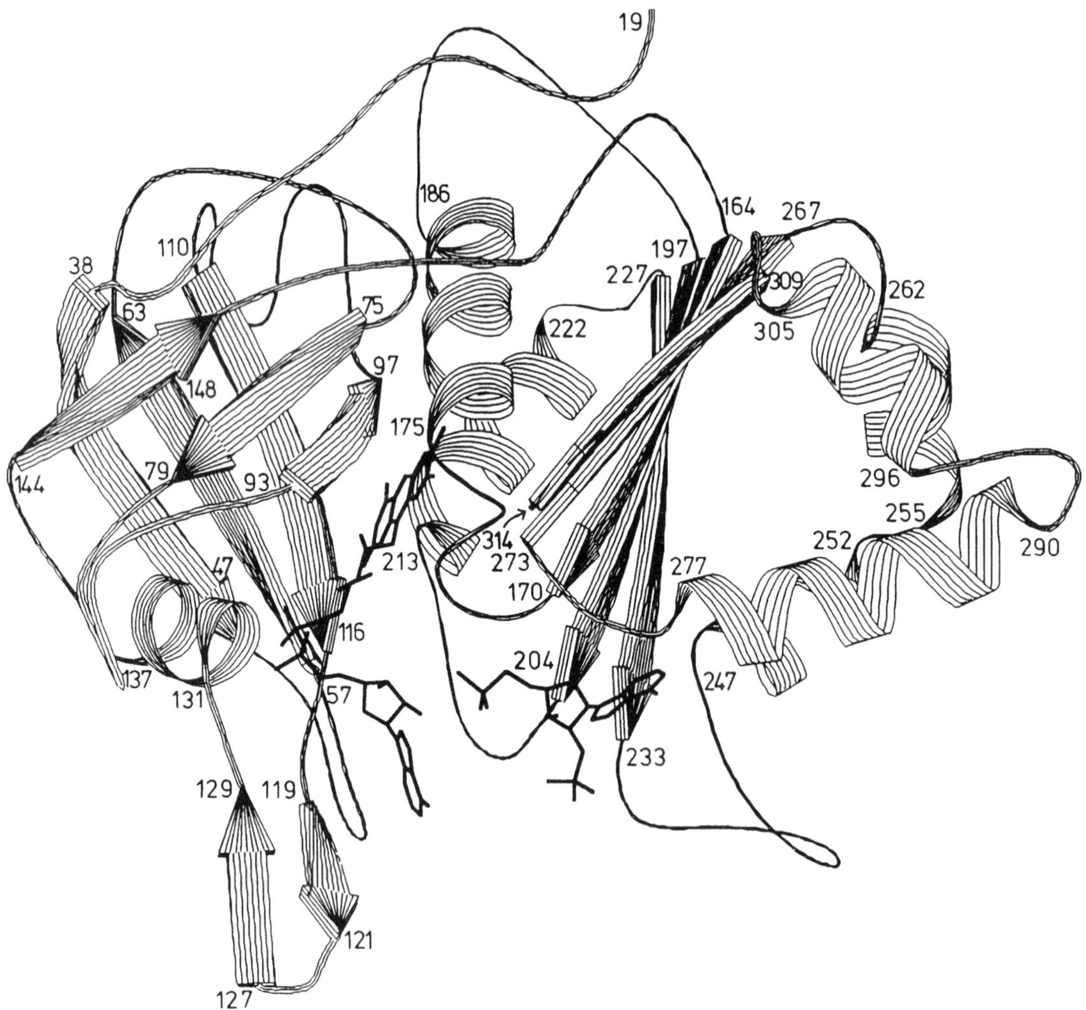

Fig. 5. Polypeptide chain folding and domain structure of spinach FNR. In this ribbon diagram, β-pleated sheet and the antiparallel hairpin are indicated by arrows and α-helices by coils. The location of the FAD prosthetic group and of the $NADP^+$ analog 2′-phospho-AMP are shown with stick structures. Taken from Karplus et al., 1991.

along with Leu^{94} and Ser^{96}, covers the 'back' (*si*-face) of the isoalloxazine ring. Only the dimethylbenzyl portion of the isoalloxazine ring system is exposed to the bulk solvent, while the rest is shielded by protein and an oriented water molecule. The pyrophosphate group is anchored by hydrogen bonds to the side chains of Arg^{93} and Ser^{133} and to three peptide amides from the α-helix running from residue 131 to 137. The adenosine portion of FAD is bound only by van der Waals contacts with the region of the protein from residue 116 to 128. These features are consistent with the properties of FNR reconstituted with FAD analogs (Zanetti et al., 1983). Residues directly involved in FAD-binding are absolutely conserved, except for Cys^{114}, which is replaced by serine in the FNR from *Cyanophora paradoxa*. However, the interaction between Cys^{114} and FAD involves the peptide backbone and not the side chain of this residue (Karplus and Bruns, 1994).

E_m values for the FAD/$FADH_2$ couple in spinach FNR near –360 mV (pH 7.0) have been obtained by several groups (Knaff and Hirasawa, 1991). Slightly more positive E_m values (pH 7.0) of –320 mV and –344 mV have been reported for the FNR's from two *Anabaena* species (Sancho et al., 1988; Pueyo et al., 1991). The FAD semiquinone state, observed in a number of studies of FNR (Knaff and Hirasawa, 1991; Zanetti and Aliverti, 1991), is not stabilized thermodynamically. One-electron E_m values for the oxidized FAD/flavin semiquinone and the flavin semiquinone/$FADH_2$ couples at pH 7.0 have been estimated to be –370 mV and –270 mV, respectively, for the *Anabaena variabilis* FNR (Sancho et al., 1988) and –377 mV and –311 mV for the *Anabaena* PCC 7119 FNR (Pueyo et al., 1991), and it has been estimated that for spinach FNR at pH 8.0 the E_m value for the FAD/flavin semiquinone couple is at least 90 mV more negative than that for the semiquinone/$FADH_2$ couple (Batie and Kamin, 1986). Equilibrium redox titrations suggest that complex formation between spinach FNR and ferredoxin shifts the E_m value of the FAD/$FADH_2$ couple by ca. +20 mV or less and shifts the E_m value of the ferredoxin [2Fe-2S] cluster by somewhere between -25 mV and –90 mV (Batie and Kamin, 1981; Smith et al., 1981). However, kinetic measurements suggest that complex formation between spinach ferredoxin and FNR may shift the E_m value of ferredoxin to a more positive value, not to the more negative values seen in equilibrium redox titrations (Walker et al., 1991). This disagreement remains unresolved. Complex formation between *Anabaena* PCC 7119 FNR and the flavodoxin from this cyanobacterium produces essentially no change in the E_m value of the FAD/$FADH_2$ couple of FNR, but does result in small shifts in the two one-electron E_m values of the FMN/$FMNH_2$ couple of flavodoxin (Pueyo et al., 1991). Although complex formation with $NADP^+$ does not appear to affect the redox properties of the FAD group of spinach FNR, the E_m value of the $NADP^+$/NADPH couple increases by ca. +40 mV when $NADP^+$ is bound to either spinach (Batie and Kamin, 1986) or *Anabaena* (Sancho and Gómez-Moreno, 1991) FNR. Interpretation of these results is complicated by the formation of a charge-transfer complex between NADPH and the enzyme (Batie and Kamin, 1986).

The FAD semiquinone state of FNR has been detected both by electron paramagnetic resonance (EPR) spectroscopy and by optical absorbance measurements (Knaff and Hirasawa, 1991; Zanetti and Aliverti, 1991). The absorbance spectrum of the semiquinone indicates that it is the neutral, protonated species rather than the unprotonated semiquinone anion (Knaff and Hirasawa, 1991; Zanetti and Aliverti, 1991). The fact that the X-ray crystal structure reveals that no positively charged groups from the protein are located in the vicinity of the N1/O2α locus of the pyrimidine subnucleus of the isoalloxazine ring system is consistent with the fact that a neutral semiquinone rather than a semiquinone anion is formed (Karplus, 1991; Zanetti and Aliverti, 1991). Support for the suggestion that the FAD semiquinone plays a role in electron transfer from reduced ferredoxin to FNR in vivo comes from the observation that the FNR semiquinone can be observed after flash illumination of intact cells of the green alga *Chlorella pyrenoidosa* (Bouges-Bocquet, 1980).

B. Substrate Binding and Mechanism

1. $NADP^+$ Binding

Although crystallographic analysis has not yet provided evidence for the manner in which the nicotinamide portion of $NADP^+$ binds to spinach FNR, it has proven possible to crystallize the protein with a fragment of $NADP^+$, 2′-phospho-5′-AMP, bound (Karplus et al., 1991; Bruns and Karplus, 1995). Although a complex of $NADP^+$ with the *Anabaena* FNR has been crystallized and the 2′,5′-ADP portion of the substrate is well defined in the difference electron density map, no electron density was observed for the nicotinamide moiety (Serre et al, 1994). As 2′-phospho-5′-AMP competes with $NADP^+$ for binding to spinach FNR (Batie and Kamin, 1986), the site of inhibitor binding revealed by the crystallographic studies is likely to accurately delineate the binding of this portion of the true substrate, $NADP^+$, to the enzyme.

The inhibitor binds at the C-terminal portion of the central β-sheet in the $NADP^+$-binding domain (Fig. 5), with the adenine sandwiched between Tyr^{246} and Leu^{274}. The side chains of Thr^{170} and Gln^{248} form hydrogen bonds with the adenine and the ribose, respectively, and hydrogen bonds with two main-chain carbonyls also contribute to binding this portion of the inhibitor. The 2′-phosphate forms hydrogen bonds with the side chains of Ser^{234}, Arg^{235} and Tyr^{246}. As the structure was determined using crystals at a pH near 4.5, the 2′-phosphate probably carried only a single negative charge instead of the double negative charge it would likely carry at physiological pH and thus it is possible that Lys^{244}, which is only 4 Å away, makes an additional hydrogen bond to the substrate at physiological pH values (Karplus et al., 1991). Lys^{244} has been implicated in $NADP^+$-binding by both affinity-labeling (Chan et al., 1985) and site-directed mutagenesis studies (Aliverti et al., 1991). Chemical modification studies on the FNR from *Anabaena* PCC 7119, have implicated Arg^{224} and Arg^{233}, the residues equivalent to Arg^{235} and Lys^{244}, respectively, in the spinach enzyme, in $NADP^+$ binding (Medina et al., 1992a). These interactions are consistent with the discrimination of the enzyme for $NADP^+$ instead of NAD^+ as the preferred substrate, as the latter lacks a 2′-phosphate group (Batie and Kamin, 1986; Karplus et al., 1991). Lys^{116}, which has been implicated in $NADP^+$ binding in the spinach enzyme by site-directed mutagenesis (Aliverti et al., 1991) and chemical modification studies (Cidaria et al., 1985), interacts with the 5′-phosphate group of the inhibitor. These residues are all highly conserved in the FNR's for which sequences are known (Medina et al., 1993; Karplus and Bruns, 1994).

2. Ferredoxin Binding

Spinach leaf ferredoxin is known to form a high affinity, electrostatically-stabilized, 1:1 complex with spinach leaf FNR (Carrillo and Vallejos, 1987; Knaff and Hirasawa, 1991; Zanetti and Aliverti, 1991). The fact that FNR can form a ternary complex, binding both ferredoxin and $NADP^+$ (Batie and Kamin, 1984a), indicates that FNR must have two separate binding sites for the two substrates. Complex formation between spinach FNR and ferredoxin appears to make both the [2Fe-2S] cluster of ferredoxin (Batie and Kamin, 1984a; Bhattacharyya et al., 1986) and the FAD of FNR (Batie and Kamin, 1984a) less accessible to solvent. A large body of evidence, obtained from chemical modification, cross-linking, substrate binding and NMR studies suggests that ferredoxin contributes the negative charges to this interaction, while FNR contributes the positive charges (Carrillo and Vallejos, 1987; Knaff and Hirasawa, 1991; Zanetti and Aliverti, 1991; Karplus and Bruns, 1994). Calculations of the surface potential of spinach FNR reveal an asymmetric charge distribution (De Pascalis et al., 1993) that leads to a large dipole moment (Jelesarov et al., 1993). The fact that a complementary fit can be made between the two domains of negative surface potential on ferredoxin (negative lobes are centered on Asp^{65}, Asp^{66}, Glu^{92}, Glu^{93} and Glu^{94} and on Asp^{26}, Glu^{29}, Glu^{30} and Asp^{34}, respectively - See Fig. 4C) and the two domains of positive surface potential on FNR (positive lobes are centered on the lysines at positions 33, 35, 85, 88, and 91 and the lysines at positions 153, 304 and 305) is consistent with these electrostatic models for plant FNR/ferredoxin interaction (De Pascalis et al., 1993; Karplus and Bruns, 1994). Complex formation between the two proteins is driven mainly by a positive entropy change ($\Delta S = 125\ Jmol^{-1}\ k^{-1}$) with ΔH being very small. The favorable entropy change appears to originate from release of oriented water molecules at the protein:protein interface (Jelesarov and Bosshard, 1994).

The M_r = 35 kDa form of cyanobacterial FNR forms complexes with both cyanobacterial ferredoxin and flavodoxin (Pueyo and Gómez-Moreno, 1991; Sancho and Gómez-Moreno, 1991; Pueyo et al., 1991; Pueyo et al., 1992). It appears likely that the binding site on FNR for ferredoxin is the same as the binding site for flavodoxin (Pueyo et al., 1992) and, as is the case for the higher plant system, positively charged residues on FNR are involved in protein:protein interactions (Sancho et al., 1990; Medina et al., 1992a; 1992b; 1992c). However, in the case of the *Anabaena* PCC 7119 FNR/ferredoxin complex, unlike the case for the complex between the two spinach proteins, electrostatic forces do not appear to be the major contributor to complex stability

(Walker et al., 1991). Additional evidence for structural differences between the plant and cyanobacterial FNR/ferredoxin complexes comes from the observation that complex formation between the *Anabaena* proteins inhibits reduction of FNR by laser flash-generated deazariboflavin semiquinone without affecting reduction of ferredoxin, exactly the opposite pattern from that observed with the spinach proteins (Walker et al., 1991).

As there are not yet any crystallographic data available on the interaction between FNR and ferredoxin, attempts to identify specific regions of the proteins that are directly involved in complex formation between the two spinach proteins have relied on a variety of indirect techniques including chemical modification (Vieira et al., 1986; De Pascalis et al., 1993; Jelesarov et al., 1993), limited proteolysis (Gadda et al., 1990), mutagenesis (Aliverti et al., 1990), cross-linking (Vieira et al., 1986; Zanetti et al., 1988), structural comparisons to bacterial phthalate dioxygenase reductase (Correll et al., 1992; 1993) and computer modeling (De Pascalis et al., 1993; Karplus and Bruns, 1994) to predict the detailed orientation of the two proteins within the ferredoxin/FNR complex. One such model which is largely consistent with the results obtained from all of these different experimental techniques, proposed by Karplus and Bruns (1994) for the spinach system, is shown in Fig. 6. In a similar model, it was calculated that the distance separating the dimethylbenzene portion of the FAD of FNR and Fe1 in the [2Fe-2S] cluster of ferredoxin within the ferredoxin/FNR complex could be as little as 4 Å (De Pascalis et al., 1993). Although chemical modification studies have provided some evidence for the involvement of a tryptophan residue in electron transfer between the FAD of FNR and the [2Fe-2S] cluster of ferredoxin (Hirasawa et al., 1995), a 4 Å separation should allow rapid electron transfer to occur without any involvement of facilitating aromatic residues. Indeed, no aromatic side chains are found between the two prosthetic groups in the model shown in Fig. 6.

Chemical modification techniques have also been used to identify specific residues that may be involved in the interactions between *Anabaena* FNR and either *Anabaena* ferredoxin or flavodoxin (Medina et al., 1992a; 1992b; 1992d). It should be kept in mind that the role of electrostatic interactions in stabilizing these complexes may be less important in the cyanobacterial system than is the case for the plant proteins (See above).

It is not known whether the binding site on ferredoxin for FNR is also involved in the docking of ferredoxin with the Photosystem I reaction center. It has been proposed, on the basis of fluorescence measurements with eosin-modified FNR, that ferredoxin can form a ternary complex with membrane-bound FNR and Photosystem I in spinach thylakoid membranes (Wagner et al., 1982). If ternary complex formation can in fact occur, the site on ferredoxin with which FNR interacts must differ from that at which ferredoxin binds to Photosystem I.

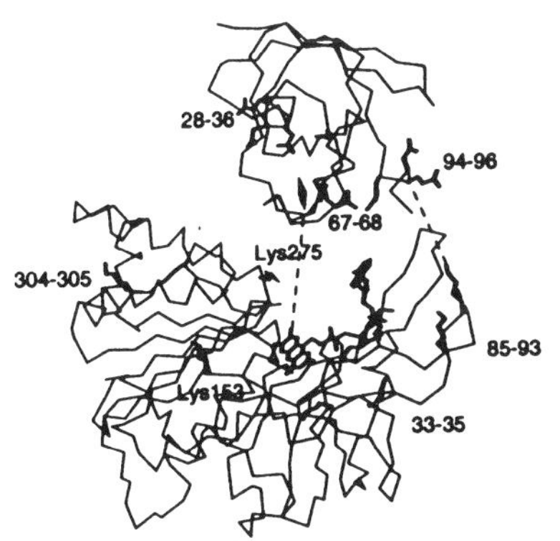

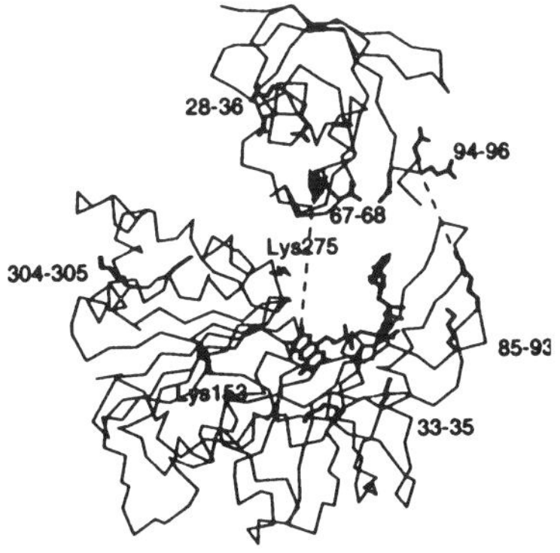

Fig. 6. Stereo view of a model for the FNR/ferredoxin complex. For the sake of clarity, the ferredoxin model has been translated ca. 10 Å above FNR. The [2Fe-2S] cluster of ferredoxin is shown as a diamond and the FAD group of FNR is shown with a line formula. Dashed lines indicate regions that may come together upon complex formation. Side chains are shown for spinach FNR residues Lys^{33}, Lys^{35}, Lys^{85}, Lys^{88}, Lys^{91}, Arg^{93}, Lys^{153}, Lys^{275}, Lys^{304} and Lys^{305} and for *Aphanothece sacrum* ferredoxin residues Asp^{28}, Glu^{31}, Glu^{32}, Asp^{36}, Asp^{67}, Asp^{68}, Glu^{94}, Glu^{95} and Asp^{96}. In order to convert the *A. sacrum* sequence numbers into the positions of the corresponding positions for spinach ferredoxin, subtract 2. Taken from Karplus and Bruns, 1994.

However, it has been argued that the hypothesis of ternary complex formation is not easily reconciled with the observed inability of a cross-linked ferredoxin/Photosystem I complex to photoreduce $NADP^+$ (Zanetti and Merati, 1987; Zilber and Malkin, 1988). One promising approach to resolving the question of whether the same site on ferredoxin is involved in both its interaction with FNR and the Photosystem I reaction will be to determine whether site-specific mutant forms of ferredoxin that have impaired abilities to reduce FNR (See below) are still able to accept electrons from Photosystem I at rates similar to those observed for wild type ferredoxin (Hervás et al., 1992).

3. Catalytic Mechanism

Stopped-flow (Batie and Kamin, 1984b) and steady-state (Masaki et al., 1982) kinetic measurements suggest that both spinach and *Spirulina platensis* FNR's function via a mechanism that involves ordered addition of substrates to form a ternary complex, with $NADP^+$ binding before reduced ferredoxin. Reduction of the FAD of FNR to the semiquinone by reduced ferredoxin is followed by dissociation of oxidized ferredoxin, binding of a second reduced ferredoxin and the subsequent reduction of the FAD to fully reduced $FADH_2$. A two-electron reduction of $NADP^+$ then occurs and is followed by release of products, with product release likely to be the rate-limiting step in the overall process (Batie and Kamin, 1984b). The extremely negative E_m value for the one-electron reduction of pyridine nucleotide (ca. –930 mV, Land and Swallow, 1968; Farrington et al., 1980) makes it very unlikely that $NADP^+$ reduction by FNR could occur in two one-electron steps. Both binding studies (Wagner et al., 1982; Batie and Kamin, 1984a) and stopped-flow kinetic measurements (Batie and Kamin, 1984b) suggest that negative cooperativity exists between the ferredoxin and $NADP^+$ binding sites on spinach FNR, with $NADP^+$ binding favoring the dissociation of oxidized ferredoxin. Flash photolysis experiments, carried out at high ionic strength, in which spinach ferredoxin was rapidly reduced by the 5-deazariboflavin radical and, in turn, was reoxidized by spinach FNR allowed determination of the second-order rate constant ($1.5 \times 10^8\ M^{-1}s^{-1}$ at pH 7.0 and 310 mM ionic strength) for reduction of FNR by reduced ferredoxin to the FAD semiquinone state and of the limiting first-order rate constant ($4000\ s^{-1}$) for electron transfer from the reduced [2Fe-2S] cluster of ferredoxin to the FAD in FNR within a transient ferredoxin/FNR complex (Bhattacharyya et al., 1986). While increasing the ionic strength significantly increases the apparent K_d for complex formation between the two spinach proteins, there is little effect of ionic strength on the rate constant for intracomplex electron transport (Walker et al., 1991). This suggests that the geometry of the initial collision between the two spinach proteins may be close to optimal for the subsequent electron transfer reaction.

Flash photolysis kinetic measurements conducted with *Anabaena* ferredoxin and FNR indicate relatively little effect of ionic strength on the apparent K_d for complex formation between the two proteins. A biphasic ionic strength dependence of the rate constant for intracomplex electron transport was observed, with the highest rate constant occurring at intermediate ionic strength (Walker et al., 1991). These observations have been interpreted in terms of a mechanism in which structural arrangements within the complex between the cyanobacterial ferredoxin and FNR facilitate electron transport. It is possible to envisage a mechanism for docking of ferredoxin and FNR in which the two proteins are first steered towards each other through the orientation of their large molecular dipole moments, a process that brings the proteins into an approximately correct orientation. Subsequently, complementary electrostatic surface potentials guide the two proteins towards an orientation of the [2Fe-2S] cluster and the FAD that is optimal for electron transfer (De Pascalis et al., 1993). Such a mechanism should increase the fraction of productive encounter complexes and would be consistent with the results of kinetic studies. If protein:protein interactions within the complex are too strong, the flexibility necessary for optimal orientation of the prosthetic groups may be absent (Walker et al., 1991).

Studies using ferredoxins altered by site-directed mutagenesis indicate that an aromatic amino acid at ferredoxin position 65 is required for optimal electron transport from reduced *Anabaena* ferredoxin to *Anabaena* FNR (Hurley et al., 1993a; 1993b). Replacement of the phenylalanine found at this position in wild type ferredoxin with either tyrosine or tryptophan had little effect on the kinetics, but replacement of phenylalanine with aliphatic amino acids decreased the rate constant by more than three orders of magnitude. A double mutant, which places an aliphatic residue at position 65 and tyrosine at the adjacent position 64, exhibits a rate constant for FNR reduction four orders less than that for wild type ferredoxin, indicating that the positioning of the aromatic moiety is important (Hurley et al., 1993b). Analysis of the kinetic data obtained with mutant ferredoxins lacking an aromatic residue at position 65 suggests that it is intracomplex electron transfer *per se* that requires an aromatic side chain at this position (Hurley et al., 1993b).

Additional experiments with mutant ferredoxins indicate that residue Glu^{94} of *Anabaena* ferredoxin provides a negative charge that is required for high rates of electron transport to *Anabaena* FNR (Hurley et al., 1993b; 1994). An E94K ferredoxin mutant has a K_d for binding to FNR that differs by only 3-fold compared to that of the wild type, while the second-order rate constant for FNR reduction is decreased by a factor of 20,000 (Hurley et al., 1993b). An E94D ferredoxin mutant, which should retain the negative charge at position 94, is essentially identical to wild type *Anabaena* ferredoxin in terms of its ability to reduce *Anabaena* FNR rapidly, while an E94Q mutant lacking a negative charge at this position shows the same slow kinetics seen with the E94K mutant even though the E_m value and the K_d for binding to FNR of this mutant are essentially identical to those of wild type ferredoxin (Hurley et al., 1994). Analysis of the kinetics of *Anabaena* FNR reduction by the E94K and E94Q ferredoxin mutants suggests that it is the rate constant for electron transport within a transient ferredoxin/FNR complex that is affected by the loss of a negative charge at position 94 (Hurley et al., 1993a; 1994). In contrast to the large effects caused by alteration of the charge at ferredoxin position 94, replacement of the negative charge at the adjacent position 95 by a positive charge, in an E95K mutant, had no effect on the kinetics of reduction of *Anabaena* FNR. The *Anabaena* heterocyst ferredoxin, which contains glutamate at position 94 but proline instead of glutamate at position 95, exhibits kinetics of FNR reduction similar to those obtained with the vegetative ferredoxin. The remarkable differences observed between removal of the negative charge at two immediately adjacent positions in ferredoxin provide evidence for a high degree of specificity and localization in the interactions between the two proteins during the electron transfer process. It should be pointed out that site-directed mutagenesis studies of changes at the equivalent position in spinach ferredoxin 1, i.e. at Glu^{95}, produce very different results. Preliminary characterization of E92Q, E92A and E92K mutant forms of spinach ferredoxin 1 indicates that these mutant ferredoxins show small enhancements in the rate of FNR-catalyzed electron transfer from NADPH to ferredoxin (Aliverti et al., 1995).

A proposal has been made for a hydride transfer mechanism from the $FADH_2$ of spinach FNR to $NADP^+$ in which the nicotinamide moiety of $NADP^+$ is bound parallel to the isoalloxazine ring of FAD in a manner similar to that observed for glutathione reductase (Karplus et al., 1991). As this postulated nicotinamide-binding site is occupied by Tyr^{314} in the unliganded structure, the side chain of Tyr^{314} would have to move out of the pocket to allow substrate binding (Karplus and Bruns, 1994). However, site-directed mutagenesis of the equivalent carboxy-terminal Tyr^{314} in pea FNR indicated that an aromatic amino acid is required at this position in order for k_{cat} for the reduction of a dye by NADPH catalyzed by FNR to be similar to the wild-type value (Oralin et al., 1993), suggesting that Tyr^{314} may play an active role in catalysis. Site-directed mutagenesis of another amino acid residue located near the FAD-binding site in spinach FNR, Ser^{272}, indicates that it too may be involved in catalysis, perhaps by stabilizing a charge-transfer complex that is a precursor to hydride transfer between flavin and pyridine nucleotide (Aliverti et al., 1993).

C. Biosynthesis and Membrane Attachment

In higher plants, FNR is encoded by a nuclear gene and synthesized on cytoplasmic ribosomes as a larger precursor protein with a cleavable leader sequence that is involved in targeting the protein for post-translational translocation across the chloroplast envelope (Grossman et al., 1982). Folding of the imported apo-FNR may involve chloroplast chaperonins (Carrillo et al., 1992). Little is known about the insertion of the FAD prosthetic group in vivo, but reconstitution of the apoprotein with FAD in vitro has been demonstrated and does not require energy input (Bookjans et al., 1978; Zanetti et al., 1982). It should be pointed out that both spinach and pea FNR's can be expressed in *E. coli* and that the FAD is properly inserted into the protein inside the bacteria (Aliverti et al., 1990; Ceccarelli et al., 1991).

A 3.4 kb promoter sequence for the single-copy gene encoding spinach leaf FNR has been characterized and shown to contain two light-responsive elements, one located within the first 118 bp upstream from the start site and the second running from nucleotide position –119 to –220 (Oelmüller et al., 1993). Neither of the fragments contains sequences similar to those of known light-responsive *cis* elements found in other plant genes. The two light-responsive regions respond differentially to phytochrome-mediated signals. Studies of chimeric fusions of these promoter regions to a reporter gene indicate that expression is not confined to tissue with functional chloroplasts in transgenic tobacco (Oelmüller et al., 1993). As mentioned above, a form of FNR different from the leaf form is found in a number of non-photosynthetic plant tissues (Suzuki et al., 1985; Hirasawa et al., 1990; Morigasaki et al., 1990a, 1990b; Green et al., 1991). It has been shown that in rice, the root FNR mRNA is induced by the addition of nitrate, but not by ammonia (Aoki and Ida, 1994), in agreement with the suggestion (see above) that the function of root FNR is to catalyze the NADPH-dependent reduction of ferredoxin so that reduced ferredoxin is available as an electron donor for nitrate assimilation in non-photosynthetic plant tissues.

Considerable uncertainty exists as to the identity of the protein(s) involved in anchoring FNR to the thylakoid membrane in vivo. It has been proposed that a 17.5 kDa intrinsic membrane protein anchors FNR to the thylakoid membrane in higher plants (Vallejos et al., 1984; Matthijs et al., 1986). A suggestion that this FNR-binding protein is identical to one of the extrinsic peptide components involved in oxygen evolution (Soncini and Vallejos, 1989), appears to have been incorrect (Berzborn et al., 1990). Evidence has been obtained for a M_r = 15 kDa protein, perhaps equivalent to the 16.5 kDa spinach peptide, involved in binding FNR to the thylakoid membrane in *Euglena gracilis* chloroplasts (Spano and Schiff, 1987). It has also been proposed that, in higher plants, FNR does not bind directly to the thylakoid membrane but that it binds instead to a M_r = 10 kDa protein, 'connectein', which in turn binds to an intrinsic protein located in the thylakoid membrane (Nozaki et al., 1985; Shin et al., 1985; Nakatani and Shin, 1991). A 200 Å^3 cavity in FNR that is lined with hydrophobic residues and is located far from the active site has been suggested as a possible site for membrane attachment (Bruns and Karplus, 1995).

As mentioned above, a M_r = 45 kDa form of FNR, with a ca. 300 amino acid-long carboxy-terminal domain that resembles plant FNR and a ca. 85 amino acid-long extension at the amino-terminus, has been found in cyanobacteria. The amino-terminal portion of *Synechococcus* PCC 7002 FNR is 78% similar to the *Synechococcus* PCC 7002 CpcD protein, a linker protein associated with the peripheral rods in phycobilisomes (Schluchter and Bryant, 1992). This finding, and the observation that the 45 kDa form of FNR can be detected in phycobilisomes, has resulted in a proposal that a substantial portion of FNR in cyanobacteria is associated with the peripheral rods of phycobilisomes at their core-distal ends (Schluchter and Bryant, 1992).

IV. Nitrite Reductase

A. Structure and Redox Properties

The six-electron reduction of nitrite to ammonia, which takes place in cyanobacteria and in the stroma of algal and plant chloroplasts, uses ferredoxin reduced in the light by Photosystem I as the electron donor and is catalyzed by the soluble enzyme nitrite reductase (Knaff and Hirasawa, 1991). Ferredoxin-dependent nitrite reductases are also present in non-photosynthetic plant tissues such as roots (Hirasawa et al., 1984; Nagaoka et al., 1984; Suzuki et al., 1985; Ogawa and Ida, 1987; Kronenberger et al., 1993). Chloroplast nitrite reductases are monomeric proteins with molecular masses near 63 kDa and contain a single siroheme and a single [4Fe-4S] cluster as prosthetic groups (Knaff and Hirasawa, 1991). It now appears very likely that a peptide of ca. 25 kDa molecular mass that was reported to be the ferredoxin-binding subunit of the ferredoxin-dependent nitrite reductases isolated from spinach leaf and from the green alga *Chlamydomonas reinhardtii* (Knaff and Hirasawa, 1991) is more likely to have been a fortuitous contaminant rather than a true subunit of the enzyme (Hilliard et al., 1991; Pajuelo et al., 1993).

Complete amino acid sequences, deduced from base sequences of the corresponding cDNA's, are available for ferredoxin-dependent nitrite reductases from four higher plants: spinach (Back et al., 1988); maize (Lahners et al., 1988); tobacco (sequences for three different isoforms are known, Kronenberger et al., 1993); and birch (Friemann et al., 1992a) and from the cyanobacterium *Synechococcus* sp. PCC 7942 (Luque et al., 1993; Suzuki et al., 1993). A partial sequence is available for nitrite reductase from Scots pine (Neininger et al., 1994a). The birch and maize enzymes are 75% and 79% identical, respectively, to spinach nitrite reductase at the amino acid level and each of the three tobacco isoforms is approximately 84% identical to the spinach enzyme. The extent of amino acid homology between the *Synechococcus* PCC 7942 and spinach nitrite reductases is 53%. Four absolutely conserved cysteines (residues 441, 447, 482 and 486 in the spinach sequence) are highly likely to function as ligands to the [4Fe-4S] cluster. Four cysteines with identical spacing are also present in the β subunit of the *E. coli* and *Salmonella typhimurium* NADPH-linked sulfite reductases (Ostrowski et al., 1989) and in the ferredoxin-linked sulfite reductase of the cyanobacterium *Synechococcus* PCC 7492 (Gisselmann et al., 1993), enzymes that catalyze the 6-electron reduction of sulfite to sulfide and contain the same siroheme and [4Fe-4S] cluster prosthetic groups as nitrite reductases. It should be pointed out that plant chloroplasts also contain a ferredoxin-dependent sulfite reductase, with the same prosthetic group content as nitrite reductase and similar catalytic properties (Krueger and Siegel, 1982a; 1982b).

A large amount of spectroscopic evidence obtained with nitrite and sulfite reductases (Krueger and Siegel, 1982b; Wilkerson et al., 1983; Day et al., 1988; Kaufman et al., 1993) and a 3 Å resolution crystal structure of the β subunit of the *E. coli* sulfite reductase (McRee et al., 1986) suggest that the two prosthetic groups of these enzymes are magnetically coupled and that a γ-sulfur from one of the cysteine ligands to the [4Fe-4S] cluster also serves as an axial ligand to the siroheme. The distance between the siroheme iron and the nearest [4Fe-4S] cluster iron in the β subunit of *E. coli* sulfite reductase is 4.4 Å and one of the cluster sulfides is in van der Waals contact with one edge of the siroheme macrocycle (McRee et al., 1986).

Resonance Raman (Ondrias et al., 1985) and EPR (Hirasawa et al., 1987) spectra are consistent with the presence of high-spin, ferric siroheme in oxidized spinach nitrite reductase and suggest that the siroheme iron is either 5-coordinate or is 6-coordinate with one weakly-bound axial ligand. ENDOR measurements indicate that the siroheme iron in the oxidized form of the β subunit of *E. coli* sulfite reductase is 5-coordinate (Cline et al., 1985b) and X-ray data indicate that the vacant sixth coordination position of the siroheme iron is close to the surface of the *E. coli* protein and freely accessible to the solvent (McRee et al., 1986). The fact that the heme-containing subunit of the *E. coli* sulfite reductase not only has the same prosthetic groups as plant nitrite reductases and shows a significant amino acid homology to the plant enzymes (Ostrowski et al., 1989; Friemann et al., 1992a; Luque et al., 1993), but also can reduce nitrite to ammonia (Siegel et al., 1982), strengthens the argument that the β subunit of the bacterial sulfite reductase is likely to be a good model for plant nitrite reductases. The absence of a second strongly-bound axial ligand to siroheme and exposure of this prosthetic group to the solvent in nitrite reductase would facilitate nitrite binding to the siroheme.

Oxidation-reduction titrations of a number of plant leaf nitrite reductases gave extremely negative values (< -500 mV) for the E_m of the [4Fe-4S] cluster in the enzymes (Knaff and Hirasawa, 1991). These values were difficult to reconcile with the considerably less negative $E_m = -420$ mV value of the [2Fe-2S] cluster in ferredoxin, the physiological electron donor to the enzyme. All of these redox titrations of nitrite reductase involved observation of EPR signals at cryogenic temperatures and many of them involved measurements made at alkaline pH and under conditions where the enzyme was likely to be present as the sulfite adduct rather than as the free enzyme (Knaff and Hirasawa, 1991; Hirasawa et al., 1994a). These measurements were thus susceptible to artifacts arising from the use of non-physiological temperatures and pH values and from the possibility that binding of a ligand to siroheme might shift the E_m value of the [4Fe-4S] cluster. More recent measurements, utilizing two different electrochemical techniques chosen to minimize these possibilities for artifacts, gave an average E_m value of -365 mV for the [4Fe-4S] cluster of spinach leaf nitrite reductase (Hirasawa et al., 1994a). These same electrochemical methods gave an average E_m value of -290 mV for the siroheme group of the spinach enzyme (Hirasawa et al., 1994), a value considerably more negative than reported earlier (Knaff and Hirasawa, 1991) but one in reasonable agreement with the -340 mV E_m value reported for the siroheme of the β subunit of *E. coli* sulfite reductase (Siegel et al., 1982). Most earlier titrations of the siroheme group of plant nitrite reductases were carried out using dithionite as a reductant, conditions where the sulfite adduct of the enzyme rather than the free enzyme was the likely species titrated. Redox titrations of the β subunit of the *E. coli* sulfite reductase have demonstrated that ligand binding can raise the E_m value for the siroheme by as much as 180 mV (Siegel et al., 1982).

B. Substrate Binding and Mechanism

Abundant spectroscopic evidence has established that nitrite is bound to the iron of the siroheme in oxidized nitrite reductases (Knaff and Hirasawa, 1991), giving a low-spin ferric siroheme form of the enzyme (Day et al., 1988). It has been demonstrated that ferredoxin binding to nitrite reductase involves a 1:1 complex of the two proteins (Mikami and Ida, 1989) and that the complex is stabilized by electrostatic interactions (Knaff and Hirasawa, 1991). The observation that a ternary complex can be formed between nitrite, the enzyme and ferredoxin indicates that there are separate binding sites on nitrite reductase for the two substrates (Hirasawa and Knaff, 1985). Negative cooperativity is observed in substrate binding by the enzyme, with the K_d for oxidized ferredoxin increasing approximately 20-fold in the presence of nitrite (Hirasawa and Knaff, 1985). Chemical modification (Hirasawa et al., 1986) and cross-linking (Privalle et al., 1985) studies suggest that the site on ferredoxin that is involved in binding to nitrite reductase is the same one involved in binding to FNR. Chemical modification studies indicate that nitrite reductase lysine and arginine residues supply the positive charges for the electrostatic interaction with ferredoxin (Hirasawa et al., 1993) and immunological studies suggest that the ferredoxin-binding site of spinach leaf nitrite reductase is antigenically related to the ferredoxin-binding site of spinach leaf FNR (Hirasawa et al., 1989). Chemical modification studies also suggest the presence of a tryptophan at the nitrite reductase-binding site for ferredoxin (Hirasawa et al., 1994b). Friemann et al. (1992a) have pointed out that a well-conserved region in plant nitrite reductases (running from residue 95 to 119 in the spinach leaf enzyme) containing a cluster of 5–6 positively charged amino acids shows some homology to a region near the amino terminus of FNR that has been implicated in the binding of ferredoxin to FNR (Zanetti et al., 1988; Aliverti et al., 1990; Gadda et al., 1990) and thus may constitute a portion of the ferredoxin-binding site of the enzyme.

Relatively few details of the mechanism by which nitrite reductase converts nitrite to ammonia are known. Recent laser flash photolysis measurements have provided support for the hypothesis that electrostatic interactions between ferredoxin and nitrite reductase play an important role in the reaction mechanism and have provided the first evidence for electron flow from the [4Fe-4S] cluster to the siroheme (Hirasawa et al., 1994). Earlier experiments using dithionite, a kinetically poor non-physiological reductant, had demonstrated that the reduced [4Fe-4S] cluster is rapidly reoxidized by nitrite (Lancaster et al., 1979). Binding of nitrite to the oxidized enzyme is relatively slow, suggesting that the reaction mechanism involves reduction of the siroheme prior to nitrite binding (Cammack et al., 1978). Evidence indicates that a ferrous siroheme-NO complex, which represents a one-electron reduction of nitrite, may be an early intermediate in the conversion of enzyme-bound substrate to ammonia (Aparicio et al., 1975; Cammack et al., 1978; Fry et al., 1980) and that hydroxylamine, which is two electrons more oxidized than the final product ammonia, may be a late intermediate (Vega and Kamin, 1977). It should also be mentioned that the nature of the isobacteriochlorin macrocycle of siroheme makes it relatively easy to oxidize the macrocycle and generate a π cation radical (Chang et al., 1981). This is also true for the siroheme in *E. coli* sulfite reductase and although no evidence has yet been obtained for a π cation radical playing a role in the mechanism of enzyme-catalyzed reduction of either nitrite or sulfite, oxidation of the isobacteriochlorin ring itself could, in principle, be a source of an 'extra' electron (in addition to the one each that can be carried by the siroheme iron and the [4Fe-4S] cluster) for the multi-electron reductions of sulfite and nitrite catalyzed by nitrite and sulfite reductases (Young and Siegel, 1988).

C. Biosynthesis

Higher plant nitrite reductases, and presumably those of algae, are coded for by nuclear genes and synthesized as larger precursors with leader sequences that target them for import into the chloroplasts (Small and Gray, 1984; Gupta and Beevers, 1987; Back et al., 1988; Lahners et al., 1988; Friemann et al., 1992a). It appears that the nitrite reductase transit sequences may be unusually short compared to other chloroplast transit sequences, only 32 amino acids long for the spinach enzyme (Back et al., 1988) and 22 amino acids for the birch enzyme (Friemann et al., 1992a). Nothing is known about the insertion of the [4Fe-4S] cluster and siroheme prosthetic groups into the apoprotein.

In all higher plants tested significant levels of nitrite reductase are only detected in the presence of nitrate. Experiments conducted with barley leaves indicate that it is nitrate, rather than nitrite, that is the regulatory species (Aslam and Huffaker, 1989). Direct measurements of mRNA levels have shown that nitrate stimulates the production of nitrite reductase mRNA in maize (Lahners et al., 1988; Kramer et al., 1989), spinach (Back et al., 1988; Seith et al., 1991), birch (Friemann et al., 1992b), tobacco (Neininger et al., 1992; Kronenberger et al., 1993) and pine (Neininger et al., 1994a), confirming earlier less direct measurements made with pea and wheat. Light, operating via a phytochrome-mediated response, also plays a role in regulating nitrite reductase levels, but the mechanism of this regulation appears to vary in different plant species. In spinach, the primary effect of nitrate is on transcription (Back et al., 1988; Seith et al., 1991), while light has no effect on the steady-state level of nitrite reductase mRNA but instead affects enzyme precursor synthesis (Seith et al., 1991). However, in both tobacco (Neininger et al., 1992) and pea (Gupta and Beevers, 1987) light does appear to play a role in the regulation of transcription. A third regulatory scenario has been reported for mustard seedlings, where the phytochrome-mediated regulation by light appears to be associated with increased transcription, with nitrate being required for translation (Schuster and Mohr, 1990). Differences have also been detected in the patterns of regulation in different tissues of the same plant (Friemann et al., 1992b; Kroneberger et al., 1993; Neininger et al., 1994a). It should also be mentioned that an as yet uncharacterized 'plastidic factor' may also be involved in the regulation of expression of the nitrite reductase gene (Mohr et al., 1992; Neininger et al., 1994b).

Initial experiments involving the use of chimeric constructs of the upstream promoter region of the spinach nitrite reductase gene with a reporter gene, followed by expression in transgenic tobacco, indicated that the regulatory element(s) responsible for both the nitrate and light responses is(are) located somewhere in a ca. 3 kb DNA fragment situated upstream from the translation start site (Back et al., 1991; Neininger et al., 1993). Deletion analysis (Rastogi et al., 1993; Neininger et al., 1994b) indicated that the basic elements required for both the nitrate-dependent and the light/phytochrome-dependent regulation of spinach nitrite reductase gene expression are located within the –200/+131 region, relative to the site for initiation of transcription. Additional positive regulatory elements exist in the –330/–200 and –1730/–1450 regions (Neininger et al., 1994b).

In *Synechococcus* PCC 7942, the gene encoding ferredoxin-dependent nitrite reductase (*nirA*) is the first gene in a six-gene operon and is co-transcribed with the other five genes, one of which codes for a ferredoxin-dependent nitrate reductase and four of which appear to code for proteins involved in the active transport of nitrate and nitrite (Suzuki at al., 1993; Luque et al., 1994). Studies with inhibitors of glutamine synthetase and glutamate synthase suggest that accumulation of a compound, the synthesis of which involves transfer of the amide nitrogen from glutamine, serves to repress transcription of the *nirA* operon (Suzuki et al., 1993).

V. Glutamate Synthase

A. Structure and Redox Properties

The two-electron reductive conversion of glutamine *plus* 2-oxoglutarate to two glutamates that occurs in cyanobacteria and in the chloroplast stroma of higher plants and algae is catalyzed, except in the case of rice, by a monomeric M_r = ca. 160 kDa enzyme that is specific for reduced ferredoxin as the electron donor (Knaff and Hirasawa, 1991). Ferredoxin-dependent glutamate synthases have also been found in plant roots (Suzuki et al., 1985; Sakakibara et al., 1992; Redinbaugh and Campbell, 1993). Plants also contain at least one pyridine nucleotide-dependent glutamate synthase that, unlike the ferredoxin-dependent enzyme, is found in the cytoplasm rather than in chloroplasts (Lea et al., 1990). The ferredoxin-dependent chloroplast glutamate synthases appear to play a key role in reassimilating the ammonia released during photorespiration (Somerville and Ogren, 1980; Kendall et al., 1986; Wallsgrove et al., 1987).

Complete amino acid sequences, deduced from the nucleotide sequence of cDNA clones, are known for the ferredoxin-dependent glutamate synthases from maize (Sakakibara et al., 1991) and the red alga *Antithamnion* sp.(Valentin et al., 1993), an essentially complete sequence (corresponding to ca. 98% of the mature form of the protein) is known for the spinach enzyme (Nalbantoglu et al., 1994) and partial sequences are known for the tobacco (Zehnacker et al., 1992) and barley (Avila et al., 1993) enzymes. The spinach and maize enzymes are 83% identical at the amino acid level (Nalbantoglu et al, 1994).

The spinach and *C. reinhardtii* enzymes are known to contain non-heme iron and acid-labile sulfide and recent EPR, resonance Raman and magnetic circular dichroism measurements conducted with the spinach enzyme have provided unambiguous evidence for the presence of a single [3Fe-4S] cluster as the only iron-sulfur center in the enzyme (Knaff et al., 1991). The NADPH-dependent *Azospirillum brasilense* enzyme has also been shown to contain a single [3Fe-4S] cluster (Vanoni et al., 1992). A comparison of the amino acid sequences of the spinach and maize enzymes to those of the α subunits of the NADPH-dependent *E. coli* and *A. brasilense* enzymes and to those of other enzymes known to have [3Fe-4S] clusters, suggests that cysteines 1132, 1138 and 1143 in the mature maize enzyme function as ligands to the [3Fe-4S] cluster (Sakakibara et al., 1991; Pelanda et al., 1993). Cysteines with identical spacings are present in the spinach ferredoxin-dependent enzyme (Nalbantoglu et al., 1994) and in the α subunits of the NADPH-dependent *E. coli* and *A. brasilense* enzymes (Gosset et al., 1989; Pelanda et al., 1993). Oxidation-reduction titrations on the spinach enzyme indicate a E_m value of –170 mV for the [3Fe-4S] cluster (Hirasawa et al., 1992).

Ferredoxin-dependent enzymes isolated from spinach (Hirasawa and Tamura, 1984), barley (Márquez et al., 1988), *C. reinhardtii* (Galván et al., 1984) and the green alga *Monoraphidium braunii* (Vigara et al., 1994) have spectra characteristic of flavoproteins. In contrast, the ferredoxin-dependent glutamate synthase of rice leaves, which appears to be a M_r = 224 kDa homodimer, contains neither flavin nor an iron-sulfur cluster (Suzuki and Gadal, 1982). Flavin analysis of both the spinach leaf (Hirasawa and Tamura, 1984) and *C. reinhardtii* (Márquez et al., 1986) enzymes indicated the presence of equimolar amounts of FAD and FMN, resulting in the proposal that the enzymes contained one mol of FAD and one mol of FMN per mol of enzyme. However, the analysis of the spinach enzyme did not include a quantitation of the total flavin content and the data obtained with the *C. reinhardtii* enzyme showed only 1.10 mol of total flavin per mol of enzyme. A ferredoxin-dependent glutamate synthase with properties similar to those of the spinach and *C. reinhardtii* enzymes, isolated from the cyanobacterium *Synechococcus* sp. PCC 6301, has been shown to contain 1 mol of FMN per mol of enzyme, but no FAD (Marqués et al., 1992). Furthermore, the extinction coefficients of the spinach and *C. reinhardtii* enzymes at the flavin maxima are comparable to that of the *Synechococcus* enzyme and, after the contribution estimated for the [3Fe-4S] cluster has been subtracted, are more consistent with the presence of only one mol of total flavin being present per mol of enzyme than with two. It should also be noted that analysis of the amino acid sequence of maize ferredoxin-dependent glutamate synthase showed a region, running from residue 1079 to residue 1136 of the mature maize protein, homologous to that of known FMN-binding domains. In contrast, no regions showing equally good homology to known FAD-binding domains could be found (Sakakibara et al., 1991). Oxidation-reduction titrations of the spinach enzyme have been interpreted in terms of equipotential FMN and FAD groups with identical E_m values of –180 mV (Hirasawa et al., 1992), but can be equally well interpreted in terms of the presence of only a single flavin. It is clear that additional investigation is needed to firmly establish the flavin content of plant leaf and algal ferredoxin-dependent glutamate synthases.

B. Substrate Binding and Mechanism

Virtually nothing is known about the mechanistic details of the reaction catalyzed by ferredoxin-dependent glutamate synthases, but some information is available on substrate binding. Complex formation between ferredoxin and glutamate synthase has been studied in detail only with the spinach leaf proteins and has been shown to be electrostatic in nature (Knaff and Hirasawa, 1991), involving negative charges on ferredoxin (Hirasawa et al., 1986) and positive charges on glutamate synthase (Hirasawa and Knaff, 1993). Both ultrafiltration (Hirasawa et al., 1989) and cross-linking (Hirasawa et al., 1991) measurements suggest that the ferredoxin:enzyme stoichiometry of the complex is 2:1. Immunological experiments suggest that the ferredoxin-binding site(s) on spinach leaf glutamate synthase is(are) antigenically similar to the ferredoxin-binding site on spinach leaf FNR (Hirasawa et al., 1989; Hirasawa et al., 1991). Similar immunological experiments suggest that the ferredoxin-binding sites of *C. reinhardtii* glutamate synthase and nitrite reductase are related (Romero et al., 1988; Gotor et al., 1990). It appears likely that the site on ferredoxin involved in binding glutamate synthase is identical to the site on ferredoxin that interacts with FNR and nitrite reductase (Hirasawa et al., 1986). No direct biochemical data have yet been obtained to identify the binding sites for glutamine and 2-oxoglutarate. However, a comparison of the sequence of maize, barley and spinach enzymes to those of the α subunit of the glutamate synthases of *E. coli* (Gosset et al., 1989) and *A. brasilense* (Pelanda et al., 1993) suggest that a *purF*-type glutamine amidotransferase domain is located within the amino-terminal 350 amino acids of the plant enzymes.

C. Biosynthesis

In higher plants, ferredoxin-dependent glutamate synthases are encoded by nuclear genes and synthesized in the cytoplasm as larger precursors with a cleavable transit sequence that targets the proteins for translocation into the chloroplast stroma (Sakakibara et al., 1991). The transit sequence of the maize enzyme is 97 amino acids long, one of the longest known for a chloroplast protein (Sakakibara et al., 1991). In contrast, the ferredoxin-dependent glutamate synthase of another photosynthetic eukaryote, the red alga *Antithamnion* sp., is encoded by a chloroplast gene (Valentin et al., 1993).

In maize (Sakakibara et al., 1991; 1992) and tobacco (Zehnacker et al., 1992) the gene for ferredoxin-dependent glutamate synthase is light-induced. In contrast to the situation for nitrite reductase, described above, the presence of nitrate has relatively little effect on levels of ferredoxin-dependent glutamate synthase in maize leaves (Sakakibara et al., 1992; Redinbaugh and Campbell, 1993). There have been conflicting reports on the effect of nitrate on ferredoxin-dependent glutamate synthase mRNA levels in maize roots (Sakakibara et al., 1992; Redinbaugh and Campbell, 1993). In tobacco, it appears that ferredoxin-dependent glutamate synthase is present in appreciable amounts only in leaves and neither the protein nor its mRNA could be detected in roots (Zehnacker et al., 1992). In spinach the level of ferredoxin-dependent glutamate synthase transcript is considerably higher in young leaves than in old leaves (Nalbantoglu et al., 1994).

VI. Thioredoxin Reductase

A. Structure and Redox Properties

Thioredoxin reductases catalyze the reduction of the disulfide bond of a single cystine in oxidized thioredoxin to two thiols, yielding two cysteines at the active site of reduced thioredoxin (Droux et al, 1987a; Buchanan, 1991; Knaff and Hirasawa, 1991). Plant leaves contain both a cytoplasmic pyridine nucleotide-dependent thioredoxin reductase and a distinct chloroplast ferredoxin-dependent thioredoxin reductase. The former, a M_r = 68 kDa flavoprotein that catalyzes the reduction of thioredoxin *h* (Florencio et al., 1988), will not be discussed in this chapter.

Chloroplast ferredoxin-dependent thioredoxin reductases (hereafter abbreviated as FTR) are soluble, heterodimeric proteins with M_r = ca. 30 kDa and are localized in the chloroplast stroma (Buchanan, 1991; Knaff and Hirasawa, 1991). It appears that FTR from the cyanobacterium *Nostoc muscorum*, unlike the chloroplast FTR's, contains one M_r = ca. 13 kDa subunit and two M_r = ca. 7 kDa subunits (Droux et al., 1987b). The M_r = ca. 13 kDa subunit of all FTR's appears to be similar, both immunologically and in size, while the size of the second subunit is different in enzymes from different sources (Droux et al., 1987b).

FTR catalyzes the reduction of two different soluble thioredoxins, both of which are located in the chloroplast stroma: Thioredoxin *f* (so-called because it serves as a regulatory reductant for *f*ructose-1,6-bisphosphatase) and thioredoxin *m* (so-called because it regulates the activity of $NADP^+$-*m*alate dehydrogenase (Buchanan, 1991). Reduced thioredoxin, in turn, reduces a regulatory cystine disulfide to two cysteines in a number of target enzymes involved in chloroplast carbon metabolism (Buchanan, 1991; Knaff and Hirasawa, 1991). Thioredoxin *f* selectively activates fructose-1,6-bisphosphatase, sedoheptulose-1,7-bisphosphatase, phosphoribulose kinase and $NADP^+$-linked glyceraldehyde-3-phosphate dehydrogenase, while thioredoxin *m* preferentially activates malate dehydrogenase, an important enzyme in C_4 plants, and deactivates glucose-6-phosphate dehydrogenase, a key enzyme in the oxidative pentose phosphate pathway (Buchanan, 1991).

Amino acid sequences are known for several plant, algal and cyanobacterial thioredoxins. Thioredoxin *m* from spinach and *C. reinhardtii* show reasonably high sequence homologies to *E. coli* thioredoxin (47% and 41% respectively). Spinach thioredoxin *f* shows much lower homology to *E. coli* thioredoxin (24%) and appears to be more closely related to animal than to bacterial thioredoxins, suggesting that the two plant chloroplast thioredoxins have different evolutionary histories (Hartman et al., 1990). Although spinach thioredoxin *f* has been crystallized (Génovésio-Taverne et al., 1991), no X-ray structure is yet available for any chloroplast thioredoxin. However the secondary structure of a *m*-type thioredoxin from *C. reinhardtii* has been determined by two-dimensional proton NMR spectroscopy (Lancelin et al., 1993) and shown to resemble that of *E. coli* thioredoxin, for which a high resolution X-ray structure is available (Katti et al, 1990). The active site sequences of spinach thioredoxins *f* and *m*, Trp-Cys-Gly-Pro-Cys-Lys, are identical to that of *E. coli* thioredoxin (Tsugita et al, 1983). Thioredoxins *m* from *C. reinhardtii* (Decottignies et al, 1990) and from the cyanobacteria *Anabaena* sp. 7119 (Gleason and Holmgren, 1988) and *Anacystis nidulans* (Muller and Buchanan, 1989) have identical active site sequences except that arginine replaces lysine as the positively-charged amino acid. However, a positive charge adjacent to the second active-site cysteine does not appear to be absolutely essential, as glutamine is found at this position in thioredoxin from the cyanobacterium *Anabaena* sp. 7120 (Gleason and Holmgren 1988). It has been demonstrated, in the case of spinach thioredoxin *f*, that Cys^{46} (the active-site disulfide nearest the amino-terminus of the protein) is responsible for the primary nucleophilic attack on the disulfide of fructose-1,6-bisphosphatase (Brandes et al, 1993).

Amino acid sequences are available for both subunits of the spinach FTR (Iwadate et al., 1992; Marc-Martin et al, 1993; Falkenstein et al, 1994; Iwadate et al, 1994), the M_r = ca. 13 kDa subunit of maize FTR (Marc-Martin et al., 1993) and the variable subunit of FTR from the cyanobacterium *Nostoc muscorum* (Muller and Buchanan, 1989). In plants, the proteins are coded for by nuclear genes and synthesized as larger precursors with cleavable transit sequences that target them to the chloroplast stroma (Marc-Martin et al., 1993; Falkenstein et al, 1994). The M_r = ca. 13 kDa subunit has been shown to contain the active-site disulfide (Droux et al., 1987a) and, recently, Cys^{54} and Cys^{84} have been demonstrated to constitute the redox-active disulfide bridge (Chow et al., 1995).

In addition to the active-site disulfide, FTR contains a single [4Fe-4S] cluster (de la Torre et al, 1982; Droux et al, 1987b; MK Johnson, DB Knaff and P Schürmann, unpublished data). The four cysteine residues that serve as ligands to the [4Fe-4S] cluster have recently been identified as Cys^{52}, Cys^{71}, Cys^{73} and Cys^{82} of the catalytic subunit (Chow et al., 1995). The cluster produces no EPR signal in the reduced state, but an EPR signal similar to that of oxidized HiPIP (High Potential Iron Protein) can be observed with the oxidized form (de la Torre et al, 1982; MK Johnson, DB Knaff and P Schürmann, unpublished data). An E_m value of +410 mV has been obtained for the [4Fe-4S] cluster of FTR isolated from the cyanobacterium *Nostoc muscorum*, using low temperature EPR spectroscopy to monitor the redox state of the cluster (de la Torre et al, 1982). A less positive E_m value of +340 mV was measured for spinach FTR using cyclic voltammetry (Salamon et al., 1995). Cyclic voltammetry, a technique that has proven useful for measuring the midpoint potentials for disulfide/dithiol couples in proteins (Salamon et al, 1992), has been used recently to obtain E_m values of –230 mV for the active-site disulfide of spinach FTR and –210 mV for the active-site disulfides of both spinach thioredoxin *f* and thioredoxin *m* (Salamon, et al., 1995). The E_m values obtained for the two spinach chloroplast thioredoxins are significantly more positive than those of –300 mV, –265 mV and –240 mV obtained previously for the thioredoxins isolated from maize (thioredoxin *m*), *E. coli* and *Saccharomyces cerevisiae*, respectively (Porqué et al, 1970; Rebeille and Hatch, 1986; Salamon et al, 1992). The E_m values for the active-site disulfides of spinach FTR and thioredoxins *f* and *m* predict thermodynamically favorable electron flow from reduced ferredoxin to FTR and then to thioredoxin. However, the much more positive E_m value measured for the [4Fe-4S] cluster in FTR makes it difficult to envisage a role for this redox couple in the electron transfer process catalyzed by the enzyme. For this reason, it has been proposed that the cluster may play a structural rather than a catalytic role in FTR (de la Torre et al, 1982). Precedents do exist for structural, rather than electron transfer roles, for [4Fe-4S] clusters in enzymes (e.g., Fu et al, 1992b).

B. Substrate Binding and Mechanism

There is essentially no information about the catalytic mechanism of FTR. No kinetic studies have been performed and key questions, such as the role of the [4Fe-4S] cluster and how the transition from the one-electron reduction of the enzyme by reduced ferredoxin to the two-electron reduction of the active-site disulfide is accomplished have yet to be addressed. However, some information about substrate binding is available. Spinach ferredoxin has been shown to form a high affinity, electrostatically stabilized 1:1 complex with spinach FTR (Hirasawa et al, 1988b). Differential chemical modification studies (De Pascalis et al, 1994) have identified five acidic amino acid residues (Asp^{34}, Asp^{65}, Glu^{92}, Glu^{93}, and Glu^{94}) and the carboxyl group of the carboxy-terminal Ala^{97} on ferredoxin that appear to be involved in binding to FTR (Fig. 4A). These residues lie largely in one of the two negative surface potential domains that surround the [2Fe-2S] cluster of ferredoxin (Fig. 4C), but do not include residues in the helical region that runs from residues 24–31, residues that lie in the second domain of negative surface potential and which appear to be involved in the interaction between ferredoxin and FNR (Fig. 4B, C, Fig. 6). The residues on FTR that provide the complementary positive charges involved in complex formation with ferredoxin have not yet been identified. The conclusion that the binding site on ferredoxin for FTR appears to be less optimally oriented with respect to the molecular dipole moment of ferredoxin than is the case for the binding site on ferredoxin for FNR (De Pascalis et al, 1994) is consistent with measurements on the relative effects of ionic strength on the two ferredoxin complexes (Droux et al, 1987b).

Relatively little is known about the binding of thioredoxin to FTR, although fluorescence quenching titrations suggest that spinach FTR and thioredoxin *f* can form a 1:1 complex and that hydrophobic interactions may play a role in the interaction between the two proteins (M Hirasawa, DB Knaff and P Schürmann, unpublished observations). A hydrophobic patch, close to the active-site disulfide of *E. coli* thioredoxin, that has been implicated in protein:protein interactions, is well conserved in spinach thioredoxin *f* (Kamo et al. 1989) and may perhaps be involved in binding to FTR.

VII. Conclusion

Recent progress has significantly expanded our knowledge of the role of ferredoxin and ferredoxin-dependent enzymes in plants, algae and cyanobacteria, as we have become aware that these proteins not only play a crucial role in light-dependent reactions but also play an important role in 'dark' metabolism. We have also begun to understand some of the intricate regulatory processes involved in expression of the genes that encode these proteins. Perhaps most impressive, particularly for biochemists and biophysicists, are the advances that have been made in beginning to understand the mechanisms of some of the electron transfer reactions in which these proteins participate. The availability of an increasing number of three-dimensional structures for ferredoxins, some at high resolution, and of structures for two FNR's, when coupled with the ability to alter the structures of both proteins by site-directed mutagenesis and to then examine individual electron transfer steps directly using rapid kinetic techniques, have opened the doors for dramatic progress in our understanding of the FNR-catalyzed electron transfer from ferredoxin to $NADP^+$, a key step of oxygenic photosynthesis. Although considerable progress has been made in defining the prosthetic group content of other ferredoxin-dependent enzymes and characterizing the interaction of these enzymes with their substrates, further advances in understanding their mechanisms will depend critically on the availability of structural information and mutagenesis systems for these enzymes as well. Such developments and direct structural information about ferredoxin/enzyme complexes from X-ray diffraction studies, of the type now available for complexes of mitochondrial (Pelletier and Kraut, 1992) and bacterial (Chen et al., 1992; 1994) electron transfer proteins, can be hoped for in the near future.

Acknowledgments

The author would like to thank Profs. Hans Bosshard, Hazel Holden and Andrew Karplus for figures used in this chapter and Profs. Karplus, Keiichi Fukuyama, Carlos Gómez-Moreno, Shoji Ida, Peter Schürmann, Gordin Tollin, Maria Vanoni and Giuliana Zanetti for access to manuscripts prior to publication. The author would also like to acknowledge extremely helpful discussions with Profs. Bosshard, Fukuyama, Karplus, Hiroshi Matsubara, Schürmann, William Thompson, Tollin, Vanoni, Keishiro Wada and Zanetti and with Dr. Masakazu Hirasawa and the assistance of Dr. Susan San Francisco with protein sequence alignments. Research in the author's laboratory was supported by grants from the U.S. Department of Energy (DE-FG05-90ER20017 and 93ER20125).

References

Alam J, Whitaker RA, Krogmann DW and Curtis SE (1986) Isolation and sequence of the gene for ferredoxin I from the cyanobacterium *Anabaena* sp. strain PCC 7120. J Bacteriol 168: 1265–1271

Aliverti A, Jansen T, Zanetti G, Ronchi S, Herrmann RG and Curti B (1990) Expression in *Escherichia coli* of ferredoxin:NADP$^+$ reductase from spinach. Eur J Biochem 191: 551–555

Aliverti A, Lübberstedt T, Zanetti G, Herrmann RG and Curti B (1991) Probing the role of Lys^{116} and Lys^{244} in the spinach ferredoxin-NADP$^+$ reductase by site-directed mutagenesis. J Biol Chem 266: 17760–17763

Aliverti A, Piubelli L, Zanetti G, Lübberstedt T, Herrmann RG and Curti B (1993) The role of cysteine residues of spinach ferredoxin-NADP$^+$ reductase as assessed by site-directed mutagenesis. Biochemistry 32: 6374–6380

Aliverti A, Hagen WR and Zanetti G (1995) Direct electrochemistry and EPR spectroscopy of spinach ferredoxin mutants with modified electron transfer properites. FEBS Lett 368: 220–224

Aoki H and Ida S (1994) Nucleotide sequence of a rice root ferredoxin-NADP$^+$ reductase cDNA and its induction by nitrate. Biochim Biophys Acta 1183: 553–556

Aoki H, Doyama N and Ida S (1994) Sequence of a cDNA encoding rice leaf ferredoxin-NADP$^+$ reductase. Plant Physiol 104: 1473–1474

Aparicio PJ, Knaff DB and Malkin R (1975) The role of an iron-sulfur center and siroheme in spinach nitrite reductase. Arch Biochem Biophys 169: 102–107

Aslam M and Huffaker RC (1989) Role of nitrate and nitrite in the induction of nitrite reductase in leaves of barley seedlings. Plant Physiol 91: 1152–1156

Avila C, Márquez AJ, Pajuelo P, Cannell ME, Wallsgrove RM and Forde BG (1993) Cloning and sequence analysis of a cDNA for barley ferredoxin-dependent glutamate synthase and molecular analysis of photorespiratory mutants deficient in the enzyme. Planta 189: 475–483

Back E, Burkhardt W, Moyer M, Privalle L and Rothstein S (1988) Isolation of cDNA clones coding for spinach nitrite reductase: Complete sequence and nitrate induction. Mol Gen Genet 212: 20–26

Back E. Dunne W, Schneiderbauer A, de Framond A, Rastogi R and Rothstein SJ (1991) Isolation of the spinach nitrite reductase gene promoter which confers nitrate inducibility on GUS gene expression in transgenic tobacco. Plant Mol Biol 17: 9–18

Barker WC, George DG, Srinivasarao, GY and Yeh LS (1992) Database of protein sequence alignments. Biophys J 61: A348

Batie CJ and Kamin H (1981) The relation of pH and oxidation-reduction potential to the association state of the ferredoxin·ferredoxin:NADP$^+$ reductase complex. J Biol Chem 256: 7756–7763

Batie CJ and Kamin H (1984a) Ferredoxin:NADP$^+$ oxidoreductase. Equilibria in binary and ternary complexes with NADP$^+$ and ferredoxin. J Biol Chem 259: 8832–8839

Batie CJ and Kamin H (1984b) Electron transfer by ferredoxin:NADP$^+$ reductase. Rapid-reaction evidence for participation of a ternary complex. J Biol Chem 259: 11976–11985

Batie CJ and Kamin H (1986) Association of ferredoxin-NADP$^+$ reductase with NADP(H). Specificity and oxidation-reduction properties. J Biol Chem 261: 11214–11223

Bayer E, Josef D, Krauss P, Hagenmaier H, Roder A and Trebst A. (1967) Abbau und resynthese des aktifzentrums von pflanzen ferredoxin. Biochim Biophys Acta 143: 435–437

Berzborn RJ, Klein-Hitpass L, Otto J, Schünemann S, Oworah-Nkruma R and Meyer HE (1990) The 'additional subunit' CF_0II of the photosynthetic ATP-synthase and the thylakoid polypeptide binding ferredoxin NADP reductase: Are they different? Z Naturforsch 45c: 60–71

Bhattacharyya AK, Meyer TE and Tollin G (1986) Reduction kinetics of the ferredoxin-NADP$^+$ reductase complex: A laser flash photolysis study. Biochemistry 25: 4655–4661

Böhme H and Haselkorn R (1988) Molecular cloning and nucleotide sequence analysis of the gene coding for heterocyst ferredoxin from the cyanobacterium *Anabaena* sp. strain PCC 7120. Mol Gen Genet 214: 278–285

Böhme H and Schrautemeier B (1987a) Comparative characterization of ferredoxins from heterocysts and vegetative cells of *Anabaena variabilis*. Biochim Biophys Acta 891: 1–7

Böhme H and Schrautemeier B (1987b) Electron donation to nitrogenase in a cell-free system from heterocysts of *Anabaena variabilis*. Biochim Biophys Acta 891: 115–120

Bookjans G, San Pietro A and Böger P (1978) Resolution and reconstitution of spinach ferredoxin-NADP$^+$ reductase. Biochem Biophys Res Comm 80: 759–765

Bouges-Bocquet B (1980) Electron and proton transfers from *P*-430 to ferredoxin-NADP-reductase in *Chlorella* cells. Biochim Biophys Acta 590: 223–233

Brandes HK, Larimer FW, Geck MK, Stringer CD, Schürmann P and Hartman FC (1993) Direct identification of the primary nucleophille of thioredoxin *f*. J Biol Chem 268: 18411–18414

Bruns CM and Karplus PA (1995) Refined crystal structure of spinach ferredoxin reductase at 1.7Å resolution: Oxidized, reduced and 2′-phospho-5′-AMP bound states. J Mol Biol 247: 125–145

Buchanan BB (1991) Regulation of CO_2 assimilation in oxygenic photosynthesis: The ferredoxin/thioredoxin system. Arch Biochem Biophys 288: 1–9

Cammack R, Rao KK, Bargeron CP, Hutson KG, Andrew PW and Rogers LG (1977) Midpoint redox potentials of plant and algal ferredoxins. Biochem J 168: 205–209

Cammack R Hucklesby DR and Hewitt EJ (1978) Electron-paramagnetic-resonance studies of the mechanism of leaf nitrite reductase. Biochem J 171: 519–526

Cammack R, Rao KK and Hall DO (1985) Ferredoxins: Structure and function of a ubiquitous group of proteins. Physiol Veg 23: 649–658

Carrillo N and Vallejos RH (1987) Ferredoxin-NADP$^+$ oxidoreductase. In: Barber J (ed) Topics in Photosynthesis, Vol 8, pp 527–560. Elsevier, Amsterdam

Carrillo N, Ceccarelli EA, Krapp AR, Boggio S, Ferreyra RG and Viale AM (1992) Assembly of plant ferredoxin-NADP$^+$ oxidoreductase in *Escherichia coli* requires GroE molecular chaperonins. J Biol Chem 267: 15537–15541

Caspar T and Quail PH (1993) Promoter and leader regions involved in the expression of the *Arabidopsis* ferredoxin A gene. The Plant J 3: 161–174

Ceccarelli EA, Viale AM, Krapp AR and Carrillo N (1991) Expression, assembly, and processing of an active plant ferredoxin-NADP$^+$ oxidoreductase and its precursor protein in *Escherichia coli*. J Biol Chem 266: 14283–14287

Chan RL, Carrillo N and Vallejos RH (1985) Isolation and sequencing of an active-site peptide from spinach ferredoxin-$NADP^+$ oxidoreductase after affinity labeling with periodate-oxidized $NADP^+$. Arch Biochem Biophys 240: 172–177

Chang CK, Hanson LK, Richardson PF, Young R and Fajer J (1981) π cation radicals of ferrous and free base isobacteriochlorins: Models for siroheme and sirohydrochlorin. Proc Natl Acad Sci USA 78: 2652–2656

Chen L, Durley R, Poliks BJ, Hamada K, Chen Z, Mathews FS, Davidson VL, Satow Y, Huizinga E, Velieux FMD and Hol WGJ (1992) Crystal structure of an electron-transfer complex between methylamine dehydrogenase and amicyanin. Biochemistry 31: 4959–4964

Chen L. Durley RCE, Mathews FS and Davidson VL (1994) Structure of an electron transfer complex: Methylamine dehydrogenase, amicyanin and cytochrome c_{551i}. Science 264: 86–90

Chow L-P, Iwadate H, Yano K, Kamo M, Tsugita A, Gardet-Salvi L, Stritt-Etter A-L and Schürmann P (1995) Amino acid sequence of spinach ferredoxin:thioredoxin reductase catalytic subunit and identification of thiol groups constituting a redox-active disulfide and a [4Fe-4S] cluster. Eur J Biochem 231: 149–156

Cidaria D, Biondi PA, Zanetti G and Ronchi S (1985) The $NADP^+$-binding site of ferredoxin-NADP reductase. Eur J Biochem 146: 295–299

Cline K, Werner-Washburne M, Lubben TH and Keegstra K (1985a) Precursors to two nuclear-encoded chloroplast proteins bind to the outer envelope membrane before being imported into chloroplasts. J Biol Chem 260: 3691–3696

Cline JF, Janick PA, Siegel LM and Hoffman BM (1985b) Electron-nuclear double resonance studies of oxidized *Escherichia coli* sulfite reductase: ^1H, ^{14}N, and ^{57}Fe measurements. Biochemistry 24: 7942–7947

Correll CC, Batie CJ, Ballou DP and Ludwig ML (1992) Phthalate dioxygenase reductase: A modular structure for electron transfer from pyridine nucleotides to [2Fe-2S]. Science 258: 1604–1610.

Correll CC, Ludwig ML, Bruns CM and Karplus PA (1993) Structural prototypes for an extended family of flavoprotein reductases: Comparison of phthalate dioxygenase reductase with ferredoxin reductase and ferredoxin. Protein Sci 2: 2112–2133

Day EP, Peterson J, Bonvoisin JJ, Young LJ, Wilkerson JO and Siegel LM (1988) Magnetization of the sulfite and nitrite complexes of oxidized sulfite and nitrite reductases: EPR silent spin S = 1/2 states. Biochemistry 27: 2126–2132

de Boer DA and Weisbeek PJ (1991) Chloroplast protein topogenesis: Import, sorting and assembly. Biochim Biophys Acta 1071: 221–253

Decottingnies P, Schmitter J-M, Jacquot J-P, Dutka S, Picaud A and Gadal PA (1990) Purification, characterization, and complete amino acid sequence of a thioredoxin from a green alga, *Chlamydomonas reinhardtii*. Arch Biochem Biophys 280: 112–121

de la Torre A, Lara C, Yee BC, Malkin R and Buchanan BB (1982) Physicochemical properties of ferralterin, a regulatory iron-sulfur protein functional in oxygenic photosynthesis. Arch Biochem Biophys 213: 545–550

De Pascalis AR, Jelesarov I, Ackermann F, Koppenol WH, Hirasawa M, Knaff DB and Bosshard HR (1993) Binding of ferredoxin to ferredoxin:$NADP^+$ oxidoreductase: The role of carboxyl groups, electrostatic surface potential, and molecular dipole moment. Protein Sci 2: 1126–1135

De Pascalis AR, Schürmann P and Bosshard HR (1994) Comparison of the binding sites of plant ferredoxin for two ferredoxin-dependent enzymes. FEBS Lett 337: 217–220

Dickey LF, Gallo-Meagher M and Thompson WF (1992) Light regulatory sequences are located within the 5′ portion of the *Fed-1* message sequence. EMBO J 11: 2311–2317

Dobres MS, Elliott RC, Watson JC and Thompson WF (1987) A phytochrome regulated pea transcript encodes ferredoxin I. Plant Mol Biol 8: 53–59

Droux M, Miginiac-Maslow M, Jacquot J-P, Gadal P, Crawford NA, Kosower NS and Buchanan BB (1987a) Ferredoxin-thioredoxin reductase: A catalytically active dithiol group links photoreduced ferredoxin to thioredoxin functional in photosynthetic enzyme regulation. Arch Biochem Biophys 256: 372–380

Droux M, Jacquot J-P, Miginac-Maslow M, Gadal P, Huet JC, Crawford NA, Yee BC and Buchanan BB (1987b) Ferredoxin-thioredoxin reductase, an iron-sulfur enzyme linking light to enzyme regulation in oxygenic photosynthesis: Purification and properties of the enzyme from C_3, C_4, and cyanobacterial species. Arch Biochem Biophys 252: 426–439

Dunham WR, Palmer G, Sands RH and Bearden AJ (1971) On the structure of the iron-sulfur complex in the two-iron ferredoxins. Biochim Biophys Acta 253: 373–384

Elliott RC, Pedersen TJ, Fristensky B, White MJ, Dickey LF and Thompson WF (1989a) Characterization of a single copy gene encoding ferredoxin I from pea. The Plant Cell 1: 681–690

Elliott RC, Dickey LF, White MJ and Thompson WF (1989b) *cis*-Acting elements for light regulation of pea ferredoxin 1 gene expression are located within the transcribed sequences. The Plant Cell 1: 691–698

Falkenstein E, Schwaewen AV and Scheibe R (1994) Full-length cDNA sequences for both ferredoxin-thioredoxin reductase subunits from spinach (*Spinacia oleracea* L.) Biochim Biophys Acta 1185: 252–254

Farrington JA, Land EJ and Swallow AJ (1980) The one-electron potentials of NAD. Biochim Biophys Acta 590: 273–276

Florencio FJ, Yee BC, Johnson TC and Buchanan BB (1988) An NADP/thioredoxin system in leaves: Purification and characterization of NADP-thioredoxin reductase and thioredoxin *h* from spinach. Arch Biochem Biophys 266: 496–507

Flügge U-F and Hinz G (1986) Energy dependence of protein transport into chloroplasts. Eur J Biochem 160: 563–570

Friedman AL and Keegstra K (1989) Chloroplast protein import. Quantitative analysis of precursor binding. Plant Physiol 89: 993–999

Friemann A, Brinkmann K and Hachtel W (1992a) Sequence of a cDNA encoding nitrite reductase from the tree *Betula pendula* and identification of conserved protein regions. Mol Gen Genet 231: 411–416

Friemann A. Lange M, Hachtel W and Brinkmann K (1992b) Induction of nitrate assimilatory enzymes in the tree *betula pendula*. Plant Physiol 90: 1214–1220

Fry IV, Cammack R, Hucklesby DP and Hewitt EJ (1980) Stability of the nitrosylsirohaem complex of plant nitrite reductase, investigated by EPR spectroscopy. FEBS Lett 111: 377–380

Fu W, Drozdzewski PM, Davies MD, Sligar SG and Johnson MK (1992a) Resonance Raman and magnetic circular dichroism studies of reduced [2Fe-2S] proteins. J Biol Chem 267: 15502–15510

Fu W, O'Handley S, Cunningham RP and Johnson MK (1992b) The role of the iron-sulfur cluster in *Escherichia coli* endonuclease III. J Biol Chem 267: 16135–16137

Gadda G, Aliverti A, Ronchi S and Zanetti G (1990) Structure-function relationship in spinach in ferredoxin-NADP$^+$ reductase as studied by limited proteolysis. J Biol Chem 265: 11955–11959

Gallo-Meagher M, Sowinski DA, Elliott RC and Thompson WF (1992) Both internal and external regulatory elements control expression of the pea *Fed-1* gene in transgenic tobacco seedlings. Plant Cell 4: 389–395

Galván F, Márquez AJ and Vega JM (1984) Purification and molecular properties of ferredoxin-glutamate synthase from *Chlamydomonas reinhardtii*. Planta 162: 180–187

Génovésio-Taverne J-C, Jetzer Y, Sauder U, Hohenester E, Hughet C, Jansonius JN, Gardet-Salvi L and Schürmann P (1991) Crystallization and preliminary X-ray diffraction studies of the spinach-chloroplast thioredoxin *f*. J Mol Biol 222: 459–461

Gisselmann G, Klausmeier P and Schwenn JD (1993) The ferredoxin:sulphite reductase gene from *Synechococcus* PCC7942. Biochim Biophys Acta 1144: 102–106

Gleason FK and Holmgren A (1988) Thioredoxin and related proteins in procaryotes. FEMS Microbiol Rev 54: 271–298

Gosset G, Merino E, Recillas F, Oliver G, Becerril B and Bolivar F (1989) Amino acid sequence analysis of the glutamate synthase enzyme from *Escherichia coli* K-12. Prot Seq Data Anal 2: 9–16

Gotor C, Pajuelo E, Romero LC, Márquez AJ and Vega JM (1990) Immunological studies of ferredoxin-nitrite reductases and ferredoxin-glutamate synthases from photosynthetic organisms. Arch Microbiol 153: 230–234

Green LS, Yee BC, Buchanan BB, Kamide K, Sanada Y and Wada K (1991) Ferredoxin and ferredoxin-NADP$^+$ reductase from photosynthetic and nonphotosynthetic tissues of tomato. Plant Physiol 96 : 1207–1213

Grossman AR, Bartlett SG, Schmidt GW, Mullet JE and Chua N-H (1982) Optimal conditions for post-translational uptake of proteins by isolated chloroplasts. J Biol Chem 257: 1558–1563

Gupta SC and Beevers L (1987) Regulation of nitrite reductase. Plant Physiol 83: 750–754

Hartman H, Syvanen M and Buchanan BB (1990) Contrasting evolutionary histories of chloroplast thioredoxins *f* and *m*. Mol Biol Evol 7: 247–254

Hase T, Kimata Y, Yonekura K, Matsumura T and Sakakibara H (1991) Molecular cloning and differential expression of the maize ferredoxin gene family. Plant Physiol 96: 77–83

Hervás M, Navarro JA and Tollin G (1992) A laser flash spectroscopy study of the kinetics of electron transfer from spinach Photosystem I to spinach and algal ferredoxins. Photochem Photobiol 56: 319–324

Hilliard NP, Hirasawa M, Knaff DB and Shaw RW (1991) A reexamination of the properties of spinach nitrite reductase: Protein and siroheme content heterogeneity in purified preparations. Arch Biochem Biophys 291: 195–199

Hirasawa M and Knaff DB (1985) Interaction of ferredoxin-linked nitrite reductase with ferredoxin. Biochim Biophys Acta 830: 173–180

Hirasawa M and Knaff DB (1993) The role of lysine and arginine residues at the ferredoxin-binding site of spinach glutamate synthase. Biochim Biophys Acta 1144: 85–91

Hirasawa M and Tamura G (1984) Flavin and iron-sulfur containing ferredoxin-linked glutamate synthase from spinach leaves. J Biochem 95: 983–994

Hirasawa M, Fukushima K, Tamura G and Knaff DB (1984) Immunochemical characterization of nitrite reductases from spinach leaves, spinach roots and other higher plants. Biochim Biophys Acta 791: 145–154

Hirasawa M, Boyer JM, Gray KA, Davis DJ and Knaff DB (1986) The interaction of ferredoxin with chloroplast ferredoxin-linked enzymes. Biochim Biophys Acta 851: 23–28

Hirasawa M, Shaw RW, Palmer G and Knaff DB (1987) Prosthetic group content and ligand-binding properties of spinach nitrite reductase. J Biol Chem 262: 12428–12433

Hirasawa M, Sung J-D, Malkin R, Zilber A, Droux M and Knaff DB (1988a) Evidence for the presence of a [2Fe-2S] ferredoxin in bean sprouts. Biochim Biophys Acta 934: 169–176

Hirasawa M. Droux M, Gray KA, Boyer JM, Davis DJ, Buchanan BB and Knaff DB (1988b) Ferredoxin-thioredoxin reductase: Properties of its complex with ferredoxin. Biochim Biophys Acta 935: 1–8.

Hirasawa M, Morrow KJ, Chang K-T and Knaff DB (1989) Circular dichroism, binding and immunological studies on the interaction between spinach ferredoxin and glutamate synthase. Biochim Biophys Acta 977: 150–156

Hirasawa M, Chang K-T and Knaff DB (1990) Characterization of a ferredoxin:NADP$^+$ oxidoreductase from a nonphotosynthetic plant tissue. Arch Biochem Biophys 276: 251–258

Hirasawa M, Chang K-T and Knaff DB (1991) The interaction of ferredoxin and glutamate synthase: Cross-linking and immunological studies. Arch Biochem Biophys 286: 171–177

Hirasawa M, Robertson DE, Ameyibor E, Johnson MK and Knaff DB (1992) Oxidation-reduction properties of the ferredoxin-linked glutamate synthase from spinach leaf. Biochem Biophys Acta 1100: 105–108

Hirasawa M, de Best J and Knaff DB (1993) The effect of lysine and arginine modifying reagents on spinach ferredoxin:nitrite oxidoreductase. Biochim Biophys Acta 1140: 304–312

Hirasawa M, Tollin G, Salamon Z and Knaff DB (1994a) Transient kinetic and oxidation-reduction studies of spinach ferredoxin:nitrite oxidoreductase. Biochim Biophys Acta 1185: 336–345

Hirasawa M, Proske PA and Knaff DB (1994b) The role of tryptophan in the reaction catalyzed by spinach nitrite reductase. Biochim Biophys Acta 1187: 80–88

Hirasawa M, Kleis-San Francisco S, Proske PA and Knaff DB (1995) The effect of N-bromosuccinimide on ferredoxin: NADP$^+$ oxidoreductase. Arch Biochem Biophys 320: 280–288

Holden HM, Jacobsen BL, Hurley JK, Tollin G, Oh B-H, Skjeldal L, Chae YK, Cheng H, Xia B and Markley JL (1994) Structure-function studies of [2Fe-2S] ferredoxins. J Bioenerg Biomembr 26: 67–88

Huisman JG, Moorman AFM and Verkley FN (1978) In vitro synthesis of chloroplast ferredoxin as a high molecular weight precursor in a cell-free protein synthesizing system from wheat germ. Biochem Biophys Res Comm 82: 1121–1131

Hurley JK, Salamon Z, Meyer TE, Fitch JC, Cusanovich MA, Markley JL, Cheng H, Xia B, Chae YK, Medina M, Gómez-Moreno C and Tollin G (1993a) Amino acid residues in *Anabaena* ferredoxin crucial to interaction with ferredoxin-$NADP^+$ reductase: Site-directed mutagenesis and laser flash photolysis. Biochemistry 32: 9346–9354

Hurley JK, Cheng H, Xia B, Markley JL, Medina M, Gómez-Moreno C and Tollin G (1993b) An aromatic amino acid is required at position 65 in *Anabaena* ferredoxin for rapid electron transfer to ferredoxin:$NADP^+$ reductase. J Am Chem Soc 115: 11698–11701

Hurley JK, Medina M, Gómez-Moreno C and Tollin G (1994) Further characterization by site-directed mutagenesis of the protein-protein interaction in the ferredoxin/ferredoxin:$NADP^+$ reductase system from *Anabaena*: Requirement of a negative charge at position 94 in ferredoxin for rapid electron transfer. Arch Biochem Biophys 312: 480–486

Ikemizu S, Bando M, Sato T, Morimoto Y, Tsukihara T and Fukuyama K (1994) Structure of [2Fe-2S] ferredoxin I from *Equisetum arvense* at 1.8 Å resolution. Acta Cryst D50: 167–174

Iwadate H, Yano K, Aso A, Kamo M, Gardet-Salvi L, Schürmann P and Tsugita A (1992) Structure of spinach ferredoxin-thioredoxin reductase. In: Murata N (ed) Research in Photosynthesis, Vol II, pp 539–542. Kluwer, Dordrecht

Iwadate H, Yano K, Kamo M, Gardet-Salvi L, Schürmann P and Tsugita A (1994) Amino acid sequence of the spinach ferredoxin-thioredoxin reductase variable subunit. Eur J Biochem 223: 465–471

Jacobson BL, Chae YK, Markley JL, Rayment I and Holden HM (1993) Molecular structure of the oxidized, recombinant, heterocyst [2Fe-2S] ferredoxin from *Anabaena* 7120 determined to 1.7-Å resolution. Biochemistry 32: 6788–6793

Jalesarov I and Bosshard HR (1994) Thermodynamics of ferredoxin binding to ferredoxin: $NADP^+$ reductase and the role of water at the complex interface. Biochemistry 33: 13321–13328

Jelesarov I, De Pascalis AR, Koppenol WH, Hirasawa M, Knaff DB and Bosshard HR (1993) Ferredoxin binding site on ferredoxin:$NADP^+$ reductase. Differential chemical modification of free and ferredoxin-bound enzyme. Eur J Biochem 216: 57–66

Kamo M, Tsugita A, Wiessner C, Wedel N, Bartling D, Herrmann RG, Aguilar F, Gardet-Salvi L and Schürmann P (1989) Primary structure of spinach-chloroplast thioredoxin *f*. Eur J Biochem 182: 315–322

Karplus PA (1991) Structure/function of spinach ferredoxin:$NADP^+$ oxidoreductase. In: Müller F (ed) Chemistry and Biochemistry of Flavoproteins, Vol 2, pp 449–455. CRC Press, Boca Raton.

Karplus PA and Bruns CM (1994) Structure-function relations for ferredoxin reductase. J Bioenerg Biomembr 26: 89–99

Karplus PA, Daniels MJ and Herriott JR (1991) Atomic structure of ferredoxin-$NADP^+$ reductase: Prototype for a structurally novel flavoenzyme family. Science 251: 60–66

Kassner RJ and Yang W (1977) A theoretical model for the effects of solvent and protein dielectric on the redox properties of iron-sulfur clusters. J Am Chem Soc 99: 4351–4355

Katti SK, LeMaster DM and Eklund H (1990) Crystal structure of thioredoxin from *Escherichia coli* at 1.68 Å resolution. J Mol Biol 212: 167–184

Kaufman J, Spicer LD and Siegel LM (1993) Proton NMR of *Escherichia coli* sulfite reductase: The unligated hemeprotein subunit. Biochemistry 32: 2853–2867

Kaufman LS, Roberts LL, Briggs WR and Thompson WF (1986) Phytochrome control of specific mRNA levels in developing pea buds: Kinetics of accumulation, reciprocity, and escape kinetics of the low fluence response. Plant Physiol 81: 1033–1038

Keegstra K and Olsen LJ (1989) Chloroplastic precursors and their transport across the envelope membranes. Ann Rev Plant Physiol 40: 471–501

Kendall AC, Wallsgrove RM, Hall NP, Turner JC and Lea PJ (1986) Carbon and nitrogen metabolism in barley (*Hordeum vulgare* L.) mutants lacking ferredoxin-dependent glutamate synthase. Planta 168 : 316–323

Kimata Y and Hase T (1989) Localization of ferredoxin isoproteins in mesophyll and bundle sheath cells in maize leaf. Plant Physiol 89: 1193–1197

Knaff DB and Hirasawa M (1991) Ferredoxin-dependent enzymes. Biochim Biophys Acta 1056: 93–125

Knaff DB, Hirasawa M, Ameyibor E, Fu W and Johnson MK (1991) Spectroscopic evidence for a [3Fe-4S] cluster in spinach glutamate synthase. J Biol Chem 266: 15050–15084

Kramer V, Lahners K, Back E, Privalle LS and Rothstein S (1989) Transient accumulation of nitrite reductase mRNA in maize following the addition of nitrite. Plant Physiol 90: 1214–1220

Kronenberger J, Lepingle A, Caboche M and Vaucheret H (1993) Cloning and expression of distinct nitrite reductases in tobacco leaves and roots. Mol Gen Genet 23: 203–208

Krueger RJ and Siegel LM (1982a) Spinach siroheme enzymes: Isolation and characterization of ferredoxin-sulfite reductase and comparison of properties with ferredoxin-nitrite reductase. Biochemistry 21: 2892–2904

Krueger RJ and Siegel LM (1982b) Evidence for siroheme-Fe_4S_4 interaction in spinach ferredoxin-sulfite reductase. Biochemistry 21: 2905–2909

Lahners K, Kramer V, Back E, Privalle L and Rothstein S (1988) Molecular cloning of complementary DNA encoding maize nitrite reductase. Plant Physiol 88: 741–746

Lancaster JR, Vega JM, Kamin H, Orme-Johnson NR, Orme-Johnson WH, Krueger RJ and Siegel LM (1979) Identification of the iron-sulfur center of spinach ferredoxin-nitrite reductase as a tetranuclear center, and preliminary EPR studies of mechanism. J Biol Chem 254: 1268–1272

Lancelin J-M, Stein M and Jacquot J-P (1993) Secondary structure and protein folding of recombinant chloroplastic thioredoxin Ch2 from the green alga *Chlamydomonas reinhardtii* as determined by ^1H NMR. J Biochem 114: 421–431

Land EJ and Swallow AJ (1968) One-electron reactions in biochemical systems as studied by pulse radiolysis. Biochim Biophys Acta 162: 327–337

Lea PJ, Robinson SA and Stewart GR (1990) The enzymology and metabolism of glutamine, glutamate and asparagine. In: Miflin BJ and Lea PJ (eds) The Biochemistry of Plants, Vol 16, pp 121–159. Academic Press, New York

Li H-M, Theg SM, Bauerle CM and Keegstra K (1990) Metal-ion-center assembly of ferredoxin and plastocyanin in isolated chloroplasts. Proc Natl Acad Sci USA 87: 6748–6752

Lubben TH, Donaldson GK, Viitanen P and Gatenby AA (1989) Several proteins imported into chloroplasts form stable complexes with groEL related chloroplast molecular chaperone. Plant Cell 1: 1223–1230

Luque I, Flores E and Herrero A (1993) Nitrite reductase gene from *Synechococcus* sp. PCC 7942: Homology between cyanobacterial and higher-plant nitrite reductase. Plant Mol Biol 21: 1201–1205

Luque I, Flores E and Herrero A (1994) Nitrate and nitrite transport in the cyanobacterium *Synechococcus* sp. PCC 7942 are mediated by the same permease. Biochim Biophys Acta 1184: 296–298

Marc-Martin S, Spielmann A, Stutz E and Schürmann P (1993) Cloning and sequencing of a corn (*Zea mays*) nuclear gene coding for the chloroplast specific catalytic subunit of ferredoxin-thioredoxin reductase. Biochim Biophys Acta 1183: 207–209

Marqués S, Florencio FJ and Candau P (1992) Purification and characterization of the ferredoxin-glutamate synthase from the unicellular cyanobacterium *Synechococcus* sp. PCC 6301. Eur J Biochem 206: 69–77

Márquez AJ, Gotor C, Romero LC, Galván F and Vega JM (1986) Ferredoxin-glutamate synthase from *Chlamydomonas reinhardtii*. Prosthetic groups and preliminary studies of mechanism. Int J Biochem 18: 531–535

Márquez AJ, Avila C, Forde BG and Wallsgrove RM (1988) Ferredoxin-glutamate synthase from barley leaves: Rapid purification and partial characterization. Plant Physiol Biochem 26: 645–651

Masaki R, Yoshikawa S and Matsubara H (1982) Steady-state kinetics of reduced ferredoxin with ferredoxin-$NADP^+$ reductase. Biochim Biophys Acta 700: 101–109

Matsubara H and Hase T (1983) Phylogenetic consideration of ferredoxin sequences in plants, particularly algae. In: Jensen U and Fairbanks DE (eds) Proteins and Nucleic Acids in Plant Systematics, pp 245–266. Springer-Verlag, Berlin

Matsubara H, Hase T, Wakabayahi S and Wada K (1980) Structure and evolution of chloroplast- and bacterial-type ferredoxin. In: Sigman DS and Brazier HAB (eds) The Evolution of Protein Structure and Function, pp 245–266. Academic Press, New York

Matthijs HCP, Coughlan SJ and Hind G (1986) Removal of ferredoxin:$NADP^+$ oxidoreductase from the thylakoid membranes, rebinding to depleted membranes, and identification of the binding site. J Biol Chem 261: 12154–12158.

Mayhew SG and Tollin G (1992) General properties of flavodoxin. In: Muller F (ed) Chemistry and Biochemistry of Flavoenzymes, Vol III. pp 389–426. CRC Press, Boca Raton

McRee DE, Richardson DC, Richardson JS and Siegel LM (1986) The heme and Fe_4S_4 cluster in the crystallographic structure of *Escherichia coli* sulfite reductase. J Biol Chem 261: 10277–10281

Medina M, Mendez E and Gómez-Moreno C (1992a) Identification of arginyl residues involved in the binding of ferredoxin-$NADP^+$ reductase from *Anabaena* sp PCC 7119 to its substrates. Arch Biochem Biophys 299: 281–286

Medina M, Mendez E and Gómez-Moreno C (1992b) Lysine residues on ferredoxin-$NADP^+$ reductase from *Anabaena* sp. PCC 7119 involved in substrate binding. FEBS Lett 298: 25–28

Medina M, Gómez-Moreno C and Tollin G (1992c) Effects of chemical modification of *Anabaena* flavodoxin and ferredoxin-$NADP^+$ reductase on the kinetics of interprotein electron transfer reactions. Eur J Biochem 210: 577–583

Medina M, Peleato ML, Mendez E and Gómez-Moreno C (1992d) Identification of specific carboxyl groups on *Anabaena* PCC 7119 flavodoxin which are involved in the interaction with ferredoxin-$NADP^+$ reductase. Eur J Biochem 203: 373–379

Medina M, Bazo IG, Fillat MF and Gómez-Moreno C (1993) Structure predictions of ferredoxin-$NADP^+$ reductase from the cyanobacterium *Anabaena* sp PCC 7119. Prot Seq Data Anal 5: 247–252

Mikami B and Ida S (1989) Spinach ferredoxin-nitrite reductase: Characterization of catalytic activity and interaction of the enzyme with substrates. J Biochem 105: 47–50

Mohr H, Neininger A and Seith B (1992) Control of nitrate reductase and nitrite reductase gene expression by light, nitrite and a plastidic factor. Bot Acta 105: 81–89

Morigasaki S, Takata K, Sanada Y, Wada K, Yee BC, Shin M and Buchanan BB (1990a) Novel forms of ferredoxin and ferredoxin-NADP reductase from spinach roots. Arch Biochem Biophys 283: 75–80

Morigasaki S, Takata K, Suzuki T and Wada K (1990b) Purification and characterization of a ferredoxin-$NADP^+$ oxidoreductase-like enzyme form radish root tissues. Plant Physiol 93: 896–901

Muller EGD and Buchanan BB (1989) Thioredoxin is essential for photosynthetic growth. J Biol Chem 264: 4008–4014

Nagaoka S, Hirasawa M, Fukushima K and Tamura G (1984) Methyl viologen-linked nitrite reductase from bean roots. Agric Biol Chem 48: 1179–1188

Nakatani S and Shin M (1991) The reconstituted NADP photoreducing system by rebinding of the large form of ferredoxin-NADP reductase to depleted thylakoid membranes. Arch Biochem Biophys 291: 390–394

Nalbantoglu B, Hirasawa M, Moomaw C, Nguyen H, Knaff DB and Allen R (1994) Cloning and sequencing of the gene encoding spinach ferredoxin-dependent glutamate synthase. Biochim Biophys Acta 1183: 557–561

Neininger A, Kronenberger J and Mohr H (1992) Coaction of light, nitrate and a plastidic factor in controlling nitrite-reductase gene expression in tobacco. Planta 187: 381–387

Neininger A , Bichler J, Schneiderbauer A and Mohr H (1993) Response of a nitrite-reductase 3.1-kilobase upstream regulatory sequence from spinach to light and nitrate in transgenic tobacco. Planta 189: 440–442

Neininger A, Seith B, Hoch B and Mohr H (1994a) Gene expression of nitrite reductase in Scots pine (*Pinus sylvestris* L.) as affected by light and nitrate. Plant Molec Biol 25: 449–457

Neininger A, Back E, Bichler J, Schneiderbauer A and Mohr H (1994b) Deletion analysis of a nitrite-reductase promoter from spinach in transgenic tobacco. Planta 194: 186–192

Nozaki Y, Tamaki M and Shin M (1985) The reconstituted $NADP^+$ photoreducing system by recombination of ferredoxin-$NADP^+$ reductase and connectein with thylakoids. Physiol Vèg 23: 627–633

Oelmüller R, Bolle C, Tyagi AK, Niekrawitz N, Breit S and Herrmann RG (1993) Characterization of the promoter from the single-copy gene encoding ferredoxin-$NADP^+$-oxido-reductase from spinach. Mol Gen Genet 237: 261–272

Ogawa M and Ida S (1987) Biosynthesis of ferredoxin-nitrite

reductase in rice seedlings. Plant Cell Physiol 28: 1501–1508

Olsen LJ, Theg SM, Selman BR and Keegstra K (1989) ATP is required for the binding of precursor proteins to chloroplasts. J Biol Chem 264: 6724–6729

Ondrias MR, Carson SD, Hirasawa M and Knaff DB (1985) Characterization of the siroheme active site in spinach nitrite reductase. Biochim Biophys Acta 830: 159–163

Oralin EG, Calcaterra NB, Carrillo N and Ceccarelli EA (1993) Probing the role of the carboxyl-terminal region of ferredoxin-$NADP^+$ reductase by site-directed mutagenesis and deletion analysis. J Biol Chem 268: 19267–19273

Ostrowski J, Wu J-Y, Rueger DC, Miller BE, Siegel LM and Kredich M (1989) Characterization of the *cysJIH* regions of *Salmonella typhimurium* and *Escherichia coli* B. J Biol Chem 264: 15726–15737

Pain D and Blobel G (1987) Protein import into chloroplasts requires a chloroplast ATPase. Proc Natl Acad Sci USA 84: 3288–3292

Pajuelo E, Borrero JA and Márquez AJ (1993) Immunological approach to subunit composition of ferredoxin-nitrite reductase from *Chlamydomonas reinhardtii*. Plant Sci 95: 9–21

Palmer, G (1973) Current insights into the active center of spinach ferredoxin and other iron-sulfur proteins. In: Lovenberg W (ed) The Iron-Sulfur Proteins, Vol 2, pp 285–325. Academic Press, New York

Pelanda R, Vanoni MA, Perego M, Piubelli L, Galizzi A, Curti B and Zanetti G (1993) Glutamate synthase genes of the diazotroph *Azospirillum brasilense*. Cloning, sequencing and analysis of functional domains. J Biol Chem 268: 3099–3106

Pelletier H and Kraut J (1992) Crystal structure of a complex between electron transfer partners, cytochrome c peroxidase and cytochrome c. Science 258: 1748–1755

Perry SE, Buvinger WE, Bennett J and Keegstra K (1991) Synthetic analogs of a transit peptide inhibit binding or translocation of chloroplast precursor proteins. J Biol Chem 266: 11882–11889

Pilon M, de Boer AD, Knols SL, Koppelman MHGM, van der Graaf RM, de Kruijff B and Weisbeek PJ (1990) Expression in *Escherichia coli* and purification of a translocation-competent precursor of the chloroplast protein ferredoxin. J Biol Chem 265: 3358–3361

Pilon M, de Kruijff B and Weisbeek PJ (1992a) New insights into the mechanism of the ferredoxin precursor into chloroplasts. J Biol Chem 267: 2548–2556

Pilon M, Weisbeek PJ and de Kruijff B (1992b) Kinetic analysis of translocation into isolated chloroplasts of the purified ferredoxin precursor. FEBS Lett 302: 65–68

Porqué PG, Baldesten A and Reichard P (1970) Purification of a thioredoxin system from yeast. J Biol Chem 245: 2363–2370

Privalle LS, Privalle CT, Leonardy NJ and Kamin H (1985) Interactions between spinach ferredoxin-nitrite reductase and its substrates. J Biol Chem 260: 14344–14350

Pueyo JJ and Gómez-Moreno C (1991) Characterization of the cross-linked complex formed between ferredoxin-$NADP^+$ reductase and flavodoxin from *Anabaena* PCC 7119. Biochim Biophys Acta 1059: 149–156

Pueyo JJ, Gómez-Moreno C and Mayhew SG (1991) Oxidation-reduction potentials of ferredoxin-$NADP^+$ oxidoreductase and flavodoxin from *Anabaena* PCC 7119 and their electrostatic complexes. Eur J Biochem 202: 1065–1071

Pueyo JJ, Revilla C, Mayhew SG and Gómez-Moreno C (1992) Complex formation between ferredoxin and ferredoxin-$NADP^+$ reductase from *Anabaena* PCC 7119. Arch Biochem Biophys 294: 367–372

Rastogi R, Back E, Schneiderbauer A, Bowsher CG, Moffatt B and Rothstein S (1993) A 330 bp region of the spinach nitrite reductase gene promoter directs nitrate-inducible tissue-specific expression in transgenic tobacco. Plant J 4: 317–326

Rebeille F and Hatch MD (1986) Regulation of NADP-malate dehydrogenase in C_4 plants: Effect of varying NADPH to NADP ratios and thioredoxin redox state on enzyme activity in reconstituted systems. Arch Biochem Biophys 249: 164–170

Redinbaugh MG and Campbell WH (1993) Glutamine synthetase and ferredoxin-dependent glutamate synthase in the maize (*Zea mays*) root primary response to nitrate. Plant Physiol 101: 1249–1255

Reith ME, Laudenbach DE and Straus NA (1986) Isolation and nucleotide sequence analysis of the ferredoxin I gene from the cyanobacterium *Anacystis nidulans* R2. J Bacteriol 168: 1319–1324

Robinson C and Ellis RJ (1984) Transport of proteins into chloroplasts. Partial purification of a chloroplast protease involved in processing imported precursor peptides. Eur J Biochem 142: 337–342

Rogers LJ (1987) Ferredoxins, flavodoxins and related proteins: Structure function and evolution. In: Fay P and van Baalen C (eds) The Cyanobacteria, pp 35–61. Elsevier, Amsterdam

Rogers WJ, Hodges M, Decottignies P, Schmitter J-M, Gadal P and Jacquot J-P (1992) Isolation of a cDNA fragment coding for *Chlamydomonas reinhardtii* ferredoxin and expression of the recombinant protein in *Escherichia coli*. FEBS Lett 310: 240–245

Romero LC, Gotor C, Márquez AJ, Forde B and Vega JM (1988) Antigenic similarities between ferredoxin-dependent nitrite reductase and glutamate synthase from *Chlamydomonas reinhardtii*. Biochim Biophys Acta 957: 152–157

Rypniewski WR, Breiter DR, Benning MM, Wesenberg G, Oh B-H, Markley JL, Rayment I and Holden HM (1991) Crystallization and structure determination to 2.5-Å resolution of the oxidized [2Fe-2S] ferredoxin isolated from *Anabaena* 7120. Biochemistry 30: 4126–4131

Sakakibara H, Watanabe M, Hase T and Sugiyama T (1991) Molecular cloning and characterization of complementary DNA encoding for ferredoxin-dependent glutamate synthase in maize leaf. J Biol Chem 266: 2028–2035

Sakakibara H, Kawabata S, Hase T and Sugiyama T (1992) Differential effects of nitrate and light on the expression of glutamine synthetases and ferredoxin-dependent glutamate synthase in maize. Plant Cell Physiol 33: 1193–1198

Sakihama N and Shin M (1987) Evidence from high-pressure liquid chromatography for the existence of two ferredoxins in plants. Arch Biochem Biophys 256: 430–434

Salamon Z, Gleason FK and Tollin G (1992) Direct electrochemistry of thioredoxins and glutathione at a lipid bilayer-modified electrode. Arch Biochem Biophys 299: 193–198

Salamon Z, Tollin G, Hirasawa M, Gardet-Salvi L, Stritt-Etter A-L, Knaff DB and Schürmann P (1995) The oxidation-reduction properties of spinach thioredoxins *f* and *m* and of ferredoxin:thioredoxin reductase. Biochem Biophys Acta 1230: 114–118

Salmeen I and Palmer G (1972) Contact-shifted NMR of spinach ferredoxin: Additional resonances and partial assignments. Arch Biochem Biophys 150: 767–773

Sancho J and Gómez-Moreno C (1991) Interaction of ferredoxin-$NADP^+$ reductase from *Anabaena* with its substrates. Arch Biochem Biophys 288: 231–238

Sancho J, Peleato ML, Gómez-Moreno C and Edmondson DE (1988)Purification and properties of ferredoxin-$NADP^+$ oxidoreductase from the nitrogen-fixing cyanobacterium *Anabaena variabilis*. Arch Biochem Biophys 260: 200–207

Sancho J, Medina M, Gómez-Moreno C (1990) Arginyl groups involved in the binding of *Anabaena* ferredoxin-$NADP^+$ reductase to $NADP^+$ and to ferredoxin. Eur J Biochem 187: 39–49

Sands RH and Dunham WR (1975) Spectroscopic studies on two-iron ferredoxin. Quart Rev Biophys 7: 443–504.

Scherer S, Alps I, Sadowski H and Böger P (1988) Ferredoxin-$NADP^+$ oxidoreductase is the respiratory NADPH dehydrogenase of the cyanobacterium *Anabaena variabilis*. Arch Biochem Biophys 267: 228–235

Schluchter WM and Bryant DA (1992) Molecular characterization of ferredoxin-$NADP^+$ oxidoreductase in cyanobacteria: Cloning and sequence of the *petH* gene of *Synechococcus* sp. PCC 7002 and studies of the gene product. Biochemistry 31: 3092–3102

Schmitz S, Schrautemeier B and Böhme H (1993) Evidence from directed mutagenesis that positively charged amino acids are necessary for interaction of nitrogenase with the [2Fe-2S] heterocyst ferredoxin (FdxH) from the cyanobacterium *Anabaena* sp., PCC7120. Mol Gen Genet 240: 455–460

Schnell DJ, Blobel G and Pain D (1991) Signal peptide analogs derived from two chloroplast precursors interact with the signal recognition system of the chloroplast envelope. J Biol Chem 266: 3335–3342

Schrautemeier B and Böhme H (1985) A distinct ferredoxin for nitrogen fixation isolated from heterocysts of the cyanobacterium *Anabaena variabilis*. FEBS Lett 184: 304–308

Schuster C and Mohr H (1990) Appearance of nitrite-reductase mRNA in mustard seedling cotyledons is regulated by phytochrome. Planta 181: 327–334

Seith B, Schuster C and Mohr H (1991) Coaction of light, nitrate and a plastidic factor in controlling nitrite-reductase gene expression in spinach. Planta 184: 74–80

Serre L, Medina M, Gómez-Moreno C, Fontecilla-Camps JC and Frey M (1991) Crystals of *Anabaena* 7119 ferredoxin-$NADP^+$ reductase. J Mol Biol 218: 271–272

Serre L, Vellieux F, Fontecilla-Camps J, Frey M, Medina M and Gómez-Moreno C (1994) Structural study of ferredoxin-$NADP^+$ reductase from *Anabaena* PCC 7119 and its complex with $NADP^+$. In: Yagi K (ed) Flavins and Flavoproteins 1993, pp 431–434. Walter de Gruyter, Berlin

Shin M, Ishida H and Nozaki Y (1985) A new protein factor, connectein, as a constituent of the large form of ferredoxin-NADP reductase. Plant Cell Physiol 26: 559–563

Siegel LM, Rueger DC, Barber MJ, Krueger RJ, Orme-Johnson NR and Orme-Johnson WH (1982) *Escherichia coli* sulfite reductase hemoprotein subunit. Prosthetic groups, catalytic parameters and ligand complexes. J Biol Chem 257: 6343–6350

Skjeldal L, Westler WM, Oh B-H, Krezel AM, Holden HM, Jacobson BL, Rayment I and Markley JL (1991) Two-dimensional magnetization exchange spectroscopy of *Anabaena* 7120 ferredoxin. Nuclear Overhauser effect and electron-self exchange cross peaks from amino acid residues surrounding the 2Fe-2S* cluster. Biochemistry 30: 7363–7368

Small IS and Gray JC (1984) Synthesis of wheat leaf nitrite reductase de novo following induction with nitrate and light. Eur J Biochem 145: 291–297

Smeekens S, van Binsbergen J and Weisbeek P (1985) The plant ferredoxin precursor: Nucleotide sequence of a full length cDNA clone. Nucleic Acids Res 13: 3179–3194

Smeekens S, Bauerle C, Hageman J, Keegstra K and Weisbeek P (1986) The role of the transit peptide in the routing of precursors toward different chloroplast compartments. Cell 46: 365–375

Smeekens S, van Steeg H, Bauerle C, Bettenbroek H, Keegstra K and Weisbeek P (1987) Import into chloroplasts of a yeast mitochondrial protein directed by ferredoxin and plastocyanin transit sequences. Plant Mol Biol 9: 377–388

Smith JM. Smith WH and Knaff DB (1981) Electrochemical titrations of a ferredoxin-ferredoxin:$NADP^+$ oxidoreductase complex. Biochim Biophys Acta 635: 405–411

Somers DE, Caspar T and Quail PH (1990) Isolation and characterization of a ferredoxin gene from *Arabidopsis thaliana*. Plant Physiol 93: 572–577

Somerville CR and Ogren WL (1980) Inhibition of photosynthesis in mutants of *Arabidopsis* lacking glutamate synthase activity. Nature 286: 257–259

Soncini FC and Vallejos RH (1989) The chloroplast reductase-binding protein is identical to the 16.5 kDa polypeptide described as a component of the oxygen-evolving complex. J Biol Chem 264: 21112–21115

Spano AJ and Schiff JA (1987) Purification, properties and cellular location of *Euglena* ferredoxin-NADP reductase. Biochim Biophys Acta 894: 484–498

Suzuki A and Gadal P (1982) Glutamate synthase from rice leaves. Plant Physiol 69: 848–852

Suzuki A, Oaks A, Jacquot J-P, Vidal J and Gadal P (1985) An electron transport system in maize roots for reactions of glutamate synthase and nitrite reductase. Plant Physiol 78: 374–378

Suzuki I, Sugiyama T and Omata T (1993) Primary structure and transcriptional regulation of the gene for nitrite reductase from the cyanobacterium *Synechococcus* PCC 7942. Plant Cell Physiol 348: 1311–1320

Suzuki S, Izumihara K and Hase T (1991) Plastid import and iron-sulfur cluster assembly of photosynthetic and nonphotosynthetic ferredoxin isoproteins in maize. Plant Physiol 97: 375–380

Takahashi Y, Hase T, Wada K and Matsubara H (1983) Ferredoxins in developing spinach cotyledons: The presence of two molecular species. Plant Cell Physiol 24: 189–198

Takahashi Y, Mitsui A, Hase T and Matsubara H (1986) Formation of iron-sulfur cluster of ferredoxin in isolated chloroplasts. Proc Natl Acad Sci USA 83: 2434–2437

Takahashi Y, Mitsui A and Matsubara H (1990a) Formation of the Fe-S cluster of ferredoxin in lysed spinach chloroplasts. Plant Physiol 95: 97–103

Takahashi Y, Mitsui A, Fujita Y and Matsubara H (1990b) Roles of ATP and NADPH in formation of the Fe-S cluster of spinach ferredoxin. Plant Physiol 95: 104–110

Theg SM, Bauerle C, Olsen LJ, Selman BR and Keegstra K (1989) Internal ATP is the only energy source required for the translocation of precursor proteins across chloroplastic membranes. J Biol Chem 264: 6730–6736

Tsugita A, Maeda K and Schürmann P (1983) Spinach chloroplast thioredoxins in evolutionary drift. Biochem Biophys Res Comm 115: 1–7

Tsukihara T, Fukuyama K, Nakamura M, Katsube Y, Tanaka N, Kakudo M, Wada K, Hase T and Matsubara H (1981) X-ray analysis of a [2Fe-2S] ferredoxin from *Spirulina platensis*. Main chain fold and location of side chains at 2.5 Å resolution. J Biochem 90: 1763–1773

Tsukihara T, Fukuyama K, Mizushima M, Harioka T, Kusunoki M, Katsube Y, Hase T and Matsubara H (1990) Structure of the [2Fe-2S] ferredoxin I from the blue-green alga *Aphanothece sacrum* at 2.2 Å resolution. J Mol Biol 216: 399–410

Valentin K, Kostrzewa M and Zetsche K (1993) Glutamate synthase is plastid-encoded in a red alga: Implications for the evolution of glutamate synthases. Plant Mol Biol 23: 77–85

Vallejos RH, Ceccarelli E and Chan R (1984) Evidence for the existence of a thylakoid intrinsic protein that binds ferredoxin-$NADP^+$ oxidoreductase. J Biol Chem 259: 8048–8051

van der Plas J, de Groot RP, Weisbeek PJ and van Arkel GA (1986a) Coding sequence of a ferredoxin gene from *Anabaena variabilis* ATCC 29413. Nucleic Acids Res 14: 7803

van der Plas J, de Groot RP, Woortman MR, Weisbeek PJ and van Arkel GA (1986b) Coding sequence of a ferredoxin from *Anacystis nidulans* R2 (*Synechococcus* PCC 7942) Nucleic Acids Res 14: 7804

Vanoni MA, Edmondson DE, Zanetti G and Curti B (1992) Characterization of the flavins and iron-sulfur centers of glutamate synthase from *Azospirillum brasilense* by absorption, circular dichroism and electron paramagnetic resonance spectroscopies. Biochemistry 31: 4613–4623

Vega JM and Kamin H (1977) Spinach nitrite reductase. Purification and properties of a siroheme-containing enzyme. J Biol Chem 252: 896–909

Vieira BJ, Colvert KK and Davis DJ (1986) Chemical modification and cross-linking as probes of regions on ferredoxin involved in the interaction with ferredoxin:NADP reductase. Biochim Biophys Acta 851: 109–122

Vigara AJ, Bes MT, Vega JM and Gómez-Moreno C (1994) Purification of Fd-glutamate synthase from *Monoraphidum braunii* and characterization of a light-dependent activity assay. J Mol Catal 89: 257–266

Vorst O, van Dam F, Oosterhoff-Teertstra R, Smeekens S and Weissbeek P (1990) Tissue-specific expression directed by an *Arabidopsis thaliana* pre-ferredoxin promoter in transgenic tobacco plants. Plant Mol Biol 14: 491–499

Vorst O, van Dam F, Weisbeek P and Smeekens S (1993) Light-regulated expression of the *Arabidopsis thaliana ferredoxinA* gene involves both transcriptional and post-transcriptional processes. Plant J 3: 793–803

Wada K, Oh-Oka H and Matsubara H (1985) Ferredoxin isoproteins and their variation during growth of higher plants. Physiol Vèg 23: 679–686

Wada K, Onda M and Matsubara H (1989) Amino acid sequences of ferredoxin isoproteins from radish roots. J Biochem 105: 619–625

Wagner R, Carrillo N, Junge W and Vallejos RH (1982) On the conformation of reconstituted ferredoxin-$NADP^+$ oxidoreductase in the thylakoid membrane. Biochim Biophys Acta 680: 317–330

Walker MC, Pueyo JJ, Navarro JA, Gómez-Moreno C and Tollin G (1991) Laser flash photolysis studies of the kinetics of reduction of ferredoxins and ferredoxin-$NADP^+$ reductases from *Anabaena* PCC 7119 and spinach: Electrostatic effects on intracomplex electron transfer. Arch Biochem. Biophys 287: 351–358

Wallsgrove RM, Turner JC, Hall NP, Kendall AC and Bright SWJ (1987) Barley mutants lacking chloroplast glutamine synthetase. Biochemical and genetic analysis. Plant Physiol 83: 155–158

Wedel N, Bartling D and Herrmann RG (1988) Analysis of cDNA clones encoding the entire ferredoxin I precursor polypeptide from spinach. Botanica Acta 101: 295–300

Wilkerson JO, Janick PA and Siegel LM (1983) Electron paramagnetic resonance and optical spectroscopic evidence for interaction between siroheme and tetranuclear iron-sulfur center prosthetic groups in spinach ferredoxin-nitrite reductase. Biochemistry 22: 5048–5054

Young LJ and Siegel LM (1988) Superoxidized states of *Escherichia coli* sulfite reductase heme protein subunit. Biochemistry 27: 5984–5990

Zanetti G and Aliverti A (1991) Ferredoxin:$NADP^+$ oxidoreductase. In: Müller F (ed) Chemistry and Biochemistry of Flavoproteins, Vol 2, pp 305–315. CRC Press, Boca Raton

Zanetti G and Merati G (1987) Interaction between photosystem I and ferredoxin. Eur J Biochem 169: 143–146

Zanetti G, Cidaria D and Curti B (1982) Preparation of apoprotein from spinach ferredoxin-$NADP^+$ reductase. Eur J Biochem 126: 453–458

Zanetti G, Massey V and Curti B (1983) FAD analogs as mechanistic and 'domain-binding' probes of spinach ferredoxin-$NADP^+$ reductase. Eur J Biochem 132: 201–205

Zanetti G, Morelli D, Ronchi S, Negri A, Aliverti A and Curti B (1988) Structural studies on the interaction between ferredoxin and ferredoxin-$NADP^+$ reductase. Biochemistry 27: 3753–3759

Zehnacker C, Becker TW, Suzuki A, Carrayol E, Caboche M and Hirel B (1992) Purification and properties of tobacco ferredoxin-dependent glutamate synthase, and isolation of corresponding cDNA clones. Planta 187: 266–274

Zilber AL and Malkin R (1988) Ferredoxin cross-links to a 22 kD subunit of Photosystem I. Plant Physiol 88: 810–814

Chapter 18

Structure Analysis of Single Crystals of Photosystem I by X-Ray, EPR and ENDOR: A Short Status Report

H. T. Witt
Max-Volmer-Institut für Biophysikalische und Physikalische Chemie, Technische Universität Berlin, Str. d. 17. Juni 135, D-10623 Berlin, Germany

Summary

Photosystem I purified from the thermophilic cyanobacterium *Synechococcus elongatus* has been crystallized. The first structural model for this reaction center was based on an electron density map at 6 Å resolution. Now, with improved crystals, a new data collection at higher structural resolution is available. This status report outlines some results on the present state of X-ray structure analysis of Photosystem I at 4.5 Å resolution. Single crystals of Photosystem I have also been analyzed by magnetic resonance spectroscopy (EPR and ENDOR). The results, presented here, provide new information on the structure of the primary electron donor (P700) and on the structures of the terminal electron acceptors, the iron-sulfur clusters F_A and F_B.

Donald R. Ort and Charles F. Yocum (eds): Oxygenic Photosynthesis: The Light Reactions, pp. 363–375.

I. Introduction

In the photosynthetic membrane of higher plants and of cyanobacteria electrons are transferred from water via numerous intermediate electron carriers to the terminal acceptor, NADP. This transfer is driven by the cooperation of two photosystems, I and II (PS I and PS II).

In PS I the photoact starts with the photooxidation of a chlorophyll *a* of the pigment 700 (P700), and in PS II with the photooxidation of a chlorophyll *a* of the pigment 680 (P680). The chlorophyll oxidations are coupled to the generation of a transmembrane electrical field. This indicates a transmembrane charge separation whereby the electrons are vectorially channeled from the excited chlorophylls at the membrane interior via different electron carriers to acceptors at the membrane exterior (for an overview see Witt, 1987). Despite these two similar primary acts, which charge the membrane electrically, the function and structure of PS I and II reaction centers are very different:

1. The acceptor at the exterior of PS II is a plastoquinone, but that of PS I is a Fe-S cluster;

2. The final electron donor in PS II is water, but in PS I it is plastocyanin or cytochrome c_6;

3. In PS I the reduction potential is extremely high (ca. –500 mV), while in PS II the oxidation potential is high (ca. +0.82 V);

4. In PS I the two main subunits contain the electron transfer chain as well as the antenna chlorophylls; in PS II the antennae are located on separate subunits;

5. There is negligible similarity in the amino acid sequences of PS I and PS II subunits.

For an understanding of the mechanism of these two different molecular machines, it is of interest to obtain information on the molecular structure of both reaction centers. However, in photosynthesis, only the structure of the reaction center of purple bacteria is known at high resolution (Deisenhofer et al., 1985). Although this reaction center is not able to oxidize water, it displays important similarities to that of PS II (Trebst, 1986; Deisenhofer and Michel, 1989). No such indirect information is, however, available for PS I, because it is related to green-sulfur bacteria (Nitschke et al., 1987; Mathis, 1990) whose reaction center structure is not known.

In the following, a short status report is given on the PS I complex, its crystallization and the present structural state at 4.5 Å resolution. Results of EPR and ENDOR structure analysis of P700 and the Fe-S clusters in PS I complexes and in single crystals at low temperature are also presented.

II. The PS I Complex of Cyanobacterium *Synechococcus elongatus*

PS I was isolated from the thermophilic cyanobacterium *Synechococcus elongatus* and purified as described in Schatz and Witt (1984), Witt et al. (1987) and Boekema et al. (1987). Isolated PS I shows a trimeric organization by analysis with electron microscopy (Boekema et al., 1987). The trimer has been separated into monomers, each of which contains one P700 and, therefore, represents one PS I reaction center (Rögner et al., 1990).

In higher plants, 12 subunits are present in PS I (for a review see Golbeck and Bryant, 1991; Golbeck, 1994; Chapter 15). Less information is available on PS I from cyanobacteria. Recently, the genes encoding 11 subunits of PS I from thermophilic *Synechococcus elongatus* were isolated and sequenced (Mühlenhoff et al., 1993; see Table 1). Seven subunits are intrinsic membrane proteins; three extrinsic subunits are located on the stromal side and the fourth is on the lumenal side. The mass of the PS I monomer deduced from amino acid sequences results in ca. 340 kDa, assuming the presence of 90 antenna chlorophylls. The isolated trimer, with a mass of ca. 1020 kDa, is the subject of the following analysis. The trimer might also be the active form in vivo (Hladik and Sofrova, 1991; Shubin et al., 1993).

The main part of the PS I electron transfer chain is suspended between the two large subunits, A and B. P700, possibly a chlorophyll *a* dimer, is located on the lumenal side and the Fe-S cluster, F_X, towards the stromal side. Between these sides two electron carriers are located: A_0, a single chlorophyll *a* and A_1, a phylloquinone (vitamin K_1). The terminal electron

Abbreviations: ENDOR – electron nuclear double resonance; EPR – electron paramagnetic resonance; F_A, F_B – Photosystem I secondary Fe-S acceptors; F_X – Photosystem I primary Fe-S acceptor; HFC – hyperfine coupling; PS – photosystem

Table 1. Subunit composition of PS I from the thermophilic cyanobacterium *Synechococcus* sp. (Mühlenhoff et al., 1993)

Subunit	Gene	Mass (kDa)			Homology**	Location
		deduced	/	apparent		
A	*psaA*	83.1	/	70	83%	transmembrane
B	*psaB*	83.0	/	70	80%	transmembrane
C	*psaC*	8.8	/	8.6	88%	stroma
D	*psaD*	15.4	/	17.3	59%	stroma
E	*psaE*	8.4	/	10.0	53%	stroma
F	*psaF*	15.1	/	15.4	50%	lumen
I*	*psaI*	4.6	/	–	41%	–
J	*psaJ*	4.8	/	3.3	49%	transmembrane
K	*psaK*	8.5	/	6.6	33%	transmembrane
L	*psaL*	15.5	/	–	47%	transmembrane
M	*psaM*	3.5	/	2.5	30%	transmembrane

* The gene of this subunit was found and sequenced, but it is uncertain whether it is expressed; it is not observed with N-terminal protein sequencing.
** By comparison with corresponding subunits from spinach, except for I and L (barley) and J and M (Marchantia)

acceptors, the Fe-S clusters, F_A and F_B, are bound to the small stromal subunit C (see review of Golbeck and Bryant, 1991).

III. Crystallization of the Trimer of PS I

In general, the solubility of proteins first increases with increasing salt concentration ('salting in'). In the subsequent range of higher salt concentrations, the solubility decreases ('salting out'). This latter range is primarily used for crystallization. Different types of salts or organic compounds are used as precipitating agents. With polyethylene glycol, 'pretty' crystals of PS I were obtained, but their X-ray diffractions were not suitable for structure analysis (Witt et al., 1987). After testing numerous sets of parameters and different agents without obtaining better crystals, we worked without any mediator for crystallization. A range of salt concentrations was used within which the protein solubility decreases with decreasing salt concentration (reverse of 'salting in'). A new crystallization point is observed in the millimolar range by decreasing the salt ($MgSO_4$) concentration of the protein solution by dialysis with water (a crystallization procedure with 'water as a precipitating agent'). The crystallization point depends slightly on the protein concentration. At that point, within seconds, large amounts of small crystals (< 0.1 mm) are formed. These were used as a last purification step. The small crystals were resolubilized by a renewed increase of the salt concentration (~50 mM). Subsequently, in a dialysis machine, the salt concentration of the protein solution was again decreased but now slowly in a controlled way: two cuvettes were used, separated by a membrane. The cuvette at the top contained the protein-salt solution, the cuvette at the bottom contained buffer plus ~10 mM $MgSO_4$. By means of a programmed pump the salt in this buffer is slowly reduced and, thereby, also the salt in the protein solution. The salt concentration is continuously monitored by conductivity measurements. In this way, conditions for crystallization can be systematically proven for different materials. In the case of the trimer of PS I, after about one day, dark green, hexagonal crystals are obtained at about 6 mM salt. These can be up to 4 mm long and are suitable for the X-ray diffraction measurements as well as EPR and ENDOR measurements at low temperature.

The photoactive crystals show a great mechanical stability; they are stable for hours when exposed to radiation of 1 Å wavelength. For estimation of the phase of diffractions, isomorphous replacement with heavy-atom compounds was used. Mercury-, platinum- and uranyl-compounds were found to be suitable. The X-ray diffraction indicates a 4 Å resolution (Witt et al., 1988).

IV. X-Ray Structure Analysis of Single Crystals of PS I at 4.5 Å Resolution

A first model of PS I was proposed on the basis of an

electron density map at 6 Å resolution published by Witt et al. (1992) and Krauß et al. (1993). An improved crystallization protocol coupled with the use of synchrotron radiation has led to significant progress in collecting data at higher resolution. In turn, an electron density map of a nominal resolution of 4.5 Å could be computed (Schubert et al., 1995). In the following sections, recent progress towards this higher resolution structure is discussed.

A. Overall Structure

Figure 1, top, shows the arrangement of two PS I trimers in the hexagonal unit cell of the crystal. The space group is $P6_3$ with dimensions of a = b = 287 Å and c = 167 Å. The view is onto the crystallographic a,b plane, which is assumed to be parallel to the membrane bilayer. The local 3-fold axis of the trimer coincides with the crystallographic 3-fold axis. The side view on one monomer (Fig. 1, bottom, left) indicates a smaller domain close to the 3-fold axis, which is possibly involved in the trimerization. Its height (40 Å) corresponds to that of membrane bilayers. The large domain is the catalytic site, because it harbors the electron transport chain (see below). The front view in Fig. 1, bottom, right, indicates an asymmetric hump on the stroma side which probably contains the C, D, and E subunits and is the docking site for ferredoxin. The 10 Å cavity at the lumen side is very probably the site for docking the immediate electron donor to PS I, either plastocyanin or cytochrome c_6. The essential elements of Photosystem I, as apparent from the electron density maps at 6 Å and 4.5 Å resolution, remain unchanged. However, a diffuse region on the lumenal side of Photosystem I (originally assigned to a disordered subunit F – dotted in Fig. 1) now shows no disorder.

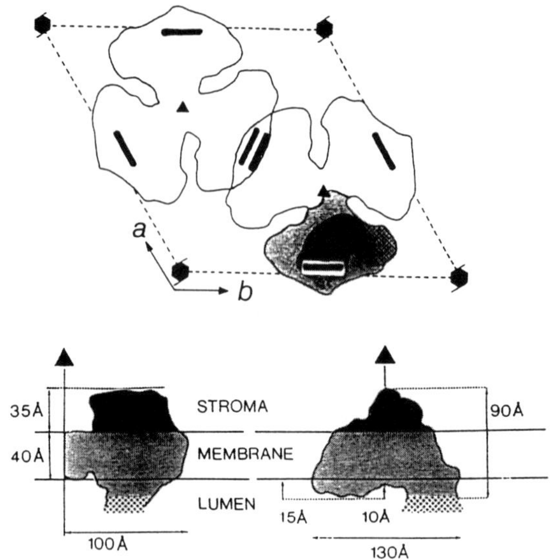

Fig. 1. Top: Packing of the PS I complexes in the crystal unit cell with symmetry elements indicated in top view onto the crystallographic a,b-plane, which is assumed to be parallel to the membrane bilayer. One monomer is shaded, with stromal part dark, membrane-embedded part light, and lumenal, less-well ordered part, dotted. The orientation of the P700 dimers at the position indicated by bold lines is perpendicular to the a,b-plane (see Sections IV, V and VI). Bottom, left: View direction along the b-axis onto one PS I monomer, which is shaded in Fig. 1, top. Bottom, right: View direction perpendicular to the b,c-plane.

B. Structure Elements

The electron density map shows a large number of tubular structures. These can be fitted with polyalanine α-helices and were attributed to 35 helices. Other prominent features are disc-like globules of 8–10 Å diameter and 4–5 Å thickness. These can be modeled with the dihydroporphyrin system of chlorophyll *a* and were attributed to 70 chlorophylls. For clarity, in Fig. 2 the helices are depicted as columns and the chlorophylls have been omitted. The view is parallel to the membrane plane onto the plane defined by the threefold and the local, pseudo-twofold axis (see below). The left part of Fig. 2, top, represents the small domain with a height corresponding to the 40 Å membrane thickness and the other part represents the large catalytic domain (see Fig. 1). Essential features of this arrangement are unchanged as compared to the 6 Å model. Yet, a number of differences are obvious. For example, while most so-called transmembrane helices could not be observed to span the membrane in the 6 Å map, the improved resolution now reveals that all such helices span the membrane completely (see Fig. 2). Secondly, in the 6 Å map 21 transmembrane helices were observed; the corresponding number in the new map is 29. Furthermore, while two symmetry-related helices parallel to the membrane plane were identified in the 6 Å map, another pair near the stromal side of the core of PS I is visible in the new 4.5 Å map. At the top, the positions of three extremely high electron densities (12 root mean-square deviations above the mean density) are indicated by cubes. They must, without doubt, represent the three Fe-S clusters, because these compounds have the highest electron density. Because two Fe-S clusters, F_A and F_B, are known to be bound to subunit C, which is exposed to the stroma, one can identify this area unambiguously as the stromal side on the electron density map. The Fe-S cluster closest to the membrane can be attributed to F_X, since this cluster coordinates with 4 cysteine residues on subunits A and B.

C. Helices Assigned to the Large Subunits A and B

By rotating Fig. 2, top, by 90° about the horizontal, the view from the stromal side onto the membrane plane becomes visible in Fig. 2, bottom. A refined inspection indicates that the shaded helices in the area A are symmetrically related with the shaded helices in B. Helix a corresponds to a′, b to b′, etc. These helices are related to each other by a local, pseudo-twofold axis. This axis, indicated by the black dot, is both parallel to the c-axis and the membrane normal, and runs through the center of the cluster F_X. Since the amino acid sequences of subunits A and B are similar, with a sequence homology of 45%, the two arrays of helices in A and B are assigned to the large subunits A and B of PS I. Helix i, in the new map, has a pseudo-symmetrical counterpart, i′. As a result, the 10 transmembrane helices e to i and e′ to i′ near the center—5 each from subunits A and B—can now be interpreted to form the central core surrounding the electron transport chain. This new arrangement now bears some similarity to that of the 10 helices of subunits L and M of the purple bacterial reaction centers.

The peripheral helices a, b, c, d, and a′, b′, c′, d′ were already observed in the 6 Å map, whereas the four helices u, t and u′, t′ were identified in the 4.5 Å map. Each of the A and B subunits is therefore assigned a total of 11 transmembrane helices. This result is in satisfactory agreement with the amino acid sequence analysis (Mühlenhoff et al., 1993) and its corresponding hydropathy plot (Mühlenhoff, 1991), which predicts 11 transmembrane regions (Mühlenhoff, 1991, Fromme et al., 1994). Helices lying parallel to the membrane are indicated by dots: two are near the lumenal side, n and n′, detected in the 6 Å map, and two are near the stromal side, s and s′, identified in the 4.5 Å map. Both pairs occur in symmetry-related fashion and are, therefore, also assigned to the A and B subunits. The unshaded helices which are not symmetry-related are assigned to the small membrane-intrinsic subunits.

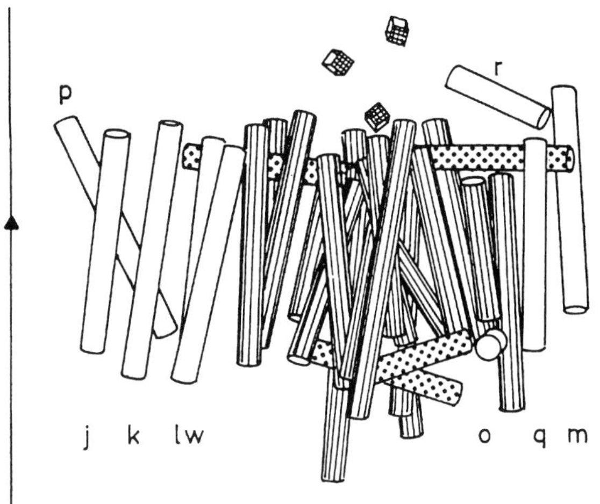

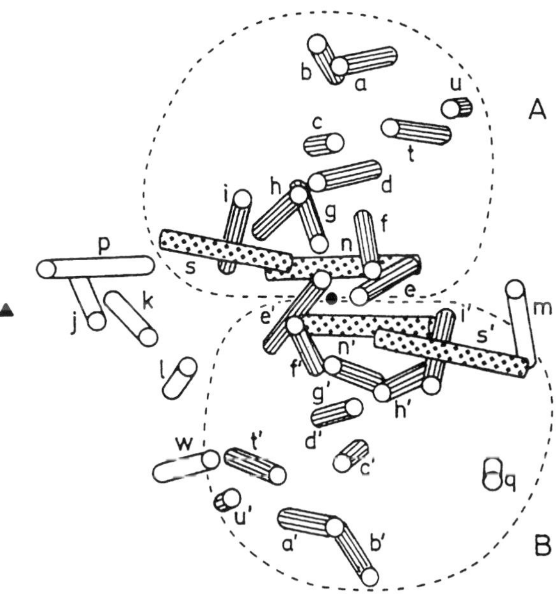

Fig. 2. Top: Model of the arrangement of helices in the PS I monomer at 4.5 Å resolution. Side view as in Fig. 1, bottom, left; stroma is on top, lumen below. The three-fold axis is marked and Fe-S clusters F_X, F_A, and F_B are indicated as cubes. Helices attributed to subunits A and B are shaded and dotted. The unshaded helices belong to the small membrane-intrinsic subunits. Bottom: Fig. 2, top, rotated 90° about the horizontal. In the resulting top view the stroma side of PS I is turned towards the viewer. (For details, see text) (Schubert et al., 1995)

D. Helices of the Small Membrane Intrinsic Subunits

The seven unshaded helices j, k, l, m, and the new ones, p, q, w, observed in the 4.5 Å map have no symmetry equivalence. Helices j, k and p are close to the 3-fold axis and form a tight bundle in the trimer. Since the trimers do not exist in mutants lacking subunit L (Chitnis et al., 1993), L must be responsible for the trimerization of PS I. This subunit contains two membrane-spanning helices (Mühlenhoff et al., 1993); therefore, two helices from j, k or p must belong to L. Helix m and the new one, q, located on the opposite side of the complex, could belong either to subunit F or J. This assignment is corroborated by electron microscopic studies, which identified the location of these subunits by deletion experiments (E. Boekema and M. Rögner, personal communication). The remaining helices must belong to subunits M, K or I.

E. The Electron Transfer Chain

Information regarding the location of the electron transfer chain within the electron density map is obtained from the position of the iron-sulfur clusters on the map. Because of the significantly higher electron density of the Fe-S clusters, their positions can be unambiguously determined (see above). The three clusters, F_x, F_A and F_B are arranged in the shape of an irregular triangle (Fig. 3). F_x is bound to subunits A and B (see above). The distance between F_A and F_B is 12 Å, the angle between the F_A-F_B axis and the c axis is 54°, but a distinction of F_A from F_B is not possible. F_A and F_B are coordinated with subunit C, which is similar to bacterial ferredoxin from *Peptococcus aerogenes*. This ferredoxin contains two 4Fe-4S clusters which are also separated by a distance of 12 Å (Adman et al., 1973; Oh-Oka et al., 1988). Close to F_A and F_B, two short α-helical stretches are observed in the 4.5 Å map (see Fig. 5, left). These show a clear resemblance to the two one-turn α-helices observed in the bacterial ferredoxin. Each cluster is coordinated to the helices through 4 cysteine residues; the two cysteine-binding motifs in the ferredoxin (cxxcxxcxxxcp) correspond to those found in the sequence of subunit C (Mühlenhoff et al., 1993). These results support the close structural similarity between the two systems. Because the atomic structure of the ferredoxin is known (Adman et al., 1976) this allows a prediction of detailed information on the position of the molecules and atoms within the core of subunit C in PS I (see Fig. 5 and Section V).

Close to the two-fold rotation axis passing through F_x (see above) at the lumen side, two contours become visible which can be assigned to two head groups of the chlorophylls of P700 (Fig. 3, left). The location of P700 is close to the cavity, which might be the docking site for the immediate electron donor of P700 (Fig. 1, bottom, right), so the position of P700 is confirmed. The chlorophyll rings are perpendicular to the membrane plane and parallel to each other with a distance of at least 4.5 Å, and a center-to-center distance is ~8 Å. In the 4.5 Å map, the two chlorophylls are connected with the helix e and e′, respectively, by a broad segment of density (Schubert et al., 1995). The latter may be assigned to a histidine side chain pointing towards the Mg atoms of the chlorophylls. Two more densities on both sides, at a distance of ca. 12 Å from P700, are assigned to additional chlorophylls, A and A′. Though their positions remain unchanged, the inclination of the chlorophyll planes had to be modified.

The most notable differences between the models derived from electron density maps at 6 Å and 4.5 Å resolution entail the positions of the primary (A_0) chlorophyll, and secondary (A_1 (vitamin K_1)) electron acceptors. The pockets of electron density attributed to these electron acceptors at 6 Å now appear to constitute part of the extended helices e and e′. Instead of these, at 4.5 Å resolution two new chlorophyll molecules became visible, related to each other by a two-fold symmetry. They are located at the same depth relative to the membrane surface as the A_0 of the map at 6 Å resolution, but are approximately 12 Å farther from the two-fold symmetry axis. One of these must be assigned to the redox active A_0. Because it is not yet known along which of the two pathways the electron transfer proceeds (perhaps even along both), A_0 cannot be identified at present. The arrangement of the chlorophylls is now similar to that of the purple bacteria reaction centers. The distances between the centers of the electron carriers are depicted in Fig. 3, right.

The position of A_1 (two vitamin K_1 molecules) is still open to debate. Two contours of electron density in the new map could accommodate the vitamin K_1 head groups. However, they are as yet indistinguishable from large protrusions emanating from α-helices that represent aromatic side chains.

F. The Antenna Chlorophylls

A unique feature of PS I is that the two subunits A and B harbor both the electron transfer chain as well as the antenna chlorophylls. A preliminary antenna

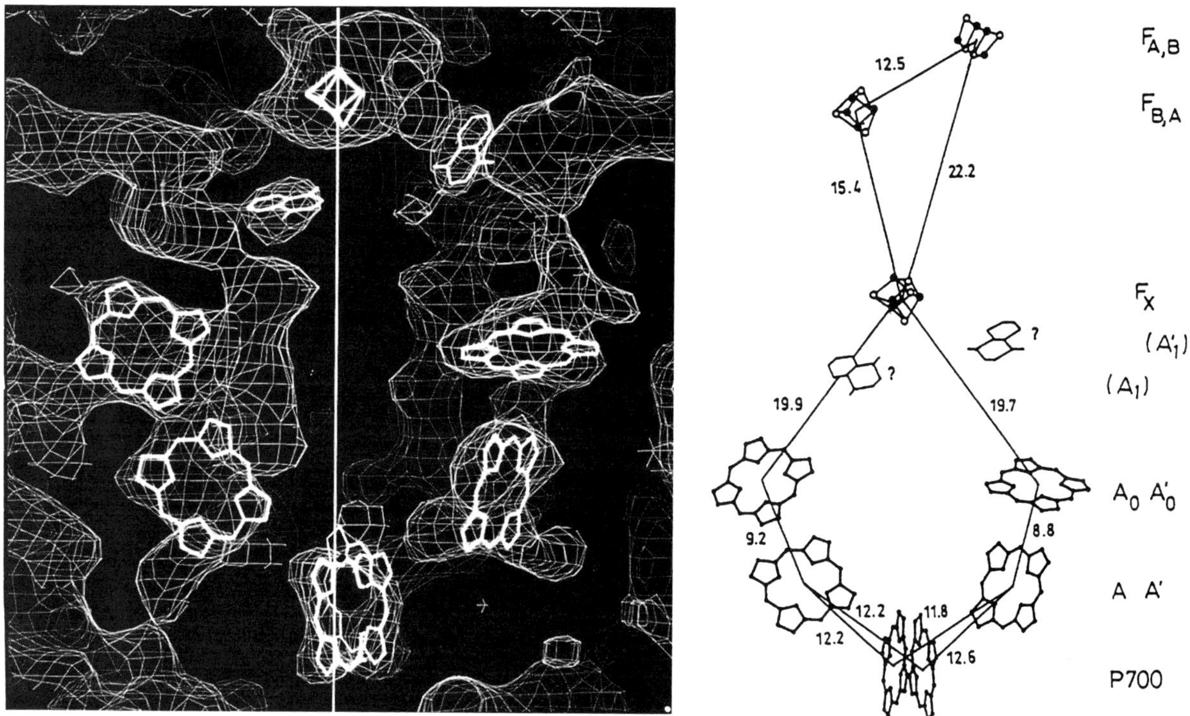

Fig. 3. Left: Section of the 4.5 Å electron density map in the region between P700 and F_X and present assignments to the electron carriers. The local twofold rotation axis is indicated. The positions of the two vitamin K_1 molecules are tentative. View direction is not directly along the plane of the rings of the chlorophylls of P700, but is rotated about the local twofold axis by 18° to make all electron carriers visible on one plane. Right: Arrangement of the geometrical centers of the electron carriers (distances in Å) (Schubert et al., 1995)

model which presently includes 64 chlorophylls is shown in Fig. 4. Although the final overall structure of the antenna assembly cannot be described yet, it is evident that there is no similarity with an arrangement of the chlorophylls recently shown for the light-harvesting complex LH 2 of purple bacteria (McDermott et al., 1995) or with the chlorophylls in the LHC II of higher plants (Kühlbrandt et al., 1994). Clearly, an independent principle was developed in PS I. The distance between the chlorophyll centers is 8–15 Å and no preferable orientation is obvious. The chlorophylls are oriented close to the periphery of subunits A and B, often with voids between the transmembrane helices. The overall arrangement of the chlorophylls is as if they are randomly positioned along the wall of an oval bowl with the bottom oriented towards the lumenal side and the large opening oriented towards the stromal side. This 'wall' may function as a storage device in which an excited state is rapidly delocalized to be trapped by P700 located towards the bottom of the device. The orientation of the antenna chlorophylls modeled at 6 Å resolution was predominantly parallel to the membrane normal. This difference is obviously due to the lower resolution and the number of chlorophylls (40) analyzed in the 6 Å map.

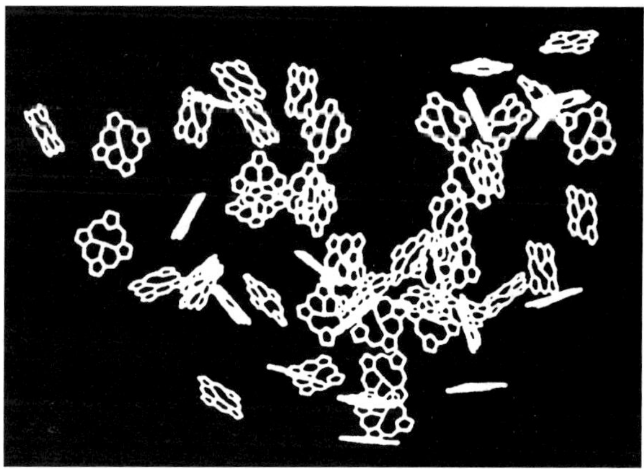

Fig. 4. Present model of the arrangement of 64 antenna chlorophylls of PS I. Side view as in Fig. 2, top (Schubert et al., 1995).

Two qualities of the crystals allow additional structure analysis by EPR and ENDOR. Since the crystals are stable at 4K—this is by no means self-evident for such large proteins—EPR measurements at low temperatures became possible, especially in the case of the Fe-S clusters. As large crystals up to 4 mm became available—a prerequisite for ENDOR measurements—analysis at low temperature with this method also became possible so as to provide refined information on the P700 dimer.

V. EPR Structure Analysis of P700 and the Fe-S clusters in PS I Complexes and Single Crystals at Low Temperature

In Photosystem I, light-induced charge separation generates a series of radical ion pairs which can be studied by EPR. The intrinsic time resolution of the method is ~10 ns so that the states $P700^{+\cdot}A_1^{-\cdot}$, $P700^{+\cdot}F_X^-$, $P700^{+\cdot}F_A^-$, and $P700^{+\cdot}F_B^-$, as well as the triplet state of the donor ($^3P700$) are observable. EPR measurements of these states yield values of the g-factors of the radicals, the dipolar coupling between the unpaired electrons and the hyperfine couplings between the electrons and the nuclei. All of these quantities are described by second rank tensors and, thus, are orientation dependent. Their orientational dependence, obtained from the EPR data, can be related to the orientation of the radical species in the reaction center if the orientation of the magnetic axes relative to the molecular axes is available. In the following, the structural information obtained from such experiments on PS I complexes at room temperature and on PS I crystals at low temperature is discussed.

A. The Primary Electron Donor P700

The idea that P700 exists as dimer was first proposed from the double band in the optical difference spectrum between P700 and $P700^{+\cdot}$ (Döring et al., 1968). This was explained on the basis of two chlorophylls causing exciton splitting of the energy levels of the dimer. Later, it was observed that the EPR spectrum of $P700^{+\cdot}$ has a linewidth narrower than that of Chl $a^{+\cdot}$ (Norris et al., 1971), which was taken as evidence for a symmetric distribution of the unpaired electron of $P700^{+\cdot}$ over two chlorophylls, the so-called special pair (see also Section VI).

Recent time-resolved EPR measurements of the triplet state, $^3P700$, at room temperature give further clear evidence that the primary donor is a dimer. The axes normal to the two chlorophyll planes are approximately parallel (Sieckmann et al., 1993). Data on oriented PS I particles (Rutherford and Sétif, 1990) demonstrate that the chlorophyll plane is oriented perpendicular to the membrane plane, as in bacterial reaction centers (Budil and Thurnauer, 1991). The fact that the EPR spectrum of $^3P700$ is that of a chlorophyll *a* monomer at low temperature (Rutherford & Sétif, 1990) indicates that the triplet excitation is localized on one half of the dimer at low temperatures.

B. The Electron Carrier Vitamin $K_1(A_1)$

The transient EPR spectra of the charge separated state $P700^{+\cdot}A_1^{-\cdot}$ are spin polarized and are very sensitive to the relative orientations of the two radicals. An analysis of the spectra at two microwave frequencies (Stehlik et al., 1989; Bittl et al., 1992; Fuechsle, et al., 1993; Snyder and Thurnauer, 1993) leads to the following important conclusions: (i) the acceptor A_1 is a phylloquinone (vitamin K_1); (ii) the phylloquinone molecule is oriented such that the in-plane axis colinear to the C=O bonds is roughly parallel to the dipolar axis joining $P700^{+\cdot}$ and $A_1^{-\cdot}$. The dipolar coupling between the electrons and thus the distance between $P700^{+\cdot}$ and $A_1^{-\cdot}$ is not significantly different from the value of 0.12 millitesla (distance ~30 Å ± 5Å) observed in *Rb. sphaeroides* (van der Est et al., 1993). This distance is expected with the X-ray analysis at higher resolution. The EPR results show no evidence for the participation of a second vitamin K_1 in electron transfer.

C. The Iron-Sulfur Acceptor F_X

Electron transfer from $A_1^{-\cdot}$ to F_X can be observed in the spin polarized EPR signals from PS I at room temperature (Bock et al., 1989). In principle, the polarized EPR spectrum of the charge separated $P700^{+\cdot}F_X^-$ state can be analyzed to give structural information. However, this is complicated by the fact that the transfer of spin polarization from $P700^{+\cdot}A_1^{-\cdot}$ to $P700^{+\cdot}F_X^-$ must be taken into account. Recent measurements on non spin-polarized F_X^- in oriented PS I particles (Guiliarelli et al., 1993) show that the $g_{xx} = g_3$ tensor axis (lowest g-value) is colinear to the membrane normal. This is in accordance with the observed spin polarized EPR, if the dipolar axis joining $P700^{+\cdot}$ and F_X^- is assumed to be colinear to the membrane normal and the c-axis. Such a $P700^{+\cdot}F_X^-$ orientation would also be consistent with the two-fold rotation axis passing through P700 and F_X in the observed X-ray structure (see Section IV).

D. The Terminal Iron-Sulfur Acceptors F_A and F_B

The charge-separated states $P700^{+\bullet}F_A^-$ and $P700^{+\bullet}F_B^-$ have been observed in the absence of spin polarization in PS I single crystals (Brettel et al., 1992) and oriented thylakoid membranes (Guigliarelli, 1993). From the thylakoid results, the relative orientation of the g-tensors of F_A and F_B with respect to the membrane normal have been determined. However, the proposed angle of 33° between the vector joining the two iron-sulfur centers and the membrane normal does not agree well with the value of 54° obtained quite accurately from the X-ray structure in Fig. 3. The PS I single crystal results provide more accurate information about the orientation of the g-tensors of F_A and F_B. By combining this information with EPR and X-ray data from smaller protein complexes such as bacterial ferredoxin, it is now possible to draw conclusions not only on the positions of the Fe-S centers as in X-ray structure analysis (Fig. 3), but on the atomic positions within the distorted 4Fe-S cubes as well.

EPR spectra of photoaccumulated F_A^- were measured at different rotation angles between the magnetic field and the c axis of the crystal. Six lines were observed for F_A^- indicating six inequivalently oriented PS I monomers in the unit cell with respect to the orientation of the g tensor axes of F_A^- (Brettel et al., 1992). This is consistent with the symmetry of the crystal (Fig. 1). From this, the orientation of the g-tensor axes relative to the crystal c axes could be evaluated with a high degree of accuracy. In recent experiments, the concentration of photoreduced F_B^- could be increased sufficiently so that the orientation of the g-tensors of the six F_B^- clusters could be obtained as well. Common transformation matrices between F_A^- and F_B^- g-tensor axes for each of the six subunits C per unit cell have been obtained. One of these is consistent with that obtained for the known X-ray structure of bacterial ferredoxin from *P. aerogenes* (Adman et al., 1976). In this way it was shown that the relative orientation of the F_A and F_B clusters is the same as in the ferredoxin (Kamlowski et al., 1995). This similarity is in accord with the result from X-ray structure analysis which indicates that the F_A-F_B distance is the same as in ferredoxin (12 Å) and that the two short α-helical stretches in PS I close to F_A and F_B, as well as their cysteine-binding motifs, correspond to the two one-turn α-helices of the ferredoxin (see Section IV).

With these data and the analogy to bacterial ferredoxin structure, it is possible to position subunit C of PS I with respect to the coordinates of F_A and F_B and all 16 atoms of these two 4Fe-4S clusters, respectively, as well as the eight cysteine residues (c) through which the clusters are coordinated with the two helices (Fig. 5). An assignment of F_A and F_B to the clusters is, however, not yet possible because of the pseudo-symmetry of both ferredoxin and subunit C. From studies on a genetically modified subunit C it was concluded that the cluster coordinated with the cysteine residues 48, 51, 54, 21 should belong to F_A, and that with Cys residues 11, 14, 17, 58 to F_B (Zhao et al., 1992).

VI. ENDOR Structure Analysis of P700 in Single Crystals of PS I at Low Temperature

Details of the electronic structure of $P700^{+\bullet}$ can be obtained from ENDOR (electron nuclear double resonance) spectroscopy. By this method, the hyperfine couplings (HFCs) between the spin of the delocalized unpaired electron in the radical and the spins of the various magnetic nuclei are determined. After assignment of the HFCs to specific nuclei, a map of the valence electron spin density distribution over the molecule is obtained.

The dimer model of P700 (see Section V) originally proposed from optical spectra (Döring et al., 1968) and EPR data (Norris et al., 1971) was supported by ENDOR spectroscopy of $P700^{+\bullet}$ (see review, Lubitz, 1991). As a result of other ENDOR studies, a monomeric Chl $a^{+\bullet}$, instead of a dimer, has been proposed for $P700^{+\bullet}$ (O'Malley and Babcock, 1984). In the following, refined ENDOR investigations are presented that are improved by measurements on single PS I crystals at low temperature.

Analysis of ENDOR spectra of monomeric Chl $a^{+\bullet}$ and $P700^{+\bullet}$ in liquid and frozen solutions (Käß et al., 1994 and unpublished data) showed that the largest HFC, which can be clearly assigned to the nuclei of the protons of the methyl group in position 12 (pyrrol ring C), is reduced by approximately 75 ± 5% in $P700^{+\bullet}$ as compared to monomeric Chl $a^{+\bullet}$. The corresponding reduction of the spin density of the unpaired electron at this position may indicate a delocalization of the electron in a dimeric species. Two other CH_3 HFCs, which were tentatively assigned

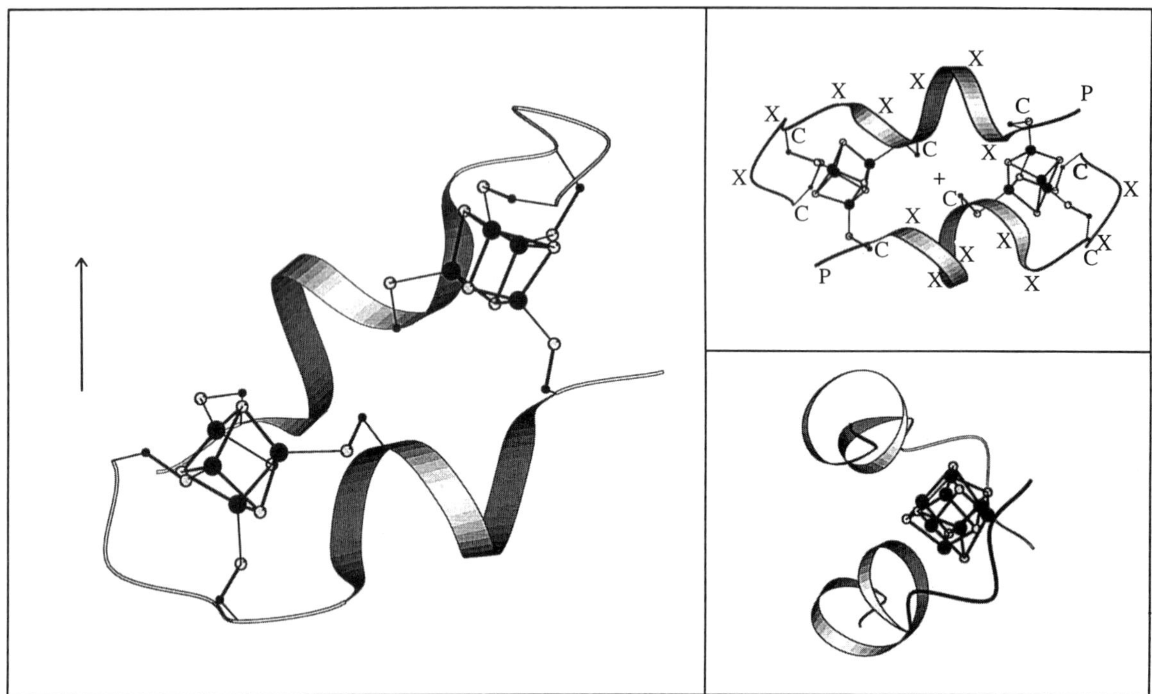

Fig. 5. Upper right: Model of the core of subunit C in the region of F_A and F_B and of the cysteine-binding pattern of the helices. View along the pseudo-twofold axis running through the cross. The model is based on the structural analogy of the bacterial ferredoxin of *Peptococcus aerogenes* (Adman et al., 1976) and subunit C as shown by X-ray and EPR-analysis of single crystals (see Sections IV and V). C denotes the cysteine residues, X other residues and P proline. Lower right: View along the line connecting the centers of F_A and F_B (Kamlowski et al., 1995). Left: Orientation of the model of the core of subunit C in PS I within the plane defined by the threefold axis and the line running through the centers of F_A and F_B. The axis F_A-F_B is tilted by 54° with respect to the threefold axis. The pseudo-twofold axis (see above) is rotated by ca. 115° around the F_A-F_B axis out of the view plane defined above.

to positions 2 and 7 in Chl *a*, showed smaller reduction factors. One of these was even larger than in monomeric Chl $a^{+\bullet}$ in vitro, indicating a spin density distribution different from that in the monomeric Chl *a* radical cation, at least within one Chl *a* half of $P700^{+\bullet}$. Furthermore, one additional smaller HFC not present in Chl $a^{+\bullet}$ appeared in the $P700^{+\bullet}$ spectra. It was, however, not clear whether this splitting belongs to the same Chl *a* or to the second Chl *a* molecule that constitutes P700.

Recently, ENDOR spectra were obtained from $P700^{+\bullet}$ in PS I single crystals at low temperature (100K) under rotation of the crystal parallel and perpendicular to the c axis (Fig. 1) (Käß et al., 1994). The measured anisotropic hyperfine couplings contain information about the orientation of $P700^{+\bullet}$ in the crystal unit cell. So far, only the larger hyperfine couplings belonging to methyl groups have been analyzed due to overlapping spectra of several sites in the unit cell; an analysis of the rotational dependence of the largest methyl proton HFC (position 12) was also performed. The rotation around the c axis shows a pattern of three cosine functions for this methyl proton HFC, which are related by 120° phase shifts. This clearly indicates that the c axis is a three-fold symmetry axis in accordance with the X-ray structure analysis (Fig. 1). For a rotation around an axis perpendicular to the c axis, according to the presence of six P700 in the unit cell, a pattern of six cosine functions is expected. However, only four have been observed. When, for rotation around this axis, the magnetic field, B_0, is parallel to c, all P700 are magnetically equivalent and all cosine functions should have a common point of intersection. This is clearly observed.

The detection of only four patterns of cosine functions shows that four of the six P700 sites in the unit cell are pairwise magnetically equivalent[1]. This pairwise equivalence of four $P700^{+\bullet}$ sites for rotation

[1] The analysis assumes an axially symmetric hyperfine tensor with the axis corresponding to the largest tensor element ($A_{\parallel}$) oriented parallel to the C-CH_3 bond axis.

perpendicular to the c axes (i.e., B_0 rotated in the a-c, b-c or d-c plane of Fig. 1, where d is the bisector of a and b) is only obtained when the C-CH_3 bond axes of the six $P700^{+\bullet}$ are in planes which are parallel to the a-c, b-c and d-c plane, respectively. Hence, the essential result is that the molecular plane of one of the potential Chl *a* of $P700^{+\bullet}$ is located perpendicular to the crystallographic a-b plane, which is proposed to also be the membrane plane (see Section IV and Fig. 1). This orientation is obviously also relevant to the second Chl *a* of $P700^{+\bullet}$, since the axes normal to both chlorophylls of the dimer are parallel (Section V). This result is consistent with the EPR analysis of ^{3}P700 in oriented PS I particles (Section V) and the X-ray analysis with respect to the orientation of the two chlorophylls assigned to P700 (see Section IV). The result is also similar to the orientation of the primary donor found in bacterial reaction centers (Allen et al., 1987).

New information also results from the phase angles of the cosine functions of the HFC of the methyl protons (position 12, ring C). Under rotation around the axis perpendicular to the two-fold axis, a phase shift indicates an angle of 33 ± 5 or 180 plus 33 ± 5 degrees between the c axis and the C-CH_3 bond axis. These are also the angles with respect to the two-fold axis which runs through P700 and F_X, because this axis is colinear with the crystal c axis (Section IV). Because the phytyl side chain has to point to the hydrophobic intrinsic part of the membrane (Fig. 1), this can be achieved only by an angle of 180 plus 33 ± 5 degrees, i.e., the CH_3 group must point towards the lumenal side (Fig. 6). In this way, the orientation of one of the Chl *a* of the dimer is unequivocally determined. The analysis of the angular dependence of the HFCs of two other methyl groups showed that they have to be assigned to positions 2 and 7 of the same Chl *a* molecule (Käß, 1995). Comparison of the HFCs of the three methyl positions (2, 7 and 12) of $P700^{+\bullet}$ with the corresponding values in monomeric Chl $a^{+\bullet}$ in vitro shows a strong localization of spin density on one Chl *a* for $P700^{+\bullet}$ (≥ 85%). The spin density assigned to the second Chl *a* of P700 varies between 0% and a maximum of 15% (Fig. 6). To clarify this, a more refined analysis of the hyperfine coupling rotation pattern of $P700^{+\bullet}$ in PS I single crystals is in progress (H. Käß et al., unpublished).

With respect to the orientation of the second Chl *a* of the dimer, one has to consider that its plane is approximately parallel to the first Chl *a* (see Sections

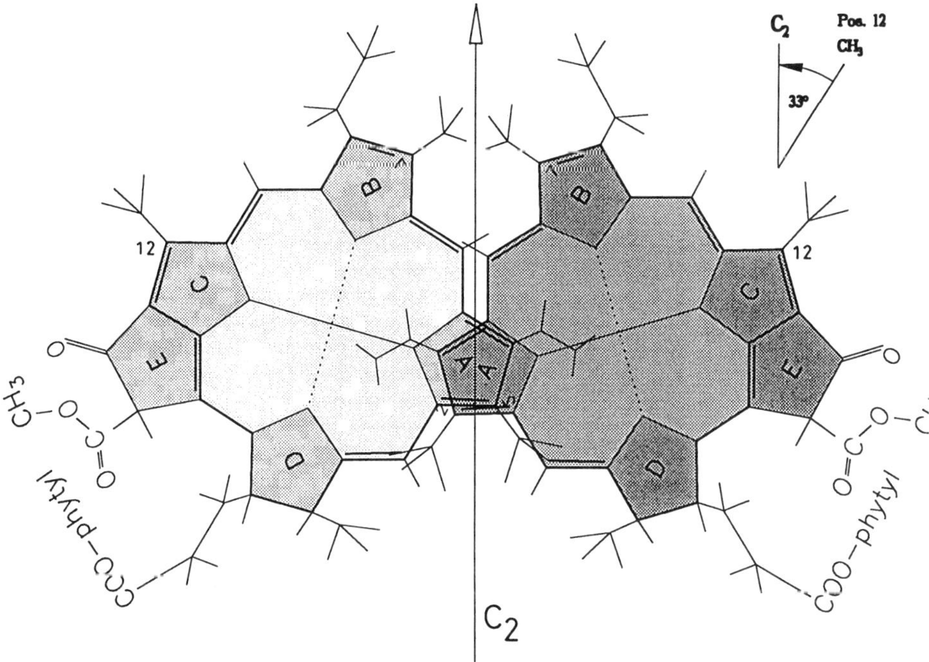

Fig. 6. Right: Circular orientation of one of the Chl *a* of the P700 dimer within its plane based on ENDOR analysis of the methyl proton hyperfine coupling in single crystals. Left: Possible orientation of the second Chl *a*. The local twofold axis c_2 is indicated. The lumenal side is located towards the top. The distribution of the electron spin density in the two Chl *a* of the $P700^{+\bullet}$ radical is ≤ 15% and ≥ 85%, respectively (Käß et al., 1994; Käß, 1995)(see Section VI).

IV and V). If it is symmetrically related to this pigment, its position should be as indicated in Fig. 6. This is the case for bacterial reaction centers.

Acknowledgments

I am very grateful to Drs. A. van der Est, A. Kamlowski and D. Stehlik for the discussion and compilation of the EPR results in Section D and to Drs. H. Käß, F. Lendzian and W. Lubitz for providing the interpretation of the ENDOR measurements prior to publication. I thank Drs. W. Saenger and N. Krauß for critical comments and Dr. P. Fromme and W. D. Schubert for their support of the compiling of the recent results at 4.5 Å resolution. This work was supported by grants from the Deutsche Forschungsgemeinschaft, Sonderforschungs-bereich 312, Teilprojekte A1, A3, A4 and D1.

References

Adman ET, Sieker LC and Jensen LH (1973) The structure of a bacterial ferredoxin. J Biol Chem 248: 3987–3996

Adman ET, Sieker LC and Jensen LH (1976) Structure of *Peptococcus aerogenes* ferredoxin. Refinement at 2 Å resolution. J Biol Chem 251: 3801–3806

Allen JP, Feher G, Yeates TO, Komiya H and Rees DC (1987) Structure of the reaction center from *Rhodobacter sphaeroides* R-26: The cofactors. Proc Natl Acad Sci USA 84: 5730–5734

Bittl R, van der Est A, Fuechsle G, Lubitz W, Stehlik D (1992) Transient EPR of photosynthetic reaction centers: P^+Q^- in Zn-substituted *Rb. sphaeroides* and Photosystem I. In: Murata N (ed) Research in Photosynthesis, Vol I, pp 461–464. Kluwer, Dordrecht

Bock CH, van der Est A, Brettel K, Stehlik D (1989) Nanosecond electron transfer kinetics in Photosystem I as obtained from transient EPR at room temperature. FEBS Lett 247: 91–96

Boekema EJ, Dekker JP, van Heel MG, Rögner M, Saenger W, Witt I and Witt HT (1987) Evidence for a trimeric organization of the Photosystem I complex from the thermophilic cyanobacterium *Synechococcus* sp. FEBS Lett 217: 283–286

Brettel K, Siekmann I, Fromme P, van der Est A and Stehlik D (1992) Low-temperature EPR on single crystals of Photosystem I: Study of the iron-sulfur-center F_A. Biochim Biophys Acta 1098: 266–270

Budil DE and Thurnauer MC (1991) The chlorophyll triplet state as a probe of structure and function in photosynthesis. Biochim Biophys Acta 1057: 1–41

Chitnis VP, Quian X, Lian Y, Golbeck JH, Nakamoto H, Dian-Lin X and Chitnis PR (1993) Target inactivation of the gene *psaL* encoding a subunit of Photosystem I of the cyanobacterium *Synechocystis* sp. PCC 6803. J Biol Chem 268: 11678–11684

Deisenhofer J and Michel H (1989) High resolution structures of photosynthetic reaction centers. EMBO J 8: 2149–2170

Deisenhofer J, Epp O, Miki K, Huber R and Michel H (1985) Structure of the protein subunits in the photosynthetic reaction centers of *Rhodopseudomonas viridis* at 3 Å resolution. Nature 318: 618–624

Döring, G, Bailey JL, Kreutz, W, Weikard, J and Witt HT (1968) The action of two chlorophyll-a_I-molecules in light reaction I of photosynthesis. Naturwiss 55: 219–224

Fromme P, Schubert WD, Krauß N (1994) Structure of Photosystem I: Suggestions on the docking sites for plastocyanin, ferredoxin and the coordination of P700. Biochim Biophys Acta 1187: 99–105

Fuechsle G, Bittl R, van der Est A, Lubitz W, Stehlik D (1993) Transient EPR spectroscopy on perdeuterated Zn-substituted reaction centers of *Rhodobacter sphaeroides* R-26. Biochim Biophys Acta 1142: 23–35

Golbeck JH (1994) Photosystem I in cyanobacteria. In: Bryant DA (ed) The Molecular Biology of Cyanobacteria, pp 319–360. Kluwer Academic Publishers, Dordrecht

Golbeck JH and Bryant DA (1991) Photosystem I. In: Lee CP (ed) Current Topics in Bioenergetics; Vol 16, pp 83–177, Academic Press, New York

Guigliarelli B, Guillaussier J, More C, Sétif P, Bottin H, Bertrand P (1993) Structural organization of the iron-sulfur centers in *Synechocystis* 6803 Photosystem I. J Biol Chem 268: 900–908

Hladik I and Sofrova D (1991) Does the trimeric form of Photosystem I reaction center of cyanobacteria in vivo exist? Photosyn Res 29: 171–175

Kamlowski A, van der Est P, Fromme P and Stehlik D (1995) Structural organization of the acceptors, A_1, F_x, F_A and F_B in Photosystem I from EPR in solution and single crystals. In: Mathis P (ed) Photosynthesis: From Light to Biosphere, Vol II, pp 29–34. Kluwer Academic Publishers, Dordrecht

Käß H (1995) Die Struktur des primären Donators P700 in Photosystem I. Thesis, TU Berlin

Käß H, Fromme P, Witt HT and Lubitz W (1994) ^1H ENDOR and ^{14}N ESEEM in single crystals of Photosystem I from *Synechococcus elongatus*. Biophys J 66: A228

Krauß N, Hinrichs W, Witt I, Fromme P, Pritzkow W, Dauter Z, Betzel Ch, Wilson KS, Witt HT and Saenger W (1993) Three dimensional structure of system I of photosynthesis at 6 Å resolution. Nature 361: 326–331

Kühlbrandt W, Da Neng W and Fujiyoshi Y (1994) Atomic model of plant light-harvesting complex by electron crystallography. Nature 367: 614–621

Lubitz W (1991) EPR and ENDOR studies of chlorophyll cation and anion radicals. In: Scheer H (ed) Chlorophylls, pp 903–944, CRC Press, Boca Raton

Mathis P (1990) Compared structure of plant and bacterial photosynthetic reaction centers. Evolutionary implications. Biochim Biophys Acta 1018: 163–167

McDermott G, Prince SM, Freer AA, Hawthornthwaite-Lawless AM, Papiz MZ, Cogdell RJ and Isaacs NW (1995) Crystal structure of an integral membrane light-harvesting complex from photosynthetic bacteria. Nature 374: 517–521

Mühlenhoff U (1991) Zur Primär- und Quartiärstruktur des Photosystem I aus dem thermophilen Cyanobakterium *Synechococcus* sp. Thesis, TU Berlin

Mühlenhoff U, Haehnel W, Witt HT and Herrmann RG (1993) Genes encoding eleven subunits of Photosystem I from the thermophilic cyanobacterium *Synechococcus* sp. GENE 127: 71–78

Nitschke W, Feiler U, Lockau W and Hauska G (1987) The photosystem of the green sulfur bacterium *Chlorobium limicola* contains two early electron acceptors similar to Photosystem I. FEBS Lett 218: 283–286

Norris JR, Uphaus RA, Crespi HL and Katz JJ (1971) Electron spin resonance of chlorophyll and the origin of signal I in photosynthesis. Proc Natl Acad Sci 68: 625–628

Oh-Oka H, Takahashi Y, Kuriyama K, Saeki K and Matsubara H (1988) The protein responsible for center A/B in spinach Photosystem I: Isolation with iron-sulfur cluster(s) and complete sequence analysis. J Biochem 103: 962–968

O'Malley PJ and Babcock GT (1984) Electron nuclear double resonance evidence supporting a monomeric nature for $P700^+$ in spinach chloroplasts. Proc Natl Acad Sci USA 81: 1098–1101

Rögner M, Mühlenhoff U, Boekema EJ and Witt HT (1990) Mono-, di and trimeric PS I reaction center complexes isolated from the thermophilic cyanobacterium *Synechococcus* sp.: Size, shape and activity. Biochim Biophys Acta 1015: 415–424

Rutherford AW and Sétif P (1990) Orientation of P700, the primary electron donor of Photosystem I. Biochim Biophys Acta 1019: 128–132

Schatz GH and Witt HT (1984) Extraction and characterization of oxygen-evolving Photosystem II complexes from a thermophilic cyanobacterium *Synechococcus* sp. Photobiochem Photobiophys 7: 1–4

Schubert WD, Klukas O, Krauß N, Saenger W, Fromme P and Witt HT (1995) Present state of the crystal structure analysis of Photosystem I at 4.5 Å resolution. In: Mathis P (ed) Photosynthesis: From Light to Biosphere, Vol II, pp 3–10. Kluwer Academic Publishers, Dordrecht

Shubin VV, Tsuprun VL, Bezsmertnaya IN and Karapetyan NV (1993) Trimeric forms of the Photosystem I reaction center complex pre-exist in the membranes of cyanobacterium *Spirulina platensis*. FEBS Lett 334: 79–82

Sieckmann I, Brettel K, Bock C, van der Est A, Stehlik D (1993) Transient EPR of the triplet state of P700 in Photosystem I: Evidence for triplet delocalization at room temperature. Biochemistry 32: 4842–4847

Snyder SW and Thurnauer MC (1993) Electron spin polarization in photosynthetic reaction centers. In: Deisenhofer J and Norris JR (eds) The Photosynthetic Reaction Center, pp 285–330, Academic Press, New York

Stehlik D, Bock CH and Petersen J (1989) Anisotropic electron spin polarization of correlated spin pairs in photosynthetic reaction centers. J Phys Chem 93: 1612–1619

Trebst A (1986) The topology of the plastoquinone and herbicide binding peptides of Photosystem II in the thylakoid membrane. Z. Naturforsch. 41c: 240–245

van der Est A, Bittl R, Abresch EC, Lubitz W, Stehlik D (1993) Transient EPR Centers of *Rhodobacter sphaeroides* R-26. Chem Phys Lett 212: 561–568

Witt HT (1987) Examples for the cooperation of photons, excitons, electrons, electric fields and protons in the photosynthesis membrane. New J Chem 11: 91–101

Witt I, Witt HT, Gerken S, Saenger W, Dekker JP and Rögner M (1987) Crystallization of reaction center I of photosynthesis. FEBS Lett 221: 260–264

Witt I, Witt HT, DiFiore D, Rögner M, Hinrichs W, Saenger W, Granzin J, Betzel Ch and Dauter Z (1988) X-Ray characterization of single crystals of the reaction center I of water splitting photosynthesis. Ber Bunsenges Phys Chem 92: 1503–1506

Witt HT, Krauß H, Hinrichs W, Witt I, Fromme P and Saenger W (1992) Three-dimensional crystals of Photosystem I from *Synechococcus* sp. and X-ray structure analysis at 6 Å resolution. In: Murata N (ed) Research in Photosynthesis, Vol I, pp 521–528, Kluwer, Dordrecht

Zhao J, Li N, Warren PV, Golbeck JH, Bryant DA (1992) Site-directed conversion of a cysteine to aspartate leads to the assembly of a [3Fe.4S] cluster in PsaC of Photosystem I. The photoreduction of F_A is independent of F_B. Biochemistry 31: 5093–5099

Chapter 19

The Cytochrome b_6f Complex—Composition, Structure and Function

Günter Hauska, Michael Schütz and Michael Büttner
Universität Regensburg, Fakultät für Biologie und Vorklinische, Medizin, Lehrstuhl für Zellbiologie und Pflanzenphysiologie, D-93053 Regensburg, Germany

Summary

The cytochrome b_6f complex universally functions in oxygenic photosynthesis as plastoquinol-plastocyanin oxidoreductase or as plastoquinol-cytochrome *c* oxidoreductase between the two light reactions of Photosystem II and Photosystem I. It is structurally and functionally related to the cytochrome bc_1 complex of respiratory organisms and comprises 4 redox centers—1 heme *c*, 2 hemes *b* and the Rieske 2Fe2S-center. These operate quinol oxidation in a concerted mechanism, called the Q-cycle, which translocates twice the protons that would be translocated by plastoquinone/quinol action alone. It is still under debate whether this efficient energy conservation is operative under all conditions. The redox centers are bound by the 3 proteins, cytochrome *f* (heme *c*), cytochrome b_6 (2 hemes *b*) and the Rieske FeS-protein, whose genes have been sequenced for numerous plants and cyanobacteria. A fourth protein (subunit IV) is universally present and corresponds to the

Donald R. Ort and Charles F. Yocum (eds): Oxygenic Photosynthesis: The Light Reactions, pp. 377–398.

C-terminal part of cytochrome *b* of the bc_1 complexes. While a high resolution 3D-structure is now available for cytochrome *f* (see Chapter 22), the 3D-structures of the Rieske FeS-protein and of cytochrome b_6 plus subunit IV still depend on predictions from the sequences. The predicted folding for cytochrome b_6 with its transmembrane arrangement of the 2 hemes is in reassuring consistency with the Q-cycle. In addition to its role as energy converter the cytochrome b_6f complex functions as a redox sensor of the plastoquinone pool in regulating the light distribution between the two photosystems.

I. Introduction

Oxidation of isoprenoid quinols by cytochrome complexes constitutes the central reaction in biological electron transport chains. It takes place in both the major energy yielding biological processes, in photosynthesis as well as in respiration, although these processes counteract each other in the overall bioenergetic cycle (Fig. 1). In the eukaryotic respiratory chain of mitochondria the cytochrome bc_1 complex functions as ubiquinol-cytochrome *c* oxidoreductase. In photosynthetic electron transport of chloroplasts the cytochrome b_6f complex likewise acts as plastoquinol-plastocyanin oxidoreductase. In photosynthetic prokaryotes the quinol oxido-reductases may function in one and the same membrane in both processes, respiration in the dark and photosynthesis in the light. Thus, as shown in Fig. 2, the cytochrome bc_1 complex of purple photosynthetic bacteria (*Rhodospirillaceae*) and the cytochrome b_6f complex of bluegreen algae (cyanobacteria) represent metabolic zwitters (Baccarini-Melandri and Zannoni, 1978; Hauska et al., 1983; Scherer, 1990). Because the oxygen in the atmosphere is thought to have originated from photosynthesis, this dual function, together with the relatedness of cytochrome bc_1- and b_6f complexes, including the primary structures of their subunits (see below), supports the current theory that aerobic respiration evolved from photosynthesis (Broda and Peschek, 1979; see addendum, however). The still controversial 'chlororespiration' of chloroplasts (Singh et al, 1992; Berger et al., 1993) possibly represents a relic of the cyanobacterial duality.

Both complexes operate very efficiently in energy conservation, the cytochrome b_6f complex at least under certain conditions (see below). According to the widely accepted 'Q-cycle'-mechanism they oxidize quinol in a coordinated manner, and 2 protons are translocated per transferred electron, twice the amount that the quinone system would translocate alone.

This chapter is intended to update our current insight into the structure and function of the cytochrome b_6f complex, as it has developed within the last decade (Hauska et al., 1983; see also Kallas, 1994). The emphasis will be on structural aspects because a review focused on function appeared recently (Hope, 1993). A parallel consideration of the related bc_1 complexes will be maintained throughout in order to present a complete story. For more details the reader is referred to recent reviews on the cytochrome b_6f complex (Cramer et al., 1991, 1994; Malkin, 1992; Anderson, 1992; Hope, 1993; Kallas, 1994), and on the cytochrome bc_1 complex from mitochondria (Beattie, 1993), and from bacteria (Knaff, 1993; Gennis et al., 1993). An extensive comparative interpretation of cytochrome *b* sequences and its functional implications, based on the solved structure of the bacterial reaction center, was provided by Degli Esposti at al. (1993).

The chloroplast cytochromes, *f* (Hill and Scarisbrick, 1951) and b_6 (Hill, 1954), were discovered more than 40 years ago. Their redox behavior led to the formulation of the Z-scheme for oxygenic photosynthesis (Hill and Bendall, 1960). Nelson and Neumann (1972) isolated a cytochrome b_6f complex and were the first to point out its analogy to the mitochondrial cytochrome bc_1 complex. Plasto-quinol-plastocyanin oxidoreductase was solubilized from thylakoids (Wood and Bendall, 1976) and later a b_6f complex of defined polypeptide composition was isolated which retained this oxidoreductase activity (Hurt and Hauska, 1981).

II. Occurrence

The classification of the cytochrome *bc* complexes into bc_1- and b_6f-types is based on two major criteria: The former are sensitive to antimycin A and other quinone antagonists, the latter have cytochrome *b* protein split into cytochrome b_6 and subunit IV.

Abbreviations: DCCD – N,N′-dicyclohexylcarbodiimide; MOA – methoxyacrylate; UHDBT – 5-n-undecyl-6-hydroxy-4,7-dioxobenzothiazole

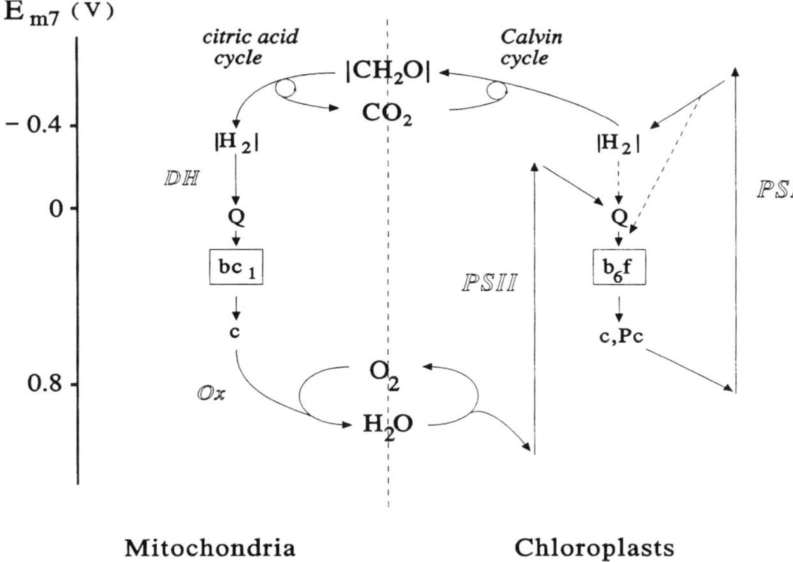

Fig. 1. Involvement of cytochrome bc_1/b_6f-complexes in the bioenergetic cycle between mitochondrial respiration and chloroplast photosynthesis. The scheme is drawn approximately to the redox potential scale on the left, except for the upward arrows in photosynthesis. The symbols represent: DH – dehydrogenase, Ox – cytochrome oxidase, PSI and PSII – Photosystem I and II, (H_2) – bound forms of hydrogen, (CH_2O) – carbohydrate, c – cytochrome *c*, Pc – plastocyanin. The dashed arrows represent cyclic electron transport around PSI.

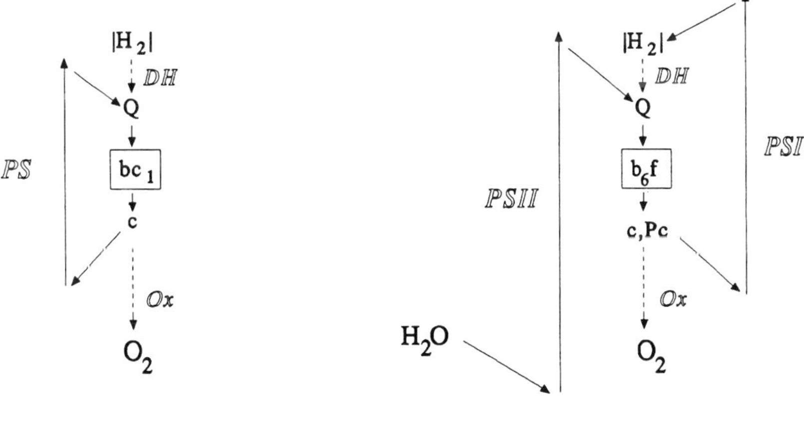

Fig. 2. Dual electron transport by the bc_1 complex in *Rhodospirillaceae* and by the b_6f complex in cyanobacteria. The symbols are as for Fig. 1. The dashed arrows represent respiration.

A. Organisms

The cytochrome b_6f complex is found in all chlorophyll *a*-containing, oxygenic organisms. It is present in the plastids of all the eukaryotic algae and plants, and in the prokaryotic cyanobacteria. It also has been detected in the prokaryotic taxon of the prochlorophyta (Greer and Golden, 1992). In view of the endosymbiont theory of eukaryotic evolution this is not surprising, nor is it surprising that the bc_1 complex is found in mitochondria and in the proteobacteria (Woese, 1987). The latter include the photosynthetic purple bacteria. More intriguing is the indication that gram-positive bacilli (Kutoh and Sone, 1988) and the related heliobacteria (Nitschke and Liebl, 1992; Sone et al., 1995), by the above criteria, may contain a b_6f-type complex. However, the division into bc_1- and b_6f-types has its limitation—the *bc* complex of *Chlorobium limicola*, a green sulfur bacterium, combines features of both (Schütz et al., 1994).

The occurrence of a cytochrome b_6f- or bc_1 complex in archaebacteria is still uncertain, although for *Sulfolobus acidocaldarius* a cytochrome *b*-like heme *a* protein has been identified which may be involved in a Q-cycle (Lübben et al., 1992). A Rieske FeS-center has been reported in *Sulfolobus acidocaldarius* (Anemüller et al., 1993). See Addendum, however.

B. Cellular Location

The cytochrome b_6f complex is localized in the membranes of chloroplasts and etioplasts (Wood and Bendall, 1976; Westhoff and Herrmann, 1988). In mature chloroplasts it is distributed over the grana and the stroma lamellae (Anderson, 1992), in contrast to the Photosystem I and II complexes, which are asymmetrically deployed between the stroma lamellae and the grana. More precisely the b_6f complex may be concentrated in the margins between grana and stroma lamellae, where it could connect Photosystem I and II (Albertsson et al., 1991).

In cyanobacteria the b_6f complexes are also membrane bound. Interestingly, however, for certain cyanobacteria it has been claimed that a cytochrome b_6f complex is located in the thylakoid membrane, whereas a cytochrome bc_1 complex is concentrated in the cytoplasmic membrane (G. A. Peschek, personal communication).

III. Isolation

The current procedures can be divided into three stages: 1) Extensive removal of peripheral proteins, including the coupling factor ATPase, from thylakoid membranes. 2) Selective solubilization of the complex by non-ionic detergents, such as alkyl glycosides. 3) Further purification via ammonium sulfate fraction–ation, or ion exchange chromatography, followed by sucrose density gradient fractionation, or molecular sieve chromatography (Hurt and Hauska, 1981, 1982a; Hauska, 1986; Black et al., 1987; Schmidt and Malkin, 1993). To preserve activity it is critical to chill carefully and to work quickly, keeping the complex from prolonged exposure to high salt, or to ion exchangers. For high concentrations, as required for EPR work, a useful final step is ammonium sulfate precipitation in the presence of cholate (Rich et al., 1987).

The isolation procedure with alkyl glycosides, originally developed for higher plants, has been applied successfully to cyanobacteria (Krinner et al., 1982; Hauska, 1986; Rögner et al., 1990; Tsiotis et al., 1992), to green algae (Schmidt and Malkin, 1993; Pierre and Popot, 1993), and to purple bacteria (Gabellini et al., 1982; Hauska, 1986; Lungdahl et al., 1986; Andrews et al., 1989; Robertson et al., 1993).

IV. Components and Structure

The cytochrome b_6f complex, like all the functionally intact, isolated cytochrome *bc* complexes, contains four metal redox centers that are bound to protein - one heme *c*, one 2Fe2S-center and two hemes *b*. In addition to protein the preparation contains lipids (and detergent) and some residual plastoquinone, a part of which may be firmly bound (Hurt and Hauska, 1982b; Robertson et al., 1993). Even the purest cytochrome b_6f complex retains about one molecule of residual chlorophyll *a* (Hurt and Hauska, 1981; Huang et al., 1994).

A. Polypeptides

The four metal redox centers are bound to three proteins: cytochrome *f*, which is a *c*-type cytochrome (*f* standing for 'folium' after its discovery in leaves (Hill and Scarisbrick, 1951)), the Rieske FeS-protein, and cytochrome *b*6, which like cytochrome *b* of bc_1 complexes is a double heme protein (Widger et al., 1984; Hauska et al., 1988). These three proteins constitute the minimum structure for a cytochrome *bc* complex as it is indeed found in bacteria (Knaff, 1993; Gennis et al., 1993). The mitochondrial bc_1 complexes contain a high number of other proteins, —eight for mammals and seven for fungi. While these do not participate in the transfer of electrons directly, they may fulfill regulatory or even unrelated functions. The two prominent, large subunits, called 'core proteins' are related to the mitochondrial signal peptidase. In plant mitochondria they are even identical with this protease (Braun et al., 1992). In contrast, the cytochrome b_6f complexes contain only one prominent additional polypeptide of about 17 kDa (Fig. 3), designated subunit IV. Together with cytochrome b_6 of 23 kDa it has high homology to the 40 kDa-cytochrome *b* of cytochrome bc_1 complexes. Subunit IV corresponds to the C-terminal part of cytochrome *b* (Widger et al., 1984; Hauska et al., 1988). Thus the b_6f complexes also could be regarded as representing the minimum structure, lacking the many supernumerary subunits of mitochondria. However, up to four small, hydrophobic and copurifying subunits of 3–6 kDa have been detected (Hurt and Hauska, 1982a; Pierre and Popot, 1993; Schmidt and Malkin, 1993), as shown in Fig. 3b, by appropriately resolving SDS-PAGE (Schägger and von Jagow, 1987). The role of these small subunits remains unknown. Possibly they are of structural importance, like a small subunit of Photosystem I (PsaL), which is responsible for trimer formation (Chitnis and Chitnis, 1993).

B. Genetic Organization

In proteobacteria (Woese, 1987) the genes for the three redox center carrying polypeptides of the minimum cytochrome bc_1 complex are organized in the *fbc*-operons, also called *pet*ABC (Knaff, 1993; Gennis et al., 1993). The gene for the Rieske FeS-protein (*pet*A) is followed by the genes for cytochrome *b* (*pet*B) and for cytochrome c_1 (*pet*C).

In contrast, for the cytochrome b_6f complex in cyanobacteria two transcription units are known, one coding for the Rieske FeS-protein and cytochrome *f*, and the other for cytochrome b_6 and subunit IV (Malkin, 1992). Recently a transcription unit in the green sulfur bacterium *Chlorobium limicola* has been discovered that lacks the gene for cytochrome c_1 or *f* from the *fbc*-operon, but has retained the gene for a Rieske FeS-protein followed by the gene for a complete cytochrome *b*. Although the latter gene is continuous, its sequence exhibits characteristics that are closer to the cytochrome b_6 plus subunit IV

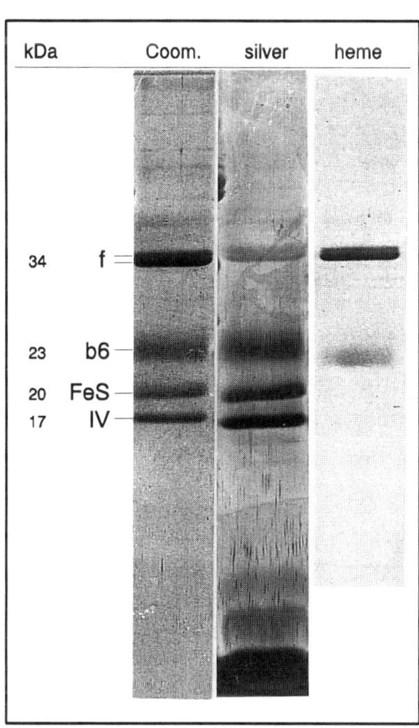

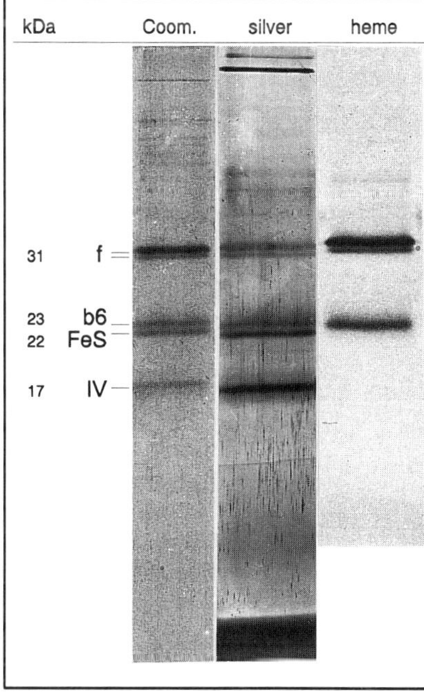

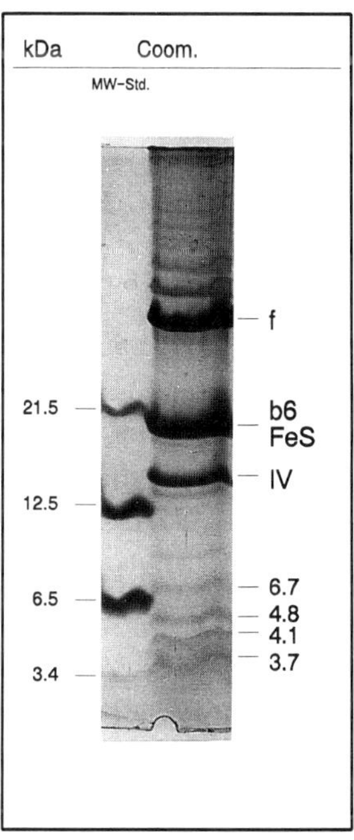

Fig. 3. Polypeptide patterns of cytochrome b_6f-complexes. SDS-PAGE of the preparations from spinach and *Anabaena variabilis* (Hauska, 1986) stained with Coomassie, silver, or for heme are shown in A and B. In C the preparation from spinach is run for resolution of small polypeptides (Schägger and von Jagow, 1987). Under these conditions cytochrome b_6 and the Rieske FeS-protein are not resolved. The numbers represent apparent M_r-values, and the symbols f, b6, FeS and IV stand for cytochromes *f* and b_6, the Rieske FeS-protein and subunit IV, respectively. The exact M_r-values of the mature proteins for spinach are 31,372 (285 residues) for cytochrome *f*, 24,038 (214 residues) for cytochrome b_6, 18,922 (178 residues) for the Rieske FeS-protein, and 17,444 (159 residues) for subunit IV (Cramer et al., 1994).

(Schütz et al, 1994). Thus it can be concluded from the phylogenetic distances (Woese, 1987), that the splitting of the gene for a cytochrome *b* into two, one for cytochrome b_6 and the other for subunit IV, must have occurred after the separation into complexes with bc_1- and b_6f-type characteristics. It is still uncertain whether *Chlorobium* contains an equivalent for cytochrome c_1 or *f*, but it is possible that yet another type of cytochrome takes over in green sulfur bacteria (Okkels et al., 1992).

Chloroplasts of higher plants have retained the cyanobacterial transcription unit for cytochrome b_6 and subunit IV on their genome (Alt et al., 1983; Heinemeyer et al., 1984), but from the other transcription unit the gene for the Rieske FeS-protein has been transferred to the nucleus (see Malkin, 1992). There it picked up a leader sequence for transport of the protein into the chloroplast (Steppuhn et al., 1987; Salter et al., 1992). The gene for cytochrome *f* remained on the plastid genome (Alt et al., 1983; Gray, 1992), which also harbors one of the small, hydrophobic subunits in a separate location (Haley and Bogorad, 1989). In green algae the situation is different, with separated genes—the ones for cytochromes *f* and *b6*, and for subunit IV on the plastome, the one for the Rieske FeS-protein on the nuclear genome (Reimann and Kück, 1989; Büschlen et al., 1991).

C. Cytochrome f

The most impressive recent progress comes from this protein. Its crystal structure has been resolved to atomic level (Martinez et al., 1994) as will be summarized in Chapter 22. For a more general review on cytochrome *f* see Gray (1992).

The heme of cytochrome *f* transfers electrons to plastocyanin or cytochrome *c* (see Figs. 5A and 6). Cytochrome *f* can be selectively reduced by ascorbate, while the reduction of the hemes in cytochrome b_6 requires the greater reducing potential of dithionite. The α-band of reduced cytochrome *f* from chloroplasts peaks at 554 nm whereas in cyanobacteria the peak is at 556 nm. The peak is asymmetric and the shoulder on the low wavelength side forms a separate peak at low temperature (Hurt and Hauska, 1981).

Cytochrome *f*, like cytochrome c_1, is a *c*-type cytochrome of some 30 kDa, and it migrates at somewhat higher M_r on SDS-PAGE (Fig. 3), depending on conditions and species (Hauska et al., 1983; Gray, 1992). In some species (spinach, bean, *Anabaena variabilis*), but not in others (pea, barley), cytochrome *f* appears as a closely spaced doublet on SDS-PAGE, as shown in Fig. 3 (Hurt and Hauska 1982a; Hauska et al., 1983). This microheterogeneity is a feature of the chloroplasts, and not an effect of isolation. Since only one gene for cytochrome *f* exists in chloroplasts (Gray, 1992), the observation either reflects heterogeneity of the gene in the plant population, or post-transcriptional events. It is different from the proteolytic clipping of the hydrophobic C-terminus during isolation of cytochrome *f* from *Brassicaceae* (Martinez et al., 1994; Gray, 1992).

The heme group is covalently bound to cysteines of the protein. The conserved binding motif CXXCH is found close to the N-terminus of cytochromes *f* and c_1. In both, the C-terminus is preceded by a pronounced hydrophobic stretch of 20 residues, which is flanked by positive charges, and is considered to fix the proteins in the membrane (see Fig. 6). Otherwise cytochromes c_1 and *f* have very little in common (Hauska et al., 1988). Cytochromes *f* are shorter at the N-terminus, but are substantially longer in the middle than cytochromes c_1. The latter possibly lacks the second, smaller globular domain found in cytochrome *f* (Martinez et al., 1994). The most significant difference between cytochromes c_1 and *f* is the 6th ligand to the heme-Fe. It is the S of a methionine in the former, but the N from the N-terminus in the latter (Martinez et al., 1994; see Chapter 22). Taken together all this may indicate that cytochromes c_1 and *f* evolved from different ancestors. The possibility that the cytochrome complex in *Chlorobium* contains neither cytochrome c_1 nor *f*, but another type of cytochrome instead (Okkels et al., 1992), is in line with this conclusion. On the other hand, a structure like cytochrome *f*, which is dominated by β-sheets, may exert relatively few mutational restrictions.

Like mitochondrial cytochromes c_1, bacterial cytochromes c_1 and cytochromes *f* are first built as precursors with N-terminal extensions, although they are not synthesized from external genes. However, they have to be translocated to the periplasmic space, or into the thylakoid, whereby the extensions are cleaved off. The final, processing protease specifically cleaves after alanine (see Gray, 1992).

D. The Rieske FeS-Protein

The Rieske FeS-protein was discovered in mitochondria by J. S. Rieske (Rieske, 1976), and functions as the primary oxidant of quinol in cytochrome *bc* complexes. It contains a 2Fe2S-center with an unusually positive redox potential, such that it can conveniently be monitored by EPR after reduction with ascorbate. The EPR-spectrum for the chloroplast protein was first reported by Malkin and Aparicio (1975). It is typical for an anisotropic, rhombic g-tensor for the unpaired electron in the reduced FeS-cluster that is characterized by the values g_z, g_y and g_x. The spectrum can be used to sensitively monitor interactions of quinones and inhibitors with the FeS-cluster (Ding et al., 1992). An apparent discrepancy between the b_6f complex and the bc_1 complex with regard to the orientation of their FeS-cluster in the membrane has been resolved (Riedel et al., 1991).

The high redox potential of the 2Fe2S-cluster is considered to be caused by N-ligation of the iron, in addition to the more typical S-ligation (Britt et al., 1991; Gurbiel et al., 1991). Indeed, each of two universally conserved peptides, CTHLGC and CPHCG, in the C-terminal half of the protein contains a histidine (Hauska et al., 1988; Gennis et al., 1993; Schütz et al., 1994). The pH-dependence of the redox potential indicates two pK-values at alkaline pH, which may result from the two histidine ligands (Nitschke et al., 1992; Link, 1994). Interestingly, two cysteines are also present in each of the conserved peptides, however, and it has been demonstrated by directed mutations of the bc_1 complexes that all four of them are required for a functional Rieske FeS-protein (Davidson et al., 1992; Graham et al., 1993; Van Doren et al., 1993a; Gennis et al., 1993), even if only two of them were involved in S-ligation of the cluster.

Over 20 sequences of Rieske FeS-proteins are known to date, including one from archaebacteria (Castresana et al., 1995) (Addendum). The sequence from *Chlorobium limicola* is now available and it represents a menaquinol-oxidizing, low-potential form (Schütz et al., 1994). Identity/similarity is high in the C-terminal half only, which carries the FeS-cluster. This has been substantiated by complementation analysis of hybrid proteins in yeast (Huang et al., 1991). The proteins of the b_6f complex have a shorter N-terminus, and migrate around 20 kDa on SDS-PAGE (Fig. 3). They accommodate one putative transmembrane helix only, which is followed by a glycine-rich cluster (Schütz et al., 1994) that could act as a molecular hinge. This cluster is missing from the somewhat larger proteins of the bc_1 complexes which have been considered to form an N-terminal hairpin structure through the membrane. However, the relative ease of dissociation from the bc complexes argues for a more peripheral binding (Geier et al., 1992; Breyton et al., 1994; de Vitry, 1994), as drawn in Fig. 4. For a more detailed discussion of predicted secondary structural elements in the Rieske FeS-protein from yeast consult Graham et al., (1993).

The Rieske FeS-proteins of chloroplasts (Steppuhn et al., 1987; Salter et al., 1992) and mitochondria are encoded in nuclear genes as precursor forms with N-terminal signal peptides, that are processed to the mature proteins on the way to the functional sites in the organelles. It came as a surprise that in mammals the signal peptide is retained as supernumerary subunit 9 in the bc_1 complex (Brandt et al., 1993). The bacterial Rieske FeS-proteins, in contrast to bacterial cytochromes c_1, are not processed beyond cleavage of the N-terminal methionine (Gennis et al., 1993). Expression in *E. coli* and *Rhodobacter sphaeroides* shows that the protein is inserted into the membrane, and the FeS-cluster is formed in the absence of the other bc_1-subunits (Van Doren et al., 1993b).

Two forms of the chloroplast Rieske FeS-proteins have been detected for spinach (Yu et al., 1994), and two different cDNA clones have been found for tobacco (Madueno et al., 1992). The meaning of this heterogeneity is unclear.

E. Cytochrome b_6 and Subunit IV

The cytochrome *b* complement is the most hydrophobic part of the *bc* complexes. It harbors two hemes, and corresponds to cytochrome b_6 plus subunit IV in b_6f complexes (Hauska et al., 1988; Malkin, 1992; Knaff, 1993; Gennis et al., 1993). The two hemes are non-covalently bound and differ in their redox potentials by about 100 mV. They are designated high- and low potential cytochrome *b*, and can be spectrally distinguished in cytochrome bc_1 complexes (von Jagow and Sebald, 1980), as well as in cytochrome b_6f complexes (Nitschke et al., 1988; Hauska et al., 1989; Kramer and Crofts, 1994).

From sequence comparisons of cytochromes *b* (Saraste, 1984; Widger et al., 1984; Hauska et al., 1988; Degli Esposti et al., 1993; Gennis et al., 1993) a transmembrane organization of the cytochrome *b*-hemes has been proposed, as depicted in Fig. 4. It was decisive for the designation of the four heme-binding histidine residues in the second and fourth

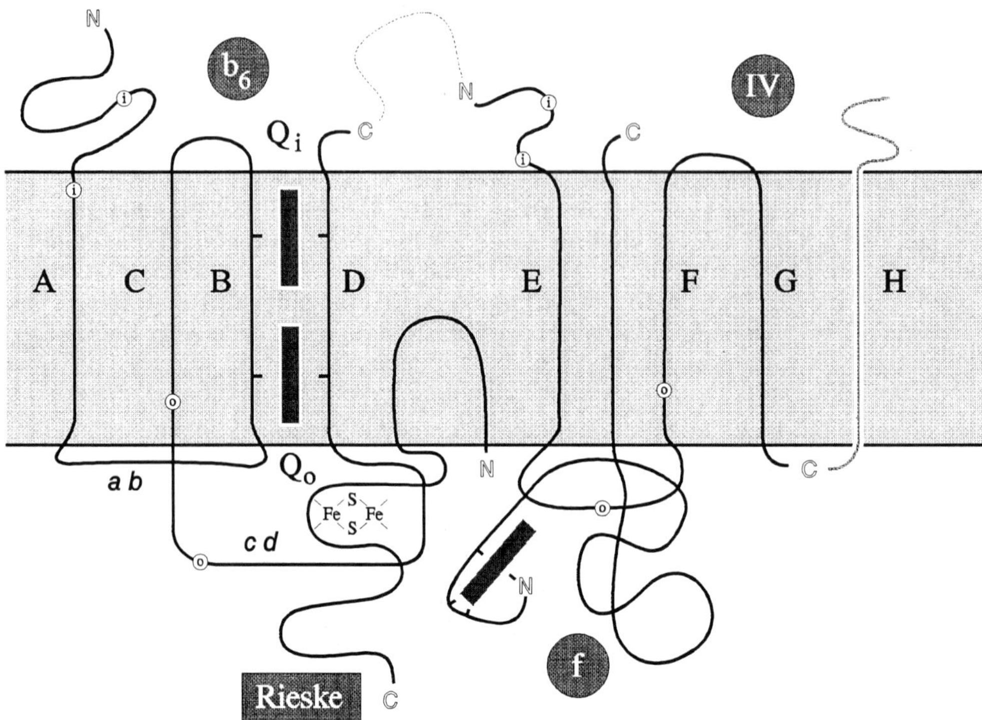

Fig. 4. Overall folding model of the cytochrome b_6f-complex in the membrane. N- and C-termini are indicated by N and C, Q_i and Q_o stand for quinone reduction and quinol oxidation sites, respectively. The 3 heme groups are represented by shaded bars. The upper part of the scheme represents the stromal surface of the membrane. The eight putative transmembrane helices are lettered A-H, the two surface helical regions 'ab' and 'cd'. Regions which fold towards Q_o or Q_i are marked by encircled 'o's' and 'i's'. The connection of cytochrome b_6 and subunit IV and the missing C-terminal part with helix H, which form a cytochrome *b* complement of bc_1-complexes are also indicated.

membrane spanning helix of cytochrome *b* that the sequence of the shorter cytochrome b_6 contains only four conserved histidines (Heinemeyer et al., 1984). Originally nine transmembrane helices were suggested for cytochrome *b*. However, the location of inhibitor resistance sites and expression experiments with fusions of cytochrome *b* truncates and alkaline phosphatase favor an arrangement with 8 transmembrane helices, which are lettered A-H in Fig. 4 (Crofts et al., 1992; Colson, 1993; Degli Esposti et al., 1993; Gennis et al., 1993). Accordingly both termini of cytochrome b_6 have been located at the stromal side of the thylakoid membrane (Szczepaniak and Cramer, 1990).

A host of cytochrome *b* sequences are known, which include a wide scope of phyla in which cytochrome *bc* complexes occur. They add up to several hundred if all the mitochondrial sequences are included. Cytochrome *b* meanwhile constitutes a marker protein for evolution (Degli Esposti et al., 1993), and at present represents the best example of a membrane protein to predict the folding structure from the helical mutability profiles through the membrane (Rees et al., 1989). Only 30 amino acid residues out of some 400 are essentially conserved (Table 1), and this number decreases further to 17 if the recently obtained sequences for *Paramecium* (Degli Esposti et al., 1993) and for *Chlorobium limicola* (Schütz et al., 1994) are included. Next to the four crucial histidines, five glycines possibly are of structural significance. The conserved aromatic residues may participate in quinone binding and/or electron transfer. In agreement with an analysis of helical hydrophobicity profiles through the membrane (Brasseur, 1988), it has been further concluded that the first four putative transmembrane helices A-D form a bundle—four conserved glycines, two each in helices A and C, providing space for the two hemes (Tron et al., 1991; Gennis et al., 1993; see also Di Rago et al., 1990b and Degli Esposti et al., 1993).

Table 1. Identical and different amino acid residues in cytochromes *b* and b_6

	Cyt *b*	b_6 + IV	*Cb. limicola*	position/function
conserved:	G33/48	37	111	A / HP-heme; Q_i-site
	G47/62	51	125	A / LP-heme
	G75/90	79	154	ab
	R79/94*	83	158	flanking B
	H82/97	86	161	B / ligand to LP-heme
	S87/102*	91	N166	B
	H96/111	100	175	B / ligand to HP-heme
	R99/114*	103	S178	flanking B
	W114/129	118	193	C
	G117/132	121	196	C / HP-heme
	G131/146	135	210	C / LP-heme; Q_o-site
	Y132/147*	136	211	C
	W142/157*	146	F221	before cd
	P155/170*	159	234	cd
	G158/173*	162	236	cd
	G167/182*	171	245	cd
	T175/190	179	253	before D
	H183/198	187	261	D / ligand to LP-heme
	P187/202*	192	266	D
	H197/212	202	276	D / ligand to HP-heme
	D229/252	35 (IV)	311	flanking/stabilizing E
	T265/288*	71	PA342	ef
	I269/292*	75	347	-"-
	PEW271/294	77	349	-"- / Q_o-site
	L282/305	88	360	-"-
	R283/306*	89	K361	-"-
	K288/311*	94	E369	flanking/stabilizing F
	G291/314	97	A372	F
different:	AF128/143	GV132	GF207	C
	S140/155	G144	A219	before cd
	T145/160	K149	Q244	cd
	V146/161	I150	V245	-"-
	- / -	T188	V262	D
	H202/217	R207	Q281	flanking D / Q_i-site

Totally conserved residues in all cytochromes *b* and b_6 + subunit IV (Hauska et al., 1988; Degli Esposti et al., 1993) are listed in the first part, and differently conserved residues in either cytochromes *b* or b_6 + subunit IV in the second. For comparison the corresponding positions in the cytochrome *b* sequence of *Chlorobium limicola* are included (Schütz et al., 1994). The position numbers for cytochrome *b* from yeast and *Rhodobacter capsulatus* are given in the first, for cytochromes b_6 + subunit IV in the second, and for *Cb. limicola* in the third column. The residues are counted including the initiator methionine and are given in single letter code. Those marked by an asterisk are essentially conserved, but differ in exceptional cases like the protozoa (Degli Esposti et al., 1993). A-F in the fourth column stand for the putative transmembrane helices A-F, ab, cd and ef for connecting loops with amphiphilic surface helices (Crofts et al., 1992; Degli Esposti et al., 1993). LP and HP denote 'low potential" and 'high-potential", Q_o and Q_i stand for quinol oxidation- and quinol reduction-sites.

This organization with the conserved four histidines and glycines is also found in the two cytochrome *b*-like heme *a* proteins from the archaebacterium *Sulfolobus acidocaldarius* (Lübben et al., 1992; see Addendum).

Table 1 also lists the few residues that differ in cytochromes *b* and b_6 plus subunit IV, but are conserved in each type. Another difference of this kind is a conserved phenylalanine in transmembrane helix B (F88 in yeast), between the two histidines in cytochrome *b* of bc_1 complexes, that was thought to be required for electron transfer between the two heme groups. It corresponds to a conserved methionine in cytochrome b_6, and this was taken as one of the indications (see below for the other one) that interheme electron transfer is not occurring in b_6f complexes (Furbacher et al., 1989). However, the phenylalanine is not completely conserved (Hauska et al., 1988; Degli Esposti et al., 1993), and site-directed replacement in *Rhodobacter sphaeroides* by isoleucine did not decrease overall electron transport (Yun et al., 1992).

Information about the contribution of individual amino acid residues to the sites of quinol/quinone interaction Q_o and Q_i (see below) in cytochrome bc_1 complexes, either from spontaneous inhibitor resistant mutants and revertants of respiratory-deficient strains (Colson, 1993), from site-directed mutants (Gennis et al., 1993), or from natural sequence variation (Degli Esposti et al., 1993) is growing rapidly. This is especially pertinent here. First, the 2D-folding of cytochrome b_6 plus subunit IV is envisioned to fold back in the third dimension, so that the amino acid residues corresponding to the inhibitor-resistances or to revertants are in close proximity with each other, as indicated in Fig. 4 by encircled 'i's' and 'o's'. Secondly, the different amino acid residues in the corresponding positions can explain why the cytochrome b_6f complex is insensitive to antimycin A and other quinone antagonists, but is sensitive to stigmatellin (Degli Esposti et al., 1993; Schütz et al., 1994). It is worth noting that the respiratory deficient point mutation G137E in yeast cytochrome *b* is suppressed by C133S, but that this revertant is resistant to mucidin, a Q_o-inhibitor (Di Rago et al., 1990a). Cytochrome b_6 carries D141 and S137 at the equivalent positions, resembling the revertant and thus explaining why the b_6f complex is mucidin insensitive. The mutant G137E is also suppressed by the more distant change N256K (Di Rago et al., 1990b; Colson, 1993), which corresponds to I62 in subunit IV (Hauska et al., 1988). To form the Q_o-pocket, helices C and F, and the interhelical loops *cd* and *ef* are in obvious topological interaction with the N-terminal regions of helices B and D, which hold the low-potential heme. It is interesting that these four regions show the highest evolutionary conservation among diverse species (Howell, 1989), possibly because the formation of the Q_o-pocket requires the interaction with the Rieske FeS-protein (Degli Esposti et al., 1993). Similarly the Q_i-pocket is formed by the N-terminal regions of the helices A and F, in addition to the C-terminal parts of B and D, which hold the high-potential heme (Fig. 4), as again suggested by inhibitor resistance and suppressor mutations (Colson, 1993). The respiration-deficient mutant S206L is suppressed by W30C (Coppée et al., 1994). The two positions correspond to I212 and C34 in cytochromes b_6 plus subunits IV (Hauska et al., 1988), again resembling the revertant.

Cytochrome b_6 and subunit IV have M_r's of about 23 and 17 kDa (Fig. 3), adding up to almost a full cytochrome *b* of 40 kDa (Fig. 4). Cytochrome b_6 ends after the 4th transmembrane helix, and contains the two heme groups. Subunit IV is essential for a functional b_6f complex, and is involved in binding plastoquinone (Li et al., 1991; Osievacz, 1992). The two most characteristic differences between cytochromes *b* of bc_1- and cytochrome b_6/subunit IV of b_6f complexes, next to the split of the protein, are the occurrence of a threonine as an additional, 14th amino acid between the 2 histidines in helix D of cytochrome b_6 (Table 1), and the lack of helix H in subunit IV (Fig. 4). Both features are found in cytochrome *b* from *Chlorobium*, although this protein is not split (Schütz et al., 1994). The additional amino acid between the histidines in the fourth transmembrane helix is a valine in this case (as it is in cytochrome b_6 from Gram-positive bacteria; see Addendum). This additional residue must influence the geometry of heme ligation (Cramer et al., 1987), and may be responsible for the lower redox potentials of the hemes in this menaquinol interacting cytochrome *b*, as well as in cytochromes b_6. Another interesting question is, how the lack of the 8th transmembrane helix that is essential for bc_1 complexes (Di Rago et al., 1993), is compensated for in b_6f complexes.

Cytochrome b_6 and subunit IV are the most hydrophobic subunits, but the interaction of the putative transmembrane, hydrophobic helices are weak. This has been concluded from their extraction at highly alkaline pH, causing electrostatic repulsion in the thylakoid membrane (Szczepaniak et al., 1991).

Cytochromes *b* are not processed from precursor forms during insertion into the membranes; only the N-terminal methionines are cleaved off.

RNA editing has been documented for the cytochrome *b* gene in protozoan mitochondria (Feagin et al., 1988) and also for the *pet*B gene of cytochrome b_6 in maize (Freyer et al., 1993). The phenomenon is not yet understood.

F. Overall Structure

A scheme for a transmembrane, monomeric arrangement of the 3 redox center-carrying polypeptides and subunit IV of the cytochrome b_6f complex is shown in Fig. 4. A high resolution 3D-structure of a cytochrome *bc* complex is still lacking, in spite of the repeated achievement of crystallization (Yu and Yu, 1993). So far only the structure of cytochrome *f* has been obtained at atomic resolution (Martinez et al., 1994; see Chapter 22).

Nearest neighbor relations of the subunits have been studied by crosslinking (Lam, 1986; Bhagvat et al., 1993). Furthermore, a point mutation of N256Y, in the vicinity of myxothiazol binding at the Q_o-site of yeast cytochrome *b*, loosens the Rieske FeS-subunit (Geier et al, 1992). Thus the loop between helices E and F (Fig. 4) is involved in binding this subunit. The amphipathic helix *cd* (Fig. 4) also seems to interact with it, as well as with cytochrome c_1. Evidence for this is that the cytochrome *b*-mutant T163F of *Rhodobacter capsulatus*, which does not assemble a bc_1 complex, is suppressed by the change A46T in the Rieske FeS-protein, or by R46C in cytochrome c_1 (Tokito et al., 1992).

In the isolated state the cytochrome b_6f complexes (Chain and Malkin, 1991; Huang et al., 1994), as well as the cytochrome bc_1 complexes appear to be dimeric. Whether this is of functional significance is still a matter of debate (Hope, 1993; Cramer et al., 1994).

V. Function

The cytochrome b_6f complex functions as plastoquinol-plastocyanin oxidoreductase in photosynthesis of chloroplasts (Fig. 1), similar to the function of the cytochrome bc_1 complex as ubiquinol-cytochrome *c* oxidoreductase in the respiration of mitochondria (Hauska et al., 1983; Cramer et al., 1991; Hope, 1993). In some algae the acceptor protein can be cytochrome *c* instead of plastocyanin, either alternatively or exclusively (Scherer, 1990). In prokaryotes these quinol oxidoreductases may function in photosynthesis as well as in respiration (Fig. 2). The question of a respiratory function in chloroplasts (chlororespiration), possibly involving the b_6f complex, is not settled (Singh et al., 1992; Berger et al., 1993). The activities of a NAD(P)H-plastoquinone oxidoreductase and of a plastoquinol oxidase are negligible in mature chloroplasts from higher plants. However, a triggering by light, or a transient function of such activities during plastid differentiation cannot be excluded.

The immediate reductant of the acceptor proteins, cytochrome *c* or plastocyanin, is cytochrome c_1 or *f*. The interaction is largely governed by electrostatic forces (Güner et al., 1993; see Chapters 21 and 22). Electron transfer from cytochrome *f* to plastocyanin is inhibited when they are crosslinked (Quin and Kostic, 1993), and it was concluded that the mutual orientation changes during electron transfer.

Cytochrome c_1 is rereduced by electrons from the Rieske FeS-center. For *Rhodobacter sphaeroides* this conclusion is based on flash experiments, in which the effect of specific inhibitors on the extent of photooxidation of cytochromes c_2 and c_1 was studied (Meinhardt and Crofts, 1982). A corresponding demonstration of electron transfer from the Rieske FeS-center to cytochrome *f* in chloroplasts is still lacking. Cytochrome *f* is photooxidized by plastocyanin/P700 with no delay. To explain these results, an operational redox potential of 370 mV for the Rieske FeS-center, when it is to reduce cytochrome *f*, was assumed (Rich et al., 1987; Hope, 1993), at least 50 mV higher than when it is to oxidize plastoquinol (as measured by oxidant-induced reduction of cytochrome b_6 (Prince et al., 1982)). The latter is close to the values obtained by redox titrations (Nitschke at al., 1992). A corresponding conformational change of the Rieske FeS-center may thus occur during turnover. Such a catalytic switch has indeed been postulated for mitochondria (Brandt and von Jagow, 1991; Brandt et al., 1992).

The turnover numbers of carefully isolated, pure cytochrome bc_1 complexes, from mitochondria and bacteria (Ljungdahl et al., 1986; Andrews et al., 1989; Robertson et al., 1993), are about 10 times higher (over 1000 s^{-1}) than those reported for cytochrome b_6f-preparations (Hauska, 1986). Whether this is an innate difference, or whether it solely reflects inactivation of the cytochrome b_6f preparation during isolation has not been resolved. The rate limitation of electron transport in chloroplasts is around 5 ms (Witt, 1971), which sets the limit for the intrinsic turnover of the cytochrome b_6f complex at 200 s^{-1}. A difficulty in comparing the rates in vitro and in membrano arises from the fact that the interaction conditions of hydrophobic substrates with the micellar state of isolates differ fundamentally from the conditions in membranes (Rich, 1984).

In addition to electrons, cytochrome *bc* complexes transfer protons through the membranes, and this is intimately linked to the redox chemistry of the quinone system.

A. Redox Potentials

The four metal centers in cytochrome bc_1/b_6f complexes operate in two pairs, two of them at relatively high, and two at relatively low redox potential (Fig. 5). The Rieske-protein carries a 2Fe2S-center with a high redox potential (about +300 mV) and is the immediate oxidant of quinol. It transfers

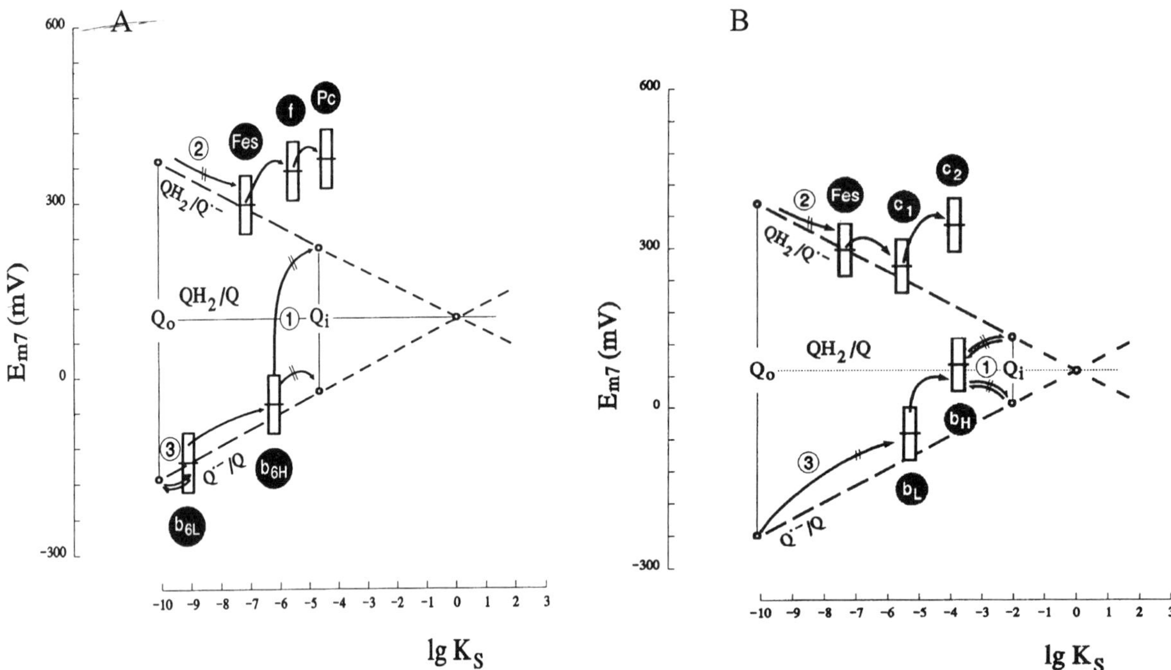

Fig. 5. Stability constant of the semiquinone and redox potentials in cytochrome *bc*-complexes. The stability constant (Ks) is given on a log (lg)-scale. Other abbreviations are as for the previous figures. A) represents the plastoquinol/b_6f-system, and B) the ubiquinol/bc_1-system. For the 4 redox centers of *bc*-complexes the redox range from 10–90% oxidation is indicated by bars. The Ks at the Q_o-site equals the value of the unbound form, for Q_i a substantially higher value was assumed, reflecting stabilization of the semiquinone by binding at this site. Three modes of inhibition are shown by numbers.

the electron to the heme *c* of cytochrome c_1 (+240 mV), and probably also of cytochrome *f* (+340 mV). The two hemes *b* of cytochrome *b* or b_6 have comparatively lower redox potentials positioned at about +90 and –50 mV for bc_1-, and about –50 and –150 mV for b_6f complexes (Fig. 5a,b). The *b*-hemes oxidize the semiquinone formed from quinol after the 1st electron transfer. The two values for cytochrome b_6 were originally obtained for the isolated b_6f complex (Hauska et al., 1983), but were recently confirmed for chloroplasts (Kramer and Crofts, 1994), in contradiction to earlier measurements (Cramer et al., 1987). The semiquinone is an unstable species over a wide pH-range, and tends to lose, as well as to take up an electron (Rich, 1984). Therefore, in this range, the redox potential of quinol/semiquinone is more positive than that of semiquinone/quinone (Fig. 5). At physiological pH, the redox potentials of the two half reactions of the quinone system in a lipid environment come close to the FeS-center plus cytochrome c_1 or *f* on one hand, and to the low-potential cytochrome b/b_6 on the other. This explains a seemingly enigmatic behavior of the cytochrome *bc* complexes. When an electron is withdrawn from cytochrome c_1 or *f*, by oxygen via the oxidase, or by light through Photosystem I, or even by addition of ferricyanide, cytochrome *b* or b_6 becomes reduced. This so-called oxidant-induced reduction of cytochrome *b* was explained, first by Wikström (1973), as a reflection of the concerted oxidation of quinol via the semiquinone intermediate. The two high potential redox centers first oxidize quinol to semiquinone, which in turn reduces the two low potential centers. In this process two protons are translocated per electron transfer through the quinol/cytochrome *bc*-system, twice that expected for quinol alone.

B. The Q-Cycle

The original explanation of oxidant-induced reduction by Wikström was carried further by Mitchell's formulation of the Q-cycle (1976), which also provides for the observed H^+/e^- ratio of 2 during one turnover of cytochrome *bc* complexes. He conceived of two sites for quinone interaction, one for the concerted quinol oxidation, called Q_o ('o' for oxidation or outside) or Q_p ('p' for positive), with a low stability of the semiquinone intermediate, and one for quinone reduction which stabilizes the intermediate semiquinone, called Q_i ('i' for inside), Q_r ('r' for reduction), or Q_n ('n' for negative). For the interplay of these two quinone reaction sites with different semiquinone stability the term 'redox seesaw' was coined. Fig. 5A shows this for the plastoquinone/b_6f-system and Fig. 5B for the ubiquinone/bc_1-system. The redox potential difference between the first and second electron transfer to the quinone reflects the stability of the intermediate semiquinone. The relation is described by the equation $(E1 - E2) = RT/nF \times \ln Ks$. In Fig. 5 it is assumed that the semiquinone is not stabilized at the site of quinol oxidation (Q_o), the stability constant being the same as in lipid solution.

For the b_6f complex and plastoquinol, the concerted reaction at Q_o is close to equilibrium (Kramer and Crofts, 1993), because the redox potentials of the FeS-center and of b_{6L} both match one of the electron transfers from the quinol-system (Fig. 5A). Consequently the semiquinone cannot be well stabilized at the Q_o-site in the b_6f complex, otherwise it could not reduce cytochrome b_{6L}. This is not so stringent for the oxidation of ubiquinol by the bc_1 complex, where the redox potential of cytochrome b_L is clearly above the one for $Q^{\cdot -}/Q$ at the Q_o-site (Fig. 5B). Rather cytochrome b_H and the Q_i-site are in equilibrium. Indeed, cytochrome b also can be reduced at this site by ubiquinol (Link et al., 1993), while cytochrome b_6 can only be reduced by plastoquinol via Q_o. This difference in the redox potential relations of the b_6f- and the bc_1-system cannot yet be rationalized.

The semiquinone intermediate at the Q_i-site is rather stable in bc_1 complexes, and can be detected by EPR (Robertson et al., 1984, 1993). For the b_6f-system $Q^{\cdot -}$ seems to be less stabilized (Fig. 5A), the documentation of any semiquinone being much more difficult (Hope, 1993). This difference at Q_i has been studied by site directed mutations of cytochrome b in *Rhodobacter*. The histidine residue H217, which is close to the ligand H212 of the high-potential heme in helix D, and corresponds to R207 in cytochrome b_6 (Table 1), is involved in forming the Q_i-pocket. More specifically, the mutants H217A (Hacker et al., 1993) and H217L do not grow photosynthetically, but the mutants H217D and H217R are photosynthetically competent (Gray et al., 1994). While H217L and H217D show drastically decreased semiquinone stability, H217R, in contrast to the expectation from an R in b_6, shows an increase in stability. Therefore, the decreased stability of $Q_i^{\cdot -}$ in the cytochrome b_6f complex is not caused by the exchange of H for R at the Q_i-site, and the structural difference must be more complex. Actually it has been shown that the R in cytochrome b_6 is accessible to trypsin, and thus the vicinity of Q_i in b_6f complexes must be exposed to the aqueous surface (Szczepaniak and Cramer, 1990), while it is thought to be buried in the membrane in bc_1 complexes (Degli Esposti et al., 1993; Cramer et al., 1994).

The fact that the Q_o-site in b_6f complexes operates close to equilibrium (Fig. 5A) accounts for the low extent of observable oxidant-induced reduction of cytochrome b_6 (Kramer and Crofts, 1993). The extent of reduction is less than one of the two hemes, the low-potential heme not being reduced unless the high-potential heme has been previously reduced (Nitschke et al., 1988). This has even led to the conclusion that interheme electron transfer in cytochrome b_6 does not occur at all (Furbacher et al., 1989), a conclusion that has been opposed by Kramer and Crofts, 1994).

The original formulation of the Q-cycle also viewed the reaction at the Q_i-site as concerted, with two different electron donors to quinone (Mitchell, 1976). The variant presented in Figs. 5 and 6 holds that only the oxidation of quinol at the Q_o-site occurs in a concerted transfer of two electrons, while reduction of the quinone at the Q_i-site occurs in two consecutive steps (Gennis et al., 1993). For clarity the full turnover of the Q-cycle is divided into two halves in Fig. 6A and B. In the first half QH_2 is formed from $Q_B^{\cdot}$ of Photosystem II, in the second from $Q_i^{\cdot -}$. In both cases two protons are bound at the stroma surface, and QH_2 from the quinone pool is oxidized at the Q_o-site on the opposite surface, by the Rieske FeS-center, to the semiquinone anion plus two protons. The semiquinone anion at this site then reduces cytochrome b_{6L} that transfers the electron through the membrane, via cytochrome b_{6H} to the quinone reduction site Q_i, which stabilizes the resulting semiquinone anion. The net effect is the translocation of four protons through the membrane per two electrons. The Q-cycle thus represents a fascinating example of how a leading reaction triggers the formation of a reactive intermediate that is channeled to work by the molecular machinery. The interactions of Q at Q_B and Q_i, and of QH_2 at Q_o from the pool follow second order kinetics (Crofts and Wraight, 1983). The transmembrane electron and proton transfer steps contribute to membrane potential formation, but are slow in the *bc* complexes (Semenov, 1993; Hope, 1993), compared to reaction centers.

The two sites of quinone interaction in the cytochrome *bc* complexes operate independently from one another, as shown convincingly for bc_1 complexes by specific inhibitors and by site directed mutants (Link et al., 1993; Gennis et al., 1993). This excludes alternative mechanisms that postulate only one site of quinone interaction leading to the assumption of different states during turnover, or of two mutually dependent sites. These alternative mechanisms are still discussed for b_6f complexes (Hope, 1993), because of two failures. Sufficiently site specific inhibitors have been lacking (see below), and in contrast to the bc_1 complex, the Q_i-site is not reversible and cannot be studied independently from the Q_o-site (see Fig. 5A in comparison to Fig. 5B). In view of the structural similarities, it would be a surprise if the mechanism of electron transfer were basically different in the two types of complexes.

The Q-cycle mechanism was also formulated for a dimer of the complexes (de Vries et al., 1983), and their structure could be dimeric indeed (see above). Careful titrations with tightly binding inhibitors are contradictory, as discussed by Hope (1993). Some results favor one inhibitor molecule per dimer, others favor one molecule per monomer as a requirement for inhibition. The latter can be interpreted that each monomeric half per se is active, and that the dimeric state is of structural rather than functional importance. The issue remains open. Furthermore, it may be necessary to extend the Q-cycle to involve a 'catalytic switch' at the Q_o-site (Brandt and von Jagow, 1991; Brandt et al., 1992; Hope, 1993).

The Q-cycle with an H^+/e^- ratio of 2 is permanently operative in the cytochrome bc_1 complexes, but there is no agreement on whether this efficiency is maintained under all conditions for the cytochrome b_6f complexes. Evidence and arguments for and against a decreased efficiency at high light intensity (= high proton potential across the membrane) have been put forward (Hope, 1993). The Q-cycle and its possible down regulation affect the estimations of how photosynthetic electron flow meets the ATP requirement in chloroplasts. The P/e_2 in photoreduction of $NADP^+$ would suffice only if the Q-cycle contributes. If it does not, the shortfall in ATP must be complemented either by cyclic electron flow, or by reuptake of oxygen via Photosystem I (Mehler reaction). The need for these complementing reactions would be even higher in C_4-plants, which require an extra ATP per fixed CO_2.

C. Inhibitors

Three modes of inhibition are discerned for electron transfer in cytochrome *bc* complexes. These were very valuable in elucidating the redox mechanism (Link et al., 1993). They are indicated by numbers in Figs. 5 and 6. Mode 1 blocks the reduction of the quinone at Q_i. As a consequence the oxidant-induced reduction of cytochrome *b* is stimulated. Mode 2 blocks electron transfer to the Rieske FeS-center, while mode 3 inhibits reduction of cytochrome b_L. Mode 1 is exerted by antimycin A, mode 2 by stigmatellin or UHDBT, and mode 3 by myxothiazol in the ubiquinol/cytochrome bc_1-system, all very efficiently. Stigmatellin actually may combine mode 2 and 3, because it also shifts the absorption of cytochrome *b*, while UHDBT does not. Unfortunately antimycin and myxothiazol do not inhibit the plastoquinol/b_6f-system, a fact that hampered the study of this system substantially. Only stigmatellin (Nitschke et al., 1989) and UHDBT (Hurt and Hauska, 1981; Riedel et al., 1991) are universally effective, both acting on the Rieske FeS-center. Interestingly MOA-stilbene, a compound related to myxothiazol, inhibits the cytochrome b_6f complex, but by mode 1, and not by mode 3 (Rich et al., 1992). Most of these inhibitory compounds in part resemble the structure of the isoprenoid quinones, and it is therefore not surprising that they are not completely specific for only one quinone interaction site, and that the relative specificity for Q_o and Q_i may be changed in b_6f compared to bc_1. In accord with this, many of these inhibitors also block the Q_B-site in the reaction centers of bacteria, or of Photosystem II (Oettmeier et al., 1985; Degli Esposti et al., 1993). MOA-stilbene may prove to be very valuable in the study of b_6f complexes, for which only alkyl-quinoline N-oxides have so far been known to act as mode 1-inhibitors. Most recent and surprising is the finding in this context, that valinomycin/K^+ also blocks electron transfer through the b_6f complex by mode 1, and does so additively to MOA-stilbene (Klughammer and Schreiber, 1993). Up to 75% of the two hemes became reduced during oxidant-induced reduction in the presence of both inhibitors, while less than half of them were reduced in the presence of only one inhibitor. The thermodynamic equilibrium at the Q_o-site (Kramer and Crofts, 1993) seems to be changed by valinomycin/MOA-stilbene.

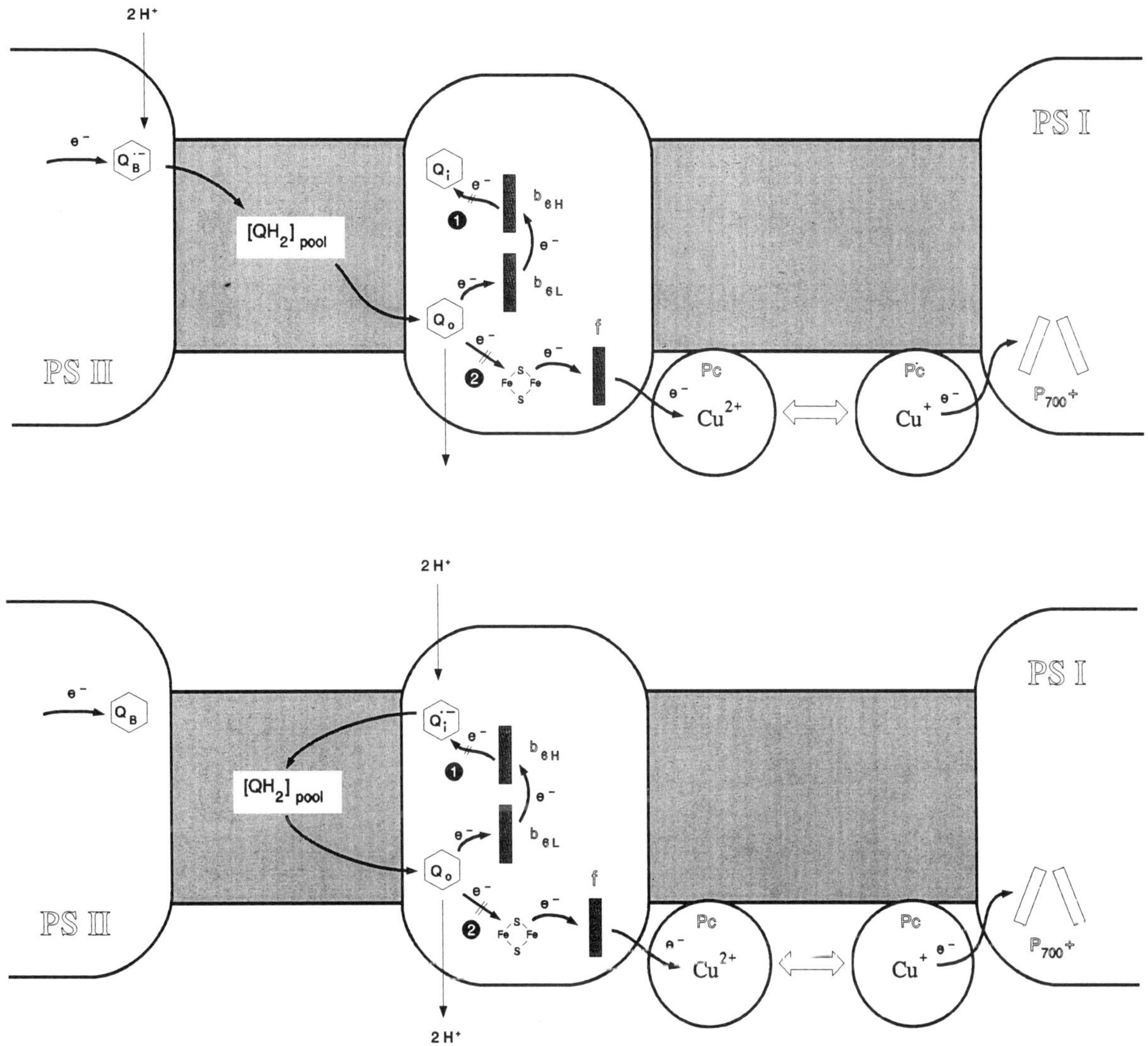

Fig. 6. The Q-cycle of chloroplasts. The schemes show the reaction centers of Photosystem II on the left and of Photosystem I on the right of the cytochrome b_6f-complex, connected by the plastoquinone pool and plastocyanin, respectively. A and B represent the 1st and 2nd half of a complete turnover of two electrons. The quinone binding sites are indicated by hexagons, and by the notations Q_B for the secondary quinone acceptor in reaction center II, and by Q_o and Q_i for the quinone oxidizing- and the quinol reducing site in the b_6f-complex, respectively. Tetraporphyrins of the hemes and the chlorophyll dimer of P700 are indicated by bars. Pc, f, b_{6L} and b_{6H} stand for plastocyanin, cytochrome *f*, and for the low- and high potential hemes of cytochrome b_6, respectively. The Rieske FeS-center is indicated by the 2Fe2S-cluster. Two modes of inhibition are shown by numbers on black circles.

Proton translocation by the cytochrome b_6f complex is inhibited by DCCD, which binds to acidic residues in the parallel helix *cd* (Fig. 4)(Wang and Beattie, 1992). The path of protons through cytochrome b/b_6 has been discussed by Beattie (1993).

VI. Regulation

Since quinol oxidation is a relatively slow reaction and often rate determining in electron transport chains (Witt, 1971), the cytochrome *bc* complexes may be involved in metabolic regulation. The possibility that they generally function as redox sensors in chloroplasts and bacteria has been discussed by Allen (1993). Moreover, the core proteins of the mitochondrial cytochrome bc_1 complexes are involved in the processing of proteins targeted for import (Braun et al., 1992), as mentioned above. A connection between this function and the electron transfer activity is not yet obvious.

It should be noted in this context that photosynthesis is efficiently shut off during gametogenesis of *Chlamydomonas reinhardtii* by specific down regulation of the cytochrome b_6f complex (Bulté and Wollmann, 1992).

Good evidence exists for the cytochrome b_6f complex of chloroplasts to act as a redox sensor, in addition to its function in energy conserving electron transport. It regulates the distribution of light excitation between Photosystem I and II (Allen, 1992). A specific kinase for the light-harvesting chlorophyll *a/b*-protein of Photosystem II (LHC-II) is activated when the plastoquinone pool is reduced (Fig. 7). The phosphorylated form of LHC-II is less closely associated with Photosystem II. Photosystem I then receives more light and reoxidizes the quinone pool. Studies with inhibitors and with mutants (Gal et al., 1987, 1990; Allen, 1992) indicated that plastoquinol is not directly activating the kinase. The degree of plastoquinone reduction is sensed by the b_6f complex, and the resulting redox- or conformational change is transferred to the kinase in some manner. Could this be linked to the 'catalytic switch' at the Q_o-site (Brandt and von Jagow, 1991; Brandt et al., 1992)? Intriguing as well is the idea that signaling via the b_6f complex required receptor-like dimerization (Cramer et al., 1994).

VIII. Conclusions and Open Questions

At the present time the following conclusions can be drawn:

1) Like all the cytochrome *bc* complexes, the cytochrome b_6f complex contains four metal redox centers on three membrane proteins—cytochrome *f*, the Rieske FeS-protein, and the double-heme protein cytochrome b_6.

2) It is plausible to assume that for quinol oxidation in the cytochrome b_6f complex the four redox centers cooperate in a universal mechanism called 'Q-cycle'. This involves a site for quinol oxidation (Q_o) and a site for quinone reduction (Q_i), which operate independently.

3) This mechanism translocates two protons through the membrane per electron transferred, double the amount expected for quinol oxidation alone.

4) The minimum structure for a functional cytochrome *bc* complex requires the three redox proteins only, as realized in bacteria. For these proteins numerous primary structures are now known. Their folding and interactions in the membrane show universal features. It is reassuring that the sequence comparison of cytochromes b/b_6 independently predicted a transmembrane orientation of the two hemes, as was postulated by the Q-cycle. Cytochrome b_6f complexes correspond to this minimal structure, with their cytochrome *b*-complement split into cytochrome b_6 and subunit IV, but also contain several very small, hydrophobic polypeptides.

5) The classification into bc_1- and b_6f-types, based on differential inhibitor sensitivities and characteristic features in the sequences, such as the split into cytochrome b_6 and subunit IV, has become less precise, since in *Chlorobium* the gene for cytochrome *b* is continuous but otherwise reveals the salient b_6f-type characteristics. During evolution the split must have occurred after the development of these characteristics.

The following central questions have yet to be solved:

1) Is the tendency of the isolated complexes to form dimers of functional, or only structural significance?

2) What is the role of the small, hydrophobic polypeptides?

3) Is the electron transfer mechanism universal indeed? Does it include a catalytic switch at the site for quinol oxidation affecting the Rieske FeS-center?

4) Why is the b_6f complex in contrast to the bc_1

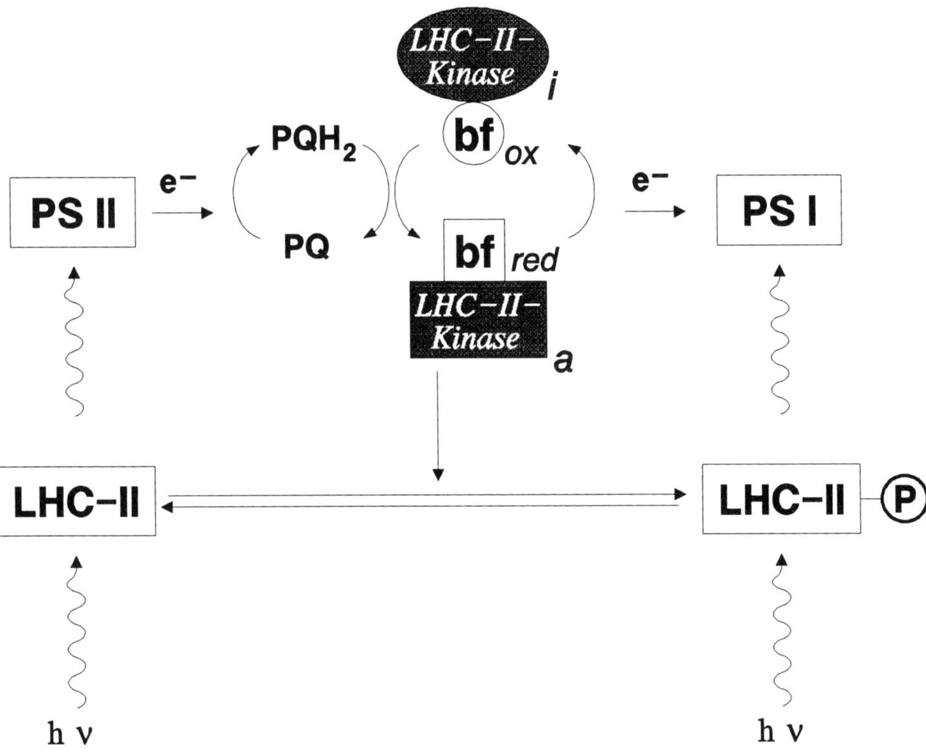

Fig. 7. Activation of the LHCII-Kinase by the cytochrome b_6f-complex. PS I, PS II, LHC-II, bf and PQ stand for Photosystem I and Photosystem II, the light harvesting complex II (chlorophyll *a,b*-protein complex), cytochrome b_6f-complex and plastoquinone, respectively. The suffixes i and a stand for 'inactive' and 'active'.

complex operating so close to equilibrium at the quinol oxidation site? Is the Q-cycle always operating in chloroplasts, or is it regulated down at high light intensity?

5) Are the predictions for the three redox proteins indeed valid for the 3D-structures in the membrane?

Undoubtedly the further elucidation of the cytochrome b_6f complexes, and of the cytochrome *bc* complexes in general, will remain in the center of bioenergetic interest for some time to come. Much is expected from atomic resolution of the 3D-structure, for which a start has been made (Martinez et al., 1994; see Chapter 22). Another promising approach has been initiated recently in the attempt to de-novo-synthesize multi-heme proteins which resemble cytochrome *b* (Robertson et al., 1994).

Acknowledgments

Work carried out in Regensburg has been supported by the Deutsche Forschungsgemeinschaft (SFB 43 C2). We are grateful to U. Brandt, T. Link and G. von Jagow/Frankfurt, B. Barquera, R. B. Gennis, D. M. Kramer and A. R. Crofts/Urbana, J. Castresana, M. Lübben and M. Saraste/ Heidelberg, W. A. Cramer/ W. Lafayette, F. Daldal and P. L. Dutton/Philadelphia, M. Degli Esposti/Bologna, J. C. Gray/Cambridge, A. B. Hope/ Adelaide, W. Nitschke/Freiburg, P. R. Rich/Bodmin, for access to unpublished work and/or stimulating discussions.

Note in Proof

Pertinent new sequence information comes from two groups.

1) Sone et al. (1995) report on a split cytochrome *b* gene for *Bacillus stearothermophilus*, which resembles the two adjacent genes for cytochrome b_6 and subunit IV of cyanobacteria and chloroplasts in the two key features discussed above—14 instead of 13 amino acids between the two histidines in transmembrane helix D, and the lack of the last hydrophobic span in subunit IV. Thus, Gram-positive bacteria indeed contain a b_6f-type complex, as suggested previously on biochemical grounds (Kutoh and Sone, 1988).

2) The first archaebacterial sequences for a Rieske FeS-protein and two cytochrome *b*-like heme *a*-proteins, together with the ones for a blue copper protein and subunits of two cytochrome oxidases, all for *Sulfolobus acidocaldarius*, have been added to the discussion (Castresana et al., 1995; Lübben, 1995). The sequence for the Rieske FeS-protein shows the two conserved, FeS-cluster binding peptides. The first reads CVHLGC instead of CTHLGC, the second remains CPHCG. They are spaced, however, by an extended stretch of amino acids. The two heme *a*-proteins show the signatures for a transmembrane arrangement like cytochrome *b* of *bc* complexes. Thirteen of the 16 universally conserved residues are present, including the four histidines and four glycines in the first four hydrophobic spans (Table 1). The conserved tripeptide PEW between the fifth and sixth transmembrane helices reads PPW and PDW, respectively. Both genes are continuous, with a C-terminal extension harboring a total of 10 or 12 putative transmembrane helices instead of eight. The two histidines in the fourth transmembrane helix are spaced by 13 amino acids, characteristic for the bc_1-type. The authors draw the most important conclusion, that a complex respiratory chain with a bc_1-type complex must have been present in the common ancestor to all living organisms, and that therefore respiration preceeded chlorophyll-based photosynthesis (Castresana et al., 1995; Lübben, 1995).

References

Albertsson P-A, Andreasson E, Svensson P and Yu S-G (1991) Localization of cytochrome f in the thylakoid membrane: Evidence for multiple domains. Biochim Biophys Acta 1098: 91–94

Allen JF (1992) Protein phosphorylation in regulation of photosynthesis. Biochim Biophys Acta 1098: 275–335

Allen JF (1993) Redox control of gene expression and function of chloroplast genomes - an hypothesis. Photosynth Res 36: 95–102

Alt J, Westhoff P, Sears BB, Nelson N, Hurt E, Hauska G and Herrmann RG (1983) Genes and transcripts for the polypeptides of the cytochrome b_6f complex from spinach thylakoid membranes. EMBO J 2: 979–986

Anderson JM (1992) Cytochrome b_6f complex: Dynamic molecular organization, function and acclimation. Photosynth Res 34: 341–357

Andrews KM, Crofts AR and Gennis RB (1989) Large scale purification and characterization of a highly active four-subunit cytochrome bc_1 complex from *Rhodobacter sphaeroides*. Biochemistry 29: 2645–2651

Anemüller S, Schmidt CL, Schäfer G and Teixeira M (1993) Evidence for a Rieske-type FeS center in the thermoacidophilic archaebacterium *Sulfolobus acidocaldarius*. FEBS Lett 318: 61–64

Baccarini Melandri A and Zannoni D (1978) Photosynthetic and respiratory electron flow in the dual functional membrane of facultative photosynthetic bacteria. J Bioenerg Biomemb 10: 109–138

Beattie DS (1993) Minireview series: The cytochrome bc_1-complex. J Bioenerg Biomemb 25: 191–273

Berger S, Ellersiek U, Westhoff P and Steinmüller K (1993) Studies on the expression of NDH-H, a subunit of the NAD(P)H-plastoquinone-oxidoreductase of higher plant chloroplasts. Planta 190: 25–31

Bhagvat AS, Blokesch A, Irrgang K-D, Salnikow J and Vater J (1993) Crosslinking of components of the cytochrome b_6f complex from spinach thylakoids with o-phthalaldehyde. Arch Biochem Biophys 304: 38–44

Black MT, Widger WR and Cramer WA (1987) Large scale purification of active cytochrome b_6f complex from spinach thylakoids. Arch Biochem Biophys 252: 655–661

Brandt U and von Jagow G (1991) Analysis of inhibitor binding to mitochondrial cytochrome *c* reductase by fluorescence quench titration. Evidence for a 'catalytic switch' at the Q_o center. Eur J Biochem 195: 163–170

Brandt U, Haase U, Schägger H and von Jagow G (1992) Significance of the 'Rieske' iron-sulfur protein for formation of the ubiquinol-oxidation pocket of mitochondrial cytochrome *c* reductase (bc_1 complex). J Biol Chem 266: 19958–19964

Brandt U, Yu L, Yu C-A and Trumpower BL (1993) The mitochondrial presequence of the iron-sulfur protein is processed in a single step after insertion into the cytochrome bc_1 complex in mammals and is retained as a subunit in the complex. J Biol Chem 269: 7603–7609

Brasseur R (1988) Calculation of the 3D-structure of *Saccharomyces cerevisiae* cytochrome *b* inserted in a lipid matrix. J Biol Chem 263: 12571–12575

Braun HP, Emmermann M, Kruft V and Schmitz UK (1992) The

general mitochondrial processing peptidase from potato is an integral part of the cytochrome *c* reductase of the respiratory chain. EMBO J 11: 3219–3227

Breyton C, de Vitry C and Popot J-L (1994) Membrane association of cytochrome b_6f subunits. The Rieske iron-sulfur protein from *Chlamydomonas reinhardtii* is an extrinsic protein. J Biol Chem 269: 7597–7602

Britt RD, Sauer K, Klein MP, Knaff DB, Kriauciunas A, Yu C-A, Yu L and Malkin R (1991) Electron spin echo modulation spectroscopy supports the suggested coordination of two histidine ligands to the Rieske Fe-S centers of the cytochrome b_6f complex of spinach and the cytochrome bc_1 complexes of *Rhodospirillum rubrum* and *Rhodobacter sphaeroides* R-26, and of beef heart mitochondria. J Biol Chem 266: 17838–17844

Broda E and Peschek GA (1979) Did respiration or photosynthesis come first? J Theor Biol 81: 201–212

Bulté L and Wollmann F-A (1992) Evidence for a selective destabilization of an integral membrane protein, the cytochrome b_6f complex, during gametogenesis of *Chlamydomonas reinhardtii*. Eur J Biochem 204: 327–336

Büschlen S, Choquet Y, Kuras R and Wollmann F-A (1991) Nucleotide sequence of the continuous and separated *petA*, *petB* and *petD* chloroplast genes in *Chlamydomonas reinhardtii*. FEBS Lett 284: 257–262

Castresana J, Lübben M and Saraste M (1995) New archaebacterial genes coding for redox proteins: Implications for the evolution of aerobic metabolism. J Mol Biol 250: 202–210

Chain RK and Malkin R (1991). The chloroplast cytochrome b_6f complex can exist in monomeric and dimeric states. Photosynth Res 28: 59–68

Chitnis VP and Chitnis PR (1993) PsaL subunit is required for the formation of Photosystem I trimers in the cyanobacterium *Synechocystis* sp. PCC 6803. FEBS Lett 336: 330–334

Colson A-M (1993) Random mutant generation and its utility in uncovering structural and functional features of cytochrome *b* in *Saccharomyces cerevisiae*. J Bioenerg Biomemb 25: 211–220

Coppée J-Y, Brasseur G, Brivet Chevilotte P and Colson A-M (1994) Non-native intragenic reversions selected from *Saccharomyces cerevisiae* cytochrome *b*-deficient mutants. Structural and functional features of the catalytic center N domain. J Biol Chem 268: 4221–4226

Cramer WA, Black MT, Widger WR and Girvin ME (1987) Structure and function of photosynthetic cytochrome bc_1 and *b6f* complexes. In: Barber J (ed) The Light Reactions, pp 447–493. Elsevier Science Publishers B.V., Amsterdam

Cramer WA, Furbacher DN, Szepaniak A and Tae G-S (1991) Electron transport between Photosystem II and Photosystem I. In: Lee CP (ed) Current Topics in Bioenergetics, Vol 16, pp 180–222. Academic Press, New York

Cramer WA, Martinez SE, Huang D, Tae G-S, Everly RM, Heymann JB, Cheng RH, Baker TS and Smith JL (1994) Structural aspects of the cytochrome *b6f* complex; structure of the lumen-side domain of cytochrome *f* J Bioenerg Biomemb 26: 31–47

Crofts AR and Wraight CA (1983) The electrochemical domain of photosynthesis. Biochim Biophys Acta 726: 149–185

Crofts A, Hacker B, Barquera B, C.H Yun and Gennis R (1992) Structure and function of the *bc*-complex of *Rhodobacter sphaeroides*. Biochim Biophys Acta 1101: 162–165

Davidson E, Ohnishi T, Atta-Asafo-Adjei E and Daldal F (1992) Potential ligands to the (2Fe-2S) Rieske cluster of the cytochrome bc_1 complex of *Rhodobacter capsulatus*. Biochemistry 31: 3342–3351

Degli Esposti M, DeVries S, Crimi M, Ghelli A, Patarnello T and Meyer A (1993) Mitochondrial cytochrome b: Evolution and structure of the protein. Biochim Biophys Acta 1143: 243–271

de Vitry C (1994) Characterization of the gene of the chloroplast Rieske iron-sulfur protein in *Chlamydomonas reinhardtii*. J Biol Chem 269: 7603–7609

de Vries S, Albracht SPJ, Berden JA, Marres CAM and Slater EC (1983) The effect of pH, ubiquinone depletion and myxothiazol on the reduction kinetics of the prosthetic groups of ubiquinol:cytochrome *c* oxidoreductase. Biochim Biophys Acta 723: 91–103

Ding H, Robertson DE, Daldal F and Dutton PL (1992) Cytochrome bc_1 complex (2Fe2S) cluster and its interaction with the ubiquinone and ubihydroquinone at the Q_o site: A double occupancy Q_o site model. Biochemistry 31: 3144–3158

Di Rago J-P, Netter P and Slonimsky PP (1990a) Pseudo-wildtype revertants from inactive apocytochrome *b* mutants as a tool for the analysis of the structure/function relationships of the mitochondrial ubiquinone-cytochrome *c* reductase of *Saccharomyces cerevisiae*. J Biol Chem 265: 3332–3339

Di Rago J-P, Netter P and Slonimsky PP (1990b) Intragenic suppressors reveal long distance interactions between inactivating and reactivating amino acid replacements generating three-dimensional constraints in the structure of mitochondrial cytochrome *b*. J Biol Chem 265: 15750–15757

Di Rago J-P, Macadre C, Lazowska J and Slonimsky PP (1993) The C-terminal domain of yeast cytochrome *b* is essential for a correct assembly of the mitochondrial cytochrome bc_1 complex. FEBS Lett 328: 153–158

Feagin JE, Shaw JM, Simpson L and Stuart K (1988) Creation of AUG initiation codons by addition of uridines within cytochrome *b* transcripts of kinetoplastids. Proc Natl Acad Sci USA 85: 539–543

Freyer R, Hoch B, Neckermann K, Maier RM and Kössel H (1993) RNA editing in *petB* of maize. Plant Journal 4: 621–629

Furbacher PN, Girvin ME and Cramer WA (1989) On the question of interheme electron transfer in the chloroplast cytochrome b_6 in situ. Biochemistry 28: 8990–8998

Gabellini N, Bowyer J, Hurt E, Melandri BA and Hauska G (1982) A cytochrome bc_1 complex with ubiquinol-cytochrome c_2 oxidoreductase activity from *Rhodopseudomonas sphaeroides* GA. Eur J Biochem 126: 105–111

Gal A, Shahak Y, Schuster G and Ohad I (1987) Specific loss of LHCII phosphorylation in the *Lemna* mutant 1073 lacking the cytochrome b_6f complex. FEBS Lett 221: 205–210

Gal A, Hauska G, Herrmann R and Ohad I (1990) Interaction between light harvesting chlorophyll-*a/b* protein (LHCII) kinase and cytochrome b_6f complex. J Biol Chem 265: 19742–19749

Geier BM, Schägger H, Brandt U and von Jagow G (1992) Point mutation in cytochrome *b* of yeast ubihydroquinone:cytochrome *c* oxidoreductase causing myxothiazol resistance and facilitated dissociation of the iron-sulfur subunit. Eur J Biochem 208: 375–380

Gennis RB, Barquera B, Hacker B, Van Doren SR, Arnaud S,

Crofts AR, Davidson , E, Gray KA and Daldal F (1993) The bc_1 complexes of *Rhodobacter sphaeroides* and *Rhodobacter capsulatus*. J Bioenerg Biomemb 25: 195–209

Graham LA, Brandt U, Sargent JS and Trumpower BL (1993) Mutational analysis of assembly and function of the iron-sulfur protein of the cytochrome bc_1 complex in *Saccharomyces cerevisiae*. J Bienerg Biomemb 25: 245–257

Gray JC (1992) Cytochrome *f*: Structure, function and biosynthesis. Photosynth Res 34: 359–374

Gray KA, Dutton PL and Daldal F (1994) Requirement of histidine 217 for ubiquinone reductase activity (Q_i-site) in the cytochrome bc_1 complex. Biochemistry 33: 723–733

Greer KL and Golden SS (1992) Conserved relationship between *psbH* and *petBD* genes: Presence of a shared upstream element in *Prochlorotrix hollandica*. Plant Mol Biol 19: 355–365

Güner S, Willie A, Millet F, Caffrey MS, Cusanovich MA, Robertson DE and Knaff DB (1993) The interaction between cytochrome c_2 and the cytochrome bc_1 complex in the photosynthetic purple bacteria *Rhodobacter capsulatus* and *Rhodopseudomonas viridis*. Biochemistry 32: 4793–4800

Gurbiel RJ, Ohnishi T, Robertson DE, Daldal F and Hoffmann BM (1991) Q-band ENDOR spectra of the Rieske protein from *Rhodobacter capsulatus* ubiquinol-cytochrome *c* oxido-reductase show two histidines coordinated to the (2Fe-2S) cluster. Biochemistry 30: 11579–11584

Hacker B, Barquera B, Crofts AR and Gennis RB (1993) Characterization of mutations in the cytochrome *b* subunit of the bc_1-complex in *Rhodobacter sphaeroides* that affect the quinone reductase site (Q_c). Biochemistry 32: 4403–4410

Haley J and Bogorad L (1989) A 4-kDa maize chloroplast polypeptide associated with the cytochrome b_6f complex: Subunit 5, encoded by the chloroplast *petE* gene. Proc Natl Acad Sci USA 86: 1534–1538

Hauska G (1986) Preparations of electrogenic, proton-translocating cytochrome complexes of the b_6f-type (chloroplasts and cyanobacteria) and bc_1-type (*Rhodo-pseudomonas sphaeroides*). Methods Enzymol 126: 271–285

Hauska G, Hurt E, Gabellini N and Lockau W (1983) Comparative aspects of quinol-cytochrome c/plastocyanin oxidoreductases. Biochim Biophys Acta 726: 97–133

Hauska G, Nitschke W and Herrmann RG (1988) Amino acid identities in the three redox center-carrying polypeptides of cytochrome bc_1/b_6f complexes. J Bioenerg Biomemb 20: 211–228

Hauska G, Herold E, Huber C, Nitschke W and Sofrova D (1989) Stigmatellin affects both hemes of cytochrome *b* in cytochrome b_6f/bc_1-complexes. Z Naturforsch 44c: 462–467

Heinemeyer W, Alt J and Herrmann RG (1984) Nucleotide sequence of the clustered genes for apocytochrome b_6 and subunit 4 of the cytochrome *bf* complex in the spinach plastid chromosome. Curr Genet 8: 543–549

Hill R (1954) The cytochrome *b* component of chloroplasts. Nature 174: 501–503

Hill R and Bendall F (1960) Function of the two cytochrome components in chloroplasts, a working hypothesis. Nature 186: 136–137

Hill R and Scarisbrick (1951) The haematin compounds of leaves. New Phytol 50: 98–111

Hope AB (1993) The chloroplast cytochrome *bf* complex: A critical focus on function. Biochim Biophys Acta 1143: 1–22

Howell N (1989) Evolutionary conservation of protein domains in the protonmotive cytochrome *b* and their possible roles in redox catalysis. J Mol Evol 29: 157–169

Huang D, Everly RM, Cheng RH, Heymann JB, Schägger H, Sled V, Ohnishi T, Baker TS and Cramer WA (1994) Characterization of the cytochrome b_6f complex as a structural and functional dimer. Biochemistry 33: 4401–4409

Huang J, Struck F, Matzinger DF and Levings CS (1991) Functional analysis in yeast of cDNA coding for the mitochondrial Rieske iron-sulfur protein of higher plants. Proc Natl Acad Sci USA 88: 10716–10720

Hurt E and Hauska G (1981) A cytochrome fb_6 complex of five polypeptides with plastoquinol-plastocyanin oxidoreductase activity from spinach chloroplasts. Eur J Biochem 117: 591–599

Hurt E and Hauska G (1982a) Identification of the polypeptides in the cytochrome b_6f-complex from spinach chloroplasts with redox center carrying subunits. J Bioenerg Biomemb 14: 405–424

Hurt E and Hauska G (1982b) Involvement of plastoquinone bound within the isolated cytochrome b_6f complex from chloroplasts in oxidant-induced reduction of cytochrome b_6. Biochim Biophys Acta 682: 466–473

Kallas T (1994) The cytochrome b_6f complex. In: Bryant (ed) The Molecular Biology of Cyanobacteria, pp 259–317. Kluwer Academic Publishers, Dordrecht

Klughammer C and Schreiber U (1993) Selective interaction of valinomycin/K^+ with the cytochrome *bf* complex. Synergistic effect with MOA stilbene on extent of cytochrome *b*563 reduction in continuous light. FEBS Lett 336: 491–495

Knaff DB (1993) The cytochrome bc_1 complexes of photo-synthetic purple bacteria. Photosynth Res 35: 117–133

Kramer DM and Crofts AR (1993) The concerted reduction of the high- and low-potential chains of the *bf* complex by plastoquinol. Biochim Biophys Acta 1183: 72–84

Kramer DM and Crofts AR (1994) Re-examination of the properties and function of the *b* cytochromes of the thylakoid cytochrome *bf* complex. Biochim Biophys Acta 1184: 193–201

Krinner M, Hauska G, Hurt E and Lockau W (1982) A cytochrome f-b_6 complex with plastoquinol-cytochrome *c* oxidoreductase activity from *Anabaena variabilis*. Biochim Biophys Acta 681: 110–117

Kutoh E and Sone N (1988) Quinol-cytochrome *c* oxidoreductase from the thermophilic bacterium PS3. Purification and properties of a cytochrome $bc_1(b_6f)$ complex. J Biol Chem 263: 9020–9026

Lam E (1986) Nearest-neighbor relationships of the constituent polypeptides in plastoquinol-plastocyanin oxidoreductase. Biochim Biophys Acta 848: 324–332

Li L-B, Zou Y-P, Yu L and Yu C-A (1991) The catalytic role of subunit IV of the cytochrome b_6f complex from spinach chloroplasts. Biochim Biophys Acta 1097: 215–222

Link TA (1994) Two pK values of the oxidised 'Rieske' (2Fe-2S) cluster observed by CD spectroscopy. Biochim Biophys Acta 1185: 81–84

Link TA, Haase U, Brandt U and von Jagow G (1993) What information do inhibitors provide about the structure of the hydroquinone oxidation of ubihydroquinone:cytochrome *c* oxidoreductase? J Bioenerg Biomemb 25: 221–232

Lübben M (1995) Cytochromes of archaeal electron transfer chains. Biochim Biophys Acta 1229: 1–22

Lübben M, Kolmerer B and Saraste M (1992) An archaebacterial terminal oxidase combines core structures of two mitochondrial complexes. EMBO J 11: 805–812

Lungdahl PO, Pennoyer JD and Trumpower BL (1986) Purification of cytochrome bc_1 complexes from phylogenetically diverse species by a single method. Methods Enzymol 126: 181–191

Madueno F, Napier JA, Cejudo FJ and Gray JC (1992) Import and processing of the precursor of the Rieske FeS protein of tobacco chloroplasts. Plant Mol Biol 20: 569–574

Malkin R (1992) Cytochrome bc_1 and b_6f-complexes of photosynthetic membranes. Photosynth Res 33: 121–136

Malkin R and Aparicio PJ (1975) Identification of a $g = 1.90$ high-potential iron-sulfur protein in chloroplasts. Biochem Biophys Res Comm 63: 1157–1160

Martinez SE, Huang D, Szczepaniak A, Cramer WA and Smith JL (1994) Crystal structure of chloroplast cytochrome *f* reveals a novel cytochrome fold and unexpected heme ligation. Structure 2: 95–105

Meinhardt SW and Crofts AR (1982) The site and mechanism of action of myxothiazol as an inhibitor of electron transfer in *Rhodobacter sphaeroides*. FEBS Lett 149: 217–222

Mitchell P (1976) Possible molecular mechanism of the protonmotive function of cytochrome systems. J Theor Biol 62: 327–367

Nelson N and Neumann J (1972) Isolation of a cytochrome b_6f particle from chloroplasts. J Biol Chem 247: 1817–1824

Nitschke W and Liebl U (1992) The cytochromes in *Heliobacillus mobilis*. In: Murata N (ed) Research in Photosynthesis, Vol 3, pp 507–510. Kluwer Academic Publishers, Dordrecht

Nitschke W, Hauska G and Crofts AR (1988) Fast electron transfer from low- to high-potential cytochrome b_6 in isolated cytochrome b_6f complex. FEBS Lett 232: 204–208

Nitschke W, Hauska G and Rutherford AW (1989) The inhibition of quinol oxidation by stigmatellin is similar in cytochrome bc_1 and b_6f complexes. Biochim Biophys Acta 974: 223–226

Nitschke W, Joliot P, Liebl U, Rutherford AW, Hauska G, Müller A and Riedel A (1992) The pH dependence of the redox midpoint potential of the 2Fe2S cluster from cytochrome b_6f complex (the 'Rieske centre'). Biochim Biophys Acta 1102: 266–268

Oettmeier W, Godde D, Kunze B and Höfle G (1985) Stigmatellin—a dual type inhibitor of photosynthetic electron transport. Biochim Biophys Acta 807: 216–219

Okkels JS, Kjaer B, Hansson Ö, Svendsen I, Lindberg-Möller B and Scheller HV (1992) A membrane-bound monoheme cytochrome *c*551 of a novel type is the immediate electron donor to P840 of the *Chlorobium vibrioforme* photosynthetic reaction center complex. J Biol Chem 267: 21139–21145

Osiewacz HD (1992) Construction of insertion mutants of *Synechocystis* sp. PCC 6803: Evidence for an essential function of subunit IV of the cytochrome b_6f complex. Arch Microbiol 157: 336–342

Pierre Y and Popot J-L (1993) Identification of two 4-kDa miniproteins in the cytochrome b_6f complex from *Chlamydomonas reinhardtii*. C R Acad Sci Paris 316: 1404–1409

Prince RC, Matsuura K, Hurt E, Hauska G and Dutton PL (1982) Reduction of cytochromes b_6 and *f* in isolated plastoquinol-plastocyanin oxidoreductase driven by photochemical reaction centers from *Rhodopseudomonas sphaeroides*. J Biol Chem 257: 3379–3381

Rees DC, DeAntonio L and Eisenberg D (1989) Hydrophobic organization of membrane proteins. Science 245: 510–513

Reimann A and Kück U (1989) Nucleotide sequence of the plastid genes for apocytochrome b_6 (*pet*B) and subunit IV of the cytochrome b_6f-complex (*pet*D) from the green alga *Chlorella protothecoides*: Lack of introns. Plant Mol Biol 13: 255–256

Rich P (1984) Electron and proton transfers through quinones and cytochrome *bc* complexes. Biochim Biophys Acta 768: 53–79

Rich PR, Heathcote P and Moss DA (1987) Kinetic studies of electron transfer in a hybrid system constructed from the cytochrome *bf* complex and photosystem 1. Biochim Biophys Acta 892: 138–151

Rich PR, Madgwick SA, Brown S, von Jagow G and Brandt U (1992) MOA-stilbene: A new tool for investigation of the reactions of the chloroplast cytochrome *bf* complex. Photosynth Res 34: 465–477

Riedel A, AW Rutherford, Hauska G, Müller A and Nitschke W (1991) Chloroplast Rieske center. EPR study on its spectral characteristics, relaxation and orientation properties. J Biol Chem 266: 17838–17844

Rieske JS (1976) Composition, structure and function of complex III of the respiratory chain. Biochim Biophys Acta 456: 195–247

Robertson DE, Prince RC, Bowyer JR, Matsuura K, Dutton PL and Ohnishi T (1984) Thermodynamic properties of semiquinone and its binding site in the ubiquinol-cytochrome c (c_2) oxidoreductase of respiratory and photosynthetic systems. J Biol Chem 259: 1758–1763

Robertson DE, Ding H, Chelminsky PR, Slaughter C, Hsu J, Moomaw C, Tokito M, Daldal F and Dutton PL (1993) Hydroubiquinone-cytochrome c_2 oxidoreductase from *Rhodobacter capsulatus*: Definition of a minimal, functional isolated preparation. Biochemistry 32: 1310–1317

Robertson DE, Farid RS, Moser CC, Urbauer JL, Mulholland SE, Pidikiti R, Lear JD, Wand AJ, DeGrado WF and Dutton PL (1994) Design and synthesis of multi-heme proteins. Nature 368: 425–432

Rögner M, Nixon PJ and Diner BA (1990) Purification and characterization of Photosystem I and Photosystem II core complexes from wildtype and phycocyanin-deficient strains of the cyanobacterium *Synochocystis* PCC 6803. J Biol Chem 265: 6189–6196

Salter AH, Newman BJ, Napier JA and Gray JC (1992) Import of the precursor of the chloroplast Rieske iron-sulphur protein by pea chloroplasts. Plant Mol Biol 20: 569–574

Saraste M (1984) Location of haem-binding sites in mitochondrial cytochrome *b*. FEBS Lett 166: 367–372

Schägger H and von Jagow G (1987) Tricine-sodium dodecyl sulfate-polyacrylamide gel electrophoresis for the separation of proteins in the range from 1 to 100 kDa. Anal Biochem 199: 223–231

Scherer S (1990) Do photosynthetic and respiratory electron transport chains share redox proteins? Trends Biochem Sci 15: 458–462

Schmidt CL and Malkin R (1993) Low molecular weight subunits associated with the cytochrome b_6f-complexes from spinach and *Chlamydomonas reinhardtii*. Photosynth Res 38: 73–81

Schütz M, Zirngibl S, le Coutre J, Büttner M, Xie D-L, Nelson N and Hauska G (1994) A transcription unit for the Rieske FeS-protein and cytochrome *b* in *Chlorobium limicola*. Photosynth Res 39: 163–174

Semenov AY (1993) Electrogenic steps during electron transfer via the cytochrome bc_1 complex of *Rhodobacter sphaeroides* chromatophores. FEBS Letters 321: 1–5

Singh KK, Chen C and Gibbs M (1992) Characterization of an electron pathway associated with glucose and fructose respiration in the intact chloroplasts of *Chlamydomonas reinhardtii* and spinach. Plant Physiol 100: 327–333

Sone N, Sawa G, Sone T and Noguchi S (1995) Thermophilic bacilli have split cytochrome *b* genes for cytochrome *b*(6) and subunit IV. First cloning of cytochrome *b* from a Gram-positive bacterium (*Bacillus stearothermophilus*). J Biol Chem 270: 10612–10617

Steppuhn J, Rother C, Hermans J, Jansen T, Salnikow J, Hauska G and Herrmann RG (1987) The complete amino-acid sequence of the Rieske FeS-precursor protein from spinach chloroplasts deduced from cDNA analysis. Mol Gen Genet 210: 171–177

Szczepaniak A and Cramer WA (1990) Thylakoid membrane protein topography. Location of the termini of the chloroplast cytochrome b_6 on the stromal side of the membrane. J Biol Chem 265: 17720–17726

Szczepaniak A, Huang D, Keenan TW and Cramer WA (1991) Electrostatic destabilization of the cytochrome b_6f complex in the thylakoid membrane. EMBO J 10: 2757–2764

Tokito MK, Gray KA, Davidson E, Park S-Y and Daldal F (1992) Structure, function and assembly of *R. capsulatus* cyt bc_1 complex. 7th EBEC Short Reports 7: 19

Tron T, Crimi M, Colson A-M and Degli Esposti M (1991) Structure/function relationships in mitochondrial cytochrome *b* revealed by the kinetic and circular dichroic properties of two yeast inhibitor resistant mutants. Eur J Biochem 199: 753–760

Tsiotis G, Lottspeich F and Michel H (1992) Isolation and characterization of cytochrome b_6f complex from *Synechocystis* PCC 6714. In: Murata N (ed) Research in Photosynthesis, Vol 2, pp 511–514. Kluwer Academic Publishers, Dordrecht

Van Doren SR, Gennis RB, Barquera B and Crofts AR (1993a) Site-directed mutations of conserved residues of the Rieske iron-sulfur subunit of the cytochrome bc_1 complex of *Rhodobacter sphaeroides* blocking or impairing quinol oxidation. Biochemistry 32: 8083–8091

Van Doren SR, Yun C-H, Crofts AR and Gennis RB (1993b) Assembly of the Rieske iron-sulfur subunit of the cytochrome bc_1 complex in *Escherichia coli* and *Rhodobacter sphaeroides* membranes independent of the cytochrome *b* and c_1 subunits. Biochemistry 32: 628–636

von Jagow G and Sebald W (1980) b-type cytochromes. Ann Rev Biochem 49: 281–314

Wang Y and Beattie DS (1992) Binding of DCCD to D155 or E166 of cytochrome b_6 in cytochrome *bf* complex isolated from spinach chloroplasts. Biochemistry 31: 8455–8459

Westhoff P and Herrmann RG (1988) Complex RNA maturation in chloroplasts: The *psbB* operon from spinach. Eur J Biochem 171: 551–564

Widger WR, Cramer WA, Herrmann RG and Trebst A (1984) Sequence homology and structural similarity between cytochrome *b* of mitochondrial complex III and the chloroplast b_6f complex: Position of the cytochrome *b*-hemes in the membrane. Proc Natl Acad Sci USA 81: 674–678

Wikström MKF (1973) The different cytochrome *b* components in the respiratory chain of animal mitochondria and their role in electron transport and energy conservation. Biochim Biophys Acta 301: 155–193

Witt HT (1971) Coupling of quanta, electrons, fields, ions and phosphorylation in the functional membrane of photosynthesis. Results by pulse spectrophotometric methods. Quart Rev Biophys 4: 365–477

Woese CR (1987) Bacterial evolution. Microbiol Rev 51: 221–271

Wood PM and Bendall DS (1976) The reduction of plastocyanin by plastoquinol-1 in the presence of chloroplasts. Eur J Biochem 61: 337–344

Yu CA and Yu L (1993) Mitochondrial ubiquinol-cytochrome *c* reductase complex: Crystallization and protein:ubiquinone interaction. J Bioenerg Biomemb 23: 259–273

Yu S-G , Romanowska E, Xue Z-T and Albertsson P-A (1994) Evidence for two different Rieske iron-sulfur proteins in the cytochrome *bf* complex of spinach chloroplast. Biochim Biophys Acta 1185: 239–242

Yun C-H, Wang Z, Crofts AR and Gennis RB (1992) Examination of the functional roles of 5 highly conserved residues in the cytochrome *b* subunit of the bc_1 complex of *Rhodobacter sphaeroides*. J Biol Chem 267: 5901–5909

Chapter 20

Basic Aspects of Electron and Proton Transfer Reactions with Applications to Photosynthesis

Lev I. Krishtalik
The Frumkin Institute of Electrochemistry, Academy of Sciences, Moscow 117 071, Russia

William A. Cramer
Dept. of Biological Sciences, Purdue University, West Lafayette, IN 47907, USA

Summary

Starting with a discussion of the dependence of the activation energy of electron transport on reorganization energy, conclusions that apply to charge transfer in photosynthetic membrane proteins are: (i) a relatively low reorganization energy is anticipated from the low dielectric constant of the protein and membrane interior; (ii) the rate of intraprotein electron transfer in theory depends on the detailed pathway, but empirically a first-order rate-distance dependence for electron transfer holds over twelve orders of magnitude in rates, a factor of ten decrease for each 1.7 Å increase in center-center separation. Most of the data apply to the photosynthetic reaction center. (iii) The theories of the distance dependence of intraprotein electron transfer do not provide an explanation for the asymmetry of electron transfer, i.e., the far greater rate through the 'L' compared to the 'M' branch of the photosynthetic reaction center. (iv) The existence of long distance intraprotein electron transfer implies that the redox poise between centers separated by as much as 20 Å must be carefully regulated in vivo

Donald R. Ort and Charles F. Yocum (eds): Oxygenic Photosynthesis: The Light Reactions, pp. 399–411.

to avoid 'promiscuous' transfer events. (v) Special aspects of intramembrane electron transfer include (a) consideration of a three layer membrane model that includes a 10–15 Å thick interfacial layer of intermediate (ε= 10–20) dielectric constant; and (b) in the case of the reaction center, a time-dependent (ps $\rightarrow \mu$s) increase in effective dielectric constant that is needed to keep the reaction activationless. (vi) Long distance trans-membrane H^+ translocation is likely to involve extended water chains, as inferred from recent atomic structure data on the *n*-side peripheral domain of the reaction center and the *p*-side peripheral domain of cytochrome *f* of the integral cytochrome b_6f complex.

I. Introduction

The purpose of this article is to provide a brief summary of (i) the foundations of the theory of charge transfer processes, and (ii) its application to selected problems in photosynthetic electron transport and proton translocation. For a more complete description of electron transfer theory, as well as its application to biological problems, the reader is referred to the following reviews (Marcus, 1964; Dogonadze, 1971; Dogonadze and Kuznetsov, 1975; Ulstrup, 1979; DeVault, 1984,1986; Marcus and Sutin, 1985; Krishtalik, 1986a; Warshel and Aqvist, 1991).

II. Medium Reorganization in Electron Transfer

Any charge transfer process inevitably involves some reorganization of the polar surroundings. A shift of charge centers results in a corresponding change in the equilibrium polarization of the medium and reorientation of both induced and permanent dipoles. The reorientation of permanent dipoles proceeds as a movement of nuclei, and this movement is many orders of magnitude slower than the movement of electrons. Therefore, according to the Franck-Condon principle, an electron transfer must occur without any change in the coordinates of the nuclei. In the absence of any fast energy input (as in optical excitation), the electron transfer takes place not only with constant nuclear coordinates, but also at constant energy.

The process is depicted schematically in Fig. 1A. The medium polarization that is relatively slow compared to electron movement is represented by a dimensionless coordinate, q. The potential energy curves describing the system energy, G, as a function of q (Fig. 1A) are assumed to have a parabolic form that corresponds to a linear response of the dielectric polarization to an electric field, and to the energy response of a harmonic oscillator displaced from its equilibrium position. At the point of intersection (q^*) of the functions for reactant and product, or initial and final energy state, the energy of the system in the initial and final states is the same, the electron energy level in both states is equal, and the electron can move under the barrier, i.e., tunnel, to the final state, due to an overlap of the electron wave functions. The activation barrier (Franck-Condon barrier), $\Delta G^{\ddagger}$, is determined by two parameters: the difference of equilibrium energies ΔG° and a non-equilibrium quantity, the solvent reorganization energy, λ_s. The latter is the energy required to bring the system at equilibrium from the initial coordinate (q_i^o) to the final (q_f^o) coordinate, while it remains on the potential energy curve of the initial state. The representation of the energy surface as a one-dimensional curve has a symbolic character. On a microscopic level there is actually a multi-dimensional surface corresponding to a large number of dipoles. Instead of an intersection point, there is a multi-dimensional intersection surface with a large number of different sets of dipole coordinates whose energies are nearly equal.

Averaging over all degrees of freedom, one obtains the free energy of activation, $\Delta G^{\ddagger}$. For parabolic potential curves,

Abbreviations: BChl – bacteriochlorophyll; BPhe – bacterio-pheophytin; D – Debye unit, unit of dipole moment; 1 D ≡ 3.3 × 10^{-30} Coulomb-meter; FTIR – Fourier transform infra-red spectroscopy; 'L' and 'M' – the two branches of the pseudo-symmetric bacterial reaction center structure, respectively containing trans-membrane helices mostly contributed by the 'L' or 'M' subunit polypeptides of the complex; LIPET – long distance intraprotein electron transfer; *n*- and *p*-sidedness – sides of the membrane with negative and positive values of the proton electrochemical potential; Q_A – primary quinone electron acceptor reduced by pheophytin and located on the 'L' branch in a pocket of the interdigitating 'M' subunit in the reaction center; Q_B – 'secondary' quinone acceptor reversibly bound on the 'M' branch in a pocket of the interdigitating 'L' subunit of the reaction center; RC – reaction center

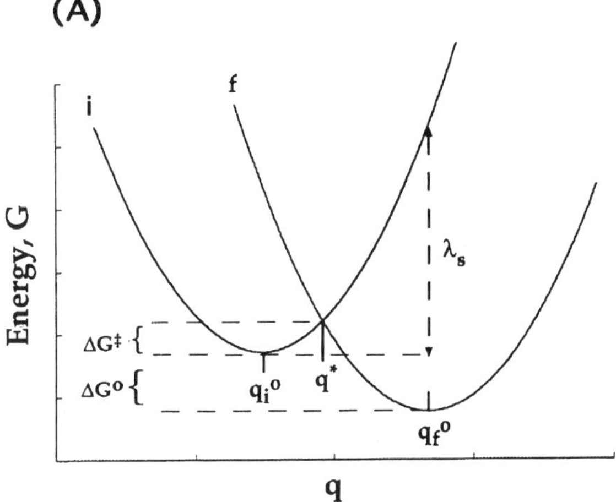

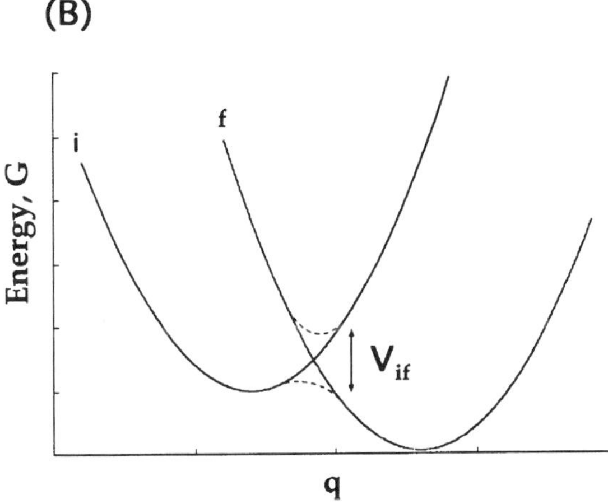

Fig. 1. (A) Multi-dimensional nuclear, or solvent, energy (G) curves, shown as parabolic functions of dimensionless normal coordinate, q, corresponding to motion of a harmonic oscillator around its equilibrium position, q^o. Definition of parameters: 'i', 'f', initial and final states; ΔG^o, free energy equilibrium difference between initial and final states at q_i^o and q_f^o; q^*, coordinate corresponding to transition state of system where energies of initial and final states are equal; $\Delta G^\ddagger$, transition activation energy $= G(q^*) - G(q_i^o)$; solvent reorganization energy, $\lambda_s = G_i(q_f^o) - G_f(q_i^o)$. (B) Energy (G) curves as a function of coordinate, q, for initial and final states (i,f) without (solid lines) and with (dotted lines) the exchange interaction energy, V_{if}.

$$\Delta G^\ddagger = \frac{(\lambda_s + \Delta G^o)^2}{4\lambda_s} \quad (1)$$

where all the quantities are the corresponding free energies: of activation ($\Delta G^\ddagger$), of solvent reorganization (λ_s), and of the electron transfer at equilibrium (ΔG^o).

This expression relates to the activation process inside a collision (reaction) complex having the optimum coordinates of all reactants. In many cases, some work (W_i) must be expended to achieve the optimum distance of approach of the reactants. The measurable activation energy is then:

$$\Delta G^\ddagger = W_i + \frac{(\lambda_s + \Delta G)^2}{4\lambda_s} \quad (1a)$$

where ΔG refers to the energy of an elementary act in the optimum reaction complex, i.e., it differs from the equilibrium energy of reaction ΔG^o of separated reactants by the effect of the work of bringing together of the initial reactants W_i, and final products, W_f.

$$\Delta G = \Delta G^o + W_f - W_i \quad (2)$$

The value of ΔG relates to the free energy change within the reaction complex. Therefore, it is independent of the concentration and the presence of reactants outside the complex. This condition is met if the reaction does not involve any change in the total number of molecules, e.g., in the case of simple electron exchange between two redox centers. However, if there is such a change, the concentration-dependent part of ΔG (transpositional entropy-dependent term or the free energy of mixing) should be excluded when analyzing the energetics of the elementary act (Krishtalik and Kuznetsov, 1986). For example, in the oxygen evolution reaction,

$$2H_2O \rightarrow O_2 + 4H^+ + 4e^-,$$

the energy required for the electron transfer is higher than the value of the standard free energy, because the latter at physiological pH is lowered by the free energy drop arising from dilution of the protons whose transfer occurs after the oxidation reaction. This changes the reaction energetics and, therefore, the kinetics (Krishtalik, 1986b, 1990).

In the framework of a model describing the medium as an infinite isotropic dielectric and the reactants as two spheres of radii, a_1 and a_2, at the distance between their centers, r_{12}, the following expression was derived for the reorganization energy (Marcus, 1956):

$$\lambda = (Ze)^2 (1/\varepsilon_{op} - 1/\varepsilon_{st}) (1/2a_1 + 1/2a_2 - 1/r_{12}) \quad (3)$$

where Ze is the charge to be transferred. The difference of reciprocal values of the optical (ε_{op}, electronic polarization) and static dielectric constants (ε_{st}) singles out the inertial, slow part of the polarization which forms the Franck-Condon barrier. The difference, $C \equiv (1/\varepsilon_{op} - 1/\varepsilon_{st})$ is often called 'the coupling constant.'

Besides the solvent (medium) reorganization energy, λ_s, which is inherent to any charge transfer process, there can be 'inner sphere' (*is*) reorganization of the reactant molecules, in which case Eqs. (1,1a) should employ the total reorganization energy, $\lambda \equiv \lambda_s + \lambda_{is}$.

III. Transition Probability

For the reorganization of the nuclear subsystem described by a harmonic oscillator, the electron transfer rate constant can be written as:

$$k_{et} = \omega/2\pi\, \kappa \exp(-\Delta G^{\ddagger}/k_B T) \quad (4)$$

where $\omega/2\pi \exp(-\Delta G^{\ddagger}/k_B T)$ defines the number of oscillations per second having sufficient energy to attain the top of the Franck-Condon barrier, ω is the radial frequency, k_B the Boltzmann constant, T the absolute temperature, and κ the transmission coefficient, i.e., the probability that reactants having an energy greater than that at the top of the barrier will form products. Considering the last problem, the transition of reactants into products means a definite transformation of the electron orbitals. This occurs because of the exchange interaction between the initial (e.g., electron at molecule A) and the final (electron at molecule B) states leading to some delocalization of the electron between A and B, with interaction energy V_{if} (Fig. 1B). The solid curves in Fig. 1B relate to the energies of the two states without exchange interaction ('ingoing' and 'outgoing' reaction channels), and the dotted curves show the effect of this interaction that causes a decrease in the barrier height. The latter two curves (adiabatic terms) are related to stationary electron orbitals corresponding to given nuclear coordinates.

In the case of a strong interaction (i.e., large value of V_{if}), the transformation of the electron orbitals upon the movement of the system along the coordinate q proceeds fast enough that at any point of this trajectory the electron cloud has time enough to obtain a stationary distribution, the latter follows the movement of nuclei adiabatically, and the system proceeds along the perturbed (lower dotted curve in Fig. 1B) adiabatic potential energy curve and, with a probability $\kappa = 1$, goes over the top of the barrier. Eq. (4) is then simplified:

$$k_{et} = \omega/2\pi \exp(-\Delta G^{\ddagger}/k_B T) \quad (4a)$$

The other limiting case, a weak interaction (i.e., small value of V_{if}), corresponds to a nonadiabatic transition in which the electron orbitals have no time for transformation during one passage of the system through the intersection point. The system has to cross the transition region many times (remaining on the ingoing channel) before the electron transfer will occur.

For non-adiabatic reactions, a frequently employed method of calculation of the rate constant utilizes the Fermi Golden Rule:

$$k_{et} = V_{if}\, FC \quad (5)$$

FC designates the Franck-Condon weighted density of states. For two (initial and final) classical harmonic oscillators of identical frequency (Marcus and Sutin, 1985),

$$FC = (4\pi\lambda k_B T)^{-1/2} \exp[-(\lambda + \Delta G)^2 / 4\lambda k_B T] \quad (6)$$

Two limiting types of degrees of freedom, classical and quantum, are characterized by the conditions $\omega >> kT$ and $\omega << kT$, respectively. The predominant path of the process for the classical mode is the transition over the barrier, while the quantum system prefers to proceed under the barrier (tunneling) (Dogonadze and Kuznetsov, 1967).

IV. Proteins As Polar Media

The protein backbone is formed by a succession of polar peptide groups whose dipole moment in free amides is close to 4 Debye (D) units ($4 \times 3.3 \times 10^{-30}$ Coulomb-meters). This value is larger in proteins than free amides because of the mutual polarization of peptide groups, especially those that are hydrogen-bonded. Dipole moments of 5–6 D, indicating increased polarity, have been both calculated (van Duijnen and Thole, 1982) and inferred from experimental data (Applequist and Mahr, 1966; Wada, 1976; Krishtalik and Topolev, 1983). Therefore, the protein is a medium with a high concentration of polar groups. However, in contrast to liquid solvents, peptide dipoles in proteins are incorporated in a relatively rigid structure, and hence their mobility and ability to reorient under the action of an external electric field is restricted.

The inability of protein dipoles to reorient in an external electric field implies a weak dielectric response, i.e., a low dielectric constant. Experimental measurements of static dielectric constants, ε, of dry proteins give values of approximately 4 (Maricíc et al., 1964; Takashima and Schwan, 1965; Pethig, 1979). Similar values have been obtained by molecular simulation, taking explicit account of peptide group mobility (Gilson and Honig, 1985; Nakamura et al., 1988; Simonson et al., 1991a,b; King et al., 1991). These considerations indicate that proteins are a highly-polar, low-dielectric medium. The high polarity influences the equilibrium energetics of enzyme reactions, while the low dielectric constant affects both equilibrium and non-equilibrium (reorganization energy) reaction parameters.

In a liquid solvent, the electric field of the dipoles fluctuates randomly around an average value equal to zero. When an ion is introduced into the solvent, its electric field will cause a preferential orientation of the dipoles, creating some reorganization of the medium and causing a non-zero dipole field. In a protein, the electric field of dipoles fluctuates around some non-zero value, this value depending upon the protein structure. Therefore, there exists a permanent time-averaged structure-dependent dipole field inside the protein. This field can partially compensate the high energy barrier involved in the transfer of a charged particle from water into the low dielectric medium of the protein. Several aspects of this problem have been reviewed (Warshel and Russel, 1984; Honig et al., 1986; Krishtalik, 1986a,1988; Harvey, 1989; Sharp and Honig, 1990; Bashford, 1991).

The low dielectric constant of proteins affects the reorganization energy of intraprotein charge transfer. For common solvents having high or intermediate dielectric constants, the inequality $\varepsilon_{op} << \varepsilon_{st}$ exists between the optical and static dielectric constants, and hence the coupling constant $C \equiv 1/\varepsilon_{op} - 1/\varepsilon_{st}$ does not vary greatly with a change of ε_S. On the contrary, the transition to a medium with a very low dielectric constant causes a large change in the value of C [Eq. (3)] and λ_s. Typical values of the optical dielectric constant are 1.8 for water and 2.2–2.4 (the usual range for amides) for proteins. For water with $\varepsilon_{st} = 80$, $C = 0.55$, whereas for a protein with $\varepsilon_{op} = 2.2 - 2.4$ and $\varepsilon_s = 4$, $C = 0.17 - 0.20$, i.e., approximately three times smaller. Therefore, from Eq. (3), a low reorganization energy associated with charge transfer reactions is typical when they occur in a protein medium (Krishtalik, 1979, 1980). These considerations are confirmed by microscopic modeling giving low reorganization energies for electron transfer in the bacterial photosynthetic reaction center (Creighton et al., 1988; Treutlein et al., 1992), for redox reactions of cytochrome *c* (Churg et al., 1983; Zheng et al., 1991), and for hydride transfer in lactate dehydrogenase (Yadav et al., 1991).

V. Proteins As an Intervening Medium in Electron Transfer

The probability of electron transfer in nonadiabatic reactions depends on the distance between reaction partners and, in principle, on their mutual orientation. The last factor may vary by one-two orders of magnitude (Cave et al., 1986). This effect gives, e.g., some preference for a parallel orientation (or, at least, orientation with an acute angle) of the planes of two reacting planar redox groups (e.g., hemes or chlorophylls). The distance dependence of long distance intraprotein electron transfer (LIPET) has been the subject of intensive investigation (e.g., reviews by Gray and Malmstrom, 1989; Bowler et al., 1990; Boxer, 1990 and Onuchic et al., 1992; as well as literature quoted in Regan et al., 1993).

It is generally accepted that the main pathway of LIPET involves a chain of covalently bonded atoms. This is the 'superexchange' mechanism employing the overlapping of the lower unfilled orbitals of the intervening bonds for the electron transfer, and the upper filled orbitals for hole transfer. Approximate estimates give a decay factor per covalent bond of approximately 0.6 per bond, which corresponds to an average value of $\beta \approx 0.8$ Å^{-1} for an exponential rate-distance (r) dependence that is proportional to exp ($-\beta$r), where β represents the average effect of the intervening medium in supporting the electron transfer (Beratan and Onuchic, 1989; Beratan et al., 1991; Betts et al., 1992). The value of $\beta = 0.8$ Å^{-1} for through-bond transfer agrees well with experimental data for electron transfer in chemical compounds with the redox centers linked via chains of σ-bonds (summarized in Bowler et al., 1990). Similar β values for several model systems have been obtained by quantum-chemical calculations (Siddarth and

Marcus, 1990; Naleway et al., 1991). This factor for a hydrogen bond is somewhat larger, and depends exponentially on the separation distance. The step with the largest decay factor is that for 'through-space tunneling.'

Different probabilities of electron transfer via covalent bonds, hydrogen bonds, and empty space result in the preference of some definite routes inside the protein. Some rules for selection of these optimal routes have been elaborated (Beratan and Onuchic, 1989; Beratan et al., 1991; Betts et al., 1992; Onuchic et al., 1992; Regan et al., 1993). Detailed quantum-chemical calculations have been carried out showing preferential paths for the electron transfer in plastocyanin (Christensen et al., 1990). In principle, the details of the bonding pattern in the protein should affect the electron transfer rate. The effective edge-edge electron transfer distance for a transfer through a mixture of covalent bonds and through-space jumps can be calculated by multiplying the number of covalent bonds in the path by 1.4 Å, and the distance of each through-space jump by 2.5 to obtain the equivalent number of covalent bonds. The experimental data indicate a particularly large rate difference between a saturated aliphatic σ-bond ($\beta \approx 0.8$ Å^{-1}) and double π-bond network (β up to 0.2 Å^{-1}) (Bowler et al., 1990). Analysis of these effects suggested that the protein interior is an inhomogeneous tunneling medium (Regan et al., 1993), and that there is a possibility of many equivalent routes of intraprotein electron transfer from donor to acceptor. It has been estimated that differences in protein structure could theoretically cause changes in electron transfer rates up to a factor of 10^4 for a fixed transfer distance (Beratan et al., 1991; Farid et al., 1993).

In contrast, using experimental data from measured electron transfer rates and midpoint oxidation-reduction potentials from reactions in (i) the bacterial photosynthetic reaction center and (ii) ruthenium-cytochrome *c* and ruthenium-myoglobin complexes (Winkler and Gray, 1992), the protein medium has been described as a homogeneous 'organic glassy solvent-like' medium with the edge-edge distance dependence of the transfer described by a single exponential (Moser et al., 1992; Farid et al., 1993) with β = 1.4 Å^{-1} and rate-axis intercept for the frequency coefficient in Eqs. (4, 4a) of 10^{13} s^{-1}, for shortest, i.e., van der Waals edge-edge distances between donor-acceptor redox pairs. The values of these parameters were obtained from values of k_{et}(max) corrected to correspond to activationless states (i.e., $-\Delta G^o = \lambda_s$, Eqs. 1, 1a) (Moser et al., 1992; Farid et al., 1993). This β = 1.4 Å^{-1} slope of the exponential was fit to the data on reactions that occur over long (ca. 20 Å) distances and over an extraordinarily wide range (12 orders of magnitude) of rates (Fig. 2). The existence of substantial local variation in β values is acknowledged in Moser et al. (1992) and Farid et al. (1993). The major point of debate between the viewpoints expressed by Moser et al. (1992) and Farid et al. (1993), compared to those by Beratan and Onuchic (1989), Beratan et al. (1991), Onuchic et al. (1992), Betts et al. (1992) and Regan et al. (1993), is whether intraprotein electron transfer reaction rates, which result from an average of individual transfer events, are altered by structure- and pathway-dependent variations in β values. In any case, some useful 'rules of thumb' derived from the average single exponential fit are that (a) the rate constant changes by approximately a factor of 10 for every factor of 1.7 Å change in distance of donor-acceptor separation; (b) the activationless rate constant is approximately 1 ns^{-1}, 1 μs^{-1}, and 1 μs^{-1}, for an edge-edge separation of 10, 15, and 20 Å (Fig. 2), respectively. (c) The main determinants of k_{et} are separation distance, r, free energy change, ΔG, and reorganization energy, λ. An empirical formula that includes the average dependence on all of these parameters is (Farid et al., 1993):

$$\log k_{et} = 15.2 - 0.61r - 3.1\,[\Delta G - \lambda]^2/\lambda \qquad (7)$$

This empirical formula does not, however, account for the asymmetry of electron transfer within the reaction center.

A. Asymmetry of Electron Transfer in the Bacterial Reaction Center

The two bacteriochlorophyll and two bacteriopheophytin molecules in the reaction center are arranged with an approximate two-fold structural symmetry, with 'L' and 'M' branches, about an axis extending from the special pair to the non-heme iron atom. In spite of this two-fold symmetry, electron transfer proceeds at a very different rate through the L and M branches, at least 10-fold faster through 'L' compared to 'M' (Michel-Beyerle et al., 1988), as shown by predominant photo-reduction of the pheophytin on the 'L' branch (Knapp et al., 1985; Maróti et al., 1985). Based on studies of the distribution of the

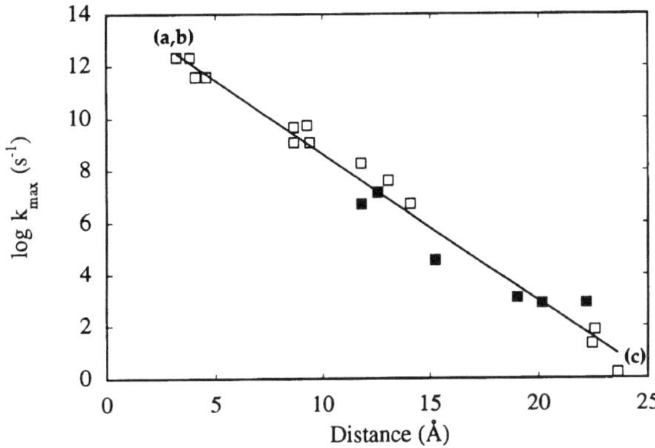

Fig. 2. Distance dependence of rate constants of electron transfer reactions in the bacterial photosynthetic reaction center (open symbols), and semi-synthetic ruthenium-cytochrome *c* and ruthenium-myoglobin. Rate constants are corrected for ΔG^o so as to be activationless. Fastest reactions a, b, are for (BChl)→ BPhe, BChl; the slowest (c) is for the recombination reaction from the anionic semiquinone of the secondary quinone acceptor, $Q_B^{\dot{-}}$→ (BChl). Average slope of exponential fit is $\beta = 1.4$ Å^{-1} (modified from Moser et al., 1992 and Farid et al., 1993).

internal electric field in the reaction center through the absorption band shifts (Stark Effect) of the intrinsic chromophores, it was concluded that there is a significant asymmetry in the effective dielectric constant on the 'L' and 'M' branches. The dielectric constant was found to be greater on the 'L' branch. Analysis of the amino acid content on the two sides indicated that the residues on the 'L' branch are more polar. It was proposed that the dielectric asymmetry is a dominant factor in determining the functional asymmetry of the electron transfer (Steffen et al., 1994). The activation energy for electron transfer in the 'L' and 'M' branches was calculated to be approximately zero and +1.6 kcal/mol, respectively (Marchi et al., 1993).

B. Problems of Control

The distance dependence, and the possibility of electron transfer over very long (ca. 20 Å) distances, is important not only from a purely quantitative point of view but also for functional purposes, e.g., for attenuation of the rates of different reactions (Moser et al., 1992; Moser and Dutton, 1992). One example is the prevention of a 'back reaction' between the semiquinone radical electron acceptors, $Q_A^{\dot{-}}$ and $Q_B^{\dot{-}}$, and the reaction center special pair bacterio-chlorophyll cation-radical of the photosynthetic reaction center, which are separated by distances of 22–24 Å (Deisenhofer and Michel, 1989; Moser et al., 1992). This distance leads to a very slow back reaction rate of 100 ms and 1 s from the $Q_A^{\dot{-}}$ and $Q_B^{\dot{-}}$ acceptor semiquinones (Feher et al., 1989) that is too slow to compete effectively with the sub-ms oxidation of $Q_A^{\dot{-}}$ by Q_B, and the millisecond oxidation of the Q_BH_2 by the cytochrome bc_1 complex.

It should also be said that negative control by the distant separation of donor-acceptor redox pairs is made more difficult by the existence of LIPET. There may be 'promiscuous' electron transfer as a consequence of electron transfer over long distances. For example, if an electron donor should not transfer an electron to a potential acceptor 15 Å away in the same protein, the transfer will occur in any case, on a μs time scale according to Fig. 2, if the acceptor is oxidized adventitiously. A possible example of such a pathway studied in vitro is that described by the 'Q cycle' (Crofts, 1985). In the prototype experiment on which the inference of this pathway is based, a small fraction of a heme of cytochrome b_6, which is initially fully oxidized in the dark, is reduced by the plastosemiquinone donor formed in a light flash (e.g., Furbacher and Cramer, 1989). It is not known, however, whether this pathway actually occurs in vivo although it is energetically feasible. This is because the redox state of the cytochrome b_6 hemes in vivo is different compared to that in isolated chloroplasts. The cytochrome b_6 hemes, which are oxidized in thylakoid membranes or chloroplasts, are kept reduced in the dark at the reducing ambient redox potential present in intact cells of the green alga, *Chlamydomonas reinhardtii* (Levine, 1969). Therefore, cytochrome b_6 in vivo is not initially able to accept an electron from the plastosemiquinone formed in a light flash, as required by the Q cycle. One then considers the possibility that the low amplitude flash-induced reduction of cytochrome b_6 seen in vitro with isolated chloroplasts is the result of 'promiscuous' electron transfer.

C. Special Aspects of Intramembrane Electron Transfer

1. The Three Layer Membrane Model

Trans-membrane electron transfer occurs within specific trans-membrane protein complexes, and hence its characteristics, including the effective midpoint potentials of the redox centers, are dependent on the specific properties of these proteins.

The relevant protein complexes have usually the form of a bundle of parallel, antiparallel, or slightly tilted α-helices in which the redox active co-factors (chlorophylls, hemes, quinones, etc.) are embedded.

Biological membranes can be simulated as a three layer structure, including a slab of low dielectric constant ($\varepsilon = 2–4$) separated on each side by a layer of intermediate polarity (lipid polar heads, hydrophilic segments of proteins; $\varepsilon = 10–20$) that separate two semi-infinite aqueous phases with $\varepsilon = 80$ (Cherepanov and Krishtalik, 1990; Krishtalik et al., 1993). The description of the system as three distinct dielectric phases with sharp boundaries between them, instead of a continuously changing dielectric, is an approximation, but better than that of neglecting the interfacial region of intermediate polarity. The equilibrium energies of the intra-membrane processes are affected by Bornian-type solvation energies of ionic species and potentials of the intra-protein electric field. The calculations predict a substantial positively directed shift (ca. 0.2 V) in the midpoint redox potential of the heme coordinated near the N-termini of two helices. This arises from the positive potential set up by α-helices at their N-termini. This model provides a possible explanation for the anomalous very positive midpoint potential of +0.4 V (Horton et al., 1976) of cytochrome *b*-559 in the Photosystem II reaction center of oxygenic photosynthesis (Krishtalik et al., 1993).

2. Activation-less Primary Charge Separation

Primary charge separation in the bacterial photosynthetic reaction center takes place with a rather low free energy gap. To ensure the irreversibility of the trans-membrane electron transfer, a larger energy gap between the initial and final electron levels is necessary. This additional substantial energy drop, $-\Delta G_{pq}$, occurs at the transition from bacteriopheophytin to the quinone acceptor, Q_A. To make this process ultrafast, and hence to impart to it a decisive preference over competing back reactions, it is desirable to make it activationless, i.e., to increase its reorganization energy up to the increased value of $-\Delta G_{pq}$. This goal may be achieved by using the final acceptor, quinone, which has a smaller effective radius and, therefore, a higher reorganization energy (Eq. 3). If nature used some porphyrin with the same redox potential as an electron acceptor in place of quinone, for example, then $-\Delta G_{pq}$ would have the same value, but λ would be much less, and the reaction would go into the so-called inverted region resulting in an increase in activation energy. The same principle is used in other photosynthetic reaction centers. In Photosystem I, the primary acceptor, a chlorophyll *a*, is rather close in energy to the excited state of the primary donor (see review by Golbeck and Bryant, 1991). A subsequent electron transfer step to a secondary acceptor is accompanied by a substantially larger energy drop, and the secondary acceptor (quinone, iron-sulfur cluster) has a substantially smaller radius than chlorophyll (Krishtalik, 1989).

In the framework of the three-layer model of the membrane (Cherepanov and Krishtalik, 1990; Krishtalik et al., 1993), the reorganization energy for the primary charge separation has been calculated (Krishtalik, 1992, 1994). The resulting value of the reorganization energy, $\lambda = 0.31 \pm 0.01$ eV, was found to be practically equal and opposite in sign to the stationary ΔG^o of this reaction which is equal to −0.31 to −0.32 eV (Woodbury et al., 1986). This result corresponds to an approximately activationless process [see Eq. (1)], and provides an explanation for the activation-less character of the primary charge separation.

These calculations were carried out using a value of the static dielectric constant $\varepsilon_{st} = 4$. However, in proteins, there exists a wide hierarchy of different relaxation times (e.g., Frauenfelder et al., 1991), and one could expect that for very short time intervals typical of electron transfer in the photosynthetic reaction center that the chromatophore surroundings will not have time to develop their response to a full extent (Woodbury and Parson, 1984; Krishtalik, 1994). This can be expressed in terms of the effective dielectric constants characteristic for different time domains. It has been shown that, for the charge separation process, the influence of the variable dielectric constant on the reaction free energy and on the reorganization energy are mutually compensating, and hence the conclusion on the activationless character of this process remains valid independently of the parameters chosen for calculations (Krishtalik, 1994). Experimentally, different ΔG values of the charge separation were found for the different relaxation times: −0.18 eV for ps (Woodbury and Parson, 1984; Woodbury et al., 1986), −0.23 eV at 10 ns (Woodbury and Parson, 1984; Woodbury et al., 1986), −0.26 eV at 100 μs (Goldstein et al., 1988) and −0.30 eV in the μs time span (Shopes and Wraight, 1987), which is close to the equilibrium value. Knowing the geometric parameters of the system, one can estimate the effective ε_{st} values corresponding to these times. Assuming a steady-state value of $\varepsilon_{st} = 4$, the effective dielectric constants were estimated to be 2.8, 3.2, and 3.4 for time intervals of ps, 10 ns, and 100 μs, respectively. In the ps time interval, the inertial polarization has not yet developed, and the dielectric constant is close to its optical value. Even after 10 ns, it is far from that corresponding to static conditions (Krishtalik, 1994).

VI. Pathways of Proton Transfer in Energy-Transducing Membranes

The focus on the central role of the electrochemical H^+ gradient in the mechanism of energy transduction in biological membranes naturally led to discussions on the mechanisms of trans-membrane H^+ translocation. The alternatives initially proposed were, (i) the proton wire, using immobilized water chains, or, (ii) a quasi-wire, using chains of discrete hydrogen-bonded amino acids (Nagle and Morowitz, 1978; Nagle and Tristram-Nagle, 1983). In either case, the average distance between H^+ donor-acceptor pairs is the length of the hydrogen bond, approximately 3.5 Å.

A. Bacteriorhodopsin—Bound H_2O

Two carboxylate amino acids (Asp85, Asp96) on opposite ends of the trans-membrane 'C' helix of the light-energy transducing integral membrane protein, bacteriorhodopsin, were implicated in the mechanism of light-induced H^+ translocation in this system (Khorana, 1988). After receiving the H^+ from the protonated retinal-Schiff base, Asp85 is closely linked to H^+ release from the membrane to the extracellular aqueous phase. Asp96 is protonated from the cytoplasmic phase and, in turn reprotonates the Schiff base (Fig. 3).

The two carboxylates together with the Schiff base provide defined intramembrane transfer sites for proton translocation (for a recent review, see Lanyi, 1993). However, because the individual proton transfer event must take place over distances $\lesssim 3.5$ Å, it is very unlikely that these two carboxylates, separated by about 16 Å along the 'C' helix, and the Schiff base are the only molecules mediating the H^+ translocation. Mutagenesis experiments indicate that Asp85 is part of a cluster including Arg82, Asp212, Tyr57, and bound water. Asp96 is similarly in a

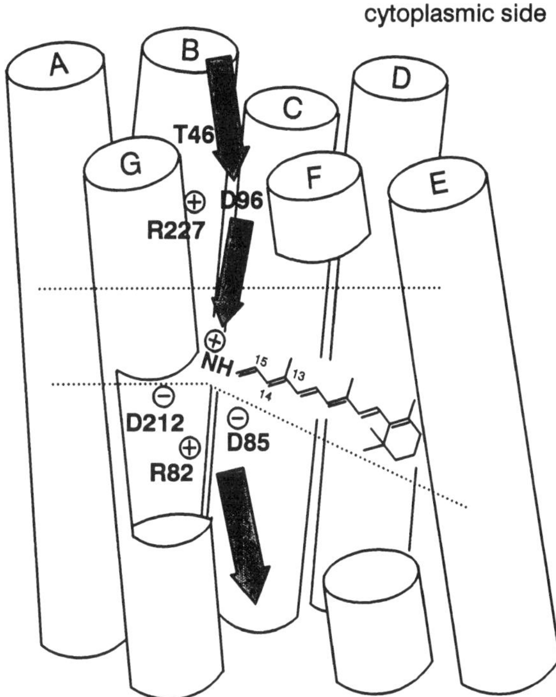

Fig. 3. Schematic structure for bacteriorhodopsin containing all-trans retinal showing residues believed to be important for proton translocation. The lipid bilayer is not shown. The H^+ pathway is shown as (i) transfer from the retinal Schiff base to Asp85, (ii) H^+ release on the extracellular side, reprotonation of the Schiff base from Asp96, and (iii) H^+ uptake from the cytoplasmic side that reprotonates Asp96 (with permission; from Lanyi, 1993).

cluster with Thr46, Ser226, Arg227, and bound water (Lanyi, 1993). Thr46 and Thr89 have been implicated by FTIR studies as a polarized hydrogen bonded pair involved in proton translocation (Rothschild et al., 1992). Evidence for several bound water molecules near the Schiff base was obtained from neutron diffraction (Papadopoulos et al., 1990), and for at least one such water molecule from hydrogen/deuterium exchange with the Schiff base (Lanyi, 1993). The role of the bound or ice-like H^+-conductive water seems especially important in filling the gaps in the vectorial H^+ translocation chain because: (i) there is relatively little information on the additional residues in the cluster forming part of a vectorially oriented chain; for both clusters described above, the additional residues influence the properties (e.g., pK) of Asp85 or Asp96, but are not essential for H^+ translocation; (ii) precedents for a vectorially-oriented bound water chain appear to exist in: (a) the pathway of H^+ uptake from the aqueous phase on the *n*-side of the membrane through the peripheral protein mass of the H-polypeptide subunit to the quinone (Q_B) binding site in the photosynthetic reaction center of *Rb. sphaeroides* (Ermler et al., 1994); (b) an intraprotein H_2O chain in cytochrome *f* in the pathway of H^+ release from the cytochrome b_6f complex on the *p*-side of the membrane (Martinez et al., 1995)—see below. The mechanism of H^+ transfer in an ordered water chain has been discussed (Nagle and Morowitz, 1978; Nagle and Tristram-Nagle, 1983; Cramer and Knaff, 1991).

B. The Bacterial Photosynthetic Reaction Center—An Extended H_2O Chain on the n-Side

Studies on the involvement of particular amino acid residues in H^+ transfer in the bacterial reaction center utilized structure information from the X-ray crystallographic analysis for the site-directed mutagenesis strategy. The physiological event that triggers H^+ uptake is the reduction of the secondary, and reversibly bound, quinone Q_B near the *n*-side of the membrane by protons originating in the *n*-side bulk aqueous phase. The net reduction of the quinone to form the quinol requires the transfer of two electrons and two protons. The electron transfer from the special pair donor, P, via the bacteriochlorophyll-bacteriopheophytin on the L-branch of the RC complex to the quinone acceptor pair Q_AQ_B, takes place in two distinguishable steps. The first electron transfer, triggered by a light flash, transiently reduces Q_A and then is transferred to the higher potential Q_B, forming an anionic non-protonated semiquinone, Q_B^-. The second light flash forms the recognizable state $Q_A^-Q_B^-$, the doubly anionic semiquinone. Two protons $H^+(1)$ and $H^+(2)$ are then transferred in succession, forming the neutral doubly protonated quinol, $Q_AQ_BH_2$.

$$PQ_AQ_B \xrightarrow{h\upsilon} P^+Q_AQ_B^-$$

$$P^+Q_AQ_B^- \xrightarrow{h\upsilon} P^+Q_A^-Q_B^- + H^+ \rightarrow P^+Q_AQ_BH^- + H^+$$

$$\rightarrow P^+Q_AQ_BH_2$$

A model for the proton transfer pathway in the region of the Q_B is shown in Fig. 4. The first proton [H^+ (1)] that is important for the transfer of the second electron to Q_B is transferred from the *n*-side aqueous phase through the peripheral protein mass of the H subunit polypeptide to Asp-L213, to Ser-L223 and then to singly reduced Q_B^- or doubly reduced Q_B^{2-}. The second proton [H^+ (2)], from the same *n*-side bulk source, is believed to be transferred to doubly reduced Q_BH^- again via Asp-L213, whose protonation is dependent upon transfer from Glu-L212 (Shinkarev and Wraight, 1993; Paddock et al., 1994).

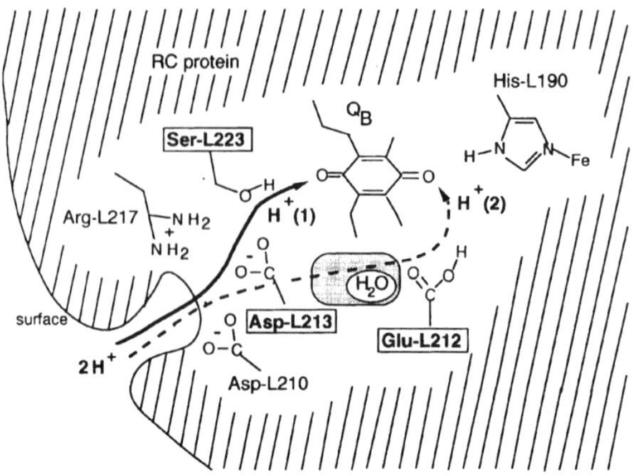

Fig. 4. Model for proton transfer from the *n*-side aqueous phase to quinone Q_B in the bacterial reaction center of *Rb. sphaeroides* (Shinkarev and Wraight, 1993; Paddock et al., 1994, from which the figure is derived, with permission). Details are discussed in the text.

A pathway from the bulk aqueous phase to Q_B or the Q_B region is suggested by a long water chain seen in the atomic model of the reaction center. The 2.6 Å X-ray structure of the *Rb. sphaeroides* reaction center complex resolved a chain of fourteen ordered, and apparently hydrogen-bonded, water molecules extending from a water accessible to the *n*-side bulk aqueous phase through the extrinsic surface peptide segments of the H subunit to the proximal ether oxygen of the quinone Q_B (Ermler et al., 1994). The degree of order in these waters is comparable to that of the surrounding amino acids. The more or less direct path to the Q_B involves twelve waters. An alternate branch from Q_B uses the same first seven waters extending from Q_B, two waters to connect to Asp210, and from there to the outside of the reaction center. This chain is suggestive of a proton-wire extending from the proton source through the shielding protein mass to the proton sink consisting in this case of the high pK reduced anionic semiquinone or doubly anionic quinol and Ser-L223,

Glu-L212. One may speculate that such a proton wire or channel is necessary for protons to find their way through the protein shield without being trapped by protonatable groups in this shield. The presence of numerous bound water molecules in the region of the Q_B niche was first noted in the 2.3 Å structure of the reaction center of *Rhodopseudomonas viridis* (Deisenhofer and Michel, 1989). It is noted, however, that the two mechanisms proposed above are not obviously complementary because the H_2O chain does not connect to Asp213, and it has been stated that there is no evidence at present about the use of this water chain for proton transfer (Ermler et al., 1994).

C. Cytochrome f – A Vectorial H_2O Chain on the p-Side

Refinement of the atomic model for the 252 residue lumen-side domain of the cytochrome *f* subunit of the cytochrome b_6f complex at a resolution of 1.96 Å revealed the presence of five internal water molecules, four of which extend 12 Å from the Histidine25 heme ligand to a solvent-accessible space within 6 Å of a Lys66 residue (see Chapter 22 by Martinez et al.), which is in the proposed docking site region for plastocyanin (Martinez et al., 1994). The existence of this vectorial water chain in cytochrome *f* suggests: (i) cytochrome *f* functions on the *p*-side of the membrane in the pathway of H^+ extrusion from the cytochrome b_6f complex; (ii) the water chain in cytochrome *f* functions as the terminal link in an inter- and intraprotein H^+ translocation pathway that starts with the oxidation of plastoquinol.

It may be suggested from these three examples that intraprotein, intramembrane extended bound water chains will turn out to provide a significant fraction of the trans-membrane pathways for H^+ translocation in energy-transducing membranes.

Acknowledgments

The collaboration of the two authors described in this article was supported by NIH Fogarty grant TW00063. This grant and that (MD9000) from the International Science Foundation provided support for studies of LIK described here, and NIH grant GM-38323 provided support for those of WAC. We thank J. Lanyi and M. Y. Okamura for communication of unpublished information, J. Hollister for skillful and dedicated work on the manuscript, and S. L. Schendel and J. B. Heymann for help with some of the illustrations.

References

Applequist J and Mahr TG (1966) The conformation of poly-L-tyrosine in quinoline from dielectric dispersion studies. J Amer Chem Soc 88: 5419–5429

Bashford D (1991) Electrostatic effects in biological molecules. Curr Op Struct Biol 1: 175–184

Beratan DN and Onuchic JN (1989) Electron tunneling pathways in proteins: Influences on the transfer rate. Photosynth Res 22: 173–186

Beratan DN, Betts JN and Onuchic JN (1991) Protein electron transfer rates set by the bridging secondary and tertiary structure. Science 252: 1285–1288

Betts JN, Beratan DN and Onuchic JN (1992) Mapping electron tunneling pathways: An algorithm that finds the minimum length/maximum coupling pathway between electron donors and acceptors in proteins. J Amer Chem Soc 114: 4043–4046

Bowler BE, Raphael AL and Gray HB (1990) Long range electron transfer in donor (spacer) acceptor molecules and proteins. Prog Inorg Chem: Bioinorg Chem 38: 259–322

Boxer SG (1990) Mechanism of long-distance electron transfer in proteins: Lessons from photosynthetic reaction centers. Ann Rev Biophys Biophys Chem 19: 267–299

Cave RJ, Siders P and Marcus RA (1986) Mutual orientation effects on electron transfer between porphyrins. J Phys Chem 90: 1436–1444

Cherepanov D and Krishtalik LI (1990) Intramembrane electric fields: A single charge, protein α-helix, photosynthetic reaction center. Bioelectrochem Bioenerg 24: 113–127

Christensen HEM, Conrad LS, Mikkelsen KV, Nielsen MK and Ulstrup J (1990) Direct and superexchange electron tunneling at the adjacent and remote sites of higher plant plastocyanins. Inorg Chem 29: 2808–2816

Churg AK, Weiss RM, Warshel A and Takano T (1983) On the action of cytochrome *c*: Correlating geometry change upon oxidation with activation energies of electron transfer. J Phys Chem 87: 1683–1694

Cramer WA and Knaff DB (1991) Energy Transduction in Biological Membranes, Chap. 2. Springer Study Edition, Springer-Verlag, New York

Creighton S, Hwang JK, Warshel A, Parson WW and Norris J (1988) Simulating the dynamics of the primary charge separation process in bacterial photosynthesis. Biochemistry 27: 774–781

Crofts AR (1985) The mechanism of ubiquinol:cytochrome *c* oxidoreductases of mitochondria and *Rb. sphaeroides*. In: Martonosi A (ed) The Enzymes of Biological Membranes, Vol 4, pp 347–382. Plenum, New York

Deisenhofer J and Michel H (1989) The photosynthetic reaction center from the purple bacterium, *Rps. viridis*. EMBO J 8: 2149–2169

DeVault D (1984) Quantum-Mechanical Tunneling in Biological

Systems, 2nd Ed, Cambridge University Press, Cambridge
DeVault D (1986) Vibronic coupling to electron transfer and the structure of the *R. viridis* reaction center. Photosynth Res 10: 125–137
Dogonadze RR (1971) Theory of molecular electrode kinetics. In: Hush NS (ed) Reactions of Molecules at Electrodes, pp 135–227. J Wiley, Intersci, London-New York
Dogonadze RR and Kuznetsov AM (1967) Effect of the changes in the first coordination sphere upon the reaction rate. Elektrokhimiya 3: 1324–1330
Dogonadze RR and Kuznetsov AM (1975) Theory of charge transfer kinetics at solid-polar liquid interfaces. Progress in Surface Science 6: 1–42
Ermler U, Fritzsch G, Buchanan SK and Michel H (1994) Structure of the photosynthetic reaction center from *Rhodobacter sphaeroides* at 2.65 Å resolution: Cofactors and protein-cofactor interactions. Structure 2: 925–936
Farid RS, Moser CC and Dutton PL (1993) Electron transfer in proteins. Curr Op Struct Biol 3: 225–233
Feher G, Allen JP, Okamura MY and Rees DC (1989) Structure and function of bacterial photosynthetic reaction centers. Nature 339: 111–116
Frauenfelder H, Sligar SG and Wolynes PG (1991) The energy landscapes and motions of proteins. Science 254: 1598–1603
Furbacher PN and Cramer WA (1989) On the question of interheme electron transfer in the chloroplast cytochrome b_6 *in situ*. Biochemistry 28: 8990–8998
Gilson M and Honig B (1985) The dielectric constant of a folded protein. Biopolymers 25: 2097–2119
Golbeck JH and Bryant DA (1991) Photosystem I. In: Lee CP (ed) Curr Topics Bioenerg, Vol 16, pp 83–177. Academic Press, Orlando
Goldstein RA, Takiff L and Boxer SG (1988) Energetics of initial charge separation in bacterial photosynthesis: The triplet decay rate in very high magnetic fields. Biochim Biophys Acta 934: 253–263
Gray HB and Malmstrom BG (1989) Long range electron transfer in multisite metalloproteins. Biochemistry 28: 7499–7505
Harvey SC (1989) Treatment of electrostatic effects in macromolecular modeling. Proteins: Structure, Function, Genetics 5: 78–92
Honig BH, Hubbell WL and Flewelling RF (1986) Electrostatic interactions in membranes and proteins. Ann Rev Biophys Biophys Chem 15: 163–193
Horton P, Whitmarsh J and Cramer WA (1976) On the specific site of action of 3-(3,4-dichlorophenyl)-1,1-dimethylurea in chloroplasts: Inhibition of a dark acid-induced decrease in midpoint potential of cytochrome *b*-559. Arch Biochem Biophys 176: 519–524
Khorana HG (1988) Bacteriorhodopsin, a membrane protein that uses light to translocate protons. J Biol Chem 263: 7439–7442
King G, Lee FS and Warshel A (1991) Microscopic simulations of macroscopic dielectric constants of solvated proteins. J Chem Phys 95: 4366–4377
Knapp EW, Fischer SF, Zinth W, Sanler M, Kaiser W, Deisenhofer J and Michel H (1985) Analysis of optical spectra from single crystals of *Rps. viridis* reaction centers. Proc Natl Acad Sci, USA 82: 8463–8467
Krishtalik LI (1979) Globule size and activation energy of an enzymatic process. Molekulyarnaya Biologiya (Moscow) 13: 577–581
Krishtalik LI (1980) Catalytic acceleration of reactions by enzymes. Effect of screening of a polar medium by a protein globule. J Theor Biol 86: 757–771
Krishtalik LI (1986a) Charge Transfer Reactions in Electrochemical and Chemical Processes. Plenum Press, New York
Krishtalik LI (1986b) Energetics of multielectron reactions. Photosynthetic oxygen evolution. Biochim Biophys Acta 849: 162–171
Krishtalik LI (1988) Charge-medium interactions in biological charge transfer reactions. In: Dogonadze RR, Kalman E, Kornyshev AA and Ulstrup J (eds) The Chemical Physics of Solvation, part C, pp 707–739. Elsevier, Amsterdam
Krishtalik LI (1989) Activationless electron transfer in the reaction center of photosynthesis. Biochim Biophys Acta 977: 200–206
Krishtalik LI (1990) Activation energy of photosynthetic oxygen evolution: An attempt at theoretical analysis. Bioelectrochem Bioenerg 23: 249–263
Krishtalik LI (1992) Intramembrane charge transfer reactions. The membrane as a dielectric medium. Molekularnaya Biologiya 26: 1377–1388
Krishtalik LI (1995) Fast electron transfers in the photosynthetic reaction centre: Effect of the time-evolution of the dielectric response. Biochim Biophys Acta 1228: 58–66
Krishtalik LI and Kuznetsov AM (1986) Energetics of the elementary act and 'configurational' electrode potential. Elektrokhimiya 22: 246–248
Krishtalik LI and Topolev VV (1983) Intraglobular electric field. The primary electric field set up by the polypeptide backbone, functional groups, and ions of the α-chymotrypsin molecule. Molekulyarnaya Biologiya (Moscow) 17: 1034–1041
Krishtalik LI, Tae GS, Cherepanov D and Cramer WA (1993) The redox properties of cytochromes *b* imposed by the membrane electrostatic environment. Biophys J 65: 1–12
Lanyi J (1993) Proton translocation mechanism and energetics in the light-driven pump bacteriorhodopsin. Biochim Biophys Acta 1183: 241–261
Levine RP (1969) A light-induced absorbance change at 564 nm in wild-type and mutant strains of *Chlamydomonas reinhardtii*. In: Metzner H (ed) Progress in Photosynthesis Research, pp 971–977. Int Union Biol Sci, Tübingen
Marchi M, Gehelen JN, Chandler D and Newton M (1993) Diabatic surfaces and the pathway for primary electron transfer in a photosynthetic reaction center. J Am Chem Soc 115: 4178–4190
Marcus RA (1956) On the theory of oxidation-reduction reactions involving electron transfer. I. J Chem Phys 24: 966–978
Marcus RA (1964) Chemical and electrochemical electron transfer theory. Ann Rev Phys Chem 15: 155–196
Marcus RA and Sutin N (1985) Electron transfers in chemistry and biology. Biochim Biophys Acta 811: 265–322
Maricíc S, Pifat G and Pravdic V (1964) Proton conductivity in solid hydrated haemoglobin. Biochim Biophys Acta 79: 293–300
Maróti P, Kirmaier C, Wraight C, Holten D and Pearlstein RM (1985) Photochemistry and electron transfer in borohydride-treated reaction centers. Biochim Biophys Acta 810: 132–139
Martinez SE, Cramer WA and Smith JL (1995) An internal H_2O chain in cytochrome *f*. Biophys J 68: 246a
Michel-Beyerle M, Plato M, Deisenhofer J, Michel H, Bixon M and Jortner J (1988) Unidirectionality of charge separation in

reaction centers of photosynthetic bacteria. Biochim Biophys Acta 932: 52–70

Moser CC and Dutton PL (1992) Engineering protein structure for electron transfer function in photosynthetic reaction centers. Biochim Biophys Acta 1101: 171–176

Moser CC, Keske JM, Warncke K, Farid RS and Dutton PL (1992) Nature of biological electron transfer. Nature 355: 796–802

Nagle JF and Morowitz H (1978) Molecular mechanisms for proton transport in membranes. Proc Natl Acad Sci, USA 75: 298–302

Nagle JF and Tristram-Nagle S (1983) Hydrogen bonded chain mechanisms for proton conduction and proton pumping. J Mem Biol 74: 1–14

Nakamura H, Sakamoto T and Wada A (1988) A theoretical study of the dielectric constant of proteins. Prot Eng 2: 177–183

Naleway CA, Curtiss LA and Miller JR (1991) Super-exchange pathway model for long distance electronic couplings. J Phys Chem 95: 8434–8437

Onuchic JN, Beratan DN, Winkler JR and Gray HB (1992) Pathway analysis of protein electron transfer reactions. Ann Rev Biophys Biomol Struct 21: 349–377

Paddock ML, Rongey SH, McPherson PH, Juth A, Feher G and Okamura MY (1994) Pathway of proton transfer in bacterial reaction centers: Role of aspartate-L213 in proton transfers associated with reduction of quinone to hydroquinone. Biochemistry 33: 734–745

Papadopoulos G, Deneher NA, Zaccai G and Büldt G (1990) Water molecules and exchangeable hydrogen ions at the active centre of bacteriorhodopsin localized by neutron diffraction. J Mol Biol 214: 15–19

Pethig R (1979) Dielectric and Electronic Properties of Biological Materials. Wiley, Chichester

Regan JJ, Risser SM, Beratan DN and Onuchic JN (1993) Protein electron transport: Single vs. multiple pathways. J Phys Chem 97: 13083–13088

Rothschild KJ, He YW, Sonar A, Marti T and Khorana HG (1992) Vibrational spectroscopy of bacteriorhodopsin mutants. Evidence that Thr-46 and Thr-89 form part of a transient network of hydrogen bonds. J Biol Chem 267: 1615–1622

Sharp KA and Honig BH (1990) Electrostatic interactions in macromolecules: Theory and applications. Ann Rev Biophys Biophys Chem 19: 301–332

Shinkarev VP and Wraight CA (1993) Electron and proton transfer in the acceptor-quinone complex of reaction centers of phototrophic bacteria. In: Deisenhofer J and Norris JR (eds) The Photosynthetic Reaction Center, Vol 1, pp 193–255. Academic Press, Orlando

Shopes RJ and Wraight CA (1987) Charge recombination from the $P^+Q_A^-$ state in reaction centers from *Rps. viridis*. Biochim Biophys Acta 893: 409–425

Siddarth P and Marcus RA (1990) Comparison of experimental and theoretical electronic matrix elements for long-range electron transfer. J Phys Chem 94: 2985–2989

Simonson T, Perahia D and Bricogne G (1991a) Intramolecular dielectric screening in proteins. J Mol Biol 218: 1859–1886

Simonson T, Perahia D and Brunger AT (1991b) Microscopic theory of the dielectric properties of proteins. Biophys J 59: 670–690

Steffen MA, Lao K and Boxer SG (1994) Dielectric asymmetry in the photosynthetic reaction center. Science 264: 810–816

Takashima S and Schwan HP (1965) Dielectric dispersion of crystalline powders of amino acids, peptides, and proteins. J Phys Chem 69: 4176–4182

Treutlein M, Schulten K, Brunger AT, Karplus M, Deisenhofer J and Michel H (1992) Chromophore-protein interactions and the function of the photosynthetic reaction center: A molecular dynamics simulation. Proc Natl Acad Sci USA 89: 75–79

Ulstrup J (1979) Charge Transfer Processes in Condensed Media; Lecture Notes in Chemistry. Springer-Verlag, Berlin

van Duijnen PT and Thole BT (1982) Cooperative effects in α-helices: An *ab initio* molecular orbital study. Biopolymers 21: 1749–1754

Wada A (1976) The α-helix as an electric macro-dipole. In: Kotani M (ed) Advances in Biophysics, Vol 9, pp 1–63. University of Tokyo Press, Tokyo

Warshel A and Aqvist J (1991) Electrostatic energies and macromolecular function. Ann Rev Biophys Biophys Chem 20: 267–298

Warshel A and Russel ST (1984) Calculations of electrostatic interactions in biological systems and in solutions. Quart Rev Biophys 17: 283–422

Winkler JR and Gray HB (1992) Electron transfer in ruthenium-modified proteins. Chem Rev 92: 369–379

Woodbury NW and Parson WW (1984) Nanosecond fluorescence from isolated photosynthetic reaction centers of *Rb. sphaeroides*. Biochim Biophys Acta 767: 345–361

Woodbury NW, Parson WW, Gunner MR, Prince RC and Dutton PL (1986) Radical pair energetics and decay mechanisms in reaction centers containing anthraquinones, naphthoquinones, or benzoquinones in place of ubiquinone. Biochim Biophys Acta 851: 6–22

Yadav A, Jackson RM, Holbrook JJ and Warshel A (1991) Role of solvent reorganization energies in the catalytic activity of enzymes. J Amer Chem Soc 113: 4800–4805

Zheng C, McCammon JC and Wolynes PG (1991) Quantum simulations of conformation reorganization in the electron transfer reactions of cytochrome *c*. Chem Phys 158: 261–270

Chapter 21

Plastocyanin: Structure, Location, Diffusion and Electron Transfer Mechanisms

Elizabeth L. Gross
Dept. of Biochemistry, The Ohio State University, 484 W. 12th. Ave. Columbus, OH 43210, USA

Summary

Plastocyanin (PC) is a 10 kDa copper protein that functions as a mobile electron carrier in photosynthetic electron transport, shuttling electrons from cytochrome *f* (Cyt *f*) in the cytochrome b_6f complex to P700 in Photosystem I (PS I). The copper atom is ligated to four residues (H37, H87, C84 and M92) in a distorted tetrahedral geometry. It is the distorted tetrahedral geometry that is responsible for the redox properties of PC. The ligands to the copper are located on two different β-strands. Thus, the protein must fold correctly to incorporate the copper and form the 'blue' copper center. Conversely, the presence of the copper stabilizes the native conformation of PC.

Donald R. Ort and Charles F. Yocum (eds): Oxygenic Photosynthesis: The Light Reactions, pp. 413–429.

PC possesses two potential binding sites for reaction partners. One is located at the H87 ligand to the heme. The other is located at Y83 which is surrounded by negatively- charged residues that are conserved in higher plant and algal PCs. The negative patches have been implicated in the binding of both Cyt *f* and PS I. In addition, Y83 has been implicated in both binding and electron transfer reactions with Cyt *f*. The possible sites of binding and electron transfer are discussed in light of the recent elucidation of the crystal structure of Cyt *f* (Martinez et al., 1994 and Chapter 22). In the case of PS I, a small positively-charged subunit of PS I (Subunit III or the *PsaF* gene product) has been implicated in PC binding. However, it has been proposed that H87 reacts directly with P700 which is located on the two large subunits that are products of the *psaA* and *psaB* genes. In addition, the structure, redox properties, and binding affinities of PC are affected by the low pH values (< 5) found in the lumen of the thylakoid upon illumination of the chloroplasts.

I. Introduction

Plastocyanin (PC) is a small (10 kD) water-soluble protein that is located in the lumen of the thylakoid (Hauska et al. 1971) where it functions as a mobile electron carrier (Selak and Whitmarsh, 1984) shuttling electrons from cytochrome *f* (Cyt *f*) in the cytochrome b_6f complex to P700 in Photosystem I. The main focus of this chapter is on the interaction of PC with its reaction partners, Cyt *f* and P700. In particular, we will discuss the interaction of PC with Cyt *f* in light of the recently published crystal structure of the lumen portion of turnip Cyt *f* (Martinez et al., 1994 and Chapter 22). However, in order to understand these interactions, it is important to understand the structure of PC and the changes that have occurred through evolution. For other reviews, see Boulter et al. (1977), Freeman (1981), Sykes (1985 and 1991) and Gross (1993). For a history of the discovery of plastocyanin, see Katoh (1995).

II. Structure

A. The Occurrence and Distribution of Plastocyanin

Plastocyanin is found in all higher plants. However, some green algae and cyanobacteria can substitute a *c* type cytochrome (Cyt *c*-553) under conditions of copper deficiency (Bohner and Boger, 1978; Ho and Krogmann, 1984). In contrast, photosynthetic bacteria use a *c* type cytochrome as the electron acceptor from cytochrome c_1. One interesting unanswered question is why photosynthetic organisms switched from using a cytochrome to using a copper protein.

Also, in true green algae and cyanobacteria, the binding sites on both Cyt *f* and Photosystem I must be able to accommodate both PC and Cyt *c*-553. It is of interest that in organisms in which PC has a positive charge, the corresponding Cyt *c*-553 also has a positive charge (Ho and Krogmann, 1984) and conversely, if the PC is negatively-charged so is the Cyt *c*-553. These results have led to the conclusion that the net charge on PC (or the cytochrome) is important for its interaction with its reaction partners.

B. Amino Acid Sequences

We will first examine the primary amino sequences of the 30 species of PC for which the primary amino acid sequences are known. A representative sample of these sequences is depicted in Fig. 1. See reviews by Boulter et al., (1977), Guss and Freeman (1983), Sykes (1985, 1991) and Briggs et al. (1990) for other compilations of PC sequences. The sequences can be divided into four classes. The first class (Class I) consists of the cyanobacteria of which only three members have been sequenced: including *Anabaena, Synechocystis* sp. PCC 6803 (Briggs et al., 1990) and *Prochlorothrix hollandica* (Arudchandran et al., 1994). The cyanobacterial PCs are the most diverse as a class and also differ significantly from the other three classes. The second class (Class II) consists of the eukaryotic algae, represented in Fig. 1 by *Scenedesmus* (Sykes, 1985) and *Enteromorpha prolifera* (Simpson et al., 1986). These PCs contain 97 amino acids. Classes III and IV are made up of PCs from higher plant species. Class III higher plant PCs are similar to those of the eukaryotic algae, in that they have only 97 amino acids. Parsley (Sykes, 1985), carrot (Shoji et al., 1985), rice (Yano et al., 1989) and

Abbreviations: CDNB – 4-chloro-3,5-dinitrobenzoic acid; Cyt *c* – cytochrome *c*; Cyt *f* – cytochrome *f*; EPR – electron paramagnetic resonance; NCBI – National Center for Biotechnology Information; NHE – normal hydrogen electrode; PC – plastocyanin; PIR – Protein Identification Resource; PS I – Photosystem I

```
Class Organism
                                       5        10        15        20        25        30
I     Anabaena         1   E T Y T V K L G S D K G L L V F E P A K L T I K P G D T V E F L
I     Synechocystis    1     N A T V K M G S D S G A L V F E P S T V T I K A G E E V K W V
II    Scenedesmus      1       A N V K L G A D S G A L V F E P A T V T I K A G D S V T W T
II    Enteromorpha     1     D A I V K L G G D D G S L A F V P N N I T V G A G E S I E F I
III   Parsley          1       A E V K L G S D D G G L V F S P S S F T V A A G E K I T F K
IV    Spinach          1       V E V L L G G G D G S L A F L P G D F S V A S G E E I V F K
IV    Poplar A         1       I D V L L G A D D G S L A F V P S E F S I S P G E K I V F K
IV    Poplar B         1       V D V L L G A D D G S L A F V P S E F S V P A G E K I V F K

                                  35        40        45        50        55        60
I     Anabaena        31   N N K V P P H N V V F D A T L N P A K S A D L A K S L S H K Q
I     Synechocystis   31   N N K L S P H N I V F A A D - - - G V D A D T A A K L S H K G
II    Scenedesmus     31   N N A G F P H N I V F D E D A V P A G V N A D A L S - - H D D
II    Enteromorpha    31   N N A G F P H N I V F D E D A V P A G V D A D A I S - - A E D
III   Parsley         31   N N A G F P H N I V F D E D E V P A G V N A E K I S - - Q P E
IV    Spinach         31   N N A G F P H N V V F D E D E I P S G V D A A K I S M S E E D
IV    Poplar A        31   N N A G F P H N I V F D E D S I P S G V D A S K I S M S E E D
IV    Poplar B        31   N N A G F P H N V L F D E D A V P S G V D V S K I S M S E E D
                                       L           N                                 N

                                65        70        75        80              85        90
I     Anabaena        62   L L M S P G Q S T S T T F P A D A P A G D Y S F Y C E P H R G A
I     Synechocystis   62   L A F A A G E S F T S T F T E P G T Y T Y       Y C E P H R G A
II    Scenedesmus     62   Y L N A P G E S Y T A K F D T A G E Y G Y       F C E P H Q G A
II    Enteromorpha    62   Y L N S K G Q T V V R K L T T P G T Y G V       Y C D P H S G A
III   Parsley         62   Y L N G A G E T Y E V T L T E K G T Y K F       Y C E P H A G A
IV    Spinach         62   L L N A P G E T Y K V T L T E K G T Y K F       Y C S P H Q G A
IV    Poplar A        62   L L N A K G E T F E V A L S N K G E Y S F       Y C S P H Q G A
IV    Poplar B        62   L L N A K G E T F E V A L S D K G E Y T F       Y C S P H Q G A
                                                                           * L     L

                                  95
I     Anabaena        91   G M V G K I T V A S
I     Synechocystis   91   G M V G K V V V E
II    Scenedesmus     91   G M V G K V I V Q
II    Enteromorpha    91   G M K M T I T V Q
III   Parsley         91   G M K G E V T V N
IV    Spinach         91   G M V G K V T V N
IV    Poplar A        91   G M V G K V T V N
IV    Poplar B        91   G M V G K V I V N
                             L
```

Fig. 1. Primary sequences of plant and algal plastocyanins. The amino acid sequences were obtained from the PIR database of NCBI. The sequences of spinach and poplar A PC can be found in Freeman (1981). Parsley and poplar B were taken from Sykes (1985) and Dimitrov et al. (1987) respectively. *Synechocystis* and *Anabaena* PCs were taken from Briggs et al. (1990) and Ho and Krogmann (1984). Residues conserved in all PCs are shaded whereas those conserved in higher plant PCs are italicized. The ligands to the copper are underlined and indicated by L. Residues in the two negative patches have double underlining and are indicated by N. Y83 is indicated by *.

barley (Nielsen and Gaussing, 1987) fall into this category. All other higher plant PCs, including poplar and spinach (Freeman, 1981), contain 99 amino acids. Compared to these PCs, Class II and III PCs have a deletion at residues #57 and #58.

Only 18 amino acid residues are conserved in all plastocyanins (shaded in Fig. 1). These conserved residues include the four ligands to the copper (H 37, H 87, C 84 and M 92) [shown in Fig. 1 underlined and indicated by L], four glycines and three invariant proline residues (P16, P36, and P86). Two of these prolines (P36 and P86) are adjacent to ligands to the copper. Note that PC contains only one cysteine (C84) and it is a ligand to the copper.

Thirty four residues are conserved in all higher plant and algal species including residues #42–44 which may be involved in the binding of Cyt *f* (see below). If only the higher plant species are considered, there are forty-five conserved residues. The residues conserved in higher plant species are shown in italics in Fig. 1.

Some higher plants, including poplar (Dimitrov et al. 1987) and parsley (Dimitrov et al., 1990), show microheterogeneity in their sequences indicating that these species contain more than one gene for PC. The two different poplar PCs (designated poplar A and poplar B) (Dimitrov et al., 1987) have been isolated and sequenced. They differ at a total of twelve positions. All of these changes are conservative except that poplar A contains an asparagine at position #76 instead of an aspartate. The difference in charge permits the two forms to be separated by anion exchange.

C. Three Dimensional Structure

1. General Features of Plastocyanin

The crystal structure has been determined for PCs from 3 species: Poplar *(Populus nigra)* (Coleman et. al., 1978; Guss and Freeman, 1983; Guss et al., 1992) and two green algae, *Enteromorpha prolifer-ata* (Collyer et al., 1990) and *Chlamydomonas reinhardtii* (Redinbo et al., 1993). Of these, poplar has been studied in the most detail since crystal structures have been determined for oxidized (Coleman et. al., 1978; Guss and Freeman, 1983), reduced (Guss et al., 1986), apo- (Garret et al., 1984), and Hg^{2+}-substituted PC (Church et al., 1986) PC.

Furthermore, the solution structures of PCs isolated from *Scenedesmus obliquus* (Moore et al., 1988), French bean (Moore et al., 1991) and parsley (Bagby et al., 1994) have been determined using NMR. A comparison of the NMR and crystal structures allows two conclusions to be drawn. First, there is essentially no difference between solution and crystal structures. Second, the major features of the three dimensional structure of PC are almost identical for all of the species studied.

The secondary structure of PC consists of two interacting β-sheets forming a flattened cylinder as shown in Fig. 2, which depicts the secondary structure of reduced poplar PC (Guss et al., 1986). The crystal structures of oxidized and reduced PC are essentially identical at neutral pH (Freeman, 1981). Residues conserved in all PCs are shown by crosshatching in Fig. 3. It can be seen that most of these residues occur in the top half of the molecule.

2. The Copper Center, apo-Plastocyanin and Protein Folding

The copper atom is coordinated to the four ligands (H37, H87, C84 and M92) in a distorted tetrahedral geometry (Fig. 2). It is the distorted tetrahedral geometry that is responsible for the deep blue color of oxidized PC, its high redox potential (+372 mV NHE; Sanderson et al., 1986) and its small EPR hyperfine coupling constant $A_{||}$ (Freeman, 1981).

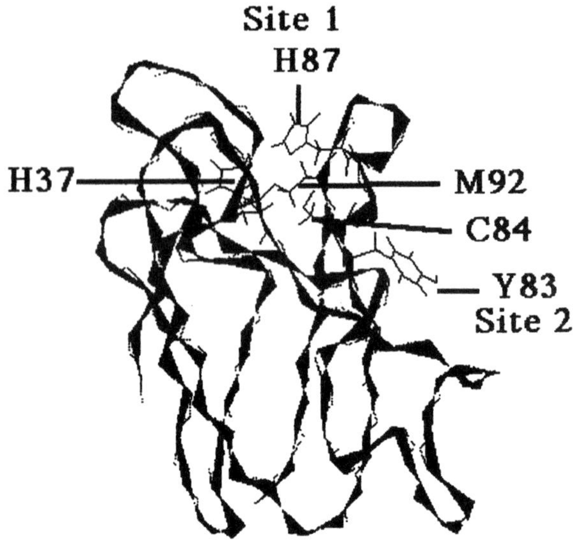

Fig. 2. Wire frame model of PC. The coordinates of reduced poplar PC (Guss et al., 1986) were obtained from the Brookhaven Protein Data Bank (Bernstein et al., 1977) and displayed using Nanovision for the MacIntosh. The ligands to the copper, Tyr 83, and the two potential binding sites are shown.

One important feature of the copper center is that the ligands of the copper are located on two different β-strands. C84, H87 and M92 are located on one strand and H37 is located on another (Fig. 2), corresponding to two different portions of the primary sequence (Fig. 1). Therefore, the protein must fold correctly in order to bind the copper atom forming the 'blue' copper center. Moreover, small changes in the geometry of the ligands may serve to fine tune the redox potential.

The question arises whether the copper is required for the correct folding of PC. PC is synthesized in the cytoplasm as a precursor (Merchant and Bogorad, 1986; de Boer and Weisbeek, 1991). It is translocated across three membranes and two separate signal sequences are removed before it arrives in the lumen of the thylakoid where copper is incorporated.

Two pieces of evidence suggest that PC can fold correctly in the absence of copper. First, Garrett et al. (1984) determined the crystal structure of apo-PC prepared by removing the Cu from preformed crystals. They found that apo-PC had a nearly native structure.

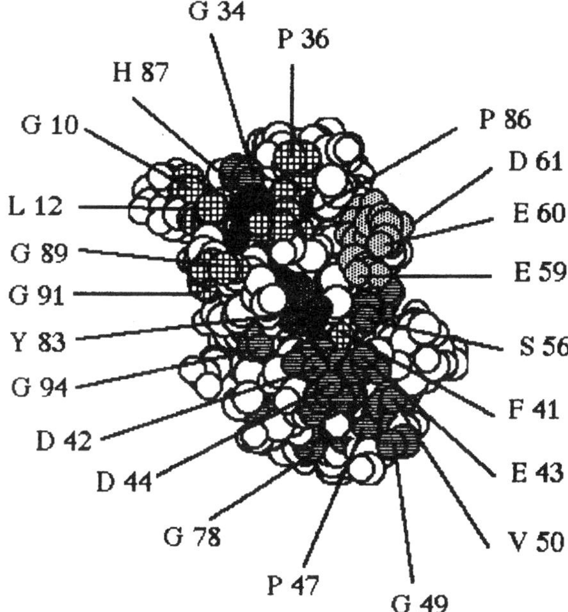

Fig. 3. Space-filling model of poplar PC showing conserved residues. H87 is shown in black and Y 83 is shown in dark gray. Residues conserved in all PCs are cross-hatched. Acidic patch B (residues #59–61) is shown in light gray. Residues conserved in all species except the cyanobacteria are designated with horizontal lines. Other conditions are as for Fig. 2.

The major difference between apo- and holo-PC was that H87 was rotated 180° about the C^{β}-C^{γ} bond and P36 was displaced slightly. They concluded that the geometry of the copper center is imposed by the polypeptide chain and that the native conformation of PC can be formed in the absence of the Cu atom. This conclusion is supported by the computer modeling studies of Skolnick and Kolinski (1990) who obtained a native secondary structure for PC using constraints such as nearest neighbor interactions, hydrophobicity and hydrogen bonds. Their results show that, when PC folds, a native structure is obtained, but they do not provide information concerning the probability of forming the native structure under in vivo conditions.

Other evidence indicates that apo-PC is unstable under low to moderate ionic strength conditions such as those observed in vivo. First, Merchant and Bogorad (1986) showed that, in *Chlamydomonas reinhardtii,* newly-synthesized and processed apo-PC is degraded in the absence of copper. In contrast, Cu-PC is much less sensitive to proteases. These results suggest that the conformation of apo-PC differs from that of Cu-PC. Second, apo-PC prepared using either cyanide (Koide et al., 1993) or the mercury/mercaptoethanol method of Scawen et al. (1975) (Gross, unpublished results) shows a non-native secondary structure. However, Koide et al., (1993) were able to preserve native secondary structure in apo-PC provided it was prepared in the presence of 0.5 M Na_2SO_4.

Another important question concerning the folding of PC involves the proline residues. Two of the three conserved proline residues (P16 and P36) are found in the cis configuration in native PC. Koide et al. (1993) found that proline isomerization occurred when apo-PC was unfolded in guanidine. HCl and that the rate limiting steps for refolding consisted of the formation of the cis-proline bonds. In vivo, proline isomerization may be particularly important for P36 which is adjacent to the H87 ligand to the copper (Fig. 3). Thus, proline isomerization may be a prerequisite for Cu incorporation.

In conclusion, the protein portion of the PC molecule influences the incorporation of the copper as well as the stability and redox properties of the copper center. Conversely, the copper center influences the stability of the PC molecule. Evidence for this is three-fold. (1) Apo-PC is more sensitive to proteases than is Cu-PC (Merchant and Bogorad, 1986). (2) Apo-PC is less stable at low ionic strength than is Cu-PC (Koide et al., 1993). (3) Reduced PC is more stable to thermal denaturation than is oxidized PC (Gross et al., 1992).

3. Potential Binding Sites for Reaction Partners

Early NMR studies of PC (Cookson et al., 1980; Hanford et al., 1980) using paramagnetic analogs of electron donors and acceptors demonstrated the existence of two potential binding sites for reaction partners (designated Sites 1 and 2).

Site 1 (shown at the top of Fig. 2) consists of the surface-exposed H87 ligand to the copper and the surrounding hydrophobic patch). H87 is the only ligand to the copper that is exposed to the surface. Electron transfer at this site occurs via an outer sphere mechanism. H87 is surrounded by a very highly conserved patch of hydrophobic residues including G10, L12, G89, P36, and P86. However, even though there are no charged residues in this region, electrostatic calculations (see below) reveal a positive electrostatic potential attributable to the copper ion which is buried below the surface in a region of low dielectric constant.

Site 2 consists of Y83 and the surrounding patches of negatively-charged residues (Figs. 3 and 4). The negative patches are designated by N and double underlining in Fig. 1. A through-bond pathway for electron transport extends from Y83 through the C84 ligand to the copper center (Betts et al., 1992; Redinbo et al., 1993). Y83 is conserved in almost all PCs except *Scenedesmus* and *Chlorella* where the tyrosine residue is replaced by a phenylalanine (Fig. 1). Y83 is designated by * in Fig. 1.

In Class IV higher plant PCs, Y83 is surrounded by two acidic patches. The center of the first patch (Patch A) (Fig. 4) consists of residues #42–44. These residues are conserved in all PCs except those isolated from cyanobacteria. In addition, all Class IV PCs have two additional negative charges: one at position #51 and the other at either position #45 or #79. Please note that, when identifying binding sites, it is important to examine the three dimensional structure of a protein as well as its sequence. Class IV PCs contain a second negative patch at residues #59–61 (Figs. 3 and 4). Both patches form β-bends that project out from the surface of the PC molecule forming part of the binding site for Cyt *f* (see below).

The second negative patch (Patch B) is located at residues #59–61 (Figs. 3 and 4). This patch is conserved in Class IV higher plant PCs but altered or absent in the other three classes. It is altered in Class II higher plant and Class III algal PCs as a consequence of the deletion of residues #57 and 58. However, in parsley, the deletion pulls #85 close to Y83 restoring one negative charge in Patch B (Fig. 4b).

Both negative patches are significantly altered in the cyanobacterial PCs. In fact, most of the differences observed between the cyanobacterial PCs and the other three classes (shown in Fig. 3 by horizontal lines) occur in the region of patch A and are located on a single strand running from D42 to V50.

If one examines Patch A (Fig. 1), one finds only one negative charge at this location. Patch B contains two positive charges rather than the three negative charges found in Class IV PCs. Interestingly, both cyanobacterial PCs sequenced to date are identical in these two regions. However, they have very different isoelectric points [pH 7.5–8.0 for *Anabaena variabilis* (Ho and Krogmann, 1984) and 5.6 for *Synechocystis* sp. PCC 6803 (Briggs et al., 1990)] and, hence different net charges at pH 7. The question arises whether it is the charge at the binding sites or the net charge on the PC molecule that controls its binding to Cyt *f* and PS I.

Site 1 and Site 2 are located 12.7 Å apart. The existence of two potential binding sites on PC has raised the speculation that Cyt *f* and P700 use different sites for binding and electron transfer.

4. The Electrostatic Field of Plastocyanin

The interaction of PC with its reaction partners is

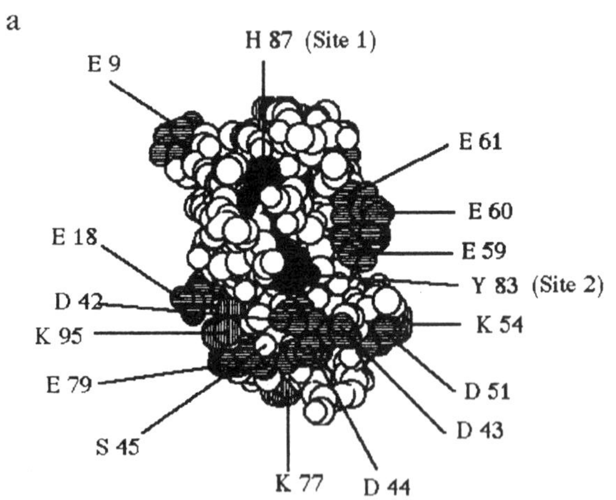

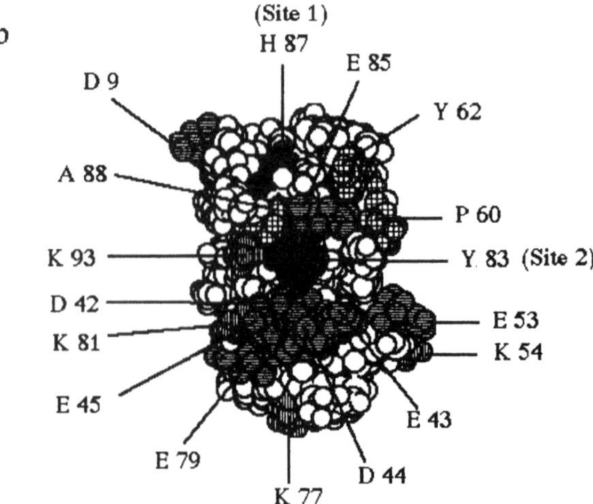

Fig. 4. Comparison of poplar and parsley PC. a. Poplar PC. b. Parsley PC. The structure of parsley PC was obtained using the molecular modeling program Insight and the molecular mechanics program Discover (Biosym). Beginning with the structure of *Enteromorpha* PC, amino acid side chains were replaced as necessary to produce the sequence of parsley PC. The final structure was produced by energy minimization. H 87 is shown in black and Y 83 in dark gray. Negative residues are depicted by horizontal lines and positive residues by vertical lines.

electrostatic in nature (see below). For this reason, it is important to be able to visualize the electrostatic field of PC. Durell et al. (1990) used the Del-Phi program of Honig (Klapper et al., 1986; Gilson et al., 1987; Gilson and Honig, 1988) to calculate the electrostatic field of PC. This program uses a macroscopic, continuum approach to numerically solve the Poisson Boltzman equation based on the crystal structure, protein dielectric constant and ionic strength of the medium. In these calculations, the dielectric constants of the medium and protein were assigned values of 80 and 2.

The calculated electrostatic field of oxidized spinach PC is shown in Fig. 5a. A region of negative potential is observed adjacent to the acidic patches at Site 2 and a region of positive potential occurs at Site 1 due primarily to the presence of the copper atom buried in a region of low dielectric constant below the surface of the protein. Reduction of PC decreased the positive potential at Site 1 (Figure 5b).

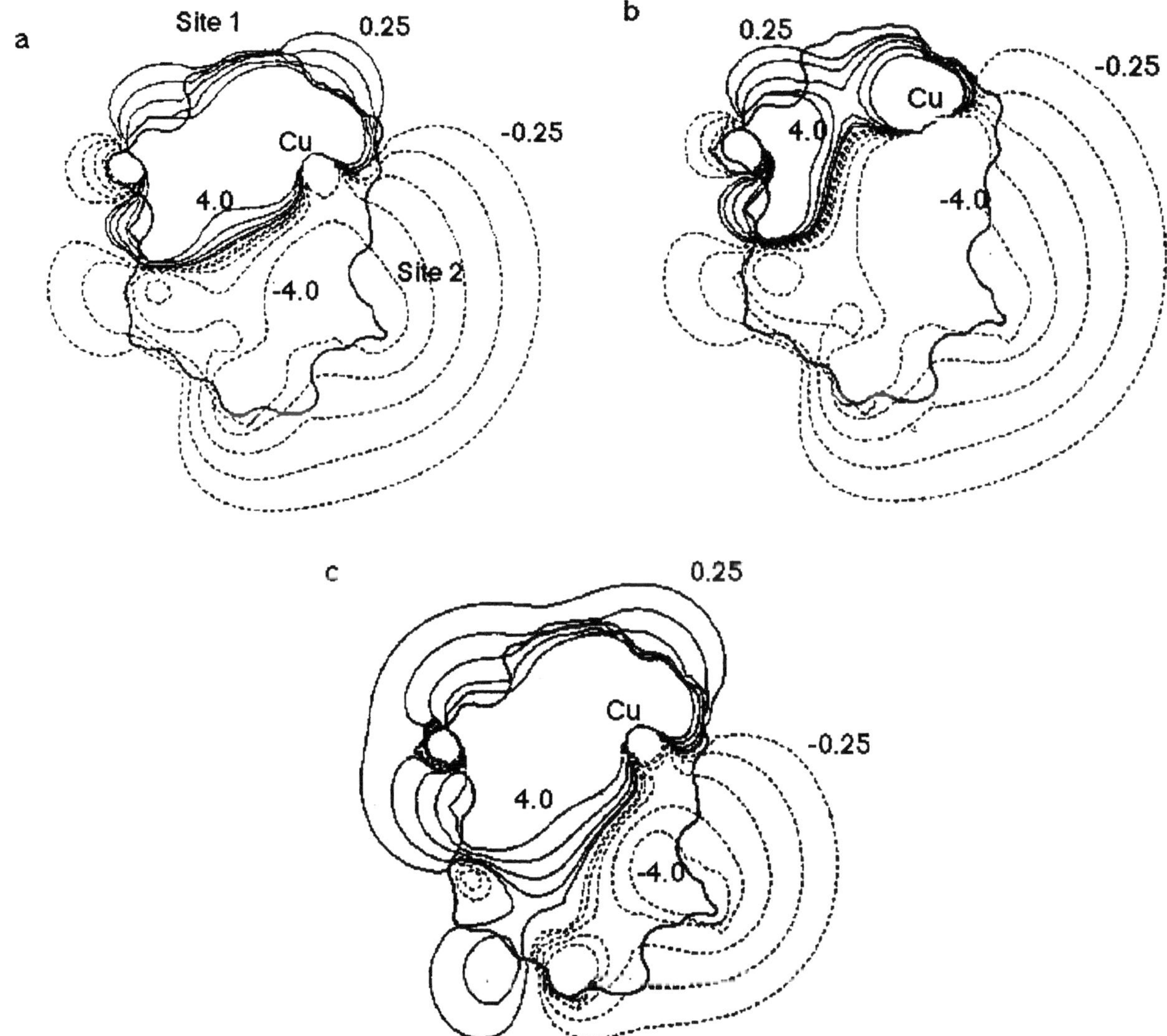

Fig. 5. The electrostatic field of PC. The electrostatic field of PC was calculated as described in Durell et al. (1990). Parameters: protein/solvent dielectric constants were 2/80, the ionic strength was 150 mM and the pH was 7 except where indicated. The contour lines are shown for 4. 0, 2. 0, 1. 0, 0. 5, 0. 25, –0. 25, –0. 5, –1. 0, –2. 0 and –4. 0 kT/e. (a) Oxidized spinach PC (b) Reduced spinach PC (c) Oxidized spinach PC (pH 5 simulation).

III. The Interaction of Plastocyanin with Cytochrome *f*

A. Interactions of Electron Carriers

The interaction of two electron carriers can be divided into three parts: (1) Approach to each other including the possible formation of a predocking complex; (2) final binding at the site where electron transfer will occur; and (3) electron transfer. The approach can be further divided into two parts: Electrostatic attraction as the mobile partner approaches from a distance and the formation of one or more predocking complexes. During the approach the interactions will initially be influenced by the net charge on the two molecules but as the two molecules approach the interactions will be determined by the charge configuration and possible hydrophobic interactions at the potential binding sites.

Computer simulations are extremely useful for examining the approach to docking as well as the precollision and final binding complexes. For example, Northrup et al. (1988) used Brownian dynamics simulations to model the docking of cytochrome *c* (Cyt *c*) with Cyt *c* peroxidase under the influence of an electrostatic field. In their simulations, Cyt *c* binding proceeded in two stages. In the first stage, Cyt *c* approached Cyt *c* peroxidase forming a large number of stable but non-specific electrostatic predocking complexes located at various points on the Cyt *c* peroxidase molecule. In the second stage, the initial binding was followed by two dimensional diffusion on the surface of the Cyt *c* peroxidase and reorientation of Cyt *c* to form the productive binding complex that resulted in electron transfer.

Furthermore, Roberts et al., (1991) studied the docking of Cyt *c* with PC and observed three different types of predocking complexes. All three predocking complexes involved Site 2 but differed both with respect to their location within the site and in the orientation of the heme. When the three predocking complexes were allowed to dock with PC, they all converged on Y 83. In the final encounter (binding) complex, the positive charges surrounding the heme on Cyt *c* formed a ring around the negative patches on PC.

Please note that, in this model, electron transfer will be inhibited if the mobile electron carrier is trapped in the initial complex by cross-linking or tight binding. In fact, Peerey and Kostic (1989) have found that covalently cross-linked complexes between PC and Cyt *c* were unable to carry out electron transfer.

There is also experimental evidence for motion within docked electron transfer complexes. NMR studies have shown that motion occurs within complexes such as the Cyt *c*-cytochrome *c* peroxidase complex (Bagby et al., 1990) and the Cyt *c*-cytochrome b_5 complex (Wendolowski et al., 1991). Thus, it is anticipated that the interaction of PC with Cyt *f* is also dynamic in nature.

B. The Electrostatic Nature of the Interaction between Plastocyanin and Cytochrome f

Several lines of evidence demonstrate the electrostatic nature of the interaction between PC and Cyt *f*. First, the rate of electron donation from Cyt *f* to PC decreases with increasing ionic strength (Niwa et al., 1980; Takabe et al., 1986; Qin and Kostic, 1992; Meyer et al., 1993) indicating that binding involves opposite charges on the two proteins. Chemical modification of PC (Takabe et al., 1984, 1986; Anderson et al., 1987) showed that the negative charges in the two acidic patches were involved in the binding of Cyt *f*. Conversely, chemical modification of both lysine (Takenaka and Takabe, 1984; Takabe et al., 1986) and arginine (Adam and Malkin, 1989) residues on Cyt *f* showed that positive charges on Cyt *f* were also involved. Finally, the cross-linking studies of Morand et al., (1989) confirmed the electrostatic nature of the interactions by cross-linking carboxyl groups on PC to lysine residues on Cyt *f*.

C. The Binding Site on Plastocyanin for Cytochrome f

1. The Negative Patches

The involvement of the negative patches has been studied using chemical modification, site-directed mutagenesis and cross-linking techniques. Chemical modification studies implicate both of the negative patches in the binding of Cyt *f*. Beoku-Betts et al. (1983, 1985) showed that chemical modification of acidic residues #42–45 of plastocyanin with Cr (III) inhibited the interaction of PC with Cyt *f*. Later, Takabe et al. (1984, 1986) and Anderson et al., (1987) modified the charge on residues in the negative patches using ethylenediamine plus a water-soluble carbodiimide. This reaction replaces a negative charge with a positive charge. Anderson et al. (1987) found that modification of a single residue in either the #42–45 or the #59–61 cluster inhibited cytochrome *f* oxidation. A much smaller effect was observed for PC modified at residue #68. These studies did not determine whether modification affected the binding of PC to Cyt *f*, electron transfer or both.

In contrast, changing D42 to an asparagine residue (D42N) using site-directed mutagenesis techniques (Nordling et al., 1991) had no effect on the interaction of PC with Cyt *f*. These results are surprising since D42 is conserved in all plant and algal PCs. Also, the electrostatic calculations of Durell et al. (1990) indicate that modification of D42 should have had an effect. Mutagenesis of additional residues in both acidic patches are required in order to clarify this question.

Morand et al., (1989) prepared a cross-linked adduct between Cyt *f* and PC. Two different types of cross-links were observed: One between residues D44 on PC and K187 on Cyt *f* (Fig. 6) and one between residue E59 or 60 on PC and an unidentified residue on Cyt *f*. Whether the cross-linked adducts are capable of intracomplex electron transfer is an open question. Peerey and Kostic (1989) showed that covalent complexes formed between PC and Cyt *c* were inactive in electron transport and proposed that motion or rearrangement of the reaction partners must occur between initial binding and final electron transfer. Qin and Kostic (1993) also failed to observe electron transfer from Cyt *f* to PC in the covalent cross linked adducts using laser flash spectroscopy with riboflavin semiquinone as the reductant. In contrast, Davis (D. J. Davis personal communication) observed rates of electron transfer from ruthenium-labeled PC to Cyt *f* within the covalent complex that were only 20% less than that observed for the non-covalent electrostatic complex. Intracomplex electron transfer was also observed by Takabe and Ishikawa (1989). The differences may reflect subtle differences in the complexes formed or in the methods of measurement. Among other factors, multiple covalent complexes are formed when PC and Cyt *f* react with the water-soluble carbodiimide (Gross et al., 1991; Qin and Kostic, 1993). The heterogeneity observed has been attributed to the covalent addition of varying amounts of N-acyl urea groups to anionic residues (Qin and Kostic, 1993).

In summary, the chemical modification and cross-linking studies show that the negative patches are important for the interaction of PC and Cyt *f*. However, these results do not prove whether electron transfer occurs via Y83 or the acid patches are only involved in binding either as an initial predocking complex or as the final binding site.

2. Tyrosine 83

Y83 has been subjected to both chemical modification and site directed mutagenesis studies. Gross et al. (1985), Anderson et al. (1985), Gross and Curtiss (1991) and Christensen et al. (1992) modified Y83 using tetranitromethane to add a nitro group adjacent to the hydroxyl. At neutral pH, this modification caused no change in the charge on Y83 and consequently, no change in activity was observed. However, addition of the nitro group lowered the pK of Y83 to pH 8.6 for reduced PC and to 8.3 for the oxidized form (Anderson et al., 1985). Thus, Y83 on oxidized PC has a net negative charge above pH 8. 3. Under these conditions, the rate of Cyt *f* oxidation by Nitro-Y83-PC was increased over that observed at neutral pH. These results were interpreted to mean that increasing the negative charge in this region of the PC molecule increases its binding affinity to Cyt *f*. Interestingly, inhibition was observed at pH values above 8.6 only for the modified PC (but not control PC)— possibly attributable to destabilization of the modified form by the extra negative charge.

Y83 has also been manipulated by site directed mutagenesis. He et al. (1991), using pea PC expressed in transgenic tobacco plants (Last and Gray, 1989), replaced Y83 with both phenylalanine (Y83F) and leucine (Y83L). Modi et al. (1992a), using spinach PC expressed in *E. coli* and using an azurin signal sequence (Nordling et al., 1990), replaced Y83 with phenylalanine (Y83F). Changes in the absorption of Cyt *f* were used to determine binding constants for complex formation allowing the investigators to distinguish between effects on binding and electron transfer. Both mutations decreased the binding constant of PC for Cyt *f* but Y83L also showed a decrease in the rate constant for electron transfer. Thus, these studies provide evidence that Y83 is involved not only in the binding of PC to Cyt *f* but also in electron transfer. They also show that an aromatic residue is required at position 83 for efficient electron transfer. All PCs have an aromatic residue at position 83 although two algal PCs, *Scenedesmus* (Fig. 1) and *Chlorella,* have a phenylalanine rather than a tyrosine at this position. He et al. (1991) also

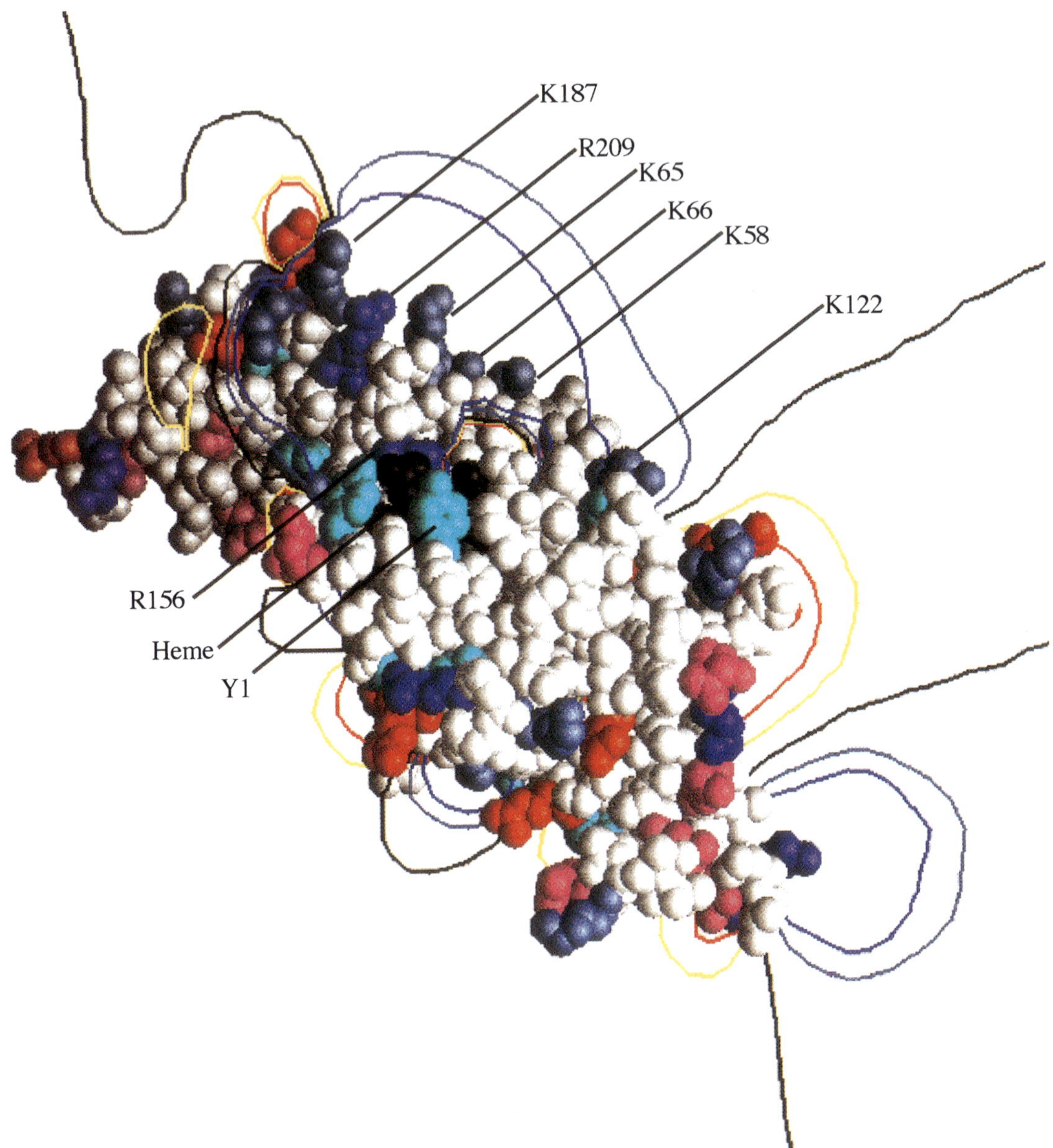

Fig. 6. A two-dimensional representation of the electrostatic potential field surrounding Cyt *f*. The crystal structure of Cyt *f* at 1. 96 Å resolution was obtained from Sergio Martinez (Martinez et al., 1994). The electrostatic map was solved for Cyt *f* using the program GRASP (Nicholls et al., 1991) The amino acid residues are color-coded as follows: arginine, dark blue; lysine, light blue; aspartate, magenta; glutamate, red, tyrosine, cyan; heme, black. The contours are coded as follows: –1, red; –0. 5, yellow; 0, black; +0. 5, light blue; +1, dark blue. All contours are in units of kT/e.

proposed that the hydroxyl of Y83 is hydrogen-bonded to a residue on Cyt *f* which would account for the difference observed between F83 and Y83.

In addition, Modi et al., (1992b) used NMR to study the binding site on PC for Cyt *f* in the electrostatic complex that is formed at low ionic strength (Qin and Kostic, 1992, 1993; Meyer et al., 1993). They concluded that the ferric ion of Cyt *f* was very close to Y83 on PC.

3. Modification of Other Residues

First, Gross et al. (1990) modified PC using 4-chloro-3,5-dinitrobenzoic acid (CDNB), a reagent which modifies amino groups, causing replacement of a positive charge by a negative charge. It was predicted that this modification would stimulate Cyt *f* oxidation. However, slight inhibition of Cyt *f* oxidation was observed for PCs modified at the α-amino of residue #1 and K54 and strong inhibition was observed for CDNB-K77 PC. It is proposed that the positive electrostatic field generated by these lysine residues is involved in steering PC to its binding site on Cyt *f*.

Second, Modi et al. (1992a) examined the effect of mutations of leucine 12 on the interaction of PC with Cyt *f*. L12 is located at Site 1 in the conserved hydrophobic patch surrounding H87. L12A showed a decrease in the binding to Cyt *f* but, surprisingly, L12N showed an increase in binding. These authors postulated a possible hydrogen bond between N12 and Cyt *f*. There is more than one possible interpretation of these data. One possible interpretation is that the binding site on PC for Cyt *f* may be larger than previously indicated including both Sites 1 and 2. Alternatively, Site 2 (i.e. the acidic patch) may be involved in the initial binding of PC with Cyt *f* and Site 1 may represent the final site at which electron transfer occurs.

D. Thc Structure of Cytochrome f and its Interaction with Plastocyanin

Cyt *f* is a 33 kD, c type cytochrome which is a component of the Cyt b_6f complex (Hauska et al., 1983; Widger, 1991; Gray, 1992). Cyt *f* is functionally analogous to Cyt c_1 of the mitochondrial and bacterial Cyt bc_1 complexes. The amino acid sequences of several Cyt *f*'s have been determined (Hauska et al., 1988; Widger, 1991; Gray, 1992). Cyt *f* is a transmembrane protein with only one membrane-spanning sequence located near the C-terminus of the protein (Willey et al., 1984). Proteolysis experiments (Willey et al., 1984) have shown that the C-terminus is located on the stromal side of the thylakoid membrane whereas the N-terminus containing the heme is exposed to the lumen where it interacts with PC. Cyt *f*s isolated from turnip leaves and other crucifers (Gray, 1992) lack the transmembrane tail, which is lost during isolation (or possibly before). Thus, these proteins can be isolated in a monomeric, water-soluble form. Cyt *f* from turnip was recently crystallized and its three-dimensional structure determined (Martinez et al., 1994; Chapter 22 of this volume).

The secondary structure of Cyt *f* (Martinez et al., 1994; Chapter 22) is unlike other c type cytochromes in that it consists of two domains (as opposed to one) and also contains a β-sheet structure as opposed to α-helix. The heme is located on the larger of the two domains, covalently bound to C21 and C24 and coordinated to H25 and the amino group of the N-terminal tyrosine residue (Fig. 6).

E. The Electrostatic Field of Cytochrome f

Cyt *f* contains several highly-conserved positively-charged residues including K65, K66, K122, K185, K187 and R209 that are depicted in Fig. 6, which shows a space-filling model of the lumenal portion of Cyt *f*. Of these, K187 is of particular interest since it can be cross-linked to D44 on PC (Morand et al., 1989). Also shown are the heme group and Y1, the N-terminus of which is ligated to the heme.

The electrostatic field of Cyt *f* was calculated using the program GRASP which graphically presents the electrostatic properties of macromolecules using an algorithm similar to that of DelPhi (Nicholls, et al., 1991). The most important electrostatic feature of Cyt *f* is the positive field contributed by K65, K66, K122, K185, K187 and R209, which will attract the negative charges on PC and draw it into the vicinity of the Cyt *f* molecule.

IV. The Interaction of Plastocyanin with P700

A. The Role of Electrostatic Forces and Divalent Cations in the Interaction between Plastocyanin and Photosystem I

As in the case of PC-Cyt *f* interactions, electrostatic forces play an important role in determining the interaction between PC and Photosystem I (PS I). At neutral pH, both PS I and PC bear a net negative charge. Under these conditions, divalent cations increase the rate of electron transfer between the two proteins, (Davis et al., 1980; Takabe et al., 1983, 1984, Ratajczak et al., 1988), presumably by screening the negative charges on PS I. The divalent cation requirement can be eliminated in several ways. When Davis et al., (1980) illuminated spinach PS I in the presence of positively-charged *Anabaena variabilis* PC or Cyt *c*-553, they found that not only was the K_m lower than for spinach PC but also that divalent cations caused inhibition rather than stimulation of the reaction. In addition, Burkey and Gross (1981a) showed that the divalent cation requirement was eliminated by changing the charge on PS I by chemical modification of carboxyl groups using ethylene-diamine and a water-soluble carbodiimide. Conversely, modification of negative charges on PC (Burkey and Gross, 1981b, 1982; Takabe et al., 1984; Anderson et al., 1987) also eliminated the divalent cation requirement.

Ratajczak et al. (1988) investigated the cation requirement for PC-P700 interaction in both intact membranes and various types of solubilized PS I particles. They concluded that divalent cations stimulated the reaction of PC with P700 in intact membranes by screening negative charges on PS I. However, divalent cations inhibited PC binding in some types of PS I particles. These authors concluded that a positively-charged subunit of PS I was involved in PC binding.

B. The Binding Site on Photosystem I for Plastocyanin

Photosystem I is a multisubunit integral membrane protein complex (Golbeck, 1987). The P700 is shared by the two large subunits (each *ca.* 83 kD) that are products of the *psaA* and *psaB* genes (Cantrell and Bryant, 1987). Recently, PS I has been crystallized and its three-dimensional structure determined at 6 Å resolution (Krauss et al., 1993). P700 is thought to be located near the lumenal surface of PS I and separated from it by two α-helices lying in the plane of the membrane. Krauss et al. (1993) proposed that PC binds at this site.

In addition to the two large subunits, one of the small subunits has been implicated in the binding of PC to P700. This subunit was designated as Subunit III by Bengis and Nelson (1985) who found that its removal from PS I particles inhibited the reaction of PC with P700. This subunit has since been identified as the product of the *psaF* gene (Sheller and Moller, 1990; Chitnis et al., 1991).

Further evidence of the involvement of this subunit comes from the cross-linking studies of Wynn and Malkin (1988) and Hippler et al. (1989) who used a water-soluble carbodiimide to link PC to PS I. They obtained a cross-linked adduct that contained both PC and Subunit III. The cross-linked adduct obtained was found to be active in $P700^+$ reduction (Hippler et al., 1989) in that it retained 50% of the amplitude of the original flash-induced P700 signal observed in control PS I. Moreover, the decay time was identical to that for control PC. This is important because the decay time represents the turnover time of active PS I centers. Although the PS I containing the cross-linked adduct was not active in steady state electron transfer (Wynn and Malkin, 1988), this is not surprising since docked PC would be locked in place and could not be replaced once oxidized.

The *psaF* gene product is positively-charged (Scheller and Moller, 1990) allowing it to interact with negatively-charged PC. These results agree with the predictions of Ratajczak et al., (1988) described above.

Other evidence suggests that Subunit III is not always required for steady state electron transfer (i. e. NADP reduction) (Scheller et al., 1990). Chitnis et al. (1991) showed that *Synechocystis* sp. PCC 6803 was able to grow photoautotrophically in the absence of Subunit III and Hatanaka et al., (1993) showed that this subunit was not required for the oxidation of Cyt *c*-553. If PC and Cyt *c*-553 are interchangeable, then this subunit should not be required for PC binding. However, there is no direct evidence for this. One possibility is that Subunit III regulates the reaction of PC with PS I but is not absolutely required. Alternatively, Mg^{2+} ions might substitute for this subunit under certain circumstances although Hatanaka et al. (1993) found that the presence of the *PsaF* gene product did not eliminate the requirement for Mg^{2+} in *Synechocystis*. It is interesting that Subunit III appears disordered in the crystal structure of PS I (Krauss et al., 1993).

C. The Binding Site on Plastocyanin for Photosystem I

The *psaF* gene product is positively-charged (Scheller and Moller, 1990), suggesting that positively-charged residues on Subunit III interact with negatively-charged residues on PC, thus implicating Site 2 as the binding site for P700. If this were true, then modification of the negative patches on PC should inhibit electron transfer from PC to P700. However, Anderson et al. (1987) observed no effect on $P700^+$

reduction in Photosystem I particles for PC modified at residues #42–45 using ethylenediamine plus a water-soluble carbodiimide. To the contrary, they observed stimulation of $P700^+$ reduction for PCs with modifications either at residues #59–61 and #68. These residues are closer to H87 (Site 1) than they are to Y83 (Site 2). However, it is uncertain whether Subunit III was present in the PS I particles used for these studies.

Further evidence implicating Site 1 comes from the work of Nordling et al. (1991) who studied the effect of two mutations (Y83H and L12E) on the interaction of PC with PS I. When they studied Y83H, located at Site 2, they found only a small effect, which could be entirely explained by a small change in redox potential for the mutant. In contrast, they observed a large effect for the L12 mutant. L12 is one of the conserved residues adjacent to H87 at Site 1. Haehnel et al. (1994) found that changing G10 in spinach PC to a leucine using mutagenesis techniques inhibited the interaction of PC with PS I. All of these studies implicate Site 1 in the binding of PS I.

These results can be reconciled as follows. PC may donate electrons to PS I using Site I for electron transfer. However, Subunit III may stabilize the complex formed between PC and P700 and hold the PC in the correct orientation for electron transfer by interacting with the negative charges at Site 2. If Subunit III is absent, divalent cations may take over its stabilizing role.

V. The Function of Plastocyanin in the Lumen of the Thylakoid at Acid pH

PC is situated in the lumen of the thylakoid where it experiences pH values of 5 or below (Rottenberg and Grunwald, 1972; Pick et al., 1974) upon illumination of the thylakoid, raising the question as to how PC functions under these conditions.

A. The Effect of pH on the Structure of Plastocyanin

The three-dimensional structure of oxidized PC has been shown to be invariant between pH 6 and 4.2 (Guss and Freeman, 1983). In contrast, reduced PC exists in two different conformations (Guss et al., 1986), one at low pH (pH 3.8) and the other which predominates at high pH (pH 7). At intermediate pH values, both forms are found in an equilibrium mixture. The conformation of the high pH form is almost identical to that of oxidized PC. However, the low pH form is redox inactive because the Cu-H87 ligand bond is broken and the histidine is rotated by 180° about the C^{β}-C^{γ} bond. Also, P36 is changed from a C-endo to a C-exo conformation.

The breaking of the H87-Cu bond is due to the protonation of H87. The pH at which protonation occurs is species dependent (Sinclair-Day et al., 1985). For example, the pK for H87 is 4.9 for spinach but 5.7 for parsley. This suggests that a large percentage of the PC may be in the low pH, inactive form under normal illumination conditions.

B. The Effect of pH on the Redox and Electron Transfer Properties of Plastocyanin

The protonation of H87 has important ramifications for the function of PC and its interaction with its reaction partners, Cyt *f* and P700 (Sykes, 1985 and references therein). The rate constant for the oxidation of reduced PC decreases as the pH is lowered because H87 becomes protonated with the resulting loss of the H87-Cu bond. In contrast, the rate constant for the reduction of oxidized PC is pH-independent over the region of interest (pH 6 to 4.2). As a consequence, there is a positive shift in the midpoint redox potential, $E^{o\prime}$.

At neutral pH values, the midpoint redox potential of Cyt *f* (365 mV; Gray, 1992) and PC (+372 mV; Sanderson et al. 1986) are almost equal, so that the driving force for electron transport is nearly zero and the electron transfer reaction is reversible under conditions in which the concentrations of the oxidized and reduced electron carriers are equal. Lowering the pH will actually increase the driving force for forward electron transfer (i. e. from Cyt *f* to PC) as a result of the positive shift in redox potential (*ca.* + 420 mV).

However, lowering the pH will also decrease the binding affinity of PC for Cyt *f* by protonating residues in the two negative patches. This is illustrated in Fig. 5c, which shows the effect on the electrostatic field of decreasing the negative charge on negative residues of PC by 50%. Beoku-Betts et al. (1985) showed that the second order rate constant for Cyt *f* oxidation decreased with decreasing pH with a pK of 5.07 ± 0.07. Thus, electrostatic interactions will be less important at physiological pH values.

The rate of electron transfer from PC to P700 is

also pH dependent. The rate of P700$^+$ reduction increases as the pH is lowered (Takabe et al., 1983, 1984) due to a decrease in the number of negative charges on both PC and PS I. The rate of electron transfer reaches a maximal value at pH 4.3 and decreases at lower pH values. The acid pH optimum for electron transport is interesting in light of the observation that reduced PC is redox inactive under these conditions as a result of the protonation of H87 (see above). One possible explanation of these contradictory results is that reduced PC becomes deprotonated, and thus reactivated, when bound to its reaction site on PS I. Deprotonation of H87 would be favored if the binding site for PC on PS I were hydrophobic in nature. This would be consistent with the known structure of PS I (Krauss et al., 1993).

The inactivation of PC at low pH may provide a control mechanism preventing a back reaction between reduced PC and oxidized Cyt *f*. At neutral pH, the back reaction (reduction of Cyt *f* by PC) is approximately equal to the forward reaction (reduction of PC by Cyt *f*) caused by the small difference between the E^{o}'s for the two molecules. At low pH, reduced PC is inactivated and the E$^{o'}$ is shifted toward +420 mV favoring the forward electron transfer reaction.

VI. Conclusions

The studies described above emphasize the dynamic nature of the interactions of PC with its reaction partners Cyt *f* and P700. The negatively-charged amino acid residues of the acid patches of PC have been implicated in the docking of PC with both Cyt *f* and P700. However, the site of electron transfer is less clear in both cases. In the case of Cyt *f*, the recent elucidation of the crystal structure should permit a final answer to this question. Molecular modeling studies and site-directed mutagenesis of both proteins will allow a detailed analysis of both the binding and electron transfer processes. However, in the case of PS I, the answer will probably have to wait for more detailed knowledge of the structure of the complex.

Acknowledgments

I wish to thank Dr. Caroline Breitenberger for helping me obtain the PC sequences from the PIR database. I also wish to thank Dr. William Cramer and Sergio Martinez for supplying me with the coordinates for cytochrome *f*. Finally, I wish to thank Robert Ross for producing the molecular model of parsley PC and Douglas C. Pearson, Jr. for helpful discussion and for Fig. 6.

References

Adam Z and Malkin R (1989) On the interaction of cytochrome *f* and plastocyanin. Biochim Biophys Acta 975: 158–163

Anderson GP, Draheim JE and Gross EL (1985) Plastocyanin conformation: The effect of oxidation state on the pK_a of nitrotyrosine-83. Biochim Biophys Acta 810: 123–131

Anderson GP, Sanderson DG, Lee CH, Durell S, Anderson LB and Gross EL (1987) The effect of ethylenediamine chemical modification of plastocyanin on the rate of cytochrome *f* oxidation and P700$^+$ reduction. Biochim Biophys Acta 894: 386–398

Arudchandran A, Seeburg D, Burkhart W and Bullerjahn GS (1994) Nucleotide sequence of the *petE* gene encoding plastocyanin from the photosynthetic prokaryote, *Prochlorothrix hollandica*. Biochim Biophys Acta 1188: 47–49.

Bagby S, Driscoll PC, Goodall, KG, Redfield C and Hill HAO (1990) The complex formed between plastocyanin and cytochrome *c*. Investigation by NMR spectroscopy. Eur J Biochem 18: 411–420

Bagby S, Driscoll PC, Haravey TS and Hill, HAO (1994) High-resolution structure of reduced parsley plastocyanin. Biochemistry 33: 6611–6622

Bengis J and Nelson N (1985) Subunit structure of Photosystem I reaction center. J Biol Chem 252: 4564–4569

Beoku-Betts D, Chapman SK, Knox CV and Sykes AG (1983) Concerning the binding site on plastocyanin for its natural redox partner cytochrome *f*. J Chem Soc Chem Commun 1983: 1150–1152

Beoku-Betts D, Chapman SK, Knox CV and Sykes AG (1985) Kinetic studies on 1:1 electron transfer reactions involving blue copper proteins. 11. Effects of pH, competitive inhibition, and chromium (III) modification on the reaction of plastocyanin with cytochrome *f*. Inorg Chem 24: 1677–1681

Bernstein FC, Doetzle TF, Williams GJB, Myer Jr EF, Brice MD, Rodgers JR, Kennard O, Shimanouchi T, and Tasumni M (1977) The protein data bank: A computer-based archival file for macromolecular structures. J Mol Biol 112: 535–542

Betts JN, Beratan DN and Onuchic JN (1992) Mapping electron tunneling pathways: An algorithm that finds the 'minimum length'/maximum coupling pathway between electron donors and acceptors in proteins. J Am Chem Soc 114: 4043–4046

Bohner H and Boger P (1978) Reciprocal formation of cytochrome Cyt *c*-553 and plastocyanin in *Scenedesmus*. FEBS Lett 85: 337–339

Boulter D, Haslett BG, Peacock D, Ramshaw JAM and Scawen MD (1977) Chemistry, function and evolution of plastocyanins. Int'l Rev Biochem 13: 1–40

Briggs LM, Pecoraro VL and McIntosh L (1990) Copper-induced expression, cloning and regulatory studies of the plastocyanin gene from the cyanobacterium *Synechocystis* sp. PCC 6803. Plant Mol Biol 15: 633–642

Burkey KO and Gross EL (1981a) Use of chemical modification to study the relationship between activity and net protein charge of the Photosystem I core complex. Biochemistry 20: 2961–2967

Burkey KO and Gross EL (1981b) Effect of carboxyl group modification on the redox properties and electron donation capability of spinach plastocyanin. Biochemistry 20: 5495–5499

Burkey KO and Gross EL (1982) Chemical modification of spinach plastocyanin: Separation and characterization of four different forms. Biochemistry 21: 5886–5890

Cantrell A and Bryant DA (1987) Molecular cloning and nucleotide sequence of the *psaA* and *psaB* genes of the cyanobacterium *Synechocystis* sp. PC 7002. Plant Mol Biol 9: 453–468

Chitnis PR Purvis D and Nelson N (1991) Molecular cloning and targeted mutagenesis of the gene *psaF* encoding subunit III of Photosystem I from the cyanobacterium *Synechocystis* sp. PCC 6803. J Biol Chem 266: 20145–20151

Christensen HEM, Conrad LS and Ulstrup J (1992) Effects of NO_2-modification of Tyr 83 on the reactivity of spinach plastocyanin with cytochrome *f*. Biochim Biophys Acta 1099: 35–44

Church WB, Guss JM, Potter JJ and Freeman HC (1986) The crystal structure of mercury-substituted poplar plastocyanin at 1. 9 Å resolution. J Biol Chem 261: 234–237

Coleman PM, Freeman HC, Guss JM, Murata M, Norris VA, Ramshaw J AM and Venkatappa MP (1978) X-ray crystal analysis of plastocyanin at 2. 7 Å resolution. Nature 272: 319–324

Collyer CA, Guss JM, Sugimura Y, Yoshizaki FY and Freeman HC (1990) Crystal structure of plastocyanin from a green alga, *Enteromorpha prolifera*. J Mol Biol 211: 617–632

Cookson DJ, Hayes MT and Wright PE (1980) NMR Studies of the interaction of plastocyanin with chromium (III) analogues of inorganic electron transfer reagents. Biochim Biophys Acta 591: 162–176

Davis DJ, Krogmann DW and San Pietro A (1980) Electron donation to Photosystem I. Plant Physiol 65: 697–702

de Boer AD and Weisbeek PJ (1991) Chloroplast protein topogenesis: Import, sorting and assembly. Biochim Biophys Acta 1071: 221–253

Dimitrov MI, Egorov CA, Donchev AA, and Atanasov BP (1987) Complete amino acid sequence of poplar plastocyanin b. FEBS Lett 226: 17–22

Dimitrov MI, Donchev AA and Egorov TA (1990) Microheterogeneity of parsley plastocyanin. FEBS Lett 265: 41–145

Durell SR, Labanowski J and Gross EL (1990) Modeling of the electrostatic potential field of plastocyanin. Arch Biochem Biophys 277: 241–254

Freeman HC (1981) Electron transfer in 'blue' copper proteins. Coord Chem 21: 29–51

Garrett TPJ, Clingeleffer DJ, Guss JM, Rogers SH and Freeman HC (1984) The crystal structure of poplar apoplastocyanin at 1. 8 Å resolution. The geometry of the copper-binding site is created by the polypeptide. J Biol Chem 59: 2822–2825

Gilson MK and Honig BH (1988) Energetics of charge-charge interactions in proteins. Proteins: Struct Funct and Genetics 3: 32–52

Gilson MK, Sharp KA and Honig BH (1987) Calculating the electrostatic potential of molecules in solution: Method and error assessment. J Comput Chem 9: 327–335

Golbeck JH (1987) Structure, function and organization of the Photosystem I reaction center complex. Biochim Biophys Acta 894: 167–204

Gray JC (1992) Cytochrome *f*: Structure, function and biosynthesis. Photosynth Res 34: 359–474

Gross EL (1993) Plastocyanin: Structure and function. Photosynth Res 37: 103–116

Gross EL and Curtiss A (1991) The interactions of nitrotyrosine-83 plastocyanin with cytochromes *f* and *c*: pH dependence and the effect of an additional negative charge on plastocyanin. Biochim Biophys Acta 1056: 166–172

Gross EL, Anderson GP, Ketchner S, and Draheim JE (1985) Plastocyanin conformation. The effect of nitrotyrosine modification and pH. Biochim Biophys Acta 808: 437–447

Gross EL, Curtiss A, Durell SR and White D (1990) Chemical modification of spinach plastocyanin using 4-chloro-3,5-dinitrobenzoic acid: Characterization of four singly-modified forms. Biochim Biophys Acta 1016: 107–114

Gross EL, Molnar S, Curtiss A, Reuter RA and Berg SP (1991) The use of monoclonal antibodies to study the structure and function of cytochrome *f*. Arch Biochem Biophys 289: 244–255

Gross EL, Draheim JE, Curtiss AS, Crombie B, Scheffer A, Pan B, Chiang C, and Lopez A (1992) Thermal denaturation of plastocyanin: The effect of oxidation state, reductants, and anaerobicity. Arch Biochem Biophys 298: 413–419

Guss JM and Freeman HC (1983) Structure of oxidized poplar plastocyanin at 1. 6 Å resolution. J Mol Biol 169: 521–563

Guss JM, Harrowell PR, Murata M, Norris VA and Freeman HC (1986) Crystal structure analyses of reduced poplar plastocyanin at six pH values. J Mol Biol 192: 361–387

Guss JM, Bartunik HD and Freeman HC (1992) Accuracy and precision in protein structure analysis. Restrained least-squares refinement of the structure of poplar plastocyanin at 1. 33 Å resolution. Acta Cryst B-48: 790–811

Haehnel W, Jansen T, Gause K, Klosgen RB, Stahl B, Michl D, Huvermann B, Kares M and Herrmann RG (1994) Electron transfer from plastocyanin to Photosystem I. EMBO J. 13: 1028–1038

Handford PM, Hill HAO, Lee RW-K, Henderson RA and Sykes AG (1980) Investigation of the binding of inorganic complexes to blue copper proteins by proton NMR spectroscopy. I. The interaction between the $[Cr (phen)_3]^{3+}$ and $[Cr (CN)_5]^{+3}$ ions and the Cu (I) form of parsley PC. J Inorg Biochem 13: 83–88

Hatanaka H, Sonoike K, Hirano M and Katoh H. (1993) Small subunits on Photosystem I reaction center complexes from *Synechococcus elongatus*. I. Is the *psaF* gene product required for oxidation of cytochrome Cyt *c*-553? Biochim Biophys Acta 1141: 45–51

Hauska GA, McCarty RE, Berzborn RJ and Racker E (1971) Partial resolution of the enzymes catalyzing photophosphorylation. VII. The function of plastocyanin and its interaction with a specific antibody. J Biol Chem 246: 3524–3531

Hauska G, Hurt E, Gabellini N and Lockau W (1983) Comparison of aspects of the quinol-cytochrome *c*/plastocyanin oxidoreductases. Biochim Biophys Acta 726: 97–133

Hauska G, Nitschke W, and Hermann RG (1988) Amino acid identities in the three redox center carrying polypeptides of the cytochrome bc_1/b_6f complexes. J Bioenerget Biomemb 20: 211–228

He S, Modi S, Bendall DA, and Gray JC (1991) The surface-exposed residue tyrosine Tyr 83 of pea plastocyanin is involved in both binding and electron transfer reactions with cytochrome *f*. EMBO J 10: 4011–4016

Hippler M, Ratajczak R and Haehnel W (1989) Identification of the plastocyanin binding subunit of Photosystem I. FEBS Lett 250: 280–284

Ho KK and Krogmann DW (1984) Electron donation to P700 in cyanobacteria and algae, an instance of unusual genetic variability. Biochim Biophys Acta 766: 310–316

Katoh S (1995) The discovery and function of plastocyanin: A personal account. Photosynth Res 43: 177–189

Klapper I, Hagstron R, Fine R, Sharp K and Honig B (1986) Focusing of electric fields in the active site of Cu-Zn superoxide dismutase: Effects of ionic strength and amino-acid modification. Proteins: Struct, Funct and Genetics 1: 47–59

Koide S, Dyson HJ and Wright PE (1993) Characterization of a folding intermediate of apoplastocyanin trapped by proline isomerization. Biochemistry 32: 12299–12310

Krauss N, Hinrichs W, Witt I, Fromme P, Pritzkow W, Dauter Z, Betzel C, Wilson KS, Witt HT and Saenger W (1993) Three-dimensional structure of system I of photosynthesis at 6 Å resolution. Nature 351: 326–351

Last DI and Gray JC (1989) Synthesis and accumulation of transgenic tobacco plants. Plant Mol Biol 12: 655–666

Martinez SE, Huang D, Szczepaniak A, Cramer WA and Smith JL (1994) Crystal structure of chloroplast cytochrome *f* reveals a novel cytochrome fold and a novel heme ligation. Structure 2: 95–105

Merchant S and Bogorad L (1986) Rapid degradation of apoplastocyanin in Cu (II) deficient cells of *Chlamydomonas reinhardtii*. J Biol Chem 261: 15850–15853

Meyer TE, Zhao ZG, Cusanovitch MA and Tollin G (1993) Transient kinetics of electron transfer from a variety of c-type cytochromes to plastocyanin. Biochemistry 32: 4552–4559

Modi S, Nordling M, Lundberg LG, Hannson O and Bendall DS (1992a) Reactivity of cytochromes *c* and *f* with mutant forms of spinach plastocyanin. Biochim Biophys Acta 1102: 85–90

Modi S, McLaughlin E, Bendall DS, He S and Gray JC (1992b) Use of NMR relaxation measurements to derive the binding site of plastocyanin in complexes with cytochrome-*f* and *c*. Bull Magnetic Res 14: 159–164

Moore JM, Case DA, Chazin WJ, Gippert GP, Havel TF, Powls R and Wright PE (1988) Three-dimensional solution structure of plastocyanin from the green alga *Scendesmus obliquus*. Science 240: 314–317

Moore JM, Lepre CA, Gippert GP, Chazin WJ, Case DA and Wright PE (1991) High-resolution solution structure of French bean plastocyanin and comparison with the crystal structure of poplar plastocyanin. J Mol Biol 221: 533–555

Morand LZ, Frame MK, Colvert KK, Johnson DA, Krogmann DW and Davis DJ (1989) Plastocyanin cytochrome *f* interaction. Biochemistry 28: 8039–8047

Nicholls A, Sharp KA and Honig B (1991) Protein folding and association: Insights from the interfacial and thermodynamic properties of hydrocarbons. Proteins 11: 281–296

Nielsen PS and Gausing K (1987) The precursor of barley plastocyanin. Sequence of cDNA clones and gene expression in different tissues. FEBS Lett 225: 159–162

Niwa S, Ishikawa H, Nikai S and Takabe T (1980) Electron transfer reactions between cytochrome *f* and plastocyanin from *Brassica komatuna*. J Biochem 88: 1177–1183

Nordling M, Olausson T and Lundberg LG (1990) Expression of plastocyanin in *E. coli*. FEBS Lett 276: 98–102

Nordling M, Sigfridsson K, Young S, Lundberg LG and Hansson O (1991) Flash-photolysis studies of the electron transfer from genetically modified spinach plastocyanin to Photosystem I. FEBS Lett 291: 327–330

Northrup SH, Boles JO and Reynolds JCL (1988) Brownian dynamics of cytochrome *c* and cytochrome *c* peroxidase association. Science 241: 67–70

Peerey LM and Kostic NM (1989) Oxidoreduction reactions involving the electrostatic and the covalent complex of cytochrome *c* and plastocyanin: Importance of the protein rearrangement for the intracomplex electron-transfer reactions. Biochemistry 28: 1861–1868

Pick U, Rottenberg H and Avron M (1974) The dependence of photophosphorylation in chloroplasts on ΔpH and external pH. FEBS Lett 48: 32–36

Qin L and Kostic NM (1992) Electron-transfer reactions of cytochrome *f* with flavin semiquinones and with plastocyanin: Importance of protein-protein electrostatic interactions and donor-acceptor coupling Biochemistry 31: 5145–5150

Qin L and Kostic NM (1993) Importance of protein rearrangement in the electron transfer reactions between the physiological partners cytochrome *f* and plastocyanin. Biochemistry 32: 6073–6080

Ratajczak R, Mitchell R and Haehnel W (1988) Properties of the oxidizing side of Photosystem I. Biochim Biophys Acta 933: 306–318

Redinbo MR, Cascio D, Choukair MK, Rice D, Merchant S and Yeates DO (1993) The 1. 5-Å crystal structure of plastocyanin from the green alga *Chlamydomonas reinhardii*. Biochemistry 32: 10560–10567

Roberts VA, Freeman HC, Olson AJ, Tainer JA, and Getzoff ED (1991) Electrostatic orientation of the electron-transfer complex between plastocyanin and cytochrome *c*. J Biol Chem 266: 13431–13441

Rottenberg H and Grunwald T (1972) Determination of ΔpH in chloroplasts. 3. Ammonium uptake as a measure of ΔpH in chloroplasts and subchloroplast particles. Eur J Biochem 25: 71–74

Sanderson DG, Anderson LB and Gross EL (1986) Determination of the redox potential and diffusion coefficient of the protein plastocyanin using optically transparent fiber electrodes. Biochim Biophys Acta 852: 269–278

Scawen MD, Hewitt EJ and James DM (1975) Preparation, crystallization and properties of *Cucurbita pepo* plastocyanin and ferredoxin. Phytochemistry 14: 1225–1234.

Scheller HV and Moller BL (1990) Photosystem I polypeptides. Physiol Plant 78: 484–494

Scheller HV, Andersen B, Okkels S, Swendsen I and Moller BL (1990) Photosystem I in barley: Subunit I is not essential for the interaction with plastocyanin. In: Baltscheffsky M (ed) Current Research In Photosynthesis Vol. II, pp 679–682. Kluwer Academic Publishers, The Netherlands

Selak MA and Whitmarsh J (1984) Charge transfer from Photosystem I to the cytochrome *b/f* complex: Diffusion and membrane lateral heterogeneity. Photochem Photobiol 39: 485–490

Shoji A, Yoshizaki F, Karahashi A, Sugimura Y, and Shimokoriyama M (1985) Seikagaku 57: 1036

Simpson RJ, Moritz RL, Nice EC, Grego B, Yoshizaki F, Sugimura Y, Freeman HC and Murata M (1986) Complete amino acid sequence of plastocyanin from a green alga, *Enteromorpha prolifera.* Eur J Biochem 157: 497–506

Sinclair-Day JD, Sisley MJ, Sykes AG, King GC and Wright PE (1985) Acid dissociation constants for plastocyanin in the Cu^{I} state. J Chem Soc Chem Commun 1985: 505–507

Skolnick J and Kolinski A (1990) Simulations of the folding of a globular protein. Science 250: 1121–1125

Sykes AG (1985) Structure and electron transfer reactivity of the blue copper protein plastocyanin. Chem Soc Rev 14: 283–315

Sykes AG (1991) Plastocyanin and the blue copper proteins. Struct Bond 75: 175–224

Takabe T and Ishikawa H (1989) Kinetic studies on a cross-linked complex between plastocyanin and cytochrome *f.* J Biochem 105: 98–102

Takabe T, Ishikawa H, Niwa S and Itoh S (1983) Electron transfer between plastocyanin and P700 in highly purified Photosystem I reaction center complex. Effect of pH, cations, and subunit peptide composition. J Biochem 94: 1901–1911

Takabe T, Ishikawa H, Niwa S and Tanaka Y (1984) Electron transfer reactions of chemically-modified plastocyanins with P700 and cytochrome *f.* Importance of local charges. J Biochem 96: 385–393

Takabe T, Takenaka K, Kawamura H and Beppu Y (1986) Charges on proteins and distances of electron transfer in metalloprotein redox reactions. J Biochem 99: 833–840

Takenaka K and Takabe T (1984) Importance of local positive charges on cytochrome *f* for electron transport to plastocyanin and potassium ferricyanide. J Biochem 96: 1813–1821

Wendoloski JJ, Matthew JB, Weber PC and Salemme FR (1991) Molecular dynamics of a cytochrome *c*-cytochrome b_5 electron transfer complex. Science 238: 794–797

Widger WR (1991) The cloning and sequencing of *Synechococcus PCC 7002 pet* CA operon: Implications for the cytochrome Cyt *c*-553 binding domain of cytochrome *f.* Photosynth Res 30: 71–85

Willey DL, Auffret AD and Gray JC (1984) Structure and topology of cytochrome *f* from pea chloroplast membranes. Cell 36: 555– 562

Wynn RM and Malkin R (1988) Interaction of plastocyanin with PS I: A chemical cross-linking study of the polypeptide that binds plastocyanin. Biochemistry 27: 5863–5869

Yano H, Kamo M, Tsugita A, Aso K and Nozu Y (1989) The amino acid sequence of plastocyanin from rice *(Oryza sativa,* subspecies japonica). Protein Seq Data Anal 2: 385–389

Chapter 22

Some Consequences of the High Resolution X-Ray Structure Analysis of Cytochrome *f*

S. E. Martinez, D. Huang, J. L. Smith and W. A. Cramer
Department of Biological Sciences, Purdue University, West Lafayette, IN 47907-1392, USA

Summary

A high resolution structure[1] is described of the lumen-side domain of cytochrome *f*. This structure is the first for a polypeptide subunit of cytochrome bc_1 or b_6f complexes, and has at least three unique or unprecedented features for a *c*-type cytochrome: (i) more than one structural domain; (ii) predominantly β-strand; and (iii) the N-terminal α-amino group is the orthogonal (6th) ligand, which is unprecedented for a heme protein.

I. Introduction

The understanding of structure/function of cytochrome *f* has been advanced by the determination of its X-ray structure at a resolution of 2.3 Å (Martinez et al., 1992,1994). The structure that was solved was that of the soluble 252-residue lumen-side domain (M_r = 28,116 with heme) of the 285-residue mature polypeptide (M_r = 31,298 in turnip chloroplasts) that spans the membrane once (Fig. 1D).

The use of the lumen-side polypeptide of the cytochrome arose from attempts to prepare intact cytochrome *f* from spinach chloroplast thylakoid membranes for crystallization. These attempts were abandoned because the intact cytochrome aggregates, as previously noted (Ho and Krogmann, 1980). Preparations of cytochrome *f* from cruciferous plants, however, were reported to be soluble and monomeric with an M_r value of 27,000 instead of 32,000–33,000 (Gray, 1978). It was suggested that the decrease in M_r value might arise from proteolysis near the C-terminus because the N-terminal sequence of cytochrome *f* from chloroplasts of the cruciferous plants was found to be identical to that of spinach and pea (Willey et al., 1984). It was inferred that the tendency of the spinach cytochrome *f* to aggregate in aqueous solu-

[1] 2.3 Å at the time of the writing of the article, now 1.96 Å.

Donald R. Ort and Charles F. Yocum (eds): Oxygenic Photosynthesis: The Light Reactions, pp. 431–437.

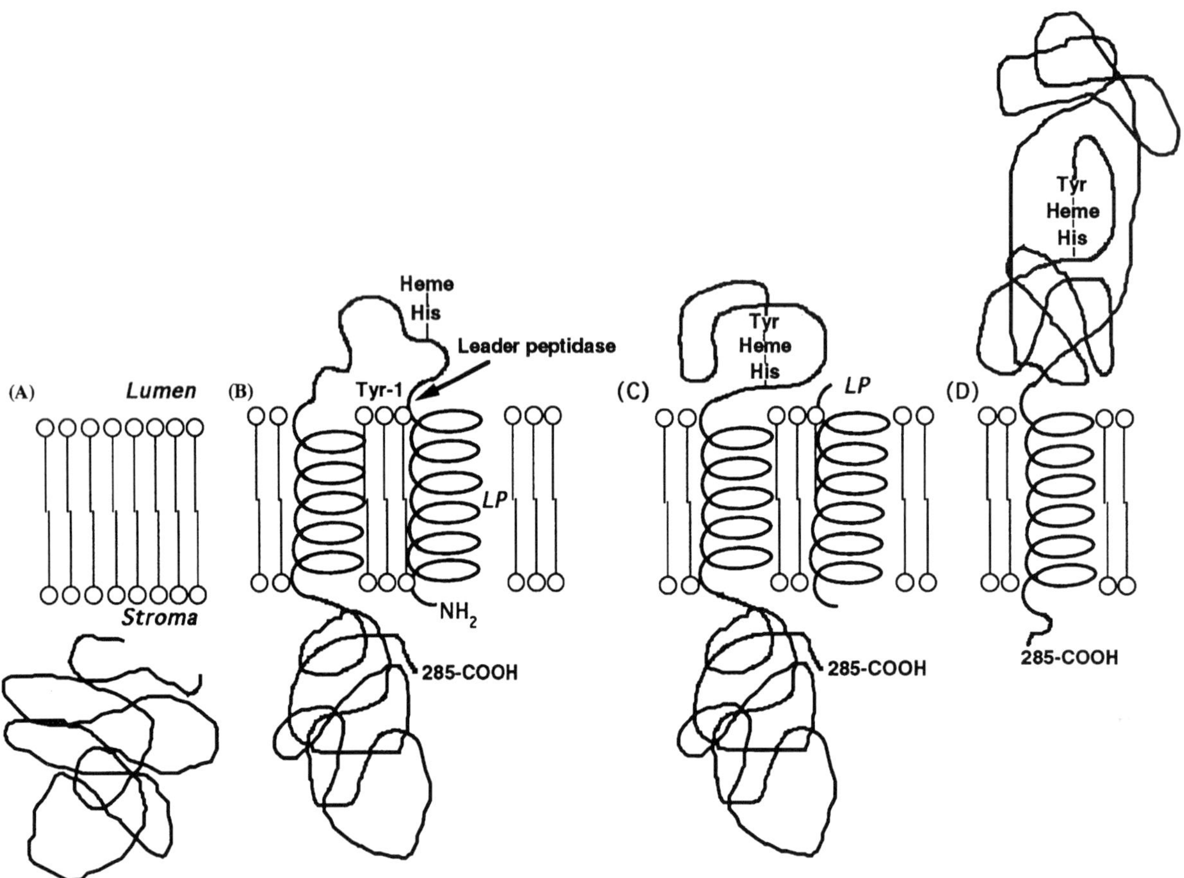

Fig. 1. Models for membrane translocation and assembly of cytochrome *f* and for the topography of the mature protein. (A-C) Model for assembly and processing. (A) Soluble precursor. (B,C) show that complete heme ligation involving the α-amino group of Tyr-1 (Martinez et al., 1994) cannot occur until the 35 residue leader peptide has been removed by the processing peptidase that is believed to reside on the lumen side of the membrane (Johnson et al., 1991; Kirwin et al., 1991). (D) Topography of mature form of the 285-residue cytochrome *f*, which has a large (250-residue) lumen-side domain, a 20-residue *trans*-membrane helix, and approximately 15-residues on the C-terminal end that protrude on the stroma side of the membrane. Rectangular box: thylakoid membrane; LP, leader peptidase, which is assumed to be degraded between the states shown in panels C and D.

tion might arise from the 20-residue hydrophobic segment (residues no. 251–270, underlined in Table 1) near the C-terminus, and that removal of the hydrophobic domain near the C-terminus might explain the water-soluble monomeric nature of the cruciferous cytochrome *f* fragment. Therefore, cruciferous cytochrome *f* was used for structural analysis.

Abbreviations: Cyt – cytochrome; ESMS – electrospray mass spectrometry; HPLC – high performance liquid chromatography; SDS-PAGE – sodium dodecyl sulfate polyacrylamide electrophoresis

II. Preparation and Crystallization of Truncated Cytochrome *f*

Cytochrome *f* purified from turnip leaves has an M_r value of 31,000 on SDS-PAGE, compared to 33,800 in the turnip thylakoid membrane (Martinez et al., 1992). C-terminal sequencing analysis, comparison with the nucleotide sequence of the cytochrome provided by J. Gray (1992), and electrospray mass spectroscopy, which yielded a peak for ~90% of the protein at a molecular weight corresponding to its true weight +31 daltons, implied a virtually pure polypeptide fragment of 252 residues terminating in a C-terminal glutamine (Table 1). The additional 31

Table 1. Amino Acid Sequence of Turnip Cytochrome *f* (Gray, 1992)[a]

```
1                                                  50
YPIFAQQNYE NPREATGRIV CANCHLASKP VDIEVQAVL PDTVFEAVVK

51                                                 100
IPYDMQLKQV LANGKKGALN VGAVLILPEG FELAPPDRIS PEMKEKIGNL

101                                                150
SFQNYRPNKK NILVIGPVPG QKYSEITFPI LAPDPATNKD VHFLKYPIYV

151                                                200
GGNRGRGQIY PDGSKSNNTV YNATAGGIIS KILRKEKGGY EITIVDASNE

201                                                250
RQVIDIIPRG LELLVSEGES IKLDQPLTSN PNVGGFGQGD AEIVLQDPLR

251                               285
VQGLLFFLGS VVLAQIFLVL KKKQFEKVQL SEMNF*
  ↑
```

[a] Twenty residues in *trans*-membrane α-helix are underlined; natural cleavage site after Q252 that occurs in preparation of turnip cytochrome *f* fragment is indicated (↑).

mass units detected by the ESMS analysis are attributed to an oxygen molecule (32 Da) that ligates the heme, which is apparently 5-coordinate after elution from reversed phase HPLC in preparation for the ESMS analysis (D. Huang, D. L. Smith, and W. A. Cramer, unpublished data).

The tentative explanation of the 'natural' proteolysis in the *Cruciferae* cytochrome *f* is the existence of a protease in chloroplasts from these plants that is activated or made accessible to the lumen-side of the cytochrome *f* during the procedures involved in its extraction from the membranes. The major steps in the extraction are (i) methyl ethyl ketone extraction of pigment and lipids, (ii) acetone precipitation of the protein in the aqueous phase, (iii) desalting on G25 and (iv) chromatography on (a) DEAE-Sephacel, (b) G100, and (c) hydroxylapatite (Gray, 1978). It is known that the cleavage occurs before the hydroxylapatite step, and presumed that it occurs during the extraction and disruption of the membrane by the methyl ethyl ketone. It is interesting that the cleavage predominantly occurs two residues into the 20-residue, hydrophobic *trans*-membrane α-helix. This implies that the protease might be related to a processing or leader peptidase; the thylakoid processing peptidase of cytochrome *f* is known to be membrane-bound (Johnson et al., 1991), and the peptidase active site for plastocyanin is on the lumen side of the membrane (Kirwin et al., 1991). The soluble turnip Cyt *f* polypeptide contains a small part (Val-251–Gln-252) of the putative hydrophobic *trans*-membrane helix. A general strategy is proposed for genetic engineering of integral redox proteins such as Cyt c_1 or the Rieske iron sulfur protein for purposes of crystallization. Including a few residues of the hydrophobic helical domain should increase the likelihood for correct folding of the protein without affecting its solubility.

The 252-residue turnip cytochrome *f* polypeptide was crystallized at 5 °C in the reduced form in the presence of dithioerythreitol using 40–42% acetone as the precipitant (Martinez et al., 1992). The space group is $P2_12_12_1$ with a = 79.2 Å, b = 81.9 Å, c = 46.3 Å, and one cytochrome *f* molecule in the asymmetric unit (Martinez et al., 1992, 1994). The structure has been solved by multiple isomorphous replacement to a resolution of 2.3 Å (Martinez et al., 1994). This is the first solved structure of a subunit of the Cyt b_6f or bc_1 complex. Compared to the intact 285-residue protein, 18 of the 20 residues of the *trans*-membrane helix and 15 residues from the C-terminal stromal end are absent. The lumen-side fragment has a normal redox potential ($E_{m7} \approx +0.37$ V) and visible absorbance spectrum, and a unimolecular rate constant for oxidation by cupri-plastocyanin = 2800 s^{-1} at pH 7, ionic strength = 0.004 (Qin and Kostic, 1992, 1993). It is of interest that the E_m is pH dependent at alkaline pH ($\Delta E_m/\Delta pH = -0.06$ V) with a pK = 8.4 (Davenport and Hill, 1952). This suggests the possibility that the cytochrome has a role in H^+ translocation.

The (24 Å × 35 Å × 75 Å) cytochrome *f* structure reveals three unprecedented features for *c*-type cytochromes, one of which is unprecedented for heme proteins in general (see III.3).

III. Unprecedented Aspects of the Structure (Fig. 2)

1. Unlike the typical structure of soluble cytochrome *c* that has one mostly α-helical domain, the elongate Cyt *f* structure is made of two domains, 'large' and 'small.' The heme is near the interdomain interface in the larger domain, the heme Fe being 45 Å from the α-carbon of the C-terminal of Arg-250, connected to the *trans*-membrane α-helix and 28 Å from that of Lys-187.

2. The major secondary structure motif is a β-strand.

3. The axial sixth heme ligand is the α-amino group of the N-terminal tyrosine residue. A number of spectroscopic studies have noted similarities between spectra of cytochrome *f* and soluble mammalian cytochrome *c* at alkaline pH (> 9) in which the methionine ligand is exchanged for a lysine (Siedow et al., 1980; Davis et al., 1988; Rigby et al., 1988; Simpkin et al., 1989). The conserved Lys-145 was proposed as the axial sixth ligand (Davis et al., 1988). However, the ε-amino group of Lys-145 is 33Å from the heme iron. The amino group function inferred from the spectroscopic studies can be fulfilled by the tyrosine α-amino group, and in one respect fulfilled even more readily because the pK of the latter (ca. 7.2) is close to neutrality.

IV. Other Major Properties of the Structure

A. Basic Patch

The small domain, consisting of residues 169–231, contains Lys-187, previously shown to cross-link to Asp-44 of plastocyanin (Morand et al., 1989), which is solvent-exposed and 28 Å from the heme Fe (Fig. 2C). Lys-187 is close to a positively charged region that includes Lys-185 and Arg-209 (turnip sequence) in the small domain, and Lys-66, Lys-65, and Lys-58 in the large domain. This set of basic residues in the small and large domains forms the most prominent basic region in the Cyt *f* polypeptide. The knowledge that such a basic region of Cyt *f* is involved in the electron transfer complex with plastocyanin (Gross, 1993; Redinbo et al., 1994) and the cross-linking between Lys-187 of Cyt *f* and Asp-44 of plastocyanin together imply that this basic region on Cyt *f* is the docking site for plastocyanin.

B. Insulation of the Heme

Unlike mammalian cytochrome *c*, the heme is insulated from the aqueous phase by an extensive shield: the front face of the heme is shielded from solvent by Tyr-1, Pro-2, Phe-4, Pro-117, and Pro-161. Of the porphyrin ring side-chains, only the vinyl methyl group of pyrrole ring C is solvent exposed (Martinez et al., 1994), compared to the entire C-D edge in cytochromes *c* and c_2. This may explain why the midpoint redox potential of Cyt *f* is approximately 0.6 V more positive than that of yeast iso-Cyt *c*, which also has His-NH_2 (Lys) ligation, at alkaline pH (Barker and Mauk, 1992).

C. Sixth Heme Ligand of Cytochrome f: Consequences for Protein Translocation and Assembly

The α-amino group of the N-terminal Tyr residue serving as the sixth heme ligand of the mature protein has cell biological consequences for the mechanism of cytochrome *f* import: complete ligation of the heme, which results in formation of the low spin coordination state and assembly of the protein in its final folded state, cannot occur until the signal peptide of the intermediate translocated preprotein has been cleaved by the processing peptidase. This peptidase is believed to be located on the lumen-side of the membrane (Johnson et al., 1991; Kirwin et al., 1991), implying that protein folding cannot be completed until translocation has proceeded so that a large part of the cytochrome including Tyr-1 and His-25 has crossed the membrane (Figs. 1B-D). In addition, ligation of the heme and associated folding may provide additional free energy needed to complete the translocation process (Pfanner and Neupert, 1990) in the transition from the state shown in Fig. 1C to that in Fig. 1D.

The use of the Tyr-1 α-amino group as the sixth heme ligand can be viewed as a control mechanism to ensure a proper delay in the timing of protein refolding and is an example of a concept in membrane protein translocation that the polypeptide to be translocated must be unfolded (Pfanner and Neupert, 1990). Although the use of amino terminal ligation thus seems to be a rational mechanism, it appears

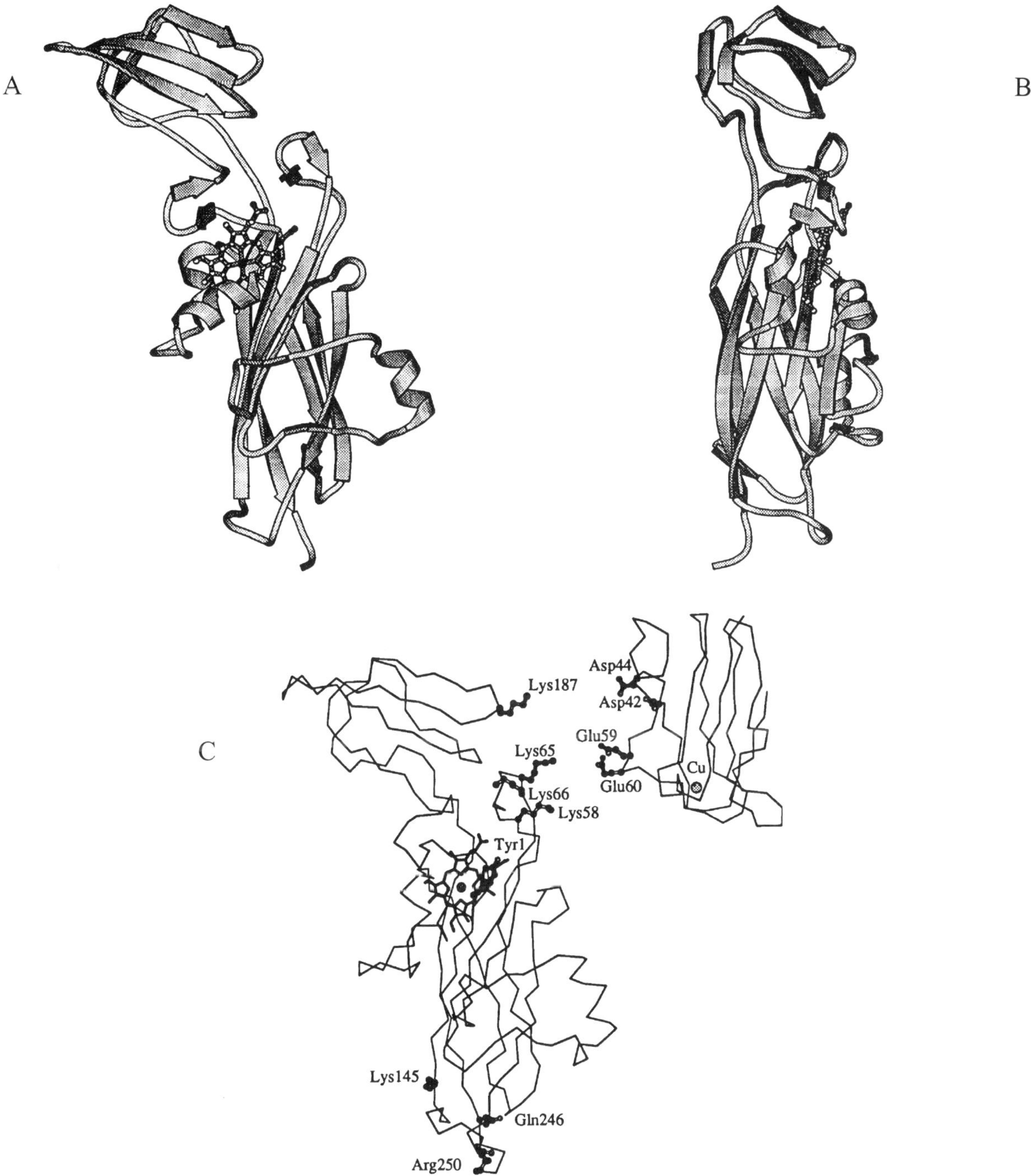

Fig. 2. (A, B) Polypeptide fold of turnip cytochrome *f* with heme shown in front view (A) and side view (B) rotated by 80° counterclockwise with respect to (A), drawn with MOLSCRIPT (Kraulis, 1991). The smaller folding domain, composed of residues 169–231, is at the top of the molecule. The large domain shown in the bottom part of the drawing contains residues 1–168 and 232–246. *n.b.*, the predominant β-strand motif. (C) Cα traces of cytochrome *f* and poplar plastocyanin in a 'pre-docking' state. The distances from the heme Fe are 28 Å, 33 Å and 45 Å, respectively, to (i) Lys-187, that can be cross-linked to Asp-44 of plastocyanin (Morand et al., 1989), (ii) Lys-145 that had been proposed to be the amine-heme ligand (Davis et al., 1988), and (iii) Arg-250, a charged residue near the end of the soluble Cyt *f* polypeptide that punctuates the lumen side of the *trans*-membrane helix. Gln-246 is the last residue toward the C-terminus in the β-strand structure; the proposed basic docking domain for plastocyanin (upper right) contains Lys-58, 65, and 66 in the large domain, and Lys-187 in the small domain. Complementary acidic residues in (Cu) plastocyanin include one or more of Asp-44, 42, and Glu-59, 60.

that it is not used by the functionally related cytochrome c_1 protein. In the nitrogen-fixing microaerobic endosymbiont, *Bradyrhizobium japonicum*, cytochromes *b* and c_1 can be made via mutagenesis as a single two-domain polyprotein connected by a 31-residue internal signal peptide (Thony-Meyer et al., 1991). Although it was shown in this work that covalent binding of the c_1' heme was required for formation of the bc_1 polyprotein, cleavage of the internal signal peptide was not. Therefore, the N-terminal α-amino group of this Cyt c_1 is involved in a peptide bond and cannot serve as the heme ligand. In addition, Cyt c_1 has a normal α-band difference spectrum in a yeast mutant blocked in cleavage of the Cyt c_1 signal peptide (Yang and Trumpower, 1994).

V. Orientation of Cytochrome *f* Relative to the Membrane Plane

EPR studies on magnetically-oriented chloroplast membranes (Bergstrom and Vannguard, 1982; Crowder et al., 1982) and dehydrated two-dimensional lattices of cytochrome b_6f complex oriented after centrifugation on Mylar sheets (Riedel et al., 1991) showed that the Cyt *f* heme has a broad (Bergstrom and Vanngard, 1982; Riedel et al., 1991) or narrow (Crowder et al., 1982) distribution of orientations relative to the membrane surface. The most probable orientation of the heme plane in both kinds of distributions would be sharply tilted at an angle of approximately 25–30° relative to the plane of the membrane. Since the heme is tilted slightly (20 Å) relative to the long axis of Cyt *f*, EPR data imply that the possible orientations of this long axis relative to the membrane plane are 45–50° or 5–10°.

VI. Evolutionary Relation between Cytochromes *f* and c_1

Because of structure-function similarities between the b_6f and bc_1 complexes, it has been thought that the respective redox-active subunits diverged in evolution from a common origin. From the low degree of identity (<20%) among five eukaryotic and bacterial Cyt c_1 sequences and one of Cyt *f*, it had been concluded that (a) the time of divergence between them was remote, or (b) that Cyt *f* had undergone a more rapid sequence evolution (Moore and Pettigrew, 1990). However, the high degree of sequence identity between Cyt *f* sequences (82% among nine sequences of higher plants (Gray, 1992)), makes the latter alternative unlikely. Another possibility, based on (i) the very low degree of sequence identity, (ii) the differences in heme ligation and resulting assembly, and (iii) different cellular origins (organelle (*f*)- vs. nuclear (c_1)-encoded genes), is that they do not have a common origin. Thus, cytochromes *f* and c_1 may have evolved convergently (Doolittle, 1994) from different gene origins leading to a protein with the same function, whose common structural features are only a *trans*-membrane anchor and the covalently bound high potential *c*-type heme (Tae et al., 1996).

VII. Electron Transfer from Cyt *f* to Plastocyanin

Electron transfer from Cyt *f* (E_m = +0.37 V) to its electron acceptor plastocyanin (E_m = +0.34–37 V) is approximately isopotential. Because the structure of plastocyanin is known (Redinbo et al., 1994), the question arises as to the nature of the structure involved in its effective docking with cytochrome *f*. The identity of the sites of interaction was indicated by covalent cross-linking of plastocyanin to cytochrome *f* using the water-soluble 1-ethyl-3-[3-(dimethylamino)-propyl]carbodiimide (EDC). As noted above, one site of linkage involved Asp-44 of plastocyanin and Lys-187 of Cyt *f*. A second linkage involved Glu-59 and/or Glu-60 to an unknown, presumably basic residue on Cyt *f* (Morand et al., 1989). Lys-187 is adjacent to a basic 'patch' of residues (Fig. 2C). The distance of the plastocyanin Asp-44 from the copper is 20 Å. With Cyt *f* in the orientation shown in Fig. 2, plastocyanin would bind through its acidic regions (residues 42–45 and 59–61 of spinach plastocyanin) to the basic region of Cyt *f* around Lys-187, or in the region near Lys-58, Cys-65, and Lys-66. In this binding geometry, the copper would be at the bottom of the plastocyanin barrel. It should be possible to obtain further information on the nature of the docking structures of Cyt *f*-plastocyanin by co-crystallization and site-directed mutagenesis.

Acknowledgments

The research described in this manuscript has been supported by grants from the USDA 9101624 (JLS and WAC), NIH GM-18457 (WAC), and the Lucille P. Markey Foundation. We thank V. Livingston and J. Hollister for careful work in the preparation of the manuscript.

References

Barker PD and Mauk AG (1992) pH-linked conformational regulation of a metalloprotein oxidation-reduction equilibrium. J Am Chem Soc 114: 3619–3624

Bergström J and Vänngard T (1982) EPR signals and orientation of cytochromes in the spinach chloroplast thylakoid membrane. Biochim Biophys Acta 682: 452–456

Crowder MS, Prince RC and Bearden A (1982) Orientation of membrane-bound cytochromes in chloroplasts, detected by low-temperature EPR spectroscopy. FEBS Lett 144: 204–208

Davenport HE and Hill R (1952) The preparation and some properties of cytochrome *f*. Proc Roy Soc London Ser B 139: 327–345

Davis DJ, Frame MK and Johnson DA (1988) Resonance Raman spectroscopy indicates a lysine as the sixth iron ligand in cytochrome *f*. Biochim Biophys Acta 936: 61–66

Doolittle RF (1994) Convergent evolution: The need to be explicit. Trends Biochem Sci 19: 15–18

Gray JC (1978) Purification and properties of cytochrome *f* from charlock. Eur J Biochem 82: 133–141

Gray JC (1992) Cytochrome *f*: Structure, function, and biosynthesis. Photosyn Res 34: 359–374

Gross EL (1993) Plastocyanin: Structure and function. Photosyn Res 37: 103–116

Ho KK and Krogmann DW (1980) Cytochrome *f* from spinach and cyanobacteria. J Biol Chem 255: 3855–3861

Johnson EM, Schabelrauch LS and Sears BB (1991) A plastome mutation affects processing of both chloroplast and nuclear DNA-encoded plastid proteins. Mol Gen Genet 225: 106–112

Kirwin PM, Elderfield PD, Williams RS and Robinson C (1991) Transport of proteins into chloroplasts: Organization, orientation, and lateral distribution of the plastocyanin processing peptidase. J Biol Chem 263: 18128–18132

Kraulis PJ (1991) MOLSCRIPT: A program to produce detailed and schematic maps of protein structures. J Appl Crystallog 24: 946–950

Martinez SE, Smith JL, Huang D, Szczepaniak A and Cramer WA (1992) Crystallographic studies of the lumen-side domain of turnip cytochrome *f*. In: Murata N (ed) Research in Photosynthesis, Vol II, pp 495–498. Kluwer, Dordrecht

Martinez SE, Huang D, Szczepaniak A, Cramer WA and Smith JL (1994) Crystal structure of chloroplast cytochrome *f* reveals a novel cytochrome fold and unexpected heme ligation. Structure 2: 95–105

Moore GR and Pettigrew GW (1990) Cytochromes *c*: Evolutionary, Structural, and Physicochemical Aspects. Springer-Verlag, Berlin

Morand LZ, Frame MK, Colvert KK, Johnson DA, Krogmann DW and Davis DJ (1989) Plastocyanin-cytochrome *f* interaction. Biochemistry 28: 8039–8047

Pfanner N and Neupert W (1990) The mitochondrial protein import apparatus. Ann Rev Biochem 59: 331–353

Qin L and Kostic NM (1992) Electron transfer reactions of cytochrome *f* with flavin semiquinones and with plastocyanin. Importance of protein-protein interactions and of donor-acceptor coupling. Biochemistry 31: 5145–5150

Qin L and Kostic NM (1993) Importance of protein rearrangement in the electron-transfer reaction between the physiological partners cytochrome *f* and plastocyanin. Biochemistry 32: 6073–6080

Redinbo MR, Yeates TO and Merchant S (1994) Plastocyanin: Structural and functional analysis. J Bioenerg Biomemb 26: 49–66

Riedel A, Rutherford W, Hauska G, Müller A and Nitschke W (1991) Chloroplast Rieske center: EPR study on its spectral characteristics, relaxation, and orientation properties. J Biol Chem 266: 17838–17844

Rigby SEJ, Moore GR, Gray JC, Godsby PMA, George SJ and Thomson AJ (1988) NMR, EPR, and magnetic CD studies of cytochrome *f*. Biochem J 256: 571–577

Siedow JN, Vickery LE and Palmer G (1980) The nature of the axial ligands of cytochrome *f*. Arch Biochem Biophys 203: 101–107

Simpkin D, Palmer G, Devlin FJ, McKenna MC, Jensen GM and Stephens PJ (1989) The axial ligands of heme in cytochromes: A near infrared magnetic circular dichroism study of cytochromes *c*, c_1, *b*, and spinach cytochrome *f*. Biochemistry 28: 8033–8039

Tae G-S, Furbacher PN and Cramer WA (1996) Evolution of cytochrome bc_1 complexes. In: Baltscheffsky H (ed) Origin and Evolution of Biological Energy Conversion. VCH

Chapter 23

An Overview of the Function, Composition and Structure of the Chloroplast ATP Synthase

Richard E. McCarty
Department of Biology, The Johns Hopkins University, 3400 N. Charles Street, Baltimore, MD 21218, U.S.A.

Summary

A concise overview of the chloroplast ATP synthase is presented, with an emphasis on its functions, regulation, composition, and structure from fluorescence energy transfer mapping. The ATP synthase may be separated into two parts: coupling factor 1 (CF_1) and CF_o. CF_1 is a large hydrophilic protein that contains the catalytic sites of the synthase. The functions of its five different polypeptides are briefly discussed. CF_o binds CF_1 and functions to conduct protons across the thylakoid membrane. Because CF_o will not be considered in detail in subsequent chapters, the composition of CF_o and some of its interactions with CF_1 are discussed. Insights into the structure and function of CF_1 as well as into subunit interactions, gleaned by structural mapping by fluorescence resonance energy transfer, are presented.

Donald R. Ort and Charles F. Yocum (eds): Oxygenic Photosynthesis: The Light Reactions, pp. 439–451.

I. Function of Complete Enzyme Complex

A. ATP Synthesis

Overall, linear electron flow in oxygenic photosynthesis is reductive:

$$H_2O + NADP^+ \rightleftharpoons ½\, O_2 + NADPH + H^+$$

At pH 7.0, the standard free energy change for this reaction is close to +220 kjoules/mol. Nonetheless, adequate free energy exists in photosynthesis that thylakoid membranes make ATP at the expense of energy released during electron transport. The reductant generated by the action of Photosystem II, Q_BH_2, has a redox potential of about 0 eV, whereas that of $P700^+$, the oxidant generated by Photosystem I activity, is approximately 0.45 eV. Thus, interphotosystem electron transfer is oxidative and exergonic (–43 kjoules/mol for one electron transfer). Part of the energy released during intersystem electron flow is conserved in the form of an electrochemical proton potential difference across the thylakoid membrane. Proton accumulation within the thylakoid vesicle is directly coupled to electron transfer from Q_BH_2 to $P700^+$ and with water oxidation. Mechanisms of H^+ and electron transfer are discussed in Chapter 20 of this volume.

The flow of protons across the membrane down their electrochemical potential ($\Delta\tilde{\mu}_{H^+}$) is exergonic. As for any ion, the electrochemical potential for protons is composed of two terms. One, the membrane potential, is electrical in nature whereas the other is essentially osmotic and expressed in terms of proton activity. The electrochemical proton potential may be expressed as

$$\Delta\tilde{\mu}_{H^+} = \Delta\psi + RT\ln[H^+]_A/[H^+]_B \qquad (1)$$

where R is the universal gas constant; T, the absolute temperature; F, Faraday's constant; $\Delta\psi$, the membrane potential; $[H^+]_A$, the concentration (activity) of protons on one side of the membrane and $[H^+]_B$, the concentration of protons on the other side. At 25 °C, equation (1) may be written (in kjoules/mol)

$$\Delta\tilde{\mu}_{H^+} = 96.3\,\Delta\psi + 5.69\,\Delta pH \qquad (2)$$

where $\Delta\psi$ is expressed in mV and ΔpH is the pH difference across the membrane. Suppose that $\Delta\psi$ is 100 mV (inside positive) and ΔpH, 2.0 units (inside acid) for a vesicular membrane preparation. The energy cost to generate this electrochemical potential is 9.63 + 11.38 = 21 kjoules/mol. The flow of protons down this electrochemical potential difference would thus liberate 21 kjoules/mol. In chloroplasts, the ΔpH term of $\Delta\tilde{\mu}_{H^+}$ is much larger than that of $\Delta\psi$ after just a second or so of illumination. For illuminated thylakoids, the $\Delta\tilde{\mu}_{H^+}$ can reach values as high as about –20 kjoules/mol (Avron, 1978). The electrochemical proton potential is the direct energy source for the endergonic synthesis of ATP from ADP and P_i (Mitchell, 1968). Proton coupled ATP synthesis in thylakoids is catalyzed by a remarkably complex enzyme known as ATP synthase, H^+-ATPase or CF_1-CF_o. The overall reaction catalyzed by the enzyme is (at pH 8.0)

$$nH^+_{in} + ADP^{3-} + HPO_3^{2-} + H^+_s \rightleftharpoons nH^+_{out} + ATP^{4-} + H_2O.$$

'In' and 'out' refer to the thylakoid lumen and the stroma, respectively. The protons so labeled are those translocated during the ATP synthase catalyzed reaction. H^+_s stands for the scalar proton that is taken up from the stroma. At pH 8, very close to one scalar proton is taken up from the stroma per ATP synthesized. A convenient in vitro monitor of photosynthetic ATP synthesis ('photophosphorylation') is to illuminate a weakly buffered thylakoid suspension and measure the pH change of the suspension resulting from scalar proton uptake.

The letter n refers to the number of protons that are translocated across the thylakoid membrane during ATP synthesis or ATP hydrolysis per ATP made or cleaved. The minimum value of n is likely to be three. Thermodynamic and steady-state kinetic approaches gave results consistent with an H^+/ATP ratio of three (Davenport and McCarty, 1981; 1984; Hangarter and Good, 1982). The phosphorylation efficiency, the amount of ATP formed per electron transferred through the electron transport chain (ATP/e^- ratio),

Abbreviations: ATP/e^- – the ratio of ATP formed to electron transferred; CF_1 – chloroplast coupling factor 1; CF_o – chloroplast F_o; DCCD – *N,N'*-dicyclohexylcarbodiimide; ECF_o – F_o from *Escherichia coli;* ECF_1 – F_1 from *Escherichia coli;* F_o – hydrophobic part of the ATP synthase; F_1 – coupling factor 1; FRET – fluorescence resonance energy transfer; H^+/ATP – protons translocated per ATP synthesized by CF_1-CF_o; H^+/e^- – protons translocated per electron transferred by the electron transport chain; Rubisco – ribulose 1,5-bisphosphate carboxylase/oxygenase; $\Delta\tilde{\mu}_{H^+}$ – electrochemical proton potential is dependent on the H^+/ATP and H^+/e^- (i.e. number of H^+ accumulated within the thylakoid vesicles for each e^- transported) stoichiometries. The maximum ATP/e^- ratio is defined by the H^+/e^- ratio divided by the H^+/ATP ratio. For linear electron transport, the H^+/e^- ratio seems to be one, at least in the steady-state. Thus, the maximum ATP/e^- ratio would be 2/3. If the Q cycle were to operate, the H^+/e^- would be higher and the ATP/e^- could in principle reach 1 (see Chapter 20 for discussion of the Q cycle).

Generally, the measured ATP/e^- ratios are significantly less than 2/3. Non-productive proton leaks (or slippage of the pumps; it is difficult to distinguish between the two possibilities), decrease ATP/e^- ratios. When mathematical corrections for measured H^+ leakage were made, the ATP/e^- ratio closely approached the predicted maximum ratio of 2/3 (Davenport and McCarty, 1984).

B. ATP Hydrolysis

The native ATPase activity of isolated thylakoids as usually prepared is very low (Avron and Jagendorf, 1959). Whereas the rate of net photophosphorylation by isolated thylakoid membranes can be as high as 5 μmol·min^{-1}·mg protein^{-1}, net ATPase rates are less than 0.05 μmol·min^{-1}·mg protein^{-1}. Much of the ATPase activity detected in untreated thylakoid membranes is likely to be catalyzed by contaminating enzymes; for example, the chloroplast envelope has ATPase activity. I have been fascinated by these simple observations for many years. Since there is virtually no ATP hydrolysis by the ATP synthase in the dark, the enzyme is clearly very inactive. Simply forming a $\Delta\tilde{\mu}_{H^+}$ across the thylakoid membranes, either artificially or by electron transport, causes a 100- to 1000-fold increase in catalytic activity. Thus, $\Delta\tilde{\mu}_{H^+}$ is required not only as the driving force for ATP synthesis, but also for activation of the synthase. Remarkably, this $\Delta\tilde{\mu}_{H^+}$-dependent activation, which likely involves intricate conformational changes in CF_1, can be complete within about 10 ms (Gräber, 1990)!

Illumination of thylakoid membranes in the presence of a disulfide bond reducing agent such as dithiothreitol allows ATP to be hydrolyzed at significant rates during subsequent dark periods (see Strotmann and Bickel-Sandkötter, 1986). Net ATP hydrolysis is not catalyzed by the reduced membranes in the light since the thermodynamic condition generated by electron transport strongly promotes ATP synthesis while inhibiting ATPase activity (Davenport and McCarty, 1986).

The inward translocation of protons is coupled to ATP hydrolysis by reduced thylakoids in the dark. An electrochemical proton potential equivalent to 150 mV or a ΔpH of 2.5 units, may be generated by ATP hydrolysis (Avron, 1978, Davenport and McCarty, 1981). This potential can actually drive intersystem electron flow in reverse, reducing electron acceptors in PS II, thereby promoting chlorophyll luminescence (Avron, 1978). ATP can also be synthesized at the expense of the $\Delta\tilde{\mu}_{H^+}$ generated by ATP hydrolysis! The reaction, often called 'ATP-P_i exchange,' is really ATP-dependent ATP synthesis. In ATP-P_i exchange, P_i is thought to be incorporated into ATP by its reaction with ADP bound to a catalytic site. The newly formed ATP can then dissociate from the enzyme. However, nucleoside diphosphate in the medium may be phosphorylated under these conditions at a much faster rate than that generated by nucleoside triphosphate hydrolysis, indicating that ATP-P_i exchange is not a true exchange (Davenport and McCarty, 1981).

As emphasized in Chapter 25, reduction of the ATP synthase has physiological significance and is involved in the intricate regulation of the activity of this enzyme. Nucleotide and subunit-subunit interactions are also important in this regulation as is modulation of the synthase by $\Delta\tilde{\mu}_{H^+}$. I think that it is rather unlikely that net ATP hydrolysis by thylakoid membranes occurs to a significant extent in vivo. In some bacteria, ATP hydrolysis by the ATP synthase is relied upon to generate the trans plasma membrane $\Delta\tilde{\mu}_{H^+}$ that is critical to the active transport of several metabolites. In thylakoids, however, the only function of the $\Delta\tilde{\mu}_{H^+}$ appears to be to drive ATP synthesis. Why waste ATP to generate a $\Delta\tilde{\mu}_{H^+}$ of no use?

How the role of reduction in chloroplast ATPase activation was discovered is interesting. Petrack and Lipmann (1961) showed that in the light, thylakoids from the cyanobacterium *Anabaena variabilis* could catalyze net ATP hydrolysis at rates as high as 184 μmol·h^{-1}·mg chlorophyll^{-1}, provided high concentrations of either cysteine or reduced glutathione were present. In these experiments, *N*-methyl-phenazonium methosulfate was added as a mediator of artificial electron transport around Photosystem I, which we recognize now, provided the $\Delta\tilde{\mu}_{H^+}$ that is required for CF_1 activation whether or not sulfhydryl reductant is present.

These results encouraged Petrack and Lipmann

(1961) to test for cysteine activation of the ATPase of spinach thylakoids. Cysteine at 80 mM was found to promote ATP hydrolysis in the light in the presence of *N*-methylphenazonium methosulfate.

We know now that reductive activation of chloroplast ATPase involves the $\Delta\tilde{\mu}_{H^+}$–dependent reduction of the disulfide bond in the γ subunit of the catalytic part of the ATP synthase (Ketcham et al., 1984). When cyanobacterial γ subunit gene sequences became available (Cozens and Walker, 1987; Curtis, 1988), it was very surprising to find that the subunit contained a single Cys residue. So why did cysteine stimulate ATPase in *Anabaena* thylakoids? I think it has little to do with the -SH group of the effector. Instead, cysteine may be a weak uncoupler – it was present at 80 mM – that decreases $\Delta\tilde{\mu}_{H^+}$ to a value that permits ATP hydrolysis even when cyclic electron flow and proton transport occur. My guess is that if a potent protonophore were added, ATPase activity would be inhibited. This guess is based on the observation that $\Delta\tilde{\mu}_{H^+}$ is required to maintain the chloroplast ATPase in active form (McCarty and Racker, 1968).

C. Nomenclature

Nomenclature of the ATP synthase and its parts should be clarified, since even some who work in the field seem confused by it. In the older literature, ATP synthase is often called 'H^+-ATPase.' Although H^+-ATPase has the advantage of indicating the proton translocating nature of the enzyme, this name might give the impression that the enzyme may operate physiologically in the hydrolytic direction in chloroplasts. Then, we have CF_1-CF_o, or variants such as CF_o-F_1, or CF_1-F_o. CF_1 stands for chloroplast factor 1 and is the only coupling factor for photophosphorylation. It was named by Vambutas and Racker (1965) by analogy to F_1, the first mitochondrial coupling factor. By definition, a coupling factor is an agent (protein in the case of CF_1) that participates in the coupling of ATP synthesis to electron flow. Removal of CF_1 uncouples ATP synthesis from electron flow. Its addition to the depleted membranes restores (couples) ATP synthesis. We know now that CF_1 contains the catalytic sites of the enzyme.

CF_o is not a coupling factor. The subscript is not zero or naught, but 'oh,' as in oligomycin. In their pioneering studies that led to the reconstitution of oxidative phosphorylation, Efraim Racker and Yasuo Kagawa (reviewed in Racker, 1976) exploited the observation that although the ATPase activity of mitochondrial membranes was very sensitive to the antibiotic, oligomycin, that of soluble F_1 was not. Detergent extracts of the inner mitochondrial membrane were found to restore oligomycin sensitivity to the ATPase activity of F_1. They called this agent F_o for factor that confers oligomycin sensitivity to F_1. F_o has come to mean those polypeptides of ATP synthases that are not part of F_1. F_o from *Escherichia coli* has just three different kinds of polypeptides, chloroplast, four and mitochondria, eight or so. Thus, CF_o stands for chloroplast F_o.

II. Composition and Function of CF_1

A. General Properties

An extrinsic membrane protein bound through CF_o to the outer surface of exposed thylakoids, CF_1 comprises about 10% of the thylakoid membrane protein (see Chapter 2, Fig. 9 and Chapter 4, Fig. 1). It is a relatively easy task to prepare 800 mg or more of CF_1 of 95% or greater purity. This is advantageous when large amounts of CF_1 are required for biophysical and nucleotide binding studies.

To remove CF_1 from thylakoids, the thylakoids are diluted into an EDTA solution of low ionic strength. The EDTA removes residual, membrane-bound, bi- and tri-valent cations. Both CF_1 and the thylakoid membrane surface are negatively charged, and without cations present to screen the charges, CF_1 dissociates from the membrane. As little as 5 mM NaCl totally prevents CF_1 release. It is difficult to remove all of the CF_1 from thylakoids by this method. Recently, I estimated the strength of the interaction between CF_1 and thylakoids. Thylakoid membranes were incubated in 0.75 mM EDTA (pH about 7) at about 40 μg chlorophyll·ml^{-1}. Filtration of a small amount of the suspension through 0.45 μm filters yielded a colorless extract that contained just 10% of the total ATPase activity. From this result, and with the reasonable assumption that the molar ratio of CF_1 to chlorophyll is 1/1000, a K_D value of about 5×10^{-10} M was calculated (R. E. McCarty, unpublished). In the procedure for CF_1 purification first developed by Binder et al. (1978), DEAE-cellulose is added directly to the thylakoid suspension. CF_1 binds to the DEAE cellulose, thereby lowering the free CF_1 concentration and, by mass action, allowing the release of more CF_1.

Once released from the membrane, CF_1 is water soluble and is also quite stable. When stored as an $(NH_4)_2SO_4$ precipitate in the presence of ATP, CF_1 will retain its capacity for activity for many months. The bane of CF_1 research is ribulose 1,5-bisphosphate carboxylase/oxygenase (Rubisco). Rubisco is everywhere. It is even a major protein in inner chloroplast envelope membrane preparations. It also contaminates CF_1 preparations and is very difficult to remove in toto, except by immunoaffinity chromatography. I know of at least several circumstances in which the Rubisco contamination of CF_1 profoundly influenced the results of experiments. Beware.

Once the Rubisco problem is obviated, CF_1 is a joy to work with. CF_1 is even more stable at room temperature than in the cold (especially in the absence of nucleotides). It can be heated to 65 °C or abused by treatment with solvents and detergents and retain activity. Provided buffers are filtered, proteolysis of CF_1 is minimal.

By itself, CF_1 is not capable of the net synthesis of ATP. As first shown by Feldman and Sigman (1982), however, ADP tightly bound to CF_1 may be phosphorylated by medium P_i to form ATP that remains tightly bound to the enzyme. CF_1 contains the nucleotide binding sites of the ATP synthase. As detailed in Chapter 24, CF_1 probably contains six nucleotide binding sites. Of these, at least two sites are noncatalytic. As isolated, CF_1 contains 1.5 to 1.9 mol of ADP per mol of enzyme. Because the ADP is very slow to dissociate, it is defined operationally, as being tightly bound. Although dissociation is slow, exchange of bound ADP with either ADP or ATP in the medium can be fast. The binding of nucleotide to one (or more) site must, therefore, promote release of tightly bound nucleotides. Two other apparently high affinity sites bind Mg^{2+}-ATP (Shapiro et al., 1991a). Occupancy of these sites, which are not catalytic, stabilizes the enzyme (Wang et al., 1993). Dissociable nucleotide binding sites, possibly two (Shapiro et al., 1991b), are also present. These sites may be catalytic.

Soluble CF_1 hydrolyzes ATP at relatively slow rates; as mentioned earlier this is also a feature of CF_1 in thylakoids. A number of treatments of CF_1 can increase its ATPase activity twenty-fold or more, as reviewed by Mills in Chapter 25. Work on soluble CF_1 has provided insights into the remarkable activation and regulation mechanisms of the membrane-bound enzyme. When assayed in the absence of oxyanions (e.g. sulfite), certain alcohols or detergents, ATP hydrolysis by activated soluble CF_1 is much more rapid in the presence of Ca^{2+} than of Mg^{2+}. In contrast, the activity of CF_1 on thylakoids is supported much better by Mg^{2+} than by Ca^{2+}. Free Mg^{2+} is a potent inhibitor of the solubilized CF_1 ATPase, whereas free Ca^{2+} is not (Hochman et al., 1976). Mg^{2+} inhibition, which can be overcome by carrying out the reaction in the presence of ethanol, HSO_3^- or octylglucoside, appears to arise from the slowing of the release of product ADP from a catalytic site(s) (Murataliev et al., 1991).

Release of product ATP, rather than phosphoanhydride bond formation, is the major energy-requiring step in ATP synthesis (Boyer, 1989). Thus, $\Delta\tilde{\mu}_{H^+}$ is used to lower the binding energy of ATP bound to the catalytic site, very likely through $\Delta\tilde{\mu}_{H^+}$-dependent conformational changes. In addition, two or more catalytic sites interact in an unusual way, alternating between high and low affinity states. In the binding change mechanism (Boyer, 1989) sites display pronounced negative cooperativity with respect to binding, but positive cooperativity with respect to catalysis. In other words, the binding of substrate to one site promotes catalysis and product release at another.

B. Subunits

As would be expected for a protein of 400,000 M_r, CF_1 is oligomeric. It consists of five different polypeptide subunits and a total of nine individual subunits. The subunits are labeled α to ϵ in order of decreasing molecular weight (Table 1; also see Chapter 24, Fig. 1). All F_1 enzymes have the unusual property of being structurally asymmetric. This asymmetry is dictated by the subunit stoichiometry of $\alpha_3\beta_3\gamma\delta\epsilon$. Interactions between the $\alpha\beta$ heterohexamer and the γ subunit are major factors in stabilizing CF_1 (Wang et al., 1993). Although the $\alpha_3\beta_3$ structure is symmetric by itself, the γ subunit with which it interacts does not show three-fold symmetry. As a result, the α and β subunits cannot interact with γ in an equivalent manner and, therefore, are in different environments. For the binding change mechanism (Chapter 24) to operate, two or more catalytic sites must sequentially become equivalent during catalysis. There is evidence that Mg^{2+}-ATP binding can accomplish this (Shapiro and

Table 1. CF_1 subunits

Subunit	M_r ($\times 10^{-3}$)[a]	Number/CF_1	Functions
α	55	3	Catalysis
β	54	3	Catalysis
γ	36	1	Regulation, proton transport, catalysis
δ	21	1	Functional binding, proton interactions
ϵ	15	1	Regulation, proton interactions

[a] Deduced from the nucleotide sequence of the genes for spinach CF_1 (Hudson and Mason, 1988).

McCarty, 1990). How asymmetry is overcome is uncertain, but the intriguing idea of 'rotational catalysis' has been advanced (Boyer, 1989). Rotational movement of the γ subunit relative to the α/β core is part of the basis of rotational catalysis. Movement of a flexible arm of the γ subunit could act as the commutator.

Why is CF_1 so complicated? In view of the H^+-ATPases of plant, yeast and fungal plasma membranes which are a single polypeptide of about 105,000 M_r, the myriad of polypeptides CF_1 contains are not required simply because it is difficult to pump protons. Certainly, all CF_1 polypeptides are required for ATP synthesis; this is probably true for CF_0 as well. I think it is possible that regulation of both activity and proton flux could account for part of the complexity of CF_1.

ATP hydrolysis at a low, but significant, rate requires only α and β, the nucleotide binding subunits of CF_1 (Gromet-Elhanan, 1992). $\alpha_3\beta_3\gamma$ is most active in ATP hydrolysis, at least 100-fold better than $\alpha\beta$. γ need not be intact for very rapid ATP hydrolysis to occur. Cleavage of the γ subunit to 14,000 and 11,000 M_r fragments that remain tenaciously bound to the $\alpha\beta$ core actually increases ATPase activity (Moroney and McCarty, 1982). γ is also involved in regulation and may be involved in proton translocation. δ is required for functional binding of CF_1 to CF_0, whereas ϵ fulfills roles in regulation as well as in proton gating.

The overall structure of CF_1 resembles that of mitochondrial F_1 (MF_1) and *Escherichia coli* F_1 (ECF_1). The genes for a large number of F_1 β subunits have been cloned and sequenced. There is striking sequence identity among the β subunits from quite different organisms. For example, the β subunit of spinach chloroplast F_1 shares an astonishing 76% sequence identity with ECF_1. The similarity of β subunits from various sources is underscored by the fact that purified β from CF_1 reconstituted ATPase activity to *Rhodospirillum rubrum* membranes from which the β subunit of its F_1 had been selectively removed (Richter et al., 1986). Interestingly, the reconstitution of ATP synthesis activity was much poorer than that of the reverse ATPase reaction.

The α subunits show less sequence homology overall, but there are regions where the sequence is well conserved. The smaller F_1 subunits seem to have been 'tailored to fit the needs' of a particular F_1. For example, only the γ subunit of CF_1 contains the thiols involved in redox regulation of CF_1. Also the ϵ subunit of ECF_1 is just 23% identical to the ϵ subunit of CF_1 and mitochondrial ϵ appears to have no counterpart in either the chloroplast or bacterial enzyme (see Jagendorf et al., 1991, for a brief review of many of these aspects).

The three larger ECF_1 subunits have been subjected to extensive mutagenesis and, especially for the β subunit, chemical modification (see for example, Futai et al., 1989). The mutagenesis approach is, of course, more difficult for CF_1. α, β and ϵ are encoded by chloroplast genes whereas γ and δ are nuclear coded. However, with the advent of chloroplast transformation in *Chlamydomonas reinhardtii*, it is possible to pursue mutagenesis studies of the δ and γ polypeptides.

Another approach is to overexpress mutated or truncated forms of CF_1 subunits in *E. coli* and investigate the activity of the expressed subunits after reconstitution. For example, purified ϵ and δ restore full activity to CF_1 deficient in either subunit and good reconstitution of ATPase activity has been achieved simply by mixing $\alpha\beta$ preparations with γ (M. L. Richter, unpublished). Although for certain overexpressed proteins, the formation of bacterial inclusion bodies can be a significant problem, for certain other proteins this behavior can actually be advantageous. For example, proteins that form inclusion bodies are readily purified. Some, including the δ and ϵ subunits of CF_1, can be renatured in full, even after treatment with sodium dodecyl sulfate or high concentrations of urea.

III. Composition and Function of CF_o

A. General Properties

By definition, CF_o is what is left of the ATP synthase after CF_1 has been removed. In another sense, CF_o is whatever copurifies with CF_1 to yield a functional ATP synthase after reconstitution. Neither definition on its own is entirely satisfying. The proteins that comprise CF_o are hydrophobic and are more difficult to purify than CF_1. Indeed, reconstitutionally active CF_o was first obtained just a few years ago (Feng and McCarty, 1990a).

There has been (in my opinion) an unfortunate tendency to focus too sharply on biochemical and biophysical aspects of CF_1. I understand why, having worked with both of the components of the ATP synthase. CF_1 is a most tractable item of study, whereas CF_o is submerged in the greasy bilayer of the thylakoid membrane.

As I have discussed, CF_1 makes ATP at the expense of the $\Delta\tilde{\mu}_{H^+}$. As the transmembrane component of the ATP synthase, it is clear that CF_o must be involved in the anchoring of CF_1 as well as, more importantly, in energy transduction.

From the mid-1960s on, there have been clues as to CF_o function. One was that removal of CF_1 greatly enhances the proton permeability of thylakoid membranes; the rebinding of CF_1 restores proton impermeability (McCarty and Racker, 1966). A second clue was that CF_o contains an extremely hydrophobic polypeptide of 8,000 M_r that is present in multiple copies (6–12). This polypeptide, sometimes called (confusingly) 'the proteolipid,' covalently binds the carboxylic acid reagent, *N,N'*-dicylcohexylcarbodiimide (DCCD). Incredibly, the substoichiometric binding of DCCD to the 8 kDa polypeptide inhibited ATP synthesis fully. In fact, the reaction of only about one in six suffices to block phosphorylation (Sigrist-Nelson et al., 1978). This extraordinary 'cooperativity' has not been fully explained.

DCCD fully offsets the enhanced proton conductance induced by release of CF_1. Said in another way, DCCD blocks proton conductance by the CF_o proton channel. More recent work (reviewed by Junge, 1989) has shown that CF_o contains a proton conductance mechanism that behaves as a remarkably effective proton channel. In addition, the channel must contain a barrier to the diffusion of cations other than protons. How such a barrier (or filter for H^+) can operate is an interesting matter for speculation. There can be no doubt, however, that CF_o functions in some way to deliver protons or the result of their transport to CF_1.

Recall that transmembranc proton potentials drive ATP synthesis, quite likely through energy ($\Delta\tilde{\mu}_{H^+}$)-dependent conformational changes in CF_1. These changes are envisioned in the binding change mechanism to promote binding of substrates and release of products. Interesting and, as yet, mostly unapproachable questions, are where within the synthase protons travel and where the $\Delta\tilde{\mu}_{H^+}$-dependent conformational changes involved in regulation of enzyme activity as well as in catalysis are initiated. There is abundant evidence that CF_1 undergoes structural rearrangements upon energization of thylakoids. The evidence ranges from energy-dependent tritium exchange (Ryrie and Jagendorf, 1972) to $\Delta\tilde{\mu}_{H^+}$-dependent enhancements of the reactivity of cys-89 of the γ subunit (McCarty and Fagan, 1973) and a lysine residue of the ϵ subunits (Komatsu-Takaki, 1993). $\Delta\tilde{\mu}_{H^+}$ itself is not sufficient to induce the change in the γ subunit that greatly increases the reaction of Cys-89 with maleimides. Thylakoids treated with DCCD generate very large transmembrane proton potentials in the light. Yet, fluorescein maleimide fails to react with the γ subunit in DCCD-treated thylakoids (Y. Evron, personal communication). The protonation of groups within either a CF_1 subunit or a CF_o polypeptide could initiate the structural changes. If, however, CF_o is involved, the group protonated would have to be on the stromal side of the membrane.

The possible connection between activation of the synthase and proton gating is rarely discussed. Under non-phosphorylating conditions, the proton leak through CF_1-CF_o is low, less than 10 μmol $H^+\cdot min^{-1}$ $mg\cdot CF_1^{-1}$ in the presence of ATP at an external pH of 8.0. DCCD, which is thought to block totally CF_o proton conductance, does not fully inhibit the thylakoid proton leak, suggesting that there are other pathways of proton transport in thylakoids. During photophosphorylation, proton flux can be greater than 150 μmol $H^+\cdot min^{-1}\cdot mg\ prot^{-1}$, assuming an H^+/ATP ratio of 3. When CF_1 is removed, the proton flux through CF_o is even faster (Junge, 1989). Clearly, CF_1 is a rather effective carburetor (blocker) of the

CF_0 proton channel, even when as under non-phosphorylating conditions, the $\Delta\tilde{\mu}_{H^+}$ can be higher than 180 mV (i.e. ΔpH > 3 units).

How might the 'proton gate' open during the onset of photophosphorylation? It seems possible that opening the gate and activation are very closely related. It may be no coincidence that the γ and ϵ subunits of CF_1 influence both proton transport and activity. CF_1 lacking ϵ fails to block the CF_0 proton channel, even though it binds normally. Reduction of the γ disulfide bond, as well as other modifications of γ, change the manner in which the synthase handles protons. Reduction improves coupling at low $\Delta\tilde{\mu}_{H^+}$ values (Mills and Mitchell, 1984; Gräber, 1990), whereas other treatments, such as reaction of Cys-89 of γ with *N*-ethylmaleimide, increase the proton leak (Y. Evron, personal communication). Thus, generation of the $\Delta\tilde{\mu}_{H^+}$ could cause changes in the γ and ϵ subunits that not only result in activation of the ATP synthase, but also, concomitant opening of the proton gate.

B. Subunits

CF_0 contains four different polypeptides, labeled in Roman numerals, I-IV (Fromme et al., 1987). Subunits I, II and III are numbered in order of decreasing molecular weight. Subunit IV is the largest CF_0 protein. Subunits I and II stain well with Coomassie blue. In contrast, subunit III can stain well by some protocols or poorly by others and the fact that subunit IV stains poorly accounts for the delay in its discovery (Fromme et al., 1987). The presence of subunit III can be confirmed by monitoring covalent [^{14}C]DCCD incorporation into subunit III. The only acidic amino acid in subunit III, Glu-65, is the target of DCCD modification.

Table 2 gives some of the properties of the CF_0 polypeptides. All F_0 complexes so far examined contain a very hydrophobic ca. 8 kDa DCCD-binding polypeptide that is predicted to contain two transmembrane helices and to assume a hairpin structure (Hudson and Mason, 1988). The loop on the hairpin is probably located on the stromal side of the membrane.

Although there is general agreement that subunit III is present in multiple copies, just how 'multiple' multiple is is still controversial. For that matter, the copy number of the three other CF_0 polypeptides is much less certain than that for CF_1 subunits. A reasonable guess, based in part on analogy with

Table 2. CF_0 subunits

Subunit	M_r[a]	Number per CF_0	Analogous ECF_0 subunit
IV	27[b]	1	a
I	17	1	–
II	16.5	1	b
III	8	6–12	c

[a] From gene sequences Hudson and Mason (1988) for I, III and IV. Hermann et al. (1993) for II.

[b] Runs as a 20,000 M_r protein on sodium dodecyl sulfate polyacrylamide gels.

E. coli F_0 and immunoprecipitation of ^{14}C-labeled CF_1-CF_0 complexes, is I_1II_1 and III_{6-12} (Süss and Schmidt, 1982). Subunit IV may be present in one copy as well. If subunit III is present in 12 copies, and the stoichiometries of the other CF_0 polypeptides are as predicted, CF_0 would have a molecular weight of about 170,000. Together with CF_1, the enzyme has a molecular weight of close to 600,000 and a total of as many as 25 individual polypeptide chains! Had I known just how complicated the ATP synthase was thirty years ago, I probably would have decided to investigate the action of a simpler enzyme; otherwise, I have no regrets about my decision.

From amino acid sequence (deduced from gene sequences), analogies to F_0 polypeptides from other coupling membranes, and immunological evidence, the topology of CF_0 polypeptide insertion into the thylakoid membrane is beginning to be elucidated. Subunits I and II likely protrude from the stromal side of the thylakoid membrane and are anchored by a single predicted transmembrane region (Hudson and Mason, 1988). The hydrophilic regions of the subunits have been proposed to be in contact with CF_1. Subunit IV probably does not play an important role in the binding of CF_1 to CF_0. Subunit IV is, however, required for DCCD-sensitive proton transport by CF_0 reconstituted liposomes (Feng and McCarty, 1990b).

Until recently, the interactions between subunits I and II and CF_1 dominated models of CF_1 binding to CF_0. It is now apparent, however, that subunit III can interact very strongly with CF_1. Through anion exchange chromatography of detergent- solubilized CF_1-CF_0, a CF_1-subunit III complex was prepared (Wetzel and McCarty, 1993a). The properties of the CF_1-III complex more closely resembled those of CF_1-CF_0 than CF_1. Even after reduction of the γ disulfide bond, the Ca^{2+}-ATPase of CF_1-III and CF_1-CF_0 remained low, suggesting that the presence of either subunit III or intact CF_0 strengthens the interaction of the inhibitory ϵ subunit. Also, the presence of either subunit III or CF_0 confers to CF_1 the ability to bind specifically to asolectin vesicles. By treatment of the vesicles that contain CF_1-III with dilute EDTA solutions, the CF_1 may be removed (Wetzel and McCarty, 1993b). Thus, vesicles containing just subunit III with a defined orientation may be prepared by a gentle method. The proton translocating properties of these vesicles have yet to be examined. With time, it should be possible to reconstitute CF_0 from purified polypeptides.

The roles of subunits I, II and III in the binding of CF_1 to CF_0 are confirmed by analysis of *C. reinhardtii* photophosphorylation mutants (Lemaire and Wollman, 1989). Both subunits I and III were required for stable binding of CF_1. In chemical cross-linking experiments, Süss (1986) found that subunit I cross-linked to β and γ, whereas subunit II cross-linked to α, β and γ. A subunit III-ϵ cross-link was also detected. Although it may be coincidental, the treatment of CF_1-III with 20% ethanol, 30% glycerol releases both the ϵ subunit and subunit III (Wetzel and McCarty, 1993b). In *E. coli*, antibodies against F_0 subunit c (analogous to CF_0-III) block ECF_1 binding to F_0 and can even displace bound ECF_1 (Deckers-Hebestreit and Allendorf, 1992).

It is interesting that subunit II in CF_1-CF_0 in situ cross-links to α, β and γ. Crystallographic evidence for mitochondrial F_1 (Petersen and Amzel, 1993; Abrahams et al., 1994), together with biochemical evidence, suggest that the α and β subunits form a staggered ring in which the α subunits are closer to the membrane than the β subunits. At least part of the γ subunit occupies the central part of the α,β hexamer (McCarty and Hammes, 1987; Abrahams et al., 1994). This structure and the cross-linking results suggest that the hydrophilic part of subunit II present on the stromal side of the membrane may extend far into CF_1.

There is very good evidence that subunit c of ECF_0 forms a hairpin structure, with two transmembrane helices forming the sides of the hairpin and a polar loop facing the F_1 side of the membrane. From ^{1}H-NMR experiments of subunit c in $CHCl_3/CH_3OH/H_2O$, it was concluded that the reaction of Asp-61 with DCCD or even protonation of Asp-61 causes changes in the polar loop region (Girvin and Fillingame, 1993). The polar loop seems less likely to protrude as far from the membrane as the larger polar domain of subunit b (analogous to subunit II). By analogy, subunit III probably assumes a structure in the thylakoid membranes similar to that of subunit c. The cross-linking between III and ϵ raises some interesting questions. Could the $\Delta\tilde{\mu}_{H^+}$-dependent changes in CF_1 be initiated in the polar loop region of subunit III and translated directly to ϵ? Can the ϵ-III association be taken as evidence that at least part of the ϵ subunit is close to the membrane?

IV. Subunit Interactions

The investigation of subunit interactions and their functions can be very satisfying. Nevertheless, it can be fraught with sometimes frustrating, albeit interesting, difficulties. A major problem in the study of multimeric proteins is that a perturbation in one part of the complex could be transmitted over tens of angstroms to other parts of the protein. I will cite two examples of this phenomenon.

For many years, it has been known that modification of the γ subunit by maleimides inhibits ATP synthesis (McCarty and Fagan, 1973). Alkylation of Cys-89 of the γ subunit by *N*-ethylmaleimide also increases proton leakiness through the enzyme. If, however, ATP is added, low proton leakage is restored (Y. Evron, personal communication). Although the alkylation of γ Cys-89 does not appear to affect ϵ binding, it has rather unusual effects on the interactions of CF_1 with nucleotides. Recall that CF_1 as isolated contains 1.5 to 1.9 mol ADP per mol of enzyme and that the ADP dissociates from the enzyme very slowly. The bound nucleotide will, however, exchange with medium nucleotide more quickly (reviewed in Strotmann and Bickel-Sandkötter, 1986). In control CF_1 in solution, exchange occurs in the presence of saturating ATP concentrations and at 20 °C with a half-time of about 11 min. Under the same conditions, the initial rate of exchange in CF_1 modified with *N*-ethylmaleimide is, astonishingly, nearly 150-fold faster.

What this very large rate acceleration means mechanistically is unclear. What is evident, however, is that the effects of *N*-ethylmaleimide modification must be transmitted over long distances. Fluorescence resonance energy transfer measurements (summarized in Shapiro et al., 1991b) show that the distance between γ Cys-89 and any of the four nucleotide binding sites so far mapped is 50 Å or greater (Fig. 1). Specifically, Cys-89 of γ and 2′,(3′),-*O*-2-4-6-trinitrophenyl-ADP bound to an exchangeable site within the α,β core is 51 Å. Exchange was monitored by following the release of the trinitrophenyl nucleotide from this site. The acceleration of exchange by the specific alkylation of γ Cys-89 must result from perturbation of the structure of γ. This structural change is passed on to the α,β core which contains the nucleotide binding sites. One could have concluded that γ Cys-89 is an essential residue. After all, it is conserved in all F_1 γ subunits for which sequence information is available, except for that from the thermophilic bacterium, PS3. I resisted this temptation and predict that a Cys-89 to Ser-89 mutation will not be deleterious. A bulkier amino acid at position 89 of γ could be inhibitory.

A second example of action over a distance is the fact that reduction of γ disulfide which is also quite distant from the nucleotide binding site (Fig. 1) greatly enhances ATPase activity, probably by overcoming inhibition by ϵ (Soteropoulos et al., 1992).

V. Structure by Fluorescence Energy Transfer Distance Mapping

Boekema and Lücken in Chapter 26 give an update on the elucidation of the structure of CF_1. Nonetheless, I will briefly discuss structure determination, mostly by fluorescence resonance energy transfer (FRET). Even if crystals of CF_1 of X-ray diffraction quality were available now, determination of the structure at atomic resolution, as was recently achieved for beef heart mitochondrial F_1 (Abrams et al., 1994), could take years. FRET distance mapping requires specific labeling and the range of distances that can be measured is limited. Also, because of the need to assume an average orientation factor (κ^2) there is a minor, but significant uncertainty in the measurements. Despite these limitations, a lot has been learned from FRET mapping about CF_1 structure that is relevant to function. Over thirty individual distances have been measured by FRET. Views of CF_1, based on FRET, are given in Fig. 1.

Confidence in FRET distance mapping for CF_1 may be derived from biochemical evidence. For example, we knew that Cys-89 and Cys-322 of γ could be readily cross-linked by bismaleimides. In the refined structure, they are positioned 8 Å apart. The ϵ subunit protects the disulfide bond between Cys-189 and Cys-205 of γ from reduction. Cys-6 of ϵ is just 24 Å from those thiols. The map is also consistent with cross-linking data (Süss, 1986). The

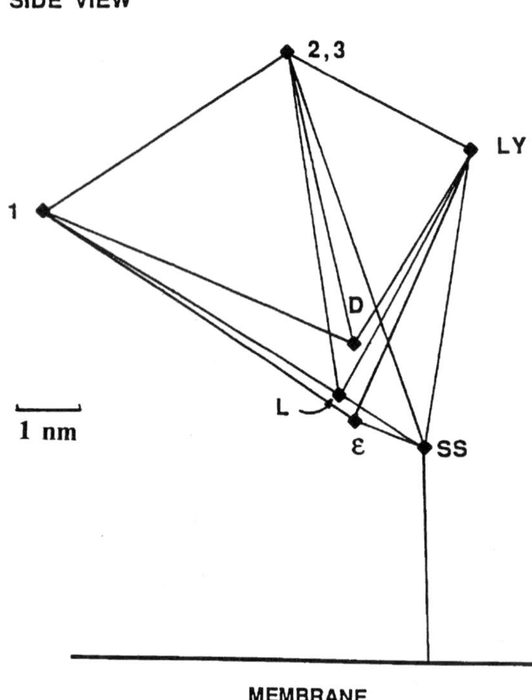

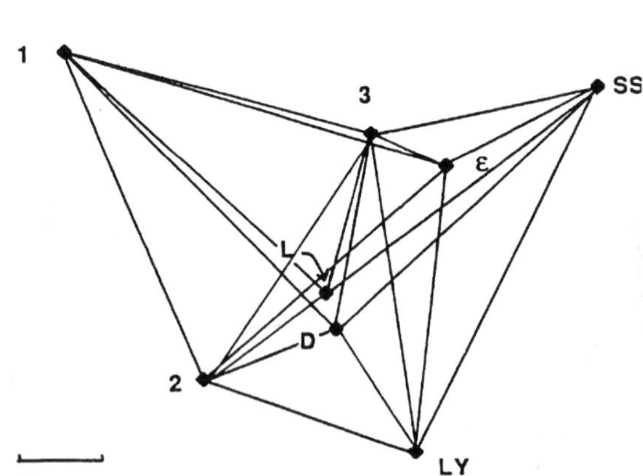

Fig. 1. Views of CF_1 as revealed by FRET. The numbers 1, 2 and 3 refer to nucleotide binding sites on CF_1 with different properties. LY stands for Lucifer yellow vinyl sulfone covalently attached to one of the three α subunits at Lys-378. L is Cys-89 of γ and D is Cys-322 of the γ subunit. SS is either Cys-199 or Cys-205, the two Cys residues of γ that participate in the disulfide bond. ϵ is Cys-6 of the ϵ subunit. To simplify the drawings, a fourth nucleotide binding site has been omitted. The distances are those fitted by least squares computer algorithm (Shapiro et al., 1992b). Where possible, the sites are connected by lines which indicate that a distance measurement has been made. In addition to the excellent agreement between the measured and fitted distances, the fact that as many as six measurements converge on a single location (e.g. Cys-6 of ϵ) gives one confidence in the approach.

central location of at least part of the γ subunit was confirmed by electron microscopy and image averaging of CF_1 (Boekema et al., 1990).

What else have we learned from FRET? As shown for the *N*-ethylmaleimide modification case, action can be felt over long distances. The γ subunit folds in such a way that the N-terminus is not too far away from the C-terminus. γ Cys-89 and Cys-322 are close to each other. The nucleotide binding sites are far away from the membrane. Could protons traverse this distance to the catalytic sites? The asymmetry of the enzyme is obvious. The fact that Lucifer yellow vinyl sulfone reacts with just one of the three α subunits allowed a novel use of FRET in which it was shown that two nucleotide binding sites switch their properties as a result of exposure of CF_1 to Mg^{2+}-ATP (Shapiro and McCarty, 1990). FRET was also used to measure the K_D for the interaction of CF_1 and thioredoxin, the natural reductant of the γ disulfide bond. Interestingly, removal of ϵ increased the extent of energy transfer without changing the apparent K_D value. This indicates that the thioredoxin can approach the γ subunit more closely when ϵ is removed. The rate of reduction of the γ disulfide by thioredoxin is also enhanced by removal of ϵ (Dann and McCarty, 1992).

These few illustrations show the usefulness of FRET. It might seem as though the partial solution of the structure of beef heart mitochondrial F_1 to 2.85Å by X-ray diffraction methods (Abrahams et al., 1994) renders the use of FRET obsolete. This is decidedly not the case. Reflections from the δ and ϵ subunits, even though they are present in the crystals, were not discerned. Moreover, only a part of the γ subunit was detected. Whether the assignments of the N- and C-terminal helices of the γ subunit of the beef heart enzyme by Abrahams et al. (1994) apply to CF_1-γ may be questioned. In CF_1 Cys residues 89 and 322 of the γ subunit may be cross-linked and FRET mapping indicates that they are close together. In the mitochondrial F_1 structure, the C-terminus of γ was proposed to be far away from the Cys residue equivalent to Cys-89 of CF_1 γ. Moreover, FRET may be used to monitor structural changes occurring over μs to ms. In contrast, an X-ray structure, however elegant, can give but one conformation of the enzyme. Although Abrahams et al. (1994) attempted to interpret their 2.85Å map in terms of a mechanism, how a single static structure can give definitive information about a dynamic mechanism is not clear.

VI. Conclusions

The chloroplast ATP synthase is a remarkable enzyme that resembles its counterparts in eubacteria and, to a lesser extent, that in mitochondria. There are, however, several aspects of CF_1-CF_0 that are unique. Among these aspects are the presence of the disulfide bond in the γ subunit and the extensive FRET map. ϵ functions are more clearly defined for CF_1 than for ECF_1 since CF_1 ϵ is not required for the binding of CF_1 to CF_0.

There are also advantages to working with the chloroplast system. Lots of CF_1 may be prepared easily from a convenient, inexpensive and edible source. Rates of ATP synthesis by thylakoid membranes can be very high and photophosphorylation is very easy to measure. Light is an inexhaustible substrate, the 'concentration' of which is readily altered. Electron transport is also readily measured since illuminated thylakoids can reduce a myriad of compounds using an abundant reductant—water.

The study of soluble CF_1 in solution has given very useful insights into the regulation and mechanism of the synthase. It is, however, more satisfying to extend the findings with this simplified model system to the ATP synthase in its natural environment — the thylakoid membrane and in intact chloroplasts and leaves or algae (Ort and Oxborough, 1992). Are there ways of monitoring the kinetics of $\Delta\tilde{\mu}_{H^+}$-dependent changes in CF_1 structure? What would happen if the γ disulfide bond could not form? Would it be possible to deplete the ATP synthase of its tightly bound nucleotide? These are but a few of the questions that will be approached in the future.

References

Abrahams JP, Leslie AGW, Lotter R and Walker JE (1994) Structure at 2.8Å resolution of F_1-ATPase from bovine heart mitochondria. Nature 230: 621–628

Avron M (1978) Energy transduction in photosynthesis. FEBS Lett 96: 225–232

Avron M and Jagendorf AT (1959) Evidence concerning the mechanism of adenosine triphosphate formation by spinach chloroplasts. J Biol Chem 234: 967–972

Binder A, Jagendorf AT and Ngo E (1978) Isolation and composition of the subunits of spinach coupling factor protein. J Biol Chem 253: 3094–3100

Boekema EJ, Xiao J and McCarty RE (1990) Structure of the ATP synthase from chloroplasts studied by electron microscopy: Location of the small subunits. Biochim Biophys Acta 1020: 49–56

Boyer PD (1989) A perspective of the binding change mechanism for ATP synthesis. FASEB J 3: 2164–2178

Cozens AL and Walker JE (1987) The organization and sequence of the genes for ATP synthase subunits in the cyanobacterium *Synechococcus* 6301. Support for an endosymbiotic origin of chloroplasts. J Mol Biol 194: 359–383

Curtis SE (1988) Structure, organization and expression of cyanobacterial ATP synthase genes. Photosynth Res 18: 223–244

Dann MS and McCarty RE (1992) Characterization of the activation of membrane-bound and soluble CF_1 by thioredoxin. Plant Physiol 99: 153–160

Davenport JW and McCarty RE (1981) Quantitative aspects of adenosine triphosphate-driven proton translocation in spinach chloroplast thylakoids. J Biol Chem 256: 8947–8954

Davenport JW and McCarty RE (1984) An analysis of proton fluxes coupled to electron transport and ATP synthesis in chloroplast thylakoids. Biochim Biophys Acta 766: 363–374

Davenport JW and McCarty RE (1986) Relationships between rates of steady state ATP synthesis and the magnitude of the proton gradient across thylakoid membranes. Biochim Biophys Acta 851: 136–145

Deckers-Hebestreit G and Altendorf K (1992) Influence of subunit-specific antibodies on the activity of the F_o complex of the ATP synthase of *Escherichia coli*. II. Effects of subunit *c*-specific polyclonal antibodies. J Biol Chem 267: 12364–12369

Feldman RI and Sigman NS (1982) The synthesis of enzyme-bound ATP by soluble chloroplast coupling factor 1. J Biol Chem 257: 1676–1683

Feng Y and McCarty RE (1990a) Purification and reconstitution of active chloroplast F_o. J Biol Chem 265: 5104–5109

Feng Y and McCarty RE (1990b) Chromatographic purification of the chloroplast ATP synthase (CF_o-CF_1) and the role of CF_o subunit IV in proton conductance. J Biol Chem 265: 12474–12480

Fromme PP, Gräber P and Salnikow J (1987) Isolation and identification of a fourth subunit in the membrane part of the chloroplast ATP synthase. FEBS Lett 210: 27–30

Futai M, Noumi T and Maeda J (1989) ATP synthase (H^+-ATPase): Results by combined biochemical and molecular biological approaches. Annu Rev Biochem 58: 111–136

Girvin ME and Fillingame RH (1993) Helical structure and folding of subunits of F_1F_o ATP synthase: ^1H NMR resonance assignments and NOE analysis. Biochemistry 32: 12167–12177

Gräber P (1990) Kinetics of proton-transport coupled ATP synthesis in chloroplasts. In: Milazzo G and Blank M (eds) Bioelectrochemistry III, pp 277–309. Plenum Press, New York

Gromet-Elhanan Z (1992) Identification of subunits required for catalytic activity of the F_1-ATPase. J Bioengr Biomembr 24: 447–452

Hangarter RP and Good ND (1982) Energy thresholds for ATP synthesis in chloroplasts. Biochim Biophys Acta 681: 396–404

Herrmann RG, Steppuhn J, Herrmann GS and Nelson N (1993) The nuclear-encoded polypeptide CF_o-II from spinach is a real, ninth subunit of chloroplast ATP synthase. FEBS Lett 326: 192–198

Hochman Y, Lanir A and Carmeli C (1976) Relations between divalent cation binding and ATPase activity in coupling factor from chloroplasts. FEBS Lett 61: 255–259

Hudson GS and Mason JG (1988) The chloroplast genes encoding subunits of the H^+-ATP synthase. Photosynth Res 18: 205–222

Jagendorf AT, Robertson N and McCarty RE (1991) Molecular biology of the chloroplast coupling factor subunits. In: Vasil I and Bogorad L (eds) The Molecular Biology of Plastids and Mitochondria, Vol 7 in the series, Cell and Somatic Mutagenesis of Plants, pp 225–254. Academic Press, New York

Junge W (1989) Protons, the thylakoid membrane and chloroplast ATP synthase. Ann NY Acad Sci 574: 268–286

Ketcham SR, Davenport JW, Warncke K and McCarty RE (1984) Role of the γ subunit of chloroplast coupling factor 1 in the light-dependent activation of photophosphorylation and ATPase activity by dithiothreitol. J Biol Chem 259: 7286–7293

Komatsu-Takaki M (1993) Energy-dependent changes in the conformation of the ATP synthase and its catalytic activity. Eur J Biochem 214: 587–591

Lemaire C and Wollman F-A (1989) The chloroplast ATP synthase in *Chlamydomonas reinhardtii*. II. Biochemical studies on its biogenesis using mutants defective in photophosphorylation. J Biol Chem 264: 10235–10242

McCarty RE and Fagan J (1973) Light stimulated incorporation of *N*-ethylmaleimide into coupling factor 1 in spinach chloroplasts. Biochemistry 12: 1503–1507

McCarty RE and Hammes GG (1987) Molecular architecture of chloroplast coupling factor 1. Trends in Biochem Sci 12: 234–237

McCarty RE and Racker E (1966) Effects of a coupling factor in its antiserum on photophosphorylation and hydrogen ion transport. Brookhaven Symposium in Biology 19: 202–214

McCarty RE and Racker E (1968) Partial resolution of the enzymes catalyzing photophosphorylation III. Activation of adenosine triphosphatase and ^{32}P-labeled orthophosphate-adenosine triphosphate exchange of chloroplasts. J Biol Chem 243: 129–137

Mills JD and Mitchell P (1984) Thiol modulation of the chloroplast proton motive ATPase and its effects on photophosphorylation. Biochim Biophys Acta 764: 63–67

Mitchell P (1968) Chemiosmotic coupling and energy transfer. Glynn Research Bodmin

Moroney JV and McCarty RE (1982) Effect of proteolytic digestion on the Ca^{2+}-ATPase activity and subunits of latent and thiol-activated chloroplast coupling factor 1. J Biol Chem 257: 5910–5914

Murataliev MB, Milgrom YM and Boyer PD (1991) Characteristics of the combination of inhibitory Mg^{2+} and azide with the F_1 ATPase of chloroplasts. Biochemistry 30: 8305–8310

Ort DR and Oxborough K (1992) In situ regulation of chloroplast coupling factor. Annu Rev Plant Physiol Plant Mol Biol 43: 269–291

Pederson PL and Amzel LM (1993) ATP synthases, structure, reaction center, mechanism and regulation of one of nature's most unique machines. J Biol Chem 268: 9937–9940

Petrack B and Lipmann F (1961) Photophosphorylation and photohydrolysis in cell-free preparations of blue-green algae. In: McElroy WD and Glass HB (eds) Light and Life, pp 621–624. Johns Hopkins Press, Baltimore

Racker E (1976) A new look at mechanisms in bioenergetics.

Academic Press, New York

Richter ML, Gromet-Elhanan Z and McCarty RE (1986) Reconstitution of the H^+ATPase of *Rhodospirillum rubrum* by the β subunit of chloroplast coupling factor 1. J Biol Chem 261: 12109–12113

Ryrie IJ and Jagendorf AT (1972) Correlation between a conformational change in the coupling factor protein and the high energy state in chloroplasts. J Biol Chem 247: 4453–4459

Shapiro AB and McCarty RE (1990) Substrate binding-induced alteration of nucleotide binding site properties of chloroplast coupling factor 1. J Biol Chem 265: 4340–4347

Shapiro AB, Huber AH and McCarty RE (1991a) Four tight nucleotide binding sites of chloroplast coupling factor 1. J Biol Chem 266: 4194–4200

Shapiro AB, Gibson KD, Scheraga HA and McCarty RE (1991b) Fluorescence resonance energy transfer mapping of the fourth of six nucleotide-binding sites of chloroplast coupling factor 1. J Biol Chem 266: 17276–17285

Sigrist-Nelson K, Sigrist H and Azzi A (1978) Characterization of the dicyclohexylcarbodiimide-binding protein isolated from chloroplast membranes. Eur J Biochem 92: 9–14

Soteropoulos P, Süss K-H and McCarty RE (1992) Modifications of the γ subunit of chloroplast coupling factor 1 alter interactions with the inhibitory ϵ subunit. J Biol Chem 267: 10348–10354

Strotmann H and Bickel-Sandkötter S (1986) Structure, function and regulation of chloroplast ATPase. Annu Rev Plant Physiol 35: 97–120

Süss KH (1986) Neighbouring subunits of CF_0 and between CF_1 and CF_0 of the soluble chloroplast ATP synthase (CF_1-CF_0) as revealed by chemical cross-linking. FEBS Lett 201: 63–68

Süss KH and Schmidt O (1982) Evidence for an $\alpha_3,\beta_3,\gamma,\delta,I,II,\epsilon,III_5$ subunit stoichiometry of chloroplast ATP synthase complex (CF_1-CF_0). FEBS Lett 144: 213–218

Vambutas VK and Racker E (1965) Stimulation of photophosphorylation by a preparation of a latent Ca^{++}-dependent adenosine triphosphatase from chloroplasts. J Biol Chem 240: 2660–2667

Wang Z-Y, Freire E, and McCarty RE (1993) Influence of nucleotide binding site occupancy on the thermal stability of the F_1 portion of the chloroplast ATP synthase. J Biol Chem 268: 20785–20790

Wetzel CM and McCarty RE (1993a) Aspects of subunit interactions in the chloroplast ATP synthase. I. Isolation of a chloroplast-coupling factor 1-subunit III complex from spinach chloroplast thylakoids. Plant Physiol 102: 241–249

Wetzel CM and McCarty RE (1993b) Aspects of subunit interactions in the chloroplast ATP synthase. II. Characterization of a chloroplast coupling factor 1-subunit III complex from spinach thylakoids. Plant Physiol 102: 251–259

Chapter 24

The Relationship Between the Structure and Catalytic Mechanism of the Chloroplast ATP Synthase

Mark L. Richter and Denise A. Mills
Department of Biochemistry, The University of Kansas, Lawrence, Kansas 66045, USA

Summary

An overview of the nucleotide binding properties and current models of the catalytic mechanism of the chloroplast ATP synthase is presented. The discussion includes consideration of the role of the small subunits of the catalytic chloroplast coupling factor 1 (CF_1) in gating the flow of protons across the membrane. Some emphasis is placed on the potential role of the ε subunit in a proton-driven activation process and an apparent role of this subunit in influencing cooperativity among the different nucleotide binding sites on both membrane-bound and isolated CF_1. Controversy over the type of nucleotide binding sites on CF_1 is discussed, together with the potential involvement of the different nucleotide binding sites in the catalytic process.

Donald R. Ort and Charles F. Yocum (eds): Oxygenic Photosynthesis: The Light Reactions, pp. 453–468.

I. Introduction

More than three decades have passed since Peter Mitchell discovered that ATP synthesis is coupled to the dissipation of a transmembrane electrochemical proton potential. Since then, many researchers have grappled with the mechanism by which this energy conserving process takes place. The fact that we have not yet arrived at a detailed coupling mechanism is due, in large part, to the challenging complexity of the ATP synthase enzyme (described in detail in the preceding chapter) and the consequent lack of detailed structural information. Nevertheless, some important steps have been taken toward achieving an understanding of the catalytic mechanism, most notably the discovery that the energy of the electrochemical potential is delivered to the active site(s) via changes in the conformational state of the enzyme (discussed in the following section). This concept challenged the earlier view of Peter Mitchell that protons traversing the membrane participated directly in ATP synthesis at the catalytic site. With the realization that the electrochemical potential plays a less direct role in ATP synthesis, work has begun to unravel the specific protein conformational events that mediate the coupling process.

Until recently, there was no detailed three-dimensional structure for the enzyme. This severely hindered the identification of the number and type of nucleotide binding sites on the enzyme. The publication of the crystal structure of the F_1-ATPase from bovine heart mitochondria at 2.8 Å resolution (Abrahams et al., 1994) determined the position of the six nucleotide binding sites. However, it still remains to be shown which of the six identified nucleotide binding sites contribute to catalysis. Some intriguing models have been put forward to explain the properties and behavior of the different nucleotide binding sites on the enzyme. These models have, however, proven difficult to test.

This chapter will focus on the role of CF_1 subunits in proton gating, a description of the nucleotide binding sites which are located on the CF_1 portion of the ATP synthase, and on some of the currently popular working models of catalytic mechanisms.

Abbreviations: AMP-PNP – adenyl-β,γ-imidodiphosphate; ANPP – 4-azido-2-nitro-phenylphosphate; CF_1 – chloroplast coupling factor 1; CF_0 – chloroplast coupling factor o; DTT – dithiothreitol; F_1 – coupling factor 1; MF_1 – mitochondrial coupling factor; TNP-ATP/ADP – 2′(3′)-(trinitrophenyl) ATP or ADP

For discussions of other relevant topics the reader is referred to several recent reviews (Nalin and Nelson, 1987; Gromet-Elhanan, 1992; Boyer, 1993) and to the accompanying chapters by R. McCarty (Chapter 23), Boekema and Lüchen (Chapter 26) and J. Mills (Chapter 25) in this volume. The structural organization of the chloroplast ATP synthase is discussed at length in Chapter 23. Our concept of the overall topology of the enzyme is shown in Fig. 1.

II. Energy Transduction via Conformational Coupling

A. *The* Binding Change *Hypothesis*

An early observation that stimulated development of the *Binding Change* hypothesis (summarized by Cross, 1992; Boyer 1993) was that the membrane-bound chloroplast ATP synthase, in the absence of net ATP hydrolysis, catalyzes a light-dependent exchange of oxygen between water and the γphosphate group on ATP bound to the enzyme, (Avron and Jagendorf, 1959; Shavit et al., 1967; Chaney and Boyer, 1969). Many subsequent studies (for example Wimmer and Rose, 1977; Hackney and Boyer, 1978; Hackney et al., 1979; Sherman and Wimmer, 1982; Sherman and Wimmer, 1983; Spencer and Wimmer, 1985; Wood et al., 1987) have elaborated on the properties of this exchange reaction with both membrane-bound and free forms of the F_1 enzymes from chloroplasts, mitochondria and bacteria. These observations led to the conclusion (Boyer, 1987) that the principle energy-dependent

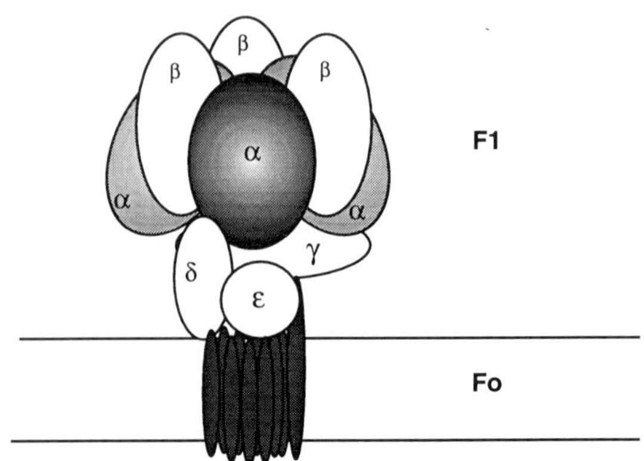

Fig. 1. Suggested organization of the subunits of the chloroplast ATP synthase.

step in ATP synthesis is not the formation of the γ phosphoanhydride bond of ATP. Instead, it was suggested that energy is required for release of product ATP and binding of substrate ADP at the catalytic site(s) during ATP synthesis. In other words, the proton potential difference ($\Delta\mu_{H^+}$) drives conformational changes in the ATP synthase that alter the binding interactions between nucleotides and the enzyme. Evidence in support of proton-driven changes in nucleotide binding affinities has come from a number of laboratories and is summarized in recent reviews by Cross (1988, 1992), Penefsky and Cross (1991) and Boyer (1993). This model has gained wide acceptance, so much so that a core element of studies currently being undertaken in this area aim at defining the conformational changes which participate in the *binding change* process.

B. Proton-driven Conformational Changes in CF_0-CF_1

Numerous studies (Roy and Moudrianakis, 1971; Strotmann et al., 1976; Magnusson and McCarty, 1976; Gräber et al., 1977; Shoshan and Selman, 1979, to name a few) have shown that illumination of thylakoid membranes leads to release of tightly-bound ADP and allows rapid exchange of medium nucleotides with nucleotides bound to CF_0-CF_1. Three different conformational states of the enzyme were proposed on the basis of such studies (Strotmann et al., 1979). In one state ADP is very tightly bound to the enzyme, while in a second state, which is induced by membrane energization, ADP is loosely bound to the enzyme. The third state is an empty ADP-binding site resulting from dissociation of nucleotide from the enzyme in the second state. Upon de-energization of the membranes ADP is incorporated to reform the tight CF_0-CF_1-ADP complex (Bickel-Sandkötter and Strotmann, 1981).

Schumann (1984) refined this model by suggesting that the proton electrochemical potential drives the enzyme between two major conformational states, from E^- which binds or exchanges nucleotides very slowly, to E^+ which binds and exchanges nucleotides rapidly. Additional, and probably more subtle, conformational changes may be induced in either of these two states upon (a) binding of nucleotide, phosphate and magnesium, (b) reduction of the γ disulfide bond (Lohse et al., 1989), or (c) further energy input from the electrochemical potential. E^+ is likely to be an activated form of the enzyme since there is a direct correlation between the kinetics of formation of this state (measured as light-induced release of tightly bound ADP) and activation of the latent ATPase activity of the ATP synthase (Shoshan and Selman, 1979; Schumann, 1981). We elaborate a little more on this model in Section IV within the context of catalytic cooperativity.

Studies by Ryrie and Jagendorf (1972) provide further evidence for the existence of proton-driven conformational changes in CF_0-CF_1 demonstrating a light-dependent exchange of ^3H_2O into the interior of CF_1. Moroney and McCarty (1982) noted that a site on the γ subunit of CF_1 becomes hypersensitive towards tryptic cleavage upon illumination of thylakoid membranes whereas in the dark this site is essentially inaccessible to trypsin. In addition, a cysteine residue on the γ subunit, which is normally buried and inaccessible to sulfhydryl modifying reagents, becomes exposed to such reagents upon illumination of the membranes (McCarty and Fagan, 1973; Moroney and McCarty, 1979). Chemical modification of this residue by N-ethylmaleimide or other sulfhydryl-directed reagents, strongly inhibits both ATP synthesis and hydrolysis.

Proton-driven conformational changes may be mediated by the ε subunit. Bakker-Grunwald and van Dam (1974) and Harris and Crofts (1978) first suggested that activation of isolated CF_1, or of membrane-bound CF_1 in the presence of DTT, may result from an interference with the ε subunit binding interaction. This idea was based on an earlier proposal (Nelson et al., 1972) that the ε subunit is an ATPase inhibitor. Later it was shown (Richter et al., 1984) that the ε subunit does indeed act as an ATPase inhibitor of isolated CF_1. Moreover, changes in the conformation of the ε subunit on membrane-bound CF_1 were observed in response to formation of the proton electrochemical potential (Richter and McCarty, 1987). Polyclonal antibodies were raised against the isolated ε subunit and shown to bind to ε on isolated CF_1 but not to ε on the latent (i.e. not active) membrane-bound form of the enzyme. Upon illumination the antibodies gained access to their binding site(s) on ε, stripped the subunit from the membranes thereby uncoupling ATP synthesis. A similar result was obtained by Komatsu-Takaki (1989) who observed a change in the reactivity/solvent accessibility of an amine residue on the ε subunit in response to formation of the proton potential difference.

C. The Role of the Small Subunits of CF_1 in Proton Gating

Several observations suggest that the γ and ε subunits of CF_1 may interact together to gate proton flow across the thylakoid membrane. Removal of the ε subunit from CF_1 does not affect the ability of the enzyme to bind specifically to thylakoid membranes previously stripped of CF_1 (Richter et al., 1984). However, the bound ε-deficient CF_1 cannot block the free flow of protons through the proton channel, nor synthesize ATP, until the ε subunit is added back to the enzyme. These studies strongly indicated that the ε subunit must be associated with the ATP synthase for proton coupling to occur.

Schumann et al., (1985) noted that removal of the ε subunit from isolated CF_1 results in exposure of the same hypersensitive tryptic cleavage site on γ that becomes exposed on the membrane-bound enzyme upon thylakoid illumination. Cleavage at this γ subunit site by trypsin renders the ε subunit completely ineffective as an ATPase inhibitor (Richter et al., 1985; Soteropoulis et al., 1992). Cleavage at this site on the membrane-bound enzyme results in uncoupling of ATP synthesis and hydrolysis with transmembrane proton movement, and in permanent activation of the ATPase activity of the enzyme (Moroney and McCarty, 1982; Larson and Jagendorf, 1986).

A close similarity between the light-driven conformation of CF_1 and the ε-deficient conformation of isolated CF_1 is further evident from the finding that removing ε from isolated CF_1 results in an increase in accessibility of the disulfide bond on the γ subunit to reducing agents in the medium, mimicking the effect of $\Delta\mu_{H^+}$ on the membrane-bound enzyme (Moroney and McCarty, 1982; Richter et al., 1985). Reversible oxidation-reduction of the disulfide bond is thought to provide a secondary level of regulation of the enzyme in vivo, and reduction of the disulfide bond in vitro partially activates the latent ATPase activity of the enzyme (see Chapter 25 for a detailed discussion of activation processes). These observations imply that (1) there is a direct binding interaction between the ε and γ subunits and that the ε subunit probably exerts its inhibitory action via the γ subunit, (2) these two subunits may act in concert to respond to the $\Delta\mu_{H^+}$ and regulate the activity of the enzyme, and (3) the ε deficient state of isolated CF_1 is equivalent (in those respects examined) to the energized state of membrane-bound CF_1. Indeed a close physical interaction between these two subunits has been demonstrated by fluorescence structural mapping studies (Richter et al., 1985).

It seems, on the basis of these observations, that at least one role of the $\Delta\mu_{H^+}$ is to remove or dislodge the inhibitory ε subunit from its binding site on CF_0-CF_1, thus activating the enzyme for both ATP synthesis and hydrolysis. Since light-induced changes in the conformation of the ε subunit, and light-induced conversion of the ATP synthase from E^- to E^+, occur simultaneously, it is quite possible that the difference between the E^- and the E^+ states results from, or is the direct result of, a change in the binding interaction between the ε subunit and one or more of the other CF_0-CF_1 subunits. At the risk of being too speculative, we suggest one possibility is that the ε subunit alternates between two conformational states, one in which it interacts primarily with CF_0, and another in which it interacts primarily with CF_1. A binding interaction between the ε subunit and subunit III of CF_0 has been shown by crosslinking studies (Suss, 1986) and by the work of Wetzel and McCarty (1993) which is discussed in Chapter 23.

It is not likely that the ε subunit dissociates from the CF_0-CF_1 complex since its presence is obligatory for proton gating and ATP synthesis (Richter et al., 1984). Furthermore, it is uncertain at this point whether the response of the ε subunit to the $\Delta\mu_{H^+}$ is part of an activation-inactivation process only, or whether it plays an integral role in the catalytic mechanism by oscillating between different conformational states during a single catalytic cycle. However, the lack of evidence for a tightly-bound ADP (the E^- state) as an intermediate during steady state catalysis (Aflalo and Shavit, 1982), and the kinetics of inhibition of ATPase activity by ADP (Schumann, 1984; Strotmann et al., 1987; Huchzermeyer, 1988a; Lohse et al., 1989), suggest that the E^+ state is probably a long-lived activated form of the enzyme.

The δ subunit, in addition to the γ and ε subunits, is also likely to assist in proton gating. Engelbrecht and Junge (1988) treated thylakoid membranes for a short time with EDTA to partially strip the CF_1 from the membranes. Under these conditions the membranes leaked protons through the exposed channels at a sufficient rate to significantly reduce ATP synthesis. Addition of excess quantities of the purified δ subunit, without any of the other CF_1 subunits, blocked the proton leak and increased the rate of ATP synthesis. This implies that the δ subunit can interact directly with subunits of CF_0 to block the free flow of protons through the proton channel. Beckers et al. (1992) showed that subunit I of CF_0 could be crosslinked to the δ subunit using the zero-length crosslinking reagent 1-ethyl-3-(dimethylaminopropyl)-carbodiimide. Crosslinking resulted in inhibition of both photophosphorylation and ATP hydrolysis without affecting formation of the transmembrane proton electrochemical potential. The results are consistent with a role for the δ subunit in the proton gating process.

Taking all of these observations together it would appear that a complex interplay between the proton channel subunits and the γ, δ and ε subunits of CF_1 may be involved in translating the energy of the $\Delta\mu_{H^+}$ into conformational changes which are propagated along to the catalytic site(s). Delineating these interactions will be an essential step in defining the coupling mechanism.

III. Nucleotide Binding Sites

A. Nucleotide Binding to CF_1

The number and type of nucleotide binding sites present on isolated CF_1 has been the subject of intense debate. Until recently, three nucleotide binding sites had been identified on both membrane-bound CF_1 (Harris and Slater, 1975; Cerione and Hammes, 1982) and isolated CF_1 (Bruist and Hammes, 1981). The tightly-bound ADP which is released upon illumination of thylakoid membranes has been proposed to reside at a non-catalytic site on CF_1, since the rate at which it was released is much slower than the onset of ATP synthesis (Bar-Zvi and Shavit, 1980; Strotmann et al., 1981; Schumann, 1981), and since medium ADP rather than the tightly-bound ADP becomes phosphorylated at the onset of illumination (Rosing et al., 1976). Alternatively, the tightly-bound ADP may be trapped at a catalytic site as the enzyme reverts to its inactive state upon de-energization. Tightly-bound nucleotides may also be stable intermediates in a reaction sequence involving two (Schumann, 1984) or three (Rosen et al., 1979) catalytic sites on the enzyme. This idea is further discussed in Section IV.

Hammes and coworkers (Carlier and Hammes, 1979; Bruist and Hammes, 1981) identified and characterized three nucleotide binding sites on isolated CF_1 which have significantly different properties from each other. One site, designated Site 1, is normally occupied by a tightly-bound ADP molecule when CF_1 is isolated. This site will exchange its nucleotide with medium nucleotides and the ADP becomes released during catalytic turnover by the enzyme. The rate of release of tightly-bound ADP was considered (Bar-Zvi and Shavit, 1982; Bruist and Hammes, 1982) too slow for Site 1 to be a catalytic site. Rather, it was presumed to have a regulatory function similar to the site that slowly releases ADP from the membrane-bound enzyme. A second site, Site 2, binds ATP only in the presence of magnesium ions. The MgATP binds very tightly to this site on the enzyme, exchanging very slowly with medium nucleotides. Nucleotide bound to this site does not dissociate during catalysis by either soluble CF_1 or membrane-bound CF_1 (Leckband and Hammes, 1987; 1988). Site 2 was, therefore, also considered to have a regulatory rather than a catalytic function. The third site, Site 3, has a much lower affinity for nucleotides than either of the other two sites since it does not retain bound nucleotides during gel filtration. Its properties were examined by titration with the fluorescent analog TNP-ATP, which it binds with an apparent K_d of 1–2 μM (Snyder and Hammes, 1984). Binding occurs in the presence or absence of magnesium ions. The lower affinity of this site makes it a prime candidate for a catalytic site since the K_m for catalysis is also in the micromolar range.

Leckband and Hammes (1987) examined the release of tightly-bound ADP from Site 1 using a quenched-flow method, and identified a lag in the onset of CaATP hydrolysis by heat-DTT activated CF_1. The lag corresponded to an activation process that appeared to be triggered by free Ca^{2+} and required binding of CaATP presumably at another site on the enzyme. Rapid mixing of activated CF_1 with CaATP caused approximately 50% of the tightly-bound ADP to be released from Site 1 within 30 s, the remaining 50% taking several minutes to dissociate. The reason for this apparent heterogeneity among sites releasing ADP is unclear. Milgrom et al., (1991) observed a similar response to binding of MgATP at putative non-catalytic sites on the enzyme, and they reasoned that binding of ATP at noncatalytic sites is prerequisite to expression of ATP hydrolytic activity. It should be noted, however, that a consensus has not been reached as to which sites are catalytic and which sites are noncatalytic.

Several more recent studies provide evidence for six nucleotide binding sites on CF_1. These involved the use of a gel filtration technique (Girault et al., 1988), equilibrium dialysis (Shapiro et al., 1991), and photoaffinity labeling (Xue et al., 1987a,b). Shapiro and McCarty (1991) utilized an HPLC method in conjunction with column centrifugal filtration to conduct a thorough analysis of all nucleotides bound to CF_1 under different conditions. Between 1.5 and 2 sites were occupied by very tightly-bound ADP on freshly isolated enzyme. Incubation with ATP in the presence of magnesium resulted in filling of 2–3 sites on CF_1 with tightly-bound MgATP, while at the same time 1–2 sites were occupied by tightly-bound ADP. Therefore as many as 4–4.5 nucleotides have been observed tightly-bound to CF_1 at one time. The term tightly-bound in this case refers to the fact that these sites stay occupied during several successive column filtration steps, and it should not be taken to infer that they cannot exchange with medium nucleotides. In our experience a site with a K_d of less than 10^{-7} M will stay occupied under these conditions. The two other sites on CF_1 have a lower affinity for nucleotides and so they presumably empty during the column filtration steps.

Shapiro and McCarty (1991) loaded radioactively labeled ATP into tight sites on CF_1 in the presence of magnesium. Labeled ADP quickly became incorporated into site 1 in exchange for the unlabeled ADP. This occurred within the first 30 min of incubation with MgATP, approximately the time it takes to pass the labeled protein through three successive centrifuge columns. Incorporation of labeled ADP, from added labeled ATP, into the other site which binds ADP tightly, takes several hours to complete. The most straightforward explanation of these results is that the labeled ATP becomes incorporated into these two sites, becomes hydrolyzed and the resulting ADP stays tightly bound at the same site. This suggests that these two sites are catalytically competent. An alternative explanation is that MgATP may become bound and hydrolyzed at a catalytic site on the enzyme and the resulting ADP subsequently exchanges into the tight ADP binding sites. Shapiro and McCarty (1991) demonstrated that medium ADP can rapidly (seconds to minutes) exchange with tightly-bound ADP. The properties of the four tight nucleotide binding sites on CF_1 suggested the presence of three more or less equivalent pairs of sites (Shapiro et al., 1991). Two of the three pairs may participate in a bi-site cooperative catalytic mechanism (see Section IV).

Studies of the mitochondrial F_1 (Cross and Nalin, 1982) identified two categories of nucleotide binding sites, three which readily exchange nucleotides with the medium and three which do not readily exchange with medium nucleotides. The three exchangeable sites are considered to be catalytic sites and the three non-exchangeable sites are considered to be non-catalytic. By analogy, it has been suggested (Xue and Boyer, 1989; Milgrom et al., 1990) that three poorly exchangeable sites on CF_1 which bind MgATP tightly are non-catalytic sites, whereas Sites 1 and 3, plus another exchangeable site, constitute catalytic sites. This model is thus in sharp contrast to that of Shapiro and McCarty (1991).

A distinction between catalytic and non-catalytic sites has also been made on the basis of their specificity for nucleotides (Kironde and Cross, 1986; Xue and Boyer, 1989). It is argued that catalytic sites bind both adenine and guanine nucleotides, whereas non-catalytic sites are adenine specific. Such a distinction may also be problematic as it was shown recently (Guerrero et al., 1990; Milgrom et al., 1991; Beharry and Bragg, 1992) that putative non-catalytic sites can also bind guanine nucleotides tightly. To complicate this picture further, it still has not been shown decisively whether more than one catalytic site operates at any given time. In fact it has been argued strongly that there is only one (Wang 1988) or maybe two (Leckband and Hammes, 1987; Berden et al., 1991; Fromme and Gräber, 1990b) active catalytic sites on F_1 enzymes, when isolated or membrane-bound.

Both the α and β subunits contain consensus amino acid sequences which are common to more than fifty different nucleotide binding proteins (Walker et al., 1982). Admon and Hammes (1987) labeled CF_1 with the photoaffinity nucleotide analog 3′-O-(4-benzoyl)benzoyl-ATP (BzATP) under conditions which led to tight binding of three BzATP molecules per CF_1. Following photolysis, most of the probe was incorporated into two peptides of the β subunit corresponding to amino acid residues 360–378 and 393–397. Similar studies were done using the analog 2-azidoATP (Xue et al., 1987a). In this case, CF_1 was treated with the nucleotide analog such that either putative catalytic or non-catalytic sites, based on the above criteria of exchangeability of catalytic sites, were labeled with the analog. Photolysis of the probe led to labeling of the β subunit in each case but at different regions of the polypeptide chain. Azido ATP bound at putative catalytic sites labeled the region of β corresponding to amino acid residues 360–378, whereas azido ATP bound at non-catalytic sites labeled the region corresponding to residues 379–390 (Xue et al., 1987b). This led to the suggestion that at least the adenine portions of both types of nucleotide binding sites are located on the β subunits. A similar result was obtained for the mitochondrial enzyme (Cross et al., 1987) using 2-azido ATP. However, the α subunits are labeled in addition to the β subunits when other photoaffinity nucleotide analogs are used (Bar-Zvi et al., 1983; Kambouris and Hammes, 1985; Admon and Hammes, 1987). The recently published crystal structure of mitochondrial F_1 (Abrahams et al., 1994) has resolved this issue. The structure, solved to 2.8 Å, clearly indicated that the nucleotide binding sites are located at each of the six interfaces formed between adjacent α and β subunits. Three sites reside on β subunits with a small structural contribution provided by the adjacent α subunits. The other three sites reside on the α subunits with a small structural contribution provided by adjacent β subunits.

A model summarizing the nucleotide binding properties of CF_1 is shown in Fig. 2. The six nucleotide binding sites are labeled 1 to 6 according to the notation of Hammes and coworkers and McCarty and coworkers. Each α or β subunit is also labeled according to the type of nucleotide occupying the β subunits of the MF_1 crystal structure as described in the figure legend. The structure was interpreted in support of the existence of three catalytic sites (sites 2, 4 and 5) and three non-catalytic sites (sites 1, 3 and 6). The β tyrosine residue equivalent to 362 in CF_1-β (#345 MF_1-β) is shown to be adjacent to the adenine ring of the nucleotide on the β subunit whereas tyrosine residue 385 (# 368 MF_1-β) is adjacent to the adenine ring of the nucleotide bound to the α subunit. This would suggest, according to the labeling studies, that the tight-binding non-catalytic site is on the α subunit and the catalytic site on the β subunit.

Identification of phosphate binding sites on CF_1 has proven difficult due to their low binding affinity. Nearly identical values were obtained for phosphate binding to isolated CF_1 using equilibrium dialysis and a rapid centrifuge column technique (Huchzermeyer, 1988b). A single phosphate binding site with an apparent K_d of 170 μM was identified on both latent and activated CF_1. This site was distinct from another site which binds the γ phosphate of ATP in activated but not latent CF_1. At low concentrations (~25 μM) the photoaffinity phosphate analog ANPP labeled the β subunit of CF_1 almost exclusively

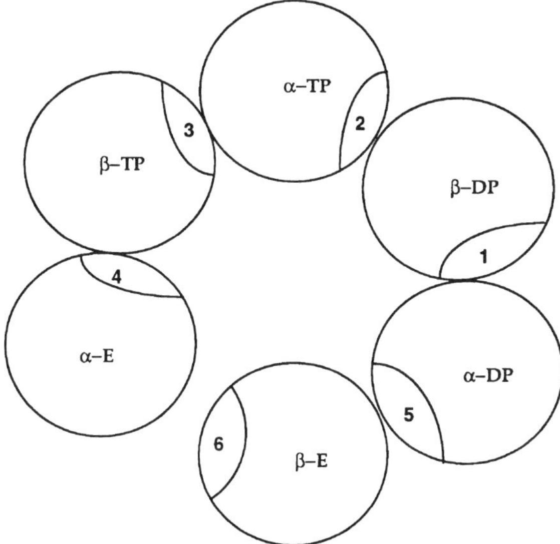

Fig. 2. Suggested organization of nucleotide binding sites on CF_1. The β subunits are defined by the nucleotide bound in the MF_1 crystal structure (Abrahams et al, 1994) with the α subunit being named according to the β nucleotide site that it structurally supports (although all of the α subunits bound AMP-PNP). β-DP had ADP bound, β-TP bound AMP-PNP and β-E had no nucleotide bound.

(Michel et al., 1992). At higher ANPP concentrations (~200 mM) two sites on the membrane-bound enzyme become labeled, one on the β subunit and another on the α subunit (G. Groth, personal communication). This is consistent with the presence of a putative phosphate-binding *p-loop* sequence in both subunits (Walker et al., 1982).

Up to six heterogeneous divalent metal cation binding sites have been identified (Hochman and Carmeli, 1981; Haddy and Sharp, 1989) on CF_1, three of which bind metals in a cooperative manner (Hiller and Carmeli, 1985). The true substrate for catalysis is believed to be Mg^{2+}ADP but the role of the metal ion has yet to be defined. Free divalent metal cations are competitive inhibitors of catalysis (Hochman et al., 1976). Recent studies using ^1H and ^{31}P nuclear relaxation measurements (Devlin and Grisham, 1990) with cobalt III tightly coordinated to the β and γ phosphates of 5′-β,γ-methylene triphosphate (i.e. $Co(NH_3)_4$AMPPCP) suggested two metal binding sites, one low affinity and one higher affinity site, both in close proximity to each other and to the nucleotide phosphates of a single bound nucleotide. Care should be taken in interpreting these data, however, since in our experience AMPPCP is a poor substitute for ATP or AMPPNP in binding to CF_1.

B. Nucleotide Binding to Isolated CF_1 Subunits

The β subunit was isolated from a preparation of the spinach CF_1 from which the two smallest subunits, δ and ε, had been removed (Richter et al., 1986). This β subunit could be reconstituted to β-deficient *Rhodospirillum rubrum* chromatophores (Philosoph et al., 1977; Khananshvili and Gromet-Elhanan, 1982) and has a single nucleotide binding site (Roux-Fromy et al., 1987; Mills and Richter, 1991) which binds the fluorescent nucleotide analog TNP-ATP with a dissociation constant of approximately 1 μM. TNP-ATP and TNP-ADP bind with equal affinities whereas TNP-AMP either does not bind or binds with a very low affinity (K_d ~mM). Binding is independent of the presence of magnesium ions. Thus the β phosphate of ATP seems to contribute mostly to the binding energy at this site. These properties of the single nucleotide binding site on the isolated β subunit are very similar to the low affinity catalytic site, Site 3, on CF_1. The specificity of the nucleotide binding site, determined as the relative binding affinity for different nucleotides through competition experiments (ATP>GTP/ITP>CTP), matches that of Site 3 and matches the nucleotide specificity for ATP hydrolysis by CF_1.

More recent studies in our laboratory (Mills et al, 1995) have identified a significant conformational change near the N-terminus of the β subunit which occurs upon ADP binding. The change is detected as an increase in fluorescence of probes located at Cys63 of β which is more than 40 Å away from the bound nucleotide (Colvert et al, 1992). A dissociation constant of 0.7 μM was measured for ADP. Interestingly, ATP can compete with ADP for binding (dissociation constant for ATP is 2 μM) but does not induce the structural change.

Two nucleotide binding sites have been identified on the β subunit isolated from the photosynthetic *Rs. rubrum* membranes (Gromet-Elhanan and Khananshvili, 1984), one site which binds ADP or ATP with apparent dissociation constants of 4.4 and 6.7 μM respectively, and a second site which binds ATP only (K_d = 6–7 μM), the binding of which is dependent on the presence of magnesium ions. Inorganic phosphate also binds in the presence, but not in the absence, of magnesium, probably at the magnesium-dependent ATP binding site. The affinities for nucleotides at both sites are sufficiently tight that the nucleotides remained bound during gel filtration (Gromet-Elhanan and Khananshvili, 1984; Khananshvili and Gromet-Elhanan, 1984). A model has been presented (Gromet-Elhanan, 1992) which suggested that each of the two sites is shared between β and α subunits such that the γ phosphate of the nucleotide interacts with one subunit while the adenine ring and ribose moiety interact mostly with the other subunit. Such an arrangement of nucleotide binding sites is consistent with the presence of a total of six sites on the enzyme, one at each of the six $\alpha\beta$ subunit interfaces. However, the crystal structure argues against a clear partitioning of the phosphate- and adenine-binding domains between α and β subunits. It also implies that the α subunit would have two nucleotide binding sites which are complementary to the two sites on the β subunit. Andralojc and Harris (1992) reported that association of the α and β subunits of *Rs. rubrum* F_1 is necessary for the formation of a tight nucleotide binding site. We (Gao et al, 1995) have shown that interaction between the γ subunit and the α and β subunits is necessary to form high affinity (tight) nucleotide binding sites on CF_1.

To date there have been no published reports on the nucleotide binding properties of isolated α subunits from *Rs. rubrum* F_1 or from CF_1. However, some preliminary data from our laboratory indicates the presence of a single nucleotide binding site on isolated spinach α subunit with very similar properties to the single site identified on the isolated spinach β subunit.

A number of models depicting the nucleotide binding domain of the β subunit of F_1 enzymes have been proposed (Duncan and Cross, 1992; Cross, 1992) based on the known structures of nucleotide binding domains of other proteins, on predictions of secondary structure, and by incorporating other information from chemical modification, metal binding and site-directed mutagenesis studies. The X-ray crystal structure of MF_1 (Abrahams et al., 1994) has identified the amino acids most closely involved in nucleotide binding, with the conserved *p-loop* being important in binding the β and γ phosphates of the bound nucleotides as predicted by the model.

Chen et al. (1992) reported over-expression in *E. coli* of the gene encoding the spinach chloroplast β subunit. The protein was solubilized from inclusion bodies and refolded by a slow dialysis procedure to yield full length β subunit with essentially identical properties, including nucleotide binding, to the native β subunit. We have recently refolded the α and γ

subunits using similar procedures and partially reconstituted all three subunits in vitro (unpublished results). Headway in this area has also been achieved with spinach subunits by Lill et al., (1993) and Chen and Jagendorf (1994) and in *Chlamydomonas reinhardtii* (Boynton et al., 1988).

IV. Catalytic Cooperativity

A. Cooperative Interactions Among Nucleotide Binding Sites

Most of the pioneering work describing cooperativity in nucleotide binding and catalysis by F_1 enzymes came from studies of the mitochondrial enzyme. This work is reviewed in depth by Cross (1981, 1992) and Penefsky and Cross (1991). Incubation of the mitochondrial F_1 with sub-stoichiometric amounts of substrate MgATP leads to its very tight binding (dissociation constant of $<10^{-12}$ M) and slow hydrolysis ($k_{cat} = 10^{-3}$ S^{-1}) at a catalytic site on the enzyme (Grubmeyer et al., 1982). At higher substrate concentrations, catalytic turnover with the concurrent release of product ADP from this site is increased by about five orders of magnitude. This apparent cooperativity, negative with respect to nucleotide binding and positive with respect to catalysis, is assumed to result from conformational changes induced by binding of nucleotide at a different site, probably another catalytic site, on the enzyme. Catalysis at the first site is referred to as 'unisite catalysis' whereas that involving more than one site under steady state conditions in the presence of higher substrate concentrations, is thought to involve at least two and possibly more than two catalytic sites and is often termed 'multisite catalysis'.

Unisite catalysis has also been shown to take place in *E. coli* F_1 (Duncan and Senior, 1985; Noumi et al., 1986) and in the F_1 from the thermophilic bacterium PS3 (Hisabori et al., 1992) but has not been reported for isolated CF_1. However, the cooperativity between Sites 1 and 3 on CF_1 (Fig. 2) is similar in many ways to unisite catalysis. Site 1 binds ADP very tightly and is considered to be catalytically competent. Site 1 releases its bound nucleotide during hydrolysis of ATP at Site 3 (or an equivalent low affinity site) (Bruist and Hammes, 1982). This is analogous to the switching from *unisite* to *multisite* catalysis seen in MF_1 except that there is no net turnover at Site 1 probably because of its lower dissociation constant than the analogous site on the mitochondrial or bacterial enzymes.

Avital and Gromet-Elhanan (1991) isolated an $\alpha\beta$ complex by lithium chloride extraction of spinach thylakoids. The complex is a mixture of $\alpha_3\beta_3$ and $\alpha_1\beta_1$ forms as determined by gel filtration (Gromet-Elhanan and Avital, 1992). The $\alpha\beta$ complex has a measurable, but very low (50–200 nmol. $min^{-1} \cdot mg^{-1}$) ATPase activity which has an uncharacteristic lack of sensitivity to the inhibitors tentoxin and azide. The relevance of this low ATPase activity to the catalytic mechanism is not clear at this time although it could represent catalytic turnover at single sites. The K_m value for the activity was not reported. However, it is clear from these experiments that, although the α and β subunit heterodimer is capable of very low rates of catalysis, the γ subunit must also be present for full cooperative catalysis by the chloroplast enzyme. An $\alpha\beta$ heterodimer was prepared from *Rs. rubrum* F_1 subunits and shown to have similar properties to the CF_1 $\alpha\beta$ complex (Andralojc and Harris, 1992).

Unisite ATP hydrolysis was reported recently for CF_o-CF_1 reconstituted into asolectin vesicles (Fromme and Gräber, 1990a) and in thylakoid membranes (Fromme and Gräber, 1990b). In order to observe the activity, the enzyme needed to be activated by membrane energization. The rate constant for the *unisite* catalysis (at sub-stoichiometric concentrations of ATP) was ~0.1 s^{-1} compared to 40–100 s^{-1} for catalysis at higher ATP concentrations. This is significantly higher than the rate constant for unisite catalysis originally reported for mitochondrial F_1 (Grubmeyer et al., 1982), but is similar to that reported for the mitochondrial enzyme under similar experimental conditions (Milgrom and Muratiliev, 1987). It was also possible to show that higher concentrations of ATP promoted turnover by several fold (Fromme and Gräber, 1990b), demonstrating the existence of catalytic cooperativity in this system. Interestingly, these workers concluded that two independent catalytic sites were responsible for the observed catalysis.

Formation of the $\Delta\mu_{H^+}$ induces release of tightly-bound ADP from the ATP synthase and also allows the rapid binding of medium ADP to occur (Schumann, 1984). Upon de-energization of the membrane the enzyme reverts to an inactive conformation that cannot bind or exchange medium ADP. The presence of ADP accelerates the inactivation process leading to formation of the very tight E-

ADP complex. Binding of phosphate and magnesium slows the inactivation process (Schumann, 1984). The basic concepts are summarized in the following scheme which is based on models presented by Shoshan and Selman (1979), Bickel-Sandkötter and Strotmann (1981), Schumann (1984), Huchzermeyer (1988a) and Fromme and Gräber (1990a; 1990b), using the E^-/E^+ notation of Schumann (1984).

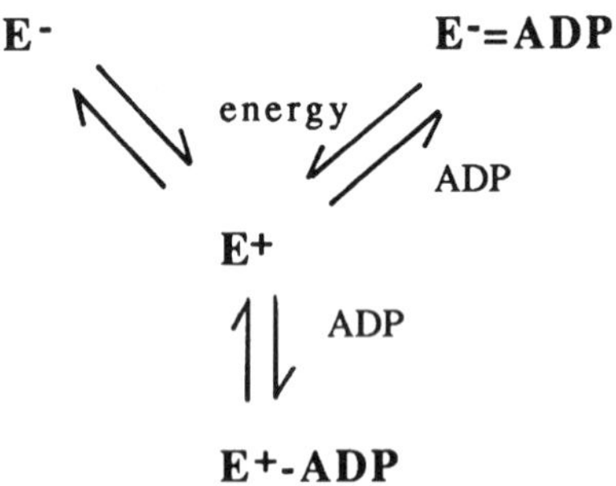

Energy is required to convert the enzyme from the inactive E^- form to the active E^+ form and probably involves a change in the position of the ε subunit. In the presence of the $\Delta\mu_{H^+}$, ADP binds to, and exchanges rapidly with, the E^+ form. In E^+-ADP the ADP is bound loosely to the enzyme and rapidly exchanges with medium nucleotides. E^+ is the catalytically active form of the enzyme. However, to observe net ATP hydrolysis the γ disulfide bond must be reduced and the $\Delta\mu_{H^+}$ lowered sufficiently to allow coupled ATP hydrolysis to take place (see chapter 25). Upon de-energization E^+ reverts to E^- slowly in the absence of ADP and more rapidly in the presence of ADP, irrespective of the oxidation state of the disulfide bond. The resulting E^- = ADP is the normal latent state of the enzyme with ADP tightly bound and non-exchangeable. If, however, the disulfide bond is in the reduced state the ATPase activity can be re-activated in the absence of a $\Delta\mu_{H^+}$ by the addition of alcohols or oxyanions (Anthon and Jagendorf, 1986; Larson and Jagendorf, 1986). Moreover, the E^+-ADP form of the enzyme can bind additional nucleotides and exhibits both negative and positive cooperativity.

B. Alternating Sites *and* Rotational Catalysis

The existence of tight and loose nucleotide binding sites, and the observed cooperativity among nucleotide binding sites on F_1 enzymes, prompted the suggestion (Kayalar et al., 1977; Cross et al., 1982) that synthesis or hydrolysis of a single ATP molecule involves participation of two or three catalytic sites which alternate between two or among three different conformational states. Cross (1992) has nicely summarized the important features of this model. In the three-site model each catalytic site is considered to undergo a transition between a tight configuration, a loose configuration and an open configuration with respect to affinity for nucleotides. Energy from the proton electrochemical potential forces F_1 to undergo a conformational change in which one of the three sites moves into the open configuration releasing product ATP while an adjacent site occupied by substrate ADP and phosphate moves into the tight configuration. In this configuration ATP is reversibly synthesized but remains tightly bound to the enzyme. These events are accompanied by loose binding of ADP and phosphate to the third site.

The idea that nucleotide binding sites alternate properties during catalysis, although it remains to be proven, attractively explains most experimental observations and so has gained considerable support. Whether two or three sites are involved in this process, however, is being hotly contested. Much of the argument for three catalytic sites comes from symmetry considerations; three pairs of $\alpha\beta$ subunits so three catalytic sites. Anywhere from one to three K_m values have been measured for CF_1 (Gresser et al., 1982; Stroop and Boyer, 1985; Zhou and Boyer, 1992) although a single rate constant is observed in oxygen exchange measurements (Kohlbrenner and Boyer, 1983).

Assuming an alternating three-site mechanism, it was suggested that the structural asymmetry required to induce different conformational states at different nucleotide binding sites resulted from the asymmetric association of the small single copy F_1 subunits, γ, δ and ε, with $\alpha\beta$ subunit pairs (Boyer and Kohlbrenner, 1981; Gresser et al., 1982). Since sites alternate sequentially through the different states it was suggested (Boyer and Kohlbrenner, 1981; Cox et al., 1984) that the $\alpha\beta$ subunits may rotate relative to the smaller subunits. A possible analogy is the proton-driven flagellar motor (Berg, 1975). This *Rotational Catalysis* model has proven extraordinarily difficult to adequately test in spite of several (published) attempts. The confusing results of these studies are summarized in the review by Cross (1992) who concludes that rotation is unlikely in isolated F_1 enzymes but likely to occur in the membrane-bound forms of the enzymes. At this stage, however, there is no compelling evidence for rotation by either form of the enzyme.

C. Site-Switching

The existence of asymmetry among the different copies of the large subunits of CF_1 was clearly demonstrated by the selective labeling of lysine 378 on one of the three α subunits following exposure of the enzyme to the probe Lucifer Yellow (Nalin et al., 1985). By monitoring fluorescence resonance energy transfer between Lucifer Yellow on α and various probes located at other sites on CF_1, the Lucifer Yellow site was mapped to the unique locus which is depicted in Fig. 1 of Chapter 23. The Lucifer Yellow site thus provided a reference point from which to follow potential changes in relative positions of nucleotide binding sites that are predicted by the *Alternating Sites* hypothesis.

Shapiro and McCarty (1990) observed a significant increase in energy transfer between the Lucifer Yellow site and TNP-ATP bound at Site 1 (Fig. 2), immediately following catalytic turnover by DTT-activated CF_1. The calculated decrease in distance between the two sites was consistent with an exchange of properties between Site 1 and Site 3, Site 3 being significantly closer to the Lucifer Yellow site in latent CF_1 (Nalin et al., 1985). It was suggested that these two nucleotide binding sites might participate in an alternating bi-site, or *site-switching*, mechanism in which they become equivalent with respect to their nucleotide binding properties during steady state catalysis. Interestingly, the increase in energy transfer induced by substrate gradually decayed when substrate was removed, implying that the enzyme relaxes into a preferred conformational state during inactivation.

The same fluorescence change was induced by binding of the non-hydrolyzable ATP analog MgAMP-PNP to Lucifer Yellow labeled CF_1, suggesting that substrate binding without catalytic turnover is sufficient to drive the conformational change involved in site switching. If this is correct, then MgAMP-PNP binding should also drive the release of tightly-bound ADP from Site 1. Such an effect was observed. Incubation of CF_1 with MgAMP-PNP resulted in tight binding of MgAMP-PNP to three sites on the enzyme with a concomitant release of the tightly-bound ADP (Shapiro et al., 1991).

The CF_1-specific inhibitor tentoxin was shown (Hu et al., 1993) to significantly reduce the rate of MgAMP-PNP-induced release of tightly-bound ADP from activated (ε-deficient) CF_1. Since tentoxin is known to be a non-competitive inhibitor of ATP synthesis/hydrolysis, it is likely that it inhibits ADP release by preventing conformational changes associated with the site to site cooperativity. This led us to test the effect of the ε subunit which is also a non-competitive inhibitor of CF_1 with respect to substrate. To do this we compared MgAMP-PNP-induced release of bound ADP (Hu et al., 1993) in latent CF_1 to CF_1 lacking the ε subunit. The disulfide bond was oxidized in both cases so that the only difference between the two forms of the enzyme was the presence or absence of the ε subunit. The result (Table 1) shows that the presence of the ε subunit greatly reduced the rate of release of ADP upon MgAMP-PNP binding. Assuming that the activated E^+ state of membrane-bound CF_1 is an ε-deficient state as suggested earlier, then this result is entirely predictable.

The fact that the ε subunit inhibits the site-switching process could explain the activating effects of CaATP (Leckband and Hammes, 1987) and MgATP (Milgrom et al., 1990). Soteropoulis et al. (1992) showed that reduction of the disulfide bond on the γ subunit significantly increases the apparent K_d for ε binding to CF_1. We predict that substrate binding will also affect ε binding. Nucleotide binding to *E. coli* F_1 has been shown to alter the conformational state of the ε subunit (Gogol et al., 1990). Nucleotide-dependent changes in the activation state of CF_1 could also explain some of the heterogeneity observed

Table 1. Inhibition of intersite cooperativity by the ε subunit

Protein	Addition	ADP Bound* (mol/mol enzyme)
CF_1	none	1.85 ± 0.10
CF_1	+ 5 mM MgAMP-PNP	1.44 ± 0.14
$CF_1(-\varepsilon)$	none	1.74 ±0.20
$CF_1(-\varepsilon)$	+ 5 mM MgAMP-PNP	0.40 ± 0.06

*Release of tightly-bound ADP was induced by incubating CF_1 or $CF_1(-\varepsilon)$ with MgAMP-PNP for 90 min. at room temperature as described by Shapiro et al., (1991). Both enzymes had been pretreated with 100 μM $CuCl_2$ to oxidize the γ disulfide bond. ADP bound was measured by HPLC following three successive centrifuge column steps to remove loosely-bound nucleotide (Soteropoulis et al., 1992). The values listed are the averages of three to four experiments.

in ADP release experiments. It is common to activate CF_1 by heating the enzyme in the presence of DTT, or by incubating the enzyme for long periods of time with high concentrations of DTT. However, Soteropoulis et al. (1992) have shown that release of the inhibitory ε subunit from reduced CF_1 does not occur until the enzyme is diluted to a very low protein concentration. It is therefore likely that the conditions of assay determine the extent of ε release and thus the extent of activation. Complete removal of the ε subunit prior to assay might provide a more homogeneous population of enzyme molecules for study.

In summary, at least two sites, and possibly two pairs of sites, have been shown to switch properties during catalysis by isolated CF_1. The two inhibitors, tentoxin and the ε subunit, allosterically block the conformational changes associated with the site-switching process.

D. A Catalytic Nucleotide *Hypothesis*

Harris (1993) recently proposed that the terminal phosphate of ATP which is bound at slowly exchanging sites (the putative non-catalytic sites) may participate directly in the catalytic process by acting as an acid/base catalyst in formation of the phosphoanhydride bond of ATP. This idea was formed from the observation that bound pyridoxal phosphate in glycogen phosphorylase (Johnson et al., 1990) catalyzes a similar reaction. In addition, there is some, although very limited, sequence homology between the phosphate binding site on glycogen phosphorylase and that on the α subunit of F_1 where the catalytic nucleotide is thought to bind.

The *catalytic nucleotide* model requires that the phosphate binding domains of catalytic and non-catalyzing (this term is used to denote the fact that these sites are involved in catalysis but do not undergo catalytic turnover) nucleotide binding sites are in close proximity to each other which is inconsistent with current structural models of F_1 (Fig. 2).

V. Concluding Remarks

We have provided an overview of several models which are currently under scrutiny as possible mechanisms of catalysis. For a model to be considered seriously it must address 1) the observed cooperativity among different nucleotide binding sites on F_1, and 2) the observed asymmetry among the nucleotide binding sites identified on the isolated enzyme and its relevance to what is happening on the membrane. Ultimately, the correct model must describe a fully reversible mechanism whereby protons can drive changes in the structure of nucleotide binding sites in one direction, and nucleotides can reverse the changes and drive protons back across the membrane. The model which has received most attention, and which has been presented in most detail, is the *Alternating Sites/Rotational* mechanism. However, definitive evidence for or against this model has not yet been provided.

Bianchet et al. (1991) have reported the crystal structure of the rat liver mitochondrial ATP synthase to a resolution of 3.6 Ångstroms. Abrahams et al. (1993, 1994) and Ishii et al. (1993) have also prepared crystals of the beef heart mitochondrial and bacterial enzymes respectively. Crystallization of the chloroplast enzyme has not yet been reported. However, the basic structure and mechanism is likely to be conserved among all of the ATP synthases so that information obtained with any one of these enzymes should be relevant to all of them. With a refined three-dimensional structure of part of the MF_1 in hand, plus the recent development of new genetic methods for manipulating the structures of the eucaryotic enzymes, we can expect much more rapid progress over the next few years in defining the catalytic mechanism of ATP synthesis.

References

Abrahams JP, Lutter R, Todd RJ, van Raaij MJ, Leslie AGW and Walker JE (1993) Inherent asymmetry of the structure of F_1-ATPase from bovine heart mitochondria at 6.5Å resolution. EMBO J 12: 1775–1780

Abrahams JP, Leslie AGW, Lutter R and Walker JE (1994) Structure at 2.8Å resolution of F_1-ATPase from bovine heart mitochondria. Nature 370: 621–628

Admon A and Hammes GG (1987) Amino acid sequence of the nucleotide binding region of chloroplast coupling factor 1. Biochemistry 26: 3193–3197

Aflalo C and Shavit N (1982) Source of rapidly labeled ATP tightly bound to non-catalytic sites on the chloroplast ATP synthase. Eur J Biochem 126: 61–68

Andralojc PJ and Harris DA (1992) Isolation and characterization of a functional $\alpha\beta$ heterodimer from ATP synthase of *Rhodospirillum rubrum*. FEBS Lett 310: 187–192

Anthon GE and Jagendorf AT (1986) Evidence for multiple effects in methanol activation of chloroplast coupling factor 1. Biochim Biophys Acta 848: 92–98

Avital S and Gromet-Elhanan Z (1991) Extraction and purification of the beta subunit and an active alpha-beta-core complex from the spinach chloroplast CF_oF_1 ATP synthase. J Biol Chem 266: 7067–7072

Avron M and Jagendorf AT (1959) Evidence concerning the mechanism of adenine triphosphate formation by spinach chloroplasts. J Biol Chem 234: 967–972

Bakker-Grunwald T and van Dam K (1974) On the mechanism of activation of the ATPase in chloroplasts. Biochim Biophys Acta 347: 290–298

Bar-Zvi D and Shavit N (1980) Role of the tight nucleotide binding in the regulation of the chloroplast ATP synthetase activities. FEBS Lett 119: 68–72

Bar-Zvi D and Shavit N (1982) Modulation of the chloroplast ATPase by tight ADP binding: Effect of uncouplers and ATP. J Bioenerg Biomemb 14: 467–478

Bar-Zvi D, Teifert MA and Shavit N (1983) Interaction of the chloroplast ATP synthetase with the photoreactive nucleotide 3′-0-(4-benzoyl)benzoyl adenosine 5′-diphosphate. FEBS Lett 160: 233–238

Beckers G, Berzborn RJ and Strotmann H (1992) Zero-length cross-linking between subunits δ and I of the H^+-translocating ATPase of chloroplasts. Biochim Biophys Acta 1101: 97–104

Beharry S and Bragg PD (1992) *E. coli* F1-ATPase can use GTP-nonchaseable bound adenine nucleotide to synthesize ATP in DMSO. Biochemistry 31: 11472–11476

Berden JA, Hartog AF and Edel CM (1991) Hydrolysis of ATP can be described only on the basis of a dual-site mechanism. Biochim Biophys Acta 1057: 151–156

Berg HC (1975) Bacterial behavior. Nature 254: 389–392

Bianchet M, Xavier Y, Hullihen J, Pedersen PL and Amzel M (1991) Mitochondrial ATP synthase: Quaternary structure of the F1 moiety at 3.6Å determined by X-ray diffraction analysis. J Biol Chem 266: 21197–21201

Bickel-Sandkötter S and Strotmann H (1981) Nucleotide binding and regulation of chloroplast ATP synthase. FEBS Lett 125: 188–192

Boyer PD (1987) The unusual enzymology of ATP synthase. Biochemistry 26: 8503–8507

Boyer PD (1993) The binding change mechanism for ATP synthase-some probabilities and possibilities. Biochim Biophys Acta 1140: 215–250

Boyer PD and Kohlbrenner WE (1981) The present status of the binding-change mechanism and its relation to ATP formation by chloroplasts. In: R Selman and Selman-Reimer S (eds) Energy Coupling in Photosynthesis pp 230–240. Elsevier, Amsterdam

Boynton JE, Gilham NW, Harris JP, Johnson AM, Jones AR, Randalf-Anderson BL, Robertson D, Klein TM, Shark KB and Sanford JC (1988) Chloroplast transformation in *Chlamydomonas* with high velocity projectiles. Science 240: 1534–1538

Bruist MF and Hammes GG (1981) Further characterization of nucleotide binding sites on chloroplast coupling factor one. Biochemistry 20: 6298–6305

Bruist MF and Hammes GG (1982) Mechanism for catalysis and regulation of adenosine 5′-triphosphate hydrolysis by chloroplast coupling factor. Biochemistry 21: 3370–3377

Carlier MF and Hammes GG (1979) Interaction of nucleotides with chloroplast coupling factor 1. Biochemistry 18: 3446–3451

Cerione RA and Hammes GG (1982) Structural mapping of nucleotide binding sites on chloroplast coupling factor. Biochemistry 21: 745–752

Chaney SG and Boyer PD (1969) Lack of detection of intermediates in the path of phosphorylative oxidation to water in photophosphorylation. J Biol Chem 244: 5773–5778

Chen GG and Jagendorf AT (1994) Chloroplast molecular chaperone-assisted refolding and reconstitution of an active multi-subunit CF_1 core. Proc Natl Acad Sci USA: 11497–11501

Chen Z, Wu I and Richter ML (1992) Over-expression and refolding of β-subunit from chloroplast ATP synthase. FEBS Lett 298: 69–73

Colvert KC, Mills DA and Richter ML (1992) Structural mapping of cysteine-63 of the chloroplast ATP synthase β subunit. Biochemistry 31: 3930–3935

Cox GB, Jans DA, Fimmel AL, Gibson F and Hatch L (1984) The mechanism of ATP synthase: Conformational change by rotation of the β-subunit. Biochim Biophys Acta 768: 201–208

Cross RL (1981) The mechanism and regulation of ATP synthesis by F_1-ATPases. Ann Rev Biochem 50: 681–714

Cross RL (1988) The number of functional sites on F_1-ATPases and the effects of quaternary structural asymmetry on their properties. J Bioenerg Biomemb 20: 395–406

Cross RL (1992). The reaction mechanism of F_oF_1-ATP synthases. In: Ernster L (ed) Molecular Mechanisms in Bioenergetics pp 317–330. Elsevier, Amsterdam

Cross RL and Nalin CM (1982) Adenine nucleotide binding sites on beef-heart F_1-ATPase. Evidence for three exchangeable sites that are distinct from three non-catalytic sites. J Biol Chem 257: 2874–2881

Cross RL, Cunningham D, Miller CG, Xue Z, Zhou J-M and Boyer PD (1987) Adenine nucleotide binding sites on beef heart F_1 ATPase: Photoaffinity labeling of β subunit Tyr 368 at a non-catalytic site and Tyr 345 at a catalytic site. Proc Natl Acad Sci USA 84: 5715–5719

Devlin CC and Grisham CM (1990) ^1H and ^{31}P Nuclear magnetic resonance and kinetic studies of the active site structure of chloroplast CF_1 ATP synthase. Biochemistry 29: 6192–6203

Duncan TM and Cross RL (1992) A model for the catalytic site of F_1-ATPase based on analogies to nucleotide-binding domains of known structure. J Bioenerg Biomemb 24: 453–461

Duncan TM and Senior AE (1985) The defective proton-ATPase of *uncD* mutants of *Escherichia coli*: Two mutations which affect the catalytic mechanism. J Biol Chem 260: 4901–4907

Engelbrecht S and Junge W (1988) Purified subunit δ of chloroplast coupling factor CF_1 reconstitutes photophosphorylation in partially CF_1-depleted membranes. Eur J Biochem 172: 213–218

Fromme P and Gräber P (1990a) Activation/inactivation and uni-site catalysis by the reconstituted ATP-synthase from chloroplasts. Biochim Biophys Acta 1016: 29–42

Fromme P and Gräber P (1990b) ATP-hydrolysis in chloroplasts: Uni-site catalysis and evidence for heterogeneity of sites. Biochim Biophys Acta 1020: 187–194

Gao F, Lipscomb B, Wu I and Richter ML (1995) In vitro assembly of the core catalytic complex of the chloroplast ATP synthase. J Biol Chem 270: 9763–9769

Girault G, Berger G, Galmiche J-M and Andre F (1988) Characterization of six nucleotide binding sites of chloroplast coupling factor 1 and one site on its purified β subunit. J Biol

Chem 263: 14690–14695

Gogol E, Johnston E, Aggeler R and Capaldi R (1990) Ligand-dependent structural variations in *E. coli* F_1-ATPase revealed by cryoelectron microscopy. Biochemistry 29: 9585–9589

Gräber P, Schlodder E and Witt HT (1977) Conformational change of the chloroplast ATPase induced by a transmembrane electric field and its correlation to phosphorylation. Biochim Biophys Acta 461: 426–440

Gresser MJ, Meyers JA and Boyer PD (1982) Catalytic site cooperativity of beef heart mitochondrial F_1-ATPase. J Biol Chem 257: 12030–12038

Gromet-Elhanan Z (1992) Identification of the subunits required for the catalytic activity of the F_1-ATPase. J Bioenerg Biomemb 24: 447–452

Gromet-Elhanan Z and Avital S (1992) Properties of the catalytic $(\alpha\beta)$-core complex of chloroplast CF_1-ATPase. Biochim Biophys Acta 1102: 379–385

Gromet-Elhanan Z and Khanashvili D (1984) Characterization of two nucleotide binding sites on the isolated, reconstitutively active β subunit of the F_0F_1 ATP synthase. Biochemistry 23: 1022–1028

Grubmeyer C, Cross RL and Penefsky HS (1982) Mechanism of ATP hydrolysis by beef heart mitochondrial ATPase. J Biol Chem 257: 12092–12100

Guerrero KJ, Ehler LL and Boyer PD (1990) Guanosine and formycin triphosphates bind at non-catalytic nucleotide binding sites of CF_1 ATPase and inhibit ATP hydrolysis. FEBS Lett 270: 187–190

Hackney DD and Boyer PD (1978) Subunit interaction during catalysis: Implications of concentration dependency of oxygen exchange accompanying oxidative phosphorylation for alternating site cooperativity. J Biol Chem 253: 3164–3170

Hackney DD, Rosen G and Boyer PD (1979) Subunit interaction during catalysis. Alternating site cooperativity in photophosphorylation shown by substrate modulation of $[^{18}O]$ ATP species formation. Proc Natl Acad Sci USA 76: 3646–3650

Haddy AE and Sharp RR (1989) Field dependence of solvent proton and deuteron NMR relaxation rates of the manganese (II) binding site of chloroplast coupling factor 1. Biochemistry 28: 3656–3664

Harris DA (1993) The 'non-exchangeable' nucleotides of F_1-F_0 ATP synthase; Cofactors in hydrolysis? FEBS Lett 316: 209–215

Harris DA and Crofts AR (1978) The initial stages of photophosphorylation: Studies using excitation by saturating, short flashes of light. Biochim Biophys Acta 502: 87–102

Harris DA and Slater EC (1975) Tightly-bound nucleotides of the energy-transducing ATPase of chloroplasts and their role in photophosphorylation. Biochim Biophys Acta 387: 335–348

Hiller R and Carmeli C (1985) Cooperativity among manganese-binding sites in the H-ATPase of chloroplasts. J Biol Chem 260: 1614–1617

Hisabori T, Muneyuki E, Odaka M, Yokoyama K, Mochizuki K and Yoshida M (1992) Single site hydrolysis of 2′,3′-O-(2,4,6-trinitrophenyl)-ATP by the F_1-ATPase from the thermophilic bacterium PS3 is accelerated by the chase-addition of excess ATP. J Biol Chem 267: 4551–4556

Hochman Y and Carmeli C (1981) Correlation between the kinetics of activation and inhibition of adenosinetriphosphatase activity by divalent metal ions and the binding of manganese to chloroplast coupling factor 1. Biochemistry 20: 6287–6292

Hochman Y, Lanir A and Carmeli C (1976) Relations between divalent cation binding and ATPase activity in coupling factor from chloroplast. FEBS Lett 61: 255–259

Hu N, Mills DA, Huchzermeyer B and Richter ML (1993) Inhibition by tentoxin of cooperativity among nucleotide binding sites on chloroplast coupling factor 1. J Biol Chem 268: 8536–8540

Huchzermeyer B (1988a) Nucleotide binding and ATPase activity of membrane bound chloroplast coupling factor (CF1). Z Naturforsch 43: 133–139

Huchzermeyer B (1988b) Phosphate binding to isolated chloroplast coupling factor (CF1). Z Naturforsch 43: 213–218

Ishii N, Yoshimura H, Nagayama K, Kagawa Y and Yoshida M (1993) Three dimensional structure of F_1-ATPase of thermophilic bacterium PS3 obtained by electron crystallography. J Biochem (Tokyo) 113: 245–250

Johnson LN, Acharya KR, Jordan MD and McLaughlin PJ (1990) Refined crystal structure of the phosphorylase-heptulose-2-phosphate-oligosaccharide-AMP complex. J Mol Biol 211: 645–661

Kambouris NG and Hammes GG (1985) Investigation of nucleotide binding sites on chloroplast coupling factor 1 with 3′-O-(4-benzoyl)benzoyl adenosine 5′-triphosphate. Proc Natl Acad Sci (USA) 82: 1950–1953

Kayalar C, Rosing J and Boyer PD (1977) An alternating site sequence for oxidative phosphorylation suggested by measurement of substrate binding patterns and exchange reaction inhibitions. J Biol Chem 252: 2486–2491

Khananshvili D and Gromet-Elhanan Z (1982) Isolation and purification of an active γ subunit of the F_0F_1-ATP synthase from chromatophore membranes of *Rhodospirillum rubrum*. J Biol Chem 257: 11377–11383

Khananshvili D and Gromet-Elhanan Z (1984) Demonstration of two binding sites for ADP on the isolated β-subunit of the *Rhodospirillum rubrum* RF_0F_1-ATP synthase. FEBS Lett 178: 10–14

Kironde FA and Cross RL (1986) Adenine nucleotide-binding sites on beef heart F_1-ATPase: Conditions that affect occupancy of catalytic and non-catalytic sites. J Biol Chem 261: 12544–12549

Kohlbrenner WE and Boyer PD (1983) Probes of catalytic site cooperativity during catalysis by the chloroplast ATPase and the ATP synthase. J Biol Chem 258: 10881–10886

Komatsu-Takaki M (1989) Energy-dependent conformational changes in the epsilon subunit of the chloroplast ATP synthase. J Biol Chem 264: 17750–17753

Larson EM and Jagendorf AT (1986) Anion stimulation of ATPase in activated spinach chloroplast coupling factor 1 (CF_1): Light activation mimic? Plant Physiol 80: S251

Leckband D and Hammes GG (1987) Interactions between nucleotide binding sites on chloroplast coupling factor 1 during ATP hydrolysis. Biochemistry 26: 2306–2312

Leckband D and Hammes GG (1988) Function of tightly bound nucleotides on membrane-bound chloroplast coupling factor. Biochemistry 27: 3629–3633

Lill H, Burkovski A, Altendorf K, Junge W and Engelbrecht S (1993) Complementation of *Escherichia coli unc* mutant strains by chloroplast and cyanobacterial F_1-ATPase subunits. Biochim Biophys Acta 1144: 278–284

Lohse D, Thelen R and Strotmann H (1989) Activity equilibria of the thiol-modulated chloroplast H^+-ATPase as a function of the proton gradient in the absence and presence of ADP and arsenate. Biochim Biophys Acta 976: 85–93

Magnusson RP and McCarty RE (1976) Light-induced exchange of nucleotides into coupling factor 1 in spinach chloroplast thylakoids. J Biol Chem 251: 7417–7422

McCarty RE and Fagan J (1973) Incorporation of N-ethylmaleimide into coupling factor 1 in spinach chloroplasts. Biochemistry 12: 1503–1507

Michel L, Garin J, Girault G and Vignais PV (1992) Photolabeling of the phosphate binding site of chloroplast coupling factor 1 with [^{32}P]azidonitrophenyl phosphate. FEBS Lett 313: 90–93

Milgrom YM and Murataliev MB (1987) Characterization of the nucleotide tight-binding sites of the isolated mitochondrial F_1-ATPase. FEBS Lett 219: 156–160

Milgrom YM, Ehler LL and Boyer PD (1990) ATP binding at non-catalytic sites of soluble CF1 is required for expression of the enzyme activity. J Biol Chem 265: 18725–18728

Milgrom YM, Ehler LL and Boyer PD (1991) The characteristics and effect on activity of nucleotide binding to non-catalytic sites of CF_1-ATPase. J Biol Chem 266: 11551–11558

Mills DA and Richter ML (1991) Nucleotide binding to the isolated β subunit of the chloroplast ATP synthase. J Biol Chem 266: 7440–7444

Mills DA, Seibold SA, Squier TC and Richter ML (1995) ADP binding induces long-distance structural changes in the β polypeptide of the chloroplast ATP synthase. Biochemistry 34: 6100–6108

Moroney JV and McCarty RE (1979) Reversible uncoupling of photophosphorylation by a new bifunctional maleimide. J Biol Chem 254: 8951–8955

Moroney JV and McCarty RE (1982) Light-dependent cleavage of the γ subunit of coupling factor 1 by trypsin causes activation of Mg^{2+}-ATPase activity and uncoupling of photophosphorylation in spinach chloroplasts. J Biol Chem 257: 5915–5920

Nalin CM and Nelson N (1987) Structure and biogenesis of chloroplast coupling factor CF_0CF_1-ATPase. Curr Top Bioenerg 15: 273–294

Nalin CM, Snyder B and McCarty RE (1985) Selective modification of an α subunit of chloroplast coupling factor 1. Biochemistry 24: 2318–2324

Nelson N, Nelson H and Racker E (1972) Partial resolution of the enzymes catalyzing photophosphorylation: Purification and properties of an inhibitor isolated from chloroplast coupling factor 1. J Biol Chem 247: 7657–7662

Noumi T, Taniai M, Kananzawa H and Futai M (1986) Replacement of arginine 246 by histidine in the β subunit of *Escherichia coli* H^+-ATPase resulted in loss of multi-site ATPase activity. J Biol Chem 261: 9196–9201

Penefsky HS and Cross RL (1991) Structure and mechanism of F_0F_1-type ATP synthases and ATPases. Advances in Enzymology and Related Areas of Molecular Biology 64: 173–214

Philosoph S, Binder A and Gromet-Elhanan Z (1977) Coupling factor ATPase complex of *R. rubrum*. J Biol Chem 252: 8747–8752

Richter ML and McCarty RE (1987) Energy-dependent changes in the conformation of the ε subunit of the chloroplast ATP synthase. J Biol Chem 262: 15037–15040

Richter ML, Patrie WJ and McCarty RE (1984) Preparation of the ε subunit and ε subunit-deficient chloroplast coupling factor 1 in reconstitutively active forms. J Biol Chem 259: 7371–7373

Richter ML, Snyder B, McCarty RE and Hammes GG (1985) Binding stoichiometry and structural mapping of the ε polypeptide of chloroplast coupling factor 1. Biochemistry 24: 5755–5763

Richter ML, Gromet-Elhanan Z and McCarty RE (1986) Reconstitution of the H^+-ATPase complex of *Rhodospirillum rubrum* by the β subunit of the chloroplast coupling factor 1. J Biol Chem 261: 12109–12113

Rosen G, Gresser M, Vinkler C and Boyer PD (1979) Assessment of total catalytic sites and the nature of bound nucleotide participation in photophosphorylation. J Biol Chem 254: 10654–10661

Rosing J, Smith DJ, Kayalar C and Boyer PD (1976) Medium ADP and not ADP already tightly bound to thylakoid membranes forms the initial ATP in chloroplast phosphorylation. Biochem Biophys Res Commun 72: 1–8

Roux-Fromy M, Neumann J-M, Andre F, Berger G, Girault G, Galmiche J-M, and Remy R (1987) Biochemical and proton NMR characterization of the isolated functional beta-subunit of coupling factor one from spinach chloroplasts. Biochem Biophys Res Commun 144: 718–725

Roy H and Moudrianakis EN (1971) Interactions between ADP and the coupling factor of photophosphorylation. Proc Natl Acad Sci USA 68: 2720–2724

Ryrie IJ and Jagendorf AT (1972) Correlation between a conformational change in the coupling factor protein and the high energy state in chloroplasts. J Biol Chem 247: 4453–4459

Schumann J (1981) Adenine nucleotide binding to CF_1 and ATPase activity of chloroplasts. In: R Selman and Selman-Reimer S (eds) Energy Coupling in Photosynthesis Research, pp 223–230. Elsevier, North Holland

Schumann J (1984) A study on the exchange of tightly bound nucleotides on the membrane-associated chloroplast ATP synthase complex. Biochim Biophys Acta 766: 334–342

Schumann J, Richter ML and McCarty RE (1985) Partial proteolysis as a probe of the conformation of the γ subunit in activated soluble and membrane-bound chloroplast coupling factor. J Biol Chem 260: 11817–11830

Shapiro AB and McCarty RE (1990) Substrate binding-induced alteration of nucleotide binding site properties of chloroplast coupling factor 1. J Biol Chem 265: 4340–4347

Shapiro AB and McCarty RE (1991) Four tight nucleotide binding sites of chloroplast coupling factor 1. J Biol Chem 266: 4194–4200

Shapiro A, Gibson KD, Scheraga H and McCarty RE (1991) Fluorescence resonance energy transfer mapping of the fourth of six nucleotide-binding sites of chloroplast coupling factor 1. J Biol Chem 266: 17276–17285

Shavit N, Skye GE and Boyer PD (1967) Occurrence and possible mechanism of ^{32}P and ^{18}O exchange reactions of photophosphorylation. J Biol Chem 242: 5125–5130

Sherman PA and Wimmer MJ (1982) Two types of kinetic regulation of the activated ATPase in the chloroplast photophosphorylation system. J Biol Chem 257: 7012–7017

Sherman PA and Wimmer MJ (1983) Kinetic effects of chemical and physical uncoupling on the energy-transducing ATPase from spinach chloroplasts. Eur J Biochem 136: 539–543

Shoshan V and Selman BR (1979) The relationship between

light-induced adenine nucleotide exchange and ATPase activity in chloroplast thylakoid membranes. J Biol Chem 254: 8801–8807

Snyder B and Hammes GG (1984) Structural mapping of chloroplast coupling factor. Biochemistry 23: 5787–5795

Soteropoulos P, Suss K-H and McCarty RE (1992) Modifications of the γ subunit of chloroplast coupling factor 1 alter interactions with the inhibitory ε subunit. J Biol Chem 267: 10348–10354

Spencer JC and Wimmer MJ (1985) Mechanisms by which reactions catalyzed by chloroplast coupling factor 1 are inhibited: ATP synthesis and ATP-H_2O oxygen exchange. Biochemistry 24: 3884–3890

Stroop SD and Boyer PD (1985) Characteristics of the chloroplast ATP synthase as revealed by reaction at low ADP concentrations. Biochemistry 24: 2304–2310

Strotmann H, Bickel S and Huchzermeyer B (1976) Energy-dependent release of adenine nucleotides tightly bound to chloroplast coupling factor CF_1. FEBS Lett 61: 194–198

Strotmann H, Bickel-Sandkötter S and Shoshan V (1979) Kinetic analysis of light-dependent exchange of adenine nucleotides on chloroplast coupling factor CF_1. FEBS Lett 101: 316–320

Strotmann H, Bickel-Sandkötter S, Franek U and Gerke V (1981) Nucleotide interactions with membrane-bound CF_1. In: R Selman and Selman-Reimer S (eds) Energy Coupling in Photosynthesis Research, pp 187–196. Elsevier, North Holland

Strotmann H, Kleefeld S and Lohse D (1987) Control of ATP hydrolysis in chloroplasts. FEBS Lett 221: 265–269

Suss K-H (1986) Stable binding interaction among subunits of the chloroplast ATP synthase (CF_1-CF_0) as examined by solid support (nitrocellulose)-subunit reconstitution-immunoblotting. FEBS Lett 199: 169–172

Walker JE, Saraste M, Runswick MJ and Gay N (1982) Distantly related sequences in the α- and β- subunits of ATP synthase, myosin, kinases and other ATP-requiring enzymes and a common nucleotide binding fold. EMBO J 1: 945–951

Wang JH (1988) Chemical modification of active sites in relation to the catalytic mechanism of F1. J Bioenerg Biomemb 20: 407–422

Wetzel CM and McCarty RE (1993) Aspects of subunit interactions in the chloroplast ATP synthase: II. Characterization of a chloroplast coupling factor 1-subunit III complex from spinach thylakoids. Plant Physiol 102: 251–259

Wimmer MJ and Rose IA (1977) Mechanism for oxygen exchange in the chloroplast photophosphorylation system. J Biol Chem 252: 6769–6775

Wood JM, Wise JG, Senior AE, Futai M and Boyer PD (1987) Catalytic properties of the F_1-adenosine triphosphatase from *Escherichia coli* K-12 and its genetic variants as revealed by 0^{18} exchanges. J Biol Chem 262: 2180–2186

Xue Z and Boyer PD (1989) Modulation of the GTPase activity of the chloroplast F_1-ATPase by ATP binding at non-catalytic sites. Eur J Biochem 179: 677–681

Xue Z, Zhou J-M, Melese T, Cross RL and Boyer PD (1987a) Chloroplast F_1ATPase has more than three nucleotide binding sites, and 2-azido ADP or 2-azido ATP at both catalytic and noncatalytic sites labels the β subunit. Biochemistry 26: 3749–3753

Xue Z, Miller CG, Zhou J-M and Boyer PD (1987b) Catalytic and noncatalytic nucleotide binding sites of chloroplast F_1ATPase. FEBS Lett 223: 391–394

Zhou J-M and Boyer PD (1992) MgADP and free Pi as the substrates and the Mg^{2+} requirement for photophosphorylation. Biochemistry 31: 3166–3171

Chapter 25

The Regulation of Chloroplast ATP Synthase, CF_o-CF_1

John D. Mills
Plant Biochemistry and Microbiology Group, Department of Biological Sciences, Keele University, Staffordshire, ST5 5BG UK

Donald R. Ort and Charles F. Yocum (eds): Oxygenic Photosynthesis: The Light Reactions, pp. 469–485.

Summary

ATP accounts for about one quarter of the energy stored by the light reactions of photosynthesis. The enzyme responsible for this energy capture is the ATP synthase, CF_0CF_1. This enzyme is regulated in a remarkable way by all of its substrates. Also, it is the only component of the light reactions known to be under redox control of the thioredoxin system. Primary control is exerted by the electrochemical proton potential difference, $\Delta\tilde{\mu}_{H^+}$. In the dark, CF_0CF_1 is catalytically inactive but becomes activated by light-dependent $\Delta\tilde{\mu}_{H^+}$. The magnitude of $\Delta\tilde{\mu}_{H^+}$ required for activation is considerably lower when CF_0CF_1 is reduced. The substrates of photophosphorylation have allosteric-like effects on activity, ADP and Mg^{2+} stabilizing the inactive states, and ATP and inorganic phosphate (Pi) favoring the active states.

Simple 4-state kinetic models can explain recent measurements in vitro of the extent of activation and the rate of catalysis of CF_0CF_1 as a function of $\Delta\tilde{\mu}_{H^+}$ and ΔG_p. However, a crucial uncertainty is the number of protons involved in both processes. In vivo, reduction by thioredoxin is reasonably understood, but short-term control of CF_0CF_1 following light/dark and dark/light transitions is not easily explained by in vitro models. The possible benefits to the plant of regulation are examined. It is argued that oxidation is a long-term mechanism designed to limit the possibility of prolonged ATP hydrolysis during darkness. Reduction maximizes the P/e_2 ratio of ATP synthesis at very low light or partially uncoupled conditions, and also maximizes the rate of photophosphorylation at high light intensity. Regulation is modeled using metabolic control theory. It is shown that CF_0CF_1 is rarely a rate-limiting step in photophosphorylation, and reduction serves to further decrease its flux control coefficient.

I. Introduction

The chloroplast ATP synthase (coupling factor, CF_0CF_1) uses the energy available in a proton electrochemical potential difference, $\Delta\tilde{\mu}_{H^+}$, to make ATP. The light reactions of photosynthesis supply the Calvin cycle with ATP and NADPH in the molar ratio 3:2. From the metabolite concentrations in chloroplasts in vivo, the free energies for ATP synthesis, ΔG_p, and $NADP^+$ reduction are around 45 and 220 kJ $mole^{-1}$ respectively (Heineke et al., 1991). CF_0CF_1 is therefore responsible for transducing one quarter of the conserved light energy.

While it is important that CF_0CF_1 operates efficiently during photosynthesis, hydrolysis of ATP must be restricted during darkness. This is accomplished by a complex regulation involving the $\Delta\tilde{\mu}_{H^+}$, thioredoxin and allosteric-like effects of *substrates*. CF_0CF_1 activity is not influenced by any other physiological effector (for example an intermediate of the Calvin cycle). This fact indicates that this enzyme is not a flux control point in photophosphorylation. In fact, it will be argued that regulation serves to decrease the flux control coefficient for CF_0CF_1 on ATP synthesis during the day, and maximize it on ATP hydrolysis at night.

II. Kinetic Studies with Thylakoids

A. The Four State Model

Figure 1 depicts a widely accepted working model for the regulation of CF_0CF_1 (after Bakker-Grunwald and van Dam, 1974). CF_0CF_1 is controlled by three inter-dependent mechanisms; i) activation by $\Delta\tilde{\mu}_{H^+}$, ii) oxidation/reduction by thioredoxin (or dithiothreitol), and iii) control by ATP, ADP, Pi and Mg^{2+}. The evidence for this scheme is as follows.

B. Early Studies

1. Regulation by Thiols and by $\Delta\tilde{\mu}_{H^+}$

Chloroplast thylakoids isolated from dark-adapted leaves do not hydrolyze ATP until given a prior illumination period in the presence of reduced thiols (Petrack et al., 1965; McCarty and Racker, 1968; Carmeli and Lifschitz, 1972). Since thiols very slowly activate the enzyme in the dark, it was proposed that illumination induces a conformational change in

Abbreviations: $\Delta\tilde{\mu}_{H^+}$ – electrochemical potential difference of H^+ across the thylakoid; ΔpH – difference in pH across the thylakoid; $\Delta\Psi$ – difference in electric potential across the thylakoid; ΔG_p – Free energy change for ATP synthesis; NEM – N-ethylmaleimide

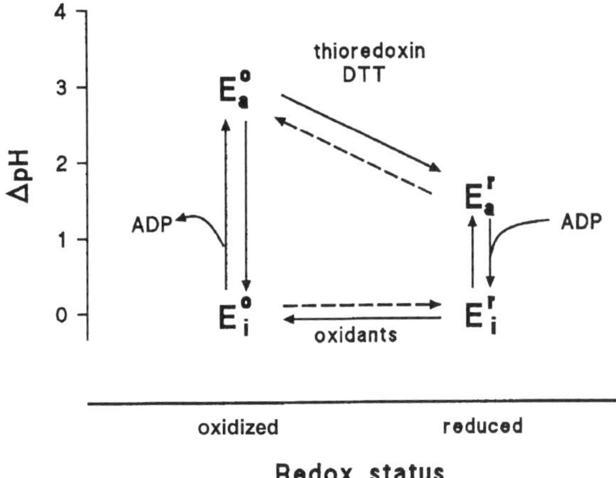

Fig. 1. Four state model for the activation and reduction of chloroplast CF_0CF_1. E_i^o, E_a^o, E_i^r and E_a^r represent the inactive oxidized, active oxidized, inactive reduced and active reduced states of CF_0CF_1, respectively. Dotted lines indicate transitions that are slow in vitro and may not occur in vivo (see text for details).

CF_0CF_1, so as to expose the redox active site (Bakker-Grunwald and van Dam, 1974). Indeed, illumination, or more specifically ΔpH, is known to induce a large conformational change in CF_0CF_1 (Ryrie and Jagendorf, 1972).

The γ subunit contains 4 cysteine (cys) residues (Moroney et al., 1984). One is surface exposed and plays no apparent role in regulation. A second (cys-89) is exposed by ΔpH and modification by NEM results in 50% inhibition of enzyme activity. The third and fourth residues are cys-199 and cys-205, located on a short insert unique to higher plants (Miki et al., 1988). It is these two residues that form the disulfide bridge in the oxidized form and are reducible by dithiothreitol (Nalin and McCarty, 1984).

Realization that ΔpH activates CF_0CF_1, aside from its thermodynamic involvement in ATP synthesis, came from observations that light-induced ATPase activity declined in the ensuing dark period (Carmeli and Lifschitz, 1972; Bakker-Grunwald and van Dam, 1974). Full activity could be restored by a second illumination in the absence of dithiothreitol. Uncouplers accelerated the dark decline of ATP hydrolysis and prevented reillumination from restoring activity. Thus ΔpH was necessary for sustained hydrolysis, even though it thermodynamically opposes the reaction.

The first indication that $\Delta\tilde{\mu}_{H^+}$ might also activate ATP synthesis came from studies by Junge (1970). Gräber et al. (1977) used light or electric pulses to drive CF_0CF_1 catalytic turnover at high rates. By studying the release of bound ADP, a measure of enzyme activation, it became clear that the rate of ATP synthesis was kinetically limited by the fraction of enzymes that were active, each active enzyme operating near its Vmax (Schlodder et al., 1982).

However, all of these experiments had been carried out with thylakoids isolated from dark-adapted leaves in which CF_0CF_1 was mostly, if not entirely, oxidized. It was shown later that preillumination of thylakoids in the presence of dithiothreitol allowed ATP to be made at much lower ΔpH values in flashing light (Harris and Crofts, 1978), acid/base transitions (Mills and Mitchell, 1982b; Junesche and Gräber, 1985) and continuous illumination (Rumberg and Becher, 1984; Mills and Mitchell, 1984; Ketcham et al., 1984). It was suggested that reduced CF_0CF_1 was activated at much lower ΔpH (see Fig. 1.), allowing either net hydrolysis or net synthesis to be observed depending on the prevailing ΔG_p.

2. Regulatory Effects of Substrates

The rates of transition among the four activity states of Fig 1. are influenced by substrates. Carmeli and Lifschitz (1972) showed that ADP accelerated the decline of ATP hydrolysis activity in darkness, but phosphate stabilized activity and overcame the ADP effect. Bakker-Grunwald and van Dam (1974) found a deactivating effect of Mg^{2+} ions on CF_0CF_1.

Deactivation correlates with tight binding of ADP. Illumination of thylakoids converts a bound ADP from a non-exchangeable to an exchangeable form (Strotmann and Bikkel-Sandkötter 1977). In a medium of low ADP concentration, this ADP dissociates with Kd in the μM range. The appearance of ATP hydrolysis correlates with release of the ADP (Schumann and Strotmann, 1981). However, the correlation is not perfect. Although a fraction of the ADP exchanges rapidly enough to precede activation as measured by ATP synthesis (Gräber et al., 1977), or hydrolysis (Du and Boyer, 1990), the remainder exchanges more slowly suggesting heterogeneity of the tightly-bound pool. Rebinding of ADP in the dark correlates with inhibition of the ATPase activity. Phosphate delays but does not prevent tight binding of ADP and thence inactivation (Schumann and Strotmann, 1981; Dunham and Selman, 1981). ATP competes with ADP with a 10 fold lower affinity, but tight binding of ATP does not inhibit activity (Strotmann and Bikkel-Sandkötter, 1977). These results suggest that inactivation of the enzyme is promoted by the binding of ADP but not ATP to the active form.

3. Models of Kinetic Regulation

Simple kinetic models have been devised that simulate many of the observations summarized above. While not intended to be completely realistic, these models help focus attention on the quantitative aspects of CF_oCF_1 regulation.

a. Role of $\Delta\tilde{\mu}_{H^+}$ in Activation

Activation by $\Delta\tilde{\mu}_{H^+}$ must involve protonation/deprotonation of acid/base groups, presumably those on CF_oCF_1. Several models have been put forward to explain this process (Schlodder et al., 1982; Mills and Mitchell, 1984; Rumberg and Becher 1984; Gräber et al., 1984). They all propose that the active state is achieved when *simultaneously*, i) groups on the enzyme of low pK_a and facing the inner thylakoid lumen (or *P* pole) are protonated and ii) groups of high Pk_a and facing out to the stroma (*N* pole) are deprotonated (see Scheme 1).

The protonation steps may be considered to be concerted (Schlodder et al., 1982; Rumberg and Becher, 1984) or sequential (Mills and Mitchell, 1984). The fraction of active enzymes, η is given by:

$$\eta = \frac{H_nE_a}{E_t} \tag{1}$$

where H_nE represents the protonated form of the enzyme.

Figure 2A shows η as a function of ΔpH according to this mechanism, and assumes *n* in each step is 3 ($n_{act} = 3$). The activation of the reduced enzyme at lower ΔpH is readily explained in terms of a decrease in K_P (defined in Scheme (1)) or K_d.

b. Role of $\Delta\tilde{\mu}_{H^+}$ in Catalysis

Usually, minimal schemes such as Scheme (2) are adopted to explain the thermodynamic effect of $\Delta\tilde{\mu}_{H^+}$ (Schlodder et al., 1982; Gräber et al., 1984), despite the considerable evidence that the catalytic mechanism is much more complex.

Figure 2B shows the expected rate of catalysis as a function of ΔpH according to this model, assuming *n* in all steps is 3 ($=n_{cat}$, the H^+/ATP ratio). The chemiosmotic hypothesis (Mitchell, 1968) states equilibrium is achieved when:

$$\Delta G_P - n_{cat}\cdot\Delta\tilde{\mu}_{H^+} = 0 \tag{2}$$

The point at which any curve in panel B cuts the x-axis (rate = 0) represents thermodynamic equilibrium. The dotted curves in Fig 2B indicate the effect of decreasing the pseudo rate constant k_2 or k_4 (Scheme (2)) by a factor of 10, which simulates an increase or

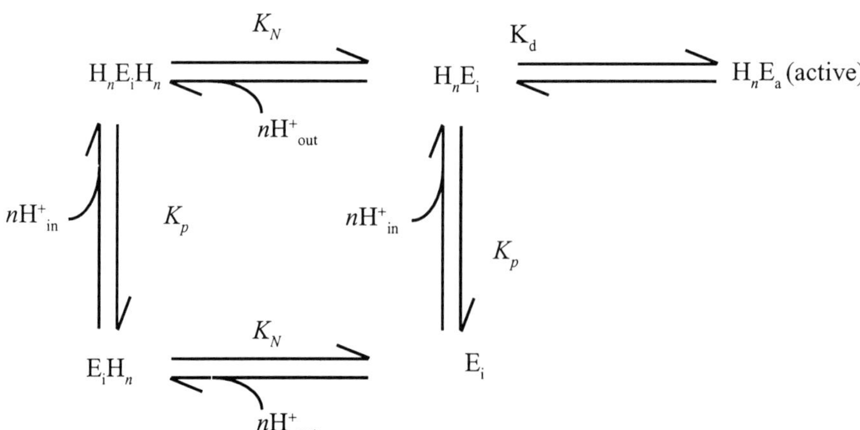

Scheme 1. Kinetic effect of $\Delta\tilde{\mu}_{H^+}$ in activation. E_i and E_a represent inactive and active states of CF_oCF_1 respectively. K_P is the apparent equilibrium constant for binding of protons to ionizable groups on the inner, lumen-facing surface of CF_oCF_1. The protonated form of the enzyme is represented as H_nE. Correspondingly, K_N is the equilibrium constant for binding of protons to groups on the outer surface, and EH_n indicates the protonated form of the enzyme.

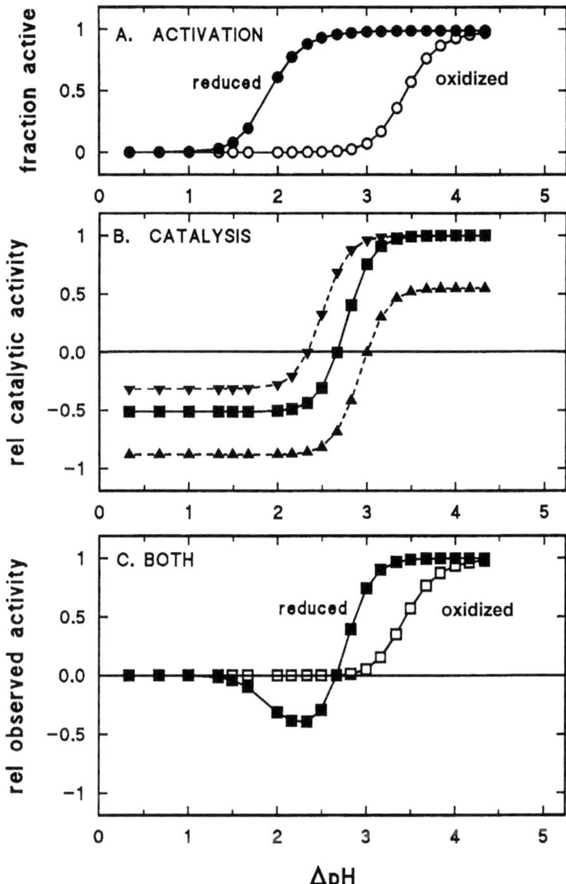

Fig. 2. Effect of ΔpH on CF_0CF_1 activation and activity. A: η, the fraction of active enzymes was calculated from Scheme 1 as described (Mills and Mitchell, 1984) assuming $K_N = 10^{-8}$ M, $K_d = 0.01$, $K_P = 1.10^{-4}$ (○, oxidized) and K_P=3.10^{-6} M (● reduced). B: Catalysis was calculated by running scheme 2 until enzyme intermediates reached steady state. No attempt was made to fit rate constants to published values. (■) K_1/K_{-1}=1.10^{-2} μM^3, k_2[ADP,Pi]/k_{-2}=10, $k_3/k_{-3}\cdot[H^+_{out}]$=1; $k_4/k_{-4}\cdot$[ATP]=10; (▼) $k_4/k_{-4}\cdot$[ATP] = 1; (▲) k_2[ADP,Pi]/k_{-2}= 1. For both A and B, H^+_{in} was varied and $H^+_{out} = 10^{-8}$ M was kept constant. C is the product of the solid curve in Fig 2B with each activation curve, Fig 2A.

decrease in ADP/ATP ratio by a factor of 10 (≡5.7 kJ mole^{-1}). As expected, the equilibrium points are shifted by 0.33 pH units.

The chemiosmotic hypothesis makes no predictions about the rate of catalysis away from equilibrium. However, standard enzyme kinetics gives rise to a linear dependence on $\Delta\tilde{\mu}_{H^+}$ in the region close to equilibrium. As $\Delta\tilde{\mu}_{H^+}$ moves further from equilibrium with ΔG_p, Vmax is approached in the direction of both ATP synthesis and hydrolysis.

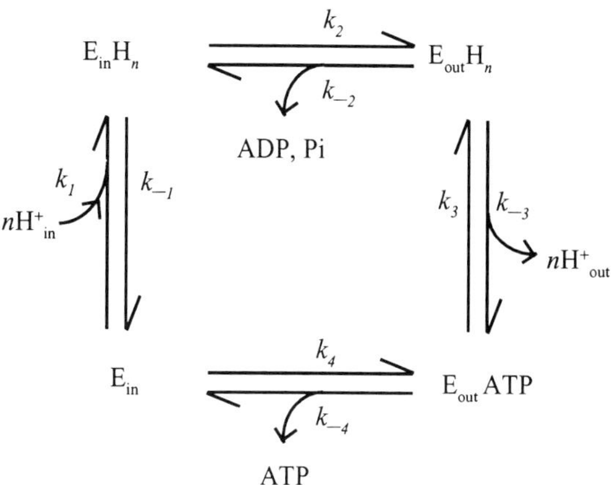

Scheme 2. Thermodynamic effect of $\Delta\tilde{\mu}_{H^+}$ on catalysis. E_{in} represents the deprotonated form of CF_0CF_1. Working clockwise round the cycle, E_{in} binds catalytic protons from the thylakoid lumen to form $E_{in}H_n$. Proton translocation to the outside then occurs and is accompanied by ADP+Pi binding. Protons are then released from $E_{out}H_n$ to the stroma. Finally, the deprotonated, outer-facing catalytic groups, E_{out} are reorientated towards the thylakoid lumen accompanied by binding/release of ATP.

c. Force-Flow Curves

Usually, it is assumed that the protonation/deprotonation events associated with activation are completely independent of those associated with catalysis (Schlodder et al., 1982; Gräber et al., 1984). The observed activity, v, is, therefore, the product of regulatory (η) and thermodynamic (w) factors:

$$v = \eta(\Delta pH, \Delta\Psi)\cdot w(\Delta pH, \Delta\Psi, ATP, ADP, Pi, Mg^{2+}) \qquad (3)$$

This is shown in Fig 2C. When the enzyme is oxidized, the curve follows that for activation at all values of $\Delta\tilde{\mu}_{H^+}$. ATP hydrolysis is not possible because CF_0CF_1 is not active at low $\Delta\tilde{\mu}_{H^+}$. However, when CF_0CF_1 is reduced, the curve follows that for catalysis at moderate to high $\Delta\tilde{\mu}_{H^+}$ because CF_0CF_1 is fully activated and controlled by conventional thermodynamic and enzyme kinetic factors. Only at low $\Delta\tilde{\mu}_{H^+}$ does the activity of ATP hydrolysis become limited by the fraction of active enzymes.

C. Recent Studies

It has been over ten years since theoretical curves such as these were first published. How well are these rather simplistic models supported by recent work?

1. Thermodynamic or Kinetic Control by $\Delta\tilde{\mu}_{H^+}$?

a. Energetic Thresholds for Activation and Catalysis

Several groups have attempted to separate the activation from the catalytic process in order to measure the respective dependence of each on $\Delta\tilde{\mu}_{H^+}$. Using continuous light to generate ΔpH, Rumberg and Becher (1984) estimated the activation of reduced CF_0CF_1 from the appearance of uncoupled ATP hydrolysis activity, and activation of oxidized enzyme from ATP synthesis. In thylakoids in which CF_0CF_1 was either fully reduced or fully oxidized, they observed energetic thresholds of ΔpH = 1.5 and ΔpH = 3 respectively. This agrees well with the work of Junesche and Gräber (1985; 1987) where estimates from acid-base transitions gave values of ΔpH = 1.5 and about ΔpH = 2.75 for reduced and oxidized CF_0CF_1. The parameters used to produce theoretical curves in Fig 2C have been chosen to approximate these data.

This difference of nearly 1.5 ΔpH units between oxidized and reduced CF_0CF_1 measures the effect of reduction on the activation profile (Fig 2A). When ATP synthesis is measured by a variety of methods, and the force-flow curves compared as in Fig 2C, the observed differences are smaller as expected, and range from 0.3 to 0.6 ΔpH units (Mills and Mitchell, 1982b; 1984; Rumberg and Becher 1984; Ketcham et al., 1984; Hangarter et al. 1987; Junesche and Gräber, 1987; 1991; Possmayer and Gräber, 1994).

b. Effects of ΔG_p

At first sight, the results quoted above appear to be in agreement. However, closer inspection reveals a number of inconsistencies, some of which are hard to reconcile.

Consider when CF_0CF_1 is oxidized. A threshold for activation of ΔpH = 3 is equivalent to 51 kJ mole^{-1} (assuming n_{act}=3). Therefore when ΔG_p>51 kJ mole^{-1}, thermodynamic control by ΔG_p should be imposed. From the number of flashes required to initiate ATP synthesis, Hangarter et al. (1987) indeed found the threshold for ATP synthesis to be increased at ΔG_p>51 kJ mole^{-1}. However as expected, ΔG_p had no effect on activation measured by the release of tightly bound ADP. In contrast, Rumberg and Becher (1984) found no effect of increasing ΔG_p from 42 to 57 kJ mole^{-1}.

If CF_0CF_1 is thermodynamically controlled at ΔG_p>51 kJ mole^{-1}, as indicated by the data of Hangarter et al (1987) it follows from the principle of microreversibility, that ATP hydrolysis, or at least significant exchange reactions should be seen at high ΔG_p. Despite considerable efforts, these reactions have never been observed at any significant rate under any condition when CF_0CF_1 is demonstrably fully oxidized (Rumberg and Becher, 1984; Junesche and Gräber, 1987).

The observed lack of ATPase activity with oxidized CF_0CF_1 might be explained if n_{act} is in fact greater than 3, if for example n_{act} = 4, CF_0CF_1 should remain under kinetic control until ΔG_p > 68 kJ mole^{-1} (though it would now be difficult to explain the results of Hangarter et al, 1987). Estimates of n_{act} vary from 2 to 4 and with redox state of CF_0CF_1 (e.g. Schlodder et al., 1982; Davenport and McCarty, 1984; Ketcham et al., 1984; Junesche and Gräber, 1985; 1987). Recent measurements with CF_0CF_1 incorporated into liposomes suggests n_{act} = 3 when CF_0CF_1 is reduced, and n_{act} = 4 when oxidized (Possmayer and Gräber, 1994). In contrast, Strelow and Rumberg (1993) estimate n_{act} = 4 for the reduced enzyme and possibly less when CF_0CF_1 is oxidized.

The difficulty is that n_{act} cannot be measured with precision. In Hill plots, a fit to n_{act} = 3 or n_{act} = 4 is hard to distinguish (Possmayer and Gräber, 1994). Theoretically, the slopes can vary without a change in n_{act} (Strelow and Rumberg, 1993). If activation and catalysis are independent, the apparent n_{act} will increase when the two thresholds overlap, because the overall activity is the product of two processes, each with an n = 3 dependency.

Now consider the reduced enzyme. A threshold for activation of reduced enzyme of ΔpH = 1.5 is equivalent to 26 kJ mole^{-1} if n_{act} = 3 (34 kJ mole^{-1} if n_{act} = 4). This is also the lowest value of ΔG_p that in practice can be generated in the laboratory. Thus when CF_0CF_1 is reduced, the thermodynamic threshold for ATP synthesis should always exceed the activation threshold. However, Hangarter et al. (1987) found from the delay in flash-dependent ATP synthesis that thermodynamic control was not imposed until ΔG_p = 45 kJ mole^{-1}.

c. Regulatory Effects of Substrates

Can some of the effects of ΔG_p be explained by

regulatory effects of ATP, ADP and Pi? Kleefeld et al. (1990) measured activation from the appearance of uncoupled ATPase activity after incubating reduced CF_0CF_1 in the presence or absence of nucleotides. They found that the threshold was $\Delta pH < 2$ with no nucleotides or with 5 μM ATP present, but was shifted up to $\Delta pH = 3$ in the presence of 5 μM ADP. Thus occupation of the 'loose' binding site by ADP increases the apparent $\Delta\tilde{\mu}_{H^+}$ needed for activation, up to that measured for the oxidized enzyme. When the substrate ratios in the ATPase assay mix were varied at constant ΔG_p = 60kJ mole^{-1}, the initial rate was progressively inhibited by increasing ADP/Pi (Strotmann et al., 1987).

Therefore when ΔG_p is adjusted by changing [ADP]/[ATP], the threshold for activation might also change. A crucial variable is the concentration of Pi. If Pi prevents deactivation at high [ADP] as it does at low [ADP] (Kleefeld et al. 1990), then it is safe to conclude that reduced CF_0CF_1 is always activated above $\Delta pH < 2$ and CF_0CF_1 is therefore always limited thermodynamically (Junesche and Gräber, 1987). If Pi protection is only partial, or is competitive with ADP (Dunham and Selman, 1981), then the energetic threshold for activation could lie anywhere between $3 > \Delta pH > 2$ depending on the relative Pi and ADP concentrations (Kleefeld et al, 1990). The relatively low level of 0.5 mM Pi used by Hangarter et al (1987) could account for the higher observed threshold of 45 kJ mole^{-1} (equivalent to $\Delta pH = 2.5$). Clearly more work is required to quantify the values for n_{act} and Pi effects on threshold energetic levels.

2. How Related are the Mechanisms of Regulation and Catalysis?

a. Protonation Reactions

To what extent do the processes of activation and catalysis share a common mechanism? Are the same acid/base groups involved? Are the protons that activate the enzyme transported across the thylakoid? In the kinetic schemes presented above, activation only requires protonation/deprotonation while catalysis additionally requires transport. However, the schemes yield very similar curves. In fact Scheme 2 has been used to describe either activation (Rumberg and Becher, 1984) or catalysis (Schlodder et al. 1982). Modeling therefore has no power to resolve these questions

In thylakoids, $\Delta\Psi$ and ΔpH are energetically equivalent for both activation and catalysis (Gräber et al., 1977; Junesche and Gräber, 1991). The simplest explanation is that the protonatable groups lie at the bottom of a proton well (Mitchell, 1968). On the lumen side of CF_0CF_1, the proton well is presumably CF_0, though the protonatable groups could be on either CF_0 or CF_1. Griwatz and Junge (1992) suggested that treatment of thylakoids with very low EDTA concentrations induced a distorted structure in which protons could be transiently trapped in a proton well on the inner side of CF_1. Proton binding was very cooperative (n = 6 in Hill plots), and the apparent pK_a was high, $pK_a = 7.3$.

It is simplest to imagine that the same protonatable groups function in activation and catalysis and that reduction of CF_0CF_1 alters their pKa. This is consistent with the observation that CF_0 channel blockers inhibit both catalysis and $\Delta\tilde{\mu}_{H^+}$-induced nucleotide release (Reimer and Selman 1978). However, it has recently been claimed that venturicidin can discriminate different pathways for activating and catalytic protons (Valerio et al., 1992). In chromatophores, ΔpH and $\Delta\Psi$ were additive for catalysis (Turina et al., 1991) but not for activation (Turina et al., 1992), suggesting different mechanisms for the two processes in photosynthetic bacteria.

Most studies have concluded that ATP synthesis depends on ΔpH, rather than the absolute pH_{in} or pH_{out} (e.g. Davenport and McCarty, 1986; Biaudet et al., 1988. Bizouarn et al, 1991). Although inconclusive, this behavior would be expected if the activation protons were translocated. Recently, using CF_0CF_1 reconstituted into liposomes, Possmayer and Gräber (1994) concluded that ATP synthesis actually depends on the absolute internal and external pH. They estimate the apparent pK_a to be 5.3 and 7.8 respectively, and suggest previous studies had not covered a wide enough range of pH. However Possmayer and Gräber (1994) interpret their data in terms of translocation of the activation protons. In fact they suggest there is only one set of groups that become protonated on the inside, cross the osmotic barrier, and are then deprotonated outside. This is certainly the minimal explanation! Until the mechanism of proton transport in CF_1 is solved, the question of a shared protonation mechanism in activation and catalysis remains speculative.

b. Substrate Binding Sites

The multiplicity of nucleotide binding sites and their

involvement in catalytic mechanisms are dealt with in Chapters 23 and 24. Several lines of work indicate that the ADP, which becomes tightly bound to CF_0CF_1 during inactivation, occupies a potentially catalytic site. For example, release of bound ADP precedes turnover (5 ms) during phosphorylation on the membrane (Gräber et al., 1977), or hydrolysis in isolated CF_1 (Du and Boyer, 1990). Labeling with 2-azido-ADP suggests the tightly-bound ADP occupies an occluded catalytic site (Zhou et al, 1988).

Inactivation can be explained in terms of the binding change hypothesis for catalysis (Feldman and Boyer, 1985; Du and Boyer, 1990). During coupled hydrolysis, ATP binding at one catalytic site promotes release of tightly-bound ADP and Pi at an alternate site. As substrate ATP becomes exhausted, the lifetime of ADP + Pi on the enzyme is longer, and Pi may be lost because its affinity is lower than ADP. The E_a.ADP form is very susceptible to inactivation in the absence of $\Delta\tilde{\mu}_{H^+}$. This sequence of events has been directly monitored by Kleefeld et al. (1990). Re-activation would involve first binding, and then transport of protons coupled to an 'abortive' turnover where the tightly-bound ADP is expelled from the catalytic site. Thus in this model, the mechanisms of activation, catalysis and their associated proton transport are very closely related.

III. Regulation in vivo

A. Regulation by Thioredoxin

In vivo, oxido-reduction of CF_0CF_1 involves the thioredoxin system. Thioredoxins are small, ubiquitous proteins of approximately 12000 Mr in which the only functional group is a redox active disulfide formed from the sequence cys-gly-pro-cys (Holmgren, 1985). Chloroplasts contain two thioredoxins, Th_f and Th_m, the latter existing in several isoforms due to a 'frayed' N-terminus (Schürmann et al, 1981). In dark-adapted intact chloroplasts, thioredoxins are almost completely oxidized, but become reduced (>75%) in the light by electron transport via ferredoxin and ferredoxin-thioredoxin reductase (Scheibe, 1981; Crawford et al., 1989).

Reduction of thioredoxins acts as dark → light signal switching chloroplast metabolism (Cséke and Buchanan, 1986). Thioredoxin can reduce a disulfide bridge in the protein structure of a number of target enzymes, thereby changing their activity. In general, enzymes of biosynthetic pathways are activated by reduction, for example fructose bisphosphatase, sedoheptulose bisphosphatase and ribulose 5-phosphate kinase of the reductive pentose phosphate pathway. In contrast, glucose-6-phosphate dehydrogenase of the oxidative pentose phosphate pathway is inactivated by reduction. Th_f was so named because fructose bisphosphatase was specific for this thioredoxin whereas malate dehydrogenase showed a higher preference for Th_m. However, selectivity is not absolute and is only observed using low concentrations of thioredoxin in vitro (Cséke and Buchanan, 1986).

1. Reduction by Thioredoxin

The activation and reduction of CF_0CF_1 in situ has been observed upon illumination of intact chloroplasts (Mills et al. 1980; Shahak, 1985), barley protoplasts (Quick and Mills, 1986), unicellular algae (Noctor and Mills, 1988; Selman-Reimer and Selman, 1988) and leaves (Morita et al., 1983; Kramer and Crofts, 1989). The appearance of ATPase activity in leaves correlates with reduction of the disulfide bridge on the gamma subunit of CF_0CF_1 (Vallejos et al., 1983). The evidence that thioredoxin functions as the physiological reductant in vivo comes mainly from kinetic studies and reconstitution studies in vitro.

Light-dependent reduction of CF_0CF_1 can be reconstituted in washed thylakoids by adding back ferredoxin, ferredoxin-thioredoxin reductase and thioredoxin (Mills et al., 1980). CF_0CF_1 shows no preference between Th_f and Th_m, and can be reduced by non-chloroplast thioredoxins (Galmiche et al. 1990). There is as yet no selective inhibitor of thioredoxin to confirm its role in vivo. Electron acceptors such as methyl viologen and phenazine methosulfate, which compete with ferredoxin for electrons from Photosystem I, block the light-dependent appearance of ATPase activity in intact chloroplasts but not in thylakoids supplemented with dithiothreitol (Mills et al., 1980; Mills and Mitchell, 1982a). Methyl viologen has therefore been used to provide evidence for an involvement of thioredoxin in activation of CF_0CF_1 in the alga *Dunaliella tertiolecta*, where it inhibits (Noctor and Mills, 1988), but against redox control of F_0F_1 in cyanobacterium *Nostoc* sp. Mac (Austin et al, 1992), where little inhibition by methyl viologen was observed.

2. Role of Oxidants in vivo

In isolated intact chloroplasts and algae, CF_0CF_1 becomes oxidized within a few minutes of terminating illumination (Mills and Mitchell, 1982a; Shahak, 1985; Selman-Reimer and Selman, 1988). If intact chloroplasts are lysed at the time illumination ceases, oxidation is much slower (Mills and Mitchell, 1982a), and thylakoids can remain reduced for hours (Bakker-Grunwald and van Dam, 1974). There appears therefore to be an oxidation system within intact chloroplasts that is diluted to ineffectiveness upon lysis. The identity of the oxidant is uncertain. Oxidation of CF_0CF_1 in lysed thylakoids is accelerated by several potentially physiological reagents such as H_2O_2 or oxidized thioredoxin (Mills and Mitchell, 1982a). A specific oxidant for CF_0CF_1, ovothiol B, has been isolated from the unicellular alga *Dunaliella salina*, but not yet reported in plants (Selman-Reimer et al., 1991).

Shahak (1985) observed an interesting effect of ΔpH on the rate of oxidation of CF_0CF_1. If intact chloroplasts were lysed during illumination, oxidants had very little effect on ATPase activity. If however the light were extinguished or uncouplers added with oxidants, then a rapid decline in activity was seen. Therefore, ΔpH inhibits oxidation of the reduced enzyme, but promotes reduction of oxidized CF_0CF_1. It was concluded that the redox active target sequence within the γ subunit is exposed in the activated, oxidized enzyme but becomes buried after its reduction (Shahak, 1985).

Similar observations have been made using methyl viologen in place of oxidants. Addition of methyl viologen to dark-adapted intact chloroplasts completely blocked reduction of CF_0CF_1, but addition in the light failed to promote oxidation (Mills and Mitchell, 1982a). Similar results were obtained using the alga *Dunaliella tertiolecta* (Noctor and Mills, 1988). Moreover, addition of methyl viologen in the light initiated an immediate and complete re-oxidation of fructose bisphosphatase, confirming that it was indeed preventing reduction of thioredoxin (Noctor and Mills, 1988). These results suggest that the natural oxidant(s) in the intact organelle can also only oxidize CF_0CF_1 from its inactive form.

B. Are in vitro Models Sufficient to Explain in vivo Regulation?

There are several observations of regulation of CF_0CF_1 in vivo that seem at odds with in vitro studies (Ort and Oxborough, 1992).

1) Using continuous light, induction of ATPase activity is much faster in intact chloroplasts than in thylakoids supplemented with thioredoxin (Mills et al., 1980; Galmiche et al., 1990; Strelow and Rumberg, 1993). In thylakoids, the slow rate of appearance of ATPase activity reflects the rate of reduction of CF_0CF_1, and depends on the concentration of added thioredoxin (Mills et al., 1980; Strelow and Rumberg, 1993). Concentrations usually employed in vitro (1–2 μM) are much lower than those found in vivo where Th_m alone is 0.1 mM (Scheibe, 1981). Direct measurements show that thioredoxin reduction is rapid in intact chloroplasts, and complete within 1 min (Crawford et al., 1989). The faster rate of CF_0CF_1 reduction in vivo may therefore be due to a higher concentration of reduced thioredoxin.

2) Re-oxidation of CF_0CF_1 in the dark can be very slow in leaves and depends on the illumination time (Kramer and Crofts, 1989; Kramer et al., 1992). It was suggested that oxidation was delayed by buffering from an unidentified redox pool. Since this delay does not seem to occur in isolated intact chloroplasts, it may reflect interactions between chloroplast and cytosol in the leaf.

3) ATP synthesis in leaves can be measured from the accelerated decay of the electrochromic shift following single turn-over flashes. Using this method, Kramer and Crofts (1989) made the remarkable observation that activation occurred after just one flash when CF_0CF_1 was poised in the reduced, inactive state by prior illumination. This suggested an energetic threshold of <50 mV (<1 ΔpH unit) in vivo. The oxidized state of CF_0CF_1 required one more flash, suggesting that the threshold was 50 mV higher. These thresholds are much lower than observed in vitro. However, as the authors point out, decay of the electrochromic shift measures ATP synthesis and therefore the energy thresholds cannot be so low. Kramer and Crofts (1989) suggested that an undetected background ΔpH of at least 1 pH unit exists. In this case, the thresholds are in line with those measured in vitro.

4) When intact chloroplasts are illuminated, the

ΔG_p increases from only 42 to 47 kJ mole^{-1} whereas thylakoids are able to phosphorylate against much higher phosphate potentials (Giersch et al., 1980). Upon darkening, CO_2 fixation ceases within 15 s, concomitant with a fall in ΔG_p back to dark levels of 42 kJ mole^{-1} (Stitt et al., 1982; Heineke et al., 1991). This corresponds to a dark [ADP]/[ATP] ratio of about 2, which is apparently in equilibrium with adenylate kinase but clearly not with CF_0CF_1. Since CF_0CF_1 remains reduced over this period (Kramer and Crofts, 1989), the in vivo enzyme must be rapidly inactivated by the collapse of $\Delta\tilde{\mu}_{H^+}$. When the same experiment is performed with thylakoids using 50 μM ATP, virtually all of the ATP is hydrolyzed before CF_0CF_1 is inactivated by the combined effect of high [ADP]/[ATP] and collapsing $\Delta\tilde{\mu}_{H^+}$ (Kleefeld et al, 1990). There are several possible explanations for these discrepancies, none of which is entirely convincing.

i) the threshold for activation of reduced CF_0CF_1 in vivo exceeds 42 kJ mole^{-1} (Ort and Oxborough, 1992). As discussed in Section IIB, this would be compatible with most measurements in vitro only at low Pi concentration (Kleefeld et al 1990), whereas Pi concentrations in vivo are high (Ort and Oxborough, 1992).

ii) the measured levels of ADP, ATP and Pi do not reflect thermodynamically active concentrations. The stroma is highly concentrated, and far from an ideal solution. Stitt et al. (1982) found that some nucleotides are bound at the time of inactivation (CF_0CF_1 binding of ATP could account for one third of the dark ATP). Thus the [ADP]/[ATP] ratio may be actually higher than it seems.

iii) the chemiosmotic hypothesis is not a good description of energetics in vivo. It has been suggested that energy coupling may be switched between delocalized and localized coupling mechanisms, by changing the concentrations of salt, Ca^{2+} (Chiang and Dilley, 1987), or low concentrations of uncouplers (Sigalat et al., 1988). A localized scheme might explain the in situ data but no quantitative kinetic schemes are available to evaluate this possibility.

Many other thermodynamic relationships are poorly understood in vivo. For example, NADPH/$NADP^+$ ratios only change by a factor of 2 between light and dark and are out of equilibrium with a number of dehydrogenases (Heineke et al, 1991). Large redox disequilibria exist across the chloroplast envelope that are still not well explained. This is clearly an area requiring further study.

C. Function of Regulation in vivo

1. Control of ATP Hydrolysis in Darkness

Only CF_0CF_1 from plants and green algae are under redox control and contain the thioredoxin target sequence in their γ subunit (Werner et al., 1990). In contrast, activation by $\Delta\tilde{\mu}_{H^+}$ is common to all photosynthetic F_0F_1. The properties of the bacterial (Turina et al., 1992) and cyanobacterial (Krab et al, 1993) enzymes seem to permanently resemble reduced CF_0CF_1, indicating that it is the oxidized state that is unique to plants. Since oxidation of CF_0CF_1 locks in the inactive state, its function must be to prevent all possibility of ATP hydrolysis during the long hours of darkness. Other coupling membranes display significant respiratory chain activity that would require an active F_0F_1 at all times.

This explanation presumes that thylakoids do not carry out oxidative phosphorylation. However, the chloroplast genome contains sequences homologous to polypeptides of NADH dehydrogenase (Fearnley and Walker, 1992). Respiratory-like activity termed chlororespiration has been documented in green algae (Bennoun, 1982), and recently in leaves where it can cause dark reduction of plastoquinone (Groom et al., 1993; Harris and Heber, 1993). However, it was from a green alga that a specific CF_0CF_1 oxidant was purified (Selman-Reimer et al., 1991), and where oxidation of the enzyme occurs rapidly in the dark (Selman-Reimer and Selman, 1988). It seems unlikely therefore that chlororespiration can be coupled to ATP synthesis.

Recently, the γ subunit from the diatom *Odontella sinensis* was shown to lack the thioredoxin target sequence (Pancic and Strotmann, 1993). From this result one can conclude either that thioredoxin control does not after all correlate with lack of respiratory activity, or that *Odontella* thylakoids can respire at significant rates. Little is known about thylakoid proteins in non-green algae, but chlororespiration in diatoms generates a dark ΔpH sufficient to quench chlorophyll fluorescence (Ting and Owens, 1993). The organization of F_0F_1 genes of this eukaryote resembles those in cyanobacteria (Pancic et al., 1992), which are known to carry out oxidative phosphorylation in the thylakoid membrane (Nichols et al, 1992). Interestingly, the organization of ATPase genes of the red alga *Antithamnion* sp. also resembles that in cyanobacteria (Kostrzewa and Zetsche, 1992). It would be instructive to examine thioredoxin control and compare this with respiratory activity of chloroplasts in the non-green algae.

2. Flux Control during ATP Synthesis

Thioredoxin control of some chloroplast enzymes is dynamic in the light, i.e. the extent of reduction is subject to 'fine-tuning' in response to changing conditions such as light intensity. A good example is malate dehydrogenase, where a high $NADPH/NADP^+$ ratio enhances the extent of reduction (Scheibe, 1987). This may promote export of reductant from the chloroplast when the rate of NADPH production exceeds consumption (malate 'valve').

Unlike malate dehydrogenase, the reduction state of CF_0CF_1 probably does not vary during the day. CF_0CF_1 in protoplasts is completely and rapidly reduced at very low light intensities (Quick and Mills, 1986). In sunflower leaves, complete reduction occurs below the compensation point for photosynthesis (Kramer et al., 1992). There are probably two reasons why CF_0CF_1 remains fully reduced in the weakest of light. The first is thermodynamic. Equilibrium redox titrations reveal that CF_0CF_1 has the most oxidizing midpoint potential of all the chloroplast enzymes so far examined (Kramer et al., 1992). Computer modeling suggests that the equilibrium redox properties alone are sufficient to explain the kinetics of oxidation and reduction of CF_0CF_1 in vivo. The second reason is kinetic. The steady state level of reduction of CF_0CF_1 represents a balance between the rates of reduction and oxidation. The work of Shahak (1985) indicates that oxidation is restricted when CF_0CF_1 is activated.

Thus the activation and thiol modulation mechanisms are synergistic. As indicated in Fig. 1, reduction of the enzyme lowers the ΔpH requirement for activation, and activation accelerates the reduction process. Since both reductants and ΔpH are only available in the light, the effect is to stabilize the active state of CF_0CF_1 in the light. Conversely, oxidation of the enzyme increases the ΔpH required for activation, and inactivation enhances the oxidation process. Darkness will favor both of these conditions. This regulatory system seems very much designed as a switch mechanism and is resistant to fine tuning.

If the function of oxidation is to prevent ATP hydrolysis in the dark, what benefit does reduction of CF_0CF_1 confer in the light? Reduction will allow ATP synthesis to proceed at lower ΔpH but this does not necessarily mean that the rate of photophosphorylation is increased relative to the oxidized enzyme. It was shown that reduction of CF_0CF_1 resulted in only a modest stimulation (20–50%) of the rate of ATP synthesis at high light intensity, and this stimulation declined as the light intensity was decreased (Mills and Mitchell, 1984). Since ΔpH is a variable during photophosphorylation, it can adjust until proton influx and efflux match. At low light intensity, it is the rate of coupled electron transport that limits ATP synthesis. Oddly, the greatest stimulation by reduction is seen under partially uncoupled conditions, where the increased activity of the reduced enzyme allows it to compete more effectively with non-phosphorylating proton efflux (Mills and Mitchell, 1984).

Kinetic models again help to shed light on this complex system. Steady-state photophosphorylation can be simulated using flux equations to describe proton influx and efflux from the thylakoid (Quick and Mills, 1987). The rate of proton influx, J_{Hin} is given by:

$$J_{Hin} = n_{et} \cdot V_E \cdot K_a / (H^+_{in} + K_a) \tag{4}$$

where V_E is the rate of uncoupled electron transport, n_{et} is the H^+/e_2 ratio, and $K_a/(H^+_{in}+K_a)$ simulates inhibition of electron transport by $\Delta\tilde{\mu}_{H^+}$. Proton efflux, J_{Hout} is determined by the phosphorylating, J_{Hp}, and non-phosphorylating, J_{Hnp} efflux as follows.

$$J_{Hp} = n_{cat} \cdot V_P \cdot \eta \tag{5}$$

$$J_{Hnp} = P_H \cdot H^+_{in} \tag{6}$$

where V_P is the turnover number of CF_0CF_1 and P_H is the passive permeability coefficient for protons. In steady state,

$$J_{Hin} = J_{Hp} + J_{Hnp} \tag{7}$$

This model can simulate steady-state photophosphorylation over a wide variety of conditions (Quick and Mills, 1987). In Fig 3A, it is used to simulate the rates of ATP synthesis for oxidized and reduced CF_0CF_1 as a function of V_E, i.e. varying the maximum rate of electron transport. The abscissa of Fig. 3 can be regarded therefore as a wide range of light intensities. When the value of P_H is small, simulating well-coupled thylakoids, reduction only slightly stimulates ATP synthesis at high light intensity and the stimulation declines at lower intensities as observed (Mills and Mitchell, 1984). Fig. 3B shows that the P/e_2 ratio is constant over most of the range and falls at very low V_E, but markedly less so when CF_0CF_1 is reduced. The coupling state of the membrane is very important. When the passive permeability of the membrane is increased 10 fold, the rates of synthesis and P/e_2 ratios are sub-optimal, but now reduction of CF_0CF_1 increases both parameters over a wider range of V_E relative to the oxidized enzyme (Figs. 3D and 3E).

The simulations in Fig 3 assume that passive proton leak is ohmic (Davenport and McCarty, 1984). Therefore, the fall in P/e_2 at very low light represents a relatively higher loss of proton efflux by non-phosphorylating pathways, a physiological 'uncoupling'. Activation of CF_0CF_1 is also associated with

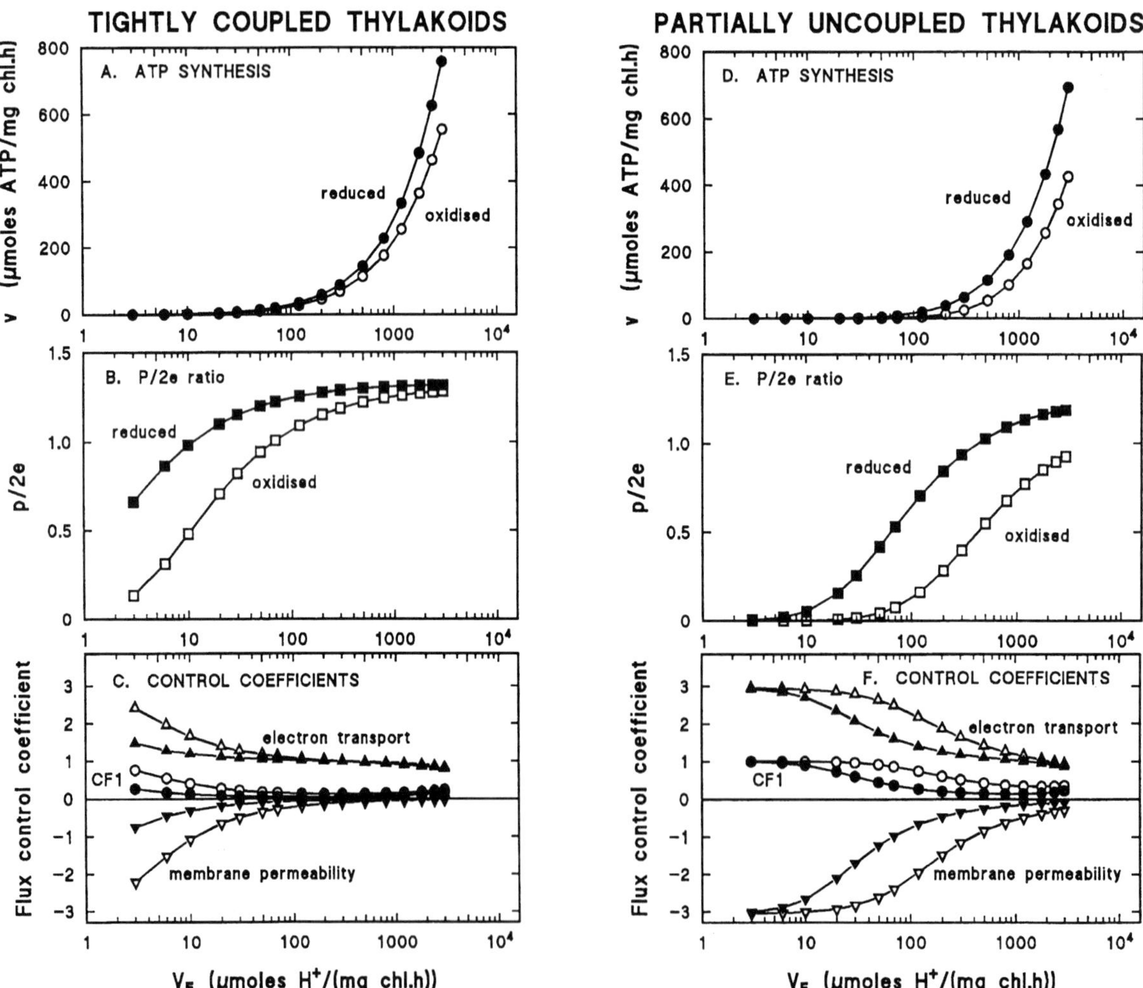

Fig. 3. Simulated rate of ATP synthesis (A,D), P/e_2 ratio (B,E) and flux control coefficients (C/F) at different V_E. Constants were set at realistic levels: $K_a = 3\cdot10^{-5}$ M, $P_H=3\cdot10^{6}$ M, and H^+_{in} was solved by iteration. The activity of CF_0CF_1 was calculated using scheme 1 for convenience assuming K_d=0.01 and $K_P = 3\cdot10^{-6}$ M (oxidized, open symbols) or $3\cdot10^{-5}$ M (reduced, closed symbols). This gives a difference in threshold activation of 0.5 ΔpH units. A,D the rate of photophosphorylation was calculated from the resulting H^+_{in} assuming $n_{cat} = 3$; B,E and hence P/e_2 assuming $H^+/e_2 = 4$; C,F flux control coefficients were calculated using equation 8 from the decrease in ATP synthesis observed upon decreasing V_E (△,▲), V_P (○,●) or P_H(▽,▼) by 1%. V_E is the rate of uncoupled electron transport, V_P is the turnover number of CF_0CF_1 and P_H is the passive permeability coefficient for protons.

an increased non-ohmic proton permeability, though the effects are reduced by nucleotides (Hangarter and Ort, 1986; Groth and Junge, 1993). Thus, as light intensity is increased, the situation may change from that in Figs. 3A,B to that of Figs. 3D,E.

Modeling indicates two benefits of reduction, a slight increase in ATP synthesis rate at high light and a marked increase in P/e_2 at low light. Hangarter et al. (1987) have observed such results in flashing light of low frequency. In fact, Gabrys et al (1994) used the increased efficiency of photophosphorylation at very low light to screen for *Arabidopsis* mutants defective in thiol modulation of CF_0CF_1. The strategy seems to have been successful, and such mutants should yield valuable information in the future.

Modeling enables one other aspect of regulation to be explored. To what extent does reduced CF_0CF_1 limit photophosphorylation, and hence photosynthesis? In terms of metabolic control theory (Kacser and Burns, 1973), the extent to which any enzyme, E limits the flux, F, in a metabolic pathway is measured by the flux control coefficient C^F_E

$$C^F_E = \frac{dF/F}{dE/E} \tag{8}$$

In the model given above, photophosphorylation is controlled by three 'enzymes', CF_0CF_1, electron transport and membrane permeability and it is a simple matter to calculate the flux control coefficients in each case (see Tager et al, 1983 for a similar analysis of oxidative phosphorylation). Figs. 3C and 3F show these values. Over most light intensities, the control coefficients are close to zero for CF_0CF_1 and membrane permeability, and close to 1 for electron transport. This means that flux is controlled (limited) by electron transport, and not by CF_0CF_1. This would be expected from the maximal activities observed for CF_0CF_1. The turnover number of CF_0CF_1 is 400 s^{-1} whereas in photophosphorylation, the rate never exceeds 150 s^{-1}, i.e. the capacity of CF_0CF_1 greatly exceeds that of electron transport (Possmayer and Gräber, 1994).

At very low light intensity, all the flux control coefficients increase. This seems surprising but is expected since the control coefficient for membrane permeability is negative, and the sum of all coefficients must be 1 (Tager et al, 1983). Again, under this condition, reduction significantly suppresses this effect, a consequence of the improved coupling. This analysis has yet to be experimentally verified.

In summary, in the dark CF_0CF_1 controls the rate of ATPase activity, and oxidation maximizes the flux control coefficient. In the light, CF_0CF_1 is not a control point of photophosphorylation (low flux control coefficient), and reduction further minimizes any such possibility at low light intensity. This analysis assumes that substrate ADP and Pi are saturating, as would be expected from the measured levels in vivo (Heineke et al, 1991). In a model of the entire Calvin Cycle, Pettersson and Ryde-Pettersson (1988) concluded that at low Pi, photosynthetic flux was controlled by CF_0CF_1. That work did not separate out the contribution of electron transport and CF_0CF_1 to ATP synthesis. Nevertheless, low Pi is one situation where control could be thrown from electron transport onto CF_0CF_1. It will be interesting to explore these possibilities theoretically and experimentally.

IV. Structural Studies

A. Role of the ε and γ subunits

Whereas catalysis is carried out by the $\alpha\beta$ subunits (Zhou et al, 1988), it is the minor subunits, particularly ε and γ that are responsible for regulation of CF_0CF_1. Understanding the roles of these subunits has come mainly from studies of isolated CF_1.

ATPase activity of isolated CF_1 is latent but can be activated by a variety of treatments. Nelson et al. (1972) originally found that addition of purified ε inhibited ATPase activity. The idea gained hold that ε is an inhibitor protein and that activation was due to dissociation of ε from CF_1. This may be true for some in vitro methods of inducing activity. Soteropoulos et al (1992) showed that simply diluting CF_1 resulted in activation of the enzyme. From the dilution factor required, they estimated a dissociation constant of 2.10^{-10} M for ε. Incubation of isolated CF_1 with antibody to ε enhances activation and causes dissociation of ε (Richter and McCarty, 1987).

Activation of CF_1 by organic solvents also involves dissociation of ε. Treatment with 30% methanol induces high MgATPase activity (Anthon and Jagendorf, 1986). When CF_1 is bound to ion exchange columns and eluted with solvents, forms depleted of ε, δ or both are found (Mitra and Hammes, 1988). The ε-depleted forms are active ATPases. Rebinding of ε to ε-depleted CF_1 results in inhibition of activity. Andralojc and Harris (1988) detected additional weak binding sites for ε by this method.

Other methods of activation can be related to effects on the γ subunit. Incubation of CF_1 with high concentrations (50 mM) of dithiothreitol for long periods (2h) partially activates CF_1 and reduces γ (Arana and Vallejos, 1982). Activation by dithiothreitol and by ε depletion are additive (Soteropoulos et al., 1992). However, the mechanism clearly involves interplay between the γ and ε subunits because reduction lowers the apparent dissociation constant of ε (Duhe and Selman, 1990). Although activation of membrane-bound CF_1 alters the conformation of ε (Richter and McCarty, 1987; Komatsu-Takaki, 1993), ε does not dissociate. In fact anti-ε antibody inhibits by uncoupling photophosphorylation (Soteropoulos et al., 1992).

One of the more complex mechanisms of in vitro activation is incubation with the protease trypsin. Schumann et al. (1985) attempted to correlate the progress of trypsin activation with the appearance of products of digestion. When CF_1 was oxidized, activation correlated best with clipping of the α-subunit, though γ and ε were also cleaved with slower kinetics. When CF_1 was reduced, activation could be correlated with cleavage of the γ subunit and the appearance of a small peptide close to the thiol-active sequence.

B. Structural Models for Regulation

Image-enhanced electron microscopy shows the six major subunits of CF_1 in a hexagonal arrangement of alternating α and β subunits, with the three β lying in a plane above the three α (Boekema and Böttcher, 1992). The minor subunits show up as an off-center mass in the central cavity. Presumably most of this central mass is contributed by the γ subunit. It was suggested that of the three gaps between the lower α subunits, δ permanently occupies one position and ε may be bound to either of the two others (Boekema and Böttcher, 1992). For an update on this view, see Chapters 24 and 26.

Recently, a high resolution structure has been published for ox heart F_1 (Abrahams et al., 1994) which depicts the α, β subunits and portions of γ. The α and β subunits are hexagonally arranged, but more symmetrically than suggested above, rather like segments of an orange. The conserved C and N terminal sequences of the γ subunit are helical, and form a long, coiled-coil through the core. The structure provides an explanation for the catalytic mechanism (Abrahams et al., 1994; see also Chapter 24).

Unfortunately, most of the central part of the γ sequence does not appear in the crystal structure. If the sequences of the spinach and ox heart γ subunits are aligned according to Werner et al (1990), and the basic structure of CF_1 is assumed to be the same as ox heart F_1, then the thioredoxin target site would occur some 25 residues from the bottom of the coiled-coil, well away from the catalytic sites and probably near F_0. As neither δ or ε, nor any part of F_0 appears in the published structure, the mechanism and regulation of proton translocation remain speculative.

Activation of CF_0CF_1 by $\Delta\tilde{\mu}_{H^+}$ could involve the following sequence of events, which are consistent with the available kinetic and structural evidence. The central region of γ makes contact with F_0, and is responsible for gating the proton current. The coiled-coil formed by the N and C terminal sequences conformationally connects the proton well to the catalytic sites. It is γ that controls the activation of the enzyme by $\Delta\tilde{\mu}_{H^+}$, transmitting the effects of protonation to the catalytic sites. Activation requires a conformational change that displaces ε from a high affinity to a low affinity binding site. When γ is oxidized, the free energy of this change is quite large, and shows up as a low pK_a for protonation of internal groups. Reduction of γ, or removal of ε, considerably lowers the activation energy, and this shows up as an increase in pK_a of the protonation sites in the proton well. A full explanation of regulation at a molecular level obviously awaits the publication of a high resolution structure for CF_0CF_1.

Acknowledgments

I am grateful to Janet Turton for correcting the manuscript, and to the Biotechnology and Biological Sciences Research Council for their financial support.

References

Abrahams JP, Leslie, AGW, Lutter R, and Walker JE (1994) Structure at 2.8 Å resolution of F_1-ATPase from bovine heart mitochondria. Nature 370: 621–628

Andralojc PJ and Harris DA (1988) Two distinct types of ε-binding site exist in chloroplast coupling factor (CF_1). FEBS Lett 233: 403–407

Anthon GE and Jagendorf AT (1986) Evidence for multiple

effects in the methanol activation of chloroplast coupling factor 1. Biochim Biophys Acta 848: 92–98

Arana JL and Vallejos RH (1982) Involvement of sulphydryl groups in the activation mechanism of the ATPase activity of chloroplast CF_1. J Biol Chem 257: 1125–1127

Austin PA, Ross IS and Mills JD (1992) Light/dark regulation of photosynthetic enzymes in intact cells of the cyanobacterium *Nostoc* sp Mac. Biochim Biophys Acta 1099: 226–232

Bakker-Grunwald T and van Dam K (1974) On the mechanism of activation of the ATPase in chloroplasts. Biochim Biophys Acta 347: 290–298

Bennoun P (1982) Evidence for a respiratory chain in the chloroplast. Proc Nat Acad Sci USA 79: 4352–4356

Biaudet P, de Kouchkovsky, F and Haraux F (1988) ΔpH-activation of the thiol-modified chloroplast ATP hydrolase. Nucleotide binding effects. Biochim Biophys Acta 933: 487–500

Bizouarn T, de Kouchkovsky Y and Haraux F (1991) Dependence of kinetic parameters of chloroplast ATP synthase on external pH, internal pH, and delta pH. Biochemistry 30: 6847–6853

Boekema EJ and Böttcher B (1992) The structure of ATP synthase from chloroplasts. Conformational changes of CF1 studied by electron microscopy. Biochim Biophys Acta 1098: 131–143

Carmeli C and Lifschitz Y. (1972) Effects of Pi and ADP on ATPase activity in chloroplasts. Biochim Biophys Acta 267: 86–95

Chiang G and Dilley RA (1987) Evidence for Ca^{2+}-gated proton fluxes in chloroplast thylakoid membranes: Ca^{2+} controls a localized to delocalized proton switch. Biochemistry 26: 4911–4916

Crawford NA, Droux M, Kosower, NS and Buchanan BB (1989) Evidence for function of the ferredoxin/thioredoxin system in the reductive activation of target enzymes in isolated intact chloroplasts. Arch Biochem Biophys 271: 223–239

Cséke C and Buchanan BB (1986) Regulation of the formation and utilization of photosynthate in leaves. Biochim Biophys Acta 853: 43–63

Davenport JW and McCarty RE (1984) An analysis of proton fluxes coupled to electron transport and ATP synthesis in chloroplast thylakoids. Biochim Biophys Acta 766: 363–374

Davenport JW and McCarty RE (1986) Relationships between rates of steady-state ATP synthesis and the magnitude of the proton activity across thylakoid membranes. Biochim Biophys Acta 851: 136–145

Du ZY and Boyer PD (1990) On the mechanism of sulfite activation of chloroplast thylakoid ATPase and the relation of ADP tightly bound at a catalytic site to the binding change mechanism. Biochemistry 29: 402–407

Duhe RJ and Selman BR (1990) The dithiothreitol-stimulated dissociation of the chloroplast coupling factor 1 ε-subunit is reversible. Biochim Biophys Acta 1017: 70–78

Dunham, KR and Selman BR (1981) Interactions of Pi with spinach CF_1. Effects on ATPase and binding activities. J Biol Chem 256: 10044–10049

Fearnley IM and Walker JE (1992) Conservation of sequences of subunits of complex I and their relationships with other proteins. Biochim Biophys Acta 1140: 105–134

Feldman RI and Boyer PD (1985) The role of tightly bound ADP on chloroplast ATPase. J Biol Chem 260: 13088–13094

Gabrys H, Kramer DM, Crofts AR and Ort DR (1994) Mutants of chloroplast coupling factor reduction in *Arabidopsis*. Plant Physiol 104: 769–776

Galmiche JM, Girault G, Berger G, Jacquot JP, Miginiac-Maslow M and Wollman E. (1990) Induction by different thioredoxins of ATPase activity coupling factor 1 from spinach. Biochimie 72: 25–32

Giersch C, Heber U, Kobayashi Y, Innoue Y, Shibata K, and Heldt HW (1980) Energy storage, phosphorylation potential and protonmotive force in chloroplasts. Biochim Biophys Acta 590: 59–73

Gräber P, Schlodder U and Witt, HT (1977) Conformational change of the chloroplast ATPase induced by transmembrane electric field and its correlation to phosphorylation. Biochim Biophys Acta 461: 426–440

Gräber P, Schlodder U and Witt, HT (1984) Mechanism of the regulation of ATP synthesis hydrolysis in chloroplasts. In: Papa S, Altendorf K, Ernster L and Packer L (eds) H^+-ATPase: Structure, Function, Biogenesis. pp 431–440. Adriatic Editrice, Bari

Griwatz C and Junge W (1992) Cooperative transient trapping of protons by a distorted chloroplast ATPase. Evidence for a proton well? Biochim Biophys Acta 1101: 244–248

Groom QJ, Kramer DM, Crofts AR and Ort DR (1993) The non-photochemical reduction of plastoquinone in leaves. Photosyn Res 36: 205–215

Groth G and Junge W (1993) Proton slip of the chloroplast ATPase: Its nucleotide dependence, energetic threshold, and relation to an alternating site mechanism of catalysis. Biochemistry 32: 8103–8111

Hangarter RP and Ort DR (1986) The relationship between light-induced increases in the H+ conductivity of thylakoid membranes and activity of the coupling factor. Eur J Biochem 158: 7–12

Hangarter RP, Grandioni P and Ort DR (1987) The effects of chloroplast coupling factor reduction on the energetics of activation and on the energetics and efficiency of ATP formation. J Biol Chem 262: 13513–13519

Harris DA and Crofts AR (1978) The initial stages of photophosphorylation. Studies using excitation by saturating short flashes of light. Biochim Biophys Acta 502: 87–102

Harris GC and Heber, U (1993) Effects of anaerobiosis on chlorophyll fluorescence yield in spinach leaf discs. Plant Physiol 101: 1169–1173

Heineke D, Riens B, Grosse H, Hoferichter P, Peter U, Flügge U-I and Heldt HW (1991) Redox transfer across the inner chloroplast envelope membrane. Plant Physiol 95: 1131–1137

Holmgren A (1985) Thioredoxin. Ann Rev Biochem 54: 237–271

Junesche U and Gräber P (1985) The rate of ATP synthesis as a function of ΔpH in normal and dithiothreitol-modified chloroplasts. Biochim Biophys Acta 809: 429–434

Junesche U and Gräber P (1987) Influence of the redox state and the activation of the chloroplast ATP synthase on proton-transport coupled ATP synthesis/hydrolysis. Biochim Biophys Acta 893: 275–288

Junesche U and Gräber P (1991) The rate of ATP synthesis as a function of ΔpH and ΔΨ catalysed by the active, reduced H^+-ATPase from chloroplasts. FEBS Lett 294: 275–278

Junge W (1970) The critical electric potential difference for photophosphorylation. Its relation to the chemiosmotic hypothesis and to triggering requirements of the ATPase system. Eur J Biochem 14: 582–592

Kacser H and Burns JA (1973) The control of flux. In: Davies DD (ed) Rate Control of Biological Processes, pp 65–104. Cambridge University Press, Cambridge

Ketcham SR, Davenport JW, Wernicke K and McCarty RE (1984) Role of the γ-subunit of the chloroplast coupling factor 1 in the light-dependent activation of photophosphorylation and ATPase activity by dithiothreitol. J Biol Chem 259: 7286–7293

Kleefeld S, Lohse D, Mansy A and Strotmann H (1990) Activation and deactivation of the thiol-modulated chloroplast H^+-ATPase during ATP hydrolysis. Biochim Biophys Acta 1019: 11–18

Komatsu-Takaki M (1993) Energy-dependent changes in the conformation of the chloroplast ATP synthase and its catalytic activity. Eur J Biochem 214: 587–591.

Kostrzewa M and Zetsche K (1992) Large ATP synthase operon of the red alga *Antithamnion sp.* resembles the corresponding operon in cyanobacteria. J Mol Biol 227: 961–970

Krab K, Bakels RHA, Scholts MJC and van Walraven HS (1993) Activation of the H^+-ATP synthase in thylakoid vesicles from cyanobacterium *Synechococcus* 6716 by $\Delta\tilde{\mu}_{H^+}$. Biochim Biophys Acta 1141: 197–205

Kramer DM and Crofts AR (1989) Activation of the chloroplast ATPase measured by electrochromic change in leaves of intact plants. Biochim Biophys Acta 976: 28–41

Kramer DM, Wise RR, Frederick JR, Alm DM, Hesketh JD, Ort DR and Crofts AR (1992) Regulation of coupling factor in field grown sunflower: A redox model relating coupling factor activity to activities of other thioredoxin-dependent enzymes. Photosyn Res 26: 213–222

McCarty RE and Racker E (1968) Partial resolution of the enzymes catalysing photophosphorylation III. Activation of ATPase and ^{32}Pi, ATP exchange in chloroplasts. J Biol Chem 243: 129–137

Miki J, Maeda M, Mukohata Y and Futai M (1988) The γsubunit of ATP synthase from spinach chloroplasts. Primary structure deduced from cloned cDNA sequence. FEBS Lett 232: 221–226

Mills JD and Mitchell P (1982a) Modulation of coupling factor ATPase activity in intact chloroplasts. Reversal of thiol modulation in the dark. Biochim Biophys Acta 679: 75–83

Mills JD and Mitchell P (1982b) Thiol modulation of CFo-CF1 stimulates acid/base dependent phosphorylation of ADP by broken pea chloroplasts. FEBS Lett 144: 63–67

Mills JD and Mitchell P (1984) Thiol modulation of the chloroplast ATPase and its effect on photophosphorylation. Biochim Biophys Acta 764: 93–104

Mills JD, Mitchell P and Schürmann P (1980) Modulation of coupling factor ATPase activity in intact chloroplasts. The role of the thioredoxin system. FEBS Lett 112: 173–177

Mitchell P (1968) Chemiosmotic coupling and energy transduction. Glynn Research Ltd, Bodmin, UK

Mitra B and Hammes GG (1988) Characterization of three-subunit chloroplast coupling factor. Biochemistry 27: 245–250

Morita S, Itoh S and Nishimura M (1983) Flash induced photophosphorylation in chloroplasts with activated ATPase. Biochim Biophys Acta 724: 411–415

Moroney JV, Fullmer CS and McCarty RE (1984) Characterization of the cysteinyl-containing peptides of the γ subunit of coupling factor 1. J Biol Chem 259: 7281–7285

Nalin CM and McCarty RE (1984) Role of a disulfide bond in the γ subunit in activation of the ATPase of chloroplast coupling factor 1. J Biol Chem 259: 7275–7280

Nelson N, Nelson H and Racker E (1972) Partial resolution of the enzymes catalysing photophosphorylation XII. Purification and properties of an inhibitor isolated from coupling factor 1. J Biol Chem 259: 7275–7280

Nichols P, Obinger C, Niederhauser H and Peschek GA (1992) Cytochrome oxidase in *Anacystis nidulans*. Stoichiometries and possible functions in the cytoplasm and thylakoid membranes. Biochim Biophys Acta 1098: 184–190

Noctor GD and Mills JD (1988) Thiol-modulation of the chloroplast ATPase. Lack of oxidation of the enzyme in the presence of $\Delta\tilde{\mu}_{H^+}$ and a possible explanation of the physiological requirement for thiol regulation of the enzyme. Biochim Biophys Acta 935: 53–60

Ort DR and Oxborough K (1992) In situ regulation of chloroplast coupling factor activity. Ann Rev Plant Physiol Plant Mol Biol 43: 269–291

Pancic PG and Strotmann H (1993) Structure of the nuclear encoded gamma subunit of CF0CF1 of the diatom *Odontella sinensis* including its presequence. FEBS Lett 320: 61–66

Pancic PG, Strotmann H and Kowallik KV (1992) Chloroplast ATPase genes in the diatom *Odontella sinensis* reflect cyanobacterial characters in structure and arrangement. J Mol Biol 224: 529–536

Petrack B, Caston A, Sheppy F and Farron F (1965) Studies on the hydrolysis of ATP by spinach chloroplasts. J Biol Chem 240: 906–914

Pettersson G and Ryde-Pettersson U (1988) A mathematical model of the Calvin photosynthesis cycle. Eur J Biochem 175: 661–672

Possmayer FE and Gräber P (1994) The pH_{in} and pH_{out} dependence of the rate of ATP synthesis catalysed by the chloroplast H^+-ATPase, CF_0CF_1, in proteoliposomes. J Biol Chem 269: 1896–1904

Quick WP and Mills JD (1986) Thiol modulation of chloroplast CFo-CF1 in isolated barley protoplasts and its significance to the regulation of carbon dioxide fixation. Biochim Biophys Acta 861: 166–172

Quick WP and Mills JD (1987) Changes in the apparent Michaelis constant for ADP during photophosphorylation are consistent with a delocalized chemiosmotic energy coupling. Biochim Biophys Acta 893: 197–207

Reimer S and Selman BR (1978) Tentoxin-induced energy independent adenine nucleotide exchange and ATPase activity with chloroplast coupling factor 1. J Biol Chem 253: 7249–7255

Richter ML and McCarty RE (1987) Energy-dependent changes in the conformation of the ε-subunit of the chloroplast ATP synthase. J Biol Chem 262: 15037–15040

Rumberg B and Becher U (1984) Multiple ΔpH control of H^+-ATP synthase function in chloroplasts. In: Papa S, Altendorf K, Ernster L and Packer L (eds) H^+-ATPase: Structure, Function, Biogenesis. pp 421–430. Adriatic Editrice, Bari

Ryrie IJ and Jagendorf AT (1972) Correlation between a conformational change in the coupling factor protein and the high energy state in chloroplasts. J Biol Chem 247: 4453–4459

Scheibe (1981) Thioredoxin$_m$ in pea chloroplasts: Concentration and redox status under light and dark conditions. FEBS Lett 133: 301–304

Scheibe R (1987) NADP$^+$-malate dehydrogenase in C3 plants: Regulation and role of a light-activated enzyme. Physiol Plantarum 71: 393–400

Schlodder E, Gräber P and Witt HT (1982) Mechanism of phosphorylation in chloroplasts. In: Barber J (ed) Topics in Photosynthesis, Vol 4, pp 105–175. Elsevier, Amsterdam

Schumann J and Strotmann H (1981) The mechanism of induction and deactivation of light triggered ATPase. In: Akoyunoglou G (ed) Photosynthesis II, Electron Transport and Phosphorylation, pp 881–892. Balaban Int Sci Serv, Philadelphia

Schumann J, Richter ML and McCarty RE (1985) Partial proteolysis as a probe of the conformation of the γ subunit in activated soluble and membrane-bound chloroplast coupling factor 1. J Biol Chem 260: 11817–11823

Schürmann P, Maeda K and Tsugita A. (1981) Isomers of thioredoxins in spinach leaves. Eur J Biochem 116: 37–45

Selman-Reimer S and Selman BR (1988) The activation and inactivation of the *Dunaliella salina* chloroplast coupling factor 1 (CF1) in vivo and *in situ*. FEBS Lett 230: 17–20

Selman-Reimer S, Duhe RJ, Stockman BJ and Selman BR (1991) L-1-N-methyl-4-mercaptohistidine disulfide, a potential endogenous regulator in the redox control of chloroplast coupling factor 1 in *Dunaliella*. J Biol Chem 266: 182–188

Shahak Y (1985) Differential effect of thiol oxidants on the chloroplast H$^+$-ATPase in the light and in the dark. J Biol Chem 260: 1459–1464

Sigalat C, de Kouchkovski Y, Haraux F and de Kouchkovski F (1988) Shift from localized to delocalized energy coupling in thylakoids by permeant amines. Biochim Biophys Acta 934: 375–388

Soteropoulos P, Süss KH and McCarty RE (1992) Modifications of the γ subunit of chloroplast coupling factor 1 alter interactions with the inhibitory ε subunit. J Biol Chem 267: 10348–10354

Stitt M, McLilley R and Heldt HW (1982) Adenosine nucleotide levels in the cytosol, chloroplasts and mitochondria of wheat protoplasts. Plant Physiol 70: 971–977

Strelow F and Rumberg B (1993) Kinetics and energetics of redox regulation of ATP synthase from chloroplasts. FEBS Lett 323: 19–22

Strotmann H and Bickel-Sandkötter S (1977) Energy dependent exchange of adenine nucleotides on chloroplast coupling factor 1. Biochim Biophys Acta 460: 126–135

Strotmann H, Kleefeld S and Lohse D (1987) Control of ATP synthesis in chloroplasts. FEBS Lett 221: 265–269

Tager JM, Wanders RJA, Groen AK, Kunz W, Bohnensack R, Kuster U, Letko G, Bohme G, Duszynski J and Wojtczak L (1983) Control of mitochondrial respiration. FEBS Lett 151: 1–9

Ting CS and Owens TG (1993) Photochemical and nonphotochemical fluorescence quenching processes in the diatom *Phaeodactylum tricornutum*. Plant Physiol 101: 1323–1330

Turina P, Melandri BA and Gräber P (1991) ATP synthesis in chromatophores driven by artificially induced ion gradients. Eur J Biochem 196: 225–229

Turina P, Melandri BA and Gräber P (1992) Activation of the H$^+$-ATP synthase in photosynthetic bacterium *Rhodobacter capsulatus*. J Biol Chem 267: 11057–11063

Valerio M, de Kouchkovsky Y and Haraux F (1992) An attempt to discriminate catalytic and regulatory proton binding sites in membrane-bound, thiol-reduced chloroplast ATPase. Biochemistry 31: 4239–4247

Vallejos RH, Arana JL and Ravizzini (1983) Changes in activity and structure of chloroplast proton ATPase induced by illumination of spinach leaves. J Biol Chem 258: 7317–7321.

Werner S, Schumann J and Strotmann H (1990) The primary structure of the γ-subunit of the ATPase from *Synechocystis* 6803. FEBS Lett 261: 204–208

Zhou JM, Xue ZX, Du ZY, Melese T and Boyer PD (1988) Relationship of tightly bound ADP and ATP to control and catalysis by chloroplast ATP synthase. Biochemistry 27: 5129–5135

Chapter 26

The Structure of the CF_1 Part of the ATP-Synthase Complex from Chloroplasts

Egbert J. Boekema
Bioson Research Institute, Biophysical Chemistry, University of Groningen, Nijenborgh 4, 9747 AG Groningen, The Netherlands

Uwe Lücken*
Fritz-Haber-Institute, Faradayweg 4-6, D-14195 Berlin, Germany

Summary

The proton ATP synthase consists of a membrane integrated part, F_0, and a hydrophilic part, F_1. F_1 is composed of five different subunits: α, β, γ, δ and ε. This chapter focuses on the chloroplast F_1 (CF_1) structure and discusses the overall shape and dimensions of CF_1, shape and size of the various subunits, subunit interactions and conformational changes in the subunit positions related to catalysis. Structural data originate mainly from X-ray diffraction and electron microscopy. Recently, the structure of F_1 from beef heart mitochondria has been determined at 2.8 Å and this structure, although not fully identical to CF_1, can be taken as a blueprint for models of the CF_1 structure.

* Present Address: Electron Optics Laboratory, Philips, 5600 MD Eindhoven, The Netherlands

Donald R. Ort and Charles F. Yocum (eds): Oxygenic Photosynthesis: The Light Reactions, pp. 487–492.

I. Introduction

The proton ATP synthase from chloroplasts, mitochondria and eubacteria consists of a membrane-integrated part, F_0, which acts as a proton channel through the membrane and a hydrophilic part, F_1, which contains the nucleotide-binding sites. The ATP synthases from different sources have a very similar structure. This is especially the case for the F_1 part. The F_1 part from spinach chloroplasts, CF_1, is composed of five different subunits: α (55.4 kDa), β (53.9 kDa), γ (35.7 kDa), δ (20 kDa) and ε (14.7 kDa). One CF_1 molecule contains three copies of the large α and β subunits, and one copy each of the smaller subunits making a total mass of nearly 400 kDa. This chapter focuses on the CF_1 structure. The lack of a 3D structure with high resolution forces us to discuss also the F_1 structure from mitochondria and eubacteria. Results obtained with related F_1's are also relevant for CF_1 structure; at the low resolution level the mitochondrial (MF_1) and CF_1 structures have a striking similarity (Boekema et al., 1990), although all 5 subunits differ in mass and amino acid sequence, resulting in a 10% smaller mass for mitochondrial F_1. Especially the two smallest subunits differ considerably, but the fact that the δ subunits of MF_1 and CF_1 have a very different amino acid sequence has often been overemphasized, because both δ subunits are functionally related (Engelbrecht and Junge, 1990).

There are three main methods to investigate the structure of a protein: X-ray diffraction, NMR and electron microscopy (EM). For F_1, the first and the third method have been widely applied. By X-ray diffraction of three-dimensional crystals the structure can be determined at atomic resolution. A structure of rat liver MF_1 at the sub-atomic level has been published (Bianchet et al., 1991). More recently, based on crystals from beef heart MF_1 (Abrahams et al., 1993) a 2.8 Å resolution structure has been determined (Abrahams et al., 1994). NMR has the same potential as X-ray diffraction but has the disadvantage that the objects studied should be smaller than 40 kDa. Thus it can only be applied to the structure elucidation of the small F_1 subunits. Results of such work have not yet been reported. With EM it is very difficult to determine a high-resolution structure, because it needs large two-dimensional (2D) crystals. With small 2D crystals a low-resolution 3D model can be obtained (Gogol et al., 1989a; Ishii et al., 1993). Low-resolution data can also be easily obtained with single-particle averaging (Boekema et al., 1986; Gogol et al., 1989b). In this chapter we will mainly discuss the outcomes of these techniques and, where necessary, extend them with results of biochemical investigations.

Abbreviations: CF_1 – chloroplast F_1; DTT – dithiothreitol; EM – electron microscopy; MF_1 – mitochrondrial F_1; NEM – N-ethylmaleimide

II. Structural Aspects of CF_1

A. Overall Shape of CF_1

If the electron density of a protein is solved by X-ray diffraction with a resolution of better than 3 Å, there is enough structural detail to define the amino acid chain backbone. If the resolution is lower, the ambiguity of fitting amino acids increases and above 5 Å only secondary elements, such as large α-helices, can be recognized. First, an X-ray study on rat liver mitochondrial F_1 was published which did not show much of the amino acid chain backbone, although a resolution of 3.5 Å resolution was claimed (Bianchet et al., 1991). But more recently, the beef heart mitochondrial ATPase structure could be solved at 2.8 Å resolution (Abrahams et al., 1994) and this enabled the determination of the amino acid chain backbone of the α- and β subunits and most of the γ subunit. We will discuss the features from both X-ray structures mainly in connection with the EM and other data on CF_1.

The maximal diameter of CF_1 parallel to the plane of the membrane is close to 110 Å, because this is the repeating distance of CF_1 in two-dimensional crystals (Boekema and Böttcher, 1992). In crystals the repeat can be determined by EM with an error of only a few Å, even if the resolution of the image features is much lower. The diameter of bovine heart MF_1 is 100 Å (Abrahams et al., 1994). This is somewhat in discrepancy with the 120 Å diameter from the rat liver MF_1 model (Bianchet et al., 1991). The height of CF_1 vertical to the membrane plane is about 83 Å. A similar height has been found for mitochondrial F_1 (Boekema et al., 1988; Abrahams et al., 1994) and *E. coli* F_1 (90 Å; Lücken et al., 1990). Average side views of F_1 complexes in CF_0F_1 (Fig. 1) and *E. coli* F_0F_1 (Lücken et al., 1990), recorded by EM, show a trapezoid form for symmetrical views (Fig. 3) of F_1.

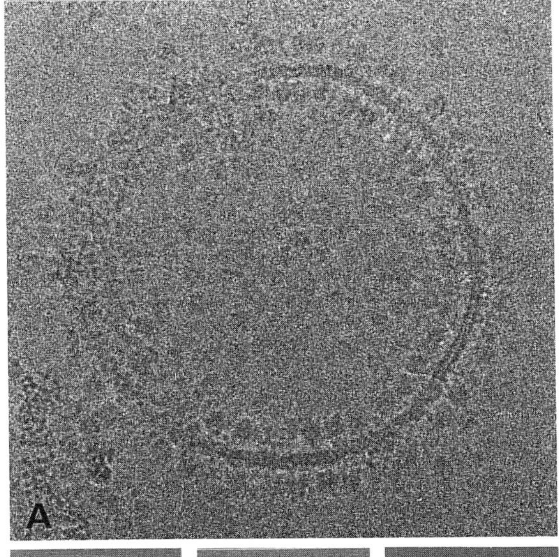

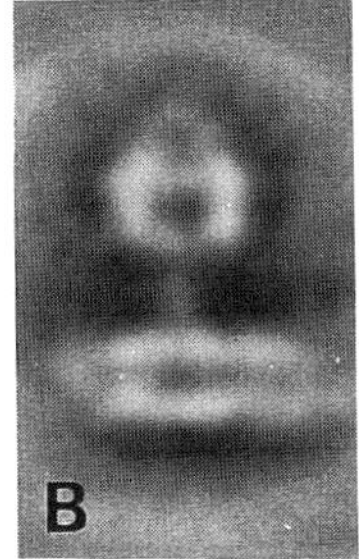

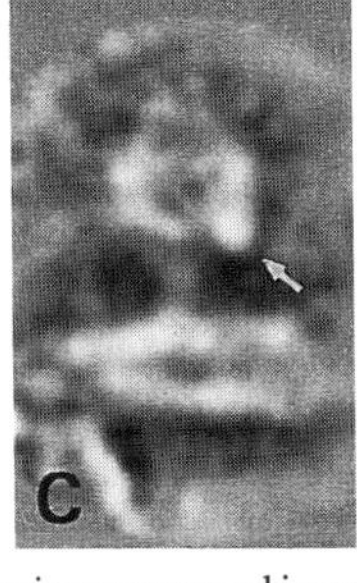

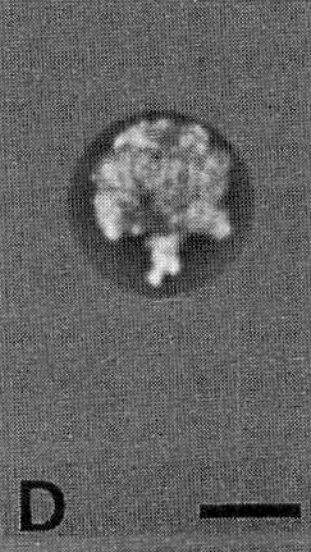

Fig. 1. Cryo-electron microscopy and image analysis of unstained F_0F_1 ATPase from spinach chloroplasts. (A) EM of F_0F_1 ATPase reconstituted in lipid vesicles (B) average image of reconstituted F_0F_1 particles (C) average image of F_0F_1 particles labeled with a gold cluster, which binds specifically to the S-S disulfide in the loop of the γ subunit. Procedure: the -SH groups in CF_1 were blocked with NEM, the S-S bond was reduced with DTT and the reduced -SH groups labeled with a monomaleimidogold label. The S-S group is in the periphery of CF_1 (position marked by an arrow), in accordance with the fluorescence energy transfer data (see chapter 23). (D) structure of the bovine heart MF_1 (Abrahams et al., 1993) shown on the same scale. For representation and comparison the electron density was low-pass filtered to enhance the low-resolution features of the map. The bar represents 100 Å.

B. Structure of the Individual Subunits

1. Shape of the α- and β-Subunits

The 2.8 Å map of MF_1 (Abrahams et al., 1994) shows that the α- and β subunits are almost identically folded: each consists of N-terminal six-stranded β-barrel, a central α-β domain containing the nucleotide-binding site and a C-terminal bundle of 7 and 6 helices, respectively. The X-ray model confirms and extends all previous work on the structure of the large subunits, such as the fact that the distribution of 21 protease-cleaved sites in the isolated α subunit is similar to that of the β subunit, thus providing experimental evidence for a similar folding topology of the two subunits (Tozawa et al., 1993).

From EM of negatively stained specimens followed by computer analysis (Boekema et al., 1992) it could be concluded that the overall shape of the isolated β subunit of spinach CF_1 is also elongated. The β subunit has dimensions of about 66–69 × 42–44 Å. This indicates that the packing of the large subunits in CF_1 and MF_1 could have a similar folding. Two independent low-resolution 3D reconstructions of bacterial F_1 from EM images also indicate that the α- and β subunits are elongated. In a model of Gogol et al. (1989a) these subunits are very elongated, e.g. 90 × 30 Å. In the model of Ishii et al., 1993 they are shown to be 70 × 30 Å.

2. Shape of the Small Subunits γ, δ and ε

By analytical ultracentrifugation experiments it was found that subunit γ from *E. coli* F_1 is rather elongated (Dunn, 1982). EM analysis of a CF_1 particle, from which subunits δ and ε were removed, showed that the diameter of the central mass was 25–28 Å, much smaller than the projected diameter of the large subunits (Fig. 2). Considering the mass of γ, 35.7 kDa, this indicates that it is an elongated subunit with a calculated length of 50 Å (Boekema et al., 1990). The X-ray map of bovine mitochondrial F_1 shows that the γ subunit is even much longer: a 90 Å long α-helix forms the center of the MF_1 structure, a second α-helix of 60 Å forms a left-handed antiparallel coiled coil with the first helix (Abrahams et al., 1994).

The chloroplast γ subunit has an interesting component, which is a flexible loop that includes a disulfide bond (McCarty and Hammes, 1987). The distance of this disulfide bond to two other sulfhydryl groups on γ, determined by fluorescence resonance energy transfer, was 44 and 47 Å, or about 38 Å, if projected in a plane parallel to the membrane. Thus this loop, which is important for ATPase activity, is really a loop protruding from the rest of the γ subunit. The position of the S-S disulfide was also investigated by cryo-EM (U. Lücken, unpublished data) using a monomaleimidogold label (Wilkens and Capaldi, 1992). Results indicate that the S-S disulfide bridge is indeed extending far from the center of CF_1 (Fig. 1).

Unfortunately, the atomic structure of δ and ε

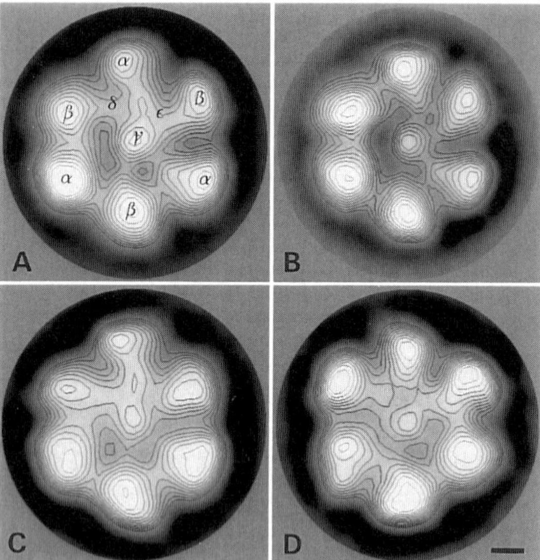

Fig. 2. Electron microscopy of negatively stained CF_1. (A) average image of holo-enzyme CF_1 in the hexagonal view (modified from Boekema et al., 1990); (B) average image of CF_1 depleted of subunits δ and ε (modified from Boekema et al., 1990); (C) and (D) average images of projections from CF_1 particles arranged upside- up and upside-down on the carbon support. In general, in negatively stained samples the lower half of the particle is more strongly embedded in the stain layer and this leads to a stronger contrast of the features of this part in the projection. In the predominant view (A and the similar view C) the α- and β subunits that are close to δ appear less pronounced, because the structure is slightly lifted up by δ. The α- and β subunits become more evenly embedded in the stain layer when the δ subunit is absent, as in Fig. B, or when δ is away from the carbon support in upside-down attached particles, as in D. The bar represents 20 Å.

could not be determined in the 2.8 Å density map (Abrahams et al., 1994). The chloroplast δ subunit has an elongated shape (Wagner et al., 1988). It is predicted to be an α-helical protein with only limited β-structure. Conserved amino acids are found in the N-terminal start of a putative amphipathic α-helix (Hoesche and Berzborn, 1993). The shape of the chloroplast ε subunit is not known. ε from *E. coli* has been crystallized (Codd et al., 1992), but the structure has not yet been reported.

C. Positions and Interactions of the Subunits

From thermal denaturation experiments it was found that the major forces that stabilize CF_1 must be between the α, β and γ subunits; subunits δ and ε have little effect on the stability (Wang et al., 1993). Thus the three largest subunits with a total of 7 copies define the shape of CF_1. Nevertheless, the smallest subunits are not totally unimportant for the overall shape. It was demonstrated that the positions of the α subunits in CF_1 changed when δ or δ plus ε were removed (Boekema and Böttcher, 1992).

Although not all interesting questions concerning the detailed structure can be answered, several conclusions concerning the arrangement of the large subunits and small subunits can be drawn:

1. The overall shape of CF_1 is asymmetric. This means that all three α- and all three β subunits 'see' a different surrounding environment. Several lines of evidence support this conclusion. For instance; the three β-β center-to-center average distances in CF_1 are 79, 77 and 73 Å (Boekema and Böttcher, 1992). Similar asymmetrical distances were found for the β subunits in *E. coli* F_1 (Wilkens and Capaldi, 1994). The 2.8 Å resolution structure of MF_1 (Abrahams et al., 1994) shows the asymmetry of the overall structure in a very detailed way. The main reason why the other X-ray map presents a symmetric F_1 structure (Bianchet et al., 1991) is the fact that a polypeptide chain can be better fitted if noisy data are 3-fold averaged. The large subunits not only 'see' a different surrounding but also have a slightly different conformation. For example: only one of the three α subunits of CF_1 reacts with Lucifer yellow which modifies lysine 378 (McCarty and Hammes, 1987). Similarly, Wilkens and Capaldi (1992) found that only one of the α subunits in *E. coli* F_1 has a cysteine group that reacts with maleimides. The inequivalent catalytic sites in CF_1 originate, of course, also from a structural difference (see chapters 24 and 25).

2. In the plane parallel to the membrane, the six copies of the large subunits alternate in position. This was first shown unambiguously by immuno-electron microscopy with monoclonal antibodies (Gogol et al., 1989b).

3. There is some evidence that the α- and β subunits in CF_1 are not exactly at the same height as shown in Fig 3. The MF_1 model from Bianchet et al. (1991) shows two slightly offset, interdigitated layers of large subunits. At the top, the β subunits protrude about 15 Å higher than the α subunits. The CF_1 data indicate a similar arrangement

(Boekema et al., 1992). However, the 2.8 Å resolution structure of MF_1 shows that only one β subunit (β_{DP}) is at a substantially higher level in the structure, whereas two α subunits are extending slightly more to the base of F_1.

4. A seventh mass seen in the center of the $\alpha_3\beta_3$ configuration (Fig. 2) is formed by the small subunits and consists mainly of subunit γ (Boekema et al., 1990; Wilkens and Capaldi, 1992).

5. The small subunits are positioned in the lower half of F_1, which is the part of F_1 closest to the membrane. This was already inferred based on biochemical evidence, but recently Ishii et al (1993) directly showed that the central mass is in the lower half of F_1.

6. Subunit α interacts with an α subunit and with subunit I, a F_0 subunit, which contributes to the stalk connecting F_1 to F_0 (Beckers et al., 1992). Although subunit δ is partly hidden within CF_1, it must also be involved in the stalk, because it forms a protrusion on F_1. In original EM images the protrusion was barely visible, but the large subunits around δ look differently embedded in projections of negatively stained specimens (Fig. 2). The protrusion is clearly visible in the structure of bovine heart MF_1 (Abrahams et al., 1993), where it consists of two α-helices in a coiled-coil making a 40 Å stem (Fig. 1D). But it needs to be confirmed if the entire protrusion consists of only the δ subunit.

7. In CF_1 the subunits γ and ε are close and interact with each other (McCarty and Hammes, 1987). Interestingly, a complex of γ and ε was purified and crystallized from *E. coli* (Cox et al., 1993).

8. According to Fig. 3, subunit γ is the central CF_1 subunit that interacts with all other F_1 subunits and with the stalk formed by F_0. Because it protrudes from MF_1, it is very likely that conformational changes occurring in the stalk of F_0F_1 can be of catalytic relevance if mediated by subunit γ. But also other CF_1 subunits may interact with the stalk and be directly influenced by changes occurring in the stalk. In *E. coli* F_1 a β subunit, different from the one which interacts with the ε subunit, binds to the b subunit of the F_0 part (Wilkens et al., 1994).

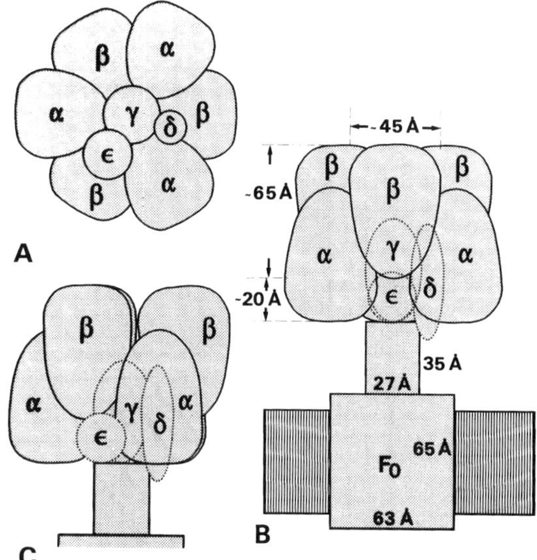

Fig. 3. Views of a model for CF_1 ATP synthase. A. The model in the hexagonal projection, parallel to the plane of the membrane, seen from the membrane. B. The model seen from aside in a 'symmetrical view'. C. The model seen from aside in an 'asymmetrical view'. The positions of 3B and 3C differ by 30°. The dimensions of the F_0 part and the stalk, symbolized as cylinders, have been determined, but the positions of the subunits within the membrane are still not fully clear.

D. Conformational Changes in Subunits and Subunit Positions

It has been shown that there is evidence for movements of subunit ε that could play a role in the catalytic mechanism and regulation of ATP synthesis/ hydrolysis (Gogol et al., 1990; Boekema and Böttcher, 1992). Wilkens and Capaldi (1994) showed that both γ and ε subunits are mobile, probably by moving as a single domain through distances as great as 10–20 Å.

III. Concluding Remarks

Further structural data are in demand, even after the determination of the MF_1 structure from beef heart mitochondria, which is a breakthrough for the field of bioenergetics. For the precise mechanism of ATP synthesis in chloroplasts it will also be necessary to solve the CF_1 structure at atomic resolution by X-ray diffraction. Only based on such a structure will all the biochemical data attain their full significance. But F_1 is not only a fixed structure, as seen by X-rays in a 3D crystal. Large rearrangements, such as the movement of the γ and ε subunit cannot be easily studied by X-ray diffraction if the conformation is unstable. Here, cryo-EM, which usually results in a much lower resolution than X-ray diffraction, will continue to be of value.

Acknowledgments

Support by the Netherlands Foundation for Chemical Research (SON) and by a grant of the European Union BIO-CT93-0078 is gratefully acknowledged.

References

Abrahams JP, Lutter R, Todd RJ, van Raaij MJ, Leslie AGW and Walker JE (1993) Inherent asymmetry of the structure of F_1-ATPase from bovine heart mitochondria at 6.5 Å resolution. EMBO J 12: 1775–1780

Abrahams JP, Leslie AGW, Lutter R and Walker JE (1994) Structure at 2.8 Å resolution of F_1-ATPase from bovine heart mitochondria. Nature 370: 621–628

Beckers G, Berzborn RJ and Strotmann H (1992) Zero-length crosslinking between subunits δ and I of the H^+-translocating ATPase of chloroplasts. Biochim Biophys Acta 1101: 97–104

Bianchet M, Ysern X, Hullihen J, Pedersen PL and Amzel, LM (1991) Mitochondrial ATP synthase. Quaternary structure of the F_1 moiety at 3.6 Å determined by X-ray diffraction. J Biol Chem 266: 21197–21201

Boekema EJ and Böttcher B (1992) The structure of ATP synthase from chloroplasts. Conformational changes of CF_1 studied by electron microscopy. Biochim Biophys Acta 1098: 131–143

Boekema EJ, Berden JA, and van Heel MG (1986) Structure of mitochondrial F_1-ATPase studied by electron microscopy and image processing. Biochim Biophys Acta 851: 353–360

Boekema EJ, Schmidt G, Gräber P and Berden JA (1988) Structure of the ATP-synthase from chloroplasts and mitochondria by electron microscopy. Zeitschrift Naturforsch. 43c: 219–225

Boekema EJ, Xiao J, and McCarty RE (1990) Structure of the ATP synthase from chloroplasts studied by electron microscopy. Localization of the small subunits. Biochim Biophys Acta 1020: 49–56

Boekema EJ, Harris D, Böttcher B and Gräber P (1992) The structure of the ATP-synthase from chloroplasts. In: Murata N (ed) Research in Photosynthesis, Vol 2, pp 645–652. Kluwer Academic Publishers, Dordrecht

Codd R, Cox G, Guss JM, Solomon RG and Webb D (1992) The expression, purification and crystallization of the ε subunit of the F_1 portion of the ATPase of *Escherichia coli*. J Mol Biol 228: 306–309

Cox GB, Cromer BA, Guss JM, Harvey, I, Jeffrey PD, Solomon RG and Webb DC (1993) Formation in vivo, purification and crystallization of a complex of the γ and ε subunits of the F_0F_1-ATPase of *Escherichia coli*. J Mol Biol 229: 1159–1162

Dunn SD (1982) The isolated γ subunit of *Escherichia coli* F_1 ATPase binds the ε subunit. J Biol Chem 257: 7354–7359

Engelbrecht S and Junge W (1990) Subunit δ of H^+-ATPases: At the interface between proton flow and ATP synthesis. Biochim Biophys Acta 1015: 379–390

Gogol EP, Lücken U, Bork T and Capaldi RA (1989a) Molecular architecture of *Escherichia coli* F_1 adenosinetriphosphatase. Biochemistry 28: 4709–4716

Gogol EP, Aggeler R, Sagermann M and Capaldi RA (1989b) Cryoelectron microscopy of *Escherichia coli* F_1 adenosinetriphosphatase decorated with monoclonal antibodies to individual subunits of the complex. Biochemistry 28: 4717–4724

Gogol EP, Johnston E, Aggeler R and Capaldi RA (1990) Ligand-dependent structural variations in *Escherichia coli* F_1 ATPase revealed by cryoelectron microscopy. Proc Natl Acad Sci USA 87: 9585–9589

Hoesche JA and Berzborn R (1993) Primary structure, deduced from cDNA, secondary structure analysis and conclusions concerning interaction surfaces of the δ subunit of the photosynthetic ATP-synthase (E.C. 3.6.1.34) from millet (*Sorghum bicolor*) and maize (*Zea mays*). Biochim Biophys Acta 1142: 293–305

Ishii N, Yoshimura H, Nagayama K, Kagawa Y and Yoshida M (1993) Three-dimensional structure of F_1-ATPase of thermophilic bacterium PS3 obtained by electron crystallography. J Biochem 113: 245–250

Lücken U, Gogol EP and Capaldi RA (1990) Structure of the ATP synthase complex (ECF_1F_0) of *Escherichia coli* from cryoelectron microscopy. Biochemistry 29: 5339–5343

McCarty, RE and Hammes GG (1987) Molecular architecture of chloroplast coupling factor 1. Trends Biol Sci 12: 234–237

Tozawa K, Miyauchi M and Yoshida M (1993) Structure of the α subunit of F_1-ATPase probed by limited proteolysis. J Biol Chem 268: 19044–19054

Wagner R, Apley EC, Engelbrecht S and Junge W (1988) The binding of eosin-labeled subunit δ to the isolated chloroplast ATPase, CF_1, as revealed by rotational diffusion in solution. FEBS Lett 230: 109–115

Wang Z-Y, Freire E and McCarty RE (1993) Influence of nucleotide binding site occupancy on the thermal stability of the F_1 portion of the chloroplast ATP synthase. J Biol Chem 268: 20785–20790

Wilkens S and Capaldi RA (1992) Monomaleimidogold labeling of the γ subunit of the *Escherichia coli* F_1 ATPase examined by cryoelectron microscopy. Arch Biochem Biophys 299: 105–109

Wilkens S and Capaldi RA (1994) Asymmetry and structural changes in ECF_1 examined by cryoelectronmicroscopy. Biol Chem Hoppe-Seyler 375: 43–51

Wilkens S, Dunn SD and Capaldi RA (1994) A cryoelectron microscopy study of the interaction of the *Escherichia coli* F_1-ATPase with subunit β dimer. FEBS Lett. 354: 37–40

Chapter 27

Light-harvesting Complexes of Plants and Algae: Introduction, Survey and Nomenclature

David John Simpson
Department of Physiology, Carlsberg Laboratory, Gamle Carlsberg Vej 10, DK-2500 Copenhagen Valby, Denmark

Jürgen Knoetzel
Institute of Cell Biology, Biochemistry and Biotechnology, University of Bremen, Postbox 330440, D-28334 Bremen, Germany

Summary

Light-harvesting complexes are pigment-proteins whose function is to absorb light energy and to transfer it rapidly and efficiently to the photosynthetic reaction centers. They bind most of the photosynthetic pigments found in higher plants, algae and photosynthetic bacteria, either non-covalently (chlorophylls and carotenoids) or via thioester bonds (phycobilins), acting as antennae in trapping and transferring excitation energy.

Our knowledge of the structure and function of these complexes has been greatly increased over the past ten years by the cloning and sequencing of the genes coding for the polypeptides, and by the determination of the crystallographic structure of the holoproteins. The chlorophyll-based membrane-intrinsic antennae of plants and algae contain 3 hydrophobic, α-helical, membrane-spanning regions and are encoded by an extended family of nuclear-encoded genes. Electron diffraction of two-dimensional crystals of LHCII has revealed the

Donald R. Ort and Charles F. Yocum (eds): Oxygenic Photosynthesis: The Light Reactions, pp. 493–506.

location and polypeptide ligands of many of the pigments. The same technique has been applied to Photosystems I and II, but not at a resolution that allows the polypeptide backbone to be fitted. Reconstitution of LHCII apoproteins with isolated pigments has shown that while carotenoids are essential, there is little specificity regarding the type of carotenoid.

The algae have developed a variety of light-harvesting antennae to absorb the green light that predominates in their environment. Red algae and cyanobacteria have large, water-soluble complexes called phycobilisomes attached to the stromal surface of their thylakoids. The X-ray crystallographic structure has revealed that the protein keeps the bilin pigment in a linear conformation, which increases its absorption in the 480–630 nm region. The different pigment-protein complexes are arranged to transfer excitation energy to the phycobilisome core and then to the membrane-associated Photosystem II reaction center. In contrast, the Chromophyta contain unusual allenic xanthophylls whose absorption maxima are red-shifted when part of a pigment-protein complex.

Additional roles have been suggested for light-harvesting complexes, including photoprotection and acting as proton channels for Photosystem II.

I. Introduction

The photosynthetic pigments of the chloroplasts of plants and algae owe their color to the absorbance of light by conjugated double bonds. They consist of carotenoids, chlorophylls (cyclic tetrapyrroles) and bilins (linear tetrapyrroles) (Fig. 1). These pigments are not free, but are associated with proteins which are located within the photosynthetic membrane (thylakoids), or bound to water-soluble proteins attached to this membrane. About half of the protein in chloroplasts and almost all of the lipid are found in the thylakoid membranes. Chl *a* and *b* comprise about 8% of the dry weight, representing a protein:Chl ratio of 6.3:1. Isolated Chl-proteins have even higher concentrations of Chl, equivalent to a concentration of up to 0.4M (Simpson, 1988). All of the Chl is bound non-covalently to the protein, since it can be removed by solvent or detergent extraction. An average plant leaf has about 70 million cells with 5 billion chloroplasts, each containing about 600 million molecules of chlorophyll (Chl), almost all of which function in light-harvesting. Chls, carotenoids and phycobilins transfer absorbed light energy to and from neighboring pigments to the primary electron donor pigments, the special pair Chl of the reaction centers of Photosystems (PS) I and II. This variety of pigments ensures that a wide spectral range of sunlight can be used to drive charge separation in the reaction center. Metallochlorins absorbing light too far into the red (e.g., bacteriochlorophyll) lack an oxidized state that is energetic enough to split water, and are not found in oxygenic photosynthetic organisms.

The light-harvesting antennae can be divided into four main types based on the characteristics of the constituent proteins and whether they are Chl- or phycobilin-antenna systems:

(i) Intrinsic, membrane-bound proteins with extensive polar regions in Chl-based systems are seen in plants and green algae, Crysophyceae, Xanthophyceae, diatoms, brown algae, Cryptophyceae, dinoflagellates, Euglenophyta and some oxygenic prokaryotes.

(ii) Water-soluble, Chl-based antenna systems are found in green photosynthetic bacteria (*Chlorobium, Prosthecochloris*) and dinoflagellates.

(iii) Water-soluble phycobilin light-harvesting systems are found in the antennae of cyanobacteria, red algae and Cryptophyceae.

(iv) Transmembrane, BChl-containing, hydrophobic proteins of Chlorobiaceae, Chloroflexaceae, Chromatiaceae and Rhodospirillaceae.

The protein component of light-harvesting complexes: (i) determines the specific binding and the spatial arrangement of the pigment molecules,

Abbreviations: BChl – bacteriochlorophyll; BPhe – bacteriopheophytin; Chl – chlorophyll; CP – chlorophyll protein; EF – exoplasmic face; ELIP – early light-induced protein; LHC – light-harvesting chlorophyll a/b binding; LHCI – light-harvesting complex of Photosystem I; LHCII – light-harvesting complex of Photosystem II; PS – photosystem; *Rb.* – *Rhodobacter; Rp.* – *Rhodopseudomonas*

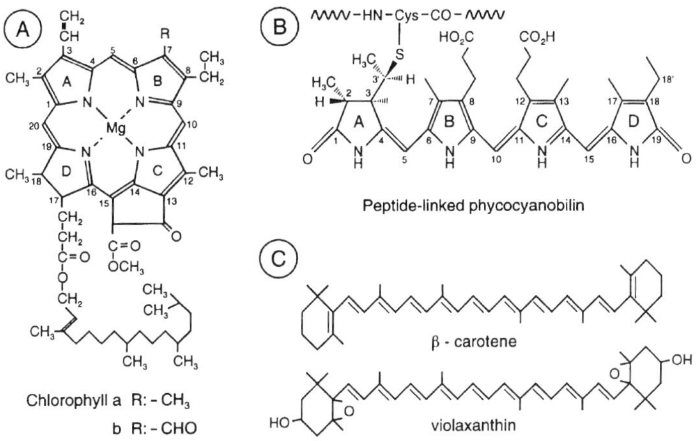

Fig. 1. Structural formulae of the light-absorbing pigments of light-harvesting proteins (chromophores). A. Chlorophyll *a* and *b* (metalloporphyrins). B. Phycobilin (linear tetrapyrrole). C. β-carotene and violaxanthin.

(ii) determines the configuration and conformation of the pigments and thus establishes their absorption and emission properties essential for light-harvesting function, and (iii) mediates the interaction with other protein components in the complex structural organization of the supramolecular antenna systems, allowing excitation energy transfer.

The two mechanisms for energy transfer between Chl molecules are delocalized exciton coupling, which takes place between similar or identical pigments at distances up to 20 Å, leading to the formation of a delocalized exciton spread over the collective pigment molecular orbitals, and Förster inductive resonance transfer which is a longer-range dipole-dipole interaction, where the acceptor is excited as the donor falls to the ground state (Sauer, 1986). An exciton state exists when two molecules are so close that it is impossible to excite one without exciting the other. A separation of 20 Å is less than twice the diameter of the porphyrin ring, and about 5 times its width. If two Chl molecules are separated by <10 Å, they can form a trap (excimer) due to orbital overlap, and the excitation energy absorbed by one of them will be quenched by the trap. The protein is thus essential to provide the correct pigment spacing and orientation for efficient absorption and energy transfer (Zuber, 1985; Huber, 1990).

Light-harvesting proteins exist as supramolecular complexes which are large enough to be seen by electron microscopy. Phycobilisomes can be seen in thin sections, while freeze-fracture reveals the membrane intrinsic components as discrete particles. The use of mutants affected in photosynthesis or pigment biosynthesis has identified the localization of PS I, PS II and LHCII, which exist as independent membrane intrinsic particles (Simpson, 1986). The particles of the protoplasmic face of stacked membranes (PFs) correspond to the mobile LHCII, which upon phosphorylation, detach from PS II and migrate laterally into the non-appressed membrane regions. The antenna systems are not only complex structures but can adapt their light-harvesting function to various conditions.

The expression of genes coding for light-harvesting polypeptides is light- and phytochrome-regulated, depending on the developmental stage of the chloroplast, the tissue and circadian rhythms (Piechulla, 1993). Regulation of light-harvesting function occurs through reversible redox-controlled LHCII protein phosphorylation and the resulting lateral migration of LHCII away from PS II leads to a redistribution of excitation energy between PS II and PS I (Allen, 1992). This chapter will introduce the membrane-intrinsic Chl *a*-proteins, the Chl *a/b*-, Chl *a/c*- and Chl *a/b/c*-proteins, as well as the water soluble phycobilin- and peridinin-proteins involved in light-harvesting. More detailed information can be found in the reviews of Bassi et al. (1990), Glazer (1983), Green et al. (1991, 1992), Jansson (1994), chapters 3,6,7 and 9 in Staehelin and Arntzen (1986) and chapters 15, 28, 29 and 30 in this book.

II. Light-harvesting Proteins: Multiple Structures

A. Antennae in Green Plants

1. LHCII

The realization that Chl and carotenoids are non-covalently bound to proteins came with the discovery, more than 25 years ago, of the plant light-harvesting Chl *a/b*-complex of PS II (LHCII) as one of the two major Chl-proteins resolved by polyacrylamide gel electrophoresis of detergent-solubilized thylakoid membranes. Thornber (1975) recognized it as a light-harvesting complex of PS II using the Chl *b*-less barley mutant *chlorina f2*, which is photochemically viable, but lacks LHCII. Since this green band, missing in *chlorina f2*, cannot contain the PS II reaction center, it must be a light-harvesting complex. As a consequence of this historical development, the term 'LHCII' describes a pigment-protein complex containing half of the Chl and about one-third of the protein of green plant thylakoids. The two major approaches to obtain structural information about the protein and the organization of the pigment molecules with respect to the polypeptide chain are the crystallization of the protein and reconstitution studies.

The first Chl-protein whose crystal structure was determined was a water-soluble BChl *a*-protein from *Prosthecochloris aestuarii* (Matthews et al., 1979). It consists of 3 identical subunits, each associated with 7 BChl *a* molecules co-ordinated through their Mg atoms via histidine residues (in 5 cases), a peptidyl carbonyl and possibly a water molecule. Histidine residues are involved in the co-ordination of BChl in the *Rhodopseudomonas capsulata* reaction center (Michel and Deisenhofer, 1988), and conserved histidine residues are the putative BChl ligands in antenna complexes of purple and green photosynthetic bacteria (Zuber, 1985). The first three-dimensional structure of membrane-intrinsic LHCII protein at 6 Å resolution shows a trimeric arrangement of monomers in a radially symmetric complex with internal symmetry within each monomer (Kühlbrandt and Wang, 1991). The trimeric structure is believed to represent the functional organization of LHCII in thylakoids, and freeze-fracture membrane particles of 8 nm diameter, which are mainly located in the stacked regions of the thylakoid membrane, have the same size and shape as fracture faces of two-dimensional crystals of LHCII (Simpson, 1986). In each monomer, there are three membrane-spanning helices, helices A and B showing a high intramolecular similarity. At 49 Å and 46 Å in length, these are the longest helices described for a membrane protein. The upper halves of the two helices and the protruding surface exposed regions, together with hook-like extensions on the stromal side of the thylakoid membrane, are conserved in other Chl *a/b*-xanthophyll proteins. LHCII contains about 8 Chl *a* and 6 Chl *b* together with 2 xanthophylls (Kühlbrandt and Wang, 1991). The 14 Chls have their chlorin rings roughly perpendicular to the membrane plane and are arranged in two layers near the surfaces of the membrane.

More recently, the structure of LHCII has been resolved to 3.4 Å, allowing the identification of some of the Chl ligands. Only 2 are histidines, while 3 are amines (2 glutamine, 1 asparagine), 3 are glutamates, 1 is co-ordinated via water and the remaining 3 identifiable Chl ligands are peptidyl carbonyls (Kühlbrandt et al., 1994). Helices A and B serve as a scaffold for packing the pigments into a small volume so that the light-harvesting proteins have an unusually large number of pigment molecules, i.e., one Chl per 15 amino acid residues, bound to a single polypeptide (Fig. 2).

Following the method for in vitro reconstitution of antenna complexes introduced by Plumley and Schmidt (1987), results have been obtained concerning the molecular basis of functional pigment-protein interactions by varying the amounts and ratios of pigments used for reconstitution, or by modifying the protein used in reconstitution experiments by deletion mutagenesis (Paulsen and Hobe, 1992; Cammarata et al., 1992). Unfortunately, the introduction of structural alterations such as deletions and substitutions in LHCII and subsequently studying the pigment-binding properties of the protein in vitro have not so far been able to identify specific pigment binding sites. Heterologous reconstitution utilizing pigments and light-harvesting proteins from spinach, pea and *Chlamydomonas* revealed no evidence for specialized binding sides for trihydroxy-α-carotene, which is interchangeable with lutein (Cammarata et al., 1992). Homologously and heterologously reconstituted green algae Chl *a/b* and Chl *a/b/c* LHC complexes bind Chls and xanthophylls in different stoichiometries, indicating great variability in xanthophyll binding specificity (Meyer and Wilhelm, 1993).

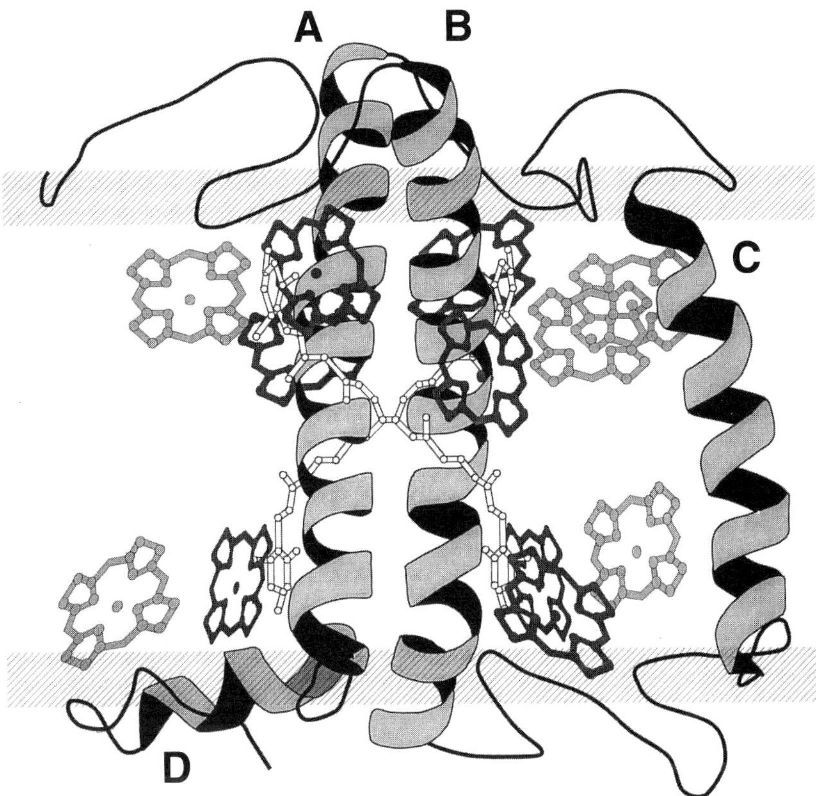

Fig. 2. Side view of the three dimensional structure of LHCII proposed by Kühlbrandt et al. (1994). The striped regions indicate the approximate boundary of the lipid bilayer, while the ribbon structures represent the α-helices. Two lutein molecules brace two of the helices, while the dark porphyrin rings indicate Chl *a* and the light porphyrin rings indicate Chl *b*. The top of the diagram is the stromal side, and is the location of the N-terminus of the polypeptide. (Reprinted with permission from Nature 367, 614–621 (1994). Copyright (1994) Macmillan Magazines Limited.)

In contrast, site-directed mutagenesis of the L and M subunits of the *Rb. capsulatus* reaction center has shown that alteration of the BChl ligand H^{M200} to leucine or phenylalanine, results in BPhe binding, producing a heterodimeric special pair (Bylina and Youvan, 1988). Interaction of BPhe with E^{L104} is responsible for the red shift of this molecule via its interaction with the C9 keto group of ring V, which is abolished when this residue is mutated to glutamine or leucine (Bylina et al., 1988). Spectral properties can also be altered by the interaction of pigment molecules with non-ligand residues, such as tyrosines in the LH2α subunit of *Rb. spheroides*. Mutation of LH2α^{44} to phenylalanine and LH2α^{45} to leucine, reveal that these tyrosines are responsible for the red shift of BChl, probably via interaction between π-π^*orbitals in aromatic residues and the tetrapyrrole ring (Fowler et al., 1992).

2. The Lhcb *Proteins*

LHCII preparations contain a mixture of different proteins which have been identified as closely related pigment-proteins encoded by a family of nuclear genes (Table 1) (Green et al., 1991, 1992). All proteins are encoded by nuclear DNA, synthesized in the cytoplasm as precursor proteins with transit sequences which direct the protein into the chloroplast. An interesting exception is *Euglena*, where cDNA clones of *Lhca* and *Lhcb* proteins encode large precursors containing several consecutive divergent polypeptides (Houlné and Schantz, 1988), which are apparently transported through the Golgi before entering the chloroplast (Schiff et al., 1991).

Methods for the preparation of LHCII yield *Lhcb1* and *Lhcb2* proteins with various amounts of LHCB3 and others. LHCB1 is very homologous to the *Lhcb2* protein, and is part of the most peripheral antenna of

Table 1. Chloroplast proteins with light-harvesting function in plants

Gene[1]	Gene product	Antenna protein	Gene	Gene product	Antenna protein
Lhca1	LHCI-730 Type I LHCI 22 kDa	Chl *a*/*b*-lutein protein Chl *a*/Chl *b*/lutein/violaxanthin/β-carotene/ neoxanthin = 1:0.4:0.17:0.11:0.02:0.005	*Lhcb1*	LHCII LHCB1 Type I LHCII LHCIIb 28 kDa	
Lhca2	LHCI-680B Type II LHCI 23 kDa	Chl *a*/*b*-lutein protein	*Lhcb2*	LHCII LHCB2 Type II LHCII LHC IIb 27 kDa	Chl *a*/*b*-lutein proteins Chl *a*/Chl *b*/lutein/neoxanthin/violaxanthin = 1:0.74:0.24:0.06:0.009
Lhca3	LHCI-680A Type III LHCI 25 kDa	Chl *a*/*b*-lutein protein	*Lhcb3*	LHCII LHCB3 Type III LHCII LHC IIb 25 kDa	
Lhca4	LHCI-730 Type IV LHCI 21 kDa	Chl *a*/*b*-lutein protein	*Lhcb4*	CP 29 LHCB4 Type II CP 29 LHC IIa 31 kDa	Chl *a*/*b*-lutein protein Chl *a*/Chl *b*/lutein/violaxanthin/neoxanthin/ β-carotene = 1:0.38:0.21:0.14:0.08:0.017
psaA	PS I-A 82 kDa	Chl *a*/β-carotene protein Chl *a*/β-carotene = 1:0.1–0.2	*Lhcb5*	CP 26 LHCB5 Type I CP 29 LHC IIc 28,29 kDa	Chl *a*/*b*-lutein protein Chl *a*/Chl *b*/lutein/violaxanthin/neoxanthin β-carotene = 1:0.52:0.23:0.08:0.06:0.016
psaB	PS I-B 82 kDa	Chl *a*/β-carotene protein Chl *a*/β-carotene = 1:0.1–0.2	*Lhcb6*	CP24 LHCB6 LHC IId 21 kDa	Chl *a*/*b*-lutein protein Chl *a*/Chl *b*/lutein/violaxanthin/β-carotene/ neoxanthin =1:0.81:0.38:0.18:0.035:traces
psbB	CP47 47 kDa	Chl *a*/β-carotene protein Chl *a*/β-carotene/lutein = 1:0.12:0.045	*psbC*	CP43 43 kDa	Chl *a*/β-carotene protein Chl *a*/β-carotene/lutein = 1:0.08:0.016

[1] The gene nomenclature presented here is now standard. There is, as yet, no consensus regarding protein nomenclature. See references in Janssen et al., 1992.

PS II (Spangfort and Andersson, 1989), where it probably exists with LHCB1 in mixed trimers. LHCB3 is the smallest pigment-protein found in LHCII, and in contrast to LHCB1 and 2, is tightly bound to the reaction center of PS II. It lacks the N-terminal phosphorylation site, and may be involved in mediating thylakoid stacking. *Lhcb4* codes for CP29, the largest of the LHC proteins, and is part of the inner antenna which connects the LHCII to the PS II reaction center. *Lhcb5* and *Lhcb6* code for the apoproteins of CP26 and CP24, respectively.

The PS II antenna is a highly complex structure and includes 40–50 Chl *a* molecules bound to CP43 and CP47 (*psbC* and *psbB*); a consensus antenna size value for the Chls organized in the *Lhcb* proteins is 230–250 Chl (Melis, 1991). Attempts to describe the topological organization are largely based on detergent fractionation studies and have resulted in several models with different protein stoichiometries (Peter and Thornber, 1991; Dainese et al., 1992b; Jansson, 1994). Most of the models are based on a dimeric PS II structure derived from detergent-isolated PS II preparations lacking LHCII (Lyon et al., 1993). The minor *Lhcb3, 4, 5* and *6* apoproteins are found in equimolar amounts (Dainese and Bassi, 1991; Peter and Thornber, 1991) and are probably monomeric. Analysis of in situ arrays of PS II particles in thylakoids, in which LHCII is also present, has led others to conclude that PS II exists as a monomer in the thylakoid membrane, surrounded by 4 LHCII complexes (Holzenburg et al., 1993). It is not known whether the number of the minor *Lhcb* proteins is 1, 1.5 or 2 copies per reaction center. Nevertheless, there appear to be at least 6–12 LHCB1 + 2 per four minor *Lhcb* proteins in PS II. The model in Fig. 3 assumes that the minor LHCB 3, 4, 5 and 6 are present in 1:1:1:1 stoichiometry and 2–4 trimers of LHCB1+2 constitute the most peripheral antenna of PS II responsible for variable antenna size (Jansson, 1994). This model with 6–12 *Lhcb* polypeptides each binding 13–15 Chl would give rise to antenna sizes of 140–240 Chl.

3. *The* Lhca *Proteins*

The light-harvesting antenna of PS I was originally found to be composed of four polypeptides from 22.5–24 kDa with a minor 10 kDa component (Mullet

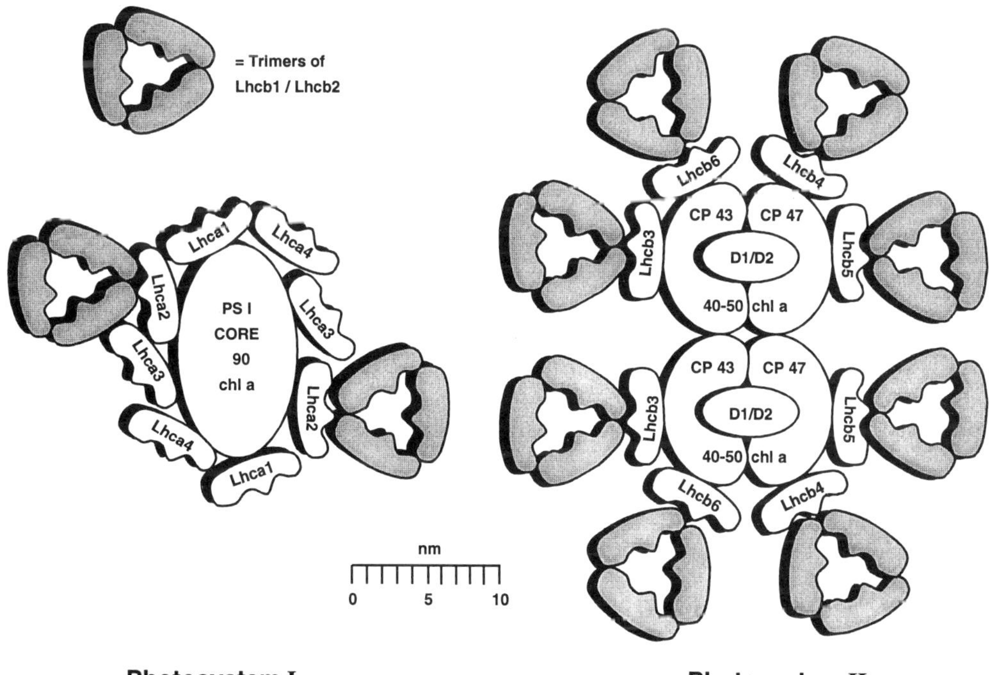

Fig. 3. Model for the organization of the Chl *a/b* proteins of the light-harvesting antennae of PS I and II in plants. The PS I core antenna of 80–100 Chl is surrounded by a monolayer of 2+2+2+2 *Lhca* polypeptides (*Lhca 1-4*) and an outer antenna of *Lhcb1/2* trimers can connect to *Lhca2* protein (Knoetzel et al., 1992). The PS II antenna has a core consisting of 40–50 Chl of CP47 and CP43, an inner antenna of *Lhcb 3, 4, 5,* and *6* polypeptides in a 1:1:1:1 stoichiometry, and *Lhcb1/2* trimers as outer antenna (Jansson, 1994).

et al., 1980) and the isolated antenna fluoresces at 730 nm at 77K. The antenna of barley Photosystem I can be fractionated into three pigment-protein complexes: a heterodimeric LHCI-730 complex composed of a 21 and a 22 kDa polypeptide and two LHCI-680 complexes with single apoproteins of 25 kDa (LHCI-680A) and 23 kDa (LHCI-680B) (Knoetzel et al., 1992). N-terminal protein sequences from the four antenna proteins from spinach, pea and barley made possible an assignment of the proteins to the four genes coding for Chl *a/b*-xanthophyll proteins of PS I (*Lhca1-4*) (Table 1) (Ikeuchi et al., 1991; Knoetzel et al., 1992).

The PS I core is surrounded by a monolayer of 8 LHC polypeptides (Boekema et al., 1990). On the basis of the finding that the four *Lhca* proteins seem to be present in approximately equal amounts and the estimate of 2 copies per reaction center (Bruce and Malkin, 1988), the model in Fig. 2 is proposed. The positions of the four *Lhca* proteins in the inner antenna, however, are highly hypothetical. LHCB1/2 trimers can be functionally attached to PS I in vitro and LHCI-680B seems to function as a connecting antenna in the transfer of excitation energy from LHCII or LHCI-730 to the reaction center (Knoetzel and Simpson, 1991). The non-pigmented subunit PSI-E co-purifies with LHCI-680 and may participate in its binding to the reaction center (Knoetzel et al., 1992). Phosphorylated LHCII may be less able to interact with LHCI than its non-phosphorylated form (Harrison and Allen, 1992).

4. The PsaA/B *Proteins*

In PS I, about half of the light-harvesting Chl is associated with the 2 large polypeptides that contain the reaction center pigment P700 and form the CPI complex. This consists of 2 homologous 82 kDa proteins which bind about 90 molecules of Chl *a* and 12–16 molecules of β-carotene. Analysis of the crystal structure of *Synechococcus* PS I at 6 Å resolution reveals the familiar trimeric structure of light-harvesting proteins and the locations of the P700 Chl *a* special pair and the subsequent electron acceptors A_0 (Chl *a*), A_1 (vitamin K_1) and F_X (a 4Fe4S cluster) have been tentatively assigned (Krauss et al., 1993; see chapter 18). Each of the polypeptides consists of 9 transmembrane helices and they are thought to be bound together via a leucine zipper located around the conserved cysteines which bind the 4Fe4S cluster that bridges the two polypeptides (Golbeck and Bryant, 1991). The Chl-binding ligands have not been identified, but the CPI complex is the counterpart of the D1/D2/CP43/CP47 complex of PS II, of which D1 and D2 are similar to the L and M subunits of the *Rp. viridis* reaction center (Michel and Deisenhofer, 1988).

5. The Early Light-induced Proteins (ELIPs)

The 13–14 and 18–19 kDa early light-inducible proteins of barley are related to the LHC family, and are synthesized transiently in the initial stages of chloroplast development, in plants exposed to light stress conditions or during recovery from photo-inhibition (Adamska et al., 1993). ELIPs are related to the *cbr* gene product of the green alga *Dunaliella bardawil,* which co-fractionates with LHCII and is co-induced with the accumulation of β-carotene under stress (Levy et al., 1992). A 22 kDa protein of PS II, the *PsbS* gene product, which is also found in cyanobacteria, shows sequence homology to the LHC proteins and ELIPs although it does not bind Chl (Kim et al., 1992).

Besides the 10 *Lhca* and *Lhcb* proteins, additional pigment-proteins exist which may have photo-protective functions. A 14–15 kDa Chl *a/b*-protein (CP14-15) with about 3 Chl *a* and 1 Chl *b* per protein was isolated from spinach PS II (Irrgang et al., 1993) and may be related to the barley LHC IIe which binds high levels of carotenoids, especially violaxanthin (Peter and Thornber, 1991). Neither protein shows cross reactivity to antibodies against other LHC proteins. The photoprotective function proposed for LHC IIe based on its high xanthophyll content (Chl *a*/Chl *b*/violaxanthin/ lutein/neoxanthin = 1:0.7:1.5:1.0:0.5) is similar to those of the ELIPs.

B. Antennae of Oxygenic Prokaryotes and Algae

Algae live in a light-limited aquatic environment with altered spectral quality of sunlight and have evolved a variety of light-harvesting pigment systems enabling them to collect blue to green light more efficiently (see Table 2).

1. Rhodophyta and Cyanobacteria

Phycobilisomes are the water-soluble light-harvesting units of cyanobacteria and the chloroplasts of red algae. They are composed of phycobiliproteins and

Table 2. Antenna proteins in oxygenic prokaryotes and algae

Systematic group/ genus	Antenna protein: Membrane-intrinsic	Antenna protein: Water-soluble	Systematic group/genus	Membrane-intrinsic
Dinophyta/ Dinophyceae *Gonyaulax*	Chl a/c_2-peridinin protein Chl a/Chl c_2/peridinin/diadinoxanthin/ dinoxanthin/diatoxanthin = 1:0.25:1.3:0.3:0.1:0.01 19 kDa	peridinin-Chl a-protein Chl a/peridinin = 1:4 32 kDa	Chlorophyta/ Chlorophyceae *Chlamydomonas*	Chl a/b-trihydroxy-α-carotene protein Chl a/Chl b/trihydroxy-α-carotene/ lutein/neoxanthin/violaxanthin = 1:0.84:0.19:0.16:0.11:0.04 LHCII: 26.5, 27, 29.5,30.5 kDa CP24, CP26, CP29: 19, 37, 35 kDa LHCI: 21, 21, 22, 23.5, 25.5, 25.8, 26 kDa
	Chl a/c_2-xanthophyll protein			
Rhodophyta *Porphyridium*	Chl a-zeaxanthin protein Chl a/zeaxanthin/β-carotene = 1:0.75:0.12 LHCI = 19.5,20,22,23,23.5 kDa	phycoerythrin/phycocyanin/ allophycocyanin 15–21 kDa Hemi-ellipsoidal phycobilisomes	Chlorophyta/ Prasinophyceae *Mantoniella*	Chl $a/b/c$*-prasinoxanthin protein Chl a/Chl b/magnesium 2,4 divinyl phaeoporphyrin a_5-monomethyl ester/neoxanthin/violaxanthin = 1:1.1:0.2:0.3:0.13:0.07 23 kDa
Heterokontophyta/ Xanthophyceae *Pleurochloris*	Chl a/c-diadinoxanthin protein Chl a/Chl c/diadinoxanthin/ heteroxanthin/vaucheriaxanthin = 1:0.22:0.26:0.15:0.13 23 kDa		Chlorophyta/ Prasinophyceae *Pseudoscourfieldia*	Chl $a/b/c$*-prasinoxanthin protein Chl a/Chl b/Chl c*/prasinoxanthin/ neoxanthin/unknown carotenoid = 1:1.5:0.3:0.9:0.4:0.1 24, 23, 26 kDa
Heterokontophyta/ Phaeophyceae (brown algae) *Acrocarpia*	Chl a/c_2-fucoxanthin protein Chl a/Chl c_2/fucoxanthin = 1:0.5:1 13–16 kDa		Chlorophyta/ Ulvophyceae *Codium*	Chl a/b-siphonaxanthin protein (siphonaxanthin, neoxanthin, siphonein, α-carotene) Chl a/Chl b = 1:1.4 26, 28, 32, 34 kDa
	Chl $a/c_1/c_2$-violaxanthin protein Chl a/Chl c_1/Chl c_2/violaxanthin = 1:0.12:0.12:0.12			Chl a/b-siphonaxanthin protein Chl a/Chl b = 1:0.6 19, 20, 22.5, 23, 24.5 kDa
Heterokontophyta/ Bacillariophyceae (diatoms) *Phaeodactylum*	Chl a/c_2-fucoxanthin protein Chl a/Chlc_2/fucoxanthin/ neofucoxanthin/diatoxanthin/ diadinoxanthin = 1:0.5:2.5:1:1:0.5:0.5 15–17.5 kDa		Chlorophyta/ Chlorophyceae *Chlorella*	Chl a/b-trihydroxy-α-carotene protein Chl a/Chl b/trihydroxy-α-carotene/ lutein/neoxanthin/violaxanthin = 1:0.8:0.15:0.15:0.07:0.06
Cryptophyta/ Cryptophyceae *Chroomonas*	Chl a/c_2-xanthophyll protein (diadinoxanthin) Chl a/Chl c_2 = 1:0.6 18, 19, 22 kDa	phycoerythrin-545 phycocyanin-645 12, 13, 19 kDa	Prochlorophyta/ Prochlorophyceae *Prochloron*, *Prochlorothrix*	Chl a/b protein Chl a/b= 2.5

linker polypeptides, the latter being required for phycobiliprotein assembly and attachment to PS II, and modulate the spectroscopic properties of biliproteins in the phycobilisome (Mörschel, 1991). A phycobilisome contains between 300–800 phycobilin pigments which are bound covalently via a thioester bond to cysteine residues of the phycobiliproteins. Their spectral properties are determined by the type of bilins, the polypeptide chain and the aggregation state of each. Phycobiliproteins consist of an α and β subunit, each carrying 2–3 covalently bound molecules of phycoerythrin or phycoerythrocyanin and phycocyanin, and form aggregates $(\alpha\beta)_n$ (n = 2 in Cryptophyceae; n = 3 or 6 in red algae and cyanobacteria). The structure of phycobiliproteins reveals that they, too, have a trimeric symmetry, each subunit consisting of an $\alpha\beta$ unit (Schirmer et al., 1985; Zuber, 1985). The protein constrains the bilins into their extended conformation, resulting in increased visible light absorption (480 – 630 nm). Phycobiliproteins form rod-like structures, five to six rod elements of these peripheral biliproteins are connected in at least two planes to a tricylindrical core of phycobiliproteins with allophycocyanin (Mörschel, 1991). The pigments are physically arranged so that excitation energy is transferred downhill from the peripheral biliproteins to the core, and detailed structural analysis in the red alga *Porphyridium cruentum* has shown that hemi-ellipsoidal phycobilisomes (PBS) transfer excitation energy via CP47/CP43 to the PS II RC (Holzwarth, 1991). Under conditions where more light is absorbed by PS I (State I), many of the phycobilisomes and EF particles are organized into parallel rows, which increases the efficiency of excitation energy transfer to PS II. In State II, the EF particles and phycobilisomes have a more random distribution, which decreases the average distance between PS II and PS I (Olive et al., 1986). In State II, there is increased energy transfer from PS II to PS I (spillover), which may be mediated by phosphorylation of linker and PS II polypeptides. It has recently been shown that the red alga *Porphyridium cruentum* contains Chl *a*-binding polypeptides functionally associated with PS I which are immunologically related to higher plant LHCI (Wolfe et al., 1994).

2. *Chlorophyta and* Prochloron

Green algae LHCs have structural characteristics similar to those of higher plants but vary in their carotenoid composition. *Chlorella* and *Chlamydomonas* bind trihydroxy-α-carotene in addition to lutein as the main xanthophylls in LHCII (Knoetzel et al., 1988). Antibodies against Chl *a/b* light-harvesting proteins of plants cross-react with 11–13 thylakoid polypeptides of *Chlamydomonas* (Plumley et al., 1993), and one *Lhcb1* gene has been cloned (Imbault et al., 1988). Immunological cross-reactivity studies identified the minor CP29, CP26 and CP24 polypeptides (Bassi and Wollman, 1991). However, the LHC family seems to be larger in *Chlamydomonas* than in green plants, and 30–35 copies of at least seven polypeptides per reaction center constitute the light-harvesting antenna of *Chlamydomonas* PS I as revealed by N-terminal sequencing and immunological analysis (Bassi et al., 1992). The similarity between the Chl-proteins of *Chlamydomonas* and higher plants justifies the use of this genetically versatile alga as a model system for photosynthesis in plants.

Chl *a/b*-proteins are not unique to eukaryotic photosynthesis, as evidenced by their occurrence in photosynthetic prokaryotes. The prokaryotic Chl *a/b*-proteins of *Prochlorothrix hollandica* or *Prochloron* sp. are different from the LHCII of green plant chloroplasts (Post et al., 1993). The pigment-protein complex co-purifies not with PS II but with PS I, has a 34–35 kDa apoprotein which can be reversibly phosphorylated, shows no immunological cross-reactivity to eukaryotic antenna structures, and thus is not likely to be an ancestor of the plant LHC.

An unusual type of LHCII, which is immunochemically distinct from those of green algae and plants and which contains not only Chl *a* and *b* but also Chl *c**, was found in the Prasinophyceae *Mantoniella squamata* and *Pseudoscourfieldia marina* (Fawley, 1993; Rhiel and Mörschel, 1993). The complexes have a lower Chl *a/b*-ratio and bind more xanthophylls than the LHCII of land plants, but the number of pigment molecules bound per polypeptide is approximately the same (Fawley, 1993). Two cDNA clones of the Chl *a/b/c*-complex of *Mantoniella squamata* share 34–38% homology with LHCII polypeptides and contain three hydrophobic α-helices with the two highly conserved regions within helices I and III common to all LHC (Rhiel and Mörschel, 1993).

3. *Chromophyta*

A limited number of light-harvesting complexes has been isolated from Crysophyceae, Xanthophyceae, diatoms, brown algae, Cryptophyceae and dinoflagellates. A common characteristic is their high xanthophyll content, and apoproteins of 15–20 kDa. The antenna systems of algae that employ Chl *a* and Chl *c* as light-harvesting Chl can be divided into three subgroups: the Chl *a/c*-fucoxanthin system of diatoms and brown algae, the Chl *a/c*-peridinin system of dinoflagellates and the Chl *a/c*-phycobilin system of Cryptophyceae. Fucoxanthin and peridinin are unusual, allenic xanthophylls and their absorption maxima are red-shifted 80–100 nm when associated with specific polypeptides.

a. Diatoms and Brown Algae

The Chl *c* containing light-harvesting polypeptides of diatoms and brown algae are also related to the LHC proteins of plants. Despite differences in pigment composition, the proteins share common structural Chl-binding domains. Polyclonal antibodies against four different apoproteins of either the Chl *a/b* antennae of PS I and PS II or CP26 react with 4–6 Chl *a/c* proteins of three diatom species (Plumley et al., 1993). Three cDNA clones encoding precursors of fucoxanthin-Chl proteins of the diatom *Phaeodactylum tricornutum* showed homology to conserved parts of the LHC of plants and green algae (Grossman et al., 1990). In an attempt to isolate native light-harvesting complexes from brown algae, seven of the Chl *a/c*-fucoxanthin proteins from *Dictyota dichotoma* were found to form a supramolecular complex which seems to be firmly bound to the thylakoid lamellae (Katoh et al., 1989).

b. Dinoflagellates

Dinoflagellates contain a membrane-intrinsic Chl *a/*c_2-peridinin protein which may be analogous to the Chl *a/c*-fucoxanthin proteins of diatoms and brown algae, and a Chl *a/*c_2-xanthophyll protein that binds approximately 60% of the total Chl c_2 (Knoetzel and Rensing, 1990). The Chl c_2 is thought to mediate excitation energy transfer from the water-soluble peridinin-Chl *a*-protein associated with PS II (Koka and Song, 1977; Boczar and Prezelin, 1987). The protein sequence of a Chl *a/c* protein from the dinoflagellate *Amphidinium* shows sequence relatedness to plant Chl *a/b* proteins (Green et al., 1992)

c. Cryptophyceae

The main light-harvesting pigments of Cryptophyceae are Chl c_2 and either phycoerythrin or phycocyanin. The phycobiliproteins of *Chroomonas* sp. and *Cryptomonas maculata* have three subunits (α_1, α_2 and β) which can probably occur as homo- and/or heterodimers organized in an uncharacterized structure in the thylakoid lumen, and not into discrete structures like the phycobilisomes of red algae which are attached to the outer thylakoid surface (Hiller and Martin, 1987; Rhiel et al., 1989). A Chl *a/*c_2-xanthophyll protein is localized in both appressed and non-appressed regions of the thylakoid membrane (Vesk et al., 1992), and a partial amino acid sequence of the protein from *Cryptomonas* shows homology to plant LHC protein (Green et al., 1992).

III. Light-harvesting Proteins: Multiple Functions

Light-harvesting proteins have other functions. Carotenoids in LHC-proteins stabilize their structure and have a photoprotective role; interconversion of the three xanthophylls—violaxanthin, antheraxanthin and zeaxanthin—is related to the safe dissipation of excess excitation energy (Siefermann-Harms, 1985; Young, 1991; see chapter 30). LHCII was shown to catalyze the epoxidation of zeaxanthin to violaxanthin (Guszecki and Krupa, 1993). The fact that minor Chl *a/b*-proteins of PS II, CP24, CP26 and CP29 bind more than 80% of the violaxanthin found in the PS II antenna, taken together with the knowledge of the distribution of the Chl spectral forms within the complexes suggests that the proteins located between the PS II core antenna and LHCII function in the dissipation of excess energy (Dainese et al., 1992a, Bassi et al., 1993; Jennings et al., 1993). The *Lhca1, 3, 4* and the *Lhcb1* (and probably *Lhcb2*) proteins covalently bind DCCD, which was correlated with short-circuiting the proton-pumping activity of PS II (Jahns and Junge, 1993). This led the authors to conclude that the proteins might have a function in funneling protons from the water-splitting complex to the lumen (see chapter 10).

Acknowledgment

J. K.'s work is supported by the Deutsche Forschungsgemeinschaft. J.K. would like to thank his four-week old son Victor for helpful evening discussions.

References

Adamska I, Kloppstech K And Ohad I (1993) Early light-inducible protein in pea is stable during light stress but is degraded during recovery at low light intensity. J Biol Chem 268: 5438–5444

Allen JF (1992) Protein phosphorylation in regulation of photosynthesis. Biochim Biophys Acta 1098: 275–335

Bassi R and Wollman FA (1991) The chlorophyll-*a/b* proteins of Photosystem II in *Chlamydomonas reinhardtii.* Planta 183: 423–433

Bassi R, Rigoni R and Giacometti GM (1990) Chlorophyll binding proteins with antenna function in higher plants and green algae. Photochem Photobiol 52: 1187–1206

Bassi R, Soen SY, Frank G, Zuber H and Rochaix JD (1992) Characterization of chlorophyll *a/b* proteins of Photosystem I from *Chlamydomonas reinhardtii.* J Biol Chem 267: 25714–25736

Bassi R, Pineau B, Dainese P and Marquardt J (1993) Carotenoid-binding proteins of Photosystem II. Eur J Biochem 212: 297–303

Boczar BA and Prezelin BB (1987) Chlorophyll-protein complexes from the red-tide dinoflagellate, *Gonyaulax polyedra* Stein. Isolation, characterization, and the effect of growth irradiance on chlorophyll distribution. Plant Physiol 83: 805–812

Boekema EJ, Wynn RM and Malkin R (1990) The structure of spinach Photosystem I studied by electron microscopy. Biochim Biophys Acta 1017: 49–56

Bruce BD and Malkin R (1988) Subunit stoichiometry of the chloroplast Photosystem I complex. J Biol Chem 263: 7302–7308

Bylina EJ and Youvan DC (1988) Directed mutations affecting spectroscopic and electron transfer properties of the primary donor in the photosynthetic reaction center. Proc Nat Acad Sci 85: 7226–7230

Bylina EJ, Kirmaier C, McDowell L, Holten D and Youvan DC (1988) Influence of an amino-acid residue on the optical properties and electron transfer dynamics of a photosynthetic reaction center complex. Nature 336: 182–184

Cammarata KV, Plumley FG and Schmidt GW (1992) Pigment and protein composition of reconstituted light-harvesting complexes and effects of some protein modifications. Photosynth Res 33: 235–250

Dainese P and Bassi R (1991) Subunit stoichiometry of the chloroplast Photosystem II antenna system and aggregation state of the component chlorophyll *a/b* binding proteins. J Biol Chem 266: 8136–8142

Dainese P, Marquardt J, Pineau B and Bassi R (1992a) Identification of violaxanthin and zeaxanthin binding proteins in maize Photosystem II. In: Murata N (ed) Research in Photosynthesis, Vol I, pp 287–290. Kluwer, Dordrecht

Dainese P, Santini C, Ghiretti-Magaldi A, Marquardt J, Tidu V, Mauro S, Bergantino E and Bassi R (1992b) The organization of pigment-proteins within Photosystem II. In: Murata N (ed) Research in Photosynthesis, Vol II, pp 13–20. Kluwer, Dordrecht

Fawley MW (1993) Structure of a prasinoxanthin-chlorophyll *a/b* light-harvesting complex of the green flagellate *Pseudoscourfieldia marina* (Micromonadophyceaea) Biochim Biophys Acta 1183: 85–90

Fowler GJS, Visschers RW, Grief GG, van Grondelle R and Hunter NC (1992) Genetically modified photosynthetic antenna complexes with blueshifted absorbance bands. Nature 355: 848–850

Glazer AN (1983) Comparative biochemistry of photosynthetic light-harvesting systems. Ann Rev Biochem 52: 125–157

Golbeck JH and Bryant DA (1991) Photosystem I. Curr Top Bioenerg 16: 83–177

Green BR, Pichersky E and Kloppstech K (1991) Chlorophyll *a/b*-binding proteins: An extended family. TIBS 16: 181–186

Green BR, Durnford D, Aebersold R and Pichersky E (1992) Evolution of structure and function in the Chl *a/b* and Chl *a/c* antenna protein family. In: Murata N (ed) Research in Photosynthesis, Vol I, pp 195–202. Kluwer, Dordrecht

Grossman A, Manodori A and Snyder D (1990) Light-harvesting proteins of diatoms: Their relationship to the chlorophyll *a/b* binding proteins of higher plants and their mode of transport into chloroplasts. Mol Gen Genet 224: 91–100

Gruszecki WI and Krupa Z (1993) LHCII, the major light-harvesting pigment-protein complex is a zeaxanthin epoxidase. Biochim Biophys Acta 1144: 97–101

Harrison MA and Allen JF (1992) Protein phosphorylation and Mg^{2+} influence light harvesting and electron transport in chloroplast thylakoid membrane material containing only the chlorophyll-*a/b*-binding light-harvesting complex of Photosystem II and Photosystem I. Eur J Biochem 204: 1107–1114

Hiller RG and Martin CD (1987) Multiple forms of a type I phycoerythrin from a *Chroomonas* sp. (Cryptophyceae) varying in subunit composition. Biochim Biophys Acta 923: 98–102

Holzenburg A, Bewley MC, Wilson FH, Nicholson WV and Ford RC (1993) Three-dimensional structure of Photosystem II. Nature 363: 470–472

Holzwarth AR (1991) Structure-function relationships and energy transfer in phycobiliprotein antennae. Physiol Plant 83: 518–528

Houlné G and Schantz R (1988) Characterization of cDNA sequences for LHCI apoproteins in *Euglena gracilis*: The mRNA encodes a large precursor containing several consecutive divergent polypeptides. Mol Gen Genet 213: 479–486

Huber R (1990) A structural basis of light energy and electron transfer in biology. Eur J Biochem 187: 283–305

Ikeuchi M, Hirano A and Inoue Y (1991) Correspondence of apoproteins of light-harvesting chlorophyll *a/b* complexes associated with Photosystem I to *cab* genes: Evidence for a novel type IV apoprotein. Plant Cell Physiol 32: 103–112

Imbault P, Wittemer C, Johanningmeier U, Jacobs JD and Howell SH (1988) Structure of the *Chlamydomonas reinhardtii cabII-1* gene encoding a chlorophyll-*a/b*-binding protein. Gene 73: 397–407

Irrgang KD, Kablitz B, Vater J and Renger G (1993) Identification, isolation and partial characterisation of a 14–15 kDa pigment binding protein complex of PS II from spinach. Biochim Biophys Acta 1143: 173–182

Jahns P and Junge W (1993) Another role of chlorophyll *a/b* binding proteins of higher plants: They modulate protolytic reactions associated with Photosystem II. Photochem Photobiol 57: 120–124

Jansson S (1994) The light-harvesting chlorophyll *a/b*-binding proteins. Biochim Biophys Acta 1184: 1–19

Jansson S, Pichersky E, Bassi R, Green BR, Ikeuchi M, Melis A,

Simpson DJ, Spangfort M, Staehelin LA and Thornber JP (1992) A nomenclature for the genes encoding the chlorophyll *a/b*-binding proteins of higher plants. Plant Mol Biol 10: 242–253

Jennings RC, Bassi R, Garlaschi FM, Dainese P and Zucchelli G (1993) Distribution of the chlorophyll spectral forms of the chlorophyll-protein complexes of photosystem antenna. Biochemistry 32: 3203–3210

Katoh T, Mimuro M and Takaichi S (1989) Light-harvesting particles isolated from a brown alga, *Dictyota dichotoma*. A supramolecular assembly of fucoxanthin-chlorophyll-protein complexes. Biochim Biophys Acta 976: 233–240

Kim S, Sandusky P, Bowlby NR, Aebersold R, Green BR, Vlahakis S, Yocum CF and Pichersky E (1992) Characerization of a spinach *psbS* cDNA encoding the 22 kDa protein of Photosystem II FEBS Lett 314: 67–71

Knoetzel J and Rensing L (1990) Characterization of the photosynthetic apparatus from the marine dinoflagellate *Gonyaulax polyedra* I. Pigment and polypeptide composition of the pigment-protein complexes. J Plant Physiol 136: 271–279

Knoetzel J and Simpson DJ (1991) Expression and organisation of antenna proteins in the light- and temperature-sensitive mutant *chlorina*-^{104}. Planta 185: 111–123

Knoetzel J, Braumann T and Grimme LH (1988) Pigment-protein complexes of green algae: Improved methodological steps for the quantification of pigments in pigment-protein complexes derived from the green algae *Chlorella* and *Chlamydomonas*. J Photochem Photobiol B Biol 1: 475–491

Knoetzel J, Svendsen I and Simpson DJ (1992) Identification of the Photosystem I antenna polypeptides in barley. Isolation of three pigment binding antenna complexes. Eur J Biochem 206: 209–215

Koka P and Song PS (1977) The chromatophore topography and binding environment of peridinin-chlorophyll *a*-protein complexes from marine dinoflagellate algae. Biochim Biophys Acta 495: 220–231

Krauss N, Hinrichs W, Witt I, Fromme P, Pritzkow W, Dauter Z, Betzel C, Wilson KS, Witt HT and Saenger W (1993) Three-dimensional structure of system-I of photosynthesis at 6 Å resolution. Nature 361: 326–331

Kühlbrandt W and Wang DN (1991) Three-dimensional structure of plant light-harvesting complex determined by electron crystallography. Nature 350: 130–134

Kühlbrandt W, Wang DN and Fujiyoshi Y (1994) Atomic model of plant light-harvesting complex by electron crystallography. Nature 367: 614–621

Levy H, Gokhman I and Zamir A (1992) Regulation and light-harvesting complex II association of a *Dunaliella* protein homologous to early light-induced proteins in higher plants. J Biol Chem 267: 18831–18836

Lyon MK, Marr KM and Furcinitti PS (1993) Formation and characterization of two-dimensional crystals of Photosystem II. J Struct Biol 110: 133–140

Matthews BW, Fenna RE, Bolognesi MC, Schmidt MF and Olson JM (1979) Structure of a bacteriochlorophyll *a*-protein from the green photosynthetic bacterium *Prosthecochloris aestuarii*. J Mol Biol 131: 259–285

Melis A (1991) Dynamics of photosynthetic membrane composition and function. Biochim Biophys Acta 1058: 87–106

Meyer M and Wilhelm C (1993) Reconstitution of light-harvesting complexes from *Chlorella fusca* (Chlorophyceae) and *Mantoniella squamata* (Prasinophyceae). Z Naturforsch 48c: 461–473

Michel H and Deisenhofer J (1988) Relevance of the photosynthetic reaction center from purple bacteria to the structure of Photosystem II. Biochemistry 27: 1–7

Mörschel E (1991) The light-harvesting antennae of cyanobacteria and red algae. Photosynthetica 25: 137–144

Mullet JE, Burke JJ and Arntzen CJ (1980) Chlorophyll-proteins of Photosystem I. Plant Physiol 65: 814–822

Olive J, M'Bina I, Vernotte C, Astier C and Wollman FA (1986) Randomization of the EF particles in thylakoid membranes of *Synechocystis* 6714 upon transition from state I to state II. FEBS Lett 208: 308–312

Paulsen H and Hobe S (1992) Pigment-binding properties of mutant light-harvesting chlorophyll-*a/b*-binding protein. Eur J Biochem 205: 71–76

Peter GF and Thornber JP (1991) Biochemical composition and organization of higher plant Photosystem II light-harvesting pigment-proteins. J Biol Chem 266: 16745–16754

Piechulla B (1993) 'Circadian clock' directs the expression of plant genes. Plant Mol Biol 22: 533–542

Plumley FG and Schmidt GW (1987) Reconstitution of chlorophyll *a/b* light-harvesting complexes: Xanthophyll-dependent assembly and energy transfer. Proc Natl Acad Sci 84: 146–150

Plumley FG, Martinson TA, Herrin DL, Ikeuchi M and Schmidt GW (1993) Structural relationships of the Photosystem I and Photosystem II chlorophyll *a/b* and *a/c* light harvesting apoproteins of plants and algae. Photochem Photobiol 57: 143–151

Post AF, Ohad I, Warner KM and Bullerjahn GS (1993) Energy distribution between Photosystems I and II in the photosynthetic prokaryote *Prochlorothrix hollandica* involves a chlorophyll *a/b* antenna which associates with Photosystem I. Biochim Biophys Acta 1144: 374–384

Rhiel E and Mörschel E (1993) The atypical chlorophyll *a/b/c* light-harvesting complex of *Mantoniella squamata*: Molecular cloning and sequence analysis. Mol Gen Genet 240: 403–413

Rhiel E, Kunz J and Wehrmeyer W (1989) Immunocytochemical localization of phycoerythrin-545 and of a chlorophyll *a/c* light harvesting complex in *Cryptomonas maculata* (Cryptohyceae) Bot Acta 102: 46–53

Sauer K (1986) Photosynthetic light reactions-physical aspects. In: Staehelin LA and Arntzen CJ (eds) Encyclopedia of Plant Physiology, Vol 19, Photosynthesis III (New Series), pp 85–97. Springer, Berlin

Schiff JA, Schwartzbach SD, Osafune T and Hase E (1991) Photocontrol and processing of LHCP II apoprotein in *Euglena*: Possible role of Golgi and other cytoplasmic sites. J Photochem Photobiol B Biol 11: 219–236

Schirmer T, Bode W, Huber R, Sidler W and Zuber H (1985) X-ray crystallographic structure of the light-harvesting biliprotein C-phycocyanin from the thermophilic cyanobacterium *Mastigocladus laminosus* and its resemblance to globin structures. J Mol Biol 184: 257–277

Siefermann-Harms D (1985) Carotenoids in photosynthesis. I. Location in photosynthetic membranes and light-harvesting function. Biochim Biophys Acta 811: 325–355

Simpson DJ (1986) Freeze-fracture studies of mutant barley

chloroplast membranes. In: Staehelin LA and Arntzen CJ (eds) Encyclopedia of Plant Physiology, Vol 19, Photosynthesis III (New Series), pp 85–97. Springer, Berlin

Simpson DJ (1988) Low temperature absorption spectroscopy of barley mutants. Gaussian deconvolution and fourth derivative analysis. Carlsberg Res Commun 53: 343–356

Spangfort M and Andersson B (1989) Subpopulations of the main chlorophyll *a/b* light-harvesting complex of Photosystem II -isolation and biochemical characterization. Biochim Biophys Acta 977: 163–170

Staehelin LA and Arntzen CJ (eds) (1986) Encyclopedia of Plant Physiology, Vol 19, Photosynthesis III (New Series), Springer, Berlin

Thornber JP (1975) Chlorophyll proteins: Light-harvesting and reaction center components of plants. Annu Rev Plant Physiol 26: 127–158

Vesk M, Dwarte D, Fowler S and Hiller RG (1992) Freeze fracture immunocytochemistry of light-harvesting pigment complexes in a cryptophyte. Protoplasma 170: 166–176

Wolfe GR, Cunningham FX, Durnford D, Green BR and Gantt E (1994) Evidence for a common origin of chloroplasts with light-harvesting complexes of different pigmentation. Nature 367: 566–568

Young AJ (1991) The photoprotective role of carotenoids in higher plants. Physiol Plant 83: 702–708

Zuber H (1985) Structure and function of light-harvesting complexes and their polypeptides. Photochem Photobiol 42: 821–844

Chapter 28

The Light-Harvesting Chlorophyll *a/b*-Binding Polypeptides and Their Genes in Angiosperm and Gymnosperm Species

Eran Pichersky
Biology Department, University of Michigan, Ann Arbor, MI 48109, USA

Stefan Jansson
Department of Plant Physiology, University of Umea, S-901 87 Umea, Sweden

Summary

The Chl *a/b*-binding (CAB) polypeptides are a set of structurally and evolutionarily related proteins found in the thylakoid membranes of chloroplasts. They form supramolecular structures known as Light-Harvesting Complexes (LHCs), where light energy is absorbed and is converted into excitation energy. There are ten distinct types of CAB polypeptides, and they are encoded by nuclear genes collectively designated as *Lhc* genes. Here we describe the characteristics of each type of CAB polypeptide. We also discuss the possible evolutionary processes that gave rise to the CAB family and some of the regulatory aspects of the patterns of expression of the *Lhc* genes.

Donald R. Ort and Charles F. Yocum (eds): Oxygenic Photosynthesis: The Light Reactions, pp. 507–521.

I. Introduction

The chlorophyll *a* and *b* pigments are the primary receptors of light energy (i.e., photons) in the photosynthetic apparatus of plants. These pigments are found in the thylakoid membranes of the chloroplasts. They are held there, apparently in precise orientation (Kühlbrandt et al., 1994), by a set of proteins collectively designated as Chl *a/b*-binding (CAB) polypeptides. The CAB polypeptides bind the majority of the photosynthetic pigments of the green plant: roughly half of the Chl *a*, all of the Chl *b*, and most, if not all, of the xanthophylls (lutein, neoxanthin, violaxanthin, antheraxanthin and zeaxanthin). The CAB polypeptides are found in several parts of the photosynthetic apparatus, and singly or in various combinations with each other they form what has been termed Light-Harvesting Complexes, or LHCs. The supramolecular light-harvesting complexes are described in an earlier chapter by Simpson and Knoetzel (chapter 27). In addition, several reviews dealing with different aspects of LHCs and CAB proteins have appeared in the last few years (Peter and Thornber, 1988; Bassi et al., 1990; Green et al., 1991; Jansson, 1994). Here we give an up-to-date description of the structural details of each of the ten recognizable types of CAB polypeptides in angiosperm and gymnosperm species, and discuss the evolution and regulation of the genes encoding them.

The first CAB polypeptide was discovered more than 25 years ago (Thornber, 1975) and for the first 20 years after this discovery considerable confusion existed as to the number of CAB polypeptides, their specific locations in the photosynthetic apparatus, and their structural relatedness to each other. These controversies have been discussed in detail elsewhere (Green, 1988; Jansson 1994). The most serious problem facing workers in the field was the difficulty in separating the various CAB polypeptides while retaining the pigment molecules (in the absence of any 'functional' assay, the demonstrated ability to bind Chl *a* and *b* constitutes the only attribute which gains a protein the designation of 'CAB' polypeptide). Additional difficulties in interpreting the data were caused by the fact that 1) all CAB polypeptides proved to migrate on SDS-PAGE in the M_r range of 20–30,000, but for a given CAB polypeptide interspecific variation in apparent molecular weight exists, 2) structural similarities among CAB polypeptides led to cross-reactivity of anti-CAB antibodies with CAB polypeptides that they were not directed against and 3) there was a virtual lack of protein sequences.

These problems have largely been solved in the last few years by the extensive isolation and characterization of genes encoding CAB polypeptides, and by the concomitant refinement of protein isolation and microsequencing techniques. These investigations have shown that each CAB polypeptide is encoded by a separate nuclear gene, and they have uncovered little evidence for post-translation modification in vivo which would lead to several polypeptides being issued from a single gene (but see further discussion below). Thus, these recent findings have led to the consensus that there are at least ten distinct 'types' of CAB polypeptides in the photosynthetic apparatus of angiosperm and gymnosperm plant species (Green et al., 1991; Jansson et al., 1992; Jansson, 1994). Much less is known about the CAB polypeptides and genes from 'lower' plants, so these will only be mentioned briefly. There is also near-unanimous agreement as to the distribution of these ten polypeptides in the various LHCs. Consequently, a nomenclature for the genes encoding the CAB polypeptides has been agreed upon (Jansson et al., 1992). This nomenclature divides the genes encoding the CAB proteins into two groups. The first consists of the *Lhca* genes (*Lhca*1, *Lhca*2, *Lhca*3, *Lhca*4), encoding CAB polypeptides associated with PS I, and the second group is the *Lhcb* genes (*Lhcb*1 through *Lhcb*6), encoding CAB polypeptides primarily associated with PS II, although the LHCB1 and LHCB2 polypeptides can also associate with PS I, serving as antenna for both photosystems. An extensive list of references reporting gene sequences was given in Jansson et al. (1992) and will not be repeated here.

Since the *Lhc* genes are found in the nucleus of the cell, and thus the mRNAs made from these genes are translated in the cytosol, the proteins encoded by the nuclear *Lhc* genes each contain an N-terminal sequence, termed a 'transit sequence', which is necessary to bring about the transfer of the precursor CAB proteins into the chloroplast. After import, the transit peptide is removed. The complete sequences of the preCAB proteins have been deduced from the nucleotide sequence of the corresponding genes.

Abbreviations: CAB – chlorophyll *a/b*-binding; Chl – chlorophyll; ELIPs – early light-induced proteins; LHC – Light harvesting complex; MSR – membrane spanning region; PS – photosystem

Because of blockage, exact location of the N-termini of several of these polypeptides is not known, and therefore their exact size is not known. This fact must be borne in mind when sequence comparisons between mature CAB polypeptides are presented.

Comparisons have clearly demonstrated that the sequences of the CAB polypeptides all show significant similarity to each other in certain parts of the proteins, indicating that they all arose from a common ancestor (however, the transit peptides of the different types of CAB polypeptides and some other parts of the proteins show little or no similarity to each other, most likely reflecting the fast evolution of these parts of the proteins). Because of the common ancestry of these proteins, they have been classified as a 'protein family' and the genes encoding them as a 'gene family' (Green et al., 1991). Recently, additional proteins have been identified as evolutionarily related to the CAB polypeptides. These are the ELIPs (Green et al., 1991) and the *psb*S gene product (Kim et al., 1992; Wedel et al., 1992). At present there is scant evidence that these distant relatives of the CAB proteins are involved in light harvesting (although this possibility has not been ruled out), so they will not be discussed further here.

II. The CAB Proteins

A. General Comments on the Structure and Function of CAB Proteins

The CAB proteins bind pigment molecules and keep them in position to ensure an efficient migration of the absorbed excitation energy into the reaction center. They are also able to adjust the functional antenna size as a response to light conditions both by dissociation/association with each other, and by changing the efficiency of energy transfer via the action of the xanthophyll cycle (Yamamoto and Bassi, Chapter 30 of this volume). At present, it is relatively clear what these proteins manage to do in co-operation, but little is known about the specific functions of each individual in the family, and this will be a major task for future research in this area.

The amino acid identity between homologous CAB proteins from different plant species is usually very high—more than 90% within the angiosperms and 80–90% when angiosperm and gymnosperm sequences are compared (Jansson and Gustafsson, 1990, 1991). Most of the differences between orthologous proteins from different plants are found in the N-terminal region. The differences among the different types of CAB polypeptides are more substantial—up to 65% sequence divergence—but two regions of the protein, the first membrane-spanning region (MSR) and its immediate N-terminal area and the third MSR and its immediate N-terminal area (see next paragraph and Fig. 1) are still highly conserved. We have chosen to present a consensus protein sequence for each member (Fig. 1), which we have deduced from all reported gene sequences. It is recommended that readers interested in protein sequences from a particular plant species consult Jansson (1994).

The structure of the *Lhcb*1 gene product has recently been solved at 3.4 Å resolution (Kühlbrandt et al., 1994). As predicted (Karlin-Neumann et al., 1985), the protein has three MSRs which are in the alpha-helix configuration (Fig. 2). Eighty percent of the polypeptide, twelve Chls and 2 luteins have been fitted, and eight, or perhaps nine, of the Chl ligands have been identified. The two lutein molecules are held in place at both ends by the polypeptide, although only one of the four ligands has been unambiguously determined; the three others have been localized to a short stretch of amino acids. The number of pigment molecules in the solved structure is not consistent with biochemical data on the pigment content that indicate eight Chl *a*, six Chl *b*, two luteins, one neoxanthin and one violaxanthin molecules were bound per polypeptide. Reconstitution experiments have also shown that all the xanthophylls are required for the creation of a stable, monomeric LHCB1 pigment/protein complex (Plumley and Schmidt, 1987), indicating that there are specific binding sites on each monomer. It thus appears that some of the pigment molecules were lost during the crystallization procedure, and the position of these molecules and their ligands are unfortunately not known.

Bearing in mind that theoretical calculations have provided evidence that all the LHC proteins would have a similar folding pattern (Green et al., 1991), we have transposed these structural elements onto the sequence alignment, and thus predicted the MSRs and pigment ligands of the other nine LHC proteins, although no empirical data of this kind are presently available. We believe that this operation is, in most cases, justified, since the sequences of these elements are remarkably conserved between the different proteins (Jansson, 1994). In addition to predicting folding patterns, comparisons between the sequences

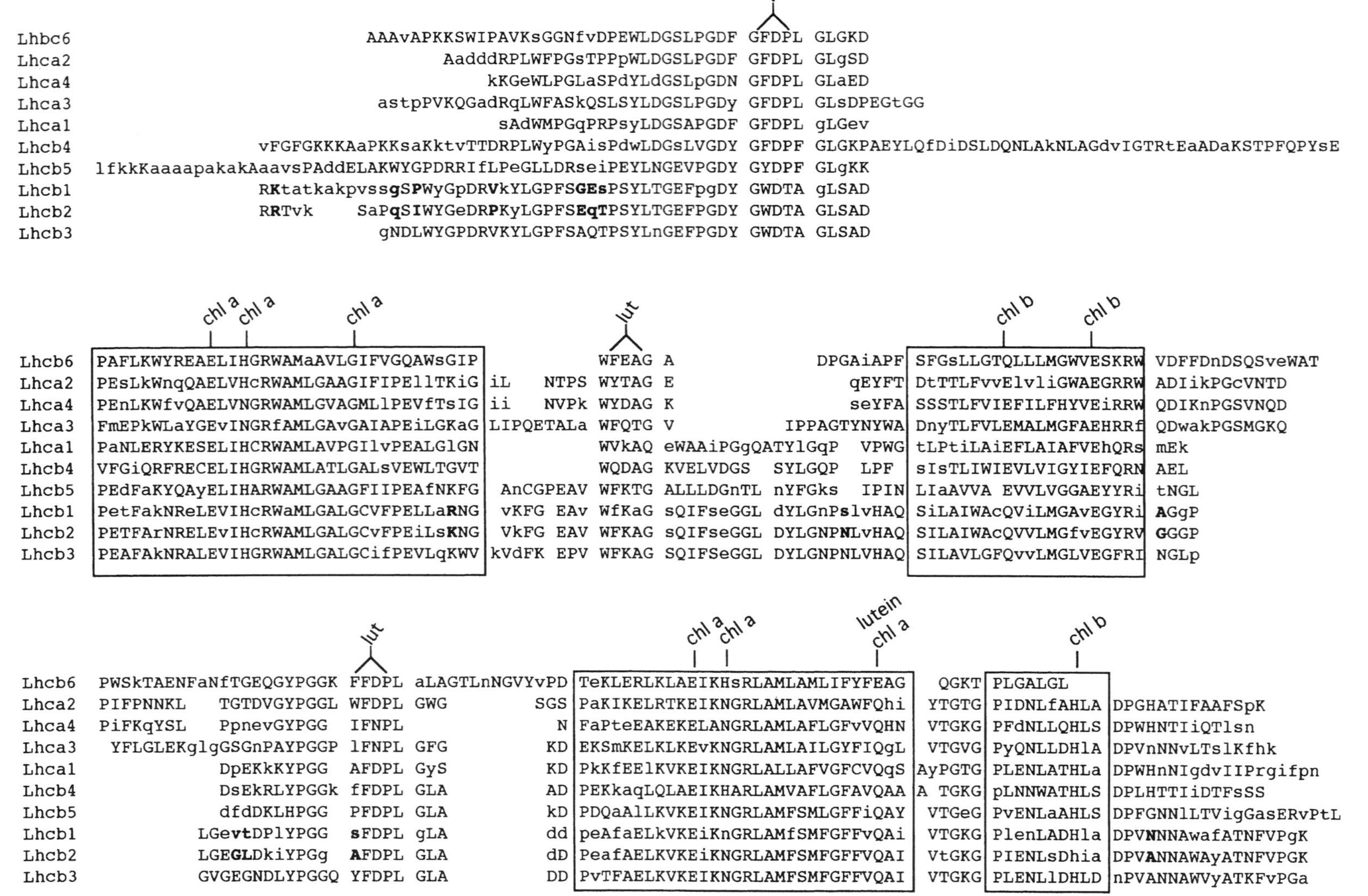

Fig. 1. Alignment of the CAB polypeptides. The actual N-termini of LHCA3, LHCB4, and LHCB5 are not known, and are thus tentative. Bold letters in the LHCB1 and LHCB2 sequences indicate residues distinctive for each polypeptide. The three MSRs, the short fourth alpha-helix (boxed areas), and the pigment ligands, which have been determined experimentally only in LHCB1, are also indicated.

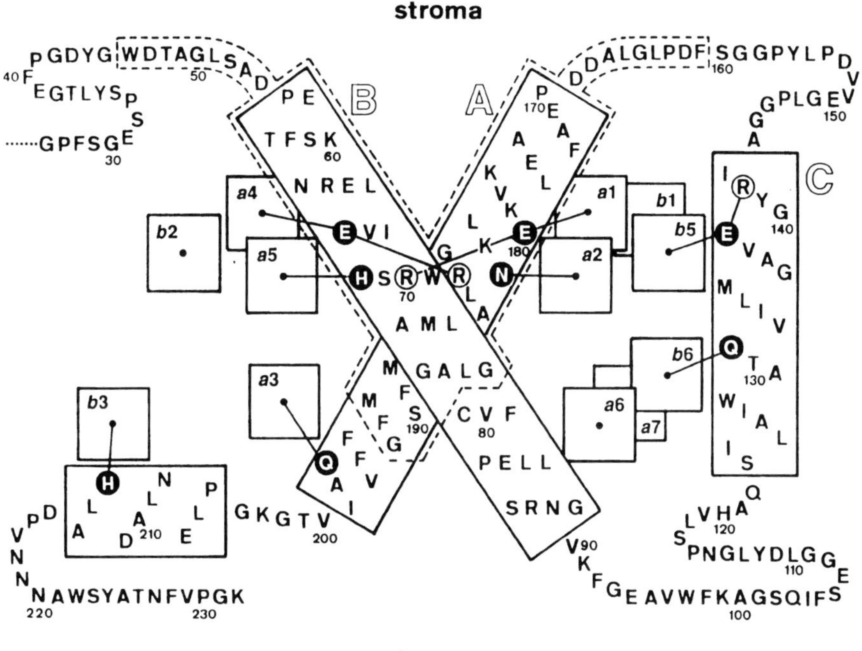

Fig. 2. A structural map of the LHCB1 protein. The three membrane-spanning alpha-helices are boxed, as is the short fourth alpha-helix on the lumen side. The Chl *a* and *b* molecules are shown as squares, and are connected to their ligands (black background) with lines. Arg (R) residues connected (with lines) to Glu (E) Chl ligands compensate for the negative charge of the Glu residues by forming ion pairs. The symmetrical regions are surrounded by a broken line (Reprinted with permission from Kühlbrandt et al., Nature 367: 614–621, copyright (1994), Macmillan Magazines Limited).

of all ten CAB proteins may yield further insights into their structure. For example, the sequence around the four lutein binding sites in LHCB1 might not immediately seem to have much in common, but if all the CAB protein sequences are compared, all four sites appear to be derivatives of a consensus sequence, WFDPL, containing one polar residue surrounded by four non-polar ones. Most common substituents are for W: G and F; for F: V, W and Y; for D: Q and K; for P: A and T; and for L: G, I and A. Since in one of these sites the lutein ligand has been identified as the polar residue Q (Fig. 1), it may be argued that the polar amino acid in the sequence motif is the lutein ligand in all sites as well.

There are four types of CAB proteins (LHCA1, LHCA2, LHCA3, LHCA4) associated with PS I, four (LHCB3, LHCB4, LHCB5, LHCB6) that are exclusively associated with PS II, and two (LHCB1, LHCB2) that may be associated with either, although they are usually thought of as constituting 'LHCII'. We begin the description of individual CAB proteins with the components of LHCI.

B. LHCA1

With 200 residues in its mature form, LHCA1 is, together with LHCA4, the shortest of the LHC proteins. Little is known about the specific functions of any of the LHCA proteins. Both at the level of overall sequence similarity (Jansson, 1994), and in the length of the loops connecting the MSRs and the position of the possible lutein-binding sites (Fig. 1), LHCA1 shows the highest similarity with LHCB4, indicating that these proteins could have homologous (but unknown) functions in PS I and PS II. It sometimes co-purifies with LHCA4 (Knoetzel et al., 1992), possibly reflecting an association of these two proteins in vivo. The unusual, strong 77 K fluorescence emission peak at 730 nm originating from the LHCA1/LHCA4 complex LHCI-730 (Bassi and Simpson, 1987) has not been attributed to specific pigment molecules bound to either of the two polypeptides. One hypothesis is that all LHCI subunits fluoresce at a long wave-length in vivo, and that absence of such emission from any sub-complexes is a consequence of pigment loss during preparation

(Ikeuchi et al., 1991). In *Chlamydomonas reinhardtii*, long wave-length fluorescence is not emitted from a single polypeptide; it is only emitted when the different LHCI subunits are assembled (Bassi et al., 1992).

Seven *Lhca*1 gene and/or cDNA sequences from five plant species—tomato, *Arabidopsis thaliana*, tobacco, barley and Scots pine—have been characterized.

C. LHCA2

The mature LHCA2 protein, which consists of 211 amino acids, can associate with LHCA3 in the LHCI-680 complex (Knoetzel et al., 1992). The protein sequences encoded by *Lhca*2, *Lhca*3 and *Lhca*4 share several features. For example, the lengths of the hydrophilic regions are very similar between the three proteins, and all have a short N-terminus and short region between MSR1 and MSR2, but a long connector between MSR2 and MSR3. If the lengths of the hydrophilic regions are considered (Fig. 1), the LHC proteins can be divided into four groups, one consisting of the LHCA1 and LHCB4 proteins, another of LHCA2, LHCA3 and LHCA4, a third group of LHCB1, LHCB2, LHCB3 and LHCB5, and the last consisting of LHCB6 alone. The four LHCA polypeptides are probably present in equimolar amounts, and both electron microscopy studies (Boekema et al., 1990) and pigment stoichiometry arguments (Jansson, 1994) indicate that there are two LHCA polypeptides of each type bound to one reaction center.

*Lhca*2 genes have been characterized from at least four plant species—tomato, petunia, Scots pine and barley.

D. LHCA3

The largest of the LHCA polypeptides, LHCA3 has an apparent molecular weight of 24–25 kDa. However, since its N-terminus is blocked by an unidentified group (Ikeuchi et al., 1991), the actual length of the mature polypeptide is not known. The sequence shown in Fig. 1 is based on the assumption that the transit peptide cleavage site is as predicted by Jansson and Gustafsson (1991). LHCA3 is composed of more basic amino acids than the other LHC proteins. It also differs from the other family members in having a six-residue insertion, G(T/P)GGF(I/M), at the beginning of the first MSR, a region which is otherwise very conserved (LHCB4 also has an insertion at the very same position, but it consists of 43 amino acids).

LHCA3 is present in the LHCI-680 pigment-protein complex (Ikeuchi et al., 1991). An oligomer of LHCA3 has been found during electrophoresis (Knoetzel et al., 1992), but it is not known whether such an oligomer occurs in vivo. Oligomeric forms of the other CAB proteins have also occasionally been found during electrophoresis. However, there are several observations which indicate that the LHCA proteins (as well as the minor LHCB proteins) are monomeric in vivo. The best evidence is probably the shape and size of the PS I complex with and without LHCA polypeptides (Boekema et al., 1990), which seems to exclude the possibility of a trimeric configuration for LHCI units (although this is the native state of LHCII, see below). Stoichiometry arguments also favor this view (Jansson, 1994).

*Lhca*3 gene sequences have been characterized from tomato, pea, Scots pine, and potato.

E. LHCA4

LHCA4 is believed to be associated with LHCA1 in the LHCI-730 complex. LHCA4 is very similar to LHCA1 both in size (199–200 amino acids) and gel mobility, but when the amino acid sequences are considered, it appears that its closest relative is LHCA2. Although the amino acid sequence identity is not remarkably higher between LHCA2 and LHCA4 than between the other LHCA proteins, the spacing between the different sequence motifs is very similar, the only difference being that LHCA4 has an extremely short sequence between the predicted lutein binding site and the third MSR. All LHCA polypeptides are stable in the absence of Chl *b*, since they are equally abundant in the chlorophyll *b*-less barley mutant chlorina-f2 (Krol et al., 1995).

*Lhca*4 genes and/or cDNAs have been characterized from tomato, *Arabidopsis thaliana*, barley, and Scots pine .

F. LHCB1

More has been written on LHCB1, the most abundant CAB protein, than about the other CAB proteins combined. The *Lhcb*1 genes and their products have been used as a model system for research into several areas of plant biology, including protein targeting into chloroplasts, light-regulated gene expression, chloroplast biogenesis, circadian rhythms, excitation balance between PS I and PS II, regulation of thylakoid ultrastructure, protein phosphorylation, and reconstitution of pigment-protein complexes. It must, however, be kept in mind that in the past, many researchers often implicitly assumed that LHCII contains the products of the *Lhcb*1 genes only, while in fact most protocols for preparation of 'LHCII' also yield not only LHCB2 as well as LHCB1, but also various amounts of LHCB3 and other LHCB proteins.

*Lhcb*1 genes have been characterized from many species, and the list of 18 species in Jansson et al. (1992) is now even longer. Multiple *Lhcb*1 genes are normally present and expressed in the same species. The products of these genes differ slightly from each other, and consequently several LHCB1 proteins from a single plant can be resolved on a high-resolution polyacrylamide gel (Bassi et al., 1990; Allen and Staehelin, 1992). The N-terminus of mature LHCB1 is blocked by an acetyl group, but tandem mass spectrometry has nevertheless identified the N-terminal amino acid as an Arg (Michel et al., 1990), yielding a mature protein of approximately 232 amino acid residues.

LHCB1 is reversibly phosphorylated at Thr3 (Michel et al., 1990). However, this amino acid is not absolutely conserved, and a nearby serine residue can also be the substrate for the kinase (Mullet, 1983). This phosphorylation has been extensively studied, since it is believed to cause the lateral movement of 'mobile LHCII' (LHCB1 and LHCB2) from PS II in the grana regions to PS I in the stroma-exposed regions of the thylakoids, thereby changing the excitation balance between PS I and PS II. It was previously believed that this process also regulated grana stacking, but since mutants lacking LHCB1 and LHCB2 are completely capable of forming normal grana stacks (Bassi et al., 1985; Murray and Kohorn, 1991), these CAB polypeptides cannot be of major importance for grana stacking.

The native form of LHCB1 and LHCB2 is the trimer (Kühlbrandt and Wang, 1991). The trimers are believed to be formed by random association of the two types of CAB proteins, and thus any possible proportion of LHCB1 and LHCB2 may be present in a given trimer (Larsson et al., 1987a, 1987b). The trimers can, however, easily dissociate during detergent treatment. The thylakoid lipid phosphatidyl glycerol (PG) containing the unusual fatty acid D^3-trans-hexadecenoic acid seems to be of particular importance for trimerization, since this fatty acid is specifically associated with trimeric LHCB1/LHCB2 (Remy et al., 1982). A *Chlamydomonas* mutant unable to synthesize this fatty acid lacks oligomeric LHCII (Garnier et al., 1990). It has recently been found that this PG is bound somewhere between amino acids 9 and 46. Removal of this region by digestion with protease completely abolishes PG binding and trimerization. However, an additional site of PG binding also occurs in the hydrophobic interior (Nussberger et al., 1993). Approximately four molecules of digalactosyl diacyl glycerol (DGDG) are also associated with each LHCB1 polypeptide, but this lipid most likely is bound to the outside of the complex and is not involved in trimerization (Nussberger et al., 1993).

Whether LHCB1 precursors can be processed at alternative sites in their amino acid sequences to give multiple mature products from a single precursor has been a matter of debate. Several in vitro studies have shown that one *Lhcb*1 gene product, when taken up by intact chloroplasts or isolated thylakoids, gave rise to polypeptides of different electrophoretic mobility (Kohorn et al., 1986; Lamppa and Abad, 1987), but many others have not detected such heterogeneity. A partially purified 'processing peptidase' has been obtained (Abad et al., 1989) which could cleave LHCB1 produced in *E. coli* at the secondary site (Abad et al., 1991). However, at present there is little evidence that such secondary processing contributes significantly to LHCB1 protein heterogeneity in vivo. Such evidence could be obtained, for example, by isolating LHCB1 proteins with the same primary sequence but with different N-termini from leaf material (but see Sigrist and Staehelin, 1994).

Reconstitution experiments with LHCB1 synthesized in *E. coli* (Plumley and Schmidt, 1987; Cammarata and Schmidt, 1992; Paulsen and Hobe, 1992) or purified from plant material (Cammarata et al., 1992) indicate that a stable pigment-protein complex cannot be formed when the C-terminus (which forms a short helix, see Fig. 2) is removed. However, an LHCB1 protein with its N-terminus and the first seven amino acids of MSR1 deleted can still reconstitute and bind apparently normal amounts of pigments (Cammarata and Schmidt, 1992; Paulsen and Hobe, 1992). This is surprising, since the crystallographic data indicate that one of the binding sites for a lutein molecule is lost in this mutant protein. The most likely interpretation of the data is that the other interactions between the lutein and the protein are strong enough to keep the pigment in place. If, however, the first Chl *a* binding site is deleted, no stable complex can be formed (Cammarata and Schmidt, 1992; Paulsen and Hobe, 1992).

G. LHCB2

LHCB2, the second component of 'LHCII', is slightly smaller than LHCB1 (approximately 228 amino acids), and is sometimes referred to as the '25 kDa LHCII protein' (Jansson et al., 1990). It is found in amounts equivalent to roughly one third of LHCB1 (Larsson et al., 1987a; Maenpaa and Andersson, 1989). The *Lhcb2* genes, formerly called Type II cab genes, are very similar to the *Lhcb1* genes, and LHCB2 mature proteins have greater than 85% sequence identity with LHCB1. They are, however, most certainly proteins of somewhat different functions, since the differences between them have been maintained throughout evolution ever since the lineages of gymnosperms and angiosperms diverged several hundred million years ago (Jansson and Gustafsson, 1990). There are 14 positions throughout the polypeptides with distinctive amino acids for each of the two types of CAB proteins (eleven on the stromal, two on the lumenal side and one within the first MSR, indicated by bold letters in Fig. 1) and it is likely that some of them contribute to the different functions of the two proteins.

At present, the only clear biochemical evidence for distinct functions for LHCB1 and LHCB2 is that phosphorylation kinetics of LHCB2 in spinach appears to be faster than that of LHCB1 (Islam, 1987; Larsson et al., 1987a), although in vitro phosphorylation experiments have shown that synthetic peptides corresponding to the LHCB1 and LHCB2 N-termini are phosphorylated at the same rate (Michel et al., 1990). The final degree of phosphorylation of the two proteins is, however, the same: one third on a molar basis (Islam, 1987). The two proteins are phosphorylated at the same residue, Thr3 (Michel et al., 1990). It is also believed that this protein is involved in the long-term acclimation to different light regimes. More specifically, the amount of this protein increases, relative to LHCB1, when the PS II antenna size is large (Larsson et al., 1987a; Maenpaa and Andersson, 1989). Phosphorylation of LHCB1 and LHCB2 regulates State 1/State 2 transitions (Mullet. 1983; Allen, 1992), which is believed to consist of a lateral movement of phosphorylated LHCB1 and LHCB2 from PS II in the grana regions into the PS I-enriched stroma-exposed regions, changing the relative antenna sizes of PS I and PS II. The phosphorylation reaction seems to be regulated by the redox state of the photosynthetic electron transport chain (Michel et al., 1990).

H. LHCB3

LHCB3 is both smaller in size (223 amino acids) and less abundant in the thylakoid membrane than either LHCB1 or LHCB2. However, the sequences are similar (greater than 70% identity), although LHCB3 is truncated at the N-terminus as compared to the other two proteins. In contrast to LHCB1 and LHCB2, LHCB3 is not phosphorylated and does not appear to leave its position at the PS II complex under any conditions. There have been conflicting reports on the native state of this protein. Some authors believe that it forms a trimer together with two LHCB1 polypeptides (Peter and Thornber, 1988), and others propose that it forms homotrimers (Dainese et al., 1992). A theory that it might be monomeric also has been put forward (Jansson, 1994).

LHCB3 has been reported to be the only LHCB protein in barley plants grown under intermittent light, suggesting that it forms the innermost antenna of PS II (Morrissey et al., 1989). However, studies using monospecific antibodies have shown that LHCB5, and not LHCB3, is the only LHCB protein present under those conditions (Krol et al., 1995). *Lhcb3* genes have been isolated from tomato, pea, barley and Brassica.

I. LHCB4

The sequence of LHCB4, the largest CAB protein, has several features in common with LHCA1 and LHCB5. Biochemical similarities between LHCB4 and LHCB5 have caused considerable confusion in the past (for discussion, see Falbel and Staehelin, 1992). The N-terminus of LHCB4 is blocked and therefore the transit peptide cleavage site is not known, although the proposal of Morishige and Thornber (1992) seems reasonable. *Lhcb4* genes were the last *Lhc* genes to be cloned, and only recently have they been isolated and characterized from barley and Arabidopsis (Morishige and Thornber, 1992; Green and Pichersky, 1993). In addition, internal protein sequences have been obtained from spinach and tomato.

The most distinctive feature of the primary structure of LHCB4 is the hydrophilic 43 amino acid insertion at the same position as the six extra amino acids of LHCA3, immediately N-terminal of the first MSR. However, the spacing between the MSRs of LHCB4 is most similar to that in LHCA1, but quite different from all the other LHC proteins.

LHCB4, LHCB5, LHCB6 and perhaps LHCB3 are thought to correspond to what has been termed the 'inner LHCII', i. e., the part of LHCII that does not disconnect from PS II after phosphorylation of 'mobile LHCII' (i.e., LHCB1 and LHCB2) (Peter and Thornber, 1991; Dainese et al., 1992; Akerlund, 1993; Jansson, 1994). Nevertheless, phosphorylation of LHCB4 has been demonstrated (Peter and Thornber, 1991), but whether it has any regulatory function remains to be demonstrated.

A protein designated CP31, which accumulates in some maize genotypes after cold stress, has been shown to be post-translational modification of LHCB4 (Hayden et al., 1988; Dainese et al., 1992). Interestingly, this protein has been reported to increase dissipation of excess energy by a zeaxanthin-independent mechanism (Dainese et al., 1992). Its accumulation correlates with the appearance of a strong 740 nm fluorescence emission peak, and this emission is likely to come from a Chl *b* molecule.

J. LHCB5

The LHCB5 protein illustrates well many of the problems related to research on the CAB proteins. It initially went undetected due to comigration with the much more abundant LHCB1 and LHCB2 proteins during gel electrophoresis. When it was finally discovered and its gene isolated, different research groups named it variously as CP29, Type I CP29, CP26 and LHCIIc. LHCB5 is N-terminally blocked and heterogeneity in apparent molecular weight is found among different plant species (Falbel and Staehelin, 1992). There is also heterogeneity within a plant, and in preparations from several species two different LHCB5 proteins separate clearly on SDS-PAGE (Bassi et al., 1990; Peter and Thornber, 1991; Allen and Staehelin, 1992). However, the three species from which *Lhcb5* genes or cDNAs have been isolated —tomato, barley, and Scots pine—have only one *Lhcb5* gene each. The mature LHCB5 protein might start close to a conserved Leu residue (Fig. 1), corresponding to a polypeptide of 245–251 amino acids, but it is also possible that the protein is 10–20 amino acids shorter.

Arvidsson et al. (1993) have reported that LHCB5 binds a Cu^{2+} ion, and they speculate that the ligand is a Cys residue located between the first MSR and the lutein binding site. They also suggest that LHCB5 might be one of the oxidases in the xanthophyll cycle.

K. LHCB6

LHCB6, more familiar as CP24 or LHCIId, is the smallest of the LHCB proteins (210 amino acids), but also the most distinctive of the CAB polypeptides. It is least related to the other proteins when amino acid sequences are compared. It also exhibits a very different pattern when the lengths of the regions connecting the MSRs are compared. It might in fact lack the short fourth helix, present on the lumenal side of other CAB polypeptides, which binds at least one Chl *b* molecule. However, the sequence LGALG, which is 100% identical to a Chl-binding site in the first MSR, is present instead at the same position and may serve the same function. The ease by which this protein loses pigments during isolation might be related to the possible absence of a fourth helix. It also appears to be unstable in the absence of Chl *b*, since it is completely absent in the Chl *b*-less barley mutant chlorina-f2 (Krol et al., 1995).

Lhcb6 DNA sequences have been characterized from tomato and spinach, and protein sequencing has identified the protein in barley.

III. The *Lhc* Genes

A. Gene Number

The first *Lhc* gene to be cloned was an *Lhcb1* gene. This was no doubt due to the presence of multiple copies in the genome and the high level of expression of this type of gene found in all plants examined; this in turn is correlated with the fact that the LHCB1 protein is the most abundant CAB polypeptide. Genes of this type have been isolated from scores of flowering plants. In cases where reliable estimates could be made, their number per haploid genome varied in angiosperms from a minimum of 5 to a maximum of 16 (McGrath et al., 1992). Not surprisingly, higher numbers were encountered in polyploid angiosperm species. In the gymnosperm Scots pine, *Pinus sylvestris*, that number was estimated to be only three.

With the exception of the *Lhcb1* genes, the other 9 types of *Lhc* genes have been extensively investigated only in tomato and Scots pine, and, to a lesser degree, in *Arabidopsis thaliana*. Nevertheless, it is clear that these genes are generally found in only one or a few copies per haploid genome. This is also roughly consistent with the observed levels of the corresponding proteins. However, it should be emphasized that there is no strict correlation of number of genes (and their level of expression, see below) with the amount of each protein observed in the cell. For example, while tomato has four *Lhcb3* genes but only two *Lhcb2* genes, the LHCB2 protein is more abundant than the LHCB3 protein under all conditions examined (Green et al., 1992).

B. Evolution of Gene Structure

Comparisons of the sequences of the ten different types of CAB proteins show significant similarity at two segments of the proteins. The first region includes the first MSR and its immediate N-terminal region, and the second highly conserved region is the third MSR and its immediate N-terminal region. These two regions together comprise about 40% of the total length of the mature polypeptide. In the other 60% of each polypeptide, the sequences may or may not show similarity to other CAB polypeptides. However, enough similarity in these regions is seen between two or more of these polypeptides to conclude that the entire set of these proteins have arisen by a process of repeated gene duplication and divergence, rather than by a process of exon shuffling. The faster-diverging regions may simply include functionally and/or structurally less essential areas, or, conversely, parts of the proteins that may have type-specific functions that have been specifically selected for that purpose.

A possible exception to the whole-gene duplication and divergence process may be the region that encodes the transit peptide in the *Lhc* genes. The transit peptides of the different CAB proteins show practically no similarity, and thus could have arisen by exon shuffling. However, it is also possible that this region has little selective constraint on its evolution, as has been shown for other transit peptides (Pichersky et al., 1986), and thus the lack of sequence similarity could simply reflect rapid divergence.

Interestingly, the two universally conserved regions of each CAB protein also show significant similarity to each other, suggesting that the prototype of the present day LHC protein was created by a head-to-tail duplication of a smaller gene (Hoffman et al., 1987; Green and Pichersky, 1994).

One of the most intriguing observations about the structure of *Lhc* genes is the location of introns. The number of introns in these genes varies from no introns in *Lhcb*1 genes to five in the *Lhcb*5, with other genes having 1, 2, 3, and 4 introns (Table 1). If the ancestral gene contained any introns, the members of the *Lhc* gene family should also contain such introns. This prediction is only partially borne out. There are only two, and possibly three, cases of shared introns, and only one of these cases clearly represents an intron ancestral to the entire family. This position is at the beginning of the first conserved region, and six different *Lhc* gene types, including the most divergent members of the family, have an intron at this position. At another position close to the beginning of the mature N-terminus coding region an intron is found in three types of CAB genes, but that region is so divergent that it is impossible to positively determine that the intron occurs at exactly the same position in all of these genes. In a third case, an intron is present in exactly the same position in both *Lhca*1 and *Lhca*4. Although sequence comparisons indicate that the two types of genes are *not* the most closely related (Jansson, 1994), there is a high level of uncertainty in these comparisons because of the high level of sequence divergence. Thus, it may still be that the two genes are the result of a gene duplication relatively later in the history of the *Lhc* gene family, and that the shared intron must have been present in their shared ancestor.

While a multitude of other introns occur in the *Lhc* genes in addition to those just mentioned, none is shared by different types of *Lhc* genes. The conclusion is inescapable that most of these introns were added later in the evolution of this gene family. This is also the most likely explanation for the shared intron between *Lhca*1 and *Lhca*4 mentioned above, which must have been inserted in their shared ancestral gene after it diverged from the rest of the family. The alternative would be that an intron in this position was ancestral to the entire family, survived until the *Lhca*1/4 ancestor diverged from the rest of the family, and then all other members of the family lost this intron independently, a not very parsimonious explanation.

On the other hand, intron loss clearly has occurred in the family. For example, this is the most parsimonious explanation for the lack of an intron in

Table 1. Introns in the *Lhc* genes

Gene	Number of introns	Contains a shared intron (see text)
*Lhca*1	3	+ (+ a 2nd shared intron)
*Lhca*2	4	+
*Lhca*3	2	+
*Lhca*4	2	+ (+ a 2nd shared intron)
*Lhcb*1	0	-
*Lhcb*2	1	-
*Lhcb*3	2	-
*Lhcb*4	1	-
*Lhcb*5	5	+
*Lhcb*6	2	+

the four types of genes that do not have the intron shared by six different *Lhc* genes. Three of the genes lacking this intron - *Lhcb*1, *Lhcb*2 and *Lhcb*3 - are closely related to each other on the gene family 'tree', and the fourth one, *Lhcb*4, may also be closely related to them (Jansson, 1994). Thus, a maximum of two independent losses (and possibly only one) may have been involved.

C. Regulation of Gene Expression

Since the *Lhc* genes were among the first plant genes to be isolated, it is not surprising that they were also among the first to be examined for their expression characteristics. By now a vast literature, which cannot be wholly cited here, exists as to the pattern and level of expression of these genes under a multitude of environmental conditions and stimuli, at various stages of development, and in the different organs and tissues of the plant. Without devoting a whole review to this topic, the highlights of *Lhc* gene regulation are listed below.

The *Lhc* genes are naturally expressed in photosynthetic tissue. This means that they are expressed not only in leaves, but also in other green tissues such as stems, sepals, and the green fruits of tomatoes, for example (the great majority of the Chl pigments in the cell are associated with the CAB proteins, thus the green appearance of a tissue is by itself a good indication of the expression of the *Lhc* genes).

Earliest investigations of the regulation of the *Lhc* genes were focused on the role of light. The embryos of angiosperms are not green, and germination and initial growth of the seedling occur in the soil under relative darkness. Later, when the seedling breaks through the soil, a 'greening' process takes place until the photosynthetic apparatus is completely formed and the plant becomes fully autotrophic. During this process, the *Lhc* genes go from being completely quiescent in the embryo to being highly transcribed in the developing young plant.

Since germination and growth of the seedling is known to be controlled by the phytochrome pathway, the effect of various types of light - red, far red, and blue - on the expression of the *Lhc* genes has been examined at some length. However, most of these investigations involved only one type of *Lhc* genes, the *Lhcb*1 genes, which, as will be remembered, are found in multiple copies in all plant species. The results can be briefly summarized as consistent with the hypothesis that the expression of the *Lhcb*1 genes is positively regulated by the phytochrome pathway. Thus, a short pulse of red light will activate *Lhcb*1 gene transcription, but a pulse of far-red light will reverse this process (Tobin and Silverthorne, 1985). However, details of the mechanism(s) of this phenomenon await further elucidation of the phytochrome signal transduction pathway itself. It has also been shown that some *Lhcb*1 genes could be activated by blue light (Palomares et al., 1991). Why there should be a difference in response to light conditions among a subset of *Lhc* genes which encode essentially identical proteins is not clear.

In conifers, the embryo has the capacity of forming Chl *a* and *b* in the dark during the early stages of development (Canovas et al., 1993). The *Lhcb1* and *Lhcb2* genes are highly expressed during these stages (Alosi and Neale, 1991), whereas their expression declines rapidly as the capacity for Chl synthesis goes down (Gustafsson et al., 1991). This indicates that Chl synthesis rather than light *per se* is a major determinant for *Lhc* gene expression in conifers, although phytochrome is also involved (Alosi and Neale, 1991). Because of the experimental difficulties of separating Chl synthesis and light stimulus in angiosperms, it is not known if conifers and angiosperms differ in this respect. In *Ginko biloba*, a gymnosperm without the capacity to synthesize Chl in the dark, Lhc gene expression does not occur in the dark (Chinn and Silverthorne, 1993), suggesting that Chl synthesis might be of fundamental importance to *Lhc* gene expression.

Extensive studies of the mode of regulation of the entire set of *Lhc* genes have been carried out only in the plant *Lycopersicon esculentum* (tomato) (Kellman et al., 1990, 1993; Piechulla et al., 1991; Piechulla, 1993). There it has been found that all the members of the family are expressed in all green tissues. In the primary photosynthetic tissue, the leaf, the steady-state levels of mRNA issuing from each of these genes were quite unequal. As expected, levels of mRNA from some *Lhcb*1 genes were 5–10 fold higher than mRNAs from Lhcb2 or *Lhcb*3 genes, and the levels of *Lhca* mRNA were even lower. However, mRNAs from some *Lhcb*1 genes (there are eight in tomato) were quite low too, comparable to that of *Lhcb*2 genes, for example.

These results are generally consistent with the observations that in the leaf, as well as in all photosynthetic tissues, all ten different types of CAB proteins are always found, and that the LHCB1 proteins are always the most abundant CAB proteins in the cell. However, data concerning both the absolute amounts of the proteins issuing from each of the *Lhc* genes, and the fraction of the total contributed by each of the *Lhc* genes are not yet well quantified. Thus, it is not yet possible to make any clear statement concerning the relationships between mRNA levels and protein abundance.

An interesting question that remains to be investigated is whether some growing conditions exist under which the relative proportion of the different CAB proteins changes, and whether under such conditions the level of transcription of the different *Lhc* genes changes as well. During leaf greening, the LHCB proteins accumulate before the LHCA proteins in both angiosperms (Anandan et al., 1993) and gymnosperms (Canovas et al., 1993). It is well established that both the Chl *a/b* ratio and the ratio between PS I and PS II can vary with growth conditions in fully green plants, but how such changes are reflected at the level of CAB polypeptide composition has not been adequately investigated. The only direct evidence for changes in CAB protein composition of green leaves is that the amounts of the LHCB1 and LHCB2 are higher in plants grown under low light (Larsson et al., 1987b). In most models, it is supposed that the other CAB proteins are constitutively bound in a fixed amount to the reaction center (Jansson, 1994), although this assumption has not been definitely proven.

The investigations in tomato have focused so far on examining the mRNA levels of the different *Lhc* genes in different organs, and under the relatively fixed conditions of these experiments no change in *relative* level of expression has been found. Thus, it has been found, for example, that the *Lhcb*1A gene is not as highly expressed in stems as it is in leaf tissue, but all the other genes also show the same relative reduction in their expression in the stem (Kellmann et al., 1990, 1993; Piechulla et al., 1991). This picture is, with few exceptions, true in the other green organs as well. Thus, while evidence suggests that the *Lhc* genes are co-ordinately regulated at the mRNA level, the mechanisms of this co-ordinate expression are at present little understood and the effects on the CAB protein composition remains to be investigated.

Another interesting phenomenon which the *Lhc* genes exhibit, and for which no mechanism has yet been found, is the coordinate *circadian* rhythm of *Lhc* expression. It has been shown by several researchers (e.g., Kloppstech, 1985), but most extensively by Piechulla and co-workers (1993), that all the *Lhc* genes show a light-entrained, circadian mode of regulation. Levels of expression are highest around mid-day, and decrease to little or no expression in the dark (however, the general level of *Lhc* gene expression in mature leaves is only a small percentage of what it is in developing leaves). Transcription begins to pick up again early the next day, even before illumination begins. If illumination is completely blocked, this oscillating pattern will be maintained in the dark for several days, although the amplitude of the transcription level will gradually decrease. The biological significance of this phenomenon is not yet clear. It is easy to rationalize that CAB proteins are constantly turned over, and the peak levels of gene transcription correspond to peak levels of demand for new CAB proteins. However, the kinetics of CAB protein turnover has been little studied and much remains to be elucidated before firm conclusions can be drawn.

Acknowledgments

The work described in this article from the authors' labs was funded by grants from the USA Department of Agriculture and the Swedish Council for Forestry and Agricultural Research. We thank Dr. Kühlbrandt for permission to reproduce Fig. 2.

References

Abad MS, Clark SE and Lamppa GK (1989) Properties of a chloroplast enzyme that cleaves the chlorophyll *a/b* binding protein precursor. Plant Physiol 90: 117–124

Abad MS, Oblong JE and Lamppa GK (1991) Soluble chloroplast

enzyme cleaves preLHCP made in *Escherichia coli* to a mature form lacking a basic N-terminal domain. Plant Physiol 96: 1220–1227

Åkerlund HE (1993) Function and organization of photosystem II. In: Sundqvist C and Ryberg M (eds): Pigment-protein Complexes in Plastids: Synthesis and Assembly. Cell Biology: A series of Monographs, pp 419–446. Academic Press, Inc., San Diego

Allen JF (1992) Protein phosphorylation in regulation of photosynthesis. Biochim Biophys Acta 1098: 275–335

Allen KD and Staehelin LA (1992) Biochemical characterization of photosystem II antenna polypeptides in grana and stroma membranes of spinach. Plant Physiol 100: 1517–1526

Alosi MC and Neale DB (1991) Light- and phytochrome-mediated gene expression in Douglas fir seedlings. Physiol Plant 86: 71–78

Anandan S, Morishige DT and Thornber JP (1993) Light-induced biogenesis of light-harvesting complex I (LHC I) during chloroplast development in barley (*Hordeum vulgare*). Plant Physiol 101: 227–236

Arvidsson PO, Bratt CE, Andreasson LE and Akerlund HE (1993) The 28-kDa apoprotein of CP-26 in PS-II binds copper. Photosynth Res 37: 217–225

Bassi R and Simpson D (1987) Chlorophyll-protein complexes of barley photosystem I. Eur J Biochem 163: 221–230

Bassi R, Hinz U and Barbato R (1985) The role of the light harvesting complex and PS II in thylakoid stacking in the chlorina-f2 barley mutant. Carlsberg Res Comm 50: 347–367

Bassi R, Rigoni F and Giacometti GM (1990) Chlorophyll binding proteins with antenna function in higher plants and green algae. Photochem Photobiol 52: 1187–1206

Bassi R, Soen SY, Frank G, Zuber H and Rochaix J-D (1992) Characterization of chlorophyll *a/b* proteins of Photosystem I from *Chlamydomonas reinhardtii*. J Biol Chem 267: 25714–25736

Boekema EJ, Wynn RM and Malkin R (1990) The structure of spinach photosystem I studied by electron microscopy. Biochim Biophys Acta 1017: 49–56

Cammarata KV and Schmidt GW (1992) In vitro reconstitution of a light-harvesting gene product: Deletion mutagenesis and analyses of pigment binding. Biochemistry 31: 2779–2789

Cammarata KV, Plumley FG and Schmidt GW (1992) Pigment and protein composition of reconstituted light-harvesting complexes and effects of some protein modifications. Photosynth Res 33: 235–250

Canovas F, McLarney B and Silverthorn J (1993) Light-independent synthesis of LHCIIb polypeptides and assembly of the major pigmented complexes during the initial stages of *Pinus Palustris* seedling development. Photosynth Res 38: 89–97

Chinn E and Silverthorne J (1993) Light-dependent chloroplast development and expression of a light-harvesting chlorophyll *a/b*-binding protein in the gymnosperm *Ginko biloba*. Plant Physiol 103: 727–732

Dainese P, Santini C, Ghiretti-Magaldi A, Marquardt J, Tidu V, Mauro S, Bergantino E and Bassi R (1992) The Organization of pigment-proteins within photosystem II. In: Murata M (ed) Research in Photosynthesis, Vol II, pp 13–20. Kluwer Academic Publishers, Dordrecht

Falbel TG and Staehelin LA (1992) Species-related differences in the electrophoretic behaviour of CP29 and CP26: An immunochemical analysis. Photosynth Res 34: 249–262

Garnier J, Wu B, Maroc J, Guyon D and Tremolieres A (1990) Restoration of both an oligomeric form of the light-harvesting antenna CP-II and a fluorescence state II/state I transition by D^3-trans-hexadecenoic acid-containing phosphatidyl glycerol in cells of a mutant of *Chlamydomonas reinhardtii*. Biochim Biophys Acta 1020: 153–162

Green BR (1988) The chlorophyll-protein complexes of higher plant photosynthetic membranes, or Just what green band is that? Photosynth Res 15: 3–23

Green BR and Pichersky E (1993) Nucleotide sequence of an *Arabidopsis thaliana Lhcb4* gene. Plant Physiol 103: 1451–1452

Green BR and Pichersky E (1994) Hypothesis for the evolution of three-helix Chl *a/b* and Chl *a/c* light-harvesting antenna proteins from two-helix and four-helix ancestors. Photosynth Res 39: 149–162

Green BR, Pichersky E and Kloppstech K (1991) Chlorophyll *a/b* binding proteins: An extended family. Trends Biochem Sci 16: 181–186

Green BR, Shen D, Aebersold R and Pichersky E (1992) Identification of the polypeptides of the major light-harvesting complex of photosystem II (LHCII) with their genes in tomato. FEBS Lett 305: 18–22

Gustafsson P, Jansson S, Lidholm J and Lundberg AK (1991) Structure and regulation of photosynthetic genes of *Pinus sylvestris* (Scots pine) and *Pinus contorta* (lodgepole pine). Forest Ecol Manag 43: 287–300

Hayden DB, Covello PS and Baker NR (1988) Characterization of a 31 kDa polypeptide that accumulates in the light-harvesting apparatus of maize leaves during chilling. Photosynth Res 15: 257–270

Hoffman NE, Pichersky E, Malik VS, Castresana C, Ko K, Darr SC and Cashmore AR (1987) A cDNA clone encoding a photosystem I protein with homology to photosystem II chlorophyll *a/b*-binding polypeptides. Proc Natl Acad Sci USA 84: 8844–8848

Ikeuchi M, Hirano A and Inoue Y (1991) Correspondence of apoproteins of light-harvesting chlorophyll *a/b* complexes associated with photosystem I to *cab* genes: Evidence for a novel Type IV apoprotein. Plant Cell Physiol 32: 103–112

Islam K (1987) The rate and extent of phosphorylation of the two light-harvesting chlorophyll *a/b* binding protein complex (LHC-II) polypeptides in isolated spinach thylakoids. Biochim Biophys Acta 893: 333–341

Jansson S (1994) The light-harvesting chlorophyll *a/b*-binding proteins. Biochim Biophys Acta 1184: 1–19

Jansson S and Gustafsson P (1990) Type I and Type II genes for the chlorophyll *a/b*-binding protein in the gymnosperm *Pinus sylvestris* (Scots pine): cDNA cloning and sequence analysis. Plant Mol Biol 14: 287–296

Jansson S and Gustafsson P (1991) Evolutionary conservation of the chlorophyll *a/b*-binding proteins: cDNAs encoding Type I, II and III LHC I polypeptides from the gymnosperm Scots pine. Mol Gen Genet 229: 67–76

Jansson S, Selstam E and Gustafsson P (1990) The rapidly phosphorylated 25 kDa polypeptide of the light-harvesting complex of photosystem II is encoded by the Type II *cab-II* genes. Biochim Biophys Acta 1019: 110–114

Jansson S, Pichersky E, Bassi R, Green BR, Ikeuchi M, Melis A, Simpson DJ, Spangfort M, Staehelin LA and Thornber JP

(1992) A nomenclature for the genes encoding the chlorophyll *a/b*-binding proteins of higher plants. Plant Mol Biol Rep 10: 242–253

Karlin-Neumann GA, Kohorn BD, Thornber JP and Tobin EM (1985) A chlorophyll *a/b*-protein encoded by a gene containing an intron with characteristics of a transposable element. J Mol Appl Genet 3: 45–61

Kellmann JW, Pichersky E and Piechulla B (1990) Analysis of diurnal expression patterns of the tomato chlorophyll *a/b* binding protein genes. Influence of light and characterization of the gene family. Photochem Photobiol 52: 35–41

Kellmann JW, Merforth N, Weise M, Pichersky E and Piechulla B (1993) Concerted circadian oscillation in transcript levels of nineteen *Lha/b* (*cab*) genes in *Lycopersicon esculentum* (tomato). Mol General Genet 237: 439–448

Kim S, Sandusky P, Bowlby NR, Aebersold R, Green BR, Vlahakis S, Yocum CF and Pichersky E (1992) Characterization of a spinach *psbS* cDNA encoding the 22 kDa protein of photosystem II. FEBS Lett 314: 67–71

Kloppstech K (1985) Diurnal and circadian rhythmicity in the expression of light-induced plant nuclear messenger RNAs. Planta 165: 502–506

Knoetzel J, Svendsen I and Simpson DJ (1992) Identification of the photosystem I antenna polypeptides in barley. Isolation of three pigment binding antenna complexes. Eur J Biochem 206: 209–215

Kohorn BD, Harel E, Chitnis PR, Thornber JP and Tobin EM (1986) Functional and mutational analysis of the light-harvesting chlorophyll *a/b* protein of thylakoid membranes. J Cell Biol 102: 972–981

Krol M, Spangfort MD, Huner NPA, Öquist G, Gustafsson P and Jansson S (1995) Chlorophyll *a/b*-binding proteins, pigment conversions, and early light-induced proteins in a Chl *b*-less barley mutant. Plant Physiol 107: 873–883

Kühlbrandt W and Wang DN (1991) Three-dimensional structure of plant light-harvesting complex determined by electron crystallography. Nature 350: 130–134

Kühlbrandt W, Wang DN and Fujiyoshi Y (1994) Atomic model of plant light-harvesting complex. Nature 367: 614–621

Lamppa GK and Abad MS (1987) Processing of a wheat light-harvesting chlorophyll *a/b* protein precursor by a soluble enzyme from higher plant chloroplasts. J Cell Biol 105: 2641–2648

Larsson UK, Anderson JM and Andersson B (1987a) Variations in the relative content of the peripheral and inner light-harvesting chlorophyll *a/b*-protein complex (LHC II) subpopulations during thylakoid light adaptation and development. Biochim Biophys Acta 894: 69–75

Larsson UK, Sundby C and Andersson B (1987b) Characterization of two different subpopulations of spinach light-harvesting chlorophyll *a/b*-protein complex (LHC II): Polypeptide composition, phosphorylation pattern and association with photosystem II. Biochim Biophys Acta 894: 59–68

Maenpaa P and Andersson B (1989) Photosystem II heterogeneity and long-term acclimation of light-harvesting. Z Naturforsch 44C: 403–406

McGrath JM, Terzaghi WB, Sridhar P, Cashmore AR and Pichersky E (1992) Sequence of the fourth and fifth photosystem II type I chlorophyll *a/b*-binding protein genes of *Arabidopsis thaliana* and evidence for the presence of a full complement of the extended CAB gene family. Plant Mol Biol 19: 725–733

Michel HP, Buvinger WE and Bennett J (1990) Redox control and sequence specificity of a thylakoid protein kinase. In: Baltscheffsky, M (ed) Current Research in Photosynthesis, Vol II, pp 747–753. Kluwer Academic Publishers, Dordrecht

Morishige DT and Thornber JP (1992) Identification and analysis of a barley cDNA clone encoding the 31-kilodalton LHCIIa (CP29) apoprotein of the light-harvesting antenna complex of photosystem II. Plant Physiol 98: 238–245

Morrissey PJ, Glick RE and Melis A (1989) Supramolecular assembly and function of subunits associated with the chlorophyll *a-b* light-harvesting complex II (LHCII) in soybean chloroplasts. Plant Cell Physiol 30: 335–344

Mullet JE (1983) The amino acid sequence of the polypeptide segment which regulates membrane adhesion (grana stacking) in chloroplasts. J Biol Chem 258: 9941–9948

Murray DL and Kohorn BD (1991) Chloroplasts of *Arabidopsis thaliana* homozygous for the ch-1 locus lack chlorophyll *b*, lack stable LHCPII and have stacked thylakoids. Plant Mol Biol 16: 71–79

Nussberger S, Dorr K, Wang DN and Kühlbrandt W (1993) Lipid-protein interactions in crystals of plant light-harvesting complex. J Mol Biol 234: 347–356

Palomares R, Herrmann RG and Oelmuller R (1991) Different blue-light requirement for the accumulation of transcripts from nuclear genes for thylakoid proteins in *Nicotiana tabacum* and *Lycopersicon esculentum*. J Photochem Photobiol B: Biol 11: 151–162

Paulsen H and Hobe S (1992) Pigment-binding properties of mutant light-harvesting chlorophyll-*a/b*-binding protein. Eur J Biochem 205: 71–76

Peter G and Thornber JP (1988) The antenna components of photosystem II with emphasis on the major pigment-protein, LHCIIb. In: Scheer H and Schneider (eds) Photosynthetic light-harvesting systems, pp 175–186. Walter de Gruyter and Co, Berlin New York

Peter GF and Thornber JP (1991) Biochemical composition and organization of higher plant photosystem II light-harvesting pigment proteins. J Biol Chem 266: 16745–16754

Pichersky E, Bernatzky R, Tanksley SD and Cashmore AR (1986) Evidence for selection as a mechanism in the concerted evolution of *Lycopersicon esculentum* (tomato) genes encoding the small subunit of ribulose-1,5-bisphosphate carboxylase/oxygenase. Proc Natl Acad Sci USA 83: 3880–3884

Piechulla B (1993) 'Circadian clock' directs the expression of plant genes. Plant Mol Biol 22: 533–542

Piechulla B, Kellmann J-W, Pichersky E, Schwartz E and Forster HH (1991) Determination of steady-state mRNA levels of individual chlorophyll *a/b* binding protein genes of the tomato *cab* gene family. Mol Gen Genet 230: 413–422

Plumley FG and Schmidt GW (1987) Reconstitution of chlorophyll *a/b* light-harvesting complexes: Xanthophyll-dependent assembly and energy transfer. Proc Natl Acad Sci USA 84: 146–150

Remy R, Tremolieres A, Duval JC, Ambard-Bretteville F and Dubacq JP (1982) Study of the supramolecular organization of light-harvesting chlorophyll protein (LHCP). FEBS Lett 137: 271–275

Sigrist M and Staehelin LA (1994) Appearance of Type 1, 2 and 3 Light-Harvesting Complex II and Light-Harvesting Complex I proteins during light-induced greening of barley (*Hordeum vulgare*) etioplast. Plant Physiol 104: 135–145

Thornber JP (1975) Chlorophyll-proteins: Light-harvesting and reaction center components of plants. Annu Rev Plant Physiol 26: 127–158

Tobin EM and Silverthorne J (1985) Light regulation of gene expression in higher plants. Annu Rev Plant Physiol 36: 569–593

Wedel N, Klein R, Ljungberg U, Andersson B and Herrmann RG (1992) The single-copy gene *psbS* codes for a phylogenetically intriguing 22 kDa polypeptide of photosystem II. FEBS Lett 314: 61–66

Chapter 29

Excitation Energy Transfer: Functional and Dynamic Aspects of *Lhc* (*cab*) Proteins

Anastasios Melis
Department of Plant Biology, 411 Koshland Hall, University of California, Berkeley, CA 94720, USA

Summary

Photosynthetic light-harvesting complexes have been the subject of extensive research because of interest in their primary function, namely the absorption of light and the transfer of excitation energy to a photochemical reaction center. As a result, a great deal of information has accumulated on the pigment co-factors, structure and folding of the corresponding proteins, molecular and supramolecular organization in the thylakoid membrane, and on their association with the photosystems. In addition to the structural-functional and assembly characteristics, this chapter reviews dynamic aspects of the light-harvesting complexes including the modulation of the light-harvesting antenna size by irradiance, the role of light-harvesting complexes in the regulation of excitation energy distribution between the two photosystems and the role they play as signal receptors for photosystem ratio adjustment in chloroplasts. Finally, the chapter examines aspects of exciton interactions between carotenoids and chlorophylls in the *Lhc* proteins and presents highlights of excitation transfer dynamics in the pigment bed of light-harvesting complexes.

Donald R. Ort and Charles F. Yocum (eds): Oxygenic Photosynthesis: The Light Reactions, pp. 523–538.

I. Introduction

In oxygenic photosynthesis, electron transport for the generation of reductant, in the form of reduced ferredoxin and NADPH, and for the formation of ATP occurs in the thylakoid membrane of chloroplasts. The overall process is highly endergonic and it requires the input of light energy. The absorption of light by photosynthetic pigments, the transfer of excitation to a photochemical reaction center molecule, and the conversion of excitation energy to chemical energy takes place in Photosystem I (PS I) and Photosystem II (PS II) in the thylakoid membrane.

Excitation energy in PS II is trapped by the photochemical reaction center chlorophyll P680 where it causes a direct endergonic charge separation between P680 and the primary electron acceptor pheophytin ($P680^{*}{\cdot}Ph \rightarrow P680^{+}{\cdot}Ph^{-}$) (see Chapters 11 and 12). This photochemical reaction generates a strong oxidant ($P680^{+}$), thermodynamically capable of driving the oxidation of H_2O molecules (Chapters 8, 9, 12).

Excitation energy in PS I is trapped by the photochemical reaction center P700, causing a direct endergonic charge separation between P700 and the primary electron acceptor chlorophyll molecule ($P700^{*}{\cdot}Chl \rightarrow P700^{+}{\cdot}Chl^{-}$) (see Chapter 16). The reduced primary electron acceptor (Chl^{-}) is a strong reductant, thermodynamically capable of transferring electrons to low-potential iron-sulfur centers in PS I, leading to the reduction of ferredoxin and $NADP^{+}$.

The efficiency of the overall process, including the transfer of excitation to a photochemical reaction center molecule and the conversion of excitation energy to chemical energy is close to 100%, consistent with the overall high quantum efficiency of photosynthesis.

Under physiological plant growth conditions, the frequency of the charge separation reactions, thus the rate of photosynthesis, depends on the rate of light absorption by a photosystem. In turn, this parameter depends on the incident light intensity and on the effective absorption cross section of the photosynthetic pigments. The intensity of photosynthetically active radiation (PAR) at the surface of the earth ranges from 0–2,000 μmol photons$\cdot m^{-2}\cdot s^{-1}$. It has been estimated that, at the maximum solar intensity of 2,000 μmol photons$\cdot m^{-2}\cdot s^{-1}$, a single chlorophyll *a* molecule will absorb approximately 10 photons per second. Obviously, this rate will be strongly attenuated under conditions of less-than-bright sunlight.

If photochemistry at PS I and PS II were to depend on the absorption of light by a single Chl molecule (e.g., the reaction center Chl molecule), then it is clear that 10 photons per second would not be able to sustain significant rates of photosynthesis. Nature, however, solved this potential limitation by organizing chlorophylls and other accessory pigments to act cooperatively in the absorption of incoming electromagnetic radiation (Emerson and Arnold, 1932a,b). The concept of the photosynthetic unit, proposed by Gaffron and Wohl (1936), postulated that distinct assemblies of photosynthetic pigment molecules serve as antennae for the collection of light and as a conducting medium for excitation migration toward a photochemical reaction center. It is now recognized that distinct chlorophyll-protein complexes are associated with the reaction center complexes of PS I and PS II and perform the functions of light absorption and excitation energy transfer to the respective photochemical reaction centers (see Chapters 27 and 28).

A. Photosystem II

Associated with the D1/D2 PS II reaction center heterodimer (the chloroplast *psbA*/*psbD* gene products (Nanba and Satoh, 1987)) are two 'core antenna' chlorophyll-proteins known as CP47 and CP43 (the organelle *psbB* and *psbC* gene products, respectively) (Green and Camm, 1981; Bricker, 1990). Together, the PS II reaction center complex, CP47 and CP43 constitute the PS II-core and they collectively bind about 37 Chl *a* molecules (Glick and Melis, 1988).

Light-harvesting in PS II is further aided by the auxiliary chlorophyll *a-b* light harvesting complex (LHC-II). The LHC-II was discovered more than 25 years ago as a major Chl containing protein band resolved in SDS-PAGE of non-thermally denatured proteins from higher plant chloroplasts (Thornber et al., 1967). Since then, considerable information has

Abbreviations: CP – chlorophyll-protein; D1 – the 32 kDa reaction center protein of Photosystem II coded by the chloroplast *psbA* gene; D2 – the 34 kDa reaction center protein of Photosystem II coded by the chloroplast *psbD* gene; LHC – the chlorophyll *a-b*-binding light-harvesting complex; P680 – the photochemical reaction center chlorophyll of Photosystem II; P700, – the photochemical reaction center chlorophyll of Photosystem I; PS – photosystem; Q_A – the primary quinone electron acceptor of Photosystem II; RC – reaction center;

accumulated on LHC-II in terms of its structure and function in the thylakoid membrane (Thornber, 1986). It was demonstrated that LHC-II contains about an equal number of Chl *a* and Chl *b* molecules (Chl *a*/Chl *b* = 1.17). When denatured, proteins of the LHC-II migrate in the 21–30 kDa region in SDS-PAGE and may contain up to six distinct subunit polypeptides (Darr et al., 1986; Peter and Thornber, 1991; Harrison and Melis, 1992; Jansson et al., 1992). Polypeptides of the LHC-II are coded for by the nuclear *Lhcb* gene family (see Chapters 27 and 28). Each *Lhcb* protein is believed to bind 13–14 Chl *a* and Chl *b* molecules (Thornber et al., 1988; Morrissey et al., 1989; Kühlbrandt et al., 1991, 1994). Collectively, the LHC-II macrocomplex may contain over 250 Chl *a* and Chl *b* molecules in the fully assembled light-harvesting antenna of PS II in chloroplasts (Melis and Anderson, 1983; Glazer and Melis, 1987).

B. Photosystem I

The photochemical apparatus of PS I contains the PS I-core antenna and the auxiliary Chl *a-b* light-harvesting antenna (LHC-I). The PS I-core antenna consists of the 82–84 kDa heterodimer protein, coded for by the organelle genes *psaA* and *psaB* (Fish et al., 1985; Kirsch et al., 1986) (under denaturing conditions on SDS-PAGE, these PS I subunits migrate in the 62–70 kDa region). Non-covalently bound to this complex is the photochemical reaction center P700 of PS I (a Chl *a* dimer), two electron acceptors including the primary acceptor A_0 (a Chl *a* molecule), and the secondary acceptor A_1 (a phylloquinone molecule). Covalently bound to the complex is the primary 4Fe-4S iron-sulfur cluster F_X. The *psaA* and *psaB* gene products also bind a total of about 100 Chl *a* molecules (Anderson et al., 1983; Glick and Melis, 1988). Associated with the protein of the *psaA-psaB* gene products is a 9 kDa protein (the organelle *psaC* gene product) containing the 4Fe-4S iron-sulfur clusters F_B and F_A. Electron transport in PS I originates in P700 (photooxidation of P700 is a strongly endergonic reaction) and proceeds through A_0, A_1, and the iron-sulfur clusters F_X, F_B and F_A (Evans, 1982). Thereafter, electrons are transferred to ferredoxin on the stromal surface of the PS I complex whereas electron donation to $P700^+$ occurs in the lumenal side of the PS I complex by soluble plastocyanin (Bruce et al., 1989; Golbeck, 1992; see also Chapters 15 and 17).

The accessory light-harvesting antenna of PS I (LHC-I) contains 80–120 Chl *a* and Chl *b* molecules. In contrast to the Chl composition of the LHC-II, the LHC-I contains more Chl *a* than Chl *b* (Chl *a*/Chl *b* = ~3.5/1) (Lam et al., 1984). Polypeptides of the LHC-I are coded for by the nuclear *Lhca* gene family (Jansson et al., 1992a; Chapters 27 and 28 of this book). When denatured, they migrate in the 19–25 kDa region in SDS-PAGE, are immunologically distinct from those of the LHC-II, and may contain up to four distinct subunits (Haworth et al., 1983; Anderson et al., 1983; Lam et al., 1984a,b; Ikeuchi et al., 1991; Knoetzel et al., 1992).

II. Size of the Chlorophyll *a-b* Light-Harvesting Antenna in Photosystem I and Photosystem II of Chloroplasts

Three different experimental approaches have been described in the literature for the determination of the functional chlorophyll antenna size for PS I and PS II. They are: (a) the spectrophotometric/kinetic approach; (b) the optical cross section approach; and (c) the biochemical approach. A brief review of the basis of each method and a summary of functional antenna size measurements is given below.

A. Spectrophotometric/kinetic Determination of Photosystem Chlorophyll Antenna Size

The spectrophotometric/kinetic method is based on the premise that the absolute size of the Chl antenna of PS II and PS I can be determined *in situ* from the known ratio of total Chl to a reaction center (Chl/PS I and Chl/PS II determined spectrophotometrically (Melis and Brown, 1980)) upon partitioning of this total Chl into distinct PS I and PS II components. The method assigns functional Chl to each photosystem in direct proportion to the rate of light utilization by each photoreaction. The kinetics of Q_A photoreduction (for PS II) and of P700 photooxidation (for PS I) have been used for the measurement of the respective rates (Thielen and van Gorkom, 1981a; Melis and Anderson, 1983; Melis, 1989). This method uses broad-band green actinic excitation, which is absorbed equally by Chl *a* and Chl *b* molecules (Ghirardi and Melis, 1984), and assumes that all Chl is functionally associated with and transfers excitation energy efficiently to a photosystem. The assumption is satisfied for plants grown under physiological conditions (Thielen and van Gorkom, 1981b; Ley and Mauzerall, 1982; Smith et al., 1990). Under these conditions, the rate of light utilization by a photosystem is directly proportional to the functional light-harvesting antenna size (Glazer and Melis, 1987; Melis, 1989).

B. Optical-cross Section Method for the Determination of Photosystem Chlorophyll Antenna Size

The optical cross section approach makes use of the probability with which a chlorophyll molecule will absorb incoming radiation of a specific wavelength. The method depends on the experimental determination of the *in vivo* absorption of a specific wavelength of light by individual pigment molecules (Mauzerall, 1986; Mauzerall and Greenbaum, 1989). Narrow-band 596 nm actinic excitation has been used in such measurements (Mauzerall and Greenbaum, 1989) because at this wavelength the extinction coefficient values of Chl *a* and Chl *b* are similar (MacKinney, 1940). The method measures the flash-saturation curve of activities that are specific to PS II (oxygen evolution (Ley and Mauzerall, 1982)) and PS I (respiratory oscillations (Greenbaum et al., 1987)), by using single-turnover flashes of light. From the flash-saturation curves, an estimate of the effective chlorophyll absorption cross section of a photosystem can be derived.

C. Biochemical Method for the Determination of Photosystem Chlorophyll Antenna Size

The biochemical method is based on the isolation of resolved PS I and PS II complexes from thylakoid membranes, followed by a direct measurement of the chlorophyll per photosystem ratio. In this case, spectrophotometric and/or immunochemical tests need to be performed in order to demonstrate the purity of these preparations. Under such conditions, quantitation of Chl provides a direct measure of the Chl antenna size associated with a particular photosystem. This approach was applied to measurements of the Chl antenna size of PS II, obtained upon isolation of resolved appressed membranes from the grana partition regions (Berthhold et al., 1981; Lam et al., 1983; Murata et al., 1984), and in measurements of the Chl antenna size of PS I, obtained upon isolation of native PS I complexes by means of detergents (Haworth et al., 1983; Anderson et al., 1983).

Table 1 provides a brief summary of the absolute chlorophyll antenna size of Photosystem I and Photosystem II in a variety of photosynthetic tissues upon utilization of the above mentioned techniques.

III. Hierarchy of *Lhcb* Protein Assembly during Chloroplast Development

The minimal functional chlorophyll antenna size of Photosystem II contains 37 Chl *a* molecules. This is believed to be the chlorophyll complement of the PS II-core complex, contained within the D1/D2 heterodimer and the CP43/CP47 Chl *a*-proteins (Glick and Melis, 1988). Further increments in the light-harvesting antenna size of Photosystem II occur during chloroplast development by addition of individual proteins and of subcomplexes of the LHC-II. It is generally assumed that three minor Chl *a-b*-proteins (CP29, CP27 and CP24, coded for by nuclear *Lhcb4, Lhcb5* and *Lhcb6* genes, respectively, see Chapters 27 and 28) act as bridges in the process of excitation energy transfer from the bulk of LHC-II (coded for by nuclear *Lhcb1* and *Lhcb2* genes) to the PS II-core ((Green and Camm, 1981; Dunahay and Staehelin, 1986; Larsson et al., 1987a; Morrissey et al., 1989; Peter and Thornber, 1991; Dainese et al., 1992b; Bratt and Åkerlund, 1992), but see (Simpson, 1990)).

The three dimensional structure of LHC-II proposed by Kühlbrandt and co-workers (Kühlbrandt and Wang, 1991; Kühlbrandt et al., 1994) infers a radially symmetric complex with 3 constituent polypeptide subunits per LHC-II subcomplex. Analysis of the bulk LHC-II organization by isoelectric focusing (Spangfort et al., 1987; Spangfort and Andersson, 1989), electron-microscopy (Staehelin, 1986; Greene et al, 1988a,b), from results obtained during the development of chloroplast thylakoids (Ghirardi and Melis, 1988; Peter and Thornber, 1991; Welty et al., 1992) and upon phosphorylation of the LHC-II (Larsson et al., 1987a,b) all suggest the occurrence of two distinct subpopulations of the LHC-II (Ghirardi et al., 1986). A 'tightly bound' subpopulation of the LHC-II (also known as LHC-II-inner) was proposed to contain about 80 Chl (*a*+*b*) molecules and to have a $\{(Lhcb1)_5(Lhcb2)_1\}$ hexameric configuration. It was suggested that the tightly bound subpopulation of the LHC-II (LHC-II-inner) binds to the PS II-core complex via two copies of the (*Lhcb3*) protein

Table 1. Absolute chlorophyll antenna sizes for PS I and PS II units. Numbers given are Chl (*a+b*) molecules functionally associated with a specific photosystem.

Plant material	Spectrophotometric/ kinetic method	Optical cross-section method	Biochemical method
spinach			
PS II	230[a]		230[b]
PS I	210[a]		200[c]
tobacco			
PS II	240[d]		
PS I	190[d]		
pea			
PS II	280[e]		
PS I	220[e]		
wild-type barley			
PS II	250[f]		
PS I	190[f]		
chlorina f2 barley			
PS II	50[f]		
PS I	150[f]		
intermittent-light barley			
PS II	37[g]		
PS I	95[g]		
Chlorella vulgaris			
PS II		390[h]	
PS II		570[i]	
PS I		540[i]	
Chlamydomonas reinhardtii			
PS II	530[j]		
PS I	240[j]		
Dunaliella salina			
PS II	560[k]		
PS I	220[k]		

References : a-Melis and Anderson 1983; b-Berthold et al., 1981, Lam et al., 1983, Murata et al., 1984; c-Anderson et al., 1983, Ortiz et al., 1984; d-Thielen and van Gorkom 1981a; e-Melis 1984; f-Ghirardi et al., 1986; g-Glick and Melis 1988; h-Ley and Mauzerall 1982; i-Greenbaum and Mauzerall 1991; j-Neale and Melis 1986; k-Guenther et al., 1988; Smith et al., 1990

(Harrison and Melis, 1992).

The second subpopulation of the LHC-II is known as 'mobile' (owing to its reversible dissociation from PS II upon phosphorylation), or LHC-II-peripheral. It was proposed to have a $\{(Lhcb1)_2(Lhcb2)_1\}_n$ configuration where $1 < n < 7$. It is believed to contain 80 Chl (*a+b*) molecules when $n = 2$, 120 Chl (*a+b*) molecules when $n = 3$, and so forth (Morrissey et al., 1989). The two LHC-II subpopulations (LHC-II-inner and LHC-II-peripheral) are probably contained within Anderson's LHCP1 (1980), Green's CPII* (Green and Camm, 1982) and Thornber's LHC-IIb (Peter and Thornber, 1988; Thornber et al., 1988; Harrison and Melis, 1992).

Studies of the LHC-II organization in Chl *b*-less and Chl *b*-deficient mutants, in which the amount of LHC-II is highly truncated, revealed peculiar properties in terms of the priority and stability of assembly for the *Lhcb* proteins. For example, the *chlorina-f2* mutant of barley (Thornber and Highkin, 1974) contains no chlorophyll *b* in its light-harvesting antenna, whereas the *chlorina-103* mutant (Simpson et al., 1985) contains approximately 10% of the chlorophyll *b* found in wild-type. The absolute chlorophyll antenna size for Photosystem II in wild-type, *chlorina-103* and *chlorina-f2* mutant was 250, 58 and 50 chlorophyll molecules, respectively (Table 1). The polypeptide composition of the LHC-II was found to be identical in the two *chlorina* mutants, with only the *Lhcb3* protein present at an equal copy number per PS II in each of the two mutants and in the wild-type barley (Harrison and Melis, 1992; Harrison et al., 1993). All other *Lhcb* proteins were substantially depleted from the light-harvesting antenna and thylakoid membrane of the mutants. Apparently, the *Lhcb3* protein assembles stably in the thylakoid membrane, in the absence of the other *Lhcb* proteins and chlorophyll *b*, and exhibits flexibility in its complement of bound chlorophylls. This is surprising in view of the general assumption that the three minor Chl *a-b*-proteins (CP29, CP27 and CP24, coded for by nuclear *Lhcb4, Lhcb5* and *Lhcb6* genes, respectively,) act as the primary structural and functional bridges in the process of excitation transfer from the bulk LHC-II to the PS II-core. These results suggested a supramolecular organization for the LHC-II in which the *Lhcb3* protein receives priority of assembly over the other *Lhcb* proteins, is in direct proximity to the PS II-core complex and serves to transfer excitation energy directly from at least a portion of the bulk LHC-II to the PS II-core unit (Harrison and Melis, 1992; Harrison et al., 1993).

In spite of these advances, uncertainty exists on the spatial organization of LHC-II-inner, LHC-II-peripheral, CP29, CP27, and CP24 in relation to the PS II-core complex. A number of models postulate obligatory transfer of excitation energy from LHC-II-peripheral through LHC-II-inner to the minor Chl *a-b*-proteins and, from there, to the PS II-core. This would require a linear arrangement of the modular LHC-II components in the thylakoid membrane (Larsson et al., 1987a; Morrissey et al., 1989; Peter and Thornber, 1991; Dainese et al., 1992b). However, other investigators have suggested a parallel rather than linear organization in which trimer or hexamer complexes of the LHC-II-inner and LHC-II-peripheral may feed excitation energy independently into the various minor Lhcb proteins (*Lhcb3*, *Lhcb4*, *Lhcb5*, and *Lhcb6*) which in turn feed the light-energy directly into the CP47 and CP43 Chl *a*-proteins of the PS II-core (Green and Camm, 1981; Hoyer-Hansen et al., 1988; Simpson, 1990; Harrison and Melis, 1992; Jansson, 1992; see also Chapter 27 of this book).

IV. Modulation of the Chlorophyll Light-harvesting Antenna Size by Irradiance

Evidence for the response of plants to the level of irradiance was first provided in the pioneering work of Björkman and coworkers (Björkman et al., 1972). The level of irradiance during plant growth defines such parameters as the relative abundance of thylakoid membrane *versus* that of the stroma phase in chloroplasts, the relative chlorophyll (Chl) content and the Chl *a*/Chl *b* ratio in thylakoids, and the photosynthetic unit size (Malkin and Fork, 1981; Melis and Harvey, 1981; Ley and Mauzerall, 1982; Lichtenthaler and Meier, 1984; Leong and Anderson, 1984a,b; Wilhelm and Wild, 1984; Neale and Melis, 1986; Larsson et al., 1987b; Sukenik et al., 1988; Fujita et al., 1989; Smith et al., 1990).

The level of irradiance during plant growth modulates the size and composition of the light-harvesting antenna of the photosystems (reviewed in Anderson, 1986; Melis, 1991). In general, low light intensity promotes larger Chl antenna size for both PS I and PS II (larger photosynthetic unit size). High-light conditions elicit a smaller Chl antenna size. This adjustment in the Chl antenna size of the photosystems comes about because of changes in the size of the auxiliary Chl *a-b* light-harvesting complex (LHC-II and LHC-I) (Leong and Anderson, 1984a, Larsson et al., 1987b, Sukenik et al., 1988, Smith et al., 1990). Mechanistically, these changes are implemented through the association of variable amounts of *Lhcb1* and *Lhcb2* proteins with the light-harvesting antenna of PS II (Morrissey et al., 1989; Mawson et al., 1994). The response appears to be highly conserved in all photosynthetic organisms examined. Recent work by LaRoche et al. (1991) suggested that changes in the abundance of LHC-II are preceded by changes in the abundance of the mRNA coding for these proteins. The regulation of these phenomena at the molecular and membrane levels is currently unknown.

The ability of a plant to modulate the chlorophyll antenna size varies widely, depending on plant species, growth conditions, and the synergism of chemical or other environmental stresses. In general, agriculturally important crop plants show limited antenna size variations as a function of light-intensity. Green algae display significant variations in Chl antenna size in response to the level of irradiance. Table 2 presents examples of the response of the chlorophyll antenna size in distinct green algae. The results show a significant attenuation in the Chl antenna size of both photosystems as a function of irradiance, with PS II displaying a greater *plasticity* in the response to light-intensity.

V. Distribution of Excitation Energy Between the Two Photosystems

The spectral distribution of solar quantum irradiance at the surface of the earth is characterized by approximately constant emission in the visible wavelength region between 450 and 700 nm and of progressively declining intensities in the violet and

Table 2. Absolute chlorophyll antenna sizes for PS I and PS II as a function of growth irradiance.
Numbers given are Chl (*a*+*b*) molecules functionally associated with a specific photosystem. For a definition of the precise range of light intensities listed as low, moderate and high, the reader should refer to the corresponding original publication. For green algae, a consensus range of light intensities is as follows: Low-light: <200 μmol photons·m^{-2}·s^{-1}; Moderate-light: 200–800 μmol photons·m^{-2}·s^{-1}; High-light: >800 μmol photons·m^{-2}·s^{-1}.

Plant material	Low-light	Moderate-light	High-light
Chlorella vulgaris			
PS II	390[h]	230[h]	140[h]
PS II	570[i]	330[i]	100[i]
PS I	540[i]	350[i]	210[i]
Chlamydomonas reinhardtii			
PS II	530[j]	450[j]	
PS I	240[j]	170[j]	
Dunaliella salina			
PS II	560[k]	—	60[k]
PS I	220[k]	—	110[k]

References : h-Ley and Mauzerall, 1982; i-Greenbaum and Mauzerall, 1991; j-Neale and Melis, 1986; k-Guenther et al., 1988; Smith et al., 1990.

ultraviolet regions (400–280 nm). This emission profile, however, is seen only by the uppermost layer of chloroplasts in the leaf of sun-adapted plants. Internal leaf chloroplasts as well as shade-adapted species and plants living in the aquatic environment will experience a substantially different light environment. For example, it is known that strong variations in light quality occur within a single leaf (Terashima and Saeki, 1983; Vogelmann, 1989, 1993), within the canopy of a single tree or within the canopy of a forest (Björkman and Ludlow, 1972), and within the aquatic environment (Kirk, 1983). Some of these environments favor absorption of light by PS I, others favor absorption of light by PS II. It is clear that such different light ecotypes will have a profound effect on the distribution of excitation between the two photosystems.

Under limiting intensity of illumination, the efficiency of photosynthesis depends on the coordinated interaction of PS II and PS I in the electron-transport chain. The quantum yield of photosynthesis in many plant species from diverse light habitats was found to be 0.106±0.001 mol O_2 evolved per mol photon absorbed (Ley and Mauzerall, 1982; Björkman and Demmig, 1987; Evans, 1987). This value is very close to a theoretical upper limit of 0.125 mol O_2 evolved per mol photon absorbed, which translates into a photosynthetic efficiency of about 85%, independent of the light climate in which plants grow. This is a remarkable feature of the photosynthetic apparatus, given the contrasting light-qualities that prevail in different plant ecosystems and the fact that substantially different pigments absorb light for PS I and PS II in the thylakoid membrane of oxygenic photosynthesis. Thus, a question is raised as to how plants may always perform photosynthesis with maximal efficiency under a variety of contrasting light-qualities.

A. State Transitions and Phosphorylation of Lhcb *Proteins*

Earlier work from many laboratories investigated the phenomenon known as 'state transitions' (Wang and Myers, 1974) and the underlying phosphorylation-dephosphorylation of the LHC-II, regulated by a redox controlled kinase (Bennett et al., 1980; Telfer et al., 1983; Larsson and Andersson, 1985; Bennett, 1991; Allen, 1992). State transitions are elicited as a 'short-term' response of the photosynthetic apparatus to a change in the proportion of excitation received by PS II and PS I. They occur promptly (on the order of minutes) whenever the balance of excitation energy distribution between the two photosystems is disturbed. For example, irradiance absorbed preferentially by PS II, and to a much lesser degree by PS I, will create unbalanced electron flow in the thylakoid membrane resulting in the reduction of the electron transport carriers functioning as inter-mediates between the two photosystems. It is believed that reduction of the plastoquinone pool (Allen et al., 1981), and/or of the *b*-type hemes in the cytochrome

b_6f complex (Coughlan, 1988; Gal et al., 1990), activates a membrane-bound kinase (Coughlan and Hind, 1986) which phosphorylates members of the *Lhcb* family of proteins. The phosphorylation site is believed to be a *Thr* or *Ser* residue near the -NH_2-terminus of the *Lhcb* protein (Mullet, 1983; Michel et al., 1988) which extends into the stromal side of the thylakoid membrane (Thornber et al, 1988). Phosphorylation of the 'mobile', or LHC-II-peripheral, component of the LHC-II causes dissociation of this subcomplex from PS II. Under the influence of the negative charge introduced upon phosphorylation (Barber, 1982), appressed membranes become partially unstacked and thus allow diffusion of the phospho-LHC-II from the grana to the stroma-exposed thylakoids. Consequently, phosphorylation of the LHC-II and dissociation from PS II brings about a significant reduction (by about 50%) in the functional Chl antenna size of PS II (Kyle et al., 1984) that tends to correct for the uneven distribution of excitation between the two photosystems. If the rate of light absorption by PS I is greater than that of PS II, then plastoquinone and the cytochrome b_6f complex become oxidized, leading to a de-phosphorylation of the LHC-II, a reaction catalyzed by a membrane bound phosphatase (Bennett, 1991).

Evidence based on 77 K Chl fluorescence emission spectra was interpreted to indicate that phospho-LHC-II became structurally and functionally associated with PS I. This hypothesis was initially attractive because the presumed association of phospho-LHC-II with PS I would tend to further diminish unbalanced distribution of excitation energy between the two photosystems. However, efforts to chemically cross-link phospho-LHC-II with PS I were unsuccessful (Bennett, personal communication). Moreover, specific efforts to demonstrate excitation energy transfer from Chl *b* to P700 in stroma thylakoid preparations that contained significant amounts of phospho-LHC-II suggested a very low efficiency of interaction between phospho-LHC-II and PS I (Telfer et al., 1986; Larsson et al., 1986; Deng and Melis, 1986; Telfer, 1987; Bassi et al., 1988).

In summary, state-transitions bring about organizational changes in the thylakoid membrane, i.e., a phosphorylation-induced reversible detachment of the LHC-II from PS II; they occur on the order of minutes and they do not involve changes in the composition of the thylakoid membrane. State-transitions provide a first-line chloroplast response to a sudden change in light-quality (Haworth et al., 1982). They result in the fine-tuning of the distribution of excitation energy between the PS I and PS II (Fork and Satoh, 1986; Ranjeva and Boudet, 1987), but they are not considered to be either a sufficient or efficient response in cases of prolonged and gross imbalance in the distribution of excitation between the two photosystems (Anderson, 1986; Anderson and Andersson, 1988; Fujita, 1992).

B. Regulation of Photosystem Stoichiometry in Chloroplasts

1. Light-quality Defines the Photosystem Stoichiometry (PS II/PS I Ratio) in Chloroplasts

Work in several laboratories provided evidence that chloroplasts respond to long-term imbalance in the distribution of excitation energy between the two photosystems by adjusting and optimizing the photosystem ratio in thylakoids. The results showed that adjustments of the photosystem ratio may be a key to the high quantum efficiency of photosynthesis under diverse light-quality conditions (Murakami and Fujita, 1988; Melis et al., 1989; Chow et al., 1990). The mechanism for the adjustment of the photosystem ratio appears to be highly conserved in nature since oxygen-evolving organisms from cyanobacteria to higher plant chloroplasts are known to possess it (reviewed in Melis, 1991). It was inferred that such a mechanism performs a highly needed function in oxygen-evolving plants (Chow et al., 1990). Table 3 provides examples of the PS II/PS I ratio in PS I-light and PS II-light grown plants, and illustrates the effect of PS II/PS I ratio adjustments on the quantum yield (Φ) of photosynthesis.

The dynamic response of the thylakoid membrane and the adjustment of photosystem stoichiometry to different light-quality conditions (Kim et al., 1993) suggested the existence of a mechanism capable of recognizing imbalance in the rate of light utilization by the two photoreactions and capable of regulating cellular metabolic activity to bring about photosystem stoichiometry adjustments. The operation of such a mechanism in chloroplasts, and the ensuing structural and functional adjustment in thylakoids, are now concepts established in the field. However, there is only a limited understanding of the actual biochemistry and molecular biology that underlie the

Table 3. Quantum Yield (Φ) of Photosynthesis in PS I-Light and PS II-Light-Acclimated *Pisum sativum* and *Synechococcus* 6301

Pisum sativum (pea)

	PS II/PS I	Φ measured under PS I-irradiance	Φ measured under PS II-irradiance
PS I-light leaves	2.5/1	0.108	0.080
PS II-light leaves	1.2/1	0.091	0.097
Difference, %	~50%	19%	21%

Absolute quantum yield units in *mol oxygen evolved per mol photon absorbed.*

Synechococcus 6301

	PS II/PS I	Φ measured under PS I-irradiance	Φ measured under PS II-irradiance
PS I-light cells	0.7/1	1.00	0.60
PS II-light cells	0.3/1	0.75	1.00
Difference, %	~60%	25%	40%

Relative quantum yield units.
From (Murakami and Fujita 1988, Melis et al., 1989, Chow et al., 1990).
Plants were allowed to acclimate for several days either under PS I-light or under PS II-light conditions. These light qualities were defined earlier (Glick et al., 1986) as light absorbed predominantly by PS I (i.e., by Chl *a* more than by Chl *b*), or by PS II (i.e., by Chl *b* or phycocyanin more than by Chl *a*).

molecular feedback mechanism for the regulation of the PS II/PS I ratio in thylakoids.

2. The Role of Lhcb Proteins as Signal Receptors for Photosystem Ratio Adjustment

There is evidence from different laboratories suggesting that photosynthetic pigments in chloroplasts are the light-quality receptor molecules and that signal transduction proceeds via the electron-transport chain in the thylakoid membrane. This was originally proposed from work in this laboratory (Melis et al., 1985; Glick et al., 1986). The identity of the primary photoreceptor was further explored in work with the Chl *b*-less *chlorina*-f2 mutant of barley (*Hordeum vulgare*) and the phycobilisome-less mutant of the cyanobacterium *Agmenellum quadruplicatum* (= *Synechococcus* PCC 7002). The research provided the following evidence :

(i) In the absence of accessory pigments (Chl *b* and phycobilins), the photosystem ratio cannot be influenced by the quality of illumination (PS I-light or PS II-light) during plant growth. Unlike their wild-type counterparts, changes in light-quality during mutant growth (PS I-light → PS II-light, and vice versa) did not elicit changes in the photosystem ratio (Kim et al., 1993).

(ii) The Chl *b*-less *chlorina f2* mutant of barley and phycobilisome-less mutant of *A. quadruplicatum* had a significantly elevated PS II/PS I ratio compared to their wild-type counterparts. Measurements of the functional light-harvesting antenna size showed that absence of Chl *b* and of the phycobilisome in the mutants resulted in smaller antenna size for both PS II and PS I. However, attenuation of the antenna size of PS II was much greater than that of PS I in the mutants. As discussed earlier (Glazer and Melis, 1987; Melis, 1991), such mutations always result in elevated PS II/PS I ratios in thylakoids. The response of the plant is a compensation for the greater attenuation of the Chl antenna size in PS II and it reflects the function of a dynamic mechanism which is directed toward restoration of the balance of light absorption between PS II and PS I in the mutants.

Other laboratories have contributed significantly in the development of these concepts. Fujita and co-workers determined that changes in light-quality during plant growth (PS I-light → PS II-light, and vice versa) alter the redox state of the intermediate electron pool between the two photosystems in a way that lowers the efficiency of electron transport and the quantum yield of photosynthesis (Murakami and Fujita, 1988; Fujita et al., 1988; Murakami and

Fujita, 1991). The balance of electron transport is released from such biased states upon adjustment of the PS II/PS I ratio (Fujita, 1992).

Anderson and co-workers (Chow et al., 1987; 1990a, Anderson and Chow, 1992) concur with the proposal that the two photosystems themselves are the predominant sensors of light-quality in triggering modulations in the PS II/PS I ratio. Further, they presented evidence to suggest that phytochrome is ineffective in modulating the photosystem composition of pea plants (Chow et al., 1987). This conclusion was strengthened in recent work by Smith et al. (1993) who used mutants deficient in phytochrome-A and showed that such plants are impaired in phytochrome-mediated responses, such as extension growth in a shade environment; however, they were fully capable of adjusting the PS II/PS I ratio in response to sun or shade conditions.

In summary, existing evidence supports the notion that signal perception for photosystem stoichiometry adjustment occurs at the thylakoid membrane level as differential sensitization of auxiliary pigments associated with PS II and PS I. The requirement for electron transport in the thylakoid membrane and the role of plastoquinone and/or of the cytochrome *b-f* complex as activators in the signal transduction pathway has not yet been investigated.

VI. Excitation Transfer Pathways within the *Lhc* Proteins

A. Exciton Interactions between Carotenoids and Chlorophylls

Exciton interactions between carotenoids and chlorophylls are complex and may take several distinct forms. This complexity is due to the presence of different types of carotenoids in the photosynthetic apparatus (Seifermanns-Harms, 1985) as well as due to the heterogeneous distribution of the various types among the Chl-proteins (Dainese et al., 1992a; also see Chapter 30 in this book). The function of carotenoids is photoprotection from the destructive effects of singlet oxygen (Frank et al., 1991), the regulated quenching of excitation energy in the chlorophyll pigment bed of photosynthesis (Demmig-Adams, 1990), and contribution to the process of light absorption and excitation energy transfer to chlorophyll (Goedheer, 1969; Öquist et al., 1980).

The type of interaction between carotenoids and chlorophyll depends strongly on the chemical configuration and on the localization of the carotenoid molecule in the pigment bed (DeCoster, 1992). For example, violaxanthin, a xanthophyll with nine conjugated C-C double bonds, is known to transfer excitation energy to chlorophyll with a time constant of 0.2–0.3 ps (Trautman et al., 1990). A light-dependent de-epoxidation of violaxanthin to zeaxanthin in situ (Yamamoto, 1979; see also Chapter 30) increases the conjugated C-C double bonds of the molecule to 11, shifts the molecule's excited state closer to that of Chl *a* (Owens et al., 1992) and results in the trapping and quenching of excitation energy in the pigment bed (Demmig-Adams and Adams, 1992). *Beta* carotenes localized in the internal Chl-proteins of the photosystems participate in light-harvesting and carotenoid to Chl *a* singlet-singlet energy transfer (Öquist et al., 1980), whereas β-carotenes in the vicinity of the photochemical reaction centers function in photoprotection via Chl *a* to carotenoid triplet-triplet energy transfer (Frank et al., 1991).

Of considerable interest is the function of fucoxanthin in the light-harvesting apparatus of certain algae (Anderson and Barrett, 1986). Fucoxanthin and other carotenoids play a significant role in light-harvesting in the aquatic environment because most of the light available in this ecotype is in the blue-green region of the spectrum (Kirk, 1983). Studies on the electronic interaction between fucoxanthin and Chl *a* suggest Förster/coulombic interactions between the electronic transition dipoles of fucoxanthin (donor) and Chl *a* (acceptor) molecules (Katoh et al., 1991; Katoh, 1992). Transfer rates with a time constant around 1 ps were reported (Trautman et al., 1990; Shreve et al., 1991).

B. Exciton Dynamics in the LHC Chlorophyll Antenna

Excitation energy transfer in the Chl antenna of the photosystems can be analyzed in terms of the in situ chlorophyll spectral forms, determined by asymmetric Gaussian deconvolution of the absorbance spectra of chloroplasts in the red region. Using this method, French et al., (1972) have discerned four major and some minor forms of Chl *a* occurring in a large number of photosynthetic species. Jennings and co-workers (Jennings et al., 1990; Zucchelli et al., 1992) employed the asymmetric Gaussian deconvolution method in combination with fluorescence emission measurements with sub-thylakoid grana particles (enriched in PS II) and with isolated and purified LHC-II. Their analyses revealed that specific fluorescence emission bands are associated with each absorbing chlorophyll spectral form and further provided information on the relative excitation transfer efficiencies between the spectrally distinct forms of Chl in the pigment bed of PS II.

A question concerning exciton transfer dynamics in the Chl antenna of the photosystems is the rate-limiting step in the kinetics of exciton decay, following absorption of light by Chl in the LHC. The kinetics of exciton decay in the process of photochemistry could be limited either by the rate of exciton diffusion in the pigment bed, or by the rate of trapping at the photochemical reaction center Chl molecule. Holzwarth and co-workers applied a global target analysis in which picosecond chlorophyll fluorescence kinetics were resolved in both time and wavelength (Beauregard et al., 1991; Holzwarth and Roelofs, 1992). This analysis provided information on the energy migration and trapping properties of PS IIα and PS IIβ in the thylakoid membrane of green plants (Karukstis, 1992) and suggested different molecular functioning for the two forms of PS II with respect to primary processes, including a slower primary charge separation in PS IIβ than in PS IIα centers (Roelofs et al., 1992). On the basis of the global target analysis, a trap-limited exciton decay model for PS II has been proposed (Holzwarth, 1992) in which the rate of exciton transfer from the Chl antenna to the reaction center P680 molecule is faster than the rate of trapping.

Energy transfer dynamics in PS I are largely defined in the domain of the LHC-I and of the PS I-core complex. Fluorescence depolarization measurements suggested an ultrafast first-time excitation transfer with an upper limit of 0.18±0.04 ps for the average single transfer step in the PS I-core Chl *a* antenna (Du et al., 1993). However, the overall pathway of excitation transfer in PS I is complicated by the function of long-wavelength absorbing pigments (C705-F735) with energy levels lower than the absorption maximum of the reaction center P700 (Satoh and Butler, 1978). It was proposed that such low-energy pigments appear in the process of cooling of photosynthetic samples to a low temperature (77 K) (Rijgersberg et al., 1979), a condition commonly applied prior to spectrophotometric measurements (Satoh and Butler, 1978; Rijgersberg et al., 1979). Time-resolved fluorescence relaxation measurements at 295 K and 77 K with isolated and purified LHC-I revealed five distinct ps excitation decay components whose relative amplitude is strongly influenced by temperature (Mukerji and Sauer, 1993). Excitation of Chl *b* resulted in a relative amplitude enhancement of the low energy emissions (F735) from the LHC-I, suggesting that Chl *b* in the LHC-I is specifically associated with the F735 emitters in PS I (Mukerji and Sauer, 1993). These results complement other measurements (Werst et al., 1992) which failed to reveal a 'funnel' arrangement of the Chl antenna in PS I, i.e., a structural and functional organization of the PS I antenna that leads to preferential or unidirectional migration of excitation toward P700. These results suggest significantly different excitation transfer dynamics in the Chl *a-b* LHC of PS I and PS II.

References

Allen JF (1992) Protein phosphorylation in regulation of photosynthesis. Biochim Biophys Acta 1098: 275–335

Allen JF, Bennett J, Steinback KE and Arntzen CJ (1981) Chloroplast protein phosphorylation couples plastoquinone redox state to distribution of excitation energy between photosystems. Nature 291: 21–25

Anderson JM (1980) P-700 content and polypeptide profile of chlorophyll-protein complexes of spinach and barley thylakoids. Biochim Biophys Acta 591: 113–126

Anderson JM (1986) Photoregulation of the composition, function, and structure of thylakoid membranes. Ann Rev Plant Physiol 37: 93–136

Anderson JM and Andersson B (1988) The dynamic photosynthetic membrane and regulation of solar energy conversion. Trends in Biochemical Sciences 13: 351–355

Anderson JM and Barrett J (1986) Light-harvesting pigment-protein complexes of algae. In: Staehelin LA and Arntzen CJ (eds) Encyclopedia of Plant Physiology, New Series, Vol., 19, pp 269–285. Springer-Verlag, Berlin

Anderson JM and Chow WS (1992) A regulatory feedback mechanism for light acclimation of the photosynthetic apparatus: Are Photosystems II and I self-regulatory sensors? In: Argyroudi-Akoyunoglou JH (ed) Regulation of Chloroplast Biogenesis, pp 475–482. Plenum Press, New York

Anderson JM, Brown JS, Lam E and Malkin R (1983) Chlorophyll *b*: an integral component of Photosystem I of higher plant chloroplasts. Photochem Photobiol 38: 205–210

Barber J (1982) Influence of surface charges on thylakoid structure and function. Ann Rev Plant Physiol 33: 261–295

Bassi R, Giacometti GM and Simpson DJ (1988) Changes in the organization of stroma membranes induced by in vivo state 1-state 2 transition. Biochim Biophys Acta 935: 152–165

Beauregard M, Martin I and Holzwarth AR (1991) Kinetic modeling of exciton migration in photosynthetic systems (1) Effects of pigment heterogeneity and antenna topography on exciton kinetics and charge separation yields. Biochim Biophys Acta 1060: 271–283

Bennett J (1991) Protein phosphorylation in green plant chloroplasts. Ann Rev Plant Physiol Plant Mol Biol 42: 281–311

Bennett J, Steinback KE and Arntzen CJ (1980) Chloroplast phosphoproteins: Regulation of excitation energy transfer by phosphorylation of thylakoid membranes. Proc Nat Acad Sci USA 77: 5253–5257

Berthold DA, Babcock GT, Yocum CF (1981) A highly-resolved, oxygen-evolving Photosystem II preparation from spinach thylakoid membranes. FEBS Lett 134: 231–234

Björkman O and Demmig B (1987) Photon yield of O_2 evolution and chlorophyll fluorescence characteristics at 77 K among vascular plants of diverse origins. Planta 170: 489–504

Björkman O and Ludlow MM (1972) Characterization of the light climate on the floor of a Queensland rainforest. Carnegie Inst Washington Yearbook 71: 85–94

Björkman O, Boardman NK, Anderson JM, Thorne SW, Goodchild DJ and Pyliotis NA (1972) Effect of light intensity during growth of *Atriplex patula* on the capacity of photosynthetic reactions, chloroplast components and structure. Carnegie Inst Washington Yearbook, 71: 115–135

Bratt CE and Åkerlund H-E (1992) The role of the chlorophyll *a/b*-binding complex CP29 in thylakoid membranes. In: Murata N (ed) Research in Photosynthesis, Vol I, pp 231–234. Kluwer Academic Publishers, Dordrecht

Bricker TM (1990) The structure and function of CPa-1 and CPa-2 in Photosystem II. Photosynthesis Research 24: 1–13

Bruce BD, Malkin R, Wynn RM and Zilber A (1989) Structural organization and function of polypeptide subunits in Photosystem I. In: Barber J and Malkin R (eds) Techniques and Developments in Photosynthesis Research, NATO ASI series, Series A, Life sciences, Vol. 168, pp 61–80. Plenum Publishing Co, New York

Chow WS, Haehnel W and Anderson JM (1987) The composition and function of thylakoid membranes from pea plants grown under white or green light with or without far-red light. Physiologia Plantarum 70: 196–202

Chow WS, Goodchild DJ, Miller C and Anderson JM (1990a) The influence of high levels of brief or prolonged supplementary far-red illumination during growth on the photosynthetic characteristics, composition and morphology of *Pisum sativum* chloroplasts. Plant, Cell and Environment 13: 135–145

Chow WS, Melis A and Anderson JM (1990b) Adjustments of photosystem stoichiometry in chloroplasts improve the quantum efficiency of photosynthesis. Proc Natl Acad Sci USA 87: 7502–7506

Coughlan SJ (1988) Chloroplast thylakoid protein phosphorylation is influenced by mutations in the cytochrome *b-f* complex. Biochim Biophys Acta 933: 413–422

Coughlan SJ and Hind G (1986) Protein kinases of the thylakoid membrane. J Biol Chem 261: 14062–14068

Dainese P, Marquardt J, Pineau B and Bassi R (1992a) Identification of violaxanthin and zeaxanthin binding proteins in maize Photosystem II. In: Murata N (ed) Research in Photosynthesis, Vol I, pp 287–290. Kluwer Academic Publishers, Dordrecht

Dainese P, Santini C, Ghiretti-Magaldi A, Marquardt J, Tidu V, Mauro S, Bergantino E and Bassi R (1992b) The organization of pigment-proteins within Photosystem II. In: Murata N (ed) Research in Photosynthesis, Vol II, pp 13–20. Kluwer Academic Publishers, Dordrecht

Darr SC, Somerville SC and Arntzen CJ (1986) Monoclonal antibodies to the light harvesting chlorophyll *a/b* protein complex of Photosystem II. J Cell Biol 103: 733–740

DeCoster B, Christensen RL, Gebhard R, Lugtenburg J, Farhoosh R and Frank HA (1992) Low-lying electronic states of carotenoids. Biochim Biophys Acta 1102: 107–114

Demmig-Adams B (1990) Carotenoids and photoprotection in plants - A role for the xanthophyll zeaxanthin. Biochim Biophys Acta 1020: 1–24

Demmig-Adams B and Adams WW III (1992) Photoprotection and other responses of plants to high-light. Ann Rev Plant Physiol & Plant Mol Biol 43: 599–626

Deng X and Melis A (1986) Phosphorylation of the light-harvesting complex II in higher plant chloroplasts: Effect on PS II and PS I absorption cross section. Photobiochem Photobiophys 13: 41–52

Du M, Xie X, Jia Y, Mets L and Fleming GR (1993) Direct observation of ultrafast energy transfer in PS I core antenna. Chemical Physics Letters 201: 535–542

Dunahay TG and Staehelin LA (1986) Isolation and characterization of a new minor chlorophyll *a-b*-protein complex (CP24) of spinach. Plant Physiol 80: 429–434

Emerson R and Arnold W (1932a) A separation of the reactions in photosynthesis by means of intermittent light. J Gen Physiol 15: 391–420

Emerson R and Arnold W (1932b) The photochemical reactions in photosynthesis. J Gen Physiol 16: 191–205

Evans JR (1987) The dependence of quantum yield on wavelength and growth irradiance. Aust J Plant Physiol 14: 69–79

Evans MCW (1982) Iron-sulfur centers in photosynthetic electron transport. In: Spiro TG (ed), Iron-sulfur Proteins, Vol. 4, pp 249–284. Wiley, New York

Fish LE, Kuck U, and Bogorad L (1985) Two partially homologous adjacent light-inducible maize chloroplast genes encoding polypeptides of the P700 chlorophyll *a* protein complex of Photosystem I. J Biol Chem 260: 1413–1421

Fork DC and Satoh K (1986) The control by state transitions of the distribution of excitation energy in photosynthesis. Ann Rev Plant Physiol 37: 335–362

Frank HA, Violette CA, Trautman JK, Owens TG and Albrecht AC (1991) Carotenoids in photosynthesis: Structure and photochemistry. Pure Appl Chem 63: 109–114

French CS, Brown JS and Lawrence MC (1972) Four universal forms of chlorophyll . Plant Physiol 49: 421–429

Fujita Y (1992) Plasticity of biological system for light-energy conversion in oxygenic photosynthesis. In: Matsunaga T, Hikuma M and Kajiwara K (eds) Proceedings of the Symposium on the Development of Opto-bio Materials. Transl Mat Res Soc Japan 10: 25–45

Fujita Y, Murakami A, Ohki K and Hagiwara N (1988) Regulation of photosystem composition in cyanobacterial photosynthetic systems: Evidence indicating that Photosystem I formation is controlled in response to the electron transport state. Plant Cell Physiol 29: 557–564

Fujita Y, Iwama Y, Ohki K, Murakami A, Hagiwara N (1989) Regulation of the size of light-harvesting antennae in response to light intensity in the green alga *Chlorella pyrenoidosa*. Plant Cell Physiol 30: 1029–1037

Gaffron H and Wohl K (1936) Zur theorie der assimilation. Naturwissenschaften 24: 81–90

Gal A, Hauska G, Herrmann R and Ohad I (1990) Interaction

between LHC-II kinase and cytochrome b_6-*f*: in vitro control of kinase activity. J Biol Chem 265:, 19742–19749

Ghirardi ML and Melis A (1984) Photosystem electron transport capacity and light-harvesting antenna size in maize chloroplasts. Plant Physiol 74: 993–998

Ghirardi ML and Melis A (1988) Chlorophyll *b*-deficiency in soybean mutants. I. Effects on photosystem stoichiometry and chlorophyll antenna size. Biochim Biophys Acta 932: 130–137

Ghirardi ML, McCauley SW and Melis A (1986) Photochemical apparatus organization in the thylakoid membrane of *Hordeum vulgare* wild-type and chlorophyll *b*-less chlorina-f2 mutant. Biochim Biophys Acta 851: 331–339

Glazer AN and Melis A (1987) Photochemical reaction centers: Structure, organization, and function. Annu Rev Plant Physiol 38: 11–45

Glick RE and Melis A (1988) Minimum photosynthetic unit size in system-I and system-II of barley chloroplasts. Biochim Biophys Acta 934: 151–155

Glick RE, McCauley SW, Gruissem W and Melis A (1986) Light quality regulates expression of chloroplast genes and assembly of photosynthetic membrane complexes. Proc Natl Acad Sci USA 83: 4287–4291

Golbeck JH (1992) Structure and function of Photosystem I. Ann Rev Plant Physiol and Plant Mol Biol 43: 293–324

Goedheer JC (1969) Energy transfer from carotenoids to chlorophyll in blue-green, red and green algae and greening bean leaves. Biochim Biophys Acta 172: 252–265

Green BR and Camm EL (1981) A model of the relationship of the chlorophyll-protein complexes associated with Photosystem II. In: Akoyunoglou G (ed) Photosynthesis III. Structure and Molecular Organization of the Photosynthetic Apparatus, pp 675–681. Balaban International Science Services, Philadelphia, PA

Green BR and Camm EL (1982) The nature of the light-harvesting complex as defined by sodium dodecyl sulfate polyacrylamide gel electrophoresis. Biochim Biophys Acta 681: 256–262

Greenbaum NL and Mauzerall DC (1991) Effect of irradiance level on distribution of chlorophylls between PS II and PS I as determined from optical cross-sections. Biochim Biophys Acta 1057: 195–207

Greenbaum NL, Ley AC and Mauzerall DC (1987) Use of a light-induced respiratory transient to measure the optical cross section of Photosystem I in *Chlorella*. Plant Physiol 84: 879–882

Greene BA, Allred DR, Morishige D and Staehelin LA (1988a) Hierarchical response of light-harvesting chlorophyll-proteins in a light-sensitive chlorophyll *b*-deficient mutant of maize. Plant Physiol 87: 357–364

Greene BA, Staehelin LA and Melis A (1988b) Compensatory alterations in the photochemical apparatus of a photoregulatory, chlorophyll *b*-deficient mutant of maize. Plant Physiol 87: 365–370

Guenther JE, Nemson JA and Melis A (1988) Photosystem stoichiometry and chlorophyll antenna size in *Dunaliella salina* (green algae). Biochim Biophys Acta 934: 108–117

Harrison MA and Melis A (1992) Organization and stability of polypeptides associated with the chlorophyll *a-b* light-harvesting complex of Photosystem II. Plant Cell Physiol 33: 627–637

Harrison MA, Nemson JA and Melis A (1993) Assembly and composition of the chlorophyll *a-b* light-harvesting complex of barley (*Hordeum vulgare* L.): Immunochemical analysis of chlorophyll *b*-less and chlorophyll *b*-deficient mutants. Photosynthesis Research 38: 141–151

Haworth P, Kyle DJ, Horton P and Arntzen CJ (1982) Chloroplast membrane protein phosphorylation. Photochem Photobiol 36: 743–748

Haworth P, Watson JL and Arntzen CJ (1983) The detection, isolation and characterization of a light-harvesting complex which is specifically associated with Photosystem I. Biochim Biophys Acta 724: 151–158

Holzwarth AR (1992) Exciton dynamics in antennae and reaction centers of Photosystems I and II. In: Murata N (ed) Research in Photosynthesis, Vol I, pp 187–194. Kluwer Academic Publishers, Dordrecht

Holzwarth AR and Roelofs TA (1992) Recent advances in the understanding of chlorophyll excited state dynamics in thylakoid membranes and isolated reaction center complexes. J Photochem Photobiol B: Biol 15: 45–62

Høyer-Hansen G, Bassi R, Honberg LS and Simpson DJ (1988) Immunological characterization of chlorophyll *a-b*-binding proteins of barley thylakoids. Planta 173: 12–21

Ikeuchi M, Hirano A and Inoue Y (1991) Correspondence of apoproteins of light-harvesting chlorophyll *a/b* complexes associated with Photosystem I to *cab* genes: Evidence for a novel Type IV apoprotein. Plant Cell Physiol 32: 103–112.

Jansson S (1992) The chlorophyll *a/b*-binding proteins: Studies on the *Lhca* and *Lhcb* genes of Scots pine. Doctoral dissertation, University of Umeå, Department of Plant Physiology

Jansson S, Pichersky E, Bassi R, Green BR, Ikeuchi M, Melis A, Simpson DJ, Spangfort M, Staehelin LA, Thornber JP (1992) A nomenclature for the genes encoding the chlorophyll *a/b*-binding proteins of higher plants. Plant Mol Biol Rep 10: 242–253

Jennings RC, Zucchelli G and Garlaschi FM (1990) Excitation energy transfer from the chlorophyll spectral forms to Photosystem II reaction centers: A fluorescence induction study. Biochim Biophys Acta 1016: 259–265

Karukstis KK (1992) Chlorophyll fluorescence analyses of Photosystem II reaction center heterogeneity. J Photochem Photobiol B: Biol 15: 63–74

Katoh T (1992) S_1 state of fucoxanthin involved in energy transfer to chlorophyll *a* in the light-harvesting proteins of brown algae. In: Murata N (ed), Research in Photosynthesis, Vol. I, pp 227–230. Kluwer Academic Publishers Dordrecht

Katoh T, Nagashima U and Mimuro M (1991) Fluorescence properties of the allelic carotenoid fucoxanthin: Implications for energy transfer in photosynthetic systems. Photosyn Res 27: 221–226

Kim JH, Glick RE and Melis A (1993) Dynamics of photosystem stoichiometry adjustment by light-quality in chloroplasts. Plant Physiol 102: 181–190

Kirk JTO (1983) Light and Photosynthesis in Aquatic Ecosystems. Cambridge University Press. New York

Kirsch W, Seyer P, and Herrmann RG (1986) Nucleotide sequence of the clustered genes for two P700 chlorophyll *a* apoproteins of the Photosystem I reaction center and the ribosomal protein S14 of the spinach plastid chromosome. Curr Genet 10: 843–855

Knoetzel J, Svendsen I and Simpson DJ (1992) Identification of the Photosystem I antenna polypeptides in barley. Isolation of

3 pigment-binding antenna complexes. Eur J Biochem 206: 209–215

Kühlbrandt W and Wang DN (1991) Three-dimensional structure of plant light-harvesting complex determined by electron crystallography. Nature 350: 130–134

Kühlbrandt W, Wang DN and Fujiyoshi Y (1994) Atomic model of plant light-harvesting complex determined by electron crystallography. Nature 367: 614–621

Kyle DJ, Kuang T-Y, Watson JL and Arntzen CJ (1984) Movement of a subpopulation of LHC-II from grana to stroma lamellae as a consequence of its phosphorylation. Biochim Biophys Acta 765: 89–96

Lam E, Baltimore B, Ortiz W, Chollar S, Melis A and Malkin R (1983) Characterization of a resolved oxygen-evolving Photosystem II preparation from spinach chloroplasts. Biochim Biophys Acta 724: 201–211

Lam E, Ortiz W and Malkin R (1984a) Chlorophyll *a/b* proteins of Photosystem I. FEBS Lett 168: 10–14

Lam E, Ortiz W, Mayfield S and Malkin R (1984b) Isolation and characterization of a light-harvesting chlorophyll *a/b* complex associated with Photosystem I. Plant Physiol 74: 650–655

LaRoche J, Mortain-Bertrand A and Falkowski PG (1991) Light-intensity-induced changes in *cab* mRNA and light-harvesting complex II apoprotein levels in the unicellular chlorophyte *Dunaliella tertiolecta*. Plant Physiol 97: 147–153

Larsson UK and Andersson B (1985) Different degrees of phosphorylation and lateral mobility of two polypeptides belonging to the light-harvesting complex of Photosystem II. Biochim Biophys Acta 809: 396–402

Larsson UK, Ögren E, Öquist G and Andersson B (1986) Electron transport and fluorescence studies on the functional interaction between phospho-LHC-II and Photosystem I in isolated stroma lamellae vesicles. Photobiochem Photobiophys 13: 29–39

Larsson UK, Sundby C and Andersson B (1987a) Characterization of two different subpopulations of spinach light-harvesting chlorophyll *a-b*-protein complex (LHC-II): Polypeptide composition, phosphorylation pattern and association with Photosystem II. Biochim Biophys Acta 894: 59–68.

Larsson UK, Anderson JM and Andersson B (1987b) Variations in the relative content of the peripheral and inner light-harvesting chlorophyll *a/b*-protein complex (LHC-II) subpopulations during thylakoid light adaptation and development . Biochim Biophys Acta 894: 69–75

Leong TA and Anderson JM (1984a) Adaptation of the thylakoid membranes of pea chloroplasts to light intensities. I. Study on the distribution of chlorophyll-protein complexes. Photosynth Res 5: 105–115

Leong TA and Anderson JM (1984b) Adaptation of the thylakoid membranes of pea chloroplasts to light intensities. II. Regulation of electron-transport capacities, electron carriers, coupling factor (CF1) activity and rates of photosynthesis. Photosynth Res 5: 117–128

Ley AC and Mauzerall DC (1982) Absolute absorption cross sections for Photosystem II and the minimum quantum requirement for photosynthesis in *Chlorella vulgaris*. Biochim Biophys Acta 680: 95–106

Lichtenthaler HK and Meir D (1984) Regulation of chloroplast photomorphogenesis by light intensity and light quality. In: Ellis RJ (ed), Chloroplast Biogenesis, pp 245–258. Cambridge University Press, Cambridge

MacKinney G (1940) Criteria for purity of chlorophyll preparations. J Biol Chem 132: 91–109

Malkin S and Fork DC (1981) Photosynthetic units of sun and shade plants. Plant Physiol 67: 580–583

Mauzerall D (1986) The optical cross section and absolute size of a photosynthetic unit. Photosynth Res 10: 163–170

Mauzerall D and Greenbaum NL (1989) The absolute size of a photosynthetic unit. Biochim Biophys Acta 974: 119–140

Mawson BT, Morrissey PJ, Gomez A and Melis A (1994) Thylakoid membrane development and differentiation: Assembly of the chlorophyll *a-b* light-harvesting complex and evidence for the origin of M_r = 19, 17.5 and 13.4 kDa proteins. Plant Cell Physiol 35: 341–351

Melis A (1984) Light regulation of photosynthetic membrane structure, organization and function. J Cell Biochem 24: 271–285

Melis A (1989) Spectroscopic methods in photosynthesis: Photosystem stoichiometry and chlorophyll antenna size. Phil Trans R Soc Lond B 323: 397–409

Melis A (1991) Dynamics of photosynthetic membrane composition and function. Biochim Biophys Acta 1058: 87–106

Melis A and Anderson JM (1983) Structural and functional organization of the photosystems in spinach chloroplasts: Antenna size, relative electron transport capacity, and chlorophyll composition. Biochim Biophys Acta 724: 473–484.

Melis A and Brown JS (1980) Stoichiometry of system I and system II reaction centers and of plastoquinone in different photosynthetic membranes. Proc Natl Acad Sci USA 77: 4712–4716.

Melis A and Harvey GW (1981) Regulation of photosystem stoichiometry, chlorophyll *a* and chlorophyll *b* content and relation to chloroplast ultrastructure. Biochim Biophys Acta 637: 138–145.

Melis A, Manodori A, Glick RE, Ghirardi ML, McCauley SW and Neale PJ (1985) The mechanism of photosynthetic membrane adaptation to environmental stress conditions: A hypothesis on the role of electron-transport capacity and of ATP/NADPH pool in the regulation of thylakoid membrane organization and function. Physiol Veg 23: 757–765

Melis A, Mullineaux CW and Allen JF (1989) Acclimation of the photosynthetic apparatus to Photosystem I or Photosystem II light: Evidence from quantum yield measurement and fluorescence spectroscopy of cyanobacterial cells. Z Naturforsch Teil C 44: 109–118

Michel H, Hunt DF, Shabanowitz J and Bennett J (1988) Tandem mass spectrometry reveals that three PS II proteins of spinach chloroplasts contain *N*-acetyl-O-phosphothreonine at their NH_2 termini. J Biol Chem 25: 1123–1130

Morrissey PJ, Glick RE and Melis A (1989) Supramolecular assembly and function of subunits associated with the chlorophyll *a/b* light-harvesting complex II (LHC-II) in soybean chloroplasts. Plant Cell Physiol 30: 335–344

Mukerji I and Sauer K (1993) Energy transfer dynamics of an isolated light-harvesting complex of Photosystem I from spinach: Time resolved fluorescence measurements at 295 K and 77 K. Biochim Biophys Acta 1142: 311–320

Mullet JE (1983) The amino acid sequence of the polypeptide segments which regulate membrane adhesion (grana stacking) in chloroplasts. J Biol Chem 258: 9941–9948

Murakami A and Fujita Y (1988) Steady state of photosynthesis in cyanobacterial photosynthetic systems before and after regulation of electron transport composition: Overall rate of photosynthesis and PS I/PS II composition. Plant Cell Physiol 29: 305–311

Murakami A and Fujita Y (1991) Steady state of photosynthetic electron transport in cells of the cyanophyte *Synechocystis* PCC 6714 having different stoichiometry between PS I and PS II: Analysis of flash-induced oxidation-reduction of cytochrome *f* and P700 under steady state of photosynthesis. Plant Cell Physiol 32: 213–222

Murata N, Miyao M, Omata T, Matsunami H and Kuwabara T (1984) Stoichiometry of components in the photosynthetic oxygen evolution system of Photosystem II particles prepared with Triton X-100 from spinach chloroplasts. Biochim Biophys Acta 765: 363–369

Nanba O and Satoh K (1987) Isolation of a Photosystem II reaction center consisting of D-1 and D-2 polypeptides and cytochrome *b*-559. Proc Natl Acad Sci USA 84: 109–112

Neale PJ and Melis A (1986) Algal photosynthetic membrane complexes and the photosynthesis-irradiance curve: A comparison of light-adaptation responses in *Chlamydomonas reinhardtii*. J Phycol 22: 531–538

Ortiz W, Lam E, Ghirardi M and Malkin R (1984) Antenna function of a chlorophyll *a/b* protein complex of Photosystem I. Biochim Biophys Acta 766: 505–509

Owens TG, Shreve AP, Albrecht AC (1992) Dynamics and mechanism of singlet energy transfer between carotenoids and chlorophylls: Light-harvesting and non-photochemical fluorescence quenching. In: Murata N (ed), Research in Photosynthesis, Vol I, pp 179–186. Kluwer Academic Publishers Dordrecht

Öquist G, Samuelson G and Bishop NI (1980) On the role of β-carotene in the reaction center chlorophyll *a* antenna of Photosystem I. Physiol Plant 50: 63–70

Peter GF and Thornber JP (1988) The antenna components of Photosystem II with emphasis on the major pigment-protein, LHC-IIb. In: Scheer H and Schneider S (eds), Photosynthetic Light-Harvesting Systems, Organization and Function, pp 175–186. Walter de Gruyter, New York

Peter GF and Thornber JP (1991) Biochemical composition and organization of higher plant Photosystem II light-harvesting pigment-proteins. J Biol Chem 266: 16745–16754

Ranjeva R and Boudet AM (1987) Phosphorylation of proteins in plants: Regulatory effects and potential involvement in stimulus/response coupling. Ann Rev Plant Physiol 38: 73–93

Rijgersberg CP, Melis A, Amesz J and Swager JA (1979) Quenching of chlorophyll fluorescence and photochemical activity of chloroplasts at low temperature. In: Chlorophyll Organization and Energy Transfer in Photosynthesis. The CIBA Foundation Symposium 61 (new series), pp 305–322. Excerpta Medica, Elsevier, North Holland

Roelofs TA, Lee C-H and Holzwarth AR (1992) Global target analysis of picosecond chlorophyll fluorescence kinetics from pea chloroplasts A new approach to the characterization of the primary processes in Photosystem II α and β units. Biophys J 61: 1147–1163

Satoh K and Butler WL (1978) Competition between the 735 nm fluorescence and the photochemistry of Photosystem I in chloroplasts at low temperature. Biochim Biophys Acta 502: 103–110

Seifermanns-Harms D (1985) Carotenoids in photosynthesis. I. Location in photosynthetic membranes and light-harvesting function. Biochim Biophys Acta 811: 325–355

Shreve AP, Trautman JK, Owens TG and Albrecht AC (1991) A femtosecond study of electronic state dynamics of fucoxanthin and implications for photosynthetic carotenoid to chlorophyll energy transfer mechanisms. Chemical Physics 154: 171–178

Simpson D (1990) The structure of Photosystem I and II. In: Baltscheffsky M (ed) Current Research in Photosynthesis, Vol II, pp 725–732. Kluwer Academic Publishers, Boston

Simpson DJ, Machold O, Høyer-Hansen G and von Wettstein D (1985) *Chlorina* mutants of barley (*Hordeum vulgare* L). Carlsberg Res Commun 50: 223–238

Smith BM, Morrissey PJ, Guenther JE, Nemson JA, Harrison MA, Allen JF, Melis A (1990) Response of the photosynthetic apparatus in *Dunaliella salina* (green algae) to irradiance stress. Plant Physiol 93: 1433–1440

Smith H, Samson G and Fork DC (1993) Photosynthetic acclimation to shade - Probing the role of phytochromes using photomorphogenetic mutants of tomato. Plant, Cell and Environment 16: 929–937

Spangfort M and Andersson B (1989) Subpopulations of the main chlorophyll *a/b* light-harvesting complex of Photosystem II: Isolation and biochemical characterization. Biochim Biophys Acta 977: 163–170

Spangfort M, Larsson UK, Anderson JM and Andersson B (1987) Isolation of two different subpopulations of the light-harvesting chlorophyll *a/b* complex of Photosystem II. FEBS Lett 224: 343–347

Staehelin LA (1986) Chloroplast structure and supramolecular organization of photosynthetic membranes. In: Staehelin LA and Arntzen CJ (eds) Encyclopedia of Plant Physiology, Photosynthesis III, Vol, 19, pp 1–83. Springer-Verlag, Berlin

Sukenik A, Bennett J and Falkowski PG (1988) Changes in the abundance of individual apoproteins of light-harvesting chlorophyll *a/b*-protein complexes of Photosystem I and II with growth irradiance in the marine chlorophyte *Dunaliella tertiolecta*. Biochim Biophys Acta 932: 206–215

Telfer A (1987) The importance of membrane surface electrical charge on the regulation of photosynthetic electron-transport by reversible protein phosphorylation. In: Biggins J (ed) Progress in Photosynthesis Research, Vol II, pp 689–696. Martinus Nijhoff Publishers, Dordrecht

Telfer A, Allen JF, Barber J and Bennett J (1983) Thylakoid protein phosphorylation during state 1-state 2 transitions in osmotically shocked pea chloroplasts. Biochim Biophys Acta 722: 176–181

Telfer A, Whitelegge JP, Bottin H and Barber J (1986) Changes in the efficiency of P700 photo-oxidation in response to protein phosphorylation detected by flash absorption spectroscopy. J Chem Soc, Faraday Trans 2, 82: 2207–2215

Terashima I and Saeki T (1983) Light environment within a leaf. I. Optical properties of paradermal sections of *Camelia* leaves with special reference to differences in the optical properties of palisade and spongy tissues. Plant Cell Physiol 24: 1493–1501

Thielen APGM and van Gorkom HJ (1981a) Quantum efficiency and antenna size of Photosystem IIα, IIβ and I in tobacco chloroplasts. Biochim Biophys Acta 635: 111–120

Thielen APGM and van Gorkom HJ (1981b) Energy transfer and quantum yield in Photosystem II. Biochim Biophys Acta 637: 439–446

Thornber JP (1986) Biochemical characterization and structure of pigment-proteins of photosynthetic organisms. In: Staehelin LA and Arntzen CJ (eds) Encyclopædia of Plant Physiology, New Series, Vol 19, pp 98–115. Springer Verlag, New York

Thornber JP and Highkin HP (1974) Composition of the photosynthetic apparatus of normal barley leaves and a mutant lacking chlorophyll *b*. Eur J Biochem 41: 109–116

Thornber JP, Gregory RPF, Smith CA and Bailey JL (1967) Studies on the nature of the chloroplast lamella. I. Preparation and some properties of two chlorophyll-protein complexes. Biochemistry 6: 391–396

Thornber JP, Peter GF, Chitnis PR, Nechushtai R and Vainstein A (1988) The light-harvesting complex of Photosystem II of higher plants. In: Stevens SE Jr and Bryant DA (eds), Light-energy Transduction in Photosynthesis: Higher Plant and Bacterial Models, pp 137–154. The American Society of Plant Physiologists, Rockville, Maryland

Trautman JK, Shreve AP, Owens TG and Albrect AC (1990) Femtosecond dynamics of carotenoid-to-chlorophyll energy transfer in thylakoid membrane preparations from *Phaeodactylum tricornutum* and N*annochloropsis* sp. Chem Phys Lett 166: 369–374

Vogelmann T (1989) Penetration of light into plants. Photochem Photobiol 50: 895–902

Vogelmann TC (1993) Plant tissue optics. Ann Rev Plant Physiol and Plant Mol Biol 44: 231–251

Wang RT and Myers J (1974) On the State 1-State 2 phenomenon in photosynthesis. Biochim Biophys Acta 347: 134–140

Welty BA, Morishige DT and Thornber JP (1992) Identity of a group of 13.5–20 kDa polypeptides whose apparent abundance decreases during plastid development. Plant Cell Physiol 33: 1049–1055

Werst M, Jia Y, Mets L and Fleming GR (1992) Energy transfer and trapping in the Photosystem I core antenna A temperature study. Biophys J 61: 868–878

Wilhelm C and Wild A (1984) The variability of the photosynthetic unit in *Chlorella*. J Plant Physiol 115: 125–135

Yamamoto HY (1979) Biochemistry of the violaxanthin cycle in higher plants. Pure Appl Chem 51: 639–648

Zucchelli G, Jennings RC and Garlaschi FM (1992) Independent fluorescence emission of the chlorophyll spectral forms in higher plant Photosystem II. Biochim Biophys Acta 1099: 163–169

Chapter 30

Carotenoids: Localization and Function

Harry Y. Yamamoto
Hawaii Institute of Tropical Agriculture and Human Resources, University of Hawaii at Manoa, 3050 Maile Way, Honolulu, Hawaii 96822, USA

Roberto Bassi
Facoltà di Scienze MM.FF.NN. a Ca' Vignal, Università di Verona, strada Le Grazie, Verona, Italy 37134

Donald R. Ort and Charles F. Yocum (eds): Oxygenic Photosynthesis: The Light Reactions, pp. 539–563.

Summary

Carotenoids are components of every pigment-protein complex in the photosynthetic apparatus of higher plants. These pigments, previously referred to as 'accessory,' are now recognized to fulfill indispensable functions in light harvesting, protection against photooxidation, and regulation of Photosystem II efficiency. The wealth of information accumulated in recent years dealing with the closely related questions of carotenoid organization and functions are summarized in this chapter. In the first section the distribution of carotenoids in the different pigment proteins is reported showing that each photosystem subunit has its characteristic composition. The organization of the different xanthophylls within the antenna complexes is discussed on the basis of recent structural and biochemical evidence. In the second section, advances in photophysical mechanisms through which carotenoids perform their classical light harvesting and protective functions are discussed. In addition, particular attention is given to discussion of the xanthophyll cycle which, in conjunction with the transthylakoid ΔpH, down-regulates Photosystem II photochemical efficiency by non-radiative dissipation of energy in the light-harvesting complexes. Down-regulation helps to keep PS II traps open, thereby helping to maintain electron transport and to protect the reaction center from photoinhibition.

I. Introduction

Carotenoids are the most widely occurring pigments in nature. In photosynthetic cells, carotenoids are localized primarily in the photosynthetic apparatus and have been termed 'accessory' pigments because they are not directly involved in the primary reaction of photosynthesis or in electron transport. The term accessory implies a secondary, nonessential role. Yet, the lack of naturally occurring carotenoidless photosynthetic organisms suggests that carotenoids are essential for survival. Their critical function is thought to be as protective agents against lethal photooxidation. This function, which is generally attributed to β-carotene, neither explains the complex and highly conserved carotenoid composition in the various photosynthetic organisms nor the heterogeneous localizations of the pigments in the photosynthetic apparatus. In this chapter, we summarize knowledge on carotenoid localization and function in the photosynthetic membranes of higher plants. In the first part, the considerable advances made on localizing the diverse carotenoids of chloroplasts in pigment-protein complexes are discussed. The carotenoids are heterogeneously and yet specifically distributed in and within the complexes. Their localizations imply structural functions in addition to photophysical ones. In the second section, the classical functions of protecting against photooxidation and light-harvesting are briefly discussed including, for the latter function, the significance of so-called 'forbidden' excited states of carotenoids. In addition, particular attention is given to a recently-recognized third function, the non-radiative dissipation of energy in LHCII to down-regulate PS II photochemical efficiency. This function, which involves the xanthophyll cycle, is exceptional among carotenoid functions for its dynamic properties. The physiology and mechanism of xanthophyll-cycle mediated energy dissipation are rapidly developing areas of research. Figure 1 shows the structures of carotenoids in higher-plant thylakoids that are dicussed in this review.

Abbreviations: A – antheraxanthin; Asc – ascorbate; Chl – chlorophyll; CP – chlorophyll–protein; DCCD – dicyclohexylcarbodiimide; Deriphat – N-lauryl β-iminodipropionate; DTT – dithiothreitol; EFs – endoplasmic face, stacked; ELIP – early light-induced protein; EM – electron microscope; HPLC – high performance liquid chromatography; LHCI – light-harvesting complex of PS I; LHCII – light-harvesting complex of PS II; MGDG – monogalactosyldiacylglycerol; NPQ – non-photochemical fluorescence quenching; NRD – non-radiative energy dissipation; OEE – oxygen evolving enhancer; PAGE – polyacrylamide gel electrophoresis; PS I – Photosystem I; PS II – Photosystem II; PFu – periplasmic face, unstacked; RC – reaction center; SDS – sodium dodecylsulphate; V – violaxanthin; Z – zeaxanthin

II. Localization

A. Photosystem I

1. Composition of Pigment-proteins

PS I is a multisubunit complex that is located in the unstacked, stroma-exposed membranes where it forms the large PFu particles (10.3 × 12.5 nm) visible by freeze fracture EM (Simpson, 1983). It is composed of a core complex (PS I core) and a light-harvesting component (LHCI). A particle containing both components can be obtained by solubilization with anionic detergents and sucrose gradient ultracentrifugation (Mullet et al., 1980; Bassi and Simpson, 1987; Nechushtai et al., 1987). Such a preparation has Chl/P700 ratio of about 210 and a Chl *a*/*b* ratio of 6–6.5 and contains at least 15 polypeptides. Similar preparations but in smaller yield can be obtained by SDS-PAGE or Deriphat-PAGE (Bassi et al., 1985; Peter and Thornber, 1991b). The PS I-LHCI complex has, besides Chl *a* and *b*, *β*-carotene, lutein and violaxanthin but little to no neoxanthin (Anderson et al., 1983a). The complex also has zeaxanthin and antheraxanthin when isolated following light treatment (Lee and Thornber, 1995). These results are consistent with previous analyses of isolated stroma lamellae enriched in PS I obtained by mechanical fractionation from dark (Henry et al., 1983) and light-treated chloroplasts (Siefermann and Yamamoto, 1976).

2. Photosystem I Core Complex

This is a Chl *a* binding complex which has an apparent size of 250 kDa in non-denaturing PAGE (Bassi et al., 1985; Bassi and Simpson, 1987; Peter and Thornber, 1991b). The complex binds P700 and can photoreduce $NADP^+$ in the presence of ferredoxin and ferredoxin-NADP reductase (Bruce and Malkin, 1988). All pigments are bound to the two major protein subunits, the products of *psaA* and *psaB* chloroplast genes, with P700 being located at the interface between the two (Golbeck, 1987). Together, the subunits bind about 90 Chl *a* (Bassi and Simpson, 1987) and 14 *β*-carotene molecules (Bengis and Nelson, 1975; Rawyler et al., 1980; Anderson et al., 1983a; Haworth et al., 1983; Siefermann-Harms, 1984). The organization of *β*-carotene in PS I is currently unknown. However, it is expected to be elucidated when the structure of PS I from *Synechococcus* is resolved from 3.8 Å diffraction crystals (Witt et al., 1987; Krauss et al., 1993; Chapter 18).

3. Light Harvesting Complex I

In higher plants, LHCI splits into two moieties, designated LHCI-680 and LHCI-730 (see Table 1 for alternative nomenclature) according to their fluorescence emission maxima at 77 K (Lam et al., 1984). Most authors agree that LHCI polypeptides are the products of four *Lhc*a genes

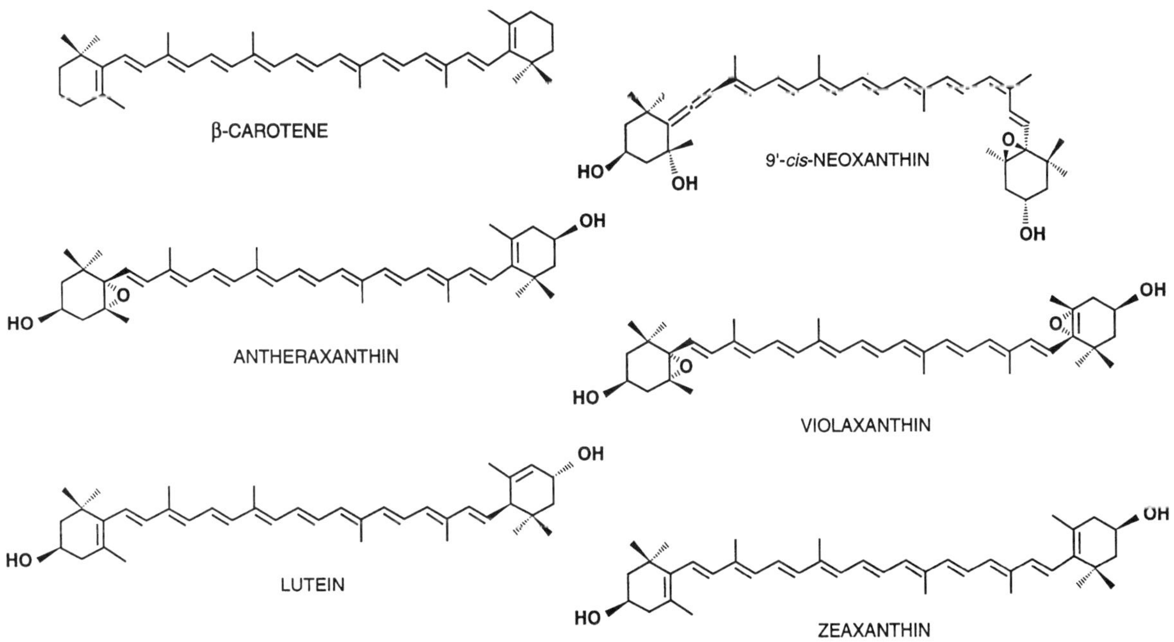

Fig. 1. Structure of carotenoid molecules in the photosynthetic apparatus of higher plants.

Table 1. Correspondence between genes encoding chlorophyll *a/b* proteins and the chlorophyll-protein complexes containing the gene product

Gene	Gene Product/Pigment-Protein Complex		
	Green et al., 1991	Thornber et al., 1991	Bassi et al., 1990
Lhca1	Type I LHCI	LHC Ib	LHCI-730
Lhca2	Type II LHCI	?	LHCI-680
Lhca3	Type III LHCI	LHC Ia	LHCI-680
Lhca4	Type IV LHCI	LHC Ib	LHCI-730
Lhcb1	Type I LHCII	LHC IIb 28kDa	LHCII
Lhcb2	Type II LHCII	LHC IIb 27kDa	LHCII
Lhcb3	Type III LHCII	LHC IIb 25kDa	LHCIIa
Lhcb4	Type II CP29	LHC IIa	CP29
Lhcb5	Type I CP29	LHC IIc	CP26
Lhcb6	CP24	LHC IId	CP24

(This table is modified from Jansson et al., 1992).

(Hoffman et al., 1987; Knoetzel et al., 1992). The first LHCI isolated was as a high molecular weight complex from *Chlamydomonas reinhardtii* (Wollman and Bennoun, 1982). In this organism the complex appears to be different from higher plants in that the fluorescence emission of the intact complex is at 705 rather than 735 nm. Moreover, the number of LHCI polypeptides is higher (7–10 vs 4–5 in higher plants) (Bassi et al., 1992). There are no rigorous measurements for the pigment-protein stoichiometry in either LHCI-730 or LHCI-680 due to pigment loss when dissociating individual proteins from the LHCI complex. Indirect evidence suggests that each LHCI polypeptide may bind 8–10 Chl molecules (Bassi et al., 1992).

The reported Chl *a/b* ratios for LHCI have ranged widely from 1.4 to 3.0 (Bassi et al., 1985; Bassi and Simpson, 1987). The lower values are probably due to some pigment proteins like LHCI-680 and CP24 being prone to loss of Chl *a* when subjected to PAGE. The number of chlorophyll molecules bound to LHCI-680 in a PS I unit is approximately 30–35 (Bassi and Simpson, 1987). The calculated carotenoid composition of LHCI based on this is shown in Table 2. Determinations on the individual pigment-proteins dissociated from the PS I-LHCI complex are qualitative (Thornber et al., 1993), confirming that violaxanthin and lutein are present while neoxanthin is absent in LHCI. The presence of β-carotene has not been reported for purified LHCI; however, since its concentration in PS I is decreased by half when LHCI is removed from a PS I-LHCI complex, β-carotene may be a component of LHCI similar to the minor pigment-proteins CP29, CP26, and CP24 of PS II (Bassi et al., 1993).

B. Photosystem II

1. Composition of Pigment-proteins

Photosystem II is located in the stacked membranes of granal chloroplasts where it forms the large EFs (11.7 × 15.5 nm) freeze-fracture particles with the core complex. The outer LHCII is arranged in the complementary fracture face (PFs) to form 9.0 × 10.3 nm particles (Simpson, 1978; Miller and Kushman, 1979). PS II is composed of a Chl *a* binding core complex and several surrounding Chl *a/b* proteins which constitute the outer antenna. The whole PS II complex can be prepared as stacked membranes, free of other thylakoid complexes (Berthold et al., 1981). When solubilized, the complex splits into the core complex and the outer antenna components that can be separated by sucrose gradient ultracentrifugation or PAGE (Bassi et al., 1987; Dainese and Bassi, 1991; Peter and Thornber, 1991a). The PS II membranes and the isolated core complex can catalyze electron transport from water to quinone analogues or other electron acceptors (for a review see Satoh, Chapter 11). Carotenoids are present in both the PS II core complex and the antenna moieties. β-carotene is bound mainly to the PS II core complex whereas the xanthophylls are components of the antenna system.

Table 2. Pigment composition of PS I proteins

	Chl *a*	Chl *b*	β-carotene	Lutein	Neoxanthin	Violaxanthin	Reference
PS I-LHCI	186	24	27	12	0-2	9	a, b, c
PS I-core	100	–[2]	14	–	–	–	b, c
LHCI*	86	24	13	24	0[1]	9	d
Lhca 1	ND	ND	ND	+	–	+	e
Lhca 2	ND	ND	ND	+	–	ND	e
Lhca 3	ND	ND	ND	+	–	ND	e
Lhca 4	ND	ND	ND	+	–	ND	e

* Calculated from the difference between PS I-LHCI complex and PS I core complex (Siefermann-Harms, 1985).
[1] The neoxanthin value was set to 0 since stroma membranes lack this pigment (Henry et al., 1983). [2] None.
a. Anderson et al. (1983a); b. Bengis and Nelson (1975); c. Setif et al. (1980); d. Siefermann-Harms (1985); e. Thornber et al. (1993)

2. Photosystem II Core Complex

The core complex binds the electron transport cofactors Mn^{2+}, P680, pheophytin, Q_A, as many as 50–55 antenna Chl *a* molecules, and carotenoids (Table 3). The two sets of cofactors are bound to distinct polypeptides, the electron transport components being bound to the D1 and D2 polypeptides while the antenna pigments are located on the two homologous CP43 and CP47 proteins. Additional subunits of the core complex include the two cytochrome *b*-559 subunits and the three oxygen-evolving enhancers (OEE).

Chlorophyll-binding proteins in PS II include the D1 and D2 polypeptides encoded by *psbA* and *psbD*, respectively. D1 and D2 are composed of 353 amino acid residues in most species; however, their apparent molecular weights as determined by SDS-PAGE are 32 and 34 kDa respectively. The isolated complex, consisting of D1 and D2 polypeptides and Cyt *b*-559, is photochemically active (Namba and Satoh, 1987). Bound pigments include 4 (Namba and Satoh, 1987) or 6 Chl *a* molecules and two pheophytins (Gounaris et al., 1990; Kobayashi et al., 1990). The number of β-carotene molecules is also under debate. One (Namba and Satoh, 1987) and two (van Dorsen et al., 1987; Gounaris et al., 1990; Kobayashi et al., 1990) molecules per RC have been reported. However, spectroscopically distinct β-carotene forms can be induced by the interaction between carotenoid molecules in dimeric RC complexes, supporting the view of one β-carotene per D1-D2-cyt *b*-559 complex (Newell et al., 1991). The organization of β-carotene in this complex can only be hypothesized on the basis of a homology with spheroidene in *Rb. sphaeroides* (Yates et al., 1988; Feher et al., 1989). If the homology holds, β-carotene should be located in D2, near the accessory Chl *a* which lies between the pheophytin and P680.

3. CP47 and CP43

In light of the unequivocal location of PS II reaction center in the D1-D2-Cyt *b*-559 complex (Nanba and Satoh, 1987), CP47 and CP43 must be considered a part of the light-harvesting system. They bind Chl *a* and carotenoids (Table 3), and are encoded by the *psbB* and *psbC* genes. These genes are located in the chloroplast genome close to the *psbA* and *psbD* genes that encode the D1 and D2 proteins. The deduced protein sequences of CP47 (508 residues) and CP43 (461 residues) are well conserved, with homology being 94 or 95% between higher plant proteins and 72 or 77% with the cyanobacterial proteins respectively (Vermaas et al., 1987). A common structure with six transmembrane helices can be hypothesized from the significant homology between the two proteins, (Holschul et al., 1984; Morris and Hermann, 1984; Bricker, 1990). The two proteins are thought to bind 20–25 Chl *a* molecules each (de Vitry et al., 1984; Satoh, 1985), though lower values have been suggested (Glick and Melis, 1989) based on functional measurements in developing plant material and mutants lacking Chl *b* or by biochemical measurements (Barbato et al., 1991). The lower values should be viewed with caution as a recent finding shows that intermittent light-grown plants not only change their subunit stoichiometry but also the number of Chl molecules per polypeptide (Marquardt and Bassi, 1993). A more complete study on pigment binding to PS II core subunits is needed.

The major carotenoid of CP43 and CP47 is β-carotene, but lutein may be present in low amounts (Table 2) (Bassi et al., 1993). Delepelaire and Chua (1979) reported that there were more β-carotene molecules in CP43 than in CP47 (probably 5 and 3 respectively). The organization of this pigment in the complex is unknown but the orientation of β-carotene

Table 3. Pigment composition of PS II proteins (molecules per polypeptide). The values refer to experimentally determined stoichiometry. Values within brackets refer to the figure obtained by assuming that 12 chlorophyll molecules bind to each pigment-protein as suggested by sequence homology to LHCII complex resolved by electron crystallography (Kühlbrandt et al., 1994).

	Chl *a*	Chl *b*	*β*-carotene	Lutein	Neoxanthin	Violaxanthin	Reference
D1/D2 heterodimer	6+ 2 phaeo.	–[1]	1(2)	–	–	–	c, d, f
CP47	25	–	2	0.4	–	–	a, c, d, f
CP43	25	–	3	1	–	–	a, b, h
PS II core	~56	–	5	–	–	–	a, b, h
CP29	6(9)	2(3)	0.1(0.15)	1.3(2)	0.5(0.7)	1(1.5)	a, h
CP26	6(8)	3(4)	0.1(0.15)	1.3(1.7)	0.4(0.5)	0.5(0.7)	a, g
CP24	3(7)	2(5)	0.2(0.5)	1.2(2.8)	–	0.5(1.3)	a, g
LHCII	7	5		2	0.4–1*	0.05–0.02	a, e
Total PS II membranes*	160	70	10	2.8	7	3.7	a

* Calculated from Bassi et al 1993.
[1] None. a. Bassi et al. (1992); b. Delepelaire and Chua (1979); c. Gounaris et al. (1990); d. Kobayashi et al. (1990); e. Kühlbrandt, et al. (1994); f. Namba and Satoh (1987); g. Peter and Thornber (1991a); h. de Vitry et al. (1984)

in CP43 and CP47 is clearly different. *β*-carotenes in CP43 are largely parallel to the plane of the membrane and those in CP47 are perpendicular. The absorption maxima of *β*-carotene in CP43 were found to be blue shifted by 10–15 nm as compared to those of CP47 (Breton and Satoh, 1987).

4. Major LHCII Complex

The light-harvesting chlorophyll *a*/*b* protein of PS II of higher plants (LHCII) is the most studied pigment protein complex and, therefore, it will be discussed in some detail (see Table 1 for alternative nomenclature). In fact, its sequence homology to many other proteins in the photosystems of higher plants and algae, as well as its photoprotective functions, makes it the most useful model for studies on the organization of carotenoids in antenna systems.

The major LHCII complex was the first Chl-protein described. It constitutes about one third of the total thylakoid protein and binds half of the total chlorophyll. This protein is mostly, if not completely, present in a trimeric form in the membranes (Peter and Thornber, 1991a). HPLC analysis of purified LHCII shows that it contains, besides Chl *a* and Chl *b*, the xanthophylls lutein, neoxanthin, and violaxanthin. The presence of *β*-carotene in LHCII has been suggested in earlier reports (Braumann et al., 1982; Lichtenthaler et al., 1982), but the analysis of a purified preparation ruled this out (Bassi et al., 1993).

While HPLC pigment analysis can accurately determine the relative pigment amounts, the pigment-protein stoichiometry cannot be determined with the same accuracy, causing intense debate. Biochemical determinations yield chlorophyll per LHCII monomer values of 7 (Ryrie et al., 1980), 12 (Dainese and Bassi, 1991), or 15 (Butler and Kühlbrandt, 1988). Image reconstruction at 6 Å resolution (Kühlbrandt and Wang, 1991) confirmed the higher value, but a recent determination at 3.4 Å supported the value of 12 or 13 (Kühlbrandt et al., 1994). On the basis of the 1.4 molar ratio between Chl *a* and *b* for a highly purified complex, 7 Chl *a*, 5 Chl *b* and two lutein molecules can be assigned to each LHCII monomer (Bassi et al., 1993; Juhler et al., 1993). Two additional pigments have been found in LHCII preparations: violaxanthin (V) and neoxanthin (N). V has always been found in variable substoichiometric amounts, i.e. ten times less with respect to lutein (Setif et al., 1980; Siefermann-Harms, 1984). The case of N is less clear since it can be found in amounts of one molecule per LHCII monomer; however, this value is variable in different preparations (Peter and Thornber, 1991a; Bassi et al., 1993), suggesting that both V and N are located more peripherally than lutein in the complex.

The recent results from Kühlbrandt's group that resolved trimeric LHCII to 3.4 Å substantiate the biochemical data. As suggested by sequence analysis, the LHCII structure exhibits three trans-membrane helices, oriented with the N-terminus extending on the stromal face while the C-terminus is in the lumenal space and includes four short amphiphilic helices parallel to the plane of the membrane (Kühlbrandt et al., 1994). The first and the third transmembrane helices (called respectively B and A) are highly conserved and homologous to one another, suggesting that the protein derives from a duplication event of an ancestral gene (see Chapter 28). Helices A and B are longer than needed for membrane spanning and extend out of the stromal side of the membrane. Moreover, they are tilted with respect to the membrane plane and cross each other in an X-shaped structure. The third helix is separate and perpendicular to the membrane plane. The structure of LHCII is shown in Fig. 2.

In the monomeric LHCII, only two carotenoid molecules are detected by electron crystallography as 30 Å long regions of density in the center of the structure, forming a cross brace between the A and B helices (Fig. 2). One of these two carotenoids, presumably lutein, connects the C-terminal on the lumenal side to the loop between helices A and C on the stromal side. The other lutein molecule symmetrically connects the N-terminus to the loop between helices C and B on the lumenal side. This has an important structural role by providing a direct and strong link between the peptide loops at both surfaces of the molecule. In the structure of LHCII, lutein molecules appear to be in van der Waals contact with 7 chlorophyll molecules, presumably Chl *a*. This is close enough for direct energy transfer, (see later section). Lutein (or the related xanthophyll Loroxanthin) seems to be essential for LHCII accumulation and stability as shown in lutein-deficient mutants (Chunaev et al., 1991). Neoxanthin and violaxanthin are not detectable, suggesting that either

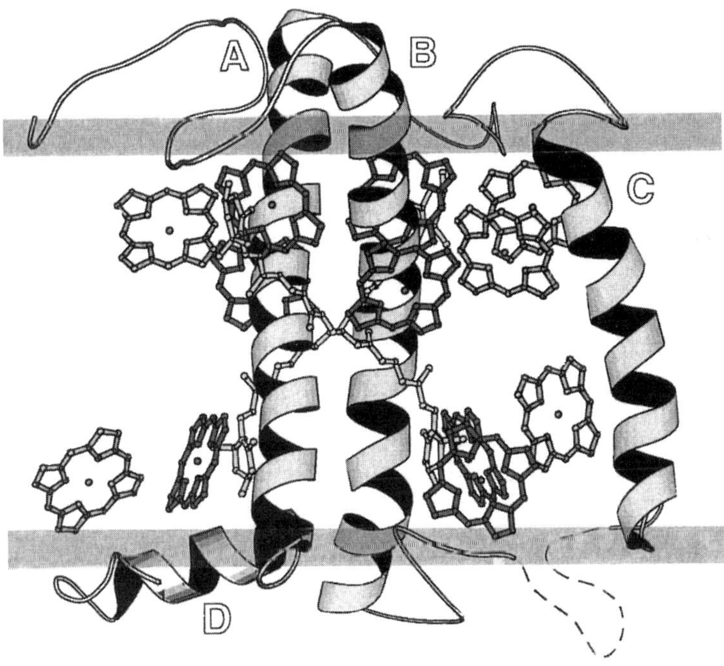

Fig. 2. Overall view of LHCII monomer as resolved from electron crystallography (Kühlbrandt et al., 1994). Lutein extends from the stromal to the lumenal surface of the molecule forming a cross brace near helices A and B.

they don't occupy specific positions in the complex or are not present in each monomer. In fact, LHCII is a heterogeneous protein. Not only can several apoproteins be resolved from the LHCII complex by denaturing electrophoresis (Spangfort and Andersson, 1989; Di Paolo et al., 1990; Sigrist and Staehelin, 1992), but also a number of LHCII subpopulations can be resolved by isoelectric focussing (Larsson et al., 1987; Bassi et al., 1988; Bassi and Dainese, 1992). This is consistent with the large number of highly homologous genes which have been found in several species (Dunsmuir, 1985; McGrath et al., 1991). These genes fall into three types known as *Lhcb*1, *Lhcb*2 and *Lhcb*3 (Table 1).

A set of interesting results was obtained from reconstitution experiments with LHCII complexes. Chl *a* and *b* were indispensable to the reassembly of denatured apoprotein, but in the absence of xanthophylls, the stability of the reconstituted complex was significantly decreased (Plumley and Schmidt, 1987; Paulsen et al., 1990; Cammarata and Schmidt, 1992). Since neoxanthin can substitute to some extent for lutein (Plumley and Schmidt, 1987), the possibility that neoxanthin (and perhaps also violaxanthin) actually occupies specific sites in some components of the heterogeneous LHCII complex cannot be excluded. Reconstitution of LHCP deletion mutants overexpressed in *E. coli* indicated that most of the N-terminal and part of the C-terminal hydrophobic regions are essential for pigment binding (Paulsen and Hobe, 1992). The significance of these effects specifically on carotenoid binding, however, was not determined. An alternative hypothesis for xanthophyll location in the photosynthetic membrane involving lipids is discussed later in the text.

5. Minor Pigment-proteins of PS II Antenna

Besides LHCII, three additional Chl *a*/*b*-xanthophyll proteins are present in the PS II complex. They are CP29, CP26, and CP24, named from their apparent mass in non-denaturing (green) gels, and are encoded by the nuclear genes *Lhcb4*, *Lhcb*5 and *Lhcb*6, respectively. They are synthesized in the cytoplasm as higher molecular-weight precursors, imported into the chloroplast and subsequently inserted into the thylakoids to form pigment-protein complexes which have a higher Chl *a*/*b* ratio than LHCII. They are present in PS II in equal amounts; however, together they bind only 15% of the total PS II chlorophyll (vs. 63% by LHCII).

The primary sequences of CP29, CP26, and CP24 show that they are highly homologous with each other and with other LHC gene products. This suggests that their molecular structures are similar to LHCII. Nevertheless, recent results show that they are distinct in their carotenoid content, most noticeably in binding more than 80% of the PS II violaxanthin. This composition strongly suggests that they have an important role in the regulation of energy transfer to the PS II core complex.

The pigment-protein complex CP29 was the first minor complex to be distinguished from LHCII (Machold et al., 1979). Its apparent mass is slightly higher (31 kDa) than that of the major LHCII polypeptides, both in denaturing and green gels (Camm and Green, 1980; Bassi et al., 1987). The protein is N-terminally blocked. Partial sequences for CP29 from spinach, maize, and tomato (Henrysson et al., 1989) show that it is coded by the *Lhcb*4 gene, isolated and sequenced by Morishige and Thornber (1991). *Lhcb*4 is the largest LHC gene, coding for a mature protein of approximately 257 amino acids. It is highly conserved between maize and barley, not only in the mature protein but also in the pre-sequence region (Morishige and Thornber, 1991). The larger size is due to a 42-residue insertion located just before the first transmembrane helix that is not present in other LHC genes. This complex has a Chl *a*/*b* ratio of 2.2–2.8 (Dainese et al., 1990; Dainese and Bassi, 1991; Peter and Thornber, 1991a), and contains lutein, violaxanthin, and neoxanthin as additional pigments (Peter and Thornber, 1991a; Bassi et al., 1993). Values of four (Barbato et al., 1989; Irrgang et al., 1991), eight (Dainese and Bassi, 1991), and ten (Henrysson et al., 1989) Chl molecules per polypeptide have been reported.

The CP26 pigment-protein complex has been described in maize and spinach as having an intermediate Chl *b* content between that of CP29 and LHCII (Bassi et al., 1987, Dunahay et al., 1987). Its pigment components include violaxanthin, lutein, neoxanthin, and Chl *a*/*b* in a 2.2 ratio (Dainese and Bassi, 1991). Lower (1.8) and higher (2.7) Chl *a*/*b* ratio values have also been reported (Barbato et al., 1989, Thornber et al., 1993). The binding of between 9 and 11 Chl molecules per polypeptide has been determined (Dainese and Bassi, 1991; Peter and Thornber 1991a). In urea gels, two closely migrating apoproteins are resolved, in similar amounts, with apparent masses of 28 and 29 kDa (Bassi et al., 1987). Both polypeptides are N-terminally blocked and therefore, the actual molecular weights and the maturation sites are not known. Antibodies against oligopeptides obtained from the *Lhcb5* gene recognize the CP26 apoprotein (Allen and Staehelin, 1992) implying that this is the coding gene. This has been confirmed by direct protein sequencing.

CP24 is the product of the *Lhcb6* gene as shown by N-terminal sequencing (Morishige et al., 1990). *Lhcb6* genes have been sequenced from tomato and spinach (Schwartz and Pichersky, 1990; Spangfort et al., 1990) coding for a 210 amino acid protein. The carotenoid content is qualitatively similar to that of CP29 and CP26 except for the absence of neoxanthin. Divergent values have been reported for the Chl *a*/*b* ratio (1.6–0.8) and the Chl to polypeptide ratio (5–13) (Dainese and Bassi, 1991; Peter and Thornber, 1991a). The latter is presumably due to the preferential loss of Chl *a* during PAGE. CP24 has a red absorption peak at 675.5 nm and fluorescence emission at 681.5 nm (Jennings et al., 1993a,b). The characteristics of the protein with respect to molecular mass, pigment composition and immunological cross-reactions are similar to the LHCI-680 component *Lhca2*, leading to the suggestion in earlier reports that they were the same protein (Bassi et al., 1987). This hypothesis was later disproved with monoclonal antibodies (Di Paolo et al., 1990) and N-terminal sequencing (Morishige and Thornber, 1991). However, a polypeptide very similar to CP24 is present in LHCI as recently found in *Chlamydomonas reinhardtii* (Bassi et al., 1992). The absence of neoxanthin is a characteristic that CP24 shares with LHCI polypeptides (Henry et al., 1983; Nechushtai et al., 1987).

Table 3 shows the pigment composition of the PS II proteins. The actual number of carotenoid (and Chl) molecules present in each pigment-protein complex is difficult to determine. Loss of pigments during isolation and/or difficulties in accurately determining protein concentration in the small amount of highly purified pigment-protein available are the chief causes for differences in the reported ratios for each complex. Although the stoichiometry for each pigment is different, it has been consistently reported that CP26, CP24 and CP29 polypeptides bind a lower number of pigment molecules than LHCII (Henrysson et al., 1989; Barbato et al., 1991; Dainese and Bassi, 1991; Irrgang et al., 1991). These results should be reconsidered in the light of the strong homology between all the LHC genes (see Pichersky and Jansson, Chapter 28 of this book) and results from electron crystallography. In fact, the amino acid residues coordinating eight out of 12–13 chlorophyll molecules that have been identified in LHCII (Kühlbrandt et al., 1994) are markedly conserved in the sequences of the *Lhcb* and *Lhca* gene products so far identified, the only exception being a histidine residue located close to the C-terminus of the LHC gene that is missing in CP24. Therefore, the number of Chl molecules for each LHC polypeptide can be estimated as close to 12. The number of carotenoids bound, assuming 12 chlorophylls per protein, is reported within brackets in Table 3. However, the unidentified Chl binding residues may not be conserved.

In the three minor Chl *a*/*b* proteins (CP24-CP29), lutein is present in more than one (probably two) molecules per polypeptide whereas neoxanthin is always present in substoichiometric amounts. The most noticeable feature in these proteins is the high violaxanthin content and/or its de-epoxidation products, antheraxanthin and zeaxanthin (Yamamoto et al. 1962), which are obtained under high-light conditions (Thayer and Björkman, 1992; Bassi et al., 1993; Ruban et al., 1994; Lee and Thornber, 1995). These pigments are components of the xanthophyll cycle, which has been implicated in the thermal dissipation of energy in the PS II antennae (see following section for further discussion). The xanthophyll-cycle pigments are present in variable and substoichiometric amounts relative to protein, depending on isolation procedure (Peter and Thornber, 1991a; Thayer and Björkman, 1992; Bassi et al., 1993; Ruban et al., 1994). It is unclear where or how violaxanthin (also neoxanthin) is bound to minor Chl *a*/*b* proteins. A tentative hypothesis can be proposed based on three observations. 1) CP26 and CP24 tightly bind between 4 and 10 times more lipids (mainly MGDG), respectively, than LHCII or PS II core complex (Tremoliere et al., 1993). 2) Violaxanthin de-epoxidase requires MGDG (Yamamoto et al., 1974). 3) The xanthophyll pool size is inversely related to the amount of Chl *a*/*b* proteins (reviewed in Demmig-Adams, 1990). It is hypothesized that xanthophyll-cycle pigments are dissolved in lipids that are tightly bound to the complexes rather than bound to specific sites within the polypeptide. Irrespective of the way xanthophyll cycle pigments bind to CP26, CP29, and CP24, it is interesting that these proteins have been consistently proposed to have a pericentral location in PS II units, thus acting as connecting antennae between the major LHCII complex and the PS II core complex (Peter and Thornber, 1991a; Harrison and Melis, 1992; Bassi and Dainese, 1992; Jansson, 1994). The location of xanthophyll cycle pigments in these proteins is consistent with their action as switches for excitation energy, which is either funneled to RCII or diverted to heat dissipation. There is also evidence that antennae proteins are not critical for xanthophyll cycle activity. The cycle is present in plants that have reduced antenna (Jahns and Krause, 1994) and in the *chlorina-f2* mutant which lacks all antennae proteins (Leverenz et al., 1992; Marquardt and Bassi, 1993; Król et al., 1995). In these cases, de-epoxidation does not appear to be related to energy dissipation.

In summary, carotenoids are present in all of the Photosystem I and II pigment-proteins. The chloroplast-encoded subunits of the PS II core are Chl *a*/*β*-carotene complexes while the nuclear-encoded subunits of the antenna system are Chl *a*/*b*/ xanthophyll complexes. Each pigment protein of the antenna system has a particular carotenoid composition in addition to the common lutein component. Lutein, in addition to a possible energy transfer function, appears to have an important structural role in the assembly of LHCII and probably also of other LHC gene products. The functional meaning of the complex topological distribution of the other xanthophylls is not clear; however, the location of xanthophyll-cycle pigments (V, A, Z) in the minor pigment-proteins CP24, CP26, and CP29, which connect the major antenna LHCII to the PS II core complex, strongly suggests that the non-radiative energy dissipation mechanism is located here. The carotenoid distribution in PS I is less clear but the data so far suggest that the situation is not very different from PS II.

C. Other Chlorophyll and/or Carotenoid Binding Proteins

Besides the relatively well-characterized polypeptides described above, recent reports suggest that other carotenoid-binding proteins may exist in the photosynthetic apparatus.

1. Psb*S*

The gene coding for a 22 kDa protein, previously reported to be important in PS II assembly (Hundal et al., 1990), has been sequenced (Wedel et al., 1992; Kim et al., 1992) and found to have homology with *Lhcb* genes. On the basis of hydrophobicity plots, *Psb*S is hypothesized to have four rather than three transmembrane helices (as other LHC genes). A recent report claims that it is a Chl *a/b*-binding subunit of Photosystem II and has a small amount of associated carotenoids (Funk et al. 1994).

2. LHCIIe and ELIPs

Following several earlier reports (Krishnan and Gnanam, 1979; Irrgang et al., 1990) the finding of a low molecular-mass Chl *a*/*b* protein, LHCIIe, has been reported which is greatly enriched in xanthophylls (Peter and Thornber, 1991a). To date, other labs have not been able to confirm this interesting report. The possibility exists that the above result may be related to the presence of ELIPs, a newly discovered class of proteins transiently synthesized during light exposure of dark grown plants (Grimm et al., 1989; Hundal et al., 1990) or after photoinhibition (Adamska et al., 1992) but also present at low levels in fully-green leaves (Droppa et al., 1987; Marquardt and Bassi, 1993). Their deduced sequence shows a molecular weight between 14 and 17 kDa (Meyer and Kloppstech, 1984). N-terminal sequencing of LHCIIe is needed to verify this hypothesis. ELIP's sequences show homology to the LHC polypeptides and to the carotenoid binding protein of green algae (Levy et al., 1993). Recent results suggest that ELIPs are binding sites for photoconvertible xanthophylls and may replace or complement a deficient xanthophyll binding capacity under conditions of light stress (Król et al., 1995). This view is supported by the cloning of an ELIP-like gene from *Dunaliella* called *Cbr*. The 17 kDa gene product is synthesized transiently during V to Z de-epoxidation and becomes associated with a LHCII fraction that thereafter contains zeaxanthin (Lerst et al., 1991; Levy et al., 1993). The isolation of ELIPs and Cbr in their putative pigment-binding form is needed to clarify this point.

3. Carotenoids in the Chloroplast Envelope

Chloroplast carotenoids, although mainly located in the photosynthetic apparatus, are not restricted to the thylakoid membranes. The envelope of plant plastids (Block et al., 1983a,b) as well as the plasma membrane of cyanobacteria (Omata and Murata, 1983) contain carotenoids. The major carotenoid in these membranes has been shown to be violaxanthin and its de-epoxidation products, although neoxanthin, lutein, and β-carotene are also present. The xanthophyll to carotene ratio is higher in the envelope than in thylakoids (6 versus 3). In plants, the two membranes of the envelope have distinct carotenoid compositions, the outer being richer in neoxanthin (Block et al., 1983b). The role of carotenoids in the envelope is unknown; the most probable hypothesis is that they are synthesized in this compartment. How the pigments are transported or transferred to the thylakoid is unknown. The level of violaxanthin in the envelope decreases coincident with light-induced de-epoxidation of violaxanthin in the thylakoid, suggestive of a dynamic link between these membranes (Siefermann-Harms et al., 1978). However, as yet, there is no direct evidence for a transport system that carries pigment from the envelope to the thylakoid. In any case, the function and organization of carotenoids in the envelope are probably different from thylakoids since the envelope has neither chlorophyll (Joyard et al., 1991; Joyard et al., 1992) nor de-epoxidase activity. The carotenoids may be freely soluble in the membrane lipids (Joyard et al., 1992) or in part, bound to outer-envelope proteins (Markwell et al., 1992). In cyanobacteria, however, carotenoids have been reported to be bound to at least two intrinsic membrane proteins of the cell envelope of *Synechocystis* (Bullerjahn and Sherman, 1986) and *Synechococcus* (Masamoto et al., 1987; Reddy et al., 1989). These are 42–45 kDa proteins which accumulate only during growth at high light intensity (Masamoto et al., 1987), suggesting that they play a role in photoprotection.

III. Function

Carotenoids have two well-recognized functions in photosynthesis. They protect against photooxidative damage and act as accessory light-harvesting pigments. The first is arguably an essential function that explains the presence of carotenoids in all photosynthetic organisms. The second is also important and possibly critical to organisms that are in environments where light available for absorption by chlorophyll is relatively low. The ability to utilize light that is poorly absorbed by chlorophyll may be important even in leaves because there is a substantial spectral gradient through the leaf cross section (Vogelman, 1993). A third function is now evident. In higher plants, zeaxanthin (Z) and antheraxanthin (A) formed by the xanthophyll cycle (Yamamoto et al., 1962) increase non-radiative dissipation of energy (NRD) in the pigment bed of PS II (Demmig et al., 1987; Gilmore and Yamamoto, 1993b). NRD reduces or down-regulates the intrinsic photochemical efficiency of PS II by diverting energy from the reaction center. Down-regulation helps to keep PS II traps open, thus preventing over-reduction of PS II that would otherwise lead to photoinhibition. Xanthophyll-related non-radiative energy dissipation can therefore also be interpreted to protect against photoinhibition by excess light (reviewed by Young, 1991; Demmig-Adams and Adams, 1992; Björkman and Demmig-Adams, 1993). In phytoplankton, diatoxanthin formed from diadinoxanthin (Hager and Stransky, 1970) may have a similar function (Demers et al., 1991; Olaizola and Yamamoto, 1994; Olaizola et al., 1994).

A. Protecting against Photooxidation

The classic experiments by Sistrom et al. (1956) demonstrated that carotenoids are essential for protecting photosynthesis against photooxidative damage. They showed that photosynthetic bacterial mutants which were devoid of colored carotenoids were highly susceptible to damage by the combined effects of light and oxygen compared to wild type. The protective function of carotenoids against photooxidation has been reviewed previously (Krinsky, 1984; Siefermann-Harms, 1987).

Excited singlet chlorophyll (^{1}Chl) can form triplet chlorophylls (^{3}Chl*) by intersystem crossing. ^{3}Chl* is a relatively long-lived species that can react with triplet oxygen ($^3O_2$) to form singlet oxygen ($^1O_2^*$), a highly oxidative species that is toxic to cells. Carotenoids with $N \geq 9$ conjugated double bonds can quench ^{3}Chl*, thus protecting against photooxidation by preventing $^1O_2^*$ formation. Carotenoids can also directly quench singlet oxygen and provide further protection against photooxidation. The ability of carotenoids to scavenge singlet oxygen, however, is thought to be diffusion limited.

The triplet-state energy of carotenoids with N = 9 is reported to be 7,340 cm^{-1}. This is below the excited triplet state of chlorophyll *a* [~10,500 cm^{-1}] and singlet oxygen [7,855 cm^{-1}] as theoretically required for a quenching function (Siefermann-Harms, 1987;

Truscott, 1990; Frank, 1993). The triplet energy levels of other carotenoids are correspondingly lower or higher depending on degree of unsaturation. The reported triplet energy level for carotenoids with N = 7 of 9,010 cm^{-1} (Frank, 1993) is lower than expected for these carotenoids that reportedly lack protective activity (Claes, 1960).

The photophysical mechanisms for quenching of chlorophyll triplets and singlet oxygen by carotenoids is incompletely understood. Quenching of triplet chlorophyll is thought to involve the Dexter electron-exchange mechanism (Truscott, 1990; Koyama, 1991). This mechanism requires donor and acceptor pairs to be close enough together for overlap of their orbitals. Evidence for close and specific structural associations of carotenoids and chlorophyll in pigment protein complexes was discussed in the previous section on localization. Bacterial reaction centers specifically bind 15-*cis* carotenoids, spheroidene, neurosporene and spirilloxanthin (Lutz et al., 1978; Koyama et al., 1990; Koyama, 1991). The *cis* carotenoids can be photo-isomerized to all-*trans* forms (Boucher and Gingra, 1984). Based on in vitro effects, it has been proposed that quenching of triplet chlorophyll involves the rapid isomerization of excited 15-*cis* triplet to the excited all-trans triplet carotenoid (Koyama et al., 1990). In higher plants, β-carotene in the reaction center is all-*trans*. All other pigments are also all-*trans* in the polyene chain except neoxanthin which is 9-*cis*. There is, however, no evidence that neoxanthin undergoes *cis-trans* isomerization or functions in protecting against photooxidation.

B. Light Harvesting

The light-harvesting function of carotenoids is readily demonstrated in brown algae and photosynthetic bacteria which have carotenoids that absorb light in spectral regions minimally absorbed by chlorophylls. In these organisms, the action spectra of photosynthesis or chlorophyll fluorescence show peaks corresponding to carotenoid absorption (Siefermann-Harms, 1985; Katoh et al., 1991). The presence of carotenoids in all light-harvesting complexes examined is presumptive evidence that energy transfer is a generalized carotenoid function. Energy transfer from lutein and peridinin to chlorophyll has been reported to be 100% efficient (Song et al. 1976; Siefermann-Harms and Ninnemann, 1982).

Light absorbance by carotenoids raises ground-state electrons to an excited singlet state. The subsequent transfer of energy to chlorophyll takes place by rapid singlet-singlet photophysical processes. The actual ability of a donor molecule to transfer energy to an acceptor molecule is complex. In addition to the relative energetic levels of donor and acceptor being appropriate, factors of internal conversion rates, extent of spectral overlap, and proximity of donor-acceptor pairs determine if transfer will occur, the rate of transfer, and the mechanism. These factors are better understood for bacterial carotenoids than for higher-plant carotenoids. This is due in part to the high degree of unsaturation of higher-plant carotenoids which made the necessary measurements difficult. Several reviews are available on the photophysics of bacterial carotenoids and short polyenes (Truscott, 1990; Frank et al., 1991; Mimuro and Katoh, 1991; Cogdell et al., 1992; Frank, 1993). Further background information can be found in two excellent earlier reviews of the carotenoid light-harvesting function (Siefermann-Harms, 1985; Cogdell and Frank, 1987).

The characteristic absorbance spectrum of carotenoids is due to the strongly dipole-dipole allowed transition from ground state (S_o or 1^1A_g) to the second excited singlet state (S_2 or 1^1B_u). In vivo, the absorbance spectra of carotenoids are strongly red-shifted compared to their spectra in hexane. This shift represents a lowering of the S_2 energy level, which has been ascribed to the mutual polarizability of the carotenoid and protein environment (Andersson et al., 1991; Cogdell et al., 1992). The S_2 state decays rapidly by internal conversion to the lower, first-excited singlet state (S_1 or 2^1A_g). The S_1 state is the so-called 'forbidden' transition because it cannot be populated from the carotenoid ground state (S_o) by single photon absorption. In contrast to S_2, the S_1 state is little affected by the surrounding environment (Andersson et al., 1991). A red-shift of the absorption spectrum thus reduces the energy gap between the S_2 and S_1 states and increases spectral overlap with an associated increase in the rate of internal conversion. Structurally, the major factor influencing the energy level of both the S_2 and S_1 state is the number of conjugated double bonds (N). In general, the energy levels decrease with increasing N (DeCoster et al., 1992) and the energy gap between the two states increases slightly with increasing N (Cogdell et al., 1992; DeCoster et al., 1992). Fluorescence from the S_1 state of pigments with $N \geq 9$ conjugated unsaturations is very low, making their detection and measurement difficult (Gillbro and Cogdell, 1989). Importantly, the energy level of the S_1 state is not highly influenced by the polarizability of the solvent (Andersson et al., 1991). Thus, the energy levels determined in vitro are thought to closely represent the levels in the pigment-protein complexes.

After excitation of a ground-state carotenoid, energy transfer to chlorophyll can follow two paths. Energy can flow directly to the Q_x level of Chl *a* (Mimuro and Katoh, 1991; Owens et al., 1992) or, following internal conversion to the low-lying S_1 state, to the Q_y level of chlorophyll *a*. The actual path taken depends on the competitiveness of the relative rates as determined by the relative energetic levels of the excited states, fluorescence lifetime of the donor, and spectral overlap of donor fluorescence and acceptor absorbance. Energy transfer through the S_1 state appears to be favored by the rapid rate of internal conversion from S_2 to S_1 (Cogdell et al., 1992).

The importance of the S_1 state in energy transfer was highlighted by the pioneering work of Thrash et al. (1979) on β-carotene. Significant advances have been made on locating the S_1 states for mini-carotenes (Andersson et al., 1992; Gillbro et al., 1993), bacterial carotenoids (DeCoster et al., 1992) and carotenes of varying unsaturation (Gillbro et al., 1993; Mimuro et al., 1993). There are, however, relatively few reports on the photophysical properties for carotenoids in higher plants and algae. Available results for plastid pigments are summarized in Table 4. The reported values are remarkably consistent although they were determined by diverse methods. The exception is β-carotene where current values for the S_1 state are consistently lower than first reported by Thrash et al. (1979). Most reports now indicate that its S_1 state is slightly above the Q_y state of Chl *a*, consistent with an energy-transfer function (Owens et al., 1992; Gillbro et al., 1993). One report locates the S_1 state of β-carotene below the Q_y state of chlorophyll *a* (Frank et al., 1994). This energetic level would exclude energy transfer from the S_1 state to Chl *a*, leaving open instead the possibility of its accepting energy from Chl *a*. Whether this explains the results of Beddard et al. (1977) that β-carotene quenches chlorophyll fluorescence in vitro is an open question. T. Gillbro (personal communications) could not confirm the effect. Fucoxanthin, a major accessory light-harvesting pigment of diatoms, has its S_1 state above Q_y of chlorophyll *a*, consistent with a light harvesting function (Katoh et al., 1991). The S_1 state

Table 4. Energy levels for singlet excited states of higher-plant carotenoids and chlorophylls

Pigment	cm^{-1} S_1 or (2^1Ag)	S_2 or (1^1Bu)	Reference
β-carotene	17,230	20,700	g
	15,000	19,700–20,800	e
	14,500		c
	13,000–14,000		a
Zeaxanthin	14,500–15,140	19,700–20,620	e
	14,200		b
Antheraxanthin	14,700		b
Violaxanthin	16,600–18,060	22,025	e
	15,200		b
Neoxanthin		18,200	
Fucoxanthin	16,100	21,500	e
	15,900	19,800	d
	Qy	Qx	
Chl *a*	14,800–15,200	17,200	a, b, d, e, f

a. DeCoster et al. (1992); b. Frank et al. (1994); c. Gillbro et al. (1993); d. Katoh et al. (1991); e. Owens et al. (1992); f. Siefermann–Harms (1985); g. Thrash et al. (1979)

of lutein in LHC chloroplasts was not detectable by picosecond transient Raman spectroscopy presumably because energy transfer from lutein to chlorophyll is faster than the detectable limits (approx. 10^{-1} ps) of the instrument (Hashimoto and Koyama, 1990). The S_1 states for lutein and neoxanthin, the characterizing xanthophylls of chlorophyll *a*/*b* binding proteins, have not been reported.

The S_1 states for V and Z measured by time-resolved absorbance spectroscopy and application of the energy-gap law (Frank et al., 1994) listed in Table 4 are in close agreement with values estimated by extrapolation from more saturated carotenoids (Owens et al., 1992). The S_1 level for violaxanthin is located just above chlorophyll *a* as expected for an energy-transfer function. Indeed, violaxanthin is the major carotenoid in the light-harvesting complexes of *Nannochloropsis*; lutein is absent (Chrystal and Larkum, 1987; Brown, 1987; Livne et al., 1992). Moreover, N. Mohanty and H.Y. Yamamoto (unpublished data), found that *Nannochloropsis* had no detectable xanthophyll-cycle activity and no significant NPQ. These findings are consistent with a light-harvesting function for violaxanthin in *Nannochloropsis*. The S_1 state for Z is below the Q_y state of chlorophyll *a*, at a level that makes it theoretically possible to accept energy from chlorophyll. This has important implications for zeaxanthin-dependent non-photochemical quenching.

The relative energetic levels of pigment pairs suggest the possibility and direction for energy transfer but not the mechanism. Because the fluorescence yield of carotenoids is low, energy transfer from carotenoids to chlorophyll is probably by the Dexter electron-exchange rather than the Förster resonance-transfer mechanism (Naqvi, 1980; Siefermann-Harms, 1985; Cogdell and Frank, 1987). This contrasts with chlorophyll-chlorophyll energy transfer which probably occurs by the Förster mechanism. The prerequisite of the Dexter mechanism is that donor and acceptor pairs, besides spectral overlap, must be close enough for orbital overlap.

Besides lutein, other xanthophylls are present in various amounts in LHC fractions. They are probably located peripherally in the pigment protein complexes, possibly in lipids that are tightly bound to the complexes as discussed in the previous section on localization. Raman-scattering spectroscopy suggests that carotenoids are present on the surface of both sides of the membranes (Picorel et al., 1992). If and how they function from these peripheral locations in light-harvesting or any other capacity is presently unknown.

C. PS II Down-regulation

1. Xanthophyll and ΔpH-dependant Energy Dissipation in LHCII.

Plants that are exposed to higher light intensities than can be used by photosynthesis must either avoid the interception of the excess light or, if absorbed, dissipate the excess energy by a harmless mechanism to avoid potential damage to the photosynthetic apparatus (Björkman and Demmig-Adams, 1993). One mechanism that accomplishes the latter is 'non-radiative' (heat) dissipation of the excess energy that is induced in the pigment bed of PS II in the presence of a trans-thylakoid ΔpH (Krause et al., 1982; Gilmore and Yamamoto, 1992a). This mechanism, which is conserved in all higher plants and many algae, effectively down-regulates PS II by diverting energy away from the reaction center, thus helping to keep the traps open (Genty et al., 1989; Genty et al., 1990) and protecting against photoinhibition (Krause and Behrend, 1986). NRD does not decrease the maximal rate of photosynthesis but instead displaces it to higher light intensity.

NRD induction can be grouped into three phases that proceed somewhat sequentially. The first phase is the induction of the trans-thylakoid ΔpH. This phase defines the 'high-energy' and rapidly-reversible nature of NRD (Briantais et al., 1979; Krause and Laasch, 1987). In the second phase, the acidic lumen and the presence of ascorbate induce de-epoxidation of violaxanthin to antheraxanthin and zeaxanthin. The third phase is the NRD mechanism. This phase is complex and poorly understood. It occurs in LHCII and is thought to involve either the direct (Demmig et al., 1987; Gilmore and Yamamoto, 1993b) or indirect effects of zeaxanthin and antheraxanthin (Horton and Ruban, 1992; see also reviews by Demmig-Adams and Adams, 1992; Pfündel and Bilger, 1994).

The three phases are easily resolved in chloroplasts by the differential effects of the inhibitors, nigericin, DTT, antimycin and dibucaine on de-epoxidation and NPQ. DTT inhibits NPQ by inhibiting de-epoxidase activity (Yamamoto and Kamite, 1972). Thus it does not inhibit NPQ if de-epoxidation has been pre-induced (Bilger and Björkman, 1990; Demmig-Adams et al., 1990; Gilmore and Yamamoto, 1991). Antimycin inhibits NPQ and an associated light-scattering increase at A_{530} (Oxborough and Horton, 1987) without inhibiting de-epoxidation (Gilmore and Yamamoto, 1990). Dibucaine inhibits lumen acidification but not NPQ (Laasch and Weis, 1989; Noctor et al., 1993). In contrast, recent results show that dibucaine can inhibit de-epoxidation by its effect on lumen acidity and thus, NPQ in dark-adapted chloroplasts (Mohanty and Yamamoto, 1995). It is concluded that whereas de-epoxidation depends on lumen acidity, NPQ depends on membrane-sequestered acidity. The three phases and the putative inhibition sites are shown schematically in Fig. 3 (Mohanty et al., 1995).

Xanthophyll-mediated NRD can respond to both short and long-term requirements for PS II down-regulation. Rapid responses are dependent on induction of ΔpH (seconds) (Gilmore and Yamamoto, 1991, 1992a) and zeaxanthin and antheraxanthin formation (minutes) (Demmig et al., 1987; Gilmore and Yamamoto, 1993b). Long-term adaptive responses to the growth environment include changes in pool size of xanthophyll-cycle pigments (Thayer and Björkman, 1990; Demmig-Adams and Adams, 1994). In general, plants that have a greater need for down-regulation because of low photosynthetic capacity or high light environments have or develop

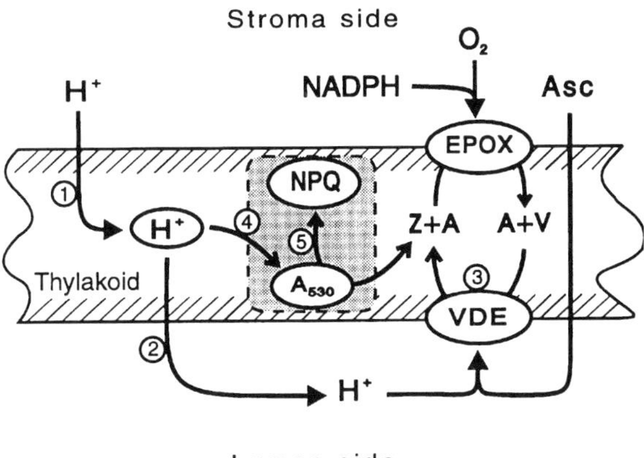

Fig. 3. Schematic of the three phases of xanthophyll-dependent non-photochemical quenching. The circled numbers indicate the putative inhibition sites of: 1, nigericin; 2, dibucaine; 3, DTT; 4 antimycin; 5, A23187. A_{530} is the light-scattering absorbance increase that reflects a conformational change, possibly the aggregation of LHCII. The three phases and the effects of the various inhibitors are described in more detail in the text and in Mohanty et al. 1995. Antheraxanthin (A) is an intermediate product of both violaxanthin de-epoxidation and zeaxanthin epoxidation. Not shown is the direct cycling of antheraxanthin through the de-epoxidase or epoxidase after the first de-epoxidation or expoxidation to ultimately form zeaxanthin or violaxanthin, respectively.

greater capacity for zeaxanthin and antheraxanthin formation (Demmig-Adams et al., 1995). Moreover, recent results suggest that sustained reductions of photochemical efficiency after exposure to temperature or light stress may not be photoinhibition but instead sustained xanthophyll-dependent NRD (Gilmore and Björkman, 1994b; Adams and Demmig-Adams 1994, 1995; Adams et al., 1995). When de-epoxidized xanthophylls are present, NRD can be sustained for long periods in the dark by ATP-hydrolysis dependent ΔpH (Gilmore and Yamamoto, 1992a).

NRD is measured indirectly as non-photochemical quenching of chlorophyll fluorescence (NPQ). Details on fluorescence parameters and various expressions of NPQ can be found elsewhere (Schreiber, 1986; van Kooten and Snel, 1990; Gilmore and Yamamoto, 1991; Krause and Weis, 1991; Björkman and Demmig-Adams, 1993). NPQ is subject to interpretation because it cannot by itself indicate the site or mechanism of quenching. Although it is well established that NPQ is increased by the presence of zeaxanthin and antheraxanthin, NPQ that is independent of zeaxanthin is also known (Bilger and Björkman, 1990; Demmig-Adams et al., 1990; Noctor et al., 1991; Gilmore and Yamamoto, 1991). This type of NPQ is reported to be small relative to xanthophyll-associated NPQ except by one laboratory that finds maximal NPQ is independent of zeaxanthin at saturating ΔpH (Noctor et al., 1989,1991). The latter findings, supported by other in-vitro evidence (Ruban et al., 1992; Mullineaux et al., 1993), have been interpreted to show that zeaxanthin sensitizes NPQ to ΔpH but otherwise is not directly involved in the quenching mechanism (Horton and Ruban, 1992). However, a recent report also shows that NPQ was significantly higher in 'light' chloroplasts with high zeaxanthin compared to 'dark' chloroplasts with negligible zeaxanthin, although quenching was still substantial in the latter (Noctor et al., 1993). Most reports show maximal NPQ depends on de-epoxidized xanthophylls, even under saturating light (see reviews by Demmig-Adams and Adams, 1992; Björkman and Demmig-Adams, 1993; also papers by Oberhuber and Bauer, 1991; Winter and Lesch, 1992; Adams and Demmig-Adams, 1994; Bilger and Björkman, 1994; Demmig-Adams and Adams, 1994; Foyer et al., 1994). In lettuce and pea chloroplasts, the relatively low levels of zeaxanthin-unrelated NRD could be accounted for by supposing that antheraxanthin quenches as effectively as zeaxanthin (Gilmore and Yamamoto, 1993a,b; Neubauer, 1993). Recent field studies appear to confirm that antheraxanthin contributes to NPQ (Adams and Demmig-Adams 1994, 1995; Demmig-Adams and Adams, 1994). The nature of high NRD in the near total absence of zeaxanthin is still unresolved. It is noted, however, that where NRD was directly dependent on zeaxanthin, ascorbate was either present by addition to isolated systems (Gilmore and Yamamoto, 1991, 1992a; Genty et al., 1992; Rees et al., 1992; also discussion in Kreiger and Weis, 1993) or, presumably, naturally present in intact leaves. ΔpH-dependent NRD that is due to inactive PS II centers, charge recombination at the reaction center and calcium-dependent donor-side inhibition have been reported (Neubauer and Schreiber, 1987; Weis and Berry, 1987; Krieger and Weis, 1993). These effects, which are de-epoxidation independent, are not observed when ascorbate is present (Genty et al., 1992). Studies reporting high levels of zeaxanthin-independent NPQ apparently were also done in the absence of ascorbate (Noctor et al., 1989,1991; Ruban et al., 1992). Thus, the presence or absence of ascorbate may dictate the NRD mechanism, at least in vitro. Further studies on the possible regulatory role of ascorbate on NRD are needed.

Two recent findings further support the view that de-epoxidized xanthophylls are essential for NPQ. First, dibucaine inhibits de-epoxidation and, concomitantly, almost all NPQ (Mohanty and Yamamoto, 1995) but has no effect on NPQ if zeaxanthin and antheraxanthin are preformed (Noctor et al., 1993; Mohanty and Yamamoto, 1995). These effects are similar to treating leaves or thylakoids with DTT before or after inducing de-epoxidation (Bilger and Björkman, 1990; Gilmore and Yamamoto, 1991). The mechanism of de-epoxidation inhibition by DTT and dibucaine, however, are different. DTT inhibits de-epoxidation by a direct effect on the de-epoxidase (Yamamoto and Kamite, 1972) whereas dibucaine inhibits the necessary lumen pH for de-epoxidation (Mohanty and Yamamoto, 1995). Second, re-epoxidation of zeaxanthin reverses NPQ even while a transthylakoid ΔpH is maintained (Gilmore et al., 1994). This result is interpreted to show that the epoxidase can effectively remove de-epoxidized pigments from quenching complexes and convert them to non-quenching forms.

2. Properties and Regulation of the Xanthophyll Cycle

Zeaxanthin and antheraxanthin are usually minor to undetectable components of dark-adapted leaves and accumulate in light due to xanthophyll (violaxanthin)-cycle activity (Yamamoto et al., 1962). The cycle operates as a transmembrane system where de-epoxidation (V →A →Z) takes place on the lumen side and epoxidation (Z →A →V) on the stroma side of the membrane (see Yamamoto, 1979 for a review). The dynamic nature of the xanthophyll cycle is exceptional for carotenoids in higher plants. No other carotenoids undergo equivalent substrate-level stoichiometric changes. De-epoxidation of violaxanthin is optimal near pH 5.0 and requires ascorbate (Hager, 1969; Yamamoto et al., 1972; Pfündel and Dilley, 1993). Epoxidase activity is optimal near pH 7.0, and requires NADPH and O_2 (Hager, 1975; Siefermann and Yamamoto, 1975b).

Violaxanthin de-epoxidase (VDE) is firmly bound to the thylakoid membrane at pH 5.0 and presumably is free in the lumen at pH 7.0 (Yamamoto et al., 1974; Hager and Holocher, 1994). Purified violaxanthin is a suitable substrate for VDE and is stoichiometrically converted to zeaxanthin only when co-suspended with MGDG (Yamamoto et al., 1974). Interestingly, purified VDE did not de-epoxidize violaxanthin bound to pigment-protein fractions of LHCII even when the latter was supplemented with MGDG (K. Hindehoffer, A. Lee, P. Thornber and H.Y. Yamamoto, unpublished). Potentially inhibitory effects of residual detergents were excluded. A possible interpretation of these results is that violaxanthin in these isolated particles does not represent the state in situ. Recently, zeaxanthin epoxidase was reported to co-purify with LHCII, implicating LHCII as the zeaxanthin epoxidase (Gruszecki and Krupa, 1993). This important observation has not yet been confirmed.

Z or Z+A and, concomitantly, NPQ increase under conditions of limiting photosynthesis (Demmig et al., 1988; Bilger and Björkman, 1990; Königer and Winter, 1991; Oberhuber and Bauer, 1991; Greer et al., 1993; Demmig-Adams and Adams, 1994; Foyer et al., 1994; Gilmore and Björkman, 1994a; Adams et al., 1995). Under such conditions, how is the necessary transthylakoid ΔpH that is required by both de-epoxidation and NPQ supported? Neubauer and Yamamoto (1992, 1993) showed that under non-CO_2 fixing conditions, the Mehler-peroxidase system (Asada and Takahashi, 1987) can generate a sufficient ΔpH to induce both de-epoxidation and NPQ. Since ascorbate is also required for de-epoxidase activity, ascorbate plays a pivotal role in modulating xanthophyll-dependent NPQ in addition to modulating the NRD mechanism as discussed above. Ascorbate is present in chloroplasts (Anderson et al., 1983b; Beck et al., 1983) and its concentration increases in plants under stress (Gillhan and Dodge, 1987; Schoner et al., 1989). It has yet to be determined to what extent, if any, the Mehler-peroxidase reaction contributes to supporting de-epoxidation and NPQ in intact plants. There is indirect evidence that cyclic electron transport possibly supports the needed ΔpH for NPQ in leaves (Gilmore and Björkman, 1994a).

Few details are known about the mechanism of the transmembrane system. Do the xanthophyll pigments migrate across the membrane? If so, do they migrate as free pigments or pigment-protein complexes? Alternatively, do the de-epoxidase and epoxidase, which presumably face opposite sides of the membrane, act on pigments that are in relatively fixed positions within the membrane core? We hypothesized earlier in this review that the xanthophyll-cycle pigments may be localized in a small amount of lipid that is tightly bound to the periphery of pigment protein complexes. This hypothesis can explain the homogeneous kinetics of de-epoxidation and the phenomenon of violaxanthin availability (Siefermann and Yamamoto, 1974, 1975a). The fraction of the violaxanthin pool that is accessible to or available for de-epoxidation increases with light intensity (Siefermann and Yamamoto, 1974). Changes in availability could be from light-induced conformational changes that release bound violaxanthin to the lipid pool. Although the xanthophyll cycle is mainly associated with PS II by activity and localization, it is not limited strictly to PS II containing membranes. Stroma lamellae (Siefermann and Yamamoto, 1976) and bundle-sheath cells (H. Y. Yamamoto and O. Björkman, unpublished) have de-epoxidase activity. Moreover, PS-1 complexes isolated following light treatments show decreased violaxanthin and increased zeaxanthin and antheraxanthin content (Thayer and Björkman, 1992; Lee and Thornber, 1995). The significance of the xanthophyll cycle in PS I-enriched membranes is unclear.

3. Mechanism of Xanthophyll-Dependent Non-Photochemical Fluorescence Quenching

The fluorescence properties of NPQ and localization of xanthophyll-cycle pigments indicate that xanthophyll-dependent NPQ takes place in the antennae of PS II (see reviews by Horton and Ruban, 1992; Björkman and Demmig-Adams, 1993). The detailed mechanism by which the presence of xanthophylls and transthylakoid ΔpH result in NRD, however, is unclear. In addition to de-epoxidized xanthophylls and transthylakoid ΔpH, there appear to be further temporal and biophysical changes that are necessary for NPQ (Oxborough and Horton, 1987; Gilmore and Yamamoto, 1991). Notably, NPQ has been correlated with a light scattering absorbance change (LSC) around 530–540 nm, also referred to as the A_{530} change (Bilger and Björkman, 1990; Ruban et al., 1993). LSC is thought to indicate a conformational change in the thylakoid membrane (Deamer et al., 1967; Murakami and Packer, 1970; Krause, 1974) and possibly aggregation of LHCII following its protonation (Horton et al., 1992; Ruban et al., 1992). Cation-proton exchange also appears to be involved in NPQ (Gilmore and Yamamoto, 1992b; Mohanty et al., 1995).

Whether de-epoxidized xanthophylls increase NPQ directly or indirectly is currently being debated. One hypothesis is that xanthophylls increase the sensitivity of LHCII to a ΔpH-induced aggregation, which results in chlorophyll 'concentration quenching', but otherwise do not participate directly in the quenching mechanism (Horton et al., 1992). Another hypothesis is that a membrane-sequestered ΔpH induces a conformational change and cation-proton exchanges to bring zeaxanthin and antheraxanthin into close proximity with chlorophyll, enabling direct quenching of excited singlet chlorophyll (Mohanty and Yamamoto, 1995). Both hypotheses include a role for the ΔpH-induced conformational change in LHCII signaled by LSC. Until recently, direct quenching by xanthophylls seemed theoretically unlikely. The excited singlet states of carotenoids, except for triplet carotenoids, were thought to be above the lowest excited state of chlorophyll *a*, thus excluding a direct quenching function. Frank et al. (1994) estimated the S_1 state of the xanthophyll cycle pigments by time-resolved absorbance decline and application of the energy-gap law. The S_1 states for antheraxanthin and zeaxanthin were found to be below the lowest excited singlet state (Q_y) of chlorophyll *a* (Table 4), in close agreement with the energetic levels estimated earlier by extrapolation (Owens et al., 1992). Owens et al. (1992) hypothesized that non-radiative energy dissipation could be due to transfer of energy from chlorophyll to the S_1 state of zeaxanthin. Frank et al. (1994) extended this hypothesis by proposing that the xanthophyll cycle functions as a 'molecular gear shift' that can either transfer energy to or accept energy from chlorophyll depending on the violaxanthin-zeaxanthin status (Fig. 4a,b). Antheraxanthin, which has an S_1 energy level between zeaxanthin and chlorophyll *a*, can facilitate energy transfer to zeaxanthin as well as quench directly as illustrated in Fig. 3c.

In summary, xanthophyll-dependent NPQ is complex, involving a series of events that can be divided into energization, de-epoxidation and quenching phases. The latter two phases depend on the transthylakoid ΔpH, de-epoxidation on the lumen acidity and quenching phase on the membrane-sequestered acidity. Ascorbate modulates de-epoxidation and possibly the quenching mechanism. Xanthophyll-dependent NPQ accounts for a major fraction or, in some cases, nearly all of energy-dependent NPQ. It functions to down-regulate PS II photochemical efficiency, thereby keeping traps open for electron transport and preventing photoinhibiton. The photophysical mechanism is not established. The direct dependency of NPQ on zeaxanthin and

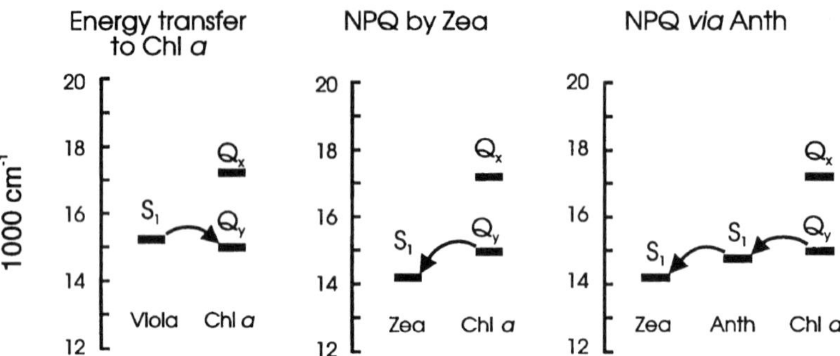

Fig. 4. Hypothetical energy-transfer paths between xanthophyll-cycle pigments and chlorophyll *a*.

antheraxanthin is consistent with direct quenching of excited singlet chlorophyll by the low-energetic S_1 states of these pigments. More direct spectroscopic evidence showing energy transfer from excited Chl to xanthophylls for in-situ quenching is needed. Inactivation of violaxanthin de-epoxidase by mutation or molecular methods may provide further clues. There are other gaps in the understanding of xanthophyll-dependent NPQ. In particular, the precise nature and role of the LSC, putatively LHCII aggregation, needs further clarification. The heterogeneous localization of pigments relative to the homogeneous de-epoxidase activity also needs study. The signals that regulate this protective mechanism at the plant-adaptation level are unknown. The xanthophyll cycle is highly conserved in higher plants. Is the cycle's role in down-regulation of PS II photochemical efficiency a sufficient explanation for its conservation in higher plants?

Acknowledgments

We thank W. Kühlbrandt (EMBL, Heidelberg), S. Jansson (Umeå), W. Adams III (Boulder), B. Demmig-Adams (Boulder), O. Björkman (Stanford), and H. Frank (Connecticut) for sharing advance copies of their work. H.Y.Y. thanks T. Gillbro for helpful discussions and D. Rockholm for technical assistance. R.B. thanks the Italian Ministry of Agriculture and Forestry for financial support. (Grant no. 349-7240-91). H.Y.Y. was supported in part by U.S.D.A. Research Initiative Competitive Grant No. 93–37100-9062.

References

Adams III WW and Demmig-Adams B (1994) Carotenoid composition and down regulation of Photosystem II in three conifer species during the winter. Physiol Plantarum 92: 451–458

Adams III WW and Demmig-Adams B (1995) The xanthophyll cycle and sustained thermal energy dissipation activity in *Vinca minor* and *Euonymus kiautschovicus* in winter. Plant Cell and Environ 18: 117–127

Adams III WW, Demmig-Adams B, Verhoeven AS and Baker DH (1995) 'Photoinhibition' during winter stress: Involvement of sustained xanthophyll cycle-dependent energy dissipation. Aust J Plant Phys 22: 261–276

Adamska I and Ohad I and Kloppstech K (1992) Involvement of the early light-inducible protein (ELIP) in mature pea plants during photoinhibition. In: Argyroudy-Akoyonoglou JH (ed) Regulation of Chloroplast Biogenesis, pp 113–118. Plenum Press, New York

Allen GF and Staehelin LA (1992) Biochemical characterization of Photosystem II antenna polypeptides in grana and stroma membranes of spinach. Plant Physiol 100: 1517–1526

Anderson JM, Brown JS, Lam E and Malkin R (1983a) Chlorophyll *b*: An integral component of Photosystem I of higher plant chloroplasts. Photochem Photobiol 38: 205–210

Anderson JW, Foyer CH and Walker DA (1983b) Light-dependent reduction of dehydroascorbate and uptake of exogenous ascorbate by spinach chloroplast. Planta 158: 442–450

Andersson PO, Gillbro T, Ferguson L and Cogdell RJ (1991) Absorption spectral shifts of carotenoids related to medium polarizability. Photochem Photobiol 54: 353–360

Andersson PO, Gillbro T, Asato AE and Liu RSH (1992) Dual singlet state emission in a series of mini-carotenes. J Luminescence 51: 11–20

Asada K and Takahashi M (1987) Production and scavenging of active oxygen in photosynthesis. *In*: Kyle DJ, Osmond CB, and Arntzen CJ (eds) Photoinhibition, pp 227–287. Elsevier Science Publishers, Amsterdam

Barbato R, Rigoni F, Giardi MT and Giacometti GM (1989) The minor antenna complexes of an oxygen evolving Photosystem II preparation: Purification and stoichiometry. FEBS Lett 251: 147

Barbato R, Race HL, Friso G and Barber J (1991) Chlorophyll levels in the pigment-binding proteins of Photosystem II. A study based on the chlorophyll to cytochrome ratio in different Photosystem II preparations. FEBS Lett. 286: 8620

Bassi R and Dainese P (1992) A supramolecular light-harvesting complex from chloroplast photosystem-II membranes. Eur J

Biochem 204: 317–326
Bassi R and Simpson D (1987) Chlorophyll-protein complexes of barley Photosystem I. Eur J Biochem 163: 221–230
Bassi R, Machold O and Simpson DJ (1985) Chlorophyll-proteins of two Photosystem I preparations from maize. Carlsberg Res Comm 50: 145–162
Bassi R, Hoyer-Hansen G, Barbato R, Giacometti GM and Simpson DJ (1987) Chlorophyll-proteins of the Photosystem II antenna system. J Biol Chem 262: 13333–13341
Bassi R, Rigoni F, Barbato R and Giacometti G (1988) Light-harvesting chlorophyll *a*/*b* proteins (LHCII) populations in phosphorylated membranes. Biochim Biophys Acta 936: 29–38
Bassi R, Rigoni F and GM Giacometti (1990) Chlorophyll binding proteins with antenna function in higher plants and green algae. Photochem Photobiol 52: 1187–1206
Bassi R, Pineau B, Dainese P and Marquardt J (1992) Carotenoid binding proteins of Photosystem II. Eur J Biochem 212: 297–303
Bassi R, Pineau B, Dainese P and Marquardt J (1993) Carotenoid-binding proteins of Photosystem II. Eur J Biochem 212: 297–303
Beck E, Burkett A and Hoffman M (1983) Uptake of L-ascorbate by intact spinach chloroplasts. Plant Physiol 73: 41–45
Beddard GS, Davidson RS and Tretheway KR (1977) Quenching of chlorophyll fluorescence by β-carotene . Nature 267: 373–374
Bengis C and Nelson N (1975) Purification and properties of the Photosystem I reaction center from chloroplasts. J Biol Chem 250: 2783–2788
Berthold D, Babcock GT and Yocum CF (1981) A highly resolved, oxygen-evolving Photosystem II preparation from spinach thylakoid membranes. EPR and electron transport-proprieties. FEBS Lett 134: 231–234
Bilger W and Björkman O (1990) The xanthophyll cycle and sustained thermal energy photoprotection elucidated by measurements of light-induced absorbance changes, fluorescence and photosynthesis in leaves of *Hedera canariensis*. Photosynth Res 25: 173–185
Bilger W and Björkman O (1994) Relationship among violaxanthin deepoxidation, thylakoid membrane conformation and nonphotochemical chlorophyll fluorescence quenching in cotton leaves. Planta 193: 238–246
Bilger W, Björkman O and Thayer S (1989) Light induced spectral absorbance changes in relation to photosynthesis and the epoxidation state of xanthophyll cycle components in cotton leaves. Plant Physiol 91: 542–551
Björkman O and Demmig-Adams B (1993) Regulation of photosynthetic light energy capture, conversion, and dissipation in leaves of higher plants. In: Schulze ED and Caldwell M (eds) Ecological Studies Vol 100, pp 14–47. Springer-Verlag, New York
Block MA, Dorne A, Joyard J and Douce R (1983a) Preparation and characterization of membrane fractions enriched in outer and inner envelope membranes from spinach chloroplasts I. Electrophoretic and immunochemical analysis. J Biol Chem 258: 13273–13280
Block MA, Dorne A, Joyard J and Douce R (1983b) Preparation and characterization of membrane fractions enriched in outer and inner envelope membranes from spinach chloroplasts II. Biochemical characterization. J Biol Chem 258: 13281–13286
Boucher F and Gingra G (1984) Spectral evidence for photo-induced isomerization of carotenoids in bacterial photoreaction center. Photochem Photobiol 40: 277–282
Braumann T, Weber G and Grimme LH (1982) Carotenoid and chlorophyll composition of light harvesting and reaction centre proteins of the thylakoid membrane. Photobiochem Photobiophys 4: 1–18
Breton J and Satoh K (1987) Orientation of the pigments in Photosystem II: Low-temperature linear-dichroism study of a core particle and of its chlorophyll-protein subunits isolated from *Synechococcus* sp. Biochim Biophys Acta 892: 99–107
Briantais J-M, Vernotte C, Picaud M and Krause GH (1979) A quantitative study of the slow decline of chlorophyll *a* fluorescence in isolated chloroplasts. Biochim Biophys Acta 548: 128–138
Bricker TM (1990) The structure and function of Cpa-1 and Cpa-2 in Photosystem II. Photosynth Res 24: 1–13
Brown JS (1987) Functional organization of chlorophyll *a* and carotenoids in the alga, *Nannochloropsis salina.* Plant Physiol 83: 434–437
Bruce BD and Malkin R (1988) Subunit stoichiometry of the chloroplast Photosystem I complex. J Biol Chem 263: 7302–7308
Bullerjahn, GS and Sherman LA (1986) Identification of a carotenoid-binding protein in the cytoplasmic membrane from the heterotrophic cyanobacterium *Synechocystis* sp. strain PCC6714. J Bacteriol 167: 396–399
Butler PJG and Kühlbrandt W (1988) Determination of the aggregated size in detergent solution of the light-harvesting chlorophyll *a*/*b*-protein complex from chloroplast membrane. Proc Natl Acad Sci USA 85: 3797–3801
Camm EL and Green BR (1980) Fractionation of thylakoid membranes with the nonionic detergent octyl-β-D-glucopyranoside. Resolution of chlorophyll-protein complex II into two chlorophyll-protein complexes. Plant Physiol 66: 428–432
Cammarata KV and Schmidt GW (1992) In vitro reconstitution of a light-harvesting gene product: Deletion mutagenesis and analyses of pigment binding. Biochemistry 31: 2779–2789
Chrystal J and Larkum AWD (1987) Pigment-protein complexes and light harvesting in Eustigmatophyte algae. In: Biggins J (ed) Progress in Photosynthesis Research, Vol II, pp 189–192. Martinus Nijhoff Publishers, Dordrecht
Chunaev A, Mirnaya O, Maslov V and Boschetti A (1991) Chlorophyll *b* and loroxanthin-deficient mutants of *Chlamydomonas reinhardtii.* Photosynthetica 25, 21: 301–307
Claes H (1960) Interaction between chlorophyll and carotene with different chromogenic groups. Biochem Biophys Res Commun 3: 585–590
Cogdell RJ and Frank HA (1987) How carotenoids function in photosynthetic bacteria. Biochim Biophys Acta 895: 63–79
Cogdell RJ, Andersson PO and Gillbro T (1992) Carotenoid singlets and their involvement in photosynthetic light-harvesting pigments. J. Photochem Photobiol B: Biol 15: 105–112
Dainese P and Bassi R (1991) Subunit stoichiometry of the chloroplast Photosystem II antenna system and aggregation state of the component chlorophyll *a*/*b* binding proteins. J. Biol. Chem 266: 8136–8142
Dainese P, Hoyer-Hansen G and Bassi R (1990) The resolution of chlorophyll *a*/*b* binding proteins by a preparative method

based on flat bed isoelectrofocusing. Photochem Photobiol 51: 693–703

de Vitry C, Wollmann FA and Delepelaire P (1984) Function of the polypeptides of the Photosystem II reaction center in *Chlamydomonas reinhardtii.* Biochim Biophys Acta 767: 415–422

Deamer DW, Crofts AR and Packer L (1967) Mechanisms of light-induced structural change in chloroplasts. Biochim Biophys Acta 131: 81–86

DeCoster B, Christensen RL, Gebhard R, Lugtenburg J, Farhoosh R and Frank HA (1992) Low-lying electronic states of carotenoids. Biochim Biophys Acta 1102: 107–114

Delepelaire P and Chua NH (1979) Lithium dodecyl sulfate/ polyacrylamide gel electrophoresis of thylakoid membranes at 4°C: Characterizations of two additional chlorophyll *a*-protein complexes. Proc Natl Acad Sci USA 76: 111–115

Demers S, Roy S, Gagnon R and Vignault C (1991) Rapid light-induced changes in cell fluorescence and in xanthophyll-cycle pigments of *Alexandrium excavatum* (Dinophyceae) and *Thalassiosira pseudonana* (Bacillariophyceae): A photo-protection mechanism. Mar Ecol Prog Ser 76: 185–193

Demmig B, Winter K, Krüger A and Czygan FC (1987) Photoinhibition and zeaxanthin formation in intact leaves. Plant Physiol 84: 218–224

Demmig B, Winter K, Krüger A and Czygan FC (1988) Zeaxanthin and the heat dissipation of excess light energy in *Nerium oleander* exposed to a combination of high light and water stress. Plant Physiol 87: 17–24

Demmig-Adams B (1990) Carotenoid and photoprotection in plants: A role for the xanthophyll zeaxanthin. Biochim Biophys Acta 1020: 1–24

Demmig-Adams and Adams III WW (1992) Photoprotection and other responses of plants to high light stress. Ann Rev Plant Physiol Plant Mol Biol 43: 599–626

Demmig-Adams B and Adams III WW (1994) Capacity for energy dissipation in the pigment bed in leaves with different xanthophyll cycle pools. Aust J Plant Physiol 21: 575–588.

Demmig-Adams B, Adams III WW, Heber U, Neimanis S, Winter K, Krüger A, Czygan F-C, Bilger W and Björkman (1990) Inhibition of zeaxanthin formation and of rapid changes in radiationless energy dissipation of dithriotreitol in spinach leaves and chloroplasts. Plant Physiol 92: 292–301

Demmig-Adams B, Adams III WW, Logan BA and Verhoeven AS (1995) Xanthophyll cycle-dependent energy dissipation and flexible Photosystem II efficiency in plants acclimated to light stress. Aust J Plant Physiol 22: 249–260

Di Paolo ML, Peruffo dal Belin A and Bassi R (1990) Immunological studies on chlorophyll-*a/b* proteins and their distribution in thylakoid membrane domains. Planta 181: 275–286

Droppa M, Ghirardi ML, Horvath G and Melis A (1987) Chlorophyll *b* deficiency in soybean mutants. II. Thylakoid membrane development and differentiation. Biochim Biophys Acta 932: 138–145

Dunahay TG, Shuster G and Staehelin LA (1987) Phosphorylation of spinach chlorophyll-protein complexes. CPII*, but not CP29, CP27, or CP24, is phosphorylated in vitro. FEBS Lett 215: 25–30

Dunsmuir P (1985) The petunia chlorophyll *a/b* binding protein genes: A comparison of *Cab* genes from different gene families. Nucleic Acids Res 13: 2503–2518

Feher G, Allen JP, Okamura MY and Rees DC (1989) Structure and function of bacterial photosynthetic reaction centres. Nature 339: 111–116

Foyer DH, Lescure J-C, Lefebvre C, Morot-Gaudry J-F, Vincentz M and Vaucheret H (1994) Adaptations of photosynthetic electron transport, carbon assimilation, and carbon partitioning in transgenic *Nicotiana plumbaginifolia* plants to changes in nitrate reductase activity. Plant Physiol 104: 171–178

Frank HA (1993) Physical and chemical properties of carotenoids. Annals of the New York Academy of Sciences 691: 1–9

Frank HA, Violette CA, Trautman JK, Shreve AP, Owens TG and Albrecht AC (1991) Carotenoids in photosynthesis: Structure and photochemistry. Pure Appl Chem 63: 109–114

Frank HA, Cua A, Chynwat V, Young A, Gosztola D and Wasielewski MR (1994) Photophysics of the carotenoids associated with the xanthophyll cycle in photosynthesis. Photosynth Res 41: 389–395

Funk C, Schröder WP, Green BR, Renger G and Andersson B (1994) The intrinsic 22 kDa protein is a chlorophyll-binding subunit of Photosystem II. FEBS Letters 342: 261–266

Genty B, Briantais JM and Baker NR (1989) The relationship between the quantum yield of photosynthetic electron transport and quenching of chlorophyll fluorescence. Biochem Biophys Acta 990: 87–92

Genty B, Harbinson J, Briantais JM and Baker NR (1990) The relationship between non-photochemical quenching of chlorophyll fluorescence and the rate of photosystem 2 photochemistry in leaves. Photosynth Res 25: 249–257

Genty B, Goulas Y, Dimon B, Peltier G, Briantais JM and Moya I (1992) Modulation of efficiency of primary conversion in leaves, mechanism involved at PS2. In: Murata N (ed) Research in Photosynthesis, Vol IV, pp 603–610. Kluwer Academic Publishers, Dordrecht

Gillbro T and Cogdell RJ (1989) Carotenoid fluorescence. Chem Phys Letters 158: 312–316

Gillbro T, Andersson PO, Liu RS, Asato AE, Takaishi S and Cogdell RJ (1993) Location of the carotenoid $2A_g$-state and its role in photosynthesis. Photochem Photobiol 57: 44–48

Gillhan DJ and Dodge AD (1987) Chloroplast superoxide and hydrogen peroxide scavenging systems from pea leaves: Seasonal variation. Plant Sc 50: 105–109

Gilmore AM and Björkman O (1994a) Adenine nucleotide and the xanthophyll cycle in leaves I. Effects of the CO_2^- and temperature-limited photosynthesis on adenylate energy charge and violaxanthin de-epoxidation. Planta 192: 526–536

Gilmore AM and Björkman O (1994b) Adenine nucleotides and the xanthophyll cycle in leaves II. Comparison of the effects of CO_2^- and temperature-limited photosynthesis on Photosystem II fluorescence quenching, the adenylate energy charge and violaxanthin de-epoxidation in cotton. Planta 192: 537–544

Gilmore AM and Yamamoto HY (1990) Zeaxanthin formation in qE-inhibited chloroplasts. In: Baltscheffsky M (ed). Current Research in Photosynthesis, Vol II, pp 495–498. Kluwer Academic Publishers, Dordrecht

Gilmore AM and Yamamoto HY (1991) Zeaxanthin formation and energy-dependent fluorescence quenching in pea chloroplasts under artificially mediated linear and cyclic electron transport. Plant Physiol 96: 636–643

Gilmore AM and Yamamoto HY (1992a) Dark induction of

zeaxanthin-dependent nonphotochemical fluorescence quenching mediated by ATP. Proc Natl Acad Sci USA 89: 1899–1903

Gilmore AM and Yamamoto HY (1992b) Zeaxanthin-dependent quenching of the variable fluorescence arising from ATP-induced reverse electron flow. In: Murata (ed) Research in Photosynthesis, Vol I., pp 255–258. Kluwer Academic Publishers, Dordrecht

Gilmore AM and Yamamoto HY (1993a) Biochemistry of xanthophyll-dependent nonradiative energy dissipation. In: Yamamoto HY and Smith CM (eds) Photosynthetic Responses to the Environment pp 160–165, American Society of Plant Physiologists, Maryland

Gilmore AM and Yamamoto HY (1993b) Linear models relating xanthophylls and lumen acidity to non-photochemical fluorescence quenching. Evidence that antheraxanthin explains zeaxanthin-independent quenching. Photosynthesis Research 35: 67–78

Gilmore AM, Mohanty N and Yamamoto HY (1994) Epoxidation of zeaxanthin and antheraxanthin reverses non-photochemical quenching of Photosystem II chlorophyll *a* fluorescence in the presence of trans-thylakoid ΔpH. FEBS Lett 350: 271–274

Glick RE and Melis A (1989) Minimum photosynthetic unit size in system I and system II of barley chloroplasts. Biochim Biophys Acta 934: 151–155

Golbeck JH (1987) Structure, function and organization of the Photosystem I reaction center complex. Biochim Biophys Acta 895: 167–204

Gounaris K, Chapman DJ, Booth P, Crystall B, Giorg LB, Klug DR, Porter G, and Barber J (1990) Comparison of the D1/D2/ cytochrome b559 reaction centre complex of photosystem two isolated by two different methods. FEBS Lett 265: 88–92

Green BR, Pichersky E and Kloppstech K (1991) Chlorophyll *a/b*-binding proteins: An extended gene family. Trends in Biochem Sci 16: 181–186

Greer DH, Laing WA and Woolley DJ (1993) The effect of chloramphenicol on photoinhibition of photosynthesis and its recovery in intact kiwifruit (*Actinidia deliciosa*) leaves. Aust J Plant Physiol 20: 33–34

Grimm B, Krause E and Kloppstech K (1989) Transiently expressed early light-inducible thylakoid proteins share transmembrane domains with light-harvesting chlorophyll binding proteins. Plant Mol Biol 13: 583–593

Gruszecki WI and Krupa Z (1993) LHCII, the major light-harvesting pigment-protein complex is a zeaxanthin epoxidase. Biochem Biophys Acta 1144: 97–101

Hager A (1969) Lichtbedingte pH-erniedriqung in einem chloroplasten-kompartiment als ursache der enzymatischen violaxanthin-zeaxanthin-umwandlung. Beziehungen zur photophosphorylierung Planta 89: 224–243

Hager A (1975) The reversible, light-induced conversions of xanthophylls in the chloroplast. Ber Deutsch Pot Ges Bd 88: 27–44

Hager A and Holocher K (1994) Localization of the xanthophyll-cycle enzyme violaxanthin de-epoxidase within the thylakoid lumen and abolition of its mobility by a (light-dependent) pH decrease. Planta 192: 581–589

Hager A and Stransky H (1970) Das carotinoidmuster und die verbreitung del lichtinduzierten xanthophyllcyclus in verschiedenen algenklassen V. Einzelne vertreter der Cryptophyceae, Euglenophyceae, Bacillariophyceae, Chrysophyceae und Phaeophyceae. Arch Mikrobiol 73: 77–89

Harrison MA and Melis A (1992) Organization and stability of polypeptides associated with the chlorophyll *a-b* light-harvesting complex of photosystem-II. Plant Cell Physiol. 33: 627–637

Hashimoto H and Koyama Y (1990) The 2^1A state of a carotenoid bound to spinach chloroplast as revealed by picosecond transient Raman spectroscopy. Biochem Biophys Acta 1017: 181–186

Haworth P, Watson JL and Arntzen CJ (1983) The detection, isolation and characterization of a light-harvesting complex which is specifically associated with Photosystem I. Biochim Biophys Acta 724: 151–158

Henry LEA, Mikkelsen JD and Lindberg-Moller B (1983) Pigment and acyl-lipid composition of Photosystem I and II vesicles and of photosynthetic mutants in barley. Carlsberg Res Commun 48: 131– 148

Henrysson T, Schroeder WP, Spangfort M and Akerlund HE (1989) Isolation and characterization of the chlorophyll *a/b* protein complex CP29 from spinach. Biochim Biophys Acta 977: 301–308

Hoffman NE, Pichersky E, Malik VS, Castresana C, Ko K, Darr SC and Cashmore AR (1987) A cDNA clone encoding a Photosystem I protein with homology to Photosystem II chlorophyll *a/b*-binding polypeptides. Proc Natl Acad Sci USA 84: 8844–8848

Holschuhl K, Bottomley W and Whitfeld PR (1984) Structure of the spinach chloroplast genes for the D2 and 44 kd reaction-centre proteins of Photosystem II and for $tRNA^{Ser}$ (UGA). Nucleic Acid Res 12: 8819–8834

Horton P and Ruban AV (1992) Regulation of Photosystem II. Photosyn Res 34: 375–385

Horton P, Ruban AV and Walters RG (1992) pH-dependent control of chloroplast light harvesting by binding of DCCD to LHCII. In: Murata N (ed) Research in Photosynthesis, Vol 1, pp 311–314. Kluwer Academic Publishers, Dordrecht

Hundal T, Virgin I, Styring S and Andersson B (1990) Changes in the organization of Photosystem II following light-induced D_1-protein degradation. Biochim Biophys Acta 1017: 235–241

Irrgang KD, Bochtel C, Vater J and Renger G (1990) A new Chl *a/b* binding protein in Photosystem II from spinach with a M_r of 14 kDa. In: Baltscheffsky M (ed) Current Research in Photosynthesis, Vol 1, pp 355–358. Kluwer Academic Publishers, Dordrecht

Irrgang KD, Renger G and Vater J (1991) Isolation, purification and partial characterization of a 30-kDa chlorophyll-*a/b*-binding protein from spinach. Eur J Biochem 201: 515–522

Jahns P and Krause GH (1994) Xanthophyll cycle and energy-dependent fluorescence quenching in leaves from pea plants grown under intermittent light. Planta 192: 176–182

Jansson S (1994) The light-harvesting chlorophyll *a/b*-binding proteins. Biochim Biophys Acta 1184: 1–20

Jansson S, Pichersky E, Bassi R, Green BR, Ikeuchi M, Melis A, Simpson DJ, Spangforth M, Staehelin LA and Thornber JP (1992). A nomenclature for the genes encoding the chlorophyll *a/b*-binding proteins of higher plants. Plant Mol Biol Reporter 10: 242–253

Joyard J, Block M and Douce R (1991) Molecular aspects of plastid envelope biochemistry. Eur J Biochem 199: 489–509

Joyard J, Block MA, Pineau B and Douce R (1992) Pigments of the plastid envelope membranes. In: Aryroudi-Akoyonoglou J (ed) Regulation of Chloroplast Biogenesis, pp 165–174. NATO-ASI series Plenum Press, New York

Juhler RK, Andreasson E, Yu SG and Albertsson PA (1993) Composition of photosynthetic pigments in thylakoid membrane vesicles from spinach. Photosynth Res 35: 171–178

Katoh T, Nagashima U and Mimuro M (1991) Fluorescence properties of the allenic carotenoid fucoxanthin: Implication for energy transfer in photosynthetic pigment systems. Photosynth Res 27: 221–226

Kim S, Sandusky P, Bowlby NR, Aebersold R, Green BR, Vlahakis S, Yocum CF and Pichersky E (1992) Characterization of spinach *psbS* cDNA encoding the 22 kDa protein of Photosystem II. FEBS Lett 314: 67–71

Knoetzel J, Svendsen I and Simpson DJ (1992) Identification of the Photosystem I antenna polypeptides in barley. Isolation of three pigment-binding antenna complexes. Eur J Biochem 206: 209–215

Kobayashi M, Maeda H, Watanabe T, Nakane H and Satoh K (1990) Chlorophyll *a* and *β*-carotene content in the D1/D2/cytochrome *b*-559 reaction center complex from spinach. FEBS Lett 260: 138–140

Königer M and Winter K (1991) Carotenoid composition and photon-use efficiency of photosynthesis in *Gossypium hirsutum* L. Grown under conditions of slightly suboptimum leaf temperatures and high levels of irradiance. Oecologia 87: 349–356

Koyama Y (1991) Structures and functions of carotenoids in photosynthetic systems. J Photochem Photobiol B: Bill 9: 265–280

Koyama Y, Takatsuka I, Kanaji M, Tomimoto K, Kito M, Shimamura T, Yamashita J, Saiki K and Tsukida K (1990) Configurations of carotenoids in the reaction center and the light-harvesting complex of *Rhodospirillum rubrum.* Natural selection of carotenoid configurations by pigment protein complexes. Photochem Photobiol 51: 119–128

Krause GH (1974) Changes in chlorophyll fluorescence in relation to light-dependent cation transfer across thylakoid membranes. Biochim Biophys Acta 333: 301–313

Krause GH and Behrend U (1986) ΔpH-dependent chlorophyll fluorescence quenching indicating a mechanism of protection against photoinhibition of chloroplasts. FEBS Letters 200: 298–302

Krause GH and Laasch H (1987) Energy-dependent chlorophyll fluorescence quenching in chloroplasts correlated with quantum yield of photosynthesis. Z Naturforsch 42c: 582–584

Krause GH and Weis E (1991) Chlorophyll fluorescence and photosynthesis: The basics. Annu Rev Plant Physiol Plant Mol Biol 42: 313–349

Krause GH, Vernotte C and Briantais JM (1982) Photoinduced quenching of chlorophyll fluorescence in intact chloroplasts and algae. Biochim Biophys Acta 679: 116–124

Krauss N, Hinrichs W, Witt I, Fromme P, Pritzkow W, Dauter Z, Betzel C, Wilson K, Witt HT and Senger W (1993) Three dimensional structure of system I of photosynthesis at 6 Å resolution. Nature 361: 326–331

Kreiger A and Weis E (1993) The role of calcium in the pH-dependent control of Photosystem II. Photosynth Res 37: 117–130

Krinsky NI (1984) Biology and photobiology of singlet oxygen. In: Bors W, Saran M, and Tait D (eds) Oxygen Radicals in Chemistry and Biology, pp 453–463. Walter de Gruyter and Co, New York.

Krishnan M and Gnanam A (1979) A basic chlorophyll-protein complex. FEBS Lett 97: 322–324

Król M, Spanfort MD, Huner NPA, Öquist G, Gustafsson P and Jansson S (1995) Chlorophyll *a*/*b*-binding proteins, pigment conversions and early-light-induced proteins in a Chl *b*-less barley mutant. Plant Physiol 107: 873–883

Kühlbrandt W and Wang DN (1991) Three-dimensional structure of plant light-harvesting complex determined by electron crystallography. Nature 350: 130–135

Kühlbrandt W, Wang DN and Fujiyoshi Y (1994) Atomic model of plant light-harvesting complex by electron crystallography. Nature 367: 614–621

Laasch H and Weis E (1989) Photosynthetic control, 'energy dependent' quenching of chlorophyll fluorescence and photophosphorylation under influence of tertiary amine. Photosynth Res 22: 137–146

Lam E, Ortiz W and Malkin R (1984) Chlorophyll *a*/*b* proteins of Photosystem I. FEBS Lett 168: 10–14

Larsson UK, Sundby C and Andersson B (1987) Characterization of two different subpopulations of spinach light-harvesting chlorophyll *a*/*b*-protein complex (LHCII): Polypeptide composition, phosphorylation pattern and association with Photosystem II. Biochim Biophys Acta 894: 59–68

Lee AL and Thornber JP (1995) Analysis of the pigment stoichiometry of pigment-protein complexes from barley (*Hordeum vulgare*). The xanthophyll cycle intermediates occur mainly in the light-harvesting complexes of Photosystem I and Photosystem II. Plant Physiol 107: 565–574

Lerst A, Levy H and Zamir A (1991) Co-regulation of a gene homologous to early-light-induced genes in higher plant and *β*-carotene biosynthesis in the alga *Dunaliella bardawil.* J Biol Chem 266: 13698– 13705

Leverenz JW, Öquist G and Wingsle G (1992) Photosynthesis and photoinhibition in leaves of chlorophyll *b*-less barley in relation to absorbed light. Physiol Plant 85: 495–502

Levy H, Tal T, Shaish A, and Zamir A (1993) Cbr, an algal homolog of plant early-light-induced proteins, is a putative zeaxanthin binding protein. J Biol Chem 268: 20892–20896

Lichtenthaler HK, Prenzel U and Kuhn G (1982) Carotenoid composition of chlorophyll-carotenoid-proteins from radish chloroplasts. Z Naturforsch 37c: 10–12

Livne A, Katcoff D, Yacobi YZ and Sukenik A (1992) Pigment-protein complexes of *Nannochloropsis* sp (Eustigmatophyceae). An algae lacking chlorophylls *b* and *c*. In: Murata N (ed) Research in Photosynthesis, Vol I, pp 203–206. Kluwer Academic Publishers, Dordrecht

Lutz M, Aalidis I, Hervo G, Cogdell R and Reiss-Husson F (1978) On the state of carotenoids bound to reaction centers of photosynthetic bacteria: A resonance Raman study. Biochim Biophys Acta 503: 287–303

Machold O, Simpson DJ and Moller BL (1979) Chlorophyll-proteins of thylakoids from wild type and mutants of barley (*Hordeum vulgare* L.). Carlsberg Res Commun 44: 235–254

Markwell J, Bruce BD and Keegstra K (1992) Isolation of a carotenoid-containing sub-membrane particle from the chloroplastic envelope outer membrane of pea (*Pisum sativum*). J Biol Chem 267: 13933–13937

Marquardt J and Bassi R (1993) Chlorophyll-proteins from maize seedlings grown under intermittent light conditions. Their stoichiometry and pigment content. Planta 191: 265–273

Masamoto K, Riethman HC and Sherman LA (1987) Isolation and characterization of a carotenoid associated thylakoid-protein from the cyanobacterium *Anacystis nidulans* R2. Plant Physiol 84: 633–639

McGrath JM, Terzaghi WB, Sridhar P, Cashmore AR and Pichersky E (1991) Sequence of the fourth and fifth Photosystem II type I chlorophyll *a/b*-binding protein genes of *Arabidopsis thaliana* and evidences for the presence of a full complement of the extended CAB gene family. Plant Mol Biol 19: 725–733

Meyer G and Kloppstech K (1984) A rapidly light-induced chloroplast protein with a high turnover coded for by pea nuclear DNA. Eur J Biochem 138: 201–207

Miller KR and Kushman RA (1979) A chloroplast membrane lacking Photosystem II. Thylakoid stacking in the absence of the Photosystem II particle. Biochim Biophys Acta 546: 481–497

Mimuro M and Katoh T (1991) Carotenoids in photosynthesis: Absorption, transfer and dissipation of light energy. Pure Appl Chem 63: 123–130

Mimuro M, Nagashima U, Nagaoka S, Takaichi S, Yamazaki I, Nishimura Y, and Katoh T (1993) Direct measurement of the low-lying singlet excited (2^1A_g) state of a linear carotenoid, neurosporene, in solution. Chem Phys Letters 204: 101–105

Mohanty N and Yamamoto HY (1995) Mechanism of nonphotochemical chlorophyll fluorescence quenching: I. The role of de-epoxidized xanthophylls and sequestered thylakoid membrane protons as probed by dibucaine. Aust J Plant Physiol 22: 231–238

Mohanty N, Gilmore AM and Yamamoto (1995) Mechanism of nonphotochemical chlorophyll quenching. II. Resolution of rapidly reversible absorbance changes at 530 nm and fluorescence quenching by the effects of antimycin, dibucaine and cation exchanger, A23187. Aust J Plant Physiol 22: 239–247

Morishige D and Thornber JP (1991) Correlation of apoproteins with the genes for the major chlorophyll *a/b* binding protein of Photosystem II in *Arabidopsis thaliana.* Confirmation for the presence of a third member of the LHCIIb gene family. FEBS Lett 293: 183–187

Morishige DT, Anandan S, Jaing JT and Thornber JP (1990) Amino-terminal sequence of the 21 kDa apoprotein of a minor light-harvesting pigment-protein complex of the Photosystem II antenna (LHCIId/CP24). FEBS Lett 264: 239–242

Morris J and Herrman RG (1984) Nucleotide sequence of the gene for the P680 chlorophyll *a* apoprotein in the Photosystem II reaction center from spinach. Nucleic Acids Res 12: 2837–2850

Mullet JE, Burke JJ and Arntzen CJ (1980) Chlorophyll proteins of Photosystem I. Plant Physiol 65: 814–822

Mullineaux CW, Pascal AA, Horton P and Holzwarth AR (1993) Excitation-energy quenching in aggregates of the LHCII chlorophyll-protein complex: A time-resolved fluorescence study. Biochim Biophys Acta 1141: 23–28

Murakami S and Packer L (1970) Protonation and chloroplast membrane structure. Jour Cell Biol 47: 332–335

Namba O and Satoh K (1987) Isolation of a Photosystem II reaction center of D-1 and D-2 polypeptides and cytochrome *b*-559. Proc Natl Acad Sci USA 84: 109–112

Naqvi K (1980) The mechanism of singlet-singlet excitation energy transfer from carotenoids to chlorophyll. Photochem Photobiol 31: 523–524

Nechushtai R, Peterson CC, Peter GF and Thornber JP (1987) Purification and characterization of a light-harvesting chlorophyll-*a/b*-protein of Photosystem I of *Lemna gibba*. Eur J Biochem 164: 345–350

Neubauer C (1993) Multiple effects of dithiothreitol on non-photochemical fluorescence quenching in intact chloroplasts: Influence on violaxanthin de-epoxidase and ascorbate peroxidase activity. Plant Physiology 103: 575–583

Neubauer C and Schreiber U (1987) The polyphasic rise of chlorophyll fluorescence upon upset of strong continuous illumination. I. Saturation characteristics and partial control by the Photosystem II acceptor side. Z Naturforsch 42c: 1246–1254

Neubauer C and Yamamoto HY (1992) Mehler-peroxidase reaction mediates zeaxanthin formation and zeaxanthin-related fluorescence quenching in intact chloroplasts. Plant Physiol 99: 1354–1361

Neubauer C and Yamamoto HY (1993) The roles of ascorbate in the related ascorbate peroxidase, violaxanthin de-epoxidase and non-photochemical fluorescence-quenching activities. In: Yamamoto HY and Smith CM (eds) Photosynthetic Responses to the Environment, pp 166–171. American Society of Plant Physiologists, Maryland

Newell WR, van Amerongen H, Barber J and van Grondelle R (1991) Spectroscopic characterization of the reaction centre of Photosystem II using polarised light: Evidence for *β*-carotene excitons in PS II reaction centres. Biochim Biophys Acta 1057: 232–238

Noctor G, Rees D and Horton P (1989) Uncoupler titrations of energy-dependent quenching of chlorophyll fluorescence in chloroplasts. In: Baltscheffsky M (ed) Current Research in Photosynthesis, Vol I, pp 627–630. Kluwer Academic Publishers, Dordrecht

Noctor G, Rees D, Young A and Horton P (1991) The relationship between zeaxanthin, energy-dependent quenching of chlorophyll fluorescence and the transthylakoid pH-gradient in isolated chloroplasts. Biochim Biophys Acta 1057: 320–330

Noctor G, Ruban AV and Horton P (1993) Modulation of ΔpH-dependent nonphotochemical quenching of chlorophyll fluorescence in spinach chloroplasts. Biochim Biophys Acta 1183: 339–344

Oberhuber W and Bauer H (1991) Photoinhibition of photosynthesis under natural conditions in ivy (*Hedera helix* L.) growing in an understory of deciduous trees. Planta 185: 545–553

Olaizola M and Yamamoto HY (1994) Short-term responses of the diadinoxanthin cycle and fluorescence yield in *Chaetoceros muelleri* (Bacillariophyceae). J Phycol 30: 606–612

Olaizola M, LaRoche J, Kolber Z and Falkowski PG (1994) Non-photochemical quenching and the diadinoxanthin cycle in a marine diatom. Photosynth Res 41: 357–370

Omata T and Murata N (1983) Isolation and characterization of the cytoplasmic membrane from the blue-green alga (Cyanobacterium) *Anacystis nidulans*. Plant Cell Physiol 24: 1101–1112

Owens TG, Shreve AP and Albrecht AC (1992) Dynamics and mechanism of singlet energy transfer between carotenoids and chlorophylls: Light harvesting and non-photochemical fluorescence quenching. In: Murata N (ed) Research in Photosynthesis, Vol 1, pp 179–186. Kluwer Academic Press, Dordrecht

Oxborough K and Horton P (1987) Characterization of the effects of antimycin A upon high energy state quenching of chlorophyll fluorescence (qE) in spinach and pea chloroplasts. Photosynth Res 12: 119–128

Paulsen H and Hobe S (1992) Pigment-binding properties of mutant light-harvesting chlorophyll-*a*/*b*-binding protein. Eur J Biochem 205: 71–76

Paulsen H, Rümler U and Rüdiger W (1990) Reconstitution of pigment-containing complexes from light-harvesting chlorophyll *a*/*b*-binding protein overexpressed in *Escherichia coli*. Planta 181: 204–211

Peter GF and Thornber JP (1991a) Biochemical composition and organization of higher plant Photosystem II light-harvesting pigment-proteins. J Biol Chem 266: 16745–16754

Peter GF and Thornber JP (1991b) Electrophoretic procedure for fractionation of Photosystem I and II pigment-proteins of higher plants and determination of their subunit composition. Methods in Plant Biochemistry 5: 195–210

Pfündel EE and Bilger W (1994) Regulation and possible function of the violaxanthin cycle. Photosynth Res 42: 89–109

Pfündel EE and Dilley RA (1993) The pH dependence of violaxanthin deepoxidation in isolated pea chloroplasts. Plant Physiol 101: 65–71

Picorel R, Bakhtiari M, Lu T, Cotton TM and Siefert M (1992) Surface-enchanced resonance Raman scattering spectroscopy as a surface topography probe in plant photosynthetic membranes. Photochem Photobiol 56: 263–270

Plumley, FG and Schmidt GW (1987) Reconstitution of chlorophyll *a*/*b* light-harvesting complexes: Xanthophyll-dependent assembly and energy transfer. Proc Natl Acad Sci USA 84: 146–150

Rawyler A, Henry LEA and Siegenthaler PA (1980) Acyl and pigment lipid composition of two chlorophyll-proteins. Carlsberg Res Commun 45: 443–451

Reddy, KJ, Masamoto, K, Sherman, DM and Sherman LA (1989) DNA sequence and regulation of the gene (*cbpA*) encoding the 42-kilodalton cytoplasmic membrane carotenoprotein of the cyanobacterium *Synechococcus* sp strain PCC 7942. J Bacteriol 171: 3486–3493

Rees D, Noctor GD, Ruban AV, Crofts J, Young A and Horton P (1992) pH Dependent chlorophyll fluorescence quenching in spinach thylakoids from light treated or dark adapted leaves. Photosynth Res 31: 11–19

Ruban AV, Rees D, Pascal AA and Horton P (1992) Mechanism of ΔpH-dependent dissipation of absorbed excitation energy by photosynthetic membranes. II. The relationship between LHCII aggregation in vitro and qE in isolated thylakoids. Biochim Biophys Acta 1102: 39–44

Ruban AV, Young AJ and Horton P (1993) Induction of nonphotochemical energy dissipation and absorbance changes in leaves. Plant Physiol 102: 741–750

Ruban A, Young AJ, Pascal AA and Horton P (1994) The effects of illumination on the xanthophyll composition of the Photosystem II light-harvesting complexes of spinach thylakoid membranes. Plant Physiol 104: 227–234

Ryrie IJ, Anderson JM and Goodchild DJ (1980) The role of the light-harvesting chlorophyll *a*/*b*-protein complex in chloroplast membrane stacking. Cation-induced aggregation of reconstituted proteoliposomes. Eur J Biochem 107: 345–354

Satoh K (1985) Protein-pigments and Photosystem II reaction center. Photochemistry Photobiology 42: 845–853

Schoner S, Foyer C, Lelandais M and Krause GH (1989) Increase in activities of scavengers for active oxygen in spinach related to cold acclimation in excess light. In: Baltscheffsky M (ed) Current Research in Photosynthesis, Vol 2, pp 483–486. Kluwer Academic Publishers, Dordrecht

Schreiber U (1986) Detection of rapid induction kinetics with a new type of high-frequency modulated chlorophyll fluorometer. Photosynth Res 261–272

Schwartz, E and Pichersky, E (1990) Sequence of two tomato nuclear genes encoding chlorophyll *a*/*b*-binding proteins of CP24, a PS II antenna component. Plant Mol Biol 15: 157–160

Setif P, Acker S, Lagoute B and Duranton J (1980) Contribution to the structural characterization of eucaryotic PS I reaction centre-II. Characterization of a highly purified photoactive SDS-CP1 complex. Photosynth Res 1: 17–21

Siefermann D and Yamamoto HY (1974) Light-induced de-epoxidation of violaxanthin in lettuce chloroplasts III. Reaction kinetics and effect of light intensity on de-epoxidase activity and substrate availability. Biochem Biophys Acta 357: 144-150

Siefermann D and Yamamoto HY (1975a) Light-induced de-epoxidation of violaxanthin in lettuce chloroplasts IV. The effects of electron-transport conditions on violaxanthin availability. Biochim Biophys Acta 387: 149–158

Siefermann D and Yamamoto HY (1975b) NADPH and oxygen-dependent epoxidation of zeaxanthin in isolated chloroplasts. Biochem Biophys Res Comm 62: 456–461

Siefermann D and Yamamoto HY (1976) Light-induced de-epoxidation in lettuce chloroplasts VI. De-epoxidation in grana and in stroma lamellae. Plant Physiol 57: 939–940

Siefermann-Harms D (1984) Evidence for a heterogeneous organization of violaxanthin in thylakoid membranes. Photochem Photobiol 40: 507–512

Siefermann-Harms D (1985) Carotenoids in photosynthesis. I. Location in photosynthetic membranes and light-harvesting function. Biochim Biophys Acta 811: 325–355

Siefermann-Harms D (1987) The light-harvesting and protective functions of carotenoids in photosynthetic membranes. Physiol Plant 69: 561–568

Siefermann-Harms D and Ninnemann H (1982) Pigment organization in the light-harvesting chlorophyll-*a*/*b* protein complex of lettuce chloroplasts. Evidence obtained from protection of the chlorophylls against proton attack and from excitation energy transfer. Photochem Photobiol 35: 719–731

Siefermann-Harms D, Joyard J and Douce R (1978) Light-induced changes of the carotenoid levels in chloroplast envelopes. Plant Physiol 61: 530–533

Sigrist, M. and Staehelin, LA (1992) Identification of type I and Type 2 light-harvesting chlorophyll *a*/*b*-binding proteins using monospecific antibodies. Biochim Biophys Acta 1098: 191–200

Simpson DJ (1978) Freeze-fracture studies on barley plastid membranes II. Wild-type chloroplasts. Carlsberg Res Commun 43: 365– 389

Simpson DJ (1983) Freeze-fracture studies on barley plastid membranes. VI. Location of the P700 chlorophyll *a*-protein 1. Eur J Cell Biol 31: 305–314

Sistrom WR, Griffiths M and Stanier TY (1956) The biology of a photosynthetic bacterium which lacks colored carotenoids. Cell Comp Physiol 48: 473–515

Song P-S, Koka P, Prezelin B and Haxo FT (1976) Molecular topology of the photosynthetic light-harvesting pigment complex, peridinin-chlorophyll-*a*-protein, from marine dinoflagellates. Biochemistry 15: 4422–4427

Spangfort M and Andersson B (1989) Subpopulations of the main chlorophyll *a/b* light-harvesting complex of Photosystem II-isolation and biochemical characterization. Biochim Biophys Acta 977: 163–170

Spangfort M, Larsson UK, Ljunberg U, Ryberg M and Andersson B (1990) The 20 kDa apo-polypeptide of the chlorophyll *a/b* protein complex CP24 – Characterization and complete primary amino acid sequence. In: Baltscheffsky M (ed) Current Research in Photosynthesis, Vol II, pp 253–256. Kluwer Academic Publishers, Dordrecht

Thayer SS and Björkman O (1990) Leaf xanthophyll content and composition in sun and shade leaves determined by HPLC. Photosynth Res 23: 331–343

Thayer SS and Björkman O (1992) Carotenoid distribution and deepoxidation in thylakoid pigment-protein complexes from cotton leaves and bundle sheath cells of maize. Photosynthesis Research 33: 213–225

Thornber JP, Morishige DT, Anandan S and Peter GF (1991) Chlorophyll-carotenoid proteins of higher plant thylakoids. In: Scheer H (ed) Chlorophylls, pp 549–585. CRC Press, Boca Raton, Florida

Thornber JP, Peter GF, Morishige DT, Gomez S, Anandan S, Kerfeld C, Welty BA, Lee A, Takeuchi TS and Preiss S (1993) Light harvesting in Photosystem I and II. Biochem Soc Trans 21: 15–18

Thrash RJ, Fang L-B and Leroi GE (1979) On the role of forbidden low-lying excited states of light-harvesting carotenoids in energy transfer in photosynthesis. Photochem Photobiol 1049–1050

Tremoliere A, Dainese P and Bassi R (1993) Heterogeneous lipid distribution among chlorophyll binding proteins of Photosystem II in maize mesophyll chloroplast. Eur J Biochem 221: 721–730

Truscott TG (1990) The photophysics and photochemistry of the carotenoids. J Photochem Photobiol B: Biol 6: 359–371

van Dorsen RJ, Breton J, Pljiter JJ, Satoh K, van Gorkom HJ and Amesz J (1987) Spectroscopic properties of the reaction center and of the 47 kDa chlorophyll protein of Photosystem II. Biochim Biophys Acta 893: 267–274

van Kooten O and Snel JFH (1990) The use of fluorescence nomenclature in plant stress physiology. Photosynth Res 25: 147–150

Vermaas WFJ, Williams JGK and Arntzen CJ (1987) Sequencing and the modification of *psbB*, the gene encoding the CP47 protein of Photosystem II in the cyanobacterium *Synechocystis* 6803. Plant Mol Biol 8: 317–326

Vogelman TC (1993) Plant tissue optics. Ann Rev Plant Physiol Plant Mol Biol 44: 231–251

Wedel N, Klein R, Ljunberg U, Andersson B and Herrman RG (1992) The single-copy gene *psbS* codes for a phylogenetically intriguing 22 kDa polypeptide of Photosystem II. FEBS Lett 314: 61–66

Weis E and Barry JA (1987) Quantum efficiency of Photosystem II of chloroplasts. Biochim Biophys Acta 894: 198–208

Winter K and Lesch M (1992) Diurnal changes in chlorophyll *a* fluorescence and carotenoid composition in *Opuntia ficus-indica*, a CAM plant and in three C3 species in Portugal during summer. Oecologia 91: 505–510

Witt I, Witt HT, Gerken S, Saenger W, Dekker JP and Rögner M (1987) Crystallization of reaction center I of photosynthesis. Low-concentration crystallization of photoactive protein complexes from the cyanobacterium *Synechococcus* sp. FEBS 221: 260–264

Wollman FA and Bennoun P (1982) A new chlorophyll-protein complex related to Photosystem I in *Chlamydomonas reinhardtii*. Biochim Biophys Acta 680: 352–360

Yamamoto HY (1979) Biochemistry of the violaxanthin cycle in higher plants. Pure and Appl chem 51: 639–648

Yamamoto HY and Kamite L (1972) The effects of dithiothreitol on violaxanthin de-epoxidation and absorbance changes in the 500-nm region. Biochim Biophys Acta 267: 538–543

Yamamoto HY, Nakayama TOM and Chichester CO (1962) Studies on the light and dark interconversion of leaf xanthophylls. Arch Biochem Biophys 97: 168–173

Yamamoto HY, Kamite L and Wang Y-Y (1972) An ascorbate-induced absorbance change in chloroplasts from violaxanthin de-epoxidation. Plant Physiol 49: 224–228

Yamamoto HY, Chenchin E and Yamada DK (1974) Effect of chloroplast lipids on violaxanthin de-epoxidase activity. In: Avron M (ed) Proceedings of 3rd International Congress on Photosynthesis, Vol III, pp 1999 –2006. Elsevier, Amsterdam

Yates TO, Komiya A, Chirino DC, Rees, JP, Allen JP and Feher G (1988) Structure of the reaction center from *Rhodobacter sphaeroides* R-26 and 2.4.1: Protein cofactor (bacterio-chlorophyll, bacteriopheophytin, and carotenoid) interactions. Proc Natl Acad Sci USA 85: 7993–7997

Young AJ (1991) The protective role of carotenoids in higher plants. Physiologia Plant 83: 702–708

Chapter 31

Chloroplast Gene Expression: Regulation at Multiple Levels

Marina K. Roell and Wilhelm Gruissem
Department of Plant Biology, 111 Koshland Hall, University of California, Berkeley, CA 94720, USA

Summary

The plastid DNA of higher plants contains genes that primarily encode photosynthetic proteins or products required for their synthesis. Expression of plastid genes resembles prokaryotic gene expression in many aspects, reflecting the plastid's probable origin as prokaryotic endosymbionts. Early theories suggested that plastid gene expression, like that of prokaryotes, was regulated principally at the level of transcription initiation, although further research led to competing theories regarding the relative importance of regulation at different levels. It is becoming apparent that regulation of transcription initiation is an important, but not exclusive, mechanism for control of plastid gene expression. Post-transcriptional control of transcript abundance has also emerged as an important level of regulation, with identification of nuclear-encoded factors that mediate processing, stabilization, and degradation of plastid transcripts. Translation initiation and elongation can also

Donald R. Ort and Charles F. Yocum (eds): Oxygenic Photosynthesis: The Light Reactions, pp. 565–587.

be specifically regulated, as can post-translational mechanisms affecting protein turnover. Thus, plastid gene expression is regulated by a combination of mechanisms that operate at multiple levels. Most of the regulatory factors affecting plastid gene expression are nuclear-encoded, illustrating the regulatory role that the nucleus plays in the control of plastid function. This allows plastid differentiation to be coordinated with the tissue-specific and developmental program of the plant. At the same time, transcription of certain nuclear-encoded plastid-localized proteins requires a plastid-generated signal; thus, expression of nuclear and plastid genes is controlled by bidirectional signaling between the two compartments. Elucidation of the mechanisms controlling the coordinated expression of the two genomes will remain a fruitful area of investigation for many years to come.

I. Introduction

Plastids of higher plants contain their own DNA, as well as the transcriptional and translational machinery required for its expression. Most of the genes located in the plastid genome encode products that are related either directly or indirectly to the chloroplast's photosynthetic function. These include genes coding either for members of photosynthetic complexes (Photosystems I and II, the cytochrome b_6f complex, ribulose-1,5-bisphosphate carboxylase (Rubisco), and ATP synthetase) or for products involved in their expression (ribosomal proteins and RNAs, tRNAs, and translation initiation and elongation factors). However, the plastid genes code for only a fraction of the proteins required for photosynthesis and protein synthesis. The remainder are encoded in the nucleus, presumably transferred from the plastid's ancestral endosymbiont as it evolved into an organelle. The nuclear-encoded chloroplast proteins must be transported post-translationally into the chloroplast, where some are assembled with the plastid-encoded proteins to form functional complexes. This division of genetic information requires the coordinated expression of nuclear and plastid genes, and provides a means by which the nucleus can regulate plastid function. The content and structure of the plastid genome and the nuclear genes encoding chloroplast proteins are covered in more detail in other chapters of this book. In this chapter we will concentrate on plastid gene expression as it relates to photosynthesis. Plastid gene expression has been reviewed recently (Igloi and Kössel, 1992; Gruissem and Tonkyn, 1993; Mullet, 1993), so here we will focus primarily on recent research.

Expression of plastid genes must be differentially regulated in a number of different situations. During development of chloroplasts from proplastids or etioplasts, Rubisco must be synthesized and assembled with its nuclear-encoded subunits, and photosynthetic complexes must be assembled and inserted into the thylakoid membranes. Differentiation of chloroplasts to chromoplasts, on the other hand, requires the disassembly of the photosynthetic complexes and accumulation of carotenoid pigments. Photosynthetic chloroplasts themselves can be altered in response to varying environmental conditions, the most notable of which is light. Some photosynthetic proteins do not accumulate until the plants are illuminated, and expression of others can be altered under low or high light conditions to optimize the rate of photosynthesis. Regulation of expression of chloroplast genes can be accomplished at a number of levels, including transcription initiation, transcript turnover, translation efficiency, and protein turnover.

Abbreviations: IR – inverted repeat; RNAP – RNA polymerase; RNP – RNA binding protein; SLF – sigma-like factors; TAC – transcriptionally active chromosome

II. Transcription

A. RNA Polymerase

The first step in gene expression is transcription by RNA polymerase (RNAP) to produce mRNA. Chloroplasts contain at least two distinct types of RNAP activities, which differ in their salt and template requirements, pH optima, and sensitivities to heparin, which prevents transcription initiation. Both are resistant to rifampicin, which binds to and strongly inhibits the bacterial RNAP β subunit (Joussaume, 1973). One plastid RNAP is a soluble complex that has been isolated from maize, spinach, and pea, and is thought to transcribe primarily the mRNA and tRNA genes (Rushlow et al., 1980; Greenberg et al., 1984), although it can also transcribe the rRNA genes to some extent (Briat et al., 1982; Rajasekhar et al., 1991). Preparations of this complex contain from five to 14 protein subunits (Gruissem and Tonkyn, 1993; Mullet, 1993). The other plastid RNAP activity, referred to as 'transcriptionally active chromosome' (TAC), is isolated as a complex that is tightly bound to and actively transcribing its DNA template. This enzyme complex, isolated from *Euglena gracilis*, spinach, and mustard, appears to preferentially transcribe the plastid rRNA operon (Gruissem et al., 1983), although it can also transcribe other genes (Reiss and Link, 1985; Little and Hallick, 1988). Characterization of these two activities continues to be the focus of intensive scrutiny, as researchers seek to identify the subunits necessary for correct initiation of transcription, the subunits' functions and coding locations, and to determine whether the RNAP complexes share common subunits.

The chloroplast genome contains four open reading frames called *rpoA, rpoB, rpoC1,* and *rpoC2*, that were identified by homology to the *E. coli* RNAP α and β subunits and the N- and C- terminal portions of the β' subunit, respectively. This splitting of the β' subunit into two peptides also occurs in the RNAP of cyanobacteria, which are more closely related to chloroplasts than is *E. coli*; the split subunits are referred to as the β' and β'' subunits. N-terminal amino acid sequencing of members of the purified soluble RNAP complex shows that four of the subunits are encoded by the *rpo* genes. Sequences from the 38, 120, and 180 kDa polypeptides in a maize plastid RNAP preparation correspond to those predicted by the sequence of the maize *rpoA, rpoB,* and *rpoC2* genes, respectively, and amino acids 2–9 of the 78 kDa peptide match the sequence from *rpoC1* (Hu and Bogorad, 1990; Hu et al., 1991). Antibodies raised against fusion proteins from the *rpoA*, *rpoB,* and *rpoC* genes (Little and Hallick, 1988; Purton and Gray, 1989) or against purified polymerase (Rajasekhar et al., 1991) can interfere with a soluble RNAP activity in vitro, and when used in Western analyses, cross-react with proteins of the expected sizes. Antibodies against the *rpo* fusion proteins do not inhibit spinach and *Euglena* TACs, suggesting that the two complexes do not share common core subunits (Little and Hallick, 1988). Alternatively, it is possible that the active site and sites of interaction between subunits of the TAC are not accessible to the *rpo* fusion protein antibodies, since the complex is already assembled and tightly bound to the template; those antibodies which bind peripheral sites on the proteins may not interfere with transcription.

Recently an RNAP activity in pea showed specificity for transcription of plastid mRNAs, and contained three subunits that are immunologically related to an RNAP activity that has specificity for plastid rRNA transcription (Lakhani et al., 1992). In this case the authors suggest that different ancillary subunits combined with a common core may bestow different specificities of the two RNAP preparations. It is uncertain how these RNAPs are related to the soluble and TAC complexes described previously.

An intriguing result was described recently regarding a novel plastid RNAP from spinach (Lerbs-Mache, 1993) in which two types of RNAP activity were separated from an active fraction. One activity contains subunits of 150, 145, 80, and 38 kDa, similar to the previously-described soluble RNAPs, while the other activity seems to be associated with a single polypeptide of 110 kDa. The latter activity could represent a single-subunit RNAP similar to that of T7 bacteriophage.

The fact that the plastid encodes RNAP subunits raises the mechanistic question of what enzyme initially transcribes these genes. *A priori*, two possibilities exist: either some level of plastid-encoded RNAP (or its mRNA) is present in the undifferentiated proplastids and is activated (or translated) at the onset of plastid development, or a nuclear-encoded complex is transported into the chloroplast to transcribe the plastid *rpo* genes. A limited amount of plastid transcripts can be translated on the small number of ribosomes present in the etioplasts, but high levels of translation require synthesis and assembly of additional ribosomes from the plastid- and nuclear-encoded components.

Several lines of evidence provide clues to the origin of the initial plastid RNAP. First, studies of the temporal pattern of transcript accumulation during development in barley show that the *rpo* and *rps* (ribosomal protein subunits) messages peak slightly earlier than those encoding subunits of the photosynthetic apparatus (Baumgartner et al., 1993). This staggered pattern of accumulation would be expected if the products of the *rpo* genes are necessary for the transcription of the photosynthetic genes, although this is not necessarily the case. Second, plastid transcription levels generally are low in undifferentiated organelles, and increase early in leaf development or light-induced chloroplast differentiation. The increase in light-induced transcription activity parallels or precedes accumulation of 16S ribosomal RNAs and mRNAs from *rpoB* and ribosomal protein genes (Baumgartner et al., 1993), indicating that the RNAP is not translated on plastid ribosomes, or that it is stored and activated post-translationally. Third, the regions upstream of the *rpo* genes do not contain promoter sequences thought to be necessary for transcription by the plastid-encoded RNAP (see below), suggesting that they may be transcribed by an RNAP of nuclear origin. Fourth, studies of the parasitic plant *Epifagus virginiana*, which has a greatly reduced plastid genome, reveal that transcription of plastid rRNA genes occurs despite the fact that the *rpoB-rpoC1-rpoC2* operon was lost (Morden et al., 1991; Feirrabend, 1992). Finally, white leaves of the albostrians mutant of barley and heat-bleached leaves of rye both bear undifferentiated plastids lacking ribosomes and, consequently, plastid translation products (including the subunits of RNAP encoded by *rpoA, rpoB, rpoC1* and *rpoC2*). In these plants transcripts of the *rpo* genes and of *rps15*, but not of genes involved in photosynthesis and related processes (*psbA, rbcL, atpI-H*), were abundantly accumulated (Hess et al., 1993). This suggests that a RNAP of nuclear origin transcribes the *rpo* genes, but not the plastid photosynthetic genes.

Considering the above evidence, it is reasonable to speculate the following model: a nuclear-encoded RNA polymerase is imported into the proplastid, where it transcribes the plastid *rpo* genes. The newly-transcribed *rpo* mRNAs are translated on the few ribosomes present in the proplastids, and the resultant RNAP transcribes genes encoding ribosomal proteins, tRNAs and photosynthetic proteins. Meanwhile the imported RNA polymerase forms the TACs, which transcribe rRNAs. The nuclear-encoded components of the ribosomes are imported and assembled with the plastid-encoded components to form more ribosomes. A positive feedback loop results in the amplification of plastid ribosomes to allow the high levels of protein synthesis required to translate the mRNAs encoding photosynthetic proteins. According to this model, the nucleus provides the signal required to initiate chloroplast development. Because a great number of chloroplast proteins are nuclear-encoded, continued coordination between nuclear and chloroplast gene expression is necessary to maintain chloroplast function in addition to initiating its development.

B. Sigma-Like Factors

One mechanism by which nuclear and chloroplast gene expression can be coordinated is through the action of the sigma-like factors (SLFs) that were identified in chloroplast extracts (Surzycki and Shellenbarger, 1976; Bülow and Link, 1988; Lerbs et al., 1988). The sigma factors of *E. coli* are proteins that interact with the core RNAP to modulate its specificity for different promoters (Helmann and Chamberlin, 1988). The SLFs from chloroplasts are proteins which, when added to *E. coli* core polymerase, can confer specificity for chloroplast promoters. An SLF from mustard was characterized by its effect on *E. coli* RNAP with respect to DNA binding, footprinting and transcript specificity for chloroplast mRNA and tRNA genes and their promoters (Bülow and Link, 1988). An SLF from spinach complements *E. coli* core RNAP and shows cross-reactivity with antiserum raised against *E. coli* sigma factor (Lerbs et al., 1988). The latter SLF, a 90 kDa peptide, is present in young, but not in mature spinach leaves, suggesting a possible role in developmental gene regulation. Additionally, three SLFs isolated from etioplasts were the same size as those previously isolated from photosynthetic leaves, although their specificities for different promoters are not the same (Tiller and Link, 1993b). The different specificities may result from different phosphorylation states of the SLFs, since treatment of the chloroplast RNAP complex with protein kinase converted it to a form with etioplast-type specificity and vice versa, treatment of the etioplast RNAP complex with phosphatase generated chloroplast-type specificities (Tiller and Link, 1993a). None of the unidentified open reading frames in the chloroplast genomes are homologous to known bacterial sigma factors, although homology among prokaryotic sigma factors themselves is low. Thus, the coding location of these SLFs is still unknown; if they are indeed encoded in the nucleus, this could provide another mechanism by which the nucleus can regulate plastid function.

Another protein factor that affects the specificity of the chloroplast RNAP was identified in maize (Jolly and Bogorad, 1980). This factor, called the S-factor, is a 27 kDa peptide which, when added to partially purified RNAP, increases specific transcription from chloroplast genes in vitro, although it has no effect on *E. coli* RNAP. Notably, this increased specific transcription is observed only when using supercoiled DNA as templates.

C. Template Topology

The sensitivity of RNAP specificity to template topography probably reflects a physiologically significant requirement, since transcriptional initiation by purified plastid RNAP becomes non-specific with linear templates (Orozco et al., 1985); correct initiation requires superhelical twisting of template DNA (Stirdivant et al., 1985). There are several reports that supercoiled templates are transcribed more efficiently by plastid RNAP than are relaxed templates (Jolly and Bogorad, 1980; Crossland et al., 1984; Stirdivant et al., 1985; Lam and Chua, 1987; Russell and Bogorad, 1987), although Link and coworkers (Link, 1984) detected discrete products from linear templates using RNAP isolated by a purification protocol different from those previously described for RNAP isolation. It is possible that the latter protocol retained some factor(s) lost in the previous methods.

Topoisomerase activity has been detected in chloroplasts from spinach (Siedlecki et al., 1983), pea (Lam and Chua, 1987; Nielsen and Tewari, 1988), and cauliflower (Fukata et al., 1991), and in whole cells of *Chlamydomonas reinhardtii* (Thompson and Mosig, 1985). Inhibitors of topoisomerase can differentially affect the expression of specific genes in chloroplasts from *Chlamydomonas*. For example, in the presence of novobiocin some chloroplast transcripts become more abundant while many others are reduced (Thompson and Mosig, 1985), which indicates that changes in topography of the plastid DNA can contribute to differential gene expression (Thompson and Mosig, 1985,1987; Lam and Chua, 1987).

D. Promoters

In keeping with the prokaryotic nature of the plastid-encoded polymerase, most of the plastid genes have sequences in their upstream regions that resemble *E. coli* promoters. These regions contain two elements, ctp1 and ctp2, which are similar to the –35 and –10 consensus sequences of *E. coli.* In fact, several of the plastid DNA sequences can direct transcription initiation of reporter genes in prokaryotic cells. The consensus for these putative promoters is 5′-TTGacc—17-19 nucleotides—TAtAat-3′, where capital letters denote invariant bases. Mutational analysis of these sequences in vitro confirms their ability to function as promoters, since deletions, insertions, or substitutions of nucleotides in this region can either increase or decrease transcription efficiency (Gruissem et al., 1986; Gruissem and Tonkyn, 1993).

The promoter can affect transcription in two ways: first, by its strength, which is a function of its DNA sequence and can be thought of as the efficiency with which the promoter recruits RNAP to initiate transcription. Stronger promoters lead to more frequent initiation, thus the different promoter strengths of different genes leads to a constant ratio of differential transcription. Comparative analysis of transcription initiation efficiency in chloroplast extracts at promoter regions from several genes shows that promoter strengths differ in vitro (Gruissem and Zurawski, 1985), which is consistent with the differential rates of transcription from different genes in vivo. For example, the transcription rates of 10 plastid genes of spinach vary up to 25- fold (Deng et al., 1987), while those of 15 barley plastid genes vary up to 300-fold (Baumgartner et al., 1993). The relative transcription rates of many plastid genes remain constant during chloroplast development in spinach (Deng et al., 1987; Deng and Gruissem, 1987) and barley (Mullet and Klein, 1987), even as the level of RNAP activity varies ten-fold (Deng and Gruissem, 1987; Baumgartner et al., 1989).

Second, regardless of strength, different promoters can be selectively utilized under varying conditions to differentially modulate the transcription of individual transcripts. For example, the transcription rates of some plastid genes can be differentially altered during chloroplast development, as in the case of *psbA* and *rbcL* in barley and maize (Klein and Mullet, 1990), *trnK* and *rpl2* in spinach (Deng et al., 1987; Deng and Gruissem, 1987), *psbD-psbC* in barley (Sexton et al., 1990; Christopher et al., 1992), *petE* in maize (Haley and Bogorad, 1990), and *rpo* and *rps* genes in barley (Baumgartner et al., 1993). Differential modulation of transcription rates is also observed in spinach in response to different light qualities (Deng et al., 1989) or photo-oxidative stress (Tonkyn et al., 1992). This selectivity could be conferred by *trans*-acting factors such as SLFs that interact with sequences in or near the promoter region.

Transcription initiation can also be regulated by alternative promoters that can be utilized under different conditions to modulate transcription rates. Identification of transcription initiation sites revealed multiple promoters for several plastid genes. For example, three promoters identified in vitro for the maize *atpB* gene correspond to three of the four transcripts observed in vivo (Chen et al., 1990). Analysis of the ORF31-*petE*-ORF42 gene cluster in maize showed that a monocistronic *petE* transcript is produced from an alternative promoter during light-induced plastid maturation in addition to the bi- and tri-cistronic transcripts that are produced constitutively from the proximal promoter (Haley and Bogorad, 1990).

The foregoing example also demonstrates that the organization of genes into operons can play a role in regulation of expression. Most plastid genes are arranged in operons of two types—those which contain genes for proteins with related function, and those for proteins with different functions. For example, the *rpoB-rpoC1-rpoC2* operon encodes subunits of RNAP, and *psbI-psbK-psbD-psbC* operon encodes members of the Photosystem II complex. Proteins from a single complex are usually required in a fixed stoichiometry, and cotranscription provides one mechanism for producing the required ratio of transcripts. On the other hand, some operons encode proteins with different functions. The *psaA-psaB-rps14* operon, for example, encodes subunits of Photosystem I and a ribosomal protein, and *psbB-psbH-petB-petD* encodes subunits of both Photosystem II and the cytochrome b_6f complex. As illustrated above in the case of the ORF31-*petE*-ORF42 gene cluster in maize, grouping of genes into an operon does not necessarily lead to fixed ratios of transcription rates of the individual genes. The pattern of transcription within an operon can be altered by a number of processes, including the use of alternative promoters and terminators. Further regulation of expression of specific genes within operons is achieved post-transcriptionally by multiple layers of control, including selective RNA processing and differential stability of the processed cistrons (see below).

E. Methylation of DNA

Regulation of plastid gene expression by methylation of the template DNA is the subject of some controversy. Tissue-specific differential methylation of plastid DNA has been detected in maize (Gauly and Kössel, 1989), and methylation has been correlated with reduced transcription in chromoplasts of tomato fruit (Kobayashi et al., 1990) and amyloplasts in sycamore suspension cultures (Ngernprasirtsiri et al., 1988; Ngernprasirtsiri, 1990). In bundle sheath and mesophyll cells of etiolated, greening, and fully green maize leaves there is an inverse correlation between the relative abundance of specific transcripts in a given cell type during greening and the methylation status of the corresponding nuclear or chloroplast genes (Ngernprasirtsiri et al., 1989). Inhibition of methylation relieves dark repression of chloroplast differentiation in *Euglena gracilis* (Fasulo et al., 1990) and restores transcriptional activity in photosynthetic genes which are usually suppressed in cultured non-photosynthetic sycamore cells (Ngernprasirtsiri and Akazawa, 1990). However, other researchers report that extensive restriction and hybridization analyses failed to reveal any difference between the chloroplast and chromoplast genomes in tomato, indicating that no developmentally related DNA methylation was detected by these methods (Marano and Carrillo, 1991). No methylation of plastid DNA was detected in maize endosperm amyloplast or leaf chloroplast (McCullough et al., 1992), or during greening, expansion, maturity and senescence of the primary leaf of barley (Tomas et al., 1992). Although methylation of plastid DNA was detected in the *rbcL, atpB,* and *psbC* genes in pea amyloplasts, this did not correlate with the level of expression of these genes (Ohta et al., 1991). Methylation of plastid DNA appears to play a role in the regulation of transcription in some conditions, although it is not clear why conflicting results were obtained for tomato chromoplasts. Further research will be needed to clarify this issue.

III. Post-transcriptional Processes

A. Overview of Transcript Processing

Chloroplast genes are typically organized into polycistronic transcripts that undergo extensive processing to yield complex sets of overlapping RNAs, as well as monocistronic tRNAs, rRNAs, and mRNAs (for review see Rochaix, 1992). The primary transcripts can be quite extensive, as in the case of one operon in *Euglena gracilis* in which 11 ribosomal protein genes, a tRNA gene, an open reading frame encoding a protein of unknown function, and at least

15 introns are transcribed as a single transcription unit (Christopher and Hallick, 1990). Processing of chloroplast primary transcripts can include removal of introns and splicing of exons, endonucleolytic cleavage between cistrons, trimming of the 5′ and 3′ ends, and RNA editing. In some cases multiple 5′ ends are generated by both alternative transcription initiation and multiple processing cleavage sites (Crossland et al., 1984; Mullet et al., 1985; Gamble et al., 1988; Sexton et al., 1990). For example, ten *psbD-psbC* RNAs synthesized in dark-grown barley share four different 5′-ends, two of which arise by transcription initiation at alternative sites, and another which is generated by 5′-processing of longer *psbD-psbC* transcripts (Sexton et al., 1990). The multiple 5′ ends may play some role in translational regulation or differential stabilization of the transcript (Crossland et al., 1984; Poulsen, 1984).

Processing of the primary mRNA transcript can proceed in a linear step-wise fashion, or it can proceed in a less-ordered manner, depending on the individual transcript (Greenberg and Hallick, 1986; Rock et al., 1987; Marion-Poll et al., 1988; Westhoff and Herrmann, 1988). The *psbB* operon, for example, includes four genes, two of which contain introns. Identification of the processing intermediates showed that splicing and cleavage can occur by many different pathways in maize (Barkan, 1988) and spinach (Westhoff and Herrmann, 1988). On the other hand, the primary transcript of the *psbB* operon of tobacco seems to be spliced very rapidly before being cleaved into several small RNA species (Tanaka et al., 1987). Similarly, in the ribosomal protein operon of *Euglena*, processing occurs by an ordered series of reactions in which splicing precedes cistron cleavage (Christopher and Hallick, 1990). The splicing of multiple introns within a single transcript can be ordered as well, as in the case of the ORF203-*rps12-rpl20* transcript of *Marchantia polymorpha*, in which the intron closest to the 5′ end is spliced first (Kohchi et al., 1988a). Processing of the ribosomal RNA occurs by a strictly ordered processing pathway, perhaps due to coupling to ribosome assembly (reviewed by Crouse et al., 1984 and Delp and Kössel, 1991).

The functional significance of this complicated transcript processing is not completely understood. While the major accumulated form of most gene transcripts is the fully processed mono- or di-cistronic RNA, complete processing of the polycistronic transcripts is not required before recruitment into polysomes (Barkan, 1988). In fact, it has been suggested that association of polycistronic mRNAs with polysomes could facilitate the processing reactions by exposing specific recognition sites in the RNA for endonucleolytic cleavage (Westhoff and Herrmann, 1988).

B. Intron Splicing

Many plastid mRNAs contain introns, yet a generalized splicing mechanism and the role that introns and their splicing may play in the regulation of plastid gene expression is not clear. Characterization of chloroplast introns (for review see Plant and Gray, 1988; Gruissem, 1989; Rochaix, 1992) revealed that most belong to the group I and group II intron classes, originally identified as self-splicing introns in *Tetrahymnena* and fungal mitochondrial genes. These introns contain sets of conserved sequences characteristic for each intron type. For example, the group II introns have the potential to form a structure with six helical domains radiating from a central core. The majority of introns from land plant chloroplasts fall into this category. A new class of intron, designated group III, was discovered in *Euglena* (Christopher and Hallick, 1989). Introns in this class have degenerate versions of the group II intron consensus boundary sequences, but lack the highly-conserved secondary structural features characteristic of group II introns. Several of the chloroplast introns are capable of self-splicing in vitro, such as the class I introns in the 23s rRNA (Herrin et al., 1990; Durrenberger and Rochaix, 1991) and *psbA* genes (Herrin et al., 1991) of *Chlamydomonas*. In the case of the 23S rRNA of *Chlamydomonas*, however, the requirement for high temperature and high [Mg2+] suggests involvement of additional splicing factors in vivo (Thompson and Herrin, 1991). Interestingly, an ORF found within the intron of the 23s rRNA gene encodes a protein with double-stranded DNA nucleolytic activity (Thompson et al., 1992) and has been implicated in intron insertion in vivo (Dujon, 1989; Durrenberger and Rochaix, 1991). The mechanism of splicing is under investigation (Mohr et al., 1993), and in some cases proceeds through formation of lariat intermediates (Kim and Hollingsworth, 1993). In some instances, splicing appears to provide a regulatory role in the accumulation of translationally competent mRNAs from intron-containing plastid genes (Deng and Gruissem, 1988; Barkan, 1989), although this is not necessarily true for all plastid introns.

C. *Trans-splicing*

In some genes, exons encoding different parts of a single protein may be located distant from each other on the plastid genome. *Trans*-splicing is then required to generate the intact mRNA encoding the entire gene product (for review see Bonen, 1993). This process has been demonstrated in chloroplasts of *Chlamydomonas* (Kück et al., 1987; Goldschmidt-Clermont et al., 1993), *Euglena* (Tessier et al., 1991), liverwort (Kohchi et al., 1988b), and tobacco (Koller et al., 1987; Hildebrand et al., 1988). Investigations of the mechanism of *trans*-splicing suggest the involvement of group II introns (Jarrell et al., 1988; Kohchi et al., 1988b) and a small chloroplast RNA encoded by the *tscA* locus (Goldschmidt-Clermont et al., 1990, 1991). Analysis of Photosystem I-deficient mutants in *Chlamydomonas* illustrates the requirement of multiple nuclear-encoded factors as well (Choquet et al., 1988; Herrin and Schmidt, 1988). *Trans*-splicing of *psaA* in *Chlamydomonas* has been suggested to play a role in post-transcriptional regulation (Goldschmidt-Clermont et al., 1993). In *Euglena* short leader sequences may be transferred from small RNAs to pre-mRNAs by *trans*-splicing (Tessier et al., 1991); these small leader sequences could affect mRNA stability or translation initiation, thus providing another possible mechanism for post-transcriptional regulation.

D. *Twintrons*

Recently, a novel type of intron was reported in *Euglena* chloroplasts that consists of an intron within an intron and has been named a 'twintron' (Copertino and Hallick, 1991). The twintrons described so far consist of group II (Copertino and Hallick, 1991), group III (Copertino et al., 1992), and mixed group II/group III (Copertino et al., 1991) introns. Splicing of internal introns in three of the four group III twintrons involves multiple 5'- and/or 3'-splice sites, and in two cases the proximal 5'-splice site can be spliced to an internal 3'-splice site, yielding alternative 'pseudo' fully spliced mRNAs (Copertino et al., 1992). One complex twintron in a *Euglena gracilis* chloroplast ribosomal protein gene contains four introns—the external intron is interrupted by an internal intron containing two additional introns. This twintron is excised by four sequential splicing reactions, two of which utilize multiple 5'- and/or 3'-splice sites (Drager and Hallick, 1993). It is not yet known whether the alternatively spliced products have any functional or regulatory significance. Twintrons have not been reported in chloroplasts of higher plants, and may be a unique feature of *Euglena*, which contains an unusually high content of intron sequences—149 introns account for 38% of the DNA in the plastid genome (Hallick et al., 1993).

E. *RNA Editing*

RNA editing occurs in a number of chloroplast genes (for review, see Kössel et al., 1993). Editing of C to U in the maize plastid *rpl2* mRNA (Hoch et al., 1991) and tobacco *psbL* mRNA (Kudla et al., 1992) converts the sequence of the first codon from ACG to AUG, thus restoring the conventional initiating codon. RNA editing of chloroplast transcripts also occurs in locations internal to the protein coding sequence, as illustrated by four C to U transitions in the maize *ndhA* mRNA, which restore amino acids that are conserved in the *ndhA*-encoded peptides of chloroplasts of other species (Maier et al., 1992).

Restoration of an initiation codon by RNA editing could provide a regulatory point for generation of translatable message, although this has not yet been shown. Evidence supporting this hypothesis is that editing of the *psbL* initiation codon and an internal codon of the *psbF* transcript is complete in RNA from etioplasts and plastids, but not in the plastid RNA of seeds or roots (Bock et al., 1993). On the other hand, editing of the plastid *psbL* gene of bell pepper occurs in the non-photosynthetic chromoplasts of ripe fruit, as well as in the chloroplasts of leaves (Kuntz et al., 1992). Therefore, it is not yet clear whether editing plays a regulatory role in plastid gene expression.

In the *psbB* and *rpoA* operons and the *rps4-rps14* gene cluster of maize, editing precedes both intron splicing and cleavage to monocistronic mRNAs. In the *psbB* operon and the *rps4-rps14* gene cluster, editing is rapid, and only completely edited transcripts are detected, while in the *rpoA* operon partially edited transcripts can be detected (Freyer et al., 1993; Ruf et al., 1994). In the former two genes, the editing occurs at internal sites in the *petB* and IRF170 coding regions, while in the latter, editing restores the initiation codon of the *rpl2* gene. It was suggested that there may be selective pressure for rapid editing of the internal positions, since unedited mRNA might be translated to produce defective proteins. The unedited *rpl2* gene, on the other hand, would not be translatable, and thus not subject to such selective pressure (Freyer et al., 1993).

The mechanism of RNA editing in plastids is likely to be complex. Sequences surrounding an editing site are not alone sufficient as determinants for the editing process in chloroplasts since, in spite of strong sequence conservation, only three of the four editing sites identified in the maize *rpoB* transcript are functional in barley (Zeltz et al., 1993). Thus it is likely that at least one *trans*-acting factor is required. If peptide factors are involved, they can not be encoded in the plastid genome, since editing can occur in the absence of chloroplast ribosomes (Zeltz et al., 1993).

F. Regulation of Transcript Turnover

The amount of transcript available for translation depends on both the rate of transcription and the rate of degradation, which together determine the abundance of a particular mRNA. Accumulation of mRNA does not always mirror the transcription rate, indicating that control of the degradation of transcripts is an important mechanism in the regulation of mRNA levels. For example, the steady state levels of *psbA* and *atpB* transcripts increase significantly during greening of etiolated spinach cotyledons, while the transcription rates, measured by a run-on transcription assay, remain relatively constant (Deng and Gruissem, 1987). The half-lives of several chloroplast genes in developing barley, measured using tagetitoxin, a selective inhibitor of chloroplast transcription, provided direct evidence that the stability of some transcripts is selectively increased and further modulated during chloroplast development (Kim et al., 1993). In mature barley chloroplasts as well, enhanced levels of *psbA* mRNA were due primarily to selective stabilization (Baumgartner et al., 1993). In contrast, differential accumulation of *psbD* mRNA relative to *rbcL* mRNA occurs as a result of light-stimulated transcription of *psbD* (Baumgartner et al., 1993). These two regulatory strategies are not mutually exclusive, and in fact, probably act in concert to optimize the expression of plastid genes. Quantitative analysis of transcription and RNA levels of 15 barley chloroplast genes indicated that transcription rates and mRNA levels vary over 300-fold, while predicted mRNA stabilities vary 30-fold (Rapp et al., 1992). Thus, the expression of a single gene can be controlled by a combined regulation of transcription initiation rate and transcript stability. A specific example is the complex interplay between transcription and mRNA stability that controls the levels of transcripts from the large ATP synthase gene cluster in spinach plastids during light-induced development. During the initial 24 h of light-induced development of an etioplast to a chloroplast, transcription decreases in conjunction with increased transcript stability. Transcriptional activity of this cluster then increases between the 24-h and later stages, with a concomitant decrease in the stability of the transcripts. As the young chloroplast matures, the transcripts from this cluster again become markedly more stable, and the transcription of this set of genes declines (Green and Hollingsworth, 1992). Thus, more than one level of gene regulation can be employed to provide a dynamic response to changing developmental or environmental conditions.

1. Cis*-elements Affecting Transcript Stability*

Most chloroplast transcripts contain inverted repeat (IR) sequences in their 3′ untranslated regions that can theoretically form stable stem-loop structures. Analysis using single- and double-strand specific nucleases has shown that such a structure forms in vitro (Stern et al., 1989). IR sequences are found at the 3′ ends of mono- and polycistronic transcripts (e.g. *rbcL, psaA-psaB-rps14*), and immediately following each coding region on a polycistronic transcript known to undergo processing into monocistronic mRNAs (e.g. *psbB-psbH-petB-petD*). These sequences are highly conserved for the same gene between evolutionarily divergent plants, but differ between genes in the chloroplast genome (Zurawski and Clegg, 1987). Such 3′ stem-loop structures participate in transcription termination in prokaryotic cells (Platt, 1986), and function as RNA processing sites or protective structures against 3′ exonucleolytic degradation in both prokaryotic and eukaryotic cells (Birchmeier et al., 1984; Belasco et al., 1985; Wong and Chang, 1986; Newbury et al., 1987; Marzluff and Panday, 1988). In vitro studies with partially fractionated chloroplast extracts (Gruissem, 1984; Gruissem et al., 1986; Orozco et al., 1986) indicate that chloroplast 3′ IR sequences do not function efficiently as transcription terminators (Stern and Gruissem, 1987). It is not likely that these extracts lack proteins required for efficient termination, since chloroplast RNAP in these extracts can efficiently terminate at certain tRNA genes that lack 3′IR sequences (Stern and Gruissem, 1987; Tonkyn and Gruissem, 1993), or at known bacterial terminator regions (Chen and Orozco, 1988). The 3′IR sequences instead appear to be *cis*-acting elements required for processing and stabilization of mRNA 3′ends. This has been demonstrated in vitro for several chloroplast mRNA 3′end precursors (Stern and Gruissem, 1987; Stern et al., 1989; Schuster and Gruissem, 1991). When the 3′ IR sequences are removed either by deletion mutagenesis (Stern and Gruissem, 1989) or endonucleolytic cleavage (Hsu-Ching and Stern, 1991a), the resulting RNAs are rapidly degraded in vitro. More direct evidence for a stabilizing function of the 3′IR sequence was obtained from experiments with transformed *Chlamydomonas* chloroplast, in which the partial or complete deletion of the *atp*B 3′ IR results in a 60 to 80% loss of the mRNA (Stern et al., 1991).

2. *Trans-acting Factors Affecting Transcript Turnover*

The presence of *trans*-acting factors affecting chloroplast mRNA turnover has been indirectly demonstrated by the identification of single nuclear mutations in *Chlamydomonas* that affect the accumulation of specific chloroplast mRNAs. Mutation of the *Chlamydomonas* nuclear *Nac2* locus specifically affects stability of the chloroplast *psbD* transcript encoding polypeptide D2 of Photosystem II (Kuchka et al., 1989). The defect in the *Nac2-26* mutant causes accelerated decay of the *psbD* mRNA, which is cotranscribed with exon 2 of the *trans*-spliced *psaA* mRNA. *Trans*-splicing of the *psaA* exon 2 occurs normally, although the *psbD*-exon 2 precursor does not accumulate in double mutants for *Nac2-26* and *trans*-splicing. This indicates that removal of *psaA* exon 2 from the dicistronic transcript is completed before the *psbD* mRNA is degraded, and that the 5′ end of the message is critical for mRNA stabilization. The *Nac2-26* does not affect the stability of other chloroplast mRNAs for Photosystem II, although the proteins do not accumulate. Other *Chlamydomonas* nuclear mutants have been described in which the chloroplast *psbB* mRNA (Jensen et al., 1986; Sieburth et al., 1991; Monod et al., 1992) or *psbC* mRNA (Sieburth et al., 1991) do not accumulate, although the transcription of their genes is not affected.

A large number of nuclear mutants in vascular plants show reduced levels of photosynthetic complexes (Metz and Miles, 1982; Miles, 1982; Somerville, 1986; Taylor, 1989), although few data have been published on the chloroplast mRNA metabolism in such mutants (Barkan et al., 1986; Kobayashi et al., 1987; Taylor et al., 1987; Gamble and Mullet, 1989; Bruce and Malkin, 1991). One maize mutant, *Hcf38*, is reported to exhibit alterations in chloroplast mRNAs. This mutant accumulates reduced levels of the *psaA*, *atpB/E*, and *petA* mRNAs, and an altered population of transcripts of the *psbB* operon (Barkan et al., 1986). Thus, in higher plants as well as *Chlamydomonas,* the nucleus encodes *trans*-acting factors that can alter the stability or processing of specific chloroplast mRNAs.

A biochemical approach has been used to identify more directly proteins that may function in the processing and stabilization of plastid mRNA and tRNA 3′ends. Experiments using chloroplast extracts in RNA mobility shift and UV crosslinking assays showed that specific RNA-protein complexes form in vitro with the different 3′ ends of the *petD, rbcL, psbA* (Stern et al., 1989), and *trnK* (Nickelsen and Link, 1989). The proteins that interact with the plastid 3′ ends fall into two general classes. The first class consists of proteins between 24 and 33 kDa, which interact with all mRNA 3′ends tested, although their affinity for the different mRNAs may vary (Stern et al., 1989; Schuster and Gruissem, 1991). Two such proteins from spinach chloroplasts, a 28 kDa RNA binding protein (28RNP; Schuster and Gruissem, 1991) and a 24 kDa RNA binding protein (24RNP; S. Abrahamson and W. Gruissem, manuscript in preparation), were purified, cloned, and their sequences deduced from the cDNAs. These proteins contain conserved RNA-binding domains (reviewed in Gruissem and Schuster, 1993), as well as non-conserved, but very acidic N-terminal domains that may have a role in protein-protein interactions. Both proteins are necessary, but neither is sufficient for mRNA 3′end processing in vitro. A cDNA encoding part of a 100 kDa RNA binding protein (100RNP) was recently cloned and sequenced, and homology with a bacterial enzyme suggests that it may possess exonucleolytic activity. Immunoprecipitation of the 100RNP also precipitates 28RNP, suggesting that these two proteins interact with fairly high affinity. This suggests a model in which the 28RNP interacts with both mRNA and a nuclease to direct specific mRNA 3′ end processing. In the absence of 28RNP, the mRNA 3′ end precursors are not processed and are degraded (Schuster and Gruissem, 1991). Once a stable mRNA 3′end has been established, binding of the 28RNP to the mRNA 3′ end may no longer be required. This is supported by results showing that most of the 28RNP in chloroplast extracts is not associated with mRNAs on polysomes (Gruissem and Tonkyn, 1993).

The second class of proteins that interact with plastid 3′ ends consists of those that interact more specifically with particular mRNA 3′ ends. For example, a 55 kDa RNA binding protein (55RNP) was identified that binds with greater affinity to the *petD* 3′ end (Hsu-Ching and Stern, 1991b) than to the 3′ ends of *psbA* or *rbcL*. The 55RNP interacts with a highly conserved duplicated sequence motif AUUYNAUU, which is located 17 to 30 nucleotides 5′and 2 to 7 nucleotides 3′of the *petD* 3′IR sequence (Stern et al., 1989). The function of this protein is not known, but it may interfere with the activity of a specific endonuclease that cleaves the *petD* mRNA 3′end at or near the 55RNP binding site. Cleavage at that site would remove the 3′ stem-loop structure, which has been shown in vitro (Stern and Gruissem, 1987; Stern et al., 1989; Adams and Stern, 1990) and in vivo (Stern et al., 1991) to be necessary for stabilization of the mRNA.

Several RNPs similar to the 24RNP and 28RNP of spinach have been isolated from tobacco chloroplasts using single-stranded DNA chromatography. These proteins range in size from 28 to 33 kDa and contain conserved RNA-binding and acidic N-terminal domains. Their function has not yet been determined, although it has been suggested that they have a role in chloroplast intron splicing (Li and Sugiura, 1990; Ye et al., 1991). cDNAs encoding similar proteins have been isolated from maize (Cook and Walker, 1992) and *Arabidopsis* (R. Hell, unpublished data) expression libraries using either specific DNA sequence motifs or DNA fragments generated by PCR using primers to the RNA-binding domains. Transgenic *Arabidopsis* plants expressing antisense to a 24RNP indicate that this protein may play a fundamental role in chloroplast function (R. Hell, unpublished data). All of the RNPs described above are nuclear-encoded, illustrating another mechanism by which the nucleus controls chloroplast functions.

3. Role of Translation in Chloroplast mRNA Stability

Several lines of investigation suggest that association with ribosomes may have a complex effect on the stability of plastid transcripts. When translation in leaf chloroplasts is blocked with lincomycin, which inhibits an early step in polypeptide synthesis on 70S ribosomes prior to polysome assembly, degradation of *psbA* and *rbcL* mRNAs is reduced. Chloramphenicol, which specifically inhibits the peptidyltransferase activity of 70S ribosomes, accelerates the decay of the two RNAs, but with different kinetics (Klaff and Gruissem, 1991). Although it cannot be ruled out that these inhibitors have additional effects, these results indicate that translation of *psbA* and *rbcL* may facilitate their turnover in a differential manner. A similar conclusion may be drawn from experiments with *Chlamydomonas* in which a nonsense mutation in *rbcL* (Spreitzer et al., 1985) or a frameshift mutation in *psaB* (Xu et al., 1993) cause premature translation termination. The affected mRNA in these mutants accumulates to increased levels, indicating that decreased association with ribosomes leads to decreased rates of degradation. Contrasting effects were observed in maize mutants that appear to have global defects in chloroplast translation. In these mutants the level of *rbcL* mRNA was reduced four-fold, although many other chloroplast mRNAs accumulated to normal levels. These results indicate that in maize the *rbcL* mRNA, but not all chloroplast mRNAs, is destabilized as a consequence of its decreased association with ribosomes (Barkan, 1993). The different results may be due to different metabolism in different plants or may reflect the different approaches used to probe the effects of reduced translation, and point out that caution must be exercised when interpreting the effects of mutations and inhibitors that may have pleiotrophic effects. A nuclear mutation in *Chlamydomonas* was described that causes reduced stability of the *atpB* mRNA and prevents its translation, while another nuclear mutation responsible for the decreased stability of the *atpA* mRNA had limited effect on the translation of this transcript (Drapier et al., 1992). Further characterization of these mutations is necessary to determine the primary target of the mutant protein and its effect upon translation of the different mRNAs. While the frameshift and nonsense mutants of *Chlamydomonas* provide strong, direct evidence for the effects of translation on transcript stability, the nuclear mutants of both *Chlamydomonas* and maize may have additional effects that have not yet been identified. Nevertheless, while the details may vary, it is clear that translation does have some effect on the stability of some plastid mRNAs. More complete characterization of the nuclear mutations will help to clarify those effects.

IV. Translation

Protein synthesis in plastids is similar to bacterial translation, occurring on 70S ribosomes that are sensitive to chloramphenicol, and mediated by similar initiation and elongation factors. The rRNAs and some ribosomal proteins are encoded in the plastid genome, and the remainder of the ribosomal proteins are encoded in the nucleus (Mache, 1990). Sequences resembling prokaryotic ribosome binding sites precede most plastid open reading frames, and translation begins with incorporation of a formyl-methionine residue. A survey of nearly 200 chloroplast-encoded protein genes showed that the general sequence requirements for translation are consistent with the prokaryotic nature of the plastid translation apparatus. In the positions from –25 to +25 (relative to the initiating AUG), A or U is the predominant nucleotide, except for C at +5 and G at +22, and in 92% of the genes, a sequence resembling the Shine-Dalgarno ribosome binding site consensus sequence (GGAGG) was present within 100 base pairs upstream of the start codon (Bonham-Smith and Bourque, 1989). The universal code is employed, although there is a clear preference for A or U in the third position, a bias that appears to be a consequence of an overall high A + T content of plastid genomes. An interesting exception to this bias occurs in the *psbA* genes of several plants. In this gene, codon use favors C over T for the third position of two-fold degenerate amino acids. In each case the only tRNA encoded by the genome is complementary to the codon ending in C. Since *psbA* is the major translation product of the chloroplast, this indicates that selection is acting on the codon use of this gene to adapt codons to tRNA availability, as previously suggested for unicellular organisms (Morton, 1993).

A. Suborganellar Localization

Ribosomes are found in both the stromal and thylakoid fractions of chloroplast extracts, although it is not yet clear whether binding of polysomes to membranes plays an immediate role in the regulation of chloroplast protein synthesis (for review see Jagendorf and Michaels, 1990). It has been reported that *psbA* and *rbcL* transcripts are present in both the stroma and thylakoid of spinach chloroplasts, and that the former mRNA in the stroma and the latter mRNA in the thylakoid fraction are not translated *in situ*. While transcripts of the *psbA* gene are not associated with ribosomes in the stroma, most transcripts of the *rbcL* gene are associated with polysomes in the stroma (Minami et al., 1988). Because *psbA* encodes a membrane-associated protein and *rbcL* encodes a soluble protein, it may be that the location of translation was a reflection of the final location of these chloroplast proteins. However, almost half of the total population of transcripts of *psbA* were detected in stromal RNA, and more than half of that of *rbcL* in thylakoid-associated RNA (Minami et al., 1988). This raises questions concerning the extent of accumulation of these transcripts in the 'non-functioning' subchloroplast fractions. The large amount of *psbA* mRNA found free of polysomes has been reported earlier (Klein et al., 1988b; Klaff and Gruissem, 1991; Gruissem and Tonkyn, 1993), and may reflect the regulation of its expression by a translational, rather than simply transcriptional, mechanism (Fromm et al., 1985; Gamble and Mullet, 1989). A survey of several membrane-associated chloroplast proteins and their mRNAs showed that transcripts of the chloroplast genes *psaA, psbB, psbC, psbD* and *petA* were found predominantly on thylakoid-bound polysomes engaged in the synthesis and the cotranslational integration of membrane proteins. In contrast, transcripts of the genes *rbcL, psbE, petD, atpA, atpB, atpE* and *atpH* were found more frequently on the free polysomes corresponding to a stroma-located translation of these mRNAs and a post-translational integration of the encoded intrinsic membrane proteins. Thus chloroplast-encoded membrane proteins can be integrated by both co- and post-translational mechanisms (Friemann and Hachtel, 1988), and while some chloroplast-encoded membrane proteins may be translated predominantly on thylakoid-associated polysomes and inserted cotranslationally into the thylakoid membrane, this is not necessarily the case for all such proteins.

In *Chlamydomonas* the partitioning of mRNA between the stromal and thylakoid fractions is equal for *rbcL* and *psbA* (Breidenbach et al., 1988b). Measurements of the rate of protein synthesis and correlation with mRNA partitioning during light cycles in differentiated (not greening) *Chlamydomonas* chloroplasts indicated that light-induced binding of polysomes to thylakoids might be an essential, but not the only, prerequisite for protein synthesis in chloroplasts (Breidenbach et al., 1988a). Similar analysis of the chloroplast-encoded elongation factor Tu (EF-Tu) showed that synthesis of EF-Tu, a soluble chloroplast protein, is associated with thylakoid membranes (Breidenbach et al., 1990) as well. The association of different mRNAs with polysomes in different suborganellar locations may vary among plants, and will probably be affected by the growth conditions as well as differences in isolation protocols. Further investigation will be necessary to elucidate the significance of the association of mRNA with polysomes of different suborganellar fractions in relation to the expression of chloroplast genes.

B. Translational Regulation

Control of gene expression at the translational level can allow an organism to adapt rapidly to changes in environmental conditions by altering the preferential translation of mRNAs without concomitant adjustments in transcript level. This can be an especially important mechanism for regulating the synthesis of highly expressed genes, when continued translation of existing mRNA under unfavorable conditions could unnecessarily consume limited metabolites. Translation of chloroplast proteins appears to be affected by ATP levels (Michaels and Herrin, 1990) and the metabolic state of the organelles (Scheibe, 1990; Reinbothe et al., 1991b), possibly through reversible phosphorylation of ribosomal proteins (Guitton et al., 1984; Posno et al., 1984). Association of regulatory proteins with ribosomes or mRNAs also could affect translation initiation or elongation. Such association would provide a mechanism for the regulation of specific genes under specific conditions.

Translational control of gene expression in plastids has been well-documented in a number of cases. For example, in *Euglena* cells transferred to darkness, synthesis of Rubisco large subunits stopped immediately, even as levels of the *rbcL* transcripts increased transiently (Reinbothe et al., 1991a). Control of Rubisco synthesis was determined to be translationally mediated during the cell-cycle in *Euglena* (Brandt et al., 1989), and during light-induction of chloroplast development in barley (Klein, 1991), *Chlamydomonas* (Breidenbach et al., 1988a), amaranth (Berry et al., 1986, 1990) and *Euglena* (Pastushenkova and Volodarskii, 1990). Expression of the plastid *psbA* gene can also be modulated at the translational level (Jensen et al., 1986; Kuchka et al., 1988). In senescing leaves of bean, translation of the *psbA* transcript remains high in spite of a 9-fold decrease in the level of the transcript (Bate et al., 1990). This is in contrast to other photosynthetic proteins, whose translation decreases with the decreasing transcript levels (Bate et al., 1990). In tobacco, *psbA* mRNA is present in all tissues, while its product accumulates tissue-specifically and in response to light. Investigations with a chimeric *uidA* gene encoding β-glucuronidase (GUS) under control of the *psbA* 5′- and 3′-regulatory regions (224 and 393 bp, respectively), integrated into the tobacco plastid genome, showed that this pattern of expression is mediated at the translational level by the *psbA* untranslated sequences (Staub and Maliga, 1993). Similarly, replacement of the *Chlamydomonas petD* gene with a *uidA-petD* fusion showed that sequences essential for translation reside in the *petD* 5′ untranslated region (Sakamoto et al., 1993). Both nuclear and chloroplast mutants of *Chlamydomonas* have been described that are unable to synthesize P6, the 43 kDa Photosystem II core polypeptide encoded by the chloroplast *psbC* gene. The chloroplast mutation (Fu34) was localized near the middle of a 550 base pair 5′ untranslated region of *psbC* where the RNA can be folded into a stem-loop structure. A second mutation near that of Fu34, in the same stem-loop region, partially restores synthesis of P6. These chloroplast mutations appear to define the target site of a nuclear factor that is involved in P6 translation (Rochaix et al., 1989). RNA binding proteins have been isolated from *Chlamydomonas* that protect a 36 base RNA fragment containing a stem-loop located upstream of the ribosome binding site of the *psbA* gene. Binding of these proteins to the *psbA* mRNA correlates with the level of translation of *psbA* mRNA observed in light- and dark-grown wild type cells and in a mutant that lacks D1 synthesis in the dark (Danon and Mayfield, 1991). One of the proteins, a 47 kDa RNP which binds to the stem-loop structure, is present at similar levels in both light- and dark-grown cells, but binds the RNA primarily when isolated from light-grown cells. This shows that protein modification may be necessary for binding (Rochaix, 1992).

Another mechanism of translational regulation is mediated by ribosome stalling at hairpin structures in a leader peptide, which leads to premature termination of translation. This mechanism, similar to attenuation of amino acid biosynthetic genes in *E. coli*, would be controlled by the presence or absence of ribosomes translating a leader peptide. The presence of potential hairpin structures in the leader region of a spinach chloroplast operon has led to the postulation that this operon could be regulated by such a mechanism. Transformation of *E. coli* with a construct in which the spinach rDNA operon leader is fused to the *galK* gene yields results consistent with a role of the leader region in a translation-mediated attenuation of the chloroplast rDNA expression (Laboure et al., 1988).

V. Post-translational Regulation

A. Cofactor Stabilization

Many photosynthetic proteins require the noncovalent attachment of prosthetic groups such as chlorophylls, carotenoids, quinones, hemes, and ions in order to function. In some cases, cofactor binding stabilizes the cognate protein, protecting it from proteolysis, thus allowing it to accumulate. Such cofactor stabilization has been well-documented in the requirement of chlorophyll for accumulation of plastid-encoded chlorophyll *a*-apoproteins (Klein et al., 1988a; Eichaker et al., 1990; Mullet et al., 1990). In the case of the Photosystem II apoprotein CP43, pulse-labeling assays revealed transiently labeled proteins that are rapidly degraded in plastids of dark-grown plants that lack chlorophyll. In contrast, CP43 synthesized in plastids from illuminated plants was stable (Mullet et al., 1990). The role of chlorophyll in this stabilization was demonstrated in vitro in extracts of etiolated barley, which do not accumulate the chlorophyll apoproteins. When chlorophyll *a* synthesis in these extracts was initiated by the addition of chlorophyll precursors, accumulation of both CP43 and another apoprotein, CP47, as well as the D1 protein of Photosystem II, was detected (Eichaker et al., 1990). In contrast to CP43, which is synthesized in its entirety and rapidly degraded in the absence of chlorophyll, only 15- to 25-kDa translation intermediates of the D1 protein are detectable in the absence of chlorophyll. It appears that chlorophyll is required for the translation of the full-length D1 protein (Mullet et al., 1990). This is consistent with the results of a toeprint analysis, in which the distribution of ribosomes on the D1 mRNA was mapped revealing that sites of ribosome pausing were consistent with the 15-28 kDa translation intermediates. Ribosome pausing may facilitate cotranslational binding of cofactors such as chlorophyll to D1, and in the absence of chlorophyll, causes the premature termination of translation. Cofactor binding is also necessary for the stabilization of plastocyanin, which is rapidly degraded in Cu(II) deficient cells of *Chlamydomonas* (Merchant and Bogorad, 1986). Studies with mutants of *Chlamydomonas* blocked at specific steps of carotenoid or chlorophyll synthesis demonstrated that carotenoids, as well as chlorophylls, have a direct role in the stabilization of certain chlorophyll apoproteins (Herrin et al., 1992). In contrast, cytochrome *f* accumulates in the thylakoid membranes of an *Oenothera hookeri* chloroplast mutant with a defect in heme binding to cytochrome *f*, indicating that cofactor binding may not be necessary for stabilization of this protein (Johnson and Sears, 1990).

B. Assembly of Protein Complexes

Many proteins of multisubunit complexes in chloroplasts are rapidly degraded when prevented from assembling. For example, a mutant of *Chlamydomonas* that carries a lesion in the *Nac2* gene is unable to accumulate the *psbD* transcript encoding D2, but synthesizes all other Photosystem II proteins normally. However, the PS II complex is completely unstable in the *Nac2-26* mutant, and all major PS II polypeptides, including the three oxygen-evolving enhancing proteins, are absent or greatly reduced in this strain because of post-translational degradation (Kuchka et al., 1989). Likewise, in a transformant of *Chlamydomonas* in which the *psaC* gene has been replaced by a non-functional gene, neither PSI reaction center subunits nor the seven small subunits belonging to PSI accumulate stably in the thylakoid membranes of the transformants. Pulse-chase labeling of cell proteins shows that the PSI reaction center subunits are synthesized normally but turn over rapidly in the transformants (Takahashi et al., 1991). Such instability of unassembled subunits has been extensively documented for Photosystem I (Girard-Bascou et al., 1987; Choquet et al., 1988; Herrin and Schmidt, 1988), Photosystem II (Bennoun et al., 1986; Erickson et al., 1986; Jensen et al., 1986; Kuchka et al., 1988; de Vitry et al., 1989; Gamble and Mullet, 1989), the cytochrome b_6f complex (Bendall et al., 1986; Olive et al., 1986; Bruce and Malkin, 1991), and the ATP synthetase complex

(Biekmann and Feierabend, 1985; Drapier et al., 1992).

C. Degradation of Chloroplast Proteins

Proteolysis is an important step in the regulation of protein levels in the chloroplast (Mishkind et al., 1985). As mentioned above, many proteins that are unable to assemble into complexes are rapidly degraded, which prevents the accumulation of non-functioning subunits, and provides one mechanism for maintaining stoichiometric amounts of subunits for a given protein complex. Proteolysis is also important in the differentiation of chloroplasts to chromoplasts, which requires the degradation of photosynthetic complexes and accumulation of carotenoid pigments. Additionally, proteolysis is critical for the removal of photosystem D1 protein that has been damaged by oxygen radicals during photosynthetic electron transport (Kyle et al., 1984; Ohad et al., 1984). Degradation of this protein is coupled to photosynthetic electron transport (Mattoo et al., 1984) and has been shown to be mediated by a membrane-associated protease (Inagaki et al., 1989; Adir et al., 1990).

Plants contain proteins that are homologous to Clp proteins (ClpA, ClpB, ClpP) (Gottesman et al., 1990; Kitagawa et al., 1991; Moore and Keegstra, 1993), which are subunits of ATP-dependent serine protease complexes. These proteins catalyze initial rate-limiting steps in proteolysis, and are present in all organisms examined to date (Gottesman et al., 1990; Goldberg, 1992). ClpA is an ATPase that activates an ATP-dependent ClpP endoprotease, and ClpB has been identified as a heat-shock protein that may regulate proteolysis, but could also protect proteins from heat-induced denaturation (Kitagawa et al., 1991; Squires et al., 1991). The plant Clp analogs could be part of the ATP-dependent serine protease complex associated with the thylakoid membrane that degrades soluble and thylakoid membrane proteins (Liu and Jagendorf, 1986). The gene for the ClpP endoprotease is encoded by the chloroplast genome (Gray et al., 1990; Maurizi et al., 1990) and is retained in the reduced plastid genome of the parasitic plant *Epifagus virginiana* (Morden et al., 1991). Homologs of ClpA (Gottesman et al., 1990) are encoded in the nucleus, and the protein is transported into the chloroplast (Keegstra et al., 1989). The nuclear location of the gene for the chloroplast regulatory ClpA homolog provides an interesting mechanism for nuclear control of proteolysis in the chloroplast. Regulation of proteolysis functions as an important, yet often neglected final step in the regulation of chloroplast gene expression.

VI. Overview of Plastid Gene Regulation

Regulation of gene expression in chloroplasts is a complex process that involves combinations of mechanisms operating at multiple levels. Different mechanisms may be dominant for particular genes under different conditions, which results in data that may at first glance appear contradictory, but in fact is a reflection of the complex regulation necessary for survival in a variable environment. For example, a plant's metabolism must be adjusted in response to changes in light, temperature, and water and nutrient availability that occur both diurnally and seasonally. It is probably not possible to completely separate the different mechanisms, since they act in concert to optimize the plant's metabolism. Additionally, different strategies may have evolved in different plants that allow them to survive in varying conditions. For these reasons it is difficult to formulate a single model for chloroplast gene expression in chloroplasts. However, a basic framework can be drafted that will take into account the possible mechanisms of gene regulation, with the understanding that different portions will be utilized in the expression of specific genes by different plants (or tissues) under different conditions (see Fig. 1). In general, it appears that most, if not all of the plastid genes are transcribed at constitutive, but differential basal levels, although for some genes transcription can be further modulated by developmental or environmental signals. Accumulation of transcripts is further regulated by complex processing and degradation pathways, and finally, protein levels can be regulated both at translation initiation and elongation, and by post-translational degradation. Plastid gene expression is affected at every level by factors provided by the nucleus: nuclear-encoded sigma-like factors can affect the specificity of plastid RNAPs during transcription initiation; transcript processing and degradation are mediated by nuclear-encoded proteins; translation initiation can be affected by nuclear-encoded factors; translation itself occurs on ribosomes that contain nuclear-encoded components; and turnover of plastid proteins is carried out by nuclear-encoded proteases

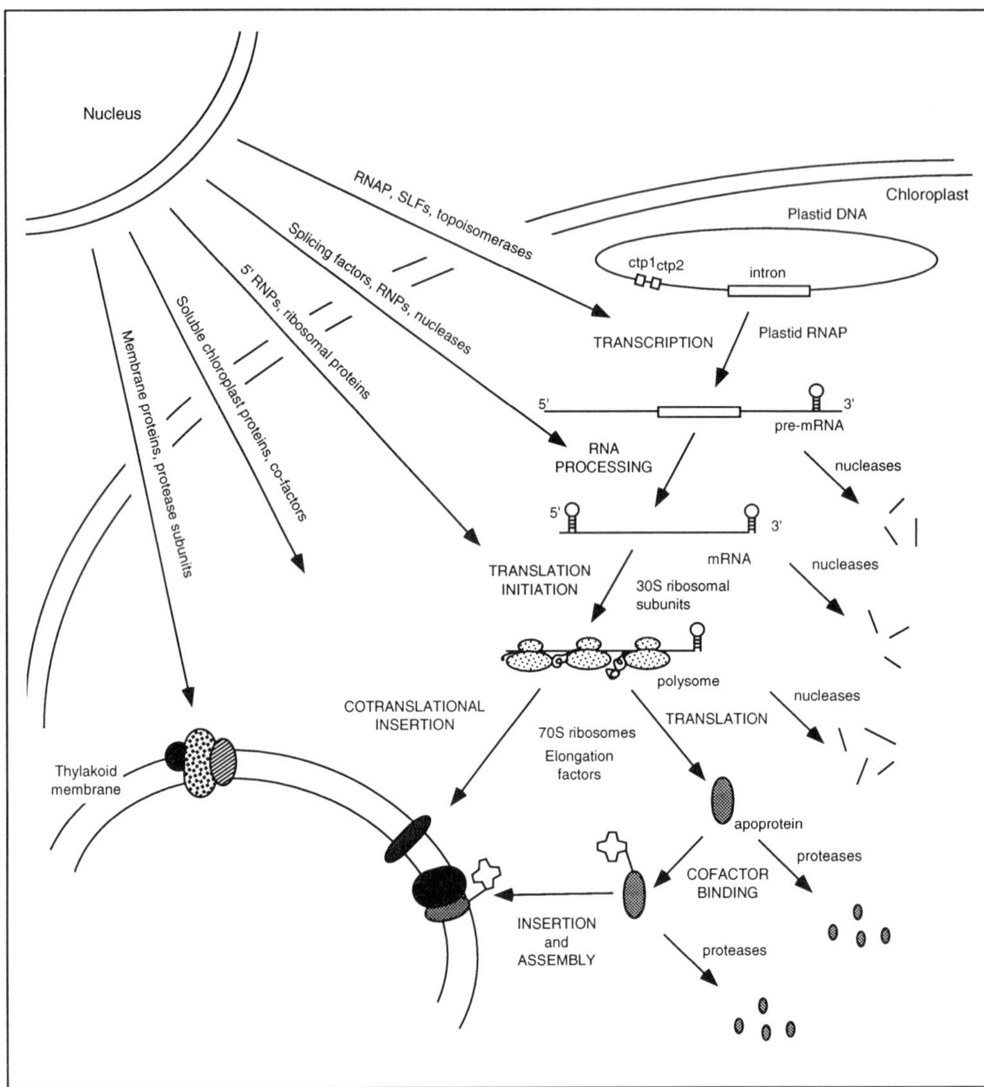

Fig. 1. Summary of regulatory processes that can affect the expression of plastid genes.

and a plastid-encoded protease that is regulated by a nuclear-encoded subunit. At the same time, expression of some nuclear-encoded plastid-localized proteins requires a plastid-generated signal (Susek et al., 1993). Thus, expression of nuclear and plastid genes is coordinated by bidirectional signaling between the two organelles. Elucidation of the mechanisms controlling this coordinated expression of the two genomes during chloroplast development and plastid differentiation will remain a fruitful area of investigation.

Acknowledgments

We thank Susan Abrahamson and Shaul Yalovsky for critical reading of the manuscript. Published and unpublished research on plastid gene expression from the authors' laboratory is supported by grants to WG from the U.S. Department of Energy and the National Science Foundation, and to MR and SA from the National Institutes of Health.

References

Adams CC and Stern DB (1990) Control of mRNA stability in chloroplasts by 3′ inverted repeats: Effects of stem and loop mutations on degradation of *psbA* mRNA in vitro. Nuc Acids Res 18: 6003–6010

Adir N, Shochat S, Inoue Y and Ohad I (1990) Mechanism of the light dependent turnover of the D1 protein. Curr Res Photosynth II: 409–413

Barkan A (1988) Proteins encoded by a complex chloroplast transcription unit are each translated from both monocistronic and polycistronic mRNAs. EMBO J 7: 2637–2644

Barkan A (1989) Tissue-dependent plastid RNA splicing in maize: Transcripts from four plastid genes are predominantly unspliced in leaf meristems and roots. Plant Cell 1: 437–445

Barkan A (1993) Nuclear mutants of maize with defects in chloroplast polysome assembly have altered chloroplast RNA metabolism. Plant Cell 5: 389–402

Barkan A, Miles D and Taylor WC (1986) Chloroplast gene expression in nuclear, photosynthetic mutants of maize. EMBO J 5: 1421–1427

Bate NJ, Straus NA and Thompson JE (1990) Expression of chloroplast photosynthesis genes during leaf senescence. Physiol Plant 80: 217–225

Baumgartner BJ, Rapp JC and Mullet JE (1989) Plastid transcription activity and DNA copy number increase early in barley chloroplast development. Plant Physiol 89: 1011–1018

Baumgartner BJ, Rapp JC and Mullet JE (1993) Plastid genes encoding the transcription/translation apparatus are differentially transcribed early in barley (*Hordeum vulgare*) chloroplast development: Evidence for selective stabilization of *psbA* mRNA. Plant Physiol 101: 781–791

Belasco JG, Beatty JT, Adams CW, von Gabain A and Cohen SN (1985) Differential expression of photosynthetic genes in *R. capsulata* results from segmental differences in stability within the polycistronic *rxcA* transcript. Cell 40: 171–181

Bendall DS, Sanguansermsri M, Girard-Bascou J and Bennoun P (1986) Mutations of *Chlamydomonas reinhardtii* affecting the cytochrome b_6f complex. FEBS Lett 203: 31–35

Bennoun P, Spierer-Herz M, Erickson J, Girard-Bascou J, Pierre Y, Delosme M and Rochaix J-D (1986) Characterization of Photosystem II mutants of *Chlamydomonas reinhardtii* lacking the *psbA* gene. Plant Mol Biol 6: 151–160

Berry JO, Nikolau BJ, Carr JP and Klessig DF (1986) Translational regulation of light-induced ribulose 1,5-bisphosphate carboxylase gene expression in Amaranth. Mol Cell Biol 6: 2347–2353

Berry JO, Breiding DE and Klessig DF (1990) Light-mediated control of translational initiation of ribulose 1,5-bisphosphate carboxylase in Amaranth cotyledons. Plant Cell 2: 795–803

Biekmann S and Feierabend J (1985) Synthesis and degradation of unassembled polypeptides of the coupling factor of photophosphorylation CF_1 in 70S ribosome-deficient rye leaves. Eur J Biochem 152: 529–535

Birchmeier C, Schumperli D, Sconzo G and Birnstiel ML (1984) 3′ editing of mRNAs: Sequence requirements and involvement of a 60-nucleotide RNA in maturation of histone mRNA precursors. Proc Natl Acad Sci USA 81: 1057–1061

Bock R, Hagemann R, Kössel H and Kudla J (1993) Tissue- and stage-specific modulation of RNA editing of the *psbF* and *psbL* transcript from spinach plastids: A new regulatory mechanism? Mol Gen Genet 240: 238–244

Bonen L (1993) *Trans*-splicing of pre-mRNA in plants, animals and protists. Faseb (Fed Am Soc Exp Biol) J 7: 40–46

Bonham-Smith PC and Bourque DP (1989) Translation of chloroplast-encoded mRNA: Potential initiation and termination signals. Nuc Acids Res 17: 2057–2080

Brandt P, Breidenbach E, Prestin B and Boschetti A (1989) Transcriptional, translational and assembly control of the large subunit of ribulose-1,5-bisphosphate carboxylase/oxygenase during the cell cycle of *Euglena gracilis*. J Plant Physiol 134: 420–426

Breidenbach E, Jenni E and Boschetti A (1988a) Synthesis of two proteins in chloroplasts and mRNA distribution between thylakoids and stroma during the cell cycle of *Chlamydomonas reinhardtii*. Eur J Biochem 177: 225–232

Breidenbach E, Jenni E, Leu S and Boschetti A (1988b) Quantification of messenger RNA in chloroplasts of *Chlamydomonas reinhardtii*: Equal distribution of messenger RNA for a soluble and a membrane polypeptide in stroma and thylakoids. Plant Cell Physiol 29: 1–8

Breidenbach E, Leu S, Michaels A and Boschetti A (1990) Synthesis of EF-Tu and distribution of its messenger RNA between stroma and thylakoids during the cell cycle of *Chlamydomonas reinhardtii*. Biochim Biophys Acta 1048: 209–216

Briat J, Dron M, Loiseaux S and Mache R (1982) Structure and transcription of the spinach chloroplast rDNA leader region. Nuc Acids Res 10: 6865–6878

Bruce BD and Malkin R (1991) Biosynthesis of the chloroplast cytochrome b_6f complex: Studies in a photosynthetic mutant of *Lemna*. Plant Cell 3: 203–212

Bülow S and Link G (1988) Sigma-like activity from mustard (*Sinapis alba* L.) chloroplasts conferring DNA-binding and transcription specificity to *E. coli* core RNA polymerase. Plant Mol Biol 10: 349–357

Chen L-J and Orozco EMJ (1988) Recognition of prokaryotic transcription terminators by spinach chloroplast RNA polymerase. Nuc Acids Res 16: 8411–8431

Chen L-J, Rogers SA, Bennett DC, Hu M-C and Orozco EMJ (1990) An in vitro transcription termination system to analyze chloroplast promoters: Identification of multiple promoters for the spinach *atpB* gene. Curr Genet 17: 55–64

Choquet Y, Goldschmidt-Clermont M, Girard-Bascou J, Kuck U, Bennoun P and Rochaix J-D (1988) Mutant phenotypes support a *trans*-splicing mechanism for the expression of the tripartite *psaA* gene in the *C. reinhardtii* chloroplast. Cell 52: 903–913

Christopher DA and Hallick RB (1989) *Euglena gracilis* chloroplast ribosomal protein operon: A new chloroplast gene for ribosomal protein L5 and description of a novel organelle intron category designated group III. Nuc Acids Res 17: 7591–7608

Christopher DA and Hallick RB (1990) Complex RNA maturation pathway for a chloroplast ribosomal protein operon with an internal tRNA cistron. Plant Cell 2: 659–671

Christopher DA, Kim M and Mullet JE (1992) A novel light-regulated promoter is conserved in cereal and dicot chloroplasts. Plant Cell 4: 785–798

Cook WB and Walker JC (1992) Identification of a maize nucleic acid-binding protein (NBP) belonging to a family of nuclear-encoded chloroplast proteins. Nuc Acids Res 20: 359–364

Copertino DW and Hallick RB (1991) Group II twintron: An intron within an intron in a chloroplast cytochrome *b*-559 gene. EMBO J 10: 433–442

Copertino DW, Christopher DA and Hallick RB (1991) A mixed group II/group III twintron in the *Euglena gracilis* chloroplast ribosomal protein S3 gene: Evidence for intron insertion during gene evolution. Nuc Acids Res 19: 6491–6497

Copertino DW, Shigeoka S and Hallick RB (1992) Chloroplast group III twintron excision utilizing multiple 5′- and 3′-splice sites. EMBO J 11: 5041–5050

Crossland LD, Rodermel SR and Bogorad L (1984) Single gene for the large subunit of ribulosebisphosphate carboxylase in maize yields two differentially regulated mRNAs. Proc Natl Acad Sci USA 81: 4060–4064

Crouse EJ, Bohnert HJ and Schmitt JM (1984) Chloroplast RNA synthesis. In: Ellis RJ (ed) Chloroplast Biogenesis, pp 83–136. Cambridge University Press, Cambridge

Danon A and Mayfield SP (1991) Light regulated translational activators: Identification of chloroplast gene specific mRNA binding proteins. EMBO J 10: 3993–4001

de Vitry C, Olive J, Drapier D, Recouvreur M and Wollman F-A (1989) Posttranslational events leading to the assembly of Photosystem II protein complex: A study using photosynthesis mutants from *Chlamydomonas reinhardtii*. J Cell Biol 109: 991–1006

Delp G and Kössel H (1991) rRNAs and rRNA genes of plastids. In: Bogorad L and Vasil IK (eds) The Molecular Biology of Plastids. Cell Culture and Somatic Cell Genetics of Plants, Vol 7A, pp 139–167. Academic Press, San Diego

Deng X-W and Gruissem W (1987) Control of plastid gene expression during development: The limited role of transcriptional regulation. Cell 49: 379–387

Deng X-W and Gruissem W (1988) Constitutive transcription and regulation of gene expression in non-photosynthetic plastids of higher plants. EMBO J 7: 3301–3308

Deng X-W, Stern DB, Tonkyn JC and Gruissem WG (1987) Plastid run-on transcription. Application to determine the transcriptional regulation of spinach plastid genes. J Biol Chem 262: 9641–9648

Deng X-W, Tonkyn JC, Peter GF, Thornber JP and Gruissem W (1989) Post-transcriptional control of plastid mRNA accumulation during adaptation of chloroplasts to different light quality environments. Plant Cell 1: 645–654

Drager RG and Hallick RB (1993) A complex twintron is excised as four individual introns. Nucleic Acids Res 21: 2389–2394

Drapier D, Girard-Bascou J and Wollman F-A (1992) Evidence for nuclear control of the expression of the *atpA* and *atpB* chloroplast genes in *Chlamydomonas*. Plant Cell 4: 283–295

Dujon B (1989) Group I introns as mobile genetic elements: Facts and mechanistic speculations - a review. Gene 82: 91–114

Durrenberger F and Rochaix J-D (1991) Chloroplast ribosomal intron of *Chlamydomonas reinhardtii*: In vitro self-splicing, DNA endonuclease activity and in vivo mobility. EMBO J 10: 3495–3502

Eichaker LA, Soll J, Lauterback P, Rüdiger W, Klein RR and Mullet JE (1990) In vitro synthesis of chlorophyll *a* in the dark triggers accumulation of chlorophyll *a* apoproteins in barley etioplasts. J Biol Chem 265: 13566–13571

Erickson JM, Rahire M, Malnoe P, Girard-Bascou J, Pierre Y, Bennoun P and Rochaix J-D (1986) Lack of D2 protein in a *Chlamydomonas reinhardtii psbD* mutant affects Photosystem II stability and D1 expression. EMBO J 5: 1745–1754

Fasulo MP, Pancaldi S, Bonora A, Bruni A and Dall'Olio G (1990) 5-azacytidine-removal of the dark repression in plastid development of *Euglena gracilis* Klebs. J Plant Physiol 137: 723–728

Feirrabend J (1992) Conservation and structural divergence of organellar DNA and gene expression in non-photosynthetic plastids during ontogenetic differentiation and phylogenetic adaptation. Bot Acta 105: 227–231

Freyer R, Hoch B, Neckermann K, Maier RM and Kössel H (1993) RNA editing in maize chloroplasts is a processing step independent of splicing and cleavage to monocistronic mRNAs. Plant J 4: 621–629

Friemann A and Hachtel W (1988) Chloroplast messenger RNAs of free and thylakoid-bound polysomes from *Vicia faba* L. Planta 175: 50–59

Fromm H, Devic M, Fluhr R and Edelman M (1985) Control of *psbA* gene expression: In mature *Spirodela* chloroplasts light regulation of 32-kd protein synthesis is independent of transcript level. EMBO J 4: 291–295

Fukata H, Mochida A, Maruyama N and Fukusawa H (1991) Chloroplast DNA topoisomerase I from cauliflower. J Biochem 109: 127–131

Gamble PE and Mullet JE (1989) Translation and stability of proteins encoded by the plastid *psbA* and *psbB* genes are regulated by a nuclear gene during light-induced chloroplast development in barley. J Biol Chem 264: 7236–7243

Gamble PE, Sexton TB and Mullet JE (1988) Light-dependent changes in *psbD* and *psbC* transcripts of barley chloroplasts: Accumulation of two transcripts maintains *psbD* and *psbC* translation capability in mature chloroplasts. EMBO J 7: 1289–1297

Gauly A and Kössel H (1989) Evidence for tissue-specific cytosine-methylation of plastid DNA from *Zea mays*. Curr Genet 15: 371–376

Girard-Bascou J, Choquet Y, Schneider M, Delosme M and Dron M (1987) Characterization of a chloroplast mutation in the *psaA2* gene of *Chlamydomonas reinhardtii*. Curr Genet 12: 489–495

Goldberg AL (1992) The mechanism and functions of ATP-dependent proteases in bacterial and animal cells. Eur J Biochem 203: 9–23

Goldschmidt-Clermont M, Girard-Bascou J, Choquet Y and Rochaix J-D (1990) *Trans*-splicing mutants of *Chlamydomonas reinhardtii*. Mol Gen Genet 223: 417–425

Goldschmidt-Clermont M, Choquet Y, Girard-Bascou J, Michel F, Shirmer-Rahire M and Rochaix J-D (1991) A small chloroplast RNA may be required for *trans*-splicing in *Chlamydomonas reinhardtii*. Cell 65: 135–143

Goldschmidt-Clermont M, Choquet Y, Girard-Bascou J, Michel F and Rochaix J-D (1993) Post-transcriptional control of chloroplast gene expression in *Chlamydomonas*, the case of *psaA trans*-splicing. In: Brennicke A and Kück U (eds) Plant Mitochondria: With Emphasis on RNA Editing and Cytoplasmic Male Sterility, pp 233–240. VCH Publishers, Inc, New York

Gottesman S, Squires C, Pichersky E, Carrington M, Hobbs M,

Mattick JS, Dalrymple B, Kuramitsu H, Shiroza T, Foster T, Clark WP, Ross B, Squires CL and Maurizi MR (1990) Conservation of the regulatory subunit for the Clp ATP-dependent protease in prokaryotes and eukaryotes. Proc Natl Acad Sci USA 87: 3513–3517

Gray JC, Hird SM and Dyer TA (1990) Nucleotide sequence of a wheat chloroplast gene encoding the proteolytic subunit of an ATP-dependent protease. Plant Mol Biol 15: 947–950

Green CD and Hollingsworth MJ (1992) Expression of the large ATP synthase gene cluster in spinach plastids during light-induced development. Plant Physiol 100: 1164–1170

Greenberg BM and Hallick RB (1986) Accurate transcription and processing of 19 *Euglena* chloroplast tRNAs in a *Euglena* soluble extract. Plant Mol Biol 6: 89–100

Greenberg BM, Narita JO, Deluca-Flaherty C, Gruissem W, Rushlow KA and Hallick RB (1984) Evidence for two RNA polymerase activities in *Euglena gracilis* chloroplasts. J Biol Chem 259: 14880–14887

Gruissem W (1984) A chloroplast transcription system from higher plants. Plant Mol Biol Rep 4: 15–23

Gruissem W (1989) Chloroplast RNA: Transcription and processing. In: Marcus A (ed) The Biochemistry of Plants, Vol 15, pp 151–191. Academic Press, San Diego

Gruissem W and Schuster G (1993) Control of mRNA degradation in organelles. In: Brawerman G and Belasco J (eds) Control of mRNA Stability, pp 329–365. Academic Press, Orlando

Gruissem W and Tonkyn JC (1993) Control mechanisms of plastid gene expression. Crit Rev in Plant Sci 12: 19–55

Gruissem W and Zurawski G (1985) Analysis of promoter regions for the spinach chloroplast *rbcL*, *atpB* and *psbA* genes. EMBO J 4: 3375–3383

Gruissem W, Narita JO, Greenberg BM, Prescott DM and Hallick RB (1983) Selective in vitro transcription of chloroplast genes. J Cell Biochem 22: 31–46

Gruissem W, Greenberg BM, Zurawski G and Hallick RB (1986) Chloroplast gene expression and promoter identification in chloroplast extracts. In: Weissbach A and Weissbach H (eds) Methods in Enzymology. Plant Molecular Biology, Vol 118, pp 253–270. Academic Press, Orlando

Guitton C, Dorne A-M and Mache R (1984) *In organello* and in vitro phosphorylation of chloroplast ribosomal proteins. Biochem Biophys Res Comm 121: 297–303

Haley J and Bogorad L (1990) Alternative promoters are used for genes within maize chloroplast polycistronic transcription units. Plant Cell 2: 323–333

Hallick RB, Hong L, Drager RG, Favreau MR, Monfort A, Orsat B, Spielmann A and Stutz E (1993) Complete sequence of *Euglena gracilis* chloroplast DNA. Nucleic Acids Res 21: 3537–3544

Helmann JD and Chamberlin MJ (1988) Structure and function of bacterial sigma factors. Annu Rev Biochem 57: 839–872

Herrin DL and Schmidt GW (1988) *Trans*-splicing of transcripts for the chloroplast *psaA1* gene. *In vivo* requirement for nuclear gene products. J Biol Chem 263: 14601–14604

Herrin DL, Chen Y-F and Schmidt GW (1990) RNA splicing *Chlamydomonas* chloroplasts. Self-splicing of 23 S preRNA. J Biol Chem 265: 21134–21140

Herrin DL, Bao Y, Thompson AJ and Chen Y-F (1991) Self-splicing of the *Chlamydomonas* chloroplast *psb*A introns. Plant Cell 3: 1095–1107

Herrin DL, Battey JF, Greer K and Schmidt GW (1992) Regulation of chlorophyll apoprotein expression and accumulation: Requirements for carotenoids and chlorophyll. J Biol Chem 267: 8260–8269

Hess WR, Prombona A, Fieder B, Subramanian AR and Borner T (1993) Chloroplast *rps15* and the *rpoB/C1/C2* gene cluster are strongly transcribed in ribosome-deficient plastids: Evidence for a functioning non-chloroplast-encoded RNA polymerase. EMBO J 12: 563–571

Hildebrand M, Hallick RB, Passavant CW and Bourque DP (1988) *Trans*-splicing in chloroplasts: The *rps12* loci of *Nicotiana tabacum*. Proc Natl Acad Sci USA 85: 372–376

Hoch B, Maier RM, Appel K, Igloi GL and Kössel H (1991) Editing of a chloroplast mRNA by creation of an initiation codon. Nature 353: 178–180

Hsu-Ching C and Stern DB (1991a) Specific ribonuclease activities in spinach chloroplasts promote mRNA maturation and degradation. J Biol Chem 266: 24205–24211

Hsu-Ching C and Stern DB (1991b) Specific binding of chloroplast proteins in vitro to the 3′ untranslated region of spinach chloroplast *petD* mRNA. Mol Cell Biol 11: 4380–4388

Hu J and Bogorad L (1990) Maize chloroplast RNA polymerase: The 180-, 120-, and 38-kilodalton polypeptides are encoded in chloroplast genes. Proc Natl Acad Sci USA 87: 1531–1535

Hu J, Troxler RF and Bogorad L (1991) Maize chloroplast RNA polymerase: The 78-kilodalton polypeptide is encoded by the plastid *rpoC1* gene. Nuc Acids Res 19: 3431–3434

Igloi GL and Kössel H (1992) The transcriptional apparatus of chloroplasts. Crit Rev Plant Sci 10: 525–558

Inagaki N, Fujita S and Satoh K (1989) Solubilization and partial purification of a thylakoidal enzyme of spinach involved in the processing of D1 protein. FEBS Lett 246: 218–222

Jagendorf AT and Michaels A (1990) Rough thylakoids translation on photosynthetic membranes. Plant Sci 71: 137–146

Jarrell KA, Dietrich RC and Perlman PS (1988) Group II intron domain 5 facilitates a *trans*-splicing reaction. Mol Cell Biol 8: 2361–2366

Jensen KH, Herrin DL, Plumley FG and Schmidt GW (1986) Biogenesis of Photosystem II complexes: Transcriptional, translational, and posttranslational regulation. J Cell Biol 103: 1315–1325

Johnson EM and Sears BB (1990) Structure and expression of cytochrome *f* in an *Oenothera* plastome mutant. Curr Genet 17: 529–534

Jolly SO and Bogorad L (1980) Preferential transcription of cloned maize chloroplast DNA sequences by maize chloroplast RNA polymerase. Proc Natl Acad Sci USA 77: 822–826

Joussaume M (1973) Mise en evidence de deux formes de RNA polymerase dependante du DNA dans les chloroplastes isoles de feuilles de Pois. Physiol Veg 11: 69–82

Keegstra K, Olsen LJ and Theg SM (1989) Chloroplastic precursors and their transport across the envelope membranes. Annu Rev Plant Physiol Plant Mol Biol 40: 471–501

Kim J-K and Hollingsworth MJ (1993) Splicing of group II introns in spinach chloroplasts (in vivo): Analysis of lariat formation. Curr Genet 23: 175–180

Kim M, Christopher DA and Mullet JE (1993) Direct evidence for selective modulation of *psbA, rpoA, rbcL* and 16S RNA stability during barley chloroplast development. Plant Mol Biol 22: 447–463

Kitagawa M, Wada C, Yoshioka S and Yura T (1991) Expression of *ClpB*, an analog of the ATP-dependent protease regulatory subunit in *Escherichia coli*, is controlled by a heat shock σ factor (σ^{32}). J Bacteriol 173: 4247–4253

Klaff P and Gruissem W (1991) Changes in chloroplast mRNA stability during leaf development. Plant Cell 3: 517–529

Klein RR (1991) Regulation of light-induced chloroplast transcription and translation in eight-day-old dark-grown barley seedlings. Plant Physiol 97: 335–342

Klein RR and Mullet JE (1990) Light-induced transcription of chloroplast genes: *psbA* transcription is differentially enhanced in illuminated barley. J Biol Chem 265: 1895–1902

Klein RR, Gamble PE and Mullet JE (1988a) Light-dependent accumulation of radiolabeled plastid-encoded chlorophyll *a*-apoproteins requires chlorophyll *a*. I. Analysis of chlorophyll-deficient mutants and phytochrome involvement. Plant Physiol 88: 1246–1256

Klein RR, Mason HS and Mullet JE (1988b) Light-regulated translation of chloroplast proteins. I. Transcripts of *psaA-psaB*, *psbA*, and *rbcL* are associated with polysomes in dark-grown and illuminated barley seedlings. J Cell Biol 106: 289–301

Kobayashi H, Bogorad L and Miles CD (1987) Nuclear gene-regulated expression of chloroplast genes for coupling factor one in maize. Plant Physiol 85: 757–767

Kobayashi H, Ngernprasirtsiri J and Akazawa T (1990) Transcriptional regulation and DNA methylation in plastids during transitional conversion of chloroplasts to chromoplasts. EMBO J 9: 307–313

Kohchi T, Ogura Y, Umesono K, Yamada Y, Komano T, Ozeki H and Ohyama K (1988a) Ordered processing and splicing in a polycistronic transcript in liverwort chloroplasts. Curr Genet 14: 147–154

Kohchi T, Umesono K, Ogura Y, Komine Y, Nakahigashi K, Komano T, Yamada Y, Ozeki H and Ohyama K (1988b) A nicked group II intron and *trans*-splicing in liverwort, *Marchantia polymorpha*, chloroplasts. Nuc Acids Res 16: 10025–10036

Koller B, Fromm H, Galun E and Edelman M (1987) Evidence for in vivo *trans*-splicing of pre-mRNAs in tobacco chloroplasts. Cell 48: 111–119

Kössel H, Hoch B, Maier RM, Igloi GL, Kudla J, Zeltz P, Freyer R, Neckermann K and Ruf S (1993) RNA editing in chloroplasts of higher plants. In: Brennicke A and Küch U (eds) Plant Mitochondria: With Emphasis On RNA Editing and Cytoplasmic Male Sterility, pp 93–102. VCH Publishers, Inc., New York

Kuchka MR, Mayfield SP and Rochaix J-D (1988) Nuclear mutations specifically affect the synthesis and/or degradation of the chloroplast-encoded D2 polypeptide of Photosystem II in *Chlamydomonas reinhardtii*. EMBO J 7: 319–324

Kuchka MR, Goldschmidt-Clermont M, van Dillewijn J and Rochaix J-D (1989) Mutation at the *Chlamydomonas* nuclear NAC2 locus specifically affects stability of the chloroplast *psbD* transcript encoding polypeptide D2 of PSII. Cell 58: 869–876

Kück U, Choquet Y, Schnedider M, Dron M and Bennoun P (1987) Structural and transcription analysis of two homologous genes for the P700 chlorophyll *a*-apoproteins in *Chlamydomonas reinhardtii*: Evidence for in vivo *trans*-splicing. EMBO J 6: 2185–2195

Kudla J, Igloi GL, Metzlaff M, Hagemann R and Kössel H (1992) RNA editing in tobacco chloroplasts leads to the formation of a translatable *psbL* mRNA by a C to U substitution within the initiation codon. EMBO J 11: 1099–1103

Kuntz M, Camara B, Weil J-H and Schantz R (1992) The *psbL* gene from bell pepper (*Capsicum annuum*): Plastid RNA editing also occurs in non-photosynthetic chromoplasts. Plant Mol Biol 20: 1185–1188.

Kyle DJ, Ohad I and Arntzen CJ (1984) Membrane protein damage and repair: Selective loss of a quinone-protein function in chloroplast membranes. Proc Natl Acad Sci USA 81: 4070–4074

Laboure A-M, Lescure A-M and Briat J-F (1988) Evidence for a translation-mediated attenuation of a spinach chloroplast rDNA operon. Biochimie 70: 1343–1352

Lakhani S, Khanna NC and Tewari KK (1992) Two distinct transcriptional activities of pea (*Pisum sativum*) chloroplasts share immunochemically related functional polypeptides. Biochem J 286: 833–841

Lam E and Chua N-H (1987) Chloroplast DNA gyrase and in vitro regulation of transcription by template topology and novobiocin. Plant Mol Biol 8: 415–424

Lerbs S, Brautigam E and Mache R (1988) DNA-dependent RNA polymerase of spinach chloroplasts: Characterization of α-like and σ-like polypeptides. Mol Gen Genet 211: 459–464

Lerbs-Mache S (1993) The 110-kDa polypeptide of spinach plastid DNA-dependent RNA polymerase: Single-subunit enzyme or catalytic core of multimeric enzyme complexes? Proc Natl Acad Sci U S A 90: 5509–5513

Li Y and Sugiura M (1990) Three distinct ribonucleoproteins from tobacco chloroplasts: Each contains a unique amino terminal acidic domain and two ribonucleoprotein consensus motifs. EMBO J 9: 3059–3066

Link G (1984) DNA sequence requirements for the accurate transcription of a protein-coding plastid gene in a plastid in vitro system from mustard (*Sinapis alba* L.). EMBO J 3: 1697–1704

Little MC and Hallick RB (1988) Chloroplast *rpoA*, *rpoB*, and *rpoC* genes specify at least three components of a chloroplast DNA-dependent RNA polymerase active in tRNA and mRNA transcription. J Biol Chem 263: 14302–14307

Liu X-Q and Jagendorf AT (1986) Neutral peptidases in the stroma of pea chloroplasts. Plant Physiol 81: 603–608

Mache R (1990) Chloroplast ribosomal proteins and their genes. Plant Sci 72: 1–12

Maier RM, Hoch B, Zeltz P and Kössel H (1992) Internal editing of the maize chloroplast *ndhA* transcript restores codons for conserved amino acids. Plant Cell 4: 609–616

Marano MR and Carrillo N (1991) Chromoplast formation during tomato fruit ripening. No evidence for plastid DNA methylation. Plant Mol Biol 16: 11–19

Marion-Poll A, Hibbert CS, Radebaugh CA and Hallick RB (1988) Processing of mono-, di- and tricistronic transfer RNAs precursors in a spinach or pea chloroplast soluble extract. Plant Mol Biol 11: 45–56

Marzluff WF and Panday NB (1988) Multiple regulatory steps control histone mRNA concentrations. Trends Biochem Sci 13: 49–56

Mattoo AK, Hoffman-Falk H, Marder JB and Edelman M (1984) Regulation of protein metabolism: Coupling of photosynthetic electron transport to in vivo degradation of the rapidly metabolized 32-kilodalton protein of the chloroplast membranes. Proc Natl Acad Sci USA 81: 1380–1384

Maurizi MR, Clark WP, Kim S-H and Gottesman S (1990) *ClpP* represents a unique family of serine proteases. J Biol Chem 265: 12546–12552

McCullough AJ, Kangasjarvi J, Gengenbach BG and Jones RJ (1992) Plastid DNA in developing maize endosperm: Genome structure, methylation, and transcript accumulation patterns. Plant Physiol 100: 958–964

Merchant S and Bogorad L (1986) Rapid degradation of apoplastocyanin in Cu(II)-deficient cells of *Chlamydomonas reinhardtii*. J Biol Chem 261: 15850–15853

Metz JG and Miles D (1982) Use of a nuclear mutant of maize to identify components of Photosystem II. Biochim Biophys Acta 681: 95–102

Michaels A and Herrin DL (1990) Translational regulation of chloroplast gene expression during the light-dark cell cycle of *Chlamydomonas*: Evidence for control by ATP/energy supply. Biochem Biophys Res Comm 170: 1082–1088

Miles CD (1982) The use of mutations to probe photosynthesis in higher plants. In: Hallick R, Edelman M and Chua N-H (eds) Methods in Chloroplast Molecular Biology, pp 75–109. Elsevier, New York

Minami E, Shinohara K, Kawakami N and Watanabe A (1988) Localization and properties of transcripts of *psbA* and *rbcL* genes in the stroma of spinach chloroplasts. Plant Cell Physiol 29: 1303–1309

Mishkind ML, Branagan AJ, Jensen KH, Plumley FG and Schmidt GW (1985) Roles of proteases in chloroplast biogenesis. Curr Topics Plant Biochem Physiol 4: 34–50

Mohr G, Perlman PS and Lambowitz AM (1993) Evolutionary relationships among group II intron-encoded proteins and identification of a conserved domain that may be related to maturase function. Nucleic Acids Research 21: 4991–4997

Monod C, Goldschmidt-Clermont M and Rochaix J-D (1992) Accumulation of chloroplast *psbB* RNA requires a nuclear factor in *Chlamydomonas reinhardtii*. Mol Gen Genet 231: 449–459

Moore T and Keegstra K (1993) Characterization of a cDNA clone encoding a chloroplast-targeted Clp homologue. Plant Mol Biol 21: 525–537

Morden CW, Wolfe KH, dePamphilis CW and Palmer JD (1991) Plastid translation and transcription genes in a non-photosynthetic plant: Intact, missing and pseudo genes. EMBO J 10: 3281–3288

Morton BR (1993) Chloroplast DNA codon use: Evidence for selection at the *psbA* locus based on tRNA availability. J Mol Evol 37: 273–280

Mullet JE (1993) Dynamic regulation of chloroplast transcription. Plant Physiol 103: 309–313

Mullet JE and Klein RR (1987) Transcription and RNA stability are important determinants of higher plant chloroplast RNA levels. EMBO J 6: 1571–1579

Mullet JE, Orozco EM and Chua N-H (1985) Multiple transcripts for higher plant *rbcL* and *atpB* genes and localization of transcription initiation site of the *rbcL* gene. Plant Mol Biol 4: 39–54

Mullet JE, Klein PG and Klein RR (1990) Chlorophyll regulates accumulation of the plastid-encoded chlorophyll apoproteins CP43 and D1 by increasing apoprotein stability. Proc Natl Acad Sci USA 87: 4038–4042

Newbury SF, Smith NH, Robinson EC, Hiles ID and Higgins CF (1987) Stabilization of translationally active mRNA by prokaryotic REP sequences. Cell 48: 297–310

Ngernprasirtsiri J (1990) DNA methylation is a determinative element of photosynthesis gene expression in amyloplasts from liquid-cultured cells of sycamore (*Acer pseudoplatanus* L.). Cell Struc Func 15: 285–293

Ngernprasirtsiri J and Akazawa T (1990) Modulation of DNA methylation and gene expression in cultured sycamore cells treated by hypomethylating base analog. Eur J Biochem 194: 513–520

Ngernprasirtsiri J, Kobayashi H and Akazawa T (1988) DNA methylation as a mechanism of transcriptional regulation in nonphotosynthetic plastids in plant cells. Proc Natl Acad Sci USA 85: 4750–4754

Ngernprasirtsiri J, Chollet R, Kobayashi H, Sugiyama T and Akazawa T (1989) DNA methylation and the differential expression of C_4 photosynthesis genes in mesophyll and bundle sheath cells of greening maize leaves. J Biol Chem 264: 8241–8248

Nickelsen J and Link G (1989) Interaction of a 3′ RNA region of the mustard *trnK* gene with chloroplast proteins. Nuc Acids Res 17: 9637–9648

Nielsen BL and Tewari KK (1988) Pea chloroplast topoisomerase I: Purification, characterization, and role in replication. Plant Mol Biol 11: 3–14

Ohad I, Kyle J and Arntzen CJ (1984) Membrane protein damage and repair: Removal and replacement of inactivated 32-kilodalton polypeptides in chloroplast membranes. J Cell Biol 99: 481–485

Ohta N, Sato N, Kawano S and Kuroiwa T (1991) Methylation of DNA in the chloroplasts and amyloplasts of the pea, *Pisum sativum*. Plant Sci 78: 33–42

Olive J, Vallon O, Wollman F-A, Recouvreur M and Bennoun P (1986) Studies on the cytochrome $b_6 f$ complex II. Localization of the complex in the thylakoid membranes from spinach and *Chlamydomonas reinhardtii* by immunocytochemistry and freeze-fracture analysis of b_6/f mutants. Biochim Biophys Acta 851: 239–248

Orozco EM, Mullet JE and Chua NH (1985) An in vitro system for accurate transcription initiation of chloroplast protein genes. Nuc Acids Res 13: 1283–1302

Orozco EM, Mullet JE, Hanley-Bowdoin L and Chua N-H (1986) *In vitro* transcription of chloroplast protein genes. In: Weissbach A and Weissbach H (eds) Methods in Enzymology, Vol 118, pp 232–253. Academic Press Inc, Orlando

Pastushenkova IA and Volodarskii AD (1990) Posttranscriptional regulation of the synthesis of ribulose bisphosphate carboxylase subunits during the light-induced development of chloroplasts in *Euglena*. Fiziol Rast 37: 116–125

Plant AL and Gray JC (1988) Introns in chloroplast protein-coding genes of land plants. Photosynth Res 16: 23–39

Platt T (1986) Transcription termination and the regulation of gene expression. Annu Rev Biochem 55: 339–372

Posno M, Van Noort M, Debise R and Groot GSP (1984) Isolation, characterization, phosphorylation and site of synthesis of *Spinacia* chloroplast ribosomal proteins. Curr Genet 8: 147–154

Poulsen C (1984) Two mRNA species differing by 258 nucleotides at the 5′ end are formed from the barley chloroplast *rbcL* gene. Carlsberg Res Commun 49: 89–104

Purton S and Gray JC (1989) The plastid *rpoA* gene encoding a protein homologous to the bacterial RNA polymerase alpha subunit is expressed in pea chloroplast. Mol Gen Genet 217: 77–84

Rajasekhar VK, Sun E, Meeker R, Wu B-W and Tewari KK (1991) Highly purified pea chloroplast RNA polymerase transcribes both rRNA and mRNA genes. Eur J Biochem 195: 215–228

Rapp JC, Baumgartner BJ and Mullet J (1992) Quantitative analysis of transcription and RNA levels of 15 barley chloroplast genes: Transcription rates and mRNA levels vary over 300-fold: Predicted mRNA stabilities vary 30-fold. J Biol Chem 267: 21404–21411

Reinbothe S, Reinbothe C, Krauspe R and Parthier B (1991a) Changing gene expression during dark-induced chloroplast dedifferentiation in *Euglena gracilis*. Plant Physiol Biochem 29: 309–318

Reinbothe S, Reinbothe C and Parthier B (1991b) Glucose repression of chloroplast development in *Euglena gracilis*. Plant Physiol Biochem 29: 309–318

Reiss T and Link G (1985) Characterization of transcriptionally active DNA-protein complexes from chloroplasts and etioplasts of mustard (*Sinapis alba* L.). Eur J Biochem 148: 207–212

Rochaix J-D (1992) Post-transcriptional steps in the expression of chloroplast genes. In: Palade GE (ed) Annual Review of Cell Biology, Vol 8, pp 1–28. Annual Reviews Inc, Palo Alto

Rochaix J-D, Kuchka M, Mayfield SP, Schirmer-Rahire M, Girard-Bascou J and Bennoun P (1989) Nuclear and chloroplast mutations affect the synthesis or stability of the chloroplast *psbC* gene product in *Chlamydomonas reinhardtii*. EMBO J 8: 1013–1021

Rock CD, Barkan A and Taylor WC (1987) The maize plastid *psbB-psbF-petB-petD* gene cluster: Spliced and unspliced *petB* and *petD* RNAs encode alternative products. Curr Genet 12: 69–77

Ruf S, Zeltz P and Kössel H (1994) Complete RNA editing of unspliced and dicistronic transcripts of the intron-containing reading frame IRF170 from maize chloroplasts. Proc Natl Acad Sci USA 91: 2295–2299

Rushlow KE, Orozco EMJ, Lipper C and Hallick RB (1980) Selective in vitro transcription of *Euglena* chloroplast ribosomal RNA genes by a transcriptionally active chromosome. J Biol Chem 255: 3786–3792

Russell D and Bogorad L (1987) Transcription analysis of the maize chloroplast gene for the ribosomal protein S4. Nuc Acids Res 15: 1853–1867

Sakamoto W, Kindle KL and Stern DB (1993) In vivo analysis of *Chlamydomonas* chloroplast *petD* gene expression using stable transformation of β-glucuronidase translational fusions. Proc Natl Acad Sci U S A 90: 497–501

Scheibe R (1990) Light/dark modulation: Regulation of chloroplast metabolism in a new light. Bot Acta 103: 327–334

Schuster G and Gruissem W (1991) Chloroplast mRNA 3′ end processing requires a nuclear-encoded RNA-binding protein. EMBO J 10: 1493–1502

Sexton TB, Christopher DA and Mullet JE (1990) Light-induced switch in barley *psbD-psbC* promoter utilization: A novel mechanism regulating chloroplast gene expression. EMBO J 9: 4485–4494

Sieburth LE, Berry-Lowe S and Schmidt GW (1991) Chloroplast RNA stability in *Chlamydomonas*: Rapid degradation of *psbB* and *psbC* transcripts in two nuclear mutants. Plant Cell 3: 175–189

Siedlecki J, Zimmermann W and Wissback A (1983) Characterization of a prokaryotic topoisomerase I activity in chloroplast extracts from spinach. Nuc Acids Res 11: 1523–1536

Somerville CR (1986) Analysis of photosynthesis with mutants of higher plants and algae. Annu Rev Plant Physiol 37: 467–507

Spreitzer RJ, Goldschmidt-Clermont M, Rahire M and Rochaix JD (1985) Nonsense mutations in the *Chlamydomonas* chloroplast gene that codes for the large subunit of ribulosebisphosphate carboxylase/oxygenase. Proc Natl Acad Sci USA 82: 5460–5464

Squires CL, Pedersen BM, Ross BM and Squires C (1991) *ClpB* is the *Escherichia coli* heat shock protein F84.1. J Bacteriol 173: 4254–4262

Staub JM and Maliga P (1993) Accumulation of D1 polypeptide in tobacco plastids is regulated via the untranslated region of the *psbA* mRNA. EMBO J 12: 601–606

Stern DB and Gruissem W (1987) Control of plastid gene expression: 3′ inverted repeats act as mRNA processing and stabilizing elements, but do not terminate transcription. Cell 51: 1145–1157

Stern DB and Gruissem W (1989) Chloroplast mRNA 3′ end maturation is biochemically distinct from prokaryotic mRNA processing. Plant Mol Biol 13: 615–625

Stern DB, Jones H and Gruissem W (1989) Function of plastid mRNA 3′ inverted repeats. RNA stabilization and gene-specific protein binding. J Biol Chem 264: 18742–18750

Stern DB, Radwanski ER and Kindle KL (1991) A 3′ stem/loop structure of the *Chlamydomonas* chloroplast *atpB* gene regulates mRNA accumulation in vivo. Plant Cell 3: 285–297

Stirdivant SM, Crossland LD and Bogorad L (1985) DNA supercoiling affects in vitro transcription of two maize chloroplast genes differently. Proc Natl Acad Sci USA 82: 4886–4890

Surzycki SJ and Shellenbarger DL (1976) Purification and characterization of a putative sigma factor from *Chlamydomonas reinhardtii*. Proc Natl Acad Sci USA 73: 3961–3965

Susek RE, Ausubel FM and Chory J (1993) Signal transduction mutants of *Arabidopsis* uncouple nuclear *cab* and *rbcS* gene expression from chloroplast development. Cell 74: 787–799

Takahashi Y, Goldschmidt-Clermont M, Soen S-Y, Franzen LG and Rochaix J-D (1991) Directed chloroplast transformation in *Chlamydomonas reinhardtii*: Insertional inactivation of the *psa*C gene encoding the iron sulfur protein destabilizes Photosystem I. EMBO J 10: 2033–2040

Tanaka M, Obokata J, Chunwangse J, Shinozaki K and Sugiura M (1987) Rapid splicing and stepwise processing of a transcript from the *psbB* operon in tobacco chloroplasts: Determination of the intron sites in *petB* and *petD*. Mol Gen Genet 209: 427–431

Taylor WC (1989) Regulatory interactions between nuclear and plastid genomes. Annu Rev Plant Physiol Plant Mol Biol 40: 211–233

Taylor WC, Barkan A and Martienssen RA (1987) Use of nuclear mutants in the analysis of chloroplast development. Dev Genet 8: 305–320

Tessier L-H, Keller M, Chan RL, Fournier R, Weil J-H and Imbault P (1991) Short leader sequences may be transferred from small RNAs to pre-mature mRNAs by *trans*-splicing in *Euglena*. EMBO J 10: 2621–2625

Thompson AJ and Herrin DL (1991) In vitro self-splicing reactions of the chloroplast group I intron Cr.LSU from *Chlamydomonas reinhardtii* and in vivo manipulation via gene-replacement. Nuc Acids Res 19: 6611–6618

Thompson AJ, Yuan X, Kudlicki W and Herrin DL (1992) Cleavage and recognition pattern of a double-strand-specific endonuclease (I-CreI) encoded by the chloroplast 23S rRNA intron of *Chlamydomonas reinhardtii*. Gene 119: 247–251

Thompson RJ and Mosig G (1985) An ATP-dependent supercoiling topoisomerase of *Chlamydomonas reinhardtii* affects accumulation of specific chloroplast transcripts. Nuc Acids Res 13: 873–891

Thompson RJ and Mosig G (1987) Stimulation of a *Chlamydomonas* chloroplast promoter by novobiocin *in situ* and in *E. coli* implies regulation by torsional stress in the chloroplast DNA. Cell 48: 281–287

Tiller K and Link G (1993a) Phosphorylation and dephosphorylation affect functional characteristics of chloroplast and etioplast transcription systems from mustard (*Sinapis alba* L.). EMBO J 12: 1745–1753

Tiller K and Link G (1993b) Sigma-like transcription factors from mustard (*Sinapis alba* L.) etioplast are similar in size to, but functionally distinct from, their chloroplast counterparts. Plant Mol Biol 21: 503–513

Tomas R, Vera A, Martin M and Sabater B (1992) Changes in protein synthesis without evidence of DNA methylation in barley chloroplasts during leaf growth and development. Plant Sci 85: 71–77

Tonkyn JC and Gruissem W (1993) Differential expression of the partially duplicated chloroplast s10 ribosomal protein operon. Mol Gen Genet 241: 141–152

Tonkyn JC, Deng X-W and Gruissem W (1992) Regulation of plastid gene expression during photooxidative stress. Plant Physiol 99: 1406–1415

Westhoff P and Herrmann RG (1988) Complex RNA maturation in chloroplasts. The *psbB* operon from spinach. Eur J Biochem 171: 551–564

Wong HC and Chang S (1986) Identification of a positive retroregulator that stabilizes mRNAs in bacteria. Proc Natl Acad Sci USA 83: 3233–3237

Xu R, Bingham SE and Webber AN (1993) Increased mRNA accumulation in *psaB* frame-shift mutant of *Chlamydomonas reinhardtii* suggests a role for translation in *psaB* mRNA stability. Plant Mol Biol 22: 465–474

Ye L, Li Y, Fukami-Kobayashi K, Go M, Konishi T, Watanabe A and Sugiura M (1991) Diversity of a ribonucleoprotein family in tobacco chloroplasts: Two new chloroplast ribonucleoproteins and a phylogenetic tree of ten chloroplast RNA-binding domains. Nuc Acids Res 19: 6485–6490

Zeltz P, Hess WR, Neckermann K, Borner T and Kössel H (1993) Editing of the chloroplast *rpoB* transcript is independent of chloroplast translation and shows different patterns in barley and maize. EMBO Journal 12: 4291–4296

Zurawski G and Clegg MT (1987) Evolution of higher-plant chloroplast DNA-encoded genes: Implication for structure-function and phylogenetic studies. Annu Rev Plant Physiol 38: 391–418

Chapter 32

Chloroplast Transformation: Current Results and Future Prospects

Jeanne M. Erickson
Department of Biology, University of California, Los Angeles, Los Angeles, CA 90024-1606, USA

Donald R. Ort and Charles F. Yocum (eds): Oxygenic Photosynthesis: The Light Reactions, pp. 589–619.

Summary

Genetic engineering of proteins is a powerful tool used in both basic and applied research. In vitro alteration of the primary structure of a gene and subsequent introduction of the mutated DNA into the genome of a living cell allows the directed manipulation of protein structure. Such an approach, often termed 'reverse genetics,' has been widely used to investigate the complex relationship between the structure of a protein and its function, and to explore the intricacies of biochemical and developmental pathways. An obvious prerequisite for genetic engineering is the ability to introduce DNA into a living cell in such a way that it is stably maintained and properly expressed in the appropriate genome of the host cell. The genome of a prokaryote is in the cell cytoplasm and generally consists of one or a few copies of a large DNA molecule. In contrast, photosynthetic eukaryotes contain three distinct genomes, each located within a subcellular organelle enveloped by one or more membranes and hence separated from the cytoplasm. The three plant cell genomes are those of the nucleus, the mitochondrion, and the plastid. The genome of the chloroplast, the plastid type found in photosynthetic cells, presents a complex genetic target because there are often hundreds of copies of the circular chloroplast DNA molecule per chloroplast, and often hundreds of chloroplasts per cell. Obtaining a plant cell in which every resident copy of a given chloroplast gene has been replaced by an engineered, mutant gene copy is an essential step in experiments involving DNA-mediated chloroplast transformation. An ideal model organism for such studies is provided by the unicellular green alga *Chlamydomonas reinhardtii*, which contains a single large chloroplast. This chapter presents current results and future prospects for chloroplast transformation, both in *Chlamydomonas* and in plants of agronomic interest.

I. Introduction

A. Transformation

Classical genetics has long been recognized as a powerful tool for studying a myriad of biological phenomena, phenomena with complexities that often stymie tidy and direct biochemical analysis. Since the discovery half a century ago that DNA is the 'transforming principle,' the material responsible for transforming the phenotype of bacterial cells (Avery et al., 1944), enormous progress has been made in understanding the structure of DNA and the biochemistry of nucleic acids. The subsequent advances in techniques for the in vivo and in vitro manipulation of DNA, combined with new techniques for introducing DNA into both prokaryotic and eukaryotic cells, opened the door to the current era of modern molecular genetics. Genetic engineering of gene products provides tremendous opportunities

Abbreviations: AAD – aminoglycoside adenylyltransferase; *C.* – *Chlamydomonas;* FdUrd – 5-fluoro-2′-deoxyuridine; GUS – β-glucuronidase; NPTII – neomycin phosphotransferase; PCR – polymerase chain reaction; PEG – polyethylene glycol; PS I – Photosystem I; PS II – Photosystem II; RFLP – restriction fragment length polymorphism

for the elucidation of biochemical pathways, and the investigation of relationships between biological structure and consequent function.

This chapter on chloroplast transformation includes a brief review of studies that have used chloroplast transformation to investigate specific biological processes. However, the main focus is on aspects of the transformation process itself, including techniques for the introduction of DNA into the chloroplasts of plant and algal cells, molecular markers which facilitate selection and screening of putative transformants, and molecular mechanisms that lead to stable integration of DNA into the chloroplast genome and segregation of homoplasmic transformant cell lines.

B. Plastids

Photosynthetic eukaryotes possess three distinct genomes, each located within a separate membrane-enveloped organelle. These three DNA-containing organelles are the nucleus, the mitochondrion and the plastid. The bulk of total DNA mass and total genetic information of the cell is in the cell nucleus; likewise, the majority of proteins found in mitochondria and plastids are encoded in the nuclear genome. However, both the mitochondrial and plastid genomes contain significant amounts of DNA. This non-nuclear DNA is replicated and expressed within those organelles, and mitochondrial and plastid gene products are essential for mitochondrial and plastid function.

In plants, plastids and mitochondria are generally inherited from the cytoplasm of the maternal egg cell and are propagated and distributed to daughter cells during cell division. Neither the process of plastid division itself nor the processes that determine whether a single cell will have one plastid or up to hundreds of plastids are well understood. There are several types of mature plastids in plants, all of which appear to develop from a common plastid progenitor, termed the proplastid or eoplast (Thompson and Whatley, 1980), and all of which appear to contain the same plastid genome. Plastid differentiation is dependent on differential expression of the plastid genome as well as differential nuclear gene expression, and is regulated by environmental and developmental signals. Many mature plastid types appear to be interconvertible.

Although plastids are unique to photosynthetic eukaryotes, there are many types of mature plastids that do not carry out photosynthesis (see Emes and Tobin, 1993, for a recent review of plastid metabolism and development). Rather, these plastid types are geared to specific metabolic functions and localized to specific tissues in multicellular plants. Amyloplasts synthesize and store starch, and are found primarily in endosperm, cotyledons, tubers, the root, and the root cap. Chromoplasts are involved in the synthesis and storage of carotenoid pigments; these plastids give color to fruits and flowers. The oil-rich elaioplasts found in the stems of some cacti and the epidermal cells of certain flowering plants, including several in the lily and orchid families, synthesize and store terpenoid compounds.

The chloroplast is the plastid type found in all eukaryotic photosynthetic cells. This includes the unicellular euglenoids, diatoms and dinoflagellates, the unicellular and multicellular algae, and all photosynthetic cells of non-vascular and vascular plants. In the latter case, photosynthesis takes place primarily in leaves, but also in stems, tendrils, runners, cotyledons, and even green fruits. Mature chloroplasts have an extensive and highly organized thylakoid membrane that contains the pigment-protein complexes of the light-driven photosynthetic electron transfer chain: Photosystem II (PS II), the cytochrome b_6f complex, and Photosystem I (PS I). PS II mediates the initial photo-oxidation of water to molecular oxygen, generating electrons for the chain, while PS I produces a reductant capable of reducing NADP. The thylakoid membrane also contains the ATP synthetase complex necessary for photophosphorylation of ADP. The ATP and NADPH generated as a result of photosynthetic electron transfer provide energy for CO_2 fixation and the intermediate carbohydrate biosynthetic pathways of the so-called 'dark reactions' of photosynthesis which take place in the stroma of the chloroplast (see Erickson, 1992, for a general review of chloroplasts and photosynthesis).

Chloroplast structure and function can vary significantly between organisms, and within the same organism under different environmental and developmental situations. There is great diversity among algal species, including differences in the number of membranes surrounding the chloroplast, the organization of the thylakoid membrane, and the pigment composition of the light-harvesting apparatus (reviewed in Hoober, 1984), as suggested by the generic categories of brown, green and red algae. This diversity may reflect independent endosymbiotic origins, and/or divergent evolutionary pathways (see M. Reith, Chapter 34). The green algae, which include *Chlamydomonas*, have chloroplasts most similar to those of vascular plants with respect to the chlorophyll pigments and the organization of the thylakoid into appressed (granal) and non-appressed (stromal) regions. However, unlike vascular plants, in which photosynthetic cells can contain up to several hundred chloroplasts, the unicellular green alga *Chlamydomonas* contains a single, large chloroplast.

Since the observation and description of chloroplasts in 1837, chloroplast structure and function have been studied extensively (Gregory, 1989). Partly because of its essential role in the photosynthetic process crucial for plant survival, and partly because green leaves provide a ready source of material for study, the chloroplast has been the plastid type most studied to date. It is, therefore, not surprising that genetic engineering of plastids has been aimed primarily at the chloroplast. Although this review focuses primarily on the chloroplast, its genome, and the DNA-mediated transformation of this organelle in algae and vascular plants, it is important to remember that future applications of genetic engineering will most reasonably be aimed at achieving gene expression in other plastid types as described above.

II. Chloroplast Molecular Biology

A. The Chloroplast Genome: Organization, Replication and Recombination

The plastid genome of all species studied to date is polyploid. A single plastid may contain 20 to 900 copies of the large, covalently closed circular DNA molecule (reviewed in Bendich, 1987). All evidence indicates that, while copy number of the circular plastid DNA molecule may vary, the genetic complexity of the plastid DNA remains the same. Fluorescent staining of chloroplast DNA in situ reveals 10 to 20 DNA clusters, or nucleoids, per chloroplast (Coleman, 1978; James and Jope, 1978; Kuroiwa et al., 1981). The nucleoids appear to be associated with membranes (reviewed in Possingham and Lawrence, 1983), including the thylakoid membrane in mature chloroplasts and the chloroplast envelope membrane in younger chloroplasts, which suggests a means for segregation of plastid DNA during plastid division (Herrmann and Possingham, 1980). Such a model is supported by the recent demonstration that a 130 kDa protein of the chloroplast inner envelope membrane binds to specific sequences of chloroplast DNA in young pea seedlings (Sato et al., 1993). Although the size of the circular DNA molecule varies among organisms, in most plants it consists of 120,000 to 165,000 base pairs (reviewed in Palmer, 1985; Wolfe et al., 1991; Harris et al., 1994). The organization of plastid genes on the circle is fairly conserved among species and includes, in all but a few cases, the presence of a large ~20,000 bp inverted repeat region. The complete nucleotide sequences of the plastid DNA molecule of a non-vascular plant (liverwort; Ohyama et al., 1986), an angiosperm (tobacco; Shinozaki et al., 1986), a gymnosperm (black pine; Wakasugi et al., 1994), a monocot (rice; Hiratsuka et al., 1989), *Euglena* (Hallick et al., 1993) and a nonphotosynthetic parasitic flowering plant (beechdrops; Wolfe et al., 1992) have been determined. Comparative analysis shows that there are over 100 conserved chloroplast genes in photosynthetic plants (Shimada and Sugiura, 1991; Sugiura, 1992) including those encoding rRNA and tRNA, ribosomal proteins, and other polypeptides involved in transcription, translation and photosynthesis. In contrast, the minimal plastid genome of the non-photosynthetic flowering parasite, approximately half the size of photosynthetic plastid genomes, contains only 42 genes (Wolfe et al., 1992). Most of these are related to plastid gene expression. Comparison of the coding content of plastid and nuclear genomes among plants and algae reveals that genes may be in transit, in recent time on an evolutionary scale, from the plastid to the nucleus (reviewed in Wolfe et al., 1991). Proposals for naming chloroplast genes have helped to standardize chloroplast nomenclature (Hallick and Bottomley, 1983; Hallick, 1989; Price et al., 1994).

Replication of chloroplast DNA appears to occur by both Cairns and rolling circle mechanisms (Kolodner and Tewari, 1975). Chloroplast DNA origins of replication have been mapped by electron microscopy (reviewed in Chiu and Sears, 1992). Although mutants of *C. reinhardtii* with elevated (Nakamura et al., 1994) or reduced (our unpublished results) amounts of chloroplast DNA have been isolated, the mechanisms controlling chloroplast DNA copy number are not yet well understood.

Homologous recombination occurs within the large inverted repeat region of the chloroplast DNA molecule (Palmer, 1983), presumably as intramolecular recombination events. Electron microscopy shows that while most DNA circles are single monomers, catenated dimers and dimeric unicircles are present (Herrmann et al., 1975; Kolodner and Tewari, 1975). Pulse-field gel electrophoresis likewise demonstrates that monomeric, dimeric, trimeric, and tetrameric forms of chloroplast DNA exist (Deng et al., 1989). The multimers may account for as much as 20% of the total chloroplast DNA, arising possibly through replication and/or intermolecular recombination. Intermolecular recombination may occur among repeated elements dispersed throughout the chloroplast DNA circle (Bowman et al., 1988; Hiratsuka et al., 1989). Homologous recombination and/or gene conversion may be involved in maintaining a homogeneous chloroplast DNA population (Birky and Walsh, 1992). The presence of a homologous recombination system within the chloroplast no doubt facilitates integration of transforming DNA into the chloroplast genome (Lemieux et al., 1981; Medgyesy et al., 1985).

B. Chloroplast Gene Expression

Chloroplast genomes share many features common to prokaryotic genomes, including the presence of polycistrons, rRNA binding sequences in the 5′ untranslated region of the mRNA, and stem-loop structures near the 3′ end of mRNAs. The expression of chloroplast genes can be regulated by transcription, post-transcriptional *cis* and *trans* splicing of split plastid genes, processing of polycistronic transcripts, mRNA stability, translation, and post-translational processing. Chloroplast gene expression is largely controlled by nuclear-encoded regulatory factors, many of which are imported into the chloroplast where they interact with specific chloroplast DNA or RNA elements. All of the above processes have been extensively reviewed (Mullet, 1988, 1993; Taylor, 1988; Boschetti et al., 1990; Mayfield, 1990; Bogorad, 1991; Rochaix, 1992; Gruissem and Tonkyn, 1993; Gillham et al., 1994; Harris et al., 1994, and Roell and Gruissem, Chapter 31). RNA editing of initiator codons and internal mRNA codons has been shown to occur not only in plant mitochondria, but also in the chloroplasts of many plants, including monocots and dicots (Hallick, 1992). For some plastid genes, a C to U edit of the transcript may be required to produce an AUG initiator codon (Neckermann et al., 1994). Post-transcriptional editing of an initiator codon was demonstrated in both leaves and fruits, showing that the RNA editing machinery is present in chromoplasts as well as chloroplasts (Kuntz et al., 1992). Other studies suggest that RNA editing precedes processing of polycistronic transcripts and intron-splicing (Ruf et al., 1994), and that the editing process can occur in the absence of functional chloroplast ribosomes (Zeltz et al., 1993).

Translational and post-translational processes may play a larger role than the other processes in determining the steady-state level of chloroplast-encoded polypeptides present in the chloroplast. The amount of protein product present in the chloroplast does not increase in direct proportion to the number of gene templates present (Hosler et al., 1989; Kindle et al., 1994). However, the latter authors showed that a substantial increase in the number of gene copies per chloroplast led to an increase in the gene product. Considerations of gene dosage are of particular importance given the polyploid nature of the chloroplast genome and the changes in chloroplast ploidy observed in plant cells.

III. Chloroplast Transformation in *Chlamydomonas*

A. Overview

1. The Chloroplast Genome of C. reinhardtii

The single chloroplast of the unicellular green alga, *Chlamydomonas reinhardtii*, contains 80 to 100 copies of the ~196,000 base pair DNA circle (reviewed in Harris, 1989). The large inverted repeat region of ~22 kb includes the 16S, 7S, 3S, 23S and 5S rRNA genes and *psb*A, which encodes the D1 reaction center polypeptide of PS II. A complete restriction map of the chloroplast DNA molecule has been published (Rochaix, 1978; see Harris, 1989). The DNA sequence data now exist for approximately 75% of the *C. reinhardtii* chloroplast genome, and sequence analysis of the remaining regions is currently underway or planned (S. Surzycki and E. Harris, personal communication). Restriction fragments representing the entire chloroplast genome have been amplified in *E. coli,* and plasmid clones containing chloroplast DNA are maintained and distributed through the *Chlamydomonas* Culture Center at Duke University in Durham, North Carolina (Boynton and Gillham, 1993). The chloroplast genomes of other species within the genus *Chlamydomonas* have been characterized also and compared to that of *C. reinhardtii* (Boudreau and Turmel, 1995).

Chlamydomonas is a heterothallic alga. Under conditions of nitrogen starvation, the haploid vegetative cells differentiate into gametes. Fusion of gametes of opposite mating types (*mt*$^+$ and *mt*$^-$ leads to formation of a diploid zygote, which then undergoes meiosis, resulting in four haploid meiotic products. Sexual inheritance of chloroplast DNA in *Chlamydomonas* is uniparental, under the control of several nuclear loci including that which determines mating type (Armbrust et al., 1993; VanWinkle-Swift et al., 1994; for reviews see Gillham et al., 1991; Sears and VanWinkle-Swift, 1994). In *C. reinhardtii*, chloroplast DNA of the *mt*$^-$ parent appears to be selectively degraded after gametes fuse, and >95% of the time a chloroplast genetic marker is inherited from the *mt*$^+$ parent. Hence, the use of *mt*$^+$ strains as transformation hosts facilitates transfer of the transformed chloroplast genome into different nuclear genetic backgrounds, and permits classical genetic analysis of chloroplast genes in transformants.

An unusual model by Sears and Van Winkle-Swift (1994) presents the concept that uniparental inheritance of the highly polyploid chloroplast genome in *Chlamydomonas* may provide a direct selective benefit to the organism itself if dispensable copies of the chloroplast DNA molecule are degraded and salvaged to provide nucleotides for DNA recombination and repair, RNA synthesis, and cell metabolism.

2. C. reinhardtii *as a Target for Chloroplast Transformation*

Chlamydomonas reinhardtii vegetative cells generally undergo two successive rounds of DNA synthesis and cell division before separation of daughter cells, which results in cells that appear to divide into four (see Fig. 1). On occasion, and at high frequency within certain genetic backgrounds, daughter cells stay together as clumps of 8 and even 16 cells before separating. Algal cells harvested from liquid culture in log phase and plated on solid medium provide foci for colony formation. These foci, or colony forming units, include single cells of different sizes containing different amounts of chloroplast DNA, as well as doublets, tetrads, and octads of cells. After transformation, the efficiency of chloroplast DNA segregation within cells of a single colony depends in part on the composition of the initial colony forming unit.

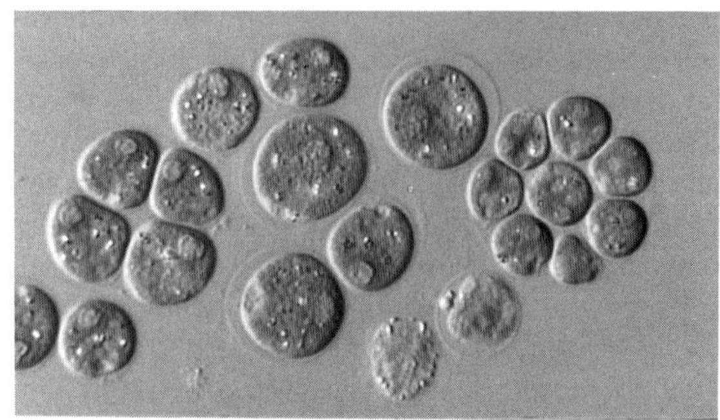

Fig. 1. A light micrograph of *C. reinhardtii* cells. Cells were harvested at mid-log stage and viewed using Nomarsky optics. (Photographed at 640× magnification).

The chloroplast itself is a cup shaped organelle that appears to lie adjacent to the plasma membrane over most of the periphery of the *C. reinhardtii* cell (Gillham, 1978), and to take up 40% to 60% of the cell volume. The diameter of a *C. reinhardtii* cell is ~5 to 10 μm, depending on the strain and age of the culture.

3. Heteroplasmy and Homoplasmy

DNA introduced into the chloroplast of a single recipient *Chlamydomonas* host cell can integrate via homologous recombination (Boynton et al., 1988) into one or more of the approximately 100 resident chloroplast DNA molecules. Integrated DNA thus replaces the homologous DNA segment of the host DNA molecule. Genes or DNA fragments with no homology to the host chloroplast DNA also can be inserted into the host genome as long as the transforming DNA construct contains some additional DNA homologous to that found in the host chloroplast. An initial transformed cell is heteroplasmic because its population of chloroplast DNA molecules is no longer homogeneous but consists of resident host DNA molecules plus the newly-formed recombinant DNA molecule(s). During vegetative algal growth, as occurs during formation of a colony from an initial single cell on solid medium, daughter cells may receive, in mixed proportions, both types of chloroplast DNA molecules. The rapid segregation of chloroplast and mitochondrial DNA molecules during vegetative growth is a well-documented phenomenon (reviewed in Birky, 1991). Factors that foster such segregation may include 1) the organization of chloroplast DNA molecules into 10 to 15 nucleoids, which are attached to the chloroplast envelope and which may segregate as a unit, 2) recombination or gene conversion between DNA molecules within or between nucleoids, and 3) relaxed replication, which allows both mother and daughter DNA molecules to be used more than once as a template during a given cycle of DNA synthesis. Eventually, daughter cells will contain only one type of chloroplast DNA molecule: either the original host molecule, or the recombinant 'engineered' chloroplast DNA. Such cells are termed homoplasmic. If all progeny are viable, a single transformant colony may contain heteroplasmic cells as well as the two types of homoplasmic cells. Transfer of a colony (~10^6 cells) into liquid medium and subsequent replating of cells usually results in secondary colony isolates that are homoplasmic. Heteroplasmy is usually an unstable condition. Exceptions to this are seen when selective pressure favors or requires the presence of two types of chloroplast DNA molecules (Goldschmidt-Clermont, 1991; Yu and Spreitzer, 1992; Liu et al., 1993a).

Transformant cell lines of interest for further study should be homoplasmic and stable. Homoplasmicity can be verified by careful molecular analysis of initial transformants. Such verification is of particular importance when mutant transformant lines function at very low levels compared to similar function in wild-type host cells; a few undetected wild-type chloroplast DNA molecules could in theory account for the low level of function observed in transformants. Use of the polymerase chain reaction (PCR) has allowed us to amplify a region from a chloroplast DNA molecule present at extremely low levels in a heterogeneous mixture of chloroplast DNAs (Fig. 2). DNA from transformants (Erickson et al., 1992) was subjected to a similar assay, and no host DNA molecules were detected (J. P. Whitelegge and J. M. Erickson, unpublished). These results indicate that such transformant cell lines are indeed homoplasmic.

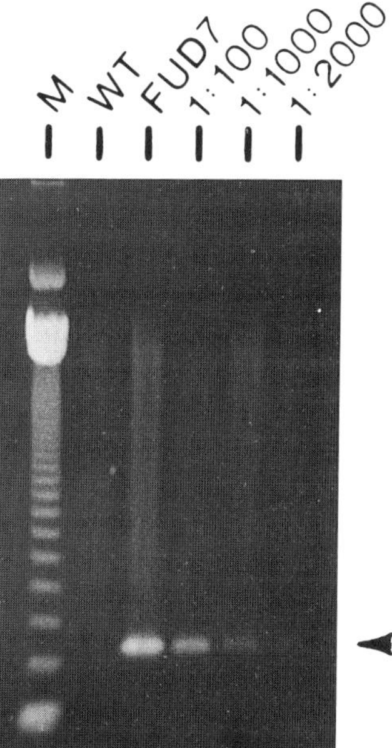

Fig. 2. Polymerase chain reaction (PCR) assay for heterogeneity in chloroplast DNA. 100 ng of total DNA isolated from wild-type *C. reinhardtii* was mixed with total DNA from *C. reinhardtii* mutant FUD7 and amplified by PCR. Five μl of each 25 μl PCR reaction was electrophoresed on a 1.4 % agarose gel and ethidium-bromide stained DNA fragments were visualized under UV light. DNA size markers (M) of the 123 bp ladder (BRL) are shown at the left. DNA templates for PCR reactions were mixed such that the ratio of mutant:wild-type chloroplast DNA was 1:100, 1:1000, or 1:2000 as indicated. The oligonucleotide primer pair permits amplification of a ~300 bp fragment of the FUD7 chloroplast DNA. Recent results show that we can detect the 300 bp fragment when FUD7 and WT DNAs are mixed at a ratio of 1:20,000 (W. Harenburg and J. M. Erickson, unpublished).

4. Other Considerations

Factors that should be considered when devising a strategy for chloroplast transformation of *Chlamydomonas* include the following: choice of selectable marker and algal host strain, age of host cell culture and method of plating, conformation of the transforming DNA molecule, method for introducing DNA into the chloroplast, length of recovery time after transformation procedure, selective medium, conditions for and time of incubation during selection, screening procedures, and isolation and maintenance of homoplasmic transformants. Several of these issues will be discussed in more detail in the following section.

B. Strategies for Transformation

1. C. reinhardtii *Selectable Marker Genes and Host Strains*

One element essential to any successful transformation is the strategy used to separate transformed cells from non-transformed host cells. The most powerful selective methods are those in which transformed cells gain the ability to grow on selective medium, while non-transformed cells die. A given marker gene must be introduced into an appropriate algal host strain. Many non-photosynthetic chloroplast mutants, lacking one or more products essential for photosynthesis, have been isolated and maintained on medium containing acetate as a sole carbon source (reviewed in Harris, 1989). Positive selection for transformants can be achieved through restoration of photosynthetic growth to such non-photosynthetic host cells after chloroplast transformation with a cloned gene encoding the missing product(s). Other positive selective markers include the genes conferring resistance to antibiotics (Harris et al., 1989; Goldschmidt-Clermont, 1991), herbicides (Johanningmeier et al., 1988; Erickson et al., 1989) and phytotoxins (Avni et al., 1992). With the elucidation of chloroplast biochemical pathways and the characterization of more *C. reinhardtii* chloroplast genes, other selective markers will no doubt become available. In the case of resistance markers, either wild-type or non-photosynthetic algal strains can serve as host.

Table 1 summarizes the most common *C. reinhardtii* chloroplast marker gene/host systems used to date. For a selective marker to be useful, the frequency of transformation must be above the frequency at which the selected phenotype arises by mutation. Although point mutations are responsible for drug and herbicide-resistance, the chloroplast transformation efficiencies in *C. reinhardtii* (10^{-4} to 10^{-5}) are generally above any background mutation rate. Obviously, the non-photosynthetic deletion mutants do not revert, and are excellent hosts in this respect.

2. Co-transformation: One or Two Transforming DNA Constructs?

In most studies involving chloroplast transformation, the selective marker gene is not the gene of interest. An important consideration is whether to transform cells with a single DNA construct in which the marker gene is physically linked to the gene of interest, or whether to use a co-transformation strategy involving two separate DNA molecules. Both approaches have been used with success (Table 1, sections I and II). The advantage of using one DNA construct is that physical linkage of the marker and the gene of interest theoretically increases the chance that selected cells will be co-transformants of interest. However, even when two markers are on the same transforming DNA construct, recombination can occur and the unselected marker is not always found in the selected transformants (Blowers et al., 1989; see Newman et al., 1990, 1992). Another consideration when using one DNA construct is that the marker gene will, of necessity, be integrated into the recipient chloroplast DNA molecule adjacent to the gene of interest. Each gene being studied must be linked to the marker gene, and the marker must not disrupt a chloroplast DNA region essential for cell viability or which affects the phenotype under investigation. In contrast, use of a co-transformation strategy, in which the marker gene is on a separate DNA molecule from the gene of interest, allows the marker to be targeted to a neutral site in the chloroplast DNA that is not forcibly adjacent to the gene of interest. Once such a site is identified (as many have been), the same marker DNA construct can be used in conjunction with any other plasmid of interest. Since co-transformation efficiencies are generally high (~10%–90%) and all putative transformant cell lines of interest must be screened for homoplasmy in any case, co-transformation with two DNA constructs is quite feasible.

A co-transformation strategy for studies of *psbA*, as shown in Fig. 3, can be used with a wild-type algal host or a chloroplast deletion mutant host. Use of the wild-type host requires screening by RFLP analysis or DNA sequence analysis (Whitelegge et al., 1992, 1995), while use of the deletion host facilitates screening via colony hybridization. Regardless of the host strain used, we usually isolate stable homoplasmic co-transformant lines from 20% to 80% of the initial transformant colonies appearing on solid medium. Researchers in other labs have seen a similar range of co-transformation frequencies for chloroplast genes (Kindle et al., 1991; Newman et al., 1991). *psbA* of *C. reinhardtii* contains four large introns and the gene spans nearly 7 kb (Erickson et al., 1984). Plasmid constructs containing intronless *psbA* genes have been engineered (Johanningmeier and Heiss, 1993; Mayfield et al., 1994; Minegawa and Crofts, 1994) to facilitate the in vitro manipulation of *psbA* coding sequences. The latter authors have linked *psbA* cDNA to an antibiotic-resistance marker (see Table 1).

3. Calculating Transformation Efficiencies

The efficiencies of chloroplast transformation are

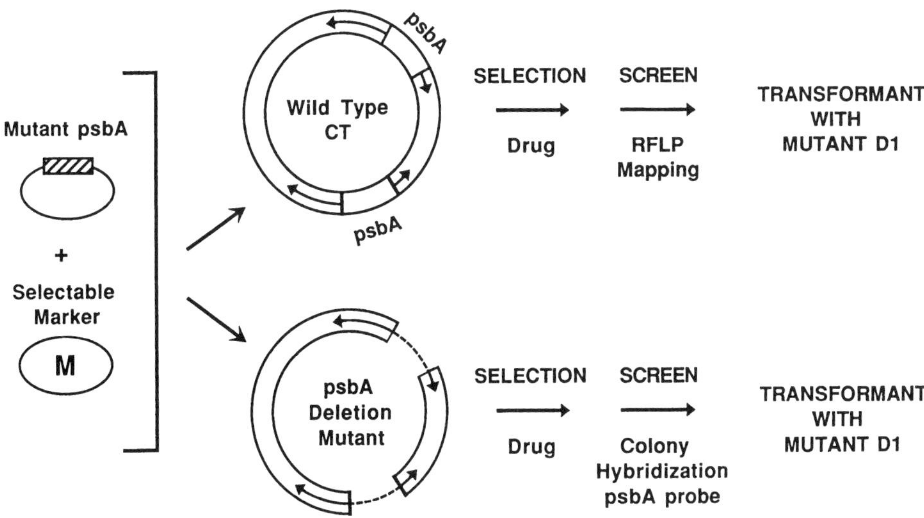

Fig. 3. Strategy for isolating transformants with a mutant D1 polypeptide. *psbA*, which encodes the D1 polypeptide, is present in the inverted repeat of *C. reinhardtii* chloroplast DNA, and thus present in two copies per DNA molecule. Wild-type cells or a chloroplast *psbA* deletion mutant can serve as algal hosts. Host cells are bombarded with microparticles coated with a mixture of the two transforming plasmid DNAs—the selectable antibiotic-resistant marker DNA and the mutant *psbA* DNA. Colonies appearing on growth medium containing antibiotic are screened initially by RFLP mapping or by colony hybridization, depending on the host strain used. Homoplasmic transformant cell lines contain mutant *psbA* genes and produce mutant D1 polypeptides. Although only one mutant copy of *psbA* may integrate initially into the chloroplast DNA of the host, the remaining wild-type copy of *psbA* most likely is altered through gene conversion or recombination (see Erickson et al., 1984).

generally presented on a per cell basis, that is, the number of cells that must be subjected to the transformation procedure in order to obtain one stable transformant cell line. Chloroplast biolistic transformation frequencies of 10^{-4} to 2×10^{-5} are reported routinely. Thus, after a single agar plate containing 2×10^{7} algal cells is bombarded with DNA-coated microparticles, from 100 to 2000 colonies are expected to appear after appropriate incubation on selective medium. Efficiencies have also been reported as the number of chloroplast transformants obtained per μg of transforming DNA used (Kindle and Sodeinde, 1994). Calculating the number of transformants per molecule of transforming DNA might also facilitate comparison between experiments using different sizes of DNA molecules. All these simple representations of transformation frequencies are purely pragmatic. Obviously, the number of transformants obtained will depend not only on the number of target cells that are hit by one or more microprojectiles, but also on the total number of chloroplasts entered by microprojectiles, the number of DNA molecules contained on a microprojectile and actually deposited within the chloroplast, the frequency with which deposited DNA is integrated into the chloroplast genome, and the probability for segregation of homoplasmic cell lines containing the selectable marker. Transformation efficiency and reproducibility may be altered when any of the above parameters change. Corrected transformation efficiencies have been reported, taking into account the estimated number of cells within the microprojectile spray pattern, and the estimated percentage of cells that survive the physical manipulation procedures associated with transformation and selection (Boynton et al., 1988; Newman et al., 1991; reviewed in Boynton and Gillham, 1993). Such corrections are aimed at determining the probability that a given DNA molecule, when introduced into the chloroplast, will actually cause stable transformation of that chloroplast genome. However, given the large number of variables discussed above, direct representations of transformation efficiencies may be the most useful means of comparing results among laboratories.

Table 1. Chloroplast Transformation in *C. reinhardtii:* Strategies, Selectable Markers and Algal Hosts

I Transformation with a single DNA construct

I.A Non-photosynthetic *C. reinhardtii* host: *C. reinhardtii* chloroplast gene as selectable marker

Basis of Selection: Hosts are acetate-requiring; chloroplast marker gene restores synthesis of missing protein, results in photoautrotrophic transformants
Selective medium and conditions: Minimal medium, incubation in light

I.A.1 Chloroplast deletion mutant as algal host strain

Deletion[a] in host	Selectable marker [b]	Features	Reference
atpB (13 kb)[†]	*atpB* (7.6 kb)	WT *atpB*	Boynton et al., 1988
atpB (2.5 kb)[†]	*atpB* (7.6 kb)	WT *atpB*	Boynton et al., 1988, 1990[F]
			Kindle et al., 1991[F]
		altered 3′ *atpB*	Kindle et al., 1994[F]
	atpB (2.9–7.6 kb)	bacterial NPTII[c]	Blowers et al., 1989
	atpB (7.6 kb)	bacterial *uidA*[d]	Blowers et al., 1990
			Klein et al., 1992
			Salvador et al., 1993
	atpB (*atpB*-INT)	ORF, 23S rRNA intron	Dürrenburger and Rochaix, 1991
		bacterial *aadA*[e]	Nickelsen et al., 1994
psbA (8.5 kb)[IR]	*psbA* (12 kb)	*psbA* * and WT	Erickson et al., 1992
	psbA (10 kb)	*psbA* herb[R]	Heiss and Johanningmeier, 1992
	psbA (6 kb)	intronless *psbA* gene	Johanningmeier and Heiss, 1993
psbA (9 kb)[IR]	*psbA* (10 kb)	*psbA* herb[R]	Boynton et al., 1992
tscA, *chlN*[fg] (2.8 kb)	*tscA* (1.8–5 kb)	bacterial *aadA*[e]	Goldschmidt-Clermont et al., 1991
	tscA (5 kb)	screen for *chlN*	Choquet et al., 1992
tscA, *chlN*[†f] (10 kb)	*tscA* (5 kb)	glass bead transformation[h]	Kindle et al., 1991[F]

I.A.2 Chloroplast point mutant as algal host strain

Point mutation in host	Selectable marker	Features	Reference
atpB (transversion)	*atpB* (7.6 kb)	WT *atpB*	Boynton et al., 1988, 1990[F]
psaB (frameshift)	*psaB* (3.8–5.8 kb)	map frameshift	Bingham et al., 1991
	psaB (3.8 kb)	*psaB* mutations	Webber et al., 1993

I.B Photosynthetic or non-photosynthetic *C. reinhardtii* host: Chemical resistance marker gene

Basis of selection: Photosynthetic and non-photosynthetic hosts are sensitive to chemical agents; chloroplast genes confer resistance to antibiotics, herbicides or chemical toxins
Selective medium and conditions: Acetate medium plus selective chemical, light or dark incubation; Minimal medium plus selective chemical, light incubation

I.B.1 *Chlamydomonas* [i] chloroplast gene as selectable marker

Selectable gene (plasmid construct)	Selection [j]	Features	Reference
[Ribosome as target for antibiotics]			
16S rRNA, nt 474*	strep	genetic and	Newman et al., 1990[F]
16S rRNA, nt 1123*	spec	molecular map of	Boynton et al., 1992[F]
23S rRNA, nt 2067* (plasmid p-183)	ery	integration sites	
16S rRNA, nt 1123* (plasmid p-228)[◊]	spec	as above	Newman et al., 1990[F]

16S rRNA*, C. *smithii* (plasmid p-229)	spec	as above	Newman et al., 1990[F]
16S rRNA, nt 1123*	spec	deletion in ORF,	Thompson and Herrin, 1991, 1994
rRNA *er-u-11** (plasmid pES 7.2)	ery	23S rRNA intron	

[PS II Q_B site as target for herbicides]

psbA, codon 264*	metribuzin	*psbA* mutations	Przibilla et al., 1991
		psbA, truncated D1	Schrader and Johanningmeier, 1992
	DCMU[k]		Newman et al., 1992[F]

[Proton ATP synthase CF1 as target for phytotoxins]

atpB, codon 83*	tentoxin	other substitutions near codon 83	Avni et al., 1992

I.B.2 Bacterial *aadA*[e] as selectable antibiotic resistance marker for chloroplast transformation

Resistance	Chloroplast gene studied	Reference
spec	*tscA*, *psaC*, ORF472;	Goldschmidt-Clermont, 1991 (see also section IIIA)
spec	*psbA* cDNA	Minegawa and Crofts, 1994[F]

II Co-transformation with two DNA constructs

Basis of selection and screen: Hosts containing wild-type or mutant chloroplast genomes are sensitive to antibiotics. Transformants selected for antibiotic resistance (as in IB) are then screened for the presence or absence of the second marker.

Host ct	Selectable marker (plasmid construct)	Resistance[j]	Second Marker	Reference
WT	16S, 23S rRNA	spec	yeast insert in *atpB*	Newman et al., 1991[F]
	(p-183)	strep	yeast insert in *rbcL*	Newman et al., 1991[F]
		strep/spec	*petA*: signal sequence	Smith and Kohorn, 1994[F]
			kan^R insert in *chlL*	Suzuki and Bauer, 1992[F]
WT	16S rRNA (p-228)[◊]	spec	altered *atpB*	Kindle et al., 1991[F]
			psbA point mutations	Erickson et al., 1992 Whitelegge et al., 1992, 1995 Roffey et al., 1994[F]
			psbA: truncated D1	Lers et al., 1992
			petD initiation codons	Chen et al., 1993[F]
			psaB point mutations	Webber et al., 1993[F]
			rbcL point mutations	Zhu and Spreitzer, 1994[F]
Δ*psbA*	16S rRNA (p-228)[◊]	spec	WT or mutant *psbA*	Erickson et al., 1992 Whitelegge et al., 1995
			altered 5′ *psbA*	Mayfield et al., 1994
**psbA*	16S rRNA (p-228)[◊]	spec	*psbA* point mutation	Roffey et al., 1991[F]
Δ*atpB*	16S rRNA (p-228)[◊]	spec	selection also for *atpB*[l]	Kindle et al., 1991[F]
			altered 3′ *atpB*	Stern et al., 1991[F] Stern and Kindle, 1993[F]

III Targeted insertional inactivation of *C. reinhardtii* chloroplast genes

III.A *aadA*[e] as a selective marker inserted into the gene of interest

Basis of selection: antibiotic sensitive host is transformed with plasmid DNA construct containing *aadA* selective marker targeted to gene of interest via flanking chloroplast DNA sequences
Selective medium and conditions: as in section IB.

Gene target	Function	Reference
chlB	Chlorophyll biosynthesis	Li et al., 1993 Liu et al., 1993b[F]
clpP	Clp protease	Huang et al., 1994[F]
petA	*cyt f*, cytochrome b_6f complex	Kuras and Wollman, 1994
petB	*cyt b6*, cytochrome b_6f complex	Kuras and Wollman, 1994
petD	subunit IV, cytochrome b_6f complex	Kuras and Wollman, 1994
psaC	PS I Fe-S polypeptide	Goldschmidt-Clermont, 1991 Takahashi et al., 1991
psbK	essential PS II gene	Takahashi et al., 1994
tscA	trans-splicing of *psaA*, essential for PS I	Goldschmidt-Clermont, 1991 Goldschmidt-Clermont et al., 1991
ycf8[m]	PS II gene, needed in adverse conditions	Monod et al., 1994
ORF 427	heteroplasmic, essential for cell viability	Goldschmidt-Clermont, 1991
ORF 712	heteroplasmic, may encode ribosomal protein 3 (Rps3)	Liu et al., 1993a[F]

III.B 16S rRNA* gene as selective marker; co-transformation with inserted gene of interest

Basis of selection: Antibiotic sensitive host is co-transformed with resistance marker construct (p-183; see I.B.1) and disrupted gene of interest.
Screen for co-transformants: antibiotic resistant colonies are screened by RFLP
Selective medium and conditions: acetate medium containing antibiotic(s), incubation in light or dark

Gene target	Disrupting element	Screen	Reference
atpB	0.48 kb yeast DNA	RFLP	Newman et al., 1991[F]
rbcL	0.48 kb yeast DNA	RFLP	Newman et al., 1991[F]
chlL	1.6 kb kan^R, ble^R	RFLP	Suzuki and Bauer, 1992[F]

IV Chloroplast expression of foreign 'reporter' genes

IV.A Transcriptional expression only

Foreign DNA	Flanking control sequences 5′	3′	Reference
aadA[e]	*rbcL*	*rbcL*	Salvador et al., 1993
uidA[d]	*atpA*	*rbcL*	Blowers et al., 1990
	atpB	*rbcL*	
	rbcL[n]	*rbcL*	
	atpB	*rbcL*	Klein et al., 1992
	16S rRNA	*rbcL*	
	atpB	*rbcL* *psaB*	Blowers et al., 1993
	rbcL[o]	*rbcL*	Salvador et al., 1993
	rbcL	*psaB*	
	atpB	*psaB*	
NPTII[c]	*rbcL* (maize)		Blowers et al., 1989

IV.B Translational expression

Foreign gene (protein)	Flanking control sequences 5′	3′	References
aadA (AAD)[e]	*rbcL*	*rbcL*	Goldschmidt-Clermont, 1991
	atpA	*rbcL*	
	psaC, *tscA*, ORF 472 insertion		
	psbD	*psbD*	Nickelsen et al., 1994
uidA (GUS)[d]	*petD*	*rbcL*	Sakamoto et al., 1993[F]

a – approximate size of deletion is indicated in parentheses
b – approximate size of chloroplast DNA used in transformation is indicated in parentheses
c – gene for neomycin phosphotransferase
d – gene for β-glucuronidase (GUS)
e – gene for aminoglycoside adenylyltransferase (AAD)
f – see Roitgund and Mets (1990), Goldschmidt Clermont et al. (1991) and Choquet et al. (1992) for identification of *tscA* and *chlN* (previously called *gidA*).
g – Deletion, duplication and inversion of chloroplast DNA in host, see Choquet et al., 1992
h – see reference for optimization of conditions and comparison with biolistic transformation
i – unless otherwise noted, all selectable markers are *C. reinhardtii* chloroplast genes
j – antibiotics: streptomycin (strep), spectinomycin (spec), erythromycin (ery)
k – diuron (DCMU)
l – Kindle et al., 1991 selected for each marker and screened for co-transformation of the unselected marker; co-transformants were found only when spectinomycin was the selected marker.
m – conserved ORF in chloroplasts; see Hallick and Bairoch (1994) for *ycf* nomenclature
n – ~1% level of transcripts compared to native *rbcL* mRNA
o – examined stability of chimeric mRNAs in light and dark, +/- *rbcL* coding region
F – pre-treatment of host cells with 5′- fluordeoxyuridine to reduce chloroplast genome ploidy
IR – deletion within chloroplast inverted repeat
† – deletion in single copy region enters one side of chloroplast inverted repeat; several deletion mutants are compared in Palmer et al., 1985
* – gene contains point mutation(s); mutant nucleotide positions (nt) are indicated for rRNA markers (see Harris et al., 1989)
◊ – p-228 is also called pCrBH4.8 (Kindle et al., 1991)

4. Conformation of the Transforming DNA Molecule

Surprisingly, relatively few experiments have been done to determine whether the structure of the transforming DNA molecule has a significant effect on the efficiency of chloroplast transformation. This may reflect, in part, the fact that early work was carried out using the gunpowder propelled biolistic transformation apparatus (reviewed in Sanford et al., 1991; see next section). Results with this apparatus were qualitatively reproducible, but the transformation efficiencies observed could vary as much as two orders of magnitude in replicate assays. Nevertheless, early reports showed that chloroplast transformation could be achieved using either single or double-stranded DNA (Blowers et al., 1989). These authors also tested linear vs. circular plasmid DNA, and found that linear DNA molecules transform with higher efficiency. Results from other laboratories suggest that circular DNA is more efficient (Kindle et al., 1991). The helium-driven biolistic transformation procedure results in higher efficiency of transformation and greater reproducibility, and studies on the form of DNA used in transformation may now yield further insight into this issue. Double-stranded covalently closed circular plasmid DNA is the most common form of transforming DNA used currently.

Transforming DNA molecules can integrate into the chloroplast genome of recipient cells through single or double crossover events, as diagrammed in Fig. 4. Integration via a single homologous recombination event appears to occur when the chloroplast host molecule has homology on only one side of the plasmid gene being used as a selective marker (Boynton et al., 1990; Kindle et al., 1991; Erickson et al., 1992). Obviously, such constructs transform these hosts much more efficiently when the DNA is circular and the entire marker construct can be integrated by a single crossover.

It is not yet clear what minimum length of

homologous DNA is required for efficient recombination, or how prevalent recombination 'hot spots' are (Newman et al., 1992). We have seen DNA replacement at fair efficiency when ~1.8 kb of homologous chloroplast DNA is present on the non-selected transforming plasmid construct, but not when the homologous region was reduced to 0.6 kb. PCR-amplified chloroplast DNA fragments have been used for successful chloroplast transformation, and are particularly useful in the engineering of site-directed mutants (A. Webber, personal communication).

5. Transgenic Chloroplast Markers and Reporters

a. Bacterial aadA

For reasons not clearly understood, it has been difficult to obtain chloroplast expression of 'foreign' genes at the level of protein product. One potential problem is the codon bias observed in most chloroplast genes, especially those of *C. reinhardtii* (Morton, 1993). In addition, lack of information on the nature of the chloroplast promoter and the effect that 5′ and 3′ untranslated regions of mRNA have on message stability and translation may have hindered many attempts at transgenic expression in chloroplasts.

Goldschmidt-Clermont (1991) obtained full expression of the bacterial *aadA* gene, encoding aminoglycoside adenylyltransferase (AAD), which confers resistance to aminoglycosides, in the chloroplast of *C. reinhardtii*. A series of plasmid constructs or cassettes, in which the bacterial *aadA* is under the control of different *C. reinhardtii* chloroplast gene 5′ and 3′ sequences, was tested and the resulting transformants were resistant to spectinomycin and streptomycin. Since this bacterial gene has no homology to chloroplast DNA, the site at which *aadA* is inserted into the chloroplast DNA molecule can be controlled by choosing the chloroplast DNA sequences that flank the *aadA* coding sequence. Such targeting allows for placement of the dominant selectable marker in a neutral site in the chloroplast DNA molecule, or for inactivation of specific chloroplast genes by *aadA* insertion (Table 1, III). Since both *aadA* mRNA and AAD protein are present in the transformants, *aadA* is a useful reporter gene for studies on transcriptional and translational signals in the 5′ and 3′ flanking regions of chloroplast genes.

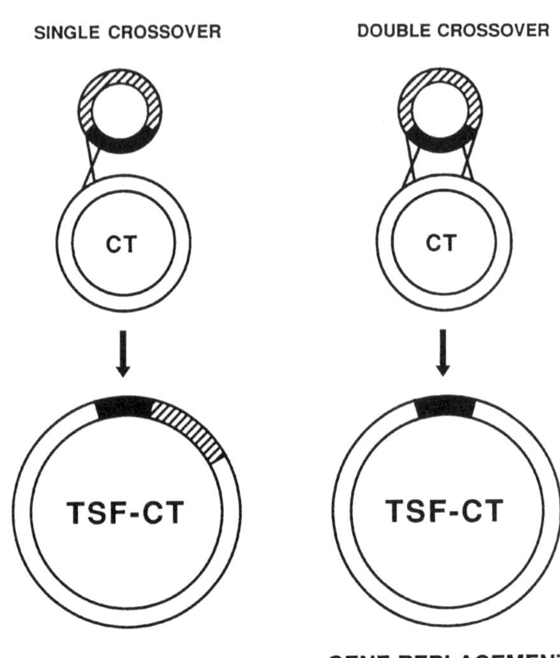

Fig. 4. Schematic representation of the integration of plasmid DNA into the chloroplast genome via homologous recombination. A single crossover event results in the incorporation of the entire transforming plasmid, including homologous chloroplast DNA (■) and vector DNA (▨), into the chloroplast DNA (CT) of the host. A double crossover event results in replacement of wild-type DNA with the homologous chloroplast DNA sequences (■) on the transforming plasmid. The two types of recombinant chloroplast DNA molecules found in transformants (TSF-CT) are shown.

b. Bacterial uidA

Chloroplast transformants of *C. reinhardtii* containing transcriptional fusions between a *C. reinhardtii* chloroplast promoter region and a bacterial gene, *uidA* (Blowers et al., 1990; Klein et al., 1992; Salvador et al., 1993) contain chimeric transcripts for the foreign gene, but no detectable gene product. *uidA* encodes *β*-glucuronidase (GUS), which can cleave artificial chromogenic substrates. A similar result was seen in *C. reinhardtii* transformants containing a transcriptional fusion between a maize chloroplast reporter and the bacterial gene for neomycin phosphotransferase (Blowers et al., 1989), and the transformants were not kanamycin-resistant. Subsequent studies showed that, when translational fusions between the chloroplast *petD* 5′ untranslated region and *uidA* were introduced into the chloroplast, transformants accumulated chimeric mRNA and showed a high level of GUS activity (Sakamoto et al., 1993). The translational fusions extended the N-terminus of GUS by 9 amino acid residues. Thus, *uidA* should be a useful reporter gene for studies of chloroplast gene expression. Such studies will no doubt lead to increased success with expression of foreign genes in the *C. reinhardtii* chloroplast.

C. General Transformation Methods

1. Biolistic Transformation

The biolistic process, in which DNA-coated microprojectiles are propelled at high velocities into eukaryotic cells under partial vacuum, was first used for stable transformation of the nuclear genome of plant cells (Klein et al., 1987). This novel method for delivering DNA into cells opened the door to chloroplast and mitochondrial transformation. In 1988, the first biolistic transformations of the chloroplast genome of the unicellular alga *C. reinhardtii* (Boynton et al., 1988), and the mitochondrial genome of yeast (Johnston et al., 1988) were reported. Since then, much progress has been made in this area. Biolistic transformation of all three genomes of *C. reinhardtii*—the nuclear (Debuchy et al., 1989; Fernandez et al., 1989; Mayfield and Kindle, 1990), the chloroplast (Boynton et al., 1988; reviewed in Boynton and Gillham, 1993) and the mitochondrial (Randolph-Anderson et al., 1993) genomes—has been reported.

In early models of the biolistic (biological ballistic) transformation apparatus, the force used for initial acceleration of the DNA-coated microparticles was provided by a gunpowder charge (Klein et al., 1987; Zumbrunn et al., 1989). More recent models use pressurized air (Oard et al., 1990) or helium (Sanford et al., 1991; Finer et al., 1992; Takeuchi et al., 1992; reviewed in Sanford et al., 1993). While the gunpowder-driven apparatus is still used with success, the advantages of pressurized air or helium include greater reproducibility, increased safety, the absence of toxic gunpowder residue on cells, and in some cases higher transformation efficiencies (Sanford et al., 1991; 1993). Regardless of how the initial force is provided, dense inorganic microparticles, generally tungsten or gold spheres, serve as microprojectiles for each apparatus described above. The DNA-coated microprojectiles are suspended in solution and placed or dried onto a larger macrocarrier. This loaded macrocarrier is forcibly launched as described above, and then caught (before impact with the host cells) in such a way that the microprojectiles pass through a stopping plate or screen and continue on at high velocity towards the target cells. Considerations for biolistic transformation, including microprojectile size, composition, and biologic toxicity, DNA-coating procedures, strength of gunpowder charge or pressure of helium, level of vacuum in chamber, and velocity of microprojectiles, are discussed in detail in a recent review (Sanford et al., 1993).

2. Other Transformation Methods

Other methods for introduction of DNA into chloroplasts have been reported. These include the polyethylene glycol-mediated stable chloroplast transformation of plant protoplasts (Golds et al., 1993; O'Neill et al., 1993), and stable chloroplast transformation of *C. reinhardtii*, achieved by vortexing cells in the presence of DNA and glass beads (Kindle et al., 1991) or by electroporation (Boynton and Gillham, 1993). These methods obviate the need for a biolistic particle delivery system, but are less efficient and often more time consuming than the biolistic method. Glass bead transformation requires the use of algal hosts lacking cell walls. Cell wall mutants can be used as hosts, or alternatively the cell wall of the host strain can be removed by treatment with autolysin (Kindle et al., 1991).

DNA was introduced into the cytoplasm or the chloroplasts of plant cells, as well as into isolated chloroplasts, using a UV-laser microbeam to make a small hole in the cell membrane and the chloroplast envelope (Weber et al., 1989). Organellar membranes opened by the laser beam close within seconds of the laser pulse, and microbeam irradiation focused on the chloroplast of *C. reinhardtii* cells did not reduce cell viability.

Another method for chloroplast transformation, currently under development, shows great promise for introducing large DNA molecules into a host. Cells are bombarded not with DNA-coated tungsten or gold microprojectiles but with globular microparticles composed of DNA itself (personal communication from Laurens Mets, University of Chicago). The quantity of DNA in a single particle can be equivalent to that found in a single human chromosome. An aerosol of the condensed DNA molecules (in helium at 1 atmosphere pressure) is expanded through a small opening into a chamber under partial vacuum that contains the target cells, and the DNA particles accelerate to speeds great enough to penetrate the cell and the chloroplast envelope membranes. One advantage of this method is that no inorganic microprojectiles are introduced into the cell. Eventually, it might be possible to introduce very large molecules of DNA—perhaps even artificial nuclear or chloroplast 'chromosomes' capable of autonomous replication—into host cells.

3. Growth of Host Cells in FdUrd

The segregation of chloroplast DNA molecules during vegetative colony growth from a single heteroplasmic cell is a well documented phenomenon, as previously discussed. Recombination, gene conversion, and stochastic events may determine whether a mutant chloroplast DNA molecule will become fixed in the population of cells or lost. One factor that affects the probability of obtaining homoplasmic chloroplast mutants, following mutagenesis or chloroplast transformation, is the number of chloroplast DNA molecules present in the initial host cell. Wurtz et al. (1977) showed a reduction in the number of chloroplast DNA molecules following treatment of *C. reinhardtii* with 5-fluorodeoxyuridine (FdUrd), and a subsequent increase in the frequency at which chloroplast mutants were recovered. FdUrd treatment of host cells increased the chloroplast transformation efficiency (Boynton et al., 1990; Kindle et al., 1991) and is used by many groups (see Table 1 footnote *F*). However, FdUrd can act as a chemical mutagen (Wurtz et al., 1979), and chloroplast deletion mutants have been recovered after FdUrd treatment (Bennoun et al., 1986). For this reason, other groups avoid treating host cells with FdUrd. In our experience, chloroplast transformation in the absence of FdUrd yields 200 to 2000 transformants per agar plate, when either photoautotrophic growth or spectinomycin-resistance (*aadA* marker) is the selected phenotype. Since we do not replate after bombardment (Fig 5) there is little or no advantage to having >2000 colonies on a plate, and our frequencies (10^{-4} to 10^{-5}) are more than adequate for most experiments.

4. Transformation Efficiency and Reproducibility

The efficiencies reported for chloroplast transformation vary depending on the transformation method, the transforming DNA, the selective strategy and the host used. However, reproducibility using a given host and DNA marker construct is generally high when the helium-driven biolistic apparatus is used (Sanford et al., 1991). A recent review (Sanford et al., 1993) discusses the relative merits of using gold *vs* tungsten particles, and suggests that coating microprojectiles with DNA may be a major source of the variability in transformation efficiency.

We find that tungsten M10 particles transform *C. reinhardtii* with greater efficiency than the 1 μm diameter gold particles (unpublished observations). Theoretically, this could be due to reduced DNA coating of the gold particles, or to the difference in particle size. Tungsten M10 particles are non-uniform in size (0.4 to 1.5 μm diameter) and shape (Sanford et al., 1993). Boynton et al. (1988) report a *C. reinhardtii* cell diameter of 10 μm and obtain good results using the 1 μm gold particles (Boynton and Gillham, 1993). The cells we use for transformation are generally smaller (5–7 μm in diameter), which could account for the greater efficiency we observe using the smaller tungsten particles.

Another factor influencing transformation efficiency is the quality of the transforming DNA. Super-coiled plasmid DNA isolated by cesium chloride (CsCl) density gradient centrifugation is commonly used for transformation (see Boynton and Gillham, 1993). In our experience, column-purified plasmid DNAs (Qiagen) and CsCl-purified DNAs transform with equal efficiencies.

D. A Brief Protocol for Chloroplast Transformation in C. reinhardtii

The general protocol given below has been used in our lab for successful transformation of the chloroplast of *C. reinhardtii,* with several dominant selectable marker/algal host systems as indicated in Table 1. Growth medium containing acetate (TAP) and minimal medium (HSM) were prepared as described (Rochaix et al., 1988). TAP-SPEC selective medium contains spectinomycin at a final concentration of 100 μg per ml of medium. Algal strains were maintained on TAP medium at 25 °C unless otherwise noted. All transformants containing mutant genes for polypeptides involved in photosynthesis are isolated and maintained in the dark to eliminate the selective advantage light gives to photoautotrophic cells.

1. Algal Host Cells

Grow algal host cells shaking gently in liquid TAP medium until the culture density reaches 2×10^6 cells per ml. Harvest cells by low speed centrifugation at room temperature and resuspend gently in TAP liquid to a density of 8×10^7 cells per ml. Spread 2×10^7 cells (0.25 ml) over the surface of a sterile nylon membrane filter [0.45 μm, 82 mm diameter, obtained from Schleicher and Schuell (Nytran 01790) or Micron Separations Inc. (Magna N04HY08250)] placed on the surface of a TAP agar plate. Prior to use, sterilize nylon membranes by autoclaving in a covered glass petri dish. Incubate plated cells ~2–3 h prior to transformation.

2. DNA and Tungsten Microparticles

Plasmid DNAs are column purified (Qiagen) from bacterial cell lysates. The DNA concentration is adjusted to 1 μg per μl in sterile 10 mM Tris-HCl (pH 8.0), 1 mM EDTA and samples are stored at 4 °C until use. M10 tungsten particles (diameter 0.4–1.5 μm, Sylvania) are washed in ethanol and coated with DNA as described by Sanford et al. (1993), except that 100 mg tungsten is washed and resuspended in 50% glycerol at 100 μg tungsten per ml. Two and one-half mg (25 μl) of washed tungsten coated with 2.5 μg of each plasmid DNA (when co-transforming with two DNA constructs) is resuspended in 24 μl of absolute ethanol, and 7 μl of this is used for each transformation or 'shot'. We coat anywhere from 2.5 to 10 mg of tungsten in one tube, keep coated particles on ice, and use them immediately or no longer than 4 h after preparation.

3. Biolistic Transformation

Biolistic transformation is achieved using the helium-driven PDS-1000/He (Du Pont, now distributed by BioRad) as instructed by the manufacturer. Coated microprojectiles are loaded onto the thin (0.06 mm) membrane macrocarriers (2.54 cm diameter, Kapton, Du Pont) as described (Sanford et al., 1993). The loaded macrocarrier is positioned 1 cm from the rupture disc (1100 PS I) and 1 cm from the stopping screen. Plates are placed on the second shelf from the bottom of the chamber, and the chamber is evacuated to ~27 to 29 inches Hg immediately prior to bombardment.

4. Selection and Screening

Bombarded cells on nylon filters are incubated on the original TAP agar plate for 24 h in complete darkness. The nylon filters then are removed and placed on petri dishes containing an appropriate selective agar medium (see Table 1). Cells are selected for antibiotic resistance by incubation on TAP-SPEC plates, in total darkness, for 10 to 14 days. Cells are selected for photoautotrophy by incubation on minimal medium in bright light for 7 to 10 days. Colonies are transferred to liquid medium and then replated to obtain secondary isolates that can be screened for homoplasmy. Screening is accomplished through colony hybridization or RFLP analysis as outlined in Fig. 3. Homoplasmic transformant cell lines are maintained on TAP agar in total darkness except during transfer to fresh medium, which is performed under low light conditions.

5. Use of Nylon Membrane Filters

Many groups plate cells in soft agar and then replate on selective medium after bombardment, as originally described (Boynton et al., 1988). However, colony isolates appearing after replating are not forcibly independent transformants when cells are incubated prior to replating. Other groups plate directly on the selective agar surface and do not let cells 'recover' under non-selective conditions prior to selection (Kindle et al., 1991). Our initial studies showed that cells plated directly on nylon membranes placed on the surface of non-selective agar plates survived procedural manipulations as well as cells plated directly on the agar surface or in soft agar (J. M. Erickson and S. Tahtakran, unpublished). Use of the membranes facilitates transfer of cells to selective medium by decreasing the time needed for transfer and reducing the opportunities for contamination. Most importantly, all transformants are of independent origin, and we can see the original distribution pattern of transformants. Figure 5 shows a typical pattern, with transformants distributed fairly evenly over the entire filter.

IV. Results of Studies Using Chloroplast Transformation in *C. reinhardtii*

A. Transcription and RNA Stability

Structural elements 5′ to gene coding sequences

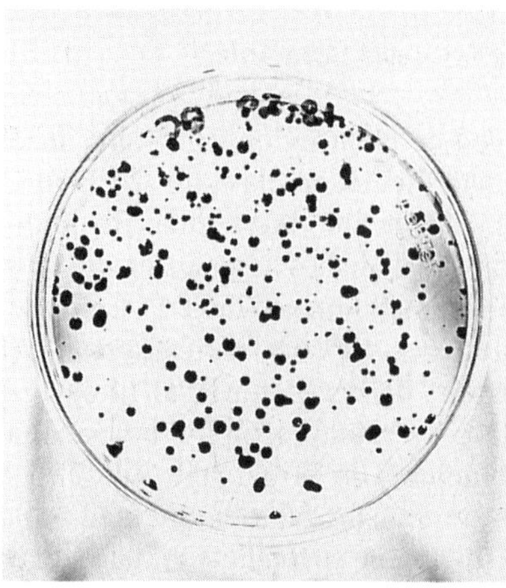

Fig. 5. Distribution of transformant *C. reinhardtii* colonies arising on selective medium after particle bombardment transformation. 2×10^7 cells wild-type algal cells (5 to 7 μm in diameter) were plated on a nylon membrane filter (82 mm in diameter) placed on solid TAP agar medium. We predict that ~5 to 15 % of the filter area is occupied by cells. Cells were bombarded with M10 tungsten microprojectiles coated with marker plasmid atpX-AAD DNA. The bacterial *aadA* gene of this plasmid confers resistance to spectinomycin (Goldschmidt-Clermont, 1991). 24 h after bombardment, the filter was transferred to agar medium containing 100 μg per ml spectinomycin and cells were incubated for 7 days. The green colonies (>500) were distributed fairly evenly over the entire filter, reflecting an even distribution of tungsten particles during bombardment. Similar patterns and efficiencies are seen when non-photosynthetic cells are transformed with an appropriate gene and transformants are selected for growth on minimal medium in high light. Using the helium-driven PDS 1000/He (Du Pont) for particle bombardment, we generally observe 200–2000 colonies per 2×10^7 cells plated, representing a transformation efficiency, on a per cell basis, of 1×10^{-5} to 1×10^{-4}.

promote transcription and in some cases affect mRNA stability. Transformants containing foreign reporter genes flanked by 5′ and 3′ regions of *C. reinhardtii* chloroplast genes (see Table 1, IV) have been used to investigate the nature of the chloroplast 'promoter' regions of *C. reinhardtii atpA* (Blowers et al., 1990; Goldschmidt-Clermont, 1991), *atpB* (Blowers et al., 1990, 1993; Klein et al., 1992; Salvador, 1993), 16S rRNA (Klein et al., 1992), *rbcL* (Blowers et al., 1990; Goldschmidt-Clermont, 1991; Salvador et al., 1993; Klein et al., 1994), *petD* (Sakamoto et al., 1993, 1994; Sturm et al., 1994), and *psbD* (Nickelsen et al., 1994). *psbD* 5′ sequences destabilize *psbD* mRNA and chimeric *psbD*-reporter transcripts in the absence of a nuclear-encoded *psbD* RNA binding protein (Nickelsen et al., 1994). *rbcL* 5′ sequences contain an element which destabilizes chimeric *rbcL*-reporter transcripts in the light, unless *rbcL* N-terminal coding sequences are added back to the chimeric gene construct (Salvador et al., 1993). Sequences well within the coding region of *rbc*L may have a dramatic positive effect on the rate of transcription from a minimal, *rbcL* promoter (Klein et al., 1994).

An inverted repeat sequence capable of forming a stem/loop structure is found in the 3′ untranslated region of chloroplast mRNAs. Controversy surrounds the role such structures play in mRNA processing and stability. The effect of altered 3′ *atpB* sequences on *atpB* mRNA stability and processing has been investigated in transformants (Stern et al., 1991; Stern and Kindle, 1993; Kindle et al., 1994). These authors conclude that the 3′ stem/loop structure stabilizes *atpB* mRNA, and that exonucleolytic processing generates the mature mRNA 3′ end. In contrast, Blowers et al., (1993) studied the effect of altered 3′ *rbcL* or *psaB* sequences on chimeric reporter transcripts. When the 3′ inverted repeat sequence was deleted, chimeric transcript size was heterogeneous but message stability was unchanged. In transformants containing multiple 3′ inverted repeat sequences, the 3′ end point of the mature mRNA occurs at the first inverted repeat sequence, which indicates that the end point is established by transcriptional termination or endonucleolytic cleavage. Therefore, the 3′ *rbcL* and *psaB* sequences may not determine chloroplast mRNA stability in vivo, but may determine the location of the 3′ end of the mRNA. It is not yet clear whether the discrepancy in conclusions reflects differences in the specific gene 3′ regions examined, or other differences in experimental design or interpretation of data.

In other studies, *psaB* mRNA was more stable when chloroplast protein synthesis was inhibited by chloramphenicol treatment or when the *psaB* mRNA contained a frameshift mutation (Xu et al., 1993). Such results indicate a link between chloroplast mRNA stability and translation. All transformants containing mutations in the 5′ untranslated region of *petD* that reduced translation of the PetD protein also showed a decrease in *petD* mRNA levels, leading Sakamoto et al. (1994) to conclude that the translatability and mRNA stability of *petD* mRNA are linked also.

An interesting sub-class of transformants containing unstable *atpB* transcripts was recovered by isolating colonies displaying robust photosynthetic growth. Characterization of these transformants showed that, although *atpB* mRNA was still unstable, a ~20–30 fold amplification of *atpB* DNA resulted in a 3–5 fold increase in *atpB* transcriptional activity and restored approximate wild-type levels of the *atpB* message and the ATP synthase β subunit (Kindle et al., 1994). These authors suggest that the amplified DNA is present as an episomal tandem repeat, which replicates independently of the chloroplast genome, and propose that the conformation of this episomal DNA may reduce transcription, thus accounting for the discrepancy between gene copy number and *atpB* transcription.

B. Introns and RNA Splicing

Several *C. reinhardtii* chloroplast genes contain introns, and several of the introns contain open reading frames (ORFs) encoding putative polypeptides. The Group I intron of the 23S rRNA gene acts as a ribozyme in vitro, as well as a mobile genetic element in vivo when both a target site and the double-strand DNA endonuclease encoded by the intron ORF are present in transformant cells (Dürrenberger and Rochaix, 1991). In vitro studies show that the 23S rRNA is capable of self-splicing and, in certain conditions, reverse-splicing; in vivo studies demonstrate that the ORF-encoded endonuclease of the intron is not required for 23S rRNA splicing (Thompson and Herrin, 1991, 1994). *psbA* Group I introns self-splice in vitro (Herrin et al., 1991). Chloroplast insertional mutagenesis studies showed that the chloroplast gene *tscA* encodes a small transcript essential for *psaA trans*-splicing (Goldschmidt-Clermont, 1991; Goldschmidt-Clermont et al., 1991). The four large introns of *psbA* have been removed in *psbA* cDNA constructs, and transformants containing 'intronless' *psbA* genes have been obtained (Johanningmeier and Heiss, 1993; Mayfield et al., 1994; Minegawa and Crofts, 1994).

C. Translation

The relationship between the initiation codon of a message and its rate of translation was examined in transformants containing *petD* initiation codons altered to AUU or AUC (Chen et al., 1993). The rate of translation of cytochrome b_6f subunit IV in these mutants was reduced 5 to 10 fold at room temperature, and more than 20 fold at elevated temperatures, and illustrates that translation initiation from these codons is temperature dependent and inefficient. Mutations introduced into the 5′ untranslated region of *psbA* alter the structure recognized by a *psbA*-specific RNA binding protein and reduce translation, leading the authors to hypothesize that this 5′ structural element may serve as a light-dependent positive regulatory signal, activated by protein binding, that increases ribosome association and translation initiation (Mayfield et al., 1994). Two classes of mutations in the 5′ untranslated regions of *petD* have been identified that act as positive regulatory elements for translation (Sakamoto et al., 1994).

The relationship between the 5′ untranslated region of the chloroplast *psbC* mRNA and two nuclear-encoded functions identified in the F34 and F64 mutants lacking the PsbC protein (P6) was further explored by Zerges and Rochaix (1994). These authors characterized chloroplast transformants containing a heterologous reporter gene under control of a wild-type or mutant 5′ *psbC* leader sequence, in algal hosts with a wild-type or mutant nuclear background. Translation of the reporter mRNA was subject to the same regulation as translation of *psbC* mRNA, which demonstrates unequivocally that the nuclear functions encoded by F34 and F64 are required for *psbC* translation.

D. Recombination

Detailed molecular and genetic analysis of transformants containing *C. reinhardtii* and *C. smithii* chloroplast markers (Newman et al., 1990, 1991, 1992) characterized the pattern of integration of transforming DNA markers into the host chloroplast and identified recombination 'hot spots' in the chloroplast inverted repeat. In most cases, homologous recombination between a circular DNA plasmid and the circular host chloroplast DNA involves two cross-over events, and plasmid vector sequences are not integrated into the chloroplast DNA molecule (Boynton et al., 1988). Exceptions to this have been seen when the selected marker of the transforming plasmid DNA is flanked on only one side by chloroplast DNA sequences homologous to the host strain (Boynton et al., 1990; Kindle et al., 1991; Erickson et al., 1992). Evidence for intramolecular recombination between copies of the inverted repeat was seen by Blowers et al. (1989) who found that transformation of an asymmetric *atpB* deletion mutant resulted in wild-type chloroplast chromosomes, even when the transforming DNA construct did not contain the DNA sequences deleted from one half of the inverted repeat. These authors conclude that wild-type DNA present in the unaltered half of the inverted repeat was recruited and duplicated, thus replacing the deleted DNA.

Introduction of a wild-type *E. coli recA* gene into the chloroplast genome increased the frequency of plastid DNA recombination with no apparent effect on cell viability or DNA repair (Cerutti et al., 1995). In contrast, these authors found that chloroplast transformants carrying mutant forms of the *E. coli recA* gene had reduced survival rates after exposure to DNA damaging agents, and were deficient in plastid DNA repair and DNA recombination. Thus, dominant negative mutations altering the bacterial RecA protein adversely affect chloroplast DNA recombination in vivo. These results indicate that the DNA recombination mechanisms in plastids and eubacteria share common features.

E. Photosynthesis

Most intrinsic trans-membrane polypeptides of the multi-subunit thylakoid membrane complexes of PS I, PS II and the cytochrome b_6f complex are encoded in the chloroplast genome. Chloroplast transformation has been used to study these polypeptides and their function in oxygenic photosynthesis. Targeted insertion of the bacterial *aadA* gene (see Table 1, III) into chloroplast genes in vivo demonstrates that among components essential for photosynthesis are the PS I polypeptide encoded by *psaC* (Takahashi et al., 1991), and the PS II polypeptide encoded by *psbK* (Takahashi et al., 1994). Interestingly, engineered mutants containing a targeted disruption in *psbI* grow photoautotrophically in dim light, but accumulate reduced levels of PS II and show an increased photosensitivity in high light (Kunstner et al., 1995). These results indicate that, while the PsbI protein is not essential for photosynthesis, it may play a role in stabilizing PS II and/or modulating PS II energy transfer. Targeted disruption of *ycf8*, a conserved chloroplast open reading frame, reveals that this ORF encodes a novel PS II polypeptide essential for PS II function under adverse conditions (Monod et al., 1994). Mutants produced by targeted disruption of *ycf*5 lack cytochrome *f* and cytochrome c_6, and transformation of the B6 chloroplast cytochrome *f* mutant with the wild-type cloned *ycf*5 restores cytochrome *f* synthesis (S. Merchant, personal communication). *ycf*5 has thus been renamed *ccsA*, reflecting its role in the synthesis of *c*-type cytochromes. Post-translational and co-translational regulation affecting biogenesis of the cytochrome b_6f complex has been studied in *C. reinhardtii* transformants containing targeted disruptions in *petA*, *petB*, or *petD* (Kuras and Wollman, 1994). Transformation was used to map the location of a chloroplast *psaB* frameshift mutation (Bingham et al., 1991).

Directed chloroplast genetic engineering in *C. reinhardtii* has been used to explore the relationship between the primary structure of an individual polypeptide and the function of the multi-subunit protein complex. Several site-directed mutations were introduced into *rbcL*, the gene encoding the large subunit of ribulose bisphosphate carboxylase/oxygenase (Zhu and Spreitzer, 1994), into *psaB*, encoding a PS I core reaction center polypeptide (Webber et al., 1993; Cui et al., 1995; Rodday et al., 1995), and into *psbA*, encoding the PS II reaction center D1 polypeptide (Przibilla et al., 1991; Roffey et al., 1991, 1994; Erickson et al., 1992; Heiss and Johanningmeier, 1992; Whitelegge et al., 1992, 1995; Xiong et al., 1994).

In wild-type *C. reinhardtii*, D1 is translated as a precursor polypeptide that contains 9 carboxy terminal residues not found in the mature D1 polypeptide. Truncated *psbA* genes, missing codons for the last nine amino acid residues of the gene, have been introduced into the chloroplast and the resulting transformants display normal PS II function (Lers et al., 1992; Schrader and Johanningmeier, 1992). These results suggest that mature D1 can assemble into PS II complexes, and that the carboxy terminal processing step itself has no other obvious function than production of mature D1.

F. Other Chloroplast Processes

Three genes whose products are required for light-independent chlorophyll biosynthesis in *C. reinhardtii* are thought to encode components of the light-independent protochlorophyllide reductase, responsible for the 'green-in-the-dark' phenotype of wild-type *Chlamydomonas*. Targeted disruption to produce mutant phenotypes confirmed the identification of *chlB* (Li et al., 1993; Liu et al., 1993b), and *chlL* (Suzuki and Bauer, 1992). *chlN* was characterized by transforming a chloroplast deletion mutant with cloned fragments that restored wild-type phenotype to the mutant host, which was yellow when grown in the dark (Choquet et al., 1992). The chloroplast gene *clpP*, encoding a Clp-like protease, has been identified and its essential function verified by insertional disruption (Huang et al., 1994). This protease is one of a few unusual proteins characterized to date that appear to undergo post-translational protein splicing.

Translocation of the cytochrome *f* protein across the thylakoid membrane was studied by engineering *petA* transformants with mutations affecting the hydrophobic core of the cytochrome *f* signal sequence (Smith and Kohorn, 1994). The identification and characterization of chloroplast and nuclear suppressors further define critical regions of the signal peptide and the translocation apparatus.

Targeted disruption of chloroplast ORF 712, which by sequence comparison appears to encode an unusual form of ribosomal protein 3, produced transformants that were stably heteroplasmic on acetate medium (Liu et al., 1993a). Since chloroplast translation is essential for cell survival, these results are consistent with the identification of ORF 712 as *rps3*. Transformants obtained following the *aadA*-mediated disruption of an otherwise unidentified polypeptide encoded by chloroplast ORF 427 were also heteroplasmic under mixotrophic growth conditions, which suggests that the ORF 427 product is likewise essential for cell viability (Goldschmidt-Clermont, 1991)

V. Plastid Transformation in Plants

A. The Plastome and the Transplastome

The plastid genome of plants presents a complex genetic target for transformation. A single plastid of a plant cell generally contains from 20 to 100 copies of the circular plastid DNA molecule. In addition, the number of plastids per cell can vary from 10–15 proplastids in a meristematic cell to several hundred chloroplasts in a leaf cell (Bendich, 1987). Thus, a given plant cell may contain a few hundred to 50,000 copies of the plastid DNA molecule. The genome of plant plastids has been called the 'plastome', and a transformed plastid genome is referred to as a 'transplastome' (Maliga, 1993) or plastid 'transgenome' (Svab and Maliga, 1993). A stable transplastomic cell line contains only transformed copies of the chloroplast DNA molecule, i.e., each plastid in a cell is homoplasmic for the transgenome.

For most applications using plastid transformation in plants, the goal is to obtain a fertile plant capable of transmitting the transplastome to the next generation. Essential steps are the a) introduction of DNA into a plastid within a plant cell, b) recombination of introduced DNA into one or more plastid DNA molecules within the plastid, c) segregation of recombinant plastid DNA molecules within the plastid during successive plastid divisions, d) segregation of homoplasmic transformed plastids within the plant cell during successive cell divisions and e) regeneration of a transplastomic cell to produce a multicellular transformed, fertile plant. The last step may require the regeneration of plants from transformed protoplasts, calli, shoots, or other plant tissues. This section will focus primarily on the process of plastid transformation, and the results obtained to date.

B. Selective Markers

Several genes that confer resistance to antibiotics are available as selectable markers for plastid transformation in plants. Nucleotide substitutions at any one of six positions in the 16S rRNA gene of *Nicotiana tabacum* result in resistance to spectinomycin or streptomycin (reviewed in Svab and Maliga, 1991). A mutant 16S rRNA gene conferring resistance to both streptomycin and spectinomycin was used as a selectable marker for plastid transformation of tobacco leaves; this is the first report of stable plastid transformation in plants (Svab et al., 1990). A mutation at any one of three different sites in the 23S rRNA gene of *Nicotiana plumbaginifolia* confers resistance to lincomycin (Cséplö et al., 1988). A mutation in the tobacco gene (*rps12*) encoding the ribosomal protein 12 confers resistance to streptomycin (Galili et al., 1989). Mutations in the 16S rRNA, 23S rRNA, and *rps12*, conferring resistance to spectinomycin, lincomycin and streptomycin, respectively, have been characterized in *Solanum nigrum* (Kavanagh et al., 1994). All three *S. nigrum* markers are located in the inverted repeat region of the *S. nigrum* chloroplast DNA, and should be useful markers for chloroplast gene recombination following plastid transformation, or after protoplast fusion with *Nicotiana*.

The bacterial gene *aadA*, which encodes aminoglycoside adenylyltransferase (AAD) and confers resistance to spectinomycin, was expressed in tobacco plastids under control of the 5′ *rrn* and 3′ *psbA* sequences (Svab and Maliga, 1993). This chimeric marker was targeted, through use of flanking sequences in the transforming plasmid construct, to the region between *rbcL* and *ORF512* in the tobacco plastid DNA, and spectinomycin-resistant, transplastomic plants were recovered. A similar plasmid construct containing the structural gene for neomycin phosphotransferase (NPTII), conferring resistance to kanamycin, was used to isolate kanamycin-resistant transplastomic plants (Carrer et al., 1993).

A third bacterial gene, *sulI*, encoding a mutant form of dihydropteroate synthase (DHPS), provides resistance to sulfonamides, and was used as a selectable marker for nuclear transformation when a transit peptide sequence targeted the DHPS to the chloroplast of transgenic tobacco leaf cells (Guerineau et al., 1990). The modified DHPS gene, under control of plastid 5′ and 3′ sequences, might provide a selectable marker for direct plastid transformation.

C. Transformation and Isolation of Transplastomic Cell Lines

1. Stable Biolistic Plastid Transformation in Tobacco Leaf Cells

Gun-powder or helium-driven biolistic transformation devices, as previously described, have been used to introduce plasmid DNA containing antibiotic resistance markers into tobacco leaf cells, and stable transplastomic plants have been recovered (reviewed in Maliga, 1993). Transformation efficiency with the targeted, chimeric *aadA* construct as an antibiotic-resistance marker (Svab and Maliga, 1993) was approximately 100 fold greater than that achieved with the 16S rRNA (Svab et al., 1990; Staub and Maliga, 1992). The increased frequency may be due to the fact that *aadA* serves as a dominant selectable marker, such that heteroplasmic cells are resistant and cell growth and division can continue during segregation. In contrast, cells heteroplasmic for rRNA resistance markers may be sensitive to the selective antibiotics, and most transformed cells may die before segregation of the transplastome can occur (Svab and Maliga, 1993). Transformation efficiency using a chimeric bacterial NPTII selectable marker (Carrer et al., 1993) was 30–40 fold less than that using the chimeric *aadA* marker.

The first green shoots appearing on selective medium after bombardment are typically heteroplasmic, while 50% of secondary shoots regenerated from a leaf of the initial shoot are homoplasmic for the transplastome (Svab and Maliga, 1993). Fertile plants were regenerated from these secondary transplastomic shoots, and selfed seed progeny were uniformly resistant to spectinomycin. In contrast, progeny of seeds obtained after reciprocal crosses to wild-type plants were resistant only when the female parent carried the transplastome. The strict maternal inheritance in tobacco will prevent uncontrolled pollen-mediated transfer of the transplastome in nature. Svab and Maliga (1993) recovered 84 spectinomycin-resistant clonal cell lines after bombarding 79 tobacco leaves. Eighty percent of these were true transformants in which *aadA* had integrated in the plastid DNA, while the remainder contained no *aadA* sequences and were assumed to be spontaneous spectinomycin resistant mutants. Due to the 100-fold lower transformation efficiency observed with the mutant rRNA and *rps12* selectable markers, spontaneous antibiotic-resistant mutants make up the vast majority of streptomycin-resistant colonies arising after bombardment with these markers.

2. Stable Polyethylene Glycol (PEG)-Mediated Plastid Transformation in Tobacco Leaf Protoplasts

Antibiotic-resistant transplastomic regenerated plants have been recovered after protoplasts prepared from tobacco leaf cells were incubated in the presence of PEG and transforming plasmid DNA containing an antibiotic-resistance marker gene (reviewed in Dix and Kavanagh, 1995). A plastid 16S rRNA gene conferring resistance to both spectinomycin and streptomycin was introduced into leaf protoplasts of *N. tabacum* (Golds et al., 1993) or *N. plumbaginifolia* (O'Neill et al., 1993), and transplastomic cell lines were recovered from antibiotic-resistant green calli. Approximately 85% of the *N. tabacum* transformant colonies selected for resistance to spectinomycin also retained the streptomycin marker, and were considered true transplastomic lines rather than spontaneous streptomycin-resistant mutants. Regenerated fertile *N. plumbaginifolia* plants showed maternal inheritance of spectinomycin resistance. Both groups obtained approximately one putative transplastomic colony for every 2×10^5 protoplasts treated.

D. Transient Expression of Foreign Genes in Plant Plastids

Transient plastidic expression of chimeric foreign genes under control of plastid promoter sequences has been observed following the biolistic delivery of bacterial *uidA* into wheat leaves and calli (Daniell et al., 1991) and bacterial *cat* (encoding chloramphenicol acetyltransferase) into cultured tobacco cells (Daniell et al., 1990). A pea promoter functioned in wheat, while a maize promoter functioned in the cultured tobacco cells (reviewed in Daniell, 1993). Addition of a pea replicon unit, which promotes autonomous chloroplast DNA replication in vitro, to the transforming plasmid construct increased the level of transient foreign gene expression observed in tobacco cells (Daniell et al., 1990). Ye et al. (1990) optimized conditions for the biolistic delivery of DNA into plastids. Results of studies on the uptake and plastidic expression of chimeric markers in plant cells after co-cultivation with *Agrobacterium* (De Block et al., 1985; Venkateswarlu and Nazar, 1991) or incubation with PEG (Spörlein et al., 1991) were inconclusive but spurred interest in plastid transformation methodology.

E. Studies of Plant Plastid Processes

1. Transcription and Translation

Tissue-specific transcription and light-regulated accumulation of GUS, under the control of tobacco *psbA* 5′ and 3′ regulatory regions, were observed in transplastomic tobacco cells selected for resistance to spectinomycin after leaves were bombarded with the chimeric *uidA* marker gene linked to a 16S rRNA selectable marker (Staub and Maliga, 1993). Further studies showed that the 5′ untranslated region of *psbA* was required for the light-regulated translation and accumulation of GUS in transgenic tobacco (Staub and Maliga, 1994a). Similar *psbA* regulatory regions have been identified in *Chlamydomonas* (Mayfield et al., 1994). These results show that D1 protein accumulation is regulated by the 5′ untranslated region of *psbA* mRNA that controls translation initiation.

McBride et al. (1994) have achieved plastid expression of a reporter *uidA* gene in transplastomic tobacco, under the control of a phage T7 RNA polymerase promoter, when a transgenic nuclear gene in the same tobacco plant encodes the phage T7 RNA polymerase expressed from a cauliflower mosaic virus 35S promoter and targeted to the chloroplast by the *rbcS* transit peptide sequence. GUS mRNA and enzyme activity were detected in all tissues examined, and were especially high in mature leaves. These authors showed that the transplastomic reporter gene, while silent in plants with a nuclear genome lacking the T7 RNA polymerase gene, was transmitted in genetic crosses and activated in the F1 progeny receiving a copy of the T7 RNA polymerase gene construct. Such results highlight the potential use of plants for controlled, high-level expression of foreign proteins encoded by plastid transgenes.

2. RNA Editing

Studies designed to investigate potential relationships between the mitochondrial and chloroplast RNA editing mechanisms showed that a chimeric plastid mRNA containing sequences derived from the petunia mitochondrial *cox*II gene is not edited in transgenic tobacco chloroplasts (Sutton et al., 1995). The portion of the *cox*II gene expressed in the chloroplast under control of the tobacco chloroplast *rps*16 or *rRNA* promoter contains 7 sites normally subject to RNA editing in petunia mitochondria. Hence, organelle-specific components may be required for RNA editing, or petunia and tobacco RNA editing mechanisms may differ. The potential role of chloroplast-encoded, trans-acting guide RNAs (gRNAs) in chloroplast RNA editing was explored in transplastomic tobacco plants containing a point-mutation in a 14-nucleotide chloroplast DNA sequence complementary to a region of the tobacco *psbL* mRNA that is normally edited by a C to U change that creates the initiator codon (Bock and Maliga, 1995). Changing the putative gRNA critical A residue to G did not abolish editing of this *psbL* codon. This 14-nucleotide putative gRNA was the only tobacco chloroplast sequence showing complementarity to the edited site of *psbL*. The authors conclude that, in the event that gRNAs help mediate plastid editing of *psbL* mRNA, these trans-acting factors may be encoded in the nuclear genome or may have limited sequence homology to the edited site.

The evolutionary conservation of RNA editing mechanisms was explored by Bock et al. (1994) who examined whether tobacco chloroplasts could perform a specific C to U editing of *psbF* mRNA normally carried out in spinach. Such an editing event changes a serine codon in the spinach gene to a conserved phenylalanine codon in spinach *psbF* mRNA, while in tobacco the *psbF* DNA already codes for phenylalanine at that position. When the endogenous tobacco gene was replaced with a modified tobacco *psbF* containing a serine codon at this site, the resulting transplastomic tobacco plants did not edit the heterologous codon. The phenotype of the mutant plants included high chlorophyll fluorescence, low chlorophyll content and slow growth, all consistent with a mutation affecting photosynthesis. Thus, spinach and tobacco chloroplasts differ in the ability to edit this particular *psb*F codon. Such studies suggest that any evolutionary loss of a requirement for this specific editing event in tobacco preceded any evolutionary loss of this specific editing capacity in tobacco chloroplasts.

3. Recombination

Studies of plastid transformation in tobacco leaf cells show that fairly long regions of homologous DNA are incorporated into the plastid DNA of the host cell. Staub and Maliga (1992) used a transforming plasmid construct containing 6.2 kb of plastid tobacco DNA, including two selectable marker genes (16S rRNA and *rps12*) and six non-selectable molecular markers. Most of the transforming DNA markers were integrated into the chloroplast DNA of transplastomic lines recovered after particle bombardment and selection; similar results were seen in *Chlamydomonas* (Newman et al., 1990). Plasmid vectors that facilitate stable, targeted integration of foreign DNA into the tobacco plastid genome have been developed (Zoubenko et al., 1994). These pRRV plasmids contain the *aad*A selective marker gene and a multiple cloning site polylinker sequence flanked on either side by 1 to 2 kb of tobacco chloroplast DNA sequence that targets the foreign gene between *trn*V and the *rsp*12/7 operon. Approximately one transplastomic cell line is recovered per sample subjected to particle bombardment. Transplastomes generated using these vectors are stable and maternally inherited.

Unusual results demonstrated that a short *psbA* 3′ sequence (383 bp) appeared to mediate homologous integration of the chimeric *aadA* reporter construct (under control of 5′*rrn* and 3′*psbA*) into the tobacco chloroplast DNA molecule (Svab and Maliga, 1993). In the transforming plasmid, the chimeric reporter gene was flanked on either side by larger regions homologous to *rbcL* (~1.6 kb) and ORF 512 (~1.3 kb), and integration events targeted the chimeric *aadA* between these two genes. However, approximately 50% of the 'transplastomic' clonal lines examined contained a minor population of transformed chloroplast DNA molecules in which *aadA* was found, not between *rbcL* and ORF 512, but between *rbcL* and the 3′*psbA* sequences. These molecules contain a duplication of a portion of 3′ *psbA* and *trnH* and a deletion of ~27 kb of the chloroplast DNA molecule. Although all DNA molecules contain *aadA*, these cell lines are technically heteroplasmic in that they contain mixed populations of chloroplast DNA molecules. It is not clear whether the heteroplasmy exists within a given plastid, between plastids in a cell, or between cells in the clonal line. If the minor class of DNA molecules is continually regenerated from the major class by secondary recombination events, it may be impossible to obtain pure homoplasmic cell lines. Recombination hot spots in the chloroplast DNA molecule of *Chlamydomonas* have been identified (Newman et al., 1992). The identification of other types of highly recombinant sequences may allow for construction of transforming plasmids that facilitate integration but avoid subsequent rearrangements of the transformed DNA molecules.

Of particular interest is the construction and use of a plasmid shuttle vector containing sequences initially isolated as an extrachromosomal tobacco plastid DNA minicircle of 868 bp, called NICE1 (Staub and Maliga, 1994b). Although the shuttle plasmid is maintained as an extrachromosomal plastid element in transplastomic tobacco plants, homologous recombination can occur between tobacco plastid DNA present in the shuttle vector and the tobacco plastid genome of the transformed plant. Thus, use of this shuttle vector facilitates the reciprocal homologous recombination of tobacco plastid DNA from the shuttle vector into the plastid genome of tobacco plants subjected to biolistic bombardment, as well as the maintenance of foreign genes on an extrachromosomal plastid element. Because the shuttle plasmids can also replicate in *E. coli*, they can be used to recover mutant DNA from tobacco plastid mutants. Such shuttle plasmids will aid in the genetic engineering of plastid genes and the characterization of plastid mutants.

VI. Future Prospects

A. Chlamydomonas

This unicellular green alga should continue to serve as a model organism for studies concerning the chloroplast. The relative ease of obtaining stable, homoplasmic chloroplast transformants, compared to the situation in vascular plants, combined with the well-developed classical genetic system and established collection of algal mutants, make *Chlamydomonas* an attractive experimental system. The alga shares many advantages of other well-exploited organisms, including yeast and bacteria, while retaining the chloroplast organization and function of a photosynthetic plant cell. The number of cells that can be readily analyzed is large enough to accommodate studies using chloroplast genetic engineering for random-mutagenesis of targeted genes.

Although significant progress has been made since the advent of chloroplast transformation, much remains to be done. In the general area of molecular biology, some basic questions remain regarding 1) chloroplast DNA replication, regulation of chloroplast DNA copy number, segregation, recombination and copy correction, 2) regulation of chloroplast transcription and mRNA stability, 3) regulation of chloroplast translational initiation and elongation, 4) links between transcription and translation, 5) RNA splicing, 6) protein splicing, and 7) molecular chaperones. In the area of photosynthesis, work will continue on biogenesis of the thylakoid membrane, assembly, stability and function of multisubunit membrane complexes, and function and regulation of stromal protein complexes involved in carbon metabolism.

At the cellular level, function is determined by information contained in three genetic compartments and by complex biochemical pathways involving transport of metabolites among cellular compartments, including the chloroplast and mitochondria. And, although not known primarily as an organism for studies on development, this heterothallic alga does undergo differentiation during gamete formation, sexual recombination, zygote formation, and spore germination. *Chlamydomonas* will no doubt be an ideal system for investigating signal transduction pathways and single-cell response to a variety of environmental cues. Since both the mitochondrial and chloroplast genomes of this alga can be transformed, *Chlamydomonas* may also provide an ideal system for studying signals between these two ATP-producing organelles.

Future applications might involve the exploitation of unicellular algae, including *Chlamydomonas*, as light-driven factories for the production of starches, fatty acids, oils, enzymes, plastics, drugs, and perhaps even hydrogen.

B. Plants

The formidable challenge of plastid transformation in plants has been undertaken and surprising progress made towards obtaining stable, transplastomic plants. Mutant plastid genes conferring resistance to antibiotics have been used to select plastid transformant cell lines from which fertile plants can be regenerated. However, screening of putative transformants has been hampered by the large percentage of resistant cells that are spontaneous antibiotic-resistant mutants and not true transformants. The use of bacterial *aadA* as a spectinomycin and streptomycin resistance marker resulted in a 100 fold increase in the efficiency at which transplastomic cell lines were recovered from bombarded tobacco leaves (Svab and Maliga, 1993). This chimeric marker holds promise for the future of basic and applied studies involving plant plastid engineering. An increase in plastid transformation efficiencies might also be achieved by reducing the chloroplast DNA content of host cells prior to transformation, either through treatment with inhibitors of chloroplast DNA synthesis (Ye and Sayre, 1990), or by using plant host cell types, such as meristematic cells, which contain fewer plastids.

Questions related to plant differentiation and development are of particular interest in studies of multicellular plants. These include studies of flowering, reproduction, and embryogenesis. Technical advances in both nuclear and plastid transformation, and an increased understanding of basic plant metabolic processes and biochemical pathways will facilitate the identification of specific targets for engineering. Potential applications of genetic engineering in plants include obtaining plants of commercial interest that have greater photosynthetic efficiency, an increased resistance to pests, and increased tolerance to adverse environmental conditions, including those related to temperature, light, moisture, and salt. Improvements in nutritional content and post-harvest quality of plant products are also desired, and could in many cases be made through plastid engineering.

Plants or plant cells in culture may be used also, in addition to algae, as factories for producing a variety of commercial products, including secondary plant metabolites. More than 50,000 such 'natural' plant chemicals may exist, some produced by only one family or species of plant. A significant world market exists for many plant chemicals such as corticosteroids, ephedrine, codeine, spearmint, vinblastine and taxol, quinine, cardamon, pyrethrins and vanilla (listed in decreasing order of estimated retail market value, as reviewed in Chrispeels and Sadava, 1994). Plants may also detoxify soils and air in enclosed spaces.

One advantage of targeting a transforming gene to the plastid when high levels of gene product are desired is the polyploid nature of the plastid genome. A given cell may thus contain thousands of template molecules for transcription. As more is learned about plastid molecular biology, including promoter characteristics, codon usage, translation, and protein stability, it may be possible to achieve and control expression of any given gene in the chloroplast.

Acknowledgments

I thank the many colleagues who kindly provided reprints and preprints of their work, and Laurens Mets for stimulating conversations.

References

Armbrust EV, Ferris PJ, and Goodenough UW (1993) A mating type-linked gene cluster expressed in *Chlamydomonas* zygotes participates in the uniparental inheritance of the chloroplast genome. Cell 74: 801–811

Avery OT, MacLeod CM and McCarty M (1944) Studies on the chemical nature of the substance inducing transformation of pneumococcal types. Induction of transformation by a desoxyribonucleic acid fraction isolated from pneumococcus type III. J Exp Med 79: 137–157

Avni A, Anderson JD, Holland N, Rochaix J-D, Gromet-Elhanan Z, and Edelman M (1992) Tentoxin sensitivity of chloroplasts determined by codon 83 of β subunit of proton-ATPase. Science 257: 1245–1247

Bendich AJ (1987) Why do chloroplasts and mitochondria contain so many copies of their genome? Bio-essays 6: 279–282

Bennoun P, Spierer-Herz M, Erickson J, Girard-Bascou J, Pierre Y, Delosme M and Rochaix J-D. (1986) Characterization of Photosystem II mutants of *Chlamydomonas reinhardtii* lacking the *psbA* gene. Plant Mol Biol 6: 151–160

Bingham SE, Xu R and Webber AN (1991) Transformation of chloroplasts with the *psaB* gene encoding a polypeptide of the Photosystem I reaction center. FEBS Lett 292: 137–140

Birky CW, Jr (1991) Evolution and population genetics of organellar genes: Mechanisms and models. In: Selander RK, Clark AG and Whitam TS (eds) Evolution at the Molecular Level, pp 112–134. Sinauer Associates, Sunderland, MA

Birky CW, Jr and Walsh JB (1992) Biased gene conversion, copy number, and apparent mutation rate differences within chloroplast and bacterial genomes. Genetics 130: 677–683

Blowers AD, Bogorad L, Shark KB, and Sanford JC (1989) Studies on *Chlamydomonas* chloroplast transformation: Foreign DNA can be stably maintained in the chromosome. Plant Cell 1: 123–132

Blowers AD, Ellmore GS, Klein U, and Bogorad L (1990) Transcriptional analysis of endogenous and foreign genes in chloroplast transformants of *Chlamydomonas*. Plant Cell 2: 1059–1070

Blowers AD, Klein U, Ellmore GS and Bogorad L (1993) Functional in vivo analyses of the 3′ flanking sequences of the *Chlamydomonas* chloroplast *rbcL* and *psaB* genes. Mol Gen Genet 238: 339–349

Bock R and Maliga P (1995) In vivo testing of a tobacco plastid DNA segment for guide RNA function in *psbL* editing. Mol Gen Genet 247: 439–443

Bock R, Kossel H and Maliga P (1994) Introduction of a heterologous editing site into the tobacco plastid genome: The lack of RNA editing leads to a mutant phenotype. EMBO J 13: 4623–4628

Bogorad L (1991) Replication and transcription of plastid DNA. In: Bogorad L, and Vasil IK (eds) The Molecular Biology of Plastids, pp 93–124. Academic Press, San Diego

Boschetti A, Breidenbach E, and Blättler R (1990) Control of protein formation in chloroplasts. Plant Sci 68: 131–149

Boudreau E and Turmel M (1995) Gene rearrangements in *Chlamydomonas* chloroplast DNAs are accounted for by inversions and by the expansion/contraction of the inverted repeat. Plant Mol Biol 27: 351–364

Bowman CM, Barker RF and Dyer TA (1988) In wheat ctDNA, segments of ribosomal protein genes are dispersed repeats, probably conserved by nonreciprocal recombination. Curr Genet 14: 127–136

Boynton JE and Gillham NW (1993) Chloroplast transformation in *Chlamydomonas*. Methods in Enzymol 217: 510–536

Boynton JE, Gillham NW, Harris EH, Hosler JP, Johnson AM, Jones AR, Randolph-Anderson BL, Robertson D, Klein TM, Shark KB, and Sanford JC (1988) Chloroplast transformation in *Chlamydomonas* with high velocity microprojectiles. Science 240: 1534–1538

Boynton JE, Gillham NW, Harris EH, Newman SM, Randolph-Anderson BL, Johnson AM and Jones AR (1990) Manipulating the chloroplast genome of *Chlamydomonas*: Molecular genetics and transformation. In: Baltscheffsky M (ed) Current Research in Photosynthesis, Vol III, pp 509–516. Kluwer Academic Publishers, Dordrecht

Boynton JE, Gillham NW, Harris EH and Newman SM (1992) Organelle genetics and transformation of *Chlamydomonas*. In: Herrmann R (ed) Cell Organelles, Advances in Plant Gene Research, Vol 6, pp 3–64. Springer-Verlag, Vienna

Carrer H, Hockenberry TN, Svab Z, and Maliga P (1993) Kanamycin resistance as a selectable marker for plastid transformation in tobacco. Mol Gen Genet 241: 49–56

Cerutti H, Johnson AM, Boynton JE and Gillham NW (1995) Inhibition of chloroplast DNA recombination and repair by dominant negative mutants of *Escherichia coli* RecA. Mol Cellular Biol 15(6): 3003–3011

Chen, X, Kindle K, and Stern D (1993) Initiation codon mutations in the *Chlamydomonas* chloroplast *petD* gene result in temperature-sensitive photosynthetic growth. EMBO J 12: 3627–3635

Chiu W-L, and Sears BB (1992) Electron microscopic localization of replication origins in *Oenothera* chloroplast DNA. Mol Gen Genet 232: 33–39

Choquet Y, Rahire M, Girard-Bascou J, Erickson J, and Rochaix J-D (1992) A chloroplast gene is required for the light-independent accumulation of chlorophyll in *Chlamydomonas reinhardtii*. EMBO J 11: 1697–1704

Chrispeels MJ and Sadava DE (1994) Plants, Genes and Agriculture. Jones and Bartlett, Boston

Coleman AW (1978) Visualization of chloroplast DNA with two fluorochromes. Exp Cell Res 114: 95–100

Cséplö A, Etzold T, Schell J and Schreier P (1988) Point mutations in the 23S rRNA genes of four lincomycin resistant *Nicotiana plumbaginifolia* mutants could provide new selectable markers for chloroplast transformation. Mol Gen Genet 214: 295–299

Cui L, Bingham SE, Kuhn M, Kass H, Lubitz W and Webber AN (1995) Site-directed mutagenesis of conserved histidines in the helix VIII domain of *psaB* impairs assembly of the Photosystem I reaction center without altering spectroscopic characteristics of P700. Biochemistry 34: 1549–1558

Daniell H (1993) Foreign gene expression in chloroplasts of higher plants mediated by tungsten particle bombardment. Meth Enzymol 217: 536–556

Daniell H, Vivekananda J, Nielsen BL, Ye GN, Tewari KK, and Sanford JC (1990) Transient foreign gene expression in chloroplasts of cultured tobacco cells after biolistic delivery of chloroplast vectors. Proc Natl Acad Sci USA 87: 88–92

Daniell H, Krishnan M, and McFadden, BF (1991) Transient expression of β-glucuronidase in different cellular compartments following biolistic delivery of foreign DNA into wheat leaves and calli. Plant Cell Reports 9: 615–619

De Block M, Schell J and Van Montagu M (1985) Chloroplast transformation by *Agrobacterium tumefaciens*. EMBO J 4, 1367–1372

Debuchy R, Purton S and Rochaix J-D (1989) The arginosuccinate lyase gene of *Chlamydomonas reinhardtii*: An important tool for nuclear transformation and for correlating the genetic and molecular maps of the ARG7 locus. EMBO J 8: 2803–2809

Deng X-W, Wing RA and Gruissem W (1989) The chloroplast genome exists in multimeric forms. Proc Natl Acad Sci USA 86: 4156–4160

Dix PJ and Kavanagh TA (1995) Transforming the plastome: Genetic markers and DNA delivery systems. Euphytica (in press)

Durrenberger F and Rochaix J-D (1991) Chloroplast ribosomal intron of *Chlamydomonas reinhardtii*: in vitro self-splicing, DNA endonuclease activity and in vivo mobility. EMBO J 10: 3495–3501

Emes MJ and Tobin AK (1993) Control of metabolism and development in higher plant plastids. Int Rev Cytol 145: 149–216

Erickson JM (1992) Photosynthesis and chloroplasts. In: Lederberg J (ed) Encyclopedia of Microbiology, Vol 3, pp 371–401. Academic Press, San Diego

Erickson JM, Rahire M and Rochaix J-D (1984) *Chlamydomonas reinhardtii* gene for the 32 000 mol. wt. protein of Photosystem II contains four large introns and is located entirely within the chloroplast inverted repeat. EMBO J 3: 2753–2762

Erickson JM, Pfister K, Rahire M, Togasaki RK, Mets L and Rochaix J-D (1989) Molecular and biophysical analysis of herbicide-resistant mutants of *Chlamydomonas reinhardtii*: Structure-function relationship of the Photosystem II D1 polypeptide. Plant Cell 1: 361–371

Erickson JM, Whitelegge J, Koo D, and Boyd K (1992) Site-directed mutagenesis of the Photosystem II D1 gene and transformation of the *Chlamydomonas* chloroplast genome. In: Murata N (ed) Current Research in Photosynthesis, Vol III, pp 421–424. Kluwer Academic Publishers, Dordrecht

Fernandez E, Schnell R, Ranum LP, Hussey SC, Silflow CD and Lefebvre PA (1989) Isolation and characterization of the nitrate reductase structural gene of *Chlamydomonas reinhardtii*. Proc Natl Acad Sci USA 86: 6449–6453

Finer JJ, Vain P, Jones MW and McMullen MD (1992) Development of the particle inflow gun for DNA delivery to plant cells. Plant Cell Reports 11: 323–328

Galili S, Fromm H, Aviv D, Edelman M and Galun E (1989) Ribosomal protein S12 as a site for streptomycin resistance in *Nicotiana* chloroplasts. Mol Gen Genet 218: 289–292

Gillham NW (1978) Organelle Heredity. Raven Press, New York

Gillham NW, Boynton JE and Harris EH (1991) Transmission of plastid genes. In: Bogorad L and Vasil IK (eds) The Molecular Biology of Plastids, pp 55–92. Academic Press, San Diego

Gillham NW, Boynton JE, and Hauser CR (1994) Translational regulation of gene expression in chloroplasts and mitochondria. Annu Rev Genet 28: 71–93

Golds T, Maliga P, and Koop H-U (1993) Stable plastid transformation in PEG-treated protoplasts of *Nicotiana tabacum*. Bio/Technology 11: 95–97

Goldschmidt-Clermont M (1991) Transgenic expression of aminoglycoside adenine transferase in the chloroplast: A selectable marker for site-directed transformation of *Chlamydomonas*. Nuc Acids Res 19: 4083–4089

Goldschmidt-Clermont M, Choquet Y, Girard-Bascou J, Michel F, Schirmer-Rahire M and Rochaix J-D (1991) A small chloroplast RNA may be required for *trans*-splicing in *Chlamydomonas reinhardtii*. Cell 65: 135–143

Gregory RPF (1989) Photosynthesis. Chapman and Hill, New York

Gruissem W and Tonkyn J (1993) Control mechanisms of plastid gene expression. CRC Reviews in Plant Sciences 12: 19–55

Guerineau F, Brooks L, Meadows J, Lucy A, Robinson C, and Mullineaux P (1990) Sulfonamide resistance gene for plant transformation. Plant Mol Biol 15: 127–136

Hallick RB (1989) Proposals for the naming of chloroplast genes. II. Update to the nomenclature of genes for thylakoid membrane polypeptides. Plant Mol Biol Reporter 7: 266–275

Hallick RB (1992) Chloroplasts. Current Opinion in Genetics and Development 2: 926–930

Hallick RB and Bairoch A (1994) Proposals for the naming of chloroplast genes. III. Nomenclature for open reading frames encoded in chloroplast genomes. Plant Mol Biol Reporter 12: S29–S30

Hallick RB and Bottomley W (1983) Proposals for the naming of chloroplast genes. Plant Mol Biol Reporter 1: 38–43

Hallick RB, Hong L, Drager RG, Favreau MR, Monfort A, Orsat B, Spielmann A and Stutz E (1993) Complete sequence of *Euglena gracilis* chloroplast DNA. Nuc Acids Res 21: 3537–3544

Harris E (1989) The *Chlamydomonas* Sourcebook. A Comprehensive Guide to Biology and Laboratory Use. Academic Press, San Diego

Harris EH, Burkhart BD, Gillham NW, and Boynton JE (1989) Antibiotic resistance mutations in the chloroplast 16S and 23S rRNA genes of *Chlamydomonas reinhardtii*: Correlation of genetic and physical maps of the chloroplast genome. Genetics 123: 281–292

Harris EH, Boynton JE and Gillham NW (1994) Chloroplast ribosomes and protein synthesis. Microbiol Rev 58: 700–754

Heiss S and Johanningmeier U (1992) Analysis of a herbicide resistant mutant obtained by transformation of the *Chlamydomonas* chloroplast. Photosyn Res 34: 311–317

Herrin DL, Bao Y, Thompson AJ and Chen YF (1991) Self-splicing of the *Chlamydomonas* chloroplast *psbA* introns. Plant Cell 3: 1095–1107

Herrmann RG and Possingham JV (1980) Plastid DNA–the plastome. Results and Problems in Cell Differentiation 10: 45–96

Herrmann RG, Bohnert H-J, Kowallik KV and Schmitt JM (1975) Size, conformation and purity of chloroplast DNA of some higher plants. Biochim Biophys Acta 378: 305–317

Hiratsuka J, Shimada H, Whittier R, Ishibashi T, Sakamoto M., Mori M, Kondo C, Honji Y, Sun CR, Meng BY, Li YQ, Kanno A, Nishizawa Y, Hirai A, Shinozaki K and Sugiura M (1989) The complete sequence of the rice (*Oryza sativa*) chloroplast genome: Intermolecular recombination between direct tRNA genes accounts for a major plastid DNA inversion during the evolution of cereals. Mol Gen Genet 217: 185–194

Hoober JK (1984) Chloroplasts. Plenum Press, New York

Hosler JP, Wurtz EA, Harris EH, Gillham NW and Boynton JE (1989) Relationship between gene dosage and gene expression in the chloroplast of *Chlamydomonas reinhardtii*. Plant Physiol 91: 648–655

Huang C, Wang S, Chen L, Lemieux C, Otis C, Turmel M and Liu X-Q (1994) The *Chlamydomonas* chloroplast *clpP* gene contains translated large insertion sequences and is essential for cell growth. Mol Gen Genet 244: 151–159

James TW, and Jope C (1978) Visualization by fluorescence of chloroplast DNA in higher plants by means of the DNA-specific probe 4′,6-diamidino-2-phenylindole. J Cell Biol 79: 623–630

Johanningmeier U and Heiss S (1993) Construction of a *Chlamydomonas reinhardtii* mutant with an intronless *psbA* gene. Plant Mol Biol 22: 91–99

Johanningmeier U, Bodner U and Wildner G (1988) Amino acid substitutions in herbicide-resistant *Chlamydomonas* mutants. In: Behrens D (ed) Technology of Biological Processes—Safety in Biotechnology—Applied Genetic Engineering. DECHEMA Biotechnology Conferences, Vol 1, pp 281–287. VHC Verlagsgesellschaft, Weinheim

Johnston SA, Anziano P, Shark K, Sanford JC and Butow R (1988) Transformation of yeast mitochondria by bombardment of cells with microprojectiles. Science 240: 1538–1541

Kavanagh TA, O'Driscoll KM, McCabe PF and Dix PJ (1994) Mutations conferring lincomycin, spectinomycin, and streptomycin resistance in *Solanum nigrum* are located in three different chloroplast genes. Mol Gen Genet 242: 675–680

Kindle KL and Sodeinde OA (1994) Nuclear and chloroplast transformation in *Chlamydomonas reinhardtii*: Strategies for genetic manipulation and gene expression. J Applied Phycology 6: 231–238

Kindle KL, Richards KL and Stern DB (1991) Engineering the chloroplast genome: Techniques and capabilities for chloroplast transformation in *Chlamydomonas reinhardtii*. Proc Natl Acad Sci USA 88: 1721–1725

Kindle KL, Suzuki H, and Stern DB (1994) Gene amplification can correct a photosynthetic growth defect caused by mRNA instability in *Chlamydomonas* chloroplasts. Plant Cell 6: 187–200

Klein TM, Wolf ED, Wu R and Sanford JC (1987) High-velocity microprojectiles for delivering nucleic acids into living cells. Nature 327: 70–73

Klein U, De Camp JD and Bogorad, L (1992) Two types of chloroplast gene promoters in *Chlamydomonas reinhardtii*. Proc Natl Acad Sci USA 89: 3453–3457

Klein U, Salvador ML and Bogorad L (1994) Activity of the *Chlamydomonas* chloroplast *rbcL* gene promoter is enhanced by a remote sequence element. Proc Natl Acad Sci USA 91: 10819–10823

Kolodner RD and Tewari KK (1975) Chloroplast DNA from higher plants replicates by both the Cairns and the rolling circle mechanism. Nature 256: 708–711

Kunstner P, Guardiola A, Takahashi Y and Rochaix JD (1995) A mutant strain of *Chlamydomonas reinhardtii* lacking the chloroplast Photosystem II *psbI* gene grows photoautotrophically. J Biol Chem 270: 9651–9654

Kuntz M, Camara B, Weil JH and Schantz R (1992) The *psbL* gene from bell pepper (*CaPS Icum annum*): Plastid RNA editing also occurs in non-photosynthetic chromoplasts. Plant Mol Biol 20: 1185–1188

Kuras R and Wollman F-A (1994) The assembly of cytochrome b_6f complexes: An approach using genetic transformation of the green alga *Chlamydomonas reinhardtii*. EMBO J 13: 1019–1027

Kuroiwa T, Suzuki T, Ogawa K and Kawano S (1981) The chloroplast nucleus: Distribution, number, size, and shape, and a model for the multiplication of the chloroplast genome during chloroplast development. Plant Cell Physiol 22: 381–396

Lemieux C, Turmel M and Lee RW (1981) Physical evidence for recombination of chloroplast DNA in hybrid progeny of *Chlamydomonas eugametos* and *C. moewusii*. Curr Gen 3: 97–103

Lers A, Heifetz PB, Boynton JE, Gillham NW and Osmond CB (1992) The carboxyl-terminal extension of the D1 protein of Photosystem II is not required for optimal photosynthetic performance under CO_2- and light-saturated growth conditions. J Biol Chem 267: 17494–17497

Li J, Goldschmidt-Clermont M and Timko MP (1993) Chloroplast-encoded *chlB* is required for light-independent protochlorophyllide reductase activity in *Chlamydomonas reinhardtii*. Plant Cell 5: 1817–1829

Liu X-Q, Huang C and Xu H (1993a) The unusual *rps3*-like orf712 is functionally essential and structurally conserved in *Chlamydomonas*. FEBS Lett 336: 225–230

Liu X-Q, Xu H and Huang C (1993b) Chloroplast *chlB* gene is required for light-independent chlorophyll accumulation in *Chlamydomonas reinhardtii*. Plant Mol Biol 23: 297–308

Maliga P (1993) Towards plastid transformation in flowering plants. Trends in Biotechnol 11: 101–107

Mayfield SP (1990) Chloroplast gene regulation: Interaction of the nuclear and chloroplast genomes in the expression of photosynthetic proteins. Curr Opinion Cell Biol 2: 509–513

Mayfield SP and Kindle KL (1990) Stable nuclear transformation of *Chlamydomonas reinhardtii* by using a *C. reinhardtii* gene as the selectable marker. Proc Natl Acad Sci USA 87: 2087–2091

Mayfield SP, Cohen A, Danon A and Yohn CB (1994) Translation of the *psbA* mRNA of *Chlamydomonas reinhardtii* requires a structured RNA element contained within the 5′ untranslated region. J Cell Biol 127: 1537–1545

McBride KE, Schaaf DJ, Daley M and Stalker DM (1994) Controlled expression of plastid transgenes in plants based on a nuclear DNA-encoded and plastid-targeted T7 RNA polymerase. Proc Natl Acad Sci USA 91: 7301–7305

Medgyesy P, Fejes E and Maliga P (1985) Interspecific chloroplast recombination in a *Nicotiana* somatic hybrid. Proc Natl Acad Sci USA 82: 6960–6964

Minagawa J and Crofts AR (1994) A robust protocol for site-directed mutagenesis of the D1 protein in *Chlamydomonas reinhardtii* : A PCR-spliced *psbA* gene in a plasmid conferring spectinomycin resistance was introduced into a *psbA* deletion strain. Photosyn Res 42: 121–131

Monod C, Takahashi Y, Goldschmidt-Clermont M and Rochaix J-D (1994) The chloroplast *ycf8* open reading frame encodes a Photosystem II polypeptide which maintains photosynthetic activity under adverse growth conditions. EMBO J 13: 2747–2754

Morton BR (1993) Chloroplast DNA codon use: Evidence for selection at the *psbA* locus based on tRNA availability. J Mol Evol 37: 273–280

Mullet JE (1988) Chloroplast development and gene expression. Annu Rev Plant Physiol Plant Mol Biol 39: 475–502

Mullet JE (1993) Dynamic regulation of chloroplast transcription. Plant Physiol 103: 309–313

Nakamura S, Sakihara M, Chibana H, Ikehara T and Kuroiwa T (1994) Mutations disturbing the condensation of plastid nucleoids in *Chlamydomonas reinhardtii*. Protoplasma 178: 111–118

Neckermann K, Zeltz P, Igloi GL, Kossel H and Maier RM (1994) The role of RNA editing in conservation of start codons in chloroplast genomes. Gene 146: 177–182

Newman SM, Boynton JE, Gillham NW, Randolph-Anderson BL, Johnson AM, and Harris EH (1990) Transformation of chloroplast ribosomal RNA genes in *Chlamydomonas*: Molecular and genetic characterization of integration events. Genetics 126: 875–888

Newman SM, Gillham NW, Harris EH, Johnson AM and Boynton JE (1991) Targeted disruption of chloroplast genes in *Chlamydomonas reinhardtii*. Mol Gen Genet 230: 65–74

Newman SM, Harris EH, Johnson AM, Boynton JE and Gillham NW (1992) Nonrandom distribution of chloroplast recombination events in *Chlamydomonas reinhardtii*: Evidence for a hotspot and an adjacent cold region. Genetics 132: 413–429

Nickelsen J, van Dillewijn J, Rahire M and Rochaix JD (1994) Determinants for stability of the chloroplast *psbD* RNA are located within its short leader region in *Chlamydomonas reinhardtii*. EMBO J 13: 3182–3191

Oard JH, Paige DF, Simmonds JA and Gradziel TM (1990) Transient gene expression in maize, rice, and wheat cells using an airgun apparatus. Plant Physiol 92: 334–339

Ohyama K, Fukuzawa H, Kohchi T, Shirai H, Sana T, Sano S, Umesono K, Shiki Y, Takeuchi M, Chang Z, Aota SI, Inokuchi H and Ozeki H (1986) Chloroplast gene organization deduced from complete sequence of liverwort *Marchantia polymorpha* chloroplast DNA. Nature 322: 272–274

O'Neill C, Horváth GV, Horváth É, Dix PJ and Medgyesy P (1993) Chloroplast transformation in plants: Polyethylene glycol (PEG) treatment of protoplasts is an alternative to biolistic delivery systems. Plant Journal 3: 729–738

Palmer JD (1983) Chloroplast DNA exists in two orientations. Nature 301: 92–93

Palmer JD (1985) Comparative organization of chloroplast genomes. Annu Rev Genet 19: 325–354

Palmer JD, Boynton JE, Gillham NW and Harris EH (1985) Evolution and recombination of the large inverted repeat in *Chlamydomonas* chloroplast DNA. In: Steinback KE, Bonitz S, Arntzen C and Bogorad L (eds) Molecular Biology of the Photosynthetic Apparatus, pp 269–278. Cold Spring Harbor Laboratory, New York

Possingham JV and Lawrence ME (1983) Controls to plastid division. Intl Rev Cytol 84: 1–56

Price C, et al. (1994) Commission on Plant Gene Nomenclature Databases: Mnemonic/Numeric 1, Plant genes: A database of plant-wide designations. Plant Mol Biol Reporter 12: S81–S88

Przibilla E, Heiss S, Johanningmeier U and Trebst A (1991) Site-specific mutagenesis of the D1 subunit of Photosystem II in wild-type *Chlamydomonas*. Plant Cell 3: 169–174

Randolph-Anderson B, Boynton JE, Gillham NW, Harris EH, Johnson AM, Dorthu M-P and Matagne RF (1993) Further characterization of the respiratory deficient *dum-1* mutation of *Chlamydomonas reinhardtii* and its use as a recipient for mitochondrial transformation. Mol Gen Genet 236: 235–244

Rochaix J-D (1978) Restriction endonuclease map of the chloroplast DNA of *Chlamydomonas reinhardtii*. J Mol Biol 126: 597–617

Rochaix J-D (1992) Post-transcriptional steps in the expression of chloroplast genes. Annu Rev Cell Biol 8: 1–28

Rochaix J-D, Mayfield, S, Goldschmidt-Clermont M, and Erickson J (1988) Molecular biology of *Chlamydomonas*. In: Shaw CH (ed) Plant Molecular Biology, a Practical Approach, pp 253–275. IRL Press, Oxford

Rodday SM, Webber AN, Bingham SE and Biggins J (1995) Evidence that the FX domain in Photosystem I interacts with the subunit *PsaC*: Site-directed changes in *PsaB* destabilize the subunit interaction in *Chlamydomonas reinhardtii*. Biochem 34: 6328–6334

Roffey RA, Golbeck JH, Hille CR and Sayre RT (1991) Photosynthetic electron transport in genetically altered Photosystem II reaction centers of chloroplasts. Proc Natl Acad Sci USA 88: 9122–9126

Roffey RA, Kramer DM, Govindjee and Sayre RT (1994) Lumenal side histidine mutations in the D1 protein of Photosystem II affect donor side electron transfer in *Chlamydomonas reinhardtii*. Biochim Biophys Acta 1185: 257–270

Roitgund C and Mets LJ (1990) Localization of two novel chloroplast genome functions: *Trans*-splicing of RNA and

protochlorophyllide reduction. Curr Genet 17: 147–153
Ruf S, Zeltz P and Kossel H (1994) Complete RNA editing of unspliced and dicistronic transcripts of the intron-containing reading frame IRF170 from maize chloroplasts. Proc Natl Acad Sci USA 91: 2295–2299
Sakamoto W, Kindle KL and Stern DB (1993) In vivo analysis of *Chlamydomonas* chloroplast *petD* gene expression using stable transformation of β-glucuronidase translational fusions. Proc Natl Acad Sci USA 90: 497–501
Sakamoto W, Chen X, Kindle KL and Stern DB (1994) Function of the *Chlamydomonas reinhardtii petD* 5′ untranslated region in regulating the accumulation of subunit IV of the cytochrome b_6f complex. Plant J 6: 503–512
Salvador ML, Klein U and Bogorad L (1993) 5′ sequences are important positive and negative determinants of the longevity of *Chlamydomonas* chloroplast gene transcripts. Proc Natl Acad Sci USA 90: 1556–1560
Sanford JC, De Vit MJ, Russell JA, Smith FD, Harpending PR, Roy MK and Johnston SA (1991) An improved, helium-driven biolistic device. Technique 3: 3–16
Sanford JC, Smith FD and Russell JA (1993) Optimizing the biolistic process for different biological applications. Meth Enzymol 217: 483–509
Sato N, Albrieux C, Joyard J, Douce R and Kuroiwa T (1993) Detection and characterization of a plastid envelope DNA-binding protein which may anchor plastid nucleoids. EMBO J 12: 555–561
Schrader S and Johanningmeier U (1992) The carboxy-terminal extension of the D1-precursor protein is dispensable for a functional Photosystem II complex in *Chlamydomonas reinhardtii*. Plant Mol Biol 19: 251–256
Sears BB and Van Winkle-Swift K (1994) The salvage/turnover/repair (STOR) model for uniparental inheritance in *Chlamydomonas*: DNA as a source of sustenance. J Heredity 85: 366–376
Shimada H and Sugiura M (1991) Fine structural features of the chloroplast genome: Comparison of the sequenced chloroplast genomes. Nuc Acids Res 19: 983–995
Shinozaki K, Ohme M, Tanaka M, Wakasugi T, Hayshida N, Matsubayashi T, Zaita N, Chunwongse J, Obokata J, Yamaguchi-Shinozaki K, Ohto C, Torazawa K, Meng BY, Sugita M, Deno H, Kamogashira T, Yamada K, Kusuda J, Takaiwa F, Kato A, Tohdoh N, Shimada H and Sugiura M (1986) The complete nucleotide sequence of the tobacco chloroplast genome: Its gene organization and expression. EMBO J 5: 2043–2049
Smith TA and Kohorn BD (1994) Mutations in a signal sequence for the thylakoid membrane identify multiple protein transport pathways and nuclear suppressors. J Cell Biol 126: 365–374
Spörlein B, Streubel M, Dahlfeld B, Westhoff P and Koop HU (1991) PEG-mediated plastid transformation: A new system for transient gene expression assays in chloroplasts. Theor Appl Genet 82: 717–722
Staub JM and Maliga P (1992) Long regions of homologous DNA are incorporated into the tobacco plastid genome by transformation. Plant Cell 4: 39–45
Staub JM and Maliga P (1993) Accumulation of D1 polypeptide in tobacco plastids is regulated via the untranslated region of the *psbA* mRNA. EMBO J 12: 601–606
Staub JM and Maliga P (1994a) Translation of *psbA* mRNA is regulated by light via the 5′-untranslated region in tobacco plastids. Plant J 6: 547–553
Staub JM and Maliga P (1994b) Extrachromosomal elements in tobacco plastids. Proc Natl Acad Sci USA 91: 7468–7472
Stern DB and Kindle KL (1993) 3′ end maturation of the *Chlamydomonas reinhardtii* chloroplast *atpB* mRNA is a two-step process. Mol Cellular Biol 13: 2277–2285
Stern DB, Radwanski ER and Kindle KL (1991) A 3′ stem/loop structure of the *Chlamydomonas* chloroplast *atpB* gene regulates mRNA accumulation in vivo. Plant Cell 3: 285–297
Sturm NR, Kuras R, Buschlen S, Sakamoto W, Kindle KL, Stern DB and Wollman FA (1994) The *petD* gene is transcribed by functionally redundant promoters in *Chlamydomonas reinhardtii* chloroplasts. Mol Cell Biol 14: 6171–6179
Sugiura M (1992) The chloroplast genome. Plant Mol Biol 19: 149–168
Sutton CA, Zoubenko OV, Hanson MR and Maliga P (1995) A plant mitochondrial sequence transcribed in transgenic tobacco chloroplasts is not edited. Mol Cell Biol 15: 1377–1381
Suzuki JY and Bauer CE (1992) Light-independent chlorophyll biosynthesis: Involvement of the chloroplast gene *chlL (frxC)*. Plant Cell 4: 929–940
Svab Z and Maliga P (1991) Mutation proximal to the tRNA binding region of the *Nicotiana* plastid 16S rRNA confers resistance to spectinomycin. Mol Gen Genet 228: 316–319
Svab Z and Maliga P (1993) High-frequency plastid transformation in tobacco by selection for a chimeric *aadA* gene. Proc Natl Acad Sci USA 90: 913–917
Svab Z, Hajdukiewicz P and Maliga P (1990) Stable transformation of plastids in higher plants. Proc Natl Acad Sci USA 87: 8526–8530
Takahashi Y, Goldschmidt-Clermont M, Soen S-Y, Franzén LG and Rochaix J-D (1991) Directed chloroplast transformation in *Chlamydomonas reinhardtii*: Insertional inactivation of the *psaC* gene encoding the iron sulfur protein destabilizes Photosystem I. EMBO J 10: 2033–2040
Takahashi Y, Matsumoto H, Goldschmidt-Clermont M and Rochaix JD (1994) Directed disruption of the *Chlamydomonas* chloroplast *psbK* gene destabilizes the Photosystem II reaction center complex. Plant Mol Biol 24: 779–788
Takeuchi Y, Dotson M and Keen NT (1992) Plant transformation: A simple particle bombardment device based on flowing helium. Plant Mol Biol 18: 835–839
Taylor WC (1988) Regulatory interactions between nuclear and plastid genomes. Annu Rev Plant Physiol Plant Mol Biol 40: 211–233
Thompson AJ and Herrin DL (1991) In vitro self-splicing reactions of the chloroplast group I intron Cr.LSU from *Chlamydomonas reinhardtii* and in vivo manipulation via gene-replacement. Nuc Acids Res 19: 6611–6618
Thompson AJ and Herrin DL (1994) A chloroplast group I intron undergoes the first step of reverse splicing into host cytoplasmic 5.8 S rRNA. Implications for intron-mediated RNA recombination, intron transposition and 5.8 S rRNA structure. J Mol Biol 236: 455–468
Thompson WW and Whatley JM (1980) Development of nongreen plastids. Annu Rev Plant Physiol 31: 375–394
VanWinkle-Swift K, Hoffman R, Shi L and Parker S (1994) A suppressor of a mating-type limited zygotic lethal allele also suppresses uniparental chloroplast gene transmission in *Chlamydomonas monoica*. Genetics 136: 867–877
Venkateswarlu K and Nazar RN (1991) Evidence for T-DNA-

mediated gene targeting to tobacco chloroplasts. Bio/ Technology 9: 1103–1105

Wakasugi T, Tsudzuki J, Ito S, Nakashima K, Tsudzuki T and Sugiura M (1994) Loss of all *ndh* genes as determined by sequencing the entire chloroplast genome of the black pine *Pinus thunbergii*. Proc Natl Acad Sci USA 91: 9794–9798

Webber AN, Gibbs PB, Ward JB, and Bingham SE (1993) Site-directed mutagenesis of the Photosystem I reaction center in chloroplasts. J Biol Chem 268: 12990–12995

Weber G, Monajembashi S, Greulich K-O and Wolfrum J (1989) Uptake of DNA in chloroplast of *Brassica napus* (L.) facilitated by a UC-laser microbeam. Eur J Cell Biol 49: 73–79

Whitelegge JP, Koo D and Erickson JM (1992) Site-directed mutagenesis of the chloroplast *psbA* gene encoding the D1 polypeptide of Photosystem II in *Chlamydomonas reinhardtii*: Changes at aspartate 170 affect the assembly of a functional water-splitting manganese cluster. In: Murata N (ed) Research in Photosynthesis, Vol II, pp 151–154. Kluwer Academic Publishers, Dordrecht

Whitelegge JP, Koo D, Diner BA, Domian I and Erickson JM (1995) *Chlamydomonas reinhardtii psbA* site-directed mutants assemble chloroplast Photosystem II reaction centers partially or totally inhibited in oxygen evolution. J Biol Chem 270: 225–235

Wolfe KH, Morden CW and Palmer JD (1991) Ins and outs of plastid genome evolution. Current Opinion in Genetics and Development 1: 523–529

Wolfe KH, Morden CW and Palmer JD (1992) Function and evolution of a minimal plastid genome from a nonphotosynthetic parasitic plant. Proc Natl Acad Sci USA 89: 10648–10652

Wurtz EA, Boynton JE and Gillham NW (1977) Perturbation of chloroplast DNA amounts and chloroplast gene transmission in *Chlamydomonas reinhardtii* by 5-fluorodeoxyuridine. Proc Natl Acad Sci USA 74: 4552–4556

Wurtz EA, Sears BB, Rabert DK, Shepherd HS, Gillham NW and Boynton JE (1979) A specific increase in chloroplast gene mutations following growth of *Chlamydomonas* in 5-fluorodeoxyuridine. Mol Gen Genet 170: 235–242

Xiong J, Hutchinson RS, Sayre RT and Govindjee (1994) Construction and characterization of a putative bicarbonate binding site mutant in the Photosystem II D1 protein of *Chlamydomonas reinhardtii*. 4th Intl Congress of Plant Mol Biol, Amsterdam, Abstract 1043

Xu R, Bingham SE and Webber AN (1993) Increased mRNA accumulation in a *psaB* frame-shift mutant of *Chlamydomonas reinhardtii* suggests a role for translation in *psaB* mRNA stability. Plant Mol Biol 22: 465–474

Ye G-N, Daniell H and Sanford JC (1990) Optimization of delivery of foreign DNA into higher-plant chloroplasts. Plant Mol Biol 15: 809–819

Ye J and Sayre RT (1990) Reduction of chloroplast DNA content in *Solanum nigrum* suspension cells by treatment with chloroplast DNA synthesis inhibitors. Plant Physiol 94: 1477–1483

Yu W and Spreitzer RJ (1992) Chloroplast heteroplasmicity is stabilized by an amber-suppressor tryptophan tRNA (CUA). Proc Natl Acad Sci USA 89: 3904–3907

Zeltz P, Hess WR, Neckermann K, Borner T and Kossel H (1993) Editing of the chloroplast *rpoB* transcript is independent of chloroplast translation and shows different patterns in barley and maize. EMBO J 12: 4291–4296

Zerges W and Rochaix JD (1994) The 5′ leader of a chloroplast mRNA mediates the translational requirements for two nucleus-encoded functions in *Chlamydomonas reinhardtii*. Mol Cell Biol 14: 5268–5277

Zhu G and Spreitzer RJ (1994) Directed mutagenesis of chloroplast ribulosebisphosphate carboxylase/oxygenase. J Biol Chem 269: 3952–3956

Zoubenko OV, Allison LA, Swab Z and Maliga P (1994) Efficient targeting of foreign genes into the tobacco plastid genome. Nuc Acids Res 22: 3819–3824

Zumbrunn G, Schneider M and Rochaix J-D (1989) A simple particle gun for DNA-mediated cell transformation. Technique 1: 204–216

Chapter 33

Regulation of Expression of Nuclear Genes Encoding Polypeptides Required for the Light Reactions of Photosynthesis

John C. Gray
Department of Plant Sciences, University of Cambridge, Downing Street, Cambridge CB2 3EA, UK

Donald R. Ort and Charles F. Yocum (eds): Oxygenic Photosynthesis: The Light Reactions, pp. 621–641.

Summary

Nuclear genes encode approximately half of the 60 or so different polypeptides involved in the light reactions of photosynthesis in higher plants. cDNAs encoding 31 different components of Photosystem II, Photosystem I, the cytochrome *bf* complex and the electron transfer chain, and ATP synthase have been isolated and their characterization has provided important structural information about the proteins. In all cases, the cDNAs encode larger precursor proteins with *N*-terminal presequences containing chloroplast-targeting information. Genomic clones have been isolated for a majority of these polypeptides and promoter analysis has defined *cis*-elements and *trans*-acting factors involved in light regulation, tissue specificity and circadian rhythmicity of gene expression. Coordination of expression of genes in the nuclear and chloroplast genomes is achieved at least in part by the control of nuclear gene expression by a 'plastidic factor' produced by functional chloroplasts.

I. Introduction

The light reactions of photosynthesis are catalyzed by approximately 60 different polypeptides located in the thylakoid membrane system of higher plant chloroplasts. These polypeptides are either arranged in integral membrane complexes (Photosystems I and II, the cytochrome *bf* complex and ATP synthase) or are soluble electron transfer components located in the thylakoid lumen (plastocyanin) or in the stroma (ferredoxin and ferredoxin-NADP$^+$ oxidoreductase). The genetic information for these polypeptides is distributed between the chloroplast and nuclear genomes, with each genome encoding approximately half of the components required for the light reactions. The circular chloroplast genome encodes 30 identified components of the light reactions, including 14 polypeptides of Photosystem II, 6 polypeptides of Photosystem I, 4 polypeptides of the cytochrome *bf* complex and 6 polypeptides of the ATP synthase. These chloroplast-encoded subunits include the reaction center polypeptides responsible for charge separation and primary electron transfer in both Photosystems I and II. The nuclear genome encodes the remainder of the polypeptides required for the light reactions. Currently cDNAs or nuclear genes encoding 31 different components of the light reactions have been identified, although it can be anticipated that genes encoding several more small polypeptides of the photosynthetic complexes will be isolated in the near future. All these different nuclear genes, and the chloroplast genes, must be correctly expressed, temporally and spatially, to ensure the synthesis and assembly of a functional photosynthetic apparatus. The aim of this article is to review our current knowledge of the regulation of expression of the nuclear genes encoding polypeptides required for the light reactions. Regulation of expression of chloroplast genes is discussed in Chapter 31 by Roell and Gruissem. The coordination of expression of the nuclear and chloroplast genomes is of fundamental importance for the formation of a thylakoid membrane system capable of carrying out the light reactions, including light harvesting, primary photochemistry and electron and proton transfer, necessary for the production of ATP and reducing power.

Abbreviations: A_0 – the first electron acceptor in the Photosystem I complex; A_1 – the second electron in the Photosystem I complex; F_A – iron-sulfur center A; F_B – iron-sulfur center B; F_X – iron-sulfur center X

II. Nuclear Genes

A. Isolation of cDNA and Genomic Clones

The isolation of cDNAs and genes encoding polypeptides required for the light reactions has been aided by two main factors: the abundance of photosynthetic proteins in the leaves of higher plants, allowing the purification of individual polypeptides in amounts sufficient for raising antibodies and for determining protein sequence, and the light-regulated expression of the genes, allowing differential screening of cDNAs produced from light- and dark-grown plants. A majority of the cDNAs for polypeptides of the light reactions were first isolated by screening expression libraries in *Escherichia coli* vectors, such as λgt11, with antibodies raised against the purified polypeptides. Protein sequencing confirmed the identity of the proteins encoded by the cDNAs and allowed the cleavage sites between the *N*-terminal presequences and the mature polypeptides to be determined. Antibodies have also been used to identify proteins synthesized on transcripts hybrid-selected by cloned cDNAs (Tittgen et al., 1986), although the approach is very labor intensive and was superseded by immunoscreening of expression libraries in λgt11. Available protein sequence has also been used to design degenerate oligonucleotides for screening cDNA libraries directly or for use as primers for the polymerase chain reaction (PCR) to amplify cDNAs. The isolation and identification of cDNAs provided hybridization probes for isolation of genomic clones and the isolation of cDNAs from other species. A compilation of nuclear genes encoding polypeptides required for the light reactions is presented in Table 1. The genes are grouped according to membrane complex or function, and are named according to the mnemonics recommended by Hallick (1989) and Jansson et al. (1992) with capitalized first letters to signify nuclear genes. The table gives the range of sizes of the precursor proteins encoded by the genes from different species, as well as the sizes of the *N*-terminal presequences, or transit peptides, and the mature polypeptides. Details of gene copy number and the number of introns in the genes are also presented, and will be referred to in Section II.B.

1. Photosystem II: Psb *and* Lhcb *Genes*

Photosystem II may superficially be regarded as made up of three assemblages of polypeptides: the reaction center core complex, the peripheral oxygen evolving complex and the light-harvesting complex. The reaction center core complex appears to be composed exclusively of chloroplast-encoded polypeptides. The peripheral oxygen-evolving complex and the light-harvesting complex are composed of nuclear gene products. The products of the *PsbO*, *PsbP*, *PsbQ*, *PsbR* and *PsbT* genes are all associated with the oxygen-evolving complex and are located on the lumenal side of Photosystem II from which they may be washed off with various aqueous media. *PsbO*, *PsbP* and *PsbQ* encode the major 33 kDa, 23 kDa and 16 kDa polypeptides of the oxygen-evolving complex. In each case cDNAs were isolated first by screening an expression library with specific antibodies (Jansen et al., 1987; Tyagi et al., 1987). *PsbR* encodes a 10 kDa polypeptide extracted from Photosystem II particles with alkaline Tris (Ljungberg et al., 1984) and *PsbT* encodes an extrinsic 3 kDa polypeptide in cotton, homologous to a 5 kDa spinach protein (Kapazoglou et al., 1995). cDNAs encoding these polypeptides were isolated by differential screening for light-induced transcripts in potato (Eckes et al., 1985) and cotton (Kapazoglou et al., 1995), and identified by reference to *N*-terminal amino acid sequence from the isolated proteins (Lautner et al., 1988; Webber et al., 1989; Ikeuchi et al., 1989). The *PsbR* gene was originally known as ST-LS1, for *Solanum tuberosum* leaf-stem specific (Eckes et al., 1986).

Genomic clones have been isolated for *PsbP* from mustard and tobacco (Merkle et al., 1990; Hua et al., 1991), *PsbR* from potato and *Arabidopsis thaliana* (Eckes et al., 1986; Gil-Gómez et al., 1991) and *PsbT* from cotton (Kapazoglou et al., 1995). Genomic clones of *PsbO* have not been reported, although a truncated *PsbO* pseudogene has been isolated and characterized from pea (R. Wales and J.C. Gray, unpublished). A genomic clone of maize *PsbQ* has been isolated and characterized, although it has been erroneously described as encoding the ferredoxin-$NADP^+$ reductase (FNR)-binding protein (Pessino et al., 1994).

The nomenclature of the genes encoding light-harvesting polypeptides has been recently clarified (Jansson et al., 1992) and the designations *Lhcb1*, *Lhcb2*, *Lhcb3*, *Lhcb4*, *Lhcb5* and *Lhcb6* were introduced to eliminate the confusion caused by the *Cab* gene nomenclature. *Lhcb1*, *Lhcb2* and *Lhcb3* encode the major light-harvesting chlorophyll *a/b*-binding proteins in the thylakoid membrane, whereas *Lhcb4*, *Lhcb5* and *Lhcb6* encode the minor components CP29, CP26 and CP24 (Jansson et al., 1992). In addition *PsbS* encodes the phylogenetically-related polypeptide CP22 (Wedel et al., 1992; Kim et al., 1992). However, in view of the recent observation that the polypeptide binds chlorophyll (Funk et al., 1994), the gene should probably be given a *Lhcb*, rather than *Psb*, designation. Details of the phylogenetic relationships among these polypeptides have been summarized by Green et al. (1991). A partial cDNA encoding a type I light-harvesting protein was first identified and isolated by hybrid-selected translation of pea leaf RNA followed by import and processing of the protein by isolated chloroplasts (Broglie et al., 1981).This cDNA was used to isolate a pea *Lhcb1* genomic clone (Cashmore, 1984) and a large number of other cDNAs by heterologous hybridization. All the *Lhcb* genes show sequence homology allowing isolation by hybridization at low stringency. *Lhcb1* genes or cDNAs have been isolated from at least 18 different plants (Jansson et al., 1992). An *Lhcb2* sequence was first isolated from *Lemna gibba* (Karlin-Neumann et al., 1985) and has subsequently been isolated from at least 8 more plants (Jansson et al., 1992). *Lhcb3* sequences have been isolated from tomato (Schwartz et al., 1991a) and barley (Brandt et al., 1992). cDNAs encoding CP29 and CP26 polypeptides were first isolated by antibody screening of expression libraries from barley and tomato (Pichersky et al., 1991; Morishige and Thornber, 1992). *Lhcb6* cDNA and genomic clones have been isolated from spinach and tomato (Spangfort et al., 1990; Schwartz and Pichersky, 1990). Apparently complete sets of *Lhcb* genes have been isolated from tomato (Green et al., 1991) and *Arabidopsis thaliana* (McGrath et al., 1992).

2. Photosystem I: Psa *and* Lhca *Genes*

Photosystem I consists of a core complex, made up of about 14 different polypeptides, and a light-harvesting chlorophyll *a/b*-binding protein complex, LHCI, made up of 4 or 5 different polypeptides. As with Photosystem II, all the redox-active groups in Photosystem I are associated with chloroplast-encoded polypeptides. The products of the chloroplast genes *psaA*, *psaB* and *psaC* provide the binding sites for P700 and the redox centers A_0, A_1, F_X, F_A and F_B. Several other small polypeptides of Photosystem I are encoded by chloroplast genes (*psaI*, *psaJ*, *psaM*) but their function is unknown. The products of the *PsaD*, *PsaE*, *PsaF*, *PsaG*, *PsaH*, *PsaK*, *PsaL* and *PsaN* genes are all associated with the Photosystem I core complex in higher plants. The *PsaD*, *PsaE*, *PsaF*, and *PsaG* cDNAs were all isolated first from spinach by antibody screening of expression libraries (Lagoutte, 1988; Münch et al., 1988; Steppuhn et al., 1988), whereas the *PsaH*, *PsaL*, *PsaK* and *PsaN* cDNAs were first isolated from barley (Okkels et al., 1989, 1991; Kjaerulff et al., 1993; Knoetzel and Simpson, 1993). There has been considerable confusion in the past over the correspondence between the original numbering system for the subunits of Photosystem I and the gene names (Okkels et al., 1992). This appears to have been caused by incorrect identification of cDNAs obtained by antibody screening of expression libraries (Okkels et al., 1992). However, alphabetic mnemonics for the gene products based on the gene names, e.g. PSI-D, PSI-E, etc., are now widely used and have largely replaced the original numbering system.

The light-harvesting chlorophyll *a/b*-binding proteins of Photosystem I are encoded by the *Lhca* genes. The nomenclature was suggested by Jansson et al. (1992) to avoid confusion caused by the use of the *Cab* nomenclature for genes encoding both Photosystem I and Photosystem II proteins. Four different *Lhca* genes have been identified to date encoding polypeptides in the range 17-24 kDa. However, another smaller polypeptide is reported to be part of the LHCI complex (Preiss et al., 1993), although cDNA or genomic sequences encoding this polypeptide have not yet been isolated. *Lhca1* cDNAs were first isolated from a tomato leaf expression library following screening with antibodies to the 24 kDa LHCI polypeptide from *Vicia faba* (Hoffman et al., 1987). An *Lhca2* cDNA was first isolated by differential screening of a petunia leaf library with cDNAs from light- and dark-grown leaves (Stayton et al., 1987). Hybridization with the *Lhca2* cDNA resulted in the isolation of an *Lhca3* gene from tomato (Pichersky et al., 1989), as well as *Lhca2* and *Lhca3* cDNAs from *Pinus sylvestris* (Jansson and Gustafsson, 1991). An *Lhca4* gene was first isolated from *Arabidopsis thaliana* using the polymerase chain reaction with oligonucleotide primers based on conserved *Lhca* sequences (Zhang et al., 1991).

3. Cytochrome bf *Complex and Electron Transfer Components:* Pet *Genes*

The cytochrome *bf* complex is structurally the simplest of the complexes of the photosynthetic electron transfer chain (see Chapter 19). It consists of four larger polypeptides, of which three contain prosthetic groups, and two or more smaller polypeptides. Cytochrome *f* (31 kDa), cytochrome *b*-563 (20 kDa), subunit IV (17 kDa) and subunit V (4 kDa) are all encoded by chloroplast genes (*petA*, *petB*, *petD* and *petG*, respectively). The Rieske iron-sulfur protein (20 kDa) is nuclear-encoded in higher plants and *PetC* cDNAs have been isolated by antibody screening of spinach and pea leaf expression libraries (Steppuhn et al., 1987; Salter et al., 1992). These cDNAs were used as hybridization probes to isolate cDNA and genomic clones from tobacco and *Arabidopsis* (Palomares et al., 1991; Madueño et al., 1992; J.S. Knight, A.R. Walker and J.C. Gray, unpublished). At least one other small subunit of the complex is likely to be nuclear-encoded; Schmidt and Malkin (1993) determined the *N*-terminal sequence of a 4 kDa polypeptide from the spinach complex and showed that it does not match any of the unidentified open reading frames in chloroplast DNA.

The *Pet* gene nomenclature is also used for the extrinsic electron transfer proteins, plastocyanin, ferredoxin and ferredoxin-NADP$^+$ reductase (FNR). A *PetE* cDNA encoding plastocyanin was first isolated by screening a plasmid cDNA library from *Silene pratensis* with a degenerate oligonucleotide encoding a conserved part of the plastocyanin sequence (Smeekens et al., 1985a). A similar approach was used to isolate pea *PetE* cDNAs (Last and Gray, 1989). Differential screening, antibody screening and hybridization with previously isolated *PetE* cDNAs were used to isolate *PetE* cDNA and genomic sequences from spinach, barley, tomato and *Arabidopsis* (Rother et al., 1986; Nielsen and Gausing, 1987, 1993; Vorst et al., 1988; Detlefsky et al., 1989).

Differential screening for cDNAs derived from RNA preferentially expressed in light-grown tissue identified *PetF* sequences encoding ferredoxin from *Silene pratensis* (Smeekens et al., 1985b) and pea (Dobres et al., 1987). Ferredoxin cDNA and genomic sequences were also isolated from spinach, *Arabidopsis*, maize and wheat (Wedel et al., 1988; Vorst et al., 1990; Hase et al., 1991; Bringloe et al., 1995).

PetH cDNAs encoding ferredoxin-NADP$^+$ oxido-reductase (FNR) were first isolated by antibody screening of expression libraries from pea and spinach (Newman and Gray, 1988; Jansen et al., 1988) and were subsequently isolated from *Mesembryanthemum crystallinum* and rice (Michalowski et al., 1989; Aoki et al., 1994). The *M. crystallinum PetH* cDNA was isolated by differential screening for salt-induced RNAs (Michalowski et al., 1989). *PetH* genomic clones have been isolated from spinach and pea (Oelmüller et al., 1993; P. Dupree, R. Sornarajah and J.C. Gray, unpublished). cDNA or genomic clones have not yet been isolated for the FNR-binding protein, which anchors the FNR protein to the stromal side of the thylakoid membrane. A reported genomic clone from maize (Pessino et al., 1994) encodes a protein that is homologous to the lumenal 16 kDa protein of the oxygen-evolving complex of Photosystem II and should therefore be regarded as a *PsbQ* gene.

4. ATP Synthase: Atp Genes

The thylakoid ATP synthase is composed of an extrinsic CF_1 complex and a membrane-embedded CF_0 complex (see Chapters 23 and 26). The CF_1 complex is composed of five different polypeptides in the stoichiometry $\alpha_3\beta_3\gamma\delta\varepsilon$. The α, β and ε subunits are encoded by the chloroplast genes *atpA*, *atpB* and *atpE*. The γ and δ subunits are nuclear-encoded. An incomplete *AtpC* cDNA encoding part of the γ subunit precursor was originally isolated by Miki et al. (1988). Thereafter genomic clones were isolated from spinach and *Arabidopsis* (Mason and Whitfeld, 1990; Inohara et al., 1991), and complete cDNA sequences were isolated from tobacco and pea (Larsson et al., 1992; Napier et al., 1992a). *AtpD* cDNAs have been isolated from spinach, tobacco and maize by antibody screening of expression libraries (Hermans et al., 1988; Napier et al., 1992b; Hoesche and Berzborn 1993), and from pea and *Sorghum bicolor* by hybridization with spinach and maize cDNAs, respectively (Hoesche and Berzborn 1992, 1993). *AtpD* genomic clones have been reported only for spinach (Bichler and Herrmann, 1990).

The CF_0 complex is composed of four different polypeptides in the probable stoichiometry I II $III_{6\text{-}12}$ IV. Subunits I, III and IV are encoded by the co-transcribed chloroplast genes *atpF*, *atpH* and *atpI*. Subunit II is nuclear-encoded and an *AtpG* cDNA has to date been isolated only from spinach (Herrmann et al., 1993).

B. Gene Structure and Organization

Complete genes encoding a majority of the components required for the light reactions have been isolated and characterized from one or more plant species (Table 1). However, details of complete *PsbO, PsbS, PsaE, PsaF, PsaG, PsaK, PsaL, PsaN, AtpD and AtpG* genes have not been published. The nucleotide sequences of the isolated genes allow comparison with cDNA sequences and the identification of introns within the genes (Section II.B.1). The isolation of complete genes is a prerequisite for promoter analysis and the identification of sequence elements required for the regulated expression of the genes (Section III.B.1).

1. Exon/Intron Structure

Introns are present in a majority of the nuclear genes encoding components required for the light reactions, although several genes are not interrupted by introns (Table 1). Introns are not present in *Lhcb1, PetE or PetF* genes from any plant species examined, which suggests that the absence of introns is evolutionarily conserved in these plant genes. Introns are also not present in the *PsbT* gene from cotton (Kapazoglou et al., 1995), in the *PsaD* genes from tobacco (Yamamoto et al., 1993) nor in the *AtpC* genes from *Arabidopsis* (Inohara et al., 1991). Single introns are present in the *Lhcb2, Lhcb4* and *Lhcb6* genes (Karlin-Neumann et al., 1985; Schwartz and Pichersky, 1990; Sagliocco et al., 1992; Green and Pichersky, 1993), whereas two introns are present in maize *PsbQ* (Pessino et al., 1994), tomato *Lhcb3* (Schwartz et al., 1991b), *PsaH* from rice and tobacco (de Pater et al., 1990; Nakamura and Obokata, 1994), tomato *Lhca3* and *Lhca4* (Pichersky et al., 1989; Schwartz et al., 1991b) and spinach *AtpC* (Mason and Whitfeld, 1990). Three introns are present in mustard and tobacco *PsbP* genes (Merkle et al., 1990; Hua et al., 1991), the *Arabidopsis PsbR* gene (Gil-Gómez et al., 1991) and tomato *Lhca1* genes (Hoffman et al., 1987). Four introns were found in the potato *PsbR* gene (Eckes et al., 1986), the tomato *Lhca2* gene (Pichersky et al., 1988) and the *Arabidopsis PetC* gene (J.S. Knight, A.R. Walker and J.C. Gray, unpublished), whereas five introns are present in the tomato *Lhcb5* gene (Pichersky et al., 1991). The pea *PetH* gene contains seven introns (P. Dupree, R. Sornarajah and J.C. Gray, unpublished), the highest number found in any of the photosynthesis genes examined to date.

In general, the majority of introns in these genes are small (73-377 bp), a common feature of introns in plant genes (Luehrsen et al., 1994), although larger introns (522-1066 bp) are present in tobacco *PsbP* (Hua et al., 1991), potato *PsbR* (Eckes et al., 1986), tomato *Lhca2* (Pichersky et al., 1988), tomato *Lhca3* (Pichersky et al., 1989), *Arabidopsis PetC* (J.S. Knight, A.R. Walker and J.C. Gray, unpublished) and pea *PetH* (P. Dupree, R. Sornarajah and J.C. Gray, unpublished). In general, the position of the introns is conserved among different species. For example, the three introns in the mustard and tobacco *PsbP* genes are in identical positions (Merkle et al., 1990; Hua et al., 1991), as are the two introns in the rice and tobacco *PsaH* genes (de Pater et al., 1990; Nakamura and Obokata, 1994). The two introns present in the 5' end of the spinach *PetH* gene (Oelmüller et al., 1993) are at exactly the same positions as the introns in the pea *PetH* gene (P. Dupree, R. Sornarajah and J.C. Gray, unpublished). Three of the introns in the potato and *Arabidopsis PsbR* genes are in identical positions but the potato *PsbR* gene contains an additional intron near the 3' end of the coding region that is not present in the *Arabidopsis* gene (Eckes et al., 1986; Gil-Gómez et al., 1991). A similar situation occurs with the spinach *AtpC* gene, which contains two large introns (Mason and Whitfeld 1990) that are not present in the *Arabidopsis AtpC* gene (Inohara et al., 1991). This indicates that some introns are an ancient feature of plant genes and that they are likely to have been present in the progenitors of higher plants. However some introns may have been gained during evolution. This is suggested for the intron in the Lemna *Lhcb2* gene which has characteristics of a transposable element (Karlin-Neumann et al., 1985).

2. Gene Copy Number

Most of the genes encoding proteins required for the photosynthetic light reactions are single copy (Table 1); that is, they are present as a single gene in a haploid complement of chromosomes. Normal diploid plants therefore contain two copies that are likely to be identical in self-pollinating in-breeding plants, but may be different (alleles) in out-breeding plants. In amphidiploid or allotetraploid plants, such as tobacco (*Nicotiana tabacum*), 'single-copy' genes will be present in four copies, and in allohexaploid plants like wheat (*Triticum aestivum*) 'single-copy' genes will be present in six copies, a pair of genes being derived from each parental species. For example, in tobacco two different genes have been detected by Southern hybridization for *PetC* (Madueño et al., 1992) and *AtpD* (Napier et al., 1992b), whereas in other plants such as pea, spinach and *Arabidopsis* they are present in a single copy (Tittgen et al., 1986; Salter et al., 1992; J.S. Knight, A.R. Walker and J.C. Gray, unpublished). A possible consequence of increased numbers of genes in alloploid plants is heterogeneity of the polypeptide products. This has been observed most often in tobacco (*Nicotiana tabacum*) which arose from the product of a cross between the progenitors of the present-day species *N. sylvestris* and *N. tomentosiformis* (Gray et al., 1974). Isoforms of the *PsbP, PsaD, PsaH* and *PsaL* gene products have been detected in *N. tabacum* and shown to be the sum of the products found in *N. sylvestris* and *N. tomentosiformis* (Hua et al., 1993; Obokata et al., 1993). These studies also detected heterogeneity in the *PsbP, PsaD, PsaE, PsaF, PsaH* and *PsaL* gene products of *N. sylvestris* and indicate the presence of two or more genes in the haploid genome of this parental species

Table 1. Nuclear genes encoding polypeptides required for the light reactions of photosynthesis. Genes are organized according to the complexes with which the gene products are associated. The range of sizes, in amino acid residues, of the precursor polypeptide, the *N*-terminal presequence and the mature polypeptide in different species is given. The gene copy number and the number of introns is given where the information is available.

Gene	Component	Precursor	Presequence	Mature	Gene copy number	Introns
Photosystem II						
PsbO	33 kDa OEC	325–333	79–85	246–248	1–9	–
PsbP	23 kDa OEC	258–268	72–82	185–186	2–4	3
PsbQ	16 kDa OEC	213–232	64–83	149–150	1	2
PsbR	10 kDa	140	41	99	1	3–4
PsbS	22 kDa	274	69	205	1	–
PsbT	3 kDa	105	77	28	–	0
Lhcb1	Type I LHCII, LHC IIb 28 kDa	264–268	33–35	231–233	5–8	0
Lhcb2	Type II LHCII, LHC IIb 27 kDa	265	36	229	2	1
Lhcb3	Type III LHCII, LHC IIb 25 kDa	265–268	42–45	223	1	2
Lhcb4	CP29, LHC IIa	290	–	–	1	1
Lhcb5	CP26, LHC IIc	286–291	–	–	1	5
Lhcb6	CP24, LHC IId	256–261	46–51	208–210	1–2	1
Photosystem I						
PsaD	PSI–D, subunit II	204–214	48–53	155–164	1–2	0
PsaE	PSI–E, subunit IV	125	34	91	2	–
PsaF	PSI–F, subunit III	231–235	77–78	154–158	2	–
PsaG	PSI–G, subunit V	143–167	44–69	98–99	1	–
PsaH	PSI–H, subunit VI	142–145	47–50	95	1–2	2
PsaK	PSI–K	132–143	44	99	–	–
PsaL	PSI–L	209–216	47	169	1	–
PsaN	PSI–N	145	60	85	–	–
LHCA1	Type I LHCI, LHC Ib	246	44–45	201–202	1–2	3
Lhca2	Type II LHCI	270–278	66	212	1	4
Lhca3	Type III LHCI, LHC Ia	273–286	32–55	231–241	1	2
Lhca4	Type IV LHCI, LHC Ib	251	–	–	1–2	2
Cytochrome *bf* complex, electron transfer components						
PetC	Rieske FeS protein	228–247	49–68	179–180	1–2	4
PetE	Plastocyanin	155–168	66–69	97–99	1	0
PetF	Ferredoxin	143–151	46–52	96–98	1–6	0
PetH	Ferredoxin–NADP$^+$ oxidoreductase	360–369	52–58	304–314	1–2	7
ATP synthase						
AtpC	CF_1 γ subunit	364–386	41–60	322–326	1–2	0, 2
AtpD	CF_1 δ subunit	247–257	60–70	187	1–2	–
AtpG	CF_0 subunit II	222	75	147	1	–

– indicates that the information is not available.

(Hua et al., 1993; Obokata et al., 1993). This has been confirmed by cDNA and gene isolation for *PsaD* and *PsaH* from *N. sylvestris* (Yamamoto et al., 1993; Nakamura and Obokata, 1994).

Several genes show differences in copy number among species that cannot be ascribed to alloploidy. For example, *AtpC* is present as a single-copy gene in spinach (Tittgen et al., 1986) but is present in two copies in *Arabidopsis thaliana* (Inohara et al., 1991), a plant with a very small genome. Tomato, which also has a relatively small genome, contains two copies of several genes, including *Lhcb6, Lhca1* and *Lhca4*, which are present in single copies in *Arabidopsis* (Green et al., 1991). Differences in gene copy number between species appear to be particularly marked with *PsbO*. This has been reported to be a single-copy gene in spinach (Tittgen et al., 1986) and low copy in potato (van Spanje et al., 1991) and tomato (Görlach et al., 1993), whereas a *PsbO* multigene family of 7-9 members has been suggested for pea (Wales et al., 1989a). *Lhcb1* appears to be present as a multigene family in all plants examined, although gene copy numbers determined solely by Southern hybridization may be inaccurate because of cross-hybridization with other *Lhc* genes. Eight *Lhcb1* genes are present in the haploid genome of tomato (Green et al., 1991) whereas five *Lhcb1* genes are present in *Arabidopsis* (McGrath et al., 1992).

The presence of two or more genes in a haploid complement of chromosomes suggests that gene duplication has occurred. Gene duplication most likely results initially in closely-linked genes at a single locus. This has been observed for *Lhcb1* in petunia, tomato, *Arabidopsis*, soybean and *Nicotiana plumbaginifolia* (Dunsmuir et al., 1983; Pichersky et al., 1985; Leutwiler et al., 1986; Castresana et al., 1987; Walling et al., 1988; McGrath et al., 1992). Two to four genes are present at each locus, arranged either in tandem as in the tomato *Cab1* locus (Pichersky et al., 1985) or as pairs of genes in an inverted orientation transcribed divergently (Dunsmuir et al., 1983; Pichersky et al., 1985; Leutwiler et al., 1986; Castresana et al., 1987; McGrath et al., 1992). A similar arrangement of divergently transcribed genes is observed for *Lhcb6, Lhca1* and *Lhca4* genes in tomato (Hoffman et al., 1987; Schwartz and Pichersky, 1990; Schwartz et al., 1991a).

The processes of gene duplication and deletion may produce truncated pseudogenes containing only parts of the original gene. This appears to have occurred in the tomato *Cab1* locus where part of a *Lhcb1* gene is present at the end of two full *Lhcb1* genes in tandem (Pichersky et al., 1985) and in a soybean *Lhcb1* gene cluster (Walling et al., 1988). It is rather surprising that pseudogenes have not been described more often in the *Lhc* extended gene family. A *PsbO* pseudogene has been isolated from pea (R. Wales and J.C. Gray, unpublished) suggesting that the presence of pseudogenes may account at least in part for the *PsbO* multigene family detected by Southern hybridization (Wales et al., 1989a).

3. Gene Organization

There have been relatively few studies of the organization of photosynthesis genes in the chromosomes of higher plants. As outlined above (Section II.B.2) multiple *Lhc* or *Cab* genes are clustered in many species, although the cluster is composed of duplicated copies of an individual gene type, e.g. *Lhcb1*, and there are no examples of clusters of copies of different genes. The different genes encoding photosynthesis components are apparently dispersed throughout the genome in higher plants. *Lhc* gene loci have been mapped in tomato (Pichersky et al., 1985, 1988; Vallejos et al., 1986; Schwartz et al., 1991a), pea (Polans et al., 1985) and *Arabidopsis* (Zhang et al., 1991; Hauge et al., 1993). In tomato, the *Lhcb1* genes are located on chromosomes 2 and 3, the *Lhcb2* genes on chromosomes 7 and 12, the *Lhca2* gene on chromosome 10 and *Lhca4* genes on chromosomes 3 and 6 (Pichersky et al., 1985, 1988; Vallejos et al., 1986; Schwartz et al., 1991a). In pea, allelic variation in *Lhcb1* genes detected by Southern hybridization was localized to a single genetic locus on chromosome 2 (Polans et al., 1985). In *Arabidopsis*, the *Lhca4* gene is near the middle of chromosome 3 (Zhang et al., 1991; Hauge et al., 1993). Map positions for the single-copy genes, *PetC, PetE, PetF* and *PetH*, were determined by examining restriction fragment length polymorphisms in F_8 recombinant inbred lines between the Landsberg and Columbia ecotypes of *Arabidopsis. PetE, PetF* and *PetH* are located on chromosome 1, whereas *PetC* is located near the top of chromosome 4 (A.R. Walker, T. Cole, K. Torney and J.C. Gray, unpublished).

III. Regulation of Gene Expression

The expression of the nuclear photosynthesis genes is regulated to ensure that appropriate amounts of the polypeptides are produced at appropriate times in appropriate parts of the plant to allow the assembly of a functional photosynthetic apparatus. Gene expression may be regulated at a number of steps required for the accumulation of the mature protein, including transcription initiation, mRNA stability, translation initiation and protein degradation. However, there is now abundant evidence for the primary regulation of expression of nuclear photosynthesis genes at the level of transcriptional initiation. In addition, there is growing evidence for regulation of some genes at the levels of mRNA stability, translation and at post-translational steps, including protein degradation.

A. Patterns of Gene Expression

The following sections will give a brief overview of our current knowledge of the patterns of expression of nuclear genes encoding components required for the light reactions of photosynthesis. Gene expression is regulated in response to a variety of external and endogenous factors. For photosynthesis genes the most obvious, and the most widely studied, factor is light, although the tissue specificity of expression has also been investigated in depth. In addition, however, the expression of many of these genes is regulated by a circadian rhythm, by sugars and plant growth regulators, such as cytokinins and abscisic acid, as well as by endogenous developmental processes.

1. Tissue Specificity

Photosynthesis takes place in the green aerial parts of higher plants, so it is obvious that genes for photosynthesis components must be expressed in these tissues. Abundant evidence for the presence of transcripts and protein products in leaves and stems, but not in roots, has been produced for many genes in many plants by western and northern blot analysis (see for example, Eckes et al., 1985; Lamppa et al., 1985; Wales et al., 1989b). More recently, with the analysis of transgenic plants containing promoter-reporter gene constructs, it has been demonstrated that this organ-specific expression is regulated primarily at the level of transcriptional initiation (Simpson et al., 1986a,b; Stockhaus et al., 1987; Ha and An, 1988; Bichler and Herrmann, 1990; Vorst et al., 1990, 1993a; Caspar and Quail, 1993; Pwee and Gray, 1993). However, even within a photosynthetic organ such as a leaf, photosynthesis genes are normally expressed only in cells containing chloroplasts or etioplasts. This has been established for several genes, including *Lhcb1*, *PsbR, PetC, PetE, PetF* and *PetH*, by expression of promoter-reporter gene constructs in transgenic plants (Stockhaus et al., 1989b; Vorst et al., 1990, 1993a; Harkins et al., 1990; Dupree et al., 1991; Caspar and Quail, 1993; Pwee and Gray, 1993). The correlation between the presence of chloroplasts and reporter gene expression in transgenic tobacco is extremely high for some genes, for example the pea *PetE* gene (Dupree et al., 1991; Pwee and Gray, 1993), but much less for some others, for example the spinach *AtpD* and *PetH* genes (Bichler and Herrmann, 1990; Oelmüller et al., 1993). A link between nuclear gene expression and the presence of functional chloroplasts has been established and has led to the hypothesis that chloroplasts produce a 'plastidic factor' that is required for the expression of a set of nuclear genes encoding polypeptides required for photosynthesis-related functions (Section IV.C).

An important feature that remains unresolved is whether tissue-specific expression of nuclear photosynthesis genes is regulated solely via the presence of chloroplasts and the 'plastidic factor' or if there are other independent tissue-specific factors regulating gene expression. At present insufficient knowledge is available about *trans*-acting factors involved in regulating tissue-specific expression of photosynthesis genes (see Section III.B.2).

2. Light Regulation

Light is required for photosynthesis, for chlorophyll synthesis and for chloroplast development. However different photoreceptors are required for each of these processes. Light absorbed by chlorophylls and carotenoids is required for photosynthesis, whereas protochlorophyllide is the photoreceptor for protochlorophyllide reductase needed for chlorophyll synthesis in most higher plants. Red and blue light, operating through the phytochrome system and the blue-light receptors, are required for chloroplast development and the expression of nuclear genes encoding photosynthesis components. These studies on nuclear gene expression have been extensively reviewed (Thompson and White, 1991; Tobin and Kehoe, 1994) and will not be discussed in detail here. The most widely used systems for studying light-regulated gene expression are illuminated dark-grown seedlings and illuminated dark-adapted leaves of older plants. However, it is becoming clear that there may be differences in mechanisms of light-regulation in seedlings and in older leaves (Thompson and White, 1991; Gallo-Meagher et al., 1992).

3. Circadian Rhythms

Diurnal variation in the abundance of transcripts of several nuclear photosynthesis genes, including *Lhcb1, PsbO, PsaD* and *PetF*, has been observed in pea, tomato, *Arabidopsis* and wheat (Kloppstech, 1985; Giuliano et al., 1988; Piechulla, 1988; Nagy et al., 1988; Millar and Kay, 1991; Bringloe et al., 1995). Extended dark or light periods showed the existence of an underlying endogenous circadian rhythm with a periodicity of about 24 hours. The amplitude and phase of the rhythm are influenced by light (Giuliano et al., 1988; Lam and Chua, 1989a; Meyer et al., 1989; Piechulla, 1989; Taylor, 1989b;

Kellmann et al., 1990), perceived through the phytochrome system (Nagy et al., 1988). These marked changes in transcript abundance were mirrored by changes in protein levels of LHCII and Rieske protein in pea leaves (Adamska et al., 1991), although protein levels did not change in tomato leaves (Piechulla, 1988).

4. Sugar Repression

Feedback regulation of photosynthesis by the accumulation of sugars in leaves appears to be due, at least in part, to the repression of nuclear gene expression. This was shown initially by repression of the expression of promoter-reporter gene constructs, including a maize *Lhcb1* promoter, by sucrose and glucose during transient expression assays in maize protoplasts (Sheen, 1990). Subsequently decreased expression of several nuclear photosynthesis genes, including *Lhcb1* and *AtpD,* in the presence of glucose or sucrose was observed in cell cultures of *Chenopodium rubrum* (Krapp et al., 1993) and rape (Harter et al., 1993) and in light-grown tobacco seedlings (Knight and Gray, 1994). Sugar repression appears to override light activation of nuclear genes and is dependent on the phosphorylation of hexoses by hexokinase (Jang and Sheen, 1994). However, in dark-grown *Arabidopsis* seedlings, sucrose only slightly delays the developmental increase in *Lhcb1* expression and eventually results in a three-fold increase in *Lhcb1* transcripts compared to seedlings without sucrose (Brusslan and Tobin, 1992).

5. Plant Growth Regulators

Several of the endogenous plant growth regulators influence chloroplast development and might be expected to affect the expression of genes encoding photosynthesis components. Cytokinins stimulate expression of *Lhcb1* genes in tobacco suspension cultures and *Lemna* fronds (Teyssendier de la Serve et al., 1985; Flores and Tobin, 1986), apparently by decreasing rates of mRNA degradation (Tobin and Turkaly, 1982). In contrast, abscisic acid decreases *Lhcb1* expression in developing soybean cotyledons (Chang and Walling, 1991) and in tomato leaves by decreasing transcription rates (Bartholomew et al., 1991).

6. Developmental Regulation

Changes in photosynthesis gene expression due to endogenous developmental processes have been observed in several species. The most widely studied changes are in the developing first leaf of cereals, which provides a developmental time-course from the youngest cells at the base to the oldest cells at the tip. Several studies have shown differences in the levels of chloroplast proteins in different sections of maize, wheat and barley leaves (Dean and Leech, 1982; Mayfield and Taylor, 1984b; Lamppa et al., 1985; Barkardottir et al., 1987). The accumulation of transcripts of the *Lhcb1, PetE* and *PetF* genes also shows differences in different leaf sections (Viro and Kloppstech, 1980; Martineau and Taylor, 1985; Lamppa et al., 1985; Barkardottir et al., 1987; Bringloe et al., 1995). However, the patterns of accumulation of the transcripts are different for different genes; in barley, *Lhcb1* transcripts accumulate in the youngest leaf sections and decline towards the leaf tip, whereas *PetE* transcripts increase towards the middle of the leaf and reach a maximum near the tip (Barkardottir et al., 1987). Developmental changes in expression of *Lhcb1* genes, independent of light, have also been observed in soybean and *Arabidopsis* seedlings (Chang and Walling, 1992; Brusslan and Tobin, 1992).

B. Promoter Analysis

The involvement of transcriptional control in the tissue-specific and light-regulated expression of nuclear genes has resulted in many studies on the promoters of these genes in an attempt to identify *cis*-elements and the *trans*-acting factors that bind to them. The first studies were carried out with promoters of *Lhcb1* genes from wheat, pea, *Arabidopsis* and tobacco (Lamppa et al., 1985; Simpson et al., 1986a,b; Ha and An, 1988; Castresana et al., 1988), but subsequently promoter analysis has been carried out on many other photosynthesis genes, including *PsbR, PsaD, PsaF, PetE, PetF, PetH* and *AtpD* (Stockhaus et al., 1987, 1989a,b; Bichler and Herrmann, 1990; Vorst et al., 1990; Flieger et al., 1993; Oelmüller et al., 1993; Vorst et al., 1993a,b; Pwee and Gray, 1993; Caspar and Quail, 1993; Bolle et al., 1994a,b; Lübberstedt et al., 1994a,b).

1. Identification of cis-elements

Two general approaches have been used, usually sequentially, to define elements required for the expression of nuclear genes. The first approach involves the production and expression of chimeric gene constructs to identify regions of the gene responsible for regulated gene expression. The identification of *cis*-acting elements has been largely dependent on the production of transgenic plants containing chimeric genes consisting of promoter elements fused to conveniently-assayed reporter genes, such as the *E. coli uidA* gene encoding β-glucuronidase (Jefferson et al., 1987). Transient assays of reporter gene activity after the introduction of chimeric genes into protoplasts by electroporation or polyethylene glycol treatment have not been used extensively for studies on nuclear genes encoding photosynthesis components because light-regulated expression has not been demonstrated in protoplasts and they are of limited use for examining tissue-specificity. Protoplasts have, however, been used to demonstrate sugar-mediated repression of *Lhcb1* gene expression in maize (Sheen 1990). Introduction of chimeric genes into organized tissues, by biolistics or electroporation, has been used for rice and *Lemna gibba* (Dekeyser et al., 1990; Luan and Bogorad, 1992; Kehoe et al., 1994). The second approach to identifying *cis*-elements uses protein-DNA binding assays, such as electrophoretic mobility shift and footprinting assays, to identify regions of the gene containing binding sites for nuclear proteins. These assays have also been used to identify the proteins binding to the *cis*-elements (Section III.C.2).

Enhancer and silencer elements have been identified in the promoters of several genes, including *Lhcb1, PsbR, PetE* and *AtpD* (Simpson et al., 1986a; Stockhaus et al., 1987, 1989b; Bichler and Herrmann, 1990; Vorst et al., 1993; Flieger et al., 1993), although their mode of action has not yet been fully established. The enhancers generally appear to contain elements required for tissue-specific and/or light regulation (Simpson et al., 1986a; Nagy et al., 1987; Stockhaus et al., 1987, 1989b; Bichler and Herrmann, 1990). In contrast a more general, non-tissue-specific enhancer, is present in the promoter of the pea *PetE* gene (J.S. Sandhu, C.I. Webster and J.C. Gray, unpublished). This highly AT-rich region may be responsible for modulating local chromatin structure by binding to HMG I(Y) proteins and the chromosome scaffold (C.I. Webster, D. Hatton and J.C. Gray, unpublished).

Promoter deletion analysis has been used to define regions necessary for light regulation (Nagy et al., 1987; Ha and An, 1988; Stockhaus et al., 1989a; Bichler and Herrmann, 1990; Millar et al., 1992; Luan and Bogorad, 1992; Vorst et al., 1993b; Lübberstedt et al., 1994a,b; Kehoe et al., 1994), tissue specificity (Ha and An, 1988; Bichler and Herrmann, 1990; Pwee and Gray, 1993), chloroplast control (Stockhaus et al., 1987, 1989b; Vorst et al., 1993a; Bolle et al., 1994b; Fisscher et al., 1994) and circadian rhythmicity (Nagy et al., 1988; Millar et al., 1992). In general this approach defines regions of the promoter necessary for certain regulatory features, but it is often difficult to separate elements required, for example, for light regulation and tissue specificity.

Sequence comparisons of regions in different genes necessary for similar regulatory features have been successful in identifying common sequence elements (Grob and Stüber, 1987; Gilmartin et al., 1990). The similarities in sequences close to the TATA box in promoters from phytochrome-regulated genes led to the identification of a putative LAMP (light-dependent amplification-mediating protein) binding site (Grob and Stüber, 1987). This sequence element was also called the I box in *RbcS* genes (Giuliano et al., 1988b) and is now usually called the GATA element (Gidoni et al., 1989; Schindler and Cashmore, 1990). Two copies of the GATA element are present in tandem between the CCAAT and TATA sequences in *Lhcb1* and *Lhcb2* genes (Kehoe et al., 1994) and are required for high level expression in the light (Castresana et al., 1988; Gidoni et al., 1989; Kehoe et al., 1994). Protein factors GAF1 (Gilmartin et al., 1990) and GA-1 (Schindler and Cashmore, 1990) bind to GATA elements and it is possible that they act by repressing expression in the dark (Castresana et al., 1988). GATA elements may also be involved in tissue-specific expression. A tetramer of a GATA element directed expression in the leaves of transgenic tobacco when fused to a minimal –90 CaMV (cauliflower mosaic virus) 35S promoter (Lam and Chua, 1989b).

Sequences now called GT-1 elements were first identified by sequence comparisons of *RbcS* promoter regions required for light-regulated expression (Gilmartin et al., 1990) and are present in the promoters of several photosynthesis genes, including *Lhcb1* and *PetE* (Schindler and Cashmore, 1990; Luan and Bogorad, 1992; Fisscher et al., 1994). A tetramer of a GT-1 element fused to a -90 CaMV 35S promoter directed light-responsive expression mainly in chloroplast-containing cells in transgenic tobacco (Lam and Chua, 1990). Studies with GT-1 elements from a pea *RbcS* gene indicated that the element acts by repressing expression in the dark (Kuhlemeier et al., 1987).

G box sequences were first identified by sequence comparisons in *RbcS* genes (Giuliano et al., 1988b) but are present in other photosynthesis genes, including *Lhcb1*, *PetH* and *PetE* (Schindler and Cashmore, 1990; Argüello et al., 1992; Luan and Bogorad, 1992; Oelmüller et al., 1993; Nielsen and Gausing, 1993) as well as in non-photosynthetic genes (Williams et al., 1992). Deletion of the G box sequence in the tobacco *Lhcb1* promoter resulted in the loss of light-inducible gene expression (Castresana et al., 1988).

Several other sequence elements including AT-1 elements (Datta and Cashmore, 1989), GC-1 elements (Schindler and Cashmore, 1990) and CA-1 elements (Sun et al., 1993) have been defined in promoters of various *Lhcb1* genes. The mnemonics indicate the predominant bases in the elements. Proteins in nuclear extracts of several plants bind to these elements (Schindler and Cashmore, 1990; Sun et al., 1993).

Although some of these promoter elements occur in similar positions in genes in different species, for example GATA elements between the CCAAT and TATA sequences in *Lhcb1* and *Lhcb2* genes (Gidoni et al., 1989; Kehoe et al., 1994), in general there appears to be little conservation of the organization of *cis*-elements in the promoters of different plants. This is most clearly illustrated by the *PetE* genes from spinach, *Arabidopsis* and pea (Bichler and Herrmann, 1990; Vorst et al., 1993a; Pwee and Gray, 1993), and the *PetF* genes from pea, *Arabidopsis* and wheat (Elliott et al., 1989; Vorst et al., 1990, 1993b; Caspar and Quail, 1993; Bringloe et al., 1995), which are markedly different in the organization of promoter elements. In addition, in some genes, such as the spinach *PetE* and *PsaF* genes, sequences in the 5' untranslated leader regions were required for transcriptional activation (Bolle et al., 1994a).

2. Identification of trans*-acting Factors*

Protein factors binding to the *cis*-elements described above have been demonstrated by electrophoretic mobility shift assays and their binding sites defined by footprinting assays. In several instances, cDNAs encoding putative *trans*-acting factors have been isolated by screening expression libraries with multimers of the binding sites (Perisic and Lam, 1992; Gilmartin et al., 1992; Schindler et al., 1992). However, because of the redundancy in the binding sites among different genes, it is not certain that the cDNA encodes the protein that regulates the expression of a particular gene.

Proteins binding to GATA elements have been identified in nuclear extracts from pea, tobacco and tomato (Schindler and Cashmore, 1990). Binding activities have been called GA-1 (Schindler and Cashmore, 1990) and GAF-1 (Gilmartin et al., 1990), but cDNAs encoding these proteins have not yet been isolated. The tobacco nuclear factor ASF2 binds GATA sequences in a petunia *Lhcb1* gene, although there appear to be slight differences in sequence specificity to GAF-1 (Lam and Chua, 1989b).

The *trans*-acting factor GT-1 contains a novel helix-helix-turn-helix motif (Perisic and Lam, 1992; Gilmartin et al., 1992). Binding activity has been extracted from the nuclei of several plants (Schindler and Cashmore, 1990; Fisscher et al., 1994). The protein acts as a repressor in the dark, rather than as an activator in the light (Kuhlemeier et al., 1987). This may be caused by dephosphorylation of GT-1 in the light, resulting in loss of DNA binding and derepression of gene expression (Fisscher et al., 1994).

The G box-binding protein, GBF, also appears to be regulated by phosphorylation and dephosphorylation (Klimczak et al., 1992). The protein is present in extracts from nuclei of both light- and dark-grown plants (Giuliano et al., 1988b), but the DNA-binding activity is stimulated by phosphorylation by a casein kinase II activity (Klimczak et al., 1992). cDNAs encoding GBF have been isolated and indicate it is a bZIP protein (Schindler et al., 1992). The protein factor ABF1 which binds to a G box-like sequence in a pea *Lhcb1* gene, probably corresponds to GBF (Argüello et al., 1992).

The protein factors AT-1 and CA-1, which bind to AT-1 and CA-1 sequences were both modulated by phosphorylation and dephosphorylation (Datta and Cashmore, 1989; Sun et al., 1993). AT-1 is active in the non-phosphorylated form and loses all binding activity on phosphorylation (Datta and Cashmore, 1989). DNA-protein complexes produced by CA-1 show different electrophoretic mobilities when CA-1 is in the phosphorylated or dephosphorylated forms (Sun et al., 1993). Reversible phosphorylation appears to modulate the DNA-binding activity of several of the *trans*-acting factors involved in the regulated expression of nuclear photosynthesis genes, and this probably is the point of interaction with the signal transduction pathways relaying information about environmental factors such as light. Sheen (1993) showed that the light-induced expression of several photosynthesis genes is inhibited by protein phosphatase inhibitors such as okadaic acid and calyculin.

C. Post-transcriptional Regulation

There are many steps at which gene expression may be regulated after the initiation of transcription. However there are few well-characterized examples of post-transcriptional regulations of nuclear genes encoding polypeptides required for the light reaction of photosynthesis. The best characterized example appears to be the light regulation of the pea *PetF* gene. Promoter analysis in transgenic tobacco demonstrated that the promoter region was not able to direct light-regulated expression of the GUS reporter gene. The sequences needed for light-regulated expression were present in the transcribed region of the gene (Elliott et al., 1989; Dickey et al., 1992) and a functioning translation start codon was necessary (Dickey et al., 1994). Light regulation of the pea *PetF* gene may be exerted by differential stability of the ferredoxin mRNA in the light and the dark. A similar explanation is also suggested for the light regulation of the pea *PetE* gene (C.A. Helliwell and J.C. Gray, unpublished). Promoter regions of the gene are not able to direct full light-regulated expression in transgenic tobacco (Pwee and Gray, 1993) whereas the transcribed region can direct full light regulation (C.A. Helliwell and J.C. Gray, unpublished). Changes in RNA stability also accounted for the increased expression of the *Lemna gibba Lhcb1* genes in the presence of cytokinin (Tobin and Turkaly, 1982).

IV. Coordination of Expression of Photosynthesis Genes

The construction and maintenance of the photosynthetic apparatus requires the coordinated expression of genes in the nuclear and chloroplast genomes. However, the coordination is relatively loose, with complexes being assembled from the available subunits and excess subunits being degraded in the chloroplasts (Leto et al., 1985; Bruce and Malkin, 1991). The expression of all the nuclear photosynthesis genes is not strictly coordinated temporally, either during greening of dark-grown seedlings or during normal development. This suggests there is little feedback control among individual nuclear genes to coordinate their expression. There does however appear to be feedback control from the chloroplast to nuclear genes encoding photosynthesis components. This may operate to prevent the expression of nuclear genes in the absence of functional chloroplasts.

A. Chloroplast Control of Nuclear Gene Expression

As outlined above (Section III.A.1), there is a strong correlation between the expression of nuclear photosynthesis genes and the presence of chloroplasts. There is now a large body of experimental evidence indicating that functioning chloroplasts are required for the high-level expression of nuclear genes for proteins required for photosynthetic processes. The earlier work has been reviewed by Taylor (1989a) and Oelmüller (1989).

In many mutant plants with defective chloroplasts the nuclear genes encoding photosynthesis-related proteins are not expressed (Taylor, 1989a). This is most clearly seen in carotenoid-deficient mutants of maize and barley grown under high light intensities, a treatment that prevents chloroplast formation (Harpster et al., 1984; Mayfield and Taylor, 1984a; Batschauer et al., 1986; Taylor, 1989a). A similar phenotype can be induced by treating wild-type plants with the carotenoid synthesis inhibitor norflurazon in the light (Reiss et al., 1983; Batschauer et al., 1986; Oelmüller and Mohr, 1986; Mayfield and Taylor, 1987). Norflurazon-treated plants showed decreased levels of transcripts of a large number of nuclear genes for photosynthesis proteins, including *PsbO, PsbP, Lhcb1, PetC, PetE* and *PetF*, but transcripts for most cytosolic proteins were unaffected (Burgess and Taylor, 1987,1988; Sagar et al., 1988). Experiments with nuclei isolated from norflurazon-treated plants demonstrated that the effect was on the rate of transcription of the nuclear genes (Ernst and Schefbeck, 1988; Sagar et al., 1988). This led to the suggestion that functional chloroplasts are required for the expression of these nuclear genes (Oelmüller, 1989; Taylor, 1989a). Studies with promoters of photosynthesis genes fused to reporter genes, such as those encoding neomycin phosphotransferase or β-glucuronidase (GUS), in transgenic plants have established the requirement for functional chloroplasts for the expression of these genes (Simpson et al., 1986b; Stockhaus et al., 1989b; Oelmüller et al., 1993; Vorst et al., 1993a; Bolle et al., 1994b; Lübberstedt et al., 1994b). The expression of chimeric genes containing the *PsbR, Lhcb1, PsaF, PetE, PetF, PetH* and *AtpC* promoters was inhibited by norflurazon in high light, even though the reporter proteins remained in the cytosol and were not targeted to the chloroplasts. These experiments provide additional evidence for an effect on transcription. The most widely accepted suggestion is that functional chloroplasts produce a positive signal, the 'plastidic factor', required for the expression of nuclear genes for photosynthesis proteins, and that production of this factor is prevented by photo-oxidation of chloroplasts (Oelmüller, 1989; Taylor, 1989a). An alternative explanation, that photo-oxidation of chloroplasts prevents the import of precursor proteins from the cytosol and this leads to feedback inhibition of nuclear gene expression (Oelmüller, 1989; Rajasekhar, 1991), is not supported by a recent study that shows no effect of norflurazon treatment on chloroplast protein import (Bolle et al., 1994b). However, these results conflict with a previous study which showed decreased protein import by chloroplasts from norflurazon-treated leaves (Dahlin, 1993).

The identity of the 'plastidic factor' is unknown, but experiments with the inhibitor tagetitoxin indicate that chloroplast transcription is necessary for expression of photosynthesis-related nuclear genes (Mathews and Durbin, 1990; Rapp and Mullet, 1991). Tagetitoxin is a specific inhibitor of chloroplast RNA polymerase in higher plants and its application to wheat and barley seedlings resulted in inhibition of accumulation of *Lhcb1* transcripts (Rapp and Mullet, 1991). There is conflicting evidence for a role for chloroplast translation in the formation of the 'plastidic factor'. Nuclear-encoded photosynthesis proteins are present in several plants with defects in chloroplast ribosomes, due either to mutations or to growth at elevated temperatures (Feierabend and Schrader-Reichhardt, 1976; Bradbeer et al., 1979). This indicated that chloroplast protein synthesis is not necessary for the production of the 'plastidic factor'. However, chloramphenicol, a chloroplast translation inhibitor, prevented the expression of *Lhcb1* genes when presented during the early stages, but not the later stages, of seedling development in mustard (Oelmüller and Mohr, 1986). Similar results have been obtained with lincomycin and erythromycin, inhibitors of chloroplast protein synthesis, in transgenic tobacco seedlings containing chimeric genes with the pea *PetE* and *PetH* promoters (R. Sornarajah and J.C. Gray, unpublished). Inhibition of chloroplast protein synthesis during the first 2-3 days of seedling growth, but not subsequently, prevented the normal light-induced expression from these promoters in transgenic plants. It has been suggested that this requirement for chloroplast protein synthesis at an early stage of development may be necessary for the formation of the chloroplast RNA polymerase, and that once the polymerase is present further protein synthesis is not necessary for the production of the 'plastidic factor' (Rapp and Mullet, 1991). There is currently no evidence for the export of RNA or protein from the chloroplast. This suggests that the transcripts giving rise to the 'plastidic factor' act within the chloroplast.

The involvement of the chloroplast glutamyl-tRNA in the synthesis of 5-aminolaevulinate, a key intermediate in chlorophyll synthesis (Schön et al., 1986), led to suggestions that glutamyl-tRNA was the essential transcript produced by the chloroplasts and that intermediates of the chlorophyll biosynthetic pathway may function as the 'plastidic factor' (Oelmüller, 1989; Susek and Chory, 1992). However recent studies with gabaculine, a potent inhibitor of 5-aminolaevulinate synthesis, appear to rule out the involvement of intermediates of the chlorophyll biosynthetic pathway (R. Sornarajah and J.C. Gray, unpublished). Gabaculine had no effect on expression from the *PetE* and *PetH* promoters in transgenic tobacco seedlings even though it inhibited the synthesis of chlorophyll in these plants. Other chloroplast transcripts, which may be involved in the production of the 'plastidic factor', have not yet been identified. It will be important to establish if the 'plastidic factor' requires the transcription of a specific chloroplast gene or all (or a subset of) chloroplast genes.

An alternative, genetic, approach may provide important information for identification of the 'plastidic factor' or the signal transduction pathway from the chloroplast to the nucleus. Susek *et al.* (1993) have identified at least three *Arabidopsis* nuclear genes necessary for coupling the expression of nuclear genes encoding photosynthesis-related proteins to the functional state of the chloroplasts. Homozygous recessive *gun* (*g*enomes *un*coupled) mutations allow expression from the *Lhcb1* promoter in plants treated with norflurazon or chloramphenicol, or in white sectors of variegated leaves produced by the chloroplast mutator (*chm*) mutation (Susek et al., 1993). The recessive nature of the *gun* mutations suggests that the signal transduction pathway normally functions to repress nuclear gene expression in the absence of chloroplast function.

Promoter analysis in transgenic plants has started

to define elements required for chloroplast control of nuclear gene expression. A far-upstream element is required for norflurazon-sensitive expression of the *Arabidopsis PetE* gene (Vorst et al., 1993a; Fisscher et al., 1994), whereas TATA-proximal elements in the pea *PetE* and *PetH* promoters appear to be involved in chloroplast control (R. Sornarajah and JC Gray, unpublished). Gel retardation assays with the TATA-proximal *PetE* and *PetH* elements and nuclear extracts of green and norflurazon-treated leaves indicate binding activity in norflurazon-treated leaves, not present in green leaves (R. Sornarajah and JC Gray, unpublished). This binding activity may represent a transcriptional repressor that prevents gene expression in the absence of the 'plastidic factor'.

V. Conclusions

Although considerable progress has been made with the isolation and characterization of cDNA and genomic clones encoding most of the components required for the light reactions of photosynthesis, much work remains for us to understand the mechanisms by which these genes are regulated. Regulation of *Lhcb1* genes has been most intensively studied but we are seemingly far from understanding how any of the regulatory mechanisms work in molecular terms. The differences in promoter organization in different genes and in different organisms provide an enormous wealth of material for future studies of gene regulation. It is to be hoped that some unifying principles will emerge to simplify the analysis of expression of all the genes encoding proteins required for the light reactions of photosynthesis.

Acknowledgments

The unpublished work from the author's laboratory cited in this review has been supported by research grants from the Science and Engineering Research Council and the Agricultural and Food Research Council of the UK.

References

Adamska I, Scheel B and Kloppstech K (1991) Circadian oscillations of nuclear-encoded chloroplast proteins in pea (*Pisum sativum*). Plant Mol Biol 17: 1055–1065

Aoki H, Doyama N and Ida S (1994) Sequence of a cDNA encoding rice (*Oryza sativa* L.) leaf ferredoxin-NADP$^+$ reductase. Plant Physiol 104: 1473–1474

Argüello G, Garcia-Hernández E, Sánchez M, Gariglio P, Herrrera-Estrella L and Simpson J (1992) Characterization of DNA sequences that mediate nuclear protein binding to the regulatory region of *Pisum sativum* (pea) chlorophyll *a/b*-binding protein gene *AB80*: Identification of a repeated heptamer motif. Plant J 2: 301–309

Barkardottir RB, Jenson BF, Kreiberg JD, Nielsen PS and Gausing K (1987) Expression of selected nuclear genes during leaf development in barley. Dev Genet 8: 495–511

Bartholomew DM, Bartley GE and Scolnik PA (1991) Abscisic acid control of *rbcS* and *cab* transcription in tomato leaves. Plant Physiol 96: 291–296

Batschauer A, Mösinger E, Kreuz K, Dörr I and Apel K (1986) The implication of a plastid-derived factor in the transcriptional control of nuclear genes encoding the light-harvesting chlorophyll *a/b* protein. Eur J Biochem 154: 625–634

Bichler J and Herrmann RG (1990) Analysis of the promoters of the single-copy genes for plastocyanin and subunit δ of the chloroplast ATP synthase from spinach. Eur J Biochem 190: 415–426

Bolle C, Sopory S, Lübberstedt T, Herrmann RG and Oelmüller R (1994a) Segments encoding 5'-untranslated leaders of genes for thylakoid proteins contain *cis*-elements essential for transcription. Plant J 6: 513–523

Bolle C, Sopory S, Lübberstedt T, Klösgen RB, Herrmann RG and Oelmüller R (1994b) The role of plastids in the expression of nuclear genes for thylakoid proteins studied with chimeric β-glucuronidase gene fusions. Plant Physiol 105: 1355–1364

Bradbeer JW, Atkinson YE, Börner T and Hageman R (1979). Cytoplasmic synthesis of plastid polypeptides may be controlled by plastid-synthesized RNA. Nature 279: 816–817

Brandt J, Nielsen VS, Thordal-Christiansen H, Simpson D and Okkels JS (1992) A barley cDNA encoding a type III chlorophyll *a/b*-binding protein of the light-harvesting complex II. Plant Mol Biol 19: 699–703

Bringloe DH, Dyer TA and Gray JC (1995) Developmental, circadian and light regulation of wheat ferredoxin gene expression. Plant Mol Biol 27: 293–306

Broglie R, Bellemare G, Bartlett S, Chua N-H and Cashmore AR (1981) Cloned DNA sequence complementary to mRNAs encoding precursors to the small subunit of ribulose-1,5-bisphosphate carboxylase and a chlorophyll *a/b* binding protein. Proc Natl Acad Sci USA 78: 7304–7308

Bruce BD and Malkin R (1991) Biosynthesis of the chloroplast cytochrome b_6f complex—studies in a photosynthetic mutant of *Lemna*. Plant Cell 3: 203–211

Brusslan JA and Tobin EM (1992) Light-independent developmental regulation of *cab* gene expression in *Arabidopsis thaliana* seedlings. Proc Natl Acad Sci USA 89: 7791–7795

Burgess DG and Taylor WC (1987) Chloroplast photooxidation affects the accumulation of cytosolic mRNAs encoding chloroplast proteins in maize. Planta 170: 520–527

Burgess DG and Taylor WC (1988) The chloroplast affects the transcription of a nuclear gene family. Mol Gen Genet 214: 89–96

Cashmore AR (1984) Structure and expression of a pea nuclear gene encoding a chlorophyll *a/b*-binding polypeptide. Proc

Natl Acad Sci USA 81: 2960–2964
Caspar T and Quail PH (1993) Promoter and leader regions involved in the expression of the *Arabidopsis* ferredoxin A gene. Plant J 3: 161–174
Castresana C, Staneloni R, Malik VR and Cashmore AR (1987) Molecular characterization of two clusters of genes encoding the type I CAB polypeptides of PS II in *Nicotiana plumbaginifolia*. Plant Mol Biol 10: 117–126
Castresana C, Garcia-Luque I, Alonso E, Malik VR and Cashmore AR (1988) Both positive and negative elements mediate expression of a photoregulated CAB gene from *Nicotiana plumbaginifolia*. EMBO J 7: 1929–1936
Chang YC and Walling LL (1991) Abscisic acid negatively regulates expression of chlorophyll *a/b* binding protein genes during soybean embryogeny. Plant Physiol 97: 1260–1264
Chang YC and Walling LL (1992) Spatial and temporal expression of Cab mRNA in cotyledons of developing soybean seedlings. Planta 186: 262–272
Dahlin C (1993) Import of nuclear-encoded proteins into carotenoid-deficient young etioplasts. Physiol Plant 87: 410–416
Datta N and Cashmore AR (1989) Binding of a pea nuclear protein to promoters of certain photoregulated genes is modulated by phosphorylation. Plant Cell 1: 1069–1077
Dean C and Leech RM (1982) Genomic expression during normal leaf development. I. Cellular and chloroplast numbers and DNA, RNA, and protein levels in tissues of different ages within a seven-day-old wheat leaf. Plant Physiol 69: 904–910
Dekeyser RA, Claes B, De Rycke RMU, Habets ME, Van Montagu M and Caplan AB (1990) Transient gene expression in intact and organized rice tissues. Plant Cell 2: 591–602
de Pater S, Hensgens LAM and Schilperoort RA (1990) Structure and expression of a light-inducible shoot-specific rice gene. Plant Mol Biol 15: 399–406
Detlefsky DJ, Pichersky E and Pecoraro V (1989) Pre-plastocyanin from *Lycopersicon esculentum*. Nucleic Acids Res 17: 6414
Dickey L, Gallo-Meagher M and Thompson WF (1992) Light regulatory sequences are located within the 5' portion of the *Fed-1* message sequence. EMBO J 11: 2311–2317
Dickey L, Nguyen T-T, Allen GC and Thompson WF (1994) Light modulation of ferredoxin mRNA abundance requires an open reading frame. Plant Cell 6: 1171–1176
Dobres M, Elliott R, Watson J and Thompson W (1987) A phytochrome regulated transcript encodes ferredoxin I. Plant Mol Biol 8: 53–59
Dunsmuir P, Smith SM and Bedbrook J (1983) The major chlorophyll *a/b* binding protein is composed of several polypeptides encoded by a number of distinct nuclear genes. J Mol Appl Genet 2: 285–300
Dupree P, Pwee K-H and Gray JC (1991) Expression of photosynthesis gene-promoter fusions in leaf epidermal cells of transgenic tobacco plants. Plant J 1: 115–120
Eckes P, Schell J and Willmitzer L (1985) Organ-specific expression of three leaf/stem specific cDNA's from potato is regulated by light and correlated with chloroplast development. Mol Gen Genet 199: 216–224
Eckes P, Rosahl S, Schell J and Willmitzer L (1986) Isolation and characterization of a light-inducible, organ-specific gene from potato and analysis of its expression after tagging and transfer into tobacco and potato shoots. Mol Gen Genet 205: 14–22
Elliott RC, Pedersen TJ, Fristensky B, White MJ, Dickey L and Thompson WF (1989) Characterization of a single copy gene encoding ferredoxin 1 from pea. Plant Cell 1: 681–690
Ernst D and Schefbeck K (1988). Photooxidation of plastids inhibits transcription of nuclear encoded genes in rye (*Secale cereale*). Plant Physiol 88: 255–258
Feierabend J and Schrader-Reichhardt U (1976). Biochemical differentiation of plastids and other organelles in rye leaves with a high-temperature-induced deficiency of plastid ribosomes. Planta 129: 133–145
Fisscher U, Weisbeek P and Smeekens S (1994) Identification of potential regulatory elements in the far-upstream region of the *Arabidopsis thaliana* plastocyanin promoter. Plant Mol Biol 26: 873–886
Flieger K, Tyagi A, Sopory S, Cseplö A, Herrmann RG and Oelmüller R (1993) A 42 bp promoter fragment of the gene for subunit III of Photosystem I (*psaF*) is crucial for its activity. Plant J 4: 9–17
Flores S and Tobin EM (1986) Benzyladenine modulation of the expression of two genes for nuclear-encoded chloroplast proteins in *Lemna gibba*: Apparent post-transcriptional regulation. Planta 168: 340–345
Funk C, Schroeder WP, Green BR, Renger G and Andersson B (1994) The intrinsic 22 kDa protein is a chlorophyll-binding subunit of Photosystem II. FEBS Lett 342: 261–266
Gallo-Meagher M, Sowinski DA and Thompson WF (1992) The pea ferredoxin 1 gene exhibits different light responses in pea and tobacco. Plant Cell 4: 383–388
Gidoni D, Brosio P, Bond-Nutter D, Bedbrook J and Dunsmuir P (1989) Novel *cis*-acting elements in petunia *cab* gene promoters. Mol Gen Genet 215: 337–344
Gil-Gómez G, Marrero PF, Haro D, Ayté J and Hegardt FG (1991) Characterization of the gene encoding the 10 kDa polypeptide of Photosystem II from *Arabidopsis thaliana*. Plant Mol Biol 17: 517–522
Gilmartin P, Sarokin L, Memelink J and Chua N-H (1990) Molecular light switches for plant genes. Plant Cell 2: 369–378
Gilmartin P, Memelink J, Hiratsuka K, Kay SA and Chua NH (1992) Characterization of a gene encoding a DNA binding protein with specificity for a light-responsive element. Plant Cell 4: 839–849
Giuliano G, Hoffman NE, Ko K, Scolnik P and Cashmore AR (1988a) A light-entrained circadian clock controls transcription of several plant genes. EMBO J 7: 3635–3642
Giuliano G, Pichersky E, Malik VS, Timko MP, Scolnik P and Cashmore AR (1988b) An evolutionarily conserved protein binding sequence upstream of a plant light-regulated gene. Proc Natl Acad Sci USA 85: 7089–7093
Görlach J, Schmid J and Amrhein N (1993) The 33 kDa protein of the oxygen-evolving complex: A multigene family in tomato. Plant Cell Physiol 34: 497–501
Gray JC, Kung SD, Wildman SG and Sheen SJ (1974) Origin of *Nicotiana tabacum* L. detected by polypeptide composition of Fraction I protein. Nature 252: 226–227
Green BR and Pichersky E (1993) Nucleotide sequence of an *Arabidopsis thaliana Lhcb*4 gene. Plant Physiol 103: 1451–1452
Green B, Pichersky E and Kloppstech K (1991) Chlorophyll *a/b*-binding proteins: An extended family. Trends Biochem Sci 16: 181–186
Grob U and Stüber K (1987) Discrimination of phytochrome

dependent light inducible from non-light inducible plant genes. Prediction of a common light-responsive element (LRE) in phytochrome dependent light inducible plant genes. Nucleic Acids Res 15: 9957–9973

Ha S-B and An G (1988) Identification of upstream regulatory elements involved in the developmental expression of the *Arabidopsis thaliana cab1* gene. Proc Natl Acad Sci USA 85: 8017–8021

Hallick RB (1989) Proposals for the naming of chloroplast genes. II. Update to the nomenclature of genes for thylakoid membrane polypeptides. Plant Mol Biol Rep 7: 266–275

Harkins KR, Jefferson RA, Kavanagh TA, Bevan MW and Galbraith DW (1990) Expression of photosynthesis-related gene fusions is restricted by cell type in transgenic plants and in transfected protoplasts. Proc Natl Acad Sci USA 87: 816–820

Harpster M, Mayfield SP and Taylor WC (1984) Effects of pigment-deficient mutants on the accumulation of photosynthesis proteins in maize. Plant Mol Biol 3: 59–71

Harter K, Talke-Messerer C, Barz W and Schäfer E (1993) Light- and sucrose-dependent gene expression in photomixotrophic cell suspension cultures and protoplasts of rape (*Brassica napus* L.). Plant J 4: 507–516

Hase T, Kimata Y, Yonekura K, Matsumara T and Sakakibara H (1991) Molecular cloning and differential expression of the maize ferredoxin gene family. Plant Physiol 96: 77–83

Hauge BM, Hanley SM, Cartinhour S, Cherry JM, Goodman HM, Koornneef M, Stam P, Chang C, Kempin S, Medrano L and Meyerowitz EM (1993) An integrated genetic/RFLP map of the *Arabidopsis thaliana* genome. Plant J 3: 745–754

Hermans J, Rother C, Bichler J, Steppuhn J and Herrmann RG (1988) Nucleotide sequence of cDNA clones encoding the complete precursor for subunit delta of thylakoid-located ATP synthase from spinach. Plant Mol Biol 10: 323–330

Herrmann RG, Steppuhn J, Herrmann GS and Nelson N (1993) The nuclear-encoded polypeptide Cfo-II from spinach is a real, ninth subunit of chloroplast ATP synthase. FEBS Lett 326: 192–198

Hoesche JA and Berzborn RJ (1992) Cloning and sequencing of a cDNA for the δ-subunit of photosynthetic ATP-synthase (E.C. 3.6.1.34) from pea (*Pisum sativum*). Biochim Biophys Acta 1171: 201–204

Hoesche JA and Berzborn RJ (1993) Primary structure, deduced from cDNA, secondary structure analysis and conclusions concerning interaction surfaces of the δ subunit of the photosynthetic ATP-synthase (E.C. 3.6.1.34) from millet (*Sorghum bicolor*) and maize (*Zea mays*). Biochim Biophys Acta 1142: 293–305

Hoffman NE, Pichersky E, Malik VS, Castresana C, Ko K, Darr SC and Cashmore AR (1987) A cDNA clone encoding a Photosystem I protein with homology to Photosystem II chlorophyll *a/b*-binding polypeptides. Proc Natl Acad Sci USA 84: 8844–8848

Hua S, Dube SK, Barnett NM and Kung S (1991) Nucleotide sequence of gene *oee2-A* and its cDNA encoding 23 kDa polypeptide of the oxygen-evolving complex of Photosystem II in tobacco. Plant Mol Biol 17: 551–553

Hua S-B, Dube SK and Kung S-D (1993) Molecular evolutionary analysis of the *PsbP* gene family of the Photosystem II oxygen-evolving complex in *Nicotiana*. Genome 36: 483–488

Ikeuchi M, Takio K and Inoue Y (1989) N-terminal sequencing of Photosystem II low-molecular-mass proteins. 5 and 4.1 kDa components of the O_2-evolving core complex from higher plants. FEBS Lett 242: 263–269

Inohara N, Iwamoto A, Moriyama Y, Shimomura S, Maeda M and Futai M (1991) Two genes, *atpC1* and *atpC2*, for the γ subunit of *Arabidopsis thaliana* chloroplast ATP synthase. J Biol Chem 266: 7333–7338

Jang J-C and Sheen J (1994) Sugar sensing in higher plants. Plant Cell 6: 1665–1679

Jansen T, Rother C, Steppuhn J, Reinke H, Beyreuther K, Jansson C, Andersson B and Herrmann RG (1987) Nucleotide sequence of cDNA clones encoding the complete '23kDa' and '16kDa' precursor proteins associated with the photosynthetic oxygen-evolving complex from spinach. FEBS Lett 216: 234–240

Jansen T, Reiländer H, Steppuhn J and Herrmann RG (1988) Analysis of cDNA clones encoding the entire precursor-polypeptide for ferredoxin: $NADP^+$ oxidoreductase from spinach. Curr Genet 13: 517–522

Jansson S and Gustafsson P (1991) Evolutionary conservation of the chlorophyll *a/b*-binding proteins: cDNAs encoding type I, II and III LHC I polypeptides from the gymnosperm Scots pine. Mol Gen Genet 229: 67–76

Jansson S, Pichersky E, Bassi R, Green BR, Ikeuchi M, Melis A, Simpson DJ, Spangfort M, Staehelin LA and Thornber JP (1992) A nomenclature for the genes encoding the chlorophyll *a/b*-binding proteins of higher plants. Plant Mol Biol Rep 10: 242–253

Jefferson RA, Kavanagh TA and Bevan MW (1987) GUS fusions: β-glucuronidase as a sensitive and versatile gene fusion marker in higher plants. EMBO J 6: 3901–3907

Kapozoglou A, Sagliocco F and Dure L (1995) PS II-T, a new nuclear encoded lumenal protein from Photosystem II. Targeting and processing in isolated chloroplasts. J Biol Chem 270: 12197–12202

Karlin-Neumann GA, Kohorn BD, Thornber JP and Tobin EM (1985) A chlorophyll *a/b*-protein encoded by a gene containing an intron with characteristics of a transposable element. J Mol Appl Genet 3: 45–61

Kehoe DM, Degenhardt J, Winicov I and Tobin EM (1994) Two 10-bp regions are critical for phytochrome regulation of a *Lemna gibba Lhcb* gene promoter. Plant Cell 6: 1123–1134

Kellmann JW, Pichersky E and Piechulla B (1990) Analysis of the diurnal expression patterns of the tomato chlorophyll *a/b* binding protein genes. Influence of light and characterization of the gene family. Photochem Photobiol 52: 35–41

Kim S, Sandusky P, Bowlby NR, Aebersold R, Green BR, Vlahakis S, Yocum CF and Pichersky E (1992) Characterization of a spinach *psbS* cDNA encoding the 22 kDa protein of Photosystem II. FEBS Lett 314: 67–71

Kjaerulff S, Andersen B, Nielsen VS, Møller BL and Okkels JS (1993) The PSI-K subunit of Photosystem I from barley (*Hordeum vulgare* L.). Evidence for a gene duplication of an ancestral PSI-G/K gene. J Biol Chem 268: 18912–18916

Klimczak LJ, Schindler U and Cashmore AR (1992) DNA binding activity of the *Arabidopsis* G-box binding factor GBF1 is stimulated by phosphorylation by casein kinase II from broccoli. Plant Cell 4: 87–98

Kloppstech K (1985) Diurnal and circadian rhythmicity in the expression of light-induced plant nuclear messenger RNAs. Planta 165: 502–506

Knight JS and Gray JC (1994) Expression of genes encoding the tobacco chloroplast phosphate translocator is not light-regulated and is repressed by sucrose. Mol Gen Genet 242: 586–594

Knoetzel J and Simpson D (1993) The primary structure of a cDNA for *psaN*, encoding an extrinsic lumenal polypeptide of barley Photosystem I. Plant Mol Biol 22: 337–345

Krapp A, Hofmann B, Schaefer C and Stitt M (1993) Regulation of the expression of *rbcS* and other photosynthesis genes by carbohydrates: A mechanism for the "sink regulation" of photosynthesis? Plant J 3: 817–828

Kuhlemeier C, Fluhr R, Green P and Chua N-H (1987) Sequences in the pea *rbcS-3A* gene have homology to constitutive mammalian enhancers but function as negative regulatory elements. Genes Dev 1: 247–255

Lagoutte B (1988) Cloning and sequencing of spinach cDNA clones encoding the 20 kDa PSI polypeptide. FEBS Lett 232: 275–280

Lam E and Chua N-H (1989a) Light to dark transition modulates the phase of antenna chlorophyll protein gene expression. J Biol Chem 264: 20175–20176

Lam E and Chua N-H (1989b) ASF-2: A factor that binds to the cauliflower mosaic virus 35S promoter and a conserved GATA motif in *Cab* promoters. Plant Cell 1: 1147–1156

Lam E and Chua N-H (1990) GT-1 binding site confers light responsive expression in transgenic tobacco. Science 248: 471–474

Lamppa GK, Morelli G and Chua N-H (1985) Structure and developmental regulation of a wheat gene encoding the major chlorophyll *a/b*-binding polypeptide. Mol Cell Biol 5: 1370–1378

Larsson KH, Napier JA and Gray JC (1992) Import and processing of the precursor form of the gamma subunit of the chloroplast ATP synthase from tobacco. Plant Mol Biol 19: 343–349

Last DI and Gray JC (1989) Plastocyanin is encoded by a single-copy gene in the pea haploid genome. Plant Mol Biol 12: 655–666

Lautner A, Klein R, Ljungberg U, Reiländer H, Bartling D, Andersson B, Reinke H, Beyreuther K and Herrmann RG (1988) Nucleotide sequence of cDNA clones encoding the complete precursor for the '10-kDa' polypeptide of Photosystem II from spinach. J Biol Chem 263: 10077–10081

Leto KJ, Bell E and McIntosh L (1985) Nuclear mutation leads to an accelerated turnover of chloroplast-encoded 48 kd and 34.5 kd polypeptides in thylakoids lacking Photosystem II. EMBO J 4: 1645–1653

Leutwiler LS, Meyerowitz EM and Tobin EM (1986) Structure and expression of three light-harvesting chlorophyll *a/b*-binding protein genes in *Arabidopsis thaliana*. Nucleic Acids Res 14: 4051–4064

Ljungberg U, Akerlund H-E and Andersson B (1984) The release of a 10-kDa polypeptide from everted Photosystem II thylakoid membranes by alkaline Tris. FEBS Lett 175: 255–258

Luan S and Bogorad L (1992) A rice *cab* gene promoter contains separate *cis*-acting elements that regulate expression in dicot and monocot plants. Plant Cell 4: 971–981

Lübberstedt T, Bolle CEH, Sopory S, Flieger K, Herrmann RG and Oelmüller R (1994a) Promoters from genes for plastid proteins possess regions with different sensitivities toward red and blue light. Plant Physiol 104: 997–1006

Lübberstedt T, Oelmüller R, Wanner G and Herrmann RG (1994b). Interacting *cis* elements in the plastocyanin promoter from spinach ensure regulated high-level expression. Mol Gen Genet 242: 602–613

Luehrsen KR, Taha S and Walbot V (1994) Nuclear pre-mRNA processing in higher plants. Prog Nucleic Acid Res Mol Biol 47: 149–193

Madueño F, Napier JA, Cejudo FJ and Gray JC (1992) Import and processing of the precursor of the Rieske FeS protein of tobacco chloroplasts. Plant Mol Biol 20: 289–299

Martineau B and Taylor WC (1985) Photosynthetic gene expression and cellular differentiation in developing maize leaves. Plant Physiol 78: 399–404

Mason JG and Whitfeld PR (1990) The γ-subunit of spinach chloroplast ATP synthase: Isolation and characterization of cDNA and genomic clones. Plant Mol Biol 14: 1007–1018

Mathews DE and Durbin RD (1990). Tagetitoxin inhibits RNA synthesis directed by RNA polymerases from chloroplasts and *Escherichia coli.* J Biol Chem 265: 493–498

Mayfield SP and Taylor WC (1984a) Carotenoid-deficient maize seedlings fail to accumulate light-harvesting chlorophyll *a/b* binding protein (LHCP) mRNA. Eur J Biochem 144: 79–84

Mayfield SP and Taylor WC (1984b) The appearance of photosynthetic proteins in developing maize leaves. Planta 161: 481–486

Mayfield SP and Taylor WC (1987) Chloroplast photooxidation inhibits the expression of a set of nuclear genes. Mol Gen Genet 208: 309–314

McGrath JM, Terzaghi WB, Sridar P, Cashmore AR and Pichersky E (1992) Sequences of the fourth and fifth Photosystem II type I chlorophyll *a/b*-binding protein genes of *Arabidopsis thaliana* and evidence for the presence of a full complement of the extended CAB gene family. Plant Mol Biol 19: 725–733

Merkle T, Krenz M, Wenng A and Schäfer E (1990) Nucleotide sequence and deduced amino acid sequence of a gene encoding the 23 kDa polypeptide of the oxygen-evolving complex from mustard (*Sinapis alba* L.). Plant Mol Biol 14: 889–890

Meyer H, Thienel U and Piechulla B (1989) Molecular characterization of the diurnal/circadian expression of the chlorophyll *a/b*-binding proteins in leaves of tomato and other dicotyledonous and monocotyledonous plant species. Planta 180: 5–15

Michalowski CB, Schmitt JM and Bohnert HJ (1989) Expression during salt stress and nucleotide sequence of cDNA for ferredoxin-NADP$^+$ oxidoreductase from *Mesembryanthemum crystallinum.* Plant Physiol 89: 817–822

Miki J, Maeda M, Mukohata Y and Futai M (1988) The γ-subunit of ATP synthase from spinach chloroplasts. FEBS Lett 232: 221–226

Millar AJ and Kay SA (1991) Circadian control of cab gene transcription and mRNA accumulation in *Arabidopsis.* Plant Cell 3: 541–550

Millar AJ, Short SR, Chua N-H and Kay SA (1992) A novel circadian phenotype based on firefly luciferase expression in transgenic plants. Plant Cell 4: 1075–1087

Morishige D and Thornber JP (1992) Identification and analysis of a barley cDNA clone encoding the 31-kilodalton LHC IIa (CP29) apoprotein of the light-harvesting antenna complex of Photosystem II. Plant Physiol 98: 238–245

Münch S, Ljungberg U, Steppuhn J, Schneiderbauer A, Nechushtai R, Beyreuther K and Herrmann RG (1988) Nucleotide sequences of cDNAs encoding the entire precursor polypeptides for subunits II and III of the Photosystem I reaction center from spinach. Curr Genet 14: 511–518

Nagy F, Boutry M, Hsu M-Y, Wong M and Chua N-H (1987) The 5' proximal region of the wheat *Cab-1* gene contains a 268-bp enhancer-like sequence for phytochrome response. EMBO J 6: 2537–2542

Nagy F, Kay SA and Chua N-H (1988) A circadian clock regulates transcription of the wheat *Cab-1* gene. Genes Dev 2: 376–382

Nakamura M and Obokata J (1994) Organization of the *PsaH* gene family of Photosystem I in *Nicotiana sylvestris*. Plant Cell Physiol 35: 297–302

Napier JA, Höglund A-S, Plant AL and Gray JC (1992a) Chloroplast import of the precursor of the gamma subunit of pea chloroplast ATP synthase. Plant Mol Biol 20: 737–741

Napier JA, Larsson KH, Madueño F and Gray JC (1992b) Import and processing of the precursor of the delta subunit of tobacco chloroplast ATP synthase. Plant Mol Biol 20: 549–554

Newman BJ and Gray JC (1988) Characterization of a full-length cDNA clone for pea ferredoxin-NADP$^+$ reductase. Plant Mol Biol 10: 511–520

Nielsen PS and Gausing K (1987) The precursor of barley plastocyanin. Sequence of cDNA clones and gene expression in different tissues. FEBS Lett 225: 159–162

Nielsen PS and Gausing K (1993) In vitro binding of nuclear proteins to the barley plastocyanin gene promoter region. Eur J Biochem 217: 97–104

Obokata J, Mikami K, Hayashida N, Nakamura M and Sugiura M (1993) Molecular heterogeneity of Photosystem I. *PsaD*, *PsaE*, *PsaF*, *PsaH*, and *PsaL* are all present in isoforms in *Nicotiana* spp. Plant Physiol 102: 1259–1267

Oelmüller R (1989) Photooxidative destruction of chloroplasts and its effects on nuclear gene expression and extraplastidic enzyme levels. Photochem Photobiol 49: 229–239

Oelmüller R and Mohr H (1986) Photooxidative destruction of chloroplasts and its consequences for expression of nuclear genes. Planta 167: 106–113

Oelmüller R, Bolle C, Tyagi AK, Niekrawietz N, Breit S and Herrmann RG (1993) Characterization of the promoter from the single-copy gene encoding ferredoxin-NADP$^+$-oxido-reductase from spinach. Mol Gen Genet 237: 261–272

Okkels JS, Scheller HV, Jepsen LB and Møller BL (1989) A cDNA clone encoding the precursor for a 10.2 kDa Photosystem I polypeptide of barley. FEBS Lett 250: 575–579

Okkels JS, Scheller HV, Svendsen I and Møller BL (1991) Isolation and characterization of a cDNA clone encoding an 18 kDa hydrophobic Photosystem I subunit (PSI-L) in barley (*Hordeum vulgare* L.). J Biol Chem 266: 6767–6773

Okkels JS, Nielsen VS, Scheller HV and Møller BL (1992) A cDNA clone from barley encoding the precursor for the Photosystem I polypeptide PSI-G: Sequence similarity to PSI-K. Plant Mol Biol 18: 989–994

Palomares R, Herrmann RG and Oelmüller R (1991) Different blue-light requirement for the accumulation of transcripts from nuclear genes for thylakoid proteins in *Nicotiana tabacum* and *Lycopersicon esculentum*. J Photochem Photobiol B: Biol 11: 151–162

Perisic O and Lam E (1992) A tobacco DNA binding protein that interacts with a light-responsive box II element. Plant Cell 4: 831–838

Pessino S, Caelles C, Puigdomenech P and Vallejos RH (1994) Structure and characterization of the gene encoding the reductase binding protein from *Zea mays* L. Gene 147: 205–208

Pichersky E, Bernatzky R, Tanksley SD, Breidenbach RB, Kausch AP and Cashmore AR (1985) Molecular characterization and genetic mapping of two clusters of genes encoding chlorophyll *a/b*-binding proteins in *Lycopersicon esculentum* (tomato). Gene 40: 247–258

Pichersky E, Tanksley SD, Piechulla B, Stayton MM and Dunsmuir P (1988) Nucleotide sequence and chromosomal location of *Cab-7*, the tomato gene encoding the type II chlorophyll *a/b*-binding polypeptide of Photosystem I. Plant Mol Biol 11: 69–71

Pichersky E, Brock TG, Nguyen D, Hoffman NE, Piechulla B, Tanksley SD and Green BR (1989) A new member of the CAB gene family: Structure, expression and the chromosomal location of *Cab-8*, the tomato gene encoding the type III chlorophyll *a/b*-binding polypeptide of Photosystem I. Plant Mol Biol 12: 257–270

Pichersky E, Subramanian R, White MJ, Reid J, Aebersold R and Green BR (1991) Chlorophyll *a/b* binding polypeptides of CP29, the internal chlorophyll *a/b* complex of PS II: Characterization of the tomato gene encoding the 26 kDa (type I) polypeptide, and evidence for a second CP29 polypeptide. Mol Gen Genet 227: 277–284

Piechulla B (1988) Plastid and nuclear mRNA fluctuations in tomato leaves - diurnal and circadian rhythms during extended dark and light periods. Plant Mol Biol 11: 345–353

Piechulla B (1989) Changes of the diurnal and circadian (endogenous) mRNA oscillations of the chlorophyll *a/b* binding protein in tomato leaves during altered day/night (light/dark) regimes. Plant Mol Biol 12: 317–327

Polans NO, Weeden NF and Thompson WF (1985) Inheritance, organization, and mapping of *rbcS* and *cab* multigene families in pea. Proc Natl Acad Sci USA 82: 5083–5087

Preiss S, Peter GF, Anandan S and Thornber JP (1993) The multiple pigment-proteins of the Photosystem I antenna. Photochem Photobiol 57: 152–157

Pwee K-H and Gray JC (1993) The pea plastocyanin promoter directs cell-specific but not full light-regulated expression in transgenic tobacco plants. Plant J 3: 437–449

Rajasekhar VK (1991). Regulation of nuclear gene expression for plastidogenesis as affected by developmental stage of plastids. Biochem Physiol Pflanzen 187: 257–271

Rapp JC and Mullet JE (1991). Chloroplast transcription is required to express the nuclear genes *rbcS* and *cab*. Plastid DNA copy number is regulated independently. Plant Mol Biol 17: 813–823

Reiss T, Bergfeld R, Link G, Thien W and Mohr H (1983) Photooxidative destruction of chloroplasts and its consequences for cytosolic enzyme levels and plant development. Planta 159: 518–528

Rother C, Jansen T, Tyagi A, Tittgen J and Herrmann RG (1986) Plastocyanin is encoded by an uninterrupted nuclear gene in spinach. Curr Genet 11: 171–176

Sagar AD, Horwitz BA, Elliott RC, Thompson WF and Briggs WR (1988). Light effects on several chloroplast components in norflurazon-treated pea seedlings. Plant Physiol 88: 340–347

Sagliocco F, Kapazoglou A and Dure L (1992) Sequence of *cab*-151, a gene encoding a Photosystem II type II chlorophyll *a/b*-binding protein in cotton. Plant Mol Biol 18: 841–842

Salter AH, Newman BJ, Napier JA and Gray JC (1992) Import of the precursor of the chloroplast Rieske iron-sulfur protein by pea chloroplasts. Plant Mol Biol 20: 569–574

Schindler U and Cashmore AR (1990) Photoregulated gene expression may involve ubiquitous DNA binding proteins. EMBO J 9: 3415–3427

Schindler U, Terzaghi W, Beckmann H, Kadesch T and Cashmore AR (1992) DNA binding site preferences and transcriptional activation properties of the *Arabidopsis* transcription factor GBF1. EMBO J 11: 1275–1289

Schmidt CL and Malkin R (1993). Low molecular weight subunits associated with the cytochrome b_6f complexes from spinach and *Chlamydomonas reinhardtii.* Photosynth Res 38: 73–81

Schön A, Krupp G, Berry-Lowe S, Kannangara G and Söll D (1986). The RNA required in the first step of chlorophyll biosynthesis is a chloroplast glutamate tRNA. Nature 322: 281–284

Schwartz E and Pichersky E (1990) Sequence of two new tomato nuclear genes encoding chlorophyll a/b-binding proteins of *CP24*, a PS II antenna component. Plant Mol Biol 15: 157–160

Schwartz E, Shen D, Aebersold R, McGrath JM, Pichersky E and Green BR (1991a) Nucleotide sequence and chromosomal location of Cab11 and Cab12, the genes for the fourth polypeptide of the Photosystem I light-harvesting antenna (LHCI). FEBS Lett 280: 229–234

Schwartz E, Stasys R, Aebersold R, McGrath JM, Green BR and Pichersky E (1991b) Sequence of a tomato gene encoding a third type of LHCII chlorophyll *a/b*-binding polypeptide. Plant Mol Biol 17: 923–925

Sheen J (1990) Metabolic repression of transcription in higher plants. Plant Cell 2: 1027–1038

Sheen J (1993) Protein phosphatase activity is required for light-inducible gene expression in maize. EMBO J 4: 1645–1653

Simpson J, Schell J, Van Montagu M and Herrera-Estrella L (1986a) Light-inducible and tissue-specific pea *lhcp* gene expression involves an upstream element combining enhancer- and silencer-like properties. Nature 323: 551–554

Simpson J, van Montagu M and Herrera-Estrella L (1986b) Photosynthesis-associated gene families: Differences in response to tissue-specific and environmental factors. Science 233: 34–38

Smeekens S, de Groot M, van Binsbergen J and Weisbeek P (1985a) Sequence of the precursor of the chloroplast thylakoid lumen protein, plastocyanin. Nature 317: 456–458

Smeekens S, van Binsbergen J and Weisbeek P (1985b) The plant ferredoxin precursor: Nucleotide sequence of a full length cDNA clone. Nucleic Acids Res 13: 3179–3194

Spangfort M, Larsson UK, Ljungberg U, Ryberg M and Andersson B (1990) The 20 kDa apo-polypeptide of the chlorophyll *a/b* protein complex CP24—characterization and complete primary amino acid sequence. In: Baltscheffsky M (ed) Current Research in Photosynthesis, Vol 2, pp 253–256. Kluwer Academic Publishers, Dordrecht

Stayton MM, Brosio P and Dunsmuir P (1987) Characterization of a full-length petunia cDNA encoding a polypeptide of the light-harvesting complex associated with Photosystem I. Plant Mol Biol 10: 127–137

Steppuhn J, Rother C, Hermans J, Jansen T, Salnikow J, Hauska G and Herrmann RG (1987) The complete amino-acid sequence of the Rieske FeS-precursor protein from spinach chloroplasts deduced from cDNA analysis. Mol Gen Genet 210: 171–177

Steppuhn J, Hermans J, Nechushtai R, Ljungberg U, Thümmler F, Lottspeich F and Herrmann RG (1988) Nucleotide sequence of cDNA clones encoding the entire precursor polypeptides for subunits IV and V of the Photosystem I reaction center from spinach. FEBS Lett 237: 218–224

Stockhaus J, Eckes P, Rocha-Sosa M, Schell J and Willmitzer L (1987) Analysis of cis-active sequences involved in the leaf-specific expression of a potato gene in transgenic plants. Proc Natl Acad Sci USA 84: 7943–7947

Stockhaus J, Schell J and Willmitzer L (1989a) Identification of enhancer elements in the upstream region of the nuclear photosynthetic gene *ST-LS1*. Plant Cell 1: 805–813

Stockhaus J, Schell J and Willmitzer L (1989b) Correlation of the expression of the nuclear photosynthetic gene ST-LS1 with the presence of chloroplasts. EMBO J 8: 2445–2451

Sun L, Doxsee RA, Harel E and Tobin EM (1993) CA-1, a novel phosphoprotein, interacts with the promoter of the *cab*140 gene in Arabidopsis and is undetectable in *det*1 mutant seedlings. Plant Cell 5: 109–121

Susek RE and Chory J (1992). A tale of two genomes: Role of a chloroplast signal in coordinating nuclear and plastid genome expression. Austr J Plant Physiol 19: 387–399

Susek RE, Ausubel FM and Chory J (1993). Signal transduction mutants of Arabidopsis uncouple nuclear *CAB* and *RBCS* gene expression from chloroplast development. Cell 74: 787–799

Taylor WC (1989a) Regulatory interactions between nuclear and plastid genomes. Ann Rev Plant Physiol Plant Mol Biol 40: 211–233

Taylor WC (1989b) Transcriptional regulation by a circadian rhythm. Plant Cell 1: 259–264

Teyssendier de la Serve B, Axelos M and Péaud-Lenoël C (1985) Cytokinin modulates the expression of genes encoding the protein of the light-harvesting chlorophyll *a/b* complex. Plant Mol Biol 5: 155–163

Thompson WF and White MJ (1991) Physiological and molecular studies of light-regulated nuclear genes in higher plants. Ann Rev Plant Physiol Plant Mol Biol 42: 423–466

Tittgen J, Hermans J, Steppuhn J, Jansen T, Jansson C, Andersson B, Nechushtai R, Nelson N and Herrmann RG (1986) Isolation of cDNA clones for fourteen nuclear-encoded thylakoid membrane proteins. Mol Gen Genet 204: 258–265

Tobin EM and Kehoe DM (1994) Phytochrome regulated gene expression. Seminars Cell Biol 5: 335–346

Tobin EM and Turkaly A (1982) Kinetin affects rates of degradation of mRNAs encoding two major chloroplast proteins in *Lemna gibba* L. G-3. J Plant Growth Regul 1: 3–13

Tyagi A, Hermans J, Steppuhn J, Jansson C, Vater F and Herrmann RG (1987) Nucleotide sequence of cDNA clones encoding the complete "33kDa" precursor protein associated with the photosynthetic oxygen-evolving complex from spinach. Mol Gen Genet 207: 288–293

Vallejos CE, Tanksley SD and Bernatzky R (1986) Localization in the tomato genome of DNA restriction fragments containing sequences homologous to the rRNA (45S), the major chlorophyll *a/b*-binding polypeptides and the ribulose 1,5-bisphosphate carboxylase genes. Genetics 11: 93–105

van Spanje M, Dirkse WG, Nap J-P and Stiekema WJ (1991) Isolation and analysis of cDNA encoding the 33 kDa precursor protein of the oxygen-evolving complex of potato. Plant Mol Biol 17: 157–160

Viro M and Kloppstech K (1980) Differential expression of the genes for ribulose-1,5-bisphosphate carboxylase and light-harvesting chlorophyll *a/b* protein in the developing barley leaf. Planta 150: 41–45

Vorst O, Oosterhoff-Teertstra R, Vankan P, Smeekens S and Weisbeek P (1988) Plastocyanin of *Arabidopsis thaliana*: Isolation and characterization of the gene and chloroplast import of the precursor protein. Gene 65: 59–69

Vorst O, van Dam F, Oosterhoff-Teertstra R, Smeekens S and Weisbeek P (1990) Tissue-specific expression directed by an *Arabidopsis thaliana* pre-ferredoxin promoter in transgenic tobacco plants. Plant Mol Biol 14: 491–499

Vorst O, Kock P, Lever A, Wetering B, Weisbeek P and Smeekens S (1993a) The promoter of the *Arabidopsis thaliana* plastocyanin gene contains a far upstream enhancer-like element involved in chloroplast-dependent expression. Plant J 4: 933–945

Vorst O, van Dam F, Weisbeek P and Smeekens S (1993b) Light-regulated expression of the *Arabidopsis thaliana ferredoxin A* gene involves both transcriptional and post-transcriptional processes. Plant J 3: 793–803

Wales R, Newman BJ, Pappin D and Gray JC (1989a) The extrinsic 33kDa polypeptide of the oxygen-evolving complex of Photosystem II is a putative calcium-binding protein and is encoded by a multi-gene family in pea. Plant Mol Biol 12: 439–451

Wales R, Newman BJ, Rose SA, Pappin D and Gray JC (1989b) Characterization of cDNA clones encoding the extrinsic 23kDa polypeptide of the oxygen-evolving complex of Photosystem II in pea. Plant Mol Biol 13: 573–582

Walling LL, Chang YC, Demmin DS and Holzer FM (1988) Isolation, characterization and evolutionary relatedness of three members from the soybean multigene family encoding chlorophyll *a/b* binding proteins. Nucleic Acids Res 16: 10477–10492

Webber AN, Packman LC and Gray JC (1989) A 10 kDa polypeptide associated with the oxygen-evolving complex of Photosystem II has a putative C-terminal non-cleavable thylakoid transfer domain. FEBS Lett 242: 435–438

Wedel N, Bartling D and Herrmann RG (1988) Analysis of cDNA clones encoding the entire ferredoxin I precursor polypeptide from spinach. Bot Acta 101: 295–300

Wedel N, Klein R, Ljungberg U, Andersson B and Herrmann RG (1992) The single-copy gene *psbS* codes for a phylogenetically intriguing 22 kDa polypeptide of Photosystem II. FEBS Lett 314: 61–66

Williams ME, Foster R and Chua N-H (1992) Sequences flanking the hexameric G-box core CACGTG affect the specificity of protein binding. Plant Cell 4: 485–496

Yamamoto Y, Tsuji H and Obokata J (1993) Structure and expression of a nuclear gene for the PSI-D subunit of Photosystem I in *Nicotiana sylvestris*. Plant Mol Biol 22: 985–993

Zhang H, Hanley S and Goodman H (1991) Isolation, characterization and chromosomal location of a new *cab* gene from *Arabidopsis thaliana*. Plant Physiol 96: 1387–1388

Chapter 34

The Evolution of Plastids and the Photosynthetic Apparatus

Michael Reith
NRC Institute for Marine Biosciences, 1411 Oxford St., Halifax, Nova Scotia, B3H 3Z1, Canada

Summary

Understanding how the diversity of present day plastids arose is the major challenge in plastid evolution. While it is clear that one or more endosymbiotic events involving a cyanobacterium and a eukaryotic host occurred, many of the subsequent details of plastid evolution are murky. In particular, three critical questions must be addressed: how many primary endosymbioses were involved; how many secondary endosymbioses (where the plastid arises from a eukaryotic endosymbiont) occurred; and, how have the light-harvesting complexes of chlorophytes and chromophytes evolved? A summary of current evidence on the number of primary endosymbioses indicates that there was only a single primary event. On the other hand, secondary endosymbioses appear to have occurred at least three times and possibly as many as five. Within this basic framework of plastid evolution, the evolution of the photosynthetic apparatus is considered by focusing on the presence/absence and coding location of photosynthetic genes as markers to further define the events of plastid evolution. In addition, the question of the origins of chlorophyte and chromophyte light-harvesting complexes is discussed. A recently described group of small cyanobacterial and rhodophyte proteins contain a critical pigment-binding, transmembrane domain that is also present in two copies in chlorophyte and chromophyte light-harvesting proteins and in the early light-induced proteins and PsbS. These similarities indicate that the cyanobacterial protein donated the critical domain for both chlorophyte and chromophyte light-harvesting proteins. The possible pathways and driving forces of the evolution of light-harvesting proteins are considered. Finally, a very basic model of the possible events and pathways of plastid evolution, with emphasis on the evolution of the photosynthetic apparatus, is presented.

Donald R. Ort and Charles F. Yocum (eds): Oxygenic Photosynthesis: The Light Reactions, pp. 643–657.

I. Introduction

Photosynthetic eukaryotes employ a variety of pigments, protein complexes and membrane structures in their plastids to capture light energy. While all plastids contain chlorophyll *a*, differences in accessory pigments and the number of surrounding membranes allow the classification of photosynthetic eukaryotes into three main types (Gray, 1992): (i) chlorophytes (Chlorophyta, Gamophyta and Euglenophyta) and metaphytes (Tracheophyta and Bryophyta), whose plastids contain chlorophyll *b* and are usually bounded by two membranes (except for Euglenophytes which have three); (ii) rhodophytes (Rhodophyta), with plastids containing phycobiliproteins and bounded by two membranes; and (iii) chromophytes, an assemblage of eight phyla that have plastids containing chlorophyll *c* and are usually bounded by four membranes. Two groups of chromophytes, the Cryptophyta and the Eustigmatophyta, contain phycobiliproteins as well as chlorophylls *a* and *c*. Another unusual chromophyte group is the Dinoflagellata, which have plastids that are usually bounded by three membranes, but in some cases only two. This amazing variety of plastid pigments, light-harvesting systems and ultrastructure presents a considerable challenge to those working in the field of plastid evolution: how have plastids evolved into so many diverse forms?

It is widely accepted that plastids arose through the endosymbiosis of a cyanobacterium with a eukaryotic cell (Gray, 1992). What is not yet clear are the details of how this process led to the diversity of present day plastids. A major issue still to be resolved is whether there was only a single, original endosymbiosis, from which all plastids evolved or whether multiple primary symbiotic events are responsible for plastid diversity. Secondary symbioses (the gain of plastids from a eukaryotic rather than prokaryotic endosymbiont) have also played an important role in the generation of plastid diversity (Douglas et al., 1991). Again, the number of these events that occurred is still to be determined. Finally, the origins of chlorophyte and chromophyte light-harvesting complexes are essentially unknown. If plastids arose from a single endosymbiotic event, what were the driving forces and mechanisms for the evolution of chlorophyte and chromophyte light-harvesting complexes?

In this review of plastid evolution, I will focus on the evolution of the photosynthetic apparatus, since this is where many of the distinguishing differences between plastids occur and since it seems most appropriate for this volume. To do this, however, a brief summary of the current evidence on the questions mentioned above is necessary. Many of these aspects of plastid evolution have been reviewed recently (Gray, 1991, 1992; Palmer, 1991; Kowallik, 1992; Douglas, 1994; Loiseaux-de Goër, 1994; Reith, 1995) and will be only briefly summarized here. On this framework of plastid evolution, I will attempt to overlay one view of the evolution of the photosynthetic apparatus and suggest a pathway for the evolution of chromophyte and chlorophyte light-harvesting complexes.

II. Plastid Evolution

A. Hypotheses of Plastid Evolution

More than 80 years ago, Mereschkowsky (1910) first suggested that plastids of different pigment composition arose from symbioses involving different photosynthetic prokaryotes containing the corresponding pigments. This hypothesis (polyphyletic origin) has since been refined by many authors (e.g. Raven, 1970; Gibbs, 1981; Whatley and Whatley, 1981). However, it was not until the discovery of the prochlorophytes, chlorophyll *a*/*b*-containing prokaryotes (Lewin and Withers, 1975), and later the brownish photoheterotrophic eubacterium, *Heliobacterium chlorum* (Gest and Favinger, 1983), that plastid progenitors other than cyanobacteria could be proposed. A relatively recent, alternate hypothesis (monophyletic origin) is that all plastids stem from a single primary endosymbiotic event and that the variations in plastid pigment and membrane composition are derived features (e.g., Cavalier-Smith, 1982; Kowallik, 1992). Key aspects of monophyletic schemes of plastid origin are the relatively simple biochemical differences between

---

Abbreviations: CAB – Chlorophyll *a*/*b* binding protein; ELIP – Early light-induced protein; FCP – Fucoxanthin chlorophyll protein; FCPC – Fucoxanthin chlorophyll *a*/*c* protein complex; HLIP – High light inducible protein; LHC – Light-harvesting complex; LHCPII – Light-harvesting chlorophyll *a*/*b*-binding proteins of Photosystem II; OEC – Oxygen-evolving complex; PBGD – Porphobilinogen deaminase; PS I – Photosystem I; PS II – Photosystem II; Rubisco – Ribulose-1,5-bisphosphate carboxylase/oxygenase; SSU – Small subunit; TMH – Transmembrane helix

chlorophylls *a*, *b* and *c* and the assumed difficulty of establishing an organelle from an endosymbiont.

A critical feature of recent schemes of both types is the necessity of secondary endosymbioses to explain differences in the number of surrounding membranes. First proposed by Gibbs (1978) on the basis of morphological studies of *Euglena*, this idea was rapidly extended to explain the origins of cryptomonads, which have since been shown to contain a remnant of the nucleus of the eukaryotic endosymbiont called the nucleomorph, and other chromophytes. Potentially, any plastid bounded by more than two membranes may have arisen through a secondary symbiosis. Recent proposals of both monophyletic and polyphyletic origins of plastids suggested the occurrence of one (Cavalier-Smith, 1982) to several (Douglas, 1992, 1994; Kowallik, 1992) secondary endosymbioses in the evolution of plastids.

Both monophyletic and polyphyletic proposals of plastid origins were initially based on ultrastructural features difficult to categorize as ancestral or derived. In the last 15 years, a significant amount of molecular data has become available that allows more definitive analyses of these proposals, primarily through phylogenetic analyses of individual genes or proteins. Data on plastid genome size, organization and content have been summarized recently (Palmer, 1991; Douglas, 1994; Loiseaux-de Goër, 1994; Reith, 1995) and will not be covered here. In the next two sections, recent evidence pertinent to the number of secondary symbioses and the monophyletic vs. polyphyletic debate are reviewed.

B. Evidence for Multiple Secondary Symbioses

Definitive evidence that secondary symbioses were involved in plastid evolution first came from phylogenetic analysis of 18S rRNA genes from the cryptomonad *Cryptomonas* Φ (Douglas et al., 1991). Two 18S genes were isolated and sequenced from this alga, both of which are expressed. One clustered specifically with 18S genes from rhodophytes while the other was located in a branch containing chlorophytes and *Acanthamoeba*. In conjunction with evidence indicating that cryptomonad and rhodophyte plastid genes are closely related, these observations indicate that the endosymbiont leading to the cryptomonad plastid was a red alga. A second noteworthy point from this study is that the *Cryptomonas* nuclear 18S gene does not group with other chromophytes, suggesting separate origins for these two groups. Similar observations were made independently in a study of 18S genes from another cryptomonad, *Pyrenomonas salina* (Eschbach et al. 1991).

More extensive rRNA phylogenies provide further insights into the possible number of secondary symbioses. An 18S rRNA phylogeny of 150 eukaryotes (Cavalier-Smith, 1993) clearly separates *Euglena* and dinoflagellates from all other photosynthetic eukaryotes (and from each other) and places cryptomonads and chromophytes (chrysophytes, phaeophytes, xanthophytes, eustigmatophytes, and diatoms) on separate branches of the tree. An additional taxon traditionally grouped with the chromophytes, the Haptophyta, also appears to be separated from the remaining chromophytes. While such trees are highly dependent on the species included and the accurate alignment of the sequences, this analysis indicates that as many as five different secondary endosymbioses may have occurred in the evolution of plastids. Based on our present knowledge of the plastid genomes of these groups, it appears that rhodophytes were the symbionts leading to cryptomonad and chromophyte plastids and perhaps those of haptophytes and dinoflagellates, although there is no molecular data on the plastid genomes of the latter groups. A chlorophyte appears to have been the predecessor of *Euglena* plastids, although 16S rRNA phylogenies indicate that the *Euglena* 16S gene is more similar to those of chromophytes than chlorophytes (Douglas and Turner, 1991).

A second line of evidence relevant to secondary symbioses comes from analyses of nuclear genes for plastid proteins in cryptomonads, chromophytes and *Euglena*. The first analysis of a chromophyte nuclear gene for a plastid protein, the fucoxanthin chlorophyll protein (FCP) of the diatom *Phaeodactylum tricornutum* (Bhaya and Grossman, 1991), demonstrated the presence of a signal peptide at the amino terminal end of the preprotein (Fig. 1). This signal peptide was capable of directing the transport of the precursor across the endoplasmic reticulum in in vitro studies using canine pancreatic microsomes. This observation is consistent with ultrastructural observations of ribosomes associated with the outermost plastid envelope membrane in algae containing plastids bounded by four membranes (Gibbs, 1962). In these algae, the outer two membranes (the chloroplast ER) are often contiguous

Fig. 1. Signal peptides and transit peptides of plastid localized proteins. A. Presequences of chromophyte (*Odontella*, *Phaeodactylum*), cryptomonad (*Chroomonas*) and euglenoid plastid proteins. Signal peptide regions are boxed. B. Transit peptides of rhodophyte plastid proteins. A pea GapA transit peptide (Brinkmann et al., 1989) is included for comparison. A conserved region of GapA transit peptides is boxed. See text for references.

with the ER or nuclear envelope. Signal peptides also occur on the precursors of the ATPase γ subunit of *Odontella sinensis*, another chromophyte (Pancic and Strotmann, 1993), and the phycoerythrin α subunit of the cryptophyte *Chroomonas* (Jenkins et al. 1990) (Fig. 1).

Studies of the *Euglena* precursors of the SSU of Rubisco (Chan et al., 1990), PBGD (Sharif et al., 1989) and LHCPII (Kishore et al., 1993) proteins have also identified signal peptide sequences. The *Euglena* LHCPII signal peptide directs the insertion of the precursor into canine microsomes (Kishore et al., 1993). In *Euglena*, chromophytes and cryptophytes, the signal peptides are the first part of a longer presequence that presumably further directs the precursor to the plastid and possibly to specific sites within the plastid. However, both the signal peptides themselves and the remainder of the presequence show considerable length and amino acid sequence differences between the *Euglena* preproteins and those of chromophytes and cryptophytes (Fig. 1A), suggesting that the targeting mechanisms directing the protein from the ER to the plastid may differ among these groups. The finding of a similar initial plastid-targeting mechanism in these three groups implies that all cells containing plastids bounded by three or four membranes will direct proteins to the plastid by way of the ER and that this may be part of a common mechanism in the establishment of plastids from secondary endosymbioses.

C. A Monophyletic Primary Origin of Plastids?

If, as suggested by the above data, all plastids bounded by three or four membranes arose through secondary endosymbioses, the question of the primary origin of plastids is reduced to choosing between monophyletic and diphyletic alternatives, assuming that rhodophytes and chlorophytes are each monophyletic groups. Until recently, phylogenetic studies were the only means for resolving this question. In general, phylogenetic analyses of plastid genes, with the exception of the *rbcLS* operon, indicated a common, cyanobacterial ancestor for rhodophyte and chlorophyte plastids (see Morden et al., 1992 for review).

In the case of *rbcL* and *rbcS*, rhodophyte (except *Cyanophora paradoxa*) and chromophyte proteins are similar to those of α and β purple eubacteria while chlorophyte and cyanobacterial proteins are closely related (Martin et al., 1992). Rather than supporting a polyphyletic origin of plastids, however, these data indicate a lateral transfer of these genes from an α or β purple bacterium into the rhodophyte/chromophyte lineage.

Other phylogenetic trees were constructed to test the hypothesis that chlorophyte plastids arose from a prochlorophyte symbiont. In both rRNA (Urbach et al., 1992) and protein trees (Palenik and Haselkorn, 1992), prochlorophytes appear to be polyphyletic within the cyanobacteria and show no specific affiliation with chlorophyte plastids. This finding indicates that the ability to synthesize chlorophyll *b* evolved multiple times or was acquired through lateral transfer. Likewise, the suggestion that chromophyte plastids evolved from *Heliobacterium chlorum* (Margulis and Obar, 1985) is not supported by phylogenetic analyses (Witt and Stackebrandt, 1988), nor by the above-mentioned data indicating an origin for these plastids through a secondary symbiosis.

While the above data fail to support the polyphyletic theory of plastid evolution, positive evidence for the monophyletic hypothesis has been more difficult to obtain. Such evidence would require the identification of shared, derived characters in plastids; that is, characters that are common among plastids but different from those of cyanobacteria. Until recently, the molecular biology of the plastids of the rhodophytes and the chromophytes was too poorly characterized to identify any such characters. The nearly complete sequencing of the plastid genomes of the rhodophytes *Porphyra purpurea* (Reith and Munholland, 1993; Reith and Munholland, unpublished data) and *Cyanophora paradoxa* (Löffelhardt and Bohnert, 1994; DA Bryant, V Stirewalt, C Michalowski, W Löffelhardt and HJ Bohnert, unpublished data) and the chromophyte *Odontella sinensis* (U Freier, B Stöbe and KV Kowallik, unpublished) in addition to the characterization of several rhodophyte nuclear genes for plastid proteins (Apt et al., 1993; Liaud et al., 1993; Zhou and Ragan, 1993) have led to the identification of two types of characters that may provide positive support for a monophyletic origin of plastids.

The first of these positive characters is the presence in chlorophyte and rhodophyte plastid genomes of clusters of genes that are physically separated in cyanobacteria. As has been discussed previously (Reith and Munholland, 1993), two groups of genes, the *psbBNH* operon and the *rpoBC1C2-rps2/tsf-atpIHGFDA* array, are physically linked in both rhodophyte and metaphyte plastid genomes but are separated in cyanobacteria. Whether these gene sets are cotranscribed (Loiseaux-de Goër, 1994) is immaterial. The relevant point is that the rearrangement of these genes into linked groups in an ancestral plastid common to all present day plastids is the most parsimonious explanation for these observations.

Interestingly, the *psbBNH* operon, which is cotranscribed with *petBD* in metaphytes, is unlinked to *petBD* in the chlorophyte *Chlamydomonas reinhardtii* (Johnson and Schmidt, 1993). The arrangement of these genes appears to have evolved from *psbB*, *psbNH* and *petBD* as separate operons in cyanobacteria to *psbBNH* and *petBD* in rhodophytes, chlorophytes and chromophytes to a single operon, *psbBNH/petBD* in metaphytes. Thus the arrangement of these genes serves as a marker for two events, the establishment of chloroplasts and the separation of the metaphytes from chlorophytes.

The second positive character linking rhodophyte and chlorophyte plastids is the mechanism by which proteins are imported into the plastid. In chlorophytes and metaphytes, nuclear-encoded, plastid proteins are synthesized with an amino-terminal extension called the transit peptide. This sequence is recognized by the plastid protein import complex which transports the preprotein through the plastid envelope and cleaves off the transit peptide. Several nuclear genes for rhodophyte plastid proteins have been isolated and characterized (Apt et al., 1993; Liaud et al., 1993; Zhou and Ragan, 1993). In all cases, an apparent transit peptide sequence with many of the same characteristics as chlorophyte transit peptides is present at the amino terminus of these proteins (Fig. 1B). Although the *Antithamnion neglectum* γ phycoerythrin transit peptide is considerably shorter than the GapA transit peptides, there is remarkable similarity between the two rhodophyte GapA transit peptides and in fact, noticeable similarity between these sequences and those of metaphytes (Fig. 1B). These observations are similar to those of metaphytes where transit peptides from the same gene in different organisms are more similar than those of two different genes in one organism (Keegstra et al., 1989). In addition, Apt et al. (1993) used the *A. neglectum* γ phycoerythrin transit peptide to direct pea Rubisco SSU into isolated pea plastids where it was assembled into the holoenzyme. The similarity between rhodophyte and chlorophyte transit peptides indicates a similar mechanism of protein import in both types of plastid. Since the plastid protein import mechanism involves a complex series of events (recognition, transport through the envelope, transit peptide processing, protein folding and assembly) and a substantial number of proteins, the most likely explanation is that both rhodophyte and chlorophyte plastids shared a common ancestor in which this system developed.

Thus, the above data indicate that only a single primary endosymbiosis occurred during the evolution of plastids but that there were multiple (perhaps as many as five) secondary endosymbiotic events. Why would secondary endosymbioses more frequently result in the establishment of plastids? An important factor may be the ease of transferring genes from the symbiont to the host. Perhaps genes are more readily transferred from eukaryote to eukaryote than from prokaryote to eukaryote since the symbiont's gene expression systems could be used directly by the host. Only the addition of appropriate signals for directing the preprotein to the chloroplast (just a signal peptide?) would be required. This question will likely remain a topic of speculation for some time.

III. Evolution of the Photosynthetic Apparatus

If one accepts the above arguments for a monophyletic primary origin of plastids followed by multiple secondary origins, it is possible to look more closely at the evolution of specific plastid components. However, since the proteins of the photosynthetic apparatus are highly conserved from cyanobacteria to metaphytes, phylogenetic analyses are not always informative. Instead, analyses of coding location (plastid or nuclear), the presence of alternative, functionally equivalent proteins in different lineages and differences in post-translational modifications reveal some of the processes and details in the evolution of the photosynthetic apparatus.

A. Core Thylakoid Membrane Complexes

1. Coding Location

The proteins of the core complexes of the thylakoid membrane (PS I core, PS II core, cytochrome b_6f complex and CF_1CF_0 ATPase) are largely encoded on the plastid genome. In all photosynthetic eukaryotes investigated, *psaABCIJM*, *psbA-F,H-L,T*, *atpABEFHI* and *petABDG* are plastid encoded, except in *Euglena*, where *psaI*, *psbM* and *petD* are nuclear and *petA* is absent (Hallick et al., 1993) and *C. paradoxa* where *atpI* is nuclear (Bryant et al., unpublished data). In addition, *atpC* and *petC* are always located in the nucleus. However, recent studies of the plastid genomes of rhodophyte and chromophyte algae revealed that the coding location of several genes of these complexes may vary among photosynthetic eukaryotes (Table 1). *atpD* and *atpG*, which are nuclear genes in chlorophytes and metaphytes, are plastid encoded in all rhodophytes (Kostrzewa and Zetsche, 1992, 1993; Reith and Munholland, 1993; Bryant et al., unpublished) and chromophytes (Pancic et al., 1992; Freier et al., unpublished) examined so far. Similarly, the chlorophyte/metaphyte nuclear genes *psaDEFKL* are plastid encoded in the rhodophyte *P. purpurea* (Reith and Munholland, 1993 and unpublished) while some of these genes have also been detected on the plastid genomes of *C. paradoxa* (Bryant et al., unpublished data), *Cryptomonas* (Douglas, 1992) and *O. sinensis* (Freier et al., unpublished) (Table 1). On the other hand, *psbM* appears to have been transferred to the nucleus in *P. purpurea* (Reith and Munholland, unpublished), but maintained in the plastid in *C. paradoxa*, chlorophytes and metaphytes. Although the data on the coding site of the core complex genes from rhodophytes and chromophytes are based on only a few species, it is clear that more genes for photosynthetic apparatus proteins were transferred to the nucleus in the chlorophyte/metaphyte lineage than in the rhodophyte/chromophyte lineage.

2. PS I Content

Although the protein constituents of the core thylakoid membrane complexes are, for the most part, conserved among all groups of photosynthetic eukaryotes, there are some minor variations in the proteins found in PS I. In chlorophytes there are three proteins, PsaG, PsaH and PsaN, that are encoded by nuclear genes and are absent from cyanobacteria (Ikeuchi, 1992; Knoetzel and Simpson, 1993). PsaG presumably arose through a duplication of *psaK* (Okkels et al., 1992) and PsaH and PsaN appear to be involved in linking the PS I LHC, which is absent in

Table 1. Coding Location of Photosynthetic Apparatus Genes

Genes	*Cyanophora*	Rhodophytes	Chromophytes[1]	Chlorophytes	Metaphytes
PS I					
psaABCIJM	C	C	C	C[2]	C
psaEF	C	C	C	N	N
psaDL	N	C	C	N	N
psaK	N	C	+	N	N
psaGHN	-	-	-	N	N
PS II					
psbA-F,H-L,T	C	C	C	C	C
psbM	C	N	+	C	C
psbRS	?	?	?	N	N
Cytochrome b_6f Complex					
petABDG	C	C	C	C[3]	C
petC	N	N	N	N	N
ATPase					
atpABEFH	C	C	C	C	C
atpI	N	C	C	C	C
atpDG	C	C	C	N	N
atpC	N	N	N	N	N
OEC					
psbO	N	N	N	N	N
psbPQ	-	-	-	N	N
psbU	N	N	N	-	-
psbV	C	C	C	-	-
Mobile Electron Carriers					
petE	-	-	-	N	N
petF	C	C	C	N	N
petH	N	N	N	N	N
petI	?	N	+	+	-
petJ	N	C	+	N	-

Symbols used are: C: chloroplast, N: nucleus, - absent, + present, but coding location unknown, ? presence or absence unknown
[1] Chromophytes and *Cryptomonas* Φ
[2] *psaM* is a nuclear gene in *Euglena*
[3] *petA* is absent and *petD* is nuclear in *Euglena*

cyanobacteria, to the reaction center (Ikeuchi, 1992). Likewise, a 4.8 kDa component was found in cyanobacteria, but not metaphytes (Koike et al., 1989a). None of these genes were detected on the *P. purpurea* or *C. paradoxa* plastid genomes. Until there is evidence to the contrary, the absence of these genes in cyanobacteria indicates that they are specific for chlorophyte PS I and thus probably evolved after the separation of the chlorophyte lineage.

3. Grana Stacking and Protein Phosphorylation

A unique feature of chlorophytes is the organization of the thylakoid membranes into grana stacks. The current evidence indicates that protein phosphorylation, particularly that at the N-terminus of LHCP II, affects membrane stacking and energy transfer between the two photosystems (Bennett, 1991; Allen, 1992). Several integral PS II proteins (D1, D2, CP43 and PsbH) have also been observed to be phosphorylated in chlorophytes, although the significance of this phosphorylation is unclear. Of these four proteins, PsbH is the most highly phosphorylated, but in cyanobacteria, including the prochlorophyte *Prochlorothrix hollandica* (Koike et al., 1989b; Mayes and Barber, 1990; Greer and Golden, 1992), and *P. purpurea* (Reith and Munholland, unpublished), the PsbH phosphorylation site, a threonine at position two, is missing due to the absence of the twelve N terminal amino acids of chlorophyte PsbHs. Interestingly, cyanobacterial PsbH still appears to be phosphorylated (Race and Gounaris, 1993), although the phosphorylation site has not yet been determined. The N-terminal extension of chlorophyte PsbH and the change in phosphorylation position appear to be another modification that evolved after the separation of the chlorophyte lineage. Whether the other three PS II proteins are phosphorylated in cyanobacteria, rhodophytes and chromophytes is still to be determined.

B. Mobile Electron Carriers

In many cyanobacteria, two pairs of proteins are involved in the transfer of electrons from the cytochrome b_6f complex to PS I and from PS I to NADPH oxidoreductase. Which member of each pair is synthesized depends on nutrient conditions. Flavodoxin (encoded by *PetI*) is made in place of ferredoxin (encoded by *petF*) under iron-limiting conditions while the synthesis of cytochrome *c*-553 (*petJ*) and plastocyanin (*PetE*) is influenced by copper levels (Morand et al., 1994; Straus, 1994). Among photosynthetic eukaryotes, there are differences in which of these proteins have been retained and the coding location of their genes (Table 1). Flavodoxin occurs in rhodophytes, chromophytes and chlorophyte algae (Fitzgerald et al., 1978; Price et al., 1991; La Roche et al., 1993) but is absent from metaphytes. Preliminary studies (Reith, unpublished data) indicate that flavodoxin is encoded by a nuclear gene, at least in rhodophytes. Ferredoxin is present in all photosynthetic eukaryotes, with one known exception, and is plastid-encoded in the rhodophyte/chromophyte lineage, but nuclear-encoded in chlorophytes. The exception is the rhodophyte *Chondrus crispus* which is deficient in ferredoxin and constitutively expresses flavodoxin (Fitzgerald et al., 1978). Whether ferredoxin is also absent from other closely related rhodophytes has not been explored. Plastocyanin appears to occur only in the chlorophyte lineage, where it is encoded in the nucleus (Merchant et al., 1990) while cytochrome *c*-553 is present in all groups except metaphytes. Again, cytochrome *c*-553 is encoded on the nuclear genome of green algae (Merchant and Bogorad, 1987), but on the plastid genome of *P. purpurea* (Reith and Munholland, 1993). This variety in the presence or absence of mobile electron carriers and their coding location provides useful markers in understanding the process of plastid evolution.

C. Oxygen-Evolving Complex

In chlorophytes, the oxygen-evolving complex (OEC) is extremely well characterized. It consists of three proteins of 33, 23 and 16 kDa, encoded by the nuclear genes *PsbO*, *PsbP* and *PsbQ* (Bricker and Ghanotakis, Chapter 8). In cyanobacteria, the 33 kDa protein is present, but the other proteins in this complex had not been identified until recently. In an elegant series of cellular fractionation studies, Shen and coworkers (Shen et al., 1992; Shen and Inoue, 1993), demonstrated that two proteins of 17 and 12 kDa are closely associated with the 33 kDa OEC protein and that the presence of these two proteins enhances oxygen evolution. N-terminal sequencing of these proteins identified the 17 kDa protein as a low-potential cytochrome, cytochrome *c*-550, that was previously characterized in several cyanobacteria (Holton and Myers, 1963; Alam et al., 1984) and the 12 kDa protein as a homologue of a 9 kDa protein from *Phormidium laminosum* that is also important for oxygen evolution (Wallace et al., 1989). Shen and Inoue (1993) conclude that these two proteins are the cyanobacterial equivalents of the chlorophyte 23 and 16 kDa proteins. In addition, they suggest that the heme group of cytochrome *c*-550 doesn't play a significant role in the water splitting process due to the high redox potential on the oxidizing side of PS II . The genes for the 9/12 kDa protein and cytochrome *c*-550 are designated *psbU* and *psbV*, respectively (R.B. Hallick, pers. comm.).

There are apparently no investigations of the OEC in rhodophytes and chromophytes, although cytochrome *c*-550 was detected in the red alga *Porphyridium cruentum* (Evans and Krogmann, 1983). Genes encoding cytochrome *c*-550 were recently detected on the plastid genomes of *P. purpurea* (Reith and Munholland, 1993), *C. paradoxa* (Bryant et al., unpublished) and *O. sinensis* (Freier et al., unpublished). These plastid-encoded proteins have apparent signal peptides at their amino termini, consistent with a lumenal location for the protein. The presence of *psbV* in these plastid genomes indicates that rhodophytes and chromophytes have a cyanobacterial-type OEC, with *psbO* and *psbU* presumably encoded in the nucleus. It appears that *psbP* and *psbQ* replaced *psbU* and *psbV* during the evolution of the chlorophytes.

D. Light-Harvesting Complexes

A key characteristic that distinguishes the different types of plastids is the composition of the pigments associated with light-harvesting antennae. In the different plastid types, these pigments are associated with rather different protein complexes: the phycobilisomes of rhodophytes (and cyanobacteria), the chlorophyll *b*-containing LHC II of chlorophytes and the fucoxanthin, chlorophyll *a/c* complex (FCPC) of chromophytes. In addition, chlorophytes have a chlorophyll *b*-containing LHC associated with PS I (LHC I). Essential to understanding plastid evolution is an understanding of how these different complexes evolved. Indeed, a monophyletic hypothesis of plastid evolution requires a mechanism whereby the loss of phycobilisomes and the gain of LHC or FCPC can be explained.

Recent molecular biological data may shed some light on the origins of the LHC and FCPC. Both LHC I and LHC II of chlorophytes are composed of proteins of the CAB family (Green et al., 1991; Pichersky and Jansson, Chapter 28). These well-studied proteins contain three transmembrane helices (TMHs), two of which (the first and third) are implicated in chlorophyll binding and have a high degree of sequence similarity. Xanthophyll carotenoids are also associated with CAB proteins and are required for LHC II assembly (Plumley and Schmidt, 1987). The unusual chlorophyll *a*, *b* and *c*-containing LHC of the primitive chlorophyte *Mantoniella squamata* is also similar to the CAB family with a high degree of sequence conservation in TMH I and III (Rhiel and Mörschel, 1993). The FCPC of *Phaeodactylum tricornutum* is also composed of a group of proteins, the FCPs, encoded by a small gene family. Like CAB polypeptides, the FCPs contain three TMHs, the first and third being highly conserved among family members (Bhaya and Grossman, 1991,1993). These two TMHs are also similar to each other as well as to the first and third TMHs of CABs, although the first TMH of FCPs is more similar to the third TMH of CABs than to the first (Fig. 2). Obviously, FCPs also bind both chlorophyll and xanthophyll.

Two other gene families in chlorophytes are similar to CAB proteins. The first of these is the ELIP (early light-induced protein) family, that was first identified as being induced during greening of barley and pea seedlings (Green et al., 1991). These proteins also contain three TMHs, the first and third showing significant similarity to those of CABs (Fig. 2). More recent studies indicate that these proteins bind xanthophylls such as zeaxanthin and their synthesis is induced under high light intensities or other conditions, such as sulfate stress or norflurazon treatment, that cause photooxidative stress (Adamska et al., 1992; Levy et al., 1993). These characteristics suggest that ELIPs are stress-induced proteins that prevent photooxidative damage. The second CAB-related protein is an intrinsic 22 kDa PS II protein encoded by *PsbS* (Kim et al., 1992; Wedel et al., 1992). This protein has four TMHs that appear to have arisen through a duplication since the first half of the protein is highly similar to the second half. In addition, the first and third TMHs are similar to the first and third TMHs of CABs (Fig. 2). PsbS is present in a wide variety of chlorophytes, and at least one cyanobacterium (Nilsson et al., 1990) and binds chlorophyll *a* and *b*, but doesn't appear to bind carotenoids (Funk et al., 1994).

All of these eukaryotic pigment-containing proteins are related by the presence of two similar TMHs that bind chlorophylls and/or xanthophylls. Genes for prokaryotic proteins containing a single copy of this pigment-binding TMH were recently identified in two cyanobacteria, *Anabaena* sp. PCC 7120 (Brahamsha and Haselkorn, 1991) and *Synechococcus* sp. PCC 7942 (Dolganov et al., 1995), and two rhodophyte plastid genomes, *P. purpurea* (Reith and Munholland, unpublished) and *Cyanidium caldarium* (Maid and Zetsche, 1992). The encoded proteins (HLIPs for high light inducible proteins) are quite small (48 to 72 aa) and are most similar to the TMHs of chlorophyte ELIPs. The *Synechococcus* HLIP is induced under high light and blue/UV light indicating a role in adaptation to high light (Dolganov et al., 1995). However, disruption of the HLIP gene does not result in particular sensitivity to high light (Dolganov et al., 1995), thus its function remains unclear.

The alignment of the TMH I and III regions of HLIPs, ELIPs, PsbS, FCP and CABs shows remarkable similarity among these proteins (Fig. 2). In particular, the motif E-X-X-N/H-G-R-X-A-M is highly conserved in both TMH regions. The recently determined crystal structure of LHC II (Kühlbrandt et al., 1994) indicates that the E and R residues in this motif form ion pairs between TMHs I and III, serving to lock these two helices together. In addition, the E

```
                           I                                  III
HLIPS            *  *** ** *       *                *  *** **
ANABAENA    FTPQAEIWNGRLAMIGFLAATLIELFSGQGFL   (FTPQAEIWNGRLAMIGFLAATLIELFSGQGFL)
SYN7942     FHDRAEKLNGRLAMIGFVALILTEVALGQGLL   (FHDRAEKLNGRLAMIGFVALILTEVALGQGLL)
PORPHYRA    FTDSAETWNGRFAMIGFMAVIFIELVTGKGLL   (FTDSAETWNGRFAMIGFMAVIFIELVTGKGLL)
CYANIDIUM   FTTGAENWNGRLAMIGFVSALVTELITGKGVL   (FTTGAENWNGRLAMIGFVSALVTELITGKGVL)
ELIPS
DSP22       DGLAPERINGRSAMIGFVAAVGVELATGRDVF    WNSDAEIWNGRFAMIGLVALAFTEYVKGGPLI
CBR         FSGAPEIINGRLAMLGFVAALGAELSTGESVL    FTPDAEMTNGRFAMIGFAAMLVYEGIQGIALF
ELIP58      SGPAPERINGRLAMVGFVAALSVEAARGGGLL    WSADAELWNGRFAMLGLVALAATEFITGAPFV
PEA         SGPAPERINGRLAMIGFVAAMGVEIAKGQGLS    MSSDAEFWNGRIAMLGLVALAFTEFVKGTSLV
FERN        RWCRPETINGRMAMVGFVWALVIDRFTGVVEK    FNAKAERWNGRLAMVRSPPSSLARWCARLPSS
PSBS
PSBS        FTKENELFVGRVAMIGFAASLLGEGITGKGIL    FTKSNELFVGRLAQLGFAFSLIGEIITGKGAL
FCP
FCP1        RLRYVEIKHGRICMLAVAGYLTQEAGIRLPGD    QKRAIELNQGRAAQMGILALMVHEQLGVSILP
CABS
MANTON      KNAEREVIHGRWAMLGVTGAWTTENGTGIPWF    ELKIKELKHCRLSMFAWLGCIFQALATQEGPI
CHLAMY      KYRELEVIHARWALLGALGILTPELLSTYAGV    ELKVKEIKNGRLAMFSCFGFFVQAIVTGKGPI
DUN         RYREIELIHARWALLGALGILTPELLSQYAGV    ELKVKEIKNGRLAMFSCLGFFVQAIVTGKGPV
EUG         RYREAEVIHARWAMLGALGVVTPELLAGNGVP    ELKVKELKNGRLAMVAMLGFYVQPLVTKAGPV
LHCII/I     KNRELEVIHCRWAMLGALGCVFPELLARNGVK    ELKVKEIKNGRLAMFSMFGFFVQAIVTGKGPL
CP29/I      KYQAYELIHARWAMLGAAGFIIPEAFNKFGAN    ILKVKEIKNGRLAMFSMLGFFIQAYVTGQGPV
LHCI/III    WLAYGEVINGRFAMLGAAGAIAPEILGKAGLI    ELKLKEIKNGRLAMLAILGYFIQALVTGVGPY
LHCI/IV     WFIQAELVNGRWAMLGVAGMLLPEVFTSIGIL    EAKEKELANGRLAMLAFLGFIVQHNVTGKGPF
```

Fig. 2. Alignment of HLIP, ELIP, PsbS, FCP and CAB TMH regions I and III. Only the most conserved regions of the TMHs are shown. The HLIP sequences are duplicated (indicated by parentheses) in the TMH III alignment for reference. Conserved amino acids are indicated by an asterisk over the alignment. References for amino acid sequences are: *Anabaena*: Brahamsha and Haselkorn, 1991; *Synechococcus* sp. strain PCC 7842: Dolganov et al., 1995; *Porphyra*: Reith and Munholland, unpublished; *Cyanidium*: Maid and Zetsche, 1992; Dsp22 (*Craterostigma planagineum*): Bartels et al., 1992; Cbr (*Dunaliella bardawil*): Lers et al., 1991; Elip58 (Barley): Grimm et al., 1989; Pea: Kolanus et al., 1987; Fern (*Onoclea sensibilis*): Raghavan and Kamalay, 1993; PsbS: Kim et al., 1992, Wedel et al., 1992; FCP (*Phaeodactylum tricornutum*): Bhaya and Grossman, 1991; Manton (*Mantoniella squamata*): Rhiel and Mörschel (1993); Chlamy (*Chlamydomonas moewusii*): Larouche et al., 1991; Dun (*Dunaliella tertiolecta*): La Roche et al., 1990; Eug (*Euglena gracilis*): Muchhal and Swartzbach, 1992; LHCII/I (tomato): Pichersky et al., 1985; CP29/I (tomato): Pichersky et al., 1991; LHCI/III (tomato): Pichersky et al., 1989; LHCI/IV (tomato): Schwartz et al., 1991.

and N/H residues bind chlorophyll *a* molecules. The conservation of this motif throughout these different protein families indicates that their TMHs I and III should be locked together in a similar fashion or, in the case of the HLIPs, that the formation of intermolecular ion pairs would produce a similar structure through HLIP dimers. In addition, two chlorophyll molecules would be expected to be bound to each TMH, although there may be other interactions that are required to stabilize these chlorophylls.

How have all these related proteins evolved? Since the regions flanking the conserved TMHs are generally not conserved between these groups of proteins, the CAB, ELIP, FCP and PsbS groups each may have arisen independently through the combination of two HLIP-like domains with one or two additional TMHs. Alternately, a cyanobacterial PsbS, in which an HLIP domain and an additional TMH were linked and then duplicated, may have been the evolutionary precursor for all the above groups. This would be consistent with the proposal by Bryant (1992) that eukaryotic light harvesting proteins evolved from a cyanobacterial ancestor. However, the overall identity between the CABs and the FCPs is only approximately 22% (Bhaya and Grossman, 1993) and several gaps must be introduced to align these proteins. Perhaps a cyanobacterial PsbS provided domains for the CAB and PsbS families but HLIPs, or another intermediate containing three TMHs, were the progenitors of ELIPs and/or FCPs. While the details of the evolution of these proteins await further characterization of cyanobacterial and algal members, the HLIPs appear to be the prokaryotic progenitor of the critical domain for the light-harvesting proteins of both the chlorophyte LHC and the chromophyte FCPC.

IV. A Model for the Evolution of the Photosynthetic Apparatus

All of the above information, although still somewhat incomplete, allows us to begin to envision some of the details in the evolution of the photosynthetic apparatus. By applying the principles of parsimony and minimizing the number of gene loss and transfer events, a model such as that in Fig. 3 can be developed. It is clear from such a model that, relatively early in the evolution of plastids, the rhodophyte and chlorophyte lineages diverged and that the chlorophyte plastids lost a number of genes or transferred them to the nucleus. Unfortunately, little is known about the plastid gene content of the more primitive chlorophytes and thus it is possible that these events may have been spread out during the evolution of chlorophyte plastids. Similarly, there are probably more loss/transfer events at the points where cryptomonads and chromophytes arise than are

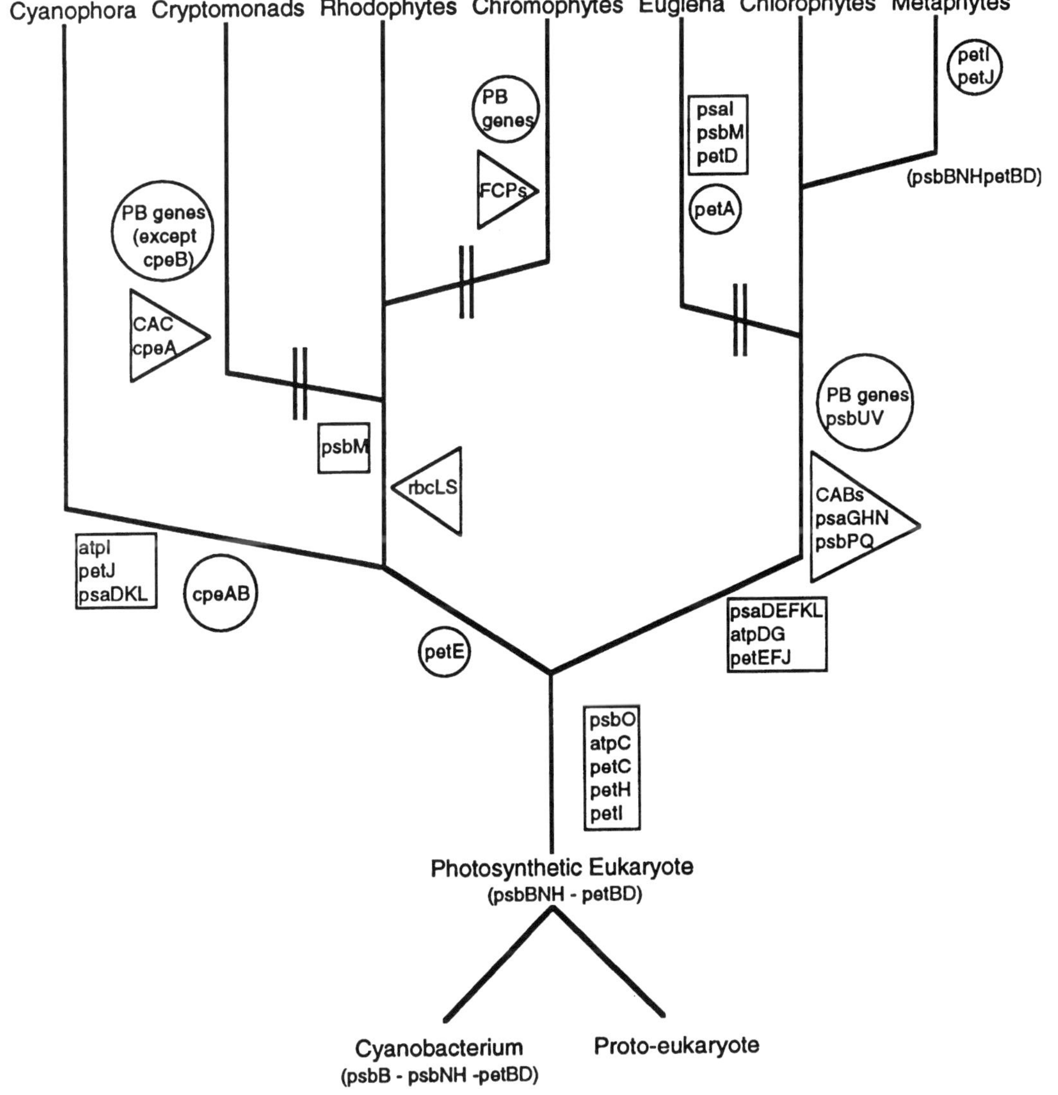

Fig. 3. A schematic for the evolution of plastids and the photosynthetic apparatus. Genes transferred to the nucleus are enclosed in boxes. Gene loss is indicated by circles. The evolution of new genes is indicated by rightward-pointing triangles. The introduction of genes is indicated by leftward-pointing triangles. Secondary endosymbioses are indicated by double vertical lines. The rearrangement of the psbB -psbNH-petBD genes from three separate operons to a single one (in metaphytes) is shown in parentheses. CAC refers to the PSII chlorophyll *a/c* binding protein of cryptomonads.

indicated in the figure. In addition, if, as indicated by Cavalier-Smith's (1993) 18S phylogeny, dinoflagellates, haptophytes and chromophytes arose through three separate secondary endosymbiosis events, then there should be two additional lines arising from the rhodophytes that result in the dinoflagellates and haptophytes.

The enigmatic alga *Cyanophora paradoxa*, which still maintains a peptidoglycan wall around its plastid, is somewhat difficult to place in the scheme of Fig. 3. *C. paradoxa* appears to be a very ancient alga and it has transferred more genes from the plastid to the nucleus than, for example, a lower rhodophyte such as *P. purpurea*. Like rhodophytes, the *C. paradoxa* plastid genome contains most of the phycobiliprotein genes, *petF* and *atpDG* (as well as several other genes not found on metaphyte plastid genomes). On the other hand, like chlorophytes and metaphytes, it has a cyanobacterial *rbcLS* operon and has apparently transferred *petJ* and *psaDKL* to the nucleus. Thus, it could potentially branch from the base of the chlorophyte/metaphyte lineage rather than from the bottom of the rhodophyte/chromophyte lineage. However, *C. paradoxa* apparently has more features in common with rhodophytes and thus branches from this lineage in Fig. 3. An alternate possibility (Löffelhardt and Bohnert, 1994) is that *C. paradoxa* arose prior to the separation of the rhodophyte and chlorophyte lineages.

Figure 3 also shows that the transfer of genes to the nucleus has been a continuous process in the evolution of plastids. While a large number of genes were probably transferred to the nucleus in the period before the divergence of the rhodophyte and chlorophyte lines, this process has continued throughout plastid evolution. This premise is further supported when genes other than those of the photosynthetic apparatus (e.g. *tufA*, genes for ribosomal proteins, genes of various biosynthetic enzymes, etc.) are considered (see Palmer, 1991; Gray, 1992, Reith, 1995). In addition, several genes have been transferred to the nucleus solely within the metaphyte lineage (see Palmer, 1991 for review).

Another subtle aspect of plastid evolution is that DNA has moved *into* chloroplasts as well as out of them (Fig. 3). The apparent lateral transfer of Rubisco genes from an α or β purple bacterium into the rhodophyte/chromophyte lineage is one indication of this. However, there is more evidence of this phenomenon in the chlorophyte/metaphyte lineage where several types of introns have been introduced. Introns are rare in cyanobacteria (Kuhsel et al., 1990) and the *C. paradoxa* (Bryant et al., unpublished) plastid genome and completely absent in the *P. purpurea* plastid genome (Reith and Munholland, 1993 and unpublished data), but occur in significant numbers in metaphyte plastids (Palmer, 1991) and are rampant in the *E. gracilis* plastid genome (Hallick et al., 1993). In addition, metaphyte introns are mostly group II introns, whereas those of chlorophytes are mostly group I introns and *E. gracilis* has both group II and group III introns. Introns thus appear to have entered the plastid genomes of the chlorophyte/metaphyte lineage several times.

The model in Fig. 3 is obviously only a beginning to our understanding of plastid evolution and will undoubtedly be modified as more information becomes available. More information on the contents of chromophyte (especially dinoflagellate and haptophyte), cryptomonad and chlorophyte plastid genomes is essential before a complete picture can emerge. The isolation of more rhodophyte and chromophyte nuclear genes is necessary to compare nuclear gene phylogenies with plastid gene phylogenies and plastid gene content analyses. The complexity and intricacy of plastid evolution has only become evident over the last few years and undoubtedly more interesting insights will unfold as data are accumulated.

Acknowledgments

I would like to thank Drs D. Bryant and K. Kowallik for communication of results prior to publication and Dr S. Douglas for critical reading of the manuscript. This is NRCC publication number 34916.

References

Adamska I, Kloppstech K and Ohad I (1992) UV light stress induces the synthesis of the early light-inducible protein and prevents its degradation. J Biol Chem 267: 24732–24737

Alam J, Sprinkle J, Hermodson MA and Krogmann DW (1984) Characterization of cytochrome c_{550} from cyanobacteria. Biochim Biophys Acta 766: 317–321

Allen JF (1992) Protein phosphorylation in regulation of photosynthesis. Biochim Biophys Acta 1098: 275–335

Apt KE, Hoffman NE and Grossman AR (1993) The γ subunit of R-phycoerythrin and its possible mode of transport into the plastid of red algae. J Biol Chem 268: 16206–16215

Bartels D, Hanke C, Schneider K, Michel D and Salamini F (1992) A dessication-related Elip-like gene from the resurrection plant *Craterostigma plantagineum* is regulated by light and ABA. EMBO J 11: 2771–2778

Bennett J (1991) Protein phosphorylation in green plant chloroplasts. Annu Rev Plant Physiol Plant Mol Biol 42: 281–311

Bhaya D and Grossman A (1991) Targeting proteins to diatom plastids involves transport through an endoplasmic reticulum. Mol Gen Genet 229: 400–404

Bhaya D and Grossman A (1993) Characterization of gene clusters encoding the fucoxanthin chlorophyll proteins of the diatom *Phaeodactylum tricornutum*. Nucl Acids Res 21: 4458–4466

Brahamsha B and Haselkorn R (1991) Isolation and characterization of the gene encoding the principal sigma factor of the vegetative cell RNA polymerase from the cyanobacterium *Anabaena* sp. strain PCC 7120. J Bacteriol 173: 2442–2450

Brinkmann H, Cerff R, Salomon M and Soll J (1989) Cloning and sequence analysis of cDNAs encoding the cytosolic precursors of subunits GapA and GapB of chloroplast glyceraldehyde-3-phosphate dehydrogenase from pea and spinach. Plant Mol Biol 13: 81–94

Bryant DA (1992) Puzzles of chloroplast ancestry. Current Biology 2: 240–242

Cavalier-Smith T (1982) The origins of plastids. Biol J Linn Soc 17: 289–306

Cavalier-Smith T (1993) Kingdom Protozoa and its 18 phyla. Microbiol Rev 57: 953–994

Chan RL, Keller M, Canaday J, Weil JH and Imbault P (1990) Eight small subunits of *Euglena* ribulose 1-5 bisphosphate carboxylase/oxygenase are translated from a large mRNA as a polyprotein. EMBO J 9: 333–338

Dolganov NAM, Bhaya D and Grossman AR (1995) Cyanobacterial protein with similarity to the chlorophyll *a/b* binding proteins of higher plants: Evolution and regulation. Proc Natl Acad Sci USA 92: 636–640

Douglas SE (1992) Eukaryote-eukaryote endosymbioses: Insights from studies of a cryptomonad alga. BioSystems 28: 57–68

Douglas SE (1994) Chloroplast origins and evolution. In: Bryant DA (ed) The Cyanobacteria, pp 91–118. Kluwer Academic Publishers, Dordrecht

Douglas SE and Turner S (1991) Molecular evidence for the origin of plastids from a cyanobacterium-like ancestor. J Mol Evol 33: 267–273

Douglas SE, Murphy CA, Spencer DA and Gray MW (1991) Cryptomonad algae are evolutionary chimeras of two phylogenetically distinct unicellular eukaryotes. Nature 350: 148–151

Eschbach S, Wolters J and Sitte P (1991) Primary and secondary structure of the nuclear small subunit ribosomal RNA of the cryptomonad *Pyrenomonas salina* as inferred from the gene sequence: Evolutionary implications. J Mol Evol 32: 247–252

Evans P and Krogmann DW (1983) Three c-type cytochromes from the red alga *Porphyridium cruentum*. Arch Biochem Biophys 227: 494–510

Fitzgerald MP, Hussain A and Rogers LJ (1978) A constitutive flavodoxin from a eukaryotic alga. Biochem Biophys Res Commun 81: 630–635

Funk C, Schröder WP, Green BR, Renger G and Andersson B (1994) The intrinsic 22 kDa protein is a chlorophyll-binding subunit of Photosystem II. FEBS Lett 342: 261–266

Gest H and Favinger JL (1983) *Heliobacterium chlorum*, an anoxygenic brownish-green photosynthetic bacterium containing a 'new' form of bacteriochlorophyll. Arch Microbiol 136: 11–16

Gibbs SP (1962) Nuclear envelope-chloroplast relationships in algae. J Cell Biol 14: 433–444

Gibbs SP (1978) The chloroplasts of *Euglena* may have evolved from symbiotic green algae. Can J Bot 56: 2883–2889

Gibbs SP (1981) The chloroplasts of some algal groups may have evolved from endosymbiotic eukaryotic algae. Ann NY Acad Sci 361: 193–207

Gray MW (1991) Origin and evolution of plastid genomes and genes. In: Bogorad L and Vasil IK (eds) Cell Culture and Somatic Cell Genetics of Plants, Vol 7A, The Molecular Biology of Plastids, pp 303–330. Academic Press, San Diego

Gray MW (1992) The endosymbiont hypothesis revisited. Int Rev Cytol 141: 233–357

Green BR, Pichersky E and Kloppstech K (1991) Chlorophyll *a/b*-binding proteins: An extended family. Trends Biochem Sci 16: 181–186

Greer KL and Golden SS (1992) Conserved relationship between *psbH* and *petBD* genes: Presence of a shared upstream element in *Prochlorothrix hollandica*. Plant Mol Biol 19: 355–365

Grimm B, Kruse E and Kloppstech K (1989) Transiently expressed early light-inducible thylakoid proteins share transmembrane domains with light-harvesting chlorophyll binding proteins. Plant Mol Biol 13: 583–593

Hallick RB, Hong L, Drager RG, Favreau MR, Monfort A, Orsat B, Spielmann A and Stutz E (1993) Complete sequence of *Euglena gracilis* chloroplast DNA. Nucl Acids Res 21: 3537–3544

Holton RW and Myers J (1963) Cytochromes of a blue-green alga: Extraction of a *c*-type with a strongly negative redox potential. Science 142: 234–235

Ikeuchi M (1992) Subunit proteins of Photosystem I. Plant Cell Physiol 33: 669–676

Jenkins J, Hiller RG, Speirs J and Godovac-Zimmerman J (1990) A genomic clone encoding a cryptophyte phycoerythrin α-subunit. Evidence for three α-subunits and an N-terminal membrane transit sequence. FEBS Lett 273: 191–194

Johnson CH and Schmidt GW (1993) The *psbB* gene cluster of the *Chlamydomonas reinhardtii* chloroplast: Sequence and transcriptional analyses of *psbN* and *psbH*. Plant Mol Biol 22: 645–658

Keegstra K, Olsen LJ and Theg SM (1989) Chloroplastic precursors and their transport across the envelope membranes. Annu Rev Plant Physiol Plant Mol Biol 40: 471–501

Kim S, Sandusky P, Bowlby NR, Aebersold R, Green BR, Vlahakis S, Yocum CF and Pichersky E (1992) Characterization of a spinach *psbS* cDNA encoding the 22 kDa protein of Photosystem II. FEBS Lett 314: 67–71

Kishore R, Muchhal US and Schwartzbach SD (1993) The presequence of *Euglena* LHCPII, a cytoplasmically synthesized chloroplast protein, contains a functional endoplasmic reticulum-targeting domain. Proc Natl Acad Sci USA 90: 11845–11849

Knoetzel J and Simpson DJ (1993) The primary structure of a cDNA for PsaN encoding an extrinsic lumenal polypeptide of barley Photosystem I. Plant Mol Biol 22: 337–345

Koike H, Ikeuchi M, Hiyama T and Inoue Y (1989a) Identification of Photosystem I components from the cyanobacterium, *Synechococcus vulcanus* by N-terminal sequencing. FEBS Lett 253: 257–263

Koike H, Mamada K, Ikeuchi M and Inoue Y (1989b) Low-molecular-mass proteins in cyanobacterial Photosystem II: Identification of *psbH* and *psbK* gene products by N-terminal sequencing. FEBS Lett 244: 391–396

Kolanus W, Scharnhorst C, Kuehne U and Herzfeld F (1987) The structure and light-dependent transient expression of a nuclear-encoded chloroplast protein gene from pea (*Pisum sativum* L). Mol Gen Genet 209: 234–239

Kostrzewa M and Zetsche K (1992) Large ATP synthase operon of the red alga *Antithamnion* sp. resembles the corresponding operon in cyanobacteria. J Mol Biol 227: 961–970

Kostrzewa M and Zetsche K (1993) Organization of plastid-encoded ATPase genes and flanking regions including homologues of *infB* and *tsf* in the thermophilic red alga *Galderia sulphuraria*. Plant Mol Biol 23: 67–76

Kowallik KV (1992) Origin and evolution of plastids from chlorophyll-*a*+*c*-containing algae: Suggested ancestral relationships to red and green algal plastids. In: Lewin RA (ed) Origins of Plastids, pp 223–263. Chapman and Hall, New York

Kühlbrandt W, Wang DN and Fujiyoshi Y (1994) Atomic model of plant light-harvesting complex by electron crystallography. Nature 367: 614–621

Kuhsel MG, Strickland R and Palmer JD (1990) An ancient group I intron is shared by eubacteria and chloroplasts. Science 250: 1570–1573

La Roche J, Bennett J and Falkowski PG (1990) Characterization of a cDNA encoding the 28.5 kDa LHCII apoprotein from the unicellular marine chlorophyte *Dunaliella tertiolecta*. Gene 95: 165–171

La Roche J, Geider RJ, Graziano LM, Murray H and Lewis K (1993) Induction of specific proteins in eukaryotic algae grown under iron-, phosphorus-, or nitrogen-deficient conditions. J Phycol 29: 767–777

Larouche L, Tremblay C, Simard C and Bellemare G (1991) Characterization of a cDNA encoding a PSII-associated chlorophyll *a/b*-binding protein (CAB) from *Chlamydomonas moewusii* fitting into neither type I nor type II. Curr Genet 19: 285–288

Lers A, Levy H and Zamir A (1991) Co-regulation of a gene homologous to early light-induced genes in higher plants and beta-carotene biosynthesis in the alga *Dunaliella bardawil*. J Biol Chem 266: 13698–13705

Levy H, Tal T, Shaish A and Zamir A (1993) Cbr, an algal homolog of plant early light-induced proteins, is a putative zeaxanthin-binding protein. J Biol Chem 268: 20892–20896

Lewin RA and Whithers NW (1975) Extraordinary pigment composition of a prokaryotic alga. Nature 256: 735–737

Liaud M-F, Valentin C, Brandt U, Bouget F-Y, Kloareg B and Cerff R (1993) The GAPDH gene system of the red alga *Chondrus crispus*: Promoter structures, intron/exon organization, genomic complexity and differential expression of genes. Plant Mol Biol 23: 981–994

Löffelhardt W and Bohnert HJ (1994) Molecular biology of cyanelles. In: Bryant DA (ed) The Cyanobacteria, pp 65–89. Kluwer Academic Publishers, Dordrecht

Loiseaux-de Goër S (1994) Plastid lineages. In: Round FE and Chapman DJ (eds) Progress in Phycological Research, Vol 10, pp 137–177. Biopress Ltd, Bristol

Maid U and Zetsche K (1992) A 16 kb small single-copy region separates the plastid DNA inverted repeat of the unicellular red alga *Cyanidium caldarium*: Physical mapping of the IR-flanking regions and nucleotide sequences of the *psbD-psbC*, *rps16*, 5S rRNA and *rpl21* genes. Plant Mol Biol 19: 1001–1010

Margulis L and Obar R (1985) *Heliobacterium* and the origin of chrysoplasts. Biosystems 17: 317–325

Martin W, Somerville CC and Loiseaux-de Goër S (1992) Molecular phylogenies of plastid origins and algal evolution. J Mol Evol 35: 385–404

Mayes SR and Barber J (1990) Nucleotide sequence of the *psbH* gene of the cyanobacterium *Synechocystis* 6803. Nucl Acids Res 18: 194

Merchant S and Bogorad L (1987) The Cu(II)-repressible plastidic cytochrome c. Cloning and sequence of a complementary DNA for the pre-apoprotein. J Biol Chem 262: 9062–9067

Merchant S, Hill K, Kim JH, Thompson J, Zaitlin D and Bogorad L (1990) Isolation and characterization of a complementary DNA clone for an algal pre-apoplastocyanin. J Biol Chem 265: 12372–12379

Mereschkowsky C (1910) Theorie der zwei Plasmaarten als Grundlage der Symbiogenesis einer neuen Lehre von der Entstehung der Organismen. Biol Zentralblatt 30: 278–303, 321–347, 353–367

Morand LZ, Cheng RH, Krogmann DW and Ho KK (1994) Soluble electron transfer catalysts of cyanobacteria. In: Bryant DA (ed) The Molecular Biology of Cyanobacteria, pp 381–407. Kluwer Academic Publishers, Dordrecht

Morden CW, Delwiche CF, Kuhsel M and Palmer JD (1992) Gene phylogenies and the endosymbiotic origin of plastids. BioSystems 28: 75–90

Muchhal US and Schwartzbach SD (1992) Characterization of a *Euglena* gene encoding a polyprotein precursor to the light-harvesting chlorophyll *a/b*-binding protein of Photosystem II. Plant Mol Biol 18: 287–299

Nilsson F, Andersson B and Jansson C (1990) Photosystem II characteristics of a constructed *Synechocystis* 6803 mutant lacking synthesis of the D1 polypeptide. Plant Mol Biol 14: 1051–1054

Okkels JS, Nielsen VS, Scheller HV and Møller BL (1992) A cDNA clone from barley encoding the precursor from the Photosystem I polypeptide PSI-G: Sequence similarity to PSI-K. Plant Mol Biol 18: 989–994

Palenik B and Haselkorn R (1992) Multiple evolutionary origins of prochlorophytes, the chlorophyll *b*-containing prokaryotes. Nature 355: 265–267

Palmer JD (1991) Plastid chromosomes: Structure and evolution. In: Bogorad L and Vasil IK (eds) Cell Culture and Somatic Cell Genetics of Plants, Vol 7A, The Molecular Biology of Plastids, pp 5–53. Academic Press, San Diego

Pancic PG and Strotmann H (1993) Structure of the nuclear encoded γ subunit of CF_0CF_1 of the diatom *Odontella sinensis* including its presequence. FEBS Lett 320: 61–66

Pancic PG, Strotmann H and Kowallik KV (1992) Chloroplast ATPase genes in the diatom *Odontella sinensis* reflect cyanobacterial characters in structure and arrangement. J Mol Biol 224: 529–536

Pichersky E, Bernatzky R, Tanksley SD, Breidenbach RB, Kausch

AP and Cashmore AR (1985) Molecular characterization and genetic mapping of two clusters of genes encoding chlorophyll *a/b*-binding proteins of *Lycopersicon esculentum* (tomato). Gene 40: 247–258

Pichersky E, Brock TG, Nguyen D, Hoffman NE, Piechulla B, Tanksley SD and Green BR (1989) A new member of the CAB gene family: Structure, expression and chromosomal location of *Cab-8*, the tomato gene encoding the type III chlorophyll *a/b*-binding polypeptide of Photosystem I. Plant Mol Biol 12: 257–270

Pichersky E, Subramanian R, White MJ, Reid J, Aebersold R and Green BR (1991) Chlorophyll *a/b* binding (CAB) polypeptides of CP29, the internal chlorophyll *a/b* complex of PSII: Characterization of the tomato gene encoding the 26 kDa (type I) polypeptide and evidence for a second CP29 polypeptide. Mol Gen Genet 227: 277–284

Plumley FG and Schmidt GW (1987) Reconstitution of chlorophyll *a/b* light-harvesting complexes: Xanthophyll-dependent assembly and energy transfer. Proc Natl Acad Sci USA 84: 146–150

Price NT, Smith AJ and Rogers LJ (1991) Relationship of the flavodoxin isoforms from *Porphyra umbilicalis*. Phytochemistry 30: 2841–2843

Race HL and Gounaris K (1993) Identification of the *psbH* gene product as a 6 kDa phosphoprotein in the cyanobacterium *Synechocystis* 6803. FEBS Lett 323: 35–39

Raghavan V and Kamalay JC (1993) Expression of 2 cloned messenger RNA sequences during development and germination of spores of the sensitive fern *Onoclea sensibilis* L. Planta 189: 1–9

Raven PH (1970) A multiple origin for plastids and mitochondria. Science 169: 641–646

Reith M (1995) Molecular biology of rhodophyte and chromophyte plastids. Annu Rev Plant Physiol Plant Mol Biol 46: 549–575

Reith M and Munholland J (1993) A high-resolution gene map of the chloroplast genome of the red alga *Porphyra purpurea*. Plant Cell 5: 465–475

Rhiel E and Mörschel E (1993) The atypical chlorophyll *a/b/c* light-harvesting complex of *Mantoniella squamata*: Molecular cloning and sequence analysis. Mol Gen Genet 240: 403–413

Schwartz E, Shen D, Aebersold R, McGrath JM, Pichersky E and Green BR (1991) Nucleotide sequence and chromosomal location of *Cab11* and *Cab12*, the genes for the fourth polypeptide of the Photosystem I light-harvesting antenna (LHCI). FEBS Lett 280: 229–234

Sharif AL, Smith AG and Abell C (1989) Isolation and characterization of a cDNA clone for a chlorophyll synthesis enzyme from *Euglena gracilis*. The chloroplast enzyme hydroxymethylbilane synthase (porphobilinogen deaminase) is synthesised with a very long transit peptide in *Euglena*. Eur J Biochem 184: 353–359

Shen J-R and Inoue Y (1993) Cellular localization of cytochrome c_{550}. Its specific association with cyanobacterial Photosystem II. J Biol Chem 268: 20408–20413

Shen J-R, Ikeuchi M and Inoue Y (1992) Stoichiometric association of extrinsic cytochrome c_{550} and 12 kDa protein with a highly purified oxygen-evolving Photosystem II core complex from *Synechococcus vulcanus*. FEBS Lett 301: 145–149

Straus NA (1994) Iron deprivation: Physiology and gene regulation. In: Bryant DA (ed) The Cyanobacteria, pp 731–750. Kluwer Academic Publishers, Dordrecht

Urbach E, Robertson DL and Chisholm SW (1992) Multiple evolutionary origins of prochlorophytes within the cyanobacterial radiation. Nature 355: 267–270

Wallace TP, Stewart AC, Pappin D and Howe CJ (1989) Gene sequence for the 9 kDa component of Photosystem II from the cyanobacterium *Phormidium laminosum* indicates similarities between cyanobacterial and other leader sequences. Mol Gen Genet 216: 334–339

Wedel N, Klein R, Ljungberg U, Andersson B and Herrmann RG (1992) The single-copy gene *psbS* codes for a phylogenetically intriguing 22 kDa polypeptide of Photosystem II. FEBS Lett 314: 61–66

Whatley JM and Whatley FR (1981) Chloroplast evolution. New Phytol 87: 233–247

Witt D and Stackebrandt E (1988) Disproving the hypothesis of a common ancestry for the *Ochromonas danica* chrysoplast and *Heliobacterium chlorum*. Arch Microbiol 150: 244–248

Zhou YH and Ragan MA (1993) cDNA cloning and characterization of the nuclear gene encoding chloroplast glyceraldehyde-3-phosphate dehydrogenase from the marine red alga *Gracilaria verrucosa*. Curr Genet 23: 483–489

Index

A

B

D

E

F

G

H

I

K

L

M

N

O

P

Q

R

S

T

U

V

W

X

Y

Z

Advances in Photosynthesis

Series editor: Govindjee, University of Illinois, Urbana, Illinois, U.S.A.

1. D.A. Bryant (ed.), *The Molecular Biology of Cyanobacteria.* 1994
   ISBN Hb: 0-7923-3222-9; Pb: 0-7923-3273-3
2. R.E. Blankenship, M.T. Madigan and C.E. Bauer (eds.): *Anoxygenic Photosynthetic Bacteria.* 1995 ISBN Hb: 0-7923-3681-X; Pb: 0-7923-3682-8
3. J. Amesz and A.J. Hoff (eds.): *Biophysical Techniques in Photosynthesis.* 1996 ISBN 0-7923-3642-9
4. D.R. Ort and C.F. Yocum (eds.): *Oxygenic Photosynthesis: The Light Reactions.* 1996 ISBN Hb: 0-7923-3683-6; Pb: 0-7923-3684-4

KLUWER ACADEMIC PUBLISHERS – DORDRECHT / BOSTON / LONDON